Rodent Societies: An Ecological & Evolutionary Perspective

Edited by Jerry O. Wolff and Paul W. Sherman

The University of Chicago Press
Chicago and London

PAUL W. SHERMAN is professor of animal behavior and a Weiss Presidential Fellow at Cornell University.

JERRY O. WOLFF was professor and chair of the biology department at the University of Memphis; he is now professor of biology at St. Cloud State University.

The University of Chicago Press, Chicago 60637
The University of Chicago Press, Ltd., London
© 2007 by The University of Chicago
All rights reserved. Published 2007
Printed in the United States of America

16 15 14 13 12 11 10 09 08 07   1 2 3 4 5

ISBN-10: 0-226-90536-5 (cloth)
ISBN-13: 978-0-226-90536-5 (cloth)
ISBN-10: 0-226-90537-3 (paper)
ISBN-13: 978-0-226-90537-2 (paper)

Library of Congress Cataloging-in-Publication Data

Rodent societies : an ecological & evolutionary perspective / edited by Jerry O. Wolff and Paul W. Sherman.
   p.   cm.
   Includes bibliographical references and index.
   ISBN-13: 978-0-226-90536-5 (cloth : alk. paper)
   ISBN-10: 0-226-90536-5 (cloth : alk. paper)
   ISBN-13: 978-0-226-90537-2 (pbk. : alk. paper)
   ISBN-10: 0-226-90537-3 (pbk. : alk. paper)
   1. Rodents—Ecology.   2. Rodents—Evolution.   3. Social behavior in animals.   4. Animal societies.   I. Wolff, Jerry.   II. Sherman, Paul W., 1949–
   QL737.R6R623   2007
   599.35'156—dc22

                                     2006021884

Rodent Societies

Contents

Conclusions

Contributors

Brandon J. Aragona
Department of Psychology
Florida State University
Tallahassee, FL 32306
aragona@psy.fsu.edu

Kenneth B. Armitage
Department of Ecology and Evolutionary Biology
University of Kansas
Lawrence, KS 66045
marmots@lark.cc.ukans.edu

J. M. Barker
Centre for the Neurobiology of Stress
Department of Life Sciences
University of Toronto at Scarborough
Scarborough, Ontario, Canada, M1C 1A4

William E. Bemis
Department of Ecology and Evolutionary Biology
Corson Hall
Cornell University
Ithaca, NY 14853
web24@cornell.edu

Nigel C. Bennett
Department of Zoology and Entomology
University of Pretoria
Pretoria, South Africa
ncbennett@zoology.up.ac.za

Manuel Berdoy
Oxford University Veterinary Services
Parks Rd, OX1 3PT
Oxford, UK
manuel.berdoy@vet.ox.ac.uk

Daniel T. Blumstein
Department of Ecology and Evolutionary Biology
University of California
621 Charles E. Young Drive South
University of California
Los Angeles, CA 90095-1606
marmots@ucla.edu

Rudy Boonstra
Centre for the Neurobiology of Stress
Department of Life Sciences
University of Toronto at Scarborough
Scarborough, Ontario, Canada, M1C 1A4
boonstra@utsc.utoronto.ca

Joel Brown
Department of Biological Sciences
University of Illinois at Chicago
Chicago, IL 60607
squirrel@uic.edu

Peter E. Busher
Center for Ecology and Conservation Biology
Boston University
Boston, MA 02215 USA
pbusher@bu.edu

Lara S. Carroll
Howard Hughes Medical Institute
University of Utah
Salt Lake City, UT 84112-5331
lara.carroll@utah.edu

J. Castillo
Centre for the Neurobiology of Stress
Department of Life Sciences
University of Toronto at Scarborough
Scarborough, Ontario, Canada, M1C 1A4

Richard C. Coss
Department of Psychology
University of California
One Shields Avenue
Davis, CA 95616-8686
rgcoss@ucdavis.edu

J. Thomas Curtis, PhD
Department of Psychology
Florida State University
Tallahassee, FL 32306
tcurtis@psy.fsu.edu

F. Stephen Dobson
Department of Biological Sciences
Auburn University
Auburn, AL 36849
fdobson@acesag.auburn.edu

Lee C. Drickamer
Department of Biological Sciences
Northern Arizona University
Flagstaff, AZ 86001
Lee.Drickamer@NAU.EDU

Luis A. Ebensperger
Departamento de Ecología
P. Universidad Católica de Chile
Centro de Estudios Avanzados
en Ecología & Biodiversidad
Casilla 114-D, Santiago, Chile
lebenspe@genes.bio.puc.cl

Christopher G. Faulkes
Queen Mary & Westfield College
School of Biological Sciences
London E1 4NS UK
C.G.Faulkes@qmul.ac.uk

Quinn E. Fletcher
Centre for the Neurobiology of Stress
Department of Life Sciences
University of Toronto at Scarborough
Scarborough, Ontario, Canada, M1C 1A4

Laurence J. Frabotta
Department of Biology
Texas A&M University
College Station, Texas 77843-3258

Bennett G. Galef
Department of Psychology
McMaster University
Hamilton, Ontario L8S 4K1
Canada
galef@mcmail.mcmaster.ca

James F. Hare
Department of Zoology
University of Manitoba
Manitoba, Canada R3T 2N2
harejf@cc.umanitoba.ca

Emilio A. Herrera
Departamento de Estudios Ambientales
Universidad Simón Bolívar
Caracas 1080-A, Venezuela
eherre@usb.ve

Warren Holmes
Department of Psychology and
Center for Ecology and Evolutionary Biology
University of Oregon
Eugene, OR 97403-5289
wholmes@darkwing.uoregon.edu

Rodney L. Honeycutt
Texas A&M University
Department of Wildlife & Fisheries Sciences
College Station, Texas 77843-2258
rhoneycutt@neo.tamu.edu

John L. Hoogland
The University of Maryland
Appalachian Laboratory
Frostburg, MD 21532
hoogland@al.umces.edu

Matina Kalcounis-Rüppell
Department of Biology
The University of North Carolina at Greensboro
Greensboro, North Carolina 27402-6170
matina_kalcounis@uncg.edu

Brian Keane
Department of Zoology
Miami University
Hamilton, OH 45011
keaneb@muohio.edu

John L. Koprowski
Wildlife and Fisheries Science
School of Renewable Natural Resources
University of Arizona
Tucson, AZ 85721
squirrel@ag.arizona.edu

Dr. Charles J. Krebs
CSIRO Sustainable Ecosystems
Canberra, A.C.T. 2601
Australia
Charles.Krebs@csiro.au

Eileen Lacey
Museum of Vertebrate Zoology
University of California
Berkeley, CA 94720
ealacey@socrates.berkeley.edu

Xavier Lambin
School of Biological Sciences
Department of Biology
University of Aberdeen
Aberdeen, Scotland
x.lambin@abdn.ac.uk

William Z. Lidicker, Jr.
Museum of Vertebrate Zoology
University of California
Berkeley, CA 94720
lidicker@socrates.Berkeley.edu

Yan Liu
Department of Psychology
Florida State University
Tallahassee, FL 32306

David W. Macdonald
Wildlife Conservation Research Unit
Department of Zoology
University of Oxford
Tubney, Oxon OX13 5QL, UK
david.macdonald@zoology.ox.ac.uk

Jill M. Mateo
Department of Comparative Human Development
Committee on Human Development
Institute for Mind and Biology
University of Chicago
Chicago, IL 60637
jmateo@uchicago.edu

Betty McGuire, PhD
Department of Ecology and Evolutionary Biology
Cornell University
Ithaca NY 14853
bam65@cornell.edu

James N. Mills
Special Pathogens Branch
Centers for Disease Control and Prevention (MS G-14)
Atlanta, GA, 30333
jumo@cdc.gov

José Roberto Moreira
Embrapa Recursos Genéticos e Biotecnologia
Parque Estação Biológica
70770-900 Brasília DF
Brazil
jmoreira@cenargen.embrapa.br

Jan O. Murie
Department of Biological Sciences
University of Alberta
Edmonton, AB T6G 2E9
Alberta, Canada
jan.murie@ualberta.ca

Eviatar Nevo
Department of Environmental and Evolutionary Biology
University of Haifa
Mount Carmel, Haifa 31905
Israel
nevo@research.haifa.ac.il

Scott Nunes
Department of Biology
University of San Francisco
San Francisco, CA 94117
nunes@ace.usfca.edu

Karen J. Nutt
School of Biological Sciences
University of Auckland
Auckland, New Zealand
knut006@ec.auckland.ac.nz

Madan K. Oli
Department of Wildlife Ecology and Conservation
University of Florida
Gainesville, FL 32611
OliM@wec.ufl.edu

Richard S. Ostfeld
Institute of Ecosystem Studies
Millbrook, NY 12545
ROstfeld@ecostudies.org

Donald H. Owings
Department of Psychology
University of California
One Shields Avenue
Davis, CA 95616-8686
dhowings@ucdavis.edu

Wayne K. Potts
Molecular Biology Program
University of Utah
Salt Lake City, UT 84112
potts@biology.utah.edu

Jan Randall
Department of Biology
University of San Francisco
San Francisco, CA 94117
jrandall@sfsu.edu

David O. Ribble
Biology Department
Trinity University
San Antonio, TX 78212
dribble@trinity.edu

S. Craig Roberts
School of Biological Sciences
University of Liverpool
Liverpool UK L69 7ZB
craig.roberts@liv.ac.uk

Diane L. Rowe
Texas A&M University
Department of Wildlife & Fisheries Sciences
College Station, Texas 77843-2258

Albrecht I. Schulte-Hostedde
Department of Biology
Laurentian University
Sudbury, Ontario, Canada
P3E 2C6
aschultehostedde@laurentian.ca

Paul W. Sherman
Department of Neurobiology and Behavior
Cornell University
Ithaca, NY 14853
pws6@cornell.edu

Robert S. Sikes
Biology Department
University of Arkansas, Little Rock
Little Rock, AR 72204
rssikes@ualr.edu

Nancy G. Solomon
Department of Zoology
Miami University
Oxford, Ohio 45056
solomong@muohio.edu

Andrew Taber
Wildlife Conservation Society
2300 Southern Boulevard
Bronx, NY 10460
ataber@wcs.org

Beatrice Van Horne
1829 Elgin Dr.
Vienna, VA 22182
bvanhorne@fs.fed.us

Zuoxin Wang
Department of Psychology
Department of Biological Sciences
Neuroscience Program
Florida State University
Tallahassee, FL 32306
zwang@darwin.psy.fsu.edu

Jane M. Waterman
Department of Biology
University of Central Florida
Orlando, FL 32816-2368
waterman@mail.ucf.edu

Jerry O. Wolff
Department of Biological Sciences
St. Cloud State University
St. Cloud, MN 56301
jowolff@stcloudstate.edu

Hannu Ylönen
Department of Biological and Environmental Science
University of Jyväskyla
FIN 40351 Jyväskyla, Finland
hylonen@dodo.jyu.fi

Acknowledgments

2 Rodney L. Honeycutt, Laurence J. Frabotta, and Diane L. Rowe: Research reported on hystricognath rodents was funded by National Science Foundation Grant DEB 9615163 to RLH.

4 Nancy G. Solomon and Brian Keane: We thank the editors for the invitation to participate in this book. Kristen Lucia, Gail Michener, Paul Sherman, and Jerry Wolff provided valuable feedback on a previous version of this manuscript. We also are grateful to Michelle Edwards, Stephanie Kortering, Samantha Lowe, Lisa Walter, and Beth Widen for assistance with references. NGS and BK were supported by NSF DEB-0316818 and NGS was supported by 1 R15 6M069409-01 during the preparation of this book chapter.

5 Lara S. Carroll and Wayne K. Potts: We would like to thank the editors for substantive comments on an earlier version of this manuscript. This manuscript was written while WKP was supported by NIH (GM39578) and NSF (IBN-9904609) grants.

6 Matina C. Kalcounis-Rüppell and David O. Ribble: We would like to acknowledge the assistance of Stacy Huff, Michelle Icenhower, and Adrian Sherman in compiling literature for our review and analysis. Robert Bradley helped us to understand Neotomine-Peromyscine systematics. Comments of Jack Millar, Maarten Vonhof, Eileen Lacey, and two anonymous reviewers improved earlier versions of this manuscript. We thank the editors for their insight, suggestions, and editorial guidance.

8 F. Stephen Dobson and Madan K. Oli: We owe special thanks to the editors for suggesting that we review rodent life histories. D. R. Broussard, P. H. Harvey, T. J. Karels, and S. C. Stearns provided excellent comments and suggestions for improvement of the manuscript. FSD's contribution to the chapter was supported by a National Science Foundation grant for research (DEB-0089473).

9 Lee C. Drickamer: I thank the many undergraduates, particularly at Williams College, graduate students and undergraduates at Southern Illinois University, and faculty colleagues at those two schools and at Northern Arizona University for their invaluable assistance, discussions, and enthusiasm. Portions of the work in my laboratory and in field settings were supported by grants from the National Institutes of Health and National Science Foundation. I thank the editors for their foresight in organizing this volume and for their diligence in putting it all together.

10 Albrecht I. Schulte-Hostedde: I thank the editors for the opportunity to contribute this chapter and for improving it with their insightful comments. Thanks to all who responded to my queries, including T. Best, S. Boutin, J. Hoogland, I. Khokhlova, H. Levenson, and R. Sweitzer. Climate data from weather stations were provided by J. Pither. A. Oey provided valuable assistance.

11 Robert S. Sikes: I am grateful for constructive comments on a previous version of this manuscript by T. G. Finley, S. Krackow, P. W. Sherman, and J. O Wolff. This work was supported in part by NSF Grant 9975445.

12 R. Boonstra, J. M. Barker, J. Castillo, and Q. E. Fletcher: The Natural Sciences and Engineering Research Council of Canada supported this research. We thank Jim Kenagy for helpful comments on an earlier draft of this chapter.

13 Scott Nunes: I thank Jerry Wolff and Steve Dobson for constructive, incisive, and helpful comments on earlier versions of this chapter.

14 F. Stephen Dobson: I owe special thanks to the editors for suggesting that I review the genetic properties of social breeding groups. Ron Chesser provided patient instruction on the gene dynamics of social breeding groups during the summers of 1993–1996, and the Savannah River Ecology Laboratory provided support via Visiting Faculty Fellowships in 1993, 1994, and 1996. In 1995, I was an Oak Ridge Institute for Science and Education Research Fellow at SREL. I especially appreciate the support of the director of SREL, M. H. Smith, during these 4 years. R. K. Chesser, N. Perrin, and M. F. Winterrowd provided excellent comments and suggestions for improvement of the manuscript. I also owe special thanks to John Hoogland, who generously made his incredible field data so freely available, and to Dave Foltz, for encouraging analyses on his and John's allozyme data. Preparation of the current manuscript was completed while I was supported by a National Science Foundation grant for research (DEB-0089473).

15 Charles J. Krebs, Xavier Lambin, and Jerry O. Wolff: We thank Peter Brown, Grant Singleton, and Alice Kenney for their comments and suggestions on the manuscript.

16 J. Thomas Curtis, Yan Liu, Brandon J. Aragona, and Zuoxin Wang: We are grateful to Christie Fowler and

Mike Smeltzer for critical reading of this manuscript. This work was partially supported by NIH grants HD48462 and HD40722 (JTC), MH67396 (BJA), MH58616 and MH67396 (ZXW)

18 Bennett G. Galef Jr.: I thank Paul Sherman for his thoughtful comments on earlier drafts of the manuscript. The author's research described here was supported for 35 years by the Natural Science and Engineering Research Council of Canada.

21 Eileen A. Lacey and Paul W. Sherman: The ideas presented in this paper reflect 25 years of collaborative interactions between the authors regarding sociality in rodents. For insights and encouragement along the way we thank R. D. Alexander, J. U. M. Jarvis, M. L. Morton, C. K. Sherman, J. S. Sherman, J. R. Wieczorek, J. O. Wolff, E. Yensen, the Animal Behavior Lunch Bunch at Cornell University, and the many graduate and undergraduate students who have assisted with our field and laboratory studies of subterranean rodents. Our work with social rodents has been supported by the National Science Foundation, the National Geographic Society, Sigma Xi, the Center for Latin American Studies at the University of California, Berkeley, and the College of Agriculture and Life Sciences at Cornell University.

22 S. Craig Roberts: I thank Morris Gosling, Paul Sherman, Jerry Wolff, and Sarah Zala for their comments on the manuscript.

23 Luis A. Ebensperger and Daniel T. Blumstein: We thank the editors for inviting us to write this chapter, and for comprehensive suggestions, which improved the original manuscript considerably. Comments by Bob Elwood and Stefano Parmigiani also are greatly appreciated. During the writing of this article, LAE was supported by the Centro de Estudios Avanzados en Ecología & Biodiversidad (FONDAP 1501-001) and by a FONDECYT grant No. 1020861.

25 Eviatar Nevo: This chapter is based on the research program of speciation and adaptive radiation in Israeli subterranean mole-rats that started in the early 1950s. The subset of aggression studies was initiated in 1975. The *Spalax ehrenbergi* superspecies research program incorporates the extensive Israeli and worldwide collaborations of active students, colleagues, and collaborators of subterranean mammals. The *Spalax* research program has been documented since 1961 in some 280 multidisciplinary scientific publications. Much of the evidence in this chapter stems from the aforementioned collaborations. We acknowledge all participants' contributions with much appreciation.

I wish to extend my deepest gratitude to my colleagues at the Institute of Evolution, whose generous and devoted help made this chapter possible: Mr. Michael Margulis and Mrs. Robin Permut.

I also appreciate the continuous financial support of the University of Haifa, Israeli Discount Bank Chair of Evolutionary Biology, the Ancell-Teicher Research Foundation for Genetics and Molecular Evolution, the Israeli Ministries of Science and Absorption, and grants from the Israel Science Foundation, BSF, the Israel Academy of Sciences, and Guggenheim Foundation.

26 Donald H. Owings and Richard G. Coss: We are indebted to Eric Charles and the editors for their careful reviews of drafts of this manuscript.

27 Daniel T. Blumstein: I thank Alexander Nikolskii for introducing me to the Russian literature on alarm communication and Luis Ebensberger for an introduction to South American rodents; neither is responsible for anticipated omissions. I thank Janice Daniel, Kim Pollard, and my reviewers—Jan Randall and Ron Swaisgood—for constructive comments on previous versions, and Erin Shelly, who found reports of genera of alarm calling rodents that I originally missed. I am *extremely* grateful to Paul Sherman, who went out of his way to improve the clarity of this chapter, and to remind me of the importance of the "limits of nepotism"—a hypothesis that is likely to explain interspecific variation in the adaptive utility of alarm communication and other nepotistic behaviors.

28 Hannu Ylönen and Joel S. Brown: We thank the editors for inviting us to join the team of *Rodent Societies*. It was a great honor. For our experiences and journeys into the fascinating world of desert rodents, voles, and the ecology of fear, we owe much to the help, instruction, and good ideas of a number of colleagues and friends, too many to mention by name, but denizens of Fennoscandia, North America, the Middle East, and Australia. We are grateful to them all.

31 Jan A. Randall: I am grateful to my Russian collaborators on the gerbil project, especially Kostya Rogovin, who kindly provided references from the Russian literature and constructed table 31.1. I thank all the students who participated in my research on kangaroo rats and gerbils over the years, and Bruce MacEvoy for his support and editing skills. He, Debra Shier, Kostya Rogovin, and Jerry Wolff provided helpful comments on the manuscript. I thank the National Science Foundation, National Institute of Health, National Geographic Society, and the Research and Development Foundation for grant support. This chapter is dedicated to John Eisenberg, who conducted the early studies on the behavior of desert rodents and encouraged me to study them.

32 Manuel Berdoy and Lee C. Drickamer: The authors would like to thank J. Galef, A. Hanson, P. Honess, G. Singleton, A. Voight, J. Webster, and S. Wolfensohn for helpful comments on this chapter.

33 David W. Macdonald, Emilio A. Herrera, Andrew B. Taber, and José Roberto Moreira: We are grateful to Ruth Feber for considerable help in the preparation of this chapter, and to our many colleagues in the WildCRU who have commented on and encouraged our research.

34 Eileen A. Lacey and Luis A. Ebensperger: We thank the editors for inviting us to write this chapter. Figures 34.1–34.4 were compiled by Karen Klitz. During the preparation of this chapter, LAE was supported by the Centro de Estudios Avanzados en Ecología and Biodiversidad (FONDAP 1501-001). The authors' ongoing studies of degus, cururos, and tuco-tucos are supported by funding from the National Science Foundation (DEB-0128857; EAL), the National Geographic Society (EAL and LAE), and by FONDECYT grant No. 1020861 (LAE).

35 Karen J. Nutt: I would like to thank the Marshall Commission and Roger Nutt for providing financial support for me while writing this chapter. Thanks also to Paula White, Eileen Lacey, and the members of both LARG and Molecular Ecology at Cambridge University for their continued moral support. I am also grateful to the editors for their kindness, patience, and constructive comments on this chapter. Thanks also to the gundis for making the study of rock-dwelling rodents a truly fascinating and worthwhile experience.

36 Chris G. Faulkes and Nigel C. Bennett: We are most grateful for financial support from the following: National Research Foundation (NCB and JUMJ), The University of Pretoria and the Mellon Foundation (NCB), The University of Cape Town and the National Geographic Society (JUMJ), and the Natural Environment Research Council (CGF). Thanks to Jenny Jarvis, Steve Le Comber, and Pippa Faulkes for proofreading and many helpful comments, and to the editors for further suggestions that greatly improved the manuscript.

39 Beatrice Van Horne: NatureServe compiled and formatted a ground squirrel custom dataset that was useful in developing the cross-species summarizations.

Last, Jerry Wolff and Paul Sherman sincerely thank all of the authors for contributing their professional expertise to *Rodent Societies* and for their diligence, patience, and good humor in working with us to complete this anthology. We also thank Janet Sherman for assistance with the index and Christie Henry, Monica Holliday, and Jennifer Howard of the University of Chicago Press and Susan Dodson of Graphic Composition for their support, encouragement, and editorial expertise in bringing this book to fruition.

Introduction

Chapter 1 Rodent Societies as Model Systems

Jerry O. Wolff and Paul W. Sherman

THE RODENTIA is the largest order of mammals, consisting of more than 2,000 species and comprising 44% of all mammals. Rodents come in a variety of body shapes, from cylindrical to spherical, and sizes, from less than 10 g to more than 66 kg. The characteristic that unites this order—and is its most conspicuous trait— is a single pair of razor-sharp incisors, which are used to gnaw food, excavate tunnels, and defend themselves. The name "rodent" derives from the Latin word *rodere,* which means "to gnaw."

No matter where you live, a rodent is probably not far away. Rodents inhabit all continents except Antarctica, and they occur in terrestrial, subterranean, arboreal, and aquatic habitats—from the high arctic tundra to equatorial rain forests, temperate bogs and swamps to hot, arid deserts, and rocky mountaintops to sandy canyon bottoms. Many species live in close association with humans. With such varied characteristics and expansive ecology, rodents provide a range of attributes that have captivated scientists and annoyed laypersons for hundreds of years.

The diversity of rodents, and the ease with which many species can be maintained in captivity, has led to their choice as model systems for observational and experimental studies in genetics, ecology, demography, physiology, and psychology. The social and reproductive behaviors of rodents also are diverse and intriguing. Although most rodents are nocturnal, a surprising amount of research has been conducted on rodent social biology. As a result, large databases are available, and these can be used to test hypotheses about the ecological and evolutionary forces that mold mammalian social and reproductive behaviors. Many

aspects of behavioral ecology are similar across species (e.g., the effects of resource distributions on mating systems [Slobodchikoff 1984; Ostfeld 1990]; the role of ecological factors in favoring group-living [Hoogland 1995; Ebensperger and Cofré 2001]; and the role of kinship in structuring social interactions [Sherman 1981a; Lacey and Wieczorek 2003]). Unifying theories developed from other taxa, such as primates, ungulates, or canids, can be experimentally tested in the field and laboratory with rodents. These options are not so readily available for the larger, wider-ranging taxa. Thus rodents are not only models for testing hypotheses developed from rodents, but they have become models for other taxa as well. The relevant information, however, is widely scattered and sometimes conflicting. Hence this volume.

Our goals were to synthesize and integrate the current state of knowledge about the social behavior of rodents, to provide ecological and evolutionary contexts for understanding rodent societies, and to highlight emerging conservation and management issues to preserve these societies. Thus we attempted to emulate *Primate Societies* (1987) and *Cetacean Societies* (2000), the two outstanding preceding volumes in this series published by the University of Chicago Press.

In selecting topics and contributors for the present volume, we first chose areas of behavioral biology and model species that we considered essential for understanding the adaptive significance of rodent social behavior generally. Then we invited contributions from researchers who have demonstrated their preeminence in illuminating these areas and in studying the focal taxa. We urged authors to use the comparative approach, and to discuss not only the behavior of their subjects but also that of ecologically similar and phylogenetically related species. We instructed authors to differentiate proximate (mechanistic) from ultimate (evolutionary) causes of behavioral phenomena, and to emphasize the usefulness of integrating results at different levels of analysis. Finally, we encouraged authors to highlight the roles of experimentation and hypothesis testing in advancing our understanding and in developing the next generation of predictive models.

All chapters were peer reviewed, sometimes by authors of related chapters, in an attempt to enhance integration. The resulting volume attests not only to the authors' expertise, good humor, and willingness to revise and reconsider their work in light of constructive criticisms and new information, but also why these contributors have made studying rodent societies their life's work. We hope that after diving between the covers of this volume, you too will be struck by rodentophilia!

We begin this anthology with a summary of the evolution, phylogeny, and biogeography of rodents, to provide an historical context and a basis for comparative analyses. We divided the subsequent 39 chapters into seven major areas that characterize rodent societies. These include sexual behavior (chaps. 3–7), life history (chaps. 8–15), behavioral development (chaps. 16–19), social behavior (chaps. 20–25), antipredator behavior (chaps. 26–28), comparative socioecology (chaps. 29–37), and conservation and disease (chaps. 38–41). Although each section focuses on a particular aspect of rodent social biology, several central themes reemerged throughout each chapter and the entire volume, enabling us to identify some universal components of rodent societies.

Universal Components of Rodent Societies

A basic understanding of behavior quickly reveals patterns that recur in various taxa. And in that rodents have been studied extensively in many areas of behavior, ecology, and evolution, authors have been able to synthesize theory with empirical studies to develop conceptual models to explain many aspects of rodent societies. It is these predictable patterns that we attempt to emphasize in this anthology.

Regarding reproductive strategies, we deviated from the conventional approach of chapters on monogamy, polygyny, and promiscuity, in favor of examining sex-specific mating strategies. Thus there are chapters on male mating strategies (Waterman, chap. 3) and female mating strategies (Solomon and Keane, chap. 4), and these focus on individual reproductive behaviors rather than the species- or population-level mating system. Our authors demonstrate how sex-specific mating and reproductive tactics maximize fitness; similar tactics also occur among males and females of other mammalian taxa. Carroll and Potts (chap. 5) use house mice as models to demonstrate the importance of female choice in selecting for genetically compatible mating partners. The ideas, mechanisms, and results are applicable to sexual selection theory in that they provide evidence for the "compatible genes" hypothesis for mate choice. In a phylogenetic analysis of breeding systems in Neotomine and Peromyscine rodents, Kalcounis-Rüppell and Ribble (chap. 6) examine a series of behavioral and ecological parameters to develop a model that predicts mating systems. Koprowski (chap. 7) also uses a comparative approach in describing alternative sexual behaviors in male tree squirrels, which reveals how male tactics are dependent in part on female tactics. All these authors remind readers of the evolutionary arms race between the sexes, and their chapters demonstrate how the behavioral strategy of one sex often is dependent on the behavior of the other sex. Mating systems are thus a consequence of alternative mating tactics used by males and females.

All animals must communicate with each other at some time, especially during intrasexual conflict and for mating and defensive purposes. Important aspects of communication that have been well studied in rodents include chemical factors in urine that accelerate and suppress reproduction (Drickamer, chap. 9), express dominance and territoriality (Roberts, chap. 22), and serve as cues of genetic compatibility in mate choice (Carroll and Potts, chap. 5). Vocal communication, especially alarm calling, has also been well studied in rodents (Blumstein, chap. 27, and Hoogland, chap. 37), as has nepotism (assisting kin) more generally (e.g., Holmes and Mateo, chap. 19).

A major theme that unites this volume is the importance of genetic relatedness in modulating the social behavior of rodents, as showcased, for example, in the chapters on general topics such as social learning (Galef, chap. 18), kin recognition (Holmes and Mateo, chap. 19), parental care (McGuire and Bemis, chap. 20), alarm calling (Blumstein, chap. 27, and Hoogland, chap. 37) as well as in chapters focused on specific taxa, including beavers (Busher, chap. 24), mole-rats (Nevo, chap. 25; Faulkes and Bennett, chap. 36), ground squirrels (Hare and Murie, chap. 29), and marmots (Armitage, chap. 30). Social groups of kin most often result from philopatry of daughters (Solomon and Keane, chap. 4), which in turn creates advantages for males that disperse to seek unrelated females for mating and to avoid inbreeding or reproduction competition in the natal site (Nunes, chap. 13). Coloniality, sociality, and especially eusociality depend on kinship (Lacey and Sherman, chap. 21) and occur in blind mole-rats (Nevo, chap. 25), North American ground squirrels (Hare and Murie, chap. 29), desert rodents (Randall, chap. 31), some groups of South American rodents (Lacey and Ebensperger, chap. 34), and African mole-rats (Faulkes and Bennett, chap. 36).

Experimentation and Hypothesis Testing

Science does not advance by gathering data that support hypotheses, but by conducting "strong inference" tests (Platt 1964) of the alternatives. Such tests involve developing competing, critical predictions from each hypothesis, and then conducting experiments or observations that attempt to falsify these predictions. The hypothesis that is left standing at the end of this process becomes the front-runner as the most likely explanation for the phenomenon at issue.

To meet our primary goals, we asked authors to emphasize their experimental and strong inferential approaches whenever possible. For example, the role of hormones and neural control of behavior (Curtis and colleagues, chap. 16) have been elucidated experimentally. The expression of hormones such as vasopressin and oxytocin are positively correlated with monogamy, which in turn affects mating behavior, paternal care, and juvenile development. Rigorous testing of alternative hypotheses is used by Carroll and Potts (chap. 5) to examine the genetic consequences of mate choice and sexual selection in house mice via the major histocompatibility complex (MHC), by Drickamer (chap. 9) to determine the role of chemical signals in accelerating or suppressing reproduction, and by Sikes (chap. 11) to demonstrate how field and laboratory experimentation was used to test hypotheses for facultative sex ratio adjustment. Sikes concludes that the theory for adaptive sex ratio adjustment in rodents has thus far exceeded the data, and that no mechanism for varying the sex ratio adaptively has been discovered. Experimentation is also used to determine what factors contribute to stress and how stress affects fitness and demography (Boonstra et al., chap. 12) and how rodents learn what foods to eat and avoid (Galef, chap. 18). McGuire and Bemis (chap. 20) use cross-fostering experiments to discern the role of paternal care in polygynous and monogamous species of voles. An experimental approach also is used to address mechanisms of reproductive suppression and self regulation (Krebs et al. chap. 15), scent marking (Roberts, chap. 22), antipredator processes (Owings and Coss, chap. 26; Ylönen and Brown, chap. 28), the genetic basis for aggression in mole-rats (Nevo, chap. 25), alarm calls (Blumstein, chap. 27; Hoogland, chap. 37), and behavior conducive to commensal living in rats and mice (Berdoy and Drickamer, chap. 32).

Proximate Mechanisms and Ultimate Causation

Why do animals do what they do? There are multiple, complementary answers to such a question, which lie at different levels of analysis (Tinbergen 1963; Sherman 1988). One type of answer addresses the mechanistic (hormonal, neuronal) causes of a behavioral phenomenon, while another addresses the ontogeny of the behavior in each individual's lifetime, still a third examines the effects of the behavior on the individual's fitness, and the fourth type of answer addresses the evolutionary history of the behavior. The first two levels are known as *proximate* and the latter two as *ultimate* explanations. Complete understanding of any behavioral phenomenon requires answers at all four levels and their integration.

Synthesis of mechanism and function is well illustrated in chapters 3 (Waterman) and 4 (Solomon and Keane), which describe how the behavior of one sex becomes a stimulus for the opposite sex to utilize a particular reproductive tactic, such as wandering versus mate guarding, or territoriality versus searching. Alternative reproductive tactics are typically functions of both social and ecological variables, as is also illustrated in chapters 6 (neotomine and peromyscine rodents, Kalcounis-Rüppell and Ribble), 7 (tree squirrels, Koprowski), 30 (marmots, Armitage), and 34 (South American hystrigonath rodents, Lacey and Ebensberger). In chapter 8, Dobson and Oli explain how reproductive effort is tied to altricial and precocial development of offspring. The presence of an opposite-sex parent is the stimulus for sex-biased natal dispersal in mice, which ultimately prevents inbreeding and intrasexual competition with male relatives (Nunes, chap. 13). Male affiliative behavior leading to monogamy can be explained proximately by neuronal and hormonal mechanisms and ultimately by the value of paternal care in contributing to survival of young (Curtis et al. chap. 16). Although density of voles and mice, or more likely number of adult females, seems to be a proximate mechanism to suppress reproduction in young females, the ultimate benefit apparently is to conserve reproductive effort by aborting embryos that would likely be killed by infanticide at birth (Krebs et al. chap. 15). Conversely, Ebensperger and Blumstein (chap. 23) discuss how the act of ejaculation inhibits infanticide by males, which in turn might be a selective factor for multi-male mating by females in some species, to confuse paternity and deter infanticide. Finally, Mateo (chap. 17) uses experimental and observational data from ground squirrels to demonstrate how learning, maturation, and ecological selective pressures intertwine to shape the ontogeny of adaptive behavior. Although we gave each author guidelines and a general subject area to cover, we found that many authors were able to go beyond a summation of data for a particular topic and synthesize two or more disciplines to create new theory with new predictions.

A Synthesis of Disciplines and New Paradigms

Many chapters included in *Rodent Societies* cover topics that occur in most behavior texts (e.g., Alcock 2005; Dugatkin 2003). However, we encouraged authors to go beyond summarizing current knowledge to develop new paradigms by combining ideas from various disciplines. For example, Kalcounis-Rüppell and Ribble (chap. 6) provide a phylogenetic analysis of life history and physiological correlates of male and female mating strategies in which they conclude that litter mass and distributional range of a species are good predictors of the occurrence of paternal care. Dobson and Oli (chap. 8) present a model of reproductive effort involving "fast and slow" life histories that is ultimately based on demography. The authors attempted to understand if developmental stage at birth might become a constraint on the ontogeny of reproductive effort and other life history variables. Albrecht Schulte-Hostedde (chap. 10) applies sexual selection and parental investment theory to investigate how ecological and energetic constraints might limit sexual dimorphism, independent of the mating system. Schulte-Hostedde presents two alternative hypotheses to illustrate how the energetics of reproduction, sexual selection, and costs of extreme climatic conditions interact to limit sexual dimorphism in chipmunks.

The idea that stress plays an important role in reproduction and population demography is well known. However, Boonstra et al. (chap. 12) provide a new synthesis of how stress affects survival and mating success. In chapter 14 Dobson proposes that social genetic structuring can occur in breeding groups in which males are polygynous and females are philopatric. He defines the conditions under which social behavior, primarily social groupings of females and behavioral influences on mating systems, influence gene dynamics. Krebs, Lambin, and Wolff (chap. 15) join forces to synthesize behavioral, ecological, and evolutionary theory to address the potential for self-regulation in rodent population dynamics. Likewise, Curtis et al. (chap.16) combine research on hormones, brain function, and social behavior as they relate to affiliative and paternal care leading to monogamy, to develop a new paradigm in neural behavior. Another synthesis of disciplines is Owings and Coss's (chap. 26) exploration of the processes that generate connections between rodent anti-predator behavior and social systems. They note that although the social and antipredatory domains are partially independent sources of natural selection, both select for phenotypes through modifications in developmental systems. Combining paradigms

from ontogeny of development, risk avoidance, kin selection, cognition, and sociality, the authors provide a novel synthesis.

Further examples of new paradigms are provided by Ylönen and Brown (chap. 28) in which they apply optimal foraging strategies and life history parameters to avoidance of predation risk. Similarly, Armitage (chap. 30) synthesizes physiology and social behavior, proposing that energetic constraints and costs of hibernation select for sociality in marmots. Conceptual and predictive models to explain the probable adaptive significance of the continuum of social systems from asociality through eusociality are provided for subterranean rodents (Lacey and Sherman, chap. 21), blind mole-rats (Nevo, chap. 25), ground squirrels (Hare and Murie, chap. 29), kangaroo rats (Randall, chap. 31), and African mole-rats (Faulkes and Bennett, chap. 36).

The scope of many of these chapters can be attributed to the wide range of species studied within a subgroup of rodents. It is this relatively large database of similar studies on phylogenetically and ecologically similar species that allowed authors to use the comparative approach to develop conceptual models to predict how extrinsic (ecological) and intrinsic (genetic) variables interact to affect the formation and structure of rodent societies.

Comparative Behavioral Ecology

The comparative method is a useful approach to develop and test evolutionary hypotheses (e.g., Barash 1989; Harvey and Pagel 1991; Komers and Brotherton 1997). This approach utilizes comparable data, often from related or similar species collected under similar conditions (to control for some environmental or social variable) and character mapping with appropriate statistical analyses (to control for phylogeny). Sufficient comparative data are available for particular groups of rodents and associated theory is developed well enough to support such studies. Therefore, we included a section on Comparative Socioecology.

Rodents are excellent models for using the comparative method to develop paradigms and test hypotheses in socioecology, as illustrated here in chapters on kinship and sociality in ground squirrels (Hare and Murie, chap. 29), the role of hibernation in the evolution of sociality in marmots (Armitage, chap. 30), environmental constraints on sociality in desert rodents (Randall, chap. 31), adaptations of commensal rats and mice (Berdoy and Drickamer, chap. 32), group living in social South American species (Macdonald et al., chap. 33; Lacey and Ebensperger, chap. 34), the uniqueness of rock-dwelling rodents (Nutt, chap. 35), social and ecological diversity of African mole-rats (Faulkes and Bennett, chap. 36), and comparison of alarm calling, multiple mating, and infanticide in three species of prairie dogs (Hoogland, chap. 37). In many of these chapters, authors have produced conceptual models that predict how environmental factors interact with social pressures to form rodent societies. It is these conceptual models that we anticipate will in many ways shape the directions of future research in animal socioecology.

Issues in Conservation and Management

Rodents typically are regarded as pests. While some populations of a few species do cause economic damage, we hope this volume, along with Feldhammer et al.'s recent (2003) anthology, will help change the general perception that rodents are vermin unworthy of interest—much less preservation. Indeed, rodents are fascinating creatures, and they have added substantially to our understanding of ecological, evolutionary, and behavioral biology.

The final section of *Rodent Societies* provides several chapters that demonstrate problems and emerging issues in rodent conservation and management. Lidicker (chap. 38) starts off with a theoretical and empirical analysis of how ecology and behavior can be applied to conservation issues to preserve threatened and endangered populations. Van Horne (chap. 39) and Hoogland (chap. 40) provide compelling evidence for the demise of ground squirrels and prairie dogs, respectively, throughout western North America. Van Horne describes factors that govern the distribution and abundance of ground squirrels. She takes a bottom-up view of the role of ground squirrels in ecosystems and describes how range management and habitat change affect population numbers and distribution. Van Horne concludes with a summary of threats to ground squirrels and actions that can be taken to ameliorate these threats. Hoogland similarly points out that whereas historically several billion prairie dogs (in 5 species of the genus *Cynomys*) once lived

in the western grasslands of North America, today, they occupy less than 1% of their former range; four of the five species are on, or under consideration for, the Federal list of threatened and endangered species. Hoogland proposes a strategy for developing sanctuaries to conserve and manage the remaining populations. Lastly, Ostfeld and Mills (chap. 41) provide an unnerving treatise of the role of social behavior and demography in the transmission of rodent-borne diseases. Although rodent-borne diseases are well controlled in North America, they do pose a threat in many parts of the world. Knowledge of social behavior and factors affecting movements and demography are essential to develop models to predict transmission rates of various diseases and the potential for outbreaks.

This section reminds us that although most research on rodents has been basic, it has increased understanding and raised concern about the declines and threats of extinction of many species of rodents. Application of rodent social biology to conservation and management is in its infancy, but this volume illustrates that knowledge of behavior may prove useful in protecting rodents and the integrity of rodent-dominated ecosystems.

We Do Not Have All of the Answers

The 42 chapters in *Rodent Societies* cover many species and varied aspects of social and reproductive behavior. Authors draw some conclusions based on the progress made in research over the past five decades. However, many intriguing and important questions remain, some of which are summarized in chapter 42. We hope that this volume will draw greater attention to the fascinating social lives of rodents and that it will help stimulate further inquiries into the extrinsic (environmental) and intrinsic (genetic) factors that mold their societies.

Chapter 2 Rodent Evolution, Phylogenetics, and Biogeography

Rodney L. Honeycutt, Laurence J. Frabotta, and Diane L. Rowe

T HERE ARE four main components to this chapter. First, we provide a detailed survey of rodent diversity at the familial level, and in this survey we summarize information on various attributes (e.g., morphology, ecology, behavior) that both contribute to this diversity and complicate the diagnosis of relationships between families and genera. Second, we provide an introduction to the paleontological and biogeographical history of rodents, and elaborate on several hypotheses regarding rodent ancestry and distribution. Third, we provide a detailed account of the phylogenetic relationships of rodent families, and in some cases genera. This account discusses phylogenetic hypotheses based on more traditional studies of morphology and paleontology, as well as recent studies employing molecular phylogenetics. Finally, we provide some examples of how a well-supported phylogeny for particular groups of rodents can be used to test hypotheses related to the evolution of complex social behavior and the evolution of life history traits.

Rodent Diversity

In terms of abundance, diversity, and distribution, the order Rodentia represents the most successful group in the class Mammalia. There are 1,135 genera and 4,629 recognized species of mammals, of which 443 genera (39%) and 2,015 species (43.5%) are rodents (Wilson and Reeder 1993). Of the 29 extant families of rodents, approximately 89% of rodent species diversity can be partitioned into five families (table 2.1 and fig. 2.1): (1) Muridae (rats and mice), (2) Sci-

uridae (squirrels), (3) Echimyidae (spiny rats), (4) Heteromyidae (pocket mice and kangaroo rats) and (5) Dipodidae (jerboas and jumping mice). The most speciose and widely distributed family is the Muridae, representing 66% of rodent taxa. Murid rodents are a significant component of mammalian diversity on most continents, and ancestral stocks of murids have experienced adaptive radiations on several continents subsequent to invasion via either land bridges or waif dispersal.

Biogeographic distribution

Except for Antarctica, representatives of the order occur on all continents and many oceanic islands. Most major biogeographic regions are characterized by unique rodent fauna as well as rodent families shared among regions. Five rodent families are endemic to sub-Saharan Africa, including Bathyergidae, Thryonomyidae, Petromuridae, Anomaluridae, and Pedetidae (table 2.2). These families represent older faunal elements dating to the Early Oligocene to Lower Miocene (Lavocat 1978). More recent immigrants to Africa include the Sciuridae (Lower Miocene), Muridae (Upper Miocene to Plio-Pleistocene), Myoxidae (Upper Miocene to Pliocene), and Hystricidae (Plio-Pleistocene). Northern Africa contains two additional families: Dipodidae (Oligocene) and currently endemic Ctenodactylidae (Miocene). Although Madagascar has been isolated from Africa since the Late Cretaceous, this island contains the endemic murid subfamily Nesomyinae, represented by nine genera and 21 species (Jansa et al. 1999). Apparently, nesomyines colonized Madagascar via over-water dispersal, but the

Table 2.1 Species and genera of recent rodents

Family	Genus	Species
Muridae	281	1,326
Sciuridae	50	273
Echimyidae	20	78
Dipodidae	15	51
Myoxidae	8	26
Capromyidae	8	20
Heteromyidae	6	59
Octodontidae	6	9
Geomyidae	5	35
Caviidae	5	14
Bathyergidae	5	12
Erethizontidae	4	12
Ctenodactylidae	4	5
Heptaxodontidae[a]	4	5
Hystricidae	3	11
Anomaluridae	3	7
Chinchillidae	3	6
Dasyproctidae	2	13
Ctenomyidae	1	38
Abrocomidae	1	3
Castoridae	1	2
Thryonomyidae	1	2
Agoutidae	1	2
Aplodontidae	1	1
Pedetidae	1	1
Petromuridae	1	1
Dinomyidae	1	1
Hydrochaeridae	1	1
Myocastoridae	1	1

[a]Recently extinct.

mainland source and the number of invasions are still debated (Simpson 1952; Dubois et al. 1996; Jansa et al. 1999). South American rodents can be subdivided into early and late colonizers. Caviomorph rodents (13 families) represent the oldest rodent radiation in South America, with the earliest fossils dating to the Late Eocene or Early Oligocene (Wyss et al. 1993). The murid subfamily Sigmodontinae represents the most diverse rodent group in South America. Timing of the sigmodontine invasion is a source of controversy, with some advocating a recent origin subsequent to the formation of the Panamanian land bridge 2 to 3 million years ago (mya; Simpson 1950; Smith and Patton 1999) and others suggesting an ancient invasion via waif dispersal in the Early Miocene (Hershkovitz 1966; Reig 1986). There are 59 species of rodents in Australia, making up 25% of the terrestrial mammalian fauna. Fifty-seven percent of the rodent species are endemic to Australia and are members of the family Muridae. This island continent experienced two invasions, one occurring approximately 10 to 15 mya (subfamily Hydromyinae; Watts and Aslin 1981) and the other 2 mya (subfamily Murinae; Watts and Kemper 1989). Both invasions were the result of waif dispersal from the Oriental biogeographic region via Wallacea. The Nearctic, Palearctic, and Oriental biogeographic regions show less endemism in terms of both rodent subfamilies and families. Like most continents, the rodent faunal elements in these regions are dominated by members of the families Sciuridae and Muridae. Two families, Geomyidae and Heteromyidae, are restricted to the New World, with their center of origin

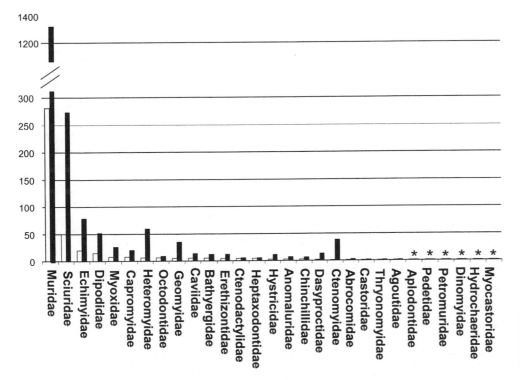

Figure 2.1 Number of genera (open bar) and species (black bar) per family of rodent. Information was compiled from Wilson and Reeder (1993). Asterisk (*) denotes monotypic families.

Table 2.2 Two classification schemes of rodents (only extant families are shown)

Simpson, 1945	Wilson and Reeder, 1993	
Categories	Categories	Distribution
Sciuromorpha	**Sciurognathi**	
Aplodontoidea	Aplodontidae	SW British Columbia (Canada) to Central California
Aplodontidae (Mountain Beaver)	Sciuridae	Worldwide except Australia and southern South America
Sciuroidea	Geomyidae	North America and Middle America
Sciuridae (Squirrels)	Heteromyidae	North America and Middle America
Geomyoidea	Castoridae	Nearctic and Palearctic (NW and NC Eurasia)
Geomyidae (Pocket Gophers)	Muridae	Worldwide except Antarctica and some oceanic islands
Heteromyidae (Pocket Mice and Kangaroo Rats)	Dipodidae	North America, Northern Africa, Eurasia, Central Asia
Castoridea	Myoxidae	Africa, Europe, and Central Asia
Castoridae (Beaver)	Anomaluridae	Central African forests
Anomaluroidea *incertae sedis*	Pedetidae	Southern African arid regions
Anomaluridae (Scaly-tailed Squirrels)	Ctenodactylidae	Northern Africa
Pedetidae (Spring Haas)	**Hystricognathi**	
Myomorpha	Erethizontidae	Nearctic, Middle America, parts of S America
Muroidea	Caviidae	S America
Cricetidae (Hamsters, New World Rats and Mice)	Hydrochaeridae	Panama and S American, tropics
Spalacidae (Mole-Rats)	Dinomyidae	Northern S America, Andean slopes
Rhizomyidae (Cane Rats)	Dasyproctidae	Mexico, Middle America, S America, tropics
Muridae (Old World Rats and Mice)	Agoutidae	Middle and South America
Gliroidea	Chinchillidae	Southern S America in C and S Andes
Gliridae (Dormice)	Capromyidae	West Indies
Platacanthomyidae (Spiny Dormice)	Octodontidae	S America in Peru, Bolivia, Chile, and Argentina
Dipodoidea	Ctenomyidae	Southern S America
Zapodidae (Jumping Mice)	Abrocomidae	S America Andes in S Peru to N Chile
Dipoididae (Jerboas)	Echimyidae	Middle America and S America, tropics
Hystricomorpha	Hystricidae	Africa and Asia
Hystricoidea	Thryonomyidae	Sub-Saharan Africa
Hystricidae (Old World Porcupines)	Petromyidae	Sub-Saharan Africa
Erethizontoidea	Bathyergidae	Sub-Saharan Africa
Erethizontidae (New World Porcupines)		
Cavioidea		
Caviidae (Guinea Pigs & Relatives)		
Hydrochaeridae (Capybara)		
Dinomyidae (Pacarana)		
Dasyproctidae (Agouti, Paca)		
Chinchilloidea (Chinchilla, Vizcacha)		
Chinchillidae (Chinchilla, Vizcacha)		
Octodontoidea		
Capromyidae (Hutia, Coypu)		
Octodontidae (Degu,		
Ctenomyidae (Tuco Tuco)		
Abrocomidae (Chinchilla Rat)		
Echimyidae (Spiny Rats		
Thryonomyidae (African Cane Rat)		
Petromyidae (African Rock Rat)		
Bathyergoidea *incertae sedis*		
Bathyergidae (African Mole-Rats)		
Ctenodactyloidea *incertae sedis*		
Ctenodactylidae (Gundis)		

and greatest diversity in the Nearctic (Wilson and Ruff 1999). The mountain beaver, family Aplodontidae, occurs exclusively in the Pacific Northwest of North America. The hystricognath rodents are represented by a single species, *Erethizon dorsatum* (New World porcupine); jumping mice (family Dipodidae) and beaver (family Castoridae) are shared between the Nearctic and Palearctic biogeographic regions.

Convergence

Changes in climate, exchanges between biogeographic regions, and rapid faunal turnover on many continents have resulted in local adaptive radiations characterized by ecomorphological specializations. As a result, a frequent theme associated with rodent radiations is convergent evolution in terms of life-history strategies, behavior, and anatomy and physiology, and with evidence of convergence emerging when one compares unrelated rodent lineages occupying different biogeographic regions. For instance, specializations for a subterranean lifestyle have evolved multiple times in rodents, as seen by the degree of convergent morphologies and physiologies displayed by the families Bathyergidae (Africa), Ctenomyidae (South America), Muridae (Palearctic and Oriental), Octodontidae (South America), and Geomyidae (North America). These specializations include a fusiform body, reduction in limb length, dorsally flattened skulls, and behavior associated with burrow construction (Stein 2000). Subterranean rodents also show physiological convergence, characterized by specializations for living in a high CO_2 environment (Buffenstein 2000). Members of several rodent families (Heteromyidae, Dipodidae, and Pedetidae) show similar adaptations for life in arid environments, especially as they relate to structure of the kidney (Schmidt-Nielsen 1964; Al-kahtani et al. 2004) and changes in the postcranial skeleton associated with saltatorial locomotion. Convergence for group living in response to similar environmental constraints is a common explanation for the evolution of complex social systems, and several behavioral studies of rodents have used comparative methods to investigate ecological features shared by unrelated species of rodents that form social groups and have similar mating systems (Lacher 1981; Faulkes et al. 1997; Blumstein and Armitage 1998a; Armitage 1999a; Ebensperger and Cofré 2000; Lacey and Wieczorek 2003). Finally, convergence and parallelism of the zygomasseteric system and dentition of rodents have exacerbated efforts to identify monophyletic groups of rodent families, making the classification of rodents difficult (Wood 1955).

Evolutionary History

Rodents first appear in the fossil record during the Paleocene approximately 55 to 60 mya, with one of the oldest families being the Paramyidae (Vianey-Liaud 1985; Hartenberger 1998). Most modern families of rodents underwent an adaptive radiation in the Paleocene/Eocene, and were well established by the Late Eocene to Early Oligocene (Vianey-Liaud 1985; Jaeger 1988). Therefore, the rodent tree of life has received little pruning in terms of its major lineages.

Issues related to divergence times

Rodents are well represented in the fossil record, and paleontologists have established divergence times for several rodent lineages. Nevertheless, recent divergence times derived from "molecular clocks" often conflict with presumably well-established paleontological dates. For instance, earliest fossils place the divergence for the murid genera *Mus/Rattus* between 12 and 14 mya (Jacobs and Pilbeam 1980; Jacobs et al. 1989). Using a molecular clock based on albumin immunology, Sarich (1985) provided an estimate of the *Mus/Rattus* divergence at 22 to 24 mya, and a date of 41 mya was obtained by Kumar and Hedges (1998) from their comparisons of eutherian mammals and estimates of amino acid replacement differences across 658 nuclear gene loci. When one considers new molecular dates for orders of mammals, the discrepancy between fossils and molecules becomes more controversial. The fossil record suggests a divergence date no older than 65 myr for most orders of placental mammals (Foote et al. 1999; Archibald 2003), whereas molecules place the origin of major placental lineages in the Cretaceous at around 100 mya (Springer 1997; Kumar and Hedges 1998; Eizirik et al. 2001). Therefore, the molecular estimates suggest gaps in the fossil record that are 50 to 60 myr earlier than the first appearance of fossil evidence for particular orders (Foote et al. 1999; Archibald 2003). How do these new molecular clocks impact estimates of rodent divergence? Based on these molecular dates, the origin of Rodentia would be 100 to 110 myr or older (Springer 1997; Cooper and Fortey 1998; Kumar and Hedges 1998), extending divergence time estimates within Rodentia beyond any known fossils. According to the estimate of Kumar and Hedges (1998), this results in a divergence time of 110 myr (Middle Cretaceous) for the separation of the two rodent suborders, Hystricognathi and Sciurognathi, and this date is 50 myr prior to the first fossil rodent or rodent ancestor.

There is at least one issue that tends to complicate recent dates derived from a molecular clock of eutherians. Rates of molecular evolution within rodents and among eutherian orders are not homogeneous, and the calibration of a molecular clock based on a single rate tends to bias divergence times, especially for groups like rodents that evolve considerably faster than other eutherian orders (Bromham et al. 2000). Therefore, a more accurate estimate is one that corrects for apparent rate heterogeneity across lineages. Two recent studies (Adkins et al. 2001; Huchon et al. 2002) attempted to correct for rate heterogeneity and derive dates based on multiple calibration points, and obtained divergence times for rodents more in line with the fossil record. For instance, Huchon et al. (2002) provided an estimate of 55.8 mya for the origin of most rodent lineages, and this estimate is congruent with fossil evidence suggesting a Paleocene/Eocene radiation for rodents (Huchon et al. 2002). Likewise, the *Mus/Rattus* divergence provided by Adkins et al. (2001) was 23 mya, a date older than the fossil record but congruent with earlier molecular dates (e.g., Sarich 1985).

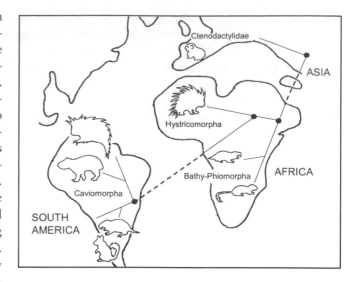

Figure 2.2 Biogeographic scenario for the evolution and dispersal of *Hystricognathi* rodents based on Huchon and Douzery (2001) and Rowe (2002). The origin of *Caviomorpha* from an African ancestor is well supported, as well as the sister-group relationship between the *Ctenodactylidae* and hystricognaths. The placement of *Hystricomorpha* is unclear.

Biogeographic Hypotheses and Phylogenetics

A combination of phylogenetics and molecular dating can be used to examine broad-scale biogeographic events coinciding with the diversification and distribution of rodents. There are several intriguing controversies surrounding rodent biogeography and the invasion of island continents that lend themselves to renewed investigations using a phylogenetic approach.

Origin of South American rodents

Existing rodent fauna of South America represent two separate invasions: an ancient African invasion of hystricognath ancestors and a more recent Nearctic invasion of murid ancestors of the subfamily Sigmodontinae. The fossil records and levels of diversity seen for both South American groups have stimulated considerable debate regarding the timing of these invasions and the dispersal pathways utilized.

The 13 families of South American hystricognath rodents are collectively placed in the Caviomorpha (Wood 1965), and their closest relatives, sometimes referred to as Phiomorpha (Lavocat 1973), reside primarily in Africa (families Bathyergidae, Petromuridae, and Thryonomyidae), with the Old World porcupines (Hystricidae) occurring in both Africa and Asia. The earliest caviomorph rodent fossils in South America appear approximately 31–37 mya (Wyss et al. 1993, 1994), and the oldest fossil phiomorphs date to 34–37 mya in Africa (Hartenberger 1998). By the Deseadan (21–27 mya) representatives of all major superfamilies and seven distinct families were present in South America

(Patterson and Wood 1982). These dates are well after the Late Cretaceous separation of South America from Africa, suggesting a trans-Atlantic, waif dispersal across an ocean expanse of up to 1700 km (fig. 2.2). As with other cases of island invasions via waif dispersal, considerable debate has focused on whether South American caviomorph rodents are the result of a single African invasion or are derived independently from multiple invasions. Earlier studies employing both immunological (Sarich and Cronin 1980) and morphological (Bugge 1985; Woods and Hermanson 1985) data suggested that caviomorphs might be the result of two to three independent invasions. To the contrary, more recent molecular data and morphological comparisons strongly support monophyly of caviomorphs and a single colonization event of South America (Luckett 1985; Nedbal et al. 1994; Adkins et al. 2001; Huchon and Douzery 2001). Recent estimates based on molecular clocks place the origin of caviomorph rodents in South America at 40–60 mya (Huchon and Douzery 2001; Rowe 2002), supporting colonization via waif dispersal during the late Paleocene to Middle Eocene.

The Sigmodontinae represents a New World subfamily of the Muridae that some taxonomic treatments divide into North American and South American groups (Hooper and Musser 1964; Carleton and Musser 1984; Musser and Carleton 1993). The earliest sigmodontines in South America appear between 2 and 3 mya (Simpson 1950), and presumably represent an adaptive radiation of a North American stock that entered South America subsequent to the formation of the Panamanian land bridge. The major controversy surrounding South American sigmodontines relates to their diversity and timing of their invasion. For instance, both Hershkovitz (1966) and Reig (1986) suggested an old invasion (early Miocene) by a sweepstakes route prior to the Panamanian land bridge, whereas Simpson (1950) and Flynn et al. (1985) argued for a more recent invasion (Plio-Pleistocene) across the land bridge, thus suggesting rapid speciation subsequent to the invasion. According to a molecular clock based on sequences of mitochondrial genes (Engel et al. 1998), dates for the origin of South American sigmodontines are intermediate between the late and early hypotheses, suggesting an origin of ancestral sigmodontines in North America (between 9 and 14.8 mya) followed by a somewhat newer adaptive radiation in South America (between 5 and 8.3 mya). In addition, the molecular phylogeny suggests that South American sigmodontines are a monophyletic group derived from a single ancestor.

Origin of the Malagasy Nesomyinae

Although Madagascar and Africa were part of Gondwana, they separated approximately 150–165 mya, with Madagascar occupying its current position since the Cretaceous (Rabinowitz et al. 1983). Large portions of the endemic mammalian fauna in Madagascar represent invasions by mainland sources derived from Africa and/or Asia. Although all rodents in Madagascar are endemic and placed in the subfamily Nesomyinae, two major controversies concerning their origin and evolution persist. First, monophyly of the Nesomyinae is disputed. Carleton and Musser (1984) placed all genera and species from Madagascar in the same subfamily but indicated that the primary basis for this classification was the unique distribution of nesomyines. They also noted the high level of morphological differences separating Malagasy genera. At least two earlier classifications also recognized Malagasy rodents as a monophyletic group (Simpson 1945; Lavocat 1978). Second, the traditional view of nesomyine biogeography advocates a single invasion from African cricetodontid rodents followed by an adaptive radiation (Simpson 1952; Lavocat and Parent 1985). A contrasting viewpoint is that nesomyine rodents are the product of multiple invasions, possibly from different colonizing sources (Jansa et al. 1999). Recent studies of Malagasy rodents have attempted to address these controversial issues with the use of molecular phylogenetics. Two studies examined a small number of nesomyines in comparison to other murid subfamilies and concluded that the subfamily was monophyletic and related to either the African subfamilies Mystromyinae or Cricetomyinae (Dubois et al. 1996; Michaux and Catzeflis 2000). Jansa et al. (1999) examined 15 of the 21 species of nesomyines, and produced a phylogenetic tree based on nucleotide sequences from the cyto-

chrome *b* gene. Their resultant phylogeny revealed a paraphyletic Nesomyinae, with some Malagasy taxa grouping closer to the subfamilies Rhizomyinae and Dendromurinae. Furthermore, an overlay of the cladogram on a map of Indo-Africa suggested that rather than Africa, Malagasy rodents are the result of a single invasion from Asia followed by an adaptive radiation in Madagascar and a secondary invasion of Africa by Malagasy lineages.

Biogeography of squirrels

Based on the fossil record, the family Sciuridae originated in North America in the Late Eocene (35 mya) and had a Holarctic distribution by the Early Oligocene (Thorington et al. 1997); ancestral stocks entered Africa, parts of southeast Asia, and South America more recently. Today, the family is the second most diverse group of rodents (table 2.1 and fig. 2.1) and is distributed worldwide, except for Madagascar, Australia, and the southern regions of South America (Hoffmann et al. 1993). As stated by Mercer and Roth (2003, p. 1570), "The geographic coherence of modern clades of squirrels suggests that dispersal over large expanses of water or desert is uncommon and that squirrel distributions may therefore be important indicators of geological and environmental history." Therefore, based on their distribution and fossil record, the family Sciuridae provides an ideal model for reconstructing patterns of continental interchange in response to vicariant events initiated during changes in climate and the formation of land bridges. Two recent molecular phylogenetic studies (Mercer and Roth 2003; Steppan et al. 2004) have established relationships among the major lineages of squirrels occurring worldwide (fig. 2.3), and estimates derived from a molecular clock show congruence between the overall phylogenetic pattern and the establishment of overland dispersal pathways linking various continents (Mercer and Roth 2003). These patterns include: (1) two separate invasions of sciurids from Asia to Africa, establishing monophyletic groups of both ground squirrels and tree squirrels, and occurring during a period when a land bridge linked these two continents; (2) single invasion of sciurids into South America subsequent to the formation of the Panamanian land bridge; (3) exchange of both flying squirrels and tree squirrels between North America and Eurasia during periods when Beringea provided a land bridge for dispersal; (4) colonization of islands during periods of lowered sea levels and connections with the mainland, such as seen for a monophyletic group of tree squirrels occupying Sundra Shelf islands in Southeast Asia. The only exception to this general pattern is the basal position of two tree squirrel lineages that display the extreme ranges of body size, the giant squirrels (*Ratufa*) from Borneo and the pygmy tree squirrels (*Sci-*

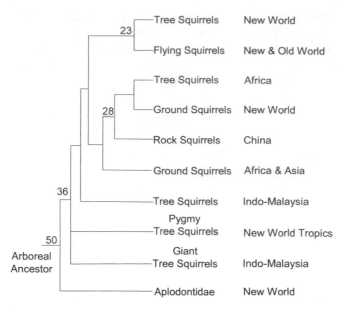

Tree Squirrels	New World
Flying Squirrels	New & Old World
Tree Squirrels	Africa
Ground Squirrels	New World
Rock Squirrels	China
Ground Squirrels	Africa & Asia
Tree Squirrels	Indo-Malaysia
Pygmy Tree Squirrels	New World Tropics
Giant Tree Squirrels	Indo-Malaysia
Aplodontidae	New World

Figure 2.3 Phylogeny of the major groups of the family *Sciuridae* based on Mercer and Roth (2003) and Steppan et al. (2004). Numbers at specific nodes represent divergence times in millions of years.

urillus) of the New World tropics (fig. 2.3). A sound explanation for the phylogenetic positions of these two lineages requires a more complex biogeographic scenario.

Phylogenetic Position of Rodents among Eutherians

With the advent of improved molecular techniques and the development of analytical methods that accommodate molecular data and increased taxon sampling, nagging questions regarding the origin and evolution of mammals, especially eutherians, have received increased attention from molecular systematists. In particular, these studies have focused primarily on ordinal level relationships (Honeycutt and Adkins 1993; Graur et al. 1996; Douzery and Huchon 2004; Waddell, Cao, et al. 1999; Stanhope et al. 1998; Madsen et al. 2001; Murphy et al. 2001a, b; Scally et al. 2001; Huchon et al. 2002; Waddell and Shelley 2003). The phylogenetic position of rodents among the eutherian radiation has received considerable attention, especially with respect to two major issues, the monophyly of and sister-group to Rodentia.

Defining the order Rodentia

The order Rodentia represents an assemblage of species characterized by diversity in morphology, physiology, ecology, and many other biological attributes. Nevertheless, all rodents share derived characteristics associated with the

dentition, skull, soft anatomy, postcranial skeleton, and jaw (Luckett and Hartenberger 1993; Hartenberger 1998). In rodents, several features of the dentition and jaw mechanism accommodate a dual jaw action that enhances both gnawing and chewing over the cheekteeth. These include: (1) a pair of continually growing incisors with enamel restricted to the labial surface and dentine on the lingual surface, thus allowing for continual sharpening (Jacobs 1984); (2) the loss of canines and a diastema separating the incisors from the cheekteeth; (3) a zygomasseteric system characterized by specializations of the masseter muscles and infraorbital foramen (Marivaux et al. 2004).

Question of rodent monophyly

In his classification of mammals, Gregory (1910; p. 323) noted that members of the order Rodentia had dentition "so striking and representative of the order" that recognition of "the modern conception of the group was reached much more rapidly than was the case in other orders." Luckett and Hartenberger (1985; p. 690) confirmed Gregory's (1910) notion of Rodentia by stating, "Virtually every aspect of comparative biology examined in this volume, from paleontological to molecular, corroborates monophyly of the order Rodentia, in relation to other Eutheria." Given these statements pertaining to rodent monophyly and the features diagnostic of the order, it is rather surprising that some molecular phylogenetic studies failed to find support for rodent monophyly (Graur et al. 1991, 1992; Li et al. 1992; D'Erchia et al. 1996; Reyes et al. 1998; Reyes et al. 2000). In all of these studies, rodents appeared either paraphyletic or polyphyletic with two unrelated groups: a clade containing hystricognath rodents, glirids, and sciurids, and another clade containing the family Muridae. The molecular markers supporting rodent polyphyly or paraphyly are inconsistent in the placement of the two rodent clades. Amino acid sequences from multiple nuclear genes and the inclusion of small numbers of taxa (Graur et al. 1991, 1992; Li et al. 1992) place guinea pigs (hystricognaths) at the base of the eutherian tree, followed by a separate lineage represented by the family Muridae and other orders of eutherian mammals. Based on the analysis of amino acid sequences of genes derived from whole mitochondrial genomes (D'Erchia et al. 1996; Arnason et al. 1997; Reyes et al. 1998; Reyes et al. 2000; Arnason et al. 2002), the Muridae resides at the base of the eutherian radiation, followed by a clade containing hystricognaths, sciurids, and glirids that is sister to other eutherians. The problem with the "mitogenomic" approach is that the placement of rodent clades varies with the taxa included and the phylogenetic method used. For instance, Lin et al.'s (2002) analysis of an increased number of mitochondrial genomes of rodents found support for rodent monophyly, thus implying the utility of complete mitochondrial genomes as a marker with increased taxon sampling. All more recent molecular studies, employing larger numbers of taxa, increased numbers of characters, and multiple nuclear genes, have obtained strong support for rodent monophyly and do not place rodent lineages at the base of the eutherian radiation (Madsen et al. 2001; Murphy et al. 2001a, b; Waddell et al. 2001; Delsuc et al. 2002; Waddell and Shelley 2003). Several problems may explain the failure of some studies to find support for rodent monophyly, including the lack of analytical approaches that adequately correct for rate heterogeneity among rodent lineages (Cao et al. 1994; Sullivan and Swofford 1997), influence on tree rooting by selection of an outgroup (Philippe 1997), and simply sampling artifact in terms of the taxa selected (Philippe 1997; Arnason et al. 2002). In addition, many rodent lineages are old and arose rapidly during the Eocene. Therefore, large numbers of both taxa and sequences may be required to split long branches and obtain support for short and more basal internodes.

Sister-group of Rodentia

The concept of the order Lagomorpha (rabbits, hares, and pikas) as the sister-group to rodents has fallen in and out of favor over the decades since Brandt (1855), and this is reflected in different classifications of eutherian mammals, especially as they relate to the placement of rabbits (Wood 1957). Simpson (1945) recognized the cohort Glires as containing the orders Rodentia and Lagomorpha (rabbits), whereas McKenna's (1975) classification placed Rodentia as *incertae sedis* in the cohort Epitheria (a group of eutherians containing all orders but Xenarthra), and the order Lagomorpha was placed in a grandorder Anagalia that included the order Macroscelidea (elephant shrews). More recent cladistic analyses based on morphology (Luckett 1985; Luckett and Hartenberger 1985; Novacek 1985; Landry 1999) provide support for a monophyletic Glires containing the orders Rodentia and Lagomorpha, and according to Novacek (1992), this relationship is strongly supported. Despite a growing consensus among paleontologists and neontologists for a monophyletic Glires, several molecular studies using either nucleotide sequences of mitochondrial genes or amino acid sequences of nuclear-encoding genes failed to find support for the monophyly of Glires (D'Erichia et al. 1996; Graur et al. 1991; Graur et al. 1996; Misawa and Janke 2003). These molecular studies have been criticized for their lack of taxon sampling (Halanych 1998; Lin et al. 2002; Delsuc et al. 2002), the use of inappropriate tree

building methods that do not consider rate heterogeneity across lineages (Sullivan and Swofford 1997), and an inaccurate rooting of the eutherian phylogeny (Douzery and Huchon 2004). In fact, most recent molecular studies, utilizing primarily nucleotide sequences of nuclear genes and more complex models of sequence evolution, have provided strong support for the monophyly of Glires to include rabbits and rodents (Waddell and Shelley 2003; Madsen et al. 2001; Murphy et al. 2001a, b).

Rodent Phylogenetics and Classification

Traditional classifications of rodents

One of the major dilemmas in rodent taxonomy pertains to the diagnosis of higher taxonomic categories, specifically superfamilies and suborders. Traditional classifications have used variants of two schemes (table 2.1), one based on characteristics of the zygomasseteric system (Brandt 1855) and the other emphasizing the lower jaw (Tullberg 1899). The classification using the zygomasseteric system groups rodent families into three suborders: (1) Sciuromorpha—characterized by a small infraorbital foramen and an expansion of the masseter muscle on the anterior part of the zygomatic arch; (2) Hystricomorpha—with an enlarged infraorbital foramen through which the masseter passes; and (3) Myomorpha—displaying an intermediate zygomasseteric system characterized by a "keyhole-shaped" infraorbital foramen penetrated by the masseter muscle. In contrast, Tullberg (1899) recognized two suborders, Hystricognathi (angle of lower jaw originating lateral to the plane of the incisor) and Sciurognathi (angle of lower jaw arising below the incisor). He further divided each suborder into tribes on the basis of the zygomasseteric system. In general, rodent families display different combinations of the zygomasseteric system and structure of the lower jaw, thus making the placement of some families difficult (e.g., Anomaluridae, Castoridae, Ctenodactylidae, Myoxidae, Pedetidae). In addition, these traits display parallel evolution among unrelated families, resulting in the inability to discern between natural groups derived from a common ancestor and unrelated forms sharing similar specializations and life history strategies (Hartenberger 1985; Jaeger 1988).

Relationships among rodent families

"Rodentology, more than any other taxonomic group, needs the collaboration of different research fields, from morphology to biochemistry, to determine evolutionary relationships" (Hartenberger, 1985, p. 25). Unraveling the evolutionary history of rodents is complicated by at least two factors: the degree to which unrelated groups show parallel changes in morphology, and the relationship between the age of particular lineages and their current distributions. As suggested by the preceding quote, a natural classification of rodents requires knowledge of phylogenetic relationships based on both morphological and molecular characters. The starting point for establishing such a classification requires a reassessment of interfamilial relationships among rodents. Recent cladistically based morphological analyses that consider both fossil and extant forms (Luckett and Hartenberger 1985; Meng 1990; Marivaux et al. 2004) and molecular analyses (Nedbal et al. 1996; Adkins et al. 2001, 2003; DeBry and Sagel 2001; Huchon et al. 2001, 2002; Montegelard et al. 2002; DeBry 2003; Jansa and Weksler 2004) have provided strong support for several rodent clades as well as several hypotheses that can be tested with the addition of taxa and characters. Although phylogenies derived from molecules and morphology are not totally congruent (fig. 2.4), they do provide strong support for several monophyletic groups, including: (1) Huchon et al.'s (1999) Ctenohystrica, containing a sister-group relationship between the Ctenodactyliae and the Hystricognathi (Phiomorpha 1 Caviomorpha); (2) the Sciuroidea, consisting of the families Sciuridae and Aplodontidae; (3) Montgelard et al.'s (2002) Anomaluromorpha, represented by Anomaluridae and Pedetidae; (4) the Myodonta, containing the superfamilies Dipodoidea and Mu-

roidea; and (5) a larger clade comprised of Dipodoidea/ Muroidea, Anomaluridae/Pedetidae, Castoridae, and Geomyoidea (Geomyidae and Heteromyidae). These two phylogenies differ in the placement of the families Castoridae, Geomyidae, and Myoxidae. They also support the notion that characteristics of both the zygomasseteric system and structure of the lower jaw have evolved independently several times throughout the rodent radiations.

Relationships among the Hystricognathi

The suborder Hystricognathi represents a monophyletic group (Luckett and Hartenberger 1993; Martin 1993; Nedbal et al. 1996; Huchon et al. 2000; Adkins et al. 2001; Huchon and Douzery 2001; Murphy et al. 2001a) containing 17 of the 28 extant families of rodents and one-tenth of the species (229 of 2021 species), and it shares a sister-group relationship with the family Ctenodactylidae (fig. 2.2; Huchon et al. 2000; Adkins et al. 2001). In terms of ecology and behavior, members of this suborder display a diversity of mating systems (monogamous, polygynous, polyandrous, and promiscuous), maintain lifestyles ranging from solitary to group living (e.g., the eusocial naked mole-rat, *Heterocephalus glaber*), and occupy a broad range of habitats. Given the interest in understanding the influence that environmental and historical constraints have on these ecological and behavioral attributes, a well-resolved phylogeny for hystricognath rodents provides an ideal interpretive framework for testing evolutionary hypotheses (Ebensperger and Cofré 2000). Therefore, we will attempt to synthesize what is known and not known about the phylogeny of hystricognaths.

Families of hystricognaths can be subdivided into three major groups (fig. 2.2): (1) Bathy-Phiomorpha (Wood 1965), representing three families (Bathyergidae, Thryonomyidae, and Petromuridae) endemic to sub-Saharan Africa; (2) Hystricomorpha (Wood 1965), consisting of Old World porcupines distributed in Africa and parts of Asia; and (3) Caviomorpha (Lavocat 1973), a South American group containing 13 of the 17 rodent families. Within the Bathy-Phiomorpha, there is strong support for a monophyletic Thryonomyoidea, comprised of the families Thryonomyidae and Petromuridae (Lavocat 1973; Nedbal et al. 1994; Huchon et al. 2001; Rowe 2002), and several studies support a monophyletic Caviomorpha (Lavocat and Parent 1985; Huchon et al. 2001; Rowe 2002). Although families of caviomorphs are subdivided into four superfamilies, Cavioidea, Chinchilloidea, Erethizontoidea, and Octodontoidea (Patterson and Wood 1982), the contents of and relationships among these superfamilies have not been adequately addressed. In addition, the placement of Hystrico-

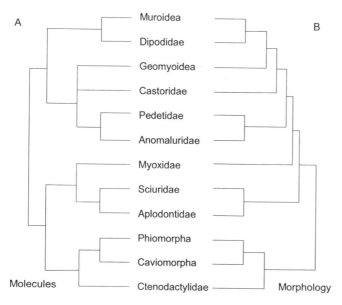

Figure 2.4 Phylogeny of rodent families (molecules versus morphology). A. Molecular-based phylogeny representing a compilation of information from several sources (Nedbal et al. 1996; Adkins et al. 2001; Montgelard et al. 2002; DeBry 2003). B. Results of a cladistic analysis of morphological data by Marivaux et al. (2004).

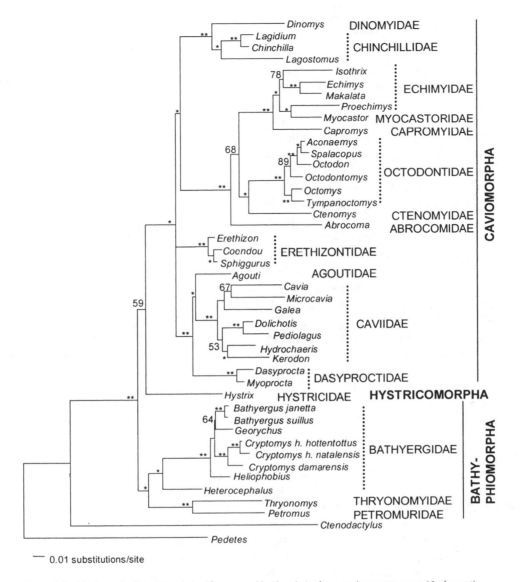

0.01 substitutions/site

Figure 2.5 Maximum likelihood tree derived from a combined analysis of two nuclear genes, exon 10 of growth hormone receptor and intron 1 of the prealbumin transthyretin gene. Bootstrap values are indicated for nodes with greater than 50% support, with the symbols * and ** representing > 90% and 100% support, respectively. Based on results from Rowe (2002).

morpha relative to the other two groups is uncertain. Recently, several molecular phylogenetic studies have advanced our knowledge of relationships among families and genera of Old World and New World hystricognaths (Nedbal et al. 1994; Huchon et al. 2001; Rowe 2002; Honeycutt et al. 2003; Ingram et al. 2004). In terms of interfamilial relationships, the monophyly of several clades are well supported, including (fig. 2.5): (1) the Bathy-Phiomorpha and the Thryonomyoidea; (2) Caviomorpha and the superfamilies Octodontoidea (families Octodontidae, Ctenomyidae, Echimyidae, Myocastoridae, Capromyidae, and Abrocomidae), Cavioidea (Caviidae, Hydrochaeridae, Agoutidae, and Dasyproctidae), and the Chinchilloidea (Chinchillidae); (3) a well-resolved phylogeny for the Bathyergidae, which places *Heterocephalus glaber* at the base of the phylogeny; (4) grouping of the Dinomyidae within the Chinchilloidea rather than the Erethizontoidea; and (5) strong support for an association between the superfamilies Chinchilloidea and Octodontoidea. Problems still persist relative to the placement of the Erethizontoidea (New World porcupines) and Hystricomorpha (Old World porcupines). Although not shown, phylogenies of genera and species are published for the families Octodontidae and Bathyergidae (Honeycutt et al. 2003; Ingram et al. 2004).

Relationships among subfamilies of Muroidea

The family Muridae has a worldwide distribution and represents the most diverse group of rodents, containing approximately 1,326 species subdivided into 17 subfamilies (Musser and Carleton 1993). The two most diverse subfamilies are the Old World Murinae (500 species) and New World Sigmodontinae (300 species). The myomorphous zygomasseteric system, loss of the upper fourth premolar, and characteristics of the first molar support monophyly of Muridae (Flynn et al. 1985). As indicated by Carleton and Musser (1984), lineages within the family show considerable parallel changes in morphology, and although monophyly of subfamilies is confirmed (Michaux and Catzeflis 2000), relationships among most of the subfamilies and many genera and species are poorly resolved. One primary explanation for this lack of resolution within and between subfamilies relates to the numerous adaptive radiations of murids on most continents, and as indicated by Michaux and Catzeflis (2000), these radiations are reflected in the "bush-like" phylogenies often obtained with molecular data.

Two recent molecular phylogenetic studies based on nuclear gene sequences provide the most thorough investigations of subfamilial relationships (Michaux and Catzeflis 2000; Jansa and Weksler 2004). Both of these studies suggest rather rapid radiations, resulting in a bush-like phylogeny for most subfamilies (fig. 2.6). Three subfamilies containing subterranean species, Rhizomyinae, Spalacinae, and Myospalacinae, represent the most basal clade, and two recent studies suggest that this group be placed in a separate family, Spalacidae (Michaux et al. 2001; Norris et al. 2004). The larger clade consists of a polytomy containing four lineages: (1) Calomyscinae; (2) a clade consisting of Malagasy Nesomyinae (not monophyletic) and African Petromyscinae, Mystromyinae, Cricetomyinae, and Dendromurinae; (3) a monophyletic Murinae sister to the subfamilies Acomyinae and Gerbillinae; and (4) a clade containing New World subfamilies Sigmodontinae, Tylomyinae, and Neotomyinae, as well as the subfamilies Arvicolinae (Old and New World) and Cricetinae (Old World).

The family Muridae is represented in the New World by members of the subfamily Sigmodontinae, which can be subdivided into North American and South American groups (Hooper and Musser 1964; Carleton and Musser 1984; Musser and Carleton 1993). North American lineages are represented by 18 genera and 124 species, subdivided into two or more tribes. South American sigmodontines presumably were derived from North American stock, and today represent one of the most speciose groups of rodents in South America, consisting of 61 genera and 299 species, subdivided into seven tribes (Musser and Carleton 1993; Weksler 2003). Comparisons of the phallus

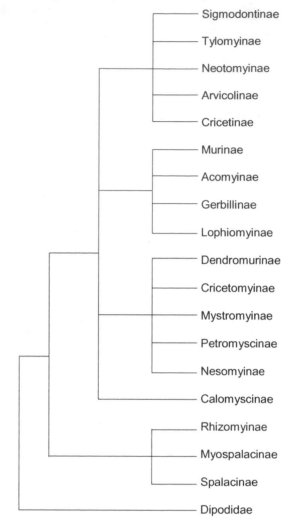

Figure 2.6   Phylogeny depicting relationships among subfamilies of muroid rodents (Jansa and Weksler 2004).

by Hooper and Musser (1964) prompted these authors to separate South American and North American sigmodontines into reciprocally monophyletic groups, whereas Weksler (2003) restricted the subfamily Sigmodontinae to only South American forms. Two recent molecular studies (Engel et al. 1998; Weksler 2003) found support for the monophyly of South American Sigmodontinae, with the genus *Sigmodon* (a group occurring in North America, Central America, and northern South America) occupying a basal position in the phylogeny. These same studies found little support for a sister-group relationship between North American neotomine/peromyscine rodents and the South American tribes of sigmodontines. However, based on the analysis of cytochrome *b* sequences, Bradley et al. (2004) did find strong support for the recognition of a monophyletic North American group, represented by four tribes, Peromyscini, Neotomini, Baiomyini, and Tylomyini. Relationships among these tribes are more equivocal.

Finally, a species level phylogeny was produced for the genus *Microtus* (Jaarola et al. 2004), a diverse group in the subfamily Arvicolinae that is distributed throughout the Holarctic (Musser and Carleton 1993; Chaline et al. 1999). This phylogeny was derived using nucleotide sequences from the mitochondrial cytochrome *b* gene and 48 of the 65 species of *Microtus*. Several clades received good support and appeared in both maximum likelihood and maximum parsimony analyses. These clades include members of the subgenus *Microtus* (containing the social voles of the *socialis* species group), the subgenus *Terricola* (Palearctic), Nearctic species, and an Asian group also supported by a separate molecular study (Conroy and Cook 2000). Deeper nodes in the phylogeny received poor support, making the diagnosis of relationships among the major clades more difficult.

Relationships among genera and species of Sciuridae

The squirrel family Sciuridae is the second most diverse group of rodents, consisting of 273 species and 50 genera, traditionally subdivided into the subfamilies Pteromyinae (flying squirrels) and Sciurinae, consisting of tree squirrels and ground squirrels (Hoffmann et al. 1993; Thorington et al. 2002). Most members of the family are diurnal and are diverse in terms of behavior and ecology, including the formation of complex social groups among various species of ground dwelling sciurids, and the occupation of a variety of habitats, ranging from deserts to grasslands and forests (Dobson 1985; Armitage 1999a). Several genera (e.g., *Spermophilus* and *Marmota*) have a Holarctic distribution, whereas others like *Cynomys* speciated in North America. As indicated by Thorington et al. (1997), the two major systematic issues are relationships among tribes and evidence for monophyly of the recognized subfamilies. For instance, there is strong morphological evidence for the monophyly of flying squirrels (Thorington 1984). Nevertheless, immunology data (Hight et al. 1974) suggest paraphyly for the subfamily Pteromyinae, with gliding evolving independently in Old World and New World forms. In addition, relationships among Old and New World groups of ground squirrels and tree squirrels in the subfamily Sciurinae are unclear. Several recent molecular studies (Harrison et al. 2003; Mercer and Roth 2003; Steppan et al. 2004; Herron et al. 2004) have contributed greatly to the resolution of the sciurid phylogeny (see fig. 2.3 for a summary of major clades). In terms of the two subfamilies, the Pteromyinae is monophyletic, suggesting a single origin for the gliding mechanism, and the Sciurinae is paraphyletic, with flying squirrels sister to New World tree squirrels. Although the tribes representing Holarctic (Marmotini) and Afro-Asian (Xerini) ground squirrels are not sister-groups, the monophyly of

each is supported (Mercer and Roth 2003; Steppan et al. 2004). In addition to resolving relationships among higher taxa of sciurids, several studies provide detailed assessments of relationships among ground squirrel genera (Steppan et al. 1999; Harrison et al. 2003; Herron et al. 2004). Three of these genera, *Cynomys*, *Marmota*, and *Spermophilus*, are of considerable interest to behavioral ecologists. Monophyly for both *Marmota* and *Cynomys* is supported (Harrison et al. 2003; Herron et al. 2004), whereas the genus *Spermophilus* appears paraphyletic, with the genera *Ammospermophilus*, *Cynomys*, and *Marmota* grouping closer to subsets of *Spermophilus*.

As previously indicated, the phylogeny for sciurids provides an interpretive framework for reconstructing the biogeographic history of the group. In addition, this same phylogeny has been used to test specific hypotheses about the ancestry of sciurids in general. For instance, Steppan et al. (2004) used a phylogenetic approach to determine whether the ancestral sciurid was arboreal or terrestrial. They defined three character states for degree of arboreality: arboreal (tree dwelling), intermediately arboreal (terrestrial with time in trees), and terrestrial (ground dwelling). Optimization of these traits on the existing phylogeny provided overall support for an arboreal ancestor. Using their phylogeny for *Marmota*, Steppan et al. (1999) tested the hypothesis that the establishment of large colonies and increased sociality evolved once, and found that such complex social systems originated two or more times. Finally, species-level studies of the genus *Cynomys* found a strong association between sister-group relationships and geographic distribution (Harrison et al. 2003; Herron et al. 2004). The following sister-group relationships in *Cynomys* were found: (1) (*Cynomys ludovicianus* + *Cynomys mexicanus*); (2) (*Cynomys gunnisoni* + *Cynomys leucurus*) + *Cynomys parvidens*.

Testing Evolutionary Hypotheses

A robust phylogeny provides an interpretive framework that allows for inferences to be drawn regarding the evolution of various traits, including morphological, behavioral, ecological, physiological, and so on. A phylogeny allows one to test for relationships between the phenotype, environment, and phylogeny (Harvey and Pagel 1991; Crespi 1996; Irwin 1996; Burda et al. 2000). Thus one can elucidate shared traits resulting from similar environmental constraints among phyogenetically distantly related lineages versus traits arising as a result of shared evolutionary history.

Comparative methods and phylogenies provide valuable tools for studying the ecology and behavior of social mammals. Specifically, a phylogeny allows one to determine whether a behavior is the result of shared common ancestry or evolved independently in response to similar environmental conditions. For instance, behavioral or ecological traits that enhance the fitness of individuals under specific environmental conditions can become established in populations, and if species occupying similar niches and environments experience similar selective pressures, one might expect evidence of convergence with respect to those traits. A phylogeny allows one to map traits and directly test for evidence of convergent evolution. Phylogenies also allow for tracing combinations of behavioral traits that may or may not map to the same node of the tree. In short, a phylogenetic perspective allows one to evaluate the function of a particular behavior as well as the evolution of that behavior.

There is a presumption that current environmental factors provide an explanation for the evolution of life-history traits associated with behavioral and ecological strategies (e.g., Dobson and Oli, chap. 8, and Kalcounis-Rüppell and Ribble, chap. 6, this volume). For instance, it has been hypothesized that the formation of complex social groups reflects a compromise between the cost of dispersal versus the cost of foregoing reproduction and staying within the natal group (e.g., Nunes, chap. 13 and Solomon and Keane, chap. 4, this volume). In similar environmental circumstances defined by limited food resources, risk of predation, and lack of suitable territory, short-term costs associated with alloparenting and other forms of reproductive altruism are offset by long-term benefits of staying within the natal group (Alexander 1974; Lacey and Sherman, chap. 21, this volume). A phylogeny provides a means of testing such predictions. In the following we provide two examples of how this phylogenetic approach can be used.

Evolution of mating systems in cavioid rodents

The superfamily Cavioidea represents a monophyletic group (Woods 1993; Nedbal et al. 1996; Huchon et al. 1999) of South American hystricognath rodents that display a diverse array of behaviors, ecological specializations, and life-history strategies. For instance, habitat use by members of this superfamily ranges from generalists to desert specialists, with the habitat occupied by the former defined by an even distribution of resources and the latter by clumped resources. Mating systems in cavioid rodents are also diverse, ranging from hierarchical promiscuity to polygyny and monogamy (Kleiman et al. 1979; Macdonald et al., chap 33, this volume). In view of differences in habitat use and mating systems observed with the Cavioidea, Lacher (1981) proposed the ecological constraints hypothesis. This hypothesis relied on the classical taxonomy of the Cavioidea,

which divides members of the family Caviidae into two subfamilies, Caviinae and Dolichotinae. Except for *Kerodon rupestris* (rock cavy), which displays a polygynous mating system and the formation of social groups, most members of the Caviinae, such as *Cavia apera* (common guinea pig), *Galea musteloides,* and *Microcavia australis* display hierarchical promiscuity and low levels of sociality. Members of the Dolichotinae, *Dolichotis patagonum* and *Pediolagus salinicola,* are socially complex and have a monogamous mating system. Given the presumed phylogenetic placement of *Kerodon rupestris,* Lacher (1981) suggested that the complex social system and mating system seen in *Kerodon* converged on that seen in the Dolichotinae, primarily in response to similar ecological constraints related to increased risk of predation and the distribution of resources. Rowe and Honeycutt (2002) used nucleotide sequences from both nuclear and mitochondrial genes to derive a molecular phylogeny for the Cavioidea. This phylogeny was used to test Lacher's (1981) original hypothesis and to examine the correlation between character evolution and characteristics of both habitat and degree of sociality. The molecular phylogeny (fig. 2.7) did not support current ideas of cavioid taxonomy, placing *Kerodon,* the capybara (*Hydrochaeris hydrochaeris*), and members of the Dolichotinae in a monophyletic clade characterized by increased sociality and complex mating systems. In fact, there was strong support for a sister-group relationship between *Kerodon* and *Hydrochaeris,* both of which are habitat specialists and have a harem-based mating system. A concentrated-changes test (MacClade 3.1; Maddison and Maddison 1992) suggested that the probability of sociality and habitat specialization being associated by chance alone was high. Therefore, the ancestor of this clade may have been social, thus allowing for the occupation of habitats characterized by patchily distributed resources and/or increased predation. This study should be expanded to include more diversity within the hystricognaths in general.

Life-history traits and social complexity

Blumstein and Armitage (1998a) devised a metric for degree of social complexity and used independent contrasts analysis to evaluate the consequences of social complexity in terms of the evolution of life-history traits (e.g., number of breeding females, time to first reproduction by females, gestation time, litter size, and survival of first offspring). Using current taxonomy as the phylogenetic framework, the following results were obtained: (1) an increase in social complexity results in fewer breeding adult females and higher survival of first-year offspring; (2) there is a correlation between age at first breeding and social complexity,

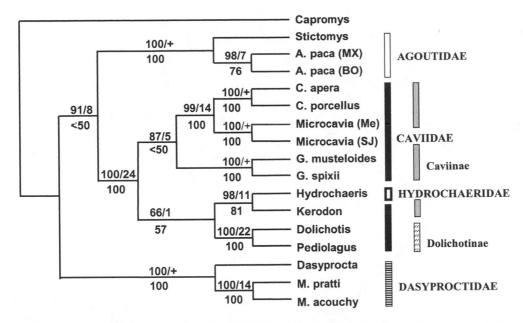

Figure 2.7 Phylogeny of the caviomorph superfamily *Cavioidea,* derived from a combined maximum parsimony analysis of one mitochondrial (12S rRNA) and two nuclear sequences (exon 10 of growth hormone receptor and intron 1 of pre-albumin transthyretin). Bootstrap values (100 replicates), followed by Bremer decay indices, are given above the nodes for the MP analysis. A positive sign (+) indicates Bremer support of greater than 20. Bootstrap values (100 replicates) below the nodes correspond to the ML analysis. Based on the results of Rowe and Honeycutt (2002).

and 60% of this correlation is phylogenetically constrained; (3) Litter size and increases in social complexity are negatively correlated, and there is no correlation between gestation time and level of social complexity. Given apparent problems with existing taxonomic treatments of some groups of sciurids, recent molecular studies on ground-dwelling sciurids provide the necessary components for the derivation of a well-resolved phylogenetic framework that can be used in future studies of social complexity and the evolution of life history traits. As an example of how this phylogeny can be used, we recoded the numerical values shown in the appendix of Blumstein and Armitage (1998a), and used MacClade 4.05 (Maddison and Maddison 1992) to map character states on the consensus phylogeny compiled from existing molecular data (fig. 2.8; Steppan et al. 1999; Harrison et al. 2003; Herron et al. 2004). The conversion from continuous to discrete characters required partitioning values into non-overlapping subsets (e.g., high complexity 0.8 to > 1, medium complexity < 0.8 to 0.4, low complexity 0.2 to < 0.4). Despite the problems with such character coding, some of the results can be compared to patterns observed by Blumstein and Armitage (1998a). For instance, the ancestral litter sizes are large (gray), whereas lineages with high social complexity show a decrease in litter size (black), which seems to be an ancestral state in all

species of *Marmota*, including the less socially complex *Marmota monax* (fig. 2.8a). In addition, the ancestral state for ground-dwelling squirrels appears to be low social complexity, with socially complex groups arising independently at least twice (fig. 2.8b). The clade containing the highly socially complex *Cynomys* reveals an early trend toward increasing complexity in the ancestor. The ancestral state for percent females breeding is low and a decrease appears ancestral to the clade containing the socially complex *Marmota flaviventris, M. olympus,* and *M. caligata,* but is possibly independently derived in the clade containing *M. caudata* and *M. marmota.* A similar pattern is seen for the *Cynomys* clade, characterized by an intermediate percentage of females breeding, except for *Cynomys leucurus,* which has secondarily derived a high percentage of females breeding.

Summary

As can be seen in the preceding sections, the order Rodentia represents a major branch of the mammalian tree of life, and unlike many branches, they have experienced little extinction, with many of the extant families dating almost to the beginning of the radiation. Rodents are excellent colo-

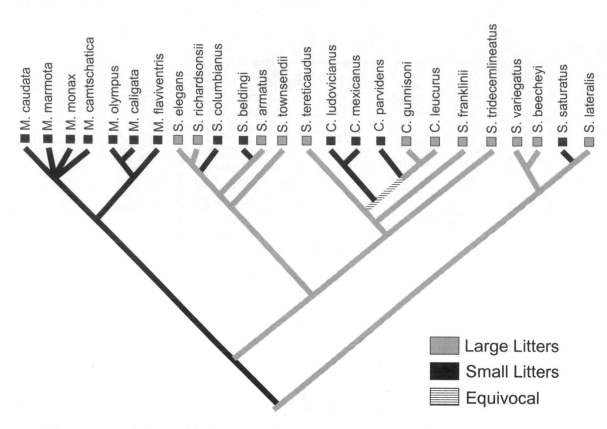

Figure 2.8a Average litter size for ground dwelling sciurids mapped using MacClade 4.05 on a composite phylogeny derived from several sources (Steppan et al. 1999; Harrison et al. 2003; Herron et al. 2004). Values for litter size were taken from Blumstein and Armitage (1998) and represent a conversion of continuous characters to discrete characters (low shown in black and high in gray).

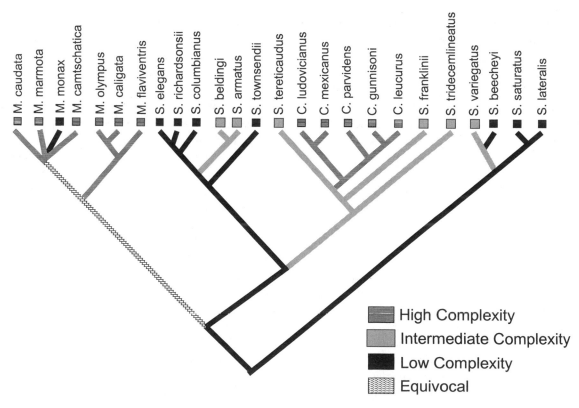

Figure 2.8b Social complexity for ground dwelling sciurids mapped using MacClade 4.05 on a composite phylogeny derived from several sources (Steppan et al. 1999; Harrison et al. 2003; Herron et al. 2004). Values for social complexity were taken from Blumstein and Armitage (1998) and represent a conversion of continuous characters to discrete character. High social complexity is stippled, intermediate complexity is gray, and low complexity is black.

nizers, and most biogeographic regions have rodent assemblages formed by continental exchanges followed by bursts of speciation. This rapid diversification and the age of many rodent lineages make the establishment of well-resolved phylogenies challenging, especially in diverse groups like the Muridae. Clearly, paleontological information and phylogenetic approaches using morphological comparisons are essential for the establishment of a natural classification. Nevertheless, to quote Hartenberger (1985; p. 25), "Paleontology has many good questions to ask, but it is unable to provide answers by itself." Obviously, questions related to the rodent radiations are probably some of the most complex in vertebrate phylogenetics, and an accurate reconstruction of rodent evolutionary history requires the use of multiple characters. Molecular data provide a wealth of markers that can be used to construct detailed phylogenies, and this is mandatory for portions of a phylogeny with short internodes. Whole genome sequences of model mammalian species, some of which are rodents, offer a ready resource of molecular markers for future phylogenetic studies on rodents. For instance, according to O'Brien et al. (1999),

there are approximately 3.2 billion nucleotide pairs in the mammalian genome, and these data provide not only nucleotide sequence information but higher order changes associated with gene rearrangements, duplications, indels, amplification of sequence blocks, and so on. Many of these markers may very well define the monophyly of groups that would otherwise be intractable using only nucleotide sequences (e.g., Pascale et al. 1990).

Although our primary research is in the area of molecular phylogenetics, we are sensitive to the needs of behavioral ecologists, who desire an accurate phylogenetic framework for testing questions related to the evolutionary process. Probably one of the best statements regarding this need was made by Blumstein and Armitage (1998a; p. 10) who said, "No published phylogenetic hypothesis includes all species of interest. We eagerly anticipate the publication of inclusive and well-supported phylogenies." Hopefully, our summary of rodent phylogenetics and evolution provides a starting point for future investigations pertaining to the evolution of complex rodent societies.

Sexual Behavior

Chapter 3 Male Mating Strategies in Rodents

Jane Waterman

CONFLICT BETWEEN THE SEXES is a driving force in the evolution of mating patterns and reproductive behavior (Alonzo and Warner 2000). Each individual strives to maximize reproductive success, and since each sex has different constraints on reproduction, males and females generally have evolved different strategies for mating. However, mating strategies are not the same thing as mating systems. A mating strategy is all the tactics used by an individual to maximize reproductive success, whereas a mating system is characteristic of a population or a species. Thus a mating system may reflect a variety of sex-specific mating tactics. Mating systems are generally defined by whether multiple mating (mating with > 1 member of the opposite sex) occurs in males, females or both. Descriptions of monogamy and polygyny, for example, assume that females mate with a single male. Within a given mating system (e.g., polygyny), males may adopt different strategies (e.g., female defense or competitive searching for mates) to try to maximize reproductive success.

Sexual selection theory predicts that when one sex (usually females) is limited in availability, the other sex competes for access to members of that sex (Trivers 1972). Thus research on mating systems has traditionally focused on the strategies of males. Male strategies for maximizing reproductive success have been viewed as a balance between the advantage of copulating with as many females as possible and the advantage of providing parental care (Clutton-Brock 1989b). Under this scenario, the mating pattern is determined by the distribution of receptive females in time and space (Emlen and Oring 1977). Thus the behavior of males is commonly the predominant factor used to describe mating patterns.

However, a male's ability to access additional females is affected by a number of factors besides the distribution of females, including competitive ability with other males, habitat structure, and perhaps most important of all, female reproductive strategies. Most research has focused on examining the environmental factors that influence male reproductive success, and often ignores how female strategies can influence male reproductive success (Ahnesjo et al. 1993; Lacey et al. 1997). Females may not often be passive participants (see Solomon and Keane, chap. 4 this volume) and females regularly multiply mate in several species that have been described as polygynous. Multiple mating by females changes the competitive landscape for males.

In rodents, males use many different strategies to successfully fertilize females, including overt conflict and defense of mates, competitive mate searching, and sperm competition (Schwagmeyer and Wootner 1985, 1986). This variety of mating strategies reflects the wide variation in spatial and temporal distribution of receptive females. Female social organization ranges from solitary, dispersed, and relatively asocial to highly clustered, colonial matrilineal groups (Michener 1983a; Wolff 1985a, 1989; Schwagmeyer 1990; Kalcounis-Rüppell and Ribble, chap. 6 this volume). Multiple mating by females, evidenced by genetic studies documenting multiple paternity, may confound the strategies used by males to maximize their reproductive success.

Among rodents, the mating tactics used by sciurids probably are the best known, because their diurnal behavior and large body size make observations easy. However, many rodent species are small, cryptic, and/or nocturnal, making mating behavior difficult to observe (Wolff 1989). Thus the mating patterns of small rodents are usually inferred by (1) lab or enclosure studies, (2) field monitoring and manipulation using traps or radiotelemetry, or (3) genetic analysis (e.g., Boonstra et al. 1993). Regardless of the ease or difficulty, the entire range of classical mating systems has been described in rodents, ranging from monogamy to polygyny to promiscuity (table 3.1). However, polygyny has been considered the primary mating system in rodents, as in mammals in general (Clutton-Brock 1989b), because gestation and lactation basically emancipate males from having to provide parental care.

In this chapter I will discuss how males gain access to females and how the strategies of rival males and females can constrain a male's reproductive success. I will address each of the primary mating systems described for rodents and discuss the constraints leading to each and the reproductive strategies used by males to maximize their reproductive success within these constraints (fig. 3.1). I will conclude that promiscuity, not polygyny, is the most common mating strategy of males. Since rodents make up 44% of all mammal species, this conclusion may represent a paradigm shift for mating systems research on mammals in general.

Monogamy

Monogamy, in which neither sex can access additional mates (Emlen and Oring 1977), is rare in mammals, occurring in less than 5% of species (Kleiman 1977). Most cases of monogamy in murid rodents have been inferred from the spatial distribution of males and females, where the home ranges of male-female pairs overlap (Wolff 1985a, 1989). Some of the other traits that have been used to infer monogamy in laboratory studies include the presence of pair-bonding, delayed sexual maturation, parental care, and the absence of sexual dimorphism, or copulatory plugs (Dewsbury 1981). Rarely, however, do studies examine if the animals are socially monogamous (in which pair bonds are formed but copulations outside the pair bond may occur) or genetically monogamous (in which extra-pair copulations do not occur outside the pair bond; Birkhead and Møller 1998).

Monogamy is especially enigmatic for males, because male reproductive success is usually a function of the number of mates he can inseminate (Trivers 1972). Why then should a male be monogamous? In rodents, two major patterns emerge. First, monogamy may occur in situations in which paternal care is critical for the survival of the offspring (*obligate monogamy;* Emlen and Oring 1977; Clutton-Brock 1989b; Ribble 2003). Paternal care could include directly caring for offspring (e.g., grooming, retrieving, huddling), as well as defending them against predators or infanticide. Second, monogamy may occur when males are unable, for various environmental reasons (e.g., widely dispersed females or resources), to gain access to more than one female during a mating season (*facultative monogamy;* Holmes 1984a; Komers and Brotherton 1997; fig. 3.2).

Obligate monogamy

Obligate monogamy has evolved in cases in which males are not emancipated from providing parental care, because male care is critical to offspring survival. Therefore, these males do not have the opportunity to seek additional mates, and their strategy for maximizing reproductive success is to maximize offspring survival. For example, in the genus *Peromyscus,* California mice (*P. californicus*) and oldfield mice (*P. polionotus*) have been documented as monogamous (Wolff 1989), both socially and genetically, and males do not mate multiply even if the opportunity arises (Foltz 1981; Ribble 2003). Male parental care (grooming, retrieving, and huddling over the offspring) appears critical to offspring survival in California mice (fig. 3.2a), especially when the ambient temperature is cold or resources are low (Gubernick and Teferi 2000; Ribble 2003). Removal of males significantly decreased pup survival, suggesting that direct paternal care and not infanticide prevention (no other males were present) is the primary function of male care in this species (Gubernick and Teferi 2000). Males also contribute anogenital stimulation during parturition, potentially aiding the birth process (Lee and Brown 2002). Since male ranges are often large and could overlap more than a single female range, female dispersion does not explain monogamy in this species (Ribble 2003). Similar advantages of direct paternal care are found in other species of monogamous rodents, such as mound-building mice (*Mus spicilegus*), in which males spend time covering young and retrieving stray pups (Patris and Baudoin 2000), and prairie voles (*Microtus ochrogaster*), in which males contribute levels of care similar to females (Oliveras and Novak 1986; McGuire and Bemis, chap. 20, this volume). Perhaps the epitome of monogamy in rodents is the American beaver (*Castor canadensis*), which is reviewed extensively in Busher, chap. 24, this volume).

Maintaining long-term pair bonds is a strategy for monogamous males that minimizes the risk of not finding a mate or mating with an infertile female. The Malagasy giant rat (*Hypogeomys antimena*) also forms monogamous long-term bonds that apparently last until one mate dies (Som-

Table 3.1 A review of the described mating systems in rodents, and if multiple mating by females has also been recorded

Described common name	Scientific name	Mating system	Multiple mating common	Litter paternity examined	Relative testes size*	References
Family Sciuridae						
White-tailed antelope squirrel	*Ammospermophilus leucurus*	non-defense polygyny			large	Belk and Duane Smith 1991
Gunnison's prairie dog	*Cynomys gunnisoni*	promiscuity	yes	yes		Travis 1993; Hoogland 1998, 2001
Black-tailed prairie dog	*C. ludovicianus*	defense polygyny	rarely	yes		Hoogland 1995
Utah prairie dog	*C. parvidens*	promiscuity	yes	yes		Hoogland 2001; Haynie et al. 2003
Yellow-bellied marmot	*Marmota flaviventris*	defense polygyny	no	yes		Armitage 1986b; Schwartz and Armitage 1980
Alpine marmot	*M. marmota*	monogamy/polyandry	yes	yes		Goossens et al. 1998
Woodchuck	*M. monax*	monogamy/polygyny			small	Meier 1983; Armitage 1986b; Allaine 2000
Olympic marmot	*M. olympus*	monogamy				Allaine 2000
Vancouver marmot	*M. vancouverensis*	monogamy/polygyny	no			Allaine 2000
Long-tailed marmot	*M. caudata*	monogamy/polyandry	yes			Allaine 2000
Hoary marmot	*M. caligata*	monogamy/polygyny	possibly			Holmes 1984a; Barash 1981
Bobak marmot	*M. bobak*	monogamy/polygyny				Allaine 2000
Menzbier's marmot	*M. menzbieri*	monogamy				Allaine 2000
Tarbagan marmot	*M. sibirica*	monogamy				Allaine 2000
Gray marmot	*M. baibacina*	monogamy				Allaine 2000
Black-capped marmot	*M. camtschatica*	monogamy				Allaine 2000
Yellow-footed bush squirrel	*Paraxerus cepapi*	promiscuity	yes			Skinner and Smithers 1990
Abert's squirrel	*Sciurus aberti*	non-defense polygyny	yes			Farentinos 1980
Arizona gray squirrel	*Sciurus arizonensis*	non-defense polygyny				Best and Riedel 1995
Eastern gray squirrel	*Sciurus carolinensis*	non-defense polygyny	yes		large	Koprowski 1993a
Fox squirrel	*Sciurus niger*	non-defense polygyny	yes			Koprowski 1993b
Eurasian red squirrel	*Sciurus vulgaris*	non-defense polygyny	yes		large	Wauters et al.1990
Uinta ground squirrel	*Spermophilus armatus*	non-defense/defense polygyny	possibly			Balph and Stokes 1963; Eshelman and Sonnemann 2000
California ground squirrel	*Spermophilus beecheyi*	defense polygyny	yes	yes	large	Dobson 1984; Boellsstorff et al. 1994
Belding's ground squirrel	*Spermophilus beldingi*	non-defense polygyny	yes	yes		Hanken and Sherman 1981; Sherman and Morton 1984
Idaho ground squirrel	*Spermophilus brunneus*	non-defense polygyny	yes	yes		Sherman 1989
European ground squirrel	*Spermophilus citellus*	non-defense polygyny	no			Millesi et al. 1998; Huber et al. 2002
Columbian ground squirrel	*Spermophilus colombianus*	defense polygyny	yes	yes		Dobson 1984; Murie 1995
Golden-mantled ground squirrel	*Spermophilus lateralis*	defense polygyny	yes		large	Phillips 1981; Kenagy pers. comm
Mexican ground squirrel	*Spermophilus mexicanus*	defense polygyny				Edwards 1946
Mohave ground squirrel	*Spermophilus mohavensis*	defense polygyny				Best 1995
Arctic ground squirrel	*Spermophilus parryii*	defense polygyny	yes	yes		Lacey et al. 1997
Richardson's ground squirrel	*Spermophilus richardsonii*	non-defense/defense polygyny	yes			Davis and Murie 1985; Michener and McLean 1996
Round-tailed ground squirrel	*Spermophilus tereticaudus*	non-defense polygyny				Dunford 1977b

(continued)

Table 3.1   (continued)

Described common name	Scientific name	Mating system	Multiple mating common	Litter paternity examined	Relative testes size*	References
Thirteen-lined ground squirrel	*Spermophilus tridecemlineatus*	non-defense polygyny	yes	yes		Schwagmeyer and Wootner 1985
Rock squirrel	*Spermophilus variegatus*	non-defense/defense polygyny	possibly			Johnson 1981
Yellow-pine chipmunk	*Tamias amoenus*	promiscuity	yes	yes	large	Schulte-Hostedde and Millar 2002, 2004; Schulte-Hostedde et al. 2003
Eastern chipmunk	*Tamias striatus*	non-defense polygyny	yes			Yahner 1978
Douglas' squirrel	*Tamiasciurus douglasii*	promiscuity	yes			Steele 1999
Pine squirrel	*Tamiasciurus hudsonicus*	promiscuity	yes			Steele 1998
Cape ground squirrel	*Xerus inauris*	promiscuity	yes		large	Waterman 1998
Northern palm squirrel	*Funambulus pennanti*	promiscuity	yes			Purohit et al. 1966
Malaysian tree squirrel	*Callosciurus caniceps*	promiscuity	yes			Tamura 1993
Formosan tree squirrel	*Callosciurus erythraeus*	non-defense polygyny	yes			Tamura 1995
Plantain squirrel	*Callosciurus notatus*	promiscuity	yes			Tamura 1993
Family Castoridae						
North American beaver	*Castor canadensis*	monogamy			small	Jenkins and Busher 1979
European beaver	*Castor fiber*	monogamy				Herr and Rosell 2004
Family Geomyidae						
Botta's pocket gopher	*Thomomys bottae*	polygyny/promiscuity	possibly		large	Eisenberg 1966; Patton and Feder 1981
Family Heteromyidae						
Desert kangaroo rat	*Dipodomys deserti*	polygyny/promiscuity				Jones 1993
Giant kangaroo rat	*Dipodomys ingens*	monogamy/polygyny/ promiscuity	yes			Randall et al. 2002
Merriam's kangaroo rat	*Dipodomys merriami*	promiscuity	yes		average	Randall 1987a; Jones 1993
Banner-tailed kangaroo rat	*Dipodomys spectabilis*	polygyny/promiscuity	yes	yes		Jones 1993; Winters and Waser 2003
Salvin's spiny pocket mouse	*Liomys salvini*	polygyny/promiscuity	yes			Jones 1993
Long-tailed pocket mouse	*Chaetodipus formosus*	polygyny/promiscuity	yes			Jones 1993
Family Muridae						
Bank voles	*Clethrionomys glareolus*	non-defense polygyny/promiscuity	yes		large	Marchlewska-Koj et al. 2003
Gapper's red-backed vole	*Clethrionomys gapperi*	promiscuity	yes		large	Heske and Ostfeld 1990
Northern red-backed vole	*Clethrionomys rutilus*	promiscuity	yes		large	Heske and Ostfeld 1990
Beach vole	*Microtus breweri*	promiscuity	yes		large	Heske and Ostfeld 1990
California vole	*Microtus californicus*	monogamy/polygyny			small	Wolff 1985a
Gray-tailed vole	*Microtus canicaudus*	polygyny/promiscuity		no	small	Wolff et al. 1994
Montane vole	*Microtus montanus*	monogamy/polygyny/ promiscuity	yes		small	Wolff 1985a; Heske and Ostfeld 1990
Tundra vole	*Microtus oeconomus*	polygyny			small	Heske and Ostfeld 1990
Prairie vole	*Microtus ochrogaster*	monogamy			small	Wolff 1985a
Meadow vole	*Microtus pennsylvanicus*	promiscuity	yes		large	Wolff 1985a
Pine vole	*Microtus pinetorum*	monogamy/ promiscuity	yes		small	Wolff 1985a; Heske and Ostfeld 1990
Water vole	*Microtus richardsoni*	promiscuity	yes			Wolff 1985a
Townsend's vole	*Microtus townsendii*	monogamy/polygyny				Lambin and Krebs 1991b

Table 3.1 (continued)

Described common name	Scientific name	Mating system	Multiple mating common	Litter paternity examined	Relative testes size*	References
Taiga vole	Microtus xanthognathus	non-defense polygyny			small	Wolff 1985a
Malagasy giant jumping rat	Hypogeomys antimena	monogamy	rare	yes	small	Sommer 2000; Sommer & Tichy 1999
Muskrat	Ondatra zibethicus	monogamy/polygyny	no		small	Marinelli et al. 1997
Djungarian hamster	Phodopus campbelli	monogamy	no			Ross 1995
Giant gerbil	Rhombomys opimus	defense polygyny	no			Randall and Rogovin 2002
Mound-building mouse	Mus spicilegus	monogamy/polygyny				Patris and Baudoin 2000; Gouat et al. 2003
Bush rat	Rattus fuscipes	promiscuity	yes			Taylor and Calaby 1988
Swamp rat	Rattus lutreolus	promiscuity	yes			Taylor and Calaby 1988
Yellow-nosed mouse	Abrothrix xanthorhinus	promiscuity	yes			Lozada et al. 1996
Long-tailed field mouse	Apodemus sylvaticus	monogamy/polygyny/ promiscuity	yes	yes	large	Bartmann and Gerlach 2001; Baker et al. 1999
Striped field mouse	Apodemus agarius	monogamy/polygyny/ promiscuity	yes	yes		Baker et al. 1999
Brants' whistling rat	Parotomys brantsii	non-defense polygyny				Jackson 1999
Bushy-tailed woodrat	Neotoma cinerea	defense polygyny	no	yes	small	Topping and Millar 1998, 1999
California mouse	Peromyscus californicus	monogamy	no	yes		Ribble 2003
Cactus mouse	Peromyscus eremicus	monogamy?				Wolff 1989
White-footed mouse	Peromyscus leucopus	monogamy/polygyny/ promiscuity	yes	yes		Wolff 1989
Deer mouse	Peromyscus maniculatus	promiscuity	yes	yes	large	Wolff 1989
Old field mouse	Peromyscus polionotus	monogamy				Foltz 1981; Wolff 1989
Family Bathyergidae						
Naked mole-rat	Heterocephalus glaber	monogamy/polyandry				Bennett et al. 2000
Family Erethizontidae						
American porcupine	Erethizon dorsatum	defense polygyny				Sweitzer and Berger 1998
Family Hystricidae						
Cape porcupine	Hystrix africaeaustralis	monogamy				Skinners and Smithers 1990
Family Caviidae						
Southern cavy	Microcavia australis	non-defense polygyny				Tognelli et al. 2001
Maras	Dolichotis patagonum.	monogamy				Taber and Macdonald 1992
Greater guinea pig	Cavia magna	promiscuity	yes			Kraus et al. 2003
Brazilian guinea pig	Cavia aperea	polygyny	no	yes		Sachser et al. 1999; Asher et al. 2004
Yellow-toothed cavy	Galea musteloides	promiscuity	yes	yes		Schwarz-Weig and Sachser 1996; Sachser et al. 1999; Hohoff et al. 2003
Family Ctenomyidae						
Talas tuco-tuco	Ctenomys talarum	defense polygyny	no	yes	small	Zenuto et al. 1999
Family Myocastoridae						
Coypu	Myocastor coypus	non-defense polygyny	no	yes		Guichon et al. 2003

SOURCE: Relative testes size is from Kenagy and Trombulak (1986), Heske and Ostfeld (1990), and Stockley and Preston (2004).
* Testes size was classified as large if the mass was greater than 1.1 and small if less than 0.9 for data from Kenagy and Trombulak (1986) and Stockley and Preston 2004 or was classified as large or small by Heske and Ostfeld (1990).

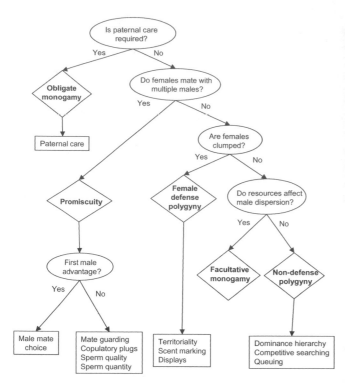

Figure 3.1 Constraints on male mating success (ovals) that lead to different mating systems (diamonds, bold) and male mating strategies (boxes).

mer 2000). Further evidence of monogamy in this species includes sexual monomorphism, small testes size, and single-male paternity (Sommer and Tichy 1999; Sommer 2000).

Monogamous pairs may also recruit helpers to provide additional parental behavior to maximize offspring survival. Most Eurasian marmots (*Marmota* spp.) are cooperative breeders, with only a single alpha pair breeding (Allaine 2000). This strategy (monogamy and cooperative breeding) may have evolved as an adaptation to harsh environments, in which group hibernation gives a thermal advantage and increases offspring survival (Allaine 2000; see also Armitage, chap. 30, this volume). Thermoregulation and cooperative maintenance of territories has been suggested as the main benefit of grouping and monogamy in Cape porcupines (*Hystrix africaeaustralis;* Corbet and van Aarde 1996).

Facultative monogamy

Monogamy has also arisen in species of rodents in which males do not contribute paternal care, but may be unable to access more than one mate because females or other resources are spatially dispersed (Kleiman 1977; Komers and Brotherton 1997; Ribble 2003). Mating patterns within

A

B

C

Figure 3.2 Examples of obligate monogamy: (a) California mouse, facultative monogamy, (b) prairie vole, and polygyny/promiscuity (c) gray-tailed voles. Photo of California mouse © David J. Gubernick; prairie vole and gray-tailed vole © J. Wolff.

these species can be quite variable, depending on local resource dispersion, with the annual number of mates per male varying greatly.

Many species of rodents have variable mating systems, ranging from monogamy to promiscuity (Wolff 1985a, 1989). Prairie voles (fig. 3.2b) have been studied extensively in the field and laboratory, and although they show many characteristics of monogamy, a large portion of males appear to be polygynous and multi-male mating has been reported (Getz et al. 1993; Carter and Getz 1993; Wolff and Dunlap 2002). Wandering males do not appear to differ phenotypically from monogamous males (Solomon and Jacquot 2002); thus, prairie voles might be a good example of facultative monogamy. Prairie vole males can form long-term bonds, and males guard females from other males (DeVries and Carter 1999). In deer mice (*Peromyscus maniculatus*), male movements and spatial overlap with females suggest that mating may range from monogamy in some cases to polygyny or promiscuity in other populations (Birdsall and Nash 1973; Merritt and Wu 1975; Kleiman 1977; Mihok 1979). Similarly, in white-footed mice (*P. leucopus*), mating systems range from monogamy to polygyny to promiscuity, depending on density and spatial distribution of females (Wolff and Cicirello 1990). In low densities, white-footed mice appeared to form monogamous bonds (Mineau and Madison 1977), whereas in a moderately dense population, females mated multiply (Wolff 1989). Wolff (1989) suggests that multiple mating in higher densities is a female strategy to reduce infanticide by males (see also Ebensperger and Blumstein, chap. 23, and Solomon and Keane, chap. 4, this volume). A similar pattern of facultative monogamy is seen in California voles (*Microtus californicus*), montane voles (*M. montanus*), and Townsend's voles (*M. townsendii*), in which mating patterns vary with density (Kleiman 1977; Wolff 1985a; Lambin and Krebs 1991b). Likewise, density of female giant kangaroo rats (*Dipodomys ingens*) influences the number of mates per male (Randall et al. 2002), with mating by both sexes ranging from a single partner at low densities to multiple partners at higher densities. Because males must defend a food territory, their mating opportunities are restricted to neighboring females (Randall et al. 2002) and they are unable to pursue a strategy of competitive searching (see the following).

Facultative monogamy in marmots appears to be influenced by the distribution and quality of resources. In yellow-bellied marmots (*Marmota flaviventris*), males that occupy marginal habitat are usually associated with a single female, whereas males in better habitats are polygynous (Armitage 1986b). Polygynous males have twice the reproductive success of monogamous males because annual reproductive success is directly linked to the number of fe-

males residing in a male's territory (Armitage 1986b, 1998). Males will expand territories to include more females if the opportunity arises, but they are constrained from moving into new territories by the need for a decent hibernaculum and sufficient resources to prepare for hibernation (Armitage 1986b). Hoary marmots (*M. caligata*) in Alaska are monogamous, most likely because the burrows females need for hibernation are too widely spaced and resources are too low for clumping of females (Holmes 1984a). Males likely do not have the opportunity to travel from one valley to another to seek additional fertile females, especially considering the period of female breeding is fairly short (Holmes 1984a). In more southern populations, when resources permit, females clump and males are polygynous (Barash 1981). Similar patterns have been observed in other species of marmots; 19% of Olympic marmots (*M. olympus*) and 50% of woodchucks (*M. monax*) were reportedly monogamous (Barash 1973; Armitage 1986b).

The distribution of resources or females, not just abundance, may also affect mating patterns. If mating systems are associated with resource dispersion, when resources are patchy, females should form groups to defend the patches, whereas when resources are uniform, females should not clump (Slobodchikoff 1984). Thus males should associate with a group of females when resources are patchy, but only with a single female when resources are uniform. When this hypothesis was tested in Gunnison's prairie dog (*Cynomys gunnisoni*), altering habitat to a more uniform distribution of resources did result in a greater number of monogamous associations (Slobodchikoff 1984; Travis and Slobodchikoff 1993; Travis et al. 1995). However density, which reflects resources, is also a strong predictor of mating patterns in this species, as monogamy occurred only at the lowest densities with the lowest patchiness. Similarly, in grey-sided voles (*Clethrionomys rufocanus*), males had greater overlap with other males when females were clumped and less so when females were dispersed (Ims 1988). However, in field voles (*Microtus agrestis*), the spatial distribution of females did not affect male spacing, whereas female density did (Nelson 1997).

Polygyny

Polygyny, in which a male monopolizes access to more than a single female, is considered the most common mating system in mammals (Clutton-Brock 1989b). Males should attempt to mate with as many females as possible to maximize their reproductive success (Trivers 1972). Many different types of polygyny have been described, but these fall into two major categories: defense and non-defense polygyny. Defense polygyny occurs if males are territorial and/or defend access to females during the breeding season, whereas lack of any form of defense is considered non-defense polygyny (Dobson 1984). Defense polygyny has also been called male territoriality, female-defense polygyny, harem-defense polygyny, resource-defense polygyny, and male-defense polygyny (Schwagmeyer 1990). Armitage (1986b) and Schwagmeyer (1990) argue that these distinctions do not reflect the real differences in reproductive behavior, and suggest that female defense polygyny is the appropriate term. Non-defense polygyny may include dominance hierarchies or competitive searching (i.e., scramble competition). Whether defense or non-defense polygyny occurs is determined by the costs of defense. The economical defensibility of mating territories is affected by (1) population density, (2) the spatial distribution of females, and (3) the temporal distribution of receptive females (Emlen and Oring 1977).

As density increases, so does the number of competitors, making territory defense uneconomical ("the density-male hypothesis," Dobson 1984). This hypothesis predicts that, for species that are not widely dispersed, non-defense polygyny will occur if density is high during the mating season, and defense polygyny will occur when densities are low (Dobson 1984). Dobson (1984) tested this hypothesis in polygynous ground-dwelling squirrels. Using only the local density of males during the mating season, the hypothesis was not supported, but when local density of both sexes was examined, the predictions of the hypothesis were supported. Dobson concluded that the cost of monopolizing females was affected not only by an increase in the number of male intruders, but also by the cost of having to sequester more females.

As discussed previously, mating patterns can be flexible, and species with male territoriality may respond to environmental changes when territoriality becomes economically expensive. Males defend territories at low densities and form dominance hierarchies at high densities in species such as mouse mice (*Mus musculus;* Davis 1958), white-footed mice, and deer mice (Wolff 1989). Grey-sided voles can switch to territoriality if the density of competitors is low (Ims 1987b).

Spatially clumped females are more economical to defend than females that are spatially dispersed, leading to territoriality. In ground-dwelling squirrels, which live in relatively open areas, female clumping and low densities are characteristic of male territoriality (Dobson 1984). However, female clumping does not result in territoriality in voles, as habitat structure may make territoriality and detection of intruders difficult for many small mammals (Ims 1988).

Temporal clumping, as well as spatial clumping, affects

male mating strategies. The timing or degree of synchrony of sexual receptivity by females has a profound effect on the operational sex ratio (OSR), the "ratio of fertilizable females to sexually active males at any given time" (Emlen and Oring 1977, 216). As female receptivity becomes more asynchronous, the OSR becomes more biased toward males, which should intensify male mating competition (Clutton-Brock and Parker 1992) and change male mating strategies. If the OSR is highly skewed toward males, then territoriality would be uneconomical, as any territorial male would have to defend against many intruders, and non-defense polygyny would be predicted. Three factors would increase this skew and make territorial defense uneconomical: a long mating season, short and asynchronous periods of receptivity by females, and an actual skew in the population sex ratio (Dobson 1984). In Richardson's ground squirrels, which typically use female defense polygyny, 50% of males in one population became nonterritorial when a late snowstorm increased asynchrony of receptive females and skewed the OSR toward males (Davis and Murie 1985).

The OSR hypothesis has been supported in many species of animals (Kvarnemo and Ahnesjo 1996); however, in rodents the support has been mixed. Randall et al. (2002) found that shifts in the OSR toward a male bias did intensify male-male competition for mates in giant kangaroo rats. Ims (1987a, 1988) proposed that the spacing system of male microtines can be predicted from the temporal distribution of females, in that males should have overlapping home ranges when females are asynchronous and exclusive home ranges when females are synchronous, which would fit the predictions of Emlen and Oring (1977). However, Agrell et al (1996) found just the opposite in male field voles (*Microtus agrestis*).

The value of the OSR as a predictor of the intensity of male competition in ground squirrels has also been questioned (Dobson 1984; Schwagmeyer and Wootner 1985; Michener and McLean 1996). Michener and McLean (1996) found more overt conflict during low OSR in Richardson's ground squirrels. Similarly, Schwagmeyer and Wootner (1985) determined that the density of female thirteen-lined ground squirrels (*S. tridecemlineatus*) affected male-male competition more than the OSR. Dobson (1984) tested the relationship between OSR and the type of mating system in ground-dwelling squirrels and rejected it as an explanation of male mating strategies. However, his calculation of OSR was based on the average number of receptive mates over a breeding season, which may be an inappropriate measure (see Michener and McLean 1996 for a discussion). A better measure of OSR would be the local OSR on a daily basis, which would more closely reflect the changes in intensity of competition among males (Waterman 1998).

Using the OSR as a predictor of mating systems by looking at its effect on only one mating strategy (overt conflict) may be too simplistic. Other factors may influence male and female mating strategies. As most species of ground-dwelling squirrels are promiscuous (table 3.1), increased intensity of sexual selection among males may be apparent in other more subtle forms of sexual selection besides overt competition (see the following).

Defense polygyny

In defense polygyny, dominance is usually site-related (Schwagmeyer 1990), where males win fights in their own area and lose outside of this area (see also Wolff et al. 1983 for *Peromyscus*). In yellow-bellied marmots, resident males have never been observed to lose an encounter with invading males that attempt to take over their territory (Armitage 1998). Similarly, in white-footed mice and deer mice residents, resident males win the majority of encounters with intruding males (Wolff et al. 1983).

Severe wounding during territorial battles has been described in a number of species in which territorial males directly defend resident females by overt conflict. In California ground squirrels (*Spermophilus beecheyi*), 93% of males were wounded during the mating season (Boellstorff et al. 1994), and severe wounding has also been reported for Columbian ground squirrels (*S. columbianus;* Steiner 1970) and Richardson's ground squirrels (*S. richardsonii;* Michener 1983a). However, less costly means of excluding other males from a territory are also used, including scent marking, patrolling, and parallel running (Schwagmeyer 1990; Roberts, chap. 22, and Busher, chap. 24, this volume).

Non-defense polygyny

In non-defense polygyny there is no site-specific dominance. Males usually roam over wide areas and access to females (or at least first access) is determined by who wins encounters. There are two important forms of pre-copulatory male-male competition in non-defense polygyny: establishing dominance hierarchies (through overt conflict or displacements) and competitive searching.

With dominance hierarchies, more dominant males should have greater reproductive success. Dominance is correlated with reproductive success in Belding's ground squirrels (*Spermophilus beldingi;* Sherman 1976) and round-tailed ground squirrels (*S. tereticaudus;* Dunford 1977b), where larger males have an advantage. Even in species in which size does not affect dominance status, more dominant individuals have greater reproductive success, as in Cape ground squirrels (*Xerus inauris;* Waterman 1998; fig. 3.3)

and bushy-tailed woodrats (*Neotoma cinerea;* Topping and Millar 1999).

Most tree squirrels, which breed very asynchronously within a local population, also form dominance hierarchies to determine access to receptive females (Koprowski, chap. 7, this volume). Males in these species gather near an estrous female and determine dominance relationships through highly agonistic interactions, but do not maintain their aggregations beyond the estrous period (e.g., tassle-eared tree squirrel, *Sciurus aberti,* Farentinos 1972; Eurasian red squirrel, *S. vulgaris,* Wauters et al. 1990; eastern gray squirrel, *S. carolinensis,* Koprowski 1993a). In eastern gray squirrels the female can be pursued by 4 to 8 males in a single mating bout, and adult male gray and fox squirrels (*S. niger*) aggressively compete for copulations during mating bouts (Thompson 1977; Koprowski 1993a,b). However, there appear to be two alternative reproductive strategies in gray squirrels that are condition dependent (Koprowski 1993a,b). The males participating in the active

Figure 3.3 Displacement interactions between two male Cape ground squirrels involves nose-nose recognition and assessment but no overt conflict (top). Males use these displacements to establish a linear dominance hierarchy within their all-male social groups (bottom). Photos by J. Waterman.

pursuit of females are older, dominant males (greater than 3 years of age), while satellite males are younger, subordinate males who remain dispersed in the area near the mating activity. Dominant males enjoy great reproductive success because they are more successful in guarding females; however, subordinate males still are able to gain some copulations if females break away from the dominant male pursuers. Koprowski (1993a) suggests these younger males are "making the best of a bad job" by promoting potential reproductive success during a time when they cannot compete with more dominant males. Male European red squirrels also have a sneaker alternative reproductive strategy, but sneaker males rarely succeeded in copulating with females (Wauters et al. 1990). In tassle-eared squirrels, females appeared to initiate copulations with subordinates, but only after copulating with the dominant male (Farentinos 1980).

Dominance is also age related in the Cape ground squirrel. Males live in all-male bands that are composed of unrelated individuals, and dominance is established using nonaggressive displacements (Waterman 1995). The most dominant male is usually the first male to copulate with a female on the day of estrus, and dominant males have the highest copulatory success (Waterman 1998; fig. 3.3). This species has extremely asynchronous receptivity in females, with estrus in females occurring unpredictably and year-round.

However, the relationship between dominance and reproductive success is not always clear (Dewsbury 1982a). In competitive searching, skill in finding receptive females is more important to male reproductive success than overt conflict. Competitive searching has also been described in European ground squirrels (*S. citellus*), Brant's whistling rats (*Parotomys brantsii*), and bushy-tailed woodrats (Millesi et al. 1998; Topping and Millar 1999; Jackson 1999). Thirteen-lined ground squirrels use competitive searching because females are widely dispersed, making it uneconomical for males to control access to more than a single female simultaneously (Schwagmeyer 1990). The influence of this wide dispersion of females on male strategies has been discussed previously under facultative monogamy, but differs here because the monogamous males defend nonbreeding territories, limiting their ability to search for females. Thirteen-lined ground squirrels are not territorial and can range over large areas (Schwagmeyer 1990). Male mobility is critical, and male mating success is mostly determined by ability to locate females. Males that obtained an above average number of mates tended to search twice the amount of area as other males (Schwagmeyer 1994). These males also spent more time with the female on the day prior to mating compared with other males (Schwagmeyer et al. 1998). Males using female defense polygyny can switch to mate search-

ing. In hoary marmots, territorial males use mate searching to increase reproductive success by gaining access to mating opportunities outside of their territories (Barash 1981), similar to Arctic and Columbian ground squirrels (Murie and McLean 1980).

Another strategy in thirteen-lined ground squirrels is queuing, in which order of access to females is determined by order of arrival (Schwagmeyer and Parker 1987). In this case, the first individual gets access to the female, and the second male waits until the first is finished. Earlier arrivals are generally heavier, as well as better at finding females, which creates an asymmetry that increases the costs of a second male trying to displace the first. Because the likelihood of finding another female is low, staying and waiting gives the best reproductive payoff to the second male.

Problems with use of the term polygyny

Polygyny has been described in many species of rodents, but depicting the primary mating system of this group as polygynous may be misleading because of female reproductive strategies. Of 33 studies that have described polygyny as the predominant mating system in rodents, only 67% actually examined if females mate multiply. Of those that have looked at female behavior, 65% documented females mating with more than one male (table 3.1). In the majority of squirrels examined, for example, females mate multiply. In Arctic ground squirrels (*Spermophilus parryii*), males defend territories to keep out other males, but territoriality does not provide exclusive access to females, although it does increase the chance of being the first mate (Lacey et al. 1997). It could be that some males are just dominant overall, and on the day of a female's estrus they can supplant or at least overrule the resident male and gain access to the female (Schwagmeyer 1990). Even with the high aggression seen among territorial male California ground squirrels, females copulated with an extremely high number of males, averaging 6.7 partners each (Boellstorff et al. 1994).

One way to document multiple mating by females is to examine the paternity of litters. Studies examining paternity suggest that behavioral data collected during mating are often a poor predictor of actual paternity (Travis et al. 1996). However, few polygynous species (where one male appears to have exclusive access to multiple females) have been tested genetically for multiple paternity. Even so, true genetic polygyny has been established in some species of rodents. Black-tailed prairie dogs (*Cynomys ludovicianus*) have little or no multiple mating by females (about 3%; Foltz and Hoogland 1981), even if more than a single male inhabits a coterie. Likewise, yellow-bellied marmots have no indication of genetic multiple paternity (Schwartz and Armitage 1980). Female gundis (*Ctenodactylus gundi*) typically live in female groups and associate with one or more males (Nutt 2003). Unimale groups were the most common association, and there was no genetic evidence of multiple paternity, even though some females did appear to mate outside of their social group. In bushy-tailed woodrats, males and females occupy large, overlapping home ranges during the breeding season, and genetic data suggest that males gain exclusive access to females during estrus with no multiple paternity within litters (Topping and Millar 1999).

One problem with using multiple paternity to document multiple mating by females is that since copulations do not always accomplish fertilization, paternity examinations necessarily underestimate the occurrence of multiple mating. Furthermore, in some species of rodents, multiple-mated females can have a lower probability of pregnancy (Dewsbury 1982c; Wolff 1989). In addition, older paternity studies underestimated the amount of multiple mating by females because of the low probability of detecting multiple paternity (Wolff 1989). However, current methods allow more resolution of genetic relationships with a greater degree of confidence (Travis et al. 1996).

Promiscuity

Promiscuity, in which both sexes mate with multiple partners, is common in the small rodents, as the frequency of multiple paternity of litters demonstrates (table 3.1). In a moderately dense population of white-footed mice, up to 30% of litters were sired by multiple males (Wolff 1989) and multiple paternity has been well-documented in deer mice (Birdsall and Nash 1973; Merritt and Wu 1975; Ribble and Millar 1996). Multiple paternity of litters was common in one population of meadow voles (*Microtus pennsylvanicus*), in which males were nonterritorial with no fixed home range (Boonstra et al. 1993). Low variability in male competitive ability (cohorts are similar because of their short life span and seasonality in breeding) and habitat that reduced the ability of males to detect rivals made it more difficult for males to monopolize females (Boonstra et al. 1993).

Most of the diurnal ground squirrels described previously as polygynous routinely have females that multiply mate. However, direct observations of their mating behaviors historically focused on male overt conflict, ignoring female behavior. Research over the past two decades has revealed that the majority of sciurid rodents are promiscuous, with both sexes pursuing strategies that often conflict with each other. Extra-pair copulations have been documented in the monogamous Alpine marmot (*Marmota marmota*; Goossens et al. 1998). Travis et al. (1996) found that in ter-

ritorial Gunnison's prairie dogs there was a very high occurrence of multiple paternity of litters; 61% of offspring were sired by extra-territory males. Travis et al. (1996) suggest that the mating system be reclassified from female defense polygyny to overlap promiscuity, as Boellstorff et al. (1994) suggested for California ground squirrels, where individual females mated with an average of six males. Similarly, in Columbian ground squirrels, once classified as female defense polygyny, females mate on average with four males, and a minimum of 16% of litters are multiply sired (Murie 1995). In Belding's ground squirrels, 78% of litters were multiply sired (Hanken and Sherman 1981), supporting the notion that polygyny is not an appropriate description of the mating system. Multiple mating and multiply sired litters have also been reported in the purportedly monogamous prairie voles (Solomon et al. 2004).

Effects of Sperm Competition on Male Strategies

Whenever females mate with more than one male, intrasexual competition can occur in the reproductive tract of the females. Sperm competition describes competition between the sperm of more than a single male for fertilization of ova within the reproductive tract of the female (Parker 1970). This sort of competition between males may be less obvious to observers, but may have profound implications for male reproductive tactics.

The degree to which sperm competition can affect the mating strategies of males depends on the effectiveness of strategies to prevent fertilization by other males. Parker (1970) identified two major male strategies when dealing with sperm competition: first, those that prevent rival males from copulating with a female (precopulatory) and second, those that increase a male's fertilization chances when dealing with a previously mated female (postcopulatory).

Countering sperm competition may reduce the opportunities to gain access to additional mates, but if the likelihood of finding another mate is low, staying and guarding a current female has a higher payoff. Another consideration is the cost of the time/energy investment of preventing subsequent re-mating, and how effective these strategies are (Parker 1970; Dewsbury 1982b; Sherman 1989; Schwagmeyer 1990). These considerations will be influenced by the current mate's opportunity to re-mate, which is affected by the time it takes for other males to find the female or the length of time between copulation and fertilization.

The influence of sperm competition on male mating strategies is also influenced by sperm precedence or mate order (nonrandom differential fertilization success; Lewis and Austad 1994). There are three possible mate order effects: a first male advantage or bias in fertilization (in which the majority of offspring are sired by the first male to copulate), a last male advantage, and no mate order advantage (Foltz and Schwagmeyer 1989). For example, first male effects have been found in thirteen-lined ground squirrels (Foltz and Schwagmeyer 1989), Belding's ground squirrels (Hanken and Sherman 1981), and in Arctic ground squirrels (Lacey et al. 1997). In thirteen-lined ground squirrels there is a 75% chance of the first male siring the offspring, while in Arctic ground squirrels first males sire 90% of the litter (Foltz and Schwagmeyer 1989; Lacey et al. 1997).

This precedence for mating (first, last) will have large impacts on the mating strategies of males. Mate-order effects can influence male searching strategies (whether to stay and copulate with a female or continue searching), the number and duration of copulations, and if the male should try to guard the female after copulation. Belding's ground squirrels, which have a first male advantage, resume searching for females as soon as copulation is completed (Sherman 1989). But if there is an extreme bias toward first males fertilizing the offspring, as in Arctic ground squirrels, subsequent males should not waste their time and energy on these females, with which they are unlikely to sire offspring (Lacey et al. 1997). If second or third mates are likely to sire offspring, then males should try to mate with any females they can. If there is a last male advantage, other mechanisms like mate guarding may be used to reduce sperm competition (Sherman 1989).

Precopulatory strategies

The most obvious precopulatory mechanism by which males can monopolize females is via overt conflict. This conflict can be through territoriality or dominance (as discussed earlier), or males can also attempt to prevent rival access to mates by mate guarding. Mate guarding is not predicted to occur if there is a first male advantage. For instance, mate guarding does not occur in Belding's ground squirrels, in which there is a clear first male advantage, but does occur in the Idaho ground squirrel (*Spermophilus brunneus*), which has a last male advantage (Sherman 1989). Mate guarding can be used by males to deter females from re-mating or to force females to accept them as a mate (Stockley 1997). This strategy can conflict with female reproductive strategies if harassment affects foraging efficiency or prevents females from mating with preferred or alternative males (Stockley 1997; Wolff and Macdonald 2004). However, females can occasionally escape guarding males, as seen in gray squirrels (Koprowski 1993a), undermining male attempts to monopolize the estrus. Mate guarding also occurs in California ground squirrels, but females were still able to successfully re-mate (Boellstorff et al. 1994).

Mate guarding can also have the disadvantage of reducing male opportunities to resume mate searching.

Another means by which a male can reduce the chances of re-mating by a female is by depositing a copulatory plug. This method has the advantage to the male that he can leave and search for new mates. Copulatory plugs have been described in many species of murids (Hartung and Dewsbury 1978) and sciurids (Murie and McLean 1980; Koprowski 1992); however, they are not always effective. In rats (*Rattus norvegicus*), 69% of copulatory plugs were dislodged with subsequent intromissions (Dewsbury 1984). In Columbian ground squirrels, 16 of 34 females had plugs and 7 were very loose (Murie and McLean 1980). Female fox squirrels and eastern gray squirrels were able to remove plugs within 30 seconds of the end of copulation, suggesting that plugs can be ineffective at restricting female re-mating opportunities (Koprowski 1992).

Thus the effectiveness of precopulatory mechanisms depends, in part, on the benefits of multiple mating to females, which will influence their incentives to seek out other mates (Wolff and Macdonald 2004; Solomon and Keane, chap. 4, Koprowski, chap. 7, this volume).

Postcopulatory strategies

If females are able to re-mate, males may pursue postcopulatory strategies. Males that fail to prevent rivals from initiating copulation with their mate can prevent fertilization by disrupting copulations. If a male gray squirrel detects a copulating pair he will attack, bite, and sometimes knock them from the tree (Koprowski 1993a). In the Cape ground squirrel, dominant males usually copulate with the female underground, where disruptions are rare (Waterman 1998). Disruptions can also occur within the female's reproductive tract, where ejaculate from one male could disrupt the sperm transport of a rival (Dewsbury et al. 1992). Ejaculate disruption can also interfere with males' own sperm transport if they re-copulate too soon, which may give incentive for males to use mate guarding or copulatory plugs to delay female re-mating.

Males may compete with rival males by altering sperm investment. Males may influence who fertilizes the ova through sperm quality or increased sperm quantity. Sperm morphology in rodents is extremely variable, in both size and length (Roldan et al. 1992). Species in which females mate with multiple males have longer sperm than in non-multiple mating species, and sperm length is positively correlated with maximum sperm velocity (Gomendio and Roldan 1991). Other factors that could influence fertilization success include sperm viability, sperm concentration, or ejaculate volume (Dewsbury 1984). Fertilization success could be influenced by the delay between competing males' copulations and timing of copulation relative to female ovulation.

The time interval between different multiple-male mating partners in thirteen-lined ground squirrels influenced the second male's ability to sire offspring (Schwagmeyer and Foltz 1990). A longer interval between matings by multiple males gave the first male a reproductive advantage for paternity. No such relationship with delay and fertilization was evident in house mice (Dewsbury 1984). In Belding's and thirteen-lined ground squirrels, which have first male advantages for siring offspring, second males may be able to get some portion of paternity by frequent re-mating with the female and increasing the duration of copulations (Sherman 1989; Schwagmeyer and Foltz 1990). These strategies of secondary males do not seem to work in Arctic ground squirrels. For a male Arctic ground squirrel the only way to increase paternity is to increase the number of females mated with first; number of matings per se does not seem to increase probability of paternity (Lacey et al. 1997).

Another postcopulatory option is to increase sperm volume (Schwagmeyer and Foltz 1990). In laboratory rats, golden hamsters (*Mesocricetus auratus*), and deer mice, the probability of fertilization is increased by larger volumes of ejaculate (Lanier et al. 1979; Dewsbury and Hartung 1980; Dewsbury 1984). Male meadow voles increase the amount of sperm in their ejaculates when they mate in the presence of another male's odors (delBarco-Trillo and Ferkin 2004). In rodents in general, there appears to be an increase in re-matings (number of ejaculates per female) with increasing potential for sperm competition (inferred by relative testes size; Stockley and Preston 2004). This increase in the number of ejaculations is suggested to increase the number of sperm transferred, but could also be related to the timing of insemination. If the exact time of ovulation is unknown, these repeated copulations could increase the likelihood of their sperm being present and viable at the time of ovulation (Parker 1984b).

Stockley and Preston (2004) used relative testes size as a measure of the intensity of sperm competition. In species in which females mate promiscuously, presumably resulting in sperm competition, males have relatively larger testes size than in monogamous or strictly polygynous species (Kenagy and Trombulak 1986; table 4.1). For North American voles, monogamous and polygynous species have small testes compared to promiscuous species (Heske and Ostfeld 1990). In the promiscuous Cape ground squirrel, with an average OSR of 10 males to a single female, testes size is nearly 20% of a male's head-body length (Waterman 1998; fig. 3.4). Males in this species masturbate to ejaculation, which may be related to intense sperm competition (Water-

A

B

Figure 3.4 In the promiscuous Cape ground squirrel, the potential for sperm competition is high and testes size is nearly 20% of a male's head-body length (a). Males will often masturbate to ejaculation (a) and (b), which may be related to intense sperm competition. Photo (a) by A. Rasa and (b) by J. Waterman.

man unpubl. data). Large testes size is presumably a mechanism for producing large volumes of sperm, increasing the probability of fertilizing eggs compared to males with less sperm volume; however, there are few data from wild populations to support this idea.

Another strategy for males to increase their fitness is to kill the young sired by another male, and thus provide themselves with a reproductive opportunity (Hrdy 1979; Ebensperger 1998c; Ebensperger and Blumstein, chap. 23, this volume). Theoretical and empirical evidence of reproductive advantages for males in committing infanticide have been discussed extensively for murid rodents (e.g., Labov et al. 1985; Ebensperger 1998c; Wolff and Macdonald 2004). Females that lose their young during early lactation typically recycle and mate again sooner than if they had nursed their young to weaning. Thus infanticide can provide a mating opportunity for males. In *Peromyscus* spp., males that had not mated in a given area (new immigrants) would kill pups, whereas resident males that had mated did not, supporting the sexual selection hypothesis for male infanticide (Wolff and Cicirello 1991). Similar observations of males killing unrelated offspring have been reported for other species of rodents and support infanticide as a male reproductive strategy (e.g., Wolff and Macdonald 2004; Ebensperger and Blumstein, chap. 23, this volume). A counter-strategy by females to male infanticide is to mate multiply with several males to confuse paternity, which in turn deters males from killing offspring (Agrell et al. 1998; Wolff and Macdonald 2004; Ebensperger 1998c; Solomon and Keane, chap. 4, this volume).

Another counterstrategy by females against infanticide is to prevent or terminate pregnancy. Dewsbury (1982c) points out the difference between pregnancy blockage, in which implantation is prevented, and the Bruce effect, where a successful fertilization is terminated by exposure to a new male. This strategy also benefits males, because females that terminate a pregnancy will re-ovulate sooner than if they carried the young to term. The Bruce effect has been demonstrated in the laboratory in many species of Muridae (e.g., Schwagmeyer 1979; Mallory and Brooks 1980; Huck 1982), in which male pheromones apparently cause abortion or resorption of embryos (Storey 1986b). The Bruce effect has even been reported in Alpine marmots, in which pregnancies were terminated with the takeover of the territory by a new male (Hacklander and Arnold 1999). However, recent evidence from experimental field studies failed to support any form of pregnancy disruption in gray-tailed voles (*Microtus canicaudus;* de la Maza et al. 1999), and only marginal or weak evidence exists in prairie voles (Mahady and Wolff 2002). De la Maza et al. (1999) suggested that multiple mating by females deters infanticide and thus functionally negates the benefits of the Bruce effect. Both field studies concluded that the Bruce effect might be an artifact of laboratory design and not an evolved reproductive strategy, at least in murid rodents.

Male Mate Choice

A major assumption in current theories of mating strategies is that males have an unlimited supply of sperm and that they are selected to maximize its use. However, Dewsbury (1982b, 1984) points out that there are physiological costs to ejaculation. In rats, reproductive behavior is reduced for a few days after satiation of mating, and prolonged copulation depletes sperm from the seminal vesicles, which can take 3 to 6 days to completely refill (Dewsbury 1984). Sperm production is limited and males must optimize ejaculate expenditure to achieve maximum reproductive payoff (Stockley 1997). Patterns of mate order may influence which females are preferred as mates, how males compete for access to those females, and how copulatory behaviors are used to achieve fertilization success (Lacey et al. 1997). Mate choice may be particularly important to males in species in which there is a first male effect and females are willing to re-mate. In thirteen-lined ground squirrels, males reject females that have already mated, and their chances of actually fertilizing the female are so low they are better off spending their energy to find new mates (Schwagmeyer and Parker 1990). Similarly, Arctic ground squirrels were able to distinguish between mated and unmated partners (Lacey et al. 1997). Male mate choice should receive greater attention as a reproductive strategy as more information on multiple paternity and mate order effects become available.

Summary

Strategies that maximize male reproductive success can vary under different population densities and distributions. Within the rodents, almost every type of mating system has been described, ranging from monogamy to promiscuity. Even within a species, variation in mating systems and mating strategies occur. Monogamy, although rare, appears to have evolved when either parental care is critical to offspring survival or when males are unable to access more than a single female because of environmental factors (fig. 3.1). Not only is the distribution of females or resources critical to a male's ability to access additional females, but actual density of both males and females appears to have a large influence. These variables (density and dispersion in space and time) also affect whether males gain access to additional females through overt conflict (territories or dominance hierarchies) or competitive searching (scramble competition). Many of the rodents described as polygynous routinely have females that multiply mate. Multiple mating by females can confound the reproductive strategies of males.

Copulatory success with females will not necessarily reflect fertilization success. Studies that examine whether a species is genetically monogamous or polygynous have shed light on the importance of sperm competition in male mating success. Sperm competition can have profound effects on the most effective mating tactics of males, and males have evolved precopulatory and postcopulatory strategies to counter it, ranging from behavioral (e.g., mate guarding) to physiological (e.g., modification of sperm investments). However, a strong male precedence in fertilization success, either as a first or last male, also may affect the resource value of a receptive female, leading males to be more selective in their choice of mates.

The mating strategies of male rodents are influenced by competition with other males and by the mating strategies of females. Mating patterns of rodents, and likely mammals in general, are dominated by promiscuous mating systems more than polygynous ones. Even when overt combat among males appears lacking, intense sexual selection among males may be occurring through less obvious forms of competition.

Chapter 4 Reproductive Strategies in Female Rodents

Nancy G. Solomon and Brian Keane

SEXUAL SELECTION involves two components: male-male competition (intrasexual selection) and female choice of mates (epigamic selection; Darwin 1871). However, studies of sexual selection have historically focused on males (e.g., aggression and dominance). The prevailing paradigm assumed that males were "eager" to mate whereas females were "comparatively passive," only mating as much as necessary to achieve fertilization of ova.

In reality, females often do more than just accept the winner of male-male contests or accept the most attractive or least objectionable opposite-sex conspecific from a number of candidates (Hrdy 1981, 1986; Andersson 1994; Lifjeld et al. 1994). An alternative to the assumption of female passivity is that females may actively solicit copulations with preferred males, seek copulations with more than one male, and employ a variety of behavioral strategies to attract the attention of preferred sexual partners (cf. Rosenqvist and Berglund 1992; Birkhead and Møller 1993). Thus female mate choice warrants more attention than it received in the earlier days of animal behavior studies (Cronin 1991).

If females are active participants in sexual interactions, differences in quality among potential mates may favor female selectivity (Trivers 1972). Numerous hypotheses have been proposed to explain mate choice by females. Females may choose partners based on heritable aspects of mating success and viability or on nongenetic benefits, such as a high-quality territory or parental care (Andersson 1994). As a result of female selectivity, males have evolved certain sexual traits or signals, which serve as indicators of male quality.

Previous studies have also ignored or deemphasized the potential for female control over processes occurring within their own bodies after mating. Interest in mechanisms regulating reproduction has provided new information on female reproductive anatomy and physiology, and suggests that there is potential for postcopulatory choice by females (i.e., cryptic female choice). Because fertilization is internal, female rodents may have partial or complete control over whether a particular male will sire some or all of their offspring.

After mating, females must successfully rear offspring, which involves providing sufficient nutrition for pre- and post-natal growth, warmth and protection from potentially infanticidal conspecifics or predators. In many rodent species, females will only breed if they have a home range that excludes competitors (Bujalska 1973; Boonstra and Rodd 1983). In most of these species, females actively defend their territories. Territorial defense has been suggested to prevent infanticide (Sherman 1980a, 1981b; Wolff 1993b; Wolff and Peterson 1998), or provide exclusive use of limited resources (Ostfeld 1985, 1990; Ims 1987a). In some species, females may disperse and bequeath the natal territory to their offspring (Harris and Murie 1984; Price and Boutin 1993).

Although many female rodents can successfully rear offspring by themselves, in other species females nest with males or live in extended family or kin groups (FitzGerald and Madison 1983; Getz et al. 1993; Burland et al. 2004; Ebensperger et al. 2004). In these groups, conspecifics participate in care of offspring born to the breeding female

(Solomon and Getz 1997). Female philopatry is prevalent in rodents and females may benefit by having kin as neighbors through the evolution of nepotistic behaviors (e.g., sharing of space; Hoogland 1995; Dalton 2000).

Evolutionary theory predicts that the female reproductive strategies should be those that result in the greatest number of offspring in succeeding generations. Among rodents, a female's reproductive success will be primarily influenced by two factors: whom she mates with and her ability to rear the offspring after fertilization. In the small percentage of rodent species in which males engage in paternal care of offspring (Gubernick and Alberts 1987; Wang, Insel, Rosenblatt, and Snowdon 1996; Gubernick and Teferi 2000; Wynne-Edwards 2003), a female's reproductive success also would depend on the quality of the male as a parent. These factors are not mutually exclusive, since a male's genetic contribution will affect the quality and quantity of the offspring a female produces and his behavior after the offspring are born may further increase their chances of survival (e.g., protection from infanticide). In this chapter, we will focus on precopulatory and postcopulatory behaviors that affect female reproductive success. We will attempt to draw conclusions after reviewing the published literature from field and laboratory studies, and then discuss areas for future research.

Female Mate Choice

Female choice of mates can occur at various stages: before copulation, during copulation, after copulation but before fertilization, and after fertilization (Birkhead and Møller 1993). Most information concerning precopulatory mate choice by female rodents is derived from laboratory studies (reviewed in Dewsbury 1988; table 4.1). Although only measurements of actual copulations are true indications of mate choice, few studies (~ 25%, N = 50 studies) report data on preferential mating. Instead, most studies report odor preferences (time spent with the odor cues from each male) or social preferences (time spent with each male) in settings in which females could not copulate with males (fig. 4.1). A few studies that have measured odor preferences (Egid and Brown 1989; Arcaro and Eklund 1998) or social preferences and mating preferences show that the former indeed reveal mating preferences (Pierce and Dewsbury 1991; Williams et al. 1992), but other studies suggest that a female's social preference is not a good indicator for her mating preference (Gubernick and Addington 1994; Derting et al. 1999).

Female rodents use many different traits in mate choice, including mating status (Pierce and Dewsbury 1991), infec-

tion status (Klein et al. 1999), dominance status (Shapiro and Dewsbury 1986), body size (Solomon 1993a), spatial ability (Spritzer et al. 2005), genetic compatibility (Lenington et al. 1994; Penn and Potts 1999), relatedness (Barnard and Fitzsimons 1988; Keane 1990b), and familiarity (Shapiro et al. 1986; Salo and Dewsbury 1995; Randall et al. 2002; Zhao et al. 2002; see Andersson 1994 for review).

Effects of female choice on reproductive success

One way of determining if female mate preferences are beneficial would be to test whether females are more likely to produce litters when paired with preferred males and whether these matings result in offspring of higher fitness. Drickamer and colleagues (2000) allowed female house mice to display social preferences for one of a pair of males that were selected at random without regard to any particular phenotypic trait. Half of the females were paired with their preferred social partner and the other half with the nonpreferred male. A greater proportion of females that mated with preferred males produced a litter (93% with preferred mates versus 71% with nonpreferred males). In addition, sons from preferred matings were dominant to sons from nonpreferred matings, and adult offspring from preferred matings built better nests, had larger home ranges, and survived longer in the laboratory and in field enclosures than offspring from nonpreferred matings (Drickamer et al. 2000). Although pregnancy rates of daughters from preferred and nonpreferred matings did not differ statistically, 86% of daughters from preferred matings experienced at least one pregnancy in field enclosures as compared to only 56% of daughters from nonpreferred matings, suggesting that there may be significant differences in lifetime reproductive success. These findings suggest that mate choice for particular males may result in multiple direct and indirect benefits to females. In the following sections we review the evidence for benefits from female mate choice in rodents.

Mating status of male

When males mate multiple times in quick succession, the later ejaculates tend to have reduced sperm counts and fertility (Huck and Lisk 1985; Huck et al. 1986; Austin and Dewsbury 1986). Therefore, the capacity to produce fertile ejaculates and to impregnate females may be temporarily limited (Dewsbury 1982b). Thus if females copulate with previously mated males, they risk the chance that they may not be impregnated or may have smaller litters.

Pierce and Dewsbury (1991) argued that monogamous females should be more concerned than polygamous females about mating with an unmated male because they

Table 4.1 List of studies examining female mate choice in rodents. Studies are categorized by traits and for each study, the female preference, if any, for particular males and the method(s) used to measure mate choice are summarized

Species	Choice	Measure used	Reference
Mating status of male			
Norway rat (*Rattus norvegicus*)	Unmated > mated	Odor preference	Krames and Mastromatteo 1973
Golden hamster (*Mesocricetus auratus*)	Unmated > recently mated	Social preference and lordosis	Huck et al. 1986
Prairie vole (*Microtus ochrogaster*)	Unmated > mated	Social and mating preference	Pierce and Dewsbury 1991
Meadow vole (*M. pennsylvanicus*)	No preference: unmated = mated	Social and mating preference	Salo and Dewsbury 1995
Montane vole (*M. montanus*)	No preference: unmated = mated	Social and mating preference	Pierce and Dewsbury 1991
Brandt's vole (*M. brandtii*)	No preference	Social preference and behavior	Yin and Fang-Ji 1998
Familiarity			
Mandarin vole (*M. mandarinus*)	Familiar > unfamiliar	Social preference and behavior	Zhao et al. 2002
Brown lemming (*Lemmus trimucronatus*)	Familiar > unfamiliar	Mating behavior	Coopersmith and Banks 1983
Giant kangaroo rat (*Dipodomys ingens*)	Less aggression and more amicable toward familiar than unfamiliar	Behavior in lab and field	Randall et al. 2002
Genetic compatibility—Relatedness			
House mice—Porton strain	Same strain > different strain = brother = control (clean bedding)	Odor preference	Gilder and Slater 1978
House mice—ICR-JCL strain	No preference sibling/non-sibling	Social preference	Hayashi and Kimura 1983
House mice—CFLP strain	No preference sibling, half-sibling, cousin, or unrelated	Odor choice, social preference	Barnard and Fitzsimons 1988
House mice—lab strain	No preference sister, cousin, or unrelated	Odor choice	Krackow and Matuschak 1991
Wild house mice (*Mus musculus*)	Non siblings > familiar and unfamiliar brothers	Social preference	Winn and Vestal 1986
Wild house mice (*M. musculus*)	Non-siblings > siblings	Social preference	Dewsbury 1988
Deermice (*Peromyscus maniculatus*)	Familiar siblings > non-siblings	Odor preference and behavior	Dewsbury 1990
White-footed mice (*P. leucopus*)	Cousins > siblings or unrelated	Odor preference and behavior	Keane 1990b
Common vole (*Microtus arvalis*)	Non-siblings > siblings	Odor preference	Bolhuis et al. 1988
Mandarin vole (*M. mandarinus*)	Unfamiliar > familiar individuals	Social preference and behavior	Fadao et al. 2000
Brandt's vole (*M. brandti*)	Social: brothers = unrelated Mating: unrelated > brothers	Social and mating preference	Yu et al. 2004
Naked mole-rat (*Heterocephalus glaber*)	Odor: no preference Social: estrus females: unfamiliar > familiar Social: reproductively inactive: no preference	Odor and social preference	Clarke and Faulkes 1999
Genetic compatibility—Intra-genomic conflict and self-promoting elements			
Wild house mice (*Mus musculus*)	Opposite haplotype > similar t-haplotype	Social preferences	Coopersmith and Lenington 1990
Wild house mice (*M. musculus*)	Opposite haplotype > similar t-haplotype	Social preferences	Lenington 1991; Lenington et al. 1994
Genetic compatibility—Immune system			
House mice—lab strain	Different MHC type than own	Odor and mating preference	Egid and Brown 1989
House mice—congenic strains	Different MHC type than own	Odor and mating preference	Arcaro and Eklund 1998
House mice—congenic strain	No preference	Odor preference	Ehman and Scott 2001
House mice—congenic strain	Preference for same strain regardless of own haplotype	Mating preference	Beauchamp et al. 1988
House mice—outbred	Different MHC type than own	Matings in semi-natural enclosures	Potts et al. 1991
Wild house mice (*Mus musculus*)	No preference	Odor and mating preference	Eklund 1997, 1999
Good genes—Resistance to infection			
House mice—lab strain	Non-parasitized > parasitized	Odor preference	Kavaliers and Colwell 1995b
House mice—congenic strains	Non-parasitized > parasitized Small N; 5 females/strain	Odor preference	Ehman and Scott 2001
House mice (*Mus musculus*)	Disease free > diseased	Odor preference	Penn et al. 1998
Wild house mice (*M. musculus*)	Non-parasitized > parasitized	Odor preference	Mihalcin 2002
Norway rat—lab strain (*Rattus norvegicus*)	Non-parasitized > parasitized (P = 0.059) No control for estrus state	Odor preference	Willis and Poulin 2000

Table 4.1   (continued)

Species	Choice	Measure used	Reference
Prairie vole (*Microtus ochrogaster*)	Odor: non-parasitized > parasitized Social: no preference	Odor and social preference	Klein et al. 1999
Good genes—Dominance status			
House mice ICR-JCL	Dominant > subordinate	Odor preference	Hayashi 1990
House mice—2nd generation from wild	Dominant > subordinate	Odor preference	Drickamer 1992a
Golden hamster (*Mesocricetus auratus*)	Dominant > subordinate	Scent marking in semi-natural enclosures	Lisk et al. 1989
Norway rat (*Rattus norvegicus*)	Dominant > subordinate Female in estrus	Social and mating preference	Carr et al. 1982
Bank vole (*Clethrionomys glareolus*)	Dominant > subordinate	Social and odor preference	Hoffmeyer 1982; Horne and Ylönen 1996
Bank vole (*C. glareolus*)	Dominant > subordinate	Odor preference and behavior	Kruczek 1997
Prairie vole (*Microtus ochrogaster*)	Dominant > subordinate	Social and mating preference	Shapiro and Dewsbury 1986
Montane vole (*M. montanus*)	No preference	Social and mating preference	Shapiro and Dewsbury 1986
Root vole (*M. oeconomus*)	Dominant > subordinate	Social preference	Zhao et al. 2003
Water vole (*Arvicola terrestris*)	Dominant > subordinate	Odor preference	Evsikov et al. 1994
Brown lemming (*Lemmus trimucronatus*)	Dominant > subordinate	Social and behavior	Huck and Banks 1982
Good genes—Spatial ability			
Meadow voles (*Microtus pennsylvanicus*)	Good spatial ability > poor spatial ability	Social preference	Spritzer et al. 2005
Other			
Snow vole (*Chionomys nivalis*)	Males with higher hematocrit volume > lower hematocrit volume More symmetric males > less symmetric	Odor preferences	Luque-Larena et al. 2003

Figure 4.1   One paradigm used in the study of female mate choice involves tethering two males in opposite ends of a three-chambered cage to prevent male-male interactions from influencing a female's choice. The female is then released into the central chamber and allowed to move freely throughout the three chambers. Time spent with each male and/or number of matings with each male can be quantified. This photo was taken at the end of a mating trial. The female is in the left-hand chamber with one of the males. She has moved all nest material from the chambers of both males into one single nest with the preferred male.

will only be mating with one male. The implication is that monogamous females have more to lose by mating with previously mated males than do females that mate with multiple males; however, this remains to be demonstrated. We do know that although female prairie voles (*Microtus och-rogaster*), a socially monogamous species, tended to prefer unmated males (as assessed through visitation and copulation; Pierce and Dewsbury 1991), so do females in multiple-mating species such as female laboratory rats (*Rattus norvegicus*, Krames and Mastromatteo 1973) and golden hamsters (*Mesocricetus auratus*, Huck et al. 1986). In contrast, montane voles (*Microtus montanus*) and meadow voles (*Microtus pennsylvanicus*), both of which mate with multiple males, showed no preference between the two types of males in similar test situations (Pierce and Dewsbury 1991; Salo and Dewsbury 1995). There are no data on the consequences of these preferences. To test Pierce and Dewsbury's hypothesis, more studies are needed from socially monogamous species to see if they differ from promiscuous species.

It has been suggested that female selectivity for a mate is related to the mating system, with females of monogamous species being more choosy than females of polygamous species (Salo and Dewsbury 1995). Aside from fertilization assurance, females of monogamous species may also be expected to prefer unmated males, to ensure that she and her offspring are the primary beneficiaries of the male's paternal effort.

Material benefits

Material benefits in the form of paternal care are rare in rodents, because most rodents are polygynous or promiscuous. In these species, males provide no paternal care, because they benefit more from mating with multiple females than by living with just one and helping to rear offspring. In socially monogamous species, the willingness and ability to provide paternal care (e.g., huddling over pups, licking pups, retrieving pups back to the nest—Gubernick and Alberts 1987; Solomon 1993b; Lonstein and DeVries 1999; McGuire and Bemis chap. 20, this volume) may be important for female reproductive success. For example, female Djungarian hamsters (*Phodopus campbelli*) and California mice (*Peromyscus californicus*) need help from their male partner to successfully rear offspring (Wynne-Edwards 1987; Gubernick and Teferi 2000). However, the relationship between female mate choice and male parental care is unknown.

Ownership of a territory may be another important benefit for females, since lactating females need a safe, secure nest site with high-quality food resources nearby. In house mice (*Mus musculus*), females mate more frequently with males that defend high-quality territories (Wolff, R. J., 1985), but whether they choose the male, the territory, or the combination is not known.

Good genes

If sexually selected traits are strongly correlated with a male's physical condition, males in good condition should exhibit traits that are considered attractive to females (Andersson 1994). Females may discriminate among males based on these characteristics, choosing males with heritable characteristics that would confer increased viability (Drickamer et al. 2000) or attractiveness (Weatherhead and Robertson 1979) to their offspring. For example, if characteristics that affect dominance are heritable, females may be predisposed to mate with dominant males so their offspring may inherit traits associated with dominance (Drickamer 1992a; Horne and Ylönen 1998). We will discuss three ways in which the quality of the genes a male possesses may be important for female mate choice.

Resistance to infection

Clayton (1991) and Zuk (1992) suggest that female choice should be influenced by the ability of males to resist a parasitic infection. Females may also benefit indirectly if resistance has a genetic basis. In the latter instance, resistance may be passed on to offspring (Potts and Wakeland 1993). Females and their offspring may benefit directly by avoiding infection, either with a parasite or disease, if females are able to discriminate between uninfected and infected males and mate less frequently with the latter (Kavaliers and Colwell 1995a, 1995b), although this benefit does not have a genetic basis.

Kavaliers and Colwell (1995a) reported that female house mice are able to discriminate between the odors of parasitized and nonparasitized males, and females avoid the odors of parasitized males (Kavaliers and Colwell 1995a). Furthermore, Penn et al. (1998) reported that female house mice are able to assess a male's infection status via odor cues present in male urine, and that infection eliminates the attractiveness of the male's odor cues. Only one study, Klein et al. (1999), goes beyond testing odor preference. In this study, investigators found that female prairie voles preferred the odors of uninfected to infected males, but there was no difference in time spent with either type of male. Although these laboratory studies suggest that male infection status may influence female attraction to males, it is not known whether this is true in nature.

Dominance status

The offspring of dominant males could have greater fitness either because dominance is genetically linked to other traits that improve survivorship or because traits that affect dominance status are heritable (Drickamer 1992a; Horne and Ylönen 1998). If that is the case, then sons of dominant males should have superior mating success due to female choice or male-male competition (Weatherhead and Robertson 1979). Females would then benefit indirectly by the increased reproductive success of their offspring (Dewsbury 1982b; Ellis 1995).

Females may also gain direct benefits from mating with dominant males because dominant males are better able to maintain high-quality territories (Wolff, R. J., 1985; Rich and Hurst 1998). Dominant males may also be better able to protect females from sexual harassment and protect their offspring from being killed by conspecific males (Agrell et al. 1998). Thus, cohabiting and mating with a dominant male may result in a direct benefit to the female through increased survival of her offspring.

Although female montane voles do not show a preference for males with high dominance rank compared to males with low dominance rank, at least in the laboratory (Shapiro and Dewsbury 1986), the results from the majority of lab studies on rodents consistently show that females prefer to socialize with and mate with dominant males (table 4.1). Also, in the field, yellow-toothed cavies (*Galea musteloides*) preferred larger males (Hohoff et al. 2003). If body size is positively correlated with dominance, these studies indicate that the social preference for dominant males by females observed in many lab studies may also be occurring in nature as well.

Spatial ability

Females may prefer males with good spatial ability to those with poor spatial ability because females that mate with males that have better spatial ability are likely to gain indirect fitness benefits for their male offspring. Because spatial ability is heritable (Upchurch and Wehner 1988, 1989; Okasanen et al. 1999), the male offspring of sires with good spatial ability may be better able to locate mates and would be favored by female choice. These indirect benefits might be particularly important during periods of low population density, when mates become more difficult to locate. Spatial ability is also likely to be important for females (and consequently all their offspring), for locating and returning to nests.

Genetic compatibility

Most good genes hypotheses assume that all females in a population benefit from mating with the same males. However, breeding studies indicate that the degree of relatedness between males and females can affect female reproductive success (Bateson 1983; Ralls et al. 1986). Thus some opposite-sex conspecifics would be better mates than others. Molecular evidence reinforces the hypothesis that the best mate varies among females within a population (Zeh and Zeh 1996). The genetic compatibility hypothesis differs from good genes hypotheses in that the fitness consequences of mating depend on the "fit" between the female's and male's genes, as opposed to just being a function of the male's genetic quality (Zeh and Zeh 1996). The good genes hypotheses predict that all females that mate with the same male achieve the same reproductive consequences,

but the genetic compatibility hypotheses predict that female reproductive success depends on the mate. Three ways in which genetic compatibility could be important are discussed below.

Inbreeding

Detrimental effects of incest have been demonstrated in a number of mammals (Wainwright 1980; Ralls et al. 1986). When paired with a closely related male, females may not mate (McGuire and Getz 1981), produce fewer litters (Krackow and Matuschak 1991), and have smaller litters at weaning (Barnard and Fitzsimons 1988; Keane 1990b) or smaller pups at weaning (Keane 1990b). Outbreeding also has costs, including breakup of coadapted gene complexes, loss of rare genes, and loss or suppression of genes required for adaptation to a local habitat (Bateson 1983). Females may also avoid mating with an unfamiliar individual because it might be very different genetically.

Females showed social preferences for nonsiblings as compared to siblings in only 3 of 9 studies (table 4.1). Although Keane (1990b) demonstrated female preferences for cousins as compared to siblings in captive white-footed mice (*Peromyscus leucopus*), this preference has not been found in laboratory or wild house mice (Barnard and Fitzsimons 1988; Krackow and Matuschak 1991). In most studies, cousins have not been included in the stimulus set, so we know very little about preference or lack of preference for individuals of intermediate degrees of relatedness.

Intragenomic conflict and self-promoting elements

One of the major factors involved in genetic incompatibility between mates is segregation distorters, which act to increase their proportional representation in gametes, that is, the ability to transmit themselves at the expense of the wild type homolog from heterozygous males (Silver 1993). The *t*-locus in house mice is one example of a well-studied segregation disorder (Lenington 1991; Silver 1993; and see Carroll and Potts, chap. 5, this volume). About 25% of wild house mice are heterozygous for the *t*-haplotype (i.e., +/*t*) (Lenington, Franks, and Williams 1988). Most *t*-haplotypes carry recessive lethal factors, which result in the death of embryos homozygous for the *t*-allele prior to birth. In addition, there are also semilethal haplotypes in wild populations (Lenington, Egid, and Williams 1988). Females heterozygous at the *t*-locus preferentially mate with homozygous wild-type males and thus avoid producing inviable *t/t* offspring (table 4.1; see also Carroll and Potts, chap. 5, this volume).

Immunologically based compatibility

The highly polymorphic major histocompatibility complex (MHC) genes control immunological recognition of self/

nonself antigens; therefore female mating preferences may be influenced by her MHC haplotype and that of her potential mate. Mate choice based on differences in MHC haplotypes can increase fitness due to decreased inbreeding (Brown 1999), because the MHC acts as an indicator of overall genetic similarity between mates. Therefore, the detection of MHC similarity can act to prevent matings between close relatives. In addition, mates with different MHC genotypes are more compatible, higher-quality mates, because their offspring will have a more heterogeneous immune response (Doherty and Zinkernagel 1975; but see Penn and Potts 1999) and females are less likely to suffer fetal mortality. How the MHC complex affects reproduction in rodents is discussed more thoroughly in Carroll and Potts, chap. 5, this volume.

In half of the published studies ($N = 6$), female house mice display odor preferences and mating preferences for males whose MHC haplotype differs from their own (table 4.1). These results were not consistently found when inbred congenetic strains of house mice were tested, nor were wild female house mice apparently using MHC haplotypes as a basis for mate choice (Eklund 1997; 1999). Carroll and Potts (chap. 5, this volume) argue that laboratory studies do not give the true picture of mate choice based on MHC haplotypes, and that seminatural enclosures are better for answering this question. When mice were released into seminatural enclosures, they did show disassortative mating due to settlement of male-female pairs and extraterritorial matings by females. Extraterritorial matings by females accounted for approximately 75% of the excess heterozygotes in the population (Carroll and Potts, chap. 5, this volume) and were primarily responsible for the deficiency of t/t homozygotes in the populations (Potts et al. 1992). When mice differed at only one MHC locus against a wild-type background, mating was random rather than dissassortative. This result strongly suggests that other genetic loci may be important in the expression of olfactory cues that affect mate choice (Carroll and Potts, chap. 5, this volume). More studies are necessary to determine whether or not MHC antigen-presenting alleles are the source of MHC-based female mate choice.

Choosing when multiple traits are involved

Although most studies of female mate choice focus on a single trait, investigators recognize the potential complexity of mate choice. A choosy individual will likely consider more than one trait when assessing the quality of potential mates (Burley 1981), and therefore the preference displayed may be a trade-off among these traits. In only a few studies have investigators designed mate choice tests so that the results of females' choices show the relative importance of various traits. Two studies show that the typical preference for dominant males can be altered if another factor is included. Mihalcin (2002) showed that the preference for dominant males versus subordinate males changed when parasite infection was also included as a factor. Mihalcin found that female house mice preferred odors from subordinate nonparasitized males to those of dominant parasitized males. Thus infection status of the male was more important than dominant status. In another study, female meadow voles spent more time with males that had good spatial ability and low dominance ranks as compared to males with poor spatial ability and high dominance ranks (Spritzer et al. 2005). These results suggest that spatial ability may be relatively more important than dominance rank to females.

Most recently, Roberts and Gosling (2003) showed that indicators of male quality and mate complementarity can both have important although different roles in female mate choice. In some strains of house mice, rate of scent marking (an indicator of male quality) was a better predictor of female preference, as measured by time spent with odors, than was MHC dissimilarity. MHC dissimilarity appeared to be important only when differences in scent marking were small. Thus in these experiments each trait signals separate, sometimes conflicting, adaptive values, forcing females to trade off between males with different combinations of traits.

These types of studies have significant implications for sexual selection theory. First, they demonstrate the importance of female choice in contributing to the maintenance and variability of sexually selected traits (Roberts and Gosling 2003). These studies also give insight into selection for the complex decision-making process of females faced with choosing among males with a combination of traits. A female's choice is likely to be determined by the relative fitness benefits resulting from choosing the best male. Finally, the impact of multiple male traits on female choice may explain some of the lack of consistency in the results of previous studies where only one trait was examined.

Multiple Mating by Females

In general, the hypotheses proposed to explain the functional significance of mating with more than one male (multi-male mating; MMM) are similar to those proposed to explain female mate choice (Schwagmeyer 1984; Keller and Reeve 1995; Yasui 1997, 2001; Jennions and Petrie 2000). These benefits of mating with multiple males include assurance of fertilization (can include a guard against sperm depletion in males and bet hedging against sterility in the first male), gaining material benefits (i.e., more paternal investment from males), getting better genes for offspring, increasing the probability of mating with high-quality males, mate compatibility, increasing the genetic variability of offspring, and preventing harassment by males or infanticide of offspring. We have discussed some of these hypotheses previously and do not review them here because there are already very good reviews in the literature (see Schwagmeyer 1984; Agrell et al. 1998; Wolff and MacDonald 2004 for more details on these hypotheses).

The observations of MMM by females might suggest that female mating preferences are more complex than has been assumed by sexual selection theory (Schwagmeyer 1984). Some of these assumptions of sexual selection theory are being reexamined with the increasing evidence for benefits to females from mating with more than one male.

Based on our literature search, the intraspecific frequency of MMM varies from 12% in red squirrels (*Sciurus vulgaris*, Wauters et al. 1990) to 79% in meadow voles (Berteaux et al.1999) and multiple paternity within litters varies from 16% in Columbian ground squirrels (*Spermophilus columbianus*, Murie 1995) to 90% in the yellow-toothed cavy (Hohoff et al. 2003). There are also a few data showing interpopulation variability in multiple paternity (table 4.2). Although little is known about temporal variation in multiple paternity within a population, multiple paternity can vary from none (0%) to a high level (50%) within the same population in different years (Schwagmeyer and Brown 1983).

Effects of multi-male mating on female reproductive success

Even though more social than solitary species are reported to have MMM (Møller and Birkhead 1989), MMM is reported just as frequently in socially monogamous mammals as in socially polygynous mammals. As predicted by the active female hypothesis, females solicit multiple matings in all species for which data are available (Wolff and MacDonald 2004). In most species, we still do not know with whom females mate. These data are critical for evaluating proposed hypotheses.

There is conflicting evidence on the effects of MMM on the probability of producing a litter. In Gunnison's prairie dogs (*Cynomys gunnisoni*), there is increased probability of producing a litter following MMM (Hoogland 1998b). An increased probability of conceiving following MMM may be due to fertility assurance, genetic incompatibility avoidance, or mate choice for good genes via sperm competition (Murie 1995; Stockley 2003). On the other hand, a decreased probability in conceiving has been detected in two species following MMM (deermice, *Peromyscus maniculatus,* and Djungarian hamsters, Dewsbury 1982b; Wynne-Edwards and Lisk 1984). Such a decrease following MMM may be the result of a strange-male-induced pregnancy block (Dewsbury 1982b; Wynne-Edwards and Lisk 1984). Finally, MMM had no effect on conception in four species (13-lined ground squirrels, *Spermophilus tridecemlineatus*, Schwagmeyer 1986; Columbian ground squirrels, Murie 1995; black-tailed prairie dogs, *Cynomys ludovicianus*, Hoogland 1995; prairie voles, Wolff and Dunlap 2002).

Although litter size tends to not to differ at birth (only 1 of 5 studies showed a significant difference), there was a nonsignificant tendency toward increased offspring survivorship for female Columbian ground squirrels that mated with more males (Murie 1995). Furthermore, female yellow-toothed cavies weaned more young if they mated with four males than with only one male, because more pups survived between birth and weaning (Keil and Sacher 1998).

Postcopulatory Cryptic Female Choice

Mate choice can occur after copulation in many species of animals. Eberhard (1996) reviewed available data on postcopulatory (cryptic) female choice. Cryptic female choice is defined as processes that occur after copulation and influence paternity in favor of particular males. This type of female choice is referred to as cryptic because it cannot be observed (Bolhuis and Giraldeau 2005). Cryptic female choice is considered to be relatively common in some animal taxa while still remaining relatively unexplored in others. Of the numerous possible mechanisms by which cryptic female choice may occur (see the following) some are not likely to occur in rodents, but we will review evidence for several that are more likely (table 4.3).

Failure to transport sperm to the site of fertilization

In rodents, male genitalia do not reach the site where ova are fertilized; females actively transport the sperm (Hunter 1975). In mammals, sperm transport is often triggered by copulation. For example, at least one preejaculatory intromission is needed to stimulate a female Norway rat (*Rattus norvegicus*) to transport sperm to the oviducts (Adler 1969). Sperm are not transported past the cervix if a female receives additional intromissions too soon after receiving the first ejaculation (Adler and Zoloth 1970). It takes a minimum of 10 minutes before sperm are completely transported to the site of fertilization (Matthews and Adler 1977). Female Norway rats solicited intromissions from

Table 4.2 Rodent species for which there is evidence that females mate with more than one male

Family	Species	Type of study	Evidence for multiple mating	Reference
Heteromyidae	Kangaroo rat (*Dipodomys ingens*)	F	Observed multiple males entering female burrow when female in estrus	Randall 1987
	Merriam's kangaroo rat (*D. merriami*)	F	Observe 1 female with > 1 male	Randall 1987
Caviidae	Yellow-toothed cavy (*Galea musteloides*)	L	90% multiple paternity (N = 10)	Hohoff et al. 2003
		L	83% multiple paternity (N = 18)	Keil et al. 1999
		L	Females mated with 2–4 males	Schwartz-Weig and Sacher 1996
Sciuridae	Fox squirrel (*Sciurus niger*)	F	Females mated with 2–4 males	Steele and Koprowski 2001
	Gray squirrel (*S. carolinensis*)	F	Females mated with 2–4 males	Steele and Koprowski 2001
	Red squirrel (*S. vulgaris*)	F	12% mated multiply (N = 26)	Wauters et al. 1990
	Gray-bellied squirrel (*Callosciurus caniceps*)	F	Females mated with 4–6 males	Tamura 1993
	Plantain squirrel (*C. notatus*)	F	Females mated with 2–4 males	Tamura 1993
	Black-tailed prairie dog (*Cynomys ludovicianus*)	F	33% copulated with > 1 male (N = 542) 32.5% multiple paternity (N = 40) 73.5% multiple paternity in multi-male coteries (N = 34 coteries)	Hoogland 1995
	Gunnison's prairie dog (*C. gunnisoni*)	F	65% copulated with > 1 male mean = 4.4 males 77% multiple paternity (N = 44)	Hoogland 1998b; Haynie et al. 2003
	Utah prairie dog (*C. parvidens*)	F	71% (N = 14) and 90% (N = 19) multiple paternity	Haynie et al. 2003
	Alpine marmot (*Marmota marmota*)	F	43% multiple paternity in multi-male groups (N = 23 litters)	Arnold et al. 1994
		F	30% multiple paternity in single male groups (N = 35)	Goosens et al. 1998
	Arctic ground squirrel (*Spermophilus parryii*)	F	Females mated with 1.9 males 8% multiple paternity (N = 12)	Lacey et al. 1997
	Belding's ground squirrel (*S. beldingi*)	F	Females mated with 1–5 males 78% multiple paternity (N = 38)	Hanken and Sherman 1981
	California ground squirrel (*S. beecheyi*)	F	All females mated multiply (N = 12) 89% multiple paternity (N = 9)	Boellstorff et al. 1994
	Columbian ground squirrel (*S. columbianus*)	F	Females mated with ~4 males 16% multiple paternity (N = 165) All females mated multiply	Murie 1995
	Idaho ground squirrel (*S. brunneus*)	F	19% multiple paternity (N = 26)	Sherman 1989
	Richardson's ground squirrel (*S. richardsonii*)	F	53% mated multiply (N = 15)	Michener and McLean 1996
	Thirteen-lined ground squirrel (*S. tridecemlineatus*)	F	Females mated with 2.3 males	Schwagmeyer and Wootner 1985
		F	0% (1979)–50% (1978) mated multiply (N = 12)	Schwagmeyer and Brown 1983
Muridae	House mouse (*Mus musculus*)	E	48% multiple paternity (N = 58)	Manning et al. 1992
	Striped field mouse (*Apodemus agrarius*)	F	80% multiple paternity (N = 10)	Baker et al. 1999
	Wood mouse (*A. sylvaticus*)	F	50% multiple paternity (N = 6)	Baker et al. 1999
		E	85% multiple paternity (N = 34)	Bartmann and Gerlach 2001
	Malagasy giant jumping rat (*Hypogeomys antimena*)	F	4% EPP (N = 48)	Sommer and Tichy 1999
	Bank vole (*Clethrionomys glareolus*)	F	22.5% obs. & 35.5% est. multiple paternity (N = 31)	Ratkiewicz and Borkowska 2000
	Red-backed vole (*C. rufocanus*)	E	12.5% multiple paternity (N = 29)	Kawata 1988
	Prairie vole (*Microtus ochrogaster*)	L	55% (N = 47) mated with >1 male	Wolff and Dunlap 2002
		F	55% multiple paternity (N = 9)	Solomon et al. 2004
	Meadow vole (*M. pennsylvanicus*)	L	79.3 copulated > 1 male	Berteaux et al. 1999
Cricetidae	Deermouse (*Peromyscus maniculatus*)	F	10% obs. and 19–43% est. multiple paternity (N = 107)	Birdsall and Nash 1973
		F	~17% multiple paternity (N = 18)	Ribble and Millar 1996
	White-footed mouse (*P. leucopus*)	F	3% (N = 29) 1987 and 19% (N = 32) 1988	Xia and Millar 1991

NOTES: E = enclosures; F = field; L = laboratory.

Table 4.3 Possible instances of cryptic female choice in rodents (adapted from Eberhard 1996)

| | Evidence for criteria | | | | |
Species	Female mate preference observed[a]	Female mate preference observed in nature[b]	Differences in male reproductive success[c]	Females mate multiply in nature[d]	Reference
Discarding sperm from current male					
House mouse (*Mus musculus*)	?	?	?	?	Leckie et al. 1973
Failure to transport sperm to fertilization site					
Norway rat (*Rattus norvegicus*)	Y	?	Y?	Y	Adler 1969; Chester and Zucker 1970; Matthews and Adler 1979; McClintock, Anisko, and Adler, 1982; McClintock, Toner, Adler, and Anisko, 1982
Remating with other males					
Golden hamster (*Mesocricetus auratus*)	?	?	Y	Y?	Carter 1973; Ogelsby et al. 1981; Huck and Lisk 1985
Norway rat (*Rattus norvegicus*)	Y	Y	?	Y	Hardy and Debold 1972; Erskine and Baum 1982; McClintock 1984
Reducing number of offspring produced after copulation					
Water vole (*Arvicola terrestris*)	Y	?	Y	?	Potapov et al.1993, cited in Eberhard 1996
Premature termination of copulation					
Golden hamster (*Mesocricetus auratus*)	Y	Y	Y	Y?	Huck and Lisk 1986
Failure to ovulate					
Norway rat (*Rattus norvegicus*)	?	?	Y	Y	Zarrow and Clark 1968
Bank vole (*Clethrionomys glareolus*)	Y	?	?	?	Clarke and Clulow 1973; Clulow and Mallory 1974
Field vole (*Microtus agrestis*)	Y	?	?	?	Clarke and Clulow 1973; Milligan 1982
Prairie vole (*M. ochrogaster*)	?	?	?	?	Gray et al. 1974
Montane vole (*M. montanus*)	Y	?	Y	?	Davis et al. 1974
Mongolian gerbils (*Meriones unguiculatus*)	Y	?	Y	Y	Ågren 1990
Failure to prepare the uterus for implantation					
House mouse (*Mus musculus*)	?	?	Y	?	Diamond 1970; McGill 1962; Land and McGill 1967
Norway rat (*Rattus norvegicus*)	Y	Y	Y	Y	McClintock 1984
Golden hamster (*Mesocricetus auratus*)	?	?	Y	Y	Huck and Lisk 1986
Deermouse (*Peromyscus maniculatus*)	Y	Y	Y	Y	Dewsbury and Baumgartner 1981; Lovecky et al. 1979
Cactus mouse (*P. eremicus*)	?	?	Y	?	Dewsbury and Estep 1975
Field vole (*M. agrestis*)	?	?	?	?	Milligan 1982
Removal of mating plugs					
Fox squirrel (*Sciurus niger*)	Y	Y	?	Y	Koprowski 1992
Gray squirrel (*S. carolinensis*)	Y	Y	?	Y	Koprowski 1992
Sometimes abort zygotes					
House mouse (*Mus musculus*)	Y	Y?	Y	Y?	Hoppe 1975; Bronson 1979; Huck 1982
Meadow vole (*Microtus pennsylvanicus*)	Y	?	Y	Y	Clulow and Langford 1971; Mallory and Clulow 1977; Webster et al. 1981; Boonstra et al. 1993
Field vole (*M. agrestis*)	Y?	?	Y	Y?	Clarke and Clulow 1973
Collared lemming (*Dicrostonyx groenlandicus*)	Y	Y?	Y	Y?	Mallory and Brooks 1978
Biased use of stored sperm					
House mouse (*Mus musculus*)	Y	—	—	Y (t-allele)	Bronson 1979; Bateman 1960, cited in Denell and Judd 1961

[a]Does female behavior towards potential mates differ among males?
[b]Does female behavior towards potential mates differ among males in nature?
[c]Do differences in female behavior towards potential mates result in differences in male reproductive success?
[d]Do females mate with multiple males in nature?

males sooner after receiving an ejaculation with a subordinate male (about 55% of females resumed mating in <10 minutes) as compared to females that received an ejaculation from a dominant male (only ~ 22% resumed mating in less than 10 minutes; McClintock, Anisko, and Adler 1982). Therefore, sperm transport was likely to be decreased when females received an ejaculate from a subordinate male.

Failure to ovulate

Most female rodents, such as Norway rats and house mice, ovulate spontaneously (i.e., on a regular cycle), while in others, such as voles, ovulation is induced (triggered) by mating (Milligan 1982). Even in rodents thought to be spontaneous ovulators, copulation can cause ovulation to occur earlier than it would have otherwise or increase ovulation rates (Zarrow and Clark 1968; Eberhard 1996). If females respond differentially to different males—for example, not all copulations result in ovulation or not all ova are ovulated after each copulation—then selection would act on male traits that increase the probability of ovulation. Increasing the likelihood of ovulation, the number of mature eggs ovulated, or the rapidity of ovulation would increase male reproductive success. Males may also benefit if ovulation coincides with the release of peak numbers of their own sperm. If this occurs, the likelihood of previous or subsequent males fertilizing most of the eggs is decreased. Although female reduction in ovulation could result in a decrease in the number of offspring produced, females may benefit by producing fewer offspring of higher quality.

The importance of copulatory stimuli in inducing ovulation has been examined in a few studies. A larger number of preejaculatory thrusts and intromissions resulted in a greater percentage of females ovulating in prairie voles (Gray et al. 1974). Similarly, Norway rats were more likely to ovulate when coital stimulation was increased, and an increased amount of stimulation resulted in a greater percentage of eggs ovulated (Zarrow and Clark 1968).

Failure to prepare uterus for implantation

In some species of rodents, follicles enter a luteal phase and produce progesterone necessary for endometrial proliferation only in response to the stimulus of copulation (table 4.3). The initiation of the luteal cycle, which results in the changes in uterine cells, is necessary for successful implantation to occur. Variation in the number and pacing of intromissions affects the likelihood that this change in uterine cells will occur. This has been demonstrated in experiments in which the amount of copulatory stimulation received by a female is manipulated. Eighty-six percent of female Norway rats receiving 10-16 intromissions had their luteal cycle triggered, as compared to 67% of females receiving 6-9 intromissions (Chester and Zucker 1970). The critical stimulus can be preejaculatory intromissions (Adler 1969; Huck and Lisk 1985) or the male ejaculatory response, which includes increased thrusting, grabbing the female with all four legs, shuddering and swelling of the penis (Land and McGill 1967; McGill et al. 1968). The most direct evidence that this mechanism can be a form of sexual selection comes from studies of McClintock, Anisko, and Adler (1982), in which dominant male Norway rats performed 6.1 intromissions as compared to 4.4 performed by subordinate males. Thus MMM may be a mechanism by which females help to prepare the uterus for implantation of blastocysts.

Removal of copulatory plugs

In a number of rodents, males deposit substances in the genital opening of females after mating; these are referred to as copulatory plugs (Dewsbury and Baumgardner 1981; Baumgardner et al. 1982). Two proposed functions of these plugs are to prevent leakage of sperm and reduce MMM. Female rodents may actively control the length of time that copulatory plugs are present in their genital tract. For example, about one-half to two-thirds of female tree squirrels have been observed to remove plugs deposited by some males but not others (Koprowski 1992). Females either removed plugs immediately after mating or left them in place for several hours. Thus females may discriminate among males by selective removal of mating plugs.

Raising the Young: Conflict and Cooperation

Female rodents continue to invest in offspring after birth. Females provide nutrition, warmth, and protection prior to weaning. All three of these forms of investment are critical in altricial species, and the latter is the most important in precocial species. These forms of investment can be enhanced by maintenance of a territory by the mother, formation of female kin clusters, or formation of cooperative breeding groups (either extended families or groups with multiple breeding females). Investment in offspring may continue even after weaning in some species by bequeathal of the natal territory to offspring or by allowing offspring to remain philopatric, that is, at the natal nest.

Female territoriality

Two hypotheses have been proposed for the function of territoriality in female rodents; both would affect reproductive success. According to the resource defense hypothesis (Ostfeld 1985; 1990), females defend their territories to provide exclusive access to food. In contrast, the pup defense hypothesis (Sherman 1980a, 1981b; Wolff 1993b; Wolff and Peterson 1998) posits that the main function of female territoriality is to protect vulnerable offspring from infanticidal conspecifics.

Considering these hypotheses with regard to the breeding season, both hypotheses predict that territory defense should be most pronounced during lactation. There are several predictions, though, that should allow us to distinguish between them.

1. The resource defense hypothesis predicts that female territorial behavior should begin to increase following fertilization, whereas the pup defense hypothesis predicts that it is not until after pups are born that female aggression to intruders should increase (Sherman 1981b).

2. According to the pup defense hypothesis, females should only behave aggressively toward potential infanticidal conspecifics (Wolff and Peterson 1998). In some species infanticide is committed by male conspecifics (e.g., house mice, Labov et al. 1985; deer mice, Wolff and Circirello 1991); female conspecifics (e.g., meadow voles, Webster et al. 1981; field voles, *Microtus agrestis,* Agrell 1995); or conspecifics of both sex (e.g., some ground squirrel species, Sherman 1981b; Michener 1983a; Vestal 1991; white-footed mice, Wolff and Cicirello 1991; bank voles, *Clethrionomys glareolus,* Ylönen et al. 1997 and some species of marmots, Brody and Melcher 1985; Blumstein 1997). Furthermore, females of some species are much more likely to kill the offspring of nonrelatives than relatives (e.g., Belding's ground squirrels, *Spermophilus beldingi,* Sherman 1981b; Columbian ground squirrels, Stevens 1998). In these species, females should be more aggressive toward intruders that are nonkin. In contrast, the food defense hypothesis predicts that females in all species would be equally aggressive toward intruders regardless of sex, and only very close kin would be allowed to feed on the territory.

3. Home range size should decrease at the onset of lactation according to the pup defense hypothesis, but may increase according to the food defense hypothesis (Boutin 1990; Wolff and Schauber 1996; Koskela et al. 1997).

Previous studies show that females do become more aggressive immediately after parturition and during lactation (Svare and Gandelman 1976; Ayer and Whitsett 1980; Sherman 1980a, 1981b; Ostermeyer 1983; Maestripieri

1992; Koskela et al. 1997), as predicted by both hypotheses. Aggression toward conspecific females is seen in some species in which females are the infanticidal sex (Madison 1980a; Koskela et al. 1997; Ylönen et al. 1997), as predicted by the pup defense hypothesis. Although intersexual aggression is uncommon in mammals, females are aggressive toward males (Agrell et al. 1998) in species in which infanticide is committed by males (vom Saal 1984; Wolff, J. O., 1985c; Mallory and Brooks 1978), or males and females (Wolff, J. O., 1985c).

Home range size in female bank voles decreased during the time between mating and lactation, at the time when the energetic demands of reproduction are increasing (Koskela et al. 1997). These data are inconsistent with the prediction of the food defense hypothesis but support the pup defense hypothesis. In addition, supplemental food did not result in decreased home ranges in field manipulations of prairie and meadow voles populations when density is controlled, suggesting that females may not be food limited as expected, according to the resource defense hypothesis (Desy et al. 1990; Fortier and Tamarin 1998). Thus existing data are more consistent with the pup defense hypothesis than with the resource defense hypothesis.

Territory bequeathal

After offspring are weaned they have three options: to disperse from their natal nest, remain philopatric and share space and resources with their mother (or parents), or remain at their natal nest after their mother/parents disperse. In most species, offspring disperse around the time of weaning (Anderson 1989), but in some, juveniles may remain and inherit some or all of their natal territory after their parents' death (Anderson 1989). In a few species, adult females have been observed to disperse when offspring are about weaning age (King 1955; Harris and Murie 1984; Price and Boutin 1993). This post-breeding dispersal may be a form of extended parental investment or parental facilitation (Brown 1987), in which parents allow their offspring to remain on their natal territory, where these offspring have access to necessary resources such as food and a nest site (Myllymaki 1977).

The territory bequeathal hypothesis makes three critical predictions:

1. Parents with weaned offspring are more likely to disperse than those with no offspring around weaning age (for example, Harris and Murie 1984; Price and Boutin 1993).

2. Offspring will remain within their natal home range after it is vacated by parents (Lambin 1997).

3. Offspring should benefit by dispersal of parents through increased chances of survival and reproduction (sensu Jones 1986).

Lambin (1997) reviewed territory bequeathal studies and concluded that most of this information was anecdotal. We reexamined the studies reviewed by Lambin (1997), eliminating those that did not pertain directly to the three critical predictions of the territory bequeathal hypothesis. Of the 16 remaining rodent studies, offspring did not remain in the natal territory for long after weaning in two studies, there was no support for the territory bequeathal hypothesis in six studies, and there was support for the territory bequeathal hypothesis in the eight (50%) remaining studies.

The best evidence in favor of the territory bequeathal hypothesis comes from studies of Columbian ground squirrels (Harris and Murie 1984), red squirrels (Price and Boutin 1993; Berteaux and Boutin 2000), and banner-tailed kangaroo rats (*Dipodomys spectabilis,* Jones 1986). Breeding dispersal by female Columbian ground squirrels was most likely to occur when a yearling daughter was still present on the natal territory (females dispersed in 11 of 17 [65%] cases when daughters were present, compared to 8 of 32 [25%] when no daughter was present). Yearling females tended to remain on the territories held by their mothers during the previous year. There was also a nonsignificant tendency for daughters to be more reproductively successful if their mother dispersed from the natal territory and they remained (8 of 11 [73%] as compared to only 2 of 6 [33%] whose mothers did not disperse). In red squirrels, more breeding females (7 of 28 [25%]) than nonbreeding females (0 of 14) dispersed. More young females survived on their natal territory after the mother had dispersed (97 of 119 [81.5%]) compared to young females that obtained another territory with a midden (56 of 96 [58.3%]) or those that did not obtain a territory with a midden (95 of 164 [57.9%]). These data support all three predictions of the territory bequeathal hypothesis.

Remain philopatric and cooperate

Some adult females and offspring remain at or near the natal nest. Hypotheses proposed to explain the occurrence of natal philopatry have emphasized costs of dispersal on the one hand and benefits of philopatry on the other (Koenig et al. 1992). Three main factors have received the most attention: (1) life-history variables—certain taxa are predisposed to remain philopatric due to life-history traits (Arnold and Owens 1998; Pen and Weissing 2000; Kokko and Lundberg 2001), (2) ecological constraints—individuals delay dispersal and remain philopatric because costs of dispersal are prohibitive (Brown 1974; Koenig and Pitelka

1981; Emlen 1982; Jarvis et al. 1994; Koenig et al. 1992), and (3) benefits of philopatry—individuals delay dispersal and remain philopatric because they gain direct or indirect fitness benefits (Stacey and Ligon 1987, 1991; Kokko and Johnstone 1999). These hypotheses are complementary, because costs and benefits, as well as life-history factors, affect dispersal (see Emlen 1994; Solomon 2003; Nunes, chap. 13, this volume for further discussion on dispersal and philopatry).

Female kin clusters

Female philopatry often results in adult female kin living in close spatial proximity to each other (Michener 1983a; Vestal and McCarley 1984; Smith 1993; Solomon 2003; fig. 4.2). The formation of female kin clusters has been documented in a variety of rodents, including numerous species of ground squirrels (Vestal and McCarley 1984; Yensen and Sherman 2003), black-tailed prairie dogs (Hoogland 1995), yellow-bellied marmots (Armitage and Schwartz 2000), gray squirrels (*Sciurus carolinensis;* Koprowski 1996), Townsend's voles (*Microtus townsendii,* Lambin and Yoccoz 1998), and house mice (Dobson and Baudoin 2002). Having kin as neighbors may be beneficial because it creates the opportunity for nepotistic interactions between individuals that may enhance their inclusive fitness. Types of potentially nepotistic behaviors that may contribute to greater reproductive success for females living in kin clusters include reduced aggression among relatives (e.g., Charnov and Finerty 1980; Sherman 1980a; Michener 1981; McClean 1982), sharing of space (e.g., Sherman 1981a; Vestal and McCarley 1984; Mappes et al. 1995; Lambin and Yoccoz 1998), cooperative defense against predators

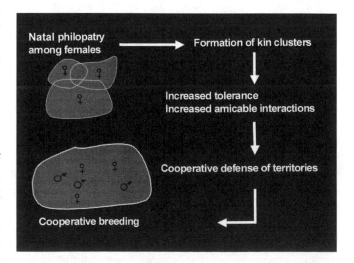

Figure 4.2   Natal philopatry by females can result in the formation of female kin clusters. Having kin as neighbors creates the opportunity for selection favoring nepotistic behaviors, which may lead to the evolution of cooperative breeding among related females. Figure courtesy of Christine R. Maher.

or conspecifics (e.g., Sherman 1980a, 1981b; Murie and Michener 1984; Wolff 1993b), and alarm calling (e.g., Sherman 1977, 1980a; Hoogland 1995).

In field experiments with bank voles (Mappes et al. 1995) and Townsend's voles (Lambin and Yoccoz 1998), survival and recruitment of juveniles was enhanced when females had kin as neighbors, whereas no fitness benefits were found for gray-tailed voles (*Microtus canicaudus*, Dalton 2000). In an experimentally manipulated population of Richardson's ground squirrels (*Spermophilus richardsonii*), females residing in kin clusters had significantly greater reproductive success compared to females in nonkin cluster groups (Davis 1984c). Having female kin as neighbors increases female reproductive success in some other species of ground-dwelling sciurids (e.g., Belding's ground squirrels, Sherman 1980a; black-tailed prairie dogs, Hoogland 1995; yellow-bellied marmots (*Marmota flaviventriis*, Armitage and Schwartz 2000). However, the effect of population kin-structure on female reproductive success still remains to be clarified for the vast majority of species in which female kin clusters occur.

Cooperative breeding

Cooperative breeding—care of young by individuals other than the genetic parents—is seen in a small percentage of rodents, many of which display natal philopatry (~ 40 species from 9 of 30 families [Solomon and Getz 1997; Blumstein and Armitage 1999; Hayes 2000]). In these species, females rear offspring in extended family groups where offspring remain after weaning, and participate in care of subsequent litters born to their mothers (figure 4.2). In about half of these cooperatively breeding species, more than one female breeds within a social group and these females may be kin but not necessarily mothers and young (Packer et al. 1992; Pusey and Packer 1994; Lewis and Pusey 1997; Ebensperger et al. 2004). Typically, all breeding females in the group provide care for young, which are reared in a communal nest (Sayler and Salmon 1971; Hayes 2000).

One of the adaptive hypotheses for alloparental care of young is that breeders benefit directly by increasing their lifetime reproductive success (Brown 1987; Solomon and Hayes, in press). Alloparents (conspecifics that participate in offspring care) may decrease the workload of breeders, which could result in more reproductive attempts per breeding season, or increase the success of individual breeding attempts by increasing the quantity or quality of pups. However, in laboratory studies, neither the presence of alloparents nor greater numbers of alloparents affected litter size at weaning in prairie voles, pine voles (*Microtus pinetorum*), or Mongolian gerbils (*Meriones unguiculatus*, Fried 1987; Solomon 1991; French 1994; Hayes and Solomon 2004). There were also no effects of living in a cooperatively

breeding group (as opposed to living solitarily) on numbers of weaned young in field studies of deer mice (Wolff 1994b), white-footed mice (Wolff 1994b), fat dormice [*Glis glis*, Pilastro et al. 1996] or yellow-bellied marmots (Van Vuren and Armitage 1994a). However, a negative relationship between number of offspring weaned and group size was found in black-tailed prairie dogs (Hoogland 1995) and social tuco-tucos (*Ctenomys sociabilis*, Lacey 2004), indicating that the presence of additional conspecifics can be detrimental rather than beneficial. In contrast, alpine marmots (*Marmota marmota*) and possibly golden marmots (*M. caudata aurea*) warm related juveniles in the hibernaculum during winter, which results in increased overwinter survival of pups (Arnold 1993; Blumstein and Arnold 1998).

Another way that alloparents may benefit the breeders is by increasing the quality of offspring produced. One measure of offspring quality is body size at weaning. Alloparental care may result in increased offspring size relative to offspring reared by a single female. Large body size at weaning may result in numerous potential benefits. Offspring that are heavier at weaning survive better than offspring that are lighter at weaning (Solomon 1991; Huber et al. 2001). Offspring that are heavier at weaning also tend to be heavier as adults. These individuals are preferred as social, and presumably mating, partners (Solomon 1993a but see Getz et al. 2004) and are likely to be able to out-compete male conspecifics for mates (Sheridan and Tamarin 1988). In some species, such as house mice, heavier offspring become sexually mature sooner than lighter individuals (Fuchs 1982). Finally, weaning weight of females affects the growth of her pups (Solomon 1994). Females that were larger at weaning had pups that grew faster prior to weaning. Increased size of weanlings was reported when prairie voles were raised with alloparents under environmentally challenging conditions (e.g., temperatures or limited food; Solomon 1991; Hayes and Solomon 2004), suggesting that cooperative breeding can increase the quality of the offspring produced by a female.

Conclusions and Future Directions for Research

The types of studies reviewed in this chapter have produced many exciting new ideas showing that female fitness plays a larger role in sexual selection and evolution of mating systems than previously thought. To make a stronger argument that a particular mechanism, whether precopulatory or postcopulatory, results in sexual selection by female mate choice, we need to show that (1) females respond differently to some males compared to other conspecific males, and that this discrimination occurs in nature, (2) females' discrimination among males leads to differences in reproductive success for these males, (3) that female preferences are associated with particular male traits, and (4) variability among males in the preferred traits has a genetic basis (Eberhard 1996).

Although there is evidence from many rodent species that females actively seek copulations with preferred males, we still need to know more about the traits preferred by females. Moreover, most mate choice studies have been conducted in the laboratory. To assess whether these data are relevant to the field (Wolff 2003c), we need field studies designed to determine preferred mating partners as compared to available choices (see Getz et al. 2004). Finally, females likely use more than one trait when assessing potential mates (e.g., Burley 1981) and observed preferences may, therefore, represent tradeoffs among the traits possessed by available mates.

The growing use of genetic markers to examine parentage in natural populations of rodents has consistently shown that litters are multiply sired in a variety of species. Numerous hypotheses have been proposed to explain why females mate with multiple males, but there are few data available to test these hypotheses. Future field studies comparing the reproductive success of females with multiply sired litters to that of females with litters sired by a single male are crucial to enhancing our understanding of the functional significance of MMM.

In addition to the male with whom a female mates, a female's reproductive success is also determined by her behavior after pups are born. Data suggest that the maintenance of a territory, tolerance of philopatric offspring, bequeathal of the natal territory to offspring by female dispersal, formation of female kin groups, and cooperative breeding are behaviors that affect female reproductive success in many rodents. However, the extent to which they affect a female's reproductive success is unknown. Furthermore, for each of these behaviors alternative hypotheses have been proposed as to how they contribute to greater female reproductive success. Although the data necessary to discriminate among these hypotheses are often difficult to obtain, it is important for future studies to attempt to do so to better understand the many questions remaining regarding how the postcopulatory behavior of female rodents may affect their fitness.

Chapter 5 Sexual Selection: Using Social Ecology to Determine Fitness Differences

Lara S. Carroll and Wayne K. Potts

THE THESIS OF this chapter is that sexual selection (usually male-male competition and female choice) operating in captive house mouse societies can be exploited to examine fitness components of genes (or other manipulated variables) that are critical for success in the wild. Since sexual selection can be studied in seminatural populations of *Mus*, we have used this simplified approach to reveal major fitness differences that were largely invisible when studying the animals with traditional laboratory approaches. The model-organism status of the house mouse provides the opportunity to merge the powerful approaches of molecular, cell, and physiological biology with the tools of field ecology. We hope our attempt to fuse these different levels of analysis in the house mouse serves as a model for developing such integrative approaches for the biological study of other social species.

There are recent compelling examples in animals, including rodents, that both intra- and intersexual selection favor mates with "good genes" and consequently produce offspring with higher fitness (Møller and Alatalo 1999; Drickamer et al. 2000; Calsbeek and Sinervo 2002; Hine et al. 2002; Drickamer et al. 2003; Gowaty et al. 2003). The ultimate measure of fitness is lifetime reproductive success, and since the most robust, pathogen-free, predator-savvy animals are those that survive to breed, mice compete for habitats that have predator-free hiding sites, and they compete for habitat with the best food resources. For this reason, in social species with intrasexual competition, sexual selection likely produces fitness outcomes approximating those of natural selection, even in seminatural population studies lacking many of the important components of natural selection. In nature, the losers of intrasexual competition are killed by starvation, predators, and disease. In our system, such animals do not breed because they are excluded from territories by competitors, and fail to gain access to mates.

Fitness measures have always been critical to questions in ecology and evolution, but increasingly they are needed to answer questions in molecular disciplines, where genes are experimentally manipulated to assay function. Perhaps the foremost category of molecular questions requiring fitness measures is in the field of functional genomics. As functional genomics programs begin to discover the power of fitness assays for revealing phenotypes, we expect such methodologies to become incorporated into the functional genomics toolbox. Conversely, insight into compelling problems at the ecological level can be gained by illuminating the mechanistic underpinnings of gene function. Reductionist scientists have committed what might be considered the ultimate reductionist act—the complete sequencing of genomes. Now we must find out what those genes do, from molecule to cell to organ to organismal fitness—what is their function in nature?

In the following pages, we relay the stories of three evolutionary puzzles that are also the stories of three natural genetic experiments whose corresponding phenotypes were either invisible or insufficiently resolved using laboratory approaches. The solution to each puzzle came only after release from the restrictive confines of the laboratory into the rich and informative context of *Mus* social structure. First, we review the role of the major histocompatibility complex (MHC) in mate choice and discuss evidence that sexual selection helps maintain the tremendous polymorphisms at these immunologically critical loci. Second, we describe an inbreeding study that shows that the magnitude of inbreeding depression is far more severe than previous laboratory measures have suggested. This work emphasizes the insight to be gained by extending any measurement of a fitness cost to include lifetime fitness. Third, we describe a partial solution to the paradoxically low frequencies of the *t* gene complex found in nature. Despite the ability to distort genetic transmission wildly in its own favor, mice heterozygous for this selfish complex of closely linked genes suffer reduced reproductive success and increased mortality, effectively crippling the spread of the *t* complex in nature. Excellent reviews have been written on each of these three topics in recent years, and our purpose is not to provide an in-depth synthesis of relevant literature. Rather, we illustrate solutions and commonalities in the study of these three distinct evolutionary genetic problems, showing how social ecology was used to ascertain fitness differences and resolve these biological puzzles.

Gene Function Is Fitness!

The ultimate function of all genes is how they contribute to fitness in nature

Mutations that decrease the fitness (reproductive success) of their bearers tend to be eliminated by natural selection, and those that increase fitness tend to become fixed. A gene with no known discernable morphological or physiological effect in the laboratory (producing a cryptic phenotype) may nevertheless be vital to fitness and must be analyzed in a context appropriate for measuring the correct components of fitness (see the following inbreeding example). Alternatively, genes with well-characterized phenotypes may have additional cryptic phenotypes (see the *t*-haplotype example in the following). Consequently, characterization of gene function will always be incomplete without fitness measurements in an ecological context. Although fitness is the ultimate target of selection, it will always be the product of particular molecular, cellular, and physiological processes. Understanding these mechanistic pathways is the focus of the mainstream molecular research program. However, this research program reaches a dead end in the absence of phenotypes. Fortunately, ecological approaches can help reveal many phenotypes that are cryptic under laboratory conditions, allowing the molecular characterization of gene function to proceed. An integrative approach combining fitness assays followed by molecular studies promises to be a powerful procedure for the discovery of gene function, particularly for genes with cryptic phenotypes.

The practical problem is how best to measure fitness. Selection is difficult to measure in the laboratory. The artificial conditions of caged breedings often produce inconsistent results and have had limited success overall in defining the mechanisms underlying reproductive differences among genotypes. Yet studies performed in the wild have their own sets of problems. Stochastic environmental conditions (weather, food, shelter, etc.) increase the variance of tested effects, adding noise to already statistically complex data sets, and the loss of subjects to dispersal and various sources of mortality confounds lifetime measures of fitness. House mouse communities are socially complex, and mating success depends on a suite of ecological and social cues for information on territory, mate, and rival quality. By providing mice with a means by which to display and utilize these socially and ecologically relevant cues while controlling extrinsic environmental factors, tests of genetic hypotheses become quite manageable. The house mouse is already the leading mammalian model organism for molecular and physiological studies. However, the ability to simulate an ecological field study within a controlled environment makes the house mouse extremely tractable for social ecology studies as well. Combining these two strengths makes the house mouse the vertebrate organism of choice for studying questions that require molecular, physiological, and ecological approaches. As we hope to demonstrate in this chapter, the massive enterprise of determining gene function demands such broad approaches.

Seminatural Populations in *Mus*

The success of seminatural *Mus* population studies is likely owed to the 10,000-year-long commensal relationship between house mice and humans that exists to this day. The highly social house mouse (*Mus musculus/Mus domesticus*) is quite willing, even enthusiastic to set up complex societies within the wire, steel, and concrete labyrinth of racquetball-court-sized artificial mouse enclosures. Our attempts to provide mice with environmental complexity without obscuring our view of them involves a considerable amount of strategically placed wire mesh, which seems to provide an important element of spatial complexity (fig. 5.1). Wire mesh partitions subdivide each enclosure into six or eight equal subsections. The same material is placed in cylinders or spirals around food corrals, and made into platforms with multiple tiers for climbing. Mice can easily scale the wire mesh and learn to maneuver quickly around and over it, but the partitioned subdivisions tend to provide dominant males with convenient territorial boundaries, and all mice have to negotiate a texturally complex three-dimensional space to cover any ground. In addition to these steel structures, mice are given unlimited food and water,

Figure 5.1 Interior view of a seminatural population enclosure for mice. The 5m by 10m enclosure is subdivided into eight subsections by 0.6 m-high hardware cloth. Each subsection contains food, water, and nest boxes.

fitness differentials of the parental generation can be determined from the analysis of these offspring.

Case 1: Mating Preferences, Odors, and MHC Diversity

In 1975, during the midst of excitement involving the recently discovered immunological role of the major histocompatibility complex, Yamazaki and his colleagues (Yamazaki et al. 1976; Yamazaki et al. 1978) made the unexpected observation that male laboratory mice preferred to mate with females carrying dissimilar MHC genes (also known as the H-2 complex in mice). Since the mice studied were genetically identical except at the MHC region, only genes residing within this large genetic cluster could be responsible for the observed mating behavior. As tiny sentries positioned on the surface of cells, MHC molecules help decipher the antigenic universe, binding bits of proteins (antigens) from self and foreign sources for presentation to the T-cells of the immune system (fig. 5.2; Zinkernagel and Doherty 1974). T-cells recognize this antigen-bound complex, targeting for destruction any cells bearing foreign antigens bound to the MHC. Without MHC molecules, T-cells would be incapable of distinguishing self from non-self, effectively rendering the immune system defenseless against pathogens, parasites, and autoimmune disease. At a cellular level, self/non-self recognition helped explain the role of MHC in the acceptance or rejection of tissue grafts,

and use wood shavings and cotton batting to make themselves comfortable in an assortment of nest boxes. Within these simulated mouse cities, the nightly activities of dozens of mice are observed and recorded under the dim red light of an artificial evening. Here, mice maneuver throughout their surroundings, strategically placing and responding to odors and displays rich in context and meaning. Territorial competition is exhaustively played out among males, usually requiring days of escalated contests before territory holders prevail. But it doesn't end there. Subordinate males continue their attempts at a coup d'état by a persistent program of insurgency. Territorial males are continually probed and challenged, and we have demonstrated that the greatest loss of fitness for some genotypes emerges during attempted overthrows rather than the initial territorial establishment (Meagher et al. 2000). Females are free to exercise choice of mates, often seeking matings from dominant males in neighboring territories. As all of these complicated social interactions play out, we obtain hundreds of litters over the course of several months. But the critical point is that these are not a random set of pups, but a highly selected set that are anticipated to be genetically superior to a set whose parents were not exposed to sexual selection. In turn, the

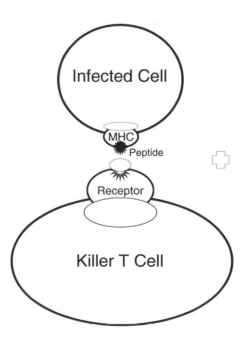

Figure 5.2 Depiction of a killer T cell receptor recognizing an MHC-bound peptide.

the phenomenon for which MHC had first been discovered (See Klein 1986 for an MHC historical account). However, only in the 1980s did the central role of MHC in immunity begin to reveal the relationship between MHC and mate choice. During this time, a reemerging interest in Darwinian theories of sexual selection, including Hamilton and Zuk's fusion of immunocompetence and good genes sexual selection theories (Hamilton and Zuk 1982), pinpointed a crucial link between immune function and reproductive behavior. These authors suggested that pathogens and parasites might help to maintain genetic variance among the secondary sex characteristics that are exploited (generally by females) for mate choice. Among the various benefits of MHC-based sexual selection, it became apparent that mate choice based on these genes could have the immediate consequence of better protecting one's offspring against infectious disease.

Rather than long tail feathers, bright plumage, or spectacular behavioral displays, secondary sex characteristics arising from the MHC are olfactorily perceived (although see Von-Schantz et al. 1996). Leinders-Zufall and coworkers (2004) have recently identified the long, mysterious source of MHC-based odorants, which arise from peptides bound to MHC molecules. These peptides are then detected by the vomeronasal organ (VNO), an olfactory accessory organ usually thought to detect soluble pheromones. Leinders-Zufall et al. (2004) applied peptides to the VNO and found sensory neurons that detected peptides with a specificity similar to that of MHC molecules themselves. If amino acids in anchor positions of the peptide were changed, the peptide would not be recognized and the sensory neurons would not fire. However, if non-anchor amino acids were mutated, the changes were ignored and the peptide was readily recognized by sensory neurons—just like MHC! This explains how mice can detect odor differences based on only a few amino acid differences at a single antigen-presenting locus (Yamazaki et al. 1983; Yamazaki et al. 1990; Carroll et al. 2002). Moreover, these highly polymorphic antigen-presenting genes, with up to 450 alleles at each locus (Adams and Parham 2001) provide tremendous underlying genetic variation, upon which sexual selection and kin selection may act. Either by choosing mates that can provide superior MHC genotypes for offspring, or by avoiding inbreeding with mates that share familial MHC odors, the extreme genetic diversity of MHC genes, coupled with allele-specific odors, provide a compelling system for allowing individuals to select among an archive of "good genes." No other candidate genes within the MHC region are quite so convincing as potential mediators of MHC-based reproductive behavior.

The origin and maintenance of MHC polymorphisms remains a puzzle. In fact, much recent experimental and theoretical work has been aimed at understanding the nature of selective forces that must operate to maintain MHC polymorphisms, which are orders of magnitude greater than polymorphisms at the majority of other vertebrate loci. Most models have focused on the role of pathogens in driving MHC diversity. Since different MHC molecules vary greatly in their binding capacities for billions of different possible protein antigens, the resulting allele-specific pathogen resistance provides one potential and powerful source of selection for diverse alleles (Doherty and Zinkernagel 1975). However, MHC-dependent mating preferences may also prove to be a powerful mechanism for driving MHC diversity. MHC-based mating preferences are not only consistent with pathogen-driven models of MHC diversity, but will tend to reinforce them, maintaining diverse alleles in the population as well as selecting for rare and novel alleles (Penn and Potts 1999).

Surprisingly, pathogen-mediated selection mechanisms have been exceedingly difficult to detect except in the presence of multiple pathogens or strains of pathogens (Thursz et al. 1997; Carrington et al. 1999; Penn et al. 2002; McClelland et al. 2003). In contrast, MHC-based mating preferences have been abundantly demonstrated in mice since Yamazaki's preliminary observations (Jordan and Bruford 1998; Penn and Potts 1999). However, laboratory studies using inbred strains have produced varied results that are somewhat difficult to interpret with respect to sexual selection theory (Manning, Potts, Wakeland, and Dewsbury, 1992). For instance, two out of ten mouse strains tested showed mating preferences for MHC-similar (assortative) rather than dissimilar (disassortative) animals (See table 1 in Jordan and Bruford 1998), somewhat inconsistent with good genes explanations of MHC-based mate choice. Moreover, most laboratory studies tested male preferences, leaving open the question of whether females, predicted to be the choosier sex due to their greater investment in offspring, also mated disassortatively.

Sexual selection in seminatural enclosures

In the first MHC study performed in seminatural enclosures, Potts et al. (1991) established three crucial details lacking from previous laboratory studies. First, MHC-based mating preferences took place in the context of *Mus* populations, laying to rest the suggestion that MHC-based mating preferences were merely a laboratory artifact. Second, as the mice were recently derived from wild populations, MHC-based mating behaviors prevailed despite the increased complexity of social cues resulting from randomly assorting wild genetic backgrounds. And third, the selection coefficient arising from nonrandom mating was clearly strong enough to maintain the allelic diversity found in surveys of wild populations, suggesting that mating preferences could indeed be the elusive source of selection maintaining MHC polymorphisms (Potts et al. 1991; Hedrick 1992). In this study, nine populations of wild-derived mice carrying four possible MHC haplotypes were maintained in seminatural enclosures until approximately 150 pups were born in each enclosure. MHC genotypes of pups born to the original founders revealed a significant 27% reduction of homozygotes from random mating expectations. Comparisons with controlled laboratory breedings suggested that abortional selection was not operating to alter genotype ratios (Alberts and Ober 1993). Moreover, genotypes of unborn litters from female founders suggested the homozygote deficiency was not due to differential mortality of neonates. Only MHC-based disassortative mating preferences could explain the pup genotype patterns. Behavioral observations also strongly suggested that these disassortative matings were primarily due to female choice, as females, not males, were observed leaving their territories to engage in extra-pair matings; these matings explained three-quarters of the heterozygote excess (Potts et al. 1992). Furthermore, communal nesting and nursing tended to occur among females with shared MHC haplotypes, providing the first experimental support for a general role of the MHC in kin recognition (Manning, Wakeland, and Potts 1992; Manning et al. 1995).

In another seminatural enclosure study, Penn and Potts (1998b) corroborated data from previous laboratory assays, showing that MHC-based mating preferences of females are established via phenotype matching. In other words, females use odor information from their rearing associates (nestmates) rather than their own MHC odors to inform mate selection (Beauchamp et al. 1988; Eklund 1997). Mating preferences of females that had been cross-fostered onto MHC-dissimilar litters as pups were compared to the preferences of in-fostered females (from MHC-similar foster litters). In direct contrast to preferences of the in-fostered control females, cross-fostered females preferred males matching their own MHC genotype over males carrying the MHC genotypes of their foster families. This study additionally established females as the choosier sex, as a male-specific preference would have yielded similar pup genotype frequencies across all populations, regardless of the female fostering treatment.

Taken together, these seminatural population experiments provide strong evidence that females learn their own MHC identity from families, that this information is used for mate choice and kin recognition (which could additionally prevent inbreeding), and that mating behavior is capable of maintaining MHC polymorphisms, even in the absence of pathogens. Yet, all this work was done using mice that differed by entire MHC haplotypes, with variation

among all classical antigen-presenting genes, as well as differences at numerous linked genes associated with the MHC region. Consequently, unambiguous evidence connecting MHC-based mating preferences to the highly polymorphic antigen-presenting genes is still lacking.

Recently, we used seminatural enclosures to study mice that differed by naturally occurring MHC mutations at a single antigen-presenting locus. On a homogeneous genetic background, odors mediated by these allelic differences are distinguishable by untrained mice (Carroll et al. 2002). However, in the context of a randomly segregating wild background, mice did not differentiate among MHC types, mating randomly rather than disassortatively (Carroll 2001). This result underscores the need for more studies to disprove or validate MHC antigen-presenting alleles as the ultimate source of MHC-based sexual selection. As our study lacked power to detect less than a 10% deficiency of homozygous pups, the result does not preclude the operation of low levels of sexual selection (even a 1% preference favoring MHC-dissimilar mates can have substantial evolutionary impact, propelling MHC-diversifying selection). However, it does suggest that other genetic loci are extremely important in the expression of odor-based cues and odor-based behavior. For example, major urinary proteins (MUPs), the most abundant proteins in mouse urine, also exhibit a high level of polymorphism, and are found at levels 100,000-fold higher than MHC protein fragments (Brennan 2001). Although unlinked and genetically unrelated to MHC, the family of genes giving rise to MUP chemical signals is also used by mice for individual recognition (Hurst et al. 2001) and is therefore one of the many possible signals competing with MHC-based odors. The rich source of odor from MUPs might easily overwhelm new odors arising from novel MHC mutations. The absence of MHC-based mating preferences for single gene differences suggests that MHC-based mating preferences may be more important in maintaining already existing MHC diversity (balancing selection), than in driving the incorporation of new alleles (diversifying selection).

Social structure and MHC-based selection in other rodents

Although genetic diversity at various MHC loci has been measured in a number of rodent species, and several of these studies have found evidence for balancing selection at MHC loci (Seddon and Baverstock 1999; Richman et al. 2001; Hambuch and Lacey 2002; Sommer 2003), very little data beyond that obtained from house mice have revealed MHC-based sexual selection. At best, there are limited data in rodents suggesting a connection between MHC diversity and social structure. For example, the endangered Malagasy giant jumping rat (*Hypogeomys antimena*) is strictly monogamous, has low MHC polymorphisms, low reproductive rates, and restricted gene flow (Sommer et al. 2002). The authors found a similar pattern of monogamy and low MHC diversity in another monogamous Malagasy rodent species, but nearly double the MHC polymorphisms in a third species with a promiscuous mating system, suggesting a correlation between type of mating system and MHC diversity. Interestingly, variation in geographic distribution did not account for variation in MHC polymorphisms among these three sympatric species (Sommer et al. 2002). Social structure may similarly influence MHC diversity of Argentine tuco-tucos (*Ctenomys*), where balancing selection for MHC Class II DQ was found to be enhanced in a social species compared to a solitary species (Hambuch and Lacey 2002). It is not clear why monogamy or solitary social structure should reduce balancing selection on MHC loci. However, even if MHC-based mating preferences are active in these systems, low polymorphism might be an unavoidable consequence of reduced gene flow increasing the effect of drift, as is found on island populations of the Australian bush rat (Seddon and Baverstock 1999). It would be worthwhile to examine the possibility that the social ecology of such rodents reduces their parasite loads, which could serve to counterbalance the loss of MHC diversity.

As MHC diversity increases in populations, so does the statistical difficulty of analyzing those populations for MHC-based mating preferences. The more alleles that are available to choose from, the more likely it is that animals will pair with MHC-dissimilar individuals by chance alone, requiring excessive power to detect statistical significance. This may be why MHC-based mating preferences have rarely been detected in wild populations, and are in fact known only for a few taxa outside of mice, including humans (*Homo sapiens*, Wedekind et al. 1995; Ober et al. 1997), Atlantic salmon (*Salmo salar,* Landry et al. 2001), and three-spined sticklebacks (*Gasterosteus aculeatus*, Reusch et al. 2001). The behavior of wild animals (especially mammals) tested in a laboratory context is often difficult to interpret biologically (Manning, Potts, Wakeland, and Dewsbury 1992). It is extremely fortunate that MHC mating preferences were initially discovered at all in inbred strains of mice, since choosy behavior contradicts years of artificial selection to create individuals that mate indiscriminately with their cage mate. Seminatural population experiments have proved successful for studying MHC-based behavioral phenomena in two of the four species for which MHC-based mating preferences are known: house mice (Potts et al. 1991) and Atlantic salmon (Landry et al. 2001). As animal biologists begin to appreciate the value of this ecological compromise between the laboratory and the field, we expect there will soon be many creative designs for adapting this powerful technique to study other biological traits in a population context.

Case 2: Fitness Consequences of Inbreeding in *Mus*

The primary cause of inbreeding depression is the expression of deleterious recessive alleles (Roff 2002), which are expressed at a higher rate in inbred individuals (Latter 1998). The deleterious effects of inbreeding have been appreciated for centuries. As schoolchildren we usually learn about it through the example of the intermarriage of royal lineages in Europe and the consequent increased incidence of genetic diseases (e.g., hemophilia) in these inbred lineages (Shaw 2001). The near universal presence of incest taboos in human societies and the myriad ways in which plants and animals avoid inbreeding (Pusey and Wolf 1996) suggest that inbreeding has been an important and persistent problem to most life forms. However, the actual fitness consequences in nature have been poorly characterized, and some authors have even suggested that inbreeding has no to little effect on animal fitness in the wild (Shields 1982). Similarly, the consequences of inbreeding in humans at the level of cousin unions have been deemed so minor that guidelines for discouraging such unions have recently been relaxed (Bennett et al. 2002). Consistent with this view are the apparent successes of naked mole-rats, a striking example of a highly inbred rodent species (Reeve et al. 1990), and black-tailed prairie dogs, which are reported to suffer no detectable fitness declines from inbreeding at the level of cousins and lower (Hoogland 1992). However, in neither of these organisms were fitnesses compared between outbred and inbred adults engaged in direct competition. As we subsequently review, by far the largest component of inbreeding-associated fitness declines occurs during adulthood. In summary, inbreeding depression is often observed, but the actual fitness costs associated with it are uncertain.

One of the possible functions of MHC-based mating preferences in house mice (described in Case 1) is the avoidance of inbreeding. To evaluate the relative importance of inbreeding avoidance on the evolution of MHC-based mating preferences, the fitness consequences of inbreeding must be known. Two major inbreeding studies had previously been conducted on house mice. From these data, the reproductive consequences of one generation of full-sib matings were estimated to be a 10% decline in litter size (Lynch 1977; Connor and Bellucci 1979). However, litter size was the only fitness-related effect measured. Thus these estimates of inbreeding in house mice only took into account the effect of embryonic lethality from inbred matings, ignoring all possible fitness defects present in the surviving inbred offspring. At the time we started designing experiments to measure these potentially important inbreeding costs in mice, studies in *Drosophila* (Charlesworth and Charlesworth 1987) were beginning to suggest that competitive conditions revealed fitness declines that were effectively invisible or underestimated in the lab.

To measure the fitness consequences for progeny of full-sib matings, Meagher et al. (2000) bred wild-caught mice to create an F2 generation derived from either outbred or full-sib matings. During these laboratory matings we found approximately the same 10% effect of inbreeding on litter size (fig. 5.3) that the two previous studies had demonstrated (Lynch 1977; Connor and Bellucci 1979). However, when we measured the adult fitness of these offspring, we found an additional 500% effect in males! Outbred males had five times more offspring than inbred males. Figure 5.3 compares the relative decline in reproductive success (fitness) of inbred animals to outbred animals for the two previous house mouse studies (which measured only litter size) and our study, which went on to measure the adult fitness of those inbred offspring (and outbred controls). The dramatic fitness declines in inbred males were due both to a 41% reduced ability to gain territories and to decreased survival. Reduced survival rates were particularly important for territorial inbred males, where 90% had died by the end of the experiment, as compared to only 24% of outbred territorial males. These results suggest that inbred males had difficulty maintaining territories, as well as gaining them. Significant fitness declines were also found for inbred females (fig. 5.3), but these declines were four times smaller than the observed male declines. We attributed these gender differences to the fact that males compete aggressively over territories, and nonterritorial (subordinate) males have little reproductive

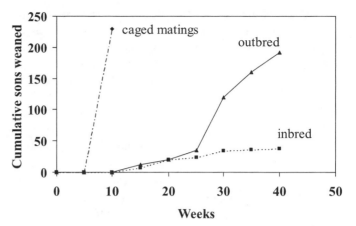

Figure 5.4 Comparison of reproductive success over time for inbred and outbred males for both caged matings and in seminatural populations. The caged matings represented equal numbers of the four possible mating types of inbred and outbred males and females.

success (< 20%). In contrast, females had no limiting resources. It remains an open question whether the fitness consequences of inbreeding in females would approach that of males if they had to compete over critical resources such as food.

Figure 5.4 shows the relative reproductive success of inbred and outbred males over time. This analysis demonstrates that the relative differences were increasing at the end of the experiment, indicating that our inbreeding depression estimates were conservative. If we had allowed the populations to continue to obtain lifetime reproductive success measures, the fitness differences between inbred and outbred animals would have been much larger, because at 40 weeks almost all (90%) of inbred territorial males were dead, whereas only 24% of outbred territorial males had died. Male lifetime fitness effects are estimated by extrapolating the slope of our curves between weeks 25–40 out to 80 weeks; outbred males would produce 25 times more offspring than inbred males. This dramatic 96% fitness decline suggests that only one generation of full-sib matings is effectively lethal to inbred sons in the face of competition from non-inbred conspecifics.

Fitness consequences of inbreeding in other rodents, measured both in lab and nature

There is only one comparable study for rodents where measures of inbreeding depression were taken both in the laboratory and in natural or seminatural population conditions. Jiménez et al. (1994) captured wild *Peromyscus leucopus* (white-footed mouse) from a study site in Illinois. This wild-derived outbred colony was maintained for two to three generations before experimental pairings of either unrelated

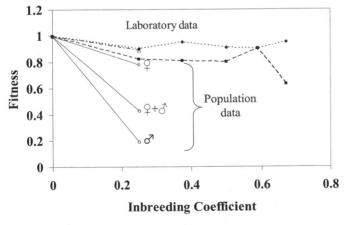

Figure 5.3 Relative fitness of inbred animals compared to outbred controls as measured by weaning success (inbred divided by outbred fitness). The laboratory data are from studies by Lynch (1977; dotted line), Connor and Belucci (1979; dashed line), and Meagher et al. (2000; dash-dotted line). The population data are all from Meagher et al. 2000; (solid lines) and are broken down by gender components and the mean of both genders (males + females).

or full-sib matings were conducted. The inbred and outbred offspring of these pairings were studied both in the laboratory and in a release and capture-release experiment. For this field experiment, 778 animals were released into the habitat where their wild ancestors had been captured a few generations earlier. Survival was estimated indirectly (trapability), and based solely on 123 mice recaptured and followed over 10 weeks. Most of the mice were never retrapped, and inbreeding was not correlated with lower survivorship (trapability) among the 663 mice that were never recaptured. Of the 123 mice recaptured at least once, inbred mice showed a 44% reduction in survival as estimated from recapture data over a 10-week period. There were no significant sex differences in estimated survival. In the laboratory, relative survivorship was measured between birth and weaning, with inbred white-footed mice showing a 6% reduction in survival. Thus the estimated fitness declines for inbreds were 7 times higher as measured in the field compared to the laboratory. These data add further support to the hypothesis that laboratory methods will often be relatively insensitive for measuring important health and vigor differences, with large fitness consequences in nature.

Case 3: The *t* Complex—What Controls the Spread of a Selfish Gene?

Around the globe, all subspecies of the house mouse are "infected" at varying frequencies with an inversion of the proximal third of chromosome 17 known as the *t* complex (Delarbre et al. 1988). When paired with a normal chromosome, recombination across this 20 centimorgan region is essentially blocked, due to four large nonoverlapping inversions, effectively linking together hundreds of loci (Artzt et al. 1982; Herrmann et al. 1986; Hammer et al. 1989). Although a heterozygote male produces *t* and + sperm in equal proportions (Silver and Olds-Clarke 1984), up to 100% of his offspring will inherit the *t* haplotype, giving this chromosomal inversion its characteristic distinction as a selfish genetic element. The extreme transmission distortion of the *t* complex is accomplished by up to five distorter genes and a linked responder gene. Protein products from the distorter genes impair wild type sperm (Silver and Remis 1987), while the responder gene protects *t*-bearing sperm from the effects of the distorter gene products (Lyon 1984; Herrmann et al. 1999). Female mice appear to transmit the *t* complex at Mendelian frequencies. Despite its excessive meiotic drive in males, the *t* complex cannot achieve fixation due to its costly effects in homozygotes. Depending on the combination of *t* haplotypes carrying linked lethality loci, homozygosity results in either complete lethality or male sterility. If

not for these dual costs balancing transmission distortion of the *t* complex, this selfish genetic element would have quickly gone to complete fixation within populations. Although other known examples of selfish genetic elements in rodents exist outside of *Mus* (Hoekstra and Hoekstra 2001), it is difficult to determine the frequency with which selfish elements evolve, because once they go to fixation (in the absence of allelic variation), they can no longer be easily detected.

The *t* paradox

Four decades of research have provided a good understanding of *t*-complex transmission distortion behavior and its underlying genetics. What remains unclear is why the *t* complex is found at frequencies across wild populations that are far lower than predicted. Despite its harmful effects in homozygotes, extreme transmission distortion through males should result in the persistence of *t*-complex haplotypes in approximately 70% of wild mice (Bruck 1957). Surveys of wild populations indicate that actual frequencies of *t*-bearing mice are far lower, ranging from 6 to 25% (Dunn and Levene 1961; Myers 1973; Figueroa et al. 1988; Lenington et al. 1988; Ardlie and Silver 1998; Huang et al. 2001; Dod et al. 2003). Much theoretical work has focused on understanding this unexpected discrepancy between observed and expected *t*-complex frequencies. Models incorporating the stochastic effects of drift show that within restricted parameters, drift together with limited migration could theoretically reduce the frequency of *t* haplotypes (Lewontin and Dunn 1960). However, house mouse populations exhibit long-range gene flow over generations, indicating that migration rates are high enough to invalidate some of these earlier stochastic models (Levin et al. 1969; Baker 1981; Berry et al. 1991). Other models have included a component of selection against heterozygotes, and these have resulted in predictions that are in fairly good accord with observed frequencies (Lewontin 1968; Levin et al. 1969; Petras and Topping 1983; Durand et al. 1997). Such heterozygote disadvantage models strongly predict the existence of a cost to heterozygotes that would counterbalance the extreme transmission distortion of the *t* complex.

Do *t* heterozygotes have a disadvantage?

Attempts to unearth this heterozygote cost have been tricky, as studies have produced greatly mixed results. The most consistent laboratory findings are from caged breedings, which tend to produce smaller litter sizes when either parent is +/*t* (Johnston and Brown 1969; Lenington et al. 1994). However, the decrease in litter size may be offset by a higher proportion of +/t animals surviving to sexual maturity in the laboratory (Dunn et al. 1958). Also perplexing are laboratory studies that found the fertility of +/t males to be both higher (Dunn and Suckling 1955) *and* lower (Levine et al. 1980) than +/+ males.

Recognizing the powerful potential of sexual selection to discriminate among genotypes, Sarah Lenington and her colleagues have done considerable work evaluating the role of mate choice in maintaining t haplotypes. The costly effects of homozygote sterility/lethality should favor individuals who avoid t-bearing mates, especially if the choosy individual already carries one t haplotype. In accord with this prediction, Lenington and her colleagues found evidence of odor preferences in both sexes for +/+ versus +/t individuals (Lenington 1983; Egid and Lenington 1985; Lenington and Egid 1985). Since the major histocompatibility complex is linked to the t complex, this obvious potential source of odor differences was tested for its contribution to t-associated odor discrimination. Odors from recombinant mouse strains carrying similar MHC haplotypes, with or without an associated t complex, were discriminated by males but not by females, suggesting that both the MHC complex and other genes within the t complex play a role in mediating t-associated odors (Egid and Lenington 1985; Lenington and Egid 1985). But despite t-associated odor avoidance behavior, short-term experiments conducted in indoor arenas revealed very little evidence for +/+ biased mating preferences (Franks and Lenington 1986), emphasizing the need to study mating preferences in an ecological context where animals assess mates on a variety of complex cues. In these short-term arenas, +/t males sired more offspring (but not significantly more) than their wild-type competitors, a result the authors attributed to higher fertility of +/t males (Dunn and Suckling 1955; Franks and Lenington 1986).

The competitive dynamics of +/+ and +/t mice were further studied by Lenington et al. (1996) in large outdoor seminatural enclosures. The results of their male dominance and survivorship analysis appear to corroborate the previous short-term population studies: +/t males were more aggressive and had slightly lower mortality than wild-type competitors (Lenington et al. 1996). Unfortunately, interpretation of these results is somewhat confounded by the experimental design, which assayed male aggression during staged encounters before release into the enclosures, and again after retrapping. Aggressive behavior during staged encounters could be influenced by the removal of important ecological and motivational cues normally present during territorial defense. Additionally, having spent part of the preceding night in traps, aggression scores might have reflected differential behavioral responses to stress between +/t and +/+ males. Finally, since reproductive data were not reported for this study, it is unknown whether aggression scores reflected actual fitness differences (Lenington et al. 1996).

Despite these multifaceted approaches to measuring selection on heterozygote t carriers, empirical support for the predicted heterozygote disadvantage remains murky. Table 5.1 outlines the major factors that have been studied for their potential effects on t frequencies. With so many different relevant factors, it is nearly impossible to integrate these data into a cohesive model that would successfully predict the fitness of t haplotypes in nature. We therefore decided to measure fitness itself.

The t complex in seminatural enclosures

Measuring the fitness of t heterozygotes requires a competitive environment in which sexual selection can operate on t-complex and wild-type genomes. Collecting long-term reproductive data is equally important for allowing estimates of lifetime fitness. The discovery of an unprecedented five-fold fitness decline in inbred relative to outbred males when measured across much of the adult lifespan (Meagher et al. 2000) clearly demonstrates the strength of large-scale seminatural experiments to reveal hidden effects of gene function (discussed previously). One year after publishing the inbreeding study, we made the fortuitous discovery of an entire natural experiment submerged within the original data set (Carroll et al. 2004). Some of the original mouse founders from the inbreeding study (Meagher et al. 2000) were found to harbor t-complex haplotypes, making it possible to test aspects of sexual selection that might serve to limit the spread of this selfish genetic complex in the wild. Because we were unaware of the presence of t haplotypes during the two generations of breeding to create founders for the inbreeding experiment, we had taken no steps to control the spread of this meiotic drive complex in our laboratory colony. The level of transmission distortion from t-bearing males during laboratory breedings was 0.88 (88% of all pups sired by +/t males inherited the t-complex version of chromosome 17), representing significant deviation from the 0.5 prediction of Mendelian inheritance. Consequently, when founder mice for the inbreeding study were selected from among the progeny of caged breeders, t frequencies had increased from 9.7% to 15.3%, a jump of 58%, but still well within the range of reported estimates from actual surveys. This fortuitous discovery allowed us to study the population dynamics of the t complex at biologically relevant initial frequencies. Since house mice are short lived and have relatively rapid generation times, ten replicate experiments, running for ten months, allowed us to

Table 5.1 Summary of tested factors that may influence frequencies of the *t* complex in nature

Type of study	Factors serving to increase *t* frequency	Reference	Factors serving to decrease *t* frequency	Reference
Models or simulations	Moderate to high migration rates	Lewontin and Dunn 1960; Levin et al. 1969	Drift	Lewontin 1962
			Homozygote and/or heterozygote disadvantage	Dunn et al. 1958
Laboratory studies or staged trials	*Gamete effects:* Up to 100% transmission of *t* gametes through males	Silver 1989		
	Fertility: Increased fertility of +/*t* males	Dunn and Suckling 1955	*Fertility:* Decreased fertility of +/*t* males	Johnston and Brown 1969
			Approximately 20% decrease in litter size of +/*t* males and females	Lenington et al. 1994
			Estrus condition[b]	Lenington and Heisler 1991
	Survivorship: Increased survival of +/*t* pups to sexual maturity	Dunn et al. 1958		
	Behavior: Increased aggression of +/*t* males	Lenington et al. 1996	*Behavior:* Male and female odor preferences for +/+ animals	Lenington 1991
Seminatural enclosure studies	*Mortality:* Higher survival of +/*t* males[a]	Lenington et al. 1996	*Mortality:* Decreased survival of +/*t* females and +/*t* dominant males	Carroll et al. 2004
	Higher survival only found in subordinate +/*t* males	Carroll and Potts, unpublished data	*Behavior:* Reduced dominance of +/*t* males	Carroll et al. 2004
			Reproduction: Decreased reproductive success of +/*t* males and females	Carroll et al. 2004

[a]It is not known whether there was a difference in survival between dominant vs. subordinate *t*-bearing males in this study.
[b]Transmission distortion appears to decrease in litters conceived during postpartum vs. cycling estrus.

measure survival and reproduction of *t*-bearing and non-*t*-bearing mice over the approximate span of a generation, providing rough estimates of lifetime reproductive success for these animals.

In contrast to the 58% increase of *t* haplotypes during two generations of caged breedings (in the absence of competition), we found that within the context of competitive populations, *t*-complex haplotypes declined dramatically. In a single generation, eight out of ten populations experienced a decline in *t* frequencies, with an overall reduction of 34.3% (Carroll et al. 2004). This result is even more striking with respect to the biased transmission rate of *t* haplotypes from *t*-bearing males in the enclosures (0.86), which was nearly equivalent to the 0.88 distortion rate measured from *t*-bearing males breeding in the laboratory. Observed *t* frequencies were 48.5% lower than predictions based on these biased transmission patterns. In analyzing the components of this *t*-haplotype decline, we found that inheriting a single *t*-bearing chromosome depressed reproductive success for both sexes, and significantly increased mortality rates for *t*-bearing females and dominant *t*-bearing males. However, the strongest component of the *t* decline emerg-

ing from our study was a significant reduction in the ability of +/*t* males to gain territories—only 32% of +/*t* males gained territories, whereas 67% of +/+ males gained territories (Carroll et al. 2004). Since female mice overwhelmingly prefer to breed with dominant males, *t*-associated impairment of male dominance helps explain why the *t* complex was found at lower-than-expected levels among enclosure pups. These data collectively suggest that selection against *t*-bearing heterozygotes in natural populations balances the opposing force of meiotic drive, and that male-male competition is the predominant form of this selection.

Though its effects are striking, the physiological mechanisms by which the *t* complex impairs heterozygote adults are unclear. Leamy et al. (2001) examined morphometric skeletal features in 109 male and female littermates of our enclosure founders (74 +/+, 31 +/*t*, and 4 *t*/*t*), and found increased fluctuating asymmetry (FA) in *t*-bearing mice but not in inbred mice. FA is the deviation from bilateral symmetry of paired morphometric characters. Presumably, perfect symmetry correlates with a high level of developmental stability and can therefore be disrupted by a variety of genetic and environmental perturbations, the *t* complex and inbreeding being two such stressors. Although the relationship between FA and fitness is unclear, greater susceptibility to such developmental perturbations will generally be costly (Møller and Swaddle 1997), especially if there is a relationship between morphology and performance. If it can be shown that FA affects any performance measure involved in male-male competition, then male territoriality and ultimately fitness can be greatly compromised by inheriting a single *t* haplotype.

Combining the enclosure study discussed earlier with prior work from other labs, it is clear that no single factor is responsible for limiting the invasion and spread of *t*-complex haplotypes in wild populations. Rather, there are a variety of selective factors, operating on both males and females to balance the extreme *t*-haplotype transmission bias from *t*-bearing males. But with so many selective components operating against the *t* complex, both in homozygote and heterozygote carriers, it is unclear why this "deleterious gene" persists, even at low levels. One solution to this apparent paradox is if sexual selection is less efficient in small populations. Deterministic models, such as those used to predict gene frequencies in the study by Carroll et al. (2004), assume random mating within infinitely large populations. However, *Mus* population sizes vary greatly, from small, single-male founder territorial units (Selander 1970), to populations containing hundreds of individuals (Ardlie and Silver 1998). Without competition, *t*-bearing animals are quite prolific—as was evident in our laboratory breeding colony. Small populations founded by one or a few *t*-bearing males are predicted to favor rapid increases in *t* frequencies by virtue of meiotic drive. Such populations may serve as the primary sources of *t* haplotypes, which can then infect neighboring populations as *t*-bearing individuals emigrate. In contrast, larger populations with extensive competition will be more effective at selecting against individuals bearing *t* haplotypes, driving down the frequencies of this selfish genetic complex. Frequency data from natural populations appear to fit such a model, where small and medium-sized populations (< 60 individuals) have a high prevalence of *t* haplotypes, and tend toward fluctuations in *t* frequency (Ardlie and Silver 1998). By contrast, *t* frequencies in large populations (> 60 individuals) tend to be much lower (avg. 3%) or are completely absent. Although this prediction of density-dependent selection has not yet been tested experimentally, seminatural populations promise a fertile approach for such studies.

Summary and Conclusions

Function is fitness! The three case examples presented here all involve major measurable fitness differentials resulting from naturally occurring genetic variation among wild (or wild-derived) house mice. However, these major fitness differences often went undetected using laboratory approaches. Recent theory and empirical work, including a meta-analysis encompassing 14 years worth of selection studies, suggest that the dominant forms of selection-driven evolution may prove to be intra-male competition and female choice (Hoekstra et al. 2001). This does not mean that parasites, predators, and other agents of natural selection are not important, but that in many species, sexual selection may be a good proxy for natural selection, because it screens for those who will command the best resources and will have the best mating success. Consequently, it becomes a good predictor of who will survive and ultimately reproduce. Seminatural competitive populations of *Mus* largely capture the social ecology of this species, and thus present a very tractable experimental system for measuring fitness, which is required for answering many questions in biology. Fortunately, using the social ecology approaches reviewed previously can help reveal many phenotypes that are cryptic under laboratory conditions, serving the functional genomics community and the molecular research program in general.

Chapter 6 A Phylogenetic Analysis of the Breeding Systems of Neotomine-Peromyscine Rodents

Matina C. Kalcounis-Rüppell and David O. Ribble

A BREEDING SYSTEM describes who copulates with whom, who contributes genes to the next generation, and is the result of the combination of female and male mating strategies, which are often conflicting (see Waterman chap. 3 and Solomon and Keane chap. 4 this volume). Under some conditions the conflict between the sexes is ultimately played out in terms of one gender monopolizing access to the other, otherwise known as polygamy (Emlen and Oring 1977). The environmental potential for polygamy (EPP) is dictated by ecological, physiological, and life-history characteristics that, in turn, have evolved within a particular phylogenetic framework (fig. 6.1). The environmental potential for polygamy depends on the degree to which multiple mates, or the resources necessary to gain multiple mates, are economically defendable (Emlen and Oring 1977). Ecological, physiological, and life-history characteristics either allow organisms to, or prevent organisms from, taking advantage of or utilizing this potential. In the case of mammals, lactation and gestation are solely the provenance of females. However, as originally indicated by Emlen and Oring (1977), emancipation from parental care duties need not necessarily lead to the evolution of polygamy. Furthermore, as pointed out in chapter 3 by Waterman and chapter 4 by Solomon and Keane, female strategies often prevent males from monopolizing matings.

This chapter focuses on the breeding systems of Neotomine-Peromyscine rodents. There are three specific objectives to this chapter. First, we describe the patterns for major Neotomine-Peromyscine clades using data collected from the literature (table 6.1). Second, we examine data

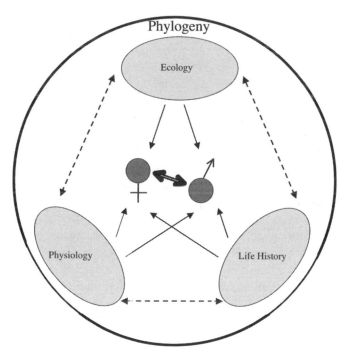

Figure 6.1 A general scenario for the determinants of breeding systems as indicated by Emlen and Oring (1977) and expanded to incorporate phylogenetic influences on ecology, physiology, and life history of the mating individuals. The breeding system (double solid line) is dictated by the influence of ecological, physiological, and life-history characteristics (single solid lines) on both males and females. In turn, ecological, physiological, and life-history characteristics are interrelated (dashed line) and have evolved within a particular phylogenetic framework.

Table 6.1 Breeding behaviors of Neotomine-Peromyscine rodent species

Taxon	Female spacing	Male spacing	Space size	Paternal care	Dispersal	References
Onychomys torridus	Solitary	Roving	M > F	Y		McCarty 1975; Horner 1961; Chew and Chew 1970; Blair 1943
Onychomys leucogaster	Solitary	Roving	M > F	N		Horner and Taylor 1968; Frank and Heske 1992; Stapp 1999
Baiomys taylori	Extensive overlap	Roving	Equitable	Y		Morrison et al 1977; Eshelman and Cameron 1987; Hudson 1974; Packard 1960; Blair 1941; Raun and Wilks 1964
Reithrodontomys humulis	Little overlap	Roving	Equitable	N		Dunaway 1968; Stalling 1997; Cawthorne and Rose 1989; Chandler 1984
Reithrodontomys fulvescens	Little overlap	Roving	M > F		Equal	Cameron and Kincaid 1982; Spencer and Cameron 1982; Packard 1968
Reithrodontomys megalotis		Polygynous	Equitable			Blaustein andRothstein 1978; Webster and Knox Jones, Jr. 1982; Fisler 1963; Fitch 1958
P. crinitus	Solitary			N		Eisenberg 1963a
P. boylii	Solitary	Roving/ polygynous	M > F	N		Ribble and Stanley 1998; Kalcounis-Rüppell and Spoon (submitted) and references therein
P. eremicus	Solitary			Y		Hatton and Meyer 1973; Lewis 1972; Eisenberg 1968
P. californicus	Solitary	Monogamous	M > F	Y	Females	Ribble 2003 and references therein
P. melanocarpus				Y		Rickart 1977; Rickart and Robertson 1985
P. attwateri			M > F			Schmidly 1974; Brown 1964
P. gossypinus	Solitary	Roving				Pournelle 1952; Wolfe and Linzey 1977; Pearson 1953
P. mexicanus	Solitary			Y		Rickart 1977; Duquette and Millar 1995a, 1995b, 1998
P. truei	Solitary	Roving	M > F	N		Hall and Morrison 1997; Ribble and Stanley 1998
P. leucopus	Solitary and gregarious	Roving/ polygynous	M > F	N	Males	Wolff 1989; Wolff and Cicirello 1989, 1991; Schug et al. 1992; Xia and Millar 1988; Xia and Millar 1989
P. polionotus	Solitary	Monogamous	M > F	Y	Equal	Blair 1951; Smith 1966; Foltz 1981
P. maniculatus	Solitary	Roving/ polygynous	M > F	N	Males	Horner 1947; Howard 1949; Ribble and Millar 1996; Wolff 1989; Wolff and Cicirello 1989, 1991
Neotomodon alstoni		Monogamous		Y		Luis et al. 2000, 2004
Neotoma albigula	Little overlap	Monogamous	Equitable	N		Boggs 1974; Batemen 1967; Macêdo and Mares 1988
Neotoma floridana	Little overlap	Monogamous	M > F	N		Rainey 1956; Fitch and Rainey 1956; Wiley 1980
Neotoma micropus	Little overlap	Roving	M > F	N	Males	Davis 1966; Wiley 1972; Raun 1966; Braun and Mares 1989
Neotoma stephensi	Solitary	Roving				Jones and Hildreth 1989; Ward 1984; Conditt and Ribble 1997
Neotoma cinerea	Little overlap	Roving	M > F		Males	Topping and Millar 1996a, 1996b, 1998
Neotoma macrotis	Solitary	Roving	M > F		Males	Matocq and Lacey 2004; Kelly 1989
Sigmodon	Solitary	Roving	M > F	N	Males	Cameron and Spencer 1985, 1981; Doonan and Slade 1995; Diffendorfer and Slade 2002
Akodon	Solitary	Roving	M > F	N	Males	Gentile et al. 1997; Suarez and Kravetz 2001; Steinmann et al. 1997; Citadino et al, 2002, 1998

NOTE: *Sigmodon* and *Akodon* are included as outgroups for comparative purposes.

from the literature on the following breeding behaviors: male spacing, female spacing, relative intersexual home range/territory size, paternal care, and juvenile dispersal patterns. We examine breeding behavior data in a phylogenetic framework to test if any phylogenetic patterns emerge in the observed variation in these breeding behaviors and if relationships occur among these behaviors. Third, we examine in a phylogenetic framework whether dietary, physiological, or life-history characteristics of the taxa are able to explain the observed variation in these breeding behaviors.

We explicitly focus on data from natural populations, although much of the information that we have on Neotomine-Peromyscine breeding systems and social behavior comes from the lab or seminatural situations. One of us recently published a phylogenetic review of monogamy and paternal care in *Peromyscus* (Ribble 2003). The current study is expanded to encompass the entire lineage of North American rats and mice within the rodent family Muridae, and includes mating strategies and taxa for which we have data from the field. Our taxonomy and phylogenetic topologies in this study reflect recent systematic work on the relationships among the genera *Peromyscus, Baiomys, Neotoma, Onychomys,* and *Reithrodontomys* (Edwards and Bradley 2002; Arellano et al. 2003; Bradley et al. 2004).

Descriptions of Breeding Systems

Peromyscini

The genus *Peromyscus* (> 50 species) has a large distribution in the Nearctic and northern Neotropics. An overview of *Peromyscus* social behavior was provided by Wolff (1989), with the majority of information coming from the deer mouse (*P. maniculatus*) and the white-footed mouse (*P. leucopus;* for discussion of the bias toward these two species see Wolff 1989). The societal structure of the deer mouse and the white-footed mouse is similar despite their varied habitats, resources, and widespread distributions. The following is from radiotelemetry studies. Males and females occupy home ranges, and in some cases home ranges are defended against conspecifics, thereby becoming territories. Maintenance of territories may be density dependent. In the wild, males and females do not nest together except in nest-boxes. Males occupy home ranges and/or defend territories to provide access to resources including food and reproductive females, whereas females occupy home ranges and/or defend territories to provide access to resources including food and space for raising their young and to pro-

tect their young from infanticide (Wolff 1993b). The pattern of overlap in home range/territories for males and females differs. In general, one male overlaps home ranges/territories of more than one female, whereas the females have home ranges or territories that are exclusive of other females. Using a polymorphic Esterase-1 locus, in a study of *P. leucopus,* Xia and Millar (1991) found that in two separate years 1 of 29 and 6 of 32 litters contained more than 2 paternal alleles, providing direct evidence that females and males were engaging in multiple mating, and they estimated that over 68% of females were involved in multiple mating. Similarly, in a study of *P. maniculatus,* Ribble and Millar (1996) found that male home ranges were significantly larger than female home ranges, and male home ranges overlapped multiple female and male home ranges. Using DNA fingerprinting they found 1 of 11 litters to be sired by two males, one of which also successfully sired litters of two more females. Additionally, 2 to 3 of 7 litters were likely sired by multiple males, based on band sharing values lower than observed among full-siblings (Ribble and Millar 1996). This frequency of multiple inseminations was similar to that found for *P. maniculatus* by using protein electrophoresis (Birdsall and Nash 1973).

In his chapter on *Peromyscus* social behavior Wolff (1989) suggested that the species diversity, habitat, and geographical variation in this genus provide a great opportunity for comparisons of social behavior. There are contrasts to the patterns of social organization in *P. maniculatus* and *P. leucopus,* and since that review, we have gained insight into the social behavior of some other *Peromyscus* species.

Relatively little is known about the behavior or ecology of the Canyon mouse (*P. crinitus*) because, as its common name implies, it inhabits rock outcrops at high elevations in western North America. Through intensive trapping and genotyping at microsatellite loci it was found that in two Utah populations of *P. crinitus* a minimum of 3 of 10 litters were multiply sired (Shurtliff et al. 2005). By genotyping all captured individuals in the population, including the potential sires, it was shown that although there is multiple mating by females, there was no instance where a male mated with more than one female (within or between litters), suggesting genetic polyandry in addition to genetic monogamy for this species. Behavioral studies were not conducted in these populations and it is not clear whether there was any nest sharing among mates; however, males assigned with high confidence of paternity were always trapped near their female mates (relative to males who did not have a high confidence of paternity; Shurtliff et al. 2005). These results suggest that home ranges of these males likely overlapped with females.

The brush mouse (*P. boylii*) is found in canyon bottoms and, in California populations, are associated with oak trees and the acorns (mast) they produce (Kalcounis-Rüppell and Millar 2002). During a two-year study when the population density was high (40–70 mice/ha), males and females did not differ in home range size, but inter- and intrasexual home-range overlap was higher during a high population density year. Males and females did not share nests and did not maintain long-term pair bonds. Although multiple mating appeared infrequent, based on behavioral observations (transfer of fluorescent powder), microsatellite analyses showed that 1 of 7 litters was sired by more than one male (Kalcounis-Rüppell 2000). At high population densities neither males nor females defended territories (Kalcounis-Rüppell 2000). These results differ from a comparable study on a *P. boylii* population at a relatively low population density in New Mexico (highest minimum number known alive: 30/2.7 ha; Ribble and Stanley 1998); where home range size was inversely related to conspecific density, females did not overlap with each other, and males had home ranges that overlapped with multiple females. Furthermore, there was a difference in home range size between the sexes, with male home ranges being almost twice as large as female home ranges. These results show population variation in social structure and highlight the importance of resource availability and population density. Indeed, during low mast years (and subsequent low population densities), *P. boylii* in California appears to have a social structure similar to that of low population density *P. boylii* in New Mexico (Kalcounis-Rüppell and Spoon [manuscript submitted for publication]).

In some species of *Peromyscus*, pairs of males and females have exclusive territories and exclusive genetic contributions to their litters, and are thus considered to be monogamous from a behavioral and a genetic standpoint. Oldfield mice (*P. polionotus*) nest in burrows that can house males, females, and offspring (50.5% of burrows contain an adult male and female; Foltz 1981). Using starch-gel electrophoresis of 5 polymorphic proteins, it was demonstrated that the males who were nesting with females were the sole sires of the litters, and when a female had at least 2 consecutive litters, the same male sired both of the litters (Foltz 1981). Home range size and dispersal distance is equal for males and females (Swilling and Wooten 2002). The California mouse (*P. californicus*) is exclusively monogamous, with DNA fingerprinting confirming that in 28 of 28 litters examined, the behavioral pairs from the field were the parents of litters (Ribble 1991). Males and females nest together during breeding and nonbreeding seasons, and maintain a pair bond permanently unless one mate dies or disappears from the grid (Ribble and Salvioni 1990).

Mean home range size is 1161 m², and does not differ between males and females (Ribble and Salvioni 1990). Dispersal is female biased (Ribble 1992). Male removal in the field has a negative effect on reproductive success, as a result of the absence of direct care of young rather than protection against infanticidal intruders (Gubernick and Teferi 2000).

The volcano mouse (*Neotomodon alstoni*) is a species that has long been taxonomically associated with *Peromyscus* (Carleton 1989), but little is known about its mating system in natural populations. In the laboratory, however, much is known about male and female parental behaviors (Luis et al. 2000; Luis et al. 2004). In captivity, males actually spend more time than females huddling, grooming, and retrieving young (Luis et al. 2000).

Lastly, the genus *Reithrodontomys* is also included in the Peromyscini clade (Bradley et al. 2004), and most of the published information from this genus is from *R. megalotis*, *R. fulvescens*, and *R. humulis*. Based on multiple captures in the same trap, *R. megalotis* is reported to be the most social of these species (Blaustein and Rothstein 1978; Cawthorn and Rose 1989), but male and female home ranges are reported to be similar in size (Fitch 1958). For *R. humulis*, there is no evidence of male care (Kaye 1961), and home ranges of males and females are similar in size and overlap extensively (Chandler 1984; Dunaway 1968; Cawthorn and Rose 1989). Male home ranges are larger than those of females in *R. fulvescens* (Packard 1968; Cameron and Kincaid 1982), but there is little evidence of territorial behavior in either sex (Packard 1968).

Neotomini

The Neotomini clade includes the genera *Neotoma* and *Onychomys* (Bradley et al. 2004). The genus *Neotoma* (approx. 20 species) is distributed from Canada to Central America. Two of these species have been studied over the long term and provide a basic understanding of their social structure in the wild. The bushy-tailed woodrat (*Neotoma cinerea*) is distributed through much of northwestern North America. Its local distribution is limited by the availability of rock outcrops for suitable nest sites (Hickling 1987), which can be up to 470 m away from foraging sites (Topping and Millar 1996a), resulting in a clumped distribution of females (Hickling 1987). Matrilineal females tend to be more closely associated in space than nonmatrilineal females, and there are fitness advantages associated with mothers and daughters coexisting in space (Moses and Millar 1994). Juvenile females were more likely to survive if they stayed on the outcrop with their mothers, and they suffered fewer reproductive failures when breeding as yearlings if their mothers were present (Moses and Millar 1994). Radiotelemetry revealed that in addition to the clumped distribution of females on rocky outcrops, females and resident males had considerable overlap of home ranges on the outcrops (Topping and Millar 1996b). DNA fingerprinting demonstrated that despite high levels of inter- and intraspecific overlap of home ranges on outcrops, 35 of 35 litters were sired by a single male; no male fathered more than one litter from a given female within or between years, resulting in low variation in reproductive success of males and females (Topping and Millar 1998).

The big-eared woodrat (*Neotoma macrotis* [formerly *N. fuscipes,* but now recognized as distinct]; Matocq 2002) occupies stick 'houses' (Linsdale and Tevis 1951) along canyon bottoms of oak woodland habitat. In a population of *N. macrotis* that has been studied for over 50 years, individuals are found in the long (580 m) and relatively narrow (26 m) riparian buffer (Matocq and Lacey 2004). Livetrapping and radiotelemetry data on this population show similar female group structure as in *N. cinerea,* with high levels of female philopatry and male-biased dispersal (Kelly 1989), suggesting matrilineal kin groups. However, using microsatellite genotyping, it was found that these were not matrilineal kin groups, because females that were grouped in space were no more related to each other than to other females in the population, and no more successful, with respect to number of pups weaned, when living in close proximity to first-order relatives (Matocq and Lacey 2004). Thus for *N. macrotis* female philopatry is not the only factor contributing to population genetic structure; factors such as habitat quality and interactions with conspecifics may be important (Matocq and Lacey 2004). Similar to *N. cinerea,* however, variation in reproductive success for males and females was low and equitable despite the occurrence of males and females having multiple partners across litters and some litters being multiply sired (Matocq 2004).

Although less intensively studied than *N. cinerea* and *N. macrotis,* the social biology of the desert woodrat *N. lepida* and the southern plains woodrat (*N. micropus*) has received some attention. The Danzante woodrat (*N. lepida latirostra*) is larger bodied than its continental counterparts, and it has larger home ranges that are more exclusive within both sexes than is displayed in continental *N. lepida* and other *Neotoma* species (Vaughan and Schwartz 1980). From trapping data, the mean home range size of *N. l. latirostra* is 0.33 ha and 0.11 ha for males and females, respectively (Vaughan and Schwartz 1980); these home ranges are larger than for populations in a Californian coastal sage (mean male home range 0.04 ha; Bleich and Schwartz 1975) and the San Gabriel Mountains (mean male home range 0.19 ha; MacMillen 1964). Home range esti-

mates from radiotelemetry in a Texas population of *N. micropus* show that like *N. lepida,* home ranges are exclusive within sexes, with males (0.19 ha) having larger home ranges than females (0.02ha); however, there was more overlap of female home ranges by male home ranges than male home ranges overlapped by female home ranges (Conditt and Ribble 1997). Additionally, there was no nest sharing among *N. micropus,* and the majority of observations of this species were of solitary animals at the nest, suggesting that *N. micropus* is relatively asocial (Conditt and Ribble 1997).

Onychomyini

The two species in the genus *Onychomys* were traditionally considered monogamous (see review in Frank and Heske 1992), but radiotelemetry studies indicate otherwise. Radiotelemetry of *O. torridus* (Frank 1989; Frank and Heske 1992) and *O. leucogaster* (Stapp 1999) indicate that males have larger ranges than females, which overlap with multiple females during the breeding season, consistent with a promiscuous mating system. No genetic studies of paternity have been conducted in either species.

Baiomyini

This monophyletic group includes *Baiomys* and *Scotinomys,* of which there is only information on *Baiomys taylori.* Early studies under laboratory conditions by Blair (1941) and Packard (1960) indicated that males will care for the young. In the field, males and females have similar size ranges, with considerable overlap within and among both sexes (Raun and Wilks 1964). No genetic paternity information is available for either species in natural populations.

Phylogenetic Analyses

Relationships among breeding behaviors

We conducted a phylogenetic comparative analysis to reconstruct ancestral character states of breeding behaviors and to test if any of these behaviors appear in the phylogeny at similar times. Furthermore, we wanted to test if there were any correlations in the appearance of these behaviors, and if the presence of one behavior influenced the appearance of others, taking into account their evolutionary history (Felsenstein 1985; Maddison 2000). Sufficient data are available (see table 6.1) for the Neotomine-Peromyscine rodents to critically evaluate mating system hypotheses. For example, various hypotheses predict that male spacing behavior will be dependent on female spacing behavior (Emlen and Oring 1977; Ostfeld 1990), or that parental care would be dependent on monogamous spacing in males (Brotherton and Komers 2003). Specifically, we examined data on male spacing, female spacing, relative intersexual home-range/territory size, paternal care, and juvenile dispersal patterns to test for phylogenetic patterns in the observed variation in these breeding behaviors and to determine if any relationships occur among these behaviors. We coded all traits as indicated in table 6.1. Female spacing patterns were scored as solitary (no overlap between home ranges), little overlap, extensive overlap, or gregarious (largely overlapping home ranges, usually accompanied with nest-sharing), based on spatial overlap during the breeding season. Species with both solitary and gregarious female spacing were scored as gregarious. Male spacing patterns were scored as monogamous, roving, polygynous, or variable if populations exhibited multiple patterns. No species has been documented to be solely polygynous; those species with polygyny have also been documented as roving. Space size was recorded as equitable or male range size being greater than female (M > F in table 6.1). If a species has been observed to exhibit male care in the laboratory, but not in the field, then they were considered nonpaternal. If a species has exhibited paternal behavior in the lab, has other life-history traits consistent with paternal care (e.g., Dewsbury 1981), and there was no conflicting information from the field, they were considered paternal (table 6.1). Where there was conflicting information from the field, we used the best evidence from the field studies to determine paternal care (e.g., we characterized *P. leucopus* as not having paternal care despite the results of Schug et al. 1992, table 6.1). Finally, dispersal of juveniles was coded as being equitable, female biased, or male biased.

Relationships between behaviors and diet, physiological, and life-history characteristics

We conducted a phylogenetic comparative analysis to test for relationships between mapped character states of breeding behaviors and ecological, physiological, or life-history characteristics, taking into account their evolutionary history (Felsenstein 1985; Maddison 2000). Specifically, we examined whether diet, physiological, or life-history characteristics of the taxa could explain the observed variation in breeding behaviors. The ecological, physiological, and life-history characters we used are shown in table 6.2. For empty cells for continuous variables (basic metabolic rate [BMR] and relative litter weight) the mean value for the genus was assumed (table 6.2).

A significant association exists between energy expenditure and diet in the wild in small mammals. Small mammals that exploit high-energy foods (vertebrates and insects) are able to spend more energy per unit mass relative to resting metabolic rates than small mammals that exploit energy-poor foods (seeds and grasses; Speakman 2000). Because Neotomine-Peromyscine rodents span this range of diets, and different costs and benefits are associated with different food resources, we hypothesized a relationship between diet and breeding behaviors. Kalcounis-Rüppell et al. (2002) demonstrate a higher energetic cost associated with mating for promiscuous males (*Peromyscus boylii*) than monogamous males (*P. californicus*), but no difference between females. Thus we predicted that roving males would have higher energy diets than solitary males, and that there would be no relationship between diet and female spacing. To test this hypothesis, we coded diet as carnivorous, insectivorous, omnivorous, granivorous, or herbivorous (table 6.2).

Ribble (2003) suggested that relative neonate and litter weight (relative to adult weight) might be correlated with mating strategies in *Peromyscus* because of the energetic cost of lactation and consequent maternal investment, which varies with offspring size and number. We hypothesized a relationship within the Neotomine-Peromyscine rodents between relative litter weight and breeding behaviors. We predicted that solitary females would have higher relative litter weights (Ribble 2003). The majority of data required to calculate relative litter weight (litter size, neonate weight (g) at birth, relative neonate weight) were from Millar (1989), with some data from Hayssen et al. (1993). Relative litter weights were calculated by dividing litter weights by adult weights (table 6.2). For the outgroups *Sigmodon* and *Akodon*, we took the average of all the species values for each genus for all variables used to calculate relative litter weights (table 6.2).

The size of the distributional range of a taxon correlates with both the ecological conditions of the range (Glazier 1980) and species life-history patterns (Glazier 1980, Brown 1995). For example, in an analysis of *Peromyscus*, Glazier (1980) found a positive correlation between geographical range and litter size, and he argued that larger geographic ranges were found in species with larger litter sizes, short life spans, and smaller body size. Since these species-level traits likely affect the distribution of organisms (Brown 1995), we wanted to test if the distribution was correlated with the behaviors we measured. To determine the species distribution area we recorded the size (km²) of the geographic ranges of all species, using the digital distribution maps of mammals of the western hemisphere (Patterson et al. 2003). To calculate species distribution areas we used the XTools extension in ArcView 3.2 (ESRI, Redlands,

Table 6.2 Ecological, physiological, and life-history characteristics of Neotomine-Peromyscine rodent species

Taxon	Species distribution area (km²)	Relative litter weight	BMR (mlO$_2$/g/ minute)	Residual BMR (mlO$_2$/g/ minute)	Feeding habit	References[a]
O. torridus	3907553.78	0.30	1.55	−0.01	Carnivorous	McCarty 1975 and references therein; Chew and Chew 1970
O. leucogaster	3907553.78	0.29	1.55*	−0.10	Carnivorous	McCarty 1978 and references therein
Baiomys taylori	1164795.22	0.45	1.95	0.01	Omnivorous	Eshelman and Cameron 1987 and references therein
R. humulis	1596620.06	0.38	2.46*	0.01	Omnivorous	Stalling 1997 and references therein
R. fulvescens	2640627.61	0.27	2.46*	0.10	Insectivorous	Spencer and Cameron 1982 and references therein
R. megalotis	5303556.43	0.44	2.46	0.07	Granivorous	Webster and Knox Jones, Jr. 1982 and references therein
P. crinitus	930352.55	0.45	1.33	−0.18	Omnivorous	Johnson and Armstrong 1987 and references therein
P. boylii	2680094.46	0.31*	2.34	0.14	Omnivorous	Kalcounis-Rüppell and Spoon (ms. submitted) and references therein
P. eremicus	1387374.30	0.29	1.47	−0.10	Omnivorous	Veal and Caire 1979 and references therein
P. californicus	158389.58	0.25	1.37	−0.04	Omnivorous	Merritt 1978 and references therein
P. melanocarpus	9334.39	0.18	1.67*	0.11		
P. attwateri	490547.28	0.31*	1.67*	0.01	Herbivorous	Brown 1964
P. gossypinus	1401995.07	0.24	1.72	0.03	Omnivorous	Wolfe and Linzey 1977 and references therein
P. mexicanus	512873.21	0.20	1.67*	0.10		
P. truei	2184728.43	0.27	1.71	0.01	Omnivorous	Hoffmeister 1981 and references therein
P. leucopus	6593854.13	0.38	1.66	−0.03	Omnivorous	Lackey et al. 1985 and references therein
P. polionotus	478832.59	0.44	1.79	−0.06	Herbivorous	Gentry and Smith 1968
P. maniculatus	14104524.90	0.40	1.74	−0.01	Omnivorous	Baker 1983
Neotomodon alstoni	51636.24	0.16	1.67*	0.09	Insectivorous	Glendinning and Brower 1990
N. albigula	1952915.28	0.13	0.74	−0.08	Herbivorous	Mačedo and Mares 1988 and references therein
N. floridana	2172842.61	0.19	0.72*	−0.06	Herbivorous	Rainey 1956; Wiley 1980 and references therein
N. micropus	1121208.66	0.13	0.72*	−0.06	Herbivorous	Braun and Mares 1989 and references therein
N. stephensi	228873.06	0.11	0.72*	−0.12		
N. cinerea	3636547.44	0.17	0.73	−0.03	Herbivorous	Smith 1997 and references therein
N. macrotis	327006.17	0.12	0.79	0.01	Herbivorous	Carraway and Verts 1991 and references therein
Sigmodon	937034.08	0.32	1.48	0.18	Omnivorous	Cameron and Spenser 1981 and references therein
Akodon	556064.75	0.49	1.70	0.02	Omnivorous	Dalby 1975

NOTES: *Sigmodon* and *Akodon* are included as outgroups for comparative purposes. Average values for genus, where specific value was not available is denoted by an asterisk (*).
[a]For feeding habit only; see text for other sources of data in this table.

CA). For outgroups *Sigmodon* and *Akodon,* we took the average of all the species for each genus as the geographic range for that genus (table 6.2).

All of our hypotheses about whether ecological, physiological, or life-history characteristics of the taxa are able to explain the observed variation in these breeding behaviors are largely based on energetic reasoning. Thus we also include basal metabolic rate (BMR) as an independent variable (table 6.2). The data on metabolic rates were from McMillan and Garland (1989). We used mass independent data in our analyses by using residual values from the predicted values from the significant relationship of body mass and mass-specific BMR for all Neotomine-Peromyscine rodents included in our analysis ($F_{1,25} = 81.18$, $P < 0.001$, $R^2 = 0.76$; logBMR = −0.32 log Body Mass + 0.67).

Phylogenetic methods

The phylogeny we used is based on the study of Bradley et al. (2004), which differentiated the Baiomyini, Neotomini, Onychomyini, and Peromyscini groups with *Sigmodon* and *Akodon* as outgroups. Not all species of interest in this study were included by Bradley et al. (2004), so we followed Edwards and Bradley (2002) for relationships among *Neotoma* and Arellano et al. (2003) for relationships among *Reithrodontomys.* Within the Peromyscini, several species of *Peromyscus* were not part of the phylogeny published by Bradley et al. (2004), so we relied on information from Stangl and Baker (1984) and Bradley (unpublished data) when necessary. We use two tree topologies that differ primarily in the placement of the Onychomyini. In the first,

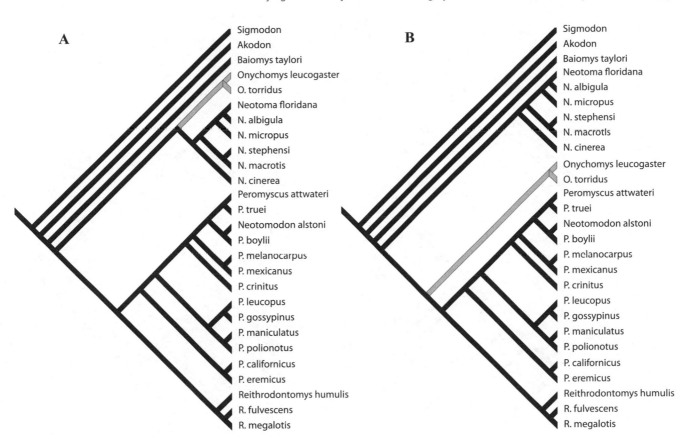

Figure 6.2 The two phylogenies used in our comparative analyses that reflect the two current hypotheses of the evolutionary relationship among the Neotomine-Peromyscine rodents. The differences are highlighted in gray. In topology A *Onychomys* is a sister taxon to Neotoma. In topology B, *Onychomys* is basal to *Peromyscus*.

(referred to as topology A) *Onychomys* is sister to *Neotoma*, and in the second, *Onychomys* (referred to as topology B) is basal to *Peromyscus* (fig. 6.2). Both topologies were used, because they reflect two current hypotheses of the evolutionary relationship among the Neotomine-Peromyscine rodents; however, current multigene data support alignment of *Onychomys* with the Peromyscini (Reeder et al. 2006).

Character states from table 6.1 and table 6.2 were mapped on the phylogeny of Neotomine-Peromyscine rodents using Mesquite, Version 1.02 (Maddison and Maddison, 2004). We made no assumptions about the evolutionary sequence in which characters changed. Continuous characters, such as geographic range and litter size and weight characteristics, were compared using Felsenstein's method of independent contrasts (Felsenstein 1985). Correlations among categorical independent characters were examined using Maddison's pairwise comparisons (Maddison 2000). To facilitate pairwise comparisons with binary categorical independent variables, spacing variables were recoded for both sexes as female spacing: solitary or not solitary, and male spacing: monogamous or not mo-

nogamous. In all cases for pairwise comparisons, we ordered categorical variables according to determination of ancestral traits that we obtained from the character trace analysis in Mesquite.

Results

Ancestral states and relationships among breeding behaviors

For all results, we use topology A to discuss and graphically describe ancestral character states. For all analyses of interrelationships among breeding behaviors, there was congruence between topology A and topology B.

The relationships between male and female spacing are indicated in the mirror phylogenies of figure 6.3a. For most Neotomine-Peromyscine rodents examined, females are solitary and males are roving, and these traits are ancestral for the clade (fig. 6.3a). This pattern is consistent with Waterman's assertion that promiscuity is common in rodents (Waterman, chap. 3). Using pairwise comparisons,

A

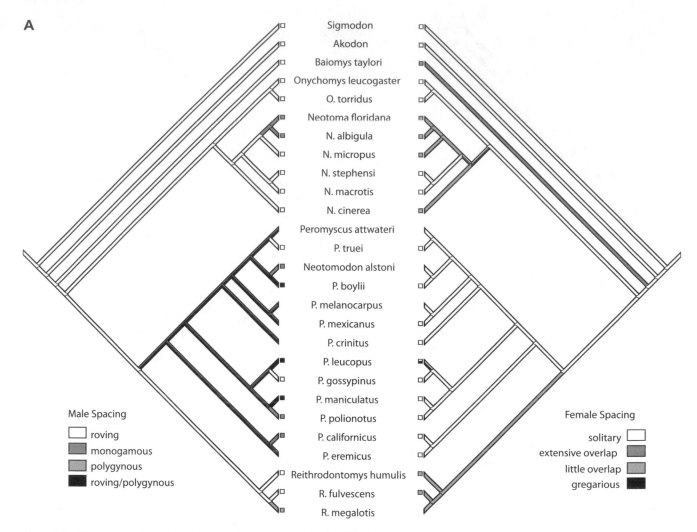

Male Spacing

□ roving
▨ monogamous
▨ polygynous
■ roving/polygynous

Sigmodon
Akodon
Baiomys taylori
Onychomys leucogaster
O. torridus
Neotoma floridana
N. albigula
N. micropus
N. stephensi
N. macrotis
N. cinerea
Peromyscus attwateri
P. truei
Neotomodon alstoni
P. boylii
P. melanocarpus
P. mexicanus
P. crinitus
P. leucopus
P. gossypinus
P. maniculatus
P. polionotus
P. californicus
P. eremicus
Reithrodontomys humulis
R. fulvescens
R. megalotis

Female Spacing

solitary □
extensive overlap ▨
little overlap ▨
gregarious ■

Figure 6.3   Mirror phylogenies showing the ancestral state reconstruction for (A) female and male spacing and (B) paternal care and male spacing in Neotomine-Peromyscine rodents. For this, and all phylogenies presented herein, the character state for each taxon is indicated in the block at the terminal end of the lineage and the origin of the character on phylogeny is indicated by the shading. No block at the terminal end of a lineage indicates insufficient data for that particular character.

we found that male spacing behavior was independent of female spacing behavior (3,096 pairings of terminal taxa with 5 pairs contrasting female behavior, $P = 0.13-0.75$) and female spacing behavior was independent of male spacing behavior (4,416 pairings of terminal taxa with 4 pairs contrasting male behavior, $P = 0.5-1.0$). In general, female spacing appears most variable among lineages of the Neotomini, whereas male spacing is most variable among the clade of Peromyscini that includes *P. californicus*, *P. eremicus*, *P. leucopus*, *P. gossypinus*, *P. maniculatus*, and *P. polionotus*. What little information we have on *Baiomys* and *Reithrodontomys* suggests these lineages have different spacing strategies.

Paternal care appears to have evolved multiple times (fig. 6.3b), consistent with the conclusion of Ribble (2003)

for *Peromyscus*. We compared the evidence for paternal care to patterns of male spacing and found monogamous males provide paternal care (8,048 pairings of terminal taxa with 6 pairs contrasting paternal care behavior, $P = 0.03-0.75$; fig. 6.3b).

By far the most common and presumably ancestral state within the Neotomine-Peromyscine rodents is for males to have larger home ranges than females, with very few taxa demonstrating equitable range sizes and no taxa with female ranges larger than male ranges (fig. 6.4a). There are relatively few data on dispersal behavior within the Neotomine-Peromyscine rodents. However, available information indicates that the ancestral condition is for natal dispersal to be male biased (fig. 6.4b). Neither home range size nor natal dispersal were related to other breeding behaviors.

B

Sigmodon
Akodon
Baiomys taylori
Onychomys leucogaster
O. torridus
Neotoma floridana
N. albigula
N. micropus
N. stephensi
N. macrotis
N. cinerea
Peromyscus attwateri
P. truei
Neotomodon alstoni
P. boylii
P. melanocarpus
P. mexicanus
P. crinitus
P. leucopus
P. gossypinus
P. maniculatus
P. polionotus
P. californicus
P. eremicus
Reithrodontomys humulis
R. fulvescens
R. megalotis

Paternal Care

☐ no care

■ care

Male Spacing

roving ☐

monogamous ▨

polygynous ▥

roving/polygynous ■

Figure 6.3 (continued)

Relationships between behaviors and diet, physiological, and life-history characteristics

For all results, we use topology A to discuss and graphically describe ancestral character states. Although there was agreement between topology A and B in the analyses of the relationships between behaviors and diet, physiological, and life-history characteristics, the congruence was not perfect. Where the two topologies differed we present both results. We found significant relationships among breeding behaviors and physiological, life history, and diet characteristics as well as trends with P values of 0.06. We treat these trends as biologically meaningful.

Among the diet, physiological, and life-history characteristics for the Neotomine-Peromyscine rodents, we found a positive correlation between species distribution area and litter size (fig. 6.5). This relationship was also significant

when placed in context of the phylogeny, using Felsenstein's method of independent contrasts (topology A: $r^2 = 0.15$, $F = 4.2$, df $= 25$, $P = 0.04$; topology B: $r^2 = 0.15$, $F = 4.6$, df $= 25$, $P = 0.04$).

Most Neotomine-Peromyscine rodents are omnivorous, but certain clades have evolved specific feeding adaptations (fig. 6.6a). For example, the *Onychomys-Neotoma* clade is derived, with *Onychomys* being carnivorous and *Neotoma* mostly herbivorous. There was a trend for male spacing behavior to be related to diet, with nonmonogamous males having a higher energy diet (i.e., carnivory and omnivory) than monogamous males (topology A: 4,416 pairings of terminal taxa with 4 pairs contrasting male spacing behavior, $P = 0.13 – 0.75$; topology B: 4,880 pairings of terminal taxa with 4 pairs contrasting male spacing behavior, $P = 0.06 – 0.75$; fig. 6.6a). There was also a trend for male spacing behavior to be related to BMR, with nonmonoga-

A

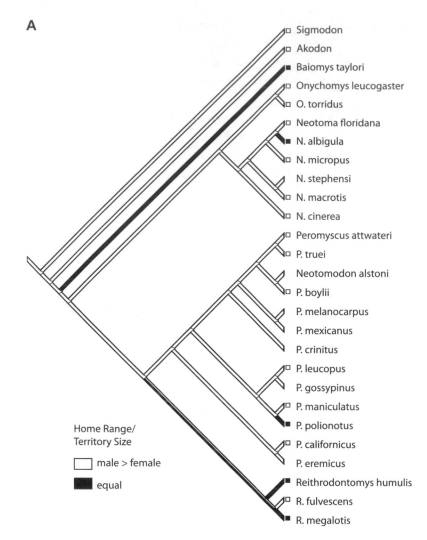

Sigmodon
Akodon
Baiomys taylori
Onychomys leucogaster
O. torridus
Neotoma floridana
N. albigula
N. micropus
N. stephensi
N. macrotis
N. cinerea
Peromyscus attwateri
P. truei
Neotomodon alstoni
P. boylii
P. melanocarpus
P. mexicanus
P. crinitus
P. leucopus
P. gossypinus
P. maniculatus
P. polionotus
P. californicus
P. eremicus
Reithrodontomys humulis
R. fulvescens
R. megalotis

Home Range/
Territory Size

□ male > female

■ equal

Figure 6.4 Phylogeny depicting the ancestral state reconstruction of (A) home range size and (B) natal dispersal in Neotomine-Peromyscine rodents.

mous males having a higher BMR than monogamous males (4,416 pairings of terminal taxa with 4 pairs contrasting male spacing behavior, $P = 0.06-0.69$; fig. 6.6b). Female spacing behavior was not related to diet but was related to BMR, with solitary females having a higher BMR than nonsolitary females (3,096 pairings of terminal taxa with 5 pairs contrasting male spacing behavior, $P = 0.03-0.5$; fig. 6-6c). Paternal care was not related to diet or BMR.

There was a trend for relative litter weight to be related to male spacing, with monogamous males having smaller relative litter weights compared to nonmonogamous males (4,416 pairings of terminal taxa with 4 pairs contrasting male spacing behavior, $P = 0.06-0.69$; fig. 6.7). There was a trend for species distribution area to be related to male spacing, with monogamous males having a relatively small species distribution area compared to nonmonogamous

males (3,096 pairings of terminal taxa with 4 pairs contrasting male spacing behavior, $P = 0.06-0.69$; fig. 6.8). Female spacing behavior was not related to species distribution area or relative litter weight. Reflecting the relationship between male spacing behavior and paternal care (fig. 6.3b), paternal care was also related to species distribution, with males displaying paternal care tending to be from taxa with smaller species distributions (8,048 pairings of terminal taxa with 6 pairs contrasting paternal care behavior, $P = 0.03-0.65$).

Discussion

For the majority of our analyses, there was strong congruence between topology A and topology B, suggesting that

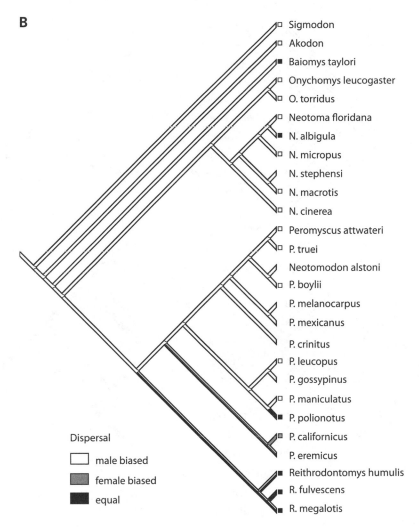

Figure 6.4 (continued)

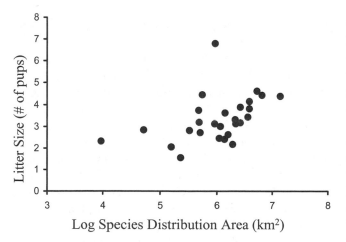

Figure 6.5 Relationship between the logarithm of species distribution area (km²) and litter size for Neotomine-Peromyscine rodents, *Sigmodon* and *Akodon*.

the *Onychomys* clade does not differ substantially from the Neotomine-Peromyscine rodents with respect to the evolution of breeding behaviors and their correlates. In general, we found that females were solitary and males were not monogamous, and these traits are ancestral for the clade. Male spacing behavior was independent of female spacing behavior, and female spacing behavior was independent of male spacing behavior. Paternal care appeared to have evolved multiple times. Monogamous males tended to provide care to offspring. The ancestral state was for males to have larger home ranges than females, with very few taxa demonstrating equitable range sizes. The ancestral state was for natal dispersal to be male biased. We found a positive correlation between species distribution area and litter size. Nonmonogamous males had a higher-energy diet (i.e., carnivory and omnivory) and a higher BMR than monoga-

A

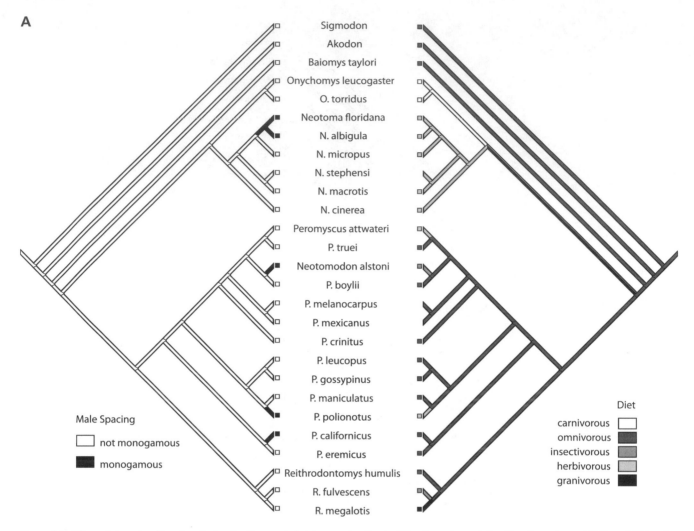

Figure 6.6 Mirror phylogenies of topology A, showing the ancestral state reconstruction for (A) male spacing and diet, (B) male spacing and BMR, and (C) female spacing and BMR in Neotomine-Peromyscine rodents. Although BMR is graphically presented as a binary variable, statistics were performed on continuous residual values.

mous males. Solitary females had a higher BMR than non-solitary females. Monogamous males had low relative litter weights and had smaller species distribution range sizes compared to males that were not monogamous. Males displaying paternal care tended to be from taxa with small species distributions.

Collectively, the influence and interrelationships of ecology, physiology, and life-history characters on the breeding behaviors of Neotomine-Peromyscine rodents from our analyses are summarized in figure 6.1 and are described as follows. We did not find significant relationships between the breeding behaviors of males and females. Males were influenced by ecological factors such as the species distribution area and feeding habits, as well as the life history characteristic of relative litter weight. Both male and female breeding behaviors were influenced by the physiological character of BMR. Aspects of ecology and life history were

interrelated. Specifically, species distribution area was related to litter size.

Our determination of ancestral states for male and female breeding behaviors provided few surprises, given our knowledge of the energetic cost of lactation in mammals (e.g., Gittleman and Harvey 1982; Thompson 1992), the differential investment in offspring by males and females (Trivers 1972), and the theory of mating system evolution (Orians 1969; Arnold and Duval 1994). Overall, females tended to be solitary, whereas males tended to be non-monogamous or roving. Males had larger home ranges than females and dispersal tended to be male biased. These results are congruent with other reviews of Peromyscine social behavior (Wolff 1989; Ribble 2003). However, exceptions to these patterns exist and probably evolved, independently, several times within the Neotomine-Peromyscine rodents.

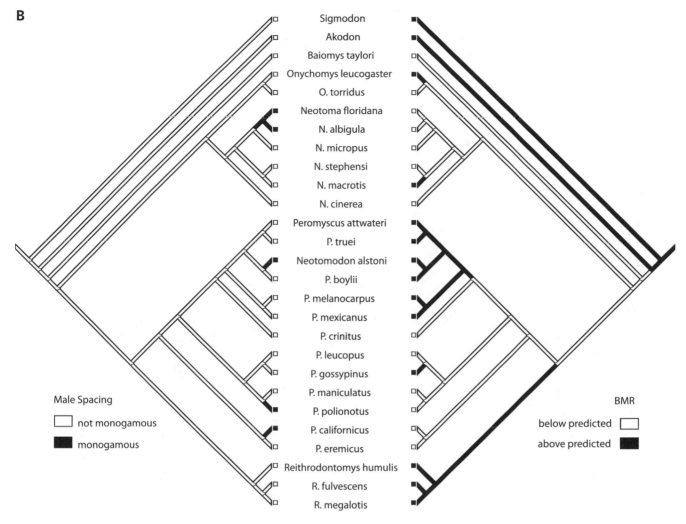

Figure 6.6 (continued)

(continued)

Our results clearly show differences in male and female mating strategies. Furthermore, ecological, physiological, and life-history characteristics influence breeding behaviors differently for males and females. The discrepancy between ecological, physiological, and life-history influences on males and females supports the contention that the breeding system of a species does need to be defined specifically in terms of male and female mating strategies, because different selective pressures have been acting differentially on the sexes within species (Reynolds 1996).

That male spacing behavior appeared to be independent of female spacing behavior suggests that reproductive success of males is not limited only by the availability of females. This is counter to the paradigm in mammalian behavioral ecology—that the reproductive success of females is limited by their ability to secure energy resources for producing and raising offspring, whereas reproductive success of males is limited by their ability to secure matings, and thus males are mainly responding to the distribution of females in space. Our results further suggest that male reproductive success is not limited by the availability of females because male spacing behavior appeared to be related to both diet and BMR, suggesting an energetic constraint to reproductive success. As expected, female spacing behavior was related to BMR, underscoring the influence that the high energetic demands of lactation impart on female breeding behavior (Thompson 1992). Although not identified by our analysis, there are other ecological factors that may affect breeding behaviors, such as pup-defense against infanticide (Wolff 1993b) and male-male competition (Bond and Wolff 1999).

Ribble (2003) suggested that relative litter weight might be correlated with the need for paternal care, and influence male mating strategies. This was not the case. There was

C

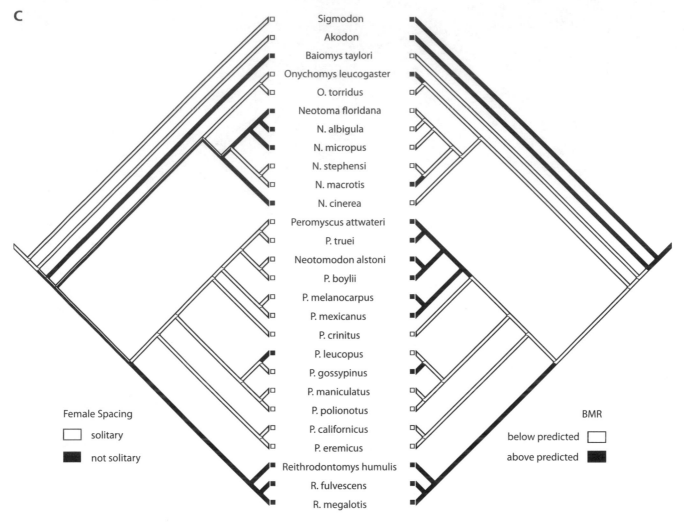

Female Spacing

☐ solitary

■ not solitary

BMR

below predicted ☐

above predicted ■

Figure 6.6 (continued)

no relationship between paternal care and relative litter weight. Although there was a relationship between male spacing behavior and relative litter weight, it was in the opposite direction to the prediction of Ribble (2003). Males of species with relatively low litter weights tended to be monogamous, suggesting that maternal investment in offspring (as measured by relative litter weight) may not necessitate male parental care.

Paternal care has evolved six times within the Neotomine-Peromyscine clade. This is consistent with the conclusions of Ribble (2003) that paternal care evolved more than once within *Peromyscus*. Comparing the evidence for male care to male spacing, we found a relationship between these two characters, with monogamously spaced males tending to provide paternal care. Thus across the entire clade, male care may be associated with monogamy, but there is little known about the social behavior of many species (e.g., *Neo-*

tomodon). These results suggest that male care may play a role in the evolution and maintenance of monogamy, and support the field experiments of Gubernick and Teferi (2000). These results are inconsistent with more global analyses (e.g., Komers and Brotherton 1997) that suggest that mammalian monogamy is not related to paternal care, but rather to female spacing and mate guarding. Whether the Neotomine-Peromyscine taxa are different from other mammals awaits more complete descriptions of social behavior from more taxa.

We found that the size of a species distributional range was related to breeding behavior and litter size. Consistent with findings of Glazier (1980), we found that litter size of Neotomine-Peromyscine rodents increased with the distributional range of a species. Furthermore, monogamous males and males that exhibited paternal care of offspring were from species with relatively small distribution ranges.

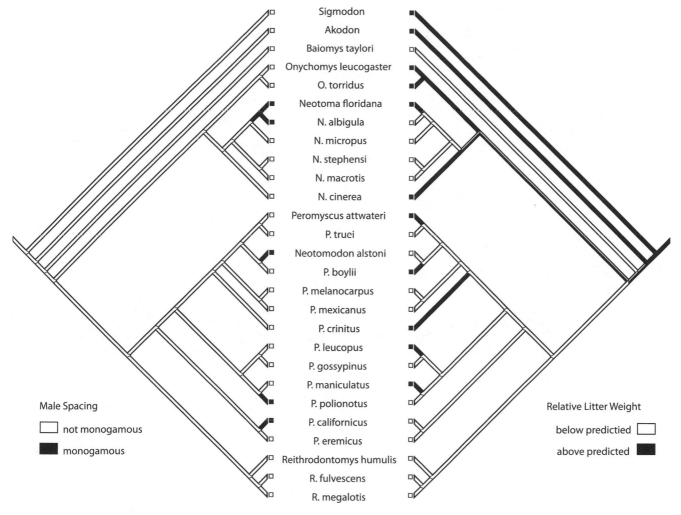

Figure 6.7 Mirror phylogenies showing the ancestral state reconstruction for male spacing and relative litter weight in Neotomine-Peromyscine rodents. Although average litter weight is graphically presented as a binary variable, statistics were performed on continuous data. Binary relative litter weight data are determined as weight being above or below the predicted values from the following significant regression line: Litter Mass = 0.14 Body Mass + 4.23 ($F_{1,23}$ = 103.64, $P < 0.001$, R^2 = 0.81).

This may be because monogamy and paternal care are a relatively specialized set of behaviors that limits the distribution of taxa exhibiting these behaviors. Furthermore, species with large litter sizes could be widely distributed because they are more successful at dispersing and colonizing new areas than species with small litter sizes. These observations support Brown's (1995) view that dispersal and social behavior affect the geographic range of species.

Relative to other groups of rodents, there has been substantial research on wild populations of species in the *Neotoma-Peromyscus* clade. In spite of this work, however, our review and analysis highlights how little we know about the breeding systems of most species. For example, our knowledge of the genetic mating system, the patterns of juvenile dispersal and recruitment, and the extent of varia-

tion among various breeding behaviors over different populations is minimal. The "molecular revolution" has largely passed by these field studies, most likely because of the difficulty in assessing behavioral interactions of nocturnal and secretive individuals in wild populations. Indeed, sampling individuals through trapping is relatively easy in this group of rodents, so we tend to know more about spacing behaviors, which are certainly suggestive of behavioral interactions and subsequent genetic mating patterns. Regardless, more fieldwork with an aim of following individuals and assessing behavior would benefit our understanding of the evolution and maintenance of breeding patterns in Neotomine-Peromyscine rodents. Clearly, we need basic information from some of the lesser-known species. One of the most exciting areas of study in the future is to examine

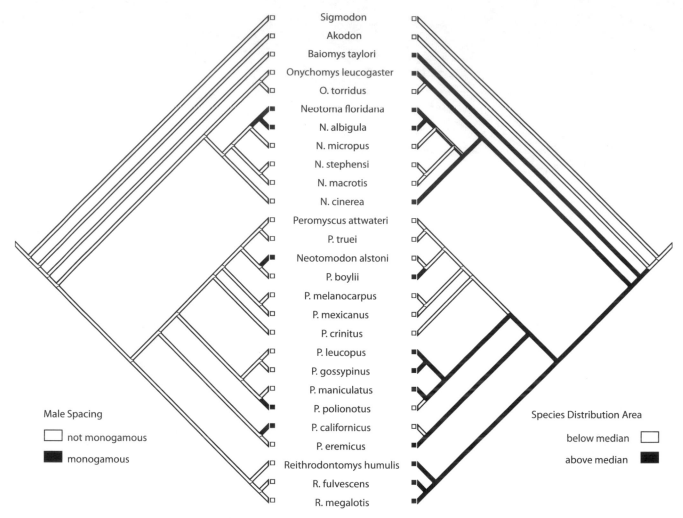

Figure 6.8 Mirror phylogenies showing the ancestral state reconstruction for male spacing behavior and species distribution area in Neotomine-Peromyscine rodents. Although species distribution area is graphically presented as a binary variable, statistics were performed on continuous data. Binary species distribution area data are determined as being above or below the average value for all represented species.

intraspecific variation in these behaviors, within some of the well-characterized species.

Summary

The purpose of this chapter was to examine, in a phylogenetic context, components of the breeding system of Neotomine-Peromyscine rodents from wild populations. First, using a review of the literature, we describe the patterns in breeding systems within this clade. Second, we examine the following breeding behaviors to test if there are any phylogenetic patterns in the observed variation in these behaviors, and if relationships exist among them: male spacing, female spacing, relative intersexual home range/ territory size, paternal care, and juvenile dispersal patterns. Third, we examine whether dietary, physiological, or life-history characteristics of the taxa explain the observed variation in these breeding behaviors. In general, we found that females are solitary and males are roving, and these traits are ancestral. Male spacing behavior is independent of female spacing behavior and female spacing behavior is independent of male spacing behavior. Paternal care has evolved multiple times, and there is a trend for monogamously spaced males to provide care. The ancestral state is for males to have larger home ranges than females, with very few taxa demonstrating equitable range sizes. Natal dispersal tends to be male biased. We found a positive correlation between species distribution area and litter size. There was a trend for nonmonogamous males to have a higher-energy diet and a higher BMR than monogamous males. Paternal care was not related to diet or BMR. Female spacing behavior was not related to diet but was related to BMR, with solitary females having a higher BMR than nonsolitary females. There was a trend for monogamous males to have smaller relative litter weights and species distribution areas compared with nonmonogamous males, and these relationships were absent in females. Our results not only demonstrate differences in male and female mating strategies, but also show that ecological, physiological, and life-history characteristics influence breeding behaviors differently for males and females. The independence of male and female spacing behaviors is counter to the paradigm in mammalian behavioral ecology that reproductive success of males is limited by their ability to secure matings and that males are mainly responding to the distribution of females in space. The independence of male and female spacing behavior, coupled with the relationships between male breeding behaviors and diet and BMR, suggests an energetic constraint to male reproductive success in Neotomine-Peromyscine rodents.

Chapter 7 Alternative Reproductive Tactics and Strategies of Tree Squirrels

John L. Koprowski

THE DISTRIBUTION OF fitness-limiting resources is often suggested to be of great importance in the evolution of social and mating systems (Emlen and Oring 1977). Male reproductive strategies often are directly related to distribution of receptive females; female distribution is influenced by food resources, nest availability, and risk of loss of juveniles (Lott 1991; Boag and Wiggett 1994a; Wolff and Peterson 1998). Given variation in resource distribution and efficacy of individuals in accessing resources, intraspecific variation in tactics used to access resources is predicted to occur.

Alternative reproductive tactics were described for a diverse array of animals (Cade 1980; Dunbar 1982; Dominey 1984; Caro and Bateson 1986; Gross 1996; Brockmann 2001). *Tactic* refers to the phenotype that results from a *strategy* (Gross 1996; Brockmann 2001). Game-theoretical approaches to understanding adaptive value of behavioral tactics provide a powerful intellectual framework and have led to important insights into evolution of alternative tactics (Maynard-Smith 1982; Parker 1984a). Alternative reproductive tactics in mammals are diverse (Dunbar 1982; Dominey 1984; Caro and Bateson 1986; Wolff, in press). The secretive nature of most rodents makes assessing reproductive behaviors challenging; however, alternative reproductive tactics are reported from a number of rodents to include ground and tree squirrels, chipmunks, and cricetids (Wolff, in press). Wolff identifies three major types of alternative tactics: (1) conditionally dependent strategies determined by resource availability, (2) mixed evolutionary stable strategies, maintained with equal payoffs through

frequency-dependent selection, and (3) "making the best of a bad job," when unable to adopt the most successful strategy. Mammalian strategies appear to be primarily of this third type, where tactics are conditional upon age, size, or dominance rank, with younger and subordinate individuals using a less successful tactic (Wolff, in press).

Unlike the vast majority of rodents, many tree squirrels are large, diurnal, and have conspicuous aboveground reproductive behaviors. Thus tree squirrels are excellent models for study of reproductive strategies and alternative reproductive tactics. In this chapter, "tree squirrel" refers to a member of the family Sciuridae that is dependent upon forest trees for food, nests, and cover, and thus have a significant arboreal component to their daily life (Gurnell 1987; Steele and Koprowski 2001). Primitive ancestors of extant Sciuridae were arboreal or semiarboreal, similar to present-day tree squirrels (Hafner 1984). Relatively unchanged over 35 million years, tree squirrels are considered to be living fossils (Emry and Thorington 1982). Herein, I briefly review the social systems of tree squirrels in which mating systems are operating, provide an overview of reproductive biology of tree squirrels, and review male and female alternative reproductive tactics.

Sociality in Tree Squirrels

Tree squirrels are often considered asocial, solitary mammals that forage for dispersed food items; however, intermediate levels of sociality are documented in numerous species (Steele and Koprowski 2001) relative to sociality of the ground-dwelling squirrels (Armitage 1981). A continuum of space-use patterns exists in the tree squirrels (Koprowski 1998). Many species of tree squirrels in the genera *Sciurus, Glaucomys,* and *Callosciurus* do not exhibit territorial defense by either sex and thus have overlapping home ranges (Gurnell 1987). Females in some species defend small exclusive core areas around nests (Havera and Nixon 1978; Tamura et al. 1989; Koprowski 1998) and *Tamiasciurus* are predominantly territorial (Smith 1968; Steele and Koprowski 2001). As a result, females are often dispersed in space. Adults of nearly all tree squirrels are known to nest in single-sex and mixed-sex groups, particularly but not exclusively in winter (Layne and Raymond 1994; Koprowski 1998; Layne 1998). In eastern gray squirrels (*Sciurus carolinensis*), nesting groups often consist of two adult females and their offspring (Koprowski 1996; Steele and Koprowski 2001). Visual cues, vocalizations, and scent marks appear important components of communication in social systems of tree squirrels (reviewed by Gurnell 1987; Steele and Koprowski 2001). In eastern gray squirrels, the most social species of tree squirrel known, natal philopatry of daughters results in formation of kin clusters of related females (Koprowski 1996). Philopatry of both males and females is uncommon in red squirrels (*Tamiasciurus hudsonicus*: Larsen and Boutin 1998); indeed, natal dispersal of offspring is typically the rule for the few species of tree squirrels studied to date (Koprowski 1998). Retention of kin in natal areas leads to formation of the fundamental social unit in ground-dwelling squirrel societies, the female-female dyad, and appears related to body size energetics (Armitage 1981, 1999a). I suggest elsewhere that the evolution of sociality in tree squirrels may involve a similar relationship, with body size energetics favoring the overwinter retention of females in their natal area (Koprowski 1998).

A General Review of Tree Squirrel Reproductive Biology: Setting the Stage

Tree squirrels typically do not breed until 1 year of age (Gurnell 1987; Koprowski 1994a, 1994b); however, in extremely good years of food availability, precocious breeding at ages as young as 5.5 months has been reported (Smith and Barkalow 1967; Koprowski 1994a, 1994b). Reproductive life is at least 8.0 years and possibly > 12.0 years (Barkalow and Soots 1975; Koprowski et al. 1988). Most tree squirrels are spontaneous ovulators (Layne 1954; Millar 1970; Gurnell 1987). Females are typically in estrus for < 1 day and likely < 8 h (Smith 1968; Dolan and Carter 1977; Koford 1982; Wells-Gosling and Heaney 1984; Koprowski

1993a, 1993b) during breeding seasons that extend over a 4 to 12 week period (Gurnell 1987; Steele and Koprowski 2001). As a result, estrous females are dispersed in both time and space. Females may attract males for several days prior to estrus (Thompson 1977). On the morning of a female's day of estrus, males congregate outside of the nest. Aggregations of males pursue the female throughout her day of estrus, usually with intense intrasexual and intersexual aggression (Farentinos 1972; Gurnell 1987; Koprowski 1993a, 1993b, 1998; Steele and Koprowski 2001). Most species of *Sciurus, Tamiasciurus,* and *Glaucomys* only produce a single litter each year after a 30–45 day gestation; however, in many species a second litter can be produced (Heaney 1984; Gurnell 1987). Some *Sciurus* from the neotropics, such as *S. granatensis,* can produce three litters per year (Heaney 1984). Modal litter size is 2 to 4 young (Heaney 1984; Gurnell 1987). No paternal care is known for any tree squirrels.

Female tactics

Tactics of females during a single day of estrus have received less attention than those of males; however, a number of behaviors executed by females are suggested to be reproductive tactics. A detailed analysis of female behavior during mating bouts suggests that females use several tactics to influence the type and outcome of competition among males (Koprowski 1998).

Evasive behavior

During mating chases, males aggregate around the female early on the morning of her day of estrus (Bakken 1959; Koprowski 1993a, 1993b). Aggression amongst males is intense and often exceeds 2,500 interactions/hour (Farentinos 1972; Koprowski 1993a). Females avoid this intense aggression and break away from males distracted by intermale interactions (Koprowski 1993a, 1993b). In eastern gray squirrels and fox squirrels (Koprowski 1993a, 1993b), the majority of copulations follow a breakaway (69.4% and 54.2% of all copulations in these species, respectively). The remainder of copulations occurs while the female is sequestered by a dominant male. Thus a majority of copulations follow evasive tactics, and suggest that factors other than male dominance are necessary for male success.

Selection of locations for mating

The location of matings does not appear to be random in eastern gray squirrels (Koprowski 1993a) or fox squirrels (Koprowski 1993b). Copulations following evasion of males occur in sites closer to the ground than those when males sequester females at the end of branches (30.7% of the

13.7 m height for sequestered copulations in *S. niger*, 55.4% of the 11.2 m mean height for sequestered copulations in *S. carolinensis;* Koprowski 1993a, 1993b). In addition, probability of being attacked by males while copulating decreased from 0.375 to 0.245 following a breakaway by females in eastern gray squirrels but not fox squirrels (Koprowski 1993a, 1993b). Females appear to actively seek locations that minimize risk of injury due to aggressive interactions by other males and falls from precarious mating sites (Koprowski 1998).

Copulatory plug removal

Immediately following copulation, females groom their genitalia (Koprowski 1993a, 1993b) and frequently remove copulatory plugs that protrude from beyond the vulva (72.6% of 22 copulations in *S. niger*, 50.0% of 26 copulations in *S. carolinensis*, Koprowski 1992). Copulatory plugs are consumed by female eastern gray squirrels (61.5% of 13 plugs) and fox squirrels (75.0% of 16 plugs) or are discarded (Koprowski 1992).

Multi-male mating

Females of all species of tree squirrels for which reproductive behavior has been studied are known to copulate on multiple occasions during a mating chase, with the vast majority of copulations occuring with different males (2 to 8 males: Farentinos 1972; Koford 1982; Tamura et al. 1988; Wauters et al. 1990; Arbetan 1993; Koprowski 1993a, 1993b; Koprowski 1998). Multiple paternity of litters has been confirmed in red squirrels and eastern gray squirrels (Arbetan 1993; David-Gray et al. 1998, 1999).

Solicitation of copulations

Female choice is a component of tree squirrel mating tactics. Female Abert's squirrels appear to actively avoid dominant males that compete for access to females, and instead may solicit subordinate males (Farentinos 1980). Evasive behavior by females through breakaways, and thus avoidance of dominant males, suggests that females (Koprowski 1993a, 1993b, 1993c) may be choosing to mate with males other than those most dominant, in addition to preferring sites with a decreased risk of injury (Steele and Koprowski 2001).

Each female tactic appears to be to the detriment of individual males participating in the mating bout; however, it likely provides advantage to estrous females (fig. 7.1). Selection of safe locations for copulations decreases risk of injury to a female that is just starting her reproductive investment (Koprowski 1993a, 1993b, 1998). Removal of copulatory plugs and multi-male mating reduces the reproductive success of individual males. Estrous females may

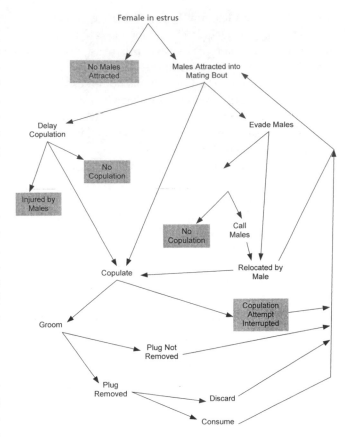

Figure 7.1 Tactics used by female tree squirrels during mating bouts, with negative fitness consequences shaded.

benefit from multiple mating in several ways, including assurance of fertilization, increased uncertainty of paternity, increased nutrition from consumption of copulatory plugs, and increased genetic variation in offspring (Koprowski 1992, 1998). At least 9 different hypotheses were proposed to explain multi-male mating (Wolff and Macdonald 2004). If assurance of fertilization or increased nutrition were the cause of multiple mating by females, one might expect females to mate with the same male frequently during mating bouts, for these benefits could be accrued simply by increasing number and duration of copulations. Because female tree squirrels rarely mate repeatedly with a male and multi-male mating is the rule, hypotheses of increased genetic variation in offspring and/or masking of paternity to deter infanticide cannot be refuted. Infanticide is reported in tree squirrels but the frequency of this behavior is not well known. Among some mammals, multi-male mating appears to be a tactic to reduce certainty of paternity and deter infanticide (Wolff and Macdonald 2004). Solicitation of copulations clearly is to the disadvantage of actively pursuing males and favors lower-ranking males. This suggests mate choice or manipulation of activities to a scramble competition where different male traits may be favored than in a contest competition.

Group nesting occurs in many species (Koprowski 1996, 1998) but does not appear unequivocally related to mating strategies (Koprowski 1998). Mixed-sex nesting groups, however, are most common during the winter breeding season in eastern gray squirrels (Koprowski 1996), fox squirrels (Adams 1976; Koprowski 1996), and Abert's squirrels (Edelman 2004). No studies have examined the relationship between frequency of nest sharing and mate choice or copulatory success; thus, the adaptive significance of mixed sex nesting groups is not yet known.

Male tactics

The spatial and temporal dispersion of estrous females provides challenges to males. The overall and instantaneous (i.e., immediate: Dearborn et al. 2001) operational sex ratios are heavily male biased, suggesting that intrasexual competition among males is high (Steele and Koprowski 2001). Indices of sexual selection reflect this intensity; variation among males in reproductive success is considerable, with a strong positive skew (Koprowski 1993a, 1993b). Distribution of copulations in tree squirrels demonstrates a strong positive skew, with a majority of copulations garnered by one or two males (Farentinos 1972; Tamura et al. 1988; Wauters et al. 1990; Koprowski 1993a, 1993b). The causes of this inequitable distribution of copulatory success in males are the multiple types of competition experienced by males due to female spatiotemporal dispersion and tactics during mating chases. Success in scramble, interference, and sperm competition likely is important in determining the reproductive success of male tree squirrels.

Scramble competition

An intense scramble competition occurs as males attempt to locate females at two different spatiotemporal scales. Receptive females must be located in the greater landscape during the breeding season if males are to increase mating opportunities. Males are attracted to estrous females from distances of nearly 1 km, and monitor females beginning at least 5 days before estrus (Thompson 1977; Steele and Koprowski 2001). Large home ranges of males relative to females (by a factor of 1.2 to 2.0: Koprowski 1998), despite a minimal or lack of sexual dimorphism (< 5%: Nixon et al. 1991; Boutin and Larsen 1993), can also be interpreted as a tactic of males to track female distribution and receptivity (Koprowski 1998). Males increase home ranges in response to decreases in female density (Kenward 1985). Home ranges of males expand during the breeding season and account for the significantly larger home ranges of

adult males than those of females (reviewed in Koprowski 1998). Male eastern gray squirrels visited 5.8 + 1.2 SD females during the 3 h immediately after emergence from their nests (n = 14: Steele and Koprowski 2001). Olfactory cues are of great import to males in the location of an estrous female (Thompson 1977; Gurnell 1987; Koprowski 1991; Steele and Koprowski 2001). Males seem to have particularly well-developed olfactory skills, and regularly visit scent mark sites (Koprowski 1993d) and odoriferous fruiting trees (L. N. Brown 1986) relative to females. Success in this geographic scale scramble competition increases potential opportunities to mate by increasing the number of mating chases in which a male is involved.

Other behaviors are also suggested to be reproductive tactics of males and serve to influence availability of receptive females. Tree squirrels are known to cannibalize conspecifics, including female (Thompson 1976) and male (Holm 1976) eastern gray squirrels, male fox squirrels (Packard 1956), and male bush squirrels (*Paraxerus cepapi*: de Villiers 1986). Only de Villiers (1986) and Holm (1976) report the actual killing of young squirrels or suggest infanticide. The rarity of observation of infanticide in the often well-concealed arboreal nests of tree squirrels makes further inference difficult; however, the ability of tree squirrels to enter estrus again in most years (Heaney 1984; Steele and Koprowski 2001) promotes speculation. Infanticide is suggested to be a male reproductive tactic, but difficulty in documenting this phenomenon obfuscates conclusions (Viljoen 1977, de Villiers 1986, but see Weissenbacher 1987). The role of infanticide as a reproductive tactic is discussed further in Ebensperger and Blumstein, chap. 23, and Hoogland, chap. 37, this volume).

Additionally, on the day of estrus, males must locate the female as she changes location among males within her home range. This scramble competition is important in determining immediate access to estrous females. Males that locate females most quickly have a high likelihood of copulating (Koprowski 1993a, 1993b, 1993d). The frequent evasive behavior of females during the mating chase provides a fitness payoff to males not equipped to succeed in interference competition, for 69.4% and 54.2% of copulations in eastern gray squirrels and fox squirrels, respectively, occur following the breakaway of a female (Koprowski 1993a, 1993b).

Interference competition

Interference competition is also important to obtain access to a female in estrus. The number of male participants in mating chases ranges between 2 and 34 individuals chasing a single female (Goodrum 1961; Farentinos 1972; Wauters et al. 1990; Arbetan 1993; Koprowski 1993a, 1993b).

Figure 7.2 Female fox squirrel (*Sciurus niger*) in estrus sequestered at the end of a branch by the dominant male using the active pursuit tactic.

Males access the female using intense aggression directed at other males and occasionally the female (Bakken 1959; Thompson 1977; Tamura et al. 1989; Wauters et al. 1990; Koprowski 1993a, 1993b); interaction rates are often excessive (Farentinos 1972; Wauters et al. 1990; Steele and Koprowski 2001). Dominance in tree squirrels is determined by an interaction between age and body size, with older, larger males dominating young, small males (Farentinos 1972; Pack et al. 1967; Benson 1980; Allen and Aspey 1986; Wauters and Dhondt 1989; Wauters et al. 1990). Older or larger males typically remain in proximity to the estrous female during a mating bout (fig. 7.2: Wauters et al. 1990; Koprowski 1993a, 1993b).

The result of shifts that occur between interference and scramble competition in each mating chase produce two tactics: active pursuit and satellite (fig. 7.3), which appear to be adopted by male eastern gray squirrels (Bakken 1959; Koprowski 1993a), fox squirrels (Koprowski 1993b), Abert's squirrels (Farentinos 1972, 1980; Koprowski, pers. observ.), Eurasian red squirrels (Wauters et al. 1990), Mexican fox squirrels (*S. nayaritensis*), and red squirrels (Koprowski, pers. observ.). Males that adopt the active pursuit tactic attempt to sequester the female and copulate while repelling other males. This tactic is adopted by 30.3% of eastern gray squirrel males and 45% of fox squirrel males (Koprowski 1993a, 1993b). Males that adopt the active pursuit tactic are the most dominant and tend to be the oldest members participating in mating chases (Koprowski 1993a, 1993b). In mating chases with few participants, all males may use this tactic; it is the only one reported for some low-density populations (Wauters et al. 1990; Koprowski 1993a, 1993b, pers. obs.). The satellite tactic is adopted by the remaining majority of males that tend to be

young and hold low dominance rank (Koprowski 1993a, 1993b). Satellite males remain in the vicinity of the estrous female but avoid active pursuit males, often foraging, feeding, or resting, and thus appearing to not be involved in the mating chase (Koprowski 1993a, 1993b; Steele and Koprowski 2001).

Why do low-ranking males remain in the vicinity of estrous females when they cannot compete with dominant males that sequester the female? The entire fitness payoff for satellite males results from the evasive behavior of females (Farentinos 1980; Koprowski 1993a, 1993b). Satellite males are unable to compete successfully in interference competition due to their low rank, and no copulations have been observed by satellites under these conditions (Koprowski 1993a, 1993b). When the female avoids dominant males in active pursuit, however, satellite males are able to copulate. As a group, satellites accrue 73.5% of the copulations in eastern gray squirrels (Koprowski 1993a) and 69.2% of the copulations in fox squirrels (Koprowski 1993b) following a breakaway, which is similar to the frequency of males that adopt the satellite tactic in the population. The tactics are equally likely to copulate first and last in a mating chase (Koprowski 1993a). Success of satellites (0.13 to 0.25 copulations/male/chase) is significantly less than that of active pursuit males (0.32 to 0.83 copulations/male/chase; Koprowski 1993a, 1993b). Measures of variation also suggest that copulatory success is most variable among satellite males (Koprowski 1993a, 1993b). As a result of this differential reproductive success per male, the benefits of actively competing for access to the estrous female are clear.

The alternative reproductive tactics of males are not the result of genetic polymorphism in the population. Males switch tactics dependent on characteristics of attending males in a mating chase (Koprowski 1993a, 1998), and in chases with few males only the active pursuit tactic may be used (Wauters et al. 1990; Koprowski, pers. observ.). Additionally, tactics do not appear to be a mixed evolutionary stable strategy for the fitness payoffs are not equivalent. Higher payoffs for active pursuit make this tactic profitable to adopt when capable of success in interference competition, while lower-ranking males "make the best of a bad job" by adopting a satellite tactic and scrounging copulations based upon female behavior (Koprowski 1993a, 1993b; Wolff, in press).

Sperm competition

Sperm competition also occurs; however, virtually nothing is known about competition at this level for tree squirrels. Copulatory plugs are common in rodents, including tree squirrels (Hartung and Dewsbury 1978; Voss 1979; Baumgardner et al. 1982; Koprowski 1992). Plugs form by coag-

Figure 7.3 Alternative reproductive tactics of male fox squirrels (*Sciurus niger*). The estrous female is located inside a cavity at the base of the tree to the left. Active pursuit males are found at the base of the tree. Satellite males are foraging in the grass to the right.

ulation of proteins in semen (Mossman et al. 1955; Mann and Lutwak-Mann 1981; Koprowski 1992). Voss (1979) suggested five potential functions for copulatory plugs: storage of sperm, prevention of sperm leakage, induction of pseudopregnancy, facilitation of sperm transport across the cervix, and prevention of subsequent intromissions, or "chastity enforcement." Copulatory plugs of tree squirrels consist of an opaque, white, rubbery acellular core; a few spermatozoa and a thin layer of epithelial cells are often found on the external surfaces after removal from the female (Koprowski 1992). Available evidence suggests that the chastity-enforcement hypothesis is the most likely explanation for rodent copulatory plugs (Voss 1979; Koprowski 1992). Additionally, males guard females for < 80 min following ejaculation in several species of tree squirrels, including fox squirrels (McCloskey and Shaw 1977; Koprowski 1993a), eastern gray squirrels (1993b), Abert's squirrels (Farentinos 1972), Eurasian red squirrels (Ognev 1940; Raspopov and Isakov 1980; Wauters et al. 1990), and red squirrels (Arbetan 1993). Unfortunately, data of sufficient detail are not available to examine the relationship between duration of guarding with measures of fitness. These tactics function to provide time that may convey advantage in sperm competition. The aggregation of the sperm of fox squirrels and flying squirrels into rouleaux (Martan et al.

1970; Martan and Hruban 1970) may also be important in sperm competition. Unfortunately, there is a dearth of knowledge about sperm competition in the tree squirrels, and order of mating effects are unknown for tree squirrels. The vigorous guarding of females after mating (Farentinos 1972; McCloskey and Shaw 1977; Thompson 1977; Tamura et al. 1988; Wauters et al. 1990; Koprowski 1993a, 1993b) and deposition of copulatory plugs (Koprowski 1992) suggest a last male advantage for most tree squirrels (Koprowski 1992), as reported for Idaho ground squirrels (*Spermophilus brunneus*), in which males tenaciously guard females after copulation (Sherman 1989).

The reproductive success of males is essentially the summation of success in these three types of competition (fig. 7.4). Older dominant, active pursuit males accrue more copulations per capita than subordinates, suggesting the relative advantage of this tactic. Young, subordinate animals that adopt the satellite tactic are less successful over the course of a breeding season, but are more successful than predicted by their low rank. The costs of extensive searching for dispersed females, monitoring of female reproductive condition, location of estrous females, intensive intermale physical contests and avoidance of these contests, and postcopulatory guarding, production of proteinaceous copulatory plugs, and sperm competition appear extensive.

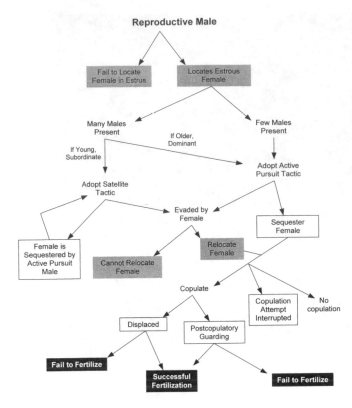

Reproductive Male

Figure 7.4 Tactics used by male tree squirrels during mating bouts with changes in competition highlighted. Gray shading indicates scramble competition, boxes indicate interference competition, and black shading indicates sperm competition.

Most males lose significant weight during the breeding season, wounding is common, and mortality related to wounding and weight loss is known to occur (Steele and Koprowski 2001).

Many tactics used by males and females are to the detriment of individuals of the opposite sex, a classic conflict between the sexes (Koprowski 1998). A common element of mating chases where male and female strategies are coincident is in the minimization of the time required for copulation. Evidence suggests that the location and duration of copulation expose the copulating pair to increased risk of injury or predation (Koprowski 1992, 1993a, 1993b). Advantage is likely gained by both sexes if copulatory times are decreased without a concomitant decrease in effectiveness of copulations.

Why Do Tree Squirrels Copulate So Quickly?

To compile data on copulatory duration in tree and ground squirrels, I used observations of uninterrupted intromissions that include pelvic thrusting, the common behavior pattern during mating in squirrels. Squirrels do not appear to

lock (classes 9 to 12 of Dewsbury 1972) during copulation (Denniston 1957; Schwagmeyer and Parker 1987; Koprowski 1993a, 1993b), and most form copulatory plugs (reviewed by Koprowski 1992). If the observer's descriptions were of sufficient detail, I used time from initial pelvic thrust to cessation and dismount to delineate duration of a single intromission. Where possible, I report duration of intromissions in natural settings, but I was limited occasionally to observations of confined pairs (table 7.1). Although many reproductive parameters may differ drastically between captive and wild populations, the few comparative data available for squirrels suggest that copulatory time is conservative (Davis 1982a; Blake and Gillet 1988; Schwagmeyer and Foltz 1990). Copulation durations of tree squirrels are nearly always shorter than those of ground squirrels, and are generally < 1 min (table 7.1).

To examine plasticity of copulatory time in tree squirrels, I also used data for fox squirrels and eastern gray squirrels collected on populations in Lawrence, Kansas, and Salem, Oregon, using the methods of Koprowski (1993a, 1993b). I recorded the duration of copulations and measures of safety (height aboveground) and competition (number of males in the mating bout, competitors present or absent) through each mating bout. If plasticity exists in copulatory duration of fox squirrels and eastern gray squirrels, then copulatory times would be longer when copulations were in safe locations (near to ground) and where mate competition was relaxed (low density of competitors, copulations in seclusion).

Duration of copulation did not vary with levels of competition. Copulatory duration is not correlated ($r = 0.02$, df $= 57$, $P > 0.50$) with the number of *S. carolinensis* males in the mating bout. Copulations last equally long when the female avoids pursuing males and copulates in seclusion with a male that has relocated her and when copulation occurs with several males nearby (*S. niger*: 18.7 ± 1.1 sec and 18.4 ± 1.0 sec, $t = 0.26$, df $= 14$, $P = 0.79$; *S. carolinensis*: 22.1 ± 0.9 sec and 21.8 ± 0.9 sec, $t = 0.40$, df $= 37$, $P = 0.69$, respectively). Duration of a copulation also was not correlated with height aboveground (*S. niger*: $r = 0.05$, df $= 18$, $P > 0.50$; *S. carolinensis*: $r = 0.16$, df $= 52$, > 0.25). At least in these two species, little plasticity exists in copulatory duration.

Among the squirrels (Mammalia: Sciuridae), some species have short copulatory times (< 30 sec, Koprowski 1993a, 1993b) while others copulate for much longer periods (1 to 33.5 min, Schwagmeyer and Foltz 1990). Koprowski (1993a) suggested that the short copulatory times of eastern gray squirrels may be advantageous and related to intense overt aggression and vulnerable location of copulations. Females of several arboreal or semi-arboreal squirrels appear to prefer to mate in seclusion (Yahner 1978a;

Table 7.1 Duration and location of copulations in the squirrels

Species	Location of mating	Sample size	Setting	Duration	Source
Arboreal squirrels					
Sciurus aberti	T	35	W	72.2 sec	Farentinos 1972
S. carolinensis	T	65	W	21.9 + 3.0 sec	Koprowski, unpublished manuscript
S. griseus	T	2	W	<30 sec	Ingles 1948; Cross 1969
S. niger	T	20	W	18.5 + 2.6 sec	Koprowski 1993b
S. vulgaris	T	5	Z	<15 sec	Raspopov and Isakov 1980
Tamiasciurus douglasii	T	10	W	1 to 25 min	Koford 1982
	G	2	W	4 and 8 min	Smith 1968
Semi-arboreal and ground-dwelling squirrels					
Tamias sibricus	G		L	3 to 135 sec	Blake and Gillet 1988
	G		W	5 to 120 sec	Blake and Gillet 1988
T. striatus	G	9	W	88 sec	Yahner 1978a
	G	3	W	1.5 to 3.0 min	Wishner 1982
Cynomys gunnisoni	G	4	W	<1 min	Fitizgerald and Lechleitner 1974
Marmota flaviventris	G		W	1 to 23 min	Armitage 1962
M. monax	G	3	L	3 to 8 min	Hoyt and Hoyt 1950
M. olympus	G		W	0.5 to 8 min	Barash 1989
Spermophilus beldingi	G		W	10 to 12 min	Sherman and Morton 1979, 1984
S. elegans	G	1	L	2.5 min	Denniston 1957
S. erythrogenys	G	1	W	1 to 2 min	Ognev 1940
S. richardsonii	G	4	W	3 to 4 min	Davis 1982a
S. tridecemlineatus	G	51	W	4.7 (1–17) min	Schwagmeyer and Brown 1983

NOTES: G = ground; T = tree; L = laboratory; W = wild; Z = zoo.

Wauters et al. 1990; Koprowski 1993a, 1993b) and may actively avoid pursuing males, to mate in some locations near the ground (Koprowski 1993a). Most ground squirrels (Sciuridae: Marmotini) mate belowground in burrows, possibly to reduce exposure to predators and/or reduce the likelihood of interruption of copulations by other males (Davis 1982a). Levels of overt aggression at the time of copulation appear reduced in most ground squirrels (e.g., Erpino 1968) and at least one species, the thirteen-lined ground squirrel (*Spermophilus tridecemlineatus*), exhibits a queuing convention, in which males wait for opportunities to copulate (Schwagmeyer and Parker 1987). A comparison of copulatory times in the Sciuridae provides insight into the potential selective pressures influencing the duration of copulation.

Several possible explanations that are *not* mutually exclusive may account for differences in copulatory times among the Sciuridae:

1. Copulatory times may have increased in ground squirrels to accrue benefits of prolonged copulation, such as increased likelihood of insemination and increased number of sperm (Schwagmeyer and Foltz 1990 and references within).

2. Copulatory times have decreased in tree squirrels due to other constraints, such as predation risk and/or interference by other males.

3. Copulatory times may be the result of within-lineage cycling evolution (Reeve and Sherman 2001).

The large aggregations, intense levels of interaction (2,500 interactions/h, Farentinos 1972; Steele and Koprowski 2001), and frequent vocalizations (Zelley 1971; Lishak 1982) characteristic of mating bouts of tree squirrels make mating activities conspicuous. Males and females in the conspicuous assemblages may be at considerable risk of predation from raptors, canids, felids, and mustelids that are known predators of tree squirrels (Gurnell 1987; Steele and Koprowski 2001). This may be one of the selective pressures for ground squirrel copulations to occur belowground (Davis 1982a). Although tree squirrels nest in cavities and leaf nests in branches of trees (however, *Tamiasciurus* use burrows in some populations: Shaw 1936; Yahner 1980), copulation rarely occurs in dens. The small size of these refuges may be too confining, or risk of interruption through the nest entrance may be too high, to permit copulation in most arboreal nests. Occupants of nests struggle to repel individuals from entering when not copulating (Steele and Koprowski 2001) suggesting that attack rates on copulating pairs might be high within nests. Predation risk is satisfactory to account for most copulations to occur in trees out of reach of many terrestrial predators. Thirteen-lined ground squirrels and Belding's ground squirrels (*Spermophilus beldingi*) are the only ground squirrels that mate primarily aboveground in open grasslands (Sherman and Morton 1979; Schwagmeyer and Foltz 1990), yet copulations are prolonged.

Another influence on copulatory time likely is the level of male-male competition and vulnerability of a copulating pair to attack by males. Competition for females is greatest in members of *Sciurus* and *Tamias* characterized by dominance polygyny or promiscuity, in which numerous males accumulate around an estrous female; copulatory times are shortest in these species. *Tamiasciurus* are intermediate, with highly variable copulatory times—perhaps because competition is limited by spatial attachment of territorial males, which decreases competition (Koford 1982). Prolonged intromission characterizes ground-dwelling squirrels with lower levels of competition around estrous females and spatial attachment of males to females (*Marmota olympus, M. flaviventris*), low adult densities and scramble competition (*M. monax, S. tridecemlineatus*), and even queuing conventions (*S. tridecemlineatus*). The short and less variable copulatory times of tree squirrels relative to ground squirrels support trends in mating systems, vulnerability, and intermale competition.

Intermale competition is great in tree squirrels, yet time to copulate is remarkably short compared to many mammals. All tree squirrels described use a nonlocking pattern of copulation. The locked pattern of copulation, in which the mechanical connection between penis and vagina holds the copulating pair together (Dewsbury 1972), likely is adaptive in species where sites for copulation are safe from predators or conspecific competitors (Dewsbury 1975); however, a copulating pair locked together may be at considerable risk if predator pressure and/or conspecific competition are intense (Langtimm and Dewsbury 1991). Nonlocking during copulation may allow increased vigilance for predators or conspecifics and permit resumption of the copulatory sequence after disruption (Langtimm and Dewsbury 1991).

If relative safety of copulatory sites from disruption influences reproductive success of males and females, then duration of copulation may also be under selective pressure. Risk of disturbance may be minimized by loss of a locked copulatory pattern (Dewsbury 1975) but may decrease further by minimizing time necessary for successful copulation. Conversely, fitness-related benefits of prolonged copulations to male rodents include increased paternity (Lanier

et al. 1979; Dewsbury and Hartung 1980; Dewsbury and Baumgardner 1981; Oglesby et al. 1981; Schwagmeyer and Foltz 1990). Reasons for increased reproductive success with prolonged copulations may be due to increased numbers of sperm, increased time guarding the female from competitors, and decreased receptivity of the female to subsequent males (Dewsbury 1984).

Among the ground squirrels, *S. beldingi* are characterized by a dominance-lek based mating system, where levels of competition in proximity of the female are the greatest reported among ground squirrels. Nearly 50% of copulation attempts are interrupted by other males, but copulatory duration is prolonged (10 to 12 min) when uninterrupted (Sherman and Morton 1979, 1984). Evidence suggests that male ground squirrels can fertilize females during copulations considerably short of maximum duration (Schwagmeyer and Foltz 1990). The great variance in ground squirrel copulatory times suggests the plasticity of this behavior. Males may attempt to maximize duration of copulation to accrue fitness benefits through increased paternity (Schwagmeyer and Foltz 1990) as a trade-off to possibility of attack. The extremely localized and high levels of overt aggression in tree squirrels such as *Sciurus* are not conducive to providing the option of prolonged copulations, as in ground squirrels.

The spatial attachment of *Tamiasciurus* males to territories precludes the higher levels of overt competition characteristic of *Sciurus* species (Koford 1982), and a queuing convention may also exist (Arbetan 1993). This reduced level of competition may result in the prolonged and highly variable copulations in this genus.

Tree squirrels usually copulate high in the canopy; copulating pairs are attacked in nearly 40% of copulation attempts, and are frequently thrown to the ground (Koprowski 1993a, 1993b). The intensity and frequency of attack, as well as the vulnerable location of copulations, may have led to selection for and/or maintenance of short copulations, forgoing potential benefits of prolonged copulation. Low variance and range in copulation times of *Sciurus* species point to some form of stabilizing selection. Chipmunks (*Tamias*) have a similar mating system to tree squirrels, but intermediate copulatory times, potentially due to the chipmunks' habit of mating on the ground. The great intraspecific variability in copulatory times of the ground squirrels suggests that strong stabilizing selection has not occurred.

These results suggest that the constraints of predation risk, male-male competition, and the female's risk of injury may present trade-offs to the benefits of prolonged copulation. Analyses of ground squirrels under varying levels of risk and competition likely will provide insight into the proximate tactics adopted by squirrels to maximize reproductive success through duration of copulations.

Summary

Reproductive strategies and alternative reproductive behaviors of tree squirrels are diverse. Females are in estrus for only a few hours on a single day of a breeding season. As a result, females are a fitness-limiting resource for males, with an availability that varies in time and space. Male fitness is related to success in scramble competition to locate mates, interference competition to access and defend mates, and sperm competition to maximize fertilization of eggs of females that mate with multiple males. Females exacerbate and manipulate the types of competition through evasive behavior, selection of sites for mating, mating with and soliciting copulations from > 1 male, and removing copulatory plugs. Conflict, both intersexual and intrasexual, has been remarkably influential in the evolution of mating systems of tree squirrels. Conflict among males appears to have been particularly important in promoting evasive behavior in females, led to reduced copulatory duration, promoted alternative reproductive tactics, and influenced postcopulatory tactics for minimizing multiple paternity. Conflict among the sexes likely acts to diminish reproductive success of any dominant individual male, as females mate with multiple males and often avoid active pursuit males. The reproductive skew among males, however, is substantial, and active pursuit males clearly are most successful. Intersexual conflict appears to increase the number of males that are able to mate due directly to the evasive behavior of females, a behavior that maintains the satellite tactic among males. The benefits to females of manipulating the types of competition remain unclear. Future research must quantify the costs and benefits of male and female tactics under different environmental and social conditions to more clearly elucidate the advantages of these behaviors in an arboreal environment.

Life Histories
& Behavior

Chapter 8 Fast and Slow Life Histories of Rodents

F. Stephen Dobson and Madan K. Oli

THE CHARACTERISTICS OF a species' life cycle, or life history, are typical patterns of growth, reproduction, and survival. Such traits vary greatly among species (Roff 1992; Stearns 1992). The study of life-history traits is important to ecology because changes in these traits reveal the ways in which populations respond to changes in environment. Life-history traits are important to the study of evolution because changes in these traits represent fitness differences of individuals, which result in genetic change within populations. Studies of the evolution of life-history traits combine ecological and evolutionary influences to reveal how particular life histories evolved and how populations can be expected to respond to future changes in the environment.

Body size has proved to be a controversial characteristic in interspecific studies of life histories. Early modeling of the evolution of life histories considered body size to be an important life-history trait (MacArthur and Wilson 1967). In this context, body size was found to be associated with a suite of life-history traits, and the suite was interpreted in an adaptive context that compared species that often exhibit high population growth rates (r-selected species) to species exhibiting relatively little fluctuation in population size (K-selected species). Subsequent research suggested that body size was not a covarying life-history trait, but a causal influence on life histories that could explain the r-K continuum (Western 1979; Western and Ssemakula 1982; Peters 1983; Schmidt-Nielson 1983; Calder 1984). Evolutionary causes of interspecific patterns of body size were not directly addressed, but processes like competition might reasonably be invoked as an explanation (e.g., Brown 1995).

Comparisons of life histories among species might reveal evolutionary patterns and environmental circumstances associated with those patterns (e.g., Gaillard et al. 1989, 2000). In searching for evolutionary patterns of life histories, Read and Harvey (1989) suggested that mortality patterns and adaptive responses to them provide the best way to identify meaningful life-history patterns. In their view, body size is an obfuscating characteristic that needs to be controlled statistically before adaptive patterns of life histories can emerge. When influences of body size are removed, associations of life-history traits are revealed that reflect adaptations to relatively greater mortality, particularly during the prereproductive period (Promislow and Harvey 1990, 1991). This analysis produces a continuum from species that have relatively shorter lives to those with longer lives, and other life-history traits like reproduction adjust to this fast-slow continuum via tradeoffs. The fast-slow continuum is an empirical observation, basically the well-known "mouse to elephant" differences in life span and related demographic rates that have been noted in mammals. The continuum remains when differences among species in adult body mass are removed; thus, body size may not be a key determinant of this aspect of life histories (Harvey and Purvis 1999).

In the present review, we consider whether the life histories of rodent species adhere to the fast-slow continuum. In studies of mammalian life histories, Oli and Dobson (2003) and Oli (2004) suggested that the fast-slow continuum might be best reflected by the ratio of fertility to age at maturity (viz., the F/α ratio). At both ends of a scale from high F/α (fast life cycles) to low F/α (slow life cycles), mammalian species of small and large body sizes occur. Within the mammals, rodents are a suitable group of species for an examination of the fast-slow continuum, in part because they are primarily herbivores and thus share aspects of their lifestyles. In addition, rodents span only a part of the range of body sizes of mammals. Thus patterns of life-history traits that follow the fast-slow continuum should be particularly evident, with or without statistically controlling possible influences of body size.

The life cycle of rodents can be adequately approximated by parameters of partial life-cycle models (Caswell 2001; Oli and Zinner 2001): juvenile and adult survival (P_J and P_A), age at maturity and at the end of the reproductive lifespan (ω), and fertility. These key elements of life history encapsulate the most salient aspects of the life cycle, and they can be used to predict population growth or evaluate whether a population is in demographic equilibrium. We estimated these key life-history traits using data from populations of rodents. We followed the approach of Gaillard et al. (1989) in applying principal component analysis (PCA) to life-history traits, and examined the outcome for evidence of a fast-slow continuum. Furthermore, we removed confounded influences of body size and phylogeny from our data, and again looked for evidence of a fast-slow continuum. Gaillard et al.'s (1989) analyses suggested that after body size and the fast-slow continuum, a tertiary mammalian life-history pattern occurred between altricial and precocial species. We also looked for evidence of the altricial/precocial pattern of life-history traits among rodent populations.

Methods

Life-history data and demographic analyses

We compiled life-history data from published sources for 43 populations of 29 species of rodents (Oli and Dobson 2003, Appendix). The database includes recently published data on Idaho ground squirrels (*Spermophilus brunneus brunneus*; Sherman and Runge 2002). For statistical analyses, we transformed variables to improve their fit to a normal distribution (after Gaillard et al. 1989). Body mass and fecundity (m) were log-transformed, age at maturity (α), and reproductive life span (ω) were calculated in months and log-transformed. Survival probabilities (P_J and P_A) were calculated as monthly rates and square root-arcsin transformed.

We evaluated the influence of body size by regressing life-history variables on log-transformed body mass. Residuals were then analyzed to investigate whether any discernible patterns in these variables persisted after statistical removal of body-size effects. We used one-way ANOVA with *family* as the main effect to evaluate the effects of phylogeny on life-history parameters. Alternative methods of evaluating phylogenetic influences on life-history traits require a well-corroborated phylogenetic tree, and this was not available for the Rodentia. We chose to limit our analyses to family because a substantial proportion of variation in mammalian life-history traits occurs among families (Stearns 1983; Promislow and Harvey 1990). Residuals were then analyzed to examine whether the pattern of covariation among life-history variables persisted after statistical removal of phylogenetic effects. It is possible that both body size and phylogeny act synergistically in their effects on life-history variables. To test this possibility, we simultaneously removed their effects using ANCOVA. Again, we analyzed residuals to determine whether patterns of covariation of life-history variables persisted.

Quantification of fast-slow continuum

We used three methods for quantifying the fast-slow continuum in rodent life histories. First, we examined loadings of life-history variables on the first principal component (PC1) of a principal components analysis. Relatively high even loadings for longevity and survival variables would reflect the fast-slow continuum. Next, we computed the ratio of fertility rate to age of first reproduction (F/α ratio; after Oli and Dobson 2003). The F/α ratio reflects the magnitude of reproduction compared to age at maturity, but it also includes the influence of adult survival (i.e., $F = P_A * m$, where m is fecundity; Oli and Dobson 2005). Because of this, we also used a similar index, the ratio of fecundity to age at maturity (viz., m/α). Third, we quantified the fast-slow continuum based on adult mortality rate. Read and Harvey (1989) argued that animals at the fast end of the continuum have high mortality rate for their body mass. Thus we examined the correlation between adult mortality rate and other life-history variables, after statistically removing the effect of body mass. We tested the adequacy of different measures of the fast-slow continuum for congruence by looking for significant associations among them.

Quantification of precociality

Data on adult female mass and weaning mass were compiled from species accounts (Mammalian Species Accounts, American Society of Mammalogists), and additional body mass data were compiled from studies reporting demographic data (Oli and Dobson 2003, Appendix) or from Silva and Downing (1995). We used the ratios of neonatal mass and of individual offspring mass at weaning to adult female body mass as measures of precociality. Larger ratios represent offspring that are closer to adult body size at birth and at the end of the major period of maternal investment, and these species might be considered more precocial, assuming that the degree of development correlates with body mass of offspring. Using this assumption, we applied correlation analysis to ask if the degree of precociality was associated with principal components of life history variables. Another variable that might reflect an aspect of the "shape" of life histories is the length of the pre-adult period compared to the length of life. We estimated this by taking the ratio of age at maturity to reproductive lifespan (viz., α/ω), and compared this ratio to principal components of the life history. Ratios that reflected the degree of development of offspring and the timing of reproductive onset within life cycles were compared to other variables using Spearman rank correlations.

Results

We analyzed life-history data for 43 populations from 29 species and 9 families of rodents. Sciuridae was the most represented family (26 populations), followed by Heteromyidae and Muridae (4 populations each), and other families were represented by 1 or 2 populations. Our sample included species from a wide range of body masses (0.019–23.8 kg) and life-history characteristics (Oli and Dobson 2003, Appendix). So our sample, although not large, included rodents with diverse patterns of life history and population dynamics.

The pattern of life histories of a population are well-described by five variables: α, ω, P_J, P_A, and m. We used fecundity (m) for this analysis rather than fertility (F), because the latter incorporates the influence of adult survival. These life-history traits all varied significantly with mean adult body mass, with between about 20% and 40% of the variation in life histories statistically accounted for by body

mass (table 8.1). We used the family taxonomic level as classes in an analysis of variance to estimate the potential influence of phylogeny on the life-history traits and body mass. Phylogeny could account for about 60% to 90% of the variation in life histories, and 90% of the variation in body mass. The latter result indicates that body mass and phylogeny were strongly confounded for the rodent populations that we examined. Thus we adjusted life-history traits statistically for both body mass and phylogeny using the general linear model, and did not consider these influences separately in further analyses. Combined influences of body mass and phylogeny were not greatly different from the same analysis with phylogeny alone, explaining between about 60% and 90% of the variation among life-history variables.

We looked for evidence of a fast-slow continuum among the life-history traits (transformed, but not adjusted for body mass and phylogeny), using a principal component analysis (table 8.2). The first two principal component (PC) axes statistically explained 84% and another 8% of the variation in traits, respectively. The first PC had strong positive loadings for α, ω, P_J, and P_A, reflecting a fast-slow continuum. Thus timing variables and survival increased or decreased together along a fast-slow continuum. A trade-off was evident with reproduction, as reflected by a negative PC1 loading for m. The second PC axis reflected changes in juvenile survival that were somewhat associated with changes in m and a trade-off with adult survival. The same general pattern was evident when adult body mass was included in the analysis, except that body mass was also strongly associated with the fast-slow continuum reflected by the first PC axis. The first three PC axes explained 77%, another 11%, and another 7%, respectively, of the variation in life-history traits and body mass. The second PC axis reflected changes in body mass that were statistically independent of PC1, and changes in juvenile survival and its associations with fecundity and adult survival occurred on

Table 8.1 Regressions (for the influence of adult body mass), analyses of variance (for the influence of family-level phylogeny), and analyses of covariance using the general linear model (for the influence of body mass and phylogeny)

Trait	Influence of adult body mass			Influence of family-level phylogeny			Influence of mass and family		
	R^2	F	P	R^2	F	P	R^2	F	P
α	0.403	27.6	<0.0001	0.816	18.9	<0.0001	0.845	20.0	<0.0001
ω	0.395	26.7	<0.0001	0.805	17.5	<0.0001	0.807	15.4	<0.0001
P_J	0.197	10.0	0.003	0.585	6.0	0.0001	0.590	5.2	0.0003
P_A	0.284	16.2	0.0002	0.879	30.9	<0.0001	0.902	33.9	<0.0001
m	0.180	9.0	0.005	0.818	19.1	<0.0001	0.839	19.1	<0.0001
Mass				0.897	37.1	<0.0001			

NOTE: Sample size in all cases is 43 populations.

Table 8.2 Factor loadings from principal component analyses

Traits	Unadjusted		Unadjusted			Adjusted for adult body mass and family-level phylogeny		
	PC1	PC2	PC1	PC2	PC3	PC1	PC2	PC3
α	0.46	0.17	0.45	0.02	0.16	0.58	0.28	0.08
ω	0.46	−0.03	0.44	0.01	−0.04	0.56	−0.28	−0.06
P_J	0.42	0.76	0.39	−0.20	0.79	0.50	0.28	−0.45
P_A	0.46	−0.36	0.43	−0.17	−0.35	0.28	−0.16	0.85
m	−0.44	0.50	−0.41	0.35	0.46	−0.11	0.86	0.26
Mass			0.32	0.90	−0.07			

NOTE: Sample size in all cases is 43 populations.

the third PC axis. For both of the previous PCAs, however, only the first principal component explained substantial variation in the life-history data.

We conducted a principal component analysis for life-history traits that were statistically adjusted for influences of body mass and family-level phylogeny (table 8.2). The first three PC axes statistically explained 42%, another 23%, and another 20%, respectively, of the variation in life-history traits. Loadings on the first PC axis revealed strong positive loadings for α, ω, P_J, and P_A, reflecting a fast-slow continuum. A trade-off of the continuum and reproduction, however, was not evident. Rather, variation in fecundity was reflected by a high positive loading for m on the second PC axis. The third PC axis reflected changes in adult survival that were somewhat associated with changes in m, and a trade-off with juvenile survival. From the statistical definition of principal components, patterns on the second

and third PC axes were statistically independent of the fast-slow continuum as reflected by the first PC.

When life-history data were not adjusted for body size and phylogeny, scores on the first PC were strongly associated with some possible measures (e.g., F/α and m/α) of the fast-slow continuum (table 8.3), but in a negative direction. Thus as PC1 scores increased and F/α and m/α decreased, life histories moved from fast to slow. Scores on PC1 were also strongly positively associated with adult mortality, as reflected by P_A. Mortality equaled 1 minus P_A, so as mortality increased, life histories moved from slow to fast. Surprisingly, the strongest single association with PC1 was with the ratio of P_A/α. Scores on the first PC were significantly negatively associated with our indices of precociality, with precocial species more likely to fall at the fast end of the continuum. For this PCA, the scores on the second PC showed no significant associations.

Table 8.3 Spearman rank correlations between principal component scores that represent the fast-slow continuum (viz., PC1), some possible proxies for the continuum, and variables that reflect developmental patterns

Index	Unadjusted		Adjusted for adult body mass and family-level phylogeny		
	PC1	PC2	PC1	PC2	PC3
F/α	−0.763***	0.191	−0.244	0.379	0.064
m/α	−0.773***	0.188	−0.257	0.367	0.0048
P_A	0.879***	−0.184	0.303	0.003	0.466*
P_A/α	0.928***	−0.009	0.365	0.080	0.414*
Neonate mass/ Adult mass	−0.746**	−0.108	−0.046	−0.261	−0.057
Weaning mass/ Adult mass	−0.589**	−0.137	−0.071	−0.009	−0.049
α/ω	−0.102	0.367	−0.129	0.449*	−0.048

NOTES: Sample size was 27 for the ratios of neonate mass and of weaning mass to adult body mass; it was 43 for all other variables.
***$P \leq 0.0001$
**$P \leq 0.001$
*$P \leq 0.01$.

When effects of body mass and phylogeny were statistically controlled, scores on the first PC axis were not significantly associated with other measures of the fast-slow continuum. Scores on the second PC (primarily reflecting fecundity, table 8.2) were significantly associated with longer subadult periods relative to the life span (viz., the α/ω ratio). Fecundity, however, was only weakly associated with the α/ω ratio ($r = 0.275$, $n = 43$, $P = 0.07$). Significant associations with the third PC (primarily reflecting variation in P_A, table 8.2) with P_A and P_A/α were likely due to the confounded influence of P_A.

Discussion

Explaining the evolution of life histories is an important goal for ecologists and evolutionary biologists. Interspecific studies attempt to evaluate the way in which life histories evolve by comparing traits of different species to aspects of the environments in which species live. Two major problems have faced such research. The first is that critical aspects of environmental influences are often difficult to identify and quantify. A possible solution to this problem is to use demographic traits as indicators of environmental changes, and then look for associated changes in life histories. This tack was taken by Murphy (1968) and later Stearns (1976) to introduce what came to be known as the bet-hedging model of life-history evolution. Harvey and his colleagues (reviewed by Harvey and Purvis 1999) also used this line of thought to argue that life-history patterns should be evolutionarily molded by mortality patterns. The primary problem with this research tactic is that demographic traits themselves are also life-history traits, so it is difficult to discern the interplay of cause and effect.

This first major problem leads us to a second problem: difficulties with drawing inferences from interspecific studies. The problem is whether inferences about adaptation can be made for species that have historical affiliations. Intraspecific studies should yield strong inferences about evolved responses of life histories to environmental conditions, but the range of variation in life histories may be narrow. Interspecific studies examine broader ranges of variation in both life histories and environments. Differences among species may be historical, however, and thus reflect evolved responses to past environments (Dobson 1985). Reconstructing past environments and discerning how ancestors responded to them is exceedingly difficult. Alternatively, differences among species may reflect current adaptation to modern environments. It is these evolutionarily maintained variations in life histories that reveal the clearest answers to our question about how life histories evolve. A possible solution to analyzing interspecific patterns of life histories is to attempt to remove influences of historical patterns by statistically adjusting for phylogeny (e.g., Miles and Dunham 1992, Martins and Hansen 1997). Such procedures also provide at least partial solutions to sampling problems associated with interspecific studies (Harvey and Pagel 1991).

Considerable interspecific research on life histories has focused on mammals. Studies of the relationship of life histories to body size used comparisons of mammalian species, and concluded that much of the variation in life histories could be explained by differences among species in size (reviewed by Harvey and Purvis 1999). Read and Harvey (1989) countered this argument by pointing out that if the influence of body mass was statistically removed from mammalian life-history traits, very similar associations among the traits remained. These associations seemed to describe a continuum of life histories, from fast traits such as high rate of reproduction, early maturity, and low survival, to slow traits such as late maturity, high survival, and low rates of repeated reproduction. This fast-slow continuum was evident not only among mammals, but within mammalian orders as well. Promislow and Harvey (1991) suggested that fast life histories evolved due to environmentally caused mortality rates that were high, and slow life histories evolved when mortality rates were low.

The purpose of our present interspecific study was to reexamine the fast-slow continuum in the order Rodentia. We improved on past analyses in two ways. First, we looked at a restricted set of life-history traits that summarize the life cycles of species. For this, we borrowed from the concept of partial life-cycle models (Caswell 2001; Oli and Zinner 2001). A partial life-cycle model can be parameterized from demographic traits that are sufficient to describe changes in population size or growth rate: age at maturity, age at last reproduction, juvenile and adult survival, and fecundity. Since these traits are sufficient to describe population dynamics, they should be key life-history characteristics. In addition, with the exception of age at last reproduction, these life-history variables have generally been used to measure the evolutionary fitness of other characteristics of organisms (Endler 1986).

Our second improvement was to borrow the idea of using principal component analysis to evaluate life-history patterns, after Stearns (1983) and Gaillard et al. (1989). Gaillard et al. (1989) used fewer traits to describe life histories, but identified the fast-slow continuum as statistically independent of body size. Principal component analysis identifies the major axis of variation in variables in a hyperspace of several axes. We can think of variables in hyperspace as similar to a rugby ball in three-dimensional space. The first principal component would run along a line drawn through the long axis of the rugby ball. If the rugby ball were "skinny" around the middle, the first component (first PC axis) would describe most of the variation in the data set, as in our data on life-history traits. Thus the first principal component summarizes traits that change together, and such traits will have high and even factor loadings on the first principal component. Factor loadings tell us which variables change the most along the principal component. The analysis also shows us whether the magnitudes of traits vary together in a positive or inverse manner, revealed by the sign of the factor loadings (viz., either the same or opposite signs). Gaillard et al. (1989) also suggested that a secondary life-history tactic to the continuum was evident in their analyses, which reflected an altricial-precocial dichotomy among species. This hypothesis can be tested by looking at PC axes after the first one, if the first axis reflects the fast-slow continuum.

Like previous studies (e.g., Western 1979; Western and Ssemakula 1982; Peters 1983; Schmidt-Nielson 1983; Calder 1984), we found that adult body mass was significantly associated with life-history traits (table 8.1). When we statistically adjusted the life-history data for family-level phylogenetic effects, however, body mass had little additional influence. Body mass varies among families of rodents, and thus influences of size and phylogeny cannot be easily separated. Thus removal of influences of body mass and phylogeny from the life-history variables was done together. Body mass and phylogeny accounted for large amounts (up to 90%) of variation in life-history variables. In particular, removal of these sources of variation left little residual variation in age at maturity, adult survival, and fecundity. This is important to keep in mind when considering residual patterns of life histories, and we expected that PCA might reflect the fast-slow continuum more poorly when influences of body size and family-level phylogeny were removed.

Loadings of life-history variables (unadjusted for body size and phylogeny) on the first principal component clearly reflected the fast-slow continuum: timing variables (α and ω) and survival were all strongly associated on the first principal component (table 8.2). In addition, reproduction varied in a negative way along this axis. These results support the idea that a fast-slow continuum exists in rodents, from species with short lives and greater reproductive investments during a relatively short period of time to species with long lives and limited but repeated investments in reproduction. A tradeoff between reproductive and somatic investments may be reflected along this axis. Our results also supported the suggestion that the fast-slow continuum may be evolutionarily influenced by mortality patterns (Read and Harvey 1989). High mortality was associated with the fast end of the continuum.

We adjusted our data statistically to control for influences of body mass and family-level phylogenetic effects. In doing this, most of the variation among populations in life-history traits was removed (see the preceding). Despite this, loadings on the first PC still indicated a fast-slow continuum, with high positive loadings for timing variables and survival, though with a weaker pattern of changes in adult survival (table 8.2). The greater strength of the loading of juvenile compared to adult survival on the first PC was in agreement with the scenario of life-history evolution presented by Promislow and Harvey (1990, 1991), that life histories evolve in response to mortality patterns, particularly among juveniles. Also in this analysis, reproduction exhibited variation that was statistically independent of the continuum, as reflected by the high positive loading of fecundity on the second PC axis. Thus reproduction did not vary in quite the same way as when the life-history traits were not adjusted for body size and phylogeny.

The second PC axis of the unadjusted life-history data accounted for such a low amount of variation in life-history traits that it may not be useful to evaluate this axis further. When adult body mass was included in the analysis, mass exhibited variations that were different from its association with the fast-slow continuum (table 8.2). In this analysis, the third PC axis exhibited a virtually identical pattern to the second PC when body mass was not included, but the third axis accounted for even less of the variation in life-history traits. When adjustments were made to the data for body mass and family-level phylogeny, substantial residual variations in life-history traits was explained by the second and third PC axes. Each axis primarily reflected changes in single life-history variables, fecundity and adult survival respectively, that were statistically independent of the fast-slow continuum. In the former case, reproduction appeared to vary independently of the fast-slow continuum. Little variation in life-history traits remained after statistical adjustment of the data for body size and phylogeny (table 8.1), however, so interpretation of these axes might be risky.

Oli and Dobson (2003, 2005) suggested that the ratio of fertility to age at maturity (F/α) might be a good index of the fast-slow continuum. This expectation was borne out by strong associations of F/α and m/α with the first principal component of the unadjusted data set (table 8.3). Surprisingly, P_A/α was very closely associated with the fast-slow continuum, as the latter was reflected by scores on the first PC. This ratio reflects mortality patterns relative to the length of prereproductive development. When the life-history variables were statistically adjusted for body mass and family-level phylogeny, however, all of these significant associations became nonsignificant. This result may have been due to removal of so much of the variation in life-history traits when statistical adjustments were made.

The mass that offspring attain at birth and at the end

of the major period of maternal investment (viz., weaning weight) should reflect how close they are to completing development to adult body size. Thus the ratios of neonatal mass and of weaning mass to adult body mass should reflect how precocial (as opposed to altricial) offspring are at the end of these periods of maternal investment. We found a significant association between our indices of precociality and scores on the first axis of the PCA for data that were not adjusted for body mass and phylogeny (table 8.3). Fast species appeared to have more precocial offspring and slow species more altricial young. When adjustments were made to PC1 scores for body mass and phylogeny, however, these patterns disappeared. Thus the role of precociality as a third major axis of life history (Gaillard et al. 1989) was not supported. In their discussion of mammalian Orders, Gaillard et al. (1989) suggested that more fully developed young (precocial species) should have slow life histories. Our measure of how close young are at birth and weaning to adult size placed precocial species (viz., those with relatively large young) at the fast end of the continuum, but only when data were not adjusted for body size and phylogeny. This result could be due to a difference between comparing Orders of mammals and species within the Rodentia, or due to our more specific definition of precocial.

The proportion of the life cycle of rodents that is spent in the prereproductive period, measured by α/ω, was not associated with the fast-slow continuum, as reflected by either unadjusted life histories or when adjustments for body mass and phylogeny were made (table 8.3). Surprisingly, this aspect of the timing of life history was most closely associated with the second PC of the adjusted (for body mass and phylogeny) data set, an axis that primarily reflected the magnitude of fecundity (though the association of α/ω and m only approached statistical significance). Rodents with long developmental periods relative to their lifespan are those with high levels of fecundity. If the first principal component of the life cycle represented the fast-slow continuum, then this pattern of long development and high fecundity is statistically independent of the continuum.

Summary

We examined whether rodent species exhibit a fast-slow continuum of life histories via analyses of life table data. We also examined whether a tertiary axis of rodent life history reflects a dichotomy between altricial and precocial species. Life histories were encapsulated by five key measures of populations: age at maturity, reproductive lifespan, juvenile and adult survival, and fecundity. Precociality was measured by how close neonatal and weaned rodents were to their adult body mass. We used principal component analyses to quantify the fast-slow continuum, both without and with statistical adjustments for body mass and family-level phylogenetic relationships. If the PCA reflected the fast-slow continuum, then high, positive factor loadings were expected for variables that reflect the tempo of life history: age at maturity, reproductive life span, and juvenile and adult survival.

We found that for rodents, the first principal component of a PCA may provide an excellent reflection of the empirically observed fast-slow continuum of mammalian species. This reflection of the fast-slow continuum appeared to hold when influences of body size and phylogeny were at least partly statistically controlled. Scores on the first PC might provide the best measure of where species occur along the fast-slow continuum, since our PCAs combined influences of the key life-history traits that make up the continuum. Rodents appear to vary in life cycle, from those that live short lives to those with long lives. Reproductive success, measured as offspring produced over a set interval time, appears to vary from high to low, respectively, with the continuum, in unadjusted data. But reproduction varied independently of the continuum in data that were statistically adjusted for body mass and phylogeny. The degree of precociality that species exhibit also appears to vary along the fast-slow continuum, with fast species exhibiting greater precociality. This pattern, however, was only significant when influences of body size and phylogeny were allowed to exhibit their natural variation.

Chapter 9 Acceleration and Delay of Reproduction in Rodents

Lee C. Drickamer

THE FIRST BEHAVIORAL and physiological responses of rodents to chemosignals were the pregnancy blocking effect for recently inseminated female mice by strange males (Bruce effect; Bruce 1959; Bruce and Parrott 1960), and the tendency for grouped females to remain anestrus (Lee-Boot effect; Andervont 1944; Lee and Boot 1955, 1956), followed by estrous synchrony when the females were exposed to a male (Whitten effect 1958, 1959; Marsden and Bronson 1964). A decade later, Vandenbergh (1967, 1969) first reported the acceleration of puberty in laboratory house mice brought about by the presence of a male or male urine. During the past thirty years, a great deal of additional information has appeared in the literature pertaining to chemosignals that influence the timing of puberty and reproduction in rodents. The house mouse (*Mus musculus*) has been the principle subject for these studies, though information has also been recorded for deer mice (*Peromyscus;* Teague and Bradley 1978; Lombardi and Whitsett 1980), voles (*Microtus;* Hasler and Nalbanov 1974; Storey and Snow 1990), and gerbils (*Meriones;* Ågren 1981). Virtually all of these effects on reproduction involve urinary chemosignals. Thus except as noted, when I refer to chemosignals, I am writing about those in urine. I also should note that where some people might choose the word *pheromone* as a label for these signals, I prefer the less specific term *chemosignal*.

These urinary compounds act as primers—that is, they generally operate via neuroendocrine mechanisms leading to physiological changes (Macdonald and Brown 1985). This means that we are dealing with substances in rodent urine that alter the physiology of any recipients, often accompanied by changes in behavior. Because the most extensive work to date has been done on house mice, deer mice, and voles, I have limited my approach here to those taxa.

My review of the urinary chemosignals that accelerate or delay estrus and puberty in female rodents has several purposes, which lead to the sections that follow. First, I will summarize information on the three major categories of urinary chemosignals accelerating or suppressing rodent reproduction. This summary is, of necessity, brief, as there are full reviews available and because my main focus is on rodent societies, not the chemistry or hormonal mechanisms that underlie these effects (table 9.1 summarizes these effects). The second section covers what is known about these chemosignal effects under field conditions. Then, I examine the issues surrounding differences between laboratory-tested phenomena and their possible occurrence in natural settings. The penultimate section covers what is known about the ecological and evolutionary perspectives regarding these reproductive chemosignals. Finally, I attempt to synthesize the current state of this field in terms of these investigations with suggestions on possible future paths of experimentation.

Rodent Chemosignals and Reproduction

Acceleration effects

Chemosignals have primarily two positive effects on reproduction. One involves accelerating the onset of puberty in females via stimulation from male presence or male urine,

Table 9.1 Summary of three chemosignal-mediated effects on reproduction in rodents

Effect	Genus	Evidence For	Against	Field tests
Acceleration of puberty	*Mus*	Vandenbergh 1967, 1969; Drickamer 1974		Massey and Vandenbergh 1981; Drickamer and Mikesic 1990
	Microtus	Batzli et al. 1977; Carter et al. 1980		
	Peromyscus	Terman 1984		
Estrus synchronization	*Mus*	Whitten 1958, 1959; Marsden and Bronson 1964		
	Microtus	Carter et al. 1980; Gavish et al. 1983		
	Peromyscus	Bronson and Marsden 1964; Marsden and Bronson 1965b		
Delay of puberty	*Mus*	Vandenbergh et al. 1972; Drickamer 1974b, 1977a		Massey and Vandenbergh 1980; Coppola and Vandenbergh 1987; Drickamer and Mikesic 1990
	Microtus	Pasley and McKinney 1973; Rodd and Boonstra 1988	Batzli et al. 1977	
	Peromyscus	Haigh 1987; Haigh et al. 1985; Terman 1968; Terman and Bradley 1981	Terman 1984; Terman 1993	Kaufman and Kaufman 1989; Terman 1993; Wolff 1992
Anestrus	*Mus*	Whitten 1959; Whitten et al. 1968; Lee and Boot 1955, 1956		
	Microtus	Pasley and McKinney 1973	Batzli et al. 1977	
	Peromyscus	Terman 1965, 1968		Terman 1993; Kaufman and Kaufman 1989
Pregnancy termination	*Mus*	Bruce 1959, 1960, 1961; Parkes and Bruce 1962; Dominic 1969	Mahady and Wolff 2002; Drickamer 1989	
	Microtus	Clulow and Clarke 1968; Stehn and Richmond 1975; Jannett 1980; Stehn and Jannett 198	Mallory and Clulow 1977; de la Maza et al. 1999; Mahady and Wolff 2002	de la Maza et al. 1999; Mahady and Wolff 2002; Wolff 2003
	Peromyscus	Bronson and Eleftheriou 1963; Kennedy et al. 1977; Haigh 1988		

NOTES: References are provided by genus for papers describing evidence for, against, and under field conditions for these three effects. All citations for and against are from laboratory studies.

and the second involves the Whitten effect, or stimulation and synchronization of estrus in adult females exposed to males or male urine (see McClintock 1983; Vandenbergh 1983; Brown 1985a; Drickamer 1986a; and Vandenbergh and Coppola 1986 for reviews).

For house mice, the acceleration of puberty is a function of both male presence and the urinary chemosignal (Vandenbergh 1967, 1969; Drickamer 1974a); puberty is accelerated most when both types of cues are present. The male urinary chemosignal is testosterone dependent, with dominant males producing more potent chemosignals than subordinate males (Vandenbergh 1969; Drickamer and Murphy 1978). The acceleration effect is seasonal in that male urine generally does not accelerate estrus or puberty in female mice during the winter nonreproductive period, but does so for the remainder of the year including the reproductive season (Drickamer 1987). The chemosignal in male urine is effective in extremely small (0.0001 cc of urine/day) doses (Drickamer 1982, 1984a) and is equally effective re-

gardless of whether treatment involves related or unrelated donors or recipients (Drickamer 1984e). The acceleration effect is blocked, with delays in puberty occurring, when young females are treated with urine from both adult males and grouped females (Drickamer 1982, 1988). Urine from pregnant and lactating females and females in estrus cause an acceleration of puberty in young female house mice (Drickamer 1983, 1986a).

The chemical substance in male urine that is responsible for acceleration of female puberty appears to be a volatile compound, but is attached to mouse urinary proteins (Vandenbergh et al. 1975, 1976; Nishimura et al. 1989; Novotny et al. 1998, 1999a; Mucignat-Caretta et al. 1995). Connections still need to be made between the role of mouse urinary proteins and male-male interactions (Hurst et al. 1998, 2001) and the urinary chemosignal from males that stimulates reproductive activity in female mice. Studies on the underlying hormonal mechanisms reveal that follicle-stimulating hormone (FSH), luteinizing hormone (LH), prolactin, and estrogen all affect reproductive activity (Bronson and Whitten 1968; Bronson and Stetson 1973; Bronson 1975; Bronson and Maruniak 1976).

In mice, urine that accelerates puberty also synchronizes estrus among grouped adults (known as the Lee-Boot and Whitten effects; Drickamer 1992, unpublished data). Many aspects of the synchronization phenomenon are similar to the acceleration of puberty. These aspects include the seasonality, testosterone-dependency, and small dosages needed for effect. The chemical nature of the compound involved appears to be the same as for puberty acceleration (Novotny et al. 1998, 1999a, 1999b)

The following is known about these acceleration and synchronization of estrous cycle effects in other rodents. For prairie voles (*Microtus ochrogaster*), activation of reproduction in young virgin females and in adult females is stimulated by a male chemosignal (Batzli et al. 1977; Carter et al. 1980, 1987; Gavish et al. 1983). There may be some intraspecific variation in this phenomenon in that prairie voles from Kansas were more responsive than those from Illinois (Roberts et al. 1998). Among voles, acceleration or activation of reproduction has also been recorded for meadow voles (*M. pennsylvanicus;* Baddaloo and Clulow 1981) and pine voles (*M. pinetorum;* Lepri and Vandenbergh 1986; Solomon et al. 1996), but not for California voles (*M. californicus;* Rissman and Johnston 1985).

For deer mice, only a few studies have been conducted to explore acceleration of female sexual maturation or synchronization of estrus in adults. For prairie deer mice (*Peromyscus maniculatus bairdii*), exposure to a male or male urine synchronizes estrus in females caged in groups (Whitten effect; Bronson and Marsden 1964; Marsden and Bronson 1964, 1965b; Bronson and Dezell 1968). Terman (1984) reported just the opposite finding; females reared in the presence of a male or male urine had smaller ovaries and fewer corpora lutea than females treated with water. There apparently have not been any studies that reported acceleration of puberty or induction of estrus in adult female deer mice following exposure to male urine. At this time, considerable work remains to be done to obtain a better picture of acceleration of reproduction in *Peromyscus*. The limited results to date provide some contradictions, and no consistent picture has emerged with regard to possible male acceleration of female puberty.

Delaying effects

Two related phenomena that characterize a delay in reproduction in rodents are that grouping among females tends to keep them in anestrus and that groups of females or urine from grouped females delays puberty in young females. I

prefer the term "delay" for both of these phenomena rather than the widely used "inhibition" and "suppression," because the latter imply an absence of estrus or puberty. Rather, this grouping or urine from grouped females tends to slow the timing of reproduction to varying degrees, but females still exhibit estrous cycles and young females attain a first vaginal estrus and puberty. The same list of reviews, cited in the previous section on acceleration effects, contains summaries of many aspects of these reproduction-delaying processes.

The disruption of estrous patterns in adult females that are grouped was first discovered by Whitten (1959) and later attributed to a urinary chemosignal (Whitten et al. 1968; McClintock 1983). The delay in sexual maturation in young mice was first reported for young females housed in groups (Vandenbergh et al. 1972) and was investigated in a series of studies exploring various facets of this phenomenon (Drickamer 1974b, 1977a, 1984a, 1984e, 1986b, 1986c; Drickamer and Assmann 1981; Coppola and Vandenbergh 1985). Based on laboratory studies, the urinary delay substance is produced by females housed in groups, is effective in very small doses, is received by the vomeronasal organ of prepubertal females, elicits the same response in females independent of the relatedness of the male, and is released by grouped females only after several weeks of caging at high density. When young females are exposed to both male urine and urine from grouped females, the latter takes precedence, except when the relative proportions involve considerably more urine from males than from grouped females (Drickamer 1982, 1988).

Limited work has been done on urinary chemosignal effects producing delays in puberty or induction of estrus in *Microtus*. For meadow voles, females housed in groups have lighter-weight ovaries and uteri than females housed alone (Pasley and McKinney 1973), suggesting that grouping delays or inhibits sexual maturation. Delay of sexual maturation for grouped females has been reported for prairie voles and California voles but not for meadow voles (Batzli et al. 1977). In contrast, Wolff et al. (2001) found no delays or inhibition of reproduction in young female meadow voles or prairie voles raised in the presence versus the absence of their mothers.

Conflicting results have been reported for *Peromyscus* regarding the effects of a urinary chemosignal on attainment of puberty. Terman (1984, 1993) reported that young female white-footed mice (*Peromyscus leucopus*) caged with an adult female reproduced at a higher rate than females caged alone. On the other hand, the presence of an adult female (Haigh et al. 1988) or soiled bedding from adult females (Bediz and Whitsett 1979; Haigh 1987; Haigh et al. 1985) delayed reproduction in young female white-footed mice. For prairie deer mice urine from females in dense laboratory populations applied to young females resulted in smaller uteri and delays in reproduction (Terman 1968a; Terman and Bradley 1981). Lower metabolic rates and altered adrenal physiology and morphology were further evidence of a physiological effect on growth and development of young females raised in the presence of adult females (Cherry et al. 2002). Lastly, Gubernick and Nordby (1992) reported delayed sexual maturation in *Peromyscus californicus* when young females were housed with both parents, though it appears that physical contact and the presence of the father rather than a urinary chemosignal may be the modulator of this effect.

Termination of pregnancy

A third type of chemosignal effect on rodent reproduction, the Bruce effect (Bruce 1959, 1960, 1961), involves the termination of pregnancy for female rodents of some species when they are exposed to a strange male or urine from a strange male early in gestation. Marchlewska-Koj (1983), Brown (1985b), Storey (1994), de la Maza et al. (1999), and Mahady and Wolff (2002) have all reviewed these effects rather extensively.

Termination of pregnancy following exposure of a recently inseminated female to a strange male or its urine was first reported for house mice, and primarily occurs prior to the implantation of the fetuses (Bruce 1959, 1960). The strongest effect occurs when exposure to the strange male is during the first 24 hours after insemination, with diminishing effects over the next 3 to 5 days until implantation occurs. The Bruce effect can be induced by exposure to male-soiled bedding (Parkes and Bruce 1962) or male urine (Dominic 1969), indicating that it is based on a chemosignal. Many facets of the Bruce effect have been explored in house mice, including the neuroendocrine pathways, tests of discrimination between sire and strange males, and varying effects of preexposure to the strange male (Parkes and Bruce 1961; Brown 1985a). The efficacy of the Bruce effect appears to vary among strains of inbred house mice (Chapman and Whitten 1969; Marsden and Bronson 1965a).

The Bruce effect occurs in a number of vole species, including field voles (*Microtus agrestis*), prairie voles, montane voles (*M. montanus*), and meadow voles (Clulow and Clarke 1968; Clulow and Langford 1971; Stehn and Richmond 1975; Jannett 1980; Stehn and Jannett 1981). Unlike the situation for house mice, in voles there appear to be interspecific differences in the timing of exposure to strange male odor, including time periods that extend well beyond the time of implantation (Stehn and Richmond 1975; Kenney et al. 1977). For example, for meadow voles, exposure to a strange male was effective in reducing pregnancy rates up to day 5 after copulation (Clulow and Langford 1971; Mallory and Clulow 1977), but for prairie voles, exposure up to 15 days after copulation induced females to abort pregnancies (Stehn and Richmond 1975; Kenney et al. 1977). There may also be differences in the efficacy of the Bruce effect that are dependent on whether the test stock is wild-caught or laboratory bred (Mallory and Clulow 1977). There are likely differences in the neuroendocrine pathways involved in the pregnancy termination, both relative to house mice and within voles, depending on when during gestation the exposure to a strange male occurs (Milligan 1980).

Prairie deer mice exhibit the Bruce effect (Bronson and Eleftheriou 1963; Bronson et al. 1964, 1969; Kenney et al. 1977; Dewsbury 1982c). The effect occurs during the first few days after insemination and involves a chemosignal cue. Lower rates of pregnancy occur when females are mated with more than one male sequentially. The effect apparently does not involve adrenocorticotrophic hormone (ACTH) or adrenal hormones, but females given doses of prolactin during exposure to strange males had pregnancy rates similar to those of unexposed females. In white-footed mice, young females housed with an adult female and a male do copulate, but pregnancy is terminated prior to implantation (Haigh et al. 1988). Additional testing on other species of *Peromyscus* might reveal a general pattern for this genus.

Field Tests

Many of the findings about chemosignals elucidated in laboratory settings are speculative with respect to their occurrence and effects in natural populations of these species. Only a few of these phenomena have been explored under field conditions and almost all such studies involve seminatural conditions where rodents are housed in some type of outdoor enclosure (table 9.1).

Delay of sexual maturation and suppression of reproduction have been reported from field populations of deer mice (reviewed by Terman 1968b, 1993; Kaufman and Kaufman 1989) and voles (reviewed by Taitt and Krebs 1985). In white-footed mice, reproductive rates seem to be reduced in some populations (Terman 1993) but not in others (Wolff 1994b). Reproduction was delayed in some young females at high densities, but only those that remained in the presence of their fathers (Wolff 1992). In voles, delayed sexual maturation and reproductive suppression have been recorded at high densities and/or peaks in population cycles (Krebs et al. chap. 15, this volume); however, one experimental study with gray-tailed voles (*M. canicaudus*) showed a minimal and short-term effect of high densities of females on reproduction of young females (Wolff et al. 2002). None of these studies involved specific tests of hypotheses about the possible roles of urinary chemosignals mediating possible delays of puberty or suppression of breeding among females.

Among actual field investigations, those on house mice involve the use of highway cloverleaf islands and structural enclosures. Urine from females in high-density populations on cloverleaf islands successfully delayed the onset of puberty in females maintained in the laboratory (Massey and Vandenbergh 1980). Urine from females in low-density populations housed at cloverleaf islands did not alter the onset of puberty. When female density was artificially inflated by adding females to cloverleaf islands, urine collected from resident females was capable of delaying puberty in young females maintained in the laboratory (Coppola and Vandenbergh 1987). Male urine collected in both high- and low-density conditions successfully accelerated puberty in female mice maintained in the laboratory (Massey and Vandenbergh 1981). These studies demonstrate that the urinary chemosignals that accelerate or delay puberty are produced in wild house mice living under natural conditions. They cannot, however, be used to indicate the possible population consequences of these signals.

A series of 0.1 ha outdoor enclosures was used to test both acceleration and delay chemosignals in a seminatural setting using introduced populations of house mice (Drickamer and Mikesic 1990). Urine from mice maintained in the laboratory was sprayed weekly on food sources and nest sites. Two enclosures were treated with urine from grouped females, two were treated with adult male urine, and two enclosures were treated with water as a control. Populations in the enclosures with male urine added attained higher densities, exhibited earlier puberty for females and more reproduction by females than mice in the control enclosures. In contrast, mice in the enclosures treated with urine collected from grouped females exhibited less reproduction, lower overall growth and densities, and less reproduction than mice in the control enclosures. In this instance, field effects on puberty and reproduction were recorded, though the urine supplies used to treat the different populations came from wild mice housed in the laboratory. Thus these two sets of investigations provide strong evidence for urinary chemosignal effects on reproduction in wild house mice, but no study has yet been conducted that involves both donors and recipients in a field setting.

Seminatural enclosures have been used effectively to study numerous aspects of rodent behavioral ecology with mixed results when compared with laboratory studies (see the following). Similar enclosures could be used to test the acceleration and delay effects of chemosignals in a field set-

ting in vole species. In an open 10-ha grassland, using experimental manipulations involving removing adults of one sex or the other, Rodd and Boonstra (1988) found that adult females play a critical role in terms of delaying the sexual maturation of young meadow voles at higher densities. In contrast, experimental manipulations of gray-tailed voles in which adult sex ratios were skewed toward males or females in large 0.2 ha enclosures produced only minimal and short-term (one week) effects of delayed sexual maturation (Wolff et al. 2002).

Laboratory Phenomena—Natural Phenomena?

One question that arises after considering the numerous laboratory studies and the available field-test data concerns whether some of the observed effects of urinary chemosignals on aspects of reproduction in rodents are laboratory artifacts (table 9.1). Do these phenomena translate to natural settings? How many of the effects are characteristic of natural populations of rodents?

Drickamer (1989) tested whether female behavior might influence pregnancy termination (the Bruce effect) in house mice. Wild female house mice were given choices between soiled bedding containing odors from the stud male and a strange male. For the first half of the 19–20 day gestation period the inseminated females exhibited a strong preference for the stud male, avoiding the odor of a strange male and the possible negative consequences. These preferences were absent in the second half of gestation, but by then, the embryos had implanted and fetuses were developing. This short study suggests that, given a chance under natural rather than caged conditions, recently inseminated females would choose to associate with the area of the stud male and could, thereby, avoid any possible pregnancy termination effects from a strange male.

More recently, Wolff (2003c) summarized studies on several rodent species that suggest that further testing of the chemosignal effects on rodent reproduction is warranted before we jump to conclusions about laboratory phenomena and their possible importance in natural populations. Field tests of the Bruce effect conducted in large seminatural enclosures failed to provide any evidence of pregnancy disruption in gray-tailed voles (de la Maza et al. 1999) or prairie voles (Mahady and Wolff 2002). An interesting side note to these two studies is that the gray-tailed vole has a promiscuous mating system, while the prairie vole has a monogamous mating system.

One hypothesis about suppression or delays in reproduction (phenomena similar to the Whitten effect) involves negative influences from the presence of predators. The mechanism of this effect would likely be the same as for intraspecific signals that delay puberty or estrus in adults. The hypothesis posits that in the presence of heavy predation, reproduction is delayed to avoid unnecessary waste of energy when pregnant females could be at greater risk of predation and progeny likely would be consumed (Cushing 1985; Ylönen 1989; Ylönen and Ronkainen 1994). However, investigations using gray-tailed voles exposed to urine and feces from mink (*Mustela vison;* Wolff and Davis-Born 1997) and bank voles (*Clethrionomys glareolus*) exposed to odors from weasels (*Mustela nivalis;* Mappes et al. 1998) failed to support this hypothesis. Interestingly, in one field study that examined the response of prairie voles and pine voles to odors of a mink predator, voles did not reduce their scent marking activity compared to control areas (Wolff 2004).

These tests raise questions about the efficacy of some of the laboratory studies designed to assess reproductive responses to chemosignals in rodents. As Wolff (2003c) notes, we need to remember that laboratory studies are critical for defining and investigating some aspects (e.g., underlying mechanisms) of these phenomena. But, we must also be vigilant about testing them in field settings. In fact, it is only under field conditions that we can judge either the true existence of the effect in natural populations or test the potential ecological and evolutionary consequences of such effects. As my concluding section will note, it is in this area that future work is needed most.

Ecological and Evolutionary Considerations

Bronson (1970, 1979a, 1979b, 1983, 1985; Bronson and Coquelin 1980) provided multiple reviews of the reproductive ecology of house mice, with some comments on deer mice. He details effects of environmental factors such as space, photoperiod, and food resources on reproduction and seasonal breeding. He discusses the possible effects on reproduction of social factors such as aggression and social signals like urinary chemosignals. Drickamer (1986a) and Vandenbergh and Coppola (1986) both examined the ecology of urinary chemosignals and their possible functions in house mice. Most of the comments on adaptive significance in these multiple papers are speculation. As noted earlier, field tests under even seminatural conditions are few in number, making it difficult to arrive at any conclusions about the ecology of the primer urinary chemosignals.

However, given that at least some of the chemosignal-mediated effects on reproduction are effective under field conditions and others remain to be tested we can proceed to ask questions like: What, if any, are the fitness conse-

quences of these urinary chemosignal effects on reproduction in various rodent species? What might be the functional effects of these various chemosignals on reproduction under natural ecological conditions? How would these systems have evolved in terms of both the production/release of the chemosignals and their reception by conspecifics? Are there any patterns that develop when these effects are examined in relation to phylogenetics? Or, do predictable patterns occur with regard to ecological settings or social systems?

The urinary chemosignals that accelerate or delay sexual maturation in individuals and, in some cases, alter female patterns of reproduction, might indirectly affect rates of population growth (Krebs et al. chap 15, this volume). At low densities and at the start of a breeding season, male urine that accelerates female puberty and onset of breeding in adult females would be predominant, leading to increases in reproduction. When density increases or other conditions change, such that there are greater risks associated with possible reproductive efforts (see the following in this section), then a combination of release of maturation-delaying chemosignal from females overrides the male urinary signal and results in delays in reproduction. I hasten to note that urinary chemosignals would only be one portion of the set of cues that are available for the rodents to process in terms of the gains from reproduction versus the possible costs and risks. The findings on urinary chemosignals need to be embedded in a larger environmental context, and one that also accounts for the changing seasons in many locations.

Drickamer (1986c) hypothesized that the active component(s) of urinary chemosignals likely originated as urinary compounds, for example, urinary proteins or components attached to urinary proteins. At some time in the past, production and/or excretion of these particular compounds could have been influenced by social, environmental, and hormonal conditions of the animal. If some advantage were gained by a conspecific receiving a message(s) represented by these urinary compounds, then natural selection could favor altering reproductive physiology to conform with the message. Once the nature of the chemical compounds and the sensory pathways and modes of action in recipient individuals are known, the fitness consequences of these phenomena can be assessed.

In addition to social factors affecting the release of these compounds that influence maturation, daylength and food resources can affect chemosignal release in house mice. Shorter daylengths and food restriction result in the release of delay chemosignal in the urine of grouped females and the lack of the puberty-accelerating chemosignal in male urine (Drickamer 1984b,d). In house mice, male urine does not accelerate reproduction during winter and urine from grouped females does not delay reproduction during summer. Taken together, these studies suggest that for female house mice, variation in the timing of puberty and reproduction is strongly influenced by a suite of factors that convey information about the adequacy of conditions for reproduction. If, as seems probable from the field tests on house mice, there are fitness consequences stemming from the influence of urinary chemosignals, then investigations of the underlying mechanisms are warranted.

A possible adaptive significance of the Bruce effect has been proposed in several reviews (e.g., Schwagmeyer 1979; Labov 1981b; Storey 1986a). The initial idea was that terminating a pregnancy because of the presence of a strange male (who has presumably displaced the stud male) is advantageous for the female in that she does not carry a litter to term that could be destroyed by infanticide. This may be true for some species. However, another idea that has gained credence is that pregnancy termination is a form of postfertilization mate selection (Dawkins 1976). Females could gain in terms of the fitness of progeny if they mate with a more aggressive and dominant male (the one who has displaced the stud male). This hypothesis may have some validity for some species with early termination of pregnancy, but for vole species, in which the Bruce effect can occur as late as day 17 of gestation, loss of a pregnancy would be a huge cost to the female. The processes involved in pregnancy termination can also be, in part, functions of other factors, such as the aggressiveness of both the female and the male(s) involved (Storey 1990). Alternatively, if the two field studies to test this phenomenon are correct and the Bruce effect does not occur in natural populations, then hypothesizing some functional significance to this response is superfluous. As de la Maza et al. (1999) and Mahady and Wolff (2002) argued, a rodent with a short lifespan cannot afford to waste pregnancies every time a strange male passes by.

It is noteworthy that pregnancy termination in house mice occurs only in the period prior to implantation of the fetuses. Since implantation occurs by about day 5 or 6 of the 19–20 day gestation period, it is evident that there is a roughly 67–75% chance that the embryos have already implanted if a male takeover effect occurs at random with respect to the stage of gestation for that female. It would be quite interesting to determine, for those species of small rodents where there is a process by which males replace each other in terms of holding particular spatial areas, whether these shifts in males take place randomly or within particular periods of time after insemination. Further, one could postulate that in those species that are polygynous, there is some likelihood that a male will depart an area in search of additional mates after a given female is inseminated. In most species of voles, males wander and do not guard their mates; consequently, strange males are not deterred from entering females' territories. It seems unlikely that females in these species would abort pregnancies following exposure to strange males. The association between mate guarding and the Bruce effect warrants further theoretical and empirical testing.

Future Explorations of Rodent Chemosignals and Reproduction

Having examined what is known about three general types of urinary chemosignals influencing rodent reproduction, I now provide some thoughts and speculations on what sorts of investigations should occur in the future. What types of findings will enable us to better understand the ecological and mechanistic aspects of these phenomena?

Comparative approach

Much of what we know in detail concerns only the house mouse, with bits and pieces of information from several other taxa, primarily *Microtus* and *Peromyscus*. Even for house mice, the complete picture regarding release and possible effects of urinary chemosignals on puberty and reproduction and/or how a reproductive response affects lifetime reproductive success has not been obtained. No tests have been conducted under any type of field conditions of urinary chemosignals from estrous, pregnant, or lactating females, all of which accelerate puberty in laboratory tests. We have no real information on the possible occurrence of the Bruce effect under natural conditions. In fact, no comparative studies have been conducted of commensal house mice that have been living in association with humans in dwellings and other human-made structures, or with feral house mice that are free-living in fields and other natural settings. All of these areas need further investigation in the species in which we know the most about the urinary chemosignals that influence reproduction. Gaps in our current understanding of these phenomena are obvious from table 9.1.

An important avenue of research would be to systematically test the effects of several urinary chemosignals on reproduction under a set of standard conditions in a laboratory, and then in natural settings consistent with the normal socioecology of each species. This experimental regime could be accomplished with a range of vole or deer mouse taxa. Large-scale studies involving several species with different social systems and from a variety of ecological settings is necessary to provide sufficient data to test the various phenomena and hypotheses for the adaptive significance of chemosignals. Given that olfactory cues are extremely important for most mammals, a series of a priori hypothesis-testing experiments with a variety of species from different trophic levels and natural histories should provide greater insight into the roles of chemosignals in mammalian societies.

Field tests and ecological issues

One of the key elements that is needed to advance our knowledge of the possible effects of urinary chemosignals on reproduction in small rodents is field experimentation. The theoretical and experimental approaches used by Wolff and colleagues (Rodd and Boonstra 1988; de la Maza et al. 1999; Mahady and Wolff 2002) were designed specifically to discern among alternative hypotheses for some behavioral phenomenon. Research must go beyond observation and description and address current paradigms experimentally to determine the functional and evolutionary significance of each trait. We do need, as advocated previously, to proceed with studies that provide comparative evidence for or against the occurrence of the chemosignal effects on reproduction in rodents, but those alone will not suffice.

Application of chemosignal treatments to experimental manipulations of density, reproductive condition, and sex ratios or age composition can be used to test various hypotheses for the functional and adaptive significance of olfactory communication. It may be possible to use chemical treatments to create animals that are incapable of producing and releasing certain chemosignals. This manipulation could provide the capacity to change densities without altering the amount of signal present. Castrated males in natural populations are a probable manipulation, to test hypotheses about density or male presence, but without chemosignal production and release.

Mechanism studies

Studies examining the underlying neural and hormonal mechanisms that regulate production and response to urinary chemosignals are in their infancy. On the hormonal side, measurements have been made of changes in blood levels of estrogens, LH, prolactin, and other hormones in association with some of the chemosignal-mediated phenomena (reviewed by Bronson and Macmillan 1983; Mc-Clintock 1983; Bronson 1989). While most of this hormone work has been done on house mice, some findings were reported for voles (e.g., Carter et al. 1986). There is ample room for additional investigations of the hormonal bases for production and release of urinary chemosignals,

as well as additional work on the mechanisms of action in recipients. Modern technologies can accelerate analyses of hormone levels. Here again, larger-scale comparative studies are necessary.

As noted earlier in this review, some findings are now available concerning the chemical composition of the various urinary chemosignals (Nishimura et al. 1989; Novotny et al. 1998, 1999; Mucignat-Caretta et al. 1995). Here also, modern technology should facilitate investigations via both better chemical analyses and rapid throughput of samples. We know virtually nothing about the chemical makeup of the chemosignals in voles or deer mice. Until such chemical information is available, we cannot construct any comparative analyses to examine questions about ecology, mating systems, or common evolutionary pathways for urinary chemosignals.

Developments in the past 20 years now make it possible to explore the underlying sensory and neural mechanisms in both recipients of chemosignal cues and also in, for example, grouped females, where estrous cycles or estrous induction are prolonged or suppressed, and where grouping results in the release of a puberty-delaying substance in the urine. One such technique, FOS immunoreactivity, has been used to examine the locations in the brain where the messages from chemical cues in urine or changes in photoperiod are sorted in prairie voles (Moffatt et al. 1995; Tubbiola and Wysocki 1997). This technique, and other immunocytochemistry procedures, can now provide specifics on the neural pathways and regions of the brain involved in the reception and processing of external cues. It should be possible to trace complete pathways for these primer chemosignals from external receptors via central nervous system processing to hormonal mechanisms, altering, for example, puberty.

Summary

Investigations of effects of chemosignals on rodent reproduction over almost 50 years fall into three main categories: those that accelerate puberty in young females, those that delay puberty in young females, and the termination of pregnancy following exposure to a strange male. These reproductive effects in response to exposure to chemosignals have been explored most extensively for house mice, but deer mice, voles, lemmings, and gerbils have also been used to study some aspects of these phenomena. While considerable information is known about chemosignals and reproduction in the laboratory for several species, sufficient information is lacking to provide a basis for comparisons across species, ecological settings, or types of mating systems. The relevance of these phenomena in natural populations of rodents is also unclear. Future work should focus on filling gaps in our knowledge about urinary chemosignals in a number of species of deer mice and voles so that comparative studies can test ecological and evolutionary hypotheses; experimental manipulations to confirm or reject predictions of various hypotheses for the functional significance of chemosignals under field conditions; and using modern techniques to thoroughly test the underlying mechanisms with regard to physiological and neural bases of these phenomena. Lastly and importantly, the production and response to chemosignals must be placed in an adaptationist paradigm such that they can be examined from an evolutionary perspective.

Chapter 10 Sexual Size Dimorphism in Rodents

Albrecht I. Schulte-Hostedde

Body size is the result of both natural and sexual selection, and may influence the demography and life history of vertebrates (Sauer and Slade 1988; Roff 1992). In particular, intraspecific variation in body size may have consequences for individual reproductive success and survival. For example, large body size is expected to enhance reproductive success of males through increased success at acquiring mates (Alexander et al. 1979; Clutton-Brock et al. 1988). For females, large size is associated with increased offspring survival through higher-quality maternal care (Ralls 1976). Selection for small body size can occur due to costs such as the time and energy required to support a large body, and the risks of predation, parasitism, or starvation associated with the rapid growth or lengthy development time required to reach large size (Blanckenhorn 2000). These costs can act to counterbalance the benefits of large size.

Sexual size dimorphism (hereafter referred to as simply sexual dimorphism) is defined as any difference in body size between males and females of the same species. Sex differences in the relationship between body size and fitness ultimately lead to different body size optima being favored by natural and sexual selection (Price 1984; Greenwood and Adams 1987; Hedrick and Temeles 1989; Andersson 1994; Blanckenhorn 2000). The sum of selection pressures acting on males and females therefore dictates the direction and magnitude of sexual dimorphism. These selection pressures include niche differentiation between the sexes, fecundity selection, intrasexual selection (male-male competition), and intersexual selection (female mate choice: Andersson 1994).

Underlying these selection pressures are genetic correlations between the sexes that may retard male and female body size from evolving apart (Lande 1980). These correlations can be powerful enough to prevent sexual dimorphism from evolving, despite differential selection for male and female body size (Merilä et al. 1998). Additionally, ecological constraints associated with locomotion and other habits may also limit the degree to which sex-specific selective forces such as sexual selection lead to sex differences in body size. Thus explanations for the evolution and maintenance of sexual dimorphism must not only consider why one sex is larger, but also why the other sex is smaller.

Rodents comprise the largest order of mammals, with over 2,000 species (Nowak 1991), and as such display a diversity of sizes, shapes, and, of particular interest, differences between the sexes. Thus rodents are an excellent phylogenetic group with which to study the selection pressures on body size that result in sexual dimorphism. The objectives of this chapter are to (1) outline several issues related to our understanding of how both body size and sex differences in body size evolve, especially of rodents (see table 10.1), (2) describe interspecific patterns of variation in sexual dimorphism of rodents, (3) examine geographic patterns of intraspecific variation in sexual dimorphism in *Tamias* spp. with respect to climatic variables, and finally (4) develop hypotheses and approaches to the study of the evolution and maintenance of sexual dimorphism, specifically using rodents.

Sexual dimorphism in mammals has been extensively studied, especially in the context of the male-biased sexual

Table 10.1 Sex-specific selective pressures that contribute to the evolution of monomorphism and sexual size dimorphism in rodents

Direction of advantage	Selective pressure	Example	Reference
Large female size	Higher fecundity	Deer mice	Myers and Masters 1983
	Better parental care	—	—
	Dominance in contests over resources	Southern flying squirrel	Madden 1974
Small female size	Early maturation and faster generation times	—	—
	Lower energetic demands for maintenance, and more efficient shunting of energy to reproduction	Yellow-pine chipmunk (?)	Schulte-Hostedde et al. 2002
Large male size	Male-male combat over females	Arctic ground squirrel	Lacey and Wieczorek 2001
Small male size	Success at scramble competition (competition in which manoeuvrability is important)	13-lined ground squirrel	Schwagmeyer 1988b
	Early maturation with more rapid reproduction	—	—

dimorphism prevalent in pinnipeds and ungulates (e.g., Alexander et al. 1979; Weckerly 1998; Loison et al. 1999), but relatively little attention has been paid to the evolution and maintenance of sexual dimorphism in rodents (but see Bondrup-Nielsen and Ims 1990; Yoccoz and Mesnager 1998). This lack of attention may be the result of the relatively subtle size differences between males and females that occur in most rodents—however, there is evidence of both male- and female-biased sexual dimorphism, a pattern that is uncommon among other mammalian orders (Ralls 1977).

Mammals are generally polygynous (Clutton-Brock 1989b), yet the mating systems of rodents are highly variable, ranging from monogamy (prairie voles [*Microtus ochrogaster;* Getz et al. 1993], California mice [*Microtus californicus;* Ribble 1991]), to polygynandry (deer mice [*Peromyscus maniculatus;* Ribble and Millar 1996], yellow-pine chipmunks [*Tamias amoenus;* Schulte-Hostedde 2004]). Sexual dimorphism is predicted to be male-biased in those species that have intense male-male competition for mates, especially if combat takes place on the ground, rather than in arboreal or aerial environments (Alexander et al. 1979; Andersson 1994). This prediction has borne true in North American voles; males had significantly longer bodies than females in polygynous species (*Microtus californicus, M. oeconomus, M. xanthognathus;* Heske and Ostfeld 1990). Nonetheless, this pattern is not universal. For example, the bushy-tailed woodrat (*Neotoma cinerea*) is highly dimorphic (males weigh approximately 30% more than females; Schulte-Hostedde et al. 2001), yet variation in male and female reproductive success is equal, and there is no genetic evidence that woodrats are polygynous (Topping and Millar 1998; 1999). Thus despite strong evidence that intense male-male competition is associated with male-biased dimorphism, this explanation cannot be universal.

Explanations for why females are larger than males tend to be more complex than explaining male-biased sexual dimorphism in rodents. Most examples of female-biased sex-

ual dimorphism are explained by the fecundity advantage afforded to large females (Andersson 1994). Indeed, in many oviparous taxa such as insects, fish, and reptiles, females are often larger than males (Andersson 1994). However there is little, if any, evidence that fecundity is correlated with body size in rodents (but see Myers and Master 1983; Dobson and Michener 1992). Ralls (1976) hypothesized that larger females were better mothers with respect to parental care, and thus selection should favor larger females. There have been few studies of female-biased sexual dimorphism in mammals, but work on rodents has suggested that the best approach to understanding the evolution of female-biased sexual dimorphism is to consider the selective pressures on both sexes (Bondrup-Nielsen and Ims 1990; Schulte-Hostedde et al. 2002). With the advent of molecular techniques for the assignment of paternity, it is possible to quantify the fitness components of both sexes, such as lifetime reproductive success, allowing sex-specific patterns to be ascertained. It is becoming clear from studies on other taxa that this approach can be fruitful when testing hypotheses related to the evolution and maintenance of sexual dimorphism (Preziosi and Fairbairn 2000; Przybylo et al. 2000).

An important consideration when testing hypotheses related to body size and sexual dimorphism is the definition of body size. Body size can be defined as the magnitude of an individual's physical structure, and two measures of body size are often used—body mass and skeletal size (e.g., Boonstra et al. 1993; Ostfeld and Heske 1993). The interpretation of intraspecific variation in body mass as an index of body size can be compromised, particularly when size dimorphism is small, because body mass can vary for two reasons. First, variation in skeletal structure may lead to large structural size and a concomitant increase in mass. Second, variation in fat reserves or muscle mass may lead to sex differences in body mass. Under the latter scenario, any observed sexual dimorphism would be due to differences in body composition rather than size. Indeed, male rodents tend to have more muscle mass than females, and thus, at the same structural size, males are heavier than females (Schulte-Hostedde et al. 2001). Additionally, female rodents gain mass during reproduction; therefore the use of mass as an index of sexual dimorphism is inappropriate during the reproductive season. The measurement of structural size components such as body length or skull dimensions is an appropriate alternative; however, it is critical that these components are measured in a repeatable fashion (Bailey and Byrnes 1990; Lougheed et al. 1991). The use of a multivariate index of body size from a factor analysis is preferred over univariate measures because the use of multiple size components in a composite index of size is more likely to accurately reflect overall structural size than a single component (Green 2001). The most appropriate index of overall body size is therefore a multivariate estimate, but when this is not possible a univariate measure of structural size, such as body length, is preferred. The use of an index of structural size avoids the problems associated with body mass.

Hand in hand with issues related to the definition of body size is how to best describe sexual dimorphism. The calculation of a dimorphism ratio is complicated by the misgivings associated with the use of ratios (Atchley et al. 1976), yet ratios have high intuitive value. Both the direction and degree of dimorphism are contained within a ratio index of size dimorphism, without the need to refer to an equation (Lovich and Gibbons 1992). Ratios for multivariate indices of body size are difficult to calculate, because such "factor scores" from factor analyses are not on a ratio scale. However, using the eigenvectors from the factor analysis to calculate the factor score of an animal of zero size, and adding the absolute value of this score to the factor scores of each individual can provide an alternative (Slattery and Alisauskas 1995). Recently, it has been suggested that the arguments against the use of ratios do not preclude their use when studying sexual dimorphism (Smith 1999), and so ratios may provide the most intuitive descriptor of the magnitude of any sex differences in body size. Nonetheless, alternative methods of analysing sexual dimorphism among species or groups have been proposed, including using residuals from a regression between the body size of one sex and the body size of the other (Ranta et al. 1994) as an index of the size of one sex relative to the size of the other sex.

Patterns of Sexual Size Dimorphism in Rodents

Evidence from some genera of rodents indicates that there is substantial variation in sexual dimorphism (Bondrup-Nielsen and Ims 1990; Heske and Ostfeld 1990; Levenson 1990; Yoccoz and Mesnager 1998). There has not been a comprehensive examination of sex differences in body size across a broad taxonomic range of rodent species.

I compiled data from published sources on structural size (body length) and/or body mass of males and females for a number of rodent species. I collected data on body mass despite misgivings regarding this metric of body size because it was often the only index of size provided. I did not compile data on body size components such as skull and pelvic characters (e.g., Lammers et al. 2001). Where geographic differences in sexual size dimorphism existed, I arbitrarily present the data from only one population. Patterns of monomorphism and sexual dimorphism are reported for 172 species of rodents (tables 10.2 and 10.3).

Male-biased sexual dimorphism

There are several broad patterns that emerge from the compiled data on sexual dimorphism among rodents. Dimorphism ratios associated with body mass tend to be higher than ratios based on body size, likely because males carry more muscle than females of the same size (Schulte-Hostedde et al. 2001). The predominant pattern among species in which dimorphism occurs is that of male-biased sexual dimorphism. In the overwhelming number of species males are larger than females, and this difference is most pronounced among the ground squirrels. The large degree of male-biased size dimorphism is likely the result of sexual selection, because the mating system of many ground squirrels involves polygyny and male-male competition (e.g., Davis and Murie 1985; Lacey and Wieczorek 2001; Hoogland 2003b). Similar patterns of male-biased dimorphism are found among desert rodents (the Heteromyidae) and fossorial rodents (e.g., the Geomyidae and the Bathyergidae). Competition occurs among male kangaroo rats (Randall 1991a; Randall et al. 2002), which may favor large male body size and the evolution of male-biased sexual dimorphism. Interestingly, males often compete through "foot-drumming" (Randall 1997), and it is not clear how patterns of foot-drumming are related to individual male size. Fossorial rodents, including solitary mole-rats and pocket gophers, have high levels of male-biased sexual dimorphism. Males may be significantly larger than females in many fossorial rodents because of the highly aggressive and xenophobic nature of intraspecific interactions (Bennett et al. 2000). The evolution of large male size and male-biased sexual dimorphism may be related to the high degree of male-male combat that occurs. The mating system of fossorial rodents has been characterized as polygynous, in which males mate with multiple females; however, it is unclear whether competition among males occurs through

Table 10.2 Common and scientific names, male and female mean body size, standard deviation (when available) and dimorphism ratio (female:male) for 49 species of the Sciuridae

Common name	Species	n (f)	n (m)	Size trait	Male	SD	Female	SD	f:m	Reference
Southern flying squirrel	Glaucomys volans	90	71	body length (mm)	125.8	5.2	129	6.4	1.03	Robins et al. 2000
Yellow-pine chipmunk	Tamias amoenus	57	37	body length (mm)	122.5	3.4	127.4	3.8	1.04	Schulte-Hostedde et al. 2002
Alpine chipmunk	T. alpinus	28	13	body length (mm)	103.9	4.3	107.3	3.5	1.03	Levenson 1990
Gray-footed chipmunk	T. canipes	21	28	body length (mm)	128.2	5.6	131.1	6.6	1.02	Levenson 1990
Gray-collared chipmunk	T. cinereicollis	17	10	body length (mm)	128.2	5.3	133.2	4.7	1.04	Levenson 1990
Cliff chipmunk	T. dorsalis	70	51	body length (mm)	123.6	8	127.5	10.1	1.03	Levenson 1990
Least chipmunk	T. minimus	204	163	body length (mm)	109.5	7.1	114.6	8.7	1.05	Levenson 1990
Long-eared chipmunk	T. quadrimaculatus	10	10	body length (mm)	131.8	5.6	138.7	6.5	1.05	Levenson 1990
Colorado chipmunk	T. quadrivittatus	21	18	body length (mm)	123.8	9	131.3	7.8	1.06	Levenson 1990
Red-tailed chipmunk	T. ruficaudus	28	24	body length (mm)	122.2	4.3	126.7	5.8	1.04	Levenson 1990
Lodgepole chipmunk	T. speciosus	16	25	body length (mm)	120.2	5.5	127.8	6.5	1.06	Levenson 1990
Townsend's chipmunk	T. townsendii	143	110	body length (mm)	141.9	8.8	146.5	7.6	1.03	Levenson 1990
Uinta chipmunk	T. umbrinus	63	78	body length (mm)	121.8	6.8	125.7	6.2	1.03	Levenson 1990
Buller's chipmunk	T. bulleri	8	11	body length (mm)	132.6	7.6	137.1	5.2	1.03	Levenson 1990
Durango chipmunk	T. durangae	17	18	body length (mm)	135.0	7.8	134.6	8.2	1.00	Levenson 1990
Merriam's chipmunk	T. merriami	40	24	body length (mm)	132.6	5.0	135.6	9.0	1.02	Levenson 1990
California chipmunk	T. obscurus	17	8	body length (mm)	125.2	5.1	128.8	4.8	1.03	Levenson 1990
Palmer's chipmunk	T. palmeri	13	16	body length (mm)	127.0	5.3	126.1	4.2	0.99	Levenson 1990
Panamint chipmunk	T. panamintinus	28	23	body length (mm)	117.9	4.6	118.6	5.4	1.01	Levenson 1990
Siberian chipmunks	T. sibiricus	29	33	body length (mm)	150.2	7.1	149.9	8.9	1.00	Levenson 1990
Sonoma chipmunk	T. sonomae	14	12	body length (mm)	136.0	5.3	140.0	7.0	1.03	Levenson 1990
Eastern chipmunk	T. striatus	45	46	body length (mm)	146.7	7.9	147.6	10.2	1.01	Levenson 1990
North American red squirrel	Tamiasciurus hudsonicus	1075	1231	body mass (g)	265	18.7	251.1	19.5	0.95	S. Boutin, personal communication
Western grey squirrel	Sciurus griseus	—	—	body mass (g)	750		960		1.28	Heaney 1984
Abert's squirrel	S. aberti	—	—	body mass (g)	589		602		1.02	Heaney 1984
Eastern grey squirrel	S. carolinensis	—	—	body mass (g)	593		593		1	Heaney 1984
Red-tailed squrirel	S. granatensis	—	—	body mass (g)	464		440		0.95	Heaney 1984
Fox squirrel	S. niger	—	—	body mass (g)	690		680		0.99	Heaney 1984
Vancouver marmot	Marmota vancouverensis	12	6	total length (mm)	695	32.8	661	39.5	0.95	Nagorsen 1987
Yellow-bellied marmot	M. flaviventris	61	38	body mass (g)	3900	43.8	2800	48.4	0.72	Armitage et al. 1976
Olympic marmot	M. olympus	—	—	body mass (g)	1900		1400		0.74	Armitage 1981
Woodchuck	M. monax	—	—	body mass (g)	3100		3080		0.99	Armitage 1981
Townsend's ground squirrel	Spermophilus townsendii	11	7	body mass (g)	259.8	16.4	173.4	22.6	0.67	Rickart 1982
Richardson's ground squirrel	S. richardsonii	22	13	body mass (g)	363	42.2	218	36.1	0.60	Michener 1984
Thirteen-lined ground squirrel	S. tridecmlineatus	—	—	body mass (g)	135	—	113	—	0.84	Armitage 1981
Round-tailed ground squirrel	S. tereticaudus	—	—	body mass (g)	145	—	100	—	0.69	Armitage 1981
Wyoming ground squirrel	S. elegans	—	—	body mass (g)	266	—	203	—	0.76	Armitage 1981
Uinta ground squirrel	S. armatus	—	—	body mass (g)	333	—	266	—	0.80	Armitage 1981
Franklin's ground squirrel	S. franklinii	—	—	body mass (g)	360	—	280	—	0.78	Armitage 1981
Columbian ground squirrel	S. columbianus	—	—	body mass (g)	492	—	270	—	0.55	Armitage 1981
California ground squirrel	S. beecheyi	—	—	body mass (g)	650	—	500	—	0.77	Armitage 1981
Arctic ground squirrel	S. parryii	—	—	body mass (g)	700	—	635	—	0.91	Armitage 1981
Golden-mantled ground squirrel	S. lateralis	—	—	body mass (g)	155	—	130	—	0.84	Armitage 1981
Belding's ground squirrel	S. beldingi	—	—	body mass (g)	220	—	218	—	0.99	Armitage 1981

Table 10.2   (continued)

Common name	Species	n (f)	n (m)	Size trait	Male	SD	Female	SD	f:m	Reference
Columbian ground squirrel	*S. columbianus*	19	7	body mass (g)	459	40	393	52	0.86	Dobson 1992
White-tailed prairie dog	*Cynomys leucurus*	17	28	body mass (g)	790.5	140	579.2	115	0.73	J.L. Hoogland, personal communication
Black-tailed prairie dog	*C. ludovicianus*	933	481	body mass (g)	727.0	94.1	692.4	94.5	0.95	J.L. Hoogland, personal communication
Gunnison's prairie dog	*C. gunnisoni*	971	903	body mass (g)	598.7	139	456.1	101	0.76	J.L. Hoogland, personal communication
Utah prairie dog	*C. parvidens*	363	286	body mass (g)	757.1	156	594.7	102	0.78	J.L. Hoogland, personal communication

NOTES: When available, sample sizes are included (f = female, m = male).

combat, sperm competition (multiple mating by females; Lacey 2000), or both. Thus the logistic difficulties of studying animals that live almost exclusively underground need to be overcome to facilitate the study of mating systems and patterns of sexual dimorphism in mole-rats and pocket gophers.

Female-biased sexual dimorphism

There are two groups in which female-biased sexual dimorphism is prevalent: the chipmunks (*Tamias* spp.) and jumping mice (*Napaeozapus insignis* and *Zapus hudsonius*). Why this pattern exists is unclear, but there is evidence that in the yellow-pine chipmunk, female-biased dimorphism may be the result of the lack of an effect of body size on male reproductive success coupled with selection for large female size with respect to both reproductive success and survival (Schulte-Hostedde et al. 2002). The mating system of most chipmunks is similar to that which is found among tree squirrels (Koprowski, chap. 7, this volume). Females become estrus in early spring and advertise their receptive state by particular vocalizations. On the day of her estrus, males aggregate around the female's burrow and chase her when she emerges (Callahan 1981). In yellow-pine chipmunks, this mating chase results in multiple paternity of the offspring, with no advantage for large males (Schulte-Hostedde et al. 2002; Schulte-Hostedde 2004). However, males with large testes, who are presumably superior at sperm competition, tend to sire more offspring than males with small testes (Schulte-Hostedde and Millar 2004).

In jumping mice, both the mating system and the relationship between body size and fitness components are unknown. Male jumping mice emerge from hibernation earlier than females (Ovaska and Herman 1988), which is consistent with ground squirrel emergence schedules, when males emerge early to compete for access to emerging females (Michener 1983b). Any inferences about the evolution of female-biased sexual dimorphism in jumping mice are hampered by a lack of information about their basic breeding patterns.

Mixed patterns of sexual dimorphism

Perhaps the most intriguing group of rodents with respect to sexual dimorphism are the voles (*Clethrionomys* and *Microtus*), because sexual dimorphism varies from female-biased to male-biased dimorphism, even among populations of the same species (Yoccoz and Mesnager 1998). This variation has stimulated much of the research on sexual dimorphism in rodents (Bondrup-Nielsen and Ims 1990; Ostfeld and Heske 1993; Yoccoz and Mesnager 1998). Alternating patterns of sexual size dimorphism may exist among voles for a number of reasons, including sex-specific variation in survival rates (Yoccoz and Mesnager 1998). Female bank voles (*Clethrionomys glareolus*) are typically larger than males, yet some alpine populations show male-biased sexual dimorphism (Yoccoz and Mesnager 1998). One hypothesis for this switch in dimorphism is that the evolution of male-biased sexual dimorphism in alpine populations of bank voles is the result of a higher survival rate. This higher survival rate could be due to an absence of predators (weasels) that selects for an increase in somatic mass and a decrease in reproductive effort (Yoccoz and Mesnager 1998). Intense sexual selection occurs when the larger males within these alpine populations compete for mates, ultimately leading to the evolution of male-biased sexual dimorphism. Other explanations are related to variation in male reproductive tactics and how these tactics might be

Table 10.3 Common and scientific names, male and female mean body size, standard deviation (when available) and dimorphism ratio (female:male) for 123 species of rodents from the Erethizontidae, Muridae, Zapodidae, Heteromyidae, Ctenomyidae, Octodontidae, Bathyergidae, Geomyidae, and Chinchillidae

Common name	Species	n (f)	n (m)	Size trait	Male	SD	Female	SD	f:m	Reference
Erethizontidae										
North American porcupine	Erethizon dorsatum	11	12	body mass (kg)	10.2	1.25	8.54	0.93	0.84	Sweitzer and Berger 1997; Sweitzer, personal communication
				vent length (cm)	78	3.46	70.9	2.98	0.91	Sweitzer and Berger 1997; Sweitzer, personal communication
Muridae										
Beach vole	Microtus breweri	25	23	body length (mm)	129.1	6.6	123.3	5.9	0.96	Heske and Ostfeld 1990
Field vole	M. agrestris	—	—	body mass (g)	30.8	—	25.9	—	0.84	Bondrup-Nielsen and Ims 1990
California vole	M. californicus aesterinus	15	16	body length (mm)	142.2	4.9	135.9	5.1	0.96	Heske and Ostfeld 1990
Long-tailed vole	M. longicaudus	6	4	body length (mm)	175	—	172	—	0.98	Smolen and Keller 1987
Singing vole	M. miurus murei	44	38	body length (mm)	123.3	4.9	115.9	5.7	0.94	Heske and Ostfeld 1990
Water vole	M. richardsoni	142	86	body mass (g)	113.7	15.8	98.9	15.7	0.87	Ludwig 1984
				body length (mm)	159	7.1	150.9	8.6	0.95	Ludwig 1984
Root vole	M. oeconomus gilmorei	42	37	body length (mm)	118.9	8.6	113.6	7.1	0.96	Heske and Ostfeld 1990
	M. oeconomus macfarlani	21	18	body mass (g)	29	5.5	24.4	5.4	0.84	Boonstra et al. 1993
Yellow-cheeked vole	M. xanthognathus	14	34	body length (mm)	159.1	10.9	153.1	8.4	0.96	Heske and Ostfeld 1990
Meadow vole	M. pennsylvanicus	43	40	body mass (g)	28.7	5.7	25.8	5.9	0.90	Boonstra et al. 1993
Woodland vole	M. pinetorum	13	5	body length (mm)	117.8	—	122.9	—	1.04	Smolen 1981
Townsend's vole	M. townsendii	58	70	body mass (g)	53.9	7.9	40.9	6.8	0.76	Boonstra et al. 1993
Northern red-backed vole	Clethrionomys rutilus dawsoni	10	11	body length (mm)	100.3	3.6	107.6	8.1	1.07	Heske and Ostfeld 1990
	C. rutilus glacialis	31	21	body length (mm)	110.9	3.9	114.5	4.3	1.03	Heske and Ostfeld 1990
	C. rutilus dawsoni	220	169	body mass (g)	24.4	3.2	23.4	5	0.96	Boonstra et al. 1993
Red-backed vole	C. gapperi	20	66	body length	97.7	0.8	98.0	1.6		Schulte-Hostedde et al. 2001
Bank vole	C. glareolus	176	70	body mass (g)	25.2	2.2	28	2.5	1.11	Bondrup-Nielsen and Ims 1990
Grey-sided vole	C. rufocanus	—	—	body mass (g)	36.4	—	42.6	—	1.17	Bondrup-Nielsen and Ims 1990
Sagebrush vole	Lagurus curtatus	10	10	body length	128	—	127	—	0.99	Carroll and Geno-ways 1980
Pouched mouse	Saccostomus mearnsi	—	—	body mass (g)	79.5	3.2	62.3	2	0.78	Keesing 1998a
Deer mouse	Peromyscus maniculatus	21	83	body length (mm)	86.5	0.7	88.2	1.6	1.02	Schulte-Hostedde et al. 2001
Volcano mouse	P. alstoni	—	—	body length (mm)	204.4	—	211.1	—	1.03	Williams et al. 1985
Canyon mouse	P. crinitus	81	79	body length (mm)	174.6	—	176.1	—	1.01	Johnson and Armstrong 1987
Yellow-nosed mouse	Abrothrix xanthorhinus	166	194	body length (mm)	80.7	5.2	82.0	7.6	1.02	Lozado et al. 1996
Longtail rice rat	Oryzomys longicaudatus	58	54	body mass (g)	37.7	8.5	27.8	6.3	0.74	Pearson 1983
				body length (mm)	101	7.0	93.2	9.2	0.92	Pearson 1983

Table 10.3 (continued)

Common name	Species	n (f)	n (m)	Size trait	Male	SD	Female	SD	f:m	Reference
South American rock rat	*Aconaemys fuscus*	4	3	body mass (g)	133	—	134	—	1.01	Pearson 1983
				body length (mm)	177	—	179	—	1.01	Pearson 1983
Long-haired grass mouse	*Akodon longipilis*	87	121	body mass (g)	38.2	5.9	36.9	6.2	0.97	Pearson 1983
				body length (mm)	104.9	7.0	105.4	7.6	1.00	Pearson 1983
Olive grass mouse	*Akodon olivaceus*	23	41	body mass (g)	28.2	3.8	25.8	4.8	0.91	Pearson 1983
				body length (mm)	95.6	5.4	93.2	6.1	0.97	Pearson 1983
Bolivian big-eared mouse	*Auliscomys micropus*	25	20	body mass (g)	72.8	12.3	72.6	12.2	1.00	Pearson 1983
				body length (mm)	133.1	9.3	131.5	5.4	0.99	Pearson 1983
Greater long-clawed mouse	*Chelemys macronyx*	11	14	body mass (g)	74.6	11.4	72.0	13.6	0.97	Pearson 1983
				body length (mm)	130.0	7.2	130.8	11.7	1.01	Pearson 1983
Long-clawed mole mouse	*Geoxus valdivianus*	8	15	body mass (g)	31.7	3.2	31.3	4.5	0.99	Pearson 1983
				body length (mm)	101.7	3.2	101.7	6.5	1.00	Pearson 1983
Chilean rat	*Irenomys tarsalis*	13	11	body mass (g)	44.4	12.9	41.9	9.0	0.94	Pearson 1983
Long-tailed mouse	*Pseudomys higginsi*	26	29	body length (mm)	131.1	—	131.7	—	0.99	Driessen and Rose 1999
House mouse	*Mus musculus*	20	20	body length (mm)	79.0	—	77.6	—	0.98	Southern 1977
Harvest mouse	*Micromys minutus*	47	72	body length (mm)	57.0	—	57.1	—	1.00	Southern 1977
Wood mouse	*Apodemus sylvaticus*	20	20	body length (mm)	87.6	—	87.7	—	1.00	Southern 1977
Yellow-necked mouse	*Apodemus flavicollis*	10	22	body length (mm)	105	—	101.6	—	0.97	Southern 1977
Musk-rat	*Ondatra zibethicus*	14	11	body length (mm)	312.7	20.6	312.5	10.1	1.00	Virgl and Messier 1992
Round-tailed muskrat	*Neofiber alleni*	108	52	body mass (g)	279	35.4	262	38.3	0.94	Birkenholz 1963
Spiny mouse	*Acomys cahirinus*	19	14	body mass (g)	43.1	4.4	36.3	4.9	0.84	Khokhlova et al. 2000
Spinifex hopping mouse	*Notomys alexis*	—	—	body mass (g)	27.9	—	33.3	—	1.19	Breed 1983
Sundevall's jird	*Meriones crassus*	30	37	body mass (g)	74.7	15.9	67.4	14	0.90	Khokhlova et al. 2000
Desert woodrat	*Neotoma lepida*	24	16	body length (mm)	174.4	7.8	159.5	6.8	0.91	Hoffmeister 1986
Eastern woodrat	*N. floridana*	27	41	total length (mm)	384.2	19.4	369.8	17.1	0.96	Birney 1973
Southern plains woodrat	*N. micropus*	23	31	total length (mm)	370	22.7	355.8	16.6	0.96	Birney 1973
Sonoran woodrat	*N. phenax*	10	10	body mass (g)	239	—	216	—	0.90	Jones and Genoways 1978
Mexican woodrat	*N. mexicana*	6	8	body length (mm)	333	—	327	—	0.98	Cornely and Baker 1986
White-throated woodrat	*N. albigula*	6	9	body mass (g)	215.6	—	162.5	—	0.75	Macedo and Mares 1988
Bushy-tailed woodrat	*N. cinerea*	40	22	body mass (g)	388	80.3	297.6	37.1	0.77	Schulte-Hostedde et al. 2001
				body length (mm)	234.6	91.3	221.6	31.1	0.94	Schulte-Hostedde et al. 2001
Capromyidae										
Nutria	*Myocastor coypus*	82	99	body mass (g)	670	79	636	116	0.95	Southern 1977
Zapodidae										
Meadow jumping mouse	*Zapus hudsonius*	64	42	body length (mm)	81.4	3.5	84.2	4.5	1.03	Whitaker 1963
Woodland jumping mouse	*Napeozapus insignis*	33	40	body length (mm)	88.8		92.1	—	1.04	Wrigley 1972
Heteromyidae										
Pacific kangaroo rat	*Dipodomys agilis*	1741	1425	body length (mm)	116.1	—	114	—	1.02	Best 1993
Desert kangaroo rat	*D. deserti*	254	204	body length (mm)	141.2	—	135.3	—	0.96	Best 1993
San Quintin kangaroo rat	*D. gravipes*	56	54	body length (mm)	130.6	—	127.1	—	0.97	Best 1993

(continued)

Table 10.3 (continued)

Common name	Species	n (f)	n (m)	Size trait	Male	SD	Female	SD	f:m	Reference
Heermann's kangaroo rat	D. heermanni	474	366	body length (mm)	121.6	—	119.4	—	0.98	Best 1993
Giant kangaroo rat	D. ingens	55	47	body length (mm)	147.6	—	144.5	—	0.98	Best 1993
San Jose island kangaroo rat	D. insularis	9	16	body length (mm)	108.2	—	97.3	—	0.90	Best 1993
Merriam's kangaroo rat	D. merriami	433	397	body length (mm)	100.6	—	99.2	—	0.99	Best 1993
Narrow-faced kangaroo rat	D. venustus	65	73	body length (mm)	128.7	—	122.9	—	0.95	Best 1988
California kangaroo rat	D. californicus	191	150	body length (mm)	119.5	—	120.5	—	1.01	Best 1988
Gulf coast kangaroo rat	D. compactus	48	29	body length (mm)	112.5	—	110.6	—	0.98	Best 1988
Texas kangaroo rat	D. elator	120	86	body length (mm)	124.3	—	124.0	—	1.00	Best 1988
Big-eared kangaroo rat	D. elephantinus	38	32	body length (mm)	129.0	—	128.9	—	1.00	Best 1988
Chisel-toothed kangaroo rat	D. microps	156	174	body length (mm)	113.5	—	111.8	—	0.98	Best 1988
Nelson's kangaroo rat	D. nelsoni	112	87	body length (mm)	128.3	—	127.1	—	0.99	Best 1988
Tipton kangaroo rat	D. nitratoides	276	200	body length (mm)	97.1	—	98.1	—	0.99	Best 1988
Ord's kangaroo rat	D. ordii	691	662	body length (mm)	114.2	—	114.0	—	1.00	Best 1988
Panamint kangaroo rat	D. panamintinus	467	385	body length (mm)	120.2	—	121.1	—	1.01	Best 1988
Phillip's kangaroo rat	D. phillipsii	93	77	body length (mm)	105.0	—	104.2	—	0.99	Best 1988
Banner-tailed kangaroo rat	D. spectabilis	296	232	body length (mm)	142.3	—	142.0	—	1.00	Best 1988
Stephen's kangaroo rat	D. stephensi	81	70	body length (mm)	115.7	—	115.8	—	1.00	Best 1988
Tehachapi pocket mouse	Perognathus alticola	20	20	body length (mm)	77.6	—	72.5	—	0.93	Best 1993
Great Basin pocket mouse	P. parvus	20	20	body length (mm)	83.6	—	78.3	—	0.94	Best 1993
Arizona pocket mouse	P. amplus	20	20	body length (mm)	70.8	—	70.1	—	0.99	Best 1993
Olive-backed pocket mouse	P. fasciatus	20	20	body length (mm)	71.2	—	72.1	—	1.01	Best 1993
Plains pocket mouse	P. flavescens	20	20	body length (mm)	72.1	—	70.4	—	0.98	Best 1993
Silky pocket mouse	P. flavus	20	20	body length (mm)	59.0	—	60.4	—	1.02	Best 1993
San Joaquin pocket mouse	P. inornatus	20	20	body length (mm)	72.6	—	72.1	—	0.99	Best 1993
Little pocket mouse	P. longimembris	20	20	body length (mm)	58.5	SD	56.9	—	0.97	Best 1993
Narrow-skulled pocket mouse	Chaetodipus artus	20	20	body length (mm)	92.1	—	85.9	—	0.93	Best 1993
Little desert pocket mouse	C. arenarius	20	20	body length (mm)	69.5	—	67.3	—	0.97	Best 1993
Bailey's pocket mouse	C. baileyi	20	20	body length (mm)	94.5	—	92.5	—	0.98	Best 1993
California pocket mouse	C. californicus	20	20	body length (mm)	88.5	—	85.3	—	0.96	Best 1993
San Diego pocket mouse	C. fallax	20	20	body length (mm)	84.6	—	82.0	—	0.97	Best 1993
Long-tailed pocket mouse	C. formosus	20	20	body length (mm)	82.7	—	79.0	—	0.96	Best 1993
Goldman's pocket mouse	C. goldmani	20	20	body length (mm)	81.4	—	83.7	—	1.03	Best 1993
Hispid pocket mouse	C. hispidus	20	20	body length (mm)	101.5	—	106.9	—	1.05	Best 1993
Rock pocket mouse	C. intermedius	20	20	body length (mm)	74.0	—	74.1	—	1.00	Best 1993
Line pocket mouse	C. lineatus	20	20	body length (mm)	74.4	—	73.0	—	0.98	Best 1993

Table 10.3 (continued)

Common name	Species	n (f)	n (m)	Size trait	Male	SD	Female	SD	f:m	Reference
Nelson's pocket mouse	C. nelsoni	20	20	body length (mm)	80.7	—	78.4	—	0.97	Best 1993
Desert pocket mouse	C. penicillatus	20	20	body length (mm)	76.6	—	75.8	—	0.99	Best 1993
Sinaloan pocket mouse	C. pernix	20	20	body length (mm)	75.2		68.9	—	0.92	Best 1993
Spiny pocket mouse	C. spinatus	20	20	body length (mm)	85.0	—	82.6	—	0.97	Best 1993
Dark kangaroo mouse	Microdipodops megacephalus	20	20	body length (mm)	65.0	—	66.8	—	1.03	Best 1993
Pale kangaroo mouse	M. pallidus	20	20	body length (mm)	85.0	—	82.6	—	0.97	Best 1993
Southern spiny pocket mouse	Heteromys australis	20	20	body length (mm)	127.5	—	120.3	—	0.94	Best 1993
Mountain spiny pocket mouse	H. oresterus	9	10	body length (mm)	159.4	—	141.1	—	0.89	Best 1993
Forest spiny pocket mouse	H. anomalus	20	20	body length (mm)	133.8	—	128.9	—	0.96	Best 1993
Desmarest's spiny pocket mouse	H. desmarestianus	20	20	body length (mm)	133.1	—	129.7	—	0.97	Best 1993
Gaumer's spiny pocket mouse	H. gaumeri	20	20	body length (mm)	125.3	—	123.4	—	0.98	Best 1993
Goldman's spiny pocket mouse	H. goldmani	20	20	body length (mm)	148.5	—	143.6	—	0.97	Best 1993
Nelson's spiny pocket mouse	H. nelsoni	20	20	body length (mm)	161.5	—	150.6	—	0.93	Best 1993
Mexican spiny pocket mouse	Liomys irroratus	20	20	body length (mm)	125.4	—	118.6	—	0.95	Best 1993
Salvin's spiny pocket mouse	L. salvini	20	20	body length (mm)	114.2	—	107.5	—	0.94	Best 1993
Painted spiny pocket mouse	L. pictus	35	27	total length (mm)	241	12.2	229.7	9.7	0.95	Genoways 1973
Panamanian spiny pocket mouse	L. adspersus	18	6	total length (mm)	265.9	11.2	249.7	7.3	0.94	Genoways 1973
	L. spectabilis	20	20	body length (mm)	109.4	—	101.5	—	0.93	Best 1993
Ctenomyidae										
Tuco-tuco	Ctenomys talarum	110	95	body mass (g)	136.2	18.8	102.3	12.6	0.76	Zenuto et al. 1999
				body length (mm)	145.99	10.7	139.2	9.35	0.95	Zenuto et al. 1999
Octodontidae										
Cururo	Spalacopus cyanus	7	8	body mass (g)	99.3	4.4	80.2	6.4	0.81	Contreras 1986
Bathyergidae										
Giant mole-rat	Cryptomys mechowi	15	18	body mass (g)	345.3	95.4	252	34	0.73	Scharf et al. 2001
Silvery mole-rat	Heliophobius argente-eocinereus	70	74	body mass (g)	190.1	58.1	162.1	47.2	0.85	Sumbera et al. 2003
				body length (mm)	155.3	17.8	148.8	15	0.96	Sumbera et al. 2003
Geomyidae										
Northern pocket gopher	Thomomys talpoides	13	13	body length (mm)	147.5	8.6	139.2	9.8	0.94	Hoffmeister 1986
Southern pocket gopher	T. umbrinus	11	20	body length (mm)	135.9	8.5	128.9	7.6	0.95	Hoffmeister 1986
Desert pocket gopher	Geomys arenarius	10	7	body length (mm)	253	—	233	—	0.92	Williams and Baker 1974
Texas pocket gopher	G. personatus	13	11	body length (mm)	275.0	—	252.9	—	0.92	Williams 1982
Chinchillidae										
Plains viscacha	Lagostomus maximus	65	66	body mass (g)	630	—	400	—	0.63	Jackson et al. 1996
				total length (mm)	753	—	641	—	0.85	Jackson et al. 1996

NOTES: When available, sample sizes are included (f = female, m = male).

size-dependent. For example, variation in patterns of dimorphism in voles may simply be because variation in mating systems occurs with larger males occurring in species and/or populations with intense male-male competition (Heske and Ostfeld 1990). Variation in spacing behaviour can also lead to variation in male mating tactics and size, and therefore can influence the evolution of sexual dimorphism (Bondrup-Nielsen and Ims 1990). To date, no study has incorporated genetic techniques to measure variation in male reproductive success and examine questions related to sexual dimorphism in voles.

Despite these general patterns of sexual dimorphism it is also clear that many species are monomorphic. Few studies have examined sex-specific patterns of selection on body size in rodents, and therefore it is not possible to determine whether the lack of dimorphism is due to a lack of differential selection on male and female body size or because of ecological or allometric constraints that limit the evolution of sexual dimorphism.

Geographic Variation in Sexual Size Dimorphism of the Chipmunks (*Tamias* spp.): Effects of Climate

Geographic variation in sexual size dimorphism has been documented in many taxa, including snakes (Pearson et al. 2002), lizards (Wikelski and Trillmich 1997), and birds (Badyaev et al. 2000), but there are few studies examining this phenomenon in mammals, specifically rodents. Geographic variation in sexual dimorphism has been related to prey size in carnivores (Dayan and Simberloff 1994), but this is unlikely to apply to rodents. Climatic variation can directly affect body size (Smith et al. 1998; Ashton et al. 2000) and the direction and magnitude of sexual dimorphism in mammals (Dobson and Wigginton 1996; Sullivan and Best 1997; Post et al. 1999). Harsh or extreme climates may place selective pressures on body size by affecting individual energy expenditure through factors such as increased thermoregulatory costs due to cold temperatures or increased foraging costs due to lower food availability. Temporal variation in climate has been suggested to affect sexual size dimorphism in tundra voles (*Microtus oeconomus*), a species in which males are heavier than females due to sexual selection, but seem to adjust body mass downward during the winter to counteract the increase in winter mortality associated with large size (Aars and Ims 2002). This process leads to temporal variation in sexual dimorphism.

The chipmunks (*Tamias* spp.) show a strong pattern of female-biased sexual dimorphism—all species with statistically significant differences in body length between the sexes are exclusively female-biased (Levenson 1990). In yellow-pine chipmunks, male body size does not influence reproductive success (Schulte-Hostedde et al. 2002), perhaps because scramble competition among males for mating opportunities favors small male size (Alexander et al. 1979). The variation in sexual dimorphism in this genus has been partly attributed to the severity of the environment, with females tending to be larger in the most severe environments ("severe" environments were those that were at high altitude and/or more northern latitude; Levenson 1990). Because large female yellow-pine chipmunks tended to have lower reproductive success than small females when rainfall was excessive, yet had greater reproductive success than small females when rainfall was average, Schulte-Hostedde et al. (2002) hypothesized that large female chipmunks were at an energetic disadvantage when environmental conditions were harsh because of their inability to meet the metabolic demands of both maintenance and lactation. The fitness advantage that large females experience under average conditions is hypothesized to occur because they are "better mothers" (Ralls 1976); either providing higher-quality parental care by protecting the offspring or higher-quality milk than small females (Schulte-Hostedde et al. 2002). Similar sex-specific responses to climate that lead to variation in sexual dimorphism have been seen in the blue tit (*Parus caerulus*), a small size-dimorphic bird (Blondel et al. 2002). Thus if large females are at an energetic disadvantage when climatic conditions are severe, then female-biased sexual dimorphism should be less pronounced in populations that experience climate extremes, in which selection should favor smaller females.

Using a qualitative assessment based on both latitude and altitude, Levenson (1990) argued that female-biased sexual dimorphism was more pronounced in populations with extreme climatic conditions, in contradiction to the suggestions of Schulte-Hostedde et al. (2002). Levenson (1990) published data on sex differences in body size from 40 populations from 11 species of chipmunks of the genus *Tamias*, almost all from the western United States. Here, I determine whether climate influences the evolution of female-biased sexual dimorphism in chipmunks, and if so, which factors are most important. I obtained the location of capture for the specimens from each of the 40 populations (H. Levenson, pers. comm.), and determined the latitude and longitude of these locations. Monthly climate normals (long-term averages; 1961–1990) were obtained from 3,044 weather stations throughout the continental United States (National Oceanic and Atmospheric Association [NOAA] 1994). For each location of capture, the closest weather station was selected.

To quantify the sex differences in body length among the 40 populations, I used the residuals from the regression of female body length on male body length for each population as an index of female size relative to male size (Ranta

Table 10.4 Description, mean and standard deviation (SD) of the 11 climatic variables used in the analysis of geographic patterns of sexual size dimorphism among 40 chipmunk populations

Variable	Description	Mean	SD
Latitude	degrees and decimal number of minutes	41.7	4.11
January min. temp.	average minimum temperature in January (°C)	−9.4	5.52
Year min. temp	average monthly within-year minimum temperature (°C)	1	4.23
July max. temp.	average maximum temperature in July (°C)	30	4.05
Year max. temp.	average monthly within-year maximum temperature (°C)	16.3	4.74
Year mean temp.	average monthly within-year mean temperature (°C)	8.7	4.37
Max. temp. range	average annual temperature range (max. July–min. Jan.) (°C)	39.4	4.57
Year total precip.	average total yearly precipitation (mm)	483.4	273.3
Max. precip.	average maximum monthly precipitation (mm)	67.2	35.32
Min. precip.	average minimum monthly precipitation (mm)	19.1	17.65
Max. precip. range	average annual precipitation range (max. month–min. month) (mm)	48.1	32.55

et al. 1994). Because these populations are from species that belong to the same genus, phylogenetic correction was not applied to the estimates of sexual size dimorphism. To determine the climatic factors that predict female size relative to male size, I conducted a forward step-wise regression using residual female size as the dependent variable, and 11 climatic and geographic variables as independent variables (table 10.4).

As expected, male and female body lengths were highly correlated (fig. 10.1; $r^2 = 0.744$, $P < 0.001$), and in the majority of populations (32/40), females were larger than males. Based on the step-wise multiple regression ($r^2 = 0.236$, $df = 4,35$, $P = 0.047$), four climatic variables appeared to be important in explaining geographic variation in sexual dimorphism. Average annual range in precipitation was positively related to relative female size ($\beta = 0.50$, partial $r = 0.39$, $P = 0.02$; fig. 10.2a), whereas average yearly total precipitation ($\beta = −0.38$, partial $r = 0.32$, $P = 0.06$; fig. 10.2b), average January minimum temperature ($\beta = −0.33$, partial $r = 0.28$, $P = 0.09$), and latitude ($\beta = −0.28$, partial $r = 0.26$, $P = 0.13$) were negatively related to relative female size. Thus female-biased sexual dimorphism was most pronounced in highly seasonal populations, and least pronounced in populations with high precipitation, cold winters, and at northern latitudes.

These patterns support the hypothesis that in extreme climates, small females have an advantage over large females with respect to reproductive success, leading to a reduction in female size and female-biased sexual dimorphism. How do these results compare with the qualitative assessment offered by Levenson (1990) on the same 40 populations? Levenson (1990) argued that populations that experienced more extreme climate had more pronounced female-biased sexual size dimorphism. This conclusion was based on a series of pair-wise comparisons that indicated that populations that were farther north or at higher elevation had

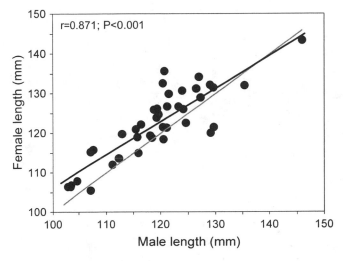

Figure 10.1 Regression between female length and male length for 40 chipmunk populations (*Tamias* spp.; male length = 11.07 + 0.885*[female length]). The gray line represents a slope of 1, in which male and female size would be equal.

higher dimorphism ratios than populations that were farther south or closer to sea level. My results indicate the opposite trend. Populations that experienced high annual rainfall and low January temperatures (i.e., more extreme climatic conditions) had less female-biased sexual size dimorphism than populations that experienced low rainfall and high January minimum temperatures. What mechanism might be at work that would lead to these patterns? If the mating system is assumed to remain the same across populations, and thus male size is likely to be optimized by sexual selection, then the most likely explanation may be an interaction between the severity of climate and the size dependence of female reproductive energetics.

These contrasting results offer a unique opportunity to present two alternative hypotheses for the evolution of female-biased sexual dimorphism in rodents. Levenson (1990)

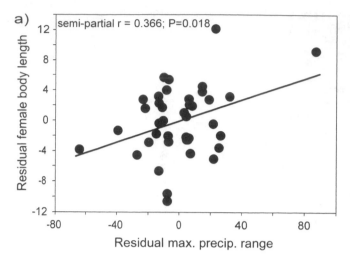

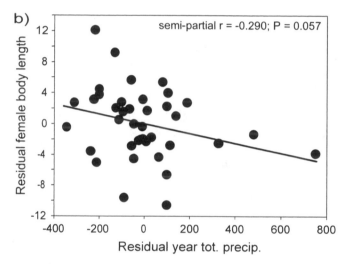

Figure 10.2 Semipartial correlation between residual female body length (female size relative to male size), and (a) residual average monthly within-year maximum temperature and (b) residual total yearly precipitation. Female-biased sexual size dimorphism increases in magnitude with increasing range in within-year maximum temperature, and declines with increased total yearly precipitation.

viding higher-quality parental care (either through higher quantity/quality of milk or better defence of the offspring) than small females (Schulte-Hostedde et al. 2002). This hypothesis predicts that females should be smaller and female-biased dimorphism less pronounced in extreme environments. The patterns I have found in the chipmunks support the latter hypothesis, but further tests on other groups of rodents should be carried out. An appropriate group may be the voles, because they are variable with respect to sexual size dimorphism. Determining the cause of the observed patterns will also require the determination of size-specific energy expenditure patterns during lactation in female rodents, perhaps with the use of doubly labelled water (Speakman 1997).

Sexual Size Dimorphism in Rodents: Developing Hypotheses and Future Directions

The study of the evolution and maintenance of sexual dimorphism is guided by the fundamental principle that selection acts on both sexes simultaneously, and any hypotheses and tests related to sexual dimorphism must take this principle into account. Thus attributing examples of male-biased sexual dimorphism in rodents to a polygynous mating system in which male size is related to mating success, while perhaps accurate, does not appreciate the complexity of the diverse factors that influence selection on body size. Because of the emphasis on male-biased sexual dimorphism, the role of selection on female size is rarely considered, yet any explanation for sexual dimorphism must consider both sexes. The mating dynamics of both sexes and female reproductive energetics can influence the evolution of sexual size dimorphism. In the following I outline these two factors and suggest future directions for the study of sexual dimorphism in rodents.

Although male-biased sexual dimorphism may evolve from a polygynous mating system, in which males compete through combat for access to females (Heske and Ostfeld 1990), other mating systems may also influence selection on body size and the concomitant evolution of sexual dimorphism. In mating systems described as scramble competition polygyny males actively seek females, and competition among males is not related to combat but rather to mobility and agility (Alexander et al. 1979). The 13-lined ground squirrel (*Spermophilus tridecemlineatus*) exhibits a mating system in which males actively seek females during the breeding season, and males that possess traits associated with mobility have high mating success (Schwagmeyer 1988a). Under these circumstances, there is little selection for large male body size, and the magnitude of sexual dimorphism is predicted to be reduced.

argued that hibernation imposed strong selection pressure on female body size, because a minimum body size might be required to survive hibernation and subsequently reproduce. Large females with higher energy reserves might be more likely to meet the costs of hibernation, reproduction, and lactation, especially when food availability is low. Thus Levenson (1990) predicted that large females should be favored in severe or seasonal environments. An alternative hypothesis predicts that large females are at a disadvantage when environmental conditions are extreme because of the energetic costs of reproduction and lactation, coupled with the high costs of somatic maintenance associated with large body size. During average conditions large females, emancipated from the prohibitively high maintenance costs associated with extreme climatic conditions, are capable of pro-

Scramble competition is also often characterized by multiple mating by females, which leads to sperm competition. Female promiscuity and the resulting multiple paternity of litters can reduce variation in male reproductive success relative to mating systems in which males compete through combat. If all males have some probability of siring offspring with a female that mates with multiple males, this will invariably decrease variation in reproductive success relative to situations in which large males can dominate small males and exclude them from any copulations with receptive females. Selection on male body size can therefore be reduced or eliminated by multiple mating by females (Wolff and Macdonald 2004). Evidence for this exists in both interspecific and intraspecific contexts. In polyandrous species of voles, males and females are the same size, and because of multiple mating and sperm competition, males had larger testes than males from polygynous species (Heske and Ostfeld 1990). Intraspecific patterns in the yellow-pine chipmunk indicate that female-biased sexual dimorphism is facilitated by low variation in male reproductive success, which is in part due to (1) scramble competition polygyny, manifested by a "mating chase" by multiple males during the female's estrus and (2) multiple mating by females, which results in high levels of multiple paternity within litters (Schulte-Hostedde 2004). If selection for large male size is relaxed, whether due to sperm competition or advantages associated with speed and agility, other factors become important in predicting the direction of sexual size dimorphism, including selection for large female body size.

Female reproductive energetics can have a profound influence on the fitness consequences of body size and the evolution of sexual dimorphism, particularly in rodents. Rodents are generally income breeders and depend on increasing ingestion rates to maintain the substantial energetic demands of lactation (Millar 1987). They are therefore susceptible to fluctuations in food resources, and these resources are often influenced by climatological variables. Variation in climate and food resources can fundamentally affect selection on female size. Litter size and offspring survival can be dramatically affected by environmental conditions (King et al. 1991; Wauters and Dhondt 1995; Neuhaus et al. 1999; Kalcounis-Rüppell et al. 2002), and these effects can be size specific, particularly if maintenance costs are high for larger females (Schulte-Hostedde et al. 2002). Geographic patterns of sexual dimorphism may in part be explained by the size-dependence of energetic costs of maintenance and reproduction.

Determining how sexual dimorphism is maintained in rodents will benefit from the study of sex-specific patterns of selection on body size in species that are monomorphic, as well as species that have both male-biased and female-biased sexual dimorphism. The application of modern techniques to answer questions related to (1) the role of male and female mating dynamics in determining the intensity of selection on body size, and (2) size-dependence of female reproductive energetics, will aid in finding general patterns and mechanisms. The use of DNA profiling to determine genetic estimates of reproductive success, levels of multiple paternity (e.g., Ribble 1991; Topping and Millar 1998) and thus the determination of the mating system, particularly in species with female-biased sexual dimorphism, should test the generality of conclusions that have been based on species with male-biased sexual dimorphism (Andersson 1994). The effects of female body size on female reproductive energetics and sexual dimorphism are rarely considered, but the use of techniques such of doubly labelled water (Speakman 1997) will help to understand how females of different size respond energetically to climatic variation, especially during the energetically expensive period of lactation. By considering both sexes and examining examples of sexual monomorphism and male-biased and female-biased sexual dimorphism, our understanding of how sexual selection operates on male and female body size in rodents will be greatly enhanced.

Summary

Sexual size dimorphism is the result of sex-specific patterns of selection on body size. These selective pressures include niche differentiation between the sexes, fecundity selection, and, especially in mammals, sexual selection. Underlying these selection pressures are genetic correlations between the sexes that tend to prevent these selective pressures from creating sex differences in body size. Rodents are highly diverse, both with respect to taxonomy and morphology, yet despite this diversity there has been little study of sexual dimorphism in rodents. A variety of selection pressures is expected to influence rodent body size. These pressures vary from allometry, sexual selection and mating systems, to climatic and other environmental influences. Unfortunately, there are few studies of these pressures in rodents, despite remarkable diversity with respect to the direction of sexual dimorphism. A review of sexual dimorphism among rodents indicates that patterns of diversity include monomorphism, male-biased sexual dimorphism, and female-biased dimorphism. Of particular interest are the ground squirrels (Sciuridae) and the kangaroo rats (Heteromyidae; both almost exclusively male-biased), the chipmunks (*Tamias* spp.) and jumping mice (Zapodidae; both almost exclusively female-biased) and the voles (*Clethrionomys* and *Microtus*; populations within species can vary from male-biased to female-biased sexual dimorphism). The selective pressures resulting in these patterns are likely as varied as the rodent taxa themselves. One understudied perspective is the effects of climatic variation on sexual size dimorphism. Evidence indicates that climate can have profound effects on the evolution of body size and sexual dimorphism. Chipmunks generally exhibit female-biased sexual dimorphism, and it has been hypothesized that climate may play a role in the evolution of this pattern of body size. Using climate normals and published data on body size of male and female chipmunks from 40 populations (11 species), I used multiple regression analysis to determine which climatic variables influence the magnitude of sexual dimorphism. The degree of female-biased sexual dimorphism declined with increasing annual rainfall and low winter temperatures, indicating that extreme climatic conditions may influence female size and reproductive energetics. Although much research on mammals has focused on male-biased sexual dimorphism and the role of sexual selection on male size, a more integrated approach examining selection on body size in both sexes will lead to a more complete understanding of the evolution of sexual dimorphism in rodents.

Chapter 11 Facultative Sex Ratio Adjustment

Robert S. Sikes

FISHER (1930) explained why sexually reproducing diploid organisms should invest equally in male and female offspring at the population level based on genetic arguments. Given this general expectation of equal investment, deviations from unity usually are considered noteworthy and beg explanation. Not surprisingly, this area of inquiry has attracted attention from researchers working with diverse species. Although numerous hypotheses that incorporate demographic and life-history parameters have been advanced to explain deviations from unity, consistent support is lacking for most vertebrate taxa (Krackow 2002). Sex ratios are a key element of life history theory (Stearns 1992) and well-deserving of attention because, as West et al. (2002, p. 122) stated, "if we cannot understand sex ratios, we cannot hope to understand most other life history traits, whose evolution usually depends upon far more complex trade-offs."

Most reviews of sex ratio adjustment (Clutton-Brock and Iason 1986; Bull and Charnov 1988; Frank 1990; Cockburn et al. 2002; Hardy 2002; Krackow 2002; West et al. 2002) note little consistency in data regarding available hypotheses. But perhaps our perception of the magnitude of biased sex ratios is overblown. If sex ratios in mammals are in fact normally distributed, then Type I statistical error will still result in a certain percentage of samples that are either female or male biased. If these male- or female-skewed ratios, representing the tails of a distribution, are then reported at a higher than expected frequency simply because they deviate from parity (Festa-Bianchet 1996; Krackow

2002), our interpretation of the frequency of skewed sex ratios becomes biased by the over-reported minority, and we feel compelled to fit these skews with explanations. Publication of sex ratio data that do not differ from unity will help counter this tendency (Duquette and Millar 1998), but only if the data constitute tests of one of the hypotheses for biased ratios. What types of studies, then, constitute appropriate tests, and what types of data are necessary if we are to make headway in eliminating alternative explanations?

Rodents are an especially attractive group for evaluating deviations of sex ratio because obtaining large samples is often feasible and because many species make suitable laboratory models for investigation of mechanisms of adjustment. Rodents typically produce litters rather than single offspring, so sex ratios potentially can be adjusted both at the level of an individual litter as well as at the population level. Further, as litter sizes for at least some taxa may be adjusted by the mother to suit environmental conditions (Slade et al. 1996), the potential exists for easy manipulation of sex ratios as well, by favoring one sex over the other within the context of litter size adjustments. This potential may be countered, however, by the short life expectancies of many species of small rodents in the wild, which in turn limits future reproductive potential. In these cases, if manipulating sex ratios requires eliminating offspring, the limited future reproductive potential of small rodents may provide too little selective advantage to offset the reduction in size of a current litter.

Hypotheses for Facultative Adjustment

Trivers-Willard hypothesis

Trivers and Willard (1973) provided theoretical justification for facultative manipulation of sex ratios by mothers in polygynously breeding mammals under certain conditions. These authors argued that mothers in poor condition or with limited resources should bias their investment toward offspring of the sex with the lowest reproductive variance. Their reasoning was that offspring of mothers in poor condition will be smaller than those of mothers in better condition and, in species in which males compete for access to females, sons from these poor mothers are less likely to breed than daughters. The Trivers-Willard hypothesis was provocative and stimulated much empirical research, as well as additional theoretical treatments on manipulating sex ratios. Follow-up work has extended incentives for parental manipulation of sex ratios to the effects of local competition for resources (Myers 1978; Silk 1983; Johnson 1988), and differential advantage of the sexes to temporal changes in resources over the course of a breeding season (Werren and Charnov 1978; Lambin 1994b,c). Each of these hypotheses focuses on a different aspect of mammalian biology that should be open to selective pressures, though not necessarily in all species.

Let us consider tests for each of the hypotheses for facultative manipulation of sex ratios in rodents in more detail, with an eye toward identifying deficiencies that limit our ability to reject explanations. As the Trivers and Willard (1973) hypothesis set the stage for this area of research, it provides a suitable starting point. The Trivers-Willard (T-W) hypothesis predicts biased investment based upon reproductive potential of offspring as influenced by maternal condition during parental investment. The hypothesis is based on three simplifying assumptions: (1) that the young's condition at the end of parental investment is correlated with the mother's condition during her investment; (2) that differences in body condition of young at the end of parental investment are maintained into adulthood; and (3) that differences in body condition affect males more than females. More specifically, Trivers and Willard predict that "natural selection must favor one or more genes that adjust the sex ratio produced by an adult female to her own condition at the time of parental investment. In taxa such as mammals where males determine sex of offspring, female control of the sex ratio must involve differential mortality by sex, either of sperm cells or of growing young during parental investment" (Trivers and Willard, 1973, p. 91). From this statement and as developed elsewhere in their argument, there are two conditions necessary for a given data

set to be consistent with the T-W hypothesis: first is the existence of differential investment by a mother between sexes, and second is that differential investment is facultative (conditional) and under maternal control. Failure to satisfy both of these conditions undermines support for the hypothesis.

Trivers and Willard's hypothesis is often misinterpreted as applying only to the secondary sex ratio (the sex ratio of a litter at birth) rather than biased investment in general, despite the fact that the authors explicitly include the entire period of postnatal investment in their model (Trivers and Willard 1973, p. 91). This interpretation perhaps is understandable given the title of their paper and the fact that most of their discussion revolves around sex ratios. Their nearly exclusive focus on sex ratios almost certainly stems from the fact that, to evaluate available data relative to their model, they made the assumption "that sex ratio at birth in mammals is a measure of the tendency to invest in one sex more than in the other" (p. 91). Nevertheless, the fundamental requirement is for differential investment between the sexes, which may or may not result in biased sex ratios. A very important consequence of this clarification is that a study need not document skewed sex ratios to support the T-W model.

The second condition of maternal control of investment is equally important, but dependent on the first. Differential allocation of resources by the mother based on maternal condition constitutes facultative adjustment in that allocation "decisions" are optional and conditional. This is a critical requirement because differential mortality or growth caused by sex differences in energy or nutrient requirements of the young would not constitute facultative adjustment and hence provide no support for the T-W hypothesis.

The Trivers-Willard hypothesis has drawn much attention from researchers, but results of studies posed as tests of the model are equivocal. One of the major difficulties in evaluating existing data is that, with few exceptions, they seldom were designed as explicit tests of the hypothesis. Instead, the T-W hypothesis, or some other explanation, is used to interpret skewed sex ratios detected in other work (recall, however, that the Trivers-Willard model is not about sex ratios per se, but differential investment in the sexes, with one potential outcome being biased sex ratios). Further, although Trivers and Willard included a consideration of polytocous species in their discussion, the potential for concurrent manipulations of litter size in such species could complicate interpretations of sex ratio adjustment (see Williams 1979).

Key studies using rodents, which may be used to evaluate the Trivers-Willard hypothesis or its assumptions, are presented in table 11.1. Among these studies, maternal condition or ability to invest in offspring was gauged as a correlate of maternal mass or age (McShea and Madison 1986; Armitage 1987a; Krackow 1997), immediate reproductive history or current litter composition in which male offspring were either assumed to be or were demonstrably more costly to produce than daughters (Gosling 1986; Krackow and Hoeck 1989; Clark et al. 1990; Millesi et al. 1999), or where resources available to the mother were altered in various ways including direct food restriction (Rivers and Crawford 1974; McClure 1981; Labov et al. 1986; Meikle and Drickamer 1986; König 1989b; Meikle and Thornton 1995; Sikes 1995, 1996a, 1996b; Moses et al. 1998; Wright et al 1988; Lamb and Aarde 2001), food augmentation (Duquette and Millar 1998, Trombulak 1991), density reduction to reduce competition (Wright et al. 1988), or separation of young (Clark et al. 1991). Fifteen of these 23 studies report either skewed sex ratios or differential investment in offspring, and hence are consistent with the predictions of the model. The remaining eight show no evidence of differential investment. Of the 15 studies consistent with predictions, only one (Gosling 1986) meets the necessary second criterion of demonstrating facultative adjustment. Consequently, the patterns of differential mortality or biased growth in the remaining 14 studies are equally consistent with nonfacultative explanations to be discussed later. As a number of these studies are of particular importance in evaluating the likelihood of Trivers and Willard-like manipulations, I will consider them in greater detail.

McClure's (1981) study of eastern woodrats (*Neotoma floridana*), in which she reported differential survival of nursing male offspring, is often cited as the only clear example of sex-biased allocation leading to differential mortality (Clutton-Brock 1991). This study is potentially very important because the experimental treatment (food restriction) was imposed postnatally, so the resulting biased sex ratios presumably resulted from postnatal maternal adjustment. Further, the sex ratio bias was coupled with a reduction in litter size in an apparent attempt by the mother to reduce overall cost of reproduction. In an effort to replicate McClure's (1981) findings, Sikes (1995, 1996b) followed the same experimental protocol of postnatal food restriction with eastern woodrats and with northern grasshopper mice (*Onychomys leucogaster*), but obtained dissimilar results. In neither of the studies by Sikes was there any evidence of sex-biased mortality or growth rates as a function of food restriction. Further, although these studies validated the first assumption of the T-W hypothesis (that offspring condition at weaning is dependent on maternal condition during parental investment), a follow-up study (Sikes 1996a) showed that the compromised offspring

condition produced by maternal food restriction was not maintained after the young were able to feed independently.

McClure's results also prompted an exceptionally well-done study of bushy-tailed woodrats (*Neotoma cinerea*) by Richard Moses and colleagues. The major criticism of McClure's work by Moses et al. (1998) was that, although she stated that male offspring of restricted mothers were often outside the nest and not nursing, no direct evidence was provided that their deaths were due to facultative adjustment. This same point had been made previously by various authors concerning tests of the T-W hypothesis (Myers 1978; Charnov 1982; Clutton-Brock 1991). To address this shortcoming, Moses et al. (1998) followed an experimental protocol of postnatal food restriction similar to that of McClure. Although these authors documented male-biased mortality among offspring of food-restricted mothers, as reported by McClure (1981), when they sought evidence that these differences were due to facultative adjustment by mothers, they found no hint of support. Instead, Moses et al. (1998) concluded that male-biased mortality occurred in the absence of maternal intervention and was most likely due to the greater energetic demand of larger male offspring in this sexually dimorphic species. Similar circumstances may explain sex-biased mortality in other sexually dimorphic species, such as horses (*Equus caballus*; Monard et al. 1997). Based on their results, Moses et al. concluded that their study failed to support the T-W hypothesis.

The descriptive study by Gosling (1986) offers the best support for the T-W hypothesis in rodents. This was a long-term study involving dissection of 5,853 female coypus (*Myocastor coypus*) from control operations in England. Of the sample, nearly 1,500 females contained litters old enough to sex and were included in subsequent analyses. Gosling documented sex-biased investment in that small litters were significantly more likely to be aborted if they were female biased than if male biased. Abortion was of entire litters, and clearly was facultative rather than simply a response to litter size, because male-biased litters tended to be retained even if small. To be sure, there was a tendency toward male-biased litters among younger mothers and among those mothers giving birth in the summer, but owing to the power afforded by an enormous sample size, age and season could be excluded as potential explanatory variables; Gosling concluded that biased investment in subsequent litters was indeed facultative and under maternal control. Gosling's documentation of facultative adjustment of maternal investment to favor one sex over the other makes this the only study of rodents to satisfy both criteria central to the T-W hypothesis. As has been pointed out repeatedly (Myers 1978; Charnov 1982; Clutton-Brock 1991; Moses et al. 1998), without solid evidence of facultative

Table 11.1 Literature using rodents relevant to Trivers and Willard's (1973) hypothesis for facultative adjustment of parental investment

Reference	Taxa	Consistent with T-W?[a]	Litter size adjustment?	Sex-ratio bias?	Bias other than sex ratio?	Facultative adjustment?[b]	Timing of condition	Type of study	Assumptions[c]
Rivers and Crawford (1974)	Mus musculus	Yes	Yes	Yes	Not tested	No	Preconception to postnatal	Manipulative	—
McClure (1981)	Neotoma floridana	Yes	Yes	Yes	Yes (weight)	No	Postnatal	Manipulative	1(yes)
Gosling (1986)	Myocastor coypus	Yes	Yes	Yes	No	Yes	Preconception/prenatal	Descriptive	1(yes), 2(yes)
Labov et al. (1986)	Mesocricetus auratus	Yes	Yes	Yes	Yes (weight)	No	Prenatal to postnatal	Manipulative	1(yes)
Meikle and Drickamer (1986)	Mus musculus	Ambiguous	No	Yes (but significant in opposite direction for some treatments)	No (tested for)	No	Preconception/prenatal	Manipulative	1(no)
McShea and Maddison (1986)	Microtus pennsylvanicus	Yes	No	Yes	Yes (weight)	No	Preconception to postnatal	Descriptive	—
Armitage (1987)	Marmota flaviventris	No	No	No (except as a function of maternal age)	Not tested	No	Preconception to postnatal	Descriptive/ manipulative	—
Wright et al. (1988)	Mus musculus	Yes	Yes	Yes	Yes (weight)	No	Preconception/ prenatal	Manipulative	1(yes)
König (1989b)	Mus domesticus	Not considered	Yes	No	Not tested	No	Postnatal	Manipulative	1(yes)
Krackow and Hoeck (1989)	Mus musculus	Yes	Yes	Yes (male biased for 2nd litters only)	Not tested	No	Preconception/ prenatal	Descriptive	—
Clark et al. (1990)	Meriones unguiculatus	Yes	No[d]	No	Yes (time nursing, but differences were among rather than within litters)	No	Prenatal	Manipulative	—
Clark et al. (1991)	Meriones unguiculatus	Yes	No	No	Yes (weight)	No	Postnatal	Manipulative	1(yes)

Study	Species								
Trombulak (1991)	*Spermophilus beldingi*	No	No	No	No (tested for)	No	Prenatal to postnatal	Manipulative	1(no)
Meikle and Thornton (1995)	*Mus musculus*	Yes	Yes	Yes	Not tested	No	Preconception to postnatal	Manipulative	1(yes), 2(yes)
Moses at el. (1995)	*Neotoma cinerea*	Ambiguous	No	Yes (but with maternal mass, not treatment)	Not tested	No	Preconception to postnatal	Manipulative	1(no)
Sikes (1995)	*Neotoma floridana*	No	Yes	No	No	No	Postnatal	Manipulative	1(yes)
Sikes (1996b)	*Onychomys leucogaster*	No	Yes	No	No	No	Postnatal	Manipulative	1(yes)
Sikes (1996a)	*Neotoma floridana Onychomys leucogaster*	No	No	No	Not tested (study on postweaning growth)	Not possible	Postnatal	Manipulative	1(yes), 2(no)
Krackow (1997)	*Mus musculus*	No	Yes	No	Not tested	No	Preconception to postnatal	Descriptive	1(yes), 2(yes)
Duquette and Millar (1998)	*Peromyscus mexicanus*	No	Yes	No	No	No	Preconception to postnatal	Manipulative	1(yes)
Moses et al. (1998)	*Neotoma cinerea*	No	Yes	Yes	Yes (weight, but tested other parameters and found no bias)	No	Postnatal	Manipulative	1(yes)
Millesi et al. (1999)	*Spermophilus citellus*	Yes	Yes (varied with date)	Yes (varied with date)	No (no bias reported during lactation)	No	Preconception to postnatal	Descriptive	—
Lamb and Aarde (2001)	*Mastomys coucha*	No	No	Yes	No (tested for)	No	Preconception/ prenatal	Manipulative	—

[a]Based on direction of sex-ratio bias or other evidence of differential investment.

[b]Was evidence presented to support that differential investment was optional and under maternal control?

[c]Which of the underlying assumptions of the Trivers-Willard hypothesis tested and were they validated?

[d]Clark et al. (1990) reported no differences in litter size but a decline in the probability that a female would breed on postpartum estrus after rearing an all-male litter as compared to those females rearing an all-female litter.

adjustments to parental investment, skewed sex ratios at weaning are equally consistent with gender-specific mortality (Werren and Charnov 1978), which is a form of nonfacultative (although potentially adaptive) litter manipulation (see the following).

A second important distinction of Gosling's work with regard to the T-W model in rodents is that differential investment was manifested *among* rather than *within* litters. This pattern is consistent with studies supporting the T-W hypothesis outside of rodents, such as in red deer (*Cervus elephas;* Clutton-Brock et al. 1986). If maternal adjustments are more likely among as opposed to within litters, then might some of the listed rodent studies provide insight concerning where to look to for additional support, even without evidence that skewed sex ratios are facultative? Work by Rivers and Crawford et al. (1974) and Wright et al. (1988) on domestic mice (*Mus musculus*), Labov et al. (1986) on golden hamsters (*Mesocricetus auratus*), and Clark et al. (1990) on Mongolian gerbils (*Meriones unguiculatus*) provides good examples. In these studies, a bias in maternal investment was reported in litters following a period of nutritional stress induced by food restriction (Rivers and Crawford 1974; Labov et al. 1986; Wright et al. 1988), or by rearing a previous litter of all-male or all-female offspring (Clark et al. 1990). But in each case the adverse condition for the mother was in place prior to conception.

Timing of stress may be a critical factor. In fact, one can quickly glean from table 11.1 that of the 15 studies reporting biased sex ratios or other forms of differential maternal investment, adverse conditions potentially eliciting differential investment were present before parturition in 12. In contrast, authors reported biased sex ratios in only 3 of 7 studies where adverse conditions were imposed postnatally, and two of these studies raise further questions. One of these is McClure's (1981), results of which have not been duplicated despite repeated attempts in several species (Sikes 1995; 1996a, 1996b; Moses et al. 1998). Another was the study by Moses et al. (1998), in which skewed ratios in bushy-tailed woodrats clearly did not result from maternal manipulation. I have argued previously, based on my inability to validate assumptions of the T-W hypotheses when stress is imposed postnatally, that application of the T-W hypothesis probably should be restricted to situations where adverse conditions are in place prior to parturition (Sikes, 1996a). The data summarized here make application of the Trivers-Willard hypothesis in response to postnatal conditions in rodents even more questionable. This conclusion is supported in part by Cameron's (2004) meta-analysis of data on T-W-like sex ratios in mammals, where she found strongest support for sex ratio adjustments when

maternal condition at conception was posed as the determining factor.

In summary, with the exception of the excellent work by Gosling (1986), unequivocal support for the Trivers and Willard hypothesis in rodents is lacking because clear evidence of facultative adjustment of reproductive investment by the mother is not provided. Given the interest that this hypothesis has generated, even with extremely limited support, any carefully designed study that addresses this critical weakness and also validates the underlying assumptions is likely to profoundly influence life-history research in mammals far into the future.

Local resource competition hypothesis

The local resource competition hypothesis (LRC) was developed primarily from work with primates (Clark 1978; Silk 1983), but has since been extended to other taxa, including ungulates (Clutton-Brock et al. 1982; Caley and Nudds 1987) and marsupials (Cockburn et al. 1985). This hypothesis holds that in species where one sex is philopatric, selection should favor the dispersing sex (usually males in mammalian systems) when resources are limited because mothers will compete with daughters for resources. Types of limiting resources are varied but include food, suitable den sights, or mates. Lambin (1994b, 1994c) provides support for this hypothesis in rodents from an observational study of Townsend's voles (*Microtus townsendii*), in which he noted female-biased sex ratios of litters in spring seasons where vole density was low (see also table 11.2). However, he concluded that sex ratios in his system were altered due to broad-scale competition caused by overall high population densities, rather than local competition as influenced by neighborhood or family size. Bond et al. (2002) presented data from a manipulative study of gray-tailed voles (*Microtus canicaudatus*) that are partially consistent with the LRC model. These investigators monitored recruitment in naturally increasing populations of voles with approximately even adult sex ratios in the spring and summer, but then manipulated adult sex ratios by adding or removing adults in the autumn. Bond et al. (2002) reported results consistent with the LRC model for early season litters, but as sex ratios later in their study were biased in the opposite direction, even though population densities and hence likelihood of local competition did not change, they concluded that their data were best explained by seasonal rather than local resource competition effects.

Local mate competition may be viewed as a special case of resource competition and potentially brings in new considerations based upon mechanisms of sex determination in some rodents. Local mate competition as proposed by

Table 11.2 Evaluation of selected literature relative to competing hypotheses of sex-ratio adjustment

Reference	Taxa	Evidence of bias?	Consistent with T-W?[a]	Consistent with FCA?[b]	Consistent with LRC?[c]	Consistent with EM?[d]
Rivers and Crawford (1974)	*Mus musculus*	Yes	Yes	—	—	Yes
McClure (1981)	*Neotoma floridana*	Yes	Yes	—	—	Yes
Gosling (1986)	*Myocastor coypus*	Yes	Yes	No	No	No
Goundie and Vessey (1986)	*Peromyscus leucopus*	Yes	No	Yes	—	Yes
Labov et al. (1986)	*Mesocricetus auratus*	Yes	Yes	—	—	Yes
Meikle and Drickamer (1986)	*Mus musculus*	Yes (but significant in opposite direction for some treatments	Partially	—	—	Yes
McShea and Maddison (1986)	*Microtus pennsylvanicus*	Yes	Yes	Yes	Yes	Yes
Armitage (1987)	*Marmota flaviventris*	No (except as a function of maternal age)	No	No	No	Yes
Wright et al. (1988)	*Mus musculus*	Yes	Yes	—	No	Yes
König (1989b)	*Mus domesticus*	No	No	—	—	No
Krackow and Hoeck (1989)	*Mus musculus*	Yes (male biased for 2nd litters only)	Yes	Yes	Yes	Yes
Clark et al. (1990)	*Meriones unguiculatus*	No	Yes	—	—	Yes
Clark et al. (1991)	*Meriones unguiculatus*	No	Yes	—	—	Yes
Trombulak (1991)	*Spermophilus beldingi*	No	No	No	No	No
Hornig and McClintock (1994)	*Rattus norvegicus*	Yes	No	No	No	Yes
Lambin (1994b,c)	*Microtus townsendii*	Yes	No	Yes	Yes	Yes
Meikle and Thornton (1995)	*Mus musculus*	Yes	Yes	—	—	Yes
Moses at el. (1995)	*Neotoma cinerea*	Yes (but with maternal mass, not treatment)	No	—	No	Yes
Sikes (1995)	*Neotoma floridana*	No	No	No	No	No
Hornig and McClintock (1994)	*Rattus norvegicus*	Yes	No	No	No	Yes
Sikes (1996a)	*Onychomys leucogaster*	No	No	Yes	No	No
Krackow (1997)	*Mus musculus*	No	No	No	No	No
Duquette and Millar (1998)	*Peromyscus mexicanus*	No	No	No	No	No
McAdam and Millar (1998)	*Peromyscus maniculatus*	Yes	No	Yes	—	Yes
Moses et al. (1998)	*Neotoma cinerea*	Yes	No	—	No	Yes
Millesi et al. (1999)	*Spermophilus citellus*	Yes (varied with date)	Yes	No	No	Yes
Lamb and Aarde (2001)	*Mastomys coucha*	Yes	No	No	No	Yes
Bond et al. 2002	*Microtus canicaudus*	Yes	No	Yes	Partially	Yes

[a]Consistent with predictions of Trivers and Willard's (T-W) 1973 hypothesis based on direction of sex-ratio bias or other evidence of differential investment?
[b]Consistent with predictions of the first cohort advantage (FCA) hypothesis?
[c]Consistent with local resource competition (LRC) hypothesis?
[d]Consistent with the extrinsic modification hypothesis (EM) for nonfacultative adjustment of parental investment?

Hamilton (1967) can occur in recently founded populations in which male offspring of founding females compete among themselves for mates. The mechanism of sex determination in wood lemmings (*Myopus schistocolor*), collared lemmings (*Dicrostonyx torquatus*), and the South American field mouse (*Akodon azarae*) has been proposed as a way to mitigate local mate competition. In these taxa, the sex-determining effects of the Y chromosome are suppressed by a modified X chromosome (X*) such that X*Y individuals are female. As a result, population sex ratios of these species are usually heavily female biased (Fredga 1988; Espinosa and Vitullo 1996). In collared lemmings, X*Y females produce both X*- and Y-bearing ova, but compensate for the loss of nonviable YY individuals by a higher ovulation rate. Wood lemmings, on the other hand, show a double disjunction during mitotic anaphase such that they produce only X* eggs (Gileva et al. 1982).

Although the mechanism of sex ratio distortion is well understood in these species, its underlying explanations are not. The stability of most mammalian populations would minimize the potential for the type of local mate competition that Hamilton envisioned in 1967, but it could occur in patchy habitats where isolated populations are prone to extinction followed by recolonization (Maynard Smith and Stenseth 1978; Werren and Hatcher 2000). However, in none of the species showing X* sex determination is there evidence of the spatial clustering and inbreeding that current models predict are necessary for such a system to evolve

(Gileva and Federov 1991; Eskelinen 1997). The role of X* chromosomes in reducing local mate competition has been further criticized based on the argument that the extent of segregation distortion of Y-bearing sperm is consistent with random mating, because selection for female-biased sex ratios in response to local mate competition would eliminate pressure for segregation distortion favoring Y sperm (Bulmer 1988). Moreover, in populations subjected to repetitive inbreeding the female-biased sex ratios tend to disappear (Jarrell 1995; Gileva 1998). Although its role in minimizing local mate competition in mammals is unclear, an evolutionary explanation for this mechanism of sex determination offers a continuing challenge.

With the exception of X* sex determination in those species listed, no proximate mechanisms for sex ratio adjustments are documented in any of the studies examining local resource competition in rodents. Further, meta-analyses of data on primates suggest that differences reported for those taxa may not deviate from random expectations, and hence provide no support for facultative adjustment of sex ratios in response to local competition (Brown and Silk 2002; but see West et al. 2002 for cautions concerning use of meta-analyses of such studies). As with data concerning the Trivers-Willard hypothesis, in the absence of mechanisms of sex ratio adjustment or evidence that skews are facultative responses by parents to local conditions, support for the local resource competition hypothesis in rodents is compromised.

First cohort advantage hypothesis

Based on his work with Virginia opossums (*Didelphis virginiana*), Wright et al. (1995) proposed that male-biased sex ratios early in the reproductive season might occur in a population if males are more likely than females to breed in their year of birth. This would be especially true given that smaller size does not affect female reproductive success as strongly as that of males in this species, and because males may have only one year for breeding due to high mortality rates. Wright et al. (1995) termed this the first cohort advantage (FCA) hypothesis. While numerous rodent studies show no seasonal changes in sex ratios (e.g., Havelka and Millar 1997), others are at least partially consistent with the FCA predictions (table 11.2). For example, Goundie and Vessey (1986) observed female-biased spring litters and male-biased fall litters in *Peromyscus leucopus* born in nest boxes.

It is important to note that the sex ratio trend in rodents typically is opposite that of opossums, in that either early litters are female biased and late season litters are male biased, or both (McShea and Madison 1986; Lambin 1994c; Bond et al. 2002). The opposite trend in ratios in rodents as compared to opossums may come about if early born females in short-season environments are able to get off a litter in the year of their birth, whereas it is unlikely for males to do the same. Alternatively, late-born males are more likely to breed the following season if maturation is delayed until the spring (Boonstra 1989). McAdam and Millar (1998) concluded that good-quality female *Peromyscus maniculatus* in the Kananaskis Valley of Alberta could maximize their inclusive fitness by producing female-biased litters early in the season.

Data from Gosling's (1986) study of naturally reproducing coypus complicates interpretation of seasonal differences of sex ratios in rodents. In this species males are 15% larger than females as adults, and thus presumably more costly to rear. Given the dimorphism in this polygynous species and the fact that single adult males are dominant to and at least partially exclude other males within the range of a group of females (Gosling 1986), it would seem that the likelihood of males breeding in their natal year is exceedingly remote. To this point there is similarity with the small-rodent systems described by McShea and Madison (1986), Lambin (1994b, 1994c) and McAdam and Millar (1998), and we would expect female-biased early litters. However, Gosling found male-biased litters in summer months, and that both litter size and mean number of female embryos per litter were significantly smaller in younger as compared to older females. In this case, younger females were producing more males than females, whereas older females produced a more even sex ratio. This pattern was a quandary for Gosling as well, but he had sufficient statistical power with his data set of ca. 1,500 dissected females to conclude that skewed sex ratios resulted from abortion of female-biased small litters in younger mothers, whereas there was no similar tendency for abortion in older mothers. The take-home message with regard to seasonal differences in sex ratios is that although seasonal differences are apparent in some rodent species, we must carefully consider and discriminate among competing explanations before we can determine the underlying cause.

Nonfacultative Adjustment?

Most hypotheses have been couched in terms of parental manipulation of investment. Framing questions in this way may constrain our interpretation of available data. As has already been discussed, without identifying a proximate mechanism for adjustment, or at least evidence that sex ratio adjustment is facultative, the existence of spatially or temporally biased ratios is equally consistent with explanations that do not rely on parental manipulation. More recent treatments have backed away from reliance on facultative differential investment. For example, Post et al. (1999) and Forchhammer (2000) proposed that biased sex ratios in dimorphic species may arise as a consequence of sex-differential viability of fetuses resulting from differences in growth rates and susceptibility to environmentally induced changes in maternal condition (the extrinsic modification hypothesis). Supporting data for nonfacultative alteration of sex ratios are also available from literature on mouse development at neurulation (Seller and Perkins-Cole 1987) and timing of mating since lights out in Norway rats (*Rattus norvegicus* Hornig and McClintock 1994).

Predictions of the extrinsic modification hypothesis are not necessarily mutually exclusive with any of the other hypotheses discussed, as its authors point out, and all may produce similar results (table 11.2). The key difference between the extrinsic modification hypothesis and all others is that alteration of sex ratios or levels of parental investment does not occur because of facultative adjustment by the parents, but rather because of differential susceptibility of the sexes to extrinsic influences. Further, even though Post et al. (1999) proposed the model as a possibility for dimorphic species, the level of dimorphism required to elicit differences in fetal viability may be subtle and may not require differences in size at all. In fact, it is entirely possible that many of the data used to support the FCA hypothesis, as well as many further studies that show seasonal differences in sex ratios at birth or weaning, but where the authors have found correlations with climatic variables rather than cohort advantage (e.g., Myers et al. 1985; Goundie and Vessey 1986; Havelka and Millar 1997) may be explained equally well by this extrinsic modification hypothesis, without invoking parental manipulation.

Proximate Mechanisms for Adjustment

Because maternal investment in each offspring increases dramatically throughout gestation and lactation, elimination of developing young to alter sex ratios would appear increasingly expensive in terms of fitness, unless litter reduction is necessary to tailor maternal output to environmental conditions (Hornig and McClintock 1996). Consequently, the earlier in the investment sequence sex ratios can be adjusted, the lower the cost for the mother. Evidence exists that timing of mating relative to ovulation can influence the probability that ova are fertilized by an X- or Y-bearing sperm (reviewed by Roberts 1978; Clutton-Brock and Iason 1986; James 1996). Specific mechanisms for these differences have not been conclusively identified, but may include differential facilitation or inhibition of X- or Y-bearing sperm in the female's reproductive tract, or preferential fertilization by one or the other (Krackow 1995;

Rorie 1999). Further, data indicate that Y-bearing sperm may swim faster than X spermatozoa, but are not as long-lived (Roberts 1978; Soede et al. 2000).

Even if equal numbers of ova are fertilized by X and Y spermatozoa, it does not follow that the sex ratio will remain even at implantation. There is a growing body of literature showing that male preimplantation embryos develop faster than females, and that there are temporal limitations to uterine receptivity to implantation (summarized by Krackow and Burgoyne 1998). As a consequence, an interaction between the asynchronous development of the embryos and uterine receptivity can produce biased sex ratios of implanted embryos (Krackow 1995; Krackow and Burgoyne 1998). Still, even if equal numbers of males and females implant, sex ratios after implantation may become skewed by male and female embryos being differentially susceptible to maternal dietary deficiencies (Rivers and Crawford 1974), or through differential fetal resorption (Krackow 1992). Cameron (2004) cited recent studies linking circulating levels of glucose and differential responses of male and female embryos to glucose as a potential mechanism for altering sex ratios. Importantly, none of these potential causes of bias described here necessarily requires parental manipulation, and hence may not constitute facultative manipulation of sex ratios. But neither do they exclude that possibility. Even if they are not facultative they may still be adaptive, depending on the environmental context.

Missing Links

Krackow (2002) argued persuasively that we should focus our efforts on identifying mechanisms for distortion of sex ratios that will prove parental manipulation and hence facultative adjustment. His conclusion underscores the fact that a key assumption of sex ratio hypotheses other than the extrinsic modification hypothesis is that observed biases arise *because* of differential investment, rather than differential investment being a result of differential offspring mortality. While differential investment in the sexes can indeed produce skews at any stage of development—up to independence—is a skewed ratio at the end of parental investment necessarily indicative of differential investment? Conversely, does differential investment necessarily alter the sex ratio? The answer to both of these questions clearly must be "NO," as distorted sex ratios may come about by differential mortality without differential investment, as per the foregoing section (see also Clutton-Brock et al. 1985; Clutton-Brock 1991). While such mortality does not necessarily constitute facultative manipulation, few studies have tested for differential investment explicitly; they simply assume that it must exist if skewed sex ratios are detected. This is faulty logic.

Is it necessary to present a proximate mechanism of manipulation to provide support for one of the hypotheses of facultative manipulation? Again, the answer is "NO," *provided there is evidence of facultative manipulation.* Even when such evidence is presented, care must be taken to be able to differentiate among competing explanations. As Cockburn et al. (2002) point out "literature is littered with studies described as 'consistent with' the prediction of one hypothesis or another." These authors go on to note that the real distinction among predictions of competing hypotheses are slight and require focus on key predictions. If this is the case, then it is the study carefully designed to discriminate among the various hypotheses that will provide the most insight.

Beyond the need to identify mechanisms and demonstrate facultative adjustment, we must also carefully consider the limitations to models of sex ratio adjustment. Pen and Weissing (2002) argue that likely explanations for relatively little progress in understanding sex ratio variation in vertebrates, as compared to haplodiploid systems, include: (1) chromosomal sex determination hindering parental manipulation; (2) longer life spans and overlapping generations, opening the door for tradeoffs between current and future reproductive effort; (3) opportunity for both parents to influence sex ratios (most male influence is likely prior to gestation); and (4) substantial costs for manipulation if sex ratios are altered at any stage of development beyond establishment of the primary sex ratio. These authors call for development of explicit models that include costs as a necessary step toward a mechanistic theory of sex allocation. In mammals, the extended period of parental investment during gestation and lactation can set the stage for parent-offspring conflict (*sensu* Trivers 1974) where parents and offspring have different sex ratios and parental allocation optima, which further complicate modeling attempts. If we are to make the most of Pen and Weissing's (2002) recommmendations, then rodents are a near-ideal mammalian system to test developing models. They have shorter life spans than many other mammalian taxa and can be studied within large enclosures or nest boxes where lifetime reproductive success can be measured. However, if the work presented by Gosling (1986) is any indication, the differences in investment of one sex relative to the other may be slight, and may require long-term data sets to provide the necessary statistical power to demonstrate empirically.

From this review it should be obvious that theories concerning facultative manipulation of sex ratios are well established. From these theories we have specific predictions and their underlying assumptions. We have data from many different taxa exhibiting skewed sex ratios. We even have a number of mechanisms that have been shown to result in skewed sex ratios in various instances. However, we lack a clear link between theory, prediction, data, and mechanism. Demonstration of facultative adjustment of parental investment, according to predictions and in line with necessary conditions of one of the facultative models of sex ratio theory, will identify systems in which to seek mechanisms for adjustment. Until such a link is established, it is more parsimonious to conclude that biased sex ratios result from causes other than facultative manipulation.

Summary

Although instances of skewed sex ratios are well documented in various rodents, our understanding of the causes of such skews is incomplete. Facultative manipulation of sex ratios in mammals is of considerable interest as this topic has been of central importance in understanding fundamental principles of other social species, most notably the hymenopteran insects. Hypotheses proposed for facultative manipulation include differential parental allocation depending upon maternal condition and cost of offspring of either sex, differential allocation to minimize competition for local resources, and advantage of favoring offspring of one sex over the other at different times within the reproductive season because of resource availability or probability of the offspring's reproduction within that same breeding season. Skewed sex ratios may also arise because of differential viability of embryos under adverse conditions experienced by the mother during gestation. Although hypotheses for sex ratio adjustment make specific predictions about the direction of skew expected, empirical data linking predictions and results are few. Most importantly, proximate mechanisms for manipulation of sex ratios have not been identified and tied to specific instances of biased ratios. A direct linking of theory, proximate mechanisms, and advantages of sex ratio adjustment over reproductive lifetime is essential before facultative sex ratio manipulation is more than an intriguing possibility in rodent ecology.

Chapter 12 The Role of the Stress Axis in Life-History Adaptations

Rudy Boonstra, J. M. Barker, J. Castillo, and Q. E. Fletcher

THE NEUROENDOCRINE SYSTEM is a major pathway in vertebrates that integrates environmental change and through which life-history decisions to reproduce, to grow, or to put energy into storage are implemented (McEwen 2001; Ricklefs and Wikelski 2002; Boonstra 2005). The goal of individuals is to maximize lifetime reproductive fitness, and the functioning of the stress axis plays a central role in the neuroendocrine system in making this happen. At the individual level, the stress axis plays a key role in allowing animals to cope with change and challenge in the face of both environmental certainty and uncertainty. At the species level, the stress axis plays a central role in evolutionary adaptations to particular ecological pressures, and an understanding of differences among species is essential to life-history adaptations. In this chapter, we discuss the basic mammalian response to stressors and how this response is modified at both the individual and species levels in response to different ecological pressures.

The stress axis is multitasking throughout the life of an organism. The stress axis is composed of the limbic system (dentate gyrus and hippocampus) and the hypothalamic-pituitary-adrenocortical axis (HPA), and is pivotal for successful adaptation for four reasons. The first three focus on common responses of individuals within a species; the fourth, on between-species differences to the basic pattern. First, the stress axis is involved in normal day-to-day activities associated with the diurnal cycle of waking, such as increased locomotion, exploratory behavior, increased appetite, and food-seeking behavior (reviewed in Wingfield and Romero 2001). Second, the stress axis permits short-term

adjustment to maintain survival in the face of acute environmental stressors. This response is the classic "flight or fight" reaction, and is a generalized response to a wide variety of stressors, such as bouts of severe weather, physical stressors such as attacks by a conspecific or a predator, or psychological stressors such as the fear of an imminent attack. Though we focus on the limbic system and the HPA axis here, it is only one part of the stress response, which includes other hormones, neurotransmitters, opioid peptides, cytokines, and brain functions (Sapolsky et al. 2000). Third, the stress axis can be permanently programmed during development because of stressors affecting the mother, and this may adapt the individual to new conditions it may experience during its lifetime (Matthews 2002). Fourth, the stress axis is subject to evolutionary modification, and equips species to succeed under different ecological contexts. Rodents span the gamut of life-history variation, with many species showing high reproductive rates, rapid development, and short life spans, while other species show the opposite traits. In the former, the stress axis functions to trade off survival for reproduction, whereas in the latter, the opposite occurs (Boonstra et al. 2001c; Boonstra 2005).

In this chapter we draw on evidence from natural populations of rodents, but supplement this with research from the laboratory, as this research is generally more in-depth (examining molecular and cellular changes), can provide a guide to indicate what is potentially occurring in nature, and is investigating areas such as neurogenesis, which field studies are only beginning to tackle. However, caution must be exercised when making extrapolations from the laboratory to the field, to avoid oversimplification and artifacts. Laboratory rodents are often less aggressive, less aware of their environment, explore less, are more social, and respond more to stressors than their natural counterparts (Künzl et al. 2003; see also Wolff 2003c).

The Stress Response

An external stressor sets off a rapid cascade of responses in vertebrates to respond to the stressor and then to reestablish homeostasis (fig. 12.1). The first line of defense, occurring within seconds of the stressor, is that the sympathetic nervous system causes the adrenal medulla to release catecholamines (epinephrine and norepinephrine) into the general circulation. The second line of defense also occurs immediately, starting with the paraventricular nucleus of the hypothalamus releasing primarily corticotropic-releasing hormone. These hormones cause the anterior pituitary to release adrenocorticotropic hormone (ACTH) into the general circulation and, within minutes, the adrenal cortex releases glucocorticoids (GCs) into the blood. In some rodents (voles and mice) the GC that is released is corticosterone, and in others (chipmunks and squirrels) it is cortisol, or a mixture of cortisol and corticosterone. The HPA axis signals the body to mobilize energy, inhibit physiological processes not required to deal with the stressor, and return the body to homeostasis after the stressor has passed. Immediate catabolic effects result in the mobilization of glucose for the muscles, the stimulation of hepatic gluconeogenesis (the breakdown of other body tissues, such as protein), and the shunting of energy resources away from peripheral tissues not needed for short-term survival. Cardiovascular tone is increased, immune function is stimulated, reproductive physiology and behavior inhibited, feeding and appetite is decreased, and cognition is sharpened (Sapolsky 2002). Under conditions where the stressor is acute, GCs exert feedback at three levels in the brain (fig. 12.1) to return the body back to the preactivation state. Key to this feedback is the intracellular GC receptors (mineralocorticoid receptors [MR] and glucocorticoid receptors [GR]) in the critical brain areas, especially the hippocampus, which regulates the overall functioning of the HPA axis (fig. 12.1; de Kloet et al. 1999).

The GC plasma carrier protein, corticosteroid binding globulin (CBG), plays a major role in allowing mammals to cope with stressors, but it changes rapidly as a function of reproduction or of chronic stressors. The mammalian body is typically buffered from the immediate impact of GCs in the blood because they are tightly bound to CBG. Only about 5 to 10% of GCs are unbound and free, and only the free GCs are biologically active (Rosner 1990). CBG is

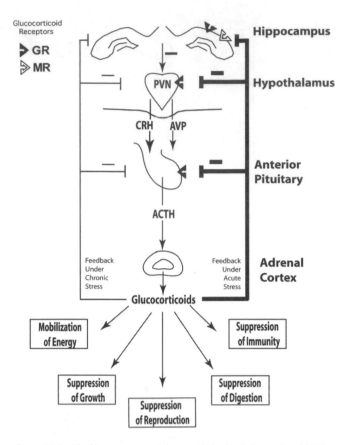

Figure 12.1   The hippocampus and the hypothalamic-pituitary-adrenal (HPA) axis, the major impacts on body processes, and the glucocorticoid (GC) feedback in the mammalian brain. The hippocampus regulates the overall functioning of the HPA. A stressor causes the hypothalamic paraventricular nucleus (PVN) to release corticotropin releasing hormone (CRH) and vasopressin (AVP), and this causes the anterior pituitary to release adrenocorticotropic hormone (ACTH). ACTH initiates the synthesis and release of glucocorticoids (GCs, corticosterone in some rodents, cortisol in others) from the adrenal cortex. GCs act at multiple sites within the body to maintain homeostasis, but because of the damaging effects of extended exposure to GCs, the HPA axis is tightly regulated through feedback (inhibition indicated by -) on glucocorticoid receptors to inhibit further HPA activity. GCs feed back on the hypothalamus and pituitary to cause a rapid inhibition of CRF release. Under conditions where the stressor is acute, feedback mechanisms operate efficiently and the system rapidly returns to normal, resulting in effects on body processes that are only short term. Under conditions where the stressor is chronic, feedback signals are weak and the system remains activated for longer periods, resulting in effects on body processes that can be long term and detrimental. Short-term effects result in suppressive impacts on body processes; long-term chronic effects result in inhibitory impacts on body processes. Glucocorticoid (GR) and mineralocorticoid receptors (MR) occur in the limbic system (hippocampus and dentate gyrus) and GRs occur in the PVN and anterior pituitary. In the brain, MRs have a higher affinity than do GRs for GCs, and at basal concentrations of cortisol, MRs are occupied whereas GRs remain largely unoccupied. During periods of stress and elevated plasma GCs there is increased occupation of GRs. Hippocampal MRs may be primarily involved in feedback regulation during basal secretion, whereas GRs become important during periods of increased GC secretion (from de Kloet et al. 1999; Matthews 2002; Sapolsky 2002).

thought to act as a reservoir of GC, so that GCs can be rapidly released in response to an environmental challenge. CBG concentrations are affected both by the stress axis and the gonadal axis. Chronic stressors, lasting for as little as 24 h, result in a marked reduction in CBG (Schlechte and

Hamilton 1987; Boonstra et al. 1998, 2001c), and hence an increase in free GCs. During pregnancy and lactation, females have higher CBG levels; breeding males have lower levels.

The stress response and the homeostatic set-point are not fixed, lifelong species-dependent characteristics, but are modified by experience, by development, and by the annual pattern of life-history changes. First, experience may alter the stress response. The stress response functions well when the stressor is acute (minutes to hours); thereafter, the negative, inhibitory effects of chronic stress become evident and intensify. Laboratory evidence in rodents indicates that the ACTH response is desensitized when the animal is repeatedly exposed to certain types of stressors (e.g., cold exposure) and not to others, but that entirely new stressors continue to elicit a typical stress response (Aguilera 1998). Under conditions of chronic stress (days to months), concentrations of free GCs increase and the normal suppressive effects of GCs grade into inhibition (fig. 12.1). The net result is potentially deleterious, affecting long-term survival and fitness through infertility, impaired resistance to disease, and inhibition of growth. Second, pre-, and postnatal periods of development are particularly vulnerable to permanent modification by stressors affecting the mother (Welberg and Seckl 2001; Matthews 2002). Offspring born to mothers who experienced a high level of stress during gestation, or offspring that experienced high levels of stress during postnatal development are programmed to have a hyperactive stress axis. An interplay also occurs between changes in the stress axis and the reproductive axis (Wartella et al. 2003) that ultimately translates into changes in adult fitness. Finally, in mammals living in seasonal environments, the annual cycle of reproduction, migration, and coping with winter may require the stress axis to be modulated in different ways at different times to optimize reproduction, survival, or both in the face of environmental challenges (Wingfield and Romero 2001). Challenges that are recurrent and predictable, such as the direct male-male aggression associated with breeding, would, if the animal did not evolve a modifying solution, inhibit reproduction.

Impact on Reproduction

The stress axis plays a key role in the entire reproductive process, as it is a transducer of how competition for resources and mates limits or augments reproduction of individuals. In turn, reproduction may result in the progressive deterioration of the stress axis with age, and thus there is a complex interaction between the stress axis and the gonadal axis (Meites and Lu 1994; McEwen 2001). Stressors cause a disruption of reproductive behavior and physiology

because of the general suppressive actions of glucocorticoids (Wingfield and Sapolsky 2003). However, the negative impacts of stress do not necessarily occur, with evolutionary adaptations allowing reproduction to proceed in spite of chronic stressors. In this section we examine some of these adaptations.

Breeding frequency in males

Organisms must successfully integrate time and space to maximize their breeding success (Southwood 1977). The breeding choices organisms must make with respect to time is whether they do it "now" or "later," and with respect to space is whether they do it "here" or "elsewhere." The dichotomy in life-history characteristics between those mammals that are semelparous and those that are iteroparous is a reflection of integrating these two dimensions. Semelparity tends to be found in those animals in which adults face low or variable probabilities of survival (Roff 1992; Stearns 1992), or in which juveniles face higher survival in one season than another (Braithwaite and Lee 1979).

Boonstra and Boag (1992) proposed a model to account for differences in the hormonal and physiological responses between species with semelparous males and those with iteroparous males. Semelparous males employ the "adaptive stress response" and trade off survival for reproduction by maximizing the energy available for a brief period of intense reproduction. This strategy results in the failure of normal feedback mechanisms of the stress axis, causing the males to die from immunosuppression, gastric ulceration, and anti-inflammatory responses. Iteroparous males employ the "homeostasis stress response," in which reproductive effort was spread out over a longer breeding season or multiple breeding seasons. This strategy results in normal feedback mechanisms of the stress axis to remain intact throughout the breeding season. Recent evidence indicates a continuum in the suite of physiological and hormonal adaptations occurring between the extremes of semelparity and iteroparity, reflecting the continuum of life histories that mammals experience (Boonstra et al. 2001c; Woods and Hellgren 2003). The only truly semelparous species are found in marsupials of the dasyurid and the didelphid families (Bradley 2003). However, partial semelparity, in which many but not all of the males die after one mating period, occurs in many mammals, including rodents. We will examine the adaptations of the stress axis in rodents under scenarios of partial semelparity and of iteroparity and discuss some of the environmental constraints that select for these life-history traits.

Partial semelparity

The arctic ground squirrel (*Spermophilus parryii*) is found throughout the alpine and arctic areas of northern North

America. They are obligate hibernators, emerging above ground from a 7–8 month hibernation in early to mid-April (Buck and Barnes 1999). Females only have sufficient time for one litter during the brief northern summer. Virtually all yearlings are reproductively mature and thus almost the entire population breeds each year. During the synchronized 2–3 wk mating period, males compete intensely for access to females, roaming widely, sustaining severe injuries, eating less, and losing more weight than females (fig. 12.2). The detrimental impact of breeding on male survival is dramatic, being immediate in some males and delayed in others. In the boreal forest, 48% of the males disappear during the mating season (Boonstra et al. 2001c), whereas in the adjacent alpine area, 28% die (Gillis 2003). The mortality rate is age dependent, with older males (i.e., those who had gone through at least two breeding seasons—i.e., > 2 yrs) bearing it differentially. Older males expend more effort on reproduction (more severe wounding and greater loss of body weight) than yearling males (Gillis 2003). For those males that do survive the mating period, summer survival is then high. However, despite apparent recovery in summer, over half of the males die while hibernating the next winter (Gillis 2003; Hubbs and Boonstra 1997), suggesting that there are long-term consequences of the severe conflict of the previous spring. Again, older males tend to survive more poorly than yearling males at this time. The net result is that about 80% of breeding male arctic ground squirrels die each year (Gillis 2003).

Breeding males in spring are chronically stressed during the mating period and have deficits in a number of areas, but not others (fig. 12.3). To assess the responsiveness of the stress axis, a brief explanation is needed to understand the hormonal-challenge protocol employed. Injections of hormones, or analogues of them that are part of the normal stress response, are used to measure an animal's stress response over a series of blood samples. Two steps were involved: the dexamethasone suppression test, followed by the adrenocorticotropic hormone (ACTH) stimulation test. Dexamethasone, a synthetic glucocorticoid agonist, should inhibit GC secretion through negative feedback mechanisms at the level of the brain by causing a reduction in ACTH release. The ACTH stimulation test is a method to directly probe the responsiveness of adrenals. Arctic ground squirrel breeding males have the highest concentrations of free cortisol as a result of the lowest CBG concentrations relative to abdominal adult males from August (Boonstra et al. 2001c). Dexamethasone resistance is modest, and thus negative feedback regulation remains largely intact (fig. 12.3). The ACTH challenge results in a rapid rise in free cortisol concentrations that exceeds those of nonbreeding males. Unlike the situation in most other species, in which both dexamethasone and ACTH inhibit testosterone secretion, the testosterone concentrations in breeding male arctic ground squirrels remain high (fig. 12.4). ACTH injections actually cause testosterone levels to increase, not decrease, reaching higher concentrations than basal levels. This pattern is unique in mammals. Stress-induced immunosuppression is pronounced, being reflected in those individuals with the poorest ability to respond to the foreign antigen challenge and the lowest number of white blood cells. Thus, the intensity of male-male competition during the mating season chronically stresses males and increases their mortality rate dramatically. However, unlike the total and immediate death of males found in the semelparous marsupials, the mortality is partial, age-dependent, and graded over time.

Iteroparity

Iteroparity occurs in most mammals, with males having multiple mating opportunities over their adult lives, either

Figure 12.2 Photographs of the same male arctic ground squirrel from an alpine site in the Ruby Mountain Range in SW Yukon in 2002. The first was taken in early April, just after the adult breeding male emerged from its hibernaculum and before the onset of the intense mating frenzy; snow still covered the alpine meadows. The second was taken three weeks later, when virtually all females had been mated and vegetation was being to appear. Note the radiocollar under its neck. He was now haggard and worn out from his mating activity and was found dead several days later. Photographs courtesy of T. J. Karels.

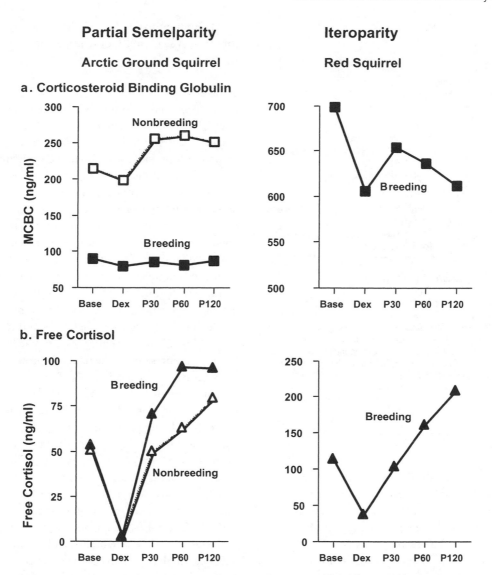

Figure 12.3 Life history variation in breeding frequency in male squirrels from the Yukon and the associated glucocorticoid plasma changes. The hormonal challenge protocol was used in both squirrels: Base levels indicate initial values, Dex indicates values 2 h after the dexamethasone injection, and P30, P60, and P120 indicate values 30, 60, and 120 min after the ACTH (adrenocorticotropic hormone) injection. In arctic ground squirrels, breeding (scrotal) males were trapped in May and nonbreeding (abdominal) males were adults trapped in August. In red squirrels, breeding males were trapped in May. Means (but not standard errors—see original publications) are presented. The data for arctic ground squirrels are from Boonstra et al. (2001) and for the red squirrel are from Boonstra and McColl (2000). Corticosteroid binding globulin measured as maximum corticosteroid binding capacity (MCBC).

during one breeding season for species that live < 1 yr (e.g., voles and mice), or over multiple breeding seasons for long-lived species (e.g., marmots). We will examine the stress axis in two wild rodents.

The meadow vole *(Microtus pennsylvanicus)* is a short-lived iteroparous small rodent found throughout grasslands of central and northern North America. Meadow voles have a promiscuous mating system, are short-lived (< 1 year), and breed continuously (ca. 3–5 litters per summer) throughout the summer and often into winter;

young born early in the breeding season can mature and breed in that season, and males are not territorial (see references in Boonstra 1994). Free GC concentrations in breeding males are usually low, with CBG concentrations always exceeding total plasma corticosterone concentrations by about 3–4 times (Boonstra and Boag 1992). In breeding males, androgen concentrations are not correlated to CBG concentrations.

The red squirrel *(Tamiasciurus hudsonicus)* is an arboreal tree squirrel whose distributional range covers the en-

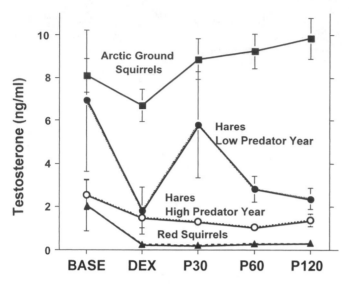

Figure 12.4 Changes in testosterone concentrations in response to the hormonal challenge protocol in adult breeding arctic ground squirrels and red squirrels (see fig.12.3 for description of hormonal challenge protocol). Changes in snowshoe hares, an iteroparous species, are included for comparison. In squirrels, males were trapped in May 1996. In hares, the males were captured in two different years: (a) when predation risk was low (1994) and the stress response was indicative of unstressed animals; and (b) when predation risk was high (1991) and the stress response was indicative of chronically stressed animals. Means are given ± SE.

tire boreal forest of North America. The red squirrel is highly territorial, asocial except during mating, has one or two litters per summer, has a sex ratio skewed toward males in the older age classes, and is long-lived (4–7 yrs; Obbard 1987). During the breeding season males have very high concentrations of cortisol and CBG (five and seven times, respectively, those of arctic ground squirrels (fig. 12.3, Boonstra and McColl 2000). Red squirrels are dexamethasone resistant, with cortisol concentrations declining only to 33% of those at baseline, whereas in nonresistant species, GC concentrations decline to ca. 5% or less. Certain rodent species are naturally dexamethasone resistant (e.g., guinea pigs [*Cavia aperea*], Keightley and Fuller 1996; and prairie voles [*M. ochrogaster*], Taymans et al. 1997), and this resistance is associated with elevated levels of GCs. Thus, red squirrels appear to be in this group. However, they do respond to ACTH with an increase in cortisol concentrations. As in most iteroparous species, the gonadal axis is very sensitive to the inhibitory effects of glucocorticoids. At the baseline levels, testosterone concentrations are negatively correlated to free cortisol concentrations (Boonstra unpublished data) and then decline markedly with the dexamethasone injection and remain low with the ACTH injection (fig. 12.4). There is no correlation between testosterone and CBG concentrations (Boonstra unpublished

data). Immunologically, breeding male red squirrels have four times the numbers of white blood cells of arctic ground squirrels.

Summary

Iteroparous males do not exhibit the symptoms of chronic stress during the breeding season, whereas partially semelparous species exhibit some of them. Iteroparous species show the following: the gonadal axis is inhibited by high GC concentrations, resulting in declines in testosterone; high testosterone concentrations do not drive down CBG levels; dexamethasone resistance, though it may occur under chronic stress conditions, is not the rule under normal conditions; and immunosuppression does not occur as a normal condition. Thus the negative feedback system continues to function well. The negative feedback system also continues to function in the partially semelparous arctic ground squirrel, but the gonadal axis becomes insensitive to the inhibitory effects of high GCs and testosterone levels remain high. The consequences are that free GCs increase. Immunosuppression is particularly pronounced in the partially semelparous species.

There are four factors that may select for a partially semelparous life history, the first being an ultimate one, and the other three being proximate ones. First, high adult mortality, particularly during the nonbreeding season (Roff 1992; Stearns 1992), may be a key factor. In the short-lived arctic ground squirrel, only 17–50% of the males from high latitudes survive winter (Hubbs and Boonstra 1997; Gillis 2003). In contrast, in the long-lived Columbian ground squirrel (*S. columbianus*) from the midlatitude mountain areas, over 90% survive winter (Neuhaus and Pelletier 2001). Second, a high degree of seasonality occurs, with the length of time during a year when reproduction is favorable being only sufficient for one litter per year. In arctic ground squirrels, 60–75% of a female's active season is occupied with pregnancy and rearing her litter. Third, mating occurs at a time that is optimal for females, but not for males. As a result, insufficient food is available for males, to sustain them or to replenish energy expended on reproduction. High GC concentrations permit the replacement of external food resources with internal body reserves through the mobilization of energy by gluconeogenesis. In arctic ground squirrels, mating in the alpine occurs when snow still covers all or most of the ground, and thus males must rely either on body stores or underground caches (Gillis 2003). Fourth, the mating system is one in which intense, direct aggression occurs among males for access to females, and male territoriality either is not present or breaks down. These latter three factors may be necessary, but not sufficient, for a partially semelparous life history, as there are examples of species in which the males are long-lived and iteroparous in spite of living under these constraints (Sherman and Morton 1984).

Socially-induced reproductive suppression

Reproductive suppression is common in many species of rodents (see reviews in Solomon and French 1997), being found both in those with social systems characterized by dominant-subordinate hierarchies (e.g., marmots and molerats) as well as those characterized by simple bonding relationships (e.g., microtines). In females, suppression could occur in any of the following ways: delay of puberty, suppression of estrus, suppression of ovulation, or failure of implantation (Faulkes et al. 1990). Direct aggression or chemical signals from conspecifics may be the causal mechanism resulting in suppression. Because of the inhibitory impact of the stress axis on the gonadal axis (Wingfield and Sapolsky 2003), the stress axis is often implicated as playing a deciding role. However, because of the reciprocal interactions between the two axes, simply showing that nonreproductive subordinates have higher GC concentrations than reproductive dominants is not sufficient to conclude that the latter inhibits the former. Testosterone acts to inhibit HPA function, and estrogen to enhance HPA function (Handa et al. 1994). Thus, in lab rodents, nonreproductive males have much higher concentrations of CBG and GCs than reproductive males, and pregnant and lactating females have higher CBG and GCs than nonreproductive females. Moreover, the response to stressors can vary with the stage of the estrous cycle (Viau and Meaney 1991) and of pregnancy (Neuman et al. 1998). In wild populations of rodents, the effects of reproduction may not have the same effect on the stress axis as in lab rats, but major differences among reproductive classes with respect to GCs are generally present (e.g., meadow voles: Boonstra and Boag 1992; degus [*Octodon degus*]: Kenagy et al. 1999; arctic ground squirrels: Boonstra et al. 2001b; Columbian ground squirrels: Hubbs et al. 2000; golden-mantled ground squirrels [*S. lateralis*]: Boswell et al. 1994; and yellow-pine chipmunks [*Tamias amoenus*]: Kenagy et al. 2000). Thus the critical requirement in studies on reproductive suppression is to compare animals of the same reproductive age under different social conditions and population densities. In addition to this caveat, a number of field studies also indicate that high levels of GCs do not invariably compromise the ability to reproduce (Creel 2001). In both territorial male alpine marmots (Arnold and Dittami 1997) and breeding male Arctic ground squirrels (Boonstra et al. 2001c) high concentrations of androgens occur, in spite of high GC concentrations.

Although surprisingly few field studies in rodents have teased out the impact of the stress axis on reproductive suppression, stress may play a crucial role. The naked mole-rat *(Heterocephalus glaber)* is a highly social species living in large underground colonies of up to almost 300 individuals, where only a single dominant female and 1 to 3 dominant males breed (see references in Faulkes et al. 1990, 1991; Faulkes and Bennett, chap. 36, this volume). Changes in reproductive hormones have been closely monitored. In nonbreeding females, lutenizing hormone (LH) from the pituitary is much lower than in breeding females, ovulation is suppressed, and progesterone levels are usually undetectable. This block is removed when these females are isolated from the colony and paired with a male (Faulkes et al. 1990). The relationship to the stress axis is not entirely clear. As expected, urinary GCs are lowest in breeding females and high in suppressed females (Faulkes and Abbott 1997), and these subordinate females are subjected to shoving matches by the dominant female. However, when the latter are removed from the colony, their GC levels remain high. These high levels may be a reflection of endocrine changes occurring as a function of endogenous reproductive changes (see previous citations on the effects of reproduction on GC levels). In males, although spermatozoa are present in both breeding and nonbreeding males, testosterone levels are much lower in nonbreeding males, and they are much less sensitive to injections of gonadotropin releasing hormone (GnRH), which is released from the hypothalamus and stimulates the pituitary to release LH (this hormone stimulates the Leydig cells of the testes to produce testosterone; Faulkes et al. 1990). Thus, both nonbreeding males and females have marked endocrine deficiencies. The link to the stress axis may be as follows: GCs have effects at the level of the brain, decreasing hypothalamic GnRH release and thus GnRH stimulated release of LH from the pituitary (reviewed in Sapolsky et al. 2000). In addition, GCs also have direct inhibitory effects on the gonads, reducing responsiveness to LH and reducing the concentrations of LH receptors.

The role of the stress axis in reproductive suppression has been most clearly worked out in alpine marmots (*Marmota marmota*; Arnold and Dittami 1997; Hackländer et al. 2003). This species is also a highly social, cooperative breeder, with one dominant pair and several subordinate offspring of up to 5 years old, not all of which are related. Reproductive suppression is complete in subordinate females, but not in subordinate males. In males, dominants predominantly attack unrelated subordinates that show higher GC concentrations and androgen suppression, but sons above a critical age and mass are left alone (Arnold and Dittami 1997). All adult females breed, but only the dominant female carries the pregnancy through to parturition. The breeding subordinates are subject to aggression by the dominant, resulting in significantly higher GC concentrations in the former and falling progesterone levels (fig. 12.5, Hack-

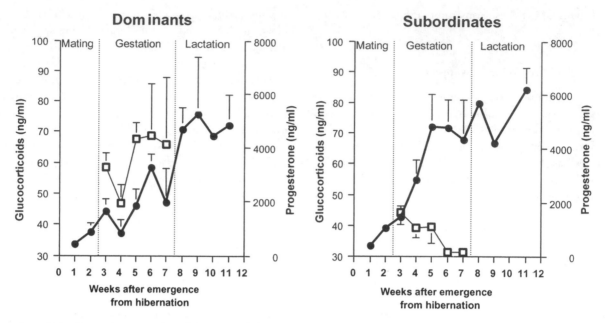

Figure 12.5 Changes in concentrations of glucocorticoids (●) and progesterone (□) in dominant and subordinate female alpine marmots during the reproductive period. Progesterone, which is needed to sustain pregnancy, was measured only during the gestation period. Parous, dominant females were those that subsequently gave birth; nonparous, subordinate females were inseminated, but pregnancy failed either before or immediately after implantation. Means are given ± SE. Adapted from Hackländer et al. (2003).

länder et al. 2003). Thus high GC concentration in the subordinates may inhibit GnRH release which results in lower progesterone levels, causing the pregnancy to terminate before or shortly after implantation (Hackländer et al. 2003).

We do not know exactly how suppression occurs in less social species, such as the voles, but recent experiments suggest that mothers do not suppress reproduction in their daughters (Wolff et al. 2001). Strangers, operating through increased social conflict at high density, may be one possible mechanism. Boonstra and Boag (1992) found that GC concentrations in the adult meadow voles were directly correlated to population density, although they did not continue the study long enough to relate this to reproductive suppression in the young. In the greater gerbil *(Rhombomys opimus)*, a more social small mammal, high density was also correlated with higher GC concentrations, probably related to increased social interactions (Rogovin et al. 2003).

Predation-induced reproductive suppression

Coping with predators is a key problem for virtually all organisms. Predators affect prey both directly (by killing them), thus potentially influencing population dynamics, and indirectly, by affecting prey behavior, foraging patterns, physiology, and reproduction, thus affecting fitness (Hik 1995; Boonstra et al. 1998; Lima 1998). Thus, predation is a major evolutionary force shaping the adaptations of the prey. In this section, we focus primarily on chronic stress that results from sustained high levels of predation

risk. There are basically two responses prey can make to chronically high predation risk: either continue to perceive it and show a chronic activation of the stress axis, resulting in the host of suppressive effects, or ignore it, at least physiologically, and get on with reproduction and the other necessities of life.

The immediate effect of predator risk on the stress axis in rodents has only been examined in the laboratory because of logistical constraints. In rats, fox odor elicits an acute stress response, causing an increase in corticosterone concentrations (e.g., Tanapat et al. 2001). When rats are chronically stressed by visual exposure to a cat (a potential predator) for 20 days, they do not habituate, even though the cat never presses home an attack (Blanchard et al. 1998). These rats show all the evidence of being chronically stressed, including higher basal corticosterone concentrations, adrenal hypertrophy, and reduced thymus weights. Some of these rats also show an enhanced stress response when challenged with an acute stressor, possibly related to a failure of the feedback system. In a laboratory study on rodents from natural populations, Eilam et al. (1999) measured both the acute stress response and the behavioral response in voles *(Microtus socialis)* and spiny mice *(Acomys cahirinus)* to owl calls. Individuals of both species showed a stress response, with higher GC concentrations. However, only the voles showed a behavioral response. Thus lack of a behavioral response is not necessarily indicative that the prey is not stressed by the threat.

The best evidence for the suppressive impact of chronic

stress of predators on natural rodent populations comes from work on ground squirrels. They are known to be extremely sensitive to predation risk, modifying their behavior both in response to direct evidence of predator presence (visual, olfactory, auditory, and tactile stimuli coming from predators), and to indirect evidence, which corresponds to the increased likelihood of encountering predators (e.g., increased foraging distance from burrows or trees, or increased visual obstructions). The only field study (to our knowledge) to measure both acute and chronic stress responses under seminatural conditions is that of Hubbs et al. (2000). Reproducing female Columbian ground squirrels were exposed to a dog (the model predator) over 8 wks. Predator-challenged females had higher levels of total and free cortisol than controls, with evidence of a heightened stress response occurring only after about one month of exposure. Nonreproductive arctic ground squirrels living in the predator-rich boreal forest exhibit evidence of chronic stress relative to those in the adjacent predator-poor alpine area (Hik et al. 2001). These squirrels exhibit lower levels of basal-free cortisol levels, dexamethasone resistance in females (but not males), reduced ability to respond to an ACTH challenge, and lower corticosteroid-binding globulin levels. Furthermore, evidence suggests that chronic physical and psychological stressors inhibit reproduction in arctic ground squirrels. In the same boreal forest as the Hik et al. (2001) study, a long-term experimental manipulation was carried out in which mammalian predators were excluded from a 1 km² area. Litter sizes and weaning rates were generally higher within the predator exclosure (Karels et al. 2000). This evidence is consistent with the hypothesis that reproduction is suppressed under conditions of chronically high predation risk and with similar findings on snowshoe hares (*Lepus americanus*; Boonstra et al. 1998). Thus, reproduction is suppressed in some species in response to the chronic stress of high predation risk.

In contrast, some species may have evolved to not be stressed by their predators. Initially, microtines (voles and lemmings) were predicted to exhibit reproductive suppression under conditions of high predation risk, particularly of weasels. This was the basis of the predator-induced breeding suppression hypothesis, which postulated that it was adaptive to delay reproduction until such time as predator density declined (Ylönen and Ronkainen 1994). Though GC concentrations were not measured, evidence in favor of this hypothesis came largely from laboratory studies using weasel odor. This odor produced suppression of reproduction in pairs of voles and delayed maturation in young females (Ylönen and Ronkainen 1994). However, most field studies using mustelid odor have failed to corroborate these findings (Wolff and Davis-Born 1997; Mappes et al. 1998), and thus predator-induced breeding suppression appears to be an artifact of the laboratory (Wolff 2003c). In contrast,

a recent field study by Fuelling and Halle (2004) reports evidence in favor of the breeding suppression hypothesis in northern Norway. We think that methodological problems (performed only for one month in mid-late August, and the possibility of a neophobic response of young born prior to the treatment avoiding traps during the treatment) call the conclusion into question. Theoretical modeling indicates that delayed reproduction is only optimal when the number of future offspring produced by not breeding exceeds that of breeding immediately (Kokko and Ranta 1996). For microtines, it may never pay to delay reproduction in the face of predation, given their short life spans and seasonal breeding.

Impact on Aging

Senescence is defined as an age-related increase in mortality rate that can be attributed to physiological deterioration (Rose 1991). Some argue that the rate of extrinsic mortality is so high in natural populations from competition, predation, parasites, and environmental stressors that animals never live long enough to experience an age-related physiological deterioration, and thus senescence would not evolve (Hayflick 2000). However, this flies in the face of theory and of most evidence from a wide variety of taxa. An age-related increase in mortality rate has been detected in many long-lived mammals (e.g., Packer et al. 1998 [*Papio anubis, Panthera leo*], Loison et al. 1999 [*Capreolus capreolus, Ovis canadensis, Rupicapra pyrenaica*]) and birds (McDonald et al. 1996 [*Aphelocoma coerulescens*]). Evidence also suggests that free-ranging animals experience age-related declines in reproduction (Packer et al. 1998, Coltman et al. 1999 [*Ovis aries*], Ericsson et al. 2001 [*Alces alces*]). Moreover, a few papers have integrated measures of survival and reproduction and have demonstrated that animals experience age-related declines in fitness (e.g., reproductive value: Newton and Rothery 1997; Møller and de Lope 1999; Ericsson et al. 2001).

In rodents, however, the evidence is mixed. Senescence was not found in five grassland species from Kansas (Slade 1995). In contrast, evidence for senescence was found in meadow voles (Boonstra and Mihok unpublished data). In the former case, uncertainty of age was a problem, whereas in the latter case, exact age was known. However, known-aged, female yellow-bellied marmots *(M. flaviventris)* apparently did not exhibit an increase in mortality rate with age but showed a total lack of reproduction in the last four years of life, suggesting reproductive senescence (Schwartz et al. 1998).

For natural populations, we do not know the nature of the physiological changes that cause an age-related increase in mortality rate, and thus we rely on evidence from laboratory populations as to the possibilities. There are three periods where the stress axis is sensitive to permanent organizational changes likely to affect age-dependent mortality: the prenatal-postnatal period, the juvenile-adult period, and periods of chronic stress. First, stressors experienced during pregnancy and the postnatal period, and a reduced level of maternal care following birth, result in nongenetic, lifetime programming of stress axis, which increases susceptibility to disease (Meaney 2001). Such offspring are more fearful, respond more strongly to stressors, and recover from them more slowly (Meaney 2001; Matthews 2002). The inherent plasticity to permit this programming may be adaptive, because it allows the environmental factors experienced by the mother to program the offspring to perform optimally for conditions that it will likely face (Meaney 2001; Welberg and Seckl 2001). For example, if the mother is in an environment where predation is unusually high, it may be beneficial for her offspring to be programmed with an extremely active HPA axis so that they are hypervigilant to predators. However, programming hypervigilant offspring means that they are likely to experience the costs of a much higher GC exposure, which may affect survival. In laboratory rodents and humans, the high levels of stress hormones that result from programming are associated with stress-related diseases in later life. There is no direct evidence from natural populations of rodents indicating that pre- or postnatal programming affects the stress axis. However, indirect evidence suggests that maternal effects may operate this way. This evidence relies on the negative reciprocal interaction between the stress axis and the reproductive axis (Wingfield and Sapolsky 2003). In cycling microtines, declining populations often show extremely low rates of survival and/or reproduction (see review in Boonstra 1994). When meadow voles from declining and low populations are brought into the lab, females and their laboratory-born progeny continue to breed poorly compared with those collected from increasing populations (Mihok and Boonstra 1992; see Sinclair et al. 2003 for comparable evidence from snowshoe hares). These intrinsic effects may be the result of stressors experienced during the decline, resulting in programming of the HPA axis that then has negative effects on the gonadal axis.

Second, the hippocampus is crucial for declarative and spatial leaning and memory, playing an important role in age-related declines of cognition (Eichenbaum 1997). The hippocampus has high concentrations of GC receptors (fig. 12.1), which perceive circulating GC concentrations, setting both basal GC concentrations and terminating the stress-related release of GCs. However, both normal aging and excessive activation of the HPA axis result in hypersecretion of GCs, causing hippocampal damage (dendritic atrophy, loss of GC receptors, and synaptic loss; Pedersen

et al. 2001). This damage impairs feedback inhibition of the HPA, leading to further damage caused by elevated GC concentration. Ultimately a positive, self-reinforcing cascade is set up, leading to progressively greater damage and a reduced ability to respond adaptively to stressors (the "glucocorticoid cascade hypothesis" Sapolsky et al. 1986). The studies of lab rodents have provided much of the evidence for our understanding of these changes in the stress response with age. Future studies should attempt to examine these changes in the natural world.

Third, long-term, chronic stressors also accelerate the rate of aging. In rats, chronic stress accelerates electrophysiological and morphological changes in the hippocampus, which have been correlated with altered HPA activity and impaired cognitive function (Pedersen et al. 2001). In addition, chronic stress in rodents causes neuron damage as well as a reduction in the GC receptor levels in the hippocampus, with the net result that negative feedback mechanisms are inhibited and the stress axis hyperactivated with age (Nichols et al. 2001). In nature, situations of long-term chronic stress are likely to be less common, though they may occur under times of high predation risk or high population density when there is competition for access to resources.

The preceding evidence all deals with changes in the stress axis that potentially occurs at the individual level to affect the rate of aging, but there is little evidence that different rodent species are programmed to age at different rates contingent on their life history. This is, however, a reasonable expectation, given that such evidence is becoming available or suggested in other groups, such as salmon and semelparous marsupial dasyurids (Finch 1990; Bradley 2003).

Impact on Neurogenesis

Conventional wisdom assumed that mammalian brains did not generate new neurons after early development (Gross 2000). However, we now know this is wrong, and that neurogenesis is a ubiquitous feature occurring in all mammals until death. Laboratory studies on rodents (primarily on mice and rats, but also on hamsters—Huang et al. 1998; eastern gray squirrels [*Sciurus carolinensis*]—Lavenex et al. 2000; and meadow voles—Galea and McEwen 1999) have played a key role in our understanding of the process, elucidating what factors increase or decrease levels of neurogenesis. However, we do not know the adaptive and functional significance of neurogenesis for animals in the natural world, nor do we know how stressors affect neurogenesis in nature (Boonstra et al. 2001a). Only three studies have examined neurogenesis in free-living rodents (Sivalingam

2002; Amrein et al. 2004; Barker et al. 2005). To put our understanding on a firm foundation, neurogenesis must ultimately be related to the ecology and evolutionary biology of species in the natural world.

In the hippocampus, neurogenesis appears to be related to spatial memory. Increased or decreased levels of neurogenesis are correlated with improved or impaired spatial memory, respectively (e.g., Lee et al. 1998). In one study involving spatial memory and neurogenesis in natural populations (Barker et al. 2005), eastern gray squirrels (a scatter-hoarding species) and yellow-pine chipmunks (a larder-hoarding species) were studied during the autumn, when memory of food storage locations would be critical. Squirrels had much higher rates of cell birth than chipmunks (as predicted), but not of early neuron survival. In the olfactory bulb, neurogenesis appears to be related to memory of odors (Carleton et al. 2003), particularly the discrimination of different odors.

Stressors at all life stages in laboratory rodents adversely affect the rate of neurogenesis. Prenatal and early postnatal stress reduces hippocampal neurogenesis in male rats (Schmitz et al. 2002), but this may be ameliorated later in life by an enriched environment and by an enhanced opportunity for learning (Ehninger and Kempermann 2003). Young rat pups show increased corticosterone concentrations and decreased rates of neurogenesis in response to the odors of adult male rats that are known to commit infanticide (Tanapat et al. 1998). In adult rats and mice, stressors produce high corticosterone concentrations that then decrease the rate of granule cell production in the hippocampus (Cameron and Gould 1994). Predator odor also inhibits neurogenesis in adult rats (Tanapat et al. 2001). Finally, stressors induced by the social situation can influence the rate of neurogenesis (Lu et al. 2003). Group housing of mice (the normal, unstressed situation in the lab) increases the number of newly generated neurons in the dentate gyrus, whereas rearing rats in isolation decreases neurogenesis, despite having no effect on endogenous GC concen-

trations. If the presence of conspecifics involves aggressive interactions, group housing can be stressful and decrease hippocampal neurogenesis (Czéh et al. 2002). Thus, rates of neurogenesis are heavily dependent on environmental conditions and appear to be readily altered, and, because of the potential implications on memory, may alter individual fitness in the natural world. We do not know whether stressors in nature will differentially affect rates of neurogenesis in wild rodents contingent on their life histories, though that is a reasonable expectation.

Summary

Coping with change is a key requirement for survival and reproduction, from both a short-term, ecological perspective and from a long-term, evolutionary perspective. The stress axis is a key control mechanism of the neuroendocrine system, which plays a central role in life-history adaptations that deal with change. At the individual level, the stress axis is integral to the regular changes associated with body function over the daily and seasonal cycles of life, and to dealing with stressors that threaten the homeostasis of the organism. However, at the individual level the axis shows a high degree of plasticity and sensitivity, and is subject to either long-term or permanent change as a result of either chronic stressors or of stressors occurring during development that permanently programs the axis at the level of the brain. At the species level, the functioning of the stress axis is not fixed, but is subject to evolutionary modification and habitat that sets the context for that modification. Differences in the functioning of the stress axis are seen most clearly in the area of reproduction (impact of mating in males and suppression of reproduction by either conspecifics or predators). We also expect, but have no evidence for at present, that differences among species in the rates of aging and in the need for neurogenesis will constrain how the stress axis responds to challenges.

Chapter 13 Dispersal and Philopatry

Scott Nunes

AT SOME POINT in their lives, nearly all organisms are faced with the "decision" of being philopatric—remaining where they are, or dispersing—moving to a new area or joining a new social group or population. This decision has important consequences for the evolutionary fitness of individuals. Moreover, because philopatry and dispersal determine the distribution of individuals, they have important consequences for the maintenance of ecosystems, the demographic and genetic structures of populations, the spread of disease, and the conservation of species (Clobert et al. 2001; Nathan 2001; Bullock et al. 2002; Ostfeld and Mills, chap. 41, this volume).

Dispersal has been extensively studied in mammals, and in particular within a broad range of rodent species. In mammalian biology, the term *dispersal* is used to describe a process in which individuals leave their home site or social group and establish residence in a new area or group (Stenseth and Lidicker 1992). In many cases, the process of dispersal is preceded by exploratory forays into potential new home areas or groups. When dispersal occurs, it includes three general stages: emigrating from the original home area or group, traversing the landscape to get to the new home, and establishing residence in the new home area or group. This movement process is potentially risky for animals. Dispersers may find themselves in unfamiliar areas, where they are vulnerable to predation (Metzgar 1967), and may encounter social resistance while attempting to immigrate into their new home areas (Wolff 1994a; Solomon 2003). Furthermore, increased vigilance and involvement in agonistic interactions during the dispersal process may

place energetic demands on individuals (Nunes and Holekamp 1996).

The dispersal process can be categorized as either natal or post-breeding. In natal dispersal, a young animal leaves its birth area before reproducing. In post-breeding dispersal, an individual leaves an area after it has successfully reproduced, and settles into a new breeding site. In mammals, natal dispersal is more common than breeding dispersal, occurring in nearly all species. Natal dispersal has a strong sex-bias in mammals, with males typically emigrating to new home areas at higher rates or over greater distances than females.

Because of its ubiquity, dispersal is an important phenomenon among taxa. Here, I evaluate dispersal among rodents, examining the ultimate and proximate causes as well as the consequences of breeding, and natal dispersal and philopatry. Because rodents are an incredibly large and diverse group of animals (Nowak 1991[vol. 1], 1991[vol. 2]), they exhibit a broad range of dispersal and philopatric behaviors, and are an excellent model system for understanding dispersal and philopatry.

Natal Dispersal

Ultimate causes of natal dispersal

In assessing the possible evolutionary forces that brought about sex-biased natal dispersal in mammals, two important questions arise. First, why do young animals leave the relative safety and familiarity of their birth areas, expose themselves to risk, and seek out and settle in new homes? Second, why is natal dispersal biased toward young males? Three explanatory hypotheses are generally proposed to explain the adaptive function of mammalian natal dispersal: young animals disperse (1) to avoid inbreeding, or in response to competition for (2) mates or (3) environmental resources (Greenwood 1980; Pusey 1987; Pusey and Wolf 1996). These hypotheses also offer possible explanations for the direction of sex bias in mammalian natal dispersal. The selection pressures that shape behavior can vary among groups of animals, and it is possible that dispersal behavior can benefit animals in different ways. Thus it is not likely that a single hypothesis can explain dispersal in all species. It is possible that in some cases more than one selective pressure shaped the evolution of dispersal behavior (Dobson and Jones 1985). Moreover, although there is a strong male bias in mammalian natal dispersal, when young females emigrate from the natal area they may do so for different adaptive reasons than do young males (e.g., Dobson 1979; Nunes et al. 1997; Gundersen and Andreassen 1998).

Inbreeding avoidance

Studies of rodents suggest that avoidance of inbreeding is an important factor in the evolution of dispersal behavior. That is, individuals leave their natal sites to find breeding areas or groups with potential mates who are unrelated to them. Inbreeding depression is a well-documented phenomenon. Although some degree of inbreeding can benefit animals by maintaining gene complexes adapted to local environmental conditions, consanguineous matings can detrimentally affect offspring by reducing heterozygosity and increasing expression of recessive lethal alleles.

Animals have a variety of behaviors for avoiding matings with close relatives (Blouin and Blouin 1989). In a study of black-tailed prairie dogs (Cynomys ludovicianus), Hoogland (1982) observed that females are less likely to become estrus when their fathers live nearby, and behaviorally avoid mating with close male relatives. Hoogland suggested that in addition to these mechanisms, natal dispersal by male prairie dogs also prevents inbreeding.

Moore and Ali (1984), however, suggested that animals rely on behaviors that are not as potentially risky as dispersal to avoid inbreeding. Because most mammals have polygynous mating systems, variance in reproductive success tends to be substantially greater for males than females. That is, males can potentially have more mating partners and father many more young than females. Thus inbreeding depression associated with consanguineous mating is likely to have a greater negative impact on the reproductive success of females than males, and reproductive fe-

males should avoid choosing male relatives as mates. In turn, males should disperse from their natal areas so as to increase the likelihood of encountering females willing to mate with them. According to Moore and Ali (1984), inbreeding avoidance is not a cause of natal dispersal, but rather natal dispersal is a consequence of females not selecting male relatives as mates.

In a study of white-footed mice (Peromyscus leucopus), Keane (1990b) observed that matings between siblings yielded smaller litters and offspring than did matings between mice with a lower coefficient of relatedness, and suggested that this was a result of inbreeding depression. He also observed that female mice discriminated between the odors of males based on their relatedness, and suggested that females might use relatedness as a criterion for selecting a mate. Keane (1990a, 1990b) suggested, as proposed by Moore and Ali (1984), that inbreeding avoidance is facilitated by the behavior of female mice, and that male mice must leave the natal area to gain access to receptive mates.

By contrast, Wolff (1992, 1993b) suggested that inbreeding avoidance was the main cause of natal dispersal in both male and female white-footed mice. He conducted removal experiments and observed that young mice dispersed at higher rates when the opposite-sex parent remained in the area. Moreover, young mice remaining in the natal area along with the opposite-sex parent were less likely to become reproductively mature than were young whose opposite-sex parent had been removed. This finding is consistent with the idea that young mice avoid mating with close relatives, and disperse to increase their chances of encountering mates to whom they are not closely related. Wolff (1992) also observed that the presence of same-sex parents did not influence dispersal by male mice, which does not support the hypothesis that competition with same-sex relatives is the motivating force for natal dispersal in white-footed mice.

In earlier work, Wolff et al. (1988) found that natal dispersal by white-footed mice decreases as population density increases, and presumably competition for environmental resources also increases. Wolff et al. (1988) surmised that resource competition is not a driving force behind natal dispersal in this species, and suggested that competition for resources may hinder dispersal by causing young mice to remain in the natal area when fewer possible home sites overall are available for settlement.

Wolff (1994a) further suggested that the male bias in natal dispersal may arise from the fact that in most mammalian species the reproductive life span is ordinarily longer in females than males. Thus the reproductive tenure of a young male who remains in his natal area is likely to overlap with that of his mother, whereas the reproductive lifespan of a young female in her natal area is not as likely to overlap with that of her father. Therefore, male but not female mice should disperse, to reduce the likelihood of inbreeding. Clutton-Brock (1989a) observed that in the few mammalian species in which natal dispersal is female biased, the reproductive tenure of males in the natal area or natal group is consistently greater than the age at which females begin reproducing.

The dispersal patterns observed by Wolff (1992) in white-footed mice have also been documented in other rodent species. Young disperse in response to the presence of the opposite-sex parent in root voles (*Microtus oeconomus;* Gundersen and Andreassen 1998) and Richardson's ground squirrels (*Spermophilus richardsonii;* Michener and Michener 1977). Moreover, in Belding's ground squirrels (*S. beldingi*) and a variety of other mammalian species, all young males emigrate from the natal area regardless of ecological or social conditions, suggesting that competition for mates or environmental resources is not driving natal dispersal in these species (Holekamp 1984, 1986; Smale et al. 1997).

Waser et al. (1986) postulated that if inbreeding avoidance were an evolutionary cause of dispersal, then in polygynous species females rather than males should disperse. They reasoned that because the maximum reproductive output of females is more limited than that of males, inbreeding depression in offspring is more costly to the fitness of females than males. Thus females should benefit more by dispersing to find unrelated mates. However, Lehmann and Perrin (2003) proposed that inbreeding avoidance is a viable ultimate explanation for male-biased natal dispersal in polygynous species if female choice is taken into account. That is, because inbreeding depression can be costly for females, natural selection should favor females who can discriminate between closely related males and more distantly related or unrelated males, and who reject as mates males who will sire inbred offspring. Lehmann and Perrin (2003) further suggested that the choosiness of females for unrelated males should be influenced by inbreeding load and the costs of finding unrelated mates. Thus according to Lehmann and Perrin (2003), inbreeding avoidance can explain sex-biased dispersal in polygynous species, whereas the direction of sex-bias can be explained by female mate choice. The occurrence of kin recognition abilities and female mate choice in a variety of rodent species (e.g., Sherman 1976; Holmes 1994, 1995; Mateo and Johnston 2000a) provide some empirical support for this idea.

Competition for mates

Dobson (1982) suggested that competition for mates may have acted synergistically with inbreeding avoidance in shaping male-biased dispersal behavior in rodents and other mammals. Because most mammalian mating systems are polygynous, competition for mates tends to be more intense among males than females. Thus Dobson (1982) suggested that males might benefit more than females from dispersing to areas with less competition for mates. Hence, dispersal may have evolved to reduce inbreeding, and competition for mates may have made dispersal more functionally adaptive for males than females in species with polygynous mating systems. Dobson (1982) in fact observed that males are the predominant dispersers in polygynous species, whereas both males and females tend to disperse in species with monogamous mating systems. However, Dobson (1979) also observed that male California ground squirrels (*S. beecheyi*) still emigrate from the natal area when adult males have been removed and competition is likely to be low. The specific importance of competition for mates as a driving force in the evolution of sex-biased dispersal remains to be elucidated. However, competition for mates does appear to have a distinct influence on some aspects of male dispersal. For example, male Columbian ground squirrels (*S. columbianus*) tend to settle in areas where the ratio of males to females is low, and consequently competition for mates might be low (Wiggett and Boag 1993b).

Competition for environmental resources

Competition for food and nest sites has been suggested to be an important ultimate cause of dispersal in female mammals. The energetic requirements of females increase dramatically during gestation and lactation, making food a vital resource for reproduction. Moreover, infanticide by conspecifics is common among mammalian species, and may be adaptively beneficial to individuals by reducing intraspecific competition that their own offspring will face in the future (Hrdy 1979; Wolff 1993a; Ebensperger and Blumstein, chap. 27, this volume). Establishing and defending territories from infanticidal conspecifics is one defense mechanism that females use to protect their young (Wolff 1993a; Solomon and Keane, chap. 4, this volume). Thus space in which to establish a territory is also a critical resource for reproduction in females. In a food supplementation experiment with California ground squirrels, Dobson (1979) observed that females were more likely to immigrate into the supplemented colony than unsupplemented colonies; however, males were equally likely to settle in both supplemented and unsupplemented colonies. By contrast, in a food provisioning experiment with Belding's ground squirrels, Nunes et al. (1997) observed that females were more likely to disperse away from areas receiving extra food, whereas emigration and immigration by males were not affected by food provisioning. They also observed that experienced adult females tended to shift their territories so that they were closer to food boxes, causing an increase in local population density. They further suggested that this increased density decreased the probability of successful competition by young females for territories during the next breeding season, prompting young females to emigrate and settle in areas with lower density. The findings of Dobson (1979) and Nunes et al. (1997) support the idea that competition for environmental resources was an important factor in the evolution of female but not male dispersal in rodents, and mammals in general.

Proximate causes of natal dispersal

A wide variety of factors have been proposed as triggers of dispersal in rodents. Changes in the habitat, demography, or population density of the home area or a nearby area may influence the future prospects of animals in either of these areas, and make it advantageous for them to transfer residence to a new area. Some of the potential triggers of dispersal are reviewed in the following.

Early developmental influences on dispersal behavior

For most rodent species, dispersal behavior is malleable and responsive to environmental variables (e.g., Lambin 1994a). However, some species exhibit rigid natal dispersal behavior, in which the probability of dispersal is not influenced by environmental conditions. These inflexible predispositions to disperse may be brought about by the action of hormones during early development (Holekamp et al. 1984; Ims 1989, 1990).

Androgens can have long-lasting influences on the brain during early development, which in turn can influence behavior later in life. Holekamp et al. (1984) and Nunes et al. (1999) suggested that androgens promote development of a predisposition to leave the natal area in young male Belding's ground squirrels. All male squirrels eventually leave the natal area, regardless of ecological and social conditions, whereas most females remain in their natal areas throughout their lives. Holekamp et al. (1984) and Nunes et al. (1999) observed that treatment of young female squirrels with androgens at the time of birth caused them to disperse at higher rates than control females, and suggested that androgens promote development of an innate drive to leave the natal area. This influence of androgens results in fairly inflexible behavior among young male squirrels, with males emigrating regardless of social conditions and resource availability in the natal area. Environmental conditions in the habitat of Belding's ground squirrels fluctuate greatly over the course of a year, but these seasonal changes are highly predictable, and this predictability might obviate the need for flexibility in the dispersal behavior of young males (Smale et al. 1997).

Androgens have also been implicated in the development of dispersal behavior in female grey-sided voles (*Clethrion-*

omys rufocanus; Ims 1989, 1990). In particular, female voles from male-biased litters tend to leave their natal areas at higher rates than do females from unbiased or female-biased litters, and their likelihood of dispersal is not influenced by social or ecological conditions. Ims (1989, 1990) proposed that female voles in male-biased litters have an increased probability of developing next to male fetuses in utero, and thus may be exposed to androgens produced by neighboring males (see vom Saal and Bronson 1978, 1980). Androgens might help organize dispersal tendencies in these females. Lambin (1994a) did not observe an effect of sex ratio on the dispersal behavior of Townsend's voles (*Microtus townsendii*), suggesting that the effects observed by Ims (1989, 1990) may not be universal among vole species. The adaptive benefit of androgenic effects on dispersal for female grey-sided voles is not fully understood, but they may help to maintain variability in the tendency to disperse (Ims 1989, 1990; vom Saal and Bronson 1978, 1980).

Bondrup-Nielsen (1993) proposed that nutrition during infancy influences dispersal in female meadow voles (*Microtus pennsylvanicus*). Female voles malnourished during early development displayed greater behavioral tendencies associated with dispersal as adults than did control females, despite being well nourished at the time they expressed these tendencies. Bondrup-Nielsen (1993) suggested that poor nutrition during early development might be an indicator of an unstable or unpredictable habitat, and might adaptively promote development of behaviors causing female voles to seek environments that are more stable.

Availability of environmental resources

In some cases, dispersal appears to be dependent on availability of vacant habitat in which to settle, and dispersal may be delayed or inhibited in the absence of obtainable, high-quality habitat with sufficient resources to support survival and reproduction (Solomon 2003). Rates of dispersal tend to decline when habitat is saturated, and competition for available space, nesting areas, or food is high (Wolff 1994a; Kokko and Lundberg 2001; Solomon 2003). For example, inhibited dispersal under conditions of high population density has been observed in field voles (*Microtus agrestis;* Sandell et al. 1991) and prairie voles (*M. ochrogaster;* McGuire et al. 1993). By contrast, dispersal rates tend to be high when nearby habitat has nest sites available for settlement (Wolff 1994a). For example, in removal studies of pocket gophers (*Thomomys talpoides*), areas from which gophers were eradicated were rapidly recolonized (Engeman and Campbell 1999). Moreover, continuous removal of gophers did not result in long-term reduction of population size (Sullivan et al. 2001). Similarly, removal of yellow-bellied marmots (*Marmota flaviventris*) from a colony resulted in increased recruitment of marmots into the colony (Brody and Armitage 1985). Removal studies with California ground squirrels also resulted in rapid recolonization of vacated habitat (Dobson 1981). However, male squirrels were less likely to immigrate into removal areas than were females, suggesting that vacant habitat is a stronger trigger of dispersal for females.

Presence of potential mates
Female prairie voles, root voles, and common voles (*M. arvalis*) may remain in a reproductively quiescent state while in the natal area (McGuire and Getz 1991; Gundersen et al. 1999; Heise and Rozenfeld 2002). However, natal female prairie voles display investigative and affiliative behavior and become reproductively mature when exposed to unfamiliar males (McGuire and Getz 1991). Moreover, female root voles preferentially settle in habitat with an experimentally high density of males, whereas male root voles are equally likely to settle in areas with few or many females. These findings suggest that female voles may be prompted to disperse by the presence of unfamiliar males in areas outside the natal site.

By contrast, Solomon (2003) observed that female prairie voles are as likely to settle in an area with potential mates as they are to settle in an area from which all males have been removed. Moreover, Lurz et al. (1997) observed that food availability was the main factor affecting female dispersal in Eurasian red squirrels (*Sciuris vulgaris*), and that male dispersal was influenced by the distribution of females. That is, females tended to settle in areas with abundant food, whereas males tended to settle in areas with potential mates. These findings suggest that although the presence of mates may influence dispersal in both males and females, availability of environmental resources may be a more important consideration for females than availability of mates.

Presence of opposite-sex parent
As noted previously, dispersal of young from the natal area can be triggered by the presence of the opposite-sex parent. This has been documented in white-footed mice (Wolff 1992), root voles (Gundersen et al. 1999), and Richardson's ground squirrels (Michener and Michener 1977). Moreover, McGuire et al. (1993) observed that emigration from the natal site was more common in prairie voles when potential mates in the natal group were parents or siblings than when they were nonrelatives, suggesting that the presence of opposite-sex siblings as well as parents might prompt dispersal (see the following).

Interactions with siblings
Jacquot and Vessey (1995) suggested that interactions with siblings trigger natal dispersal in white-footed mice. They observed that female mice dispersed farther when they were from large litters, or when they had many brothers. Jacquot and Vessey (1995) also observed that male mice with more than one littermate sister dispersed greater distances than did males with fewer than two sisters. They proposed that their results are consistent with the hypotheses that inbreeding avoidance is an important determinant of dispersal for both male and female mice. Bollinger et al. (1993) also observed in meadow voles that the presence of siblings can influence dispersal behavior. Voles released onto experimental plots with same-sex siblings were more likely to disperse from the plots than were voles released with nonsiblings. McGuire et al. (1993) observed that dispersal was more common in prairie voles when potential mates were siblings or other close relatives than when they were nonrelatives. Ribble (1992) similarly observed that dispersal distances increase with number of siblings in male mice (*Peromyscus californicus*) and with number of sisters in female mice, and suggested that competitive interaction among siblings might enhance dispersal tendencies.

Attainment of sufficient preparation
Although dispersal is risky, mechanisms exist for minimizing risks. For example, all male Belding's ground squirrels eventually leave their natal areas. However, they delay dispersing until they achieve a threshold body mass and sufficient body fat to allow them to meet the potential physical and energetic demands of dispersing (Nunes and Holekamp 1996; Nunes et al. 1998). Nunes et al. (2004) further observed that young males who engage in social play at high rates early during the juvenile period have better motor coordination and are more likely to disperse by the end of the juvenile summer than are males who play at low rates. They proposed that adequate motor development may be a requisite for dispersal in young male squirrels. Bekoff (1977) and Thompson (1998) further proposed that interaction during play bouts with siblings or conspecifics at the same stage of development might reveal information about a young animal's behavioral proficiency and competitive ability. Individuals might then use this information in determining when they are physically, behaviorally, and socially prepared for the potential risks and demands of dispersing.

Timing of natal dispersal

In many species, natal dispersal occurs near the time of reproductive maturity. However, in other species, dispersal from the natal area occurs long before the attainment of reproductive competence (Smale et al. 1997). Early dispersal is most likely to occur when individuals begin mating at the onset of reproductive maturity and changes in environmental conditions are highly predictable, allowing individuals to reliably anticipate and prepare for their first breeding season via early settlement in the new home area. In the following I assess three possible adaptive advantages of early dispersal in male Belding's ground squirrels, a species in which most males disperse between 7 to 10 weeks of age but do not begin mating until they are two years old.

First, early dispersal by young male squirrels may reduce the social resistance they encounter as they establish residence in their new home areas. Males who disperse as juveniles leave home at the end of the breeding season, when adults are fattening for hibernation and agonistic interactions are rare. By contrast, males who disperse as yearlings leave home during the following breeding season, when aggression is high among adult males as they compete with each other for mates and among reproductive females as they defend maternal territories.

Second, early dispersal may provide an energetic or size advantage at the time of breeding. According to this hypothesis, early dispersal allows young males to increase growth or acquisition of energy reserves, possibly by avoiding social resistance as they settle in their new home areas (see the preceding). If young males are able to time dispersal so that agonistic interactions with residents in their new home are minimal, they not only can reduce energy expenditure, but also increase time available for foraging. Preliminary work with Belding's ground squirrels suggests that young males may in fact gain an energetic advantage by dispersing early. Males who disperse as juveniles tend to be larger at two years of age than are males who disperse as yearlings (fig. 13.1). Because size is correlated with fighting success, and fighting success is correlated with mating success, this size difference may parlay into increased reproductive success at the beginning of the reproductive lifespan of early dispersers.

Finally, early dispersal by young males may allow them to establish social relationships with potential mates. In many species, females are selective in choosing mates, preferring some males over others, and might be more likely to mate with males if they have had the opportunity to fully assess them. Early dispersal might also provide males with the opportunity to become familiar with potential mates as well as other males in their new home area, which may be beneficial to them as they compete for mating opportunities during the breeding period.

Consequences of natal dispersal

Genetic structure of populations

The development of new techniques in molecular genetics has contributed considerably to our understanding of the relationship between dispersal and the genetic structure of populations (e.g., Dobson, chap. 14, this volume). Molecular techniques have been used to provide support for the hypothesis that dispersal by male rodents and male mammals facilitates gene flow within and between populations. Ehric and Stenseth (2001) used mitochondrial and microsatellite DNA analysis to evaluate the genetic structure of Siberian lemming (*Lemmus sibiricus*) populations and observed extensive genetic variation. Moreover, local groups of females were more genetically similar than local groups of males, suggesting that gene flow is more prominently facilitated by male than female dispersal.

Genetic studies of other rodent species have also indicated that dispersal is important in maintaining the genetic diversity of populations. In black-tailed prairie dogs, long-established colonies tend to be more genetically similar than newly formed colonies, presumably because gene flow facilitated by dispersal has counteracted founder effects in older colonies (Roach et al. 2001). Santos et al. (1995) observed that genetic diversity decreased in populations of root voles when opportunities for dispersal were limited. Limited opportunities for dispersal have also been linked to reduced gene flow and reduced genetic diversity in Malagasay jumping rats (*Hypogeomys antimena*) in Madagascar (Sommer et al. 2002).

In a study of North American collared lemmings (*Dicrostonyx groenlandicus*), Ehrich et al. (2001a, 2001b) suggested that the degree of genetic similarity among distantly separated populations could only be explained by long-distance dispersal. They noted that lemmings are occasionally observed on sea ice, and suggested that transport by ice may facilitate long-distance dispersal. Dispersal distances over a kilometer have been observed in white-footed mice, yellow-bellied marmots, and Belding's ground squirrels

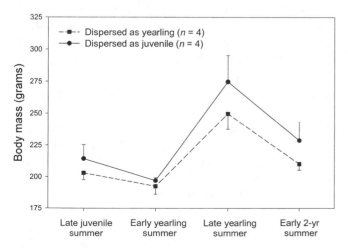

Figure 13.1 Body mass of male Belding's ground squirrels evaluated according to their age at dispersal. The same eight males are represented at each time point. Early dispersal appears to be associated with a body mass advantage early in males' second summer, when they begin mating.

(Sherman 1976; Salsbury and Armitage 1994; Bowman et al. 1999; Maier 2002), raising the possibility that periodic long-range dispersal is important in maintaining genetic relatedness among distantly separated populations. Telfer et al. (2003) observed that populations of water voles (*Arvicola terrestris*) living on islands separated by distances large enough to prohibit regular long-distance dispersal had substantially lower levels of genetic diversity than did large, continuous populations on the mainland.

Conservation of species

Dispersal also has important implications for the conservation of species. Dispersal facilitates gene flow between populations, and can be important in maintaining genetic diversity within populations. Maintaining gene flow is especially important in small populations, which are more likely than large populations to become inbred, because they provide fewer options for choosing mating partners. Dispersal may in fact counteract the stochastic tendency for alleles to be lost in small populations (Aars and Ims 2000).

Barriers to dispersal can hinder gene flow, and thus decrease genetic variability within a population and increase genetic differentiation among populations (e.g., Santos et al. 1995; Landry and Lapointe 2001). In the long term, changes in the genetic structure of populations can decrease their viability and lead to local population extinctions (e.g., Aars et al. 2001; Roach et al. 2001). Habitat destruction associated with human activity can cause continuous populations to become fragmented. Moreover, inhospitable terrain between fragments of undisturbed habitat can impede dispersal. For example, root voles in experimentally subdivided populations suffered increased predation when they attempted to disperse across inhospitable area separating habitat patches (Aars et al. 1999). Introduction of exotic predatory species can also reduce large populations to disconnected fragments and hinder dispersal, as was observed by Aars et al. (2001) among water vole populations in Scotland after introduction of the American mink (*Mustela vison*). Thus maintaining dispersal routes in fragmented habitat can be vital to the continued existence of populations in the area.

Use of dispersal corridors—narrow areas continuous with and connecting larger habitat areas—for dispersal has been observed in undisturbed populations of Columbian ground squirrels transferring between distant colonies (Wiggett and Boag 1989). Preserving dispersal corridors in fragmented habitat has been suggested to beneficially augment gene flow in a variety of rodent species, including root voles (Andreassen et al. 1998; Aars and Ims 1999; Ims and Andreassen 1999; Andreassen and Ims 2001), gray-tailed voles (*Microtus canicaudus*; Davis-Born and Wolff 2000),

and Piute ground squirrels (*Spermophilus mollis*; Antolin et al. 2001). Dispersal corridors can also enhance gene flow by allowing individuals to make excursions to distant habitat to mate with unfamiliar individuals. Such mating excursions have been suggested to occur in root voles and Japanese wood mice (*Apodemus argenteus*; Ohnishi et al. 2000).

Breeding Dispersal

Breeding dispersal occurs after successful reproduction, when an adult animal departs its breeding area and moves to a new one. Breeding dispersal has not been studied as extensively as has natal dispersal. However, three general hypotheses have been proposed to explain adaptive functions of breeding dispersal.

First, adult animals may leave their home areas to avoid mating with their offspring in the future. For example, in Belding's ground squirrels, mating is dominated by a small number of males during the breeding season (Sherman 1976). These males typically disperse from the breeding area despite their high reproductive success. Males who dominate mating are typically the largest and most successful at winning fights with other males, and presumably are good competitors. Thus these males are most likely not dispersing in response to competition for mates or environmental resources. Rather, Sherman (1976) proposed that males who have mated with several females will be fathers of a large proportion of females in the next generation, and disperse so that their reproductive tenure in the area will not overlap with that of their daughters.

Second, adult animals may disperse so that their offspring can inherit their territories. Territorial bequeathal has been observed in red squirrels (*Tamiasciurus hudsonicus*; Berteaux and Boutin 2000), Columbian ground squirrels (Harris and Murie 1984), and white-footed mice (Wolff and Lundy 1985). Berteaux and Boutin (2000) suggested that breeding dispersal by adult female red squirrels is a form of parental investment. Adult females are better able than young to establish themselves in vacant territories. They observed that by leaving their territories, mothers allowed some of their young to remain on the natal site, which increased survival of the young. Female red squirrels who engaged in breeding dispersal tended to be older and more experienced than nondispersing females, and emigrated when food resources were abundant, thus minimizing the potential costs of dispersal.

However, Lambin (1997) suggested that territorial bequeathal might be adaptive only when young inherit a critical resource such as a midden, as has been observed in red squirrels (Berteaux and Boutin 2000), or a burrow system, as has been observed in Columbian ground squirrels (Harris and Murie 1984). Demand for these resources typically exceeds supply, they are energetically expensive to construct, and they are vital to survival and reproductive success. Thus increased inclusive fitness may occur when these resources are relinquished to offspring. By contrast, Lambin (1997) suggested that territorial bequeathal is unlikely to be adaptive when inherited resources are available elsewhere or are not critical for survival and reproduction, as is the case in a variety of vole species. Lambin (1997) found no data supporting territorial bequeathal in Townsend's voles. Similarly, McGuire et al. (1993) observed that young prairie voles rarely remain in the natal site and become reproductive after their parents disappear from the area.

Third, adults may disperse to increase the quality of their territory with regard to food resources or mates. In a study of Eurasian red squirrels, Lurz et al. (1997) observed that breeding dispersal was highest among females in unstable, low-quality habitat. Only females with a continuously available supply of food resources were faithful to their breeding territories. Adult male squirrels also engaged in breeding dispersal. However, their movements tracked those of adult females. Thus whereas adult female squirrels dispersed to increase their access to food resources, males dispersed to maintain access to potential mates.

Philopatry

Ultimate causes of philopatry

It has been suggested that most female mammals are philopatric because the costs of dispersal and benefits of philopatry are greater for females than males. Mammalian reproductive strategies typically involve investing a great deal of care in a small number of young. Because embryonic and fetal development occur internally in mammals, males can abandon females after mating, to optimize their reproductive success by seeking additional mating opportunities. By contrast, females optimize reproductive success by maximizing care of young to enhance the probability that they will survive to reach reproductive maturity. The costs in females of caring for young include not only lactation, which can be energetically expensive, but also the costs of defending young against predators and conspecifics. Residing in familiar terrain may help females acquire or defend food resources and defend territories. Moreover, females may benefit by remaining in the natal site near female relatives, who are not likely to attempt infanticide.

Wolff (1994a) proposed an alternative hypothesis for female philopatry, which is predicated on the idea that inbreeding avoidance is an important driving force behind natal dispersal. He suggested that because of the intense competition for mates among polygynous males, they typically have shorter reproductive longevity than females. Thus by the time a young female reaches reproductive maturity, the probability is low that her father will still be in the area and siring offspring, and she need not disperse to avoid potentially mating with him. By contrast, males are likely to reach reproductive maturity while their mothers are present in the natal area, and disperse to avoid potential inbreeding (see the preceding). Wolff (1994a) also suggested that in polygynous species, males often wander over large areas in search of mates. Thus even if a young female remains in her natal area, there is a high probability that she will encounter unrelated males as they wander through her home area.

Perrin and Lehmann (2001) suggested that female mate choice facilitated by kin recognition might have contributed to the evolution of female philopatry. According to this idea, because inbreeding depression is potentially costly to the fitness of females, natural selection favored females who could recognize closely related males and who rejected them as mates. This choosiness by females allowed them to avoid inbreeding without emigrating from the natal area. However, it is also possible that kin recognition is a consequence rather than cause of philopatry, and might have been favored by natural selection because it facilitates cooperative interactions among kin.

Habitat constraints may be an underlying cause of philopatry in some species (Solomon 2003). This possibility has been extensively examined in African mole-rats, subterranean rodents native to sub-Saharan Africa (Spinks et al. 1999; Hazell et al. 2000; Spinks, Bennett, and Jarvis 2000, Spinks, Jarvis, and Bennett 2000; Molteno and Bennett 2002; Faulkes and Bennett, chap. 36, this volume). Mole-rat species range from solitary to highly social, and the social species predominate in arid regions where food resources are patchy. Mole-rats subsist in large part on geophytyic plants, and randomly dig to locate patches of these plants. The amount of energy an individual expends to locate a food patch can thus be substantial. Moreover, digging is limited to brief periods of rainfall during which the ground is softened. Philopatry and delayed dispersal are common among social mole-rats. The aridity food-distribution hypothesis (AFDH) has been proposed to explain the high degree of philopatry and sociality in various species of mole-rats. According to this hypothesis, foraging is risky and energetically expensive for individuals, and a solitary individual may perish before locating a food patch. By remaining philopatric and associating with conspecifics, mole-rats can act cooperatively to find food patches, thereby increasing their chances of survival.

Numerous studies have provided support for the AFDH in various mole-rat species (e.g., Spinks et al. 1999; Hazell et al. 2000; Spinks, Bennett, and Jarvis 2000; Spinks, Jarvis, and Bennett 2000; Moteno and Bennett 2002). However, Burda et al. (2000) questioned the validity of this hypothesis. They suggested that the AFDH explains large group size in mole-rats inhabiting arid regions, but does not explain philopatry. Instead, they propose that extensive philopatry in mole-rats results from a monogamous mating system and high coefficient of relatedness among individuals. Cooperative breeding occurs in social mole-rat colonies, with nonbreeding individuals contributing to the care of offspring produced by a breeding pair. If the degree of relatedness is high among individuals in a colony, nonbreeders can enhance their inclusive fitness by caring for related young. Burda et al. (2000) also suggested that monogamy has led to cooperative breeding and high degrees of philopatry in other species, such as prairie voles (e.g., Solomon 1991).

Proximate causes of philopatry and delayed dispersal

Social and physical barriers to dispersal

Social barriers to dispersal can promote facultative philopatry or delayed dispersal. For example, in many species of rodents, rates of dispersal decline sharply as population density increases and the availability of suitable habitat in which to settle declines (see Wolff 1994a). Under crowded conditions, residents of an area typically act aggressively toward prospective immigrants to deter them from settling. By contrast, dispersal rates are often high when population density is low and habitat is available for settlement. Physical barriers to dispersal can also increase tenure in the natal area. For example, dispersal is opportunistic in mole-rat species inhabiting arid regions, occurring when rain-softened ground facilitates digging and encounters with unfamiliar individuals or different groups (e.g., Spinks, Bennett, and Jarvis 2000; Spinks, Jarvis, and Bennett 2000). Moreover, Dobson et al. (1998) observed high rates of philopatry and short dispersal distances in a small lagomorph, the plateau pika (*Ochotona curzoniae*) in areas with heavy predation, and suggested that potential predation risks associated with dispersing through unfamiliar habitat might influence dispersal distance and tendencies for philopatry in small mammals.

Inheritance of territories and absence of parents

As earlier noted, adult females of some species leave their maternal territories, allowing their young to occupy them and remain philopatric. Also, as noted previously, the absence of the opposite-sex parent may prompt young to remain in the natal area (Michener and Michener 1977; Wolff 1992; Gundersen and Andreassen 1998).

Consequences of philopatry

Cooperation among kin living in proximity to each other may be a driving force behind philopatry or a consequence of it (see the preceding). Regardless of which is responsible for the other, there is typically a close association between philopatry and cooperation. Remaining in proximity to close kin throughout life provides the opportunity for cooperative interactions and the evolution of nepotistic behavior, which can increase the inclusive fitness of individuals (Sherman 1977, 1981a). Altruistic interactions depend on the ability to discriminate between closely related and more distantly related or unrelated individuals (Perrin and Lehmann 2001; Lehmann and Perrin 2002), and this ability has been documented in some rodents (e.g., Holmes 1994, 1995; Mateo and Johnston 2000a; Holmes and Mateo, chap. 19, this volume).

The behavior of female Belding's ground squirrels during the breeding period illustrates ways in which philopatry can facilitate cooperation (Sherman 1980a). Infanticide is common in Belding's ground squirrels (fig. 13.2), as it is in most mammals (Hrdy 1979; Sherman 1981b). Mother and daughter squirrels tend to set up maternal territories near each other and act cooperatively in defending the territories against infanticidal conspecifics. Early in the breeding season, much of the squirrels' habitat is covered with snow, and space in which to establish maternal territories is limited (fig. 13.2). During gestation, rates of aggressive behavior are high among female squirrels as they compete with each other for space in which to set up a territory. Rates of aggressive behavior are significantly higher among adult females that are establishing a territory next to a yearling daughter than among females with no yearling daughters (fig. 13.3). This increased aggression among females with a daughter nearby raises the possibility that the behavior of adult females may enhance their daughters' chances of acquiring a maternal territory. Adult females with territories near yearling daughters also exhibit more intense vigilant behavior than do adult females with no yearling daughters (fig. 13.4), which may reflect helping their daughters to defend their maternal territories.

Cooperation is also observed in other rodent species. For example, female black-tailed prairie dogs communally nest with close female relatives (Hoogland 1995). Moreover, female meadow voles, common mole-rats (*Cryptomys hottentotus*), Damaraland mole-rats (*Cryptomys damarensis*), and naked mole-rats (*Heterocephalus glaber*) remain in the natal area in a nonreproductive state and aid in rearing young of close relatives (e.g., Solomon 1991; Spinks et al. 1999; Burda et al. 2000; Molteno and Bennett 2002).

Armitage (1981 and chap. 30, this volume) suggested that philopatry allows for the evolution of complex social

Figure 13.2 (A) Infanticide during the lactation period of Belding's ground squirrels. Reproductive females aggressively defend territories against conspecifics that may attempt to abduct and kill their young. (B) Space constraints early in the active season of Belding's ground squirrels when females are pregnant. Much of the squirrels' habitat is covered by snow, and females aggressively compete for space available for establishing maternal territories.

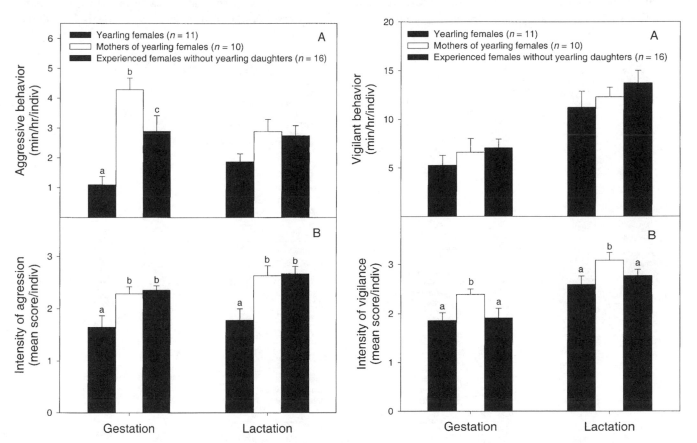

Figure 13.3 Possible assistance from mothers in establishing maternal territories among female Belding's ground squirrels. Aggressive behavior among yearling females occurs at lower (A) rates and (B) intensity than it does in experienced females. During gestation, when females establish maternal territories, squirrels with yearling daughters setting up a territory nearby exhibit higher rates of aggression toward unrelated squirrels than do experienced females without yearling daughters. Intensity scores for aggressive interactions were assigned as follows: threat—1; chase < 5 m—2; chase > 5 m—3; attack—4. Data were evaluated with ANOVA, and post-hoc comparisons were performed with Tukey's tests. Different lower case letters indicate significant differences ($P < 0.05$) between groups. Two yearling females represented in the graph had the same mother.

Figure 13.4 (A) Rates and (B) intensity of vigilant behavior among female Belding's ground squirrels during the breeding cycle. Although rates of vigilance did not differ among females, vigilant behavior tended to be more acute among females with yearling daughters nearby. Intensity scores for vigilant behavior were assigned as follows: resting alert—1; posting (standing on hind legs)—3; alarm calling—5. Females were awarded an additional point if vigilant behavior occurred on an elevated post such as a rock or fallen tree. Data were evaluated with ANOVA, and post-hoc comparisons were performed with Tukey's tests. Different lower case letters indicate significant differences ($P < 0.05$) between groups. Two yearling females represented in the graph had the same mother.

systems. For example, growth and development tend to occur over a longer time period in large-bodied compared to small-bodied sciurids, and dispersal and sexual maturation tend to be delayed in many large-bodied species, resulting in prolonged residence in the natal area. Armitage (1981) noted that kin selection favors cooperative interactions among relatives living together, which can promote complex social organization. An association between large body size, delayed maturity and dispersal, and complex social systems has been observed in a variety of sciurids, including yellow-bellied marmots, Gunnison's prairie dogs (*Cynomys gunnisoni*), and black-tailed prairie dogs (Armitage 1981; Downhower and Armitage 1981; Rayor 1985; Garrett and Franklin 1988; Blumstein and Armitage 1999). Michener (1983a) also suggested that philopatry and cooperation among female kin were important in the evolution of sociality in ground-dwelling sciurids.

Philopatry and delayed dispersal are also associated with complex social organization in common mole-rats, Dama-raland mole-rats, and naked mole-rats. These species have complex social organization, and individuals of the species act cooperatively to forage and raise young (Solomon 1991; Spinks et al. 1999; Burda et al. 2000; Molteno and Bennett 2002).

Conclusions

Rodent studies have greatly contributed to our understanding of the ultimate and proximate causes and consequences of dispersal and philopatry (figs. 13.5 and 13.6). Although competition for mates and environmental resources undoubtedly influence dispersal behavior, inbreeding avoidance appears to provide the most general evolutionary influence favoring sex-biased natal dispersal. Inbreeding depression is potentially costly (Ralls et al. 1988; Lacy et al. 1993), and leaving the natal area to find unrelated mates seems to be a common strategy among rodents for enhanc-

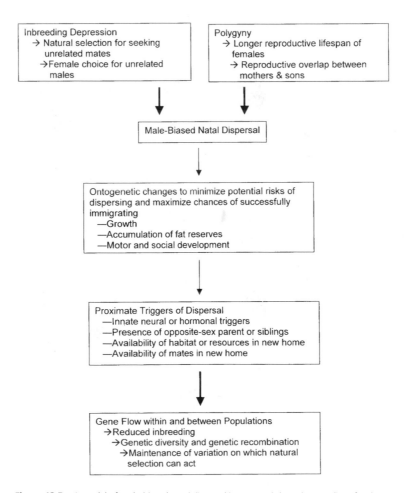

Figure 13.5   A model of male-biased natal dispersal in mammals based on studies of rodents.

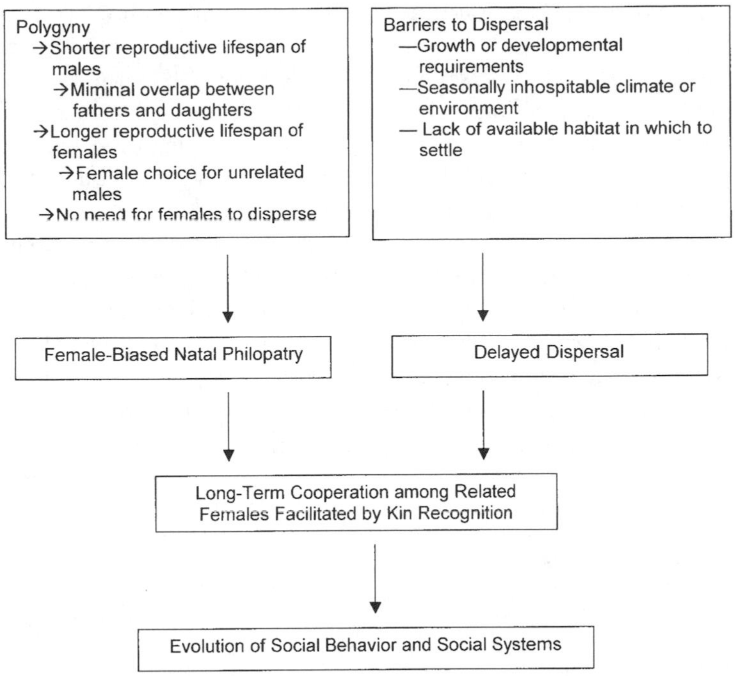

Figure 13.6 A model of delayed dispersal and female-biased philopatry in mammals based on studies of rodents.

ing fitness. Because of the high turnover rate of adult males, young females typically are not exposed to their fathers, and consequently can breed on or near their natal site with unrelated males. By contrast, the reproductive life span of mothers commonly overlaps with that of their sons, and females show a preference for unrelated males as mates. As a consequence, young males commonly emigrate from the natal site prior to mating. The timing of dispersal corresponds with growth and development and the best probability of successful immigration. Failure to emigrate from the natal site due to habitat saturation or some socioecological factor(s) often results in delayed sexual maturation if young males and females remain in the presence of opposite-sex parents.

Both philopatry and dispersal have important consequences for individuals, populations, and species. Remaining at the natal site allows for cooperation with close kin, which can be a foundation for developing complex social behavior and social systems. An increasingly important consequence of dispersal is its maintenance of genetic diversity within populations by facilitating gene flow. Loss and fragmentation of habitat associated with human activity has presented barriers to dispersal and gene flow for many populations, putting some populations in danger of becoming extinct (see Lidicker, chap. 38, this volume). Future studies on rodents and other animals that integrate dispersal

and population genetics will thus be critical in conserving threatened populations and species.

Summary

Dispersal is a ubiquitous behavior among rodents and mammals in general, and involves departure from the birth area or group (natal dispersal) or site of reproduction (breeding dispersal) and settlement in a new area or group. Natal dispersal is more prevalent than breeding dispersal, and tends to be sex-biased, with males more commonly emigrating from the birth area and females typically exhibiting philopatry, or faithfulness to the natal area throughout life.

Breeding dispersal may allow individuals to avoid mating with their young, bequeath territories to offspring, or settle in better habitat. Much of the available data suggests that avoidance of inbreeding was a primary driving force in the evolution of natal dispersal, and that polygyny and greater selectivity in mate choice by females was an important ultimate cause of the male bias in mammalian natal dispersal. Evolutionary causes for natal dispersal among ordinarily philopatric females are primarily related to competition for environmental resources such as food or nest sites. Because dispersal mediates the movement of alleles within and between populations, it has important consequences for the genetic structure as well as the conservation of populations and species.

A variety of proximate triggers for natal dispersal have been observed. In some species, a predisposition to leave the natal area arises very early in life, due to long-term effects of hormones or nutrition on the developing individual. However, in most species, dispersal behavior is more flexible and may be influenced by factors such as the availability of environmental resources, the presence of potential mates or the opposite-sex parent, or interactions with siblings. Individuals typically do not emigrate until they are prepared to cope with the potential risks of dispersing. In most species, natal dispersal occurs near the time individuals reach reproductive maturity, but may occur earlier when changes in environmental conditions are predictable and individuals begin mating at the onset of reproductive maturity.

A variety of hypotheses have been proposed to explain the prevalence of female philopatry in rodents and other mammals. Among these are the ideas that the costs of dispersal and benefits of philopatry are greater for females, and that females do not need to disperse to avoid inbreeding. Philopatry may occur among eusocial rodents such as mole-rats due to environmental constraints on dispersal and finding food, or when high coefficients of relatedness favor cooperation among kin living in the same area. The evolution of complex social systems among rodents and mammals in general has been suggested to be predicated upon philopatry and cooperation among kin living near each other.

Chapter 14 Gene Dynamics and Social Behavior

F. Stephen Dobson

Social groups of related females typify several mammalian species. These kin-based groups are found particularly in primates and rodents, and occur in other orders as well. Kin groups are associated with particular mating and dispersal patterns, usually polygynous mating, philopatry of females in the natal social group, and male dispersal away from the natal group (Greenwood 1980; Dobson 1982). Groups of closely related individuals are a necessary condition for the evolution of cooperative behavior via kin selection (Hamilton 1964), and thus mating and dispersal patterns may influence the evolution of social behavior. Kin groups may also create genetic structure within a population, and thus social evolution may depend on genetic substructuring of populations (Chesser 1998). In general, when individuals in social groups are more closely related, genetic substructuring within the population is greater, and so is the potential for social evolution.

The purpose of this chapter is to examine both theory and empirical studies that examine the coevolution of social groups, particularly kin-based groups, and genetic substructure of local and regional populations. The fundamental question is: how much do mating systems, dispersal patterns, and, especially, social group structure influence genetic properties of populations? Basset et al. (2001) used simulations to study two models of genetic substructuring of local populations, and found substantial influences of mating and dispersal patterns on estimates of gene dynamics. Under polygynous mating and sex-biased dispersal, their simulations revealed substantial genetic differentiation among groups within a local population. My goal is to examine population-genetic theory concerning the gene dynamics of kin-based social groups and point out some of the differences between the two current models that have been used to study some aspects of gene dynamics. I will also examine empirical evidence in rodents and other small mammals that focus on whether kin-based social-genetic substructuring occurs within local populations.

Models of Gene Dynamics

How might we determine the degree of genetic substructure that is associated with kin-based groups? The answer to this question was provided by Sewell Wright (1965, 1969, 1978). He devised measures of genetic properties, or gene dynamics, of populations that were structured into geographical subpopulations. These metrics compared rates of inbreeding within and among subpopulations to expectations from random mating (F-statistics), and also quantified the rate at which inbreeding accumulates (termed effective population size). The rate of inbreeding compared to what would be expected if mating were random among the members of a subpopulation was termed F_{IS}. The rate of inbreeding compared to that expected if all the members of a population were mating randomly was called F_{IT}. The degree of genetic differentiation among subpopulations was termed F_{ST}. I, S, and T indicate population levels of individuals, subpopulations, and the total regional population, respectively. Finally, effective population size was defined as a function of the relative rate of change in the inbreeding coefficient from one generation to the next (viz., $N_e = 1/(2\Delta F)$, the inbreeding effective population size; Crow and Denniston 1988).

Although Wright's measures of gene dynamics were designed for geographically distinct subpopulations that occur within a regional landscape, they can be applied to more local levels of population structure. In fact, F-statistics are hierarchical, so that several levels of population structure may be considered at once (reviewed by Chesser et al. 1996). Thus dividing a geographic subpopulation into social groups creates an additional level of population substructure, and additional F-statistics can be defined. The rate of inbreeding compared to what would be expected if mating were random among the members of the subpopulation is still F_{IS}, as noted previously. But the rate of inbreeding compared to that expected if the members of the social group were mating randomly can be defined as F_{IL}, where L indicates the lineage of the social group. By the same token, the degree of genetic differentiation among the social groups can be called F_{LS}. Since effective population size was based on the change in the average inbreeding coefficient from one generation to the next, it should be the same value at any level of hierarchical organization for populations, subpopulations, and social groups that are in genetic equilibrium.

Measures of gene dynamics reflect not only patterns of inbreeding in populations, but degree of kinship among individuals as well. Cockerham (1967, 1969, 1973) pointed out that F-statistics are closely allied to genetic correlations among individuals and can be calculated from them. The inbreeding coefficient, F, is the genetic correlation of an individual's parents. In a closed population, inbreeding accumulates over time, and thus F will increase steadily over generations. F can be averaged over all the individuals in a population to yield the average rate of inbreeding. If the value of F is compared between parents and offspring, the rate of change in F can be measured and used to estimate inbreeding effective population size (e.g., Blackwell et al. 1995). When applied to empirical data, however, this estimate assumes that the population is in genetic equilibrium, a condition that may seldom be met in nature (Dobson et al. 2004).

Genetic correlations, or gene correlations, can be measured between pairs of individuals (viz., dyads). Individuals to compare may be chosen from geographically distinct subpopulations, different social groups, or within social groups (Chesser 1991a, 1991b). The dyadic gene correlation is also called the co-ancestry of the individuals, because genetic correlations are measured by descent, preferably from a pedigree. Co-ancestry is closely linked to kinship. In the absence of inbreeding, co-ancestry is half the degree of kinship between two individuals. When there is inbreeding, however, co-ancestry takes it into account, thus measuring all sources of genetic identity by descent. So co-ancestry, like the inbreeding coefficient, increases over generations. With pedigree data, co-ancestry of any dyad (individuals i and j) can be measured as:

$$\theta_{i,j} = \frac{1}{4}(\theta_{S_iS_j} + \theta_{S_iD_j} + \theta_{S_jD_i} + \theta_{D_iD_j}) \tag{1}$$

where S represents sires and D is for dams. If this value is averaged among all the dyads within a social group or subpopulation, the average gene correlation of the social group or subpopulation can be calculated.

It is probably best to use θ as the average co-ancestry within a social group, and use another symbol for the average co-ancestry within a geographic subpopulation. Chesser (1991a, 1991b) defined α as the average co-ancestry of a geographic subpopulation or colony of smaller social groups. It (α) is averaged over all dyads of individuals that are in different social groups, but within the geographic subpopulation. At the next higher level of hierarchical population structure, co-ancestry of dyads of individuals from different subpopulations could be defined and given a symbol, such as δ (fig. 14.1). With the previously cited gene correlations, it is then possible to define F-statistics at several levels of population structure, for societies that are based on family social groups of related females:

$$F_{IL} = \frac{F - \theta}{1 - \theta} \qquad F_{IS} = \frac{F - \alpha}{1 - \alpha} \qquad F_{IT} = \frac{F - \delta}{1 - \delta}$$

$$F_{LS} = \frac{\theta - \alpha}{1 - \alpha} \qquad F_{ST} = \frac{\alpha - \delta}{1 - \delta} \tag{2}$$

These F-statistics define the gene dynamics of a population with three levels of substructure: individuals within social groups, social groups within geographical subpopulations, and subpopulations within a region. The values of F_{IL}, F_{IS}, and F_{IT} compare the inbreeding coefficient to what would be expected if mating were random at three levels of population structure: the social group, the subpopulation, and the regional population. F_{LS} indicates the degree of genetic differentiation of social groups within a subpopulation and F_{ST} is the degree of genetic differentiation of subpopulations within the regional population.

F_{IL}, F_{IS}, and F_{IT} vary between -1 and 1. A significant negative value indicates less inbreeding than expected if mating were random within the social group, subpopulation, and regional population, respectively. The sorts of things that often lead to lower inbreeding than expected are natal dispersal out of the group, subpopulation, or population, and behavioral avoidance of mating with close relatives. Dispersal can occur at any level of population structure, though rates of movement should become lower as larger geographic areas are considered. Avoiding close relatives as mates, while they are still in spatial proximity, should occur at the level of social groups or small local subpopulations. Positive values of these F-statistics occur when the inbreeding coefficient is greater than expected under ran-

REGIONAL POPULATION

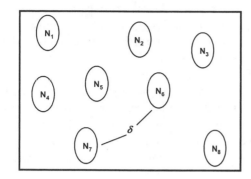

LOCAL POPULATION: N₈

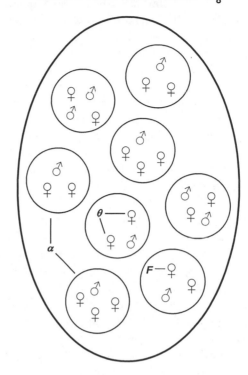

Levels of Gene Correlations:

Individual – *F*
Social Group – *θ*
Local Population – *α*
Regional Population – *δ*

Figure 14.1 Hierarchical gene dynamics. A regional population is depicted, with an expanded view of a local subpopulation. Within the subpopulation, several social breeding groups are depicted. Single examples of genetic correlations are shown: the inbreeding coefficient (*F*) for a female, co-ancestry (*θ*) within a breeding group for two females, the gene correlation between individuals in two different social groups (*α*), and the gene correlation between two subpopulations (*δ*).

dom mating. This is most likely at broad geographic scales, especially at spatial scales that are greater than maximum dispersal distances. At very large spatial scales, limitations on dispersal distances ensure that mating will be nonrandom and more localized than if individuals in the large geographic population mated randomly.

When considered on a spatial scale, F_{IL} should be negative at the level of social groups, since dispersal out of social groups usually occurs by one sex or the other (Greenwood 1980; Dobson 1982). As the spatial scale increases, estimates of F_{IS} will approach zero. When F_{IS} is not significantly different from zero, mating may in fact be nonrandom. For example, black-tailed prairie dogs have strongly negative F_{IL} values, and F_{IS} at the colony (viz., subpopulation) level can be close to zero (Dobson et al. 1997). But mating is probably not random at either level of population structure, because females in social groups (female kin-based families called *coteries*) are most likely to mate with the dominant coterie male (i.e., females make nonrandom mating choices). The fact that F_{IS} is zero occurs because at the colony level, the inbreeding coefficient (*F*) and co-ancestry among coteries (*α*) are equal. So the issue of whether there is truly a level of population structure at which matings are random is a genetic red herring. As the spatial scale becomes broader,

F_{IT} values are likely to be positive due to geographic isolation of subpopulations (Chesser 1983; Daley 1992).

F_{LS} and F_{ST} measure genetic differentiation among social groups, such as coteries, and subpopulations, such as colonies, respectively. Theoretically, these values range between zero and one. A value of zero would indicate no genetic differences among social groups or subpopulations, and a value of one would indicate complete genetic differences (viz., each subpopulation fixed for alternative alleles at all loci, an unlikely event). Note that equation (1) implies that often $\theta > \alpha > \delta$. Basically, as the spatial scale increases, our estimates of co-ancestry at each level of structure should decline. The values of F_{LS} and F_{ST} depend on the relative values of θ and α, and α and δ, respectively. In general, we might expect F_{LS} to often be somewhat greater than F_{ST}, as Chesser (1983) found for coteries and colonies of black-tailed prairie dogs. This indicates that more genetic differentiation occurs at the level of the social group than among geographically disparate subpopulations (but see Patton and Smith 1990).

At genetic equilibrium, it is unlikely that F_{LS} and F_{ST} will often exceed 0.25, despite the theoretical maximum of 1.0. F_{LS} can be used to examine why this is so. As can be seen from equation (2), the value of θ is the limiting value of F_{LS}, such that F_{LS} cannot be greater than θ. Since θ is an average genetic correlation, we can consider what its maximum value should typically be. If a social group was made up of full siblings, the average degree of relatedness would be 0.50, and the average genetic correlation in the absence of inbreeding would be 0.25. However, there are two important caveats to this generalization. One is that unlike the simple measure of kinship, co-ancestry accounts for inbreeding. Thus θ can be greater than 0.25 if there is considerable inbreeding, as may occur in the plateau pika (where $F_{LS} \cong 0.29$; Dobson and Smith et al. 2000). The second consideration is that it is unlikely that females will all be full sisters in a polygynous social group, even with only one breeding male. Rather, some mixture of full and half siblings is likely, and θ among offspring will converge on 0.17 (= 1/6; Dobson and Chesser et al. 2000). Thus in the absence of inbreeding, co-ancestries among philopatric females in matrilineal social groups should fall to near 1/6, or lower if there is more than one breeding male.

Chesser (1991a, 1991b) envisioned a buildup of genetic relatedness among philopatric females from two sources of genetic correlation. First, polygyny may often result in a single male mating with all the females in a breeding group. In this case, offspring from unrelated mothers are half siblings. This source of genetic correlation can be stronger than the second source, which is that breeding females may be related, such as full or half sisters, through their mother, depending on whether they have the same father. In general, increasing the number of successfully breeding adult males in a group will lower the gene correlations among daughters. Increasing the number of adult females in a group, however, has little effect on average gene correlations of daughters (Dobson and Chesser et al. 2000). Of course, variation among breeding groups in their gene correlations may well be increased by sampling events in small groups, or due to demographic events as individuals age (e.g., Dobson and Chesser et al. 1998).

Estimating Gene Dynamics

There are several ways that gene dynamics can be estimated at the level of social breeding groups and at higher levels of hierarchical population structure. F-statistics may be estimated from pedigrees using equations (1 and 2). Pedigrees provide the soundest measure of gene dynamics, because the gene correlations F, θ, α, and δ can be accurately measured. A good example of this technique comes from John Hoogland's long-term study of black-tailed prairie dogs (Dobson and Chesser 1997, 1998, 2004). Building a pedigree requires that both paternity and maternity be known. Pedi-

gree data might be most available from captive breeding programs in zoological parks and for domestic animals. Cases of multiple paternity, however, indicate that observations of matings alone cannot identify sires. Biochemical techniques of paternity assessment provide a possible solution to this problem (e.g., Westneat and Sherman 1997; Goossens et al. 1998).

If biochemical data are available, F-statistics can also be estimated from heterozygosity of selectively neutral alleles at variable genetic loci, under the assumption of genetic equilibrium (Slatkin 1987). In a hierarchically structured population, Wright's (1969) F-statistics would be:

$$F_{IL} = \frac{\overline{H}_L - \overline{H}_I}{\overline{H}_L} \quad F_{IS} = \frac{\overline{H}_S - \overline{H}_I}{\overline{H}_S} \quad F_{IT} = \frac{H_T - \overline{H}_I}{H_T}$$
$$F_{LS} = \frac{\overline{H}_S - \overline{H}_L}{\overline{H}_S} \quad F_{ST} = \frac{H_T - \overline{H}_S}{H_T}$$

$$(3)$$

H is heterozygosity of variable genetic loci at the indicated level of population structure, and average levels of heterozygosity are calculated among individuals, lineage groups, and subpopulations. Several statistical software packages are available for calculating F-statistics from allelic scores at polymorphic loci (e.g., Goudet 1995). Of course, gene correlations and levels of heterozygosity are related:

$$\overline{H}_I = 1 - F \quad \overline{H}_L = 1 - \theta \quad \overline{H}_S = 1 - \alpha \quad H_T = 1 - \delta$$
$$(4)$$

A final way of estimating F-statistics for lineage groups is to apply Chesser's (1991a, 1991b; Chesser et al. 1993) breeding-group model of gene dynamics. This model draws on the sorts of data collected in studies of "behavioral ecology," and was specifically designed for polygynous species that exhibit natal philopatry by females (Sugg et al. 1996). The breeding-group model is thus particularly appropriate for studying the gene dynamics of the more social rodent species. An extension of the model takes multiple paternity into account (Sugg and Chesser 1994). The basic idea behind the model is that gene dynamics can be estimated from patterns of mating, dispersal, and demography. Chesser derived mating parameters that quantify the degree of polygyny, the probability that offspring share the same mother, and the degree of multiple paternity (both within a litter and among offspring born in different years). He combined these mating parameters with demographic data such as the number of males and females in social groups, the number of social groups, mean female breeding success, and natal dispersal rates of males and females. These and a few other variables are then entered into transition matrices that are used to predict changes in gene correlations (viz., F, θ, and

α) from generation to generation. The model is iterated for several generations, until genetic equilibrium is reached. The resulting changes in gene correlations can be used to estimate F-statistics and effective population size (see the following).

Effective population sizes can also be estimated in three different ways: from pedigrees, from demographic models (*sensu* Nunney and Campbell 1993), and from biochemical data that reflect variable genetic loci. If pedigrees are available, effective sizes can be estimated from any of the different gene correlations (viz., F, θ, α, and δ). Just as inbreeding accumulates over time in a finite-sized population, the other gene correlations also increase over time. If a population is in equilibrium and is ideal (viz., one with no selection, no mutation, no dispersal, equal numbers of males and females), then all of the gene correlations increase at the same rate (Chesser et al. 1993; Sugg and Chesser 1994). Thus, the relative change from generation to generation in any of the gene correlations provides an estimate of effective population size:

$$N_e = \frac{1}{2\Delta F} \qquad N_e = \frac{1}{2\Delta\theta} \qquad N_e = \frac{1}{2\Delta\alpha} \qquad N_e = \frac{1}{2\Delta\delta} \qquad (5)$$

Relative changes in gene correlations (e.g., ΔF) are changes in average genetic correlations from the parental generation (e.g., F') to the offspring generation (F), relative to residual genetic variation (e.g., $1 - F$). Since these estimates assume genetic equilibrium, which may not typically be met in nature (Dobson et al. 2004), it may be necessary to estimate average changes in gene correlations over several generations.

Another way to estimate effective population size is to estimate changes in genetic correlations from Chesser's breeding-group model (Chesser et al. 1993; Sugg and Chesser 1994). The model estimates F-statistics from patterns of matings, demography of the local population, and dispersal patterns. In turn, F-statistics can be used to estimate effective population size. Nunney and Campbell (1993) have termed such models "demographic methods," because they predict genetic patterns that should occur given the behavior patterns and demography exhibited by local populations. F-statistics may fluctuate less over time than gene correlations, and thus may produce more robust estimates of effective population size (e.g., Dobson et al. 1998; Dobson et al. 2004). An equation for estimating effective population size using F-statistics from the breeding group model (Chesser et al. 1993; Sugg and Chesser 1994) is:

$$\hat{N}_e = \frac{4s - 3\hat{F}_{IS} - 1}{6(\hat{F}_{LS} - \hat{F}_{IS})} \qquad (6)$$

In this equation, s is the number of social breeding groups, and the "hats" emphasize that the effective size and F-statistics are estimated values. Basset et al. (2001) indicate that this equation should be applied to estimates of F-statistics based on offspring, before their dispersal from the natal area (see the following). It is important to note that equation (6) is an estimate of effective size that is made under simplifying assumptions, and more accurate and direct measures are available (Chesser et al. 1993; Sugg and Chesser 1994).

An alternative demographic model for estimating effective size of local populations that exhibit group structure was developed by Nunney (1999). This model has arbitrary groups, and does not incorporate information about genetic correlations (viz., influences of kinship and inbreeding). Moreover, the model requires additional information on mating patterns that may be difficult to gather, such as the variance in male reproductive success. Nunney's (1999) model is more appropriate for some species than the breeding-group model. Basset et al. (2001) used computer simulations to show that for monogamous species and in cases with equal dispersal of males and females, Nunney's (1999) model performed much better than the breeding-group model, which greatly overestimated effective population size. For polygynous species with sex-biased dispersal, however, both models were appropriate. Thus, studies of monogamous species should use the arbitrary-group model, even though some of the parameters will be hard to measure.

A final way to estimate effective population size is from biochemical estimates of F-statistics. Nunney and Campbell (1993) termed methods for estimating effective population size from various molecular markers (e.g., allozyme alleles, microsatellite DNA alleles) "genetic methods," to differentiate them from demographic methods. Alleles at neutral loci can be used to estimate gene dynamics, including effective population sizes, under the assumption of genetic equilibrium (reviewed by Slatkin 1987). For socially structured subpopulations, estimates of F-statistics can be made from equations (3) and substituted into equation (6) or another suitable equation, such as:

$$N_e = \frac{H_T}{4\mu(1 - H_T)} \qquad (7)$$

where μ is the mutation rate (Hartl and Clark 1997). In a local population that is structured into groups, but is otherwise ideal, Nunney (1999) points out that effective size can be estimated as:

$$N_e = \frac{N_T}{(1 + F_{IL})(1 - F_{LS})} \qquad (8)$$

where N_T is the census size of potentially breeding adults in the population. This equation also requires the assumption that all social groups are equally successful in contributing offspring to the next generation. For additional equations under a variety of assumptions, see Chesser et al. (1993), Sugg and Chesser (1994), and Nunney (1999).

Basset et al. (2001) argued that a distinction should be made between predispersal estimates of gene dynamics, which include individuals that have not yet dispersed from their natal site to their first breeding area, and postdispersal estimates of gene dynamics for adults that are capable of breeding. In many species, offspring disperse away from the natal site, like the predominant juvenile male dispersal common in many mammals (e.g., Greenwood 1980; Dobson 1982). Such dispersal can generally be expected to move individuals away from their philopatric kin. In addition, individuals that disperse to new homes may most often encounter unrelated individuals. Thus, gene correlations should generally be lower after, as compared to before, dispersal events, especially gene correlations measured within the range of dispersal movements. Consequently, F-statistics at the level of social breeding groups should also be lower following dispersal. When females become breeding adults in their natal group, and males disperse widely, greater genetic correlations among females are expected than among males and between males and females. This difference in genetic patterns between adult males and females can influence estimated values of both F-statistics and the resulting effective population sizes. Predispersal gene dynamics can also be predicted, and should generally reflect the genetic patterns of young individuals. Thus, it is critically important to distinguish pre- and postdispersal estimates and the expected gene dynamics that they reflect.

Empirical Studies of Mammals

The breeding-group model is most appropriately applied at the level of population structure where matings occur, such as lineage or family groups within colonies. At larger geographic scales, such as colonies within a geographic region, classical F-statistics are appropriate. Evidence of population-genetic structure within mammals has been demonstrated at a variety of spatial scales (e.g., Storz 1999). However, if social groups are not carefully identified, the level of population substructure at which matings occur may be missed (Sugg et al. 1996; Dobson 1998). When this level is not sampled, Chesser's (1991a, 1991b; Chesser et al. 1993) breeding model and Nunney's (1999) arbitrary-group model may lead to misleading inferences. Thus, in looking for evidence of the influence of social groups on genetic

substructuring of populations, knowledge of the behavioral ecology of a species, especially mating and dispersal patterns, is very important.

Studies of a variety of mammals have quantified the degree of substructuring of populations attributable to social groups. In a seminal and prescient paper on social groups in yellow-bellied marmots (*Marmota flaviventris*), Schwartz and Armitage (1980) found significant genetic differentiation among small colonies (table 14.1). These social rodents live in kin-based groups, exhibit a polygynous mating system, and males disperse while females are philopatric (reviewed by Armitage 1999b). Thus, the marmots are particularly appropriate for asking whether social groups create genetic substructure within local populations. Allozyme data from primarily postdispersal adults and subadults of both sexes were used to estimate F-statistics. F_{LS} was significant, indicating genetic differentiation among social groups, maintained by a polygynous mating system and female-biased philopatry. F_{ST} values were not calculated, since social groups were studied within a single local subpopulation.

A study of gene dynamics of pocket gophers (*Thomomys bottae*) identified genetic structure at the level of local polygynous groups (Patton and Feder 1981). Allozyme data were collected from the population of breeding adults of both sexes, and used to calculate postdispersal F-statistics. This species likely exhibits male-biased dispersal from the natal site. Significant genetic differentiation was found among the polygynous groups, as well as among populations within the region, among regions within broader geographic ranges, and among geographic ranges with the species range (Patton and Smith 1990). F_{LS} was lower than F_{ST}, but it still indicated significant social genetic structure of pocket gophers within a local population. F-statistics at the broader levels of genetic hierarchy likely indicated the influence of geographic variation due to the limited dispersal abilities of these fossorial rodents (Daly and Patton 1990; Steinberg and Patton 2000). Patton's studies are notable for examining several levels of population substructure, particularly at geographic scales.

VanStaaden et al. (1994) examined the gene dynamics of the moderately social Richardson's ground squirrel (*Spermophilus richardsonii*). This species is promiscuous, with multiple mating by both males and females (Michener and McLean 1996). As in more highly polygynous species, males typically exhibit greater reproductive success than females, due to a strong female bias in the adult sex ratio. Thus, genetic correlations among offspring may be high, on average, despite possible multiple paternity and dispersal by both sexes (though females are often philopatric; Michener and Michener 1977; Davis 1984c). Although matrilines of closely related and spatially contiguous adult females occur,

Table 14.1 Gene dynamics of selected rodent species, and one species of pika (order Lagomorpha) that were studied with the breeding-group model of Sugg and Chesser (1994)

Species	F_{LS}	F_{IL}	F_{IS}	F_{IT}	F_{ST}	Dispersal pattern	Notes	References
Marmota flaviventris	0.07	−0.09	−0.07			Post	allozyme	Schwartz and Armitage 1980
Thomomys bottae	0.07	0.14	0.03	0.28	0.26	Post	allozyme	Patton and Feder 1981; Patton and Smith 1990
Spermophilus richardsonii	0.05	−0.40	−0.34			Post	allozyme, females only	vanStaaden 1994
Cynomys ludovicianus	0.23	0.11	0.32	0.40	0.10	mixture	allozyme	Chesser 1983
Cynomys ludovicianus	0.16	−0.18	0.01			model	model	Sugg et al. 1996
Cynomys ludovicianus	0.16	−0.18	0.00			model	model	Dobson et al. 1997, 2004
Cynomys ludovicianus	0.19	−0.23	0.00			Pre	pedigree	Dobson et al. 1997, 2004
Cynomys ludovicianus	0.08	−0.08	0.00			Post	pedigree, both sexes	Dobson et al. 1997
Cynomys ludovicianus	0.03	−0.04	−0.01			Post	pedigree, males only	Dobson et al. 1998a
Cynomys ludovicianus	0.12	−0.13	0.01			Post	pedigree, females only	Dobson et al. 1998a
Cynomys ludovicianus	0.17	−0.21	−0.01			mixture	allozyme	Dobson et al. 1997, 2004
Cynomys ludovicianus	0.25	−0.38	−0.03			Pre	allozyme	Dobson and Chesser et al. 1998
Cynomys ludovicianus	0.16	−0.21	−0.02			Post	allozyme	Dobson and Chesser et al. 1998
Ochotona curzoniae	0.30	−0.37	0.04			model	model; single paternity	Dobson and Smith et al. 2000
Ochotona curzoniae	0.28	−0.34	0.04			model	model; multiple paternity	Dobson and Smith et al. 2000

identification of breeding groups was difficult. Consequently, vanStaaden et al. (1994) examined genetic variation among matrilines, which was likely a level of genetic structure just above the level at which matings occurs. Based on postdispersal data from 6 allozyme loci, they found significant genetic differentiation among matrilines (table 14.1). This study demonstrated potential for genetic substructuring in semisocial species that do not exhibit spatially overlapping females in strongly cohesive family groups.

By far the most complete investigation of gene dynamics of socially structured kin-groups comes from black-tailed prairie dogs (*Cynomys ludovicianus*). Chesser (1983) examined genetic variation at several levels of population structure, including coterie breeding groups, using allozyme alleles. He found significant genetic differentiation among social groups and among geographically distinct colonies (table 14.1). Surprisingly, there were greater genetic differences among social breeding groups than across considerable geographic distances. Inbreeding *F*-statistics (viz., F_{IX}, where *X* can be *L*, *S*, or *T*) at the level of breeding group, colony, and regional population were positive and increasing in magnitude. The positive value of F_{IS} contrasts with other allozyme studies of the same species that found significantly or slightly negative values (Foltz and Hoogland 1983; Daley 1992). So while Chesser's (1983) study indicated significant genetic differences among social groups, it also raised questions for further research: how could genetic differences among social groups match or surpass genetic differentiation over geographic ranges, and should inbreeding *F*-statistics exhibit a general pattern of increase?

These questions were answered in large part by Sugg et al. (1996) and Dobson et al. (1997). Sugg et al. (1996) used published data from Hoogland's (1995) long-term study of the behavioral ecology of black-tailed prairie dogs to estimate gene dynamics, using the breeding-group model (Chesser 1991a, 1991b; Chesser et al. 1993; Sugg and Chesser 1994). Sugg and Chesser's (1994) "multiple-paternity" version of the breeding-group model revealed that the combination of philopatric females and male polygyny resulted in coteries of related adult females, and consequently significant genetic differentiation occurred among coterie breeding groups (table 14.1). Surprisingly, Sugg et al. (1996) also found that effective population size was slightly greater than the size of the adult census population (figure 14.2). Dobson et al. (1997) compared estimates from the breeding-group model, allozyme data, and pedigrees in a 10-year study of black-tailed prairie dogs. All three methods of estimating F-statistics yielded very similar estimates, thus confirming results of the model. Effective population size was also estimated from more detailed data (Dobson et al. 2004), using all three methods, and again estimates of effective population size were slightly greater than the census population size (fig. 14.2).

Further work with black-tailed prairie dogs revealed differences that might be expected from predispersal and postdispersal estimates of F-statistics (table 14.1; Dobson and Chesser et al. 1998). F_{LS} was higher in predispersal (19%) than postdispersal (8%) estimates for both pedigrees, and for the allozyme data set (25% and 16%, respectively). Greater genetic differentiation is to be expected from predispersal estimates, because they are for the breeding-group young, most of whom are full or half siblings.

Genetic differentiation among breeding groups was also significant using separate estimates of F_{LS} for adult females (producing post-dispersal estimates), using pedigree data (12%). Adult males exhibited a much lower F_{LS} (3%). The difference in values for adult males and females reflects the ubiquitous natal dispersal of males and the close kinship of females. As might be expected, F_{IL} estimates were closer to zero in postdispersal estimates, though they were still significantly negative, and all F_{IS} values were close to zero. These three F-statistics are linked to each other by the formula: $(1 - F_{IS}) = (1 - F_{LS})(1 - F_{IL})$. Thus, F-statistics tend to change together. In spite of differences in F_{LS} between pre- and postdispersal estimates, it is clear that the social structure of breeding groups creates genetic structure that persists over generations.

Deviations of the inbreeding coefficient (F) from the expectation of random mating must be evaluated with caution. Patton and Feder (1981) estimated F_{IL} at a substantial positive value at the level local groups of pocket gophers (table 14.1), but felt that their estimate was biased by a substantial Wahlund effect (caused by pooling social breeding groups; see also Sugg et al. 1996). Nonetheless, an expected increase in the value of F_{IX} at increasing spatial levels seems reasonable (Dobson et al. 1997). Within breeding groups, dispersal from the natal area may often result in negative F_{IL}

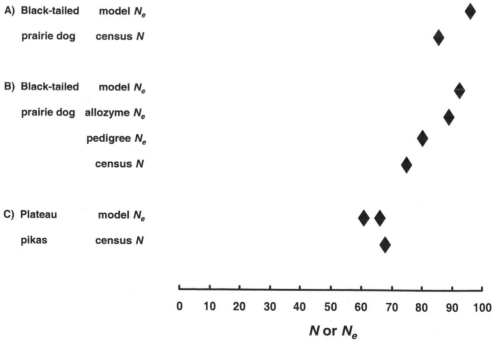

Figure 14.2 Estimates of effective population sizes and census population sizes (numbers of potentially breeding adults). (A) From Sugg et al.'s (1996) estimates from results presented in Hoogland's (1995) book on black-tailed prairie dogs. (B) From Dobson et al.'s (2004) empirical data on black-tailed prairie dogs. (C) From Dobson et al.'s (2000) study of Tibetan plateau pikas under two assumptions: complete single paternity (left hand estimate) and complete multiple paternity (estimate to the right) within pika family breeding-groups.

values. F_{IS} should approach zero as the area sampled is expanded, resulting in a balance between natal dispersal from breeding groups and a spatial scale at which the genetic influence of dispersal distances becomes limited geographically. For example, dispersal movements were much greater within the colony of black-tailed prairie dogs that Hoogland (1995) studied than the low rate of immigration to the colony from other colonies (Garrett and Franklin 1988; Dobson et al. 1997). At larger spatial scales, F_{IT} may become strongly positive, as geographic isolation results in genetic differentiation of regional subpopulations, as Wright (1969) envisioned. Relationships between inbreeding, heterozygosity, and spatial scale raise interesting questions that deserve further research.

Estimates of effective population size for local populations made up of breeding groups are higher than estimates from traditional models, which suggest effective sizes at about half the census size or less (Nunney 1993; Nunney and Elam 1994; Frankham 1996). Nunney (1999) criticized Sugg et al.'s (1996) estimate for black-tailed prairie dogs because the breeding-group model assumes nonoverlapping generations, and generations overlap in prairie dogs. Thus, he argued that the estimate may have been biased upward. However, in a study of Tibetan plateau pikas (*Ochotona curzoniae*), Dobson and Smith et al. (2000) applied the breeding-group model to behavioral and demographic results for a species that has extremely limited overlap of generations. Families of these highly social pikas exhibit a variety of mating systems, but on average are slightly polygynous (Smith and Wang 1991). They also exhibit male-biased dispersal, though either sex may be philopatric (Dobson and Smith et al. 1998). In spite of the potential for high rates of inbreeding, F_{IL} was strongly negative and F_{LS} indicated strong genetic differentiation among family groups (table 14.1). Effective population size was slightly different depending on whether multiple paternity occurred (unknown, though multiple copulations occurred; F. S. Dobson and A. T. Smith, personal observations). Estimates of effective size were slightly less than the size of the census population (fig. 14.2), and appeared much higher than expected from classical models of gene dynamics that do not take social structure into account (Dobson and Smith et al. 2000).

Gene dynamics of nonrodent species with social breeding have been reviewed by Storz (1999). The purpose of his review was to evaluate whether social breeding groups could lead to speciation, which seemed unlikely for most mammals. However, significant genetic differences among social breeding groups were found in several species. In particular, red howling monkeys (*Alouatta seniculus*) exhibited significant breeding-group structure, based on F-statistics

estimated from allozyme data from a mixture of predispersal and postdispersal individuals of both sexes and several ages. F_{LS} averaged about 0.14, F_{IL} about -0.22, and F_{IS} about -0.04 (Pope 1992). Red howlers exhibit polygyny and predominant male dispersal, often with related males dispersing together (reviewed by Pope 1992, 1998). Interestingly, troops of six other primate species, such as macaques (genus *Macaca*), exhibited generally low but substantial genetic differences among troops ($F_{LS} = 0.04 - 0.08$; Storz 1999). Two other primate species exhibited insignificant genetic differentiation among troops. Finally, spatial groups of two species of artiodactyls exhibited significant genetic differentiation, perhaps due to social breeding groups. With few exceptions and for estimates based on postdispersal estimates of F-statistics, Storz (1999) found substantial influences of social breeding groups on gene dynamics in species exhibiting polygynous or promiscuous mating systems and predominant male dispersal patterns.

Summary

In this review, I have examined whether social groups composed of philopatric female kin create genetic structure within local populations. An understanding of the hierarchical nature of gene dynamics and correct interpretations of F-statistics are necessary to address this question. Thus I've discussed the hierarchical nature of gene dynamics, and of the F-statistics that Wright (1965, 1969, 1978) used to describe them. These F-statistics, particularly F_{LS}, can be used to indicate whether social breeding groups create a level of genetic structure within local populations. For a few species of rodents and other mammals, there does appear to be significant genetic structure at the level of social breeding groups. It could hardly be otherwise, since philopatry among females will create clusters of female kin, and kin exhibit genetic correlations that are spatially concentrated. As Chesser (1991a, 1991b) pointed out, these genetic correlations are a form of genetic structure that can be measured with F-statistics and linked to higher levels of spatial genetic structure. Further studies of mammals will reveal the generality of sociogenetic structure in populations of highly social mammals. Studies of less social species, such as the groundbreaking work of vanStaaden et al. (1994) on Richardson's ground squirrels, are needed to discern whether sociogenetic substructuring extends beyond highly social species, like black-tailed prairie dogs. In addition, this review provides a working hypothesis for future studies of relationships between the inbreeding coefficient, F, and its expectation for randomly mating populations at different spatial scales, as measured by hierarchical F-statistics.

I have also pointed out that estimates of effective population size that include effects of breeding groups appear higher than estimates from models that do not include breeding-group population substructure. Inbreeding effective population size reflects changes in loss of gene diversity due to accumulation of inbreeding over generations, a property that is not directly reflected by the *F*-statistics. Divergence in estimates of effective population size from different models may occur due to differences in methods (e.g., Chesser et al. 1993; Sugg and Chesser 1994; Nunney 1999), whether estimates use predispersal or postdispersal data, or from estimating different types of effective size (e.g., inbreeding and variance effective sizes; Crow and Denniston 1988). Effective population sizes are not hierarchical, but they may vary from place to place and with increasing spatial scale. Nonetheless, social breeding groups may slow the loss of genetic diversity from populations. This phenomenon deserves further study, as it is of substantial theoretical and practical importance (e.g., Sugg and Chesser 1994; Chesser et al. 1996; Dobson and Zinner 2003).

Chapter 15 Social Behavior and Self-Regulation in Murid Rodents

Charles J. Krebs, Xavier Lambin, and Jerry O. Wolff

MURID RODENTS have been a favorite group for studies of both ecological and behavioral questions, and the broad outlines of both social behavior and population dynamics are now available for many species. What is lacking is the bridge between rodent social organization and rodent population dynamics. Two polar views can be recognized. The first view is that social behavior is interesting and important to study but has nothing to do with population dynamics. Population changes are believed to be driven by predators, disease agents, and food supplies, without any need to consider how social behavior might be involved as a destabilizing influence. At most, this first view considers that social factors may contribute as a stabilizing agent to setting some notional carrying capacity. This view can be traced back to many authors like David Lack (1954) and is widely supported in the current literature (e.g., Turchin 2003). We will refer to it as the dominant current view. The alternate view is that social behavior is an important component of population dynamics because of its potential impacts on variation in birth and death rates and dispersal. This view arose from the early work of John Christian, David E. Davis, and Dennis Chitty in the 1950s. The paradox of this polarization is that in spite of excellent evidence in many rodents about the impact of social behavior on birth, death, and dispersal rates, almost no one believes that these impacts translate into population dynamics, beyond contributing to seasonal dynamics.

Social behavior, including territoriality, dispersal, reproductive suppression, and infanticide has been studied extensively in the laboratory and field in many species of rodents. In this chapter, we consider the current theory regarding social behaviors and then show how they are al-

tered by changes in density that could affect rates of population growth. We draw heavily on examples from *Microtus* and *Peromyscus* in North America and Europe because of the extent of experimental work involving hypothesis testing in field and laboratory studies. We will not review all aspects of social behavior; rather, we limit our discussion to behaviors that show the greatest potential for self-regulation of fluctuating populations.

Avian ecologists had already suggested in the 1920s and 1930s that territorial behavior could limit population density (Howard 1920; Nice 1937; Hensley and Cope 1951). Mammal ecologists were slower to accept that social behavior might impact population dynamics, and the first approach was through physiology. Hans Selye in 1937 suggested that crowding in rodents could lead to physiological stress mediated through the adrenal gland, and that stress could reduce reproductive output as well as increase death rates. David E. Davis and John Christian (1956) did the first field experiments to show that aggressive social interactions could reduce Norway rat population size, and since these early experiments, many authors have contributed studies that evaluate the role of social behavior in affecting population events. How might social processes affect rodent population dynamics?

Mechanisms of Population Regulation via Social Behavior

Social behavior can affect population dynamics via five different mechanisms: limiting the size of the breeding population, control of the timing of sexual maturation, infanticide, control of dispersal, and direct aggression (= interference competition). Wolff (2003) has reviewed these aspects of the social ecology of rodents and has concluded that social interactions play little role in regulating or stabilizing rodent populations. We will not review the detailed aspects of these social interactions, which are covered well in Wolff, but we wish to review here the dominant current view that social interactions are rarely relevant to population dynamics.

The self-regulation hypothesis proposes that individual differences in spacing behavior influence reproductive performance and subsequent population trends either through genetic or maternal effects (Krebs 1978, 1996; Oli and Dobson 2001). Wolff (1997) provided an evolutionary argument for how various aspects of social behavior could lead to intrinsic population regulation. For social behavior to limit the growth rate of a population, it must decrease fecundity, or at least juvenile recruitment, or decrease survival rates. Decreased juvenile recruitment could result from decreased litter size, decreased juvenile survival, fewer females breeding, or a delay in time to sexual maturation of young females. There is little or no evidence that litter size is affected by social interactions other than might be affected by decreased nutritional state of females due to limited food resources. However, social stress at high density can delay the onset of sexual maturation, which slows population growth (Oli and Dobson 1999). A decrease in number of females breeding also could be affected by territoriality or some form of reproductive suppression. In the following we follow the basic conceptual arguments of Wolff (1997) and review the major determinants of social behavior that could regulate populations.

Female territoriality

The key behaviors in rodents that might impact on population density can be broadly classed as spacing behavior. If individual rodents maintain a personal or group space, then clearly the density of that population will reflect this spacing. Spacing is most readily thought of as resulting from direct physical aggression, but this mental image must be broadened to include spacing by avoidance behavior as well as spacing by direct physical interactions. We first ask if spacing behavior could limit population density (Watson and Moss 1970).

Social systems of rodents are variable and flexible and appear to depend to a great extent on the distribution of females. In most rodents, individuals or groups of related females defend territorial space against unrelated females. Territories vary considerably in size, from as little as 25 m² in high-density populations of microtines to several hectares in larger species and/or at lower densities (Wolff 1985a). For most grassland rodents that weigh less than 100 g, ter-

ritories are typically 50 to 150 m², whereas forest-dwelling species often occupy territories of several hundred square meters (Wolff 1985a, 1989). During the breeding season, territories are relatively exclusive with respect to unrelated females, but often overlap and are shared with daughters or sisters (e.g., *M. townsendii, M. canicaudus, P. leucopus*). Home range size and daily movements often are related to resource availability; however, territoriality in female rodents may not be solely based on defense of a food resource as proposed by Ostfeld (1985). According to the food-defense hypothesis, the distribution, abundance, renewability, and type of food should determine whether females defend territories or share space with other adult females (Ostfeld 1985). An alternative, but not exclusive, hypothesis is that females defend territories to protect their offspring against infanticide from conspecific females (Wolff 1993b). Species of rodents that hoard food such as seeds in a central larder seem to defend this food source; however, green vegetation, perishable, or nonstorable and other widely scattered food may not be defensible (Wolff and Peterson 1998), and whether they are limiting during the breeding season is equivocal (Taitt and Krebs 1981; Ostfeld 1985; Lambin and Krebs 1993; Wolff 1993b).

Infanticide is common among female rodents and is hypothesized to be a form of reproductive competition, in which perpetrating females kill offspring to eliminate competitors and gain access to breeding sites (Sherman 1981b; Wolff and Peterson 1998). The fact that female aggression that leads to territoriality is associated with lactation and the breeding season (Maestripieri 1992; Wolff 1993b) and does not occur in the middle of the nonbreeding seasons supports this latter hypothesis (Wolff and Peterson 1998). However, considering that territories become established when vegetation greens up, several weeks before the first litter of the year is conceived, is consistent with a dual influence of the need to secure access to food resources and social space. The infanticide hypothesis is also applicable to other taxa of mammals and is associated with altricial young that require a burrow, tunnel system, or protected den site for successful rearing of offspring. In that rodents fit these needs, females compete for this limited offspring-rearing space. Competition for breeding space contributes to territoriality in female rodents (Solomon and Keane, chap. 4, this volume).

If breeding male or female rodents defend a territory, the potential exists for spacing behavior to limit population density. The larger a territory that is defended, the lower the population density, and the immediate question arises as to what determines territory size. There has been an ongoing argument in the bird literature between those who interpret territory size as a consequence of population density and those who interpret it as a cause:

territory size → population density
population density → territory size

The only way to test these two views is to experimentally manipulate territory size, typically by manipulating aggression (e.g., Watson and Jenkins 1968; Moss et al. 1994; Mougeot, Redpath, Leckie, and Hudson 2003; Mougeot, Redpath, Moss, Matthiopoulos, and Hudson 2003), but few of these kinds of experiments have been done on wild rodents (Gipps et al. 1981; Taitt and Krebs 1982), in part because they are technically difficult.

Experiments that demonstrated the role of spacing behavior in herbivorous *Microtus*, (*townsendii, agrestis, pennsylvanicus*) have been performed near the onset of spring reproduction, when the greening of the vegetation triggers the transition from reproductive quiescence to the breeding season. Defense of space during that seasonal bottleneck in food availability is necessary for early reproduction and also secures space where offspring may later become established when vegetation availability no longer limits access to reproduction, but social space is at a premium. In *Microtus agrestis* for instance, over-wintered females defend exclusive territories, but spring-born females breeding in the year of their birth are more tolerant (Agrell 1995; Pusenius et al. 1998).

If spacing behavior limits spring *Microtus* population size, a paradox is that in the absence of any notable interannual variation of food availability that characterizes seed-eating murids (e.g., *Peromyscus* and *Apodemus*), *Microtus* outbreaks occur and are preceded by much higher densities of breeding females in spring than occurs in most years. In these years of high density, either spacing behavior is unable to reduce population size to the normal value, or individuals in high-density populations adopt an alternative spacing behavior, and their reproductive potential is not too adversely affected (since the population continues to grow to outbreak densities). The possibility that kinship among over-wintered females, and hence the risk of infanticide, differs in those years, and hence allows a different mode of spacing behavior, is considered in the following.

Does territoriality limit population size?

If females defend breeding spaces (territories) and these are limited, then potentially territoriality could limit the size of the breeding population. One view is that territory size is not fixed; rather, it shrinks and expands like an elastic disk, with changes in density and intruder pressure (Wolff 1989; Wolff and Schauber 1996). If this is the case, territoriality would only limit populations when territories have shrunk to their minimum size. How small can a territory become? Studies with voles in enclosures show that without dispersal, territory size shrinks to only a few square meters, with

Figure 15.1 Gray-tailed vole and aerial view of the vole enclosure research facility at Oregon State University used for studies on behavioral aspects of population growth. Photos by J. O. Wolff.

considerable overlap with neighboring females (e.g., Boonstra and Krebs 1977; Wolff and Schauber 1996; fig. 15.1). The fact that rodent populations in enclosures reach exceedingly high densities suggests that territoriality in itself (in the absence of potential dispersal) is insufficient to stop population growth (Boonstra and Krebs 1977; Wolff and Schauber 1996; fig. 15.2). The real question is, does territoriality create surplus individuals that are not breeding? An alternative is that factors other than intruder pressure contribute to setting territory size, as shown experimentally with red grouse (Moss et al. 1994; Mougeot, Redpath, Leckie, and Hudson 2003).

Does territoriality create surplus individuals?

We define *surplus* animals as those that are of reproductive age, but are not breeding. Following the lead of early experiments on songbirds (Hensley and Cope 1951), rodent ecologists used removal experiments to measure the number of surplus rodents. Krebs (1966) reports one early experiment on *Microtus californicus* in California. These and many other removal experiments have shown that when

Figure 15.2 The Fence Effect demonstrated here with Townsend's voles indicates that without dispersal, voles reach very high densities (left side) and do not intrinsically stop population growth, compared to unenclosed populations (right side), where dispersal is not deterred by a fence. Photo by C. J. Krebs.

you remove breeding adult rodents from an area, a flood of surplus individuals colonize the removal site, and in many cases bring the population density of the removal site back to the control density (e.g., Krebs et al. 1978). These removal experiments raise many issues that are relevant to rodent pest control. Where do these surplus animals come from? What is their fate if a removal experiment is not taking place? Do the surplus animals differ in age, sex, or size from resident animals? Many of these questions have been discussed by Anderson (1989) and Cockburn (1988). Clearly, if we accept the standard Darwinian principles, each of these individuals is attempting to maximize its own fitness, and our explanations of these results must fit in with contemporary evolutionary theory. Removal experiments to assess surplus individuals have been criticized in some species, since adjacent territory owners may shift their home ranges into the evacuated area (Schieck and Millar 1987). This criticism will affect the quantitative measurement of surplus animals, but it does not eliminate them. Schieck and Millar (1987) and Clinchy et al. (2001) have shown that surplus animals immigrate into unmanipulated areas as well as local residents shifting their home ranges into the evacuated area. A limitation of these field studies is that we do not know what the reproductive fate of immigrants into removal areas would have been if they had not moved. The assumption is apparently made that if they remained as residents they would become animals of low or zero fitness.

Given that we have surplus individuals, the second question is whether these individuals can breed when given the opportunity either in their home site or in the colonization area. The assumption is typically made that they would not achieve reproductive maturity in their home site, but this has not been tested adequately on individuals. Our results with removal experiments on *Microtus* voles have shown that there is no impediment to breeding in surplus voles that colonize removal areas, once the residents have been artificially removed from the area (Krebs et al. 1978; Myers and Krebs 1971). The impact of adult females on the suppression of maturation of young females has been studied particularly well in *Clethrionomys* voles (Bujalska 1970; Gilbert et al. 1986; Kawata 1987). The conclusion to date is that if there are surplus individuals in a rodent population, they are capable of breeding if social controls of maturation are relaxed, either in a resident population or in a newly colonized site.

Does territoriality affect recruitment?

A key question in rodent population dynamics is what controls recruitment into the trappable population. Rodents are model systems of species with very high reproductive rates coupled with high death rates, and the question we need to answer is what happens to all the young produced in a rodent population. A second question is whether that loss rate is constant or is influenced by prevailing or past density. The general finding in rodent-trapping studies is that a low fraction of the young produced ever recruit into the breeding population (Adler et al. 1987). The assumption is usually made that predators, diseases, bad weather, and other environmental factors control the survival of juveniles in their first few weeks of life.

Adult rodents can limit the recruitment of juveniles, and this can be another critical bridge between social behavior and population dynamics. If adults can limit recruitment

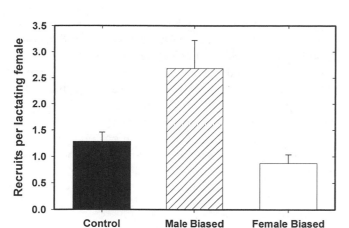

Figure 15.3 The average recruitment of juvenile *Microtus townsendii* for the summer breeding seasons of 1972 and 1973 on control (unmanipulated), male-biased (80% adult females removed) and female-biased (80% adult males removed) areas. Recruitment in this population is clearly controlled by adult females. Recruitment measured as the number of juveniles live-trapped at 2–5 weeks of age per pregnancy. Error bars are 95% confidence limits. (Data from Redfield et al. 1978, table 6).

of juveniles, we must ask if both sexes are involved or only one. Redfield et al. (1978) carried out biweekly sex-specific removal experiments on *Microtus townsendii* and *M. oregoni* for two years. The results are summarized in figure 15.3 which shows that adult females were the key to understanding why recruitment is restricted. Recruitment was 109% higher on the area with a more male-biased sex ratio, compared with a control area, and 32% lower on the area with a more female-biased sex ratio. The same pattern was found in a more carefully controlled experiment with the

same basic design with *Microtus canicaudus* by Wolff et al. (2002). Boonstra (1978) showed with a weaned juvenile introduction experiment that young juvenile *Microtus townsendii* would survive well if introduced at 3 weeks of age into a field from which all the adults had been removed, but few survived introduction into a field with only adult females present or both adult sexes present. The implication is that adult females directly kill strange juveniles or drive them out of the area, thereby limiting recruitment locally.

At the present we do not have a good quantitative analysis of the rate of loss during the first few weeks of life of nestling rodents. For *Microtus* voles, a few estimates of neonate survival are available. McShea and Madison (1989) found an average survival to recruitment of 31% in *Microtus pennsylvanicus* from 132 litters of radio-collared females. Boonstra and Hogg (1988) by contrast found an average 57% survival of neonates from 43 litters of the same species in an enclosure study. Lambin and Yoccoz (1998) found a neonate weekly survival of 70–75% over the first two weeks of life in 325 litters of *Microtus townsendii*, which translates into 0.42 to 0.54 survival to the usual trappable age of three weeks (fig. 15.4). In *Microtus canicaudus* only 1–3 juveniles are caught per pregnancy, which represents 17–50% of a modal litter size of 6 (Wolff and Schauber 1996; unpublished). Given the life history of murids, even small differences in neonate survival have profound implication for predicted population growth rate.

What happens to all these neonates? We do not know whether most of these young die in the nest or just after they leave the nest. It seems unlikely that dispersal is a primary cause of loss in the first three weeks of life (since dispersal

Figure 15.4 (A) Radiotelemetry is used to track offspring to show that females often nest near their mothers. (B) Longworth traps are used for mark-recapture studies on behavior and demography. (C) Although litter size is typically 4-6 pups, only 1-2 young per litter ever enter the population. (D) Typical grassland habitat for voles (*Microtus*) such as this study area of X. Lambin near Vancouver, Canada. Photos by X. Lambin.

usually occurs at a later age); there seem to be only three possible agents of loss: predation in the nest, infanticide, and bad weather (e.g., flooding). Considering that survival from birth to recruitment is usually inversely density dependent (Wolff et al. 2002; Krebs 2003) the limiting factors appear to be intrinsic rather than extrinsic. It is clear from a demographic perspective that variation in the nestling survival rate could account for a large portion of the variation in the per-capita rate of increase of rodent populations. One observation that has gained prominence in recent years is that infanticide is prevalent in rodent societies (Ebensperger and Blumstein, chap. 23, this volume), and could account for considerable neonate mortality.

Infanticide

Infanticide is common among male and female rodents and has been studied extensively in the laboratory and less so in the field (Agrell et al. 1998). Infanticide is committed by females to gain breeding space and by males to provide a breeding opportunity. Theoretically, the incidence of infanticide should change with density, but differ for the two sexes (Wolff 1995; table 15.1). At low densities, males move over large areas and will be more likely to commit infanticide in areas where they have not mated and have low confidence of paternity. At high densities, however, males are confined to smaller areas, have many females with which to mate, and should be less likely to commit infanticide within their resident home ranges or territories. One could speculate that females, on the other hand, should not commit infanticide at low densities, because breeding space would not be limited and thus competition is sufficiently

low that all females can acquire breeding space. At high densities, however, breeding space should be limited, competition intense, and the rate of infanticide high as females compete for limited offspring-rearing space. Thus infanticide is likely to occur at high and low densities, but differ for the two sexes (Wolff 1995); however, this has not been tested experimentally. The problem is that infanticide is most difficult to study in field populations, and we badly need information on its frequency in relation to population increases and declines.

The Bruce effect

Another factor that could decrease the rate of population growth is pregnancy disruption, commonly referred to as the Bruce effect. The Bruce effect is a form of pregnancy disruption in mammals in which exposure of a female to an unknown male results in pre- or postimplantation failure. If a female's pregnancy is disrupted following exposure to unfamiliar males intruding into her territory, this potentially could affect the rate of juvenile recruitment, especially when populations are high and many males are moving through female territories. Some form of pregnancy disruption has been reported in the laboratory for at least 12 species of rodents, including seven of the genus *Microtus* (reviewed in Mahady and Wolff 2002). However, two field studies have failed to support the relevance of this phenomenon. In field experiments with gray-tailed voles (de la Maza et al. 1999) and prairie voles (Mahady and Wolff 2002), 100% turnover of resident males and exposure to strange males every 10 days failed to detect any significant pregnancy disruption. In gray-tailed voles, no differences occurred in pregnancy rates, interbirth intervals, or juvenile recruitment between treatment and control populations. In prairie voles, nulliparous females took slightly longer to initiate first breeding in treatment populations and experienced poorer juvenile recruitment, but this effect was minimal and did not occur in parous females. The decreased juvenile recruitment in prairie voles could have been due to infanticide when young pups were exposed to strange males. In wild populations of voles and other rodents, females are commonly exposed to strange males, and it is questionable whether selection would favor any form of pregnancy termination following this exposure. Also, in most species of rodents females mate promiscuously and mated males are inhibited from committing infanticide (Agrell et al. 1998). Thus females should not need to sacrifice current pregnancies. The high rates of pregnancy and births at predictable intervals in most rodent populations make it seem unlikely that the Bruce effect has a marked effect on population growth or demography.

Table 15.1 Predicted density-dependent effects on various aspects of the social behavior of rodents (after Wolff 2003)

	Low density	High density
Territoriality	Large, widely spaced, mutual avoidance, low aggression, vacant space available	Territories small, considerable overlap, aggression high
Dispersal and philopatry	All males disperse relatively far, females disperse close to natal site, dams might bequeath maternal site to daughters	Delayed emigration, sons and daughters remain on natal site, extended families, cooperative and communal breeding of females
Age at sexual maturity	Sons and daughters mature at young age	Delayed sexual maturation for both sexes, cooperative and communal breeding for some species (see text)
Infanticide	High for males, low for females	Low for males, high for females

Reproductive suppression

The high intrinsic rate of growth of rodent populations is due to a great extent to high fecundity and early breeding of young females. In most species of rodents, young females become sexually mature and can breed shortly after weaning, as young as 20 days of age. Females exhibit postpartum estrus and breed at regular intervals as short at 21 days. Life expectancy is short, often less than 4 months; however, a given female can be expected to produce about 20 offspring in an average lifetime, so that under ideal conditions the population could double every 5 weeks.

A major factor that determines the size of the breeding population is the rate at which young females become sexually mature and experience their first pregnancy. In some species of rodents, such as *Clethrionomys*, each female needs its own individual territory to breed; that is, young females do not breed on their mothers' territories, and sexual maturation is retarded at high population densities (Gilbert et al. 1986). It is not known how common this pattern is because we have relatively few data on the relatedness of females in wild populations and position of nests within territories for most species of rodents (see kinship following). However, it is well documented for several species of *Microtus* and *Peromyscus* that daughters at times breed on their maternal site, often communally and cooperatively with other female relatives (e.g., Lambin 1994b; Wolff 1994b, Wolff et al. 1988). In *Clethrionomys* and other species in which females require individual territories, territoriality can limit the size of the breeding population and consequent rate of population growth. Females that do remain on their natal site beyond the time of normal reproductive maturity would be reproductively inhibited until they obtain a territory of their own. However, for species in which daughters are not reproductively inhibited by their mothers, and breed on shared maternal territories or even in the same nest, such as most *Microtus* and *Peromyscus*, territoriality would have a minimal effect on the size of the breeding population. The key question is what factors permit related females to breed in shared territories, and in particular whether the prevalence of these communal breeding groups is density dependent.

Resource or reproductive competition within family groups appears to be minimal for much of the breeding season but may be critical at the start of breeding in spring (Lambin and Krebs 1993; Andreassen and Ims 2001). Unrelated females are expected to commit infanticide as a form of competition for breeding territories (Wolff 1997; Wolff and Peterson 1998). At high densities, pregnancy rates typically remain high; however, juvenile recruitment declines with an increase in density (Wolff et al. 2002; Krebs 2003).

The mechanism for reproductive suppression of young females is difficult to detect in the field, but field observations with other mammals and laboratory experiments with rodents demonstrate that unrelated adult females typically inhibit young females from breeding (Drickamer 1984e; Wolff 1997; Krebs 2003).

Dispersal and philopatry

Dispersal has been studied extensively in rodents, primarily from a population regulation point of view. Most dispersal involves juveniles or young adults emigrating from the natal site and immigrating to a territory or home range area that will be the adult breeding site. The general dispersal pattern for rodents, as it is in most mammals, is for young males to disperse from their natal site and for daughters to be philopatric and remain on or near their natal site (Boonstra et al. 1987; Wolff 1993a, 1994b). Young males leave the natal site to find vacant space and unrelated breeding-age females. However, dispersal distance of young males is highly variable and dependent in part on the degree of habitat saturation and the availability of vacant territories or mating partners (Lambin et al. 2001). Since inbreeding avoidance appears to be an important function of emigration of young males from their natal site, dispersal distances are probably determined in part by the proximity of related females around the natal area, encounter rates with unrelated females, and competitive interactions of resident males (Wolff 1993a, 1994b; Andreassen and Ims 2001; Lambin 1994b; Lambin et al. 2001).

Young females typically attempt to establish breeding space close to their natal site. At low densities when adjacent space is available, young females establish individual home ranges or territories in close access to the natal site. In some cases, mothers appear to bequeath their natal site to daughters and move a short distance and establish a new territory or nesting site that overlaps the previous natal site (Wolff and Lundy 1985; Lambin and Krebs 1991a, 1993; but see Lambin 1997); however, the spatial pattern of mothers and daughters is not well understood for most species. At very high densities, mothers and daughters often nest within meters of each other or sometimes communally and cooperatively in the same nest (e.g., Wolff 1994b; Lambin and Yoccoz 1998; Solomon and Getz 1997).

If resident adults cause new recruits to disperse, and dispersal is costly in terms of survival in unfamiliar habitats, dispersal could be a process that is involved in population regulation. The most striking experimental argument for the role of dispersal in population regulation has been the fence effect (Krebs et al. 1969; Boonstra and Krebs 1977; Wolff and Schauber 1996; fig. 15.2. Ostfeld (1994) sug-

gested that the fence effect was an artifact of predator exclusion, but this conclusion is incorrect, particularly since
Boonstra's site was on an island with no mammalian predators that might be restricted by a fence.

Two general problems have plagued efforts to evaluate
the role of dispersal in population limitation. First, estimates of dispersal rate and distance are difficult to obtain.
Removal areas measure some components of dispersal but
may bias the quantitative results (Schieck and Millar 1987).
Radiotelemetry studies of dispersal are more promising,
but sample size problems and scale issues complicate interpretations (Beacham 1980; Gillis and Krebs 1999). A new
combined approach of direct capture-mark-recapture with
microsatellite genotyping to identify parents and offspring
in different populations and hence infer who disperses and
how far shows considerable promise when a large proportion of putative parents can be sampled genetically (Telfer
et al. 2003). When applied to water voles (*Arvicola terrestris*), dispersal rates were more than two times greater for
females and three times greater for males relative to estimates based on capture recapture. Second, if dispersal is to
contribute to population regulation, it must somehow be
related to population density. Many studies have suggested
that dispersal rate is inversely density dependent, with maximum rates at low density and minimal rates at high density (Gaines and McClenaghan 1980; Wolff 1997; Andreassen and Ims 2001). The low dispersal rates at high density
are supposedly due to a social fence of territorial neighbors
that deter immigration, resulting in philopatry of sons and
daughters remaining in their natal site (Wolff 1997). If this
generalization that dispersal rate is inversely density dependent continues to hold, any dispersal impacts on population
regulation or limitation will have to be achieved by indirect
means. One way to achieve this would be for dispersal to be
selective for certain phenotypes that have different types of
spacing behavior (Krebs 1985). Whether or not this occurs
in rodents is unclear at present.

Does dispersal regulate population growth?

The characteristic dispersal pattern in rodents is for males
to disperse relatively long distances and for females to settle
in territories near their natal site. However, in continuous
habitats at high densities when all breeding space is occupied by territorial males or females, young juveniles are
deterred from emigrating from their natal site by a social
fence of aggressive territorial owners inhibiting immigration (Wolff 1994b; Lambin et al. 2001). This social fence
acts as a negative density-dependent factor, reducing the
rate of dispersal (Andreassen and Ims 2001). Thus the rate
of dispersal in territorial species is inversely density dependent (Wolff 1997; Lambin et al. 2001). As density increases,
the rate of dispersal decreases, resulting in extended fami-

lies as sons and daughters remain on their natal sites past
the time of normal dispersal and sexual maturation. This
delayed emigration from the natal site can inhibit sexual
maturation of young females by direct competition with
their mothers (Gundersen and Andreassen 1998), or act
as a mechanism to avoid inbreeding with male relatives
(Wolff 1997; Lambin et al. 2001). In patchy environments,
or those in which individual movements are not deterred by
neighbors, dispersal should not be delayed and may, in fact,
help to stabilize or regulate density within the patch.

Alternatively, philopatry in female rodents may cause a
delay in density dependence that would destabilize density.
Female small mammals are highly philopatric, and breeding
females may therefore be surrounded by their philopatric
relatives under some circumstances. If female voles depress
the survival of offspring of nonkin females only, and do not
influence the survival of offspring of their female kin, time-
delayed density dependence in the regulation of vole numbers by social behavior would be the result (Lambin and
Krebs 1991a). The time delay occurs because the previous
pattern of recruitment and mortality in a population gives
rise to female kin-clusters. Kin-clusters are formed following successful reproduction, philopatric recruitment of females, and high survival; they decay with mortality and
immigration. If juvenile survival and recruitment are less
affected by female density in kin-structured populations,
such populations could temporarily escape the social consequences of high density (Lambin and Krebs 1991a, 1993;
Lambin and Yoccoz 1998).

Kinship effects

If spacing behavior can affect population size at the onset of
the breeding season and the recruitment rate of young animals, as well as their rate of sexual maturation, we need to
find out more information about the rules that govern spacing behavior in rodents. Darwinian arguments about inclusive fitness would suggest that for a start relatives should
respond differently to one another than they should to
strangers. This simple idea spawns several questions about
how relatives might recognize one another, and how familiarity might substitute for genetic relatedness (see Holmes
and Mateo, chap. 19, this volume), but the first question
we need to answer is whether or not there is a genetic structure of relatives in field populations. The formation of kin
groups, cooperation among kin, and having kin for neighbors should be beneficial for group defense and reduce the
incidence of infanticide by neighboring females (Charnov
and Finerty 1980; Wolff 1995). Lambin and Krebs (1991a)
suggested that, since females controlled recruitment in voles,
changes in female relatedness might have a significant impact on population dynamics. Specifically, they pointed out

Table 15.2 Impact of kinship on reproduction and nestling survival and summer population growth rates in seven studies

| Study | Reproduction and nestling survival | Summer population growth rate (λ) per month | | | Parous females used? | Experimental area (ha) |
		Observed	Predicted for kin	Predicted for non-kin		
Kawata 1987	No data	Spring only, unequal numbers			Yes	0.50
Boonstra and Hogg 1988	3.5 vs 2.2 young per litter	Unequal numbers	1.43	1.28	Mixed	0.15
Lambin et al. 1993, 1998	3.15 vs. 2.54 young per litter	No data	1.72	1.48	Yes	0.50
Sera and Gaines 1994	No effect	No data			No	0.09
Mappes et al. 1995	3.0 vs. 1.7	2.73 vs 1.61	1.69	1.10	Yes	0.50
Pusenius et al. 1998	2.96 vs. 0.93 young in July and August	No data	1.35	1.04	Yes	0.25
Dalton 2000	No effect	Unequal numbers			No	0.20

NOTES: For this analysis we used simple standard age-based Leslie matrices (c.f. Lambin and Yoccoz 1998) to explore the potential impact of changes in demographic parameters associated to changes in female relatedness on the dynamics of these vole populations.

that any impact of prevailing kinship on present demographic rate ought to generate a positive feedback on population growth as present kin-clusters are formed following past successful reproduction, philopatric recruitment of females, and high survival. Table 15.2 lists 7 partial tests of this idea that kinship affects population growth rates in voles by manipulating the size of matrilineal kinship groups in fenced or open populations. In four of seven studies there was higher reproductive output and juvenile survival, leading to the prediction of higher population growth rates in kin groups. For example, Lambin and Yoccoz (1998) found that relatives nested closer to one another than did unrelated females in *Microtus townsendii,* and pup survival in the nest was improved in the first two weeks of life when relatives were nearby (fig. 15.4). In addition, adult female survival at the start of the breeding season was higher for kin group females than for unrelated females. All experiments that failed to detect any effect of kinship on juvenile survival used nulliparous females, suggesting securing space for their offspring may contribute to the infanticidal behavior of female voles. If these kinds of kinship effects are significant, populations with matrilines should grow faster than those with only unrelated females. These kinds of studies need to be repeated and carried out on other rodent species before we will know how general this kinship model is in natural populations.

Summary and Conclusions

Rodent social behavior has been studied extensively, but whether it impacts population dynamics has been questioned. Social behavior is affected significantly by population density, and the question is whether this is a circular causal system. Infanticide, dispersal, sexual maturation, and direct aggression are all potential processes that are a part of spacing behavior in rodents that may cause large changes in survival, reproduction, and movements. There are several mechanisms by which social behavior can affect birth, death, and dispersal rates and thereby changes in population density. In any natural population of rodents, social behavior will operate in a matrix of extrinsic mechanisms like predation, and disentangling the relative contribution of specific factors can be done only with carefully designed experiments. We need to know both the frequency and strength of these processes in rodents, and how they change with population abundance. Recruitment in rodent populations may be limited by the presence of breeding adults, and the kin structure of populations may affect the potential rate of population growth. Further experimentation with more species is needed to uncouple the potential interaction of social behavior and extrinsic processes like predation and food supplies in limiting population abundance in natural populations of murid rodents.

Behavioral
Development

Chapter 16 Neural Regulation of Social Behavior in Rodents

J. Thomas Curtis, Yan Liu, Brandon J. Aragona, and Zuoxin Wang

SOCIAL BEHAVIOR arises from a complex interplay of numerous and often-competing sensory stimuli, the physiological and motivational states of the participants, and the ages and genders of the individuals involved. Overlying the internal responses to a social encounter are a variety of external factors, such as the context in which the encounter occurs, the time of year, environmental conditions, and the outcomes of previous social interactions. To further complicate the situation, each individual in a social encounter must be able to adjust its own actions depending on the responses of other animals. Given such complexity, a detailed understanding of the neural basis of social behavior would seem next to impossible. Nonetheless, considerable progress has been made. By examining individual components of social behavior while controlling for other variables, researchers have begun to parcel out the contributions of specific brain regions and neurochemicals to specific aspects of social behavior.

A comprehensive survey of the hormonal and neural control of all of the different types of rodent social behavior is beyond the scope of a single book chapter. Here, we will concentrate on the neuroanatomical and neurochemical substrates that underlie rodent social structures, with a particular focus on social bonds. Behavior, like all other aspects of a species' natural history, is subject to natural selection. Since the ultimate test of a behavioral repertoire is reproductive success, it is perhaps appropriate to focus on mating systems in addressing the neural control of social behavior. By focusing on mating systems we are able to place a variety of social behaviors into a firm ecological context,

since behaviors such as aggression likely derive from the mating system.

Rodents are a diverse group of creatures that inhabit a wide variety of ecological niches. As might be expected of such diversity, rodents display a wide range of mating systems. Males and females of many species often have different mating strategies (Waterman, chap. 3, Solomon and Keane, chap. 4, this volume). However, some environmental conditions require extensive cooperation between the sexes for reproductive success (Kleiman 1977). In these cases, the mating strategies of the two sexes may converge and a monogamous mating system may arise. Only about 3% of mammalian species have been categorized as being monogamous (Kleiman 1977). Within the rodent order, monogamy has arisen several times (Kleiman 1977), in some cases within genera in which other species are not monogamous. This polyphyletic origin to mating systems has presented opportunities for detailed comparative studies of the neural control of social behaviors and how the brains differ between closely related species with different social structures.

Comparative Models

Over the past two decades much research on social behavior has focused on two genera: *Microtus* and *Peromyscus*. Species within *Microtus* or *Peromyscus* often display very similar nonsocial behaviors, such as activity and feeding patterns (Madison 1985), but differ significantly in social interactions and mating systems (Dewsbury 1987; Bester-

Meredith et al. 1999). Most species exhibit a promiscuous mating system. Promiscuous species show little in the way of social ties, typically defend individual territories, and the female usually is the sole caretaker of pups. A few species, however, display characteristics of monogamy such as shared parental care, shared nests even beyond the breeding season, and selective aggression against strangers but not toward the partner. These animals form strong pair bonds with their mate, which are manifested by a preference for social contact with the partner even when other conspecifics are available.

Such species differences in social behavior have been exploited in comparative studies that allow differences in social organization to be correlated with neuroanatomical and neurochemical differences between species. Similarities in nonsocial behaviors suggest that differences found in the brain are more likely to be related to social behavior. Studies using *Peromyscus* and *Microtus* have identified a number of brain regions and neurochemical systems that are critically involved in the central control of socially relevant behaviors but that differ between species with differing mating systems. Although we will concentrate on these genera, we are in no way minimizing the contributions arising from studies using other rodent species. In many cases, work on rats, hamsters, and other species of mice (numerous strains derived from *Mus musculus*) has laid the groundwork for studies in *Peromyscus* and *Microtus,* and we will refer often to such work to provide context for findings in these latter species.

Research in Juveniles

Species differences associated with social behavior arise early in development, even among closely related species. In general, pups of promiscuous species such as the meadow voles (*M. pennsylvanicus*) and montane voles (*M. montanus*) show more rapid development when compared to pups from monogamous pine voles (M. *pinetorum*) and prairie voles (*M. ochrogaster*) (McGuire and Novak 1984; Nadeau 1985; McGuire and Novak 1986; Prohazka et al. 1986). Relative to monogamous species, promiscuous vole species display more advanced neuromuscular development at five days of age, and become independent earlier. Pups from promiscuous vole species eat solid food as early as 8 days of age and wean at 13–14 days, while pups of monogamous vole species are not weaned until about 1 week later (McGuire and Novak 1984). Similarly, among *Peromyscus,* pups of a promiscuous species, the white-footed mouse (*P. leucopus*), open their eyes earlier and wean earlier than do pups of monogamous California mice (*P. californicus*) (reviewed by Layne [1968]).

Behavioral differences reflecting the various social structures also are reflected to some extent in the play behavior of juvenile rodents (Pellis et al. 1989). This is not surprising, since juvenile play behavior may serve to prepare relevant brain circuitry for appropriate adult social behavior (Cooke et al. 2000). In fact, rats (*Rattus norvegicus*) that are deprived of opportunities to engage in play when young display deficits in social behavior as adults (van den Berg et al. 1999). Young prairie voles, which are highly social as adults, display a greater propensity for intimate contact and mutual grooming than do young meadow voles, which are rather asocial as adults (Wilson 1982a). Juveniles of highly social vole species also exhibit more complex play behavior (Pellis and Iwaniuk 1999), and the structure of play differs from that of asocial species (Pellis et al. 1989; Pierce et al. 1991). In play fighting, a passive defense posture is adopted by social species, while a more aggressive defense posture is adopted by nonsocial species (Pellis et al. 1989). Interestingly, the differences in play appear to reflect differences in precopulatory behavioral patterns of adults of each species (Pellis et al. 1989; Pierce et al. 1991).

The differences in juvenile behaviors suggest that there are differences in the central nervous system early in development among rodents with differing social systems. Indeed, several studies have shown that brain development may be delayed in monogamous voles. Allometric relationships are ratios between pairs of measures of an animal, and these ratios may change during development. Vole species with differing mating systems display different allometric relationships between brain mass and body mass during development. Promiscuous vole species switch from an immature allometric growth pattern to an adult pattern earlier in development than do the monogamous voles (Gutierrez et al. 1989), suggesting that brain development is delayed in monogamous voles. These species differences in brain growth may be attributable to the proliferation of new cells. Indices of cell proliferation in the cerebrum suggest that the brains of monogamous pine voles are still undergoing considerable mitotic activity at 5 days postnatally. At the same age, however, mitotic activity is significantly reduced in meadow voles and in other non-pair-bonding species such as rats and mice (Gutierrez et al. 1989). In the same study, monogamous vole species were also found to display a greater increase in cell proliferation in the cerebellum between 2 and 5 days of age compared to that in promiscuous voles, again suggesting that brain development is delayed in monogamous species. This difference may account for the more advanced neuromuscular development displayed by promiscuous vole pups (Prohazka et al. 1986).

In addition to differences in brain growth, the development of neurochemical systems differs between species with differing social systems. For example, brain derived neurotrophic factor (BDNF) is important for the proliferation of neurons as well as for their survival and growth. In some brain areas the promiscuous meadow vole displays adult patterns of BDNF expression at about 2 weeks of age, while the monogamous prairie vole does not show adult patterns until at least 3 weeks of age (Liu, Fowler et al. 2001). It is interesting to note that the timing of the switch to adult patterns of BDNF expression to some extent parallels the timing of weaning and independence in each species. Monogamous prairie voles and promiscuous montane voles also differ in temporal and regional expression of the gene for receptors that bind the neurochemicals vasopressin or oxytocin (Wang and Young 1997; Wang, Young et al. 1997), which, in adults, are critical for social memory and/or for the formation of social attachments in monogamous species (Dantzer et al. 1988; Williams et al. 1994; Wang et al. 1998).

Collectively, these observations demonstrate clear species differences in the ontogeny of the brain that may be important for species-specific social structures. However, it is not clear whether such differences are driven by nature or nurture. This issue typically is addressed in cross-fostering studies. Monogamous California mice, cross-fostered as pups to promiscuous white-footed mice, display some behavior patterns typical of their foster parents as adults (Bester-Meredith and Marler 2001; Bester-Meredith and Marler 2003), and the behavioral differences are correlated with changes within the brain (Bester-Meredith and Marler 2001; Bester-Meredith and Marler 2003). In studies in which pups of a promiscuous vole species were cross-fostered to monogamous parents or in-fostered to conspecific, promiscuous parents, the fostered pups showed a slight preference for the species to which it was fostered (McGuire and Novak 1987) and displayed parental behaviors at a level closer to that of their fostering parents (McGuire 1988). These results suggest that, at minimum, environmental factors can interact with genetics to influence the social behavior of rodents.

Research in Adults

Research in juveniles has provided important information about the development of brain structures and systems that are critical for social function. However, in some cases, the behavioral manifestations of developmental differences seen in juveniles do not occur until sexual maturity. Thus a thorough understanding of the neural control of social behavior also requires examination of the central nervous system in adults.

The formation and maintenance of social attachments between individuals appears to involve primarily two brain

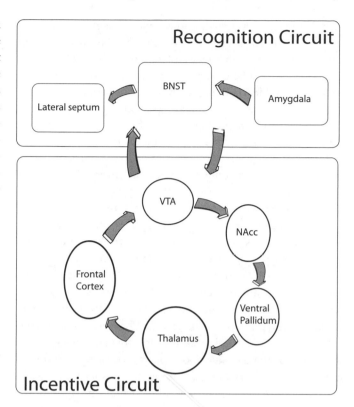

Figure 16.1 The regulation of social attachment appears to involve primarily two brain circuits. The recognition circuit may mediate the ability to distinguish individuals and thus allow context-appropriate social behaviors to be expressed. Information about the identity of another individual from the recognition circuit may modify (or even initiate) responses within the incentive circuit. Feedback from the incentive circuit may then dictate the direction (approach or avoidance, aggressive or nonaggressive, other behaviors) and intensity of the interactions. The incentive circuit was adapted from Insel 2003. BNST—bed nucleus of the stria terminalis, VTA—ventral tegmental area, Nacc—nucleus accumbens.

circuits (fig. 16.1). The first circuit is comprised of portions of the amygdala, the bed nucleus of the stria terminalis (BST), and the lateral septum. This circuit may serve as a recognition circuit, allowing appropriate social responses to be displayed upon encountering another individual. The second circuit is centered on the nucleus accumbens and includes the ventral tegmental area, ventral pallidum, certain thalamic nuclei, and portions of the cortex (Insel 2003). This circuit may serve to convey incentive value in social interactions. Examination of these two brain circuits illustrates several ways in which species differences in the central nervous system are correlated with species-specific social and mating systems.

The Amygdala-BST-Lateral Septum Circuit

In humans, social interactions can elicit a range of emotional responses. Whether rodents experience analogous "feelings" is unknown, but human responses suggest that a good place to start examining rodent social behavior is the amygdala, the "emotional brain." The amygdala has been implicated in a variety of socially relevant functions including sexual behavior, affiliative behavior, social memory, fear, and learned helplessness (Dominguez et al. 2001; also see Kling and Brothers 1992 for extensive review). Damage to the amygdala can alter the structure of play behavior by juvenile rats (Daenen et al. 2002) and, in fact, in adults, the amygdala is reduced in size in animals that were deprived of play as pups (Cooke et al. 2000). Lesions targeting particular subnuclei of the amygdala show that the medial portion of the amygdala is involved in mediating affiliative behavior in voles (Kirkpatrick et al. 1994). The findings from lesion studies are supported by observations that the amygdala is activated during the early stages of social attachment formation (Curtis and Wang 2003; Cushing et al. 2003). Interestingly, when female voles are exposed to males, the rate at which new cells are added to the amygdala increases (Fowler et al. 2002). Whether these new cells play a role in social behavior is currently being investigated.

The amygdala is an important site for the integration of a variety of sensory inputs. Among the sensory input reaching the amygdala is pheromonal information from the vomeronasal organ (VNO). Such information is important in mediating maternal behavior, pair bonding, and sexual behavior. For example, male mice in which the VNO is impaired fail to increase testosterone levels after exposure to a female and display deficits in sexual behavior (reviewed by Keverne 2002). Under natural circumstances female voles do not experience estrous cycles and require 24 to 48 hours of exposure to a male to induce sexual receptivity (Carter et al. 1987). Such reproductive activation does not occur in females from which the VNO has been removed (Lepri and Wysocki 1987; Curtis et al. 2001). Further, even if sexual receptivity is artificially induced and mating occurs, normally monogamous prairie voles do not form pair bonds after VNO lesions (Curtis et al. 2001), suggesting that pheromonal input is important in mate recognition. The importance of VNO input also is apparent after mating. For example, maternal behavior by female rats in which VNO input has been eliminated can be altered to such an extent that pup survival is compromised (Brouette-Lahlou et al. 1999).

The involvement of the amygdala in social behavior appears to be mediated, at least in part, via projections to the lateral septum, either directly or indirectly via the BST (Caffe et al. 1987). Consistent with inclusion in this pathway, the BST and lateral septum also have been implicated in a number of rodent social behaviors (Wang, Smith et al. 1994; Liu et al. 2001b). But how is information conveyed within the amygdala-BST-lateral septum circuit? The neuropeptide vasopressin has been shown to affect a variety of social behaviors.

Vasopressin and social behavior

Vasopressin is probably most widely known for its peripheral effects. Vasopressin synthesized within the hypothalamus is released via the pituitary and acts as a potent vasoconstrictor, and plays a critical role in body fluid regulation via effects at the level of the kidney. However, in addition to its peripheral effects, vasopressin can also act within the brain. For example, centrally administered vasopressin induces grooming and changes in core body temperature (Drago et al. 1997). Within the central nervous system the majority of vasopressin innervation is found in the amygdala-BST-lateral septum circuit (de Vries and Miller 1998). This extrahypothalamic vasopressin system is sexually dimorphic in rodents. Castration of neonatal male rats produces a pattern of vasopressin innervation similar to that seen in females (de Vries and Miller 1998), suggesting that this dimorphism is regulated by perinatal exposure to gonadal hormones (Wang et al. 1993; de Vries and Miller 1998; Axelson et al. 1999).

Central administration of vasopressin also produces effects on social behavior, such as facilitation of maternal behavior in rats (Pedersen et al. 1982), and induction of selective aggression (Winslow et al. 1993), paternal behavior (Wang, Ferris et al. 1994), and the formation of partner preferences (Winslow et al. 1993; Cho et al. 1999) in monogamous voles. In some cases, the effects of central vasopressin are species-specific. For example, in monogamous prairie voles, central administration of vasopressin induces aggression (Young et al. 1997), whereas the same treatment in promiscuous montane voles does not alter aggression (Young et al. 1997). If vasopressin contributes to social behavior, one might then expect that the vasopressin systems would differ among species with differing social structures. Indeed, there appear to be relationships between the density of vasopressin innervation and/or number of vasopressin receptors and species-specific social structures. The distributions of vasopressin fibers in the brain differ between monogamous and promiscuous species within both *Microtus* and *Peromyscus*. However, between genera, the distribution of vasopressin fibers differs in opposite directions. Males of a monogamous *Peromyscus* species, the California mouse, display a higher density of vasopressin immunoreactive staining in the BST than does the promiscuous white-footed mouse (Bester-Meredith et al. 1999; Bester-Meredith and Marler 2001). In *Microtus*, the opposite pattern is found: monogamous species display less vasopressin innervation in BST than do promiscuous species (Wang 1995). Vasopressin receptor densities also differ between species with differing social structures (fig. 16.2). Again, however, although there are species differences within each genus, a consistent correlation between vasopressin receptors and social structure is not found. For example, in the lateral septum, mo-

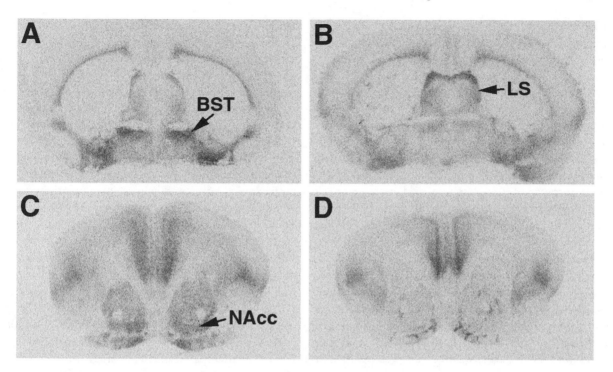

Figure 16.2 Although structurally very similar, the brains of monogamous and nonmonogamous voles differ in the densities and distribution patterns of receptors for many neurotransmitters. Panels A and B show the densities of vasopressin receptors in the bed nucleus of the stria terminalis (BST) and lateral septum (LS) in monogamous prairie voles (A) and nonmonogamous montane voles (B). Such differences in vasopressin receptor density have been correlated with species-specific social structures and patterns of aggression. Panels C and D show the densities of oxytocin receptors in nucleus accumbens (NAcc) in prairie (C) and montane (D) voles. Oxytocin receptor activation plays a critical role in pair bond formation by monogamous voles.

nogamous *Peromyscus* species have a higher density of vasopressin receptors than do promiscuous species (Insel et al. 1991; Bester-Meredith et al. 1999), exactly opposite to the pattern seen for vasopressin receptors in the lateral septum in monogamous and promiscuous *Microtus* species (Insel et al. 1994). There is a consistent correlation between vasopressin receptor densities and mating system in only one brain region, the ventral pallidum (Bester-Meredith et al. 1999), and this region has recently become the focus of several studies (cf. Pitkow et al. 2001).

Vasopressin and aggression

It has been suggested that one aspect of mating systems, species-specific aggression patterns, actually may be a better predictor of vasopressin innervation than is the mating system per se (Bester-Meredith et al. 1999). Within species, differential vasopressin innervation has been associated with individual differences in aggressiveness (Compaan et al. 1993; Everts et al. 1997). In rats, there is a negative correlation between individual aggression and vasopressin fiber density in the lateral septum (Everts et al. 1997). The negative correlation between aggression and vasopressin fiber density is also seen in other species. Aggressive mice have a lower density of vasopressin fibers in the BST than do nonaggressive mice (Compaan et al. 1993), and parental male

prairie voles, which are more aggressive than are sexually naive males, have lower vasopressin fiber density in the lateral septum relative to virgin males (Bamshad et al. 1993). Interestingly, no change in vasopressin fiber density is seen in male meadow voles after the birth of pups (Bamshad et al. 1993). This difference may reflect the fact that, after mating, monogamous prairie voles display extensive parental and nest- and mate-guarding behaviors that are not seen in promiscuous meadow voles.

Consistent with our basic premise that many social behaviors derive from the mating system, in some species mating can produce fundamental changes in social behaviors, including aggression. Male prairie voles are very different animals before and after mating (Winslow et al. 1993; Insel et al. 1995; Gammie and Nelson 2000). Sexually naive male prairie voles display little aggression when exposed to a novel male (Winslow et al. 1993). However, after 24 hours of mating, these males become less fearful and more aggressive (Insel et al. 1995). Even the pattern of agonistic behavior changes: attack bites are added to the pre-mating repertoire of defensive and threat-type behaviors (Insel et al. 1995). The transition from defense to attack appears to be mediated by vasopressin. Blockade of vasopressin receptors prior to mating blocks mating-induced aggression, while treatment with vasopressin induces aggression in the absence of mating (Winslow et al. 1993). Changes in aggression after mating are not limited to males. Sexually experienced breeder female prairie voles display more aggression and less affiliative behavior than do sexually naive females (Bowler et al. 2002), and postpartum female common voles (*M. arvalis*) become more aggressive toward males as pups develop (Heise and Lippke 1997). Whether the increases in aggression in females are attributable to mating or to gestational or postpartum changes are unknown, but changes in vasopressin gene expression after the birth of pups are known to occur (Wang et al. 2000).

The importance of vasopressin in mediating aggressive behavior suggests involvement of the amygdala-BST-lateral septum circuit. This notion is supported by studies in a variety of species. For example, agonistic behavior activates the amygdala-BST-lateral septum circuit in male Syrian hamsters (*Mesocricetus auratus*; Kollack-Walker and Newman 1995) and female house mice (*Mus musculus*; Gammie and Nelson 2001). Similarly, in a resident/intruder test, previously mated male prairie voles had elevated expression of the *c-fos* gene in the amygdala, BST, and lateral septum (Wang, Hulihan et al. 1997). Since *c-fos* expression is an indicator of neuronal activation, these observations suggest that this circuit is activated during aggression in voles as well. These results show a consistent pattern of involvement of this circuit associated with aggression, regardless of species or gender. Interestingly, in monogamous voles, activation of this system occurred only in response to a stranger, not after reexposure to the familiar partner, suggesting that aggression can be modified by familiarity (Wang, Hulihan et al. 1997).

Vasopressin and social recognition

It is apparent that the amygdala-BST-lateral septum circuit is important for a variety of social behaviors, suggesting a common factor that is involved in all of these behaviors. The ability to respond appropriately in social encounters depends to a large extent on being able to recognize individuals. We suggest that the amygdala-BST-lateral septum circuit is critical for social recognition. How important is social recognition in structuring interactions between rodent conspecifics? Although few rodent social systems appear to be structured on a hierarchical basis, for those, such as mole-rats (*Cryptomys damarensis*) that display such a social system (Gaylard et al. 1998), individual recognition likely is an important attribute. Individual recognition seems to be necessary in species that form pair bonds.

Rodents can distinguish among individuals. Evidence from Syrian hamsters suggests that the loser of an aggressive encounter can identify the individual that defeated him (Lai and Johnston 2002). Prairie voles with lesions of the amygdala have deficits in mate recognition (Demas et al.

1997), suggesting that the amygdala is involved in social recognition. Mice with impaired VNO function display deficits in the ability to discriminate sex (Stowers et al. 2002), and this structure also plays an important role in social recognition in rats (Bluthe and Dantzer 1993). Experiments in rats also suggest that the lateral septum plays an important role, mediated by vasopressin, in social memory (Dantzer et al. 1988; Bluthe and Dantzer 1993; Everts and Koolhaas 1999). The lateral septum may be critical for mate recognition in monogamous voles as well. In a number of studies on pair bonding, neurochemical manipulations have been tested for effect both before and after the formation of pair bonds. In most cases, treatments after pair bond formation are ineffective at blocking the expression of pair bonds. The single exception appears to be the effects of vasopressin in the lateral septum. Vasopressin infused into the lateral septum, presumably mimicking natural release from cells in the BST, facilitates partner preference formation by prairie voles (Liu et al. 2001b). However, administration of a vasopressin receptor blocker into the lateral septum, either before or after pair bond formation, impairs the ability of prairie voles to express a partner preference (Liu et al. 2001a). One possible explanation for these results is that the treatment interferes with mate recognition.

If the suggestion that the amygdala-BST-lateral septum circuit plays a fundamental role in organizing rodent social interactions by mediating individual identification is valid, this circuit then should interact with other brain regions involved in specific social behaviors. Further, connections within and arising from this circuit should be able to influence social interactions by modifying behaviors to ensure that responses are appropriate for the social context. An important test of such hypotheses is to show that this circuit can influence behaviors mediated by other brain regions. Flank-marking by Syrian hamsters occurs in response to stimuli associated with conspecifics (Johnston 1992) and the frequency and location of flank-marking may be modified by individual recognition (Ferris and Delville 1994). This behavior appears to be mediated via the anterior hypothalamus since injections of vasopressin into this area stimulate flank-marking (Ferris et al. 1999). Of interest here is that the anterior hypothalamus receives afferent input from the lateral septum, and stimulation of the septum also gives rise to flank marking (Irvin et al. 1990). These results show that the extrahypothalamic vasopressin can affect social behavior via connections to other brain regions.

Of course, the vasopressin system does not regulate behavior in the absence of other neurochemical systems. For example, in the lateral septum vasopressin interacts with oxytocin to regulate pair bond formation (Liu et al. 2001b). Within both the anterior hypothalamus and the ventrolateral hypothalamus, vasopressin enhances aggression, and these effects are antagonized by serotonin (Delville et al. 1996; Ferris et al. 1997) or galanin (Ferris et al. 1999). In addition to direct effects on behavior, vasopressin effects may modify, or be modified by, the effects of other neurotransmitter systems. Cells immunoreactive for tyrosine hydroxylase, an enzyme involved in the biosynthesis of catecholamines, are found in the BST of Siberian hamsters (*Phodopus sungorus;* Shi and Bartness 2000). Noradrenergic projections from the brainstem interact with vasopressin within the BST to modulate fear responses (Onaka and Yagi 1998) and norepinephrine within the olfactory bulbs interacts with vasopressin to mediate social recognition (Dluzen et al. 1998). Finally, vasopressin release in the lateral septum may modulate the release of dopamine (Ishizawa et al. 1990), and activation of vasopressin receptors in the ventral pallidum may modify responses associated with the mesolimbic dopamine system (Pitkow et al. 2001).

Dopamine and social behavior

There is a long history of research into the effects of central dopamine on behavior. Such research has implicated dopamine in responses to stress (Dunn 1988; Abercrombie et al. 1989), in mediating conditioned preferences (Kivastik et al. 1996) and the rewarding effects of food intake (Azzara et al. 2001), and in the control of mating behavior (Becker et al. 2001). More recently, researchers have begun to examine the role of dopamine in mediating social behavior (Mitchell and Gratton 1992; Mermelstein and Becker 1995; Tidey and Miczek 1996; Keer and Stern 1999; Lorrain et al. 1999) and, in particular, in pair bond formation (Gingrich et al. 2000; Aragona et al. 2003a). To date, much of this latter work has been directed toward the role of nucleus accumbens dopamine in social attachment. Mating induces dopamine release in the nucleus accumbens (Gingrich et al. 2000) and facilitates pair bond formation in prairie voles (Williams et al. 1992), suggesting a connection between dopamine release and pair bonding. Dopamine released within the nucleus accumbens is thought to be involved in reward processing, and indeed has been found to be of critical importance in the formation and maintenance of pair bonds in both sexes of prairie voles (Gingrich et al. 2000; Aragona et al. 2003a). Early work in this system identified activation of one kind of dopamine receptor, the D_2 subtype, as being a critical step in pair bond formation (Wang et al. 1999; Gingrich et al. 2000). These studies, together with a later report (Aragona et al. 2003a), also produced evidence for gender-specific effects of dopamine on pair bonding; the same doses of dopamine agonists that induced a preference for the familiar partner in female voles were ineffective in males. In addition, more recent work has provided details that make it apparent that D_2 receptor activa-

tion is just one aspect in a complex set of neurochemical interactions within nucleus accumbens during the formation and maintenance of pair bonds.

As mentioned previously, mating induces dopamine release in the nucleus accumbens in voles, and such release is shown to be important in pair bond formation (Wang et al. 1999; Gingrich et al. 2000; Aragona et al. 2003a). However, mating also induces dopamine release in the nucleus accumbens in species that do not form pair bonds (Pfaus et al. 1990; Mermelstein and Becker 1995). Why, then, does dopamine induce pair bonds in only some species? The answer may lie in interactions between dopamine and other neurochemical systems. For many years it has been known that oxytocin plays a critical role in the formation of bonds between adults (Williams et al. 1994), just as it does in the formation of bonds between mother and offspring (Carter 1998). In both *Microtus* and *Peromyscus,* monogamous and promiscuous species differ in the distribution of oxytocin receptors within the brain (fig. 16.2; Insel et al. 1991; Insel and Shapiro 1992). Within nucleus accumbens, monogamous vole species display a much higher density of oxytocin receptors than do promiscuous species (Insel and Shapiro 1992), and activation of these receptors acts in concert with the D_2 dopamine receptors to produce pair bonds. When either D_2 or oxytocin receptors are blocked no pair bonds are formed (Liu and Wang 2003). Thus the combination of mating-induced dopamine release with species-specific patterns of oxytocin activation may partially explain the variety of rodent mating systems.

There also is indirect evidence for a connection between oxytocin/dopamine interactions and pair bonding. Auto- and allogrooming play important roles in many social interactions and may facilitate the transfer of socially relevant information (Ferkin et al. 2001). Like pair bond formation, grooming behavior is to some extent mediated by an interaction of dopamine and oxytocin in the nucleus accumbens (Drago et al. 1986). It was found that non-pair-bonded male prairie voles groomed more frequently than did pair-bonded males (Wolff et al. 2002), providing further evidence of oxytocin-dopamine interaction in pair-bonding.

The nucleus accumbens contains more than just D_2 and oxytocin receptors: other dopamine receptor subtypes are expressed as well. The D_1 dopamine receptor subtype was originally described as playing no role in pair bonding (Wang et al. 1999). This is probably true in terms of pair bond *formation,* but increasing evidence suggests that D_1 dopamine receptors may be critical for pair bond *maintenance.* Male prairie voles that remain with their female partner for 2 weeks display an important change within the nucleus accumbens (Aragona et al. 2003b). In these voles, the density of D_1 dopamine receptors is substantially greater than that seen in non-pair-bonded voles. Further, activation of D_1 dopamine receptors impairs the formation of pair bonds induced either by mating or by D_2 receptor activation (Aragona et al. 2003b). The increase in D_1 receptors in pair-bonded animals may prevent the formation of a second pair bond, which in turn may serve to maintain a monogamous life strategy. It would be interesting to examine whether a similar reorganization occurs in species that display serial monogamy or whether there are sex differences in species such as Mongolian gerbil (*Meriones unguiculatus*) that appear to display sex-specific types of social bonds (Starkey and Hendrie 1998). Finally, it also would be of interest to learn whether there are individual differences in the regulation of D_1 receptor expression in pair-bonded voles. Although considered to be a monogamous species, some prairie voles can form a second pair bond (Pizzuto and Getz 1998). Differences in the ability to increase D_1 receptors may account for the small percentage of monogamous voles that form new pair bonds after losing a mate.

Corticosterone and social behavior

Vasopressin, oxytocin, and dopamine all have been implicated in social attachment in voles and, although there are sex differences in sensitivity to these neurochemicals, the direction of effects is similar in both sexes. This is not the case when the effects of the stress hormone corticosterone are examined. Monogamous prairie voles display basal circulating levels of corticosterone that are as much as ten times higher than those found in promiscuous vole species or in rats (Hastings et al. 1999). Nonetheless these voles are capable of further, stress-induced increases in corticosterone (Taymans et al. 1997). Interestingly, the effects of stress on pair bonding are sexually dimorphic in monogamous voles. In males, the effects of stress, presumably including increased circulating corticosterone, enhance the formation of pair bonds (DeVries et al. 1996). Conversely, adrenalectomy, which reduces circulating corticosterone, inhibits pair bonding (DeVries et al. 1996). In females, the opposite pattern is found; adrenalectomy enhances pair bonding, whereas stress reduces pair bond formation (DeVries et al. 1995; DeVries et al. 1996).

How might corticosterone affect pair bond formation? One possibility may be via interaction with the vasopressin system. Adrenalectomy reduces the density of vasopressin receptors in the lateral septum and BST, an effect that is reversed by hormone replacement (Watters et al. 1996). Corticosterone actions are mediated by two types of glucocorticoid receptors, high-affinity Type I receptors and low-affinity Type II receptors, and activation of the two receptor subtypes can produce differing effects (de Kloet et al. 1993). Interestingly, hormone replacement using aldosterone, which acts primarily on Type I receptors, reversed

adrenalectomy effects on vasopressin receptor density only in the BST. Dexamethasone treatment, which acts on Type II receptors, restored vasopressin receptor densities in both the lateral septum and the BST (Watters et al. 1996). These results show that changes in circulating corticosterone levels have the potential to significantly alter vasopressin-induced responses. Given the sexual dimorphism in the extrahypothalamic vasopressin system, it is possible that the sex-specific effects of corticosterone are secondary to its effects on vasopressin activity.

Corticosterone also can interact with the dopamine system. Glucocorticoid receptors are found on dopamine cells within the ventral tegmental area (VTA). It has been shown that stress alters excitatory glutamate receptors on dopaminergic cells in the VTA (Saal et al. 2003). Importantly, the stress-induced changes in VTA were blocked by glucocorticoid receptor antagonists (Saal et al. 2003). Since the VTA is the primary source of dopamine input to the nucleus accumbens (Schoffelmeer et al. 1995), these results suggest that glucocorticoid receptor activation in VTA could impact dopamine release in nucleus accumbens. Direct effects of glucocorticoid receptor activation within nucleus accumbens also are possible. For example, there is a direct correlation between corticosterone levels and dopamine transporter (DAT) activity in the shell portion of nucleus accumbens (Sarnyai et al. 1998), the subregion most strongly implicated in pair bonding (Aragona et al. 2003b). Since corticosterone levels are lower in voles that are paired (DeVries et al. 1995; DeVries et al. 1997), it is possible that DAT activity also is decreased, reducing clearance of dopamine from the synapse, and thus potentiating the effects of released dopamine. The net result of this decrease in DAT function may alter the rewarding aspects of contact with the partner. Sex differences in the distribution of glucocorticoid receptors, in the basal levels of D_1 receptors, or in the glucocorticoid/DAT interaction could explain the sex differences in responses to glucocorticoid in pair bond formation. In males the potentiated dopamine effect may be rewarding, and in females, aversive. Thus interaction between the corticosterone and dopamine systems may in part explain the sex-specific effects of stress on pair bond formation.

Synthesis

A recent review (Insel 2003) outlined a circuit that may mediate the rewarding aspects of social interaction. This circuit, involving the mesolimbic dopamine system, may be critical to an assessment of the incentive value associated with another individual, but may not account for one important aspect of social behavior, the recognition of another individual. In this regard, interplay between the vasopressin and dopamine systems may have an important impact on social behavior: the dopamine incentive system dictates the intensity of the interaction, the vasopressin recognition circuit dictates with whom the individual interacts. Are there direct connections between the recognition and incentive circuits? The answer appears to be yes. For example, there are projections from both lateral septum (Zahm et al. 2001) and BST (Georges and Aston-Jones 2002) to the VTA, a major source of dopamine to nucleus accumbens as well as to other brain regions associated with social attachment. In fact, electrical stimulation of the BST activates the vast majority of dopamine neurons in the VTA (Georges and Aston-Jones 2002). The amygdala and prefrontal cortex also may control dopamine release via direct inputs to nucleus accumbens (Carr and Sesack 2000; Howland et al. 2002). Similarly, there are efferent projections from the dopamine incentive circuit to the amygdala-BST-lateral septum circuit (cf. Hurley et al. 1991). We propose that the extrahypothalamic vasopressin system interacts with the mesolimbic dopamine reward circuit by mediating social recognition and thus modifying responses within the reward pathway.

How might these two systems interact? It is well established that nucleus accumbens dopamine is elevated in response to novelty, including exposure to another individual (Damsma et al. 1992; Noguchi et al. 2001). However, when a familiar situation is encountered, dopamine release in the nucleus accumbens, especially in the shell portion, is attenuated relative to that in earlier encounters (Bassareo et al. 2002). It is unlikely that the individuals comprising a pair remain together at all times. Evidence for sex-specific predation risk, even in monogamous species (Sommer 2000), suggests that members of a pair are at times separated. Social recognition has the potential to most greatly affect nucleus accumbens dopamine during reunion after a separation, and it is in this circumstance that a recognition circuit based on the vasopressin system may play a role in pair bonding. Since pair-bonded animals have more D_I receptors in the nucleus accumbens and activation of these receptors interferes with pair bonding, dopamine release at the wrong time within the nucleus accumbens could disrupt an existing attachment. However, recognition of the partner via the amygdala-BST-lateral septum circuit may inhibit dopamine release in the nucleus accumbens, thus precluding activation of the D_I receptors and preserving the pair bond. Such recognition would not be afforded to strangers, and the novelty-induced elevation of nucleus accumbens dopamine would then activate the increased D_I receptors, producing an aversive response to an unfamiliar individual. Such a response could in turn feed back on the extrahypothalamic vasopressin system to produce aggression toward

the stranger. In addition, the fact that D_I activation may produce an aversive response could also activate the hypothalamic-pituitary-adrenal axis, altering circulating levels of corticosterone. Corticosterone can alter function in both the vasopressin and dopamine systems, in conjunction with the fact that the corticosterone system differs in monogamous species from that in nonmonogamous species, suggesting that feedback via the corticosterone system may play a critical role in coordinating the actions of the incentive and recognition systems during social encounters. Certainly, these scenarios need to be tested in further studies.

Future Directions

Research over the past few decades has provided a strong understanding of the basic neuroanatomical substrates underlying social behavior in rodents and has begun to reveal how various neurochemical systems interact to mediate social interactions. Nonetheless, there is considerable work still to be done. Even within species, there are often subtle differences in social behavior between populations (Roberts et al. 1998; Cushing et al. 2001; Wolff and Dunlap 2002) that may be influenced by local environmental conditions. Exactly how such differences are mediated is unknown, but the fact that there are population-specific patterns of behavior suggests a genetic basis. The role of genetics in producing different social structures is just beginning to be examined. Young et al. (1999) have shown that differences in the promoter region of the vasopressin receptor gene can produce species-specific patterns of receptor expression that are correlated with social structure. In fact, expression of the vasopressin receptor gene from monogamous voles in mice can affect the social behavior of the mouse (Young et al. 1999). This line of research has the potential to provide considerable insight into the evolution of rodent social structure.

Although much is known about the effects of vasopressin, oxytocin, dopamine, and corticosterone, little is known about how these chemicals interact with each other and with other neurochemical systems to mediate social responses. For example, studies on stress responses and drug addiction have shown that neurotransmitters such as glutamate and GABA can significantly affect the activities of the mesolimbic dopamine system (Enrico et al. 1998; Takahata and Moghaddam 1998). What role such neurochemicals may play in social behaviors such as pair bonding has barely been addressed.

Finally, there is considerable work yet to be done examining the role of perinatal and exogenous influences on the central control of social behavior. Elevated stress hormones during development, perinatal exposure to vasopressin, or oxytocin, or to altered gonadal hormone levels all have been shown to impact social behaviors (Axelson et al. 1999; Stribley and Carter 1999; Catalani et al. 2000; Lonstein and De Vries 2000a; Kramer et al. 2003). Even substances consumed by the dam can affect the social behavior of offspring (Kelly and Tran 1997). Given the recent evidence that a variety of manmade chemicals in the environment can mimic the effects of endogenous substances, the study of social behavior in rodents may become even more important by providing a means to study the effects of anthropogenic substances on biologically important behaviors.

Summary

Social behavior in rodents is regulated by complex interactions between a number of brain regions and by a variety of neurotransmitter systems within the central nervous system. The combined behavioral output of these systems must be capable of responding appropriately to a wide variety of stimuli, not the least of which are the responses of the individuals with which an animal interacts. Research over the past 20 years has provided a basic framework upon which our current understanding of the neural basis of social behavior rests. Two loosely defined systems appear to interact to modulate a large percentage of social interactions. The first system may be defined as a "recognition circuit," and is responsible for distinguishing between individuals such that appropriate behavioral responses, in some cases based on past interactions, may be displayed. The second circuit, an "incentive circuit," may serve to determine the intensity of the interaction. Together these circuits may act to determine the valence and/or intensity of the interaction, that is, approach or avoidance, aggressive versus passive behavior, and so forth. Ongoing research is attempting to elaborate central changes underlying the formation and maintenance of social bonds, the effects of perinatal influences on adult social behavior, and the role of genetics in determining species- and individual-specific social behaviors.

Chapter 17 Ontogeny of Adaptive Behaviors

Jill M. Mateo

THE BEHAVIORAL BIOLOGY of rodents significantly influences their life-history strategies and reproductive success, and development creates much
of the variation in these traits on which natural selection
can act. In this chapter I present an overview of the ontogeny of rodent behavior, including pre- and postnatal effects
of social experiences, stress, and seasonality on the expression of developing phenotypes. I focus on several factors
modulating both the processes and outcomes of behavioral
development, such as maternal influences, observational
learning, hormonal effects, and ecological conditions contributing to individual and population differences in social
behaviors. I also discuss in detail, from a comparative approach, how social systems and life-history parameters mediate the development, plasticity, and functional significance
of a suite of survival behaviors. Examples are drawn from
studies of both captive and free-living animals, with a focus
on comparative research with ground-dwelling squirrels.
The framework presented here integrates both proximate
and ultimate levels of analysis, to understand how processes
of development influence, and are influenced by, social and
environmental contexts (summarized in fig. 17.1).

Behavioral development, long the realm of comparative
psychology, is emerging as an area of interest among evolutionary biologists and ecologists because of their interests
in phenotypic plasticity and the effects of experiences on
not only morphological and physiological traits but also
behavioral traits. Recent research on parental effects—the
ways in which parents' phenotypes influence the development of their offspring—has also invigorated investigations
into the processes and outcomes of behavioral development.

Compelling evidence shows that multiple phenotypes can be
favored when animals experience spatial or temporal variation in selection pressures or competition within one environment, with alternative phenotypes differentially favored
depending on frequency distributions of current traits (reviewed by Stamps 2003). Thus multiple developmental processes would be expected when individual differences are
favored, either between or within environments.

Detailed field studies of behavioral development in rodents are uncommon. In many species, adults are difficult
to view and follow, and their young are often reared in in-
accessible burrows or nests. Social interactions among parents and offspring or among littermates, so influential in
later adult relationships, are therefore unknown to us. Many
discoveries about early social interactions, such as paternal
behaviors in hamsters or voles, communal nesting, and social influences on timing of puberty, resulted from observations in captivity (e.g., Vandenbergh 1983; Manning et al.
1992; Jones and Wynne-Edwards 2000). In the lab or field
enclosure, researchers can observe a developing animal's
social and physical environment from birth (or earlier)
through adulthood, revealing the potential effects of litter
size, hormones, olfactory cues, sensory stimulation, diet,
and parental behaviors on emerging behavioral phenotypes.
However, observers must be cautious in generalizing results
from the lab to development in nature, where a young animal experiences the full range of social and physical stimulation and interacts with and modifies those environments
in ways that cannot be replicated in captivity (e.g., Wolff
2003c).

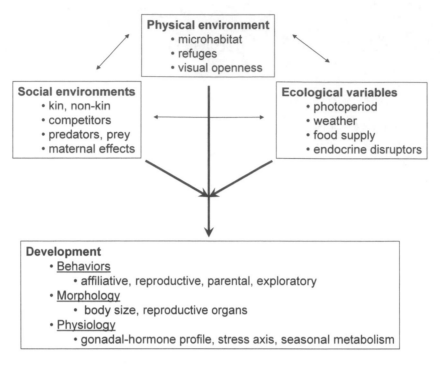

Figure 17.1 Relationships among social, physical, and ecological factors that can influence rodent development in a context-dependent fashion. Each factor can also moderate reciprocally the effects of the other factors on development, such as when the local microhabitat and current weather impact the potential for social influences on behavioral development.

General Issues in Behavioral Development

In an epigenetic approach to development, ontogeny is viewed as a series of interactions between an organism and its environment (Lehrman 1970; Johnston 1987). In addition to this epigenetic or interactionist view of development, some researchers have viewed development as a series of ontogenetic adaptations, with each stage functionally complete for that period of development and contributing to later stages (Williams 1966; Galef 1981b; Owings and Loughry 1985; Alberts 1987). These two approaches to the study of development should be considered complementary rather than competitive, however, as their interpretations focus on different levels of analysis (Tinbergen 1963; Sherman 1988). That is, proximately, what experiential factors influence the development of traits, and ultimately, are the behaviors exhibited at each stage of development adaptive for that stage? Note that we should not think of young animals as small adults, but as organisms with age-specific needs and adaptations. Indeed, young have very different selective pressures than do adults, and thus what is functional for one age group may not be for another.

I will first discuss some examples of social, hormonal, and environmental influences on rodent behavioral development (fig. 17.1), and next discuss in detail one behavioral

system—anti-predator strategies—that can be affected profoundly by interactions with parents, heterospecifics, and physical environments. The richness of the subject at hand forces me to be selective in the topics I present here. More in-depth treatment of social development in other contexts can be found throughout this volume, including territoriality (Krebs et al., chap. 15), natal dispersal (Nunes, chap. 13), the altricial-precocial continuum (Dobson and Oli, chap. 8), timing of puberty (Drickamer, chap. 9), and the specific roles of parenting styles on offspring development (McGuire and Bemis, chap. 20).

Prenatal environmental and social influences on development

Development, of course, necessarily begins at conception, and the uterine environment of developing fetuses influences their later morphology, physiology, and behavior. Sensory and perceptual development in rodents begins *in utero*. The onset of sensory function across development is remarkably consistent in birds and mammals, with the perceptual senses showing the following order of onset: tactile, vestibular, chemical, auditory, and visual. In altricial rodents, the latter two systems may not be functioning until after birth, and all systems continue to develop and form cortical connections during postnatal life (Alberts 1984; see also Gottlieb 1981). Thus nongenetic maternal influences begin prenatally through a variety of sensory experiences, such as when fetuses shift positions as the mother grooms, travels or sleeps, 'smell' what the mother eats, and experience acute changes in glucocorticoids when the mother is stressed by an agonistic social interaction or a predator encounter (see also Boonstra et al. chap. 12, this volume). Because perceptual development begins early, learning via some modalities can also begin before birth. For instance, fetuses can learn odors in the amniotic fluid that may later influence food preferences or social recognition of kin (Hepper 1987; Terry and Johanson 1996; Galef chap. 18, this volume). Thus prenatal sensory experiences can influence later behaviors, including kin recognition, mate choice, learning, filial preferences, and mother-infant interactions (Ronca and Alberts 1995).

Circulating hormones during gestation can influence later morphology, physiology, and behavior. For example, in some polytocous species (litter size > 1) prenatal exposure to gonadal steroid hormones can profoundly alter reproductive life histories. Individuals gestating between two males (2M) experience higher androgen levels than those between two females (2F), and these intrauterine position (IUP) effects have consequences in adulthood. In house mice, 2M females have delayed sexual maturation, reduced fecundity, fewer litters, are aggressive and maintain large territories, and are less sexually attractive to males, relative to 2F females. 2M males have decreased sexual activity but are more aggressive and more paternal than 2F males. In addition, 2M female mice and Mongolian gerbils tend to give birth to male-biased litters (and vice versa for 2F females), thus creating intergenerational transmission of IUP effects (reviewed in Clark and Galef 1995; Ryan and Vandenbergh 2002). Behavioral effects of IUP have been documented for only a few species of laboratory animals, although some field observations have been made of lab-born animals with known IUPs released on highway islands. Observations of these animals suggest that IUP does affect reproductive and social behaviors and perhaps, ultimately, fitness (Zielinski et al. 1992). However, it is not clear how generalizable these effects are to other species, or to animals born in the wild. Fortunately, for some species IUP can be determined without caesarian deliveries, as the ano-genital distance (AGD) reflects the degree of prenatal androgen exposure and can be used as a proxy for IUP (e.g., Vandenbergh and Huggett 1995). Thus by measuring AGDs, field researchers may be able to evaluate the consequences of prenatal experiences for adult social behaviors and reproductive success.

How might IUPs relate to a species' particular ecology? If the local environment favors one sex over another (short-term sex-ratio biasing; Fisher 1930) because males, say, are the rarer sex, or if a female in a polygynous species is in exceptionally good condition and would be favored to overproduce sons (Trivers and Willard 1973), then daughters would more often gestate between males and be partially masculinized as adults. Those daughters may also overproduce sons themselves (as 2M females tend to produce male-biased litters), continuing to bias the population toward males. If that daughter is in exceptionally good condition herself, this would be beneficial, but if she is in moderate or poor condition, then selection would not favor her sons over her daughters, and she may experience a fitness loss. Excess males would also be costly if the local population is no longer female-biased. Thus the immediate effects of IUP may, over time, influence secondary sexual characteristics, sex ratios, and reproductive success, depending on original maternal condition and local demographics.

Perinatal environmental and social influences on development

For fossorial or semifossorial animals reared in burrows, developing young do not *directly* experience a broad range of physical and social environments. Instead, parents can convey environmental variation to their offspring indirectly. A young animal can experience the effects of food shortages, climate changes (temperature, day length), or social instability before venturing out of the nest. As an example, when species inhabit a wide range geographically, photoperiodic cues (day length) can be important indicators of upcoming seasonal changes and signal appropriate times for reproductive efforts. The photoperiod experienced by females during pregnancy is transmitted hormonally via melatonin to young during gestation and lactation. In voles and hamsters, maternal melatonin influences offspring growth rate, fat deposition, pelage, and sexual maturation. Young born in the spring or early summer mature quickly and may start breeding that year, whereas those born in the late summer remain prepubertal often until the following spring, decreasing energetic needs until reproduction begins again. Thus perinatal melatonin from mothers adaptively primes young for somatic and reproductive growth appropriate for the time of year in which they are born (for an excellent review of environmental cueing of photoperiod see Lee and Gorman 2000; see also Nelson 1991 for reproductive priming via 6-MBOA, a secondary plant compound found in newly grown vegetation).

Without ever seeing daylight a young animal can apportion its somatic and reproductive efforts to best suit its future fitness potential. Although photoperiod is commonly thought to trigger breeding condition in seasonal species, social and dietary cues also influence development or regression of reproductive organs (Goel and Lee 1996; Demas and Nelson 1998; Lee and Gorman 2000). Regardless of the particular cue signaling season or day length, perinatal priming of development can have major effects on an animal's survival and reproduction, allowing it to accelerate puberty and mating if conditions are right, or to delay development and wait for more appropriate breeding conditions without wasting energy on reproductive efforts that are unlikely to succeed.

In lab rodents, maternal stress toward the end of pregnancy can influence the hypothalamic-pituitary-adrenal (HPA) functioning of her offspring (Maccari et al. 1995; Barbazanges et al. 1996), with offspring of stressed mothers showing heightened stress responses compared with young of nonstressed females. Early postnatal stress tends to decrease HPA activity and reactivity to novel objects and facilitates some forms of learning (Levine 1994; Maccari et al. 1995). However, severe stress results in maladaptive HPA axes, with animals showing inappropriate responses to novel or stressful situations, impaired learning, and altered social behaviors (e.g., McEwen and Sapolsky 1995; Lupien and McEwen 1997). Thus perinatal exposure to stress hormones, during either gestation or lactation, can fine tune sensitivity and efficacy of the HPA axis and the behaviors relating to it, producing, for example, animals with more neural receptors, lower responsiveness to stressors, and improved learning capabilities. If the mother experiences chronic stress during gestation or lactation, whether due to social instability in the group, scarce food resources, or repeated predation attempts, then this fine-tuning of her offsprings' HPA axes may help them to manage similar experiences if they remain in that environment. It is unclear to what extent adult rodent HPA systems are plastic, and thus whether animals experiencing very different environments later as adults can up- or down-regulate their glucocorticoid receptors and stress-related behaviors to show responses appropriate for those new environmental conditions.

Postnatal social influences on development

After parturition, young directly experience their physical and social environments, and social stimulation from parents and siblings can have profound effects on development. For example, rat pups actively participate in huddling with their littermates, maintaining contact and shifting position within the huddle to regulate its temperature. This thermal, tactile, and olfactory contact with littermates leads to subsequent social preferences for littermates over unfamiliar agemates (Alberts 1978; Brunjes and Alberts 1979). In rats, variation in maternal behaviors, such as nursing postures and rates of licking and grooming of pups, influences the

development of several traits in their young, as offspring of high licker/groomers are less fearful and have smaller stress responses than those of low licker/groomers. Cross-fostering studies indicate these effects on offspring are due to postnatal maternal handling rather than inherited traits. Daughters of high-licking and grooming mothers become high-licking and grooming mothers themselves, thus transmitting variation in parental behavior nongenetically across generations (Meaney 2001). Note that these long-term effects of handling and grooming result from the normal range of species-specific maternal care, rather than extreme versions of neglectful or attentive mothers. However, as yet it is not clear whether these licker/groomer phenotypes exist among free-living rodent mothers, and whether the stress-response outcomes described earlier would be observed. Before searching for this phenomenon in the field, we first need to determine if these licker/groomer effects are observed in other laboratory species, especially those that are less inbred.

When parental investment includes parent-offspring interactions, a juvenile's social behaviors may be influenced by and thus resemble its parents' (Bateson 1982). The process of parental influence on the development of offspring social behavior, or any other behavioral repertoire, can range from direct to indirect (Mateo and Holmes 1997). Parents have a direct influence when they orient their behavior toward their young, such as by leading their young to a food source or preventing them from interacting with particular conspecifics. For a parent's influence to be considered direct, its behavior must change qualitatively or quantitatively as a function of its offspring's presence. At the other end of the continuum, parents have an indirect influence when their normal behavior inadvertently affects juvenile behavior, but is not directed specifically toward their young. Adults are thus incidental models of behavior and juveniles are inadvertent observers. That is, how a parent behaves is not contingent on whether its offspring are present. Examples of indirect influences include parents' own escape responses when a predator appears, their food preferences, or their territorial responses to intrusions. Production of alarm calls (vocal signals emitted in response to predators) by adults can be an example of a direct or indirect influence, depending on whether the likelihood of calling is contingent on the presence of the adults' young (e.g., Cheney and Seyfarth 1990; Hoogland 1995).

Direct and indirect parental influences on juvenile behavior share at least four characteristics. First, influence can have an immediate effect, such as when the response of an adult to a predator evokes almost simultaneously a response by a juvenile, or it can have a delayed effect, such as when an immature animal observes an adult's response to an alarm call but does not show its own responses until it can locomote independently. Second, neither process of influence implies complex mental states or awareness of juveniles' abilities by adults (cf. Cheney and Seyfarth 1990). Third, juveniles may be passive observers or recipients of adult actions; that is, young do not have to actively seek out adults and copy their behavior. Fourth, "direct" and "indirect" describe *processes* rather than *outcomes* of influence, and do not necessarily differ in the importance they play in juvenile behavioral development (Mateo and Holmes 1997). These processes of parental influence on juvenile ontogeny are potentially common among rodent species, particularly those with extended parental care, as will be discussed in the next section.

Development of Survival Behaviors

As a detailed example of the intertwining effects of social and environmental factors on development of adaptive behaviors, consider the suite of survival skills exhibited by an animal. Many programs of research have focused on the development of antipredator behavior to identify the specific factors influencing the expression of unfolding traits. Some investigators have proposed that survival behaviors, such as alarm calls or predator-avoidance tactics, are preprogrammed or innate, so that the behaviors emerge independent of environmental input (Tinbergen 1953; Bolles 1970; Magurran 1990; Curio 1993). Others have argued that antipredator behaviors are acquired or learned, through either direct or indirect experiences with predators and conspecifics (Vitale 1989; Cheney and Seyfarth 1990; Mateo 1996a). Nature-nurture dichotomous thinking still persists in the current literature despite acknowledgments by many that behavior is not innate or acquired per se, but instead develops epigenetically, through interactions between the organism and the series of environments it encounters throughout its life span (Lehrman 1970; Gottlieb 1976; Johnston 1987). Recall that with this approach, behaviors of young animals are not considered impoverished versions of adult responses, but instead are ontogenetic adaptations, functional in their current stage of development (Williams 1966; Alberts 1987).

A behavioral system that required each juvenile to learn independently how to recognize and respond to predators would consume time and energy and be prone to fatal errors in learning (Darwin 1859; Bolles 1970). Reliance on experience, whether it be practice or learning, may at first appear less than optimal, given the vulnerability of young to predators, but a flexible developmental program might be beneficial when predator contexts vary among age groups

or among populations, favoring plasticity of species-typical behaviors (Johnston 1982; Shettleworth 1998). For example, animals that are sympatric with predatory snakes need to develop antisnake behaviors, whereas animals that do not encounter snakes obviously do not need to add (or maintain) these behaviors into their antipredator repertoires (e.g., Coss and Owings 1985; Coss et al. 1993). Such responses, then, would not necessarily be present upon first encounter with predators, but would be acquired rapidly with additional experience.

Ground-dwelling squirrels are vulnerable to both aerial (e.g., hawks, eagles, gulls) and terrestrial (e.g., coyotes, weasels, martens, venomous snakes) predators. Despite the obvious survival advantage of avoiding capture by a hunting predator, young need to learn from which animals to flee, to which warning calls to respond, and in what manner. Belding's ground squirrels (*Spermophilus beldingi*; fig. 17.2)

Figure 17.2   Adult female Belding's ground squirrel with two of her juvenile offspring. The left juvenile is engaged in a nasal investigation of its mother, smelling her oral glands, which provide information about individual and kin identity. Photo by J. M. Mateo.

have two types of alarm calls, which elicit two different behavioral responses by listeners. Whistles, which are single, nonrepetitive high-frequency notes, are elicited by fast-moving, typically aerial, predators and result in evasive behaviors such as running to or entering a burrow, and scanning the area only after reaching safety. Trills, which are composed of a series of five or more short notes, are elicited by slow-moving, primarily terrestrial predators, and usually cause others to post (a bipedal stance accompanied by visual scanning), with or without changing location (Sherman 1985; Mateo 1996a). When juvenile *S. beldingi* first emerge aboveground at about 1 month of age, they do not discriminate behaviorally between these calls, or even among alarm calls and other conspecific and heterospecific vocalizations. It takes about a week for juveniles to learn to respond selectively to alarm calls and to show the correct responses to each type of call, during which time up to 60% of juveniles may disappear, many to predation. This learning is facilitated by experience hearing the calls as well as observations of adults' reactions. Juveniles attend specifically to their mother's responses over those of other females, and tend to adopt her particular response style (e.g., more reactive, stay alert longer; Mateo 1996a; Mateo and Holmes 1997).

Some ground-dwelling squirrels show a significant degree of behavioral plasticity in their survival behaviors. I define plasticity as the modifiability of behavior as a result of specific experiences or environmental conditions (e.g., Miller 1981). The plasticity of alarm-call responses in *S. beldingi* likely reflects temporal and spatial variation in their habitats and predators (e.g., meadow versus forest, open versus closed habitat; presence or absence of snakes or weasels; natal or breeding areas). By acquiring responses that are appropriate for a given microhabitat, *S. beldingi* can optimize both their foraging and antipredator efforts, allowing juveniles (and adults) to gain adequate body weight before hibernation without expending energy on unnecessary vigilance (Mateo and Holmes 1999a). Experiences prior to natal emergence, such as hearing alarm calls or exposure to maternal stress hormones, can have long-term effects on the response repertoires juveniles exhibit (e.g., likelihood of responding, duration of alert behaviors). Yet similar experiences after emergence do not alter juvenile responses significantly, suggesting that preemergent experiences prime young to acquire antipredator behaviors appropriate for the local predator environments into which they will soon emerge (Mateo 1995; Mateo and Holmes 1999b). This developmental plasticity is not surprising given that predation can be such a potent selective agent on both juveniles and adults.

As an example of plasticity in survival behaviors, juvenile *S. beldingi* reared at the edge of meadows are more reactive to alarm calls and remain alert longer than those

from the center of a meadow (Mateo 1996a), which may reflect increased vulnerability to predators near the edge (Elgar 1989). The center-edge effect demonstrates plasticity in the expression of alarm-call responses by both mothers and juveniles according to spatially and temporally changing contexts. Juveniles may develop location-dependent responses by monitoring their mother's behavior and using it as a model for their own (Mateo and Holmes 1997). Females often nest near their mothers in subsequent years (J. M. Mateo, unpublished data), so adopting location-specific responses (e.g., increased vigilance by the meadow's edge) would be favored. Mothers' reactions, which serve as models for juvenile responses, may reflect the mothers' own vulnerabilities (indirect parental influence), or may be a form of maternal care, becoming more vigilant if they locate their natal burrow, and thus their offspring, in a dangerous area (edge) and less vigilant if in a safer region (center; direct parental influences). However, conditioning studies show that adult *Spermophilus* can acquire new antipredator responses (Mateo 1995; Shriner 1999), so if individuals disperse to new areas they can adjust their repertoires to respond appropriately for their new local predator environments.

What mechanisms might mediate this behavioral plasticity? As mentioned previously, juvenile ground squirrels are preyed on by many species, and juvenile emergence tends to draw predators. Young likely experience stress during this time, either chronically or acutely. In addition, juveniles are exposed directly to maternal stress hormones, both pre- and postnatally. In lab rodents the influence of glucocorticoids on learning and memory has an inverted-U-shaped function. Very low or very high levels of corticoids can lead to hypo- or hyperarousal and poor selective attention to input, and thus impair consolidation of new memories. Moderate levels of corticoids are optimal for attention to and consolidation of memories (reviewed in Lupien and McEwen 1997). It follows then that stress-hormone levels around the age of natal emergence may facilitate rapid learning of appropriate responses, especially in locations with high predation pressure. Indeed, adult corticoid levels vary by location and juvenile levels are elevated significantly around the time of natal emergence (J. M. Mateo, unpublished data). These trends are consistent with my hypothesis that HPA activity reflects geographic variation in stress due to predation pressure and mediates learning of antipredator behaviors (and other survival behaviors) by developing ground squirrels.

Rates of development and acquisition of behaviors

If newly emergent juveniles are highly susceptible to predation (e.g., Sherman and Morton 1984; Coss et al. 1993;

pers. obs.) then selection should favor the early acquisition of behaviors that would help young survive a predator encounter. Yet newly emergent ground squirrels do not reliably respond to alarm calls, and when they do they tend to remain in place above ground instead of running to safety. Juveniles also do not post, which would allow them to locate the predator visually and monitor its movements (e.g., Mateo 1996a). In contrast, competent, adultlike defensive behavior is present in garter snakes (*Thamnophis* spp.) at birth, and responses change very little during early life (Herzog and Burghardt 1986; see also Magurran 1990; Curio 1993). Why, then, do juvenile *S. beldingi* not emerge with these skills, which would help them to both detect and escape from predators at a time when they are most vulnerable?

The combined needs to develop in the relative safety of the natal burrow and to emerge above ground and begin foraging constrain the timing of natal emergence. Overwinter survival in some ground-dwelling squirrels is dependent upon the acquisition of adequate body fat prior to hibernation, and more than 60% of juveniles do not survive their first winter (Barash 1973; Murie and Boag 1984; Sherman and Morton 1984; pers. obs.). As a result, juveniles of hibernating species emerge less physically developed (i.e., in motor- or sensory-perceptual systems) than juveniles of nonhibernating species, but demonstrate rapid growth rates after emergence (Clark 1970; Morton and Tung 1971; Maxwell and Morton 1975; Koeppl and Hoffmann 1981; Rickart 1986). For instance, *S. beldingi* juveniles gain an average of 2 g/day after birth until they emerge at ~27 days of age at 20% of adult female (mid-season) body weight; they gain an average of 4 g/day during the 25 days following natal emergence (Mateo 1995). In comparison, California ground-squirrel young, which do not hibernate, gain 3 g/day during their ~ 45-day preemergent period, and emerge at 30% of adult-female weight (using *S. beecheyi* data in Tomich 1962). Thus *S. beldingi* have a short active season, emerge at an early age, and as a result may spend less time with their mothers and littermates, compared with *S. beecheyi*. These interspecific differences in developmental rates as a function of ecological conditions could have significant effects on mother-offspring interactions and social development. Studies of free-living rodents therefore present a unique opportunity to examine in detail how the intertwining effects of latitude, elevation, and climate on active-season length and pre- and post-weaning growth rates influence behavioral development.

S. beldingi ontogeny entails a compromise between prolonged development in the natal burrow prior to emergence and the need to begin foraging above ground to gain weight for successful overwintering. As a result, juveniles must attain the motor and perceptual capabilities needed for responses to complex alarm calls during the postemergent period, despite their susceptibility to predation. However, newly emergent juveniles' limited repertoires, prior to gaining this competence, may be sufficient for them to reduce their risk of predation, particularly if they are still near the natal burrow and if the mother is nearby to ward off predators (Mateo 1996a). Newly emergent juveniles typically freeze in response to alarm calls rather than respond like adults. This leaves them vulnerable to predators, but freezing may actually increase their chances of survival. A juvenile that seeks refuge in a burrow is safe from visual detection by a predator, yet a juvenile that remains above ground may learn to discriminate between stimuli faster by observing the responses of others (while eliminating movement cues; see Mateo and Holmes 1997). For newly emergent juveniles then, there is a trade-off between escape responses and information gathering (fig. 17.3). Further, if the likelihood of an individual actually encountering a predator is small, then the benefits of remaining above ground and observant can outweigh the costs of prolonged exposure.

Therefore, selection has favored early natal emergence by Belding's ground squirrels over prolonged preemergent maturation, with the costs of limited motor capabilities offset by the benefits of observational learning and rapid acquisition of discriminating responses. Consequently, the development of juvenile *S. beldingi* antipredator repertoires, particularly the patterns of responses that require motor coordination, may reflect neural and physical development, whereas behavioral discrimination among calls is influenced by experience with the auditory stimuli and observations of conspecifics (Mateo 1996a, 1996b; Mateo and Holmes 1997). In contrast, juvenile *S. beecheyi* have a longer period of preemergent development (see earlier comments), and perhaps as a result they show well-formed, adultlike, behavioral responses to rattlesnakes despite not having acquired adult levels of venom resistance (Coss 1991a; see also Owings and Coss, chap. 26, this volume). These findings illustrate the importance of viewing juvenile behaviors as a series of adapted phenotypes appropriate for each stage of development, rather than impoverished approximations of adult behavior (Williams 1966; Galef 1981b; Owings and Loughry 1985; Alberts 1987).

Other Issues in the Study of Behavioral Development

Length of parental dependence

There is much variation across vertebrates in the degree to which parents protect their young from predators. This variation is, in part, correlated with the length of time

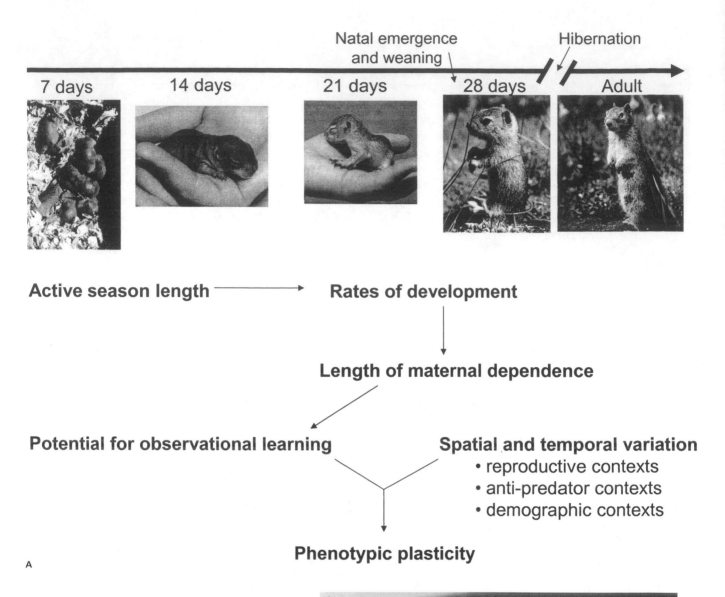

Natal emergence and weaning

Hibernation

7 days   14 days   21 days   28 days   Adult

Active season length ⟶ **Rates of development**

Length of maternal dependence

Potential for observational learning   **Spatial and temporal variation**
- reproductive contexts
- anti-predator contexts
- demographic contexts

Phenotypic plasticity

A

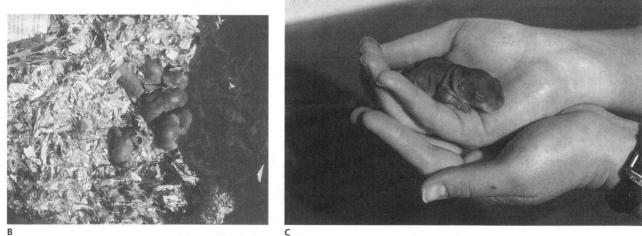

B   C

Figure 17.3 Developmental timeline of *S. beldingi,* from birth through natal emergence and overwintering to adulthood. Flow diagram highlights the potential cascading influences of abiotic factors on developmental rates and maternal dependence, which, together with variable contexts and observational learning, can in turn mediate the expression of behavioral, morphological, or physiological plasticity. Photos by J. M. Mateo.

(continued)

Figure 17.3 (continued)

young are dependent on their parents. In species with extended dependency on parental care, young have many opportunities to observe and learn from their parents' behaviors (e.g., antipredator, foraging, and social behaviors) before having to exhibit these behaviors themselves. Conversely, species with little or no parental care must acquire behavior without assistance from adults. Thus the length of dependence on parents will influence the developmental process of offspring's behavioral repertoires, and extended parental care may allow for variable pathways of behavioral development, including social facilitation of responses (Heyes and Galef 1996). For example, as discussed earlier, day-old garter snakes (*Thamnophis* spp.) that receive virtually no parental care respond defensively to predator models upon first exposure (Herzog and Burghardt 1986). In contrast, experiences with their mothers may significantly modify the responses of young vervet monkeys (*Cercopithecus aethiops*) to both predators and alarm calls, as they maintain close proximity to their mothers for at least 2 years (Cheney and Seyfarth 1990). Brown (1984) studied the development of antipredator behavior in two bass species differing in the length of time the male parent guards his young. Fry with limited paternal care (*Ambloplites rupestris*) showed predator-avoidance behavior sooner than those with extended care (*Micropterus salmoides*). Mexican jays (*Aphelocoma ultramarina*) with prolonged associations with experienced adults exhibit mobbing behavior at a later age than less social scrub jays (*A. coerulescens*) that fledge at an earlier age (Culley and Ligon 1976). Rates of behavioral development in rodents may also reflect variation in the length of parental care, which in turn could influence the development of offspring's response repertoires, while extended parental care may favor variable pathways of behavioral development, including social facilitation of responses (Heyes and Galef 1996). Comparisons among animals that differ in active-season lengths, growth rates, or periods of dependence upon mothers provide an opportunity to examine how selection has favored plasticity of behaviors as a function of environmental influences on developmental rates, and how these resulting individual differences relate to fitness.

Rates of development in ground-dwelling squirrels may also impact the length of maternal care and potential for social interactions. Juveniles of some nonhibernating species, which have slower growth rates than those that hibernate (Clark 1970; Morton and Tung 1971; Pizzimenti and McClenaghan 1974; Maxwell and Morton 1975; Koeppl and Hoffmann 1981; Rickart 1986) and subsequently longer associations with their mothers, may exhibit more social facilitation (Heyes and Galef 1996) of behavioral development than juveniles with limited growth periods. As an example, *S. beecheyi* can have long active seasons and extended co-occurrence of mothers and their young (up to 12 months; Dobson and Davis 1986), and thus maternal behavior may directly or indirectly influence the antipredator behaviors of *S. beecheyi* juveniles more so than responses of *S. beldingi* juveniles, which have only a 3–4 month developmental period prior to autumnal immergence (Morton and Tung 1971; Maxwell and Morton 1975; pers. obs.). Experiences with mothers may also affect differentially the development of behavior in species with similar active seasons but different growth rates (e.g., *S. beldingi* and *S. mollis;* Morton and Tung 1971; Rickart 1986), or in populations with varying active-season lengths (M. T. Bronson 1979; Joy 1984; Dobson and Davis 1986). In addition to physiological (Morton and Tung 1971; Maxwell and Morton 1975) and social (Armitage 1981) adaptations to the length of the growing period, then, selection may favor more experience-dependent behaviors in slowly maturing species than in species with accelerated growth (fig. 17.3A–F). This possibility has not yet been examined systematically in rodents.

Development across the lifespan

Development does not end at puberty—it continues throughout the life span—yet few researchers have examined aging in rodents in natural contexts. This is largely because animals die from extrinsic causes such as starvation or predation well before they senesce. It is important to note that senescence is not the same as aging. Aging is the postmaturation decline in survivorship and fecundity (potential for reproduction) that accompanies advancing age. Senescence is the bodily changes that cause the decline. Lab rodents have been used as models for human aging, but this work rarely considers ecological or evolutionary correlates of aging. There are some exceptions, however, including recent work examining how behavioral syndromes (see the following) influence health outcomes and longevity in rats. Juvenile rats that are neophobic continue to be so throughout their lives, have increased stress responses and more tumors, and die at an earlier age than neophilic rats (Cavigelli and McClintock 2003). This accelerated aging or early death does not necessarily reduce lifetime fitness, as neophobic animals in the field may actually experience better health during their reproductive prime, and reduced exposure to predators.

Some rodent species are well suited for studying senescence, in particular to test evolutionary predictions that selection against it would be expected in species with low rates of mortality (e.g., threat of predation) or high fecundity (Williams 1957). For example, fossorial naked mole-rats (*Heterocephalus glaber*) are protected from climatic changes and most predators, and breeding females can produce very large litters. As expected by senescence theory, then, *H. glaber* can live at least 10 years in the wild and 26 years in captivity, a remarkably long period for a small rodent (< 35 g; Braude 2000; Sherman and Jarvis 2002). Broussard et al. (2003) examined age-related investment in somatic development and reproduction in female Columbian ground squirrels (*S. columbianus*), which must allocate energy to both somatic and reproductive investments during short active seasons. Yearlings that emerge from hibernation at low weight tend to invest more in continued growth and weight gain rather than reproduction, whereas some old females (> 6 years) show a significant decline in reproductive output relative to other age classes.

As yet, however, there is scant evidence of reproductive senescence in other free-living rodents. Given the variation in ecological niches that rodents inhabit, comparative analyses of senescence could take advantage of ensuing variation in extrinsic sources of mortality, such as presence of pathogens, predators, and nest sites, similar to Austad's (1993) work on aging in mainland and island populations of opossums. Although rodents are relatively short lived compared with other mammals, researchers could fruitfully explore the selective potential for senescence and aging in sympatric species, for example, or within a species across populations spread along gradients of elevation or latitude (e.g., Dobson and Davis 1986). Further, there has been little focus on changes in social behaviors as animals age. Some older female rodents will shift their territories to make room for their reproductive daughters, occasionally moving to less-optimal spaces (Sherman 1976; Berteaux and Boutin 2000; pers. obs.). Whether this reflects an intentional bequeath of territory or a lack of competitive ability in an older animal is unclear, as is the net influence on her lifetime reproductive success, particularly when considering the indirect benefits she would receive if her daughters' reproductive success is improved by inheriting her old site (see Lambin 1997 for a treatment of territory bequeathal).

Endocrine disruptors

Endocrine disruptors, chemicals in the environment that bind to steroid-hormone receptors and act as agonists or antagonists of endogenous hormones, have been the recent focus of both field and laboratory research with rodents. Even at very low levels (e.g., 1 part per trillion) these disruptors can have profound effects on reproductive functioning, particularly if exposure occurs during fetal development, such as when estrogen mimics are present during the organizational phase of gonadal development. Endocrine disruptors such as dioxin, PCBs, and DDT can interfere with the development of gonads, external genitalia, and other sexually dimorphic structures, impede gamete production, and alter behaviors such as parenting, aggression, and territoriality (Palanza et al. 1999). Although rodents are currently used as models for how endocrine disruptors influence human development, the effects of synthetic chemicals on development and reproduction in free-living small mammals are likely to be the focus of future investigations, perhaps involving collaborations among behavioral ecologists, conservation biologists, and environmental toxicologists.

Integration of levels of analysis

"Ecological developmental biology" (also known as "eco-devo") examines how gene expression is mediated by environmental factors such as climate, day length, population density, and predators to influence developing phenotypes. Although this approach has been around for decades, we are now seeing it integrated into our studies on behavioral plasticity in natural contexts (see Gilbert 2001 for a detailed discussion). A separate, yet related, approach to behavioral development focuses on individual differences in the outcomes of developmental processes. "Behavioral syndromes" are consistent between-individual differences in behaviors across a variety of situations, such as levels of aggression in foraging, predator, and mating contexts (Sih et al. 2004). This suite of correlated traits is of interest from a proximate perspective (e.g., the developmental and physiological mechanisms underlying expression of the traits) as well as an ultimate perspective, because the syndrome probably represents life-history parameters that were positively or negatively selected in temporally or spatially variable environments, rather than traits that evolved independently. Studies of behavioral syndromes necessarily bridge areas of psychology and biology, with a focus on the expression of individual differences in context. "Evolutionary traps" refer to situations in which animals encounter drastically different environments (due to human disturbance) and exhibit behaviors based on previously reliable environmental cues, behaviors that may now be maladaptive. Phenotypic plasticity, particularly of behavioral traits, may allow these individuals to "escape" the traps, thereby adjusting their behaviors according to the new environments in which they find themselves. Emergence from hibernation by yellow-bellied marmots (*Marmota flaviventris*) in central North America occurs earlier than it did decades ago due to a warming climate. As a result, marmots become active well before snowmelt and risk starvation until food becomes available, unless they can change their seasonal activity patterns and emerge later in the spring (Inouye et al. 2000). As an example, leatherback turtles (*Dermochelys coriacea*) can adjust their food search strategy to avoid consuming plastic bags that resemble jellyfish, the turtles' typical prey (Schlaepfer et al. 2002). Studies of ecological development, behavioral syndromes, and ecological traps allow connections between population-based ecological approaches to behaviors (e.g., foraging, predator-prey interactions) and developmental studies of individual variability, learning, and plasticity. Such integrated studies could then fruitfully link early social and environmental experiences and behavioral plasticity with reproductive outcomes, life-history parameters, and fitness, and contribute to the growing interest in interdisciplinary research spanning multiple levels of analysis.

Summary

One of the hallmarks of rodent behavioral development is its sensitivity to the social and physical contexts in which ontogeny occurs. From a functional perspective, such sensitivity makes adaptive sense if one views juvenile behavioral patterns as ontogenetic or age-specific adaptations and recognizes that the factors that influence survival and reproduction in one environment will often differ from those in another environment. Further, behavioral ecologists should not think of young animals as small adults, but as organisms with age-specific needs and adaptations. Indeed, young have very different selective pressures than do adults, and thus what is functional for one age group may not be for another.

Development influences almost all aspects of rodent biology, particularly social behaviors. This chapter presents a necessarily broad survey of how anatomical, hormonal, and behavioral development are affected by an animal's physical and social environments through the life span, from the prenatal period to death. For example, maternal stress, gestational photoperiod, and androgen exposure from adjoining fetuses can cause variation in aggression, parenting behaviors, and learning abilities well into adulthood. Social and physical environments after birth continue to influence the expression of behaviors, as an animal's needs and vulnerabilities adjust with increasing age. Natural selection will favor such plasticity when an animal experiences spatial or temporal changes in microhabitat, food availability, predation pressure, social dynamics, and reproductive status.

When an emerging trait is studied descriptively, researchers tend to focus on particular aspects of the experiences of the developing individual in isolation from other experiences, deflecting attention to the interactive nature of epigenesis. When traits are examined in context, however, reciprocal associations between an animal's internal and external environments can be seen, as well as how social cues, stress and gonadal hormones, and abiotic factors feed back onto one another (fig. 17.1). This chapter highlights the need for more behavioral studies integrating levels of analysis as well as levels of biological organization. Recent interdisciplinary research offers encouraging signs that the study of behavioral development is poised to truly link the mechanisms and functions of emerging adaptive behaviors.

Chapter 18 Social Learning by Rodents

Bennett G. Galef Jr.

MANY OF THE THINGS that animals, especially young animals, need to learn, they need to learn rapidly. A fledging bird or weanling mammal venturing from the site where it was born and reared by adult kin has to learn to avoid predators before being eaten by one. The individuial has to learn to select an adequate diet before its internal reserves of any critical nutrient are exhausted and without ingesting harmful amounts of toxins.

A naive young animal faced with such problems should take advantage of opportunities that interactions with adults provide. Adults have surely learned to avoid predators and to find both appropriate substances to ingest and appropriate locations in which to seek refuge. Most important, any adult with whom a juvenile interacts is feeding, avoiding predators, and navigating about the environment in which the juvenile is struggling to achieve independence. To the extent that a juvenile can use the behavior of adults to guide development of its own behavioral repertoires, it should be able to acquire adaptive responses to environmental demands without incurring all of the costs of individual trial-and-error learning.

Formal models (e.g., Laland et al.1996) predict that dependence on social learning should evolve in environments that are neither too stable (where unlearned responses would be more valuable) nor too rapidly changing (where copying the behavior of others could lead to errors and individual learning would be most advantageous). And, in species that forage from a central location, as do most rodents, such models indicate that information exchange would be most valuable when foods are patchy in distribution and ephemeral, and naive individuals would be un-

likely to stumble upon rich feeding sites by chance (e.g., Waltz 1982). Unfortunately, theoretical approaches to the study of social learning have not yet had much impact on empirical work in the area (for review, see Galef and Giraldeau 2002), though increasing numbers of investigators are attempting to integrate theoretical and empirical approaches (e.g., Dewar 2004; Noble et al. 2001).

Behavioral processes supporting social learning range from relatively simple (e.g., local enhancement, where attention of one animal is focused on an aspect of the environment by the behavior of others [Thorpe 1956]) to cognitively complex (e.g., imitation, learning motor patterns by observing others' behavior [Galef 1988b; Whiten and Ham 1992]). Local enhancement of feeding site selection appears to be common in rodents, though apparently not of sufficient intrinsic interest to have provoked much study. Imitation has attracted considerable laboratory study but seems relatively rare in animals, and may well be nonexistent in rodents.

Observation of animals living free in undisturbed habitats has been important in calling attention to potential socially learned behaviors (i.e., behaviors the development of which is likely to have been influenced by interaction with conspecifics). However, although field observations have provided strong circumstantial evidence that some behaviors are learned socially, observation per se has not proven sufficient in most cases to conclude that social interaction is important in behavioral development. For example, chimpanzees (*Pan troglodytes*) living in East and West Africa dip for ants using both different tools and different methods of removing ants from tools (McGrew 1974; Boesch 1996).

However, whether such "traditions" reflect (1) differences in the genotypes of chimpanzees living on opposite sides of the continent (Morin et al. 1994), (2) differences in the environments with which individuals interact and about which they learn independently (Humle and Matsuzawa 2002), or (3) social influences on behavioral development (Whiten et al. 1999) is not known. Even after decades of observational study, it is not clear whether different patterns of ant dipping observed in different free-living chimpanzee populations involve social learning at all.

Identification of socially learned behaviors in populations of free-living rodents that are often small, nocturnal, shy, and subterranean is, if anything, even more difficult than in human-habituated chimpanzees, whose behavior and social interactions are relatively easily observed. Field studies are necessary to identify behaviors that rodents may learn socially. However, experiments carried out under controlled conditions are necessary to draw strong conclusions concerning the behavioral processes responsible for the development of suspect behaviors (Galef 1984, 1996c). Field experiments to determine whether any free-living animals actually learn any behavior socially are generally lacking, but needed (for review and further discussion, see Galef, 2004).

Field observations of some common rodent species reveal marked differences in the behavior of populations living in different areas that are at least superficially similar to the geographic variability in behavior observed in chimpanzees (Whiten et al. 1999) and orangutans (van Schaik et al. 2003). For example, Norway rats (*Rattus norvegicus*) living on the banks of ponds in a fish hatchery in West Virginia catch fingerlings and eat them (Cottam 1948). On the island of Norderoog in the North Sea, members of the same species frequently stalk and kill sparrows and ducks (Steiniger 1950), though they have not been reported to do so elsewhere. Colonies of black rats (*Rattus rattus*) thrive in the pine forests of Israel by removing scales from pinecones and eating the seeds that the scales conceal, a behavior not reported in other populations of black rats (Terkel 1996). Members of only some of the many colonies of Norway rats living along the banks of the Po River in Italy dive into the river and feed on mollusks inhabiting the river bottom (Gandolfi and Parisi 1972, 1973).

Almost all laboratory investigations of social learning in rodents have focused on feeding behaviors of one type or another in murids. Indeed, only six laboratories worldwide (those of Galef [Canada], Heyes [UK], Laland [UK], Terkel [Israel], Valsecchi [Italy] and Holmes and Mateo [USA]) have carried out sustained investigations of any aspect of social learning in any rodent. Four of these six laboratories have worked primarily with a single genus (*Rattus*) and five of the six within a single behavioral domain (foraging).

The other domain that has been extensively investigated is social learning of predator avoidance in a sciurid, the Belding's ground squirrel (see Mateo, chap. 17, and Holmes and Mateo, chap. 19 this volume).

The paucity of data on social learning in rodents makes my task here comparatively easy. In the space available, I can both describe, albeit briefly, five of the six major research programs mentioned previously (the sixth is discussed at length in the chapters by Mateo and Holmes and Mateo) and provide a fairly comprehensive set of references to studies in rodent social learning carried out in the last 20 years.

Pinecone Stripping by Roof Rats (*Rattus rattus*)

Roof rats living in the pine forests of Israel, but not those living elsewhere, strip the scales from pinecones and eat the seeds that the scales protect. This foraging behavior permits the rats to occupy a niche that is occupied in other parts of the world by tree squirrels; sciurids are not present in the Middle East (Aisner and Terkel 1992).

Laboratory investigation of pinecone stripping by wild-caught rats reveals that they must take advantage of the physical structure of pinecones to gain more energy from pine seeds than is expended in removing seeds from cones. To exploit pinecones efficiently, rats must first strip the scales from the base of a cone, then remove the remaining scales in succession as they spiral around the cone's shaft to its apex (Terkel 1996; fig. 18.1).

Studies of the development of the energetically efficient pattern of stripping scales from pinecones revealed that only 6 of 222 hungry laboratory-reared wild rats that were given access to a surplus of pinecones for several weeks independently learned the efficient method of feeding on them. The remaining 216 animals either ignored the cones or gnawed at them in ways that did not lead to a net energy gain (Zohar and Terkel 1995).

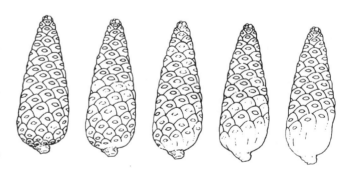

Figure 18.1 Pinecones in different stages of opening with the number of rows of stripped scales increasing from left to right (Terkel 1996, by permission of Elsevier).

Pups gestated by dams that stripped pinecones efficiently, but reared by foster mothers that did not know how to strip pinecones failed to develop the efficient technique (Aisner and Terkel 1992). However, more than 90% of pups learned to open cones properly if reared by a foster mother that stripped cones efficiently while her charges were present. Clearly, some aspect of postnatal interaction between a dam stripping scales from pinecones and the young that she rears permits transmission of the efficient means of feeding on pinecones from one generation to the next (Aisner and Terkel 1992; Zohar and Terkel 1992). Additional experiments showed that 70% of young rats with experience completing the stripping of cones that had been started appropriately by efficient adults (or by experimenters using pliers to imitate the pattern of scale removal used by proficient adult rats) themselves became efficient strippers of pinecones (Terkel 1996).

Terkel's (1996) observations of rats in the laboratory indicated that when a roof-rat mother opens pinecones by stripping scales and eating exposed seeds, her young gather around her and attempt to obtain seeds. As the young grow older, they snatch partially opened cones from their mother and continue the stripping process themselves. A mother rat's feeding activities thus appear to facilitate acquisition of pinecone stripping by her offspring in two ways: first, by focusing attention on pinecones as potential food sources (local enhancement), and later by providing young with partially opened pinecones that guide development of feeding on them (Terkel 1996). Simply watching an adult open pinecones without the opportunity to exploit pinecones started by an adult left young unable to open pinecones for themselves. Imitation seems an unlikely explanation of the behavioral process supporting transmission of the behavior from mother to young.

Poison Avoidance by Norway Rats (*Rattus norvegicus*) and Socially Learned Food Preferences in Rodents

Wild rats are social rodents that live in colonies that vary in size from a few to many hundreds of members. Each colony inhabits a burrow system from which colony members emerge to forage and to which they return between foraging bouts. The comings and goings of successful and unsuccessful foragers from a central location, where colony members can interact, provide opportunities for exchange of information about foods that would be of use to individuals both in finding food in natural environments and circumventing humans' attempts at rodent control.

When rodent control operatives attempt to use the economically efficient method of placing permanent poison-bait stations in rat-infested areas they have great initial success with rats eating ample amounts of poison and dying in large numbers. However, later bait acceptance is poor and colonies soon return to their initial sizes (Steiniger 1950). Permanent baiting stations fail because young rats born to adult colony members that have survived their first ingestion of a poison bait and learned not to eat that bait never even taste the bait for themselves (Steiniger 1950).

Young wild rats' total avoidance of foods that adults of their colony have learned to avoid eating is a robust phenomenon that can easily be brought into the laboratory (Galef and Clark 1971b). We captured adult wild rats in southern Ontario and established them in small groups in 2 m² laboratory enclosures. For 3 hr each day, we provided each of our laboratory colonies with two easily distinguished, nutritionally adequate foods.

To begin a typical experiment we introduced sublethal concentrations of toxin into one of these two foods. Our rats soon learned to avoid the poisoned bait and for weeks thereafter would not eat the food that had been poisoned even when we gave them uncontaminated samples of it.

After we had trained our colonies we waited for young to be born and grow to weaning age. When these young began to eat solid food we observed them on closed-circuit television and recorded the frequency with which they ate each of the two foods that we placed daily in their enclosure: one that adult colony members were eating and the other that adults had learned to avoid.

We found that while pups remained with the adults of their colony, they ate only the food that those adults were eating, and completely avoided the alternative that the adults were avoiding. Even after we removed pups from their natal enclosures, housed them individually, and offered them the same foods that we had made available in their colony enclosure, the pups continued for several days to prefer the food that the adults of their colony had eaten (fig. 18.2).

To determine whether such social learning of a food preference could result in a tradition that lasted for generations, we created two types of colonies (Galef and Allen 1995). We used a poisoning technique to teach all four members of each colony of one type not to eat a horseradish-flavored food (wasabi) and all four members of each colony of the other type not to eat a cayenne-pepper-flavored food. Once colony members had learned what not to eat, we offered all members of both types of colony a choice between cayenne pepper-flavored food and horseradish-flavored food for 3 hr/day. Each day, immediately after we fed a colony, we removed one of its members and replaced that member with a naive rat.

After 4 days we had replaced all the original members of each colony. For 10 days thereafter, we replaced with a

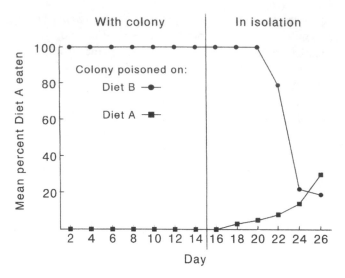

Figure 18.2 Mean number of times that juvenile Norway rats ate Diet A as a percentage of the total number of times juveniles ate both Diet A and Diet B during daily 3-hr feeding periods. Left panel: days when juveniles were with adults, Right panel: days when juveniles were moved to individual cages. Abscissa: number of days since pups started feeding on solid food.

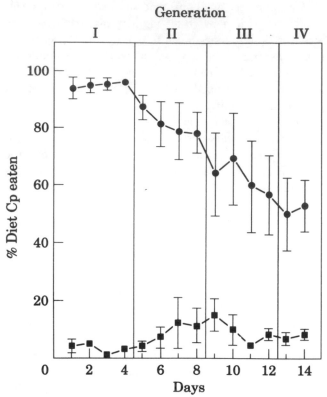

Figure 18.3 Mean ± SE amount of cayenne pepper-flavored diet (Diet Cp) ingested as a percentage of total amount eaten by subjects offered both Diet Cp and wasabi-flavored diet (Diet W) in enclosures where founding colony members ate only Diet Cp (●) or only Diet W (■). On day 1, enclosures contained only founding colony members, on days 2 to 4, both founding colony members and replacement subjects, and on days 5 to 14, successive generations of replacement subjects (Galef and Allen 1995; by permission of the American Psychological Association).

naive rat the individual in each colony that had been there longest, and we kept doing so day after day. As can be seen in figure 18.3, even after replacements of replacements of original colony members had been replaced, we still saw huge effects of the food preferences learned by original colony members (Galef and Allen 1995). The longevity of such traditions of food choice is affected by a number of factors, including colony size, rate of replacement of colony members, and number of hours each day that colony members have access to foods (Galef and Allen 1995; Galef and Whiskin 1997, 1998).

My students and I have studied how the feeding patterns of adult rats influence the food choices of young that interact with them (for reviews see Galef 1977, 1988a, 1996a, 1996b). Such social influences on food choice start before rats are born and extend throughout life. For example, a rat fetus exposed to a flavor while still in its mother's womb by injection of that flavor into its dam's amniotic fluid will, when grown, drink more of a solution containing the injected flavor than will control rats lacking such prenatal experience (Smotherman 1982). More realistically, feeding a food with a strong flavor (garlic) to a pregnant rat enhances the postnatal preference of her young for the odor of garlic (Hepper 1988).

Evidence from several laboratories indicates that flavors of foods that a lactating female rat eats affect the flavor of her milk. Exposure to food flavors in milk, a very simple sort of social learning, increases pups' preferences at weaning for foods that their mother ate (Galef and Henderson 1972; see also Galef and Sherry 1973: Bronstein et al. 1975; Martin and Alberts 1979).

When weanlings leave the safety of their natal nest to feed on solid food for the first time, they use visual cues to locate an adult at a distance and then feed with that adult (Galef and Clark 1971a). Even an anesthetized adult rat placed near one of two otherwise identical feeding sites makes the occupied site far more attractive to pups than the unoccupied one, and young pups both visit and eat more frequently at the former than the latter (Galef 1981a).

Adult rats need not be physically present at a feeding site to cause young to prefer to eat there. When leaving a feeding site to return to their burrows, adult rats deposit scent trails (Calhoun 1962a; Telle 1966) that direct young rats seeking food to locations where the adults ate (Galef and Buckley 1996). Also, when feeding, adult rats deposit olfactory cues both in the vicinity of a food source (Galef and Heiber 1976; Galef 1981a; Laland and Plotkin 1991, 1992) and on foods they are eating (Galef and Beck 1985). These residual odors attract pups and, like the presence of an adult rat at a feeding site, cause young rats to prefer marked sites to unmarked ones.

Availability of social information as to what foods are best to eat can have profound consequences for rats in environments where ingesting the most palatable foods does not lead to a nutritionally adequate diet. We placed young rats in enclosures where they had access continuously for 7 days to four different foods (Beck and Galef 1989; Galef et al. 1991). Three of these (cinnamon-, cocoa- and thyme-flavored foods) were relatively palatable, but low in protein, while one (nutmeg-flavored food) was relatively unpalatable, but protein rich. The pups failed to solve even this apparently trivial foraging problem, lost weight, and would surely have died of protein deficiency had we not terminated the experiment after 1 week. By contrast, pups that shared their enclosures with adult rats that we had trained to eat the relatively unpalatable, protein-rich, nutmeg-flavored food grew at almost the same rate as control pups offered just the protein-rich diet.

Terkel's roof rats were able to invade pine forests because of their ability to learn socially how to efficiently exploit pinecones. Similarly, Norway rats could invade an environment where needed nutrients are present only in relatively unpalatable foods because they can learn socially how to select an adequate diet under such circumstances once one of their number has learned to do so (Galef 1991).

Norway rats can also influence the food choices of conspecifics by interacting with them at a distance from a feeding site. After a naive rat (an observer) interacts with a conspecific (a demonstrator) that has recently eaten a food unfamiliar to the observer, the observer exhibits substantial enhancement of its preference for whatever food its demonstrator ate (Galef and Wigmore 1983; Posadas-Andrews and Roper 1983; Strupp and Levitsky 1984). Such effects are relatively independent of the genetic or prior social relationship of demonstrator and observer (Galef et al. 1998) but depend to some extent on the rats' previous feeding history (Dewar, 2004). Laboratory studies using procedures similar to those used with Norway rats have provided evidence of increased preference of observers for the food preferences of their respective demonstrators in a number of other rodent species (table 18.1).

In rats, both food-related odors escaping from the digestive tract of a demonstrator and the scent of bits of food clinging to its fur and vibrissae allow naive conspecifics to identify what foods others have recently eaten (Galef et al. 1985; Galef et al. 1990; Galef and Whiskin 1992). Enhancement of food preference of observers depends on their experiencing food odors together with other olfactory stimuli that are normally emitted by live conspecifics (Galef et al. 1985; Galef and Stein 1985; Galef et al. 1988; Heyes and Durlach 1990; Stetter et al. 1995). For example, rats exposed to pieces of cotton batting that are dusted with a food and moistened with distilled water do not develop a

Table 18.1 Rodent species in which increased preference of observers for foods eaten by conspecific demonstrators has been found in laboratory experiments

Species	Reference
House mouse (*Mus domesticus*)	Choleris et al., 1997; Valsecchi and Galef, 1989; Valsecchi et al. 1989
Mice (*Mus musculus*)	Valsecchi et al. 1993
Mongolian gerbil (*Meriones unguiculatus*)	Valsecchi et al. 1996; Galef et al. 1998
Roof rats (*Rattus rattus*)	Chou et al. 2000
Spiny mice (*Acomys cahirinus*)	McFayden-Ketchum and Porter 1989
Pine voles (*Microtus pinetorum*)	Solomon et al. 2002
Belding's ground squirrel (*Spermophilus beldingi*)	Peacock and Jenkins 1988; Sherman, personal communication
Golden hamster (*Mesocricetus auratus*)	Lupfer et al. 2003
Dwarf hamster (*Phodopus campbelli*)	Lupfer et al. 2003

preference for the food dusted on the cotton batting. However, rats subsequently prefer that food if exposed to it on either the head of an anesthetized conspecific or a piece of cotton batting moistened with carbon disulfide, which is a constituent of normal rat breath (Galef and Stein 1985; Galef et al. 1988).

Effects of exposure to a recently fed rat on conspecifics' food choices are powerful (Galef et al. 1984; Richard et al. 1987). Observer rats that are first taught to avoid a food by following its ingestion with an injection of toxin, then placed with a conspecific that has eaten the food to which the observer rats learned an aversion, abandon their aversion to the food they ate before being injected with toxin. Most rats that interact with conspecifics who were fed a food adulterated with cayenne pepper, a spice that is inherently unpalatable to rats, subsequently prefer peppered diet to unadulterated diet (Galef 1986; Galef 1989). Such effects of social interaction on food choice are also enduring and, under some circumstances, can be seen more than a month after social learning took place (Galef and Whiskin 2003).

Although social exposure to an odor has profound effects on rats' subsequent preferences for foods, identical experiences have no effect on rats' odor preferences in other contexts. For example, rats that have interacted with a conspecific that has eaten cinnamon-flavored diet prefer cinnamon-flavored food, but show no enhancement of their preference for cinnamon-scented nest materials or cinnamon-scented nest sites (Galef and Iliffe 1994).

Surprisingly, rats that will readily acquire preferences for foods socially do not learn aversions to foods from their fellows. Indeed, rats generally show increased preferences for foods after interacting with sick or unconscious rats that have eaten them (Galef et al. 1983; Galef et al. 1990). We surmise that, because rats encountering toxic foods are likely to eat them only once (Garcia et al. 1974), the probability that naive rats will be induced to eat toxic foods by conspecifics is low. Consequently, failure to discriminate between conspecifics that have eaten toxic and safe foods has little cost. Results of an evolutionary simulation of the effects on survival of discrimination between sick and healthy conspecifics when learning socially about foods (Noble et al. 2001) are consistent with such an explanation.

Norway Rats Diving for Mollusks

Gandolfi and Parisi (1972) reported that most members of some colonies of Norway rats living along the banks of the Po River in Italy dive to feed on mollusks that inhabit the river bottom, whereas no members of nearby colonies with equal access to mollusks do so. Nieder et al. (1982) observed such mollusk predation by small groups of rats that they confined in a large (22 × 10 m) outdoor enclosure built over a narrow branch of the Po River. Although observations in both the enclosure and the wild suggested that social learning of some sort might have been involved in diffusion of mollusk predation through rat populations, the data were not conclusive.

In a laboratory experiment undertaken to examine the role of social learning in the development of diving behavior in Norway rats, we placed second and third generation laboratory-bred female wild rats captured in Ontario together with their offspring in enclosures with separate nesting and diving areas connected by meter-long tunnels (Galef 1980). In the diving area, subjects could retrieve pieces of chocolate from beneath 15 cm of water in an aquarium.

We found that adults that had not been trained to dive for food never dove, even if housed with rats that we had trained to dive for food by placing chocolate squares in empty aquaria and, over a period of weeks, gradually increasing water levels to 15 cm. However, approximately 20% of juvenile wild rats reared in the enclosures in which adults never dove spontaneously dove for food. And juveniles were as likely to learn to dive whether their dam regularly dove and retrieved chocolates from under water or never did so. Such results suggest that observation of diving conspecifics does not, in itself, induce rats to dive.

In a subsequent study, we raised young wild rats in an enclosure where they had to swim 60 cm to reach food. When introduced into enclosures connected to a diving area, where food was available under 15 cm of water, more than 90% of rats trained only to swim spontaneously dove for food. The finding that swimming rats are effectively diving rats limits the potential role of social learning in diffu-

sion of diving behavior through a population. If rats learn to swim independently, and if swimming rats dive, then social learning could serve only to direct rats to dive in one area rather than another. However, development of swimming might itself be socially influenced. If so, then social learning might indirectly facilitate the spread of diving behavior by facilitating the spread of swimming behavior. However, wild rat pups that were reared by dams that either swam or did not swim 1.7 m down an alley to reach food did not differ in the age at which they started to swim, and all swam to food before they were 40 days old (Galef 1980).

The findings of high frequencies of willingness to swim to obtain food and to dive in wild rat pups suggests that (unless there is a relevant genotypic difference between Norway rats in Canada and Italy), all rats living along the Po River may know how to dive for mollusks, but that they do not dive unless they have to. In the laboratory, rats that reliably dove for food when food was available ashore for only 3 hr/day stopped diving when given ad libitum access to the same food on land, even if the food available ashore was considerably less palatable than that available under water (Galef 1980).

Taken together, the laboratory results offer little support for the hypothesis that variation in the frequency of diving observed among colonies of rats living along the Po River results from a socially learned tradition present in some colonies but not others. To the contrary, the laboratory data suggest that all rats may know how to dive for food, but will do so only when adequate food is not available on land.

In retrospect, some observations made in the field are consistent with the hypothesis that availability of food on land may be the major determinant of whether members of rat colonies living along the banks of the Po River feed on submerged mollusks. For example, Gandolfi and Parisi (1973, p. 69) report that in those locations where mollusk predation is observed, mollusks "represent one of the main sources, if not the main source of food for rats." Parisi and Gandolfi (1974, p. 102) suggest further that "the time dedicated by rats to mollusk capture depends greatly on the availability of other foods."

Naked Mole-Rats Recruitment of Colony Mates to Food Sources

Naked mole-rats (*Heterocephalus glaber*) are eusocial, subterranean rodents that, in nature, travel underground through an extensive labyrinth of tunnels to feed on patchily distributed bulbs and tubers. Observation of naked mole-rats kept in artificial burrow systems that mimic, albeit on a vastly reduced scale, their natural burrows, revealed that while returning to the nest after finding a new source of food individuals give a special vocalization. Upon arrival at the nest, returning successful foragers on a new food source sometimes wave the food around (Judd and Sherman 1996).

In laboratory experiments, colony mates preferred the site where the initial forager had found food, and would bypass alternative sites containing the same type of food. Such recruits preferred to use the same tunnel that the successful forager had used, even when the burrow system had been rearranged so that recruits had to turn in the opposite direction from the original scout to reach the same location.

Recruits also preferred tunnels that the initial forager had recently traversed to tunnels traversed by other colony members that were carrying the same type of food that the initial forager had carried. Such preference disappeared if tunnel segments that the initial forager had traversed were cleaned. Taken together, the results offer strong support for the hypothesis that naked mole-rats follow each other's odor trails to food, thus facilitating location of the widely dispersed foods exploited by colonies of naked mole-rats (Faulkes 1999).

Norway Rats Digging for Food

Laland and Plotkin (1990, 1992) examined social effects on the frequency with which rats dig for buried pieces of food. They discovered that the probability of observer rats digging for buried food increased if they saw demonstrator rats digging for food, and that observer rats, after learning socially to dig for food, could serve as demonstrators for other observers, that could, in turn, become demonstrators. Such chaining of socially learned behavior was first demonstrated by Curio et al. (1978) in investigations of the development of predator recognition by European blackbirds (*Turdus merula*). Curio et al.'s technique captures some features of diffusion of socially learned behaviors through free-living populations of animals. However, it does not provide opportunity for individual learning of alternative rewarding behaviors by animals in the test situation, and presence of such alternatives can be important determinants of whether socially learned behaviors will be maintained long enough in individuals to be transmitted to others. For example, the longevity of socially induced food preferences in rat colonies of the type studied by Galef depended critically on the number of hours a day that foods were present in a colony enclosure. When foods were available for 2 hr/day, preferences lasted far longer than when food was available 24 hr/day, and rats could more easily learn for themselves the relative value of available diets (Galef and Allen 1995; Galef and Whiskin 1997, 1998).

Norway Rats' Social Learning of Arbitrary Behaviors

There is a large literature concerning social effects on the bar-pressing and maze-running behavior of Norway rats and house mice (reviewed in Zentall 1988; Denny et al. 1988). I shall discuss here only the one major research program concerned with the learning of arbitrary behaviors developed since these reviews were published.

Much early work on social learning in rodents was conducted by those trained in experimental animal psychology, and was concerned with the question of whether animals could learn by imitation, with imitation defined narrowly as "learning to do an act by seeing it done" (Thorndike 1898, p. 50). Thus defined, imitation differs from other sorts of social learning in that it involves learning to produce a behavior by observing the behavior of others rather than learning about the environment by observing the behavior of others (Heyes 1993). For example, if I watch someone open a screw-top jar and eat from it, I might learn to grasp the jar with one hand and apply rotational pressure with the other. This would be imitation. Alternatively, I might learn by watching that the jar can be opened and then use trial-and-error processes to acquire the appropriate motor patterns to open the jar. This would be a nonimitative form of social learning. Discussions of such distinctions are extensive in the literature and have been reviewed by Galef (1988b) and Whiten and Ham (1992). Heyes's (1993) empirical work, described here, is the most compelling examination of imitation in rodents to date.

Heyes used the "two-action method" or "bidirectional control" procedure, in which demonstrators direct one of two patterns of behavior toward a single target to control for many alternative explanations of apparent imitative behavior that had plagued earlier attempts to demonstrate imitation learning in animals (see Zentall 1988). Observer rats were given their first opportunity to push a joystick left or right immediately after observing a conspecific demonstrator push the same joystick either left or right. Heyes found that observers given access to their demonstrators' joystick tended to push it in the same direction relative to their own body axes as had their demonstrators (Heyes et al.1992; Heyes and Dawson 1990), even if the observers were given food rewards for pushing the joystick in either direction. A variety of control procedures provided data consistent with the view that observers were copying the motor behavior of their respective demonstrators, using the orientation of their own bodies as a referent (for review, see Heyes 1996).

However, subsequent studies by Mitchell et al. (1999) and Campbell and Heyes (2002) showed that if, after the demonstrator used the joystick and before the observer did so, the joystick was rotated 180 degrees about its main axis, the direction in which the observer pushed the joystick was reversed. Apparently, demonstrators were depositing attractive odor cues on the side of the joystick against which they pushed, and these residual cues biased observers to push on the same side of the joystick as had their respective demonstrators. Heyes's experiments thus join a long line of failed attempts to find evidence of learning by imitation in rodents. Indeed, there is currently no convincing evidence of imitation learning in any rodent (or any nonhuman mammal other than dolphins and apes). This absence of laboratory evidence of imitation is somewhat surprising given the numerous examples of such learning in birds (reviewed in Zentall 2004). The reasons for the peculiar phylogenetic distribution of the ability to imitate remain obscure, though Moore (1996) has proposed that a capacity to imitate has evolved in vertebrates three times, sometimes based on vocal imitation and sometimes not.

Miscellany

In addition to the sustained research programs focused on social learning in rodents discussed earlier, there have been numerous isolated papers describing instances of social learning in rodents. Those of which I am aware are listed in table 18.2. The phenomena described in these papers are worthy of further exploration, so that their repeatability can be established and the behavioral processes supporting those that are reliable can be examined. Unfortunately, constraint on the space available here makes detailed description of each of these phenomena impossible. The interested reader is referred to the original reports.

Summary

The last 20 years have seen tremendous progress in understanding a handful of instances of apparent social learning in rodents. Obviously, it is too early to attempt generalizations as to which rodents learn socially and which learned behaviors of rodents are modified by social interactions. Less than one-half of 1% of rodent species has been examined even once in a social-learning paradigm and, even in those two species (Norway rats and house mice) that have been studied most frequently, focus has been almost entirely on social influences on foraging behavior (but see Mateo, chap. 17, and Holmes and Mateo, chap. 19, this volume). Essentially nothing is known of the role of social learning by rodents in mating, predator avoidance, predatory behavior, parental care, and so on.

We know from field observations that even congeneric rodent species can differ profoundly in the role that social cues play in various aspects of their behavior. For example, Telle (1966) observed that although both *R. rattus* and *R. norvegicus* tend to move about their territories on scent-marked runs, *R. norvegicus* attacks only those unfamiliar individuals encountered on a run, while *R. rattus* attacks unfamiliar individuals in areas between runs. Even within a species, differences in genotype (Kogan et al. 1997), hormonal state (Fleming et al. 1994), nutritional level (Galef et al. 1991), or rearing conditions (Levy et al. 2003) can affect the magnitude of social influences on learning. Such reports suggest that generalizations across species will be hard won.

Indeed, it is difficult to predict just how useful comparative approaches to the study of social learning in rodents

Table 18.2 Further instances of social learning in rodents

Species	Behavior	Reference
Mongolian gerbil	intake of novel food	Forkman, 1991; Tachiban, 1974
Golden hamster (*Mesocricetus auratus*)	chain pulling	Previde and Poli, 1996
Mouse[a]	swinging door opening	Collins, 1988
Grasshopper mouse (*Onychomys leucogastor*)[a]	cricket killing	Kemble, 1984
Red squirrels (*Tamiascurus hudsonicus*)	nut opening	Weigl and Hanson, 1980
Norway rat[a]	mouse killing	Flandera and Novakova, 1974
	avoiding flame	Lore, Blanc, and Suedfeld, 1971
	avoiding shock prod	White and Galef, 1998
	alcohol ingestion	Hunt, Lant, and Carroll, 2000; Hunt, Holloway, and Scordalakes, 2001; Honey and Galef, 2004
Prairie dog (*Cynomys gunnisoni*)[b]	alarm-call dialects	Perla and Slobodchikoff, 2002

[a]Findings my laboratory have tried to repeat without success.
[b]Authors do not discuss as an instance of social learning.

will prove to be. In all rodent species, whether solitary or social, altricial or precocial, young spend considerable time interacting with their mothers. Consequently, all juvenile rodents have an opportunity to learn about foods, harborage sites, predators, and other environmental factors from at least one other member of their species. Even European rabbits (*Oryctolagus cuniculus*; lagomorphs, not rodents) that interact with their dams for only a few minutes each day learn from her what foods to eat (Altbacker et al. 1995). Given that all rodents have opportunities to learn socially about at least some aspects of their environment, comparisons among species from different ecological situations or with different degrees of sociality may prove less informative than might be hoped.

Naive rodents generally seem to be able to learn where, when, and how to engage in a variety of behaviors from interactions with either knowledgeable conspecifics or changes that they have made in a shared environment. However, the behavioral processes underlying such social learning appear to be simple ones (e.g., local enhancment, environmental shaping), and to date, little or no reliable evidence of imitation or of teaching has been found in rodents. Of course, the observation that diffusion of behaviors through rodent populations rests on relatively simple behavioral mechanisms should not lessen our appreciation of the potentially important role that social learning can play in the development of adaptive behavioral repertoires of rodents.

Chapter 19 Kin Recognition in Rodents: Issues and Evidence

Warren G. Holmes and Jill M. Mateo

Rarely in the history of biology has a domain of empirical knowledge followed so closely and fruitfully upon an abstract theoretical idea. —E. O. Wilson (*Kin Recognition: An Introductory Synopsis*, 1987)

THE "ABSTRACT THEORETICAL IDEA" to which Wilson refers is, of course, W. D. Hamilton's (1964) theory of inclusive fitness, which specifies the conditions under which an allele would change frequency in a population due to its effects on its bearer's reproduction and the reproduction of its bearer's collateral relatives like siblings, nieces, and nephews. Hamilton knew that if kinship were to mediate social relationships, whether cooperative or competitive, then individuals must possess the means to identify kin, and he outlined four proximate means by which kin could be identified (Hamilton 1964). The importance of parent-offspring recognition and its relevance to discriminative parental care was well known prior to Hamilton's theorizing, but in the late 1970s papers began to appear on the recognition of collateral kin, and various species of rodents were often the study organisms described in these papers (reviewed in Blaustein et al. 1987; Dewsbury 1988; Holmes 2004).

We have three aims in this chapter. First, we will examine some of the conceptual and methodological issues that characterize the kin-recognition literature. Second, we will use a model-systems approach to review some of the primary empirical findings on rodent kin recognition, such as the sensory basis of recognition and the role of early expe-

rience in the ontogeny of recognition. Finally, we will outline some questions and issues about rodent kin recognition that we believe should be pursued in future studies. We discuss kin discrimination as a process that facilitates nepotism (preferential treatment of kin), which historically has been the focus of research on rodent kin recognition. We also note, however, that the issues we address apply to mate choice, as animals evaluate their relatedness to potential mates and avoid the deleterious effects of extreme inbreeding and outbreeding.

Conceptual and Methodological Issues

The meaning of kin recognition

Throughout most of the 1970s and 1980s, kin recognition referred to an internal physiological process that was inferred from kin discrimination, the differential treatment of conspecifics based on unambiguous and reliable (over evolutionary time) correlates of genetic relatedness (Holmes and Sherman 1982, 1983). The correlates of genetic relatedness are often phenotypic attributes borne by conspecifics, but contextual cues such as spatial locations or time spent in proximity to a mate (e.g., Davies et al. 1992; Alberts 1999) may also correlate with relatedness and underlie differential treatment.

It is useful to distinguish between *direct* (kin) recognition, which depends on phenotypic attributes like odors that are borne by individuals (hereafter, "kin labels") and *indirect* (kin) recognition, which depends on contextual cues like the location of a nest or territory (Waldman 1987). Likewise, Hepper (1986) distinguished between conspecific cues and environmental cues. Some authors (Barnard 1990; Tang-Martínez 2001) have argued that recognition based on indirect contextual cues is not kin recognition, despite Hamilton's (1964) observation that spatial cues can be correlates of relatedness and mediate differential treatment of kin. When behavioral ecologists are primarily interested in the fitness consequences of behavior, then kin recognition will include differential treatment mediated by direct and indirect contextual cues (Sherman et al. 1997), but when more proximate neurophysiological or cognitive questions are being addressed then kin-recognition studies are more likely to focus on cues borne by individuals (direct recognition).

In a provocative and controversial paper entitled "Do Animals Really Recognize Kin?", Grafen (1990) argued that "kin recognition" should only be used when the mechanism mediating recognition had been specifically selected to assess genetic similarity and that, under this definition, many putative examples of kin recognition were mediated by mechanisms that had evolved to solve other recognition problems, such as recognizing group members or conspecifics. In many group-living organisms, however, group members are close kin, so that kin-selected recognition abilities would operate in these groups (Sherman et al. 1997). In contrast, Grafen's argument may protect against the erroneous inference that if kin treat each other differentially then they must have evolved to do so because of nepotistic benefits. However, it is unusual in behavioral ecology to define terms based on both their proximate controls and their functional outcomes. Some investigators have accepted Grafen's (1990) arguments (e.g., Barnard 1990; Hurst and Barnard 1995), but his views have been challenged (Blaustein et al. 1991; Byers and Beckoff 1991; Stuart 1991; Sherman et al. 1997), and most empiricists studying kin recognition today operationalize kin discrimination as differential treatment based on correlates of genetic relatedness, as the term was used in the 1970s and 1980s, a tradition that we follow in this chapter.

Components of the recognition process

Kin recognition requires (1) an *actor* who tries to make a discrimination, (2) a *kin template* or internal representation against which an actor compares an "unrecognized" conspecific's phenotype, and (3) a *recipient,* the unrecognized individual whose identity an actor seeks to establish (Holmes and Sherman 1983; Liebert and Starks [2004] review these and other terms that have been central in the kin recognition literature). Some conceptual clarity was brought

to the kin-recognition process when investigators acknowledged explicitly that it comprises two (Beecher 1982; Sherman and Holmes 1985; Gamboa et al. 1986) or three (Waldman et al. 1988; Reeve 1989; Sherman et al. 1997) distinct components. First, the *production* or *expression component* refers to the nature of recipients' kin labels that make them distinguishable by actors, including when and how recipients develop these labels (Tsutsui 2004). The proximate source of kin labels may be endogenous, produced by individuals themselves, and/or exogenous, acquired from the environment, and both endogenous and exogenous labels may be inheritable (Gamboa et al. 1986). Second, the *perception component* focuses on when and how an actor acquires a kin template (reviewed in Mateo 2004), which we define below. The perception component also includes an actor's sensory detection of a recipient's kin labels and the subsequent comparison that an actor makes between its kin template and a recipient's labels. When people refer to a "kin recognition mechanism" they are often alluding to the perception component. Third, the *action component* was introduced in a seminal paper by Reeve, who defined it as "the determinants of the action taken by an actor that has calculated a particular degree of similarity between its template and a recipient's phenotype." (Reeve 1989, p. 408).

By introducing the action component (reviewed in Liebert and Starks 2004), Reeve sought to make explicit that an actor that detects a particular match between its template and a recipient's phenotype might behave differently depending on the social and ecological context, tolerating a recipient in one situation and rejecting the same recipient in another. In our view, the production and perception components encapsulate the proximate basis of the kin-recognition process, whereas the action component emphasizes the adaptive response that may follow *after* a discrimination has been made (fig. 19.1). If Hamilton's (1964) rule (r × b > c) is satisfied, nepotism may follow after kin are recognized, but if the rule is not satisfied nepotism will not occur, even if conspecifics are recognized as kin. In other words, we suggest that it is important to separate two categories of behavior: first, the *differential treatment* of kin (e.g., fitness-neutral behaviors such as proximity during foraging), which emerges from the production and perception components and comprise the recognition process, and second, the *preferential treatment* of close kin or *agonistic treatment* of distant kin or nonkin, which emerges from the action component (fig. 19.1). A fitness-neutral bioassay, like differential olfactory investigation of kin and nonkin, might reveal that kin discrimination occurs, whereas a bioassay like food sharing (Porter et al. 1980) or infanticide (Sherman 1981b; Hoogland 1985) might or might not re-

Nepotism and the Components of the Kin-Recognition Process

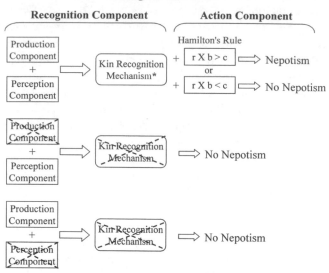

* Inferred if differrential treatment of kin is detected.

Figure 19.1 Kin recognition requires (1) the production of unique kin labels (the production component) and (2) the ability to perceive such labels and compare them with an internal representation according to some sort of matching rule (the perception component). These two requirements comprise the recognition component of the kin-recognition process. Kin recognition mechanism can facilitate nepotism, the action component, but only if Hamilton's (1964) rule is satisfied. (An X in a box indicates the absence of a functioning component.) It is thus useful to distinguish between differential treatment of kin (kin recognition) and preferential treatment of kin (nepotism), because the former will not inevitably result in the latter. Kin recognition may also mediate mate choice, which is not considered in this diagram.

veal kin discrimination, depending on the inclusive fitness costs and benefits of the behavior (Beecher 1991). If food sharing or infanticide were viewed as the action component then the absence of differential treatment of kin could mean that the actor was unable to discriminate kin from nonkin *or* that the actor's fitness interests were better served by failing to treat kin and nonkin differently. That is, a behavioral assay that is likely to affect an actor's inclusive fitness may obfuscate studying differential treatment (kin discrimination) separately from preferential treatment (nepotism). We believe that including the action component as part of the kin-recognition process entangles a proximate mechanism and the functional behavior it generates (see Griffin and West 2003). However, we certainly believe that complete explanations for kin recognition require analyses of proximate *and* ultimate factors (Holmes and Sherman 1982; Blaustein et al. 1991; Mateo 2002).

Experimental methods and research designs

Experimental studies of kin recognition raise several methodological issues (Gamboa et al. 1991; Todrank and Heth

2001) and here we highlight a few. Kin-recognition abilities are often inferred from field data that describe variation in social interactions with kinship (Hoogland 1995; Cheney and Seyfarth 1999; Smith et al. 2003; Griffin and West 2003). However, field studies that assess kin recognition with fitness-neutral assays (i.e., not involving kin favoritism) are rare (Michener 1973a; Waldman 1982; Sun and Müller-Schwarze 1997), and most experimental studies have been conducted on laboratory-born, reared, and tested animals (reviewed in Dewsbury 1988; Mateo 2003), which raises some questions about ecological validity. For example, if kin labels depend, in part, on diet, then maintaining captive animals on the same diet may minimize differences in labels that might be important to discrimination in nature, leading to false negatives. In spiny mice (*Acomys cahirinus*), for instance, both genotype and diet contribute to kin labels, and if diets are changed so that members of a sibship are maintained on different diets, then weanlings' preference to huddle with their siblings disappears (although their ability to discriminate may remain; Porter et al. 1989). If captive rearing interferes with the developmental process that produces species-typical kin-recognition abilities, then one solution is to livetrap field-reared animals and study their discrimination during brief tests in captivity, as Hare (1992; 1994) did to study littermate recognition in Columbian ground squirrels (*Spermophilus columbianus*). Another alternative is to conduct parallel studies in the laboratory and field, comparing the recognition abilities that are displayed in the two environments (Holmes and Sherman 1982; Gamboa et al. 1986; Blaustein and Waldman 1992).

Understanding kin-recognition mechanisms at the proximate level lends itself to experimental study because the two factors that often mediate recognition can be manipulated by cross-fostering. These two factors include *prior association,* direct interactions between individuals that result in learned familiarity with each other's phenotypes, and *genetic relatedness,* the factors besides prior association that correlate reliably with genes identical by descent, like phenotypic similarity among kin. Various cross-fostering designs have been used to study how prior association and genetic relatedness mediate kin recognition (Todrank and Heth 2001; Mateo and Holmes 2004). Typically, an infant is taken from its genetic mother at birth and transferred to a foster mother that is unrelated to the infant. One or more infants may be fostered and infants are often exchanged reciprocally between two litters. The effects of cross-fostering on recognition abilities can be complex (Todrank and Heth 2001) because when more than one member of a litter is transferred to a foster litter the transferee will be exposed to (1) a foster parent, (2) foster siblings, (3) a genetic sibling that was transferred with it, and (4) its own phenotype, all of which could influence subsequent recognition abilities (Mateo and Holmes 2004).

For example, there is considerable interest in whether an individual can use its own phenotype as a standard against which other phenotypes are compared to assess kinship (Hauber and Sherman 2001; Todrank and Heth 2003; Tsutsui 2004). If cross-fostering is used to test a self-matching hypothesis then only one member of a litter can be transferred to a foster litter, and none of the individuals in the foster litter, including the foster mother, can be related to the transferee whose self-matching ability is being studied. This transfer regimen ensures that any recognition of unfamiliar kin by the transferee is based on self-inspection rather than on comparison with the foster family's phenotypes (Mateo and Johnston 2000a). Moreover, cross-fostering does not address whether in utero experience influences the ontogeny of kin recognition (Hepper 1987; Robinson and Smotherman 1991), although embryo transfer can be used to address in utero effects on discrimination abilities (e.g., Isles et al. 2001).

Many kin-recognition studies on rodents have used either a paired-encounter design, in which the social interactions of two conspecifics are observed for a short period of time in an unfamiliar environment (e.g., Michener 1974; Grau 1982; Ferkin and Rutka 1990), or a single-subject design, in which a phenotypic cue from a conspecific (e.g., an odor) is presented to a subject and its response is assessed (e.g., time spent sniffing; Hepper 1987; Murdock and Randall 2001; Mateo 2002). A benefit of the paired-encounter design is that it simulates situations in nature when an actor encounters another conspecific and engages in recognition behavior. A weakness of the design is that it may not be possible to separate one individual's discrimination abilities from another's (Holmes 1986b). If, for example, A can recognize B as a sibling then A might rarely behave agonistically toward B. However, if B fails to recognize A as a sibling then B might routinely direct agonism toward A, which could elicit more agonism from A, thus hiding from the investigator A's ability to recognize B. Asymmetrical recognition abilities may be common in parent-offspring interactions at some points in development (Holmes 1990; Insley 2001), although the absence of observable behavioral discrimination on the part of young may be due to motor limitations rather than recognition abilities. Because the adaptive problems faced by parents and young differ, it would be valuable to determine if recognition is symmetric or if only one member of the parent-offspring dyad can recognize its relative (Gustin and McCracken 1987; Beecher 1991; Insley 2001), which may require use of single-subject designs.

The single-subject design is especially useful for examining the sensory basis of kin recognition. This is because cues from one modality can be presented in isolation from others (Todrank et al. 1998; Beecher et al. 1989; Parr and de Waal 1999). In rodents, olfactory cues seem to be the primary mediator of kin recognition (Halpin 1986; table 1 in Mateo 2003). Besides standard odor-preference tests (e.g., Ferkin and Rutka 1990; Clarke and Faulkes 1999), the habituation-discrimination technique has helped answer questions about odor-based kin recognition. The basic procedure entails recording a test animal's response (e.g., sniffing time) during repeated presentations of an odor—the habituation stimulus—followed by the presentation of a new odor, the test stimulus. Investigation time typically decreases across repeated presentations (habituation) and increases when a new odor is presented, if the actor can discriminate between the original and the new odor (Johnston 1993; Mateo 2002; Todrank and Heth 2003). The habituation technique has been valuable in studies of the production component of kin labels (e.g., Todrank et al. 1998) because it reveals whether test animals perceive odors from related individuals as being similar, which could facilitate recognition by phenotype matching (see the following). The technique also can demonstrate whether animals perceive odors of their distant kin (e.g., cousins) as different from nonkin (Todrank and Heth 2003), which could help explain why nepotistic treatment is (or is not) limited to certain classes of kin (Mateo 2002). When using the habituation technique, it is important to appreciate that a test animal's experience prior to the study itself may influence its response. For example, an individual that discriminates between odors of its full-siblings and paternal half-siblings might not be distinguishing differences in odor that correlate with kinship per se, but rather between familiar and unfamiliar odors, if the individual was reared with siblings and had never encountered paternal half-siblings.

Laboratory studies enable experimental control, but they usually lack the contextual features that may affect discrimination performance. For example, when juvenile (young-of-the-year) Belding's ground squirrels (S. beldingi, fig. 19.2) were tested in the laboratory, they appeared unable to discriminate between littermates and nonlittermates during (1) paired-encounter tests when agonistic interactions were assessed, (2) food-sharing tests, and (3) sleeping-partner preference tests (Holmes, unpublished data). However, when another set of juveniles was observed in large, outdoor enclosures, complete with natural vegetation and buried nestboxes, juveniles readily distinguished between littermates and nonlittermates (Holmes 1994).

In rodents, social experience during early development, often at specific times, is crucial to the ontogeny of kin recognition. For example, juvenile Columbian ground squirrels interact differentially with their littermates and nonlittermates during the first few weeks after they emerge aboveground from their natal nest (Waterman 1986, 1988).

Figure 19.2 Three juvenile Belding's ground squirrels that have come above ground for the first time from their underground natal nest. Juveniles can discriminate between their littermates and nonlittermates, based primarily on olfactory cues, and they can also recognize some classes of unfamiliar kin when they are first encountered. Photo by J. M. Mateo.

However, Hare (1992, 1994) suggested that this discrimination ability is transitory and that older juveniles fail to distinguish between littermates and nonlittermates, although the results of his behavioral tests with temporarily captive juveniles are difficult to interpret because of the large range in ages tested (e.g., 9–61 days after natal emergence). In many rodent species, individuals' social worlds expand as young leave their natal environment and begin to encounter distantly-related and unrelated conspecifics (Nunes, chap. 13 this volume). We are unaware of any longitudinal rodent kin-recognition studies that have assessed an individual's discrimination abilities at various ages, before and after different kinds of social experience. Such studies would be valuable because they would help address developmental questions like whether experience at a particular age (e.g., just prior to weaning or dispersal, when litters begin to dissolve) has a long-term impact on kin recognition.

Rodent Kin Recognition—Empirical Evidence

In this section, we describe and analyze what is known about rodent kin-recognition abilities at the proximate and ultimate levels of analysis. Because this literature is vast (see reviews in Blaustein et al. 1987; Dewsbury 1988; Halpin 1991; Mateo 2003, 2004), we concentrate on a few well-studied species.

Behavioral ecologists in North America have made detailed studies of social behavior in three genera of ground-dwelling squirrels (*Cynomys, Marmota,* and *Spermophilus*). These animals are diurnal, readily observable due to their size and preference for open habitats and because individuals can be uniquely marked for long-term identification

(Murie and Michener 1984). Field studies in the 1970s (e.g., Michener 1973a; Barash 1974c; Dunford 1977b; Sherman 1977) found that social interactions often varied with kinship in several ground-dwelling squirrels, and Michener's (1983a) review of sociality in this group emphasized that social organization was based on genetic relatedness between females, with mother-offspring groups as a common social unit (reviewed in Hare and Murie, chap. 29 this volume). Accordingly, various investigators (e.g., Sheppard and Yoshida 1971; Michener 1974; Davis 1982b) began to investigate kin-recognition abilities, especially in the genus *Spermophilus*, hereafter "ground squirrels" (reviewed by Schwagmeyer 1988a; Mateo 2003; Holmes 2004). We use "littermate" rather than "sibling" in our discussion of ground squirrel kin recognition because multiple paternity is common in *Spermophilus* (see references in Lacey and Wieszorek 2001; Solomon and Keane, chap. 4, this volume) so that litters routinely comprise full- and maternal half-siblings.

Kin recognition in Belding's ground squirrels

The kin-recognition abilities of Belding's ground squirrels (fig. 19.2) have received considerable empirical attention after field studies (Sherman 1977, 1980a, 1981a, 1981b, 1985) showed that females displayed nepotistic behavior in the context of antipredator warning calls (Sherman 1977, 1985) and cooperative defense of unweaned young by related females to prevent infanticide (Sherman 1980a, 1981a, 1981b). Cross-fostering procedures and a combination of laboratory and field experiments revealed that (1) yearlings discriminate between familiar littermates (young born in the same litter) and unfamiliar nonlittermates, (2) female yearlings discriminate between unfamiliar (reared apart) female littermates and unfamiliar female nonlittermates, and (3) mothers discriminate between their own familiar young and alien young borne and reared by other mothers (fig. 19.3). These results suggested that two recognition mechanisms, prior association and phenotype matching, are both involved in these instances of differential treatment (Holmes and Sherman 1982, 1983). We examine these mechanisms in detail because they also mediate kin recognition in several other species (see the following).

Prior association

In this recognition mechanism, an individual learns the phenotypic traits of conspecifics with which it interacts directly (e.g., nestmates), stores them in its memory, and later distinguishes between familiar and unfamiliar conspecifics. If early rearing environments comprise only kin (e.g., only full siblings share a nest) or if individuals have opportuni-

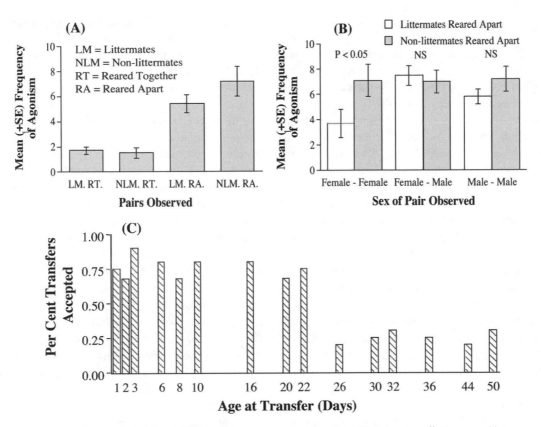

Figure 19.3 Laboratory (A and B) and field (C) data on littermate recognition (A and B) and mother-offspring recognition (C) in Belding's ground squirrels. Laboratory paired-encounter tests with cross-fostered animals revealed that pairs comprising familiar (reared together) young were significantly less agonistic than pairs comprising unfamiliar (reared apart) young (A). However, when unfamiliar pairs were categorized by sex-of-pair, pairs of unfamiliar littermate females were significantly less agonistic than pairs of unfamiliar nonkin females, which demonstrate that female *S. beldingi* can recognize their unfamiliar kin (B). Finally, free-living mothers readily retrieved into their natal burrow unrelated young placed at their burrow entrance, but only until their own young reached about 25 days of age, which coincides with when mothers' own young emerge aboveground from their natal burrow (C). From Holmes and Sherman (1982).

ties to interact only with kin before they encounter other conspecifics, then "familiarity" is a reliable proxy for relatedness (Bekoff 1981). In *Spermophilus*, young develop in underground nests, one litter and one mother per nest, which provides a developmental-ecological explanation (West et al. 2003) for the effects of prior association on littermate recognition in *S. beldingi* and *S. parryi* (Holmes and Sherman 1982), *S. richardsonii* (Sheppard and Yoshida 1971; Davis 1982b; Hare 1998b), *S. lateralis* (Holmes 1995; Mateo 2002) and *S. tridecemlineatus* (Holmes 1984b), and mother-offspring recognition in *S. richardsonii* (Michener 1971, 1974) and *S. beldingi* (Holmes 1984b). In various other species of rodents, both littermate recognition (e.g., Kareem and Barnard 1982; Sun and Müller-Schwarze 1997; Todrank et al. 1998; Mateo and Johnston 2000b) and mother-offspring recognition (e.g., Hepper 1987; Polan and Hofer 1998; Yamazaki et al. 2000) are mediated by the prior association mechanism.

We emphasize that the familiarity produced by prior association is a *descriptive* term that only indicates whether in-

dividuals have encountered each other previously, and thus has limited explanatory value. For example, the descriptor does not identify a mechanism like perceptual learning or associative learning, which might generate familiarity (Hepper 1986; Alexander 1990), it does not indicate when during development familiarity is established (Jackel and Trillmich 2003), nor does it suggest whether familiarity, once established, is relatively permanent (Kendrick et al. 2001; Insley 2000) or requires repeated exposure to former associates to be maintained (Porter and Wyrick 1979; Paz y Miño C. and Tang-Martínez 1999a). Following prolonged social isolation, for instance, *S. beldingi* retain the ability to recognize previously familiar kin, but appear unable to recognize previously familiar nonkin (Mateo and Johnston 2000b). Demonstrating that learned familiarity mediates differential treatment is a necessary first step in understanding how kin are recognized by prior association, but, as we note earlier in this paragraph, many questions must be answered to achieve a full understanding of how this mechanism operates (Porter et al. 1984; Holmes 2001).

Phenotype matching

In this recognition mechanism, an individual learns something from its own phenotype or some of the features of the kin with which it is reared, and acquires a kin template[1] that it later uses as a prototype against which other phenotypes are compared. Kin templates are often acquired during early development in a particular context, such as a natal nest in which only siblings reside. Depending on the match or overlap between the phenotype being assessed and the individual's template (Getz, 1981; Lacy and Sherman 1983), a conspecific may be recognized as a relative. Phenotype matching may explain recognition of *unfamiliar* kin like paternal half-siblings, which has been documented in ground squirrels (Holmes 1986a; Mateo 2002), laboratory mice (Kareem and Barnard 1982), and some primates (Wu et al. 1980; Widdig et al. 2001; Smith et al. 2003), father-offspring recognition when males do not associate exclusively with their young during early development (Buchan et al. 2003) and recognition between *familiar* young that develop together in litters comprising unequally related young due to multiple mating by females or communal nesting (Hauber and Sherman 2001).

Phenotype matching requires a detectable correlation between genotypic and phenotypic similarity, and in many rodents this correlation is reflected in the olfactory labels shared by close relatives (Porter et al. 1983; Hepper 1987; Sun and Müller-Schwarze 1997, 1998a). If the odors of close kin are more similar than those of distant kin (Todrank and Heth 2003) then olfactory investigation time should vary with kinship. For instance, when juvenile *S. beldingi* (fig. 19.2) were presented dorsal-gland and oral-gland odors from their more and more distantly-related kin, juveniles increased their investigation time (fig. 19.4), presumably because the odors of more distant kin were perceived as more unfamiliar (not matching the templates as well) than those of their more closely related relatives (Mateo 2002). (Across taxonomic groups and modalities, animals generally attend to novel stimuli longer or more strongly than to familiar stimuli; Johnston 1981; Halpin 1986; Stoddard 1996). The genotypes of close kin become more similar if

1. The meaning of "template" is not well developed in the kin-recognition literature of behavioral ecology (although see Sherman et al. 1997), which may help explain why investigators' views differ about the relevance of templates to the prior association and phenotype-matching mechanisms. In our treatment of phenotype matching, we use *template* like cognitive psychologists use *prototype* to refer to a set of features that are most commonly present in all members that belong to a particular category (Smith and Medin 1981; Estes 1994; Hampton 1995). Prototypes arise when organisms experience individual instances within a category and abstract the common attributes of those instances, storing them as a generic representation of the common attributes of the category as a whole. In contrast, exemplars arise when organisms experience individual instances within a category and store those individual instances rather than abstracting the commonalities among them. In our view, templates, or more properly, prototypes, are central to the phenotype-matching process, whereas exemplars are central to the prior association mechanism (and individual recognition).

inbreeding occurs, which, in group-living species, may make it difficult for individuals to distinguish close kin from other group members. For example, inbreeding is common (although not exclusive [Braude 2000]) in colonies of naked mole-rats (*Heterocephalus glaber*) and the mean coefficient of relationship is > 0.8 (Reeve et al. 1990). This may contribute to why female naked mole-rats rely on familiarity to recognize kin and do not appear to assess genetic similarity by phenotype matching (Clarke and Faulkes 1999, based on mate-choice preference tests (see also Ciszek 2000).

Both the prior association and phenotype-matching mechanisms require that kin labels, the production component of the recognition process, differ among classes of kin. In ground squirrels, odors from oral and dorsal glands are involved in social discrimination (Kivett et al. 1976; Harris and Murie 1982), and in *S. beldingi* oral and dorsal secretions differ across at least four classes of kin, which makes it possible to discriminate these kin labels in controlled tests (Mateo 2003). Just as selection has molded the perception component of the recognition process, selection is also expected to have molded the production component, so that whether ground squirrels and other rodents can recognize a particular category of kin may depend on discernable variation in the production component (Beecher 1991; Reeve 1989; Sherman et al. 1997). Thus like the perception component (Mateo 2004), selection acting on the production component may limit nepotistic behavior.

For example, *S. beldingi* females do not treat distant kin, like aunts or cousins, nepotistically (Sherman 1980a, 1981a). One possible functional explanation is that Hamilton's (1964) cost/benefit criteria are not met when members of these classes interact (fig. 19.1). Alternatively, nepotism may not occur between distant kin because actors cannot recognize them as relatives. However, we know that distant female kin in *S. beldingi* do produce olfactory kin labels, and that other females discriminate among these labels in controlled tests (fig. 19.4), which suggests that a failure to satisfy Hamilton's criteria for nepotism rather than limitations in the recognition component explains why distant kin are not recipients of nepotism (Mateo 2002). Thus a system for nepotism and one for kin recognition can evolve separately. That is, selection can act independently on the recognition component(s) and the action component (fig. 19.1), which calls into question the commonly held belief that the absence of nepotism must be due to the absence of a kin-recognition mechanism.

Kin recognition in other ground-dwelling squirrels

Based on some of the early work on ground squirrel kin recognition, investigators began to seek parallels between interspecific differences in social organization and recognition abilities (e.g., Michener 1983a; Holmes 1984b; Schwag-

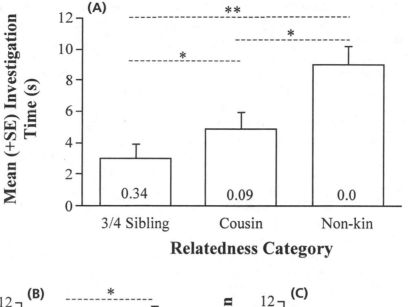

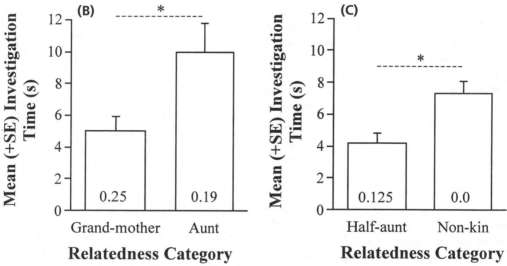

Figure 19.4 The time spent by *Spermophilus beldingi* subjects investigating the odors of subjects' unfamiliar (never encountered) relatives during olfactory preference tasks. (A) Investigation of oral-gland odors of unfamiliar kin, collected from subjects' 3/4th sibling (offspring of two sisters mated with the same male), cousins (offspring of a subject's mother's brother) and nonkin. (B) Investigation of oral-gland odors collected from subjects' grandmothers and aunts. (C) Investigation of dorsal-gland odors collected from subjects' half aunt (mother's half sister) and nonkin. Numbers inside bars are estimated coefficients of relatedness (due to mixed paternity within litters) between subjects and odor donors. Horizontal bars and asterisks represent significant differences in investigation times (* $p < 0.05$, ** $p < 0.01$). Increased investigation time as relatedness of the odor donor to the subject decreases indicates that odors of distant kin are perceived as less familiar to subjects than odors of close kin. Significant differences in investigation of odors from various kin classes indicate discrimination of those odors, and thus that the subject can recognize the difference between, for example, its distant kin and nonkin (figures 19.4a and 19.4c).

meyer 1988a). For example, *Spermophilus* social organization forms a continuum ranging from relatively asocial species (e.g., *S. tridecemlineatus*) to relatively social species (e.g., *S. richardsonii, S. beldingi*, and *S. parryi*), and it was thought that this variation could be used to predict interspecific differences in recognition abilities and mechanisms (table 2 in Sherman and Holmes 1985). More recent work, however, calls into question some of the expected links between sociality and kin-recognition abilities (Hare and Murie 1996), as seen, for example, in a comparison between

recognition abilities in a relatively social species, *S. beldingi*, with a relatively asocial species, *S. lateralis*. Despite these differences in sociality, juveniles in both species display similar social preferences for littermates over nonlittermates and are able to recognize unfamiliar kin when they are first encountered (Holmes 1995). Considering the production component, both species produce odors that can be used to discriminate among classes of kin like siblings, paternal half-siblings, and grandmothers, but only *S. beldingi* behaves nepotistically to kin other than offspring. Kin labels in *S. lateralis* may have functional value in the context of mate choice, but they appear to have no value in the context of nepotism (Mateo 2002). Current evidence for ground squirrels and other rodents as well provides little support for the belief that simple differences in social organization can be used to predict fine-grained differences in recognition abilities (Mateo 2003). This contrasts with the situation in some birds, in which colonial species have a more complex and elaborate recognition system than noncolonial species (e.g., Beecher 1991).

The recognition behavior of some ground-dwelling squirrels appears paradoxical in light of evolutionary theory (Hamilton 1964) or, as Hoogland (1995) writes about prairie dogs, "The nondiscriminatory aspects of kin recognition among prairie dogs seems puzzling" (p. 215). Both black-tailed prairie dogs (*C. ludovicanus*; Hoogland 1986, 1995) and Gunnison's prairie dogs (*C. gunnisoni*; Hoogland 1996b) distinguish between kin and nonkin, as revealed by differences in the frequency of amicable interactions (*C. ludovicanus*) and the likelihood of giving alarm calls (*C. gunnisoni*). However, field data show that prairie dogs fail to treat close kin and more distant kin differently despite the fitness benefits that would seem to accrue if such discrimination were made. For example, female *C. ludovicanus* treat distant kin, such as female full ($r = 1/8$) and half ($r = 1/16$) first cousins, just as amicably as they treat their daughters ($r = 1/2$) and full ($r = 1/2$) sisters (Hoogland 1995). To evaluate kin discrimination *independently* of kin favoritism, it would be useful to use a neutral measure like differences in olfactory investigation to determine whether prairie dogs can discriminate between close and more distant kin, even though nepotism is not directed differentially toward these kin classes in nature. Prairie dogs provide a reminder that close kin may be intense competitors (Hoogland 1985) and thus inappropriate targets for nepotism, because they do not satisfy Hamilton's rule. Accordingly, it is risky to assume that the absence of nepotism implies an inability to recognize kin (fig. 19.1; Beecher 1991), even though there is often a positive association between kin recognition and nepotism (e.g., Griffin and West 2003).

The kin-recognition abilities of Columbian ground squirrels may also be unusual, at least according to some investigators (Hare 1992; Hare and Murie 1996). *S. columbianus* are the most social members of the genus *Spermophilus* (Michener 1983a) and display kin-differential behavior in nature (Waterman 1988; King 1989a, 1989b; Wiggett and Boag 1992). Juveniles, for example, encounter both their littermates and unrelated juveniles during the first 3 weeks after emerging from their natal burrow, and play more often with their littermates than with nonlittermates (Waterman 1986), which indirectly indicates that juveniles can discriminate between these two kin classes. Hare (1992),

however, disputed the claim that juvenile *S. columbianus* could recognize their littermates and, following Grafen (1990), argued that what appears to be kin (littermate) recognition may be an artifact of colony member (familiar) versus noncolony member (unfamiliar) discrimination. Hare's recognition studies included juveniles of unreported ages (Hare 1992) or juveniles that had been aboveground at least 4 weeks before recognition tests were done (Hare and Murie 1996). We believe that to reveal the kin-recognition abilities of juvenile *S. columbianus*, including how these might change ontogenetically, it would be valuable to use a neutral bioassay (e.g., olfactory investigation) and conduct a longitudinal study, beginning soon after natal emergence and continuing throughout the juvenile summer, to better assess their skills in social recognition. That juveniles play preferentially with littermates (Waterman 1986) suggests that kin can be recognized even if nepotism is not common in adult *S. columbianus* (Hare and Murie, chap. 29 this volume).

Sex differences in kin recognition

Ground squirrel social organization is based on female kin groups; females are nepotists, whereas males are not (Michener 1983a; Murie and Michener 1984). Thus it seemed that kin-recognition abilities might differ between the sexes, which was initially confirmed in *S. beldingi* (Holmes and Sherman 1982). In paired-encounter tests, females discriminate between unfamiliar (reared apart from) littermates and unfamiliar nonlittermates, whereas males do not (fig. 19.3b), and females, but not males, discriminate between unfamiliar paternal half-siblings and unfamiliar nonkin (Holmes 1986a). In subsequent studies, however, using odor-discrimination tests rather than paired-encounter tests that monitored agonistic interactions, both females and males produced kin labels and both sexes displayed similar discrimination abilities in response to these labels (Mateo and Johnston 2000a, 2000b; Mateo 2002, 2003). We cannot explain the apparent disjunction between sex differences in nepotism and the absence of such differences in kin discrimination. However, we encourage investigators to analyze their recognition data for a sex effect whenever nepotism is sex limited, and to recall that kin recognition might serve one function in females (e.g., mediating nepotism) and another in males (e.g., mediating mate choice).

Pressing Questions and Issues

We now examine some intriguing issues surrounding rodent kin recognition, and explain why pursuing them will pay dividends for understanding recognition at both the proximate and ultimate levels of analysis.

The production component—Odors as cues to relatedness

Although it is widely recognized that odors can mediate kin recognition in Rodentia (Mateo 2003), little is known about *how* odors convey the information that allows animals to recognize their relatives. Odors serve multiple functions in rodents, such as indicating an animal's individual identity, sex, relatedness, genetic quality, health, reproductive status, or even location (Brown and MacDonald 1985; Halpin 1986; Penn and Potts 1999). Assessment of relatedness via odors may be facilitated by the major histocompatibility complex (MHC), a large group of genes involved in vertebrate immune function, which also influence odor production. These highly variable genes, some having over 100 alleles per locus, regulate recognition of self versus non-self tissues and pathogens. Because of the large number of alleles involved, the likelihood of two random conspecifics having the same MHC haplotypes is very low, but since relatives, by definition, share many genes in common, the MHC can serve as an accurate correlate of relatedness, which could be important to nepotism. In addition, the extreme variability of the MHC is thought to facilitate an individual's ability to resist a broad range of rapidly evolving parasites and pathogens, and therefore could be used in mate-choice contexts to reduce inbreeding and promote heterozygosity in offspring if adults prefer MHC-dissimilar mates (Penn 2002). Besides coding for glycoproteins, the MHC also influences odor production, likely through an interaction between MHC by-products and bacteria on gland surfaces (leading to unique secretions) or in gastrointestinal tracts (creating unique urine and fecal odors; Schellinck et al. 1997). Thus families share similar MHCs and similar odors (Boyse et al. 1991; Brown and Eklund 1994), which means that MHC-derived odors could serve as kin labels, which may be learned from family members during early development (Yamazaki et al. 1988, 2000; Penn and Potts 1998b).

Laboratory rodents (rats and mice) can discriminate between the odors of highly inbred animals that are genetically identical except for *one* MHC allele (Beauchamp et al. 1985; Brown et al. 1990). In mate-choice trials, they often prefer to mate with MHC-dissimilar individuals, although preference varies depending on the sex of the choosing individual and the methods used in the trials (Manning et al. 1992a). Thus besides conferring immune protection, the MHC may facilitate mate choice (Potts et al. 1991; Carroll and Potts, chap. 5, this volume). It is unknown, though, whether MHC-derived odors mediate recognition, kin preferences, mate choice, and reproductive success in animals that are *not* inbred. We note, however, that Manning et al. (1992b) studied the nesting habits of female house mice resulting from crosses between wild-caught mice and inbred strains, and found that females tended to nest communally with MHC-similar females. Conesting reduced infanticide compared with single-nesting females, and resulted in greater direct- and indirect-fitness benefits for conesting females (Manning et al. 1995; Dobson et al. 2000). Further, MHC-based odors of inbred rodents can be influenced by diet and health status (Schellinck et al. 1997; Ehman and Scott 2001), yet we do not know if free-living animals attend to genetic differences among odors that also reflect foraging habits and infectious conditions. Field data on MHC-based odors, including how they develop and their effects on social interactions, are needed to test functional hypotheses about the role of these odors in nepotism and mate choice.

Much of the research on the production component in rodent kin recognition has emphasized MHC-based odors, but recent work has shown that another urinary compound warrants attention because it influences chemical identity signals and may be involved in social recognition. Major urinary proteins (MUPs) are large, abundant, nonvolatile compounds that bind and carry the more volatile odor molecules that are traditionally thought to mediate recognition. The large molecules, which are costly to synthesize, extend the longevity of an odor (Hurst et al. 1998), lengthening its functional life, which may be especially important for animals that cannot demarcate and defend all areas of their territory simultaneously. MUPs may also be involved in signaling individual identity, as a male house mouse excretes urine with 10–12 different genetically encoded MUPs, which can be used by other mice to discriminate among individuals. MUPs appear to express distinct cues themselves, as the nonvolatile compounds are sufficient for individual recognition (Nevison et al. 2003; Thom and Hurst 2004). Thus in mice, and perhaps other rodents, MUPs and MHC by-products contribute to the kin-distinct urine odors (Hurst et al. 2001; Thom and Hurst 2004).

Production cues from other modalities

Most examples of rodent kin recognition implicate conspecific discrimination via olfactory cues (Mateo 2003), in part because odors vary with genetic relatedness and thus offer evolutionarily reliable cues to assess kinship (see also Todrank and Heth 2003). In theory, relatedness could be assessed through other modalities, such as vibrational signaling (Randall 2001), visual-behavioral cues (Michener 1973a; Terrazas et al. 2002), or vocalizations (Hare 1998a), but if such cues were to mediate recognition by phenotype matching then these production cues must covary with relatedness. Nonolfactory kin labels might be important if odor differences were unreliable or inaccurate indicators of relatedness, such as when a nest of social insects contains multiple matrilines but all workers acquire similar odors upon eclosion (see Keller 1997). Nonolfactory cues may also be used to detect kinship when animals do not produce or perceive olfactory labels that are sufficiently variable to support discrimination (e.g., visual similarities in zebra finches; Burley and Bartels 1990). Future work should address the potential for other nonolfactory modalities to mediate recognition, such as ultrasonic vocalizations used by bats in mother-offspring recognition (e.g., Balcombe 1990; Scherrer and Wilkinson 1993). In addition, most kin-recognition research has focused on direct mechanisms of recognition, yet there may be instances when indirect spatial cues mediate differential treatment of kin. Mother-offspring recognition, for example, is mediated by spatial cues in many burrow-nesting rodents during early stages of lactation when pups are confined to their natal burrow (Holmes 1990), but as young become mobile spatial cues will no longer correlate reliably with relatedness. It would be valuable to investigate how mothers make the transition from offspring recognition based on indirect spatial cues to direct kin labels, including the question of when during early ontogeny young begin to produce their distinctive olfactory kin labels and when mothers learn these labels and begin to respond differentially to them.

Kin recognition and cooperative breeding

Cooperative breeding, which has been variously defined (Solomon and French 1997), refers to a social structure in which individuals help care for offspring produced by other conspecifics, and is often explained at the ultimate level by ecological constraints and Hamilton's inclusive-fitness hypothesis (Emlen 1991; Jennions and Macdonald 1994). Cooperative breeding raises intriguing questions about kin recognition (Lewis and Pusey 1997; Komdeur and Hatchwell 1999; Griffin and West 2003), and we believe that this is especially true for mammals when cooperative breeding is characterized by communal nesting, which occurs when two or more females raise their young together in a common nest or burrow. For example, prior association mediates mother-offspring recognition in many species (Holmes 1990), yet in communal nesting if infants' kin labels do not develop until after litters mix, or if mothers do not learn the kin labels of their own offspring before litters mix, then mother-offspring recognition could be jeopardized, which might lead to misdirected maternal care (König 1989b). Communal nesting occurs in several species of rodents (reviewed by Solomon and Getz 1997; Hayes 2000), and we suggest that it offers one of the most fruitful avenues for future research on rodent kin recognition. For example, communal nesting provides the ecological and social circumstances under which self-referent phenotype match-

ing rather than prior association might mediate mother-offspring recognition (Holmes and Sherman 1982; Hauber and Sherman 2001), and any study that revealed mother-offspring discrimination in a communal nester would help shed light on the functional significance of communal nesting (see the following), which has remained elusive (Packer et al. 1992; Lewis and Pusey 1997; Hayes 2000).

Perhaps the greatest paradox of mammalian communal nesting is that communal nursing, the sharing of milk with young produced by another female, often is associated with it (König 1997; Hayes 2000). Given the physiological costs of lactation (König et al. 1988), behavioral ecologists expect that under most but not all circumstances mothers should avoid suckling offspring other than their own, which we refer to as "alien offspring." Communal nursing offers a valuable empirical opportunity for proximate and ultimate explanations to intersect and to enhance our understanding of mammalian communal breeding (Lewis and Pusey 1997; Griffin and West 2003). For example, Packer et al. (1992) and Roulin (2002) summarized several functional hypotheses to explain why mothers sometimes nurse aliens, including kin selection and reciprocity, as well as a nonfunctional, misdirected parental-care hypothesis. The last explanation is invoked frequently, particularly when conesting females are known to be unrelated, with the conclusion that the costs of communal nursing were not large or predictable enough to outweigh the benefits of enhanced offspring survival that sometimes occur in communal nesting (e.g., Manning et al. 1995), or that the benefits of preventing milk theft did not outweigh the costs of vigilance needed to thwart it (Packer et al. 1992). Black-tailed prairie dogs nurse their own infants exclusively until young first emerge, but when juveniles from different litters begin to share burrows after natal emergence mothers nurse communally. Investigators suggest that mothers may not be able to discriminate between their own and alien juveniles when both occupy a common burrow, or that the cost of trying to discriminate does not outweigh the cost of sharing milk (Hoogland et al. 1989). We suggest that results from mother-offspring discrimination tests would immediately help narrow the field of possible functional explanations for communal nursing in a particular species. For example, if it were shown that mothers *were* able to recognize their young, or to discriminate among infants of varying relatedness, then one could rule out the misdirected parental explanation and begin to explore functional hypotheses for communal nursing. In short, in the absence of detailed knowledge about the costs and benefits of communal nursing and pup retrieval, it would be wrong to conclude that mothers cannot recognize their own offspring simply because they retrieve and nurse alien young.

Evidence that communal nursing occurs is often used to

infer something about kin recognition (e.g., mothers cannot recognize their own offspring), but we caution that an appropriate bioassay must be used to distinguish between maternal care, which may or may not be directed differentially toward own and alien young, and maternal discrimination abilities, which should be assessed with a neutral measure such as differential olfactory investigation. For example, in laboratory tests communally-nesting female house mice (*Mus musculus domesticus*) retrieve age-matched own and alien pups with equal frequencies (Manning et al. 1995), which suggests that mothers may be unable to distinguish between their own and alien young. Similarly, captive female *S. beldingi* retrieve alien pups into their own nests as often as their own young, which suggests an inability to discriminate. However, after bringing pups back into the nests, mothers sniff and handle alien pups longer than their own pups, indicating that mothers can discriminate between the two categories of pups and that mothers are motivated to tend to both (Holmes 1990). This means that to test maternal discrimination abilities more directly, an operant procedure or one like olfactory habituation (see the preceding) should be used to reveal abilities that may be masked by maternal motivation.

What we do know about mother-offspring recognition abilities in communal nesting species is based on indirect evidence rather than on studies designed explicitly to test for recognition (Lewis and Pusey 1997). For example, we know from field work that in maras (*Dolichotis patagonum*) multiple females often share warrens, and that after young come aboveground mothers reject nursing attempts by aliens, but we know nothing about preemergent mother-offspring recognition (Taber and MacDonald 1992a). To draw ecologically valid inferences from laboratory studies of communal breeding, females should be housed together (e.g., in pairs) so that "nestmates" include young from multiple litters. In a mother-offspring recognition study of communally-breeding degus (*Octodon degus*), for example, females lived together in pairs, and laboratory tests showed that mothers could discriminate between the odors of their own young and their co-nestmate's young, despite mothers being housed together since both of their litters were born (Jesseau 2004). The paucity of empirical work on kin recognition in communal species stresses the need for research on both the proximate and ultimate levels of analysis, to explain maternal care of alien young and the role of mother-offspring recognition in this unusual form of parental care. Earlier, we emphasized the distinction between differential treatment and preferential treatment of kin (fig. 19.1), and we suggest that it is especially important to make this distinction for communal breeders because, depending on the cost and benefits experienced by females, mothers may or may not nurse their own and alien young selectively, which may or may not indirectly indicate whether mother-offspring discrimination is possible.

Does the absence of preferential treatment indicate a lack of recognition?

Whether the proximate or ultimate aspects of kin recognition are addressed, the assay chosen to assess discrimination abilities is crucial. We suggest that there are two general categories of kin-recognition studies based on the types of assays chosen. A *direct-inference study* is one in which a (presumably) fitness-neutral assay like olfactory investigation or time-in-proximity (Smale et al. 1990; Todrank et al. 1998; Mateo 2002) is used to infer that discrimination occurs. An *indirect-inference study* is one in which discrimination abilities are inferred from behavior that is either nepotistic, behavior that reflects mate choice, or behavior that is likely to have clear phenotypic costs (e.g., fighting or other forms of agonism; Barnard and Fitzsimons 1989; Keane 1990b; Sera and Gaines 1994). Both kinds of studies can reveal kin-discrimination abilities. However, when differential treatment is not detected, the conclusion that discrimination is not possible is better supported (but not verified) by results from a direct-inference study than an indirect-inference study. This is because in an indirect-inference study the phenotypic costs that actors might incur or the benefits they might gain may be such that Hamilton's rule (1964) is not satisfied, so actors should not adjust their behavior based on differences in kinship (Beecher 1991).

For instance, despite the prevalence of studies demonstrating phenotype matching in several rodents (Mateo 2003), most evidence for kin recognition in voles indicates that prior association mediates sibling recognition in *Microtus* because only kin that are reared together are treated as relatives (Boyd and Blaustein 1985; Ferkin and Rutka 1990; Sera and Gaines 1994; Berger et al. 1997; Paz y Miño C. and Tang-Martínez 1999c). The absence of differential treatment of unfamiliar relatives by voles during paired-encounter tests has been interpreted by some to demonstrate an inability to recognize unfamiliar kin (e.g., Gavish et al. 1984; Paz y Miño C. and Tang-Martínez 1999b, 1999c). This may be a premature conclusion, as these studies used indirect-inference designs. The key questions are: do voles produce cues that vary with relatedness (the production component), and if so, are they able to use this information to identify kin (the perception component)? Direct-inference studies on voles' discrimination abilities would be valuable because they could reveal whether kin-recognition abilities exist, despite not being expressed in all contexts. When bank voles (*Clethrionomys glareolus*), for instance, were presented two unfamiliar, anesthetized adult males in olfactory preference tests, females spent more time investigating their unfamiliar genetic father than an unfamiliar, unrelated adult male, demonstrating daughter-father recognition despite the absence of any prior association between daughters and their fathers (Kruczek and Golas 2003). Before concluding that voles cannot recognize their unfamiliar relatives, recall that the lack of differential treatment does not necessarily indicate the absence of recognition abilities (Gamboa et al. 1991), and we suggest that recognition abilities are more likely to be detected with direct- rather than indirect-inference studies. Until a number of direct-inference studies have been conducted on vole kin recognition, it will be unclear whether the apparent lack of kin-biased behaviors via phenotype matching in *Microtus* is due to an inability to discriminate among kin or because Hamilton's (1964) rule is not satisfied, resulting in no preferential treatment of kin that, under some circumstances, can be recognized. Indeed, the results of some recent cross-fostering studies suggest the value of a direct inference study and the potential for phenotype matching to facilitate recognition of unfamiliar kin. For example, mandarin voles (*Microtus mandarinus*) display social preferences for their unfamiliar siblings over unfamiliar nonsiblings that approach statistical significance (Fadao et al. 2000), and demonstrate mate preferences for unfamiliar nonsiblings over unfamiliar siblings (Fadao et al. 2002), a mate preference that also seems to exist in Brandt's voles (*M. brandtii;* Yu et al. 2004).

Summary

Behavioral ecologists have been very active in pursuing theoretical and empirical work on kin recognition, largely in response to Hamilton's (1964) development of inclusive fitness theory, and various species of rodents have figured prominently in the growth of this body of knowledge. In this chapter, we have addressed some of the conceptual and methodological issues surrounding kin recognition, reviewed part of the model-system work that has been done on ground squirrel (*Spermophilus*) kin recognition, and identified important questions for future research. We believe that there are good reasons for separating differential treatment of kin, which can reveal discrimination abilities, from preferential treatment of kin, which may or may not follow from kin recognition, depending on the costs and benefits of kin favoritism and the recognition context. This distinction is important because a lack of differential treatment of kin (e.g., nepotism, agonism, avoidance of inbreeding) does not necessarily indicate a lack of a recognition mechanism. Selection will operate separately on the proximate bases of recognition (e.g., recognition cues, perceptual abilities) as well as the ultimate functions of recognition (e.g., nepotism, mate choice). We also suggest that in empirical studies of discrimination abilities, fitness-neutral bioassays like differences in olfactory investigation are more appropriate for studying kin recognition than are bioassays that have clear fitness consequences like food sharing, cooperative care, or differences in agonistic behavior (e.g., kin favoritism).

Early work on ground-squirrel kin recognition revealed that two proximate mechanisms, prior association and phenotype matching, explained many instances of kin recognition, and that these mechanisms also operate in many other taxa, including several species of rodents. Prior association mediates recognition of previously encountered kin, such as siblings that develop together in a nest or burrow isolated from other conspecifics. In contrast, phenotypic matching mediates recognition of never-before-encountered kin, such as paternal half-siblings meeting for the first time, or unequally related individuals that shared a common early rearing environment. A species' ecological niche and social system will influence both the evolution of these recognition mechanisms as well as their use across various social contexts.

In future studies, we encourage investigators to examine kin recognition in cooperative and communal breeders, because when multiple individuals share a common rearing environment, intriguing questions arise as to whether and how different categories of kin might be recognized by various individuals.

Social Behavior

Chapter 20 Parental Care

Betty McGuire and William E. Bemis

W E DESCRIBE WAYS in which male and female rodents invest in their young and consider factors that might influence the level of care that parents provide. Some of these factors pertain to the young (e.g., degree of development at birth, gender, number of offspring), others to the parents (e.g., experience, concurrent pregnancy, other mating opportunities), and still others concern aspects of the physical environment in which the parental care is displayed (e.g., from small cages to seminatural environments in the laboratory to field conditions). We next describe the impact of parental care on the survival and growth of offspring, and conclude the chapter with a preliminary analysis of the evolution of parental care in voles (*Microtus* and closely related species).

Throughout the chapter we focus on studies evaluating care shown by parents toward their *own* young from birth through weaning and beyond. By their very nature, such studies require monitoring parent-offspring interactions for several weeks. We largely omit studies in which an individual's parental responsiveness is evaluated by short-term tests of exposure to pups. Individuals tested in such experiments often are not parents or are tested with unfamiliar young. While our goal was to make a broad phylogenetic survey of parental care in rodents, such care is best described for small species of Sciurognathi, particularly for species of Muridae such as rats, mice, voles, gerbils, and hamsters. Comparatively little information is available for species within Hystricognathi. Finally, because most studies on rodent parental care have been conducted in laboratory environments, such studies constitute our major data source. We include relevant field observations whenever possible.

Forms of Parental Behavior

Parental behaviors are characterized as either *direct* or *indirect* (Kleiman and Malcolm 1981). Direct parental care includes behaviors that have an immediate physical impact on offspring and their survival; in rodents, such behaviors include nursing (and feeding), grooming, transporting (most often retrieving), and huddling with young. With the obvious exception of nursing, males can exhibit all forms of direct parental care. Males of some species show levels of direct parental behavior comparable to those of females, while males of other species show little or no direct parental behavior (see reviews by Elwood 1983; Dewsbury 1985; Brown 1993). Indirect parental care includes behaviors that may be performed by parents while away from the young; these behaviors do not involve direct physical contact with offspring but still affect offspring survival, although perhaps not immediately. In rodents, indirect forms of parental care include acquiring and defending critical resources, building and maintaining nests and burrows, caring for pregnant or lactating females, and defending offspring against conspecifics or predators. Working definitions for typical direct and indirect parental behaviors based on comparative studies of voles under seminatural conditions are shown in table 20.1 (see McGuire and Novak 1984, 1986). While

Table 20.1 Categorization of parental behaviors displayed by voles (*Microtus* spp.) under seminatural conditions (modified from McGuire and Novak 1984, 1986)

Behavioral category	Definition
Direct parental behavior	
Nursing	Contact with at least one pup involving nipple attachment
Huddling	Contact with at least one pup with or without nipple attachment
Grooming pup	Licking pup
Retrieving pup	Grasping in mouth pup that has left the nest and carrying it back to the nest
Indirect parental behavior	
Nest building	Gathering, shredding, and arranging nest material
Runway building	Clearing, building, and maintaining runways
Food caching	Carrying food to a new location where it is stored for later use
Spatial location related to parenting	
In natal nest	In nest that contains young; sometimes used as an overall measure of direct care

not a specific category of parental behavior, "time in nest" often provides an estimate of overall direct parental care (table 20.1).

Direct care

Nursing
Mothers of altricial young are the sole source of early nutrition for their offspring (Alberts and Gubernick 1983), but mothers of precocial young are not. Precocial young typically supplement their diet of milk with solid food within a few days of birth, although they continue to consume milk for many weeks (Kleiman 1974; Gosling 1980). Nursing postures adopted by female rodents vary from crouching over pups to sitting or lying next to them (Kleiman 1974; Drewett 1983). Whatever form nursing takes, lactation is energetically costly to female mammals (Hanwell and Peaker 1977; Stapp et al. 1991); lactation also presents challenges to water balance, although female rodents recover some of the water and electrolytes lost in milk by consuming the urine of their pups during anogenital grooming (Alberts and Gubernick 1983). In addition to nutrients, milk contains antibodies that young rodents cannot produce on their own until a few weeks after birth (Brambell 1970).

Feeding
When young rodents can eat solid food, mothers of altricial species may bring food to the nest (e.g., woodchucks, *Marmota monax*; Barash 1974b) while mothers of precocial spe-

cies may allow young to take food from their mouths (e.g., green acouchi, *Myoprocta pratti*; Kleiman 1974). Males, too, feed juveniles: muskrat (*Ondatra zibethicus*) males provision juveniles at the home burrow (Marinelli and Messier 1995) and white-footed mice (*Peromyscus leucopus*) males accompany weaned young on foraging trips (Schug et al. 1992).

Grooming
Beginning at birth, pups are groomed extensively by mothers and sometimes by fathers (e.g., spiny mouse, *Acomys cahirinus*; Dieterlen 1962; Djungarian hamster, *Phodopus campbelli*, and Siberian hamster, *P. sungorus*; Jones and Wynne-Edwards 2000; prairie vole, *Microtus ochrogaster*; McGuire et al. 2003). Parental grooming of the anogenital region stimulates urination and defecation in young pups until they are about 2 weeks old (Capek and Jelinek 1956; Rosenblatt and Lehrman 1963). In many species, grooming of offspring continues well beyond when it is physiologically necessary for the young; it is likely that this grooming functions in the maintenance of parent-offspring bonds (Kleiman 1974; Libhaber and Eilam 2004).

Retrieving, huddling, socialization, and shoving
Other common forms of direct parental care include retrieving offspring to the nest (or transporting them to another location) and huddling with young, a behavior that provides thermoregulatory benefits. Some direct parental interactions with offspring have been described as play or socialization (e.g., green acouchi, Kleiman 1974; hoary marmot, *Marmota caligata*; Holmes 1984a). Finally, naked mole-rat (*Heterocephalus glaber*) parents shove pups around the nest; this behavior encourages pups to flee from danger and to avoid dangerous situations in the future (Stankowich and Sherman 2002).

Indirect care

Constructing and defending burrows and nests
Burrow or nest construction and maintenance by male and female parents have been described in the field for muskrats (Marinelli and Messier 1995) and for many species in the laboratory (e.g., plateau mouse, *Peromyscus melanophrys*; Ferkin 1987; volcano mouse, *Neotomodon alstoni*; Luis et al. 2000). Territorial defense against conspecifics, especially around the nest, is common for most if not all rodents, and helps to defend food resources or to protect young from infanticidal conspecifics (Sherman 1981a; Wolff 1993b; Hoogland 1995; Wolff and Peterson 1998).

Small size limits most species from defending young against predators, although male and female prairie voles reacted aggressively to shrews in the vicinity of the natal nest and effectively prevented predation on their pups in seminatural environments (Getz et al. 1992). Parental defense against small predators also has been reported for free-living black-tailed prairie dogs (*Cynomys ludovicianus*, Hoogland 1995) and naked mole-rats (Lacey and Sherman 1991). Several species warn their young with antipredator calls (e.g., Belding's ground squirrel, *Spermophilus beldingi*; Sherman 1981b; hoary marmot; Holmes 1984a; black-tailed prairie dog; Hoogland 1995; yellow-bellied marmot, *Marmota flaviventris*; Blumstein et al. 1997; Blumstein, chap. 27 this volume). Finally, coypus (*Myocastor coypus*) may delay parturition in response to the threat of predation (Gosling et al. 1988).

Food caching

Pine vole parents (*Microtus pinetorum*) carry food to specific locations where they store it in a pile for later use, and males bring food directly to the natal nest (McGuire and Novak 1984; Oliveras and Novak 1986). This behavior by male pine voles may constitute male provisioning of food for lactating females.

Attendance at the nest

A behavior recently reported for social voles (*Microtus socialis guentheri*), termed "forced babysitting," does not fit the traditional categorization of direct or indirect care, but is relevant to parental behavior. One parent, typically the male, aggressively drags the other back to the nest to remain with the pups while it leaves the nest (Libhaber and Eilam 2002). Male and female prairie voles coordinate arrivals and departures at the natal nest such that young are rarely left unattended, but the aggressive dragging of forced babysitting does not occur (McGuire and Novak 1984).

Factors that Influence Parental Behavior

In this section, we discuss seven factors that influence parental behavior. As in previous sections, most research on these topics emphasizes sciurognaths and is laboratory based.

Degree of development of young at birth

The degree of development of young varies along a continuum from altricial species typical of the Sciurognathi to precocial species typical of the Hystricognathi. Altricial species are born naked with closed eyes and ears, have poor sensory and locomotor abilities for several days after birth, and are confined to a nest. In contrast, precocial species have longer gestation periods, and their young are often fully furred, with open eyes and ears at birth, and can locomote almost immediately (Kleiman 1972; Weir 1974). Al-

tricial young rely exclusively on milk for about the first 2 weeks of life and then gradually begin to consume solid food; in contrast, precocial young may supplement milk with solid food immediately after birth (Kunkele and Trillmich 1997). Although most sciurognaths are altricial, exceptions occur (e.g., spiny mouse; Dieterlen 1962; Porter and Doane 1978).

Species with altricial young tend to give birth from a sitting or lying position and young are expelled in front of the female (e.g., Djungarian hamster and Siberian hamster; Jones and Wynne-Edwards 2000; prairie vole; McGuire et al. 2003). In several species with precocial young, parturient females assume a standing position and expel the young behind their body (e.g., spiny mouse; Dieterlen 1962; guinea pig, *Cavia porcellus*; Kunkel and Kunkel 1964; green acouchi; Kleiman 1972). Kleiman (1972) and Dieterlen (1962) noted that standing postures during parturition are assumed by other mammals with precocial young, such as ungulates, and suggested that standing during parturition may be an adaptation to the delivery of large, well-developed offspring. However, at least one species of rodent with precocial young, the cuis (*Galea musteloides*), gives birth from a sitting or lying position (Rood 1972); more data are needed to confirm that differences in birth position correspond to degree of development of young. Maternal aggression during and after parturition is characteristic of species with precocial young (Kleiman 1972) and of those with altricial young (Dewsbury 1985; McGuire et al. 2003), and likely functions to deter infanticidal conspecifics (Maestripieri and Alleva 1990; Wolff and Peterson 1998).

Rodents with precocial young often exhibit lower levels of nest building and pup retrieval than do species with altricial young (Kleiman 1974). For example, some species with precocial young do not build a nest (e.g., guinea pig; Kunkel and Kunkel 1964) or build only a temporary nest to which they retrieve young for only a few days after parturition (green acouchi; Kleiman 1972). Mothers of precocial young typically use maternal contact calls to inform their highly mobile offspring of their location and to induce following (Kleiman 1972). In contrast, females with altricial young often build an elaborate nest before parturition and maintain the nest through the preweaning period (Norway rat, *Rattus norvegicus*; Denenberg et al. 1969; meadow vole, *Microtus pennsylvanicus*; pine vole; and prairie vole; McGuire and Novak 1984). Typically, altricial young first venture from the natal nest a few days after their eyes have opened, and these brief forays trigger initially frequent retrieval back to the nest by mothers (and fathers in some species). In several species of voles, for example, eyes open 10–12 days postpartum, and a day or two later pups begin to make brief trips from the natal nest; retrieval by mothers occurs frequently during the early forays but then declines in frequency over the next few days, and eventually stops (McGuire and Novak 1984, 1986).

Despite the apparent independence of precocial offspring, mothers significantly influence their physiology and behavior (Hennessy 2003). For example, many precocial young start to eat solid food when only a day or two old and can survive without milk by 1 or 2 weeks, but nursing frequently continues for several weeks or months (Rood 1972; Kleiman 1972, 1974; Makin and Porter 1984). Mothers of precocial young also continue to groom their offspring well beyond the time when such grooming functions to stimulate urination or defecation (Kleiman 1972, 1974). The continued association between mothers and young, exemplified by prolonged nursing and grooming, is apparently beneficial to both (Kleiman 1974).

Litter size

Mothers spend more time in the nest with small than with large litters (see table 20.2, column In nest). One explana-

Table 20.2 Maternal behavior in relation to litter size among selected rodents

Species	Study design	Total care[a]	In nest	Age at weaning	Nurse	Groom pup	Nest build	Attack[b]	Reference
Norway rat (*Rattus norvegicus*)	A	S							Seitz 1958
	A		S	S					Grota and Ader 1969; Ader and Grota 1970; Grota 1973
	A		S	S	S	S			Leigh and Hofer 1973
	A					X	X (quality)		Fuemm and Driscoll 1981
House mouse (*Mus musculus*)	A		S		S (days 1–14); L (days 15–20)	X	X		Priestnall 1972
	A							L	Maestripieri and Alleva 1990
Wild house mouse (*Mus domesticus*)	N	S (days 1–4)		L					König and Markl 1987
Social vole (*Microtus socialis*)	N		X	X		X			Libhaber and Eilam 2004
Bank vole (*Clethrionomys glareolus*)	A							L	Jonsson et al. 2002
Deer mouse (*Peromyscus maniculatus*)	N			L					Millar 1979
Gerbil (*Meriones unguiculatus*)	A		S		X	L	S		Elwood and Broom 1978
Golden hamster (*Mesocricetus auratus*)	A		S[c]				S (quality)		Scott 1970
	A		S				X		Guerra and Nunes 2001
Desert woodrat (*Neotoma lepida*)	N			L					Cameron 1973
Guinea pig (*Cavia porcellus*)	A & N				L				Stern and Broner 1970
European ground squirrel (*Spermophilus citellus*)[d]	N			L					Millesi et al. 1999

NOTES: S = higher in small litters; L = higher in large litters; X = no difference with respect to litter size (modified from Mendl [1988: Tables III, IV] and updated with additional information). A = artificial manipulation of litter size, N = natural litters. Unless otherwise indicated, studies were conducted in the laboratory and patterns of behavior measured by duration, frequency of occurrence, or percent of the observation period spent performing the behavior. Empty cells indicate that the behavior was not recorded in the study.
[a]Rating of maternal quality which included categories such as response to opening of cage, reluctance to leave litter, retrieval of young, and quality of nest (Seitz 1958); sum of time percentages per observation period for nurse, groom pup, and nest build (König and Markl 1987).
[b]Attacks by mother directed at unfamiliar male conspecifics.
[c]Represents time spent by mother in bodily contact with young.
[d]Field study; age at weaning assessed by condition of mothers' teats, not behavioral observations of mothers, and young and yearling females excluded.

tion is that mothers of large litters may need to spend more time outside the nest foraging to meet nutritional demands associated with providing milk to a large number of young (Priestnall 1972; Mendl 1988). Another explanation is that mothers of small litters must help pups maintain their body temperature, whereas pups in larger litters may not require such help because they can huddle with more littermates (Priestnall 1972; Mendl 1988). Other evidence suggests, however, that littermates are relatively ineffective at maintaining a warm nesting environment (Webb et al. 1990). Finally, mothers of large litters may experience problems with hyperthermia, and so spend more time away from pups, dissipating heat (see Jans and Leon 1983 regarding maternal hyperthermia). Given the diversity of rodents, the absence of detailed behavioral studies, and the possibility that more than one of these explanations may apply to a case, we cannot establish a single cause for the observed pattern.

A second general pattern to emerge is that mothers wean large litters later than small litters (see table 20.2, column Age at weaning). Mothers may wean offspring when they reach a certain minimum weight, and this weight is achieved earlier in small than in large litters (Cameron 1973; König and Markl 1987).

No clear patterns emerge concerning litter size and the time that mothers spend nursing, grooming young, or nest building (see table 20.2, columns Nurse, Groom pup, Nest build). This inconsistency is chiefly because there are few comparable studies and criteria for behavioral categories differ across studies. Two reports indicate that maternal aggression toward unfamiliar male conspecifics increases with litter size (see table 20.2, column Attack). This result supports predictions that risks to mothers of defending young should not change with litter size, but that benefits of such defense should increase with number of offspring (Maestripieri and Alleva 1990; Jonsson et al. 2002).

Much less is known about litter size variation and paternal care, but some observations are available. For example, male gerbils (*Meriones unguiculatus*) exhibited more frequent pup grooming and body contact when litters were large but more nest building when litters were small (Elwood and Broom 1978). Paternal care in social voles is essentially independent of litter size (Libhaber and Eilam 2004).

Artificially manipulating litter size creates complications that make interpretation of results difficult. Mothers in nature may adaptively adjust their number of offspring according to their own condition, abilities, and prevailing environmental conditions, so measures of parental behavior may not vary with litter size. In contrast, females that naturally give birth to small litters and then have their litter size experimentally increased may be greatly challenged. Further, litter augmentation would be a very unusual situation under natural conditions. Although reductions in litter size would not be unusual in field populations, experimental reduction to very small litter sizes (e.g., one or two pups) may be extreme and may disrupt milk production or thermoregulation in the nest. Finally, pups can display fidelity to certain nipples (e.g., in prairie and pine voles; McGuire 1998; McGuire and Sullivan 2001) and transferring pups between litters, even on day 1 postpartum, can disrupt development of such preferences. Some researchers have teased apart the adjustments that nursing mothers make using methods to change offspring food demand (by manipulating their access to solid food) without changing litter size, but this approach only works for precocial species such as guinea pigs (Laurien-Kehnen and Trillmich 2003). Additional studies of unmanipulated litters of different sizes are needed.

An intriguing cross-species example of litter size variation is reported in naked mole-rats (Sherman et al. 1999). While most rodents have mean litter sizes equal to about one-half the number of mammae (Gilbert 1986), naked mole-rat queens raise litters on average equal to the number of mammae (about 12). Extremely large litters—up to 28 and 27 offspring—are reported in field and laboratory colonies, respectively. Sherman et al. (1999) state that such large litters are possible because offspring take turns nursing at the same nipple and colony members feed and protect the queen.

Gender of offspring

Differential investment by mothers in male and female offspring has been examined in the context of parental investment theory, especially as it relates to mating systems (Trivers 1972; Trivers and Willard 1973; Sikes, chap. 11 this volume). In polygynous species (where variance in reproductive success is typically greater for males than females) mothers in good condition are predicted to bias their investment toward sons, while mothers in poor condition should invest in daughters. Sex-biased parental investment may be reflected in the sex ratio of offspring produced or in different amounts of care shown to sons and daughters during the postnatal period (for reviews see Clutton-Brock et al. 1981; Clutton-Brock and Iason 1986; Cockburn et al. 2002). There is some evidence that female rodents adaptively manipulate the sex ratios of their litters during the prenatal period (e.g., coypus; Gosling 1986; golden hamsters, *Mesocricetus auratus*; Labov et al. 1986; house mice, *Mus musculus*; Krackow and Hoeck 1989; Krackow 1997). Here, we focus on differential parental investment during the postnatal period.

Some studies examined postnatal maternal investment in male and female offspring under ad libitum food conditions. For example, Gosling et al. (1984) found differential investment in male and female offspring in the polygynous coypu;

male offspring spent more time than female offspring sucking from the highest-yielding teats, although this pattern appeared to result from the behavior of young and not from the mother's active promotion or discouragement of particular offspring from sucking from specific teat locations. Clark et al. (1990) found that female gerbils rearing all-male litters were much more likely than females rearing all-female litters to be in the nest with young and to have pups attached to their nipples. Norway rat mothers and house mouse mothers spend more time licking the anogenital region of male than female pups, and also show enhanced nursing and nest building when rearing all-male litters (Moore and Morelli 1979; Richmond and Sachs 1984; Alleva et al. 1989). Rat mothers appear to use olfactory cues to discriminate the gender of their offspring (Moore 1981) and the specific chemosignal comes from the preputial glands of pups (Moore and Samonte 1986).

Other studies of differential postnatal parental investment compared mothers with unrestricted access to food to mothers whose food was restricted during lactation; greater investment by food-restricted mothers in female offspring was predicted. Food-restricted eastern woodrats (*Neotoma floridana*) invested more in female offspring, as evidenced by higher mortality and reduced growth of male offspring (McClure 1981). Female-biased investment also is reported for food-restricted golden hamster mothers (Labov et al. 1986). In contrast, Sikes (1995, 1996b) found no evidence of sex-biased maternal investment in food-restricted eastern woodrats and northern grasshopper mice (*Onychomys leucogaster;* also see Sikes chap. 11 this volume). Finally, recent evidence indicates that male-biased mortality in polygynous species may occur independently of parental discrimination. Moses et al. (1998), working with bushy-tailed woodrats (*Neotoma cinerea*), suggested that male-biased mortality in offspring of food-restricted mothers might reflect the greater energetic demands of male offspring, resulting from sexual selection for faster growth and greater body size. Comparative data for rodents on differential postnatal maternal investment in male and female offspring remain equivocal and the topic requires further study.

Concurrent pregnancy

Postpartum mating in some groups of rodents results in concurrent pregnancy and lactation (Gilbert 1984). Although less costly than lactation, pregnancy imposes energetic costs (e.g., bank voles, *Clethrionomys glareolus;* Kaczmarski 1966). Levels of maternal care by pregnant females are generally lower than in nonpregnant females; such differences arise late in lactation as birth of the new litter approaches (wild house mouse, *Mus domesticus;* König and Markl 1987; Norway rats; Rowland 1981; Wuensch and

Cooper, 1981; but see McGuire 1997 for red-backed voles, *Clethrionomys gapperi,* and Krackow and Hoeck 1989 for house mice). Lower levels of maternal care by pregnant females could result from the increased energetic demands faced by such females. Other options for pregnant females include diverting energy from young in utero or from themselves (Oswald and McClure 1987). All studies noted were conducted in laboratory conditions with abundant food nearby, and no temperature stresses.

Social environment

Many laboratory studies have examined effects of social experience on rodent parental behavior (McGuire 1988; Lehmann and Feldon 2000; and reviews by Dewsbury 1985; Brown 1993; Kinsley 1994). Here, we discuss how presence of other males or mating opportunities influences paternal care, and review studies that examine how the composition of a social group influences parental behavior.

Males often disproportionately increase reproductive success by seeking additional matings rather than by providing paternal care (Trivers 1972; Clutton-Brock 1989b). Thus paternal care in rodents should decrease as mating opportunities increase, and this has been found in the field for two normally monogamous species, hoary marmots (Barash 1975a) and muskrats (Marinelli and Messier 1995). In the laboratory, parent-offspring interactions in polygynous meadow voles were studied in a 2.4 by 1.2 m enclosure (Storey et al. 1994). Introduction of an estrous female did not significantly reduce the time fathers spent in the nest with their pups, even though many fathers mated with the introduced females. The enclosure's size may have made it easy for males to mate with estrous females without significantly reducing their time in the natal nest (Storey et al. 1994). Difficulties observing paternal care in the field and the need to provide extensive space in the laboratory make it challenging to study the relationship between paternal care and mating opportunities in rodents.

Field and laboratory studies show that the presence of one parent can influence care shown by the other parent. Paternal presence correlates with decreased maternal behavior in species such as rock cavies (*Kerodon rupestris,* studied in laboratory cages; Tasse 1986), Norway rats (studied in an outdoor pen; Calhoun 1962a), gerbils (studied in laboratory cages; Elwood and Broom 1978), red-backed voles (studied in seminatural laboratory environment; McGuire 1997), and muskrats (studied in the field; Marinelli and Messier 1995). The muskrat example is interesting because free-living polygynous males only provided care to young of their first mate, and these primary females displayed lower levels of maternal behavior than did secondary females, who compensated for the lack of male assistance by increasing their investment (Marinelli and Messier 1995). When male rodents provide care for young, decreased maternal behavior in the presence of males has been interpreted as evidence of reduced maternal workload. When males are present but do not care for offspring, decreased maternal behavior has been attributed to disruption caused by paternal presence; increased maternal care in such circumstances is rare. Unchanged levels of maternal behavior in the presence of fathers are reported for wild house mice (studied in laboratory cages; König and Markl 1987), prairie voles (studied in seminatural laboratory environments; Wang and Novak 1992; Wilson 1982b), meadow voles (studied in seminatural laboratory environments; Storey et al. 1994) and collared lemmings (*Dicrostonyx richardsoni*, studied in laboratory cages; Shilton and Brooks 1989).

Maternal response is a major factor influencing the level of paternal care. For example, females of biparental species frequently exclude males from the natal nest during parturition and for about a day thereafter, but subsequently permit males to fully interact with young (gerbils; Elwood 1975; southern grasshopper mice, *Onychomys torridus;* McCarty and Southwick 1977b; spiny mice; Porter et al. 1980; white-footed and deer mice, *Peromyscus spp.;* Wolff and Cicirello 1991; prairie voles; McGuire et al. 2003). In other species, female aggression toward mates may extend throughout the preweaning period (meadow voles; McGuire and Novak 1984; montane voles, *Microtus montanus;* McGuire and Novak 1986; white-footed mice; Xia and Millar 1988; but see Wolff and Cicirello 1991); in such species, male interactions with young occur primarily after weaning. Indeed, increases in male-offspring interactions with pup age have been reported for meadow voles (Oliveras and Novak 1986; Storey and Snow 1987) and montane voles (McGuire and Novak 1986) in the laboratory, and postweaning paternal care of young has been reported in a natural population of white-footed mice (*P. leucopus;* Schug et al. 1992) and deer mice, *P. maniculatus* (Wolff and Cicirello 1991). The duration of maternal aggression toward fathers is not necessarily consistent within a species; for example, female red-backed voles vary in the intensity and duration of aggressive behavior toward fathers, and this produces variation in levels of paternal care (McGuire 1997).

Few studies have examined experimentally how maternal presence affects paternal behavior (apparently polygynous taxa, such as spiny mice, Makin and Porter 1984 and collared lemmings, Shilton and Brooks 1989; and polygynous meadow voles, Storey et al. 1994). Maternal removal is problematic because it interferes with suckling and pup nutrition. Such removal is less problematic in species with precocial young, such as the spiny mouse, for which Makin and Porter (1984) conducted near-daily observations of parents from day 1 to day 23 postpartum. On any given day,

they observed pairs with their young and then temporarily removed mothers to observe paternal behavior. Males huddled with their offspring more when the mother was absent than when she was present, again confirming that females regulate interactions between fathers and offspring.

Juveniles also can affect care shown by parents to a younger litter. Norway rat mothers attack juveniles before and after the new litter is born, but juveniles still spend time in the nest with neonates (Gilbert et al. 1983). Maternal care in the presence of such juveniles is similar to that displayed in their absence (Grota and Ader 1969; Gilbert et al. 1983). Female spiny mice nest with their mate and juveniles from the previous litter both before and after a new litter is born, but keep both males and juveniles away from the nest on the day of parturition (Porter et al. 1980). Under seminatural conditions, female meadow voles aggressively exclude juveniles from the nest containing the new litter (Wang and Novak 1992); increased nest defense resulted in greater maternal workload compared to females rearing pups without juveniles present. In contrast, female prairie voles allow juveniles in the natal nest and experience reduced maternal workload in the presence of juveniles if the father is also present (Wang and Novak 1992). Presence of juveniles also may reduce paternal workload in prairie voles, but this is the only species for which data exist (Wang and Novak 1992).

Levels of paternal care positively correlate with paternity (Westneat and Sherman 1993), but this topic has received little attention in rodents, with most studies focusing on infanticidal rather than paternal behavior. An exception concerns work with meadow voles (Storey and Snow 1987). In one experiment, males spent less time in the nest with their mate's pups when another adult male was housed in a nearby wire enclosure. A second experiment compared levels of nest attendance by males housed with their mate and pups to that of males housed with a female rearing young of another male; time spent with pups was much higher for fathers than for nonfathers. Thus reduced paternal care in meadow voles correlates with uncertain paternity.

Parenting experience

Rodents can gain parenting experience by helping to care for a younger litter in their social group (alloparental care) or by caring for their own young in successive litters. In the laboratory, alloparental experience results in enhanced reproductive performance, pup growth, and development in gerbils (Salo and French 1989), but the effects of alloparental experience in other species are slight and often mixed (Solomon and Getz 1997). For example, adult prairie voles with alloparental experience did not differ in their parental behavior from those without alloparental experience, but their pups developed slightly faster (Wang 1991). In still other species, such as naked mole-rats, it is not known whether alloparental experience affects subsequent parental behavior and success in rearing young (Lacey and Sherman 1997).

Reproductive experience can influence neuroendocrine physiology of female rodents. For example, experience causes changes in the endogenous opioid system that mediates olfactory-based interactions between mother and offspring (Kinsley 1994). Parity also can influence maternal behaviors, although effects range greatly. No effect of parity is reported for captive female wild house mice (König and Markl 1987), deer mice, or white-footed mice (Hartung and Dewsbury 1979). Social environment may determine whether parity affects maternal behavior in Norway rats. When rearing young in the absence of males, multiparous and primiparous females did not differ in nursing, nest building, and retrieving (Moltz and Robbins 1965), or in the overall time spent with litters (Grota 1973). However, when rearing young in the presence of males, multiparous females more effectively switched between neonatal care and mating during postpartum estrus; such females also more effectively retrieved pups (Gilbert et al. 1984). Experience also enhances pup retrieval in other species (house mice; Cohen-Salmon 1987; golden hamsters; Swanson and Campbell 1979). In a particularly striking example, multiparous female prairie voles spend more time caring for offspring than do primiparous females, yielding more rapid physical development and a higher survival rate of young (Wang and Novak 1994).

Few studies examine the effects of parity on paternal behavior. Prior parenting experience had no effect on paternal behavior in prairie voles (Wang and Novak 1994) or white-footed mice (Hartung and Dewsbury 1979). Minor changes in a few behaviors are reported for other species, such as deer mice (Hartung and Dewsbury 1979) and Norway rats (Brown 1986a). Thus at this time, previous experience as a male caregiver appears to have little or no effect on paternal care of rodents.

Physical environment

Most rodents care for young in underground burrows or covered nests, making it difficult to observe parent-offspring interactions in the field, particularly before weaning. Direct field observations are available for some relatively large diurnal species (hoary marmots; Barash, 1975a; black-tailed prairie dogs, Hoogland 1995) and for at least one small diurnal rodent, the striped mouse (*Rhabdomys pumilio*; Schradin and Pillay 2003). Even though observations inside the nests of striped mice are not reported from the field, time spent at the nest by parents has been recorded, and

parental interactions with offspring have been studied once young are old enough to venture from the nest (Schradin and Pillay 2003). Other field studies of parent-offspring interactions involve indirect measures such as trap associations and patterns of space use revealed by radiotelemetry (Schug et al. 1992).

Testing environments in laboratory studies range from small cages with little or no cover to seminatural environments with extensive space and cover. Patterns of nesting and parental care can vary with the size and complexity of the testing environment. For example, when families of meadow and montane voles were studied in small cages in which nesting material was the only source of cover, no sex differences were apparent in the amount of time parents spent in the nest with pups (fig. 20.1a; Hartung and Dewsbury 1979). In contrast, when these same species were studied in seminatural environments that provided substantial space and hay cover, females of both species aggressively excluded males from natal nests; males nested separately and spent very little time with young pups (fig. 20.1b; McGuire and Novak 1984, 1986; Oliveras and Novak 1986). The latter results are consistent with reports from natural populations of female-only care of young and separate nesting by adult males and females during most of the breeding season (Madison 1980b; Jannett 1980). Studies on parent-offspring interactions in white-footed mice reveal a very similar pattern; whereas males in small cages show substantial pup care (Hartung and Dewsbury 1979), males in larger enclosures are excluded from nests by females (Xia and Millar 1988), and the latter pattern is more consistent with what is known about nesting and space use by males and females in natural populations (Wolff and Cicirello 1991). Female aggression toward strange males might be interpreted as defense of young, but the reasons why females aggressively exclude mates from the natal nest remain unclear.

Conflicting findings, such as those described previously, may indicate that the pup care reported for some species in small cages is a laboratory artifact (McGuire and Novak 1984, 1986; Wolff 2003c). Alternatively, males of these species have the potential to display paternal behavior, and may do so under certain conditions (Dewsbury 1985). For example, free-living male and female meadow voles sometimes nest together during colder months (Madison et al. 1984), and thus paternal care is possible under conditions of late autumn or winter breeding. Indeed, male meadow voles housed under short day lengths displayed longer grooming and huddling bouts with young than did males housed under long day lengths (Parker and Lee 2001). Additional data on parent-offspring interactions at low temperatures are needed from field or seminatural environments to confirm facultative paternal care in this species.

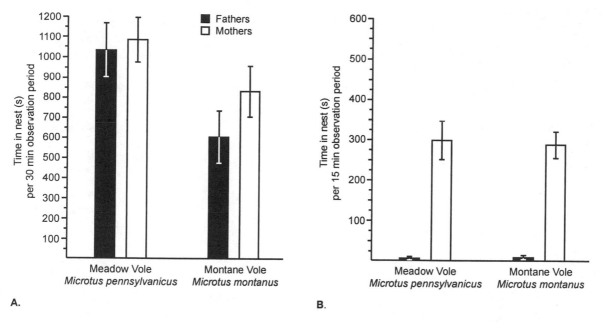

Figure 20.1 Comparison of time in the nest for both parents for meadow voles (*Microtus pennsylvanicus*) and montane voles (*M. montanus*) as a function of testing environment. A. Time in nest (total time per 30 min observation period) in small cages; data from Hartung and Dewsbury (1979). B. Time in nest (total time per 15 min observation period) in seminatural environments; data from McGuire and Novak (1984, 1986).

In addition to seasonal variation in patterns of parental care within a species, there may also be geographic variation. Inexperienced male and female prairie voles from an Illinois population displayed higher levels of parental responsiveness (time spent huddling) when presented with two unfamiliar pups than did individuals from a Kansas population (Roberts et al. 1998). The authors considered the Illinois site to have abundant resources and the Kansas site to have scarce resources, but did not discuss geographic variation in parental behavior with respect to differing resource levels. Geographic variation in parental behavior also characterizes meadow voles. Male meadow voles from Manitoba and Ontario nest with females and young and display substantial pup care even when housed in large enclosures (Storey and Snow 1987); these results differ from the separate nesting and low pup care displayed by males from Massachusetts when provided with similar space (McGuire and Novak 1984; Oliveras and Novak 1986). Increased pup care by meadow vole males from Canada correlates with harsher environmental conditions, particularly colder temperatures (Storey and Snow 1987).

Characteristics of the physical environment could either directly influence level of male parental behavior or indirectly influence it through effects on the behavior of females. The latter scenario seems more likely. For example, a testing environment's size could influence the ability of females to defend the nest against entry by males and this, in turn, influences male interactions with young. In small cages, mothers may be unable to defend the nest against entry by males. Thus males may spend time in the nest with young, especially if nest material is the only cover available in the cage. In contrast, the increased space available in seminatural environments may enable females to more effectively exclude males from the natal nest. Additionally, with cover available throughout the testing environment, separate nesting by males is more likely, and this results in limited male interaction with young pups. Paternal interaction with offspring may increase once pups are capable of leaving the natal nest; increased paternal interaction with increasing pup age has been reported for meadow voles (Oliveras and Novak 1986; Storey and Snow 1987) and white-footed mice (Schug et al. 1992). Similarly, the reported seasonal and geographic variation in paternal behavior may actually represent variation in the tolerance of females to fathers in the natal nest (Storey et al. 1994; Roberts et al. 1998).

Effects of Paternal Presence on Survival and Growth of Offspring

Effects of maternal separation on growth, development, physiology, and behavior of young rodents are reviewed by Lehmann and Feldon (2000), Levine (2001), and Braun et al. (2003). Our focus is paternal presence, particularly laboratory studies that compare pups reared by both parents with pups reared by mothers alone in testing environments ranging from standard to seminatural to challenging

Table 20.3 Effects of paternal presence on offspring

Species	Testing condition[†]	Survival	Growth	Eyes open	Reference
House mouse (*Mus musculus*)		X	X		Priestnall and Young 1978
	S	X	X		Wright and Brown 2000
	C(E)	+	X		
Southern grasshopper mouse (*Onychomys torridus*)	S	X	X		McCarty and Southwick 1977a
Gerbil (*Meriones unguiculatus*)	S	X	X	+	Elwood and Broom 1978
Meadow vole (*Microtus pennsylvanicus*)	S	X	+		Storey and Snow 1987
	SN	X			
	S	−, X			McGuire et al. 1992
	SN	X	X	−	Wang and Novak 1992
Prairie vole (*Microtus ochrogaster*)	S	X, X[a]			McGuire et al. 1992
	SN	X	X	X	Wang and Novak 1992
	C(P)	+			Getz et al. 1992
	F	X			Getz and McGuire 1993
Red-backed vole (*Clethrionomys gapperi*)	SN	X	X	X	McGuire 1997
Collared lemming (*Dicrostonyx richardsoni*)	S	X	X	X	Shilton and Brooks 1989
	C(MR)	X	X	X	
Muskrat (*Ondatra zibethicus*)	F	X			Marinelli et al. 1997
California mouse (*Peromyscus californicus*)	S		+		Dudley 1974
	C(MR)	+	+	+	
	C(WE)	+			
	S	X	X		Gubernick et al. 1993
	C(T)	+	X		
	C(E)	+	X		
	C(E)	+			Cantoni and Brown 1997
	F	+			Gubernick and Teferi 2000
	S		X	X	Vieira and Brown 2003
Djungarian hamster (*Phodopus campbelli*)	S	+	+		Wynne-Edwards and Lisk 1989
Siberian hamster (*Phodopus sungorus*)	S	X	X		Wynne-Edwards and Lisk 1989
Striped mouse (*Rhabdomys pumilio*)	C(T)	X	+[b]		Schradin and Pillay 2005b
Green acouchi (*Myoprocta pratti*)	S, SN[c]	+			Kleiman 1970

[†]NOTES: plus (+) = enhanced or accelerated by father; minus (−) = depressed or delayed by father; X = no significant effect of father. Empty cells indicate that the variable was not recorded in the study. S = standard laboratory conditions (cage, *ad libitum* food and water, warm ambient temperature); SN = seminatural environment (increased space and cover, *ad libitum* food and water, warm ambient temperature); C = challenging laboratory conditions (animals are subjected to some form of stress, including exercise for food [E], cold ambient temperatures [T], mother removed periodically [MR], pups weaned early [WE], or predator present [P]); F = field conditions.
[a]Mother primiparous, mother multiparous.
[b]Mice from desert habitat; father presence did not affect growth of young from grassland habitat.
[c]Mix of small cages, large cages and rooms; no distinction made in the data.

(see table 20.3 for definitions of testing environments; also see Brown 1993). Offspring survival, growth, and age at eye opening are the most frequently measured variables.

Male presence has little or no effect on the survival, growth, and development of most species studied in standard or seminatural environments (table 20.3). Positive effects of paternal presence on offspring survival, and to a lesser extent on offspring growth, are reported for challenging environments (table 20.3). Although challenging environments mimic some stresses under which rodents live in the field, these environments do not reflect the complexity of stresses and interactions faced by free-living rodents. For example, challenging laboratory environments do not include a variety of ground and aerial predators and potentially infanticidal conspecifics. Given the physical and so-cial stresses confronted by free-living rodents, male presence probably has an even more positive impact in the field, but only three studies have examined this. Paternal presence significantly enhanced offspring survival in free-living California mice (*Peromyscus californicus;* Gubernick and Teferi 2000), and the authors suggest that this resulted from direct care of young rather than protection against infanticidal intruders. However, paternal presence had no impact on offspring survival in free-living prairie voles (Getz and McGuire 1993) or muskrats (Marinelli et al. 1997). The lack of effect in these two species may be related to the high quality and quantity of food available in the habitat for the study population of prairie voles (Getz and McGuire 1993) and the low population density (and hence low risk of infanticide) in the muskrat study (Marinelli et al. 1997).

Evolution of Paternal Behavior in Voles

Males of some arvicoline species exhibit care behaviors unreported in other species of *Microtus*. These behaviors are routinely observed in laboratory environments and are consistent with field data of males sharing a nest with females and young. Do these behaviors represent a derived condition within *Microtus*, and if so have they evolved more than once? Although relationships among arvicoline species remain unclear (Hinton 1926; Anderson 1985; Musser and Carleton 1993; Martin et al. 2000; Conroy and Cook 2000), we have comparable behavioral data for four species of *Microtus* and two outgroups: the sagebrush vole (*Lemmiscus curtatus*) and red-backed vole (McGuire and Novak 1984, 1986; Hofmann et al. 1989; McGuire 1997). These data

permit us to make a preliminary phylogenetic analysis based on behavioral data (fig. 20.2).

Our analysis groups prairie and pine voles, a clade that is not reported in either morphological or molecular phylogenetic studies of arvicolines. Of more interest is our recovery of the subgenus *Mynomes*, which includes *Microtus pennsylvanicus* and *M. montanus*, a group typically recovered in morphological and molecular studies (e.g., Musser and Carleton 1993; Conroy and Cook 2000). An interesting pattern in our analysis is the loss of paternal behavior in subgenus *Mynomes* (Character 7). Interpreting this as a loss (rather than three separate gains of paternal behavior) also is supported by the observation that montane and meadow vole males facultatively spend time in the natal nest, depending on the testing environment (fig. 20.1).

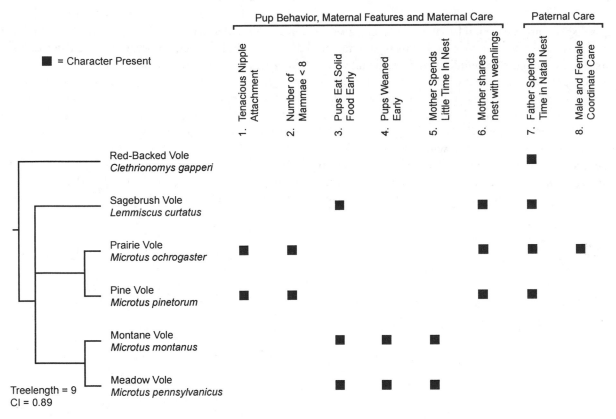

Figure 20.2 Phylogeny of parental behaviors in species of arvicolines for which comparable behavioral data are available (sources are McGuire and Novak 1984, 1986; Hofmann et al. 1989; McGuire 1997). The tree is based on an exhaustive search (PAUP*4.0b10, Swofford 2003; all characters treated as unordered; red-backed vole specified as outgroup). This single most parsimonious tree has a length of 9 and an unscaled consistency index (CI) of 0.89. Character definitions: 1. Tenacious Nipple Attachment. Pups cling so tightly to nipples that they remain attached even when mother moves. 2. Number of Mammae < 8. Number varies from four (pine vole) to six (prairie vole) to eight (other species). Character 3. Pups Eat Solid Food Early. The day that pups first eat solid food varies from 13 or 14 days postpartum (meadow, montane, and sagebrush voles) to 15 or 17 days (red-backed, prairie, and pine voles). Character 4. Pups Weaned Early. Weaning varies from 13 or 14 days postpartum (meadow and montane voles) to 19, 20, 21, or 23 days (red-backed, prairie, pine, and sagebrush voles). Character 5. Mother Spends Little Time In Nest. Montane and meadow vole females spend 45% and 50% of their time in the natal nest during first ten days postpartum. Values for pine, sagebrush, prairie, and red-backed vole females are 65%, 75%, 81%, and 81%. Character 6. Mother Shares Nest with Weanlings. Sagebrush, prairie, and pine vole females continue to nest with young after weaning; red-backed, montane, and meadow vole females nest separately from weaned young. Character 7. Father Spends Time in Natal Nest. Red-backed, pine, prairie, and sagebrush vole males spend time in the natal nest (27%, 32%, 63%, and 69% respectively); montane and meadow vole males spend less than 1% of their time in the natal nest. Character 8. Male and Female Coordinate Care. Of six species studied, only prairie voles coordinate their visits to the natal nest so that pups are rarely left alone.

To further study the evolution of paternal care, it is first important to increase the number of taxa for which comparable behavioral data are available. Such data could readily be collected for additional species of *Microtus* as well as other genera (e.g., *Synaptomys, Dicrostonyx, Arvicola, Ondatra*). Second, we need molecular and morphological character data for all species studied behaviorally. Finally, with increasing knowledge of the comparative neurobiological bases of vole social behavior (e.g., Wang and Insel 1996), such data should be incorporated into phylogenetic data matrices and analyses. Such a three-pronged approach has the potential to make arvicolines a central example for understanding the evolution of parental behaviors in rodents (Curtis et al., chap. 16, this volume).

Summary

Rodents display direct parental behaviors (nursing, grooming, retrieving, and huddling) and indirect parental behaviors (food caching, nest building, and defending young against conspecifics and predators). Except for nursing, fathers can perform all of these, and in some species fathers show levels of care comparable to mothers. Species with precocial young typically exhibit lower levels of nest building and pup retrieval than species with altricial young. Litter size affects aspects of maternal behavior. Mothers of large litters spend less time in the natal nest and wean their offspring later than do mothers of small litters. Also, maternal aggression toward male conspecifics increases with litter size. Differential prenatal and postnatal investment in male and female offspring is documented for some species, but the topic requires further study. Mothers that are pregnant while caring for a litter show lower levels of maternal care than mothers that are not pregnant, but such differences appear only late in lactation. In several species, paternal presence is associated with decreased maternal care; this may result from a decreased maternal workload (if males care for young) or disruption (if males do not contribute to care). Mothers regulate paternal interactions with young by excluding fathers from the natal nest; such exclusion typically occurs on the day of parturition, but may extend throughout the preweaning period. Male rodents decrease care when paternity is uncertain and when their mating opportunities increase, although data are limited. Experience, gained either by alloparental care or by caring for young in successive litters, has little effect on parental behavior. In contrast, characteristics of the physical environment can dramatically influence the level of care. Paternal presence increases survival and growth of offspring in challenging but not in standard or seminatural laboratory environments; field data are limited and conflicting. A preliminary phylogenetic analysis of parental behaviors in six species of voles suggests that paternal behavior may have been present in the ancestor of *Microtus* and subsequently lost in the subgenus *Mynomes* (= *M. pennsylvanicus* + *M. montanus*). Future phylogenetic, field, and seminatural studies of parental behavior are needed, especially for species of Hystricognathi.

Chapter 21 The Ecology of Sociality in Rodents

Eileen A. Lacey and Paul W. Sherman

WHETHER CONSPECIFICS LIVE ALONE or in groups has important implications for numerous aspects of behavior, including the nature and intensity of cooperative as well as competitive interactions (Brown 1987; Koenig et al. 1992; Lacey and Sherman 1997; Hayes 2000; Hoogland, chap. 37 this volume). As a result, determining why groups occur is a central goal of many studies of rodent behavior. The selective pressures favoring sociality in rodents, however, are not well understood. Although numerous benefits of group living have been proposed (Alexander 1974; Hoogland and Sherman 1976; Emlen 1984; Solomon and Getz 1997; Blumstein and Armitage 1999; Danchin and Wagner 1997), only a handful of studies of free-living rodents have rigorously tested these hypotheses, and even fewer have systematically explored the effects of multiple selective factors on social structure (Ebensperger 2001a; Ebensperger and Cofré 2001). To facilitate understanding of this important aspect of social organization, we review the occurrence of group living among rodents, including processes of group formation and the associated patterns of kin structure. We then consider the conceptual approaches that have been employed to explain these phenomena. Using studies of social, subterranean rodents as a starting point, we develop a general conceptual model of the ecological factors thought to promote group living. Our hope is that the resulting integrative framework for exploring ecological correlates of sociality will stimulate future empirical studies of this key component of rodent societies.

Group Living and Sociality

Sociality is typically defined as group living (Alexander 1974; Lee 1994). As this statement implies, the tendency for conspecifics to live in groups provides the foundation for many of the elaborate forms of social interaction observed among animals (Alexander 1974; Lacey and Sherman 1997). This definition, however, is deceptively simple given that groups may vary dramatically in size, structure, and degree of cohesion (Krause and Ruxton 2002; Safran et al., in press). Within species, the tendency to form groups may differ among populations in response to ecological conditions (Nevo et al. 1992; Jarvis et al. 1994; Spinks et al. 2000a, 2000b; Nevo, chap. 25 this volume; Macdonald et al., chap. 33 this volume), and individuals may shift between a solitary and a social existence during the course of their lifetime, including from one round of reproduction to the next (Wolff 1994b; Solomon and Getz 1997). Thus it is not possible to characterize species or even individuals as solitary versus social without considering both the environments in which they occur and the timing of their behavior relative to key life-history events such as natal dispersal, mating, and parental care.

In practice, social groups often are identified on the basis of the spatial and social interactions among conspecifics that occur during the breeding period. Although groups may form only briefly (e.g., in response to the presence of a predator), most definitions of sociality emphasize interactions that persist for a significant portion of an individual's lifetime, such as one or more rounds of reproduction (e.g., Wilson 1975; Jennions and Macdonald 1994). Spatially, members of a group are expected to show considerable overlap, including, in some cases, sharing of a nest or den site (Andersson 1984; Jennions and Macdonald 1994; Lacey 2000). Behaviorally, interactions within groups are expected to differ markedly from those between groups, with the former being more likely to include affiliative, cooperative, and nepotistic activities (Hoogland 1995; Solomon and French 1997; Blumstein and Armitage 1997a; Nevo, chap. 25 this volume). Neither criterion provides an absolute indicator of social structure (Krause and Ruxton 2002) but, taken together, information regarding spatial and social relationships provides a reasonable means of identifying group-living taxa.

One reason that it is difficult to provide a precise definition of sociality is that "solitary" and "social" are not discrete alternatives but, rather, endpoints along a continuum of spatial and social interactions among conspecifics (Jennions and Macdonald 1994; Sherman et al. 1995; Lacey 2000; Krause and Ruxton 2002; Lacey and Sherman, 2005). Numerous intermediate patterns of spatial overlap and social cohesion are expected and, indeed, many rodent species appear to fall somewhere between these extremes. For example, in multiple species of ground-dwelling sciurids, females overlap spatially with one another, although each maintains an area of exclusive use that includes her nursery burrow (Michener 1983a; Sherman 1980a, 1981a; Hoogland 2003a and chap. 37 this volume; Hare and Murie, chap. 29 this volume; Yensen and Sherman 2003). Because females do not share burrows or nests, they do not exhibit the same type of sociality found in naked mole-rats or prairie voles (*Microtus ochrogaster;* Sherman et al. 1991; Bennett and Faulkes 2000; Solomon and Getz 1997), but neither are they solitary like woodchucks (*Marmota monax;* Barash 1989; Armitage 2000), pocket gophers (*Thomomys* spp.; Nevo 1979), or blind mole-rats (*Spalax ehrenbergi;* Nevo, chap. 25 this volume). Thus rather than struggling to achieve a single, comprehensive definition of sociality, it seems more appropriate to identify criteria that are relevant to the specific conceptual issues and taxa under study (Krause and Ruxton 2002; Lacey and Sherman, 2005).

Distribution of sociality in rodents

Group living is widespread within the Rodentia, occurring in at least 70 species representing 39 genera and 18 families, including 4 subfamilies of murids. Undoubtedly, this list underestimates the actual occurrence of group living in rodents. While social structure is relatively well documented in some lineages such as bathyergid mole-rats (Sherman et al. 1991; Bennett and Faulkes 2000; Faulkes and Ben-

nett, chap. 36 this volume) and ground-dwelling sciurids (Murie and Michener 1984; Barash 1989; Armitage 2000 and chap. 30 this volume; Hoogland 2003a and chap. 37 this volume; Hare and Murie, chap. 26 this volume), the social systems of many other rodent taxa (e.g., echimyid spiny rats, thryonomyid cane rats) are unknown and, hence, many examples of group living may be unreported. One objective of this chapter is to stimulate research on the social behavior of these poorly known taxa.

Despite the paucity of behavioral data for many species, it is clear that sociality is not evenly distributed among rodent families. For example, while group living occurs in the majority of bathyergid mole-rats (Bennett and Faulkes 2000), sociality has never been reported among geomyid pocket gophers (Lacey 2000). Similarly, group living appears to be widespread in the family Octodontidae, but is rare among members of the sister family Ctenomyidae (Lacey and Ebensperger, chap. 34 this volume). More generally, sociality is particularly prevalent among hystricognath rodents (Burda 1990; Bennett and Faulkes 2000); 72% of families in the suborder Hystricognathi include at least one social species, versus 46% of families in the suborder Sciurognathi.[1] While these figures represent only crude estimates, these apparent biases in the taxonomic distribution of group living are intriguing and warrant further investigation. Multiple factors may contribute to the prevalence of sociality in some rodent lineages, including the production of precocial young (Burda 1990), the risk of predation on highly visible, diurnal animals (Jarman 1974), and the use of safe, expansible burrows (Alexander et al. 1991). None of these hypotheses has been rigorously tested, but a single causal explanation seems unlikely. For example, even among subterranean hystricognaths—species that share the tendency to produce precocial young and to live in underground burrows—the prevalence of sociality varies markedly among families (Lacey and Ebensperger, chap. 34 this volume). As a critical first step toward identifying the selective pressures that have favored group living in some rodent lineages, a better understanding of the nature of rodent sociality is needed, and thus we begin with an overview of group structure in these animals.

Characterizing social groups

Sociality occurs in myriad forms. Underlying this diversity, however, are several general elements of group structure

---

1. We have used the taxonomy of Wilson and Reeder (1993), which recognizes two suborders of rodents. A version of this reference (Wilson and Reeder, 2005) contains a substantially revised taxonomy that includes five suborders of rodents. In this revised taxonomic scheme, the Hystricognathi remain largely intact as the new Suborder Hystricomorpha, with the four other suborders pulled from the Sciurognathi.

that provide the basis for comparative studies of this phenomenon. One axis that is often used to characterize social species is the reproductive structure of groups (Brown 1987; Keller and Reeve 1994). Specifically, societies are frequently divided into those in which all group members reproduce (i.e., egalitarian, low-skew, or plural-breeding groups) versus those in which reproduction is limited to a single member of each sex (i.e., despotic, high-skew, or singular-breeding groups; Vehrencamp 1982; Brown 1987; Clutton-Brock 1998a, 1998b; Reeve et al. 1998). Although the differences between these breeding structures are likely continuous rather than dichotomous (Sherman et al. 1995; Lacey and Sherman, 2005), distinguishing between plural- and singular-breeding groups provides a useful heuristic that has important implications for patterns of cooperation and conflict among group members, as well as for the occurrence of behavioral and morphological specializations among conspecifics (Vehrencamp 1982; Sherman et al. 1995; Emlen 1996; Lacey and Sherman, 2005).

A second axis that often is used to characterize social species is the kin structure of groups. Kinship within groups typically arises due to natal philopatry (Ims 1989; Koenig et al. 1992; Jarvis et al. 1994; Emlen 1995; Nunes, chap. 13 this volume) and hence, kin structure often can be inferred from data indicating which animals remain in their natal group. In general, natal philopatry among mammals is female biased (Greenwood 1983; Dobson 1983; Brody and Armitage 1985; Ims 1990; Hoogland 1995; Solomon 2003); unlike many group-living birds (Koenig and Dickinson 2005), social mammals are rarely characterized by exclusively male natal philopatry (Solomon 2003; Nunes, chap. 13 this volume). Because natal philopatry by females predominates, groups in many mammal species consist primarily of female kin (e.g., Michener 1983a; table 1). As a result, females in these species should receive greater indirect fitness benefits from assisting group mates (Hamilton 1964; Emlen 1997; Reeve 1998) and, accordingly, nepotism (Sherman 1977; Hoogland 1995; Holmes and Mateo, chap. 19 this volume) and cooperation should be more prevalent among females. In contrast, in species in which both sexes remain in the natal area and groups consist of multiple adult females and males, kinship and cooperation should be more equitably distributed between the sexes.

Clearly, the process by which a group forms has important implications for kin structure and hence, the fitness consequences of sociality. Among rodents, natal philopatry appears to be the predominant mode of group formation (Solomon 2003). Although groups in some mammal species form when unrelated individuals aggregate to avoid predators (e.g., Thomson's and Grant's gazelle; FitzGibbon 1990) or to gain access to critical resources (e.g., river otters; Blundell et al. 2002; elephant seals; Le Boeuf and Laws

1994), we know of no social rodents in which groups do not arise primarily due to natal philopatry. The prevalence of philopatry has two important implications for rodent social structure. First, most rodent groups are composed primarily, if not exclusively, of close kin. Second, because natal philopatry is typically female biased, multifemale groups should be more common than multimale groups. Indeed, rodent groups containing multiple adult males appear to be rare (Nutt 2003), while groups containing multiple adult females are common (Solomon 2003; Nunes, chap. 13 this volume).

Philopatry and breeding structure

Among social rodents, philopatry and breeding structure appear to be closely related. Specifically, while plural-breeding groups tend to be characterized by female-only natal philopatry, singular-breeding groups frequently include philopatric animals of both sexes. Among social species of rodents for which appropriate data are available, this association between breeding structure and pattern of philopatry is significant ($N = 23$ species, $G = 17.3$, $P < 0.0001$; table 21.1). A similar, significant relationship between breeding structure and philopatry is obtained when analyses are restricted to a single species per genus to minimize the potentially confounding effects of shared evolutionary history ($N = 11$ species, Fisher's Exact $P = 0.0002$; table 21.1). Two apparent exceptions to this pattern are the California mouse (*Peromyscus californicus*), which is singular breeding with male natal philopatry (Ribble 1991; 1992), and the montane vole (*Microtus montanus*), which is also singular breeding but, at high densities, is characterized by female natal philopatry (Jannett 1978).

For three genera (*Microtus, Cryptomys, Marmota*), data on group structure and dispersal patterns are available for ≥ 4 species (table 21.1). All *Cryptomys* studied to date are singular breeding, and both sexes are philopatric. *Microtus* and *Marmota* exhibit considerable interspecific variation in social structure but, in general, covariation between breeding structure and philopatry within each genus is the same as that evident among genera. This variation among closely related species suggests that these aspects of social structure reflect species- or even population-level variation in environmental conditions, rather than constraints imposed by shared phylogenetic history (Reeve and Sherman 2001). Comparative studies of these four genera should be particularly informative regarding the ecological factors favoring a given dispersal pattern and breeding structure.

Why do patterns of philopatry and breeding structure covary? Ecological conditions play a significant role in determining which individuals remain in their natal area (Emlen 1982; Chepko-Sade and Halpin 1987; Koenig et al.

Table 21.1 List of social rodent species for which data on breeding structure and pattern of philopatry are available

Family	Species	Breeding structure	Philopatric sex	Reference
Muridae:				
Arvicolinae	Prairie vole (*Microtus ochrogaster*)[a]	S	M, F	Getz et al. 1993; McGuire et al. 1993
	Pine vole (*Microtus pinetorum*)	S	M, F	Fitzgerald and Madison 1983; Solomon et al. 1998
	Montane vole (*Microtus montanus*)	S	F	Jannett 1978
	Townsend's vole (*Microtus townsendii*)	P	F	Lambin and Krebs 1991
	Grey-sided vole (*Clethrionomys rufocanus*)	P	F (?)	Saitoh 1989
Gerbillinae	Mongolian gerbil (*Meriones unguiculatus*)[a]	S	M, F	Agren et al. 1989b
	Great gerbil (*Rhombomys opimus*)[a]	P	F	Rogovin et al. 2003
Murinae	Fat dormouse (*Myoxus glis*)[a]	P	F	Pilastro et al. 1996
Sigmdontinae	White-footed mouse (*Peromyscus leucopus*)[a]	P	F	Wolff 1994b
	Deer mouse (*Peromyscus maniculatus*)	P	F	Wolff 1994b
	California mouse (*Peromyscus californicus*)	S	M	Ribble 1992; Solomon and Getz 1997
Castoridae	Beaver (*Castor canadensis*)[a]	S	M, F	Brady and Svendson 1981; Patenaude 1983
Bathyergidae	Naked mole-rat (*Heterocephalus glaber*)[a]	S	M, F	Sherman et al. 1991; Bennett and Faulkes 2000
	Damaraland mole-rat (*Cryptomys damarensis*)[a]	S	M, F	Jarvis and Bennett 1993; Bennett and Faulkes 2000
	Common mole-rat (*Cryptomys hottentotus*)	S	M, F	Burda 1990; Jarvis and Bennett 1990
	Giant mole-rat (*Cryptomys mechowi*)	S	M, F	Burda and Kawalika 1993
	Masona mole-rat (*Cryptomys darlingi*)	S	M, F	Gabathuler et al. 1996
Ctenomyidae	Colonial tuco-tuco (*Ctenomys sociabilis*)[a]	P	F	Lacey et al. 1997; Lacey and Wieczorek 2004
Sciuridae	Black-tailed prairie dog (*Cynomys ludovicianus*)[a]	P	F	Hoogland 1995
	Yellow-bellied marmot (*Marmota flaviventris*)[a]	P	F	Armitage 1991
	Hoary marmot (*Marmota caligata*)	S	M, F	Armitage 2000
	Olympic marmot (*Marmota olympus*)	S	M, F	Armitage 2000
	Alpine marmot (*Marmota marmota*)	S	M, F	Arnold 1990; Armitage 2000
	Long-tailed marmot (*Marmota caudata*)	S	M, F	Armitage 2000

NOTES: Taxonomy follows Wilson and Reeder 1993. For murid rodents, subfamilies are indicated. S = singular breeding (typically one breeding male and one breeding female per group). P = plural breeding (multiple breeding females and/or males per group).
[a]Taxa selected for analyses employing only one species per genus were selected for inclusion in these analyses. Typically, the best-studied species in each genus was selected for inclusion in these analyses.

1992), suggesting that philopatry and associated patterns of kin structure arise due to factors that are extrinsic to social groups. Breeding structure, in turn, reflects individuals' efforts to maximize fitness within the framework imposed by kin structure and limited opportunities for dispersal. For example, the inbreeding avoidance hypothesis (Wolff 1994a; Pusey and Wolf 1996) argues that animals that remain in their natal group may refrain from reproducing to avoid incest and the associated cost of inbreeding depression. Accordingly, in species characterized by female-only natal philopatry (i.e., low probability of father-daughter or brother-sister matings), most females in a group reproduce, while in species characterized by philopatry by both sexes, typically only one female per group reproduces. Thus breeding structure appears to vary in response to the adaptive consequences of reproducing with the suite of potential partners (e.g., kin versus nonkin) generated by patterns of natal philopatry.

Reproductive skew theory (Keller and Reeve 1994; Reeve 1998; Reeve et al. 1998) also suggests that the breeding structure of social groups will be related to opportunities for individuals to disperse and to breed outside of their na-

tal group. Concessions models of reproductive skew predict that, as the difficulty of leaving the natal area increases, dominant breeding animals will be required to concede fewer direct fitness benefits to retain subordinates in their natal group (Reeve and Keller 1995). Thus the degree of skew and the prevalence of singular breeding should be positively related to the difficulty of natal dispersal. Tug-of-war models of reproductive skew (Clutton-Brock 1998a, 1998b) are less explicit concerning expected relationships between dispersal and biases in direct fitness. If, however, increased philopatry leads to greater competitive asymmetries among group members (e.g., the formation of a larger number of competitively asymmetric parent-offspring pairs), then tug-of-war models should also predict that singular breeding will be more prevalent in species in which philopatry is common (Reeve et al. 1998).

These hypotheses assume that philopatry by both sexes indicates that natal dispersal is more constrained than it is when only one sex remains in the natal area. As the probability of successfully leaving the natal area decreases, species may shift from female-only to male-and-female philopatry. Intraspecific variation in social structure provides a valuable opportunity to determine whether these patterns of philopatry represent quantitatively different endpoints along a continuum of dispersal options or qualitatively different routes to sociality that produce distinct patterns of kinship and reproductive success. Temporal variation in social structure has been reported for some group-living rodents, but these examples appear to reflect primarily changes in group size, rather than group structure. For example, prairie voles may occur as lone females, male-female pairs, or singular-breeding groups composed of multiple adults of both sexes (Getz et al. 1993; McGuire and Getz 1998). Similarly, white-footed mice (*P. leucopus*) and deer mice (*P. maniculatus*) may occur as lone females or plural-breeding multifemale groups (Wolff 1994b). None of these species, however, appears to switch from plural-breeding groups with female-only philopatry to singular-breeding groups with philopatry by both sexes, suggesting that these differences in kin and breeding structure reflect qualitatively different responses to selective pressures favoring group living.

Conceptual Approaches to Sociality

The adaptive bases for group living have long puzzled evolutionary biologists, particularly given that individuals appear to incur two unavoidable fitness costs (increased competition, increased exposure to pathogens) but receive no automatic fitness benefits when living with conspecifics (Alexander 1974; Brown and Brown 1996; Krause and Ruxton 2002). Consequently, studies of social species have traditionally focused on identifying benefits intrinsic to group living, such as increased predator protection, increased access to limited resources, and improved foraging via cooperation among group mates (Alexander 1974). Because these analyses have frequently relied upon mean group- or population-level estimates of adaptive benefits (Safran et al., in press), they have not always considered the kin structures of groups or individual variation in the adaptive consequences of sociality.

In contrast, studies of cooperatively breeding vertebrates have emphasized the individual-level fitness benefits of remaining in the natal group (Emlen 1982, 1991; Sherman et al. 1995; Koenig and Dickinson 2004). Given the prevalence of natal philopatry among group-living rodents, this conceptual approach has been widely applied to these animals (Jarvis et al. 1994; Solomon 2003; Lacey and Wieczorek 2004; Hare and Murie, chap. 29 this volume, Armitage, chap. 30 this volume, and Lacey and Ebensperger, chap 34 this volume). On the one hand, ecological constraints models of philopatry assert that groups will form

when physical or social conditions raise the costs of dispersal to the point that, on average, individuals achieve greater fitness by remaining in their natal group than by dispersing and attempting to breed elsewhere (Emlen 1982; Koenig et al. 1992). Factors frequently identified as potential constraints on dispersal include the availability of suitable habitat, potential mates, and resources required for successful reproduction (Emlen 1982; Solomon 2003). On the other hand, benefits of philopatry models (Stacey and Ligon 1991) argue that it is the intrinsic fitness benefits of living with conspecifics that lead individuals to remain in their natal area. Such benefits include increased predator detection, cooperative foraging, and the inheritance of breeding sites —a list that closely parallels the "classic" benefits of group living identified by Alexander (1974).

It is now generally accepted that ecological constraints and benefits of philopatry models represent complementary approaches to the same basic problem, namely identifying the net costs and benefits associated with remaining at home rather than dispersing (Lacey and Sherman 1997; Mumme 1997; fig. 21.1). Despite the emergence of this more synthetic conceptual framework, a dichotomous approach to studies of natal philopatry and the formation of social groups remains common (Safran et al., in press). In part, the persistence of this dichotomy reflects the difficulty of identifying the adaptive consequences of complex behavioral traits such as philopatry. To simplify this task, many studies focus on individual causal factors that, alone, may act primarily as constraints on dispersal or benefits of philopatry. Clearly, one challenge for future studies of sociality is to integrate these approaches into a comprehensive conceptual framework for assessing the fitness consequences of natal philopatry and group living.

Fitness Consequences of Sociality in Rodents

Quantifying the direct fitness consequences of sociality may yield critical insights into the nature of the selective factors favoring group living. Individuals may gain indirect fitness benefits from their interactions with group mates but, typically, these benefits accrue only after groups form (Emlen 1991; Lacey and Sherman 1997), suggesting that the adaptive consequences of group living per se are better understood by considering the direct fitness tradeoffs associated with living and breeding alone versus within a group. At one extreme, in species in which direct fitness increases with group size, sociality is probably maintained by a net balance toward intrinsic benefits associated with group living (Hoogland and Sherman 1976; Brown and Brown 1996; Safran et al., in press). At the other extreme, in species in

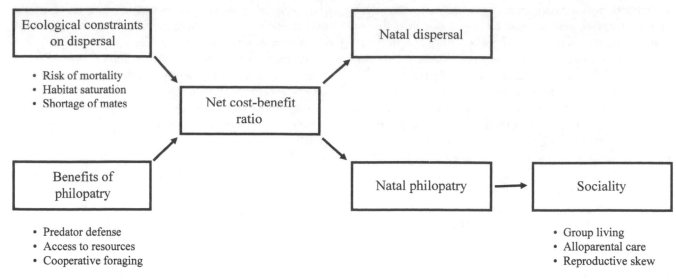

Figure 21.1 Schematic representation of the relationships between ecological constraints and benefits of philopatry models to explain natal philopatry and group living.

which direct fitness decreases with group size, individuals are likely forced to live together due to the prevalence of ecological or other extrinsic factors that limit dispersal and independent breeding (Lacey 2004; Safran et al., in press). In between are species in which the relationship between group size and fitness is more complex and may be nonlinear (e.g., yellow-bellied marmots, *Marmota flaviventris;* Armitage and Schwartz 2000). Although the inclusive fitness consequences of group living may differ somewhat due to the effects of indirect fitness gains, we believe that relationships between group size and direct fitness provide valuable information regarding the factors that promote philopatry and, hence, sociality (see also Jennions and Macdonald 1994).

Data regarding the direct fitness consequences of sociality are available for only a few species of group-living, communally nesting rodents. In several of these taxa, the per capita number of offspring produced does not appear to vary with group size (e.g., dormice, *Glis glis;* Pilastro et al. 1994; white-footed and deer mice; Wolff 1994b). In these species, tenure in the natal group tends to be brief, and individuals may switch from living in a group to living alone (or vice versa) between successive rounds of reproduction (Armitage 1991; Marin and Pilastro 1994; Wolff 1994b). These data suggest that the fitness consequences of solitary versus group life do not differ greatly, with the result that individuals are behaviorally flexible and can alter their social setting in response to small changes in environmental conditions.

In two of the remaining species for which data are available, direct fitness is negatively correlated with group size. In black-tailed prairie dogs (*Cynomys ludovicianus*), the number of young reared to weaning declines with group size (Hoogland 1995), indicating that females pay a direct fitness cost to remain in their natal group (coterie). Similarly, direct fitness is negatively correlated with group size in the colonial tuco-tuco (*Ctenomys sociabilis;* Lacey 2004), a communally nesting rodent from southern Argentina. In both of these species, females frequently remain in their natal group for life (Hoogland 1995; Lacey and Wieczorek 2004), implying that dispersal is sufficiently difficult and dangerous that natal philopatry is favored even though it entails an apparent direct fitness cost.

No communally nesting rodents are known to exhibit a consistently positive relationship between group size and per capita direct fitness. Although intrinsic benefits to group living have been suggested for at least one plural-breeding social rodent, the degu (*Octodon degus;* Ebensperger and Wallem 2002; Ebensperger and Bozinovic 2000), data regarding the direct fitness consequences of sociality in this species are not available. Among yellow-bellied marmots, net reproductive rate increases with group size in smaller groups, but this relationship becomes negative at larger group sizes (Armitage and Schwartz 2000). Finally, in some singular-breeding species such as naked mole-rats, the direct fitness of reproductive individuals may increase with group size, but the strong reproductive skew within groups dictates that nonbreeding individuals experience a decrease in direct fitness by remaining in their natal group (Lacey and Sherman 1997). An important implication of these data is that rodent groups form primarily due to the net effects of extrinsic (ecological) constraints on natal dispersal. While detailed field studies of multiple social species are required to verify this interpretation, the realization that for most individuals direct fitness declines with group size suggests a compelling conceptual framework for exploring the ecological reasons for group living in rodents.

Sociality in Subterranean Rodents

As the preceding sections of this chapter reveal, sociality is a complex phenomenon that likely evolves in response to multiple selective pressures. Given this complexity and the diverse array of rodent societies reviewed in this volume, developing a comprehensive model for understanding sociality in these animals is challenging. Hence, as a starting point for our efforts, we focus on analyses of sociality in subterranean rodents—those species in which individuals spend virtually their entire lives in underground burrows that, typically, they excavate themselves (Nevo 1979; Lacey et al. 2000). Subterranean species have played a significant role in studies of rodent sociality for the past 2 decades (Sherman et al. 1991; Lacey 2000; Faulkes and Bennett, chap. 36 this volume), and data obtained from these animals have substantially influenced thinking on a number of behavioral issues, including the importance of extrinsic determinants of social structure and the nature and taxonomic distribution of eusociality (Sherman et al. 1991; Lacey and Sherman 1997, 2005). While most of the more than 120 rodent species identified as subterranean appear to be solitary (Nevo 1979; Reig et al. 1990), group living has been documented for members of two genera of bathyergid mole-rats (Jarvis and Bennet 1991; Bennett and Faulkes 2000), one species of ctenomyid (Lacey et al. 1997), and the one species of octodontid that is truly subterranean (*Spalacopus cyanus;* Reig 1970). Each of these families represents a distinct evolutionary origin for subterranean sociality and, hence, comparative studies of these taxa provide important opportunities to assess the generality of adaptive hypotheses regarding group living (Bennett and Faulkes 2000; Lacey and Wieczorek 2003; Lacey and Ebensperger, chap. 34 this volume). Although the remainder of this chapter emphasizes subterranean taxa, we believe that the conceptual framework developed here for understanding sociality in subterranean species is applicable to all group living rodents.

Overview of subterranean sociality

The best known of the social subterranean rodents are the African mole-rats of the family Bathyergidae (Bennett and Faulkes 2000). This family includes the naked mole-rat (*Heterocephalus glaber*), which is one of the most social mammals known (Jarvis 1981; Sherman et al. 1991; Bennett and Faulkes 2000; Faulkes and Bennett, chap. 36 this volume). Naked mole-rats live in groups of 200 to 300 individuals, with a mean group size of $\sim$ 80 animals (Braude 1991; Lacey and Sherman 1997). Groups form due to natal philopatry by both sexes, such that colonies are typically extended families. A colony can exist in the same area for many years. Although natal dispersal occurs occasionally (O'Riain et al. 1996; Braude 2000), it is apparently rarely successful. Within an established colony, only one female and one to three males reproduce; reproductive partners are typically close kin (Reeve et al. 1990). Smaller, nonbreeding mole-rats participate in foraging and colony maintenance while larger nonbreeders engage in colony defense and care of the offspring of breeders (Lacey and Sherman 1991, 1997). The striking parallels between the social system of the naked mole-rat and those of social insects have led to the assertion that cooperative breeding and eusociality form a continuum of societal structures (Sherman et al. 1995; Lacey and Sherman 2005). At the same time, studies of naked mole-rats have revealed that both extrinsic (ecological) factors that promote philopatry and intrinsic (genetic) factors that favor cooperation with kin (Hamilton 1964) are important in maintaining such societies (Andersson 1984; Reeve et al. 1990; Jarvis et al. 1994; Bennett and Faulkes 2000).

All species in the bathyergid genus *Cryptomys* studied to date also live in family groups (Bennett and Faulkes 2000; Faulkes and Bennett, chap. 36 this volume). Although the social structures of these species are generally similar to that of naked mole-rats, none exhibits the extremes of colony size or behavioral specialization found in *H. glaber*. Within *Cryptomys*, the Damaraland mole-rat (*C. damarensis*) is the most social. This species lives in groups of 2–40 individuals (modal group size of $\sim$12) that form due to natal philopatry by both sexes (Bennett and Faulkes 2000). Natal dispersal, however, does not appear to be as severely constrained as in *H. glaber,* and the prevalence of philopatry varies among populations (Spinks et al. 2000b). Within a colony of *C. damarensis*, only a single male and female breed; the remaining group members assist the breeders in colony maintenance and defense activities, as well as in rearing young. Unlike naked mole-rats, colonies of *C. damarensis* do not tend to persist beyond the lifetime of a breeding pair and reproductive partners are not typically close kin. Colony sizes for the other species of *Cryptomys* that have been studied are smaller and behavioral specialization among family members is minimal (Bennett and Faulkes 2000; Spinks et al. 2000a, 2000b).

In contrast to social African mole-rats, groups of colonial tuco-tucos (*Ctenomys sociabilis*) consist of two to six closely related females and a single, unrelated male (Lacey and Wieczorek 2004). Approximately two-thirds of yearling females are still resident in their natal burrow system, where they remain for the duration of their life. In contrast, all males disperse from their natal area at the end of their juvenile season and adult males disperse between breeding seasons (Lacey and Wieczorek 2004). Within a group of *C. sociabilis,* all females reproduce and all individuals engage in activities such as burrow excavation, foraging, and alarm calling to predators (E. A. Lacey, unpublished data). Although allonursing has been observed in the lab, it has not yet been documented among free-living animals. In short, this species exhibits none of the within-group reproductive or behavioral specializations that characterize social bathyergids. Moreover, *C. sociabilis* is the only species of ctenomyid for which quantitative evidence of group living has been obtained; although anecdotal accounts of burrow sharing exist for several other species of tuco-tucos, the majority of ctenomyids are apparently solitary (Reig et al. 1990; Lacey 2000; Lacey and Ebensperger, chap. 34 this volume).

The final independently evolved example of subterranean sociality is the cururo (*Spalacopus cyanus*). This species has been studied less extensively than the colonial tuco-tuco or social African mole-rats, but available data suggest that a group consists of two to fifteen adults, including, in at least some populations, multiple adult males (Reig 1970; Lacey, Ebensperger, and Wieczorek, unpublished data). Natal philopatry has been detected for both sexes, as has dispersal between groups by adult males. The relative frequency of dispersal versus philopatry for each sex, however, is unknown. Within a group, multiple females show evidence of reproduction (Begall et al. 1999; Lacey, Ebensperger, and Wieczorek, unpublished data), although the distribution of direct fitness among group mates is unknown. As in colonial tuco-tucos, all group members appear to engage in burrow excavation and alarm calling, with no evidence of behavioral specialization among individuals.

Comparing these taxa, it appears that colonial tuco-tucos differ from social African mole-rats and cururos in that groups of *C. sociabilis* never contain more than a single adult male. Thus while colonial tuco-tucos are female kin groups that arise due to female-biased natal philopatry, groups of social mole-rats and cururos contain multiple adult males and females and arise due to philopatry by both sexes. Only the social African mole-rats, however, are known to be singular breeders in which direct fitness is restricted to a small subset of colony members and numerous nonbreeders provide alloparental care to the offspring of reproductive animals.

Adaptive bases for subterranean sociality

Social African mole-rats, colonial tuco-tucos, and cururos all form groups composed of multiple adults that live together in a single burrow system. The adaptive bases for burrow sharing have been examined most thoroughly for bathyergid mole-rats (Jarvis et al. 1994; Lacey and Sherman 1997; Bennett and Faulkes 2000; Faulkes and Bennett, chap. 36 this volume). The ecological factors thought to favor sociality in these animals are summarized by the aridity food-distribution hypothesis (AFDH), which proposes that the energetic cost of burrowing through hard soils to locate patchily distributed but locally abundant food resources is the primary selective factor favoring group living. Specifically, in arid regions characterized by difficult-to-excavate soils, the only time that it is energetically feasible for the animals to expand their burrows is when the soil has been recently softened by rain. Because the roots and tubers that the animals feed on are patchily distributed, a single individual would be unable to tunnel far enough to locate a new food supply before drying of the soil again renders the cost of tunnel excavation prohibitive. By living together and working cooperatively to excavate new tunnels, the animals are able to locate sufficient food resources to survive until the next rain.

Three lines of evidence have been used to evaluate the AFDH. First, analyses of the geographic distribution of social bathyergids suggest that while group-living species can occupy arid habitats, solitary taxa are limited to more mesic regions (Jarvis et al. 1994), as expected if sociality is necessary in areas characterized by high energetic costs of burrow excavation. Second, within species, comparative studies of two populations of the common mole-rat (*C. hottentotus*) have revealed that natal philopatry is more common in an arid than in a mesic habitat (Spinks et al. 2000a, 2000b). Third, long-term field studies of Damaraland mole-rats indicate that larger groups are better able to survive extended droughts, suggesting that sociality is an adaptive response to harsh, arid conditions (Faulkes and Bennett, chap. 36 this volume).

Although the AFDH appears to be consistent with behavioral and ecological patterns among African mole-rats, it does not seem to explain sociality in colonial tuco-tucos. Comparative studies of the habitats occupied by a population of this species and a population of the syntopic but solitary Patagonian tuco-tuco (*C. haigi*) revealed no significant differences in rainfall or food resource distributions (Lacey and Wieczorek 2003). Although significant differences in soil penetrability (a proxy for hardness) were detected, it was the soils occupied by the solitary species that were more difficult to excavate. This difference in soil conditions appeared to result from interspecific differences in habitat use. While the group-living *C. sociabilis* occurs primarily in patchily distributed areas of mesic grassland known as *mallines, C. haigi* occurs in these patches as well as in the intervening areas of arid steppe habitat. This difference in specialization creates very different spatial distributions of suitable habitat for these species, leading Lacey and Wieczorek (2003) to suggest that the difficulty of dispersing between patchily distributed mallín habitats is an important ecological variable favoring sociality in *C. sociabilis*.

The adaptive bases for sociality in cururos have not been explored. *S. cyanus* occupies a wide range of habitats in Chile, including arid mediterranean grasslands and mesic montane meadows. This intraspecific habitat variation provides an ideal opportunity to assess the ecological bases for sociality. Preliminary comparative data from two populations of *S. cyanus* indicate that soils are harder to penetrate at an arid coastal site (Parque Nacional Fray Jorge) than at a mesic montane site (Santuario de la Naturaleza Yerba Loca; Lacey, Ebensperger, and Wieczorek, unpublished data). Although food resource distributions at these localities have not been quantified, the habitat at Fray Jorge is, at least superficially, more similar to areas inhabited by social mole-rats than the habitat at Yerba Loca, which closely resembles the mesic mallínes inhabited by colonial tuco-tucos. Given these parallels, it is intriguing that preliminary mark-recapture data suggest that, contrary to the predictions of the AFDH, groups of *S. cyanus* are larger at Yerba Loca (Lacey, Ebensperger, and Wieczorek, unpublished data).

In summary, the two subterranean examples of sociality that have been studied in detail yield different pictures of the ecological factors that favor group living. At a general level, these arguments are similar in suggesting that environmental conditions constrain dispersal, thereby favoring natal philopatry and the formation of social groups. Resource distributions also appear to be important in both taxa, although the nature of these resources (food versus suitable habitat) and the spatial scales over which they are patchily distributed (within versus between burrows) may differ between social mole-rats and colonial tuco-tucos. Data from cururos are not yet sufficient to determine whether the ecology of group living in this species parallels that of either mole-rats or tuco-tucos, or whether this example of subterranean sociality reflects a third, distinct suite of environmental conditions.

An Integrated Ecological Model for Sociality in Rodents

If the ecological bases for group living differ between the two best-studied subterranean taxa—African mole-rats and colonial tuco-tucos—is it possible to develop a general explanation for sociality among subterranean species, let

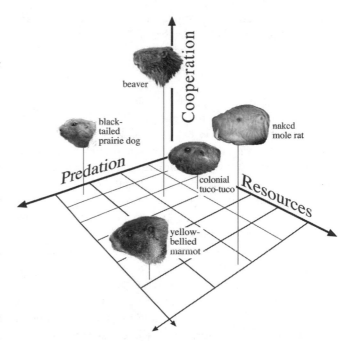

Figure 21.2 Three-dimensional conceptual model of the ecological bases for sociality. The three primary ecological factors thought to influence natal philopatry and group living are indicated. Postulated locations for several well-studied species are shown to indicate how data for these species would be represented in this model.

alone all rodents? We believe that such an explanation is possible. Here, we propose a three-dimensional model of the ecological factors that affect rodent social systems (fig. 21.2). This model is derived from comparative analyses of the social, subterranean species reviewed previously in this chapter, but should be applicable to semi-subterranean or surface-dwelling rodents as well as to facultatively versus obligately social species. Our intent is to provide a general framework for future studies that explore the ecological bases for group living in rodents. The axes included in our model are as follows:

1. Access to critical resources

Access to critical food resources is an integral component of the AFDH proposed for African mole-rats (Jarvis et al. 1994; Lacey and Sherman 1997; Bennett and Faulkes 2000). Although the reasons why colonial tuco-tucos are restricted to mesic, mallín habitats are unknown, these habitat patches themselves represent important resources, the distribution of which is thought to favor natal philopatry and group formation in *C. sociabilis* (Lacey and Wieczorek 2003). More generally, conspecifics may be required to live together in order to have access to limited, patchily distributed resources (Alexander 1974; Safran et al., in press), implying that this axis represents a logical component of any ecological model of social behavior.

2. Predation

Predation has frequently been invoked as a selective pressure favoring group living (Hoogland and Sherman 1976; Hoogland 1979a, 1981; Krause and Ruxton 2002). Surprisingly, however, predation on social subterranean rodents has received little attention, despite compelling anecdotal evidence (see the following) that these animals are preyed upon by multiple species. In particular, the AFDH does not consider predation, an omission that we address by including predation as one of three axes in our model.

Predation is often difficult to quantify. For example, even though social African mole-rats are known to be preyed upon by birds, mammals, and snakes (Jarvis and Sherman 2002), no data are available regarding the frequency of this predation. Jarvis et al. (1994) indicate that, qualitatively, predation (i.e., the identities of predatory species) does not appear to differ between solitary and social species of mole-rats, but this observation does not preclude quantitative differences in predation pressure that may favor sociality in some species but not in others. Behavioral studies of naked mole-rats indicate that colony members exhibit a conspicuous, cooperative response to threats within their burrow systems (Lacey and Sherman 1991), implying that, over evolutionary time, these animals have regularly encountered predators in their burrows. Collectively, these lines of evidence suggest that predation may be an important component of mole-rat ecology.

Similarly, predation has not been considered as part of comparative studies of the ecology of solitary and social tuco-tucos. Building from the AFDH, these analyses have focused on the same suite of ecological factors identified as important to bathyergid mole-rats (Lacey and Wieczorek 2003). Colonial and Patagonian tuco-tucos are preyed upon by foxes (*Dusicyon australis*) as well as by several species of birds (e.g., barn owls, *Tyto alba;* caracaras, *Milvago chimango;* Lacey and Wieczorek 2003). Data regarding predation are comparable to those for African mole-rats; although the suite of predators affecting *C. sociabilis* and *C. haigi* is the same, it is possible that quantitative differences in predation contribute to the differences in social structure between these species. *C. sociabilis* routinely gives alarm calls in response to predators, suggesting that, evolutionarily, predation pressure has been sufficient to favor a specific behavioral response. In sum, despite the lack of quantitative information regarding differences in predation on solitary versus social subterranean rodents, we believe that predation is an important ecological factor that should be included in our model.

3. Cooperation as a response to ecological challenges

The adaptive benefits of cooperation may vary according to the nature of the ecological factors that favor group living. For example, if (as postulated by the AFDH) temporal constraints on the excavation of new burrows (due to soil hardness in dry seasons) are a primary factor underlying sociality among African mole-rats, then cooperative digging by these animals should be favored because it increases the probability of locating additional food resources during the limited period when burrow excavation is possible. In contrast, if the difficulty of dispersing from one mallín patch to another is the primary ecological factor favoring sociality in colonial tuco-tucos, it seems unlikely that cooperation among burrow mates would provide individuals with an advantage in responding to this environmental challenge; although the animals could disperse in groups, this would not require that they share a burrow system, nor does it seem likely to increase the chances of locating a new patch of suitable habitat. This difference between social African mole-rats and colonial tuco-tucos suggests that the ability of cooperative behavior to resolve ecological challenges may be an important component of social structure.

At first glance, this axis appears to differ from the previous two in that it represents a consequence, rather than a cause of group living. Cooperative solutions to ecological challenges, however, may be critical to the persistence of groups that form due to extrinsic environmental conditions (Alexander 1974; Brown and Brown 1996). Indeed, in some species, cooperation may function as a form of group augmentation (Kokko et al. 2001) that contributes significantly to the maintenance of sociality. At a minimum, the degree of cooperation among group mates provides an important basis for comparing different social systems and, thus, inclusion of this axis in our model should facilitate distinctions between highly cooperative societies (e.g., naked mole-rats) and societies in which individuals live together but do not engage in elaborate forms of cooperation (e.g., colonial tuco-tucos).

The axes that form the basis for our model closely parallel the benefits generally ascribed to group living (Alexander 1974: Hoogland and Sherman 1976; Krause and Ruxton 2002; Safran et al., in press). Although it is implicit in many previous treatments of this subject that sociality likely reflects the combined effects of multiple environmental factors, our model makes this explicit by treating each of the relevant variables as one of three distinct factors contributing to sociality. Heuristically, the model can be viewed as a three-dimensional graph, with the distance from the origin along the x-, y-, and z-axes increasing as the strength of each selective factor (access to resources, predation pressure, benefits of cooperation) increases (fig. 21.2). Plotting data from different populations or species of group-living rodents should yield a multidimensional cloud of points. Populations or species occurring in the inner portion of the cloud (i.e., near the origin) should be those in which individuals receive only modest fitness benefits from sociality, while members of those in the outer reaches of the cloud should gain substantially from group life.

The precise shape of this cloud of points is not yet known, but should be informative. For example, if the axes in this model are not truly independent (e.g., benefits to cooperation are influenced by predation pressure or access to resources), then clustering of points in different regions of the graph may reveal previously undetected relationships among different ecological variables. At the same time, comparing the distribution of points for the two general types of social groups identified previously in this chapter may help to determine whether plural-breeding species with female philopatry are qualitatively distinct from singular-breeding species in which both sexes are philopatric. Finally, it may be possible to relate the axes of this model and the relative positions of different species to the typical direct fitness benefits associated with group living.

As an example of how this model should work, we expect the placement of African mole-rats and colonial tuco-tucos on the 3-dimensional graph to differ markedly. Given that naked and Damaraland mole-rats appear to cooperate extensively in response to ecological challenges, we predict that they will be positioned far from the origin on the corresponding axis. In contrast, colonial tuco-tucos are not highly cooperative and, hence, they would be located much closer to the origin of this axis. As a result, even if the effects of resource distributions and predation are comparable for colonial tuco-tucos and naked mole-rats, these species will occur in quite different regions of the graph. This difference, in turn, may have important implications for other aspects of the social systems of these species, including the degree of reproductive and behavioral specialization among group mates.

Extending these ideas to other group-living rodents, available data suggest that predation is an important factor favoring sociality in black-tailed prairie dogs (Hoogland 1995), indicating that this species should be positioned far from the origin on this axis, but much nearer to the origin for the other two axes of the model (fig. 21.2). In contrast, beavers are thought to form groups due largely to benefits associated with the cooperative construction of lodges and larders (Busher, chap. 24 this volume), with the result that this species should be much further from the origin along the cooperation axis than along the predation or resources axes. Finally, both resource distributions and predation are thought to influence yellow-bellied marmots (Blumstein and Armitage 1999; Armitage, chap. 30 this volume), leading to

our placement of this species far from the origin on the corresponding axes of the model.

Given the difficulty of quantifying each of the selective pressures identified by this model, it remains, for the moment, a conceptual construct for exploring the effects of different ecological variables on rodent societies. The strength of this approach is that it brings together several of the key selective factors thought to promote sociality in one comparative, at least potentially quantifiable framework. As studies of rodent social systems continue, we anticipate that quantitative comparisons of these variables will increase, leading to a more refined understanding of how ecological variables contribute to the behavioral variation evident among these animals. Although our discussion has focused on the adaptive bases for group living among subterranean species, we expect that these ideas can be extended to include other rodent societies. At a minimum, we hope that our discussion of rodent sociality will stimulate further studies of this diverse and fascinating order of mammals.

Summary and Future Directions

Determining why conspecifics live in groups and how those groups are structured with respect to kinship and reproduction is critical to understanding the diversity of rodent social systems portrayed in this volume. In many species of rodents, social groups form due to natal philopatry, with the result that groups are composed of close kin of one or both sexes. The reproductive structure of groups appears to covary with patterns of philopatry and kinship; while plural breeding is typical in species with female-only philopatry, singular breeding predominates in species in which both sexes remain in their natal area. Philopatry is expected to occur when the confluence of intrinsic benefits of group living and extrinsic constraints on dispersal dictate that individuals will achieve greater fitness by remaining in their natal group than by dispersing and attempting to breed elsewhere. Although data are available for only a few species, the generally negative relationship between direct fitness and group size implies that, among rodents, sociality may be strongly influenced by extrinsic (ecological) constraints on dispersal.

Numerous factors influence an individual's decision to remain in the natal area and, accordingly, conceptual models of sociality should consider multiple selective pressures that promote group living. Our comparative analyses of social subterranean rodents suggest that access to critical resources, predation pressure, and the use of cooperative interactions to overcome ecological challenges form the basis for a multidimensional model of the ecology of group living. Specifically, each of these parameters can be viewed as one axis in a 3-dimensional graph of social structure, with the distance from the origin reflecting the strength of the selective pressure exerted by each factor. By plotting populations or species on these axes, it should be possible to assess the relative importance of the different factors promoting sociality as well as to identify higher-order patterns of social structure, such as ecological distinctions between singular- and plural-breeding rodent societies.

Clearly, one challenge that must be met by future studies of rodent social systems is to document behavioral and environmental variables in ways that allow quantitative comparisons of the type intended by our three-dimensional model. At present, efforts to develop a general conceptual framework for sociality in rodents are hampered by a lack of quantitative data relevant to comparisons of social structures or environmental conditions. Developing metrics for behavioral and environmental variables that can be used across taxa will substantially improve our ability to identify general behavioral patterns, as will efforts to test alternative adaptive hypotheses for each species studied. Although our discussion of rodent sociality has focused largely on subterranean species, we believe that the conceptual framework and graphical model outlined here can be extended to include all rodents. Ideally, our discussion of these ideas will encourage considerable further investigation of the ecology of sociality in these animals.

Chapter 22 Scent Marking

S. Craig Roberts

THE SCENT MARKS OF RODENTS have been cast as an olfactory equivalent of the elaborate and colorful train of the peacock (*Pavo cristatus;* Penn and Potts 1998a). This is a helpful analogy, illustrating the importance of scent marking in rodent sexual selection. Just as peahens prefer males with the showiest trains and gain fitness benefits through mating with them (Petrie et al. 1991; Petrie 1994), so female rodents use scent marks of males when choosing mates (as, indeed, do females of many other mammals). However, the analogy tells only part of the story, for scent marking is also inextricably linked with competition over resources and mating opportunities, usually between males. In this sense, scent marking resembles, for example, the roars of red deer stags (*Cervus elaphus;* Clutton-Brock and Albon 1979; Clutton-Brock et al. 1979), on the basis of which potential combatants assess their relative competitive ability and decide whether to challenge an opponent physically. In rodents, as in most mammals, scent marking is a means by which individuals assess the competitive ability of opponents (Gosling 1982, 1990; Gosling and Roberts 2001a). This may occur remotely, before an encounter occurs, or in conjunction with further assessment face to face. While there is variability in, and some debate about, the mechanisms involved, there is little doubt that scent marking is a fundamental component of territorial behavior and of advertising dominance status within social hierarchies.

The benefits of being chosen as a mate or controlling access to mating opportunities account for most, if not all, scent-marking behavior. Evidence from across mammals suggests that scent marking initially evolved as a compo-

nent of competitive behavior between same-sexed individuals (usually males, although females often scent mark) and that it subsequently became used in mate choice (usually by females; Gosling and Roberts 2001a). The possibility remains, however, that scent marks of males, in some cases, are signals directed specifically to females (Gosling and Roberts 2001a).

In this chapter, I review the wealth of recent rodent studies in light of the view that scent marks are signals of status. Having described the principal glandular sources and behaviors involved, I summarize the evidence that marking is involved in intrasexual competition and mate choice. The evolution of scent marking depends, as in all signals, on the reliability of information that marks contain, and I outline some ways rodents keep signals honest. These ways include major mechanisms by which information carried in scent marks is transmitted to receivers, and key processes (e.g., signal cost, individuality, memorability) that are prerequisites for various mechanistic and functional explanations for scent marking. Finally, I describe some new research directions that may become a focus for the future, including the need for more field studies to validate and test many of the ideas discussed here, which have been largely driven from the laboratory.

Scent Sources and Scent-Marking Behavior

Rodent scent marks emanate from a variety of glandular sources (table 22.1; reviewed in Brown 1985b; Halpin 1985; Macdonald 1985). Urine and anal gland secretion are the

Table 22.1 Distribution of odor sources in major rodent families

Family	OR	HG	MG	EG	DG	VG	PG	CG	AG	PR	UR	PP	VG	Other sources
Aplodontidae	+										+			
Sciuridae	+				+				+	+	+			
Geomyidae	+													
Heteromyidae	|		+	+	+	+					+	+		
Castoridae							+		+		+	+		Castor gland
Pedetidae									+	+				
Cricetidae	+	+	+	+	+	+	+	+	+		+	+	+	Clitoral gland; flank gland; hip gland; neck gland
Spalacidae											+			
Muridae	+	+	+	+		+	+	+	+		+	+	+	Cheek gland
Caviidae									+	+	+			Coccygeal gland; chin gland
Hydrochoeridae									+		+			Morrillo
Dasyproctidae									+		+			
Chinchillidae											+			
Capromyidae									+		+			
Octodontidae											+			
Erethizontidae											+			

SOURCE: Adapted from Brown (1985b), Halpin (1985), and Macdonald (1985).
NOTES: OR = oral glands, lips, saliva; HG = harderian gland; MG = meibomian gland; EG = ear gland; DG = dorsal gland; VG = ventral (or mid-ventral) gland; PG = pedal gland (including plantar glands); CG = caudal gland; AG = anal gland; PR = perineal gland; UR = urine; PP = preputial gland; VG = vaginal secretion.

commonest odor sources. Most species use at least two sources, while others have several. For example, Libyan jirds (*Meriones libycus*) use urine and oral, gular, palmar, plantar, abdominal, preputial, and clitoral gland secretions (Djeridane 2002).

Why do certain species use multiple sources of scent? Differences in number of scent sources at the family level may be partially explained by variation in species-richness, but may also reflect strength of selection on signaling. In *Microtus* species, for example, the number of distinct sources correlates with degree of sociality (Ferkin 2001). Information available in different glands may be additive, though there may also be some redundancy. Lai et al. (1996) compared responses of Djungarian hamsters (*Phodopus campbelli*) to same- or opposite-sex scents. Females investigated male urine and mid-ventral gland secretion (MVGS) more than female scents from the same sources, suggesting redundancy in urine and MVGS at least in terms of gender recognition (similar patterns with different odor sources occur in meadow voles, *Microtus pennsylvanicus;* Ferkin and Johnstone 1995). In contrast, male hamsters respond to different female odors depending on their reproductive status (Lai et al. 1996). Mouth and urine odors were only attractive during postpartum estrus, and attractiveness of vaginal odor peaked at estrus, while MVGS was most attractive immediately before parturition. These temporal differences indicate additive information in different odors and suggest that together they provide a more precise record of individual condition than does one source alone.

Scent deposition takes a variety of forms and specialized behavior patterns. Urination and anal dragging are the most common application behaviors. That urine marking is communicatory, not simply eliminatory, is illustrated by behavioral differences associated with social rank. Dominant laboratory mice (*Mus musculus*) deposit urine in numerous small spots, subordinates typically creating large pools (Desjardins et al. 1973; Bishop and Chevins 1987). South American maras (e.g., *Dolichotis patagonum*) forcibly project urine sprays toward conspecifics (Taber and Macdonald 1984). Kangaroo rats (*Dipodomys* spp.) deposit dorsal gland secretion by rubbing themselves in sandbathing sites (Randall 1981, 1987b). Beavers (*Castor* spp.) actually create marking sites, earth mounds (Aleksiuk 1968), on which they place castoreum, a mixture of castor and anal gland secretion and urine. Artificially constructed mounds elicit normal behavioral responses when presented with castoreum, but not without it (Schulte 1998; Rosell et al. 2000).

Almost all scent-marking studies document variation according to at least one, and often to all, of the following factors: age, sex, physical condition, and season. As a generalization, scent marking is more frequent when animals are adult, male, and dominant or territorial, especially during breeding. This variation is not particular to rodents; the same applies to most mammals (Gosling and Roberts 2001a) and some other taxa (e.g., Moore et al. 1995). Careful documentation of these qualitative differences is the key to understanding the evolution and function of scent marking.

Function of Scent Marking

Functional paradigms

Historically, a variety of functional explanations have been proposed for scent marking (reviewed by Gosling 1982, 1990). Examples include the idea that marks aid in self-orientation within territories (Kleiman 1966; Walther 1978) or in monitoring resource use by providing information about previous visits to feeding sites (Henry 1976; Harrington 1981; Rozenfeld et al. 1994). Such explanations largely depend on levels of exclusive use of space or resources that are unwarranted given actual observations, and cannot readily account for qualitative differences in marking associated with sex, age, status, and season. Similarly, marking behavior in monogamous species has been interpreted as functioning in pairbond maintenance (e.g., Peters and Mech 1975) but can also be explained by general principles that apply equally across mating systems, such as intrasexual competition over mates (Gosling 1982; Roberts and

Dunbar 2000). More promising alternatives included proposals that scent marks deter or intimidate territorial intruders (Hediger 1949; Geist 1965), but these suggestions are not supported by observations, especially since intruders do not usually retreat upon finding a mark.

Gosling instead proposed that scent marking by resource holders provides a means of competitor assessment, signaling fitness costs of trespassing to receivers. These costs are a product of the probability that the signaler will return and of its relative competitive ability (Gosling 1982, 1990; Gosling and Roberts 2001a). On detecting scent marks, receivers have three main options: either withdraw from an area immediately, remain but withdraw on encountering the owner, or remain to further assess the owner, perhaps even deciding to mount an ownership challenge (fig. 22.1). Which option receivers take will be influenced by the potential costs signaled in the mark, the value of the marked resources, the costs of injury, and the scale of assessment error. This may be why responses of receivers to scent marks are so variable. Signalers also benefit because receiver re-

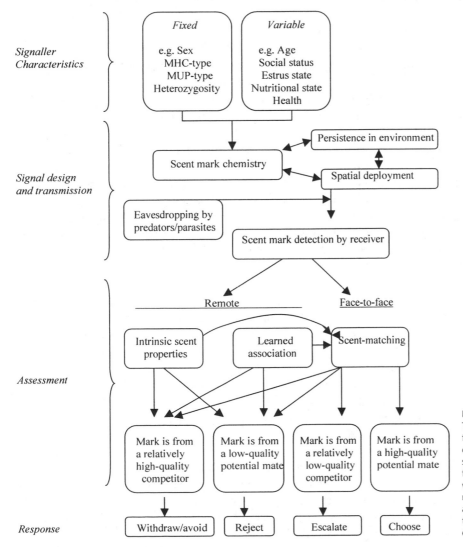

Figure 22.1 Summary of scent marking processes. The assessment and response phases are simplified for illustrative purposes, but in reality responses are complicated by many factors including the value to same-sex receivers of the marked resource, costs of fighting over it and the probability that the receiver will meet the signaller. If receivers are potential mates, responses may vary according to, for example, sexual receptivity, genetic dissimilarity, and the suitability of the signaller in comparison to others already encountered.

sponses reduce the number and intensity of potentially costly fights to which they are exposed. The formulation of scent marks as status signals builds on conventional competitor assessment and game theory (e.g., Parker 1974; Maynard Smith 1982, 1996), providing a clearly stated, unifying theoretical paradigm for understanding scent marking alongside well-developed models of visual and acoustic signaling.

This paradigm can easily be extended to incorporate signal reception by mates (Gosling and Roberts 2001a). Should females be receivers, the nature of the decisions involved may differ, but the reliability or "honesty" (Zahavi 1975) of the signal will still apply. Crucial to the idea that scent marking is an honest signal are findings that male scent-marking rates, associated gland sizes (Horne and Ylönen 1998), and attractiveness to females (Drickamer 1992a) are heritable. Signal costs are discussed more fully later in this chapter.

Despite the growing consensus that scent marks are status signals, the scent-marking literature remains replete with a variety of functional interpretations and subject to a fair amount of debate. Failure to take full account of the complexities of scent-marking behavior can lead to premature rejection of status signaling as a function of marking (see Gosling and Roberts 2001b). Confusion also exists between function and mechanism. For example, Sun and Müller-Schwarze (1998b) conclude that scent matching is the function of marking behavior in beavers, rather than a mechanism mediating a territorial defense function. Scent marks are also sometimes asserted to function as signals of individual identity (e.g., Wolff et al. 2002). Individually specific scent properties may be a necessary precondition for various functional explanations, but they do not provide an explanation of the fitness benefits gained from scent marking. Similarly, the idea that the function of scent marking is self-advertisement (that is, purely communicating presence in an area; e.g., Thomas and Kaczmarek 2002; Thomas and Wolff 2002) does not explain qualitative variation in marking behavior and responses to marks among individuals of different status.

Self-advertisement and other characteristics of marking behavior (e.g., scent individuality, countermarking, overmarking, scent masking) are integral properties or processes involved in status advertisement but are not functions in themselves. Nonetheless, there has been gradual movement over the past two decades toward a consensus view that scent marks are status signals allowing competitor and mate assessment. The following sections outline pertinent evidence arising from rodent studies.

Scent marks and male intrasexual competition

There is overwhelming evidence that scent marking is involved in intrasexual competition among males (reviews in Ralls 1971; Johnson 1973; Brown and Macdonald 1985; Gosling 1990; Gosling and Roberts 2001a). Evidence pointing to this association includes (1) correlations between social status and both scent marking and responses to marks, (2) links between frequency of scent marking and strength of intrasexual competition, (3) nonrandom deposition of scent marks within territories, (4) correlations between qualitative differences in scent chemical composition and social status, and (5) demonstration that marking and glandular development are often androgen dependent.

1. Individuals of elevated social status (i.e., dominant or territorial males) typically mark at higher frequencies than low-status males. For example, this occurs in laboratory mice (Bishop and Chevins 1987; Gosling et al. 2000), house mice (*Mus domesticus;* Hurst 1990a, 1990c), capybaras (*Hydrochaeris hydrochaeris;* Herrera and Macdonald 1994) and bank voles (*Clethrionomys glareolus;* Rozenfeld et al. 1987). Dominant male mice are quick to overmark the marks of other males (Hurst 1990b). Rates of scent marking by young mice are the best predictor of dominance in later life (Collins et al. 1997).

Correlations are also evident between social status and responses to scent marks. Males of several species of rodents avoid scent-marked substrates, especially when they are of low competitive ability (Gosling et al. 1996a, 1996b; Lai and Johnston 2002; Luque-Larena et al. 2002c) or the scent is from dominant males (Summerlin and Wolfe 1972; Jones and Nowell 1989; Hurst et al. 1994). Males avoid prolonged fights with males whose scent suggests they are territory owners (Gosling and McKay 1990; Hurst et al. 1994; Luque-Larena et al. 2001). In addition, male mice that were defeated in interactions display prolonged inhibition of urine marking compared with nondefeated controls (Lumley et al. 1999).

2. Scent-marking effort is associated with levels of intrasexual competition. Simulated territorial intrusions by males stimulate increased marking in male blind mole-rats (*Spalax ehrenbergi;* Zuri et al. 1997), beavers (Rosell et al. 2000; Rosell and Bjorkoyli 2002), and alpine marmots (*Marmota marmota;* Bel et al. 1995). Dominant male mice increase marking frequency as subadults within their territory mature (Hurst 1990b).

Increased marking frequencies also influence investment in the glandular structures that produce secretions. Among dominant male mice, those smaller than their subordinates scent mark at higher frequencies and develop absolutely larger preputial glands than relatively large dominants, indicating they have to work harder to establish and maintain social status due to their relative size disadvantage (Gosling et al. 2000). Similarly, gland weights of male house mice housed with another male for 2 weeks grow to a size almost double that of isolated males (Bronson and Marsden 1973). Males housed adjacent to intact mice develop larger glands than controls housed next to castrates, whereas glands of males housed next to females become smaller (Hayashi 1986). Regular exposure to unfamiliar scent also increases gland size in receivers (Hayashi 1990). Unaggressive strains of mice have smaller glands than aggressive strains (Yamashita et al. 1989). In bank voles, dominance is correlated with preputial gland size (Gustafsson et al. 1980), while in coypus (*Myocastor coypus*) anal gland sizes are predicted by numbers of male but not female recruits into a wild population (Gosling and Wright 1994).

In addition to absolute size, differences in glandular structure exist between individuals of different status. Preputial glands of dominant male mice were well-developed with acini at different stages of maturation, many cytoplasmic organelles, and healthy oval-shaped nuclei. In contrast, subordinates had less developed glands with fewer cellular organelles and shrunken, lobulated nuclei (Brain et al. 1983).

3. Territorial males deposit marks where they are more likely to intercept intruders. Spatial clustering of marks toward territorial boundaries is often found in mammals (Gosling 1981; Gosling and Roberts 2001a), including rodents (e.g., Bel et al. 1995). In blind mole-rats, experimentally manipulating intrusion pressure induces spatial shifts in marking effort (Zuri et al. 1997). In beavers, marks are clustered toward territorial boundaries and upstream of lodges, reflecting the direction of emigration from natal territories (Rosell et al. 1998; Schulte 1998). In addition, more mounds are formed along densely populated large rivers than small ones, probably reflecting intrusion pressure from potential immigrants (Ulevicius and Balciauskas 2000). Individuals also mark along paths and at burrow entrances (e.g., Banks and Banks 1979; Ferron and Ouellet 1989b; Rozenfeld et al. 1994; Blumstein and Henderson 1996; Brady and Armitage 1999). However, an absence of well-defined spatial patterns is not necessarily evidence against a territorial role for marking, because factors that are spatially heterogeneous (e.g., resources, intrusion pressure) may also influence marking economics (Roberts 1997; Roberts and Lowen 1997; Gosling and Roberts 2001b).

4. Chemical differences exist in the scent marks of males of different status. The best evidence comes from house mice, where chromatographic comparisons reveal quantitative differences in sixteen urinary compounds between dominant and subordinate males (Harvey et al. 1989). These characteristic differences can arise within 7 days of status establishment: concentrations of urinary dihydrofurans, ketones, and acetates decrease in subordinate urine, while 2-(*sec*-butyl)-4,5-dihydrothiazole and two sesquiterpenic compounds, alpha- and ß-farnesene, increase in dominant urine. The farnesenes are the two most prominent constituents of preputial gland secretions (Novotny et al. 1990), which as previously described are closely linked to dominant behavior.

5. If scent marking is involved in intrasexual competition between males, it would be surprising if the secretion-producing glands were not androgen dependent. Evidence for androgen dependence is suggested by correlations between mass of testes and scent glands, such as the morrillo of adult capybaras (Herrera 1992). Similarly, endogenous testosterone levels of adult male gerbils are correlated with ventral gland size and frequency of scent marking (Clark et al. 1990, 1992b).

Androgen dependence has been conclusively demonstrated by castration and androgen restoration. In house mice, prolonged investigation of an area is inhibited by scent marks. Castration of scent markers eliminates this effect, while testosterone treatment reestablishes it (Jones and Nowell 1973; Sawyer 1980). These behavioral effects are probably due to changes in the chemical constituents of urine in relation to testosterone levels (Novotny et al. 1984; Harvey et al. 1989). The four principal compounds of dominant males (farnesenes, dihydrothiazole and dehydro-*exo*-brevicomin) are not present in urine of castrates but are restored by testosterone treatment (Harvey et al. 1989). In Mongolian gerbils (*Meriones unguiculatus*), atrophy of preputial and abdominal sebaceous glandular tissues is induced by castration and restored by testosterone (although there is no effect on oral, gular, palmar, and plantar glands; Djeridane 2002). Similar effects on inhibition of marking behavior are known in Mongolian gerbils (Arkin et al. 2003), Long-Evans rats (*Rattus norvegicus*: Matochik and Barfield 1991), European ground squirrels (*Spermophilus citellus*; Millesi et al. 2002), and tree shrews (*Tupaia belangeri*; Holst and Eichmann 1998).

Scent marks and female choice

Male scent marks influence females in two main ways. First, chemical constituents of marks elicit physiological, or priming, responses in females (Driekamer chap. 9 this volume). Second, females use information in marks during mate choice. As the higher-investing sex, females should choose males in relation to their mate quality (Bateman 1948; Trivers 1972), and the inherent costs involved ensure that scent marking is a reliable indicator of male condition (Gosling and Roberts 2001a).

Females respond selectively to male odors and positively to males of high status or quality. For example, the scent of well-nourished males attracts more interest than the scent of poorly nourished males (Ferkin et al. 1997). Females prefer the odor of dominant males to that of subordinate ones in laboratory mice (Parmigiani et al. 1982), rats (Carr et al. 1982), bank voles (Hoffmeyer 1982; Horne and Ylönen 1996; Kruczek and Pochron 1997), and water voles (*Arvicola terrestris*; Evsikov et al. 1995). In laboratory mice, this preference was greater when dominants regularly encountered new rivals than when dominants were exposed to a single intact subordinate, while the latter, in turn, were more attractive than males housed with castrated subordinates (Hayashi 1990; see also Scott and Pfaff 1970; Hayashi and Kimura 1978). Preputial gland removal nullifies female preferences for dominants (Hayashi 1990). Neurons in the olfactory bulb, preoptic area, and lateral hypothalamus respond differentially to intact and castrate odors (Scott and Pfaff 1970). The chemical basis for these preferences is also known: volatile chemicals associated with dominance (as described earlier in this chapter) are attractive to females (Jemiolo et al. 1985, 1989).

Females prefer males whose pattern of odor deposition indicates high quality or resource-holding potential (RHP), such as males whose territories contain only the owner's marks (Rich and Hurst 1998), and those that countermark intruder scent (Johnston et al. 1997a; Rich and Hurst 1999). If females prefer high-status males, and these males invest more in marking, we would also expect females to use marking frequency as an indicator of male quality. In gerbils, intrauterine position correlates with circulatory testosterone levels (Clark et al. 1992b), larger scent-marking glands, and higher marking rates, which females also prefer (Clark et al. 1992a). Marking rate also predicts female preference in laboratory and house mice (Roberts and Gosling 2003; Zala et al. 2004).

Studies finding no indication of female preference based on marking frequency (e.g., Thomas 2002; Mech et al. 2003) emphasize that frequency is a proximate indicator of quality that is modulated by social conditions and that correlates with other aspects of marking behavior, including scent chemistry. This may explain why artificially increasing apparent marking rates by collecting and presenting to females many scent marks of a particular male may not successfully enhance his attractiveness (e.g., Thomas 2002). Greater success in such efforts may be achieved by altering marking behavior and scent composition more indirectly,

and realistically, through manipulation of social environment or status. For example, in wild-derived house mice (Zala et al. 2004) and in harvest mice (*Mus minutus*; Roberts and Gosling 2004), exposure of males to female or male odor, respectively, increases both male marking rate and scent attractiveness.

Female preferences for the odor of familiar males may also be linked to RHP, since the prevalence of a male's marks reflect dominance or territorial residency (Roberts and Gosling 2004). Preferences for familiar odors are known in prairie voles (*Microtus ochrogaster*; Newman and Halpin 1988), house mice (Heise and Hurst 1994), hamsters (*Mesocricetus auratus*; Lis et al. 1990; Tang-Martinez et al. 1993), kangaroo rats (*Dipodomys merriami*; Randall 1991b), and harvest mice, in which familiarity also reduces male-directed aggression (Roberts and Gosling 2004).

Other studies are consistent with the idea that females prefer males to have costly sexual displays because they indicate health and lower parasite loads (Hamilton and Zuk 1982; Penn and Potts 1998a). Female house mice distinguish odors of parasitized and unparasitized males (Kavaliers and Colwell 1992, 1995a; Kavaliers et al. 2003), and infection reduces odor attractiveness (Penn et al. 1998; Klein et al. 1999; Willis and Poulin 2000). Discrimination of healthy mates is improved when male marking rates are artificially increased (Zala et al. 2004).

Preferences are also mediated by genetic differences at the major histocompatibility complex (MHC), which codes for proteins involved in immune response and thus pathogenic resistance (chap. 5 in this volume). MHC-disassortative mating preferences (Yamazaki et al. 1976, 1979; Potts et al. 1991) benefit females because they increase offspring heterozygosity (e.g., Penn 2002). However, expression of MHC preferences is modulated by, and may trade off against, preferences for other male qualities. In congenic mouse strains, MHC haplotypes are associated with differential investment in scent marking, influencing female preferences more than genetic complementarity under defined circumstances. The interaction between these two qualities suggests a mechanism for maintaining hypervariability in both (Roberts and Gosling 2003).

Scent marking by females

Marking among females has received less research attention than in males, largely because females mark less often and less intensively (e.g., Johnson 1975; Holst and Eichmann 1998; Wolff et al. 2002). However, there is evidence that female marking is associated with intrasexual competition and enhancing opportunities for mating with high-quality mates. Reproductively active female house mice scent mark at higher frequencies than nonbreeders and preferentially countermark the marks of breeding females (Hurst 1990c). Female hamsters increase scent marking in response to female more than to male scent (Johnston 1977). Rates of aggression and scent marking are typically low in female bank voles housed together, but rise around parturition, when unfamiliar female odors trigger increased aggressiveness and scent marking (Rozenfeld and Denoel 1994). Common vole (*Microtus arvalis*) mothers mark more intensively than daughters, suggesting a link between frequency and status (Heise and Rozenfeld 1999). Female Syrian hamsters are more aggressive in the presence of their own odor than in a blank arena (Fischer and McQuiston 1991) and scent marking appears to be influential in determining low-range overlap in female bank voles (Ziak and Kocian 1996).

Female marking may also be directed toward males rather than female competitors. Female house mice mark at higher rates in the presence of intact males than castrates (Maruniak et al. 1975). When female hamsters mark in the presence of males (Johnston 1977), marks stimulate male aggression and could increase chances of mating with high-quality males (Fischer and Brown 1993). Increased marking frequency during estrus in Long-Evans rats is consistent with advertisement of reproductive status to males (Matochik et al. 1992).

Scent-Marking Mechanisms and Processes

Scent marks are signals that are usually transmitted in the absence of receivers, perhaps without targeting a particular receiver, and usually detected a relatively long time afterward, in the signaler's absence (Gosling 1982). In this regard, scent marking is unique among social signals. How then do receivers react to scent marks that they encounter? Reactions are expected to be highly variable between individuals. Assuming that the receiver is a competitor, responses will be influenced by factors that include the relative competitive abilities of receiver and signaler, the value to each of the marked resource, the costs of fighting over it, and the probability that the two will meet (Gosling and Roberts 2001a). If receivers are potential mates, responses could be influenced by condition-dependent cues of quality (which are likely to be the same that signal RHP to competitors), female receptivity, genetic dissimilarity, and the suitability of the signaler in comparison to others already encountered.

Three main mechanisms that account for how receivers use information available in scent marks have been proposed (Gosling 1990; Gosling and Roberts 2001a; see fig. 22.1). First, intrinsic properties of the marks may reveal information about the signaler. Second, receivers may have learned and remembered the signaler's identity from past encounters, whose outcome influences current behavior. Third, receivers may memorize the odor of the mark, forming a template (Sherman et al. 1997) with which to compare odors of individuals they meet subsequently, with a match between template and odor implying that the individual is the signaler. The first two mechanisms do not require the receiver to meet the signaler, while in the latter case the response is delayed until after an encounter (fig. 22.1).

These mechanisms are nonexclusive and unlikely to be species specific. Thus receivers may have previous experience of the signaler, but can update this information (template updating; Sherman et al. 1997) using current scent properties and ultimately confirm assessment through face-to-face matching, although not all options will necessarily always be used. For example, matching may become more necessary at high population densities, because of increased fighting costs, potential for mistaken identity, and higher turnover of dominant or territorial males. In contrast, potential for learned associations between marks and signaler quality is higher in temporally stable networks. One consequence of learned associations occurs between established territorial neighbors, in which familiarity reduces signaling costs along shared boundaries (the "dear enemy phenomenon"; e.g., Rosell and Bjorkoyli 2002).

Of these mechanisms, scent matching is the most accurate because it potentially takes account of all available information. Early evidence for scent matching comes from a study in which smearing urine from an unknown male mouse onto one member of a male pair resulted in increased aggression within the pair, whereas aggression was reduced if the odor came from a familiar individual (Mackintosh and Goddard 1966). Scent matching has been experimentally demonstrated; intruding males fight less with males whose odor matches the substrate odor (simulated territory owners) than when the substrate is marked by a third male (fig. 22.2; Gosling and McKay 1990). Similarly, snow vole males spend less time exploring scent-marked areas of odor matching than nonmatching males (Luque-Larena et al. 2001). Scent matching also occurs in mate choice contexts (Steel 1984).

Whichever mechanism is prevalent in a particular case, selection should act on signal design to optimize reception, accuracy, and reliability. An excellent review of the design of chemical signals, including scent marks, is provided by Alberts (1992), while Gosling and Roberts (2001a) have reviewed behavioral adaptations to increase signal efficacy across mammals. In rodents, much recent research has focused on a number of key processes that are central to our understanding of scent-marking mechanism and function, and some of these are outlined in the following sections.

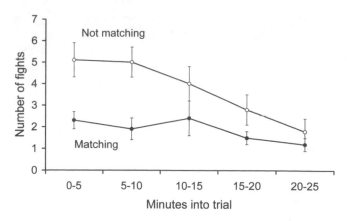

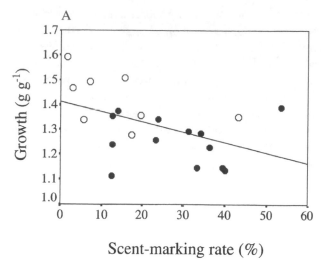

Figure 22.2 Experimental evidence for competitor assessment by scent matching. The mean number of fights (± standard error) in successive 5-min periods of trials where the scent marks on the substrate of the experimental arena either match or do not match the odor of the resident. Twenty-one male mice were tested in both conditions. Fewer fights were recorded when the scents matched. Redrawn from Gosling and McKay 1990.

Signal cost

As in all animal signals, scent marking should be costly if it is used in assessment of quality; otherwise it would be unreliable and susceptible to cheats (Zahavi 1975). Although marking can account for significant proportions of an animal's time budget, only recently has there been any quantification of the energetic costs involved. In rodents, major urinary proteins (MUPs) are synthesised in the liver and excreted in urine, their sole apparent function being in chemical signaling (Nevison et al. 2003). Average urinary protein concentrations in house mice are 30 mg ml^{-1}, almost all of which are MUPs (Beynon et al. 2001). In terms of protein turnover, house mice synthesize almost their entire liver weight every 24 hours (h), a substantial energetic cost (Beynon et al. 2001). Consistent with gender differences in marking behavior, MUP expression is androgen dependent (Knopf et al. 1983). Two studies have estimated MUP concentration in urine, finding it to be two to three (Beynon et al. 2001) or even five to twenty (Flower 1996) times higher in males than females.

In view of these levels of protein synthesis and secretion, individual variation in scent-marking investment could carry significant metabolic costs. Indeed, in male mice, scent-marking rates are inversely correlated with weight gain in young mice (fig. 22.3; Gosling et al. 2000). In males housed in pairs, dominants that are smaller than their subordinates mark at higher rates than relatively large dominants, grow more slowly, and are consequently more susceptible to dominance reversals (fig. 22.3). Reduced dominance tenure could have critical fitness costs in short-lived species. These results raise the possibility that different investment in scent marking represents alternative mating strategies, where

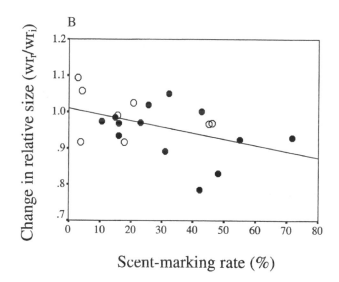

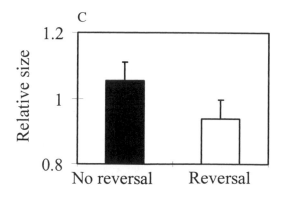

Figure 22.3 Life history costs of investment in scent marking by male mice. (A) In dominant males, mean growth rates between the ages of 9 and 25 weeks are inversely correlated with scent-marking rate ($P = 0.011$). Males were housed with another, subordinate male; closed circles denote the dominants are smaller than their subordinate, open circles denote larger dominants. (B) Dominant males that scent mark at higher rates become progressively smaller compared to their subordinate ($P = 0.038$). (C) Initially dominant males that incurred a dominance reversal were smaller, at the point of reversal, than males that maintained status throughout the experiment ($P = 0.037$). From Gosling et al. (2000).

low-marking, larger males adopt sneak-breeding or waiting strategies (Gosling et al. 2000). Gosling et al.'s study used an outbred laboratory strain (TO) with MUP concentrations of 10–11 mg ml^{-1} (Nevison et al. 2000), but these energetic costs could be even more significant in wild mice, where MUP concentrations are three times higher (Beynon et al. 2001).

Individual recognition

Individual differences in rodent odors appear to be universal (reviews in Halpin 1986 and Voznessenskaya et al. 1992). Differences are documented from Norway rats (Carr et al. 1970a), laboratory mice (Bowers and Alexander 1967), Mongolian gerbils (Dagg and Windsor 1971; Halpin 1976), chipmunks (*Tamias striatus*; Keevin et al. 1981), prairie voles (Newman and Halpin 1988), cavies (*Cavia aperea*; Martin and Beauchamp 1982), the tuco-tuco (*Ctenomys talarum*; Zenuto and Fanjul 2002), Damaraland mole-rats (*Cryptomys damarensis*; Jacobs and Kuiper 2000), red squirrels (*Tamiasciurus hudsonicus*; Vaché et al. 2001), and golden hamsters (Johnston and Rasmussen 1983; Tang-Martinez et al. 1993). Individual odors may also be recognized across species (e.g., Beauchamp et al. 1985; Johnston and Robinson 1993; Todrank and Heth 1996).

In hamsters, Johnston and Bullock (2001) demonstrated across-odor habituation to different odors from the same individual, indicating that scent from multiple sources potentially reveals individual identity. The vomeronasal organ (VNO) apparently aids discrimination, although VNO removal eliminates this ability only in males and only for certain odor sources (Johnston and Peng 2000).

As odors are influenced by environmental factors such as diet (Ferkin et al. 1997) and stress (Carr et al. 1970b; Kavaliers and Ossenkopp 2001; Marchlewska-Koj et al. 2003), the individual signal component must be discriminable over time. This is assured if odors are at least partially genetically determined. Evidence for a genetic component comes from observations that odor chemical profiles are more similar within closely related species (Heth and Todrank 2000; Heth et al. 2002) and among closely related individual beavers (Sun and Müller-Schwarze 1998a).

Two genetic regions are principal candidates for the basis of individual odors, owing to their polymorphic nature and expression in scent marks. The first is the MHC, known in mice as H-2. Mice and rats discriminate between individuals differing only at MHC (Yamazaki et al. 1979; Brown et al. 1987), even between mice carrying single MHC gene mutations (Yamazaki et al. 1990, 1991; Bard et al. 2000). Discrimination is mediated by varying proportions of volatile carboxylic acids in urine (Singer et al. 1997) and

influences preferences for mates (e.g., Yamazaki et al. 1976; Potts et al. 1991; Roberts and Gosling 2003) and nestmates (Manning et al. 1992).

The second region contributing to individuality is the polymorphic multigene family coding for MUPs (Beynon and Hurst 2003). While MUPs are known to extend the active life of scent marks (Hurst et al. 1998), recent evidence suggest they also have a more fundamental role. Males respond differently to the odor of brothers with different MUP expression but not to those of the same MUP type (Hurst et al. 2001). Countermarking responses depend on having direct contact with urine, suggesting that these involatile signal components are themselves important in individual recognition (Humphries et al. 1999; Nevison et al. 2003). Whether and how MHC and MUP genetic components interact in forming unique odor signatures remain to be addressed.

Memory

Like the ability to recognize individuals, the ability to remember scent mark properties is a key requirement for adaptive responses. That animals remember marks is implicit in many studies investigating marking behavior where responses are linked to previous experience. One common example is where female preference tests between males follow exposure to their marks (e.g., Johnston et al. 1997a; Johnston and Bhorade 1998). Preferences based on this information last for 48h in voles (Ferkin et al. 2001). Similarly, avoidance responses of subordinates to odors of dominant male mice suggest memory for odors and signaler's relative quality (Carr et al. 1970b). These kinds of responses can be directly employed to study memory. For example, Lai and Johnston (2002) showed that males could recognize, remember and avoid odor of a male that defeated them in both the short term (30 minutes after fighting) and the long term (1 week later). Other research uses habituation-dishabituation techniques (fig. 22.4). Flank gland odors are remembered for at least 10 days in hamsters (fig. 22.4; Johnston 1993) and up to 4 weeks in guinea pigs (Beauchamp and Wellington 1984). Perhaps most impressive, Belding's ground squirrels (*Spermophilus beldingi*) can remember and discriminate between familiar versus unfamiliar, and kin versus nonkin, odors after over-winter hibernation (Mateo and Johnston 2000).

Interpreting patterns of marks

Animals often scent mark near, or on top of, marks of conspecifics. This is generally termed *countermarking* (overmarking is a form of countermarking in which the second mark is placed directly over the first mark). In particu-

A

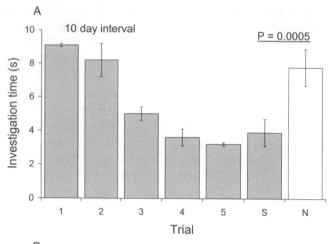

B

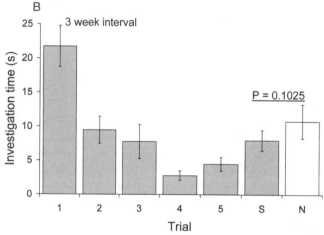

Figure 22.4 Testing memory for individual odors using habituation-discrimination techniques. In five successive trials 15 min apart, the mean number of seconds (± SE) that male hamsters investigated flank gland scent of another male decreases as habituation to the odor occurs. In a sixth trial, responses to scent from the same male (S) and a novel male (N) are compared (discrimination test). (A) Successful discrimination was recorded when the interval between the test trial and the last habituation trial was 10 days, indicating memory for the initially presented odor. (B) There is no difference in time spent investigating each scent when this interval was 3 weeks. Redrawn from Johnston (1993).

lar, resource holders (territorial or dominant males) always countermark the marks of intruders or subordinates, because these may represent a challenge for the resource and/or because it introduces ambiguity into subsequent assessment by receivers (Gosling 1982). Receivers may glean information about relative qualities of local signalers through marking patterns deposited by different individuals at different times. In essence, then, countermarking should be seen as occasional reactive scent marking to maintain the integrity of a marking network previously established and maintained on a more proactive basis.

In an early experiment in this area (reviewed by Johnston 2003), Johnston et al. (1994) showed that, following habituation to simulated hamster overmarks, novel scents are investigated more than top scents in overmarks, sug-

gesting that hamsters had habituated to the top scent. In contrast, the novel and the bottom scents were investigated equally. Johnston et al. suggested that information in the bottom scent might be masked by the top scent. However, since countermarks often do not completely cover existing marks, this experiment was repeated using partially overlapping marks (Johnston et al. 1995) with similar results. This indicates that, rather than being masked, bottom scents simply fail to attract the interest of receivers, perhaps because individuals whose scent is overmarked are apparently relatively unthreatening (see also Woodward et al. 1999).

Further work showed that these effects on receiver behavior were not explained by the area of top and bottom scents available for investigation, since top scents were investigated less even when they occurred in smaller quantities than bottom scents (Wilcox and Johnston 1995). Nor can age differences between the top and bottom marks explain the responses: when Wilcox and Johnston (1995) exposed males to two nonoverlapping scents varying in age by 20–45s (the same interval between artificial deposition of mark and overmark in previous experiments), the males habituated to both scents. This indicates that small differences in age between the two scents are insufficient in themselves to permit discrimination. In mice, however, short intervals between scent depositions (30 s) did not elicit response differences to top- and bottom-scent donors, perhaps because of scent blending while the secretion was still wet (Rich and Hurst 1999).

In view of these findings, what is the mechanism that accounts for the difference in receivers' responses to top and bottom scent? The answer appears to lie in the spatial configuration of marks. Response differences to top-scent donors only occur if a region of overlap exists between the marks. If artificial marks and overmarks are made to appear as though one overlaps the other, even though it doesn't, the response difference is still recorded (Johnston and Bhorade 1998). Similar results have been found in voles (Johnston et al. 1997a, 1997b; Ferkin et al. 1999). These results suggest that rodents are very adept at interpreting these spatial patterns and that the resulting responses may confer fitness benefits.

Mate preference tests show that females prefer top-scent males in hamsters (Johnston et al. 1997a), meadow voles (Johnston et al. 1997b; Ferkin 1999), and house mice (Rich and Hurst 1998, 1999). If countermarking has fitness effects, it should also be modulated by relatedness between potential competitors (and possibly mates). Indeed, in prairie voles, scent marks of siblings received fewer overmarks than marks of unrelated individuals (Kohli and Ferkin 1999). However, despite this evidence, it is likely that patterns of overlapping scent comprise only part of the information that receivers use, and this is illustrated by Leonard et al.'s (2001) study of gonadectomised meadow voles, in which females take into account both mark position and testosterone titers.

The overmarking studies by Johnston, Ferkin, and colleagues cited previously were designed to interpret the consequences of overmarking to artificially created overmarks and not the frequency or actual placement of overmarks per se. In contrast, a series of studies designed to quantify whether meadow voles and prairie voles overmark found that overmarking occurred less often than expected by chance (e.g., Thomas and Wolff 2002; Mech et al. 2003). These authors concluded that voles attempt to avoid overmarking, perhaps to retain individual identity (see also Thomas and Wolff 2003). Thus even though voles can discriminate top- from bottom-scent donors, and even show a preference for one over the other, overmarking in voles may not be an adaptive or sexually selected trait.

Eavesdropping

Another potential cost of scent marking is that signals may be intercepted by individuals other than the intended receivers, to the signaler's disadvantage. For example, young males may use information in marks to monitor the status of local resource-holding males, with a view to challenging poor males for the resource, or females may use the same information to avoid poor-quality mates (e.g., Rich and Hurst 1999). In addition to conspecific eavesdroppers, signalers could also alert predators to their presence, location, and movements, which could carry particularly high costs. Diurnal avian predators like the kestrel (*Falco tinnunculus*; Viitala et al. 1995), rough-legged buzzard (*Buteo lagopus*; Koivula and Viitala 1999), and great grey shrike (*Lanius excubitor*; Probst et al. 2002) can detect vole scent marks (*Microtus, Clethrionomys*) and focus hunting effort in densely marked areas (see also Koivula and Korpimäki 2001). Detection is mediated by ultraviolet (UV) reflectance of proteins in the marks. Kestrels discriminate between age and sex classes, preferring male field voles (*Microtus agrestis*) over females and juveniles (Koivula, Viitala, and Korpimäki, 1999), apparently using differences in UV reflectance between classes (Koivula, Koskela, and Viitala, 1999a).

Predation rates by terrestrial predators like mustelids may be higher than predation by raptors (Koivula and Korpimäki 2001). Here, prey odor, rather than mark visibility, may be the important cue to prey availability. Preferences of least weasels (*Mustela nivalis nivalis*) for odors of different reproductive categories of bank voles in the laboratory did not reveal the same kind of discrimination as found in raptors, although weasels preferred vole odors over the clean arm of a Y-maze (Ylönen et al. 2003). In the field, however, areas with artificially elevated scent-mark densities were

hunted more intensively and vole survival was lower, suggesting that marking density attracts greater hunting effort (Koivula and Korpimäki 2001).

Males that invest more in scent marking, through high protein concentrations in marks and/or through marking at high rates, may thus be at higher risk than females and low-investing males. Males may therefore be expected to invest less in scent marking in predator-rich areas, even temporarily reducing marking at times of high risk. Evidence for this hypothesis includes the finding that exposure to weasel odor causes reduction in hamster flank gland size, among a number of physiological effects (Zhang et al. 2003). The idea was further tested by Roberts et al. (2001), who used sib-sib comparisons to examine the degree to which male mice of known signaling investment countermarked scent marks of an unfamiliar individual in the presence or absence of predator odor (urine of ferrets, *Mustela putorius furo*). Under simulated predation risk, all males approached the competitor's marks more slowly, although high-frequency markers approached more quickly than low-frequency markers and spent more time in the vicinity of the competitor's marks. Only high-investing males significantly reduced overmarking of the competitor's scent in the presence of predator odor. These results suggest there is a unique danger inherent to scent marking at high frequencies and that high-investing males were prepared to accept increased costs of intrasexual competition to reduce the risk of predation.

In contrast, a recent study of marking by prairie voles and woodland voles (*Microtus pinetorum*) found no evidence of reduced marking in response to odor of minks (*Mustela vison*) and bullsnakes (*Pituophis melanoleucus*; Wolff 2004). Marking rates were tested both in large enclosures and in the laboratory. Some methodological differences exist between the laboratory component of this study and that of Roberts et al. (2001) that could account for the different results. In the latter study, use of sibling comparisons controlled for potentially genetic differences in marking effort (see Collins et al. 1997), the cage environment was relatively complex with reduced visibility (i.e., perceptually more dangerous), and marking was measured in response to competitor scent (counter marking), rather than marking in a blank area. However, more work is clearly needed, both in the laboratory and the field, to determine the potential sensitivity of scent-marking effort to predation risk.

Conclusions and Future Directions

One reason rodents have made such an impact on our understanding of scent marking is because of their amenability to laboratory studies. Recently, however, efforts have been made to validate some findings in the field, with mixed success (Mahady and Wolff 2002; Wolff 2003c, 2004), although some excellent examples of transferability of results from lab to field exist. Notable among these are observations that MHC-disassortative mating preferences discovered in the laboratory (Yamazaki et al. 1976) influence mating patterns in free-ranging mice (Potts et al. 1991). Thus while controlled conditions in the laboratory offer unique opportunities for research and should remain important in the future, Wolff's call for field-validation of laboratory findings (Wolff 2003c) must be heeded to improve confidence in results obtained from laboratory studies.

Studies of scent-marking behavior in rodents are opening up a range of new research directions. Two notable examples are (1) approaches in neuroscience that are revealing neural pathways involved in scent marking and olfactory perception, and (2) application of our understanding of scent marking in animal welfare and conservation biology.

Increasing research effort focuses on the neural control of marking behavior. A comprehensive review is beyond the scope of this chapter, but some examples are provided here. One focus has been the role of vasopressin (VP) in regulating social behavior in general and scent marking in particular. Microinjection of VP into several areas of the brain stimulates flank-gland marking in Syrian hamsters, while lesions of the same areas inhibit it (Hennessey et al. 1992; Albers and Bamshad 1998). VP-containing neuronal cells and fibers in the neurohypophyseal system and several extrahypothalamic areas are sexually dimorphic and androgen dependent, and control scent marking in a sex-specific manner (Dantzer and Bluthe 1992). Galanin, which antagonizes postsynaptic action of other neurotransmitters, also blocks VP-induced flank marking, suggesting that endogenous galanin may be an inhibitory force in scent-marking behavior (Ferris et al. 1999). Norepinephrine has a similar, dose-dependent effect (Whitman et al. 1992). Other research combining lesion and behavioral studies shows that the parahippocampal region is important in individual odor discrimination (Petrulis et al. 2000), while the fimbria-fornix and medial amygdala are important for regulating investigation of odor and scent marking but not individual discrimination (Petrulis and Johnston 1999; Petrulis et al. 2000). Finally, studies using rodents as models for describing neural circuitry in the mammalian olfactory system (e.g., Belluscio et al. 2002) also complement our understanding of the perceptual mechanisms involved. For example, when mice sniff conspecifics, individual neurons in the accessory olfactory bulb vary in activity depending on the sex and genetic strain of the other mouse, suggesting that populations of neurons may become tuned to recognize specific individuals (Luo et al. 2003). Such approaches hold great promise for the future.

Knowledge about the role of scent marking in modulating social behavior is also being used in designing new approaches to problems in animal welfare and conservation. In the laboratory, the link between scent marks and aggression raises the possibility of adapting husbandry practices to reduce aggressive behavior (Olsson et al. 2003). For example, transferring odor cues during routine cage cleaning may be one way to reduce postcleaning aggression peaks, but research to date is still inconclusive. In one study the transfer of urine-impregnated sawdust reduced aggression (Gray and Hurst 1995), while it intensified in another (Van Loo et al. 2000). Transfer of nesting material, on the other hand, appears to reduce aggression (Van Loo et al. 2000) and concentrations of stress-indicating hormones (Van Loo et al. 2003). In the field of conservation, Sutherland and Gosling (2000) highlighted the potential for increased understanding of processes underlying mate choice, including scent marking, in overcoming behavioral incompatibility and extreme intrapair aggression in captive breeding programs. In harvest mice, Roberts and Gosling (2004) manipulated sexual signaling characteristics of relatively unattractive males to increase their allure to females. Manipulating the degree of familiarity also influenced female preferences and reduced the amount of aggression between mates upon pairing. Once again, rodent studies are a useful model in which to test such ideas with a view to application in other, often more threatened, species such as giant pandas (*Ailuropoda melanoleuca;* Swaisgood et al. 2000).

Summary

Scent marking is an important feature of rodent social behavior. Scent marks are status signals used in assessment by receivers, who are usually same-sexed conspecifics or potential mates but could also be "eavesdroppers" (including predators). Status information carried by scent marks includes resource-holding potential (e.g., territory ownership), social status, health, and hormonal, nutritional, and reproductive condition. Scent markers benefit by reducing contest frequencies, maintaining social status, or attracting mates, and they invest heavily to ensure that their system of marks is maintained and that information carried in them is unambiguous. Receivers are expected to respond adaptively to this information. Because receivers vary in competitive ability or reproductive status, and because the costs and benefits of responding to marks in particular ways vary between individuals at different times, responses to scent marks also vary greatly.

Rodent studies have largely shaped our understanding of scent marks as signals of status. Status signaling is a theoretical paradigm that explains almost all marking behavior across taxa and provides a unifying framework, grounded in evolutionary theory, within which to study scent marking alongside other signals in the visual and acoustic modalities.

Chapter 23 Nonparental Infanticide

Luis A. Ebensperger and Daniel T. Blumstein

Male marmot 100 moved into the Grass Group. Male 69 seemed to oppose 100's sudden entry, but the females of the group appeared to accept 100. Before male 100 moved in there were 9 healthy marmot pups crawling around the Grass Group's main burrows. Within two weeks there was one injured marmot pup limping around—apparently avoiding marmot 100. The injured pup did not survive hibernation. (Blumstein 1993:14)

A female invaded an adjacent coterie territory and entered a burrow containing a recently emerged, healthy juvenile. The marauder emerged 5 minutes later with a distinctly bloody face, and then showed licking the front claws [behavior]. Several minutes later the disoriented juvenile emerged with fresh, severe wounds on the face and neck. The juvenile disappeared a few days later. (Hoogland 1995:134)

INFANTICIDE CAN STRIKE quickly and may have profound demographic consequences (Sherman 1981b; Hoogland 1995; Blumstein 1997). Nonparental infanticide, the killing of infants by conspecifics other than the parents, occurs in a variety of vertebrate and invertebrate taxa (Hausfater and Hrdy 1984; Elgar and Crespi 1992; Parmigiani and Vom Saal 1994; Van Schaik and Janson 2000). Among mammals, infanticide has been reported in primates, terrestrial and marine carnivores, artiodactyls, cetaceans, lagomorphs, perissodactyls, and tree shrews (Ebensperger 1998b). More recent additions to the literature include reports of infanticide in banded mongooses (*Mungos mungo,* Cant 2000), bottle-nose dolphins (*Tur-*

siops truncatus, Patterson et al. 1998), giant otters (*Pteronura brasiliensis,* Mourão and Carvalho 2001), hippos (*Hippopotamus amphibius,* Lewison 1998), plains zebras (*Equus burchelli,* Pluháček and Bartoš 2000), sportive lemurs (*Lepilemur edwarsi,* Rasoloharijaona et al. 2000), and suricates (*Suricata suricatta,* Clutton-Brock et al. 1998). Infanticide has been noted in the wild or under laboratory conditions in two species of hystricognath rodents and 35 species of sciurognath rodents (table 23.1). Despite the difficulty of observing and quantifying infanticide in these typically semifossorial and often nocturnal species, we know a considerable amount about the proximate regulation, evolution, and function of infanticide in rodents. Understanding the causes and consequences of infanticide in rodents provides a basis for developing and testing alternative hypotheses for the functional significance of infanticide in mammals generally.

Several field-based studies that recorded the frequency of infanticide by rodents have concluded that infanticide is a major source of juvenile mortality (Sherman 1981b; Agrell et al. 1998; Hoogland 1995; Blumstein 1997). Other seminatural and field-based studies have reached similar conclusions indirectly by showing a significant impact of adult female density on juvenile recruitment (Labov et al. 1985; Mappes et al. 1995). These studies show that the removal of breeding females usually increases the survival of resident juveniles in deer mice (*Peromyscus maniculatus;* Galindo and Krebs 1987), golden hamsters (*Mesocricetus auratus;* Goldman and Swanson 1975), gray-tailed voles (*Microtus canicaudus;* Wolff et al. 2002), and meadow voles (*Microtus pennsylvanicus;* Rodd and Boonstra 1988). In contrast,

Table 23.1 Summary of reports of nonparental infanticide in rodents

Family	Species	Common name	MN	MC	FN	FC	I	Sources
Caviidae	*Galea musteloides*	Yellow-toothed cavy				1	X	Künkele and Hoeck 1989
Hydrochaeridae	*Hydrochaeris hydro-chaeris*	Capibara				1		Da Cunha-Nogueira et al. 1999
Muridae	*Acomys cahirinus*	Spiny mouse		1		1		Porter and Doane 1978; Makin and Porter 1984
	Apodemus sylvaticus	European wood mouse		1		1		Wilson et al. 1993
	Clethrionomys glareolus	Bank vole	1	1	1	1		Ylonen et al. 1997
	Dicrostonyx groenlandicus	Collared lemming		1		2		Mallory and Brooks (1978), 1980
	Glis glis	Dormouse			1?			Pilastro et al. 1996
	Lemmus lemmus	Norwegian lemming		1?		1		Arvola et al. 1962; Semb-Johansson et al. 1979
	Meriones unguiculatus	Mongolian gerbil		1		2		Elwood 1977, 1980; Elwood and Ostermeyer 1984b
	Mesocricetus auratus	Golden hamster				1		Goldman and Swanson 1975; Marques and Valenstein 1976
	Microtus agrestis	Field vole				1	X	Agrell 1995
	Microtus brandtii	Brant's vole		1		1		Stubbe and Janke 1994
	Microtus californicus	California vole		1?				Lidicker 1979a; Heske 1987
	Microtus pennsylvanicus	Meadow vole	1	1	1	1		Louch 1956; Caley and Boutin 1985; Ebensperger et al. 2000
	Microtus ochrogaster	Prairie vole		1		2		Roberts 1994 (cited in Carter and Roberts 1997)
	Mus musculus/ domesticus	House mouse (lab stocks)		2		1		Gandelman 1972; Svare and Mann 1981; Parmigiani et al. 1989; Perrigo et al. 1993
	Mus musculus/ domesticus	House mouse (wild stocks)		1		1		Southwick 1955; Perrigo et al. 1993; Vom Saal et al. 1995; Jakubowski and Terkel 1982; Soroker and Terkel 1988
	Neotoma lepida	Desert woodrat				1?		Flemming 1979
	Ondatra zibethicus	Muskrat	?		?			Errington 1963; Caley and Boutin 1985
	Peromyscus californicus	California mouse		1		1		Gubernick 1994
	Peromyscus leucopus	White footed mouse	1	1	2	2		Wolff 1986; Wolff and Cicirello 1991
	Peromyscus maniculatus	Deer mouse	1	1	2	2		Wolff and Cicirello 1991
	Phodopus campbelli	Djungarian hamster		1		1		Gibber et al. 1984
	Phyllotis darwini	Leaf-eared mouse		1		1		D. Bustamante, R. Nespolo, and L.A. Ebensperger, unpublished ms
	Rattus norvegicus	Norway rat		2		1		Calhoun 1962; Jakubowski and Terkel 1985a
Sciuridae	*Cynomys gunnisoni*	Gunnison prairie dog	1?			1		Fitzgerald and Lechleitner 1974
	Cynomys ludovicianus	Black-tailed prairie dog	1			2		Hoogland 1985, 1995
	Cynomys parvidens	Utah prairie dog	1					Hoogland (chap. 37, this volume)
	Marmota caligata	Hoary marmot	1			1		T. Karels, unpublished ms
	Marmota caudata	Golden marmot	1					Blumstein 1997
	Marmota flaviventris	Yellow-bellied marmot	1?			1		Armitage et al. 1979; Brody and Melcher 1985
	Marmota marmota	Alpine marmot	1					Coulon et al. 1995
	Paraxerus cepapi	Tree squirrel	1					de Villiers 1986

Table 23.1 (continued)

Family	Species	Common name	MN	MC	FN	FC	I	Sources
	Spermophilus armatus	Utah ground squirrel	?		?			Balph 1984; Eshelman and Sonnemann 2000
	Spermophilus beecheyi	California ground squirrel			1			Trulio et al. 1986; Trulio 1996
	Spermophilus beldingi	Belding's ground squirrel	1		1			Sherman 1981b
	Spermophilus colum- bianus	Columbian ground squirrel	1		2			Steiner 1972; Balfour 1983; Waterman 1984; Hare 1991; Stevens 1998
	Spermophilus parryii	Arctic ground squirrel	1					Steiner 1972; Holmes 1977; McLean 1983; Lacey 1992
	Spermophilus richardsonii	Richardson's ground squirrel				1		Michener 1973b
	Spermophilus townsendii	Townsend's ground squirrel	1?					Alcorn 1940
	Spermophilus tridecem- lineatus	Thirteen-lined ground squirrel	1					Vestal 1991

NOTES: MN = male infanticide observed in nature; MC = male infanticide observed in captivity; FN = female infanticide observed in nature; FC = female infanticide observed in captivity; I = studies where individuals of the opposite sex were not examined; ? = indicate uncertainties in the database. Numbers in the MN, MC, FN, and I columns are used to indicate when one sex is more infanticidal than the other (i.e., 2 > 1). Species for which infanticide was reported but the infanticidal sex was not specified are listed, but the sex of the infanticidal animal was left blank.

juvenile recruitment per pregnancy has been shown to decrease under wild conditions as the number of adult female (but not male) gray-tailed voles sharing a patch increases (Wolff and Schauber 1996). In the laboratory, litter mortality in prairie voles (*Microtus ochrogaster*) is more negatively affected by the presence of additional females than males (Hodges et al. 2002). These observations are consistent with the assertion that infanticide by females is the mechanism for reduced recruitment of juveniles.

We view infanticide as potentially adaptive (e.g., Hrdy 1979; Sherman 1981b; Hoogland 1995), and we review functional hypotheses and evidence about the current adaptive utility of infanticide in rodents. Males and females are considered separately when infanticide serves different functions in each sex. We also address some consequences of infanticide on behavioral counter-strategies and demography.

Explanations of Infanticide: Hypotheses and Evidence

Nonadaptive explanations

As Hrdy (1979) and Sherman (1981b) pointed out, historically infanticide was considered aberrant because it was inconceivable that such a behavior could be adaptive (e.g., Fox 1968). Formally, infanticide could be neutral or maladaptive (i.e., pathological) during conditions of high density (Southwick 1955; Louch 1956; Calhoun 1962b), it

could be an accidental occurrence of dominance disputes (Rijksen 1981; Campagna et al. 1988), or result from disturbances in physical or social environments (e.g., habitat reduction coupled to high density conditions; Curtin and Dolhinow 1978; Ciani 1984).

However, four lines of evidence make it unlikely that the nonadaptive hypothesis is a general explanation for rodent infanticide. First, most studies claiming that infanticide is not an adaptive trait come from confined populations kept under seminatural conditions in which the identity of killers, and the precise circumstances (i.e., the possibility of evaluating potential benefits), of infanticide are not recorded (Southwick 1955; Calhoun 1962b; Semb-Johansson et al. 1979). Second, explanations of infanticide based on overcrowding, per se, may not be relevant because in the field, infanticide is apparently unrelated to local density (Dobson 1990; Wolff and Cicirello 1991; Hoogland 1995). Moreover, infanticide could be adaptive under conditions of high density if resources are limited. Third, there is no evidence in rodents that infanticide is accidental (e.g., pups simply get in the way of fighting adults; Sherman 1981b; Hoogland 1995). Fourth, individuals that commit infanticide do so under predictable circumstances and exhibit a number of context-specific traits. For instance, black-tailed prairie dogs engage in a specific type of self-cleaning following infanticide (Hoogland 1995). In the rest of this review we focus on potentially adaptive explanations of infanticide.

Adaptive explanations

Hypothesis 1. Direct acquisition of nutritional resources
In rodents, infanticide by males and by females has evolved together, a finding consistent with the hypothesis that infanticide originally evolved as a foraging strategy (Blumstein 2000). If infanticide initially evolved as a foraging strategy, subsequent functions of infanticide must be viewed as exaptations. Juveniles are easy prey, and infanticide may enable killers to obtain nutritious food resources (Hrdy 1979; Sherman 1981b). This hypothesis predicts that infanticide should be followed by cannibalism, and that it might be more frequent among energetically stressed individuals.

Support for the predation hypothesis is provided by field studies showing that cannibalism is negatively correlated with food availability (Holmes 1977). Most species in which infanticide and cannibalism have been noted are those with diets that normally include some animal matter (Sherman 1981b; Elwood 1992). For instance, adult rodents occasionally prey on the infants or adults of other rodents (DeLong 1966; Rood 1970; Ewer 1971; Paul and Kupferschmidt 1975; Wolff 1985c; Elwood and Ostermeyer 1986), and these same species are infanticidal.

Females

Females from 5 of 10 well-studied rodent species have been observed killing and cannibalizing pups (table 23.2). Among these, 69% ($n = 13$) of female deer mice (Wolff and Cicirello 1991) and from 67% ($n = 18$) to 100% ($n = 10$) of female white-footed mice (*Peromyscus leucopus*; Wolff and Cicirello 1989, 1991) that kill and cannibalize pups are either pregnant or lactating. Among sciurid rodents, most female black-tailed prairie dogs (*Cynomys ludovicianus*; 78%, $n = 65$; Hoogland 1985, 1995), California ground squirrels (*Spermophilus beecheyi*; 100%, $n = 36$; Trulio 1996), and Columbian ground squirrels (*Spermophilus columbianus*; 100%, $n = 7$; Stevens 1998) that committed nonparental infanticide did so while nursing their own young. Perpetrators typically consumed their victims, suggesting that they were obtaining nutritional benefits at a time of energetic stress. Interestingly, in laboratory pup-retrieval experiments, female Richardson's ground squirrels (*Spermophilus richardsonii*) that were virgins or nonparous sometimes cannibalized the young (Michener 1973b).

Predation is not a current universal function of infanticide by female rodents. Cannibalism has not been recorded in some female microtines, including collared lemmings (*Dicrostonyx groenlandicus*; Mallory and Brooks 1978, 1980), Norway lemmings (*Lemmus lemmus*; Arvola et al. 1962), bank voles (*Clethrionomys glareolus*; Ylönen et al. 1997), and field voles (*Microtus agrestis*; Agrell 1995), and it oc-

curs only rarely in yellow-bellied marmots (*Marmota flaviventris*; Armitage et al. 1979; Brody and Melcher 1985) and Belding's ground squirrels (*Spermophilus beldingi*; Sherman 1981b). In the laboratory, most female meadow voles kill (73%, $n = 11$) and consume (75%, $n = 8$) alien pups when they are pregnant, but they stop killing and consuming pups when they are lactating and/or not breeding (Ebensperger et al. 2000).

Males

Males from 9 of 11 well-studied species have been observed to kill and cannibalize pups (table 23.2), including Mongolian gerbils (*Meriones unguiculatus*; Elwood and Ostermeyer 1984a), meadow voles (Ebensperger et al. 2000), Norway rats (*Rattus norvegicus*; Paul and Kupferschmidt 1975), Belding's ground squirrels (Sherman 1981b), thirteen-lined ground squirrels (*Spermophilus tridecemlineatus*; Vestal 1991), Townsend's ground squirrels (*Spermophilus townsendii*; Alcorn 1940), Utah prairie dogs (*Cynomys parvidens*; Hoogland chap. 37 this volume), and yellow-bellied marmots (Armitage et al. 1979). As might be expected, food deprivation increases the frequency of infanticide and cannibalism in male gerbils (Elwood and Ostermeyer 1984a), Norway rats (Paul and Kupferschmidt 1975), house mice (*Mus musculus/domesticus*; Svare and Bartke 1978, but see the following), and common voles (*Microtus arvalis*; Litvin et al. 1977).

Obtaining energy is thus a common function of infanticide by male rodents. However, not all males eat the young they kill. Small proportions of male deer mice (2 out of 6) and white-footed mice (1 out of 8) ate pups after killing them (Wolff and Cicirello 1989, 1991). Thus although cannibalism does occur in some species under some circumstances by both males and females, it is not universal and does not totally explain the current motivation or functional significance of infanticide in all species or situations.

Hypothesis 2: Acquisition of space and other
physical resources
Infanticide also may provide the perpetrator, or its offspring, increased access to potentially limited resources such as food, nesting sites, or space by eliminating current or future competitors for those resources (Rudran 1973; Hrdy 1979; Sherman 1981b). In such cases, infanticide is expected to be more prevalent under conditions when resource quality varies considerably, or when resources are extremely limited (Butynski 1982). This hypothesis would also be supported by observations of individuals that commit infanticide by selectively killing the sex of young that will be competitors for the critical resource, and then taking over the resources of their victims' mother. This expectation as-

Table 23.2 Predictions of hypotheses posed to explain rodent infanticide and species where evidence supports or rejects them

Hypothesis	Main predictions	Supportive studies	Unsupportive studies
Direct acquisition of nutritional resources	Killers must consume their victims.	Females: *P. leucopus, P. maniculatus, C. ludovicianus, M. flaviventris, S. beecheyi*	Females: *D. groenlandicus, L. lemmus C. glareolus, M. agrestis*
		Males: *M. unguiculatus, R. norvegicus, M. pennsylvanicus, C. parvidens, S. beldingi, S. tridecemlineatus, S. townsendii*	Males: *P. leucopus, P. maniculatus*
	Infanticide and cannibalism common when food abundance is low, or when experimentally food-deprived.	Males: *M. unguiculatus, R. norvegicus, M. musculus-domesticus*	
	Infanticide and cannibalism common in pregnant and lactating females.	Females: *P. leucopus, P. maniculatus, C. ludovicianus, S. beecheyi, S. columbianus*	Females: *M. pennsylvanicus*
	Motivational and neurological basis of infanticide should resemble that of predatory attack.		Males: *M. musculus/domesticus*
Indirect acquisition of space and other physical resources	Infanticide more common when per capita availability of resources is low.		Females: *P. leucopus, P. maniculatus*
	Resources previously used by individuals losing litters should be taken over by killers.	Females: *S. beldingi*	
	Infanticide should be directed toward infants of the sex most likely to become competitors for the perpetrator or its offspring.		Females: *S. beecheyi*
Insurance against misdirecting parental care	Infanticide should be common in females before and after lactating their own litters.	Females: *M. auratus, M. musculus-domesticus, M. unguiculatus, R. norvegicus*	
	Infanticide by breeding females should occur when nonfilial offspring cannot be confused with own.	Females: *C. ludovicianus, S. beldingi, S. columbianus*	Females: *S. beecheyi*
	Infanticide should be common in breeding females whose nests are clumped.	No information available	
	Infanticide more frequent among species with precocial as opposed to altricial offspring.	No information available	
Acquisition of mates	Infanticidal males should not kill offspring they have sired.	Males: *M. musculus/domesticus, A. cahirinus, P. maniculatus, P. leucopus, M. pennsylvanicus*	
	The elimination of offspring should shorten the interbirth period of the victimized females.	Males: *M. musculus/domesticus, D. groenlandicus, R. norvegicus*	
	Infanticidal males should mate with and sire the subsequent offspring of the mother whose litter was killed.	Males: *M. musculus/domesticus*	

sumes that adult marauders are able to determine the sex of potential victims prior to killing them.

Females

Female Belding's ground squirrels apparently commit infanticide to obtain access to a critical resource—a burrow site that is safe from predation. In this species, females that lose their young to coyotes and badgers move to safer areas and attempt to kill young there. Indeed, 70% of females (*n* = 20) losing their litters to predators or conspecifics moved to new sites as compared with 33% of females that did not lose their litters. Nonresident adult females were responsible for 42% of observed infanticide. Infanticidal female Belding's ground squirrels seldom (9%, *n* = 8) consumed their victims (Sherman 1981b). In most cases perpetrators established nest burrows the subsequent year near their victim's natal burrow (Sherman 1981b).

Infanticide to reduce competition for space has also been suggested in white-footed mice and deer mice. Females of both species are territorial against other females, the most common perpetrators of infanticide (Wolff and Cicirello

1989, 1991). While functions are not mutually exclusive, female white-footed mice that are pregnant or lactating usually consume their victims (Wolff and Cicirello 1989, 1991). Data showing increased access by infanticidal females (or their offspring) to the territories of their victimized females are required to support the hypothesis that infanticide in these mice is a form of resource competition.

Among sciurids, such as black-tailed prairie dogs, California ground squirrels, and Belding's ground squirrels, females do not direct their infanticidal attacks selectively toward female pups (Sherman 1981b; Hoogland 1995; Trulio 1996), as would be expected from the pattern of female philopatry (Greenwood 1980; Dobson 1982). Sherman (1981b) suggested that this lack of sex-specificity was because it is more important for females to kill entire litters rapidly than to spend time sexing their victims, especially in a dark burrow. Alternatively, and what often may be the case, females kill pups as a form of direct competition with territorial females and as a means of acquiring the burrow/nest site immediately and therefore must kill all offspring and not just the philopatric sex.

The theoretical framework provided by the resource-competition hypothesis seems appropriate for exploring causal associations between infanticide and communal nesting/breeding. Infanticide is one of a series of mechanisms by which individuals may suppress reproduction in others. Females of communally breeding species might use infanticide to prevent breeding by less dominant females, and thus control the partitioning of reproduction within the group (Johnstone and Cant 1999). The observations that nursing females kill pups within the same nesting group (*Glis glis;* Pilastro et al. 1996), pups of less dominant females (house mouse, Palanza et al. 1996), and pups of the same burrow (coterie) system (black-tailed prairie dog; Hoogland 1995) support this scenario. However, lactating females do not kill pups in other communally breeding species such as meadow voles (Ebensperger et al. 2000) and Norway rats (Menella et al. 1990; Schultz and Lore 1993). The conditions under which infanticide functions as a mechanism of reproductive suppression in rodents are unclear. One aspect that requires further elucidation is the relatedness between perpetrators and victims, especially in communally nesting species or those species in which females nest close together.

Males

Two studies have examined the resource-competition hypothesis to explain infanticide by males. McLean (1983) and Lacey (1992) invoked competition for resources to explain infanticide committed by immigrant male Arctic ground squirrels (*Spermophilus parryii*). McLean (1983)

recorded 10 cases of infanticide, all of which were perpetrated by immigrant adult males. Male marauders did not cannibalize their victims, but became resident in the area after the killings. Lacey (1992) found that females who lost their litters to infanticidal males dispersed and did not mate with the killers. McLean (1983) suggested that males of this species kill infants to decrease competition for food. Lacey (1992) suggested that infanticide by male Arctic ground squirrels resulted from competition for burrow systems whereby males took over female burrows, destroyed their litters, and remained there until the next breeding season. In short, both studies have provided valuable, but still preliminary, insights into the function of male infanticide in Arctic ground squirrels. More generally, the role of resource limitation on male infanticide in rodents remains to be assessed. The function of male infanticide in sciurids (e.g., see Hoogland chap. 37 this volume), and other seasonally breeding rodents, is particularly puzzling since sexual selection seems unlikely in this case (see the following).

Hypothesis 3: Insurance against misdirecting parental care
Sherman (1981b) and Elwood and Ostermeyer (1984b) suggested that individuals sometimes commit infanticide to avoid "adopting" or otherwise providing parental care to unrelated offspring. If so, infanticide should be committed mostly by the sex that bears the primary costs of adoption (Pierotti 1991). Among mammals, lactation is the most energetically costly phase of parental care (Trillmich 1986; Gittleman and Thompson 1988), and thus females should be the ones that benefit most by committing infanticide. An additional prediction from this hypothesis is that infanticide should be more frequent in species where nests of breeding females are spatially clumped (which increases the opportunity for unrelated pups to steal milk). This hypothesis does not require that victims be consumed.

Evidence in support of the misdirected parental care hypothesis is largely circumstantial. Among species in which females are infanticidal, both laboratory and field studies show that lactating females will indeed adopt and/or provide parental care to unrelated infants (table 23.2). This is the case in spiny mice (*Acomys cahirinus;* Porter and Doane 1978), Norway lemmings (De Kock and Rohn 1972), meadow voles (McShea and Madison 1984; Sheridan and Tamarin 1986), house mice (Sayler and Salmon 1971; König 1989a, 1994b), desert woodrats (*Neotoma lepida;* Fleming 1979), white-footed mice (Hawkins and Cranford 1992; Jacquot and Vessey 1994), deer mice (Hansen 1957; Hawkins and Cranford 1992; Millar and Derrickson 1992), black-tailed prairie dogs (Hoogland et al. 1989), Belding's ground squirrels (Sherman 1980a), Columbian ground squirrels (Hare 1991), yellow-bellied marmots (Armitage and Gurri-Glass 1994), and in the yellow-toothed cavy (*Galea musteloides;* Künkele and Hoeck 1995). The biological meaning of such adoption, particularly in laboratory studies, remains to be properly evaluated. Ultimately, knowing how frequently adoption occurs in nature is essential.

Other laboratory observations provide more direct evidence for the misdirected care hypothesis. Specifically, observations have shown that female golden hamsters (Richards 1966), house mice (McCarthy and Vom Saal 1985; Soroker and Terkel 1988; Manning et al. 1995; but see Palanza et al. 1996), Mongolian gerbils (Elwood and Ostermeyer 1984b), and Norway rats (Peters and Kristal 1983) kill unrelated young when they are sexually inexperienced, pregnant, or after weaning their own litters, but rarely when they are lactating. This makes sense, because lactating females of these altricial species are those most likely to make mistakes because pups of their own are available. In the house mouse (Sayler and Salmon 1971; Ostermeyer and Elwood 1983; Manning et al. 1995; but see Palanza et al. 1996) and the cavy (Künkele 1987; cited in Künkele and Hoeck 1989), lactating females in the laboratory adopt and nurse alien pups of similar age to their own, but may attack infants that do not match the age of their own young. Thus lactating female house mice and cavies seem to kill infants that potentially could steal milk, but only at times when they can recognize pups as not their own.

Among sciurids in which lactating females kill infants, the deaths occur before young mingle (Sherman 1981b; Hoogland 1985; Hare 1991). An exception to this is in the California ground squirrel in which most victims of infanticide by lactating females are postemergent infants (Trulio 1996). Elwood (1992) suggested that committing infanticide to prevent adoption could be expected in precocial rather than altricial species; in the former, infants are mobile and may attempt to nurse from nonrelatives. However, too few precocial rodents have been studied to evaluate this prediction. Under seminatural conditions, breeding females of precocial capybaras (*Hydrochaeris hydrochaeris*) kill pups of unfamiliar females (Da Cunha-Nogueira et al. 1999), female yellow-toothed cavies kill infants that do not match the age of their own offspring (Künkele and Hoeck 1989), and female maras (*Dolichotis patagonum*) are aggressive toward alien pups that attempt to nurse from them (Taber and Macdonald 1992a). However, infanticide by females has not been observed in the similarly precocial degu (*Octodon degus;* Ebensperger 2001b). Although some evidence exists to support the predictions of the misdirected care hypothesis, the theory has not been well developed and sufficient empirical and phylogenetic data are lacking for a thorough test of its application to rodents in general.

Hypothesis 4: Acquisition of mates

Hrdy (1977b, 1979) suggested that males might kill infants to destroy another male's offspring and cause females to return to reproductive readiness. Key predictions of this "sexual selection" hypothesis are: (1) infanticidal males should not kill offspring they have sired; (2) the elimination of offspring should shorten the interbirth period of the victimized females; and (3) infanticidal males should mate with and sire the subsequent offspring of the mother of the infant(s) that were killed (Hrdy 1979; Sommer 1994).

Sexual selection has been invoked to explain infanticide by males in several species of murid rodents (Vom Saal and Howard 1982; Huck 1984; Wolff and Cicirello 1989, 1991; Elwood 1992). There is considerable evidence from laboratory studies demonstrating the existence of mechanisms enabling males to target unrelated young and avoid killing their own offspring, including direct recognition of pups (house mouse, Paul 1986; spiny mouse, Makin and Porter 1984; deer mouse, El-Haddad et al. 1988), use of indirect cues such as association with previous sexual partners (house mouse, Huck et al. 1982; meadow vole, Webster et al. 1981), location of pups (McCarthy and Vom Saal 1986a), or inhibition of male pup killing due to recent mating and cohabitation with a female (Mongolian gerbil, Elwood 1977, 1980; house mouse, Elwood 1985, 1986, Elwood and Kennedy 1991, Palanza and Parmigiani 1991; meadow vole, Webster et al. 1981; Djungarian hamster, *Phodopus campbelli,* Gibber et al. 1984; spiny mouse, Makin and Porter 1984; McCarthy and Vom Saal 1986b; Brown 1986b; and Norway rat, Jakubowski and Terkel 1985b, Mennella and Moltz 1988).

In nature, infanticide by male white-footed mice and deer mice is typically committed by individuals who are recent immigrants. Thus they are unlikely to have sired any offspring in the area (Wolff and Cicirello 1989, 1991). In seminatural (captive) populations of house mice, infanticide is committed by territorial males outside their own territories, and by nonterritorial males that have not sired any offspring (Manning et al. 1995). Again, these male rodents kill infants they are unlikely to have sired.

The second requirement of the sexual selection hypothesis—that infanticide reduces the interbirth period of the females—is supported in murid but not sciurid rodents. Captive male collared lemmings (Mallory and Brooks 1978), meadow voles (Webster et al. 1981), house mice (Vom Saal and Howard 1982; McCarthy and Vom Saal 1986b; Coopersmith and Lenington 1996), and Norway rats (Mennella and Moltz 1988) that are introduced into the cage of an unfamiliar female and her neonates attack and kill the pups. If the infanticidal males are allowed to stay and mate with the victim's mother, they produce offspring sooner than males that do not eliminate the female's original litter (Mallory and Brooks 1978; Webster et al. 1981; Vom Saal and Howard 1982; McCarthy and Vom Saal 1986b; Mennella and Moltz 1988).

Embryonic implantation in rats, house mice, and Mongolian gerbils is delayed by lactation, and by the presence of infants (Mantalenakis and Ketchel 1966; Elwood and Ostermeyer 1984b; Mennella and Moltz 1988). As the number of suckled pups decreases, females subsequently produce larger litters (Elwood and Ostermeyer 1984b). Thus by killing pups, males not only shorten the interbirth interval but also increase the female's subsequent litter size (Elwood and Ostermeyer 1984b).

Only one study with rodents has attempted to look at subsequent mating by infanticidal males (Manning et al. 1995). In seminatural enclosures, male house mice sired the subsequent litters of victimized females after committing infanticide, which supports the sexual selection hypothesis.

The sexual selection hypothesis cannot be a general explanation of infanticide by males in sciurids and other seasonally breeding mammals. In most, but not all, (e.g., de Villiers 1986) sciurids in which infanticide by adult males has been recorded, the females become estrus only once per year and the elimination of their litters does not cause them to resume their sexual receptivity until the next breeding season (Sherman 1981b; McLean 1983; Hoogland 1985; Vestal 1991; Coulon et al. 1995; Blumstein 1997). Thus males cannot increase their opportunities to reproduce in the short term by killing a female's litter (Hiraiwa-Hasegawa 1988). Moreover, models show that a year-long lag between the death of a female's offspring and her next conception may make infanticide untenable as a male reproductive strategy (Chapman and Hausfater 1979; Hausfater 1984). Nonetheless, there is a possibility that infanticidal males of seasonally breeding rodents increase their reproduction during the following breeding season, because reproductive failure one year increases a female's chance of success in the following year (e.g., in black-tailed prairie dogs [Hoogland 1985], Richardson's ground squirrels [Michener 1978], and Alpine marmots [*Marmota marmota,* Hackländer and Arnold 1999; Coulon et al. 1995]), but not in golden marmots (*M. caudata;* Blumstein 1997).

The results from the previously mentioned studies provide strong evidence that infanticidal murid males avoid killing offspring they have sired, and that the elimination of offspring may shorten the interbirth period of the victimized females. The critical prediction that infanticidal males should mate with and sire the subsequent offspring of the mother of the infants has been scarcely examined and clearly more tests, ideally involving different species, are needed.

Counterstrategies to Infanticide

Several behavioral and physiological mechanisms have been implicated as counterstrategies to infanticide, including the direct attack of potential perpetrators (either by single individuals or by group coalitions); the avoidance of infanticidal animals; multiple mating; territoriality; or the early termination of pregnancy (Hrdy 1979; Hausfater 1984; Agrell et al. 1998; Ebensperger 1998b).

The frequency and intensity of agonistic behavior by female rodents typically increases during late gestation and lactation. Reports of greater aggression by breeding females under natural conditions exists for hoary (*Marmota caligata*) and Olympic marmots (*M. olympus*), Columbian ground squirrels, grey squirrels (*Sciurus*), red squirrels (*Tamiasciurus hudsonicus*), and yellow-pine chipmunks (*Tamias amoenus*), among sciurid rodents, wood rats, and jumping mice (*Zapus*), and among murid species (Ostermeyer 1983; Maestripieri 1992). Observations of maternal aggression among animals in large pens include Hystricognath species, such as green acouchis (*Myoprocta pratti*) and Bahaman hutias (*Geocapromys ingrahami*). One explanation for such heightened aggression is that it serves to protect offspring from infanticidal conspecifics (Svare 1977; Paul 1986; Huck 1984; Parmigiani 1986). In European wood mice (*Apodemus sylvaticus;* Wilson et al. 1993) females selectively chase and attack the conspecific gender that is most likely to kill preweaned pups. Female house mice and meadow voles are more likely to attack and direct more harmful bites toward males that are infanticidal than toward noninfanticidal and less aggressive males of the same reproductive status (Parmigiani, Sgoifo, and Mainardi 1988; Parmigiani, Brain, Mainardi, and Brunoni 1988; Elwood et al. 1990; Storey and Snow 1990).

The key expectation—that maternal aggression should result in a higher likelihood of infant survival has been harder to document. A number of laboratory studies have shown that maternal aggression reduces the likelihood of infanticide (bank voles, Ylönen and Horne 2002; deer mice and white-footed mice, Wolff 1985c; golden hamsters, Giordano et al. 1984; house mice, Maestripieri and Alleva 1990; vom Saal et al. 1995; meadow voles, Storey and Snow 1987; Norway rats, Takushi et al. 1983; Flannelly and Flannelly 1985; and woodrats, Fleming 1979). However, other studies found that females were only able to delay, but not prevent, infanticide under laboratory or seminatural conditions (collared lemming, Mallory and Brooks 1980; European wood mice, Wilson et al. 1993; house mice, Brooks and Schwarzkopf 1983; Parmigiani, Sgoifo, and Mainardi 1988; Parmigiani et al. 1989; Elwood et al. 1990; Palanza and Parmigiani 1994; Palanza et al. 1994; Manning et al. 1995;

Ebensperger 1998a; and Norway rats, Erskine et al. 1978; Mennella and Moltz 1988), even if infanticidal males are not artificially confined with the female and her pups (Ebensperger 1998a). We suspect that such a delay is probably sufficient for females to prevent infanticide under more natural conditions. However, some field studies also suggest that mothers cannot completely protect their litters from infanticide (Arctic ground squirrel, McLean 1982, 1983; Richardson's ground squirrel, Michener 1983a; and yellow-bellied marmot, Armitage et al. 1979). If this is generally so, the whole topic could set the stage for future studies that would consider these male-female aggressive interactions as a coevolutionary arms race.

A second mechanism that lactating females may employ is avoiding infanticidal males (Hrdy 1974, 1977b; Butynski 1982; Sommer 1987). However, demonstrating that females leave an area to avoid infanticidal males is a difficult task because individuals can move to a different area for other reasons, including a better food supply or better nest availability. Results from two studies support this prediction. Female Arctic ground squirrels and Alpine marmots moved their litters to new locations when their territories were taken over by foreign males, who might commit infanticide (McLean 1983; Coulon et al. 1995). In at least Alpine marmots, females successfully weaned their infants after moving them (Coulon et al. 1995). Clearly, future studies need to consider other valid explanations simultaneously as to why breeding females may change their location.

A third mechanism by which individuals may prevent infanticide is by forming coalitions that cooperate to repel infanticidal conspecifics (Hrdy 1977b). Two types of evidence provide support for this mechanism in rodents. Female house mice communally nest with other female relatives (Wilkinson and Baker 1988); and in the laboratory, females nesting in pairs are successful in attacking and repelling male and female intruders (Parmigiani 1986; Maestripieri and Rossi-Arnaud 1991). As a result, in enclosed populations, infanticide occurs in single-mother nests twice as often as in communal nests (Manning et al. 1995). Sires also may participate in the direct defense of litters (Pflanz 2002), and male-female pairs of house mice are effective in repelling intruders (Palanza et al. 1996). Whereas related female Belding's ground squirrels live in close proximity and successfully defend their litters by cooperatively chasing away conspecific intruders (Sherman 1980a), pairs of female Arctic ground squirrels are rarely successful in chasing away infanticidal males (McLean 1983).

A fourth mechanism to prevent infanticide is defending a territory such that potential intruders are kept away from vulnerable infants (Sherman 1980a, 1981b; Wolff 1993b). The pup-defense hypothesis has been invoked to explain fe-male territoriality among rodents (Sherman 1980a, 1981b; Webster and Brooks 1981; McLean 1983; Michener 1983a; Brooks 1984; Wolff 1993b), and mammals in general (Wolff and Peterson 1998). Supporting evidence is that the intensity of female territoriality generally increases during pregnancy, peaks during early to mid lactation, and decreases after the weaning of infants (Sherman 1980a, 1981b; Ostermeyer 1983; Maestripieri 1992), and that female territoriality is more intense close to the females' nest site rather than in the periphery of their territories (Wolff et al. 1983; Murie and Harris 1994).

Further support for the hypothesis that female territoriality functions to prevent infanticide among rodents includes studies on three species that show a fit between the identity of infanticidal intruders and the target of territoriality. Thus both males and females may commit infanticide among Belding's ground squirrels (Sherman 1980a, 1981b), black-tailed prairie dogs (Hoogland 1985, 1995), and wild house mice (Soroker and Terkel 1988), and as expected, both male and female conspecifics are excluded from the territory of lactating females (Sherman 1981b; Chovnick et al. 1987; Hoogland 1995). In Arctic ground squirrels, male rather than female territoriality is suggested to prevent infanticide by other males (McLean 1983). However, a mismatch between the identity of infanticidal intruders and the target of territoriality occurs in at least five other species. Male rather than female Alpine marmots are infanticidal, but female territoriality is directed against other females rather than males (Arnold 1990a; Coulon et al. 1995). Further, although male white-footed mice (Wolff 1985b; Wolff and Cicirello 1991), deer mice (Wolff 1985b; Wolff and Cicirello 1991), meadow voles (Madison 1980b; Ebensperger et al. 2000), and European wood mice (Wolton 1985; Wilson et al. 1993) can be as infanticidal as females, they are not excluded from the territory of the females. Such discrepancies may be explained, to some extent, if the females use more than one strategy to deal with different types of individuals (i.e., territoriality against females, multiple mating against males). However, discrepancies also may occur if female territoriality serves different functions in different species.

According to the pup-defense hypothesis, and under a similar amount of intruder pressure, the risk of infanticide should increase with a decrease in territory size, or with the intensity of territorial defense. One study has assessed this prediction directly with supportive results. In Belding's squirrels, the size of a lactating female's territory is inversely correlated with the probability of losing infants to infanticide (Sherman 1981b). Further indirect evidence comes from field studies of voles in which neonate survival and juvenile recruitment decrease as density of adult females increases concomitant with an overlap in female territories (Rodd and Boonstra 1988; Schauber and Wolff 1996; Wolff et al. 2002).

More subtle ways by which individuals are suggested to prevent infanticide include mating with several males and pregnancy termination. By mating with multiple males, females may confuse paternity of their litters and "persuade" males to tolerate their young once born (Hrdy 1974, 1977b, 1979). This hypothesis has been frequently suggested as an evolved mechanism in female rodents (and other mammals) to prevent male infanticide (Wolff 1993b; Agrell et al. 1998; Wolff and Macdonald 2004). The promiscuity hypothesis is well supported by several laboratory studies documenting an inhibition of male pup killing due to recent mating with a female (Mongolian gerbil, Elwood 1977, 1980; house mouse, Elwood 1985, 1986; Elwood and Kennedy 1991; Palanza and Parmigiani 1991; meadow vole, Webster et al. 1981; Djungarian hamster, Gibber et al. 1984; spiny mouse, Makin and Porter 1984; McCarthy and Vom Saal 1986b; Brown 1986b; and Norway rat, Jakubowski and Terkel 1985b; Mennella and Moltz 1988). In addition, one study has supported the expectation that infanticidal male rodents should not kill the offspring of previous sexual partners. Male house mice that were introduced into the cage of either their previous mate, or a strange female, were more likely to kill pups in the cage of the strange female, even if it contained foster pups actually fathered by the test male (Huck et al. 1982). Nevertheless, several other studies have failed to replicate these results (Brooks and Schwarzkopf 1983; McCarthy and Vom Saal 1986b; Parmigiani 1989; Elwood and Kennedy 1991).

According to the promiscuity hypothesis, a relationship is expected between multiple mating and the risk of infanticide. Two studies have addressed such an expected relationship, one in the field and the other in the lab. Pregnant female water voles that moved their nest location into the home range of a new male mated with that male; but pregnant females that stayed within their original male's range did not exhibit additional mating (with the presumed resident male) once they were pregnant (Jeppsson 1986). In a lab study with field voles, Agrell et al. (1998) found that when males were close together females mated with both of them and nested between them; however when males were far apart, females mated with the dominant male and nested near him (Agrell et al. 1998). These two cases are suggestive that females assess the potential for infanticide and use multimale mating as a deterrent tactic. Assuming that mating activity involves costs to females (e.g., increased susceptibility to predators), we might expect that females will associate and mate preferentially with infanticidal rather than noninfanticidal males. The observation that female meadow voles and house mice did not prefer infanticidal

over noninfanticidal conspecific males as social or potential mating partners is inconsistent with the female promiscuity hypothesis (Ebensperger 1998d).

Wolff and Macdonald (2004) recently provided correlative support for the promiscuity hypothesis. By using examples from across sciurid and murid rodents (and from nonrodent mammals) they found that in species in which males commit infanticide, females mate with multiple males. In contrast, they recorded that multimale mating by females is not frequent in species in which male infanticide does not occur. A further analysis of their data controlling for phylogeny (e.g., Blumstein 2000) will provide a more complete test of this hypothesis.

Male-induced pregnancy disruption (also referred to as "pregnancy block," "Bruce effect." or "abortion") was initially observed in house mice, and occurs when recently inseminated females are exposed to an unfamiliar male (or to his odor), which may prevent implantation and cause a return to estrus 4–5 days later (Bruce 1959, 1960). Among other potential functions, pregnancy disruption may prevent waste of additional investment on infants that will likely be killed by invading or strange males (Hrdy 1979; Schwagmeyer 1979; Labov 1980, 1981b; Mallory and Brooks 1980). In support of this hypothesis, dominant male house mice are more infanticidal than subordinate males (Huck et al. 1982; Elwood 1986), and female encounters with dominant males are more likely to cause pregnancy disruption than encounters with subordinate males (Huck 1982; but see Labov 1981a). Infanticidal male house mice are more likely to induce pregnancy block than noninfanticidal males, which suggests an ability of females to evaluate differences in the risk of infanticide on their litters should pregnancy not be interrupted (Huck 1984; Elwood and Kennedy 1990). In golden hamsters, females are more infanticidal than males (Marques and Valenstein 1976), and pregnancy block is caused more frequently by females than males (Huck et al. 1983; Huck 1984). Apparently, females can use odor, as well as behavioral (e.g., level of aggression) cues from conspecifics to make this discrimination (Storey 1986b; Storey and Snow 1990; de Catanzaro et al. 1995).

Only two field studies have attempted to test the pregnancy disruption hypothesis and both found no or limited support for it. De la Maza et al. (1999) experimentally exposed breeding female gray-tailed voles to treatments in which males were removed and replaced by either socially unfamiliar males or females. In response to this manipulation, the researchers found no differences in intervals between parturitions, in the frequency of pregnancies, or in juvenile recruitment. Gray-tailed voles are promiscuous (Wolff et al. 1994) and males are infanticidal (J. Wolff, unpublished) and thus should fit predictions of the pregnancy disruption hypothesis. In a similar study with a population of prairie voles in outdoor enclosures, Mahady and Wolff (2002) replaced resident males with unfamiliar males every 10 days. They reported that 7 of 33 (21%) nulliparous females did not conceive during the study, but whether this was due to pregnancy failure or disruption of pairbonding in this monogamous species is not known. Production of second litters and breeding by parous females were not affected by exposure to strange males. Certainly, more field studies are needed to test the validity of the Bruce effect or pregnancy disruption hypothesis as a counterstrategy to infanticide, but at least these two field studies with two *Microtus* species do not provide strong support that pregnancy disruption occurs regularly or is an adaptive response to exposure to strange males, at least in this taxon.

Among sciurids, most takeovers by male Alpine marmots (62%, n = 21) occur after the mating period or before the end of lactation (Hackländer and Arnold 1999). Interestingly, female breeding is reduced after these male takeovers despite clear signs of pregnancy early in the season, and females failing to reproduce right after these takeovers increase their chance of breeding in the following year (Hackländer and Arnold 1999). Nonetheless, male takeovers in other populations of Alpine marmots seem to occur mostly (75%, n = 20) when juveniles are already born (King and Allainé 2002). Taken together, these field studies provide only moderate support for the idea that pregnancy disruption is a strategy to prevent losses to infanticide. Moreover, predators and other potentially stressful factors also may

cause pregnancy disruptions in female rodents (de Catanzaro and MacNiven 1992), which suggests that pregnancy disruption may indeed be part of a more general strategy to prevent the waste of energy in producing offspring likely to be lost.

Finally, socially subordinate individuals may suppress breeding as a strategy to avoid wasting energy and resources on litters that are likely to be eliminated by more dominant females within the group (Agrell et al. 1998). Such may be partially the case of subordinate females of Alpine marmots that achieve copulations and become pregnant within their social units, but only the dominant females give birth (Hackländer and Arnold 1999; King and Allainé 2002). Formal phylogenetic analyses may shed light onto the evolutionary relationships between the occurrence of within-group infanticide, social living, and breeding suppression.

Concluding Remarks

Functions of infanticide

The functional significance of infanticide in rodents is complex and cannot be explained by any one single hypothesis. Each hypothesis has its own assumptions, predictions, and tests (fig. 23.1; table 23.2). In some species individuals obtain nutritional benefits from infanticide (table 23.2). In some cases, nutritional benefits are gained by females (e.g.,

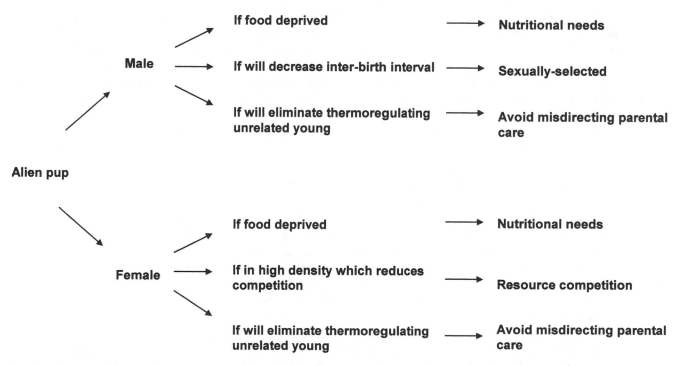

Figure 23.1 Key conditions and the subsequent benefits from infanticidal behavior reported to occur in male and female rodents. Not all benefits are equally well supported (see text).

deer mouse, the white-footed mouse, the black-tailed prairie dog, the California ground squirrel, the Columbian ground squirrel), while in other species males may gain nutritional resources (e.g., Mongolian gerbil, meadow vole, Belding's ground squirrel).

In a few species, infanticide is a mechanism of resource competition (table 23.2). The most compelling evidence supporting the idea that individuals commit infanticide to avoid misdirecting parental care to unrelated offspring comes from the infanticidal behavior of female pinnipeds, which react aggressively and bite unrelated pups that attempt to steal milk from them (e.g., Reiter et al. 1981; Bruemmer 1994). Evidence for this hypothesis among rodents is limited to associations between the breeding condition of killers and the timing of infanticide (table 23.2). Moreover, this hypothesis might explain why a female would kill a pup that wandered into her burrow, but it would not explain why a female would travel a long way from her nest burrow, enter another female's burrow, and kill young in there (as in Belding's ground squirrels and prairie dogs).

The sexual selection hypothesis in which males kill infants they have not sired as a means of reproducing with the victims' mother seems well supported in primates and African lions (reviewed in Ebensperger 1998b), but is less clear in rodents, especially sciurids. The possibility that sexually selected infanticide takes place among male rodents (particularly Muridae) is supported by laboratory studies showing that male rodents are prevented from killing their own infants (table 23.2). Nonetheless, studies that measure fitness benefits in terms of increased mating opportunities or of a reduced latency for the females to bear offspring of infanticidal males under wild or more seminatural conditions (e.g., Manning et al. 1995) are needed. Studies of sexually selected infanticide by males of seasonally breeding species also deserve further study, particularly in terms of increased chances of killers to mate with the victimized females and whether reproductive success of victimized females increases during the following breeding season.

We encourage future investigators to design studies that will simultaneously evaluate multiple functional hypotheses and their specific predictions (e.g., table 23.2). Experimental studies under natural conditions and/or those that accurately represent the social and physical environment of species are needed to discern among alternative hypotheses. Indeed, quantifying the incidence of infanticide among wild populations is a difficult task, and there are serious ethical issues with experimental studies of infanticide (Elwood 1991). However, using traps "baited" with pups (e.g., Wolff and Cicirello 1991; Ylönen et al. 1997), recording characteristic behaviors and external signs given by the perpetrators (e.g., Hoogland 1995), and potentially employing other innovative techniques while being careful to avoid pain and suffering of experimental subjects will allow future investigators to control for various social and ecological variables to test the adaptive significance of the various hypotheses for infanticide. Moreover, there is a need for future comparative studies to test the various hypotheses for the adaptive significance of infanticide.

Consequences of infanticide

Overall, the nature of the mechanisms by which parents should attempt to prevent infanticide has been controversial, and deserves further study. In particular, we believe that some important current controversies will be solved if future studies consider three major issues. First, we need information from animals whose behavior is recorded under realistic conditions of space, habitat heterogeneity, and density. This is critical to fully appreciate the meaning of any results within an evolutionary context. We acknowledge that these are difficult data to acquire in nature. Secondly, alternative hypotheses should be stated a priori, and strong inferential tests devised. The behaviors that have been suggested to be counterstrategies in rodents have other hypothesized functions as well. Multiple mating by females has many hypothesized functions (Jennions and Petrie 2000; Fedorka and Mousseau 2002). Maternal aggression could be a mechanism used by dams to assess quality of males as future mates (Parmigiani et al. 1989; Parmigiani et al. 1994). Territory defense (as opposed to defending nests and the space immediately nearby) by female rodents might be directed toward defending physical resources as well as pups (e.g., Sherman 1981b; Ostfeld 1990). Considering the great cost to females of losing their offspring, and the apparently high incidence of infanticide in natural populations, natural selection has likely favored several defensive strategies by females to protect their young in the evolutionary arms races within and between the sexes.

Summary

Nonparental infanticide, the killing of infants by conspecifics other than the parents, occurs in a variety of vertebrate and invertebrate taxa. In rodents, infanticide has been noted in the wild or under laboratory conditions in 2 species of hystricognaths and 35 species of sciurognaths. Our review supports the hypothesis that nonparental infanticide is adaptive in rodents. However, its functional significance seems complex and cannot be explained by any one single hypothesis. In some cases, nutritional benefits are gained by females, while in other species males may gain nutritional resources. In a few species, infanticide is a mechanism of resource competition. Evidence supporting the idea that individuals commit infanticide to avoid misdirecting parental care to unrelated offspring among rodents is rather limited. The sexual selection hypothesis in which males kill infants they have not sired as a means of reproducing with the victims' mother remains unproven in rodents; studies that measure fitness benefits in terms of increased mating opportunities or of a reduced latency for the females to bear offspring of infanticidal males under wild or more seminatural conditions are strongly needed. The nature of the mechanisms by which parents prevent infanticide has been controversial, and future studies need to consider two critical issues. First, information is needed from animals whose behavior is recorded under realistic ecological conditions. Second, alternative hypothesis should be stated a priori: the behaviors that have been suggested to be counterstrategies in rodents have other hypothesized functions as well. Overall, we encourage future investigators to design studies that will simultaneously evaluate multiple functional hypotheses and their specific predictions.

Chapter 24 Social Organization and Monogamy in the Beaver

Peter Busher

Beaver Natural History

The Eurasian beaver (*Castor fiber*) and the North American beaver (*Castor canadensis*) are the only two extant members of the family Castoridae. The Eurasian beaver is restricted to the Palearctic biogeographical region, while the North American beaver, which is native to and widespread in the Nearctic region, was introduced into the Palearctic region (Europe and Russia) in the past 100 years (Nolet and Rosell 1998; Hartman 1999). North American beavers have also been introduced into Tierra del Fuego, Argentina (Lizarralde 1993).

The two species are morphologically similar (Lavrov 1983; Jenkins and Busher 1979; Sieber et al. 1999) and have comparable habitat requirements and behavior (Wilsson 1971; Jenkins and Busher 1979; Hill 1982; Hodgdon and Lancia 1983; Müller-Schwarze and Sun 2003). For example, both species are primarily nocturnal, semiaquatic (living along all types of freshwater systems), strictly herbivorous, and typically live in family groups containing an adult pair, young of the year (kits) and young from the previous year (yearlings). Beavers of both species may construct lodges made of mud and sticks and dams that create ponds, which stabilize the environment. Beavers living on large lakes and rivers do not build dams; introduced North American beavers in Russia are more active in construction behaviors than Eurasian beavers (Danilov and Kan'shiev 1983; Danilov 1995). Both species communicate by tail slapping, scent marking, and through vocalizations and posture. While the two species are similar in size and appearance they do have different numbers of chromosomes (*C. canadensis* $2n = 40$; *C. fiber* $2n = 48$) and do not reproduce with each other (Lavrov and Orlov 1973; Jenkins and Busher 1979).

The Beaver Social Group and Mating System: An Overview

The two beaver species are in the small percentage of mammals (3% to 5%) and even smaller percentage of rodents that form monogamous pair bonds (Kleiman 1977; Dunbar 1984; fig. 24.1). Two types of monogamy are generally considered: (1) genetic (exclusive) monogamy, when the male and female confine their mating exclusively to the same partner, and (2) social (biparental) monogamy, when the partners maintain a close association after fertilization and care for the young, but mating may not be exclusive (Barlow 1988; Westneat and Sherman 1997; Reichard 2003).

When male care is critical for reproductive success monogamy has been considered obligate; when it is less critical and males will readily mate with additional females it has been considered facultative (Kleiman 1977; Wittenberger and Tilson 1980; Clutton-Brock 1989b). However, viewing monogamy as either obligate or facultative, while useful, has been questioned (Komers and Brotherton 1997; Smith 1997; Sun 2003). For example, the dik-dik, a dwarf antelope, is considered to exhibit facultative monogamy, yet while the male provides no parental care he also does not attempt to mate with additional available females (Brotherton and Rhodes 1996; Komers and Brotherton 1997). Additionally, monogamy in a broad array of mammal groups

Figure 24.1 A mated pair of Eurasian beavers (*Castor fiber*). The male is on the left with the right eartag and the female is on the right with the left eartag. The pair had been together at least eight years when the picture was taken, and were 15-16 years old. These beavers were reintroduced into the Netherlands in 1988. Photograph and age data provided by Frank Rosell.

was found to evolve in the absence of male parental care and in species where females occupied small, exclusive ranges (Komers and Brotherton 1997).

Beavers appear to exhibit both social and genetic monogamy since they mate exclusively with one partner (although in most cases this has yet to be confirmed by DNA analysis), have a high degree of biparental care of the young, and the pair bond is maintained for multiple years. Beavers are also considered to exhibit obligate monogamy based on the extent of male parental care and lack of behavioral dimorphism (Kleiman 1977; Sharpe and Rosell 2003). Many aspects of beaver ecology and behavior support the formation and maintenance of the monogamous pair bond. Among these are territorial defense, relatively slow maturation of young (approximately 2 years to sexual maturity), and presence of older family members living with the family group (Kleiman 1977).

A beaver group defends a territory, which can be a length of shoreline on a large river or lake or an entire section of a smaller stream (including ponds created by dam construction), and this territory includes their lodge(s), dams, and winter food cache. In many northern latitudes there is reduced plant productivity during the winter when the ponds and streams freeze, restricting beaver mobility and access to food. In this type of climate a monogamous mating system may help ensure access to a mate (breeding generally takes place in January–February and the estrus period is short). Beavers illustrate all the criteria for long-term pair bonds (Wilson 1975), and the beaver mating system represents an ideal model for investigating the evolution of monogamy.

Past reviews of monogamy in beavers have relied primarily on a few studies of the Eurasian species (Eisenberg 1966; Kleiman 1977), but more recently a number of re-

searchers in North America have investigated beaver social behavior. These studies have generally examined the ecological constraints, primarily habitat quality (Buech 1987; Smith 1997), the amount of parental care that influences beaver social organization (Svendsen 1980b; Busher 1980; Busher and Jenkins 1985; Buech 1987; Woodard 1994), or both (Sun 2003). Rosell and Pedersen (1999) summarized the social behavior of the Eurasian species. However, at present no comprehensive analysis of the social organization of both beaver species has been undertaken.

This chapter will review the past and current knowledge of the social organization of the two extant beaver species and critically evaluate the beaver social group in light of the evolutionary and ecological constraints that shape the mating system.

The Beaver Social Group

Typical group composition

The beaver social group is commonly called a colony, regardless of the fact that it does not fit the precise sociobiological definition of a colony (Wilson 1975). The idea of a beaver colony is believed to have originated with the European fur trappers and explorers (Morgan 1868; Taylor 1970). However, modern quantitative studies document the beaver social group as being a family group.

In one of the first quantitative studies of the beaver social group Bradt (1938) analyzed forty-two groups in Michigan that were completely trapped out and the beavers assigned to sex and age classes. He reported that 59.5% (25/42) consisted of an adult pair (male and female) with or without young and 45% (19/42) consisted of an adult pair with at least one litter. This led Bradt to conclude that "A typical beaver colony is shown . . . to consist of an individual family, including the parents, the kits, and often the yearlings born the previous year." Throughout the beaver's range in North America authors have consistently documented the family group as the unit of social organization, and have found the mating system to be strictly monogamous (Hall 1960; Taylor 1970; Hodgdon 1978; Busher 1975, 1980, 1987; Svendsen 1980b, 1989; Buech 1987; Payne 1982; Patenaude 1983; Peterson and Payne 1986; Wheatley 1994, 1997; Woodard 1994). Data on the composition of family groups with adult pairs and groups with adult pairs and young from selected studies illustrate the variation among populations as well as the general pattern of groups with adult pairs (fig. 24.2).

Studies of Eurasian beavers in Russia, The Netherlands, and Sweden also support the definition of the beaver social group as a family, with the adults exhibiting mating exclu-

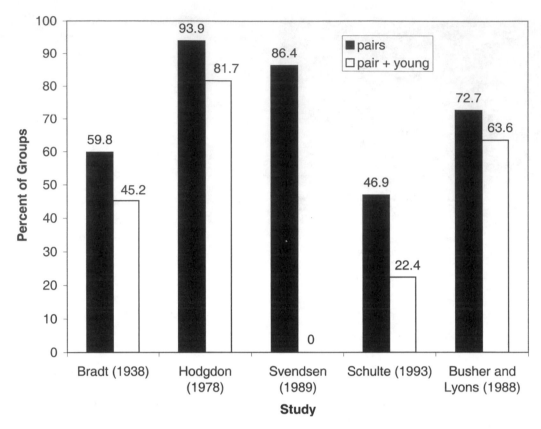

Figure 24.2 Percent of beaver groups that consisted of a mated pair of adults and the percent of those groups that consisted of a mated pair of adults with young. Svendsen (1989) did not report the percentage of groups consisting of a pair plus young. The Busher and Lyons data are from an unpublished research report for The Center for Field Research–Earthwatch (October 1988).

sivity and defending a territory against other beavers (Wilsson 1971; Djoshkin and Safonov 1972; Nolet and Rosell 1994; Hartman 1994, 1995; Rosell and Pedersen 1999).

The extended family group

While the majority of studies of both exploited and unexploited populations document the family as the most encountered social group, variations in group composition have been reported. Additional older individuals (2-year-olds or older) have been observed living in family groups (Hay 1958; Rutherford 1964; Novakowski 1967; Leege and Williams 1967; Taylor 1970; Bergerud and Miller 1977; Hodgdon 1978; Payne 1982; Busher et al. 1983; Smith 1997; Busher and Lyons 1999; Müller-Schwarze and Schulte 1999; fig. 24.3). Most authors believed that these additional adult beavers are older offspring that have delayed dispersal, which usually occurs the spring or summer when they are 2 years old (Jenkins and Busher 1979; Müller-Schwarze and Sun 2003).

In high-density populations from California to Canada between 22.2% to 87.5% (mean = 38.2%) of families had additional adults, while in low-density populations the range was 0.0% to 12.5% (mean = 5.2%). In high-density populations young adult beavers are less likely to disperse from the natal colony and more likely to remain living with their family (Müller-Schwarze and Schulte 1999; see also Nunes, chap. 13 this volume). Delayed dispersal appears to

Figure 24.3 A group of subadult Eurasian beavers (*Castor fiber*) in Norway, feeding together near the lodge. These individual are 2–3 years old and have not dispersed from the family territory, most likely due to high population density in the area. Photograph and data provided by Orsi Bozér.

be influenced by poor habitat quality and the lack of suitable dispersal sites (Hodgdon 1978; Brooks et al. 1980; Smith 1997; Busher and Lyons 1999).

Thus a beaver colony is typically a family with as many as three different related age groups of individuals present; both species exhibit this type of social group. Variations in retention of older, nonreproductive family members, as well as summer wandering by family members, have also been reported. Density and habitat quality appear to be the proximate factors influencing the composition of a beaver family.

Evidence of polygamous mating

Few studies have reported more than one reproductive female in a family group. Bergerud and Miller (1977) in Newfoundland found that three of twenty-seven (11.1%) groups had more than one adult male and three of twenty-seven (11.1%) groups had more than one adult female. However, only one of twenty-seven (3.7%) groups had two lactating females. Brooks et al. (1980) reported that 11.1% of the groups studied in Massachusetts had more than one adult female and 11.1% had more than one adult male, but no group had more than one lactating female. Schulte (1993) found 18.4% of his groups had additional adults, but also reported no additional lactating females. However, in this same population a few years later 8.57% (three of thirty-five families) had two breeding females living with one male (Sun 2003). Busher et al. (1983) reported multiple lactating adult females in 87.5% (seven of eight groups over 2 years) of the groups in a small isolated population of beavers in Nevada. They also observed more movement among group territories by adult males, adult females, and 2-year-olds, but not by yearlings. This population had an extremely high density, with close spacing between colonies, and there was little if any additional available habitat into which maturing beavers could disperse. Wheatley (1993), in Manitoba, also found one case of two lactating females living within the same lodge.

Thus while a few authors have reported more than one lactating adult female in a family, these represent a very small percentage of studies, and polygynous mating is far from frequent in beavers, as suggested by Sun (2003). In fact, throughout North America, while the group composition may range from single adults (or dispersing young adults) to mated pairs, to families with one or two generations of young to other combinations of a single adult and young, incidences of polygyny have rarely been observed. Given the large latitudinal (and elevational) distribution of beavers and the various habitat conditions (ranging from large lakes and rivers to small ponds and streams) where beavers are found, greater variety in the mating system could

be expected. However, the consistent social group observed is the family (or extended family), with monogamous mating between adults. The situation appears to be the same for the Eurasian beaver (Wilsson 1971; Rosell and Pedersen 1999; Sharpe and Rosell 2003).

Only under conditions of high density does the strictly monogamous pair bond become compromised; and further studies of beavers living in high-density populations are required to adequately evaluate the influence of density on family composition and the pair bond.

Duration of the pair bond

The general pattern in both species is for the pair bond to exist for multiple years and that the adults are closely associated during this time. Pair bond duration in eight beaver pairs over a 19-year period in a California population ranged from 1–6 years, with a mean of 3.6 years (Taylor 1970). A 12-year study in Ohio found the average duration of the bond to be 2.5 years for twenty-six pairs (Svendsen 1989). In California, a 3-year study reported that three of four pairs remained together for 3 years (Busher 1980), while a longer-term study recorded a mean duration for thirteen pairs of 1.9 years (Woodard 1994). A study in New York reported the average pair bond duration to be 4.3 years, with the longest duration 9 years (Sun 2003). Only one case of desertion was documented in these studies (a female leaving a male), with most pair bonds dissolving due to documented or presumed death of a mate. Since mating generally occurs in winter, maintaining a pair bond would ensure access to a mate, especially at latitudes or altitudes where mobility is restricted by ice and snow.

The cost of reproduction provides support for the long-term maintenance of the pair bond. For example, female beavers (introduced *Castor canadensis*) in Finland that reproduced in 2 consecutive years produced more offspring (as measured by fetus production) than females who only mated in 1 of the 2 years (Ruusila et al. 2000). Maintaining the pair bond over a series of years should increase the potential for mating, encourage reproduction in consecutive years, and increase the lifetime reproductive success of the female. Additional long-term studies of marked populations are needed, and could add to our understanding of the pair bond duration and the conditions under which they terminate.

Social Organization within the Family

Our understanding of beaver social behavior has been advanced by many recent studies (Tevis 1950; Schramm 1968; Wilsson 1971; Hodgdon and Larson 1973; Hodgdon 1978;

Lancia 1979; Hodgdon and Lancia 1983; Busher 1975, 1980, 1983a, 1983b; Busher and Jenkins 1985; Patenaude 1983; Patenaude and Bovet 1983, 1984; Buech 1987, 1995; Woodard 1994; Sharpe and Rosell 2003). Since the evolution and maintenance of the monogamous family group involves balancing reproductive advantages for both the adult male and female, it is imperative to examine their behavioral roles in the family group. As is the case with all mammals, the adult female invests heavily in her young during the relatively long period of gestation (approximately 100 days) and subsequent period of nursing (6–8 weeks). However, it is critical to examine how the other family members, especially the adult male, invest in the group.

Beaver social behavior: An overview

Comparisons between studies of marked beavers (where the age and sex of each beaver is known) are difficult, due to habitat and climatic differences at research sites or methodological differences. Additionally, due to the goal of an individual project, some specific behaviors may be over- or underrepresented. Regardless of differences between studies, it is possible to find a general pattern of beaver behavior within a family group.

Beaver behaviors can be categorized as those involved with personal maintenance (cutting food and feeding, grooming, and swimming associated with feeding), colony maintenance (construction of dams, lodges, and burrows, food caching, and bringing food to the lodge in the summer), and social interactions. Subadult beavers, especially kits, generally engage in more personal maintenance behaviors than adults, while the adults perform more colony maintenance behaviors. The degree of social interaction declines as beavers mature. Nondispersing two-year-old beavers have behavioral patterns more similar to adults and yearlings than to kits (Hodgdon 1978; Busher 1980, 1983b; Woodard 1994).

Within the social group the most common dominance hierarchy is based on age, with adults dominant to younger animals (Busher 1983a; Hodgdon and Lancia 1983). Only Hodgdon and Larson (1973) reported a hierarchy in which the adult female was dominant to the adult male, and this was based on data from one family in one season. However, no other study found any evidence of the adult female being dominant to the adult male (Svendsen 1980b; Busher 1980, 1983a; Brady and Svendsen 1981; Buech 1987, 1995). Female dominance over males in captive Eurasian beavers has been observed during the formation of the pair bond. However, after the pair formed no sexual dominance was observed (Wilsson 1971).

Monogamy in beavers has been linked with female aggression (Wittenberger and Tilson 1980), yet this was based on data from only two studies (Wilsson 1971; Hodgdon and Larson 1973). However, there is little if any evidence from other behavioral studies that female aggression and or dominance exists in beaver families; in most groups the adult male and female are codominant. It may be true that female aggression and dominance can facilitate the formation of the pair bond, but adult female dominance within an existing family has not been adequately documented. Studies designed to test hypotheses relating to pair bond formation and colony organization based on female dominance would provide additional insight into this problem.

Social behavior and maintenance of the mating system

While the importance of biparental investment in evolving and shaping monogamous mating systems has recently been questioned (Brotherton and Rhodes 1996; Brotherton and Komers 2003), the construction activities and social organization of the beaver may require increased levels of biparental care. One of the major questions regarding beaver social organization and the mating system involves adult male investment in the family group.

A study of the North American beaver in Minnesota revealed that during the active period (generally 1800 hours–0600 hours) adult females invested 6.5 hours daily to parental care and accounted for 58% of the total care during the open-water season. Adult males allocated 5.2 hours of time to behaviors associated with parental care and contributed 42% of the total care (Buech 1987). In the Eurasian beaver, pairs of adult male and female beavers were only different in the amount of time they allocated to travel; their overall time budgets were not significantly different (Sharpe and Rosell 2003). These data are consistent with hypotheses regarding parental investment and monogamy (Trivers 1972). Additional data suggest that adult males invest in the family primarily by the performance of three major types of behavior: (1) construction behaviors; (2) scent marking and territorial behavior; and (3) alarm behaviors (Busher 1980; Hodgdon and Lancia 1983; Woodard 1994).

Construction behaviors

Beavers physically alter their environment through complex construction activities, which represent a major source of investment in the territory. Beavers invest in a specific area by constructing dams, which create ponds, the construction of lodges, and, in northern latitudes, by the construction of a winter food cache (Jenkins and Busher 1979; Busher and Hartman 2001). Both adults in a family group have been observed actively performing construction behaviors, although data from different studies are not consistent (table 24.1). Because of methodological differences in data collection be-

Table 24.1 Comparison of construction behaviors performed by adult male and female beavers

Reference	Lodge work		Dam work		Dam inspection		Food cache		Canal work	
	Male	Female	Male	Female	Male	Female	Male	Female	Male	Female
Hodgdon (1978)[a]	11.4	31.3	20.5	44.3	37.3	24.4	25	29.2	28.4	34
Busher (1980)[b]	0.5	2.0	3.3	2.4	17.4	13.8	0.5	0.3	n.a.	n.a.
Buech (1987)[c]	8.5	1.1	n.a.	n.a.	n.a.	n.a.	2.6	5.3	n.a.	n.a.
Woodard (1994)[c]	0.04	0.05	9.77	2.45	0.7	0.6	n.a.	n.a.	n.a.	n.a.

NOTE: n.a. = not available.
[a]Hodgdon (1978) reported observations as the proportion of the behavior performed by each age-sex class.
[b]Busher (1980) reported observations as the frequency of occurrence of a behavior in relation to all behaviors.
[c]Buech (1987) and Woodard (1994) reported observations in the form of time budgets.

tween studies it is difficult to generalize about sexual differences in construction behavior. However, adult females have been observed to be more active than males in lodge work, while adult males have been observed to be more active in other construction behaviors (Hodgdon 1978; Busher 1980; Buech 1987; Woodard 1994).

An additional problem is that beavers in different habitats may not be involved in all construction activities. For example, beavers living on large lakes and on large rivers do not construct dams, and may not construct lodges (Buech 1987; Breck et al. 2001), and North American beavers living at more southern latitudes or in areas not subjected to freezing do not generally construct food caches. Food cache construction may also be variable within a population. Goran Hartman, studying the Eurasian beaver in Sweden, found that 33% (thirteen of thirty-nine) of families in the same geographical area did not build a food cache (G. Hartman, personal communication). However, what is evident is that the adult male in a family group is actively involved in construction behaviors; this represents a major source of male parental investment.

Territorial behavior and scent communication

Beavers mark a territory by the construction of scent mounds (fig. 24.4). These scent mounds consist of piles of mud and vegetation, onto which beavers deposit the secretions from their castor sacs and anal glands (Wilsson 1971; Hodgdon 1978; Jenkins and Busher 1979; Svendsen 1980a; Schulte 1993; Rosell and Nolet 1997). Castoreum, the fluid from the castor sacs, is produced by urine passing through the sacs and picking up the scent, and hundreds of compounds have been identified in castoreum (Svendsen and Huntsman 1988; Müller-Schwarze 1992; Tang et al. 1993, 1995). The secretion of the anal gland is oily, insoluble in both water and ethanol, and has a strong odor (Svendsen 1978; Schulte and Müller-Schwarze 1999). The anal gland

secretion has been used to differentiate between the sexes and between the two species (Schulte et al. 1995; Rosell and Sun 1999).

Scent mounds are usually constructed near feeding sites, active lodges, construction areas, and margins of territories (Wilsson 1971; Hodgdon 1978; Rosell et al. 1998). Although scent mounds do not appear to keep transient beavers out of existing territories (Wilsson 1971; Hodgdon 1978) they are most likely used to advertise the occupation of an area (Aleksiuk 1968; Taylor 1970; Schulte and Müller-Schwartz 1999).

Captive female Eurasian beavers usually marked first and more frequently than males (Wilsson 1971), while in a North American beaver population adults accounted for 74.3% of all scent marking, and males of all age classes scent marked more than females of the same age class (Hodgdon 1978). In the same North American population, when presented with unfamiliar castoreum and anal gland secretion, adult males had a significantly stronger response to un-

Figure 24.4 A typical scent mound constructed by the North American beaver (*Castor canadensis*), in Massachusetts. The mound consists of small sticks, leaves, and mud, and a beaver will deposit scent on the top of the mound. Photograph by Peter Busher.

familiar male castoreum than did adult females. Males responded at a greater distance, hissed for a longer period of time, marked more frequently, and spent more time investigating the unfamiliar scent. The response by both adult males and females was weaker to unfamiliar than to familiar female castoreum. The adult males also showed a slightly stronger response than adult females to unfamiliar anal gland secretion of both sexes, but the difference was not significant (Hodgdon 1978).

The territorial function of scent mounds has been supported by data obtained from recent field experiments. In North America, adult males had a higher response to nonneighbor castoreum than to castoreum from neighbors or to castoreum from their own offspring (Schulte 1998). A similar study of the Eurasian beaver obtained similar results, as family members responded strongly to unfamiliar scent and showed territorial behavior when confronted with unfamiliar scent (Rosell et al. 2000). Numerous other studies have been conducted in the past 20 years, often using field experimental techniques, to better understand the functional role of scent marking in beavers (for reviews of these studies see Schulte and Müller-Schwarze 1999; Rosell and Pedersen 1999; Müller-Schwarze and Sun 2003). The consensus from these studies is that beavers build and maintain scent mounds as a means of marking their territories and that the adult male is actively involved in this activity. Castoreum appears to be more important than anal gland secretion in territorial marking, and castoreum from unfamiliar males elicits a stronger response. Whether adult male scent marking behavior is a resource based territorial defense or a mate-guarding strategy is difficult to determine and it probably functions as both.

Alarm behaviors

Alarm behaviors are used for territorial defense and in response to potential predators. Beavers patrol their ponds by purposefully swimming along the perimeter, usually when they first emerge in the evening. They actively sniff unfamiliar scent and slap their tails on the surface of the water in response to unfamiliar stimuli (Wilsson 1971; Hodgdon 1978; Jenkins and Busher 1979). Tail-slapping behavior appears to be a direct response to a disturbance in the territory cued by sound, scent, or visual stimuli (fig. 24.5).

Hodgdon (1978) observed 896 tail-slapping incidents and 1,617 individual slaps by marked beavers. Adult females were involved in 26.9% of the incidences and accounted for 26.4% of the tail slaps. Adult males were involved in 15.7% of incidences, but accounted for 23.5% of the slaps. Hodgdon noted that older animals tended to slap more than younger animals, and found that males of all age classes slapped more per incident than females and that a

Figure 24.5 An adult North American beaver (*Castor canadensis*), tail slapping in response to an observer. Photograph by Peter Busher.

slap by either an adult male or an adult female elicited the same response from other family members. Busher (1980) found no difference in tail slapping between adult males and females and reported that kits slapped more than any other age class. Woodard (1994) also found no difference between the adults in tail slapping and, although she did not observe any kits involved in this behavior, she reported that yearling and 2-year-old beavers slapped more than adults. Buech (1987) does not separate tail slapping from other alarm behaviors; however, he reported that adult males allocated more time to these alarm behaviors than adult females during all months of observation.

The behavior of patrolling a territory may be a function of the specific habitat. Buech (1987), studying beavers living on a large lake, did not record this as a separate behavior, while Busher (1980) and Woodard (1994) did observe it in beavers on a small stream. Adult males have been reported to be more active in patrolling their territory than adult females or any other age class, and patrolling activity generally increases with age (Busher 1980; Brady and Svendsen 1981). Woodard (1994), while reporting that adult males and females were similar in time allocated to patrolling, found that yearlings and 2-year-olds were more active in this behavior than adults. Adult males in pairs of Eurasian beavers living on large rivers were significantly different from adult females in the time they allocated to travel (Sharpe and Rosell 2003). There appears to be variation in patrolling behavior, which is likely due to differences among individual beavers, the composition of the family group, the habitat, and the density of the population. However, adult males are at least as active as adult females in this behavior.

Other behavioral observations of male investment and care

Most data on beaver behavior are obtained while the animals are outside the rest sites during the activity period.

However, observations of adult males have also been made inside lodges and add insight into male investment. The adult male was observed being as active as the adult female in mutual grooming within the lodge, although the adult female was more apt to initiate mutual grooming with the adult male (Patenaude and Bovet 1984). During parturition the adult male and yearlings were actively involved in preparation of the lodge, and physically formed a protective enclosure around the newborn beavers. However, the adult male was less active in ingestion of the placenta than the adult female, and less active in licking the newborn beavers (Patenaude and Bovet 1983).

Ecological and Evolutionary Considerations and the Beaver Family Group

Reviews of monogamy in mammals by Eisenberg (1966), Kleiman (1977), Wittenberger and Tilson (1980), and Clutton-Brock (1989b) have all summarized conditions under which specific mammal groups exhibit a monogamous mating system. Additionally, the recent volume edited by Reichard and Boesch (2003) examines the evolution of social monogamy in numerous animal groups. The two beaver species are often cited as examples of socially (if not sociogenetic) monogamous mammals because they defend territories, invest in their territories by the construction of dams and lodges, have young with an extended period of maturation, and the adult male exhibits a relatively high degree of both direct and indirect parental care.

The specific type of mating system is an outcome of the individual reproductive interests of both the adult male and female, and these interests in reproduction are usually different (See also Waterman, chap. 3 and Solomon and Keane, chap. 4 this volume). Thus what is reproductively advantageous for one sex will pose constraints on how the other sex will mate. Three preconditions that may be necessary for the evolution of monogamy are: (1) the female must gain unique benefits from the pair bond, (2) the female must ensure and be certain of male fidelity, and (3) males must remain with the female, and that males will gain no reproductive advantage by deserting the female (Wittenberger 1979; Wittenberger and Tilson 1980).

Five hypotheses for the evolution of monogamy that are appropriate for beavers are: (1) indispensable, nonshareable male parental care is critical for female reproductive success, (2) shareable (not indispensable) male parental care is important for female reproductive success, and the female gains no advantage through a polygynous mating system, (3) female aggression constrains the male from additional mating, (4) monogamy evolved as a male mate-guarding strategy, and (5) female dispersion does not allow a male to monopolize more than one female (Wittenberger 1979; Wittenberger and Tilson 1980; Clutton-Brock 1989b; Brotherton and Rhodes 1996; Brotherton and Manser 1997; Komers and Brotherton 1997). Hypotheses 1 and 2 involve the importance and degree of male parental care, while hypotheses 4 and 5 relate to male mate-guarding and female dispersion. These hypotheses, selected beaver life-history traits, and supporting references are summarized in table 24.2.

Hypothesis 1

This hypothesis, which stresses the critical role of male parental care, is the traditional hypothesis suggested to account for beaver monogamy (Eisenberg 1966; Kleiman 1977). The primary argument is that male care is indispensable and that females will have less reproductive success without male parental investment. Data on subadult mortality rates for transplanted beavers and nontransplanted populations tend to support this hypothesis, since subadult mortality in transplanted beavers was found to be 100%, but only 50% or lower in nontransplanted animals (Henry and Bookhout 1969; Gunson 1970; Taylor 1970; Payne 1984; McKinstry and Anderson 2002). Lower subadult mortality may result from protection provided in the parent colony, and relocation appears to disrupt the family structure. A time-budget analysis of a Norwegian Eurasian beaver population found no behavioral dimorphism between mated adult males and females, and male care was considered critical for successful reproduction (Sharpe and Rosell 2003).

However, a time budget model of parental investment in North American beaver families living on a large lake in Minnesota did not support hypothesis 1. Male investment was not considered "required," and the model suggested that a female could potentially raise at least one kit by herself (Buech 1987). Additionally, the model indicated that females living in high-quality habitats with preexisting structures (lodges and dams) and possibly help from younger animals could successfully raise more than one offspring per litter. Most studies of beaver behavior indicate that adult males invest (both directly and indirectly) in the family area, and that male care is important in maintaining the mating system.

Hypothesis 2

The second hypothesis, which states that it is more advantageous for a female to be monogamous because the cost of polygyny is too high (Verner 1964; Orians 1969) was favored by Buech (1987) for beavers in northern latitudes. The extended period of parental care and the investment in lodges and dams make determining the polygyny threshold

Table 24.2 Alternative hypotheses for the evolution of monogamy in beavers

Hypothesis	Supporting life history characteristics	Selected references
1. Indispensable, non-shareable male parental care is critical for female reproductive success. Implies exclusive mating.	Precocial young, long gestation and weaning periods, long development to sexual maturity, territorial behavior, food-caching, alarm behaviors, complex, construction behaviors, pair is ecologically isolated from other potential mates during reproductive period over much of range.	Wilsson 1971; Kleiman 1977; Sharpe and Rosell 2003; Hodgdon 1978; Jenkins and Busher 1979; Hodgdon and Lancia 1983; Busher and Hartman 2001 provide reviews of beaver behavior.
2. Shareable (not indispensable) male parental care is important for female reproductive success and the female gains no advantage in a polygynous mating system.	Same as above. Behavior of yearling and two-year-old beavers in the family group may provide some of the same benefits as a polygynous system.	Wittenberger and Tilson 1980; Buech 1987, 1995. Also refer to reviews of behavior above.
3. Female aggression constrains the male from additional mating.	Female dominance during pair bond formation or within existing family group; female territorial behavior; agonistic encounters with nonfamily members.	Wilsson 1971; Hodgdon and Larson 1973; Wittenberger and Tilson 1980; Buech 1987.
4. Monogamy evolved as a male mate-guarding strategy.	Male territorial behavior, scent marking, alarm behavior, female territorial behavior.	None specifically for beavers. However, see Brotherton and Rhodes 1996; Brotherton and Manser 1997; Komers and Brotherton 1997; Brotherton and Komers 2003.
5. Female dispersion does not allow a male to monopolize more than one female. Females occupy relatively small, isolated territories.	Female dispersal patterns, female territorial behavior, scent marking, female aggression and dominance.	None specifically for beavers. However, see Brotherton and Manser 1997; Komers and Brotherton 1997. Sun et al. 2000 and Hartman 1995 may provide support.

SOURCE: Hypotheses 1–3 are summarized from Wittenberger and Tilson (1980). Hypotheses 4–5 are summarized from Brotherton and Rhodes (1996), Brotherton and Manser (1997), and Komers and Brotherton (1997). Specific cases and references are discussed in the text.

difficult. Additionally, a polygynous mating system would mean more beavers living in a defined area with greater resource use, which would require the beaver group to move more often. In a polygynous system the costs of moving and new construction would be greatly increased (Buech 1987; Sun 2003).

However, Smith (1997), also studying populations at northern latitudes, suggested that polygyny can occur in high density, poor-quality habitat populations, and related it to the "choosy generalist" foraging behavior exhibited by beavers. In this population, once the preferred food (aspen) was depleted beavers fed on a variety of other species that were readily available. Smith believed that in most habitats there is sufficient alternative food to support additional breeding females in a territory, and noted that beavers within a territory are often related, and that overall fitness would be increased if related females reproduced in the same territory. While this remains largely untested, it is interesting that while few instances of polygynous mating have been documented, all of the cases have occurred in northern-latitude populations.

Hypothesis 3

The third hypothesis states that female aggression prevents males from acquiring additional mates and has been sug-

gested to play at least a partial role in the evolution and maintenance of monogamy in beavers (Wittenberger and Tilson 1980; Buech 1987). Female aggression has been cited as a potential proximal mechanism for maintaining the pair bond (Wittenberger and Tilson 1980; Buech 1987). However, female dominance within a family has only been demonstrated in one study (Hodgdon and Larson 1973), and all other studies indicate that adult males and females are codominant (see the earlier section on social behavior). Only anecdotal evidence exists that female aggression helps maintain the pair bond. It is difficult to find much support for this hypothesis, and studies designed to investigate the role female aggression might play in maintaining the pair bond should be conducted.

Hypothesis 4

The evolution of monogamy as a male mate-guarding strategy (Brotherton and Rhodes 1996; Brotherton and Komers 2003) has not been specifically tested in beavers. However, data on beaver territorial behavior, time budgets, and timing and duration of estrus may support this hypothesis. Since beavers evolved in northern latitudes and breeding occurs when mobility is restricted by environmental conditions, male roving costs would be high and the potential for extra-pair copulations low. Additionally, estrus lasts only

12–24 hours in beavers (Wilsson 1971; Doboszynska and Zurowski 1983) and this, coupled with reduced mobility, may have selected for mate-guarding as a way to ensure reproduction (Sun 2003). However, an adult male does not always closely attend the adult female (Hodgdon 1978; Busher 1980), and the interaction rate during the activity period between adults is low (Hodgdon 1978; Busher 1983a; Woodard 1994). Male mate-guarding as the ultimate factor influencing beaver monogamy is intriguing, and more studies specifically designed to test this hypothesis are required.

Hypothesis 5

Data exist that support the role of female dispersion in the evolution of monogamy in beavers. The general pattern is for 2-year-old beavers to disperse from their parent colony during the spring or summer. While most dispersing beavers of both sexes move to sites near the parent colony, a significant difference in average dispersal distances between male and female beavers has been documented. In a population in New York, female beavers dispersed an average of 10.15 km while males dispersed an average of 3.49 km (Sun et al. 2000). Adequate data to quantify the dispersion of females are not available for beavers, but relatively wide spacing would provide support for the female-dispersion hypothesis if females were isolated in exclusive territories (Brotherton and Manser 1997; Komers and Brotherton 1997). Additional support for this hypothesis comes from the general pattern of range expansion by a reintroduced Eurasian beaver population in Sweden. Beavers spread along a very flat dispersal front from the introduction sites, with long dispersal distances (Hartman 1995, 2003). These data generally support the hypothesis that monogamy in beavers may have been influenced by female dispersal into relatively isolated territories and the subsequent mating with a single male. Mate-guarding and investment in the territory by construction activities and defense of food resources would solidify and help to maintain the pair bond.

Summary and Additional Considerations

For beavers, the ultimate ecological factors in the evolution of the pair bond may have been their foraging and construction behaviors, and the fact that they choose large prey items (trees). This behavior selected for large body size, which in turn selected for a longer developmental period. The extended length of care provided within the family group increased the need for and amount of parental investment by both sexes, but especially by the male. Since the male invests heavily in the family area through construction

and defense activities, it is less advantageous for the male to desert the female after reproduction.

Female dispersion and subsequent male mate-guarding behavior may be the ultimate social factors in the evolution of monogamy in beavers. Sexually mature females disperse alone from the natal territory and may disperse farther than males. Given the short estrus period, once a pair bond forms it would be advantageous for the male to engage in mate-guarding to ensure reproduction.

However, the complexity of the environment, the elaborate construction behaviors, the central place foraging, and food storage make it difficult to separate out the causal agents. No other socially monogamous rodent shows the same degree of construction activities or foraging and food storage behavior. Territorial behavior by both sexes of beavers may have evolved from a mate-guarding strategy, but it also may be a resource defense strategy (both food and the physical family area). However, the available evidence suggests that male mate-guarding behavior was important in the evolution of monogamy in beavers.

Problems and Future Research

Throughout this chapter I have attempted to indicate areas of concern where data were equivocal or inconsistent. Here I summarize a few major problems and suggest potential areas of research. Since both species of beaver have large geographical ranges and live in a variety of freshwater habitats it is difficult to generalize about social behavior, in that social systems might vary with habitat conditions (Busher and Jenkins 1985; Buech 1987, 1995; Sharpe and Rosell 2003). For example, a comparative behavioral study of beavers living along a large river in the southeastern United States or the Central Valley of California with those living at more northern latitudes may provide insight into the evolution of the beaver social group. Certainly river-dwelling beavers should engage in less construction behavior and partition the shoreline resource differently than beavers on small streams or in ponds. If male investment is critical for the reproductive success of the female, how does a male on a large river system invest in the family? Additionally, if movement is not restricted by ice or food availability, what mechanisms keep the male from deserting his mate for additional mating opportunities? Perhaps in these situations female aggression could restrict the ability of the adult male to seek out other mates; alternatively, the pair bond could be maintained by mate-guarding. Additional data on female dispersion and male access to multiple versus single mating partners would provide a better understanding of the role of male mate-guarding in beaver monogamy. Specific socioecological conditions could stimulate more individual plasticity in beaver mating behavior and the mating system (Smith 1997; Sun 2003).

Many aspects of beaver life history make them an ideal species for testing hypotheses regarding the evolution and maintenance of mating systems. Considerable theory and data are available for specific behaviors associated with the structure and function of the beaver social group. However, further experimental field tests are required to discern among alternative hypotheses for the evolution and maintenance of the beaver mating and social system.

Chapter 25 Evolution of Pacifism and Sociality in Blind Mole-Rats

Eviatar Nevo

Spalacidae: Eurasian Subterranean Rodents

The Spalacidae are Eurasian, primarily East Mediterranean subterranean rodents (fig. 25.1a, b) that are highly adapted for life underground (Ognev 1947; Topachevskii 1969; Savić and Nevo 1990; Nevo 1991, 1999; Nevo et al. 2001). Morphologically, these rodents are cylindrical, powerful, heavy-bodied animals with short limbs, claws, and projecting incisors (fig. 25.2a–c). The head and body length ranges from 130 to 310 mm with a minute stubby tail, which is not visible externally. Average weights range from less than 100 to 570 g (fig. 25.1d). Body sizes vary geographically with climate, soil type, and habitat productivity. Pelage color varies from dark brown to yellowish gray and is related to soil color, suggesting differential above-ground predation (Heth, Beiles et al. 1988). White stripes may occur on the head. The pelage is dense, soft, and imperfectly reversible. Bristly, apparently tactile, facial hairs

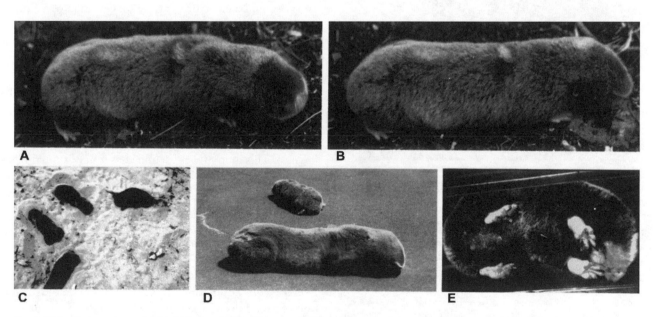

Figure 25.1 *Spalax* mole rats. (a), (b), (e) *Spalax ehrenbergi* superspecies in Israel. Note the distinct nose pad, no external eye and pinna, small ear opening, and short limbs. Photo K. Rybalko. (c) *Spalax ehrenbergi* superspecies, *2n* = 60, new species from El Alamein, North Africa (Nevo et al. 1991a–c, 1992). Animals behave socially as a group. (d) Comparison of the largest *S. microphthalmus* from the Ukraine and *S. ehrenbergi 2n* = 60, now named *Spalax judaei,* from Israel; the Ukraine *Spalax* is 5-fold larger. (e) *Spalax*— a ventral view. Photo E. Nevo.

form stiff keels of margin rhinarium imparting a broad wedge shape to the head. The dental formula is 1/1, 0/0, 0/0, 3/3 = 16. The incisors are long and procumbent, and are used in chisel-tooth digging (fig. 25.2a); their enamel thickness varies with soil type. The cheek teeth are rooted (not ever-growing) with Z- or S-shaped enamel patterns, which are species- and soil-specific (Butler et al. 1993).

Spalacids are distinguished from all other rodents (including other subterranean rodents) by the absence of any external eye opening (figs. 25.1a,b; 25.2a–c). The subcutaneous, vestigial eyes are not used in vision but rather in photoperiodic perception; they have become essentially circadian eyes (Sanyal et al. 1990; De Jong et al. 1990; Cooper et al. 1993a, 1993b; Nevo 1998, 1999). The pinnae are rudimentary (figs. 25.1b, 25.2c), but the middle ear ossicles (Burda et al. 1989, 1990), particularly the cochlea (Bruns et al. 1988), are uniquely structured among mammals. As such, they are adapted to underground low frequency and rich-repertoire, short-distance vocal communication (Heth et al. 1986; Nevo et al. 1987; Heth, Frankenberg et al. 1988b). Seismic (vibrational) communication by head thumping on the burrow ceiling (Heth et al. 1987; Rado et al. 1987; Nevo et al. 1991) is a major underground communication modality among these solitary territorial animals (Nevo 1961; Nevo et al. 1991), and varies among species (Heth et al. 1991). Adaptive morphological differentiations exist in the body and head skin (hairy skin, vibrissal fields, buccal ridge, and rhinarium [Klauer et al. 1997]). Olfaction is an important communication modality (Nevo and Heth 1976) in reproduction through sexual pheromones (Nevo et al. 1987; Menzies et al. 1992; Todrank and Heth 1996) and in food identification (Heth et al. 1992, 1996; Heth and Todrank 1995). Tactile cues are important, but as yet unquantifiable (Burda et al. 1990).

Mole-rats are confined most of their lifetime to sealed underground tunnels (Nevo 1961). They are chisel-tooth diggers, using fore and hind feet to push the earth out ahead of or behind them, thus packing and bulldozing excavated earth with their broad and flat heads to form the external mounds (fig. 25.3a–e). Specialized jaws and strong muscles aid the teeth in loosening the soil. Incisor (Flynn et al. 1987) and molar (Butler et al. 1993) structures vary with species, soil, and food.

Population Biology

Spalacids are solitary, territorial, and aggressive (Nevo 1991, 1999; Nevo et al. 1975, 1986; Guttman et al. 1975). Aggression is polymorphic within sexes, populations, and species, each involving militant, intermediate, and pacifist behavioral phenotypes (Nevo et al. 1986, see the following; fig 25.2a–c). Pacifism increases toward the Negev Desert in Israel and culminates in total fixation in a pacifistic species bordering the Sahara Desert in Egypt and North Africa (Nevo et al. 1991; Nevo, Simson, Heth, Redi, and Filippucci 1991; Nevo et al. 1992; fig 25.1c, see the following). This animal was identified as a new animal based on morphological, behavioral, chromosomal, and allozyme evidence (Nevo, Simson, Heth, Redi, and Filippucci 1991).

Territory size (range: 60–200 m^2) varies with the species, population, habitat, sex, and age (see Nevo 1979; Savić and Nevo 1990 and references therein). Population density varies with habitat productivity, from 0.1 to 23/ha (Nevo 1979). Populations abound with adults; the proportion of subadults and juveniles is low and variable between and within species. Sex ratio is skewed to different degrees

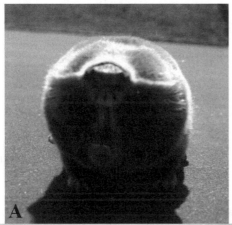
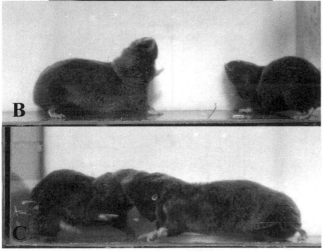

Figure 25.2 Aggressive behavior in *Spalax ehrenbergi:* (a) frontal view and exposed incisors, (c) head-on combat with biting, (b) combat with a dominant militant (left) and a submissive pacifist (right).

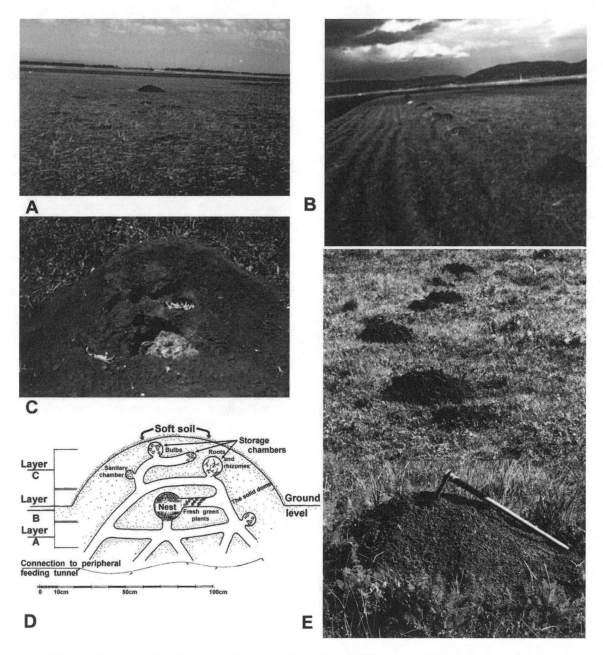

Figure 25.3 Breeding mound of the blind mole rat, *Spalax carmeli, 2n* = 58, *Spalax ehrenbergi* superspecies in Israel: (a) A breeding mound surrounded by regular nutritional mounds; (b) linear alignment of breeding mounds rising on a slightly raised ridge to avoid ground water level; (c) cross section of a natural breeding mound (note the nest in the center and storage chambers on the periphery); (d) diagrammatic section showing three layers: nest, storage chambers with bulbs, corms, roots, and sanitary chamber(s) (from Nevo 1961; photo E. Nevo); (e) small nutritional mounds radiating from a big breeding mound, with a hoe on top for scale (Photo K. Rybalko).

toward females. Populations may be isolated, semi-isolated, or even continuous in the main ranges (Nevo 1979, and, for example, in Israel, Nevo et al. 1982, fig. 25.4).

Recruitment is low. Breeding usually starts in the second year; females usually have one litter per year, which is often their sole litter in life, although they average three (ranging from 1 to 6; Dukel'skaya 1935; Ognev 1947; Nevo 1961,

1999; Savić 1973, 1982). Maximal recorded lifespan in captivity was 15 years (Savić and Nevo 1990), but the average lifespan in nature is about 3 years.

Burrow systems and dimensions vary with age, sex, habitat, productivity, soil, climate, and other environmental factors (Ognev 1947; Nevo 1961; Savić 1973, 1982; Heth 1989). Burrow systems range from 30 to 250 m total length;

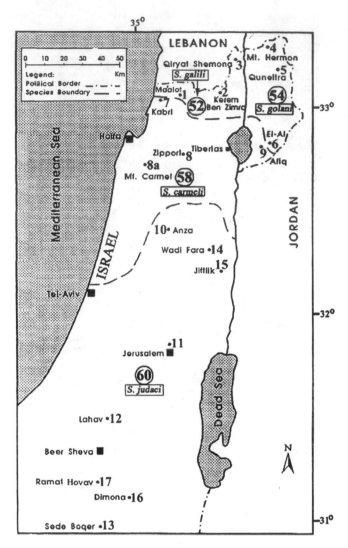

Figure 25.4 Geographic distribution of the four species belonging to the *Spalax ehrenbergi* superspecies in Israel, separated by narrow hybrid zones. The abbreviations of the 18 populations and their ecogeographic nature appear in parentheses: (C = Central; M = Marginal; NHZ = Near Hybrid Zone; SI = Semi-Isolate and I = Isolate), are as follows: *Spalax galili* (2n=52) –1=Ma'alot (NHZ); 2 = Kerem Ben Zimra (C); and 3 = Qiryat Shemona (M); *Spalax golani* (2n=54) – 4 = Mt. Hermon (M); 5 = Quneitra (C) and 6 = El-Al (NHZ) *Spalax carmeli* (2n=58) – 7 = Kabri (NHZ) 8 = Zippori, (C); 8a = Mt. Carmel, (C); and 9 = Afiq, (M; NHZ); and *Spalax judaei* (2n=60)-10 = Anza, (NHZ); 11 = Jerusalem (M); 12 = Lahav (C); 13 = Sede Boqer (I); 14 = Wadi Fara (SI); 15 = Jiftlik (SI); 16 = Dimona, (I); and 17 = Ramat Hovav (SI). (Compiled from Nevo et al. 1993).

depths of feeding tunnels are 10 to 410 cm; tunnel diameters are 5 to 12 cm; 10 to 200 mounds are formed per individual system; and up to 3 tons of soil are displaced (Topachevskii 1969; Nevo 1979; Savić 1982; Heth 1989, 1991). The superficial foraging burrows are connected to a deeper tunnel system or an elevated complex breeding mound involving nest, storage, and sanitary chambers (Nevo 1961, 1999; Savić 1973, 1982; fig. 25.3c–e).

Geographical and Ecological Distribution

The Eurasian family Spalacidae lives in East Europe, West Asia (including Asia Minor), the Near East, and North Africa (fig. 25.1 in Nevo et al. 2001). The distribution of recent and fossil Spalacidae is illustrated in figures 25.1–25.5 in Savić and Nevo (1990), and the ecogeographical distribution of recent Spalacidae is given in table 25.1 in Savić and Nevo (1990). The Spalacidae live in altitude from below sea level to an elevation of 2,600 m in the Balkans (Savić and Soldatović 1979) and Asia Minor (Nevo 1961). Spalacids range in Mediterranean and steppic habitats, inhabiting grassy and mountainous steppes, semideserts, and desert steppes. These mole-rats may penetrate open biota in agricultural plots. However, basically, they are steppic subterranean rodents whose evolutionary history, speciation, and adaptation are intimately linked to increasingly arid environments, primarily in the Near East (Nevo 1991, 1999; Nevo et al. 2001), North Africa (Nevo et al. 1991, 1992, 1994, 1995; Nevo, Simson, Heth, Redi, and Filippucci 1991), and Asia Minor (Kivanç 1988).

In Israel, the *S. ehrenbergi* superspecies *S. galili*, *S. golani*, *S. carmeli*, and *S. judaei* (2n = 52, 54, 58, and 60, respectively, fig. 25.4) range from mesic to xeric regions, but they are restricted by the 100 mm isohyet (Nevo 1961) and rarely colonize true deserts (Nevo 1989; Nevo et al. 1997). In Asia Minor, *Spalax* extend from mesic habitats near the Aegean Sea (2n = 38) and the Mediterranean Sea (2n = 56) to xeric habitats in central Anatolia (2n = 60, 62; Nevo et al. 1994, 1995; Sözen and Kivan, 1998). In Jordan, *Spalax* speciated into four species, all 2n = 60, varying in karyotype morphology, and across mountain ranges from Gilead through Amon, Moav, and Edom to Ras-e-Naqb, near Aqaba (Ivanitskaya and Nevo 1998; Nevo et al. 2000). Spalacids occupy most soil types, including open primary habitats of scrub in plains, hills, and mountains and in secondary cultivated fields where they may become serious pests (Nevo 1961).

Phylogeny and Systematics

The phylogeny and systematics of the family have been largely intractable. This is true from the familial to the specific level, and no consensus has been reached (reviewed by Ognev 1947; Topachevskii 1969; Corbet 1978; Carleton and Musser 1984; see Savić and Nevo 1990). Molecular phylogenetics based on single molecules (albumin, transferring; Sarich 1985), haemoglobin (Kleinschmidt et al. 1985), crystalline (Hendriks et al. 1987), and ribonuclease (Jekel et al. 1990) all suggest an early evolutionary divergence time for Spalacidae from the muroid-cricetoid stock 40 to 45 million years ago (mya), which was in middle Eocene times. However, DNA-DNA hybridization, based on entire genome comparisons, suggests an evolutionary divergence time of 19 mya (Catzeflis et al. 1989; Szalay et al. 1993). The oldest fossil spalacid known is *Heramys eviensis* from the lower Miocene of Greece, 25 mya (Hofmeijer and De Bruijn 1985), now considered late Oligocene. These estimates taken together suggest an Oligocene (probably late Eocene–early Oligocene) age to the origin of the Spalacidae (Nevo 1999).

To avoid confusion, and until a thorough taxonomic revision of the Spalacidae is conducted, including chromosomal, molecular-genetic, and fossil data, Savić and Nevo (1990, and their table 1) used only one generic name, *Spalax,* involving eight superspecies. Likewise, the four Israeli species were described under *Spalax* (Nevo et al. 2001). Recently, between fourteen and twenty new species, based on karyotypes and allozyme evidence, have been described in Asia Minor, almost doubling the number of extant species in the Spalacidae to about sixty (Nevo et al. 1994, 1995, 2001). Recently, four new species were discovered in Jordan, which also increased the number of species (Nevo et al. 2000).

Chromosomal Speciation

Chromosomal sibling species, or allospecies, based on Robertsonian changes (primarily fissions) and/or pericentric inversions, are generally widespread in the Spalacidae (Nevo 1999 and references therein). Notably, classical species of this family involve many cryptic sibling species or allospecies based on Robertsonian fissions (Nevo 1991, 1999; Nevo et al. 2001). The eight classical species of mole-rats from Russia, Ukraine, Balkans, Asia Minor, Israel, and North Africa actually include more than sixty species ($2n = 38$–62; NF = 72–124; Savić and Nevo 1990 references therein and table 2; Nevo, Simson, Heth, Redi, and Filippucci 1991; Nevo et al., 1994, 1995, 2000). Some (like the $2n = 60$) of the *S. ehrenbergi* superspecies in Israel, Egypt, and Jordan (Nevo 1991; Nevo et al. 1994, 1995) are late Pleistocene and recent in origin. Based on multidisciplinary studies involving natural hybridization, gene flow, pre- and post-zygotic reproductive isolation, and even morphological differentiation in middle ear ossicles (Burda et al. 1989) and baculum (Simson et al. 1993) in the *S. ehrenbergi* superspecies in Israel, most of the described karyotypes seem to be distinct biological species.

Restricted mobility, spatial isolation, and numerous small isolates in spalacids permitted their relatively rapid chromosomal evolution (Wahrman et al. 1969a, 1969b, 1985) and parapatric speciation (Nevo 1989, 1991). The initiation of speciation by Robertsonian chromosomal mutations and/or pericentric inversions and postmating reproductive isolation was complemented gradually by premating isolating mechanisms comprising olfaction, vocalization, seismic communication (reviewed in Nevo 1990, 1991) and middle ear ossicles (Burda et al. 1989), leading to the budding of new species adapted to increasing aridity, primarily in the periphery of the distribution (Nevo et al. 1997). If $2n = 38$ in west Turkey occurs near the origin of the Spalacidae, then 2n increases in the three major gradients of increasing aridity: the Balkans (up to $2n = 62$), Russia and the Ukraine (up to $2n = 62$), and the Near East and North Africa (up to $2n = 60$, and even buds of $2n = 62$; Nevo 1991, 1999).

Spalax ehrenbergi superspecies in Israel represent an evolutionary model of active speciation, studied multi- and interdisciplinarily (reviewed in Nevo 1991, 1999; Nevo et al. 2001). Adaptive speciation to four climatic regimes of the four species *Spalax galili, S. golani, S. carmeli,* and *S. judaei* ($2n = 52, 54, 58,$ and 60, respectively) was demonstrated at all organizational levels, from genetics through morphology, physiology, and behavior (Nevo 1991, 1999). Our recent finding of four presumable species in Jordan associated with varied climates (Ivanitskaya and Nevo 1998; Nevo et al. 2000) needs further substantiation. The monograph by Nevo et al. (2001) elaborates the speciation dynamics of the *S. ehrenbergi* superspecies in Israel, including natural hybridization (Nevo and Bar El 1976).

Population Structure and Dynamics, Home Range, and Territoriality in Subterranean Mammals

The optimal area hypothesis assumes that the home range of animals—the area they know and patrol (Burt 1943)—is large enough to yield an adequate supply of energy. The home ranges of subterranean mammals are generally also their exclusive and defended territories (Brown and Orians 1970), except for brief periods during the breeding season when multiple occupancies by both sexes occur. (The social subterranean animals are the exception, where colonies contain from several individuals to several hundreds of individuals [e.g., Sherman et al. 1991; Burda 1999; Lacey 2000; Bennett and Faulkes 2000].) This pattern is found in pocket gophers, mole-rats, some tuco-tucos, and moles (Nevo 1999). Territories, once established and used for one breeding season, remain largely fixed for life (except for minor boundary changes). Exceptions usually involve subadults living in marginal habitats (Howard and Childs 1959;

Godfrey and Crowcraft 1960; Miller 1964). In general, territories of same-sex individuals do not overlap, whereas partial overlap occurs between male-female territories (Howard and Childs 1959; Wilks 1963).

In accordance with the predictions of the optimal area hypotheses, the territory sizes of subterranean mammals vary with age, sex, body size, habitat, population density (Nevo 1999), and diet. (For patterns of territoriality across subterranean mammals see Nevo [1999, fig. 15.1n–p]). The territories of subadults are considerably smaller than those of adults, and females' territories are smaller than those of males. The size and shape of the territory are more constant at high densities and more variable at low densities. Finally, ranges of insectivorous species are significantly larger than those of herbivorous subterranean mammals (Nevo 1999, fig. 15.10). Average territory size for *Spalax microphthalmus* in Russia was 150 m² (Dukel'skaya 1935), for *S. leucodon* in Yugoslavia, 452 m² (range: 194–1,000 m²; Savić 1973). The larger territories of insectivores, when compared with herbivores of the same size, strongly implicate resources and energy in the selection for territory size (McNab 2002; Brown and Orians 1970; Nevo 1999, fig. 15.10).

Spacing patterns may shift from individual-territorial to colonial as environmental conditions associated with changes in food density occur (Brown and Orians 1970; Reig 1970). Coloniality occurs, though infrequently, in subterranean mammals—for example, in *Ctenomys peruanus* (Pearson 1959; Genelly 1965), *C. minutus, Spalacopus cyanus* (Reig et al. 1970), *Cryptomys hottentotus,* and several other species of African *Cryptomys* (Burda and Kawalika 1993) and *Heterocephalus glaber* (Hill et al. 1957; Sherman et al. 1991; Jarvis et al. 1994; Jarvis and Sherman, 2002). *Heterocephalus* also evolved an adaptive hierarchic caste structure (Jarvis 1981; Sherman et al. 1991; Jarvis et al. 1994; Jarvis and Sherman 2002; Nevo 1999, fig. 7.7b–f). Some cases (e.g., *Heterocephalus and Spalacopus*) are associated with unfavorable climatic and/or resource conditions, and their sociality was explained by the aridity-food distribution hypothesis (see Burda [1989] and Foulkes and Bennett, chap. 36 this volume).

Thus, territoriality and coloniality can be viewed as responses to spatiotemporal changes of exploitable resources. Both patterns are consistent with a time-and-energy-budget model (Brown and Orians 1970). The extreme case in spalacids will be discussed subsequently.

Territory, Size, and Population Density in the *Spalax ehrenbergi* Superspecies

The home ranges of individuals in the *S. ehrenbergi* superspecies, like those of most other subterranean mammals,

are exclusive, defended territories (Nevo 1961, 1979, 1999). During the breeding season (usually in December–January), for a brief period, multiple occupancies by both sexes may occur. Multiple occupancy may also occur during the raising of the young by females before weaning and dispersion (from mid-January to mid-March; Nevo 1961). Territories in *S. ehrenbergi,* once established and used for one breeding season, usually remain fixed for life except for minor boundary changes. In general, territories of the young are established in peripheral areas (Heth 1989; 1991), and territories usually do not overlap. During the breeding season, males leave their territories and usually dig long, straight tunnels, visible by a straight row of mounds, in search of the territories and breeding mounds of females. Copulation occurs inside the large female breeding mounds.

Consistent with the predictions of the optimal area hypothesis, the territory size of *S. ehrenbergi* superspecies varies with age, sex, body size, habitat, population density, and diet; that is, territory size corresponds to the metabolic demands associated with animal size (Nevo 1961, 1979, 1999; Nevo et al. 1982; Heth 1989). The territories of subadults are considerably smaller than those of adults, and the territories of females are smaller than those of males. Territory size among the species varies with productivity of the region; it is smaller in highly productive regions (in the ranges of the species *S. galili, 2n = 52,* and *S. golani, 2n = 54,* increasing southward to that of the *S. carmeli, 2n = 58* species, and climaxing in that of the *S. judaei, 2n = 60* species; Nevo 1979, 1991; Nevo et al. 1982). Average territory sizes in square meters for the four species of *S. ehrenbergi* were measured in two ways: (1) direct estimate of territory size derived from mound distribution (an underestimate), and (2) indirect estimate of territory size derived from division of the area inhabited by mole-rats by number of territories (an overestimate). For *Spalax galili, S. golani, S. carmeli,* and *S. judaei,* the estimates are (in m²) 60.7 (226.3), 55.3 (246.7), 56.3 (301.7), and 102.7 (329.7) for the results of method 1 and method 2 (in parentheses). Preliminary results from radiotracking a female *S. galili (2n = 52)* from the ecologically marginal population of Qiryat Shemona resulted in a territory size of 63 m² (Kushnirov et al. 1998), suggesting the greater accuracy of method 1.

Population density and structure in the four species of *S. ehrenbergi* have been extensively studied across the superspecies range along 1,057 km of road transects (Nevo et al. 1982; fig. 47 in Nevo et al. 2001). The results indicated the following: (1) The overall (but underestimated) number of mole-rats in their range of 15,500 km² in Israel amounts to 1.6 to 2 million individuals. (2) Population density per km² for *S. galili, S. golani, S. carmeli,* and *S. judaei* is 140, 177, 101, and 91, respectively, decreasing southward toward the desert. (3) Populations are largely continuously distributed in their main ranges corresponding to the isolation by distance model, but become semi-isolated and isolated in the marginal habitat at the southern border of *S. judaei* (fig. 25.4; and fig. 47 in Nevo et al. 2001).

These results contrast with the commonly held view that the population structure of subterranean mammals always involves numerous geographically partially isolated demes across their ranges. Moreover, the results suggest that clinal differentiation characterizes the main distributional ranges and speciation may primarily originate in the peripheral, small isolates where the initial fixation of spontaneous chromosomal mutations may take place by natural selection (Nevo et al. 1992; see speciation theory, chapter 23 in Nevo 1999).

Aggression Patterns in Speciation and Adaptation of Blind Subterranean Mole-Rats in Israel

Speciation requires reproductive isolation and ecological compatibility (Mayr 1963). Yet, while species-specific behavioral signals are well-known reinforcers of reproductive isolation, the role of aggression as a factor in species formation was poorly known. Aggressive behavior is common and adaptive *within* many animal species, chiefly in spacing out individuals, but its evolutionary significance *between* species is known primarily as an ecological effect rather than as a determinant of speciation, i.e., as a premating reproductive isolating mechanism (Wilson 1975).

The objective of long-term studies on aggression in the four species of *S. ehrenbergi* superspecies was to evaluate the evolutionary significance of interspecific aggression as a factor during final stages of speciation (Nevo et al. 1975; Guttman et al. 1975). Can species recognition be reinforced by natural selection through high levels of aggression at a stage when both reproductive isolation and ecological compatibility are still incomplete? Likewise, is the parapatric distribution among the four allospecies due to aggressive behavior? To solve these problems, we followed earlier suggestions (Nevo 1969) and investigated aggression patterns in mole-rats belonging to the *S. ehrenbergi* superspecies as a model of active and prolific speciation.

The evolutionary significance of interspecific aggression as a factor in speciation was tested among three species of the actively speciating superspecies, *Spalax ehrenbergi*. Laboratory experiments testing intra- and interspecific aggression were conducted on 48 adult animals from 10 populations comprising three species: *S. galili*, *S. carmeli*, and *S. judaei*. Twelve agonistic, motivational-conflict, and territorial-behavioral variables were recorded during 72 combats involving homo- and heterospecific encounters between opponents. Analysis of the data matrix was carried

out by the nonmetric multivariate Smallest Space Analysis (SSA-II; Nevo et al. 1975; Guttman et al. 1975). Results indicated that (1) aggression patterns, involving agonistic conflict and territorial variables, are higher in heterospecific encounters than in homospecific ones; and (2) aggression is higher among contiguous species ($2n = 58–60$, and $2n = 52–58$) than among noncontiguous ones ($2n = 52–60$). Both (1) and (2) suggest that high interspecific aggression appears to be adaptively selected at final stages of speciation in mole-rats as a premating, isolating mechanism, which reinforces species identification and establishes parapatric distributions among the evolving species. This conclusion was substantiated in our later studies (fig. 25.5d).

Adaptive aggressive behavior within and between species in *S. ehrenbergi* superspecies

Resource competition in subterranean mammals generally takes the form of intra- and interspecific aggression (Vaughan and Hansen 1964; Grant 1972; Nevo 1979; see references in Nevo 1999). The results are usually the following.

1. Solitariness accompanied by strong territoriality and competition within species, mediated by aggressive behavior between individuals (or between colonies in social mole-rats; see Reeve and Sherman in Sherman et al. 1991) in defense of their exclusive territories.
2. Competitive exclusion between species, also mediated by aggression, resulting in parapatric, that is, contiguous distributions primarily between evolving species (Nevo 1979; see fig. 25.4).

Species-specific aggression

Aggressive behavior augments other behavioral patterns (activity, exploratory, and habitat selection patterns) in optimizing the energetic balance of *S. ehrenbergi* by determining territorial size and accessible resources (Nevo 1999; Nevo et al. 1975, 1986). Aggression also functions as a premating isolating mechanism (Nevo 1990). All four species and populations of *S. ehrenbergi* are genetically polymorphic for aggression, each involving trimodal behavioral phenotypes: pacifists, intermediates, and militants (fig. 25.5a–b). Aggressive behavior displayed geographic gradients where pacifism increased and militarism decreased southward, as follows: *S. galili* > *S. golani* > *S. carmeli* > *S. judaei* ($2n = 52 > 54 > 58 > 60$, species, respectively; see fig. 25.5d). Aggression intensity was correlated with climatic determinants as well as with productivity and territory size. Aggression increases with competition for resources, as evidenced in highly dense northern populations

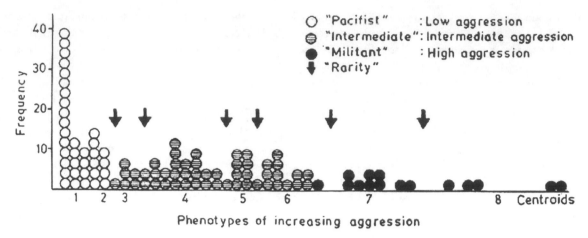

Figure 25.5a Discriminant analysis of eight categories of aggressive behavior ("total aggression") based on all four species displaying multi-peak variation of aggression in the *Spalax ehrenbergi* superspecies. Frequency of animals appears on the Y-axis. The analysis is the basis of differential polymorphism among species described in figure 25.5b and figure 25.5c (from Nevo et al. 1986).

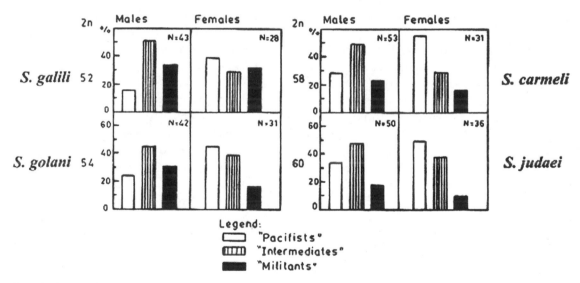

Figure 25.5b Genetic polymorphism in patterns of aggressive behavior in both sexes of each species of the *Spalax ehrenbergi* superspecies in Israel: *Spalax galili*, 2n = 52; *S. golani*, 2n = 54; *S. carmeli*, 2n = 58, and *S. judaei*, 2n = 60. The Y-axis represents percentage of animals in each of the three behavioral phenotypes (from Nevo et al. 1986).

of *S. galili, 2n = 52* and *S. golani, 2n = 54* (Nevo et al. 1982). Conversely, aggression decreases southward partly in *S. carmeli*, the *2n = 58* species, but particularly in the *S. judaei*, the *2n = 60* species, presumably due to the low available resources—hence, sparser populations, in the more xeric southern habitats (Nevo 1985; Nevo et al. 1986; fig. 25.4; and fig. 47 in Nevo et al. 2001). In the latter, climatic selection presumably operates to minimize overheating as well as water and energy expenditure (Nevo and Shkolnik 1974). This pattern supports the optimal activity hypothesis where selection maximizes fitness by optimizing net energy gain per unit activity (Nevo et al. 1982b, in Nevo 1999). Within populations, natural selection resulting from *intra*population conflicts among aggressive behavioral phenotypes, militants, intermediates, and pacifists lead to

genetic equilibria. The equilibria comprise a mixture of behavioral phenotypes (fig. 25.5b). This mix is evolutionarily stable according to the game theory definition of Maynard-Smith and Price (1973) due to the ecological heterogeneity of the habitat (Nevo et al. 1986). Militants dominate rich vegetational patches while pacifists are pushed to peripheral, poorer vegetational patches (Nevo, unpublished observations). However, these equilibria are presumably based on flexible and dynamic changes in the proportion of behavioral phenotypes in accordance with regional and local variations in ecological diversity caused by macro- and microclimatic variations.

The active ecological speciation and adaptation of *S. ehrenbergi* and its adaptive radiation into increasingly arid environments (also increasingly arid environments within

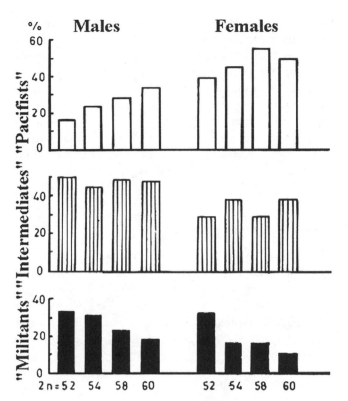

Figure 25.5c Geographic clinal variation in the three major behavioral aggressive phenotypes across the four species of the *Spalax ehrenbergi* superspecies in Israel: *Spalax galili,* 2n = 52; *S. golani,* 2n = 54; *S. carmeli,* 2n = 58, and *S. judaei,* 2n = 60, in each sex separately (from Nevo et al. 1986).

species) is intimately associated with behavioral genetic variables of activity, exploration, habitat choice, and aggression patterns. These behavioral patterns contribute collectively to the optimized energetic balance between and within species; hence, increased fitness (Nevo 1999; Nevo et al. 2001).

Evolutionary patterns and theory of aggression in *Spalax*

Aggressive behavior (fig. 25.2) is largely innate and genetically polymorphic (fig. 25.5a–b) in solitary, territorial subterranean mole-rats (Nevo et al. 1986). The pattern of aggression between individuals of the same species is qualitatively similar, but quantitatively different from aggressive patterns between individuals of different species. The proximate cues eliciting aggressive behavior in blind mole-rats, where visual stimuli do not operate, are seismic, olfactory, auditory, and tactile cues—apparently in this order (Nevo 1990, 1999; Nevo et al. 2001). The ultimate interacting determinants causing variation in the pattern and intensity of aggressive behavior among individuals, populations, and species of *Spalax* include genetic, ecological, demographical, life-history, and ethological factors, as follows.

Genetically, aggression generally increases with genetic relatedness both within and between species (fig. 25.5d). However, within populations the levels of aggression vary

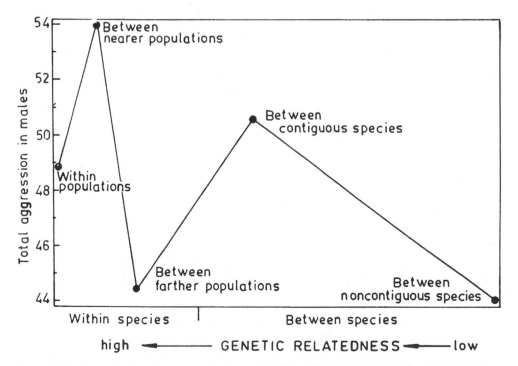

Figure 25.5d The relationship between intermale "total aggression" and genetic relatedness within and between species of the *Spalax ehrenbergi* complex. "Total aggression" is the sum of the number of attacks, bites, head-ons, and sniffings with open mouth. Genetic relatedness was estimated by geographical distance; i.e., we assume that the further apart the populations and species, the lower their genetic relatedness.

with season, sex, and neighborhood. *Ecologically*, aggression increases regionally and locally with optimal habitats and with the correlated diversity and productivity of the environment (fig. 25.5c). *Demographically*, aggression increases with population density. *Life-historically*, aggression is higher in males than in females and in the breeding than in the nonbreeding season. *Physiologically*, aggression decreases in climatically xeric and harsher environments (fig. 25.5c). Finally, *ethologically*, intermale aggression is highest between neighbors within populations and between contiguous rather than noncontiguous species (fig. 25.5c).

Can highly territorial, solitary, and aggressive *Spalax* evolve sociality?

Our extensive work on the *Spalax ehrenbergi* superspecies revealed four species (Nevo et al. 2001) in which aggression plays a significant role in both speciation and adaptation (Nevo et al. 1986). *Spalax* are strongly territorial, solitary, and innately polymorphic for aggression (fig. 25.5a–b). However, the levels of aggression decrease gradually toward the Negev Desert in Israel, presumably to minimize overheating, water, and energy expenditure and to prevent thermal death (fig. 25.5c). It was therefore a mixed surprise and expectation when we encountered total pacifistic behavior in Egyptian isolates of the *S. ehrenbergi* superspecies. These Egyptian isolates separated presumably 10,000–25,000 years ago (Lay and Nadler 1972) from the Israeli complex and were described by us as a new species (Nevo, Heth, and Pratt 1991, Nevo, Heth, and Simson 1991; Nevo et al., 1992). This extraordinary discovery supported our ecological evolutionary theory of aggression in *Spalax*, which predicts that aggression should decrease in the desert (Nevo et al. 1986), also catalyzing social evolution (Nevo et al. 1992). Here, I describe field and lab experiments with the new pacifistic species of the *Spalax ehrenbergi* superspecies.

Predominating pacifism is a prerequisite for social evolution in *Spalax*

We were able to test and verify the prediction of increasing pacifism toward the desert when we obtained 14 mole-rats from Burj-El-Arab in northern Egypt, in May 1989 (Nevo et al. 1992). The mole-rats all arrived in one wooden crate (120 × 70 × 40 cm) instead of each animal being caged separately, the usual routine with Israeli and Eurasian *Spalax* species. Furthermore, the mole-rats displayed atypical, non-*Spalax*-like social interactions, with no aggression, rather amicable and pacifistic behavior (fig. 25.1c). By contrast, even pacifist-behavioral phenotypes of *S. ehrenbergi*

from Israel must be kept in separate cages to prevent them from killing each other.

I suggest that this pacifistic behavior evolved by natural selection in the Egyptian isolates, as an adaptation for survival in the environmentally harsh, northern, marginal Sahara Desert. Overheating, loss of water, and excessive energy investment may be fatal in the desert. We tested this unique behavior of the Egyptian mole-rats and compared it to a control experiment with Israeli *S. ehrenbergi*, $2n = 60$, now named *S. judaei*, the least aggressive species (Nevo et al. 1992).

Geographical location and habitat

The two *Spalax* isolates (Burj-El-Arab and El-Hammam, 25 km apart, near El Alamein) are located in agricultural valleys in the xeric Mediterranean zone of northern Egypt fringing the Sahara Desert (fig. 1 in Nevo et al. 1992). Phytogeographically, the region is xeric-Mediterranean, surrounded by Saharo-Arabian vegetation.

Experimental animals

We carried out two field expeditions to the northern Egyptian localities (fig. 1 in Nevo et al. 1992) and conducted three sets of experiments and observations in 1989–1991. All observations, experimental designs, and detailed results appear in Nevo et al. (1992).

In general, behavioral phenotypes in Israel displayed distinct clinal, geographical variation in both sexes: pacifists increased southward in males—16%, 24%, 28%, 34%, as well as in females—39%, 45%, 55%, 50%, whereas militants decreased southward in males—33%, 31%, 23%, 18%, and in females—32%, 16%, 16%, 11%, for $2n =$ 52, 54, 58, and 60, respectively. The Egyptian population showed 100% pacifism and 0% militarism.

Experimental results

All experiments in which Egyptian animals were put together, either in small plastic containers in the field for hours, or in larger aquaria in the laboratory (for up to 12 days) resulted in pacifistic behavior and amicable interactions between animals (fig. 25.1c). The Egyptian mole-rats were characterized by huddling, tactile contact, touching, and warming each other in cohesive aggregates without aggression. Some pushing and screaming before settling down occurred, but the occurrence of contact cohabitation was in strong contrast to usual behavior of *Spalax*. Moreover, animals were easily handled by us and caressed freely like pets, an inconceivable act with other known *Spalax* species. We observed aggressive behavior on just two occasions.

One male was aggressive at capture but later displayed pacifistic behavior like the others. Another male displayed 1 hour of aggression out of several days of pacifistic behavior. In sum, the pacifistic behavior was tested five times in small containers for hours and five times in aquaria for 1 week and more. In all cases, pacifistic behavior was the rule, with only the two short-term aggressive exceptions.

Under similar experimental conditions, we conducted a control experiment with eight Israeli *S. ehrenbergi* of $2n = 60$ (*S. judaei*), a species with a relatively high level of pacifism (Nevo et al. 1986). Immediately upon insertion into the aquarium, the Israeli mole-rats started fierce fighting involving attacks, biting, and excited agonistic behavior. This encounter resulted in wounding six out of eight animals during the 10-minute experiment, which was terminated to prevent mortality. The control experiment represents the usual behavior obtained in more than 1,500 pairwise combats of mole-rats from twelve populations (Nevo et al. 1986). The Egyptian pacifistic behavior is outside the range of all 1,500 tests. The probability of obtaining such an out-of-range result in three different samples, by pure chance, is $p < 10^{-9}$. Thus a dramatic and significant behavioral difference exists between Israeli and Egyptian *S. ehrenbergi* mole-rats, separated by the Sinai Desert. The Egyptian mole-rats displayed amicable aggregative behavior well beyond the normal range of Israeli mole-rats. The results and statistical procedure appear in table 1 of Nevo et al. (1992).

Social evolution in *Spalax*

In all of our field and laboratory observations and experiments the northern Egyptian mole-rats behaved pacifistically. A tendency to aggregate also occurs, and this pattern suggests a potential for future evolution of social behavior. In contrast, however, our field observations suggest that almost all animals are solitary. Aggregative behavior may depend on certain food or temperature conditions. In our laboratory experiments, aggregation was maximal in open aquaria, where heat dissipation was larger than in the maze or field. Thus it appears that while this animal usually lives solitarily, it may allow neighbors to enter its territory at times of low temperature. Critical experiments could determine which stressor as well as any other causes may lead to aggregation and thereby possibly open the way to a social-like phase and future social evolution.

Total pacifism in the Egyptian *Spalax* isolates in the harsh environment of the Sahara Desert is understandable as an ecophysiological adaptive pattern leading to a metabolic economy (Nevo et al. 1986; Ganem and Nevo 1996). We suggest that the Egyptian *S. ehrenbergi* was selected for pacifist behavior in the Sinai Desert where aggressive phenotypes were eliminated and total pacifism evolved. A desert habitat presumably favors pacifist behavioral phenotypes to minimize overheating, water, and energy expenditures (Nevo and Shkolnik 1974), and the emergence of sociality in *Spalax* may also involve defense against potential predators. The evolution of pacifism in North African *Spalax* may follow the aridity–food distribution hypothesis of Jarvis et al. (1994), suggesting that social evolution is correlated with harsh environment. Thus ecological factors and kin selection may have interacted in social evolution in these species. In brief periods after rains the animals must cooperate to locate food patches. By living in groups, arid zone mole-rats can utilize better burrowing conditions. According to this hypothesis, subterranean rodents that inhabit xeric areas with dispersed patchy food and unpredictable rains can neither expand their tunnel systems nor disperse from them.

Social behavior in subterranean mammals is relatively rare, occurring primarily among some South American ctenomyids (Lacey 2000), but particularly in the truly social endemic African bathyergids (Jarvis and Bennett 1990; Jarvis et al. 1994; Sherman et al. 1991; Burda 1999; Bennett and Faulkes 2000). We have now found some convergence in social evolution. Future critical studies are necessary to test the nature, pattern, and degree of social evolution in the field for all Egyptian and Libyan isolates of *Spalax ehrenbergi* in North Africa.

Recent Speciation of the *Spalax ehrenbergi* Superspecies in the El-Hammam Isolate in Northern Egypt

We propose that the small northern Egyptian isolate of *S. ehrenbergi* from El- Hamman, near El Alamein, separated, presumably 10,000–25,000 years ago, from the Israeli complex (Lay and Nadler 1972) and evolved into a new species (Nevo, Simson, Heth, Redi, and Filippucci 1991). Our conclusion is based on the following evidence: (1) *Karyotype:* the diploid number of the El-Hammam mole-rats is $2n = 60$, but the karyotype differs in morphology, as deduced from banding, from that of the Israeli $2n = 60$. (2) *Genetic distance:* The Nei genetic distance of the El-Hammam Egyptian isolate from the Israeli $2n = 60$ is $D = 0.061$, larger than the genetic distances among the four Israeli species (average 0.035; range 0.002–0.056). (3) *Morphology:* The Egyptian mole-rats differ in cranial and bacular morphology from the Israeli *Spalax judaei*, $2n = 60$, and (4) *Behavior:* The El-Hammam population has evolved a totally pacifistic behavior displaying at least initial patterns of social evolution.

We propose that the El-Hammam Egyptian isolate is a

recent speciating derivative of the *Spalax ehrenbergi* superspecies, which possibly evolved because of the barrier of the Sinai Desert from the Israeli mole-rats during Holocene or Pre-Holocene times, i.e., 10,000–20,000 years ago.

Prospects

It is now a true challenge to explore all isolates of the *Spalax ehrenbergi* superspecies in North Africa and examine their genetic, chromosomal, and behavioral patterns. These future studies could highlight the extraordinary evolutionary processes of *adaptation* and *speciation* in marginal desert environments and elucidate the origin and evolution of sociality in the spalacids from solitary, territorial, and aggressive behavior of Israeli *S. ehrenbergi*.

Summary

The spalacids are Eurasian, primarily East Mediterranean subterranean rodents well adapted for life underground *ecologically, genetically, morphologically, physiologically,* and *behaviorally*. They are solitary, territorial, and aggressive. The Spalacidae originated 40–45 million years ago in Asia Minor and spread into increasingly steppic environments into the Balkans, Ukraine-Russian, and Near-East and North African steppes, budding new species, primarily by chromosomal speciation, with $2n = 38–62$ positively correlated with aridity. The Near Eastern branch of *Spalax ehrenbergi* superspecies consists of at least twelve allospecies, four of which have been studied extensively in Israel as an evolutionary model of adaptive climatic radiation and peripatric speciation: *Spalax galili*, $2n = 52$, in mesic, cool Upper Galilee; *S. golani*, $2n = 54$, in semixeric, cool Golan Heights; *S. carmeli*, $2n = 58$, in mesic, warm, central Israel; and *S. judaei*, $2n = 60$ in warm, dry Samaria, Judea, and northern Negev Desert.

Aggression in the *Spalax ehrenbergi* superspecies is polymorphic within sexes, populations, and species involving *militant, intermediate,* and *pacifist* behavioral phenotypes. Pacifism increases toward the Negev Desert in Israel and culminates in total fixation in a newly described pacifistic species with $2n = 60$ bordering the Sahara Desert in Egypt and North Africa. The new North African species was identified on morphological, behavioral, chromosomal, and allozyme grounds. In contrast to all other spalacids, this species is pacifistic and presumably evolved sociality in accordance with the aridity-food distribution hypothesis. Also a desert habitat presumably selects for pacifist behavioral phenotypes to minimize overheating, water, and energy expenditure. Social evolution may be driven by harsh ecological conditions. Thus, adaptive radiation of spalacids in North Africa stressed by desert ecology led to speciation coupled with social evolution. Future studies in North African spalacid isolates could elucidate the extraordinary evolutionary processes of adaptation and speciation in marginal desert environments.

Antipredator
Behavior

Chapter 26 Social and Antipredator Systems: Intertwining Links in Multiple Time Frames

Donald H. Owings and Richard G. Coss

Social and Individual Dynamics of a Predator-Prey Episode

The following scenario is a composite of observations made by members of our labs on what happens between California ground squirrels (*Spermophilus beecheyi*) and northern Pacific rattlesnakes (*Crotalus oreganus*). We present it to illustrate many of the themes of this chapter.

A northern Pacific rattlesnake has left its hibernation site in central California and has begun hunting for pups in a colony of ground squirrels. By crawling through the colony and interacting with a series of adult squirrels, the snake discovers the burrow system of a female (call her 9G), who stands her ground and even confronts the serpent (fig. 26.1a), behavior more typical of females with young than without. After interacting with 9G, the snake terminates the encounter and settles itself into an ambush coil in the mouth of one of the peripheral entrances to her burrow system (fig. 26.1b).

It is the season when rattlesnakes are active, and many squirrels already appear nervous, but they become more so as this rattlesnake passes through the colony. 9G and many other squirrels pause frequently to produce the snake-typical signal of tail flagging if they encounter a snake-like object (fig. 26.1c) or near burrow entrances or tall grass where snake risk is high. This flagging is contagious, inducing tail flagging and other signs of snake concern in squirrels that have not directly contacted the snake.

Over the next several hours, three adult squirrels from adjacent home ranges that do not have pups at risk come across the snake as it lies in ambush. On separate occasions lasting a few minutes, each of these adults exhibits several cycles of cautious close approach, extension of their heads in elongate investigative postures, tail fluffing as their stress-response system is activated, backing away from the snake, tail flagging, and reapproach. The third squirrel is attracted by the antisnake activity of the second, who relinquishes its position when the third initiates investigation of the snake. The second remains nearby monitoring for awhile, but then drifts away. The snake responds only with tongue-flicking.

None of these squirrels spends as much time with the snake as 9G already has, and certainly not as much time as 9G is about to. The third adult's antisnake behavior catches 9G's attention, and 9G's approach supplants the third squirrel (fig. 26.1d). After a few cautious bouts of investigation, 9G begins to throw substrate (loose soil and leaf litter) at the snake with her forepaws, and the snake strikes defensively, a maneuver that our squirrel easily evades. Two of 9G's pups pop their heads out of a burrow entrance about 8 meters away, and 9G breaks away to join them.

The snake begins to crawl toward 9G and her pups, and both 9G and a pup move to meet the snake. The pup reaches the snake before 9G does, and actually begins to confront the snake using most of the same motor patterns as its mother (fig. 26.1e). The snake strikes the pup, and the pup reels to a third burrow entrance where 9G joins it. Trailing the pup to the same entrance, the snake clears the substrate-throwing mother (fig. 26.1f) out of the way with a defensive bite, locates the dead pup and begins to swallow it. Even though 9G was hit and injured by the snakebite, she is still able to walk. At this time, almost 4 hours after her first contact with the snake, the mother begins to move her

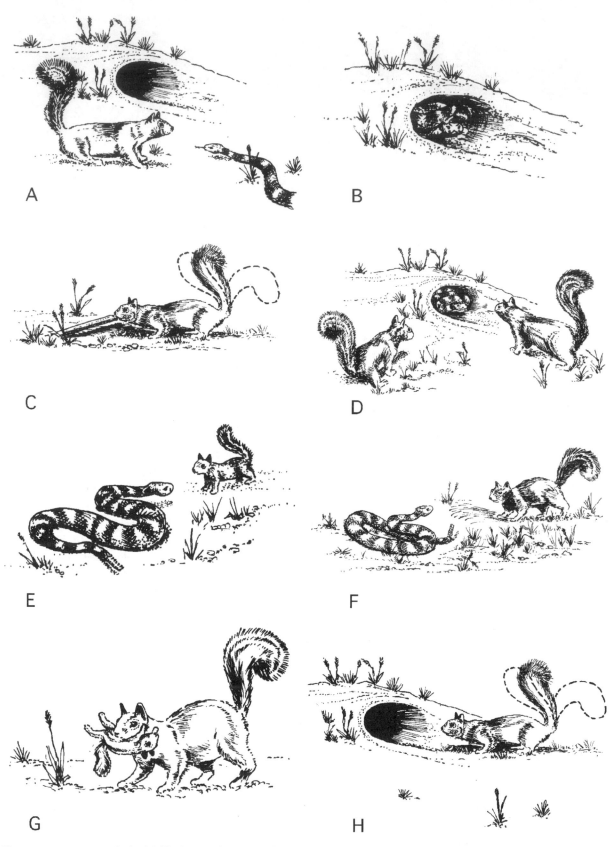

Figure 26.1 Composite episode of California ground squirrels confronting a northern Pacific rattlesnake hunting for pups. Ground squirrel mother confronts the rattlesnake (a), which pauses but then enters a nearby burrow and coils itself in an ambush posture (b). Other squirrels are vigilant after seeing the snake as characterized by this tail-flagging adult, who cautiously inspects a snake-like stick (c). The mother supplants another snake-directed adult and inspects the snake (d). The snake leaves the burrow and one of the mother's pups confronts the snake in an adult-like manner (e), an event terminated when the snake strikes and envenomates the pup. As the snake trails the pup to a burrow, the mother throws loose soil and leaf litter in an apparent effort to deter the snake (f) and is also bitten. While the snake consumes her pup, the mother hobbles on her injured leg while she transfers her other pups to a distant burrow (g). Having recovered from her snakebite, the mother tail flags the next day as she cautiously inspects burrows around the snake-encounter site (h).

other three pups (fig. 26.1g). Hobbling on her injured leg, she carries her pups one by one to a burrow about 25 meters away, where the resident male accepts them.

The next day 9G returns to the site of her encounter, exhibiting little evidence of impairment from her snakebite as she tail flags and cautiously examines burrow entrances (fig. 26.1h). But the snake has moved deep into 9G's former burrow system, and she does not enter and locate it there. The snake resumes hunting 2 days later, continuing in that area for the next 6 days, but captures no more pups among this alerted kinship cluster.

The Variety of Links between Social and Antipredator Systems

As this squirrel-snake episode illustrates, social and antipredator systems are linked in a variety of ways. We will discuss several categories of such connections, including shared components and linkages between means (e.g., social behavior) and ends (e.g., avoiding predation).

Shared components at the level of motor patterns

Many motor patterns are sufficiently broad in their applicability that they are obviously usable in both social and antipredator contexts, including locomotion, biting, pouncing, parrying, elongate investigatory postures, and freezing in quadrupedal and bipedal postures. Other motor patterns, however, require more extensive observations before their use in both contexts becomes evident; these include substrate throwing and alarm calling.

Substrate throwing with fore or hind paws is often used in the context of burrow excavation and repair, which is perhaps its original functional context. But the proximate and ultimate processes that make behavioral systems opportunistic have added this activity to a defensive context, often called defensive burying (Terlecki et al. 1979; De Boer and Koolhaas 2003). Defensive burying is typically hypothesized to serve an antipredator function (Dell'omo et al. 1994), an idea inspired by observations that substrate throwing is used by ground-dwelling sciurids to deal with snakes—for example, by entombing them in a burrow (Merriam 1901; Coss and Owings 1978; Halpin 1983; Hersek 1990). But substrate throwing is used for additional antipredator purposes, including harassing snakes above ground (fig. 26.1f; Owings and Coss 1977; Halpin 1983), as well as simply getting a snake's attention (Hennessy and Owings 1988). More broadly yet, substrate throwing is also used in social contexts to harass and even entomb conspecific interlopers (Armitage and Downhower 1970; Levy 1977; personal observations of wild *Rattus norvegicus*).

To our knowledge, no work has systematically explored how the form of substrate throwing has been modified with the acquisition of these social and antipredator functions. For example, the most common way to move earth during excavation is to kick it out behind the animal with the hind paws. It is possible that substrate throwing with the forepaws, as California ground squirrels and rock squirrels (*Spermophilus variegatus*) do, had its evolutionary origins in burrow plugging as an antisnake defense below ground, a context in which it might be important to continue monitoring the snake by facing it. This hypothesis, if true, begs the question of why black-tailed prairie dogs (*Cynomys ludovicianus*) typically (but not always) use their hind paws to kick substrate at snakes (Halpin 1983; Shier, personal communication).

Alarm calling is another class of behavior that has proven to be functionally diverse. Ground-dwelling sciurid alarm calling provided some of the earliest and best evidence of nepotistic behavior (Dunford 1977a; Sherman 1977; Hoogland 1983), and subsequently also yielded evidence that some alarm calling is self-interested rather than nepotistic (Sherman 1985; Owings et al. 1986). More recently, attention has turned to questions about what predator-related information is available in these calls (Owings and Leger 1980; Slobodchikoff et al. 1991; Blumstein and Arnold 1995; Blumstein and Armitage 1997a; Blumstein 1999a).

But some species also often use these calls socially, a rarely explored phenomenon (but see Smith et al. 1976, 1977; Leger et al. 1980; Owings and Leger 1980). At least eight species of ground-dwelling sciurids use alarm calls in sexual and agonistic contexts (Owings and Hennessy 1984). Such social use of these calls is common in at least some species, and might be expected to dilute the antipredator function of these calls over evolutionary time. For example, at least one tree squirrel species (Formosan squirrel, *Callosciurus erythraeus*) uses the same type of call in both postcopulatory and terrestrial-predator alarm contexts, as California ground squirrels do (Boellstorff et al. 1994); these tree-squirrel calls do not differ structurally, and do evoke the same suppression of conspecific movement during playbacks (Tamura 1995).

Two questions are raised by the idea that the antipredator function of calls can be diluted by using the same call type in social contexts. First, are conspecifics or the predator the primary targets of these antipredator calls? The dilution idea assumes conspecifics to be the primary targets. If, however, the predator is the primary target (Owings and Hennessy 1984; Hersek and Owings 1993), social use of calling may not interfere with its antipredator function. Second, how finely differentiated is the call system? In California ground squirrels, for example, detailed structural analyses of the same call types in social and antipredator contexts has uncovered significant structural differences (Leger et al. 1980; Owings and Leger 1980). Perhaps use of these calls in both contexts has favored structural differentiation to minimize the impact of each type of use on the functionality of the other system. If squirrels distinguish social and antipredator variants, then dilution of antipredator function may not be a problem, even if conspecifics are primary targets. However, Tamura's (1995) research on Formosan squirrels suggests that this species does not distinguish calls used in the two contexts. Playback studies have not yet ascertained whether these differences have communicative significance for California ground squirrels.

A second under-explored feature of such calling is that many sciurids and other species also call tonically (as in Schleidt 1973), that is, repetitively throughout an encounter with a predator or conspecific and even into the aftermath of encounters (Betts 1976; Smith et al. 1977; Boellstorff et al. 1994; Warkentin et al. 2001). These rhythmically repeated calling patterns regulate the behavior of conspecifics or heterospecifics by fostering a sustained state in targets—for example, of vigilance or immobility, rather than by eliciting a discrete response as initial calls often do (Owings et al. 1986; Loughry and McDonough 1988). Tonic antipredator calling may be functionally distinct; for example, working selfishly when initial calls function nepotistically even though initial callers and tonic callers are typically the same individuals (Owings et al. 1986). More work needs to be dedicated to exploring what temporally extended regulatory problems callers "solve" with these temporally extended patterns of vocal communication. Tonic antipredator patterns appear in part to be adaptations to predators that pose a more temporally extended threat—for example, because they lie in ambush for extended periods (Hanson 2003).

Shared components at the level of animal-environment integration

The stress-response system
Even though social and antipredator systems are typically functionally distinct, they are linked causally, because both conspecifics and predators can be sources of stress. Male California ground squirrels, for example, fight intensely with other males during the approximately 2-week breeding season, resulting in extensive wounding (Owings et al. 1979; Boellstorff et al. 1994). Similarly, California ground squirrel females expose themselves to substantial stress during their single day per year of estrus, soliciting courtship from multiple males, intensely resisting all males after they have mounted, but copulating with many of them (Boellstorff et al. 1994; unpublished observations). Later in the reproductive season, females also experience stress from

predators; for example, as rattlesnakes move in to prey on the developing young (our earlier scenario; Hennessy and Owings 1988). Mammalian and avian predators also endanger both pup and adult California ground squirrels, inducing stress not only through actual encounters but also indirectly through the antipredator vocalizations these predators elicit (Leger and Owings 1978; Owings and Virginia 1978; Leger et al. 1979; Owings and Leger 1980). Finally, social and predatory threats can at times be generated by the same individual; some female California ground squirrels and some members of other sciurid species are infanticidal, and may kill multiple offspring of other females (Sherman 1981b; Hoogland 1985; Trulio et al. 1986; Lacey 1991; Lacey 1992; Trulio 1996; Ebensperger and Blumstein, chap. 23 this volume).

Rodents typically cope with such threats of injury or other forms of harm by making physiological and behavioral adjustments that can increase their ability to escape or resist a threat (Koolhaas et al. 1999). These adjustments are supported by several different physiological systems, including the hypothalamic-pituitary-adrenal (HPA) stress-response system (Sapolsky 1992; Francis and Meaney 1999; Perrot-Sinal et al. 1999), and neurotransmitter/neuromodulatory systems based on the morphine-like peptides (endorphins) and the benzodiazepine-GABA neurotransmitter system (Kavaliers 1988; Edwards et al. 1990).

Activation of the HPA system contributes to many physiological and behavioral adjustments that facilitate coping with emergencies. Releasing glucose stores, inhibiting glucose storage, and increasing breathing rate, heart rate, and blood pressure all fuel physical efforts to escape or resist a threat. Blocking inflammation maintains joint mobility even when the joint has been injured, and so facilitates performance of the actions of offense and defense (Sapolsky 1992). Boosting dermal immune function can enhance resistance to pathogens that may invade cuts or punctures (Dhabhar and McEwen 1999). The effects of psychological changes complement those of these physiological adjustments. Increasing vigilance enhances the probability that sources of danger will be detected. Redirection of the focus of learning and memory systems to emotion-laden events and places facilitates learning that provides emotional labels for environmental situations; this in turn enhances anticipation of and preparation for danger (Francis and Meaney 1999).

Stress-induced analgesia is mediated by at least two neural systems: an endorphin-based neurotransmitter and neuromodulator system (Lester and Fanselow 1985; Kavaliers 1988; Hendrie 1991) and a benzodiazepine-GABA neurotransmitter system (Edwards et al. 1990). These systems appear to work in complementary time frames when activated by perception of a predator such as a weasel (Kavaliers 1988). For example, the benzodiazepine-GABA system initiates brief analgesia in white-footed mice (*Peromyscus leucopus*) after a brief exposure to an out-of-view short-tailed weasel (*Mustela erminea*), but a longer predator exposure induces a more tonic, endorphin-mediated analgesia. Activation of analgesia may allow continued defensive action when an individual has sustained a painful injury, either from a predator or a conspecific (Kavaliers 1988; Rodgers and Randall 1986). These two analgesic mechanisms may be parts of a broader defensive system supported by a midbrain structure called the periaqueductal gray (Rodgers and Randall 1987; Dielenberg and McGregor 2001; Castilho et al. 2002).

In summary, social and antipredator systems are typically functionally distinct, but they generate similar demands on the organism. These include the need to mobilize energy for vigorous activity, focus attention and learning on emotionally salient events and places, and mitigate pain and infection. These demands are met through adaptive properties of the stress-response system.

Perceptual affordances

As conceived by Gibson (1979/1986), an *affordance* is a perceived behavioral implication of the physical properties of some environmental feature in relation to the individual and its context. For example, a physical property such as brushy microhabitat has context-dependent meaning; it affords safety from aerial predators for organisms of a given size, but danger from ambushing terrestrial predators. Rodents treat brush differently depending on the current source of danger, preferring brushy habitat when threatened by raptors, but not when endangered by terrestrial predators (Sherman 1985; Kotler et al. 1993; Hanson and Coss 1997). Similarly, burrows are refuges from aerial predators but not necessarily from digging predators such as badgers (Balph 1961; Knopf and Balph 1969; MacWhirter 1992).

Conspecifics are social sources of affordances. Signals are often designed through proximate and ultimate processes to serve as cues for affordances; for example, the nepotistic antipredator vocalizations of ground-dwelling sciurids are associated with situations that afford danger and therefore refuge-seeking and further assessment (Dunford 1977a; Sherman 1977; Hoogland 1983; Owings et al. 1986). However, many socially generated affordances are not designed as such, but are exploited through active assessment by the perceiver (Otte 1974; Hennessy et al. 1981; Owings and Hennessy 1984; note that active assessment extends beyond eavesdropping on signals intended for others by also including extraction of cues from nonsignaling behavior). An example of the latter is the ability to pick up cues about an unfolding but unseen threat by using the perceptual orientation and behavior of other individuals. For instance, an individual typically directs its gaze to gather information rather than to inform others, but others may use gaze direction to localize the source of a disturbance. Belding's ground squirrels appear to locate sources of danger by exploiting such socially generated affordances, looking in the same direction as an alarm-calling conspecific who is watching a predator not visible to the gaze follower (Sherman 1977). The availability of such incidental affordances is often cited as a source of benefits derived from social living, an idea that has received support from the evidence that detection of predators can be a socially collective process (Hoogland 1981, 1995; discussed in more detail later).

Many socially generated affordances are available as by-products of the fact that behavior tends to be matched to situational demands. California ground squirrels, for example, adjust both their signaling and nonsignaling behavior to the different levels of urgency of threat posed by raptorial, mammalian, and reptilian predators. These adjustments involve trading off between the conflicting demands of minimizing conspicuousness to the predator and extracting cues from the predator about the actual danger it poses. Raptors present the most urgent threat, and squirrels minimize their conspicuousness by whistling only once or a few times, retreating to a burrow entrance, and remaining close to the ground (Owings and Hennessy 1984; Sherman 1985). Mammalian predators are intermediate in urgency of danger, and squirrels also retreat to a burrow, but then opt for more assessment and less inconspicuousness by surveying from a promontory or bipedal stance while chattering and then calling repeatedly (Owings and Hennessy 1984; Sherman 1985). Snakes pose the least urgent threat, and squirrels behave much as described in our model episode, confronting the snake in a way that augments their capacity to pick up assessment cues and simultaneously increases their conspicuousness to the snake (Owings and Hennessy 1984). Humans can judge the level of urgency of danger from such distinctive differences in behavior, and ground squirrels also appear to be able to do so (Leger and Owings 1978; Hersek and Owings 1993).

Processes for use of space

Squirrel 9G's behavior in our model episode provides a good example of a connection between social and antipredator domains with regard to the use of space. While dealing with the snake, she also faced the social task of managing her pups. And she rescued her young from the snake by transferring them to the burrow of a male willing to receive them (unpublished observations; this is not an uncommon pattern, but we do not have data on how the male's relationship with the female affects his receptivity to such burrow transfers). 9G's direct movement to the male's burrow with her pups indicated that she already knew where that burrow was. This knowledge was also needed earlier, during breeding, because estrous female California ground squirrels travel to the burrows of males to mate, running a gauntlet of other males eager to mate with them (unpublished data, as well as Boellstorff et al. 1994).

Knowledge of the spatial layout and nature of the resources in one's surroundings has utility in many domains, including both social and antipredator contexts. The benefits of such spatial ability have been a source of natural selection in some rodent species, shaping processes for learning, remembering, and using the home range (Gaulin and Fitzgerald 1989). Nevertheless, different contexts, such as predation, social interaction, and foraging may well differ in the demands they place on spatial knowledge. For example, the typical contexts for the study of spatial knowledge are low-urgency ones, such as finding, storing, and retrieving food. In these situations, some ground-dwelling sciurids have exhibited remarkably precise cognitive-mapping abilities (Devenport and Devenport 1994; Devenport et al. 2000). When a food source was moved with its conspicuous visual landmark a mere 1.5 meters from the learned location, animals initially ignored the nearby moved landmark and searched for the food in the previous location. Similar spatial abilities may be used in low-urgency social situations, such as keeping track of the home burrows of other members of a kin cluster (Boellstorff and Owings 1995). Animals also exhibit evidence of cognitive mapping under high urgency; alarmed sciurids use cues about their distance and direction from refugia, often turning up to 180 degrees before initiating a run to a refuge after an antipredator call (Leger et al. 1979), or varying their decision to run or call with their distance from a burrow (Noyes and Holmes 1979; Sherman 1985; Bonenfant and Kramer 1996). However, when social contexts are as urgent as most predatory ones—for example, during agonistic episodes—such high social and predatory urgency may not provide the time or attention needed for detailed use of cognitive mapping, and so may force the application of additional methods. These may include relying more heavily on learned landmarks close to the destination ("beacons"; Collett et al. 1986; Dinero, personal communication) or on well-practiced movements along often-used routes (Stamps 1995).

Spatial learning can be useful in ways that go beyond cognitive representations of the layout of resources. For example, laboratory rats learn not only the spatial location of food rewards, but also the appropriate motor activity, a direction of turning that will lead them to food (Restle 1957). Stamps (1995) has argued that many species may depend heavily on such motor learning aspects of spatial learning, practicing movements over regularly used trails much as humans refine their foot-racing performance for hurdles

events by running the track repeatedly. This learning may prove especially useful when rapid locomotion through cluttered surroundings is needed to deal with a conspecific or predatory adversary that has not had the same route-specific practice. Repeated use of the same route can also increase an individual's sensitivity to change along a trail, thereby enhancing the chances of detecting a conspecific or heterospecific behind a bush, for example, that wasn't there before.

Remembered cognitive, perceptual and motor knowledge of surroundings are not the only sources of effective spatial patterning of behavior. Animals may also exploit spatially varying environmental features on the basis of immediately recognized affordances rather than familiarity. Unobscured vision, for example, indicates openness and therefore affords vulnerability, motivating rats to make contact with visibility-obscuring vertical surfaces (positive thigmotaxis, e.g., Martínez et al. 2002), a response that is intensified under fear motivation (Kelley 1985). Similarly, burrowing rodents are strongly disposed to seek refuge in dark burrow-like places, a disposition that we regularly exploit in laboratory experiments by making a specific place dark where we want our squirrels to go (Rowe and Owings 1978; Rowe et al. 1986). California ground squirrels are cognizant of nearby burrows and vertical surfaces such as rocks, and vary their levels of foraging and vigilance with the proximity of these environmental features (Leger et al. 1983). Similarly, dispersing yearling Columbian ground squirrels (*Spermophilus columbianus*) follow cutbanks, roads, and game trails into new terrain, a choice that appears to guide them to new ground squirrel habitat at higher-than-chance rates (Wiggett and Boag 1989; Wiggett et al. 1989).

Means-end links: The social domain provides means for dealing with predation

Perhaps the most familiar link between social and antipredator behavior comes from the fact that rodent social systems provide a means for dealing with the threat of predation (Waterman and Fenton 2000). Our discussion of antipredator calls and exploitation of social affordances has already touched on the social domain as a means for dealing with predators. The availability of such social affordances has generated selection for increased sociality. For example, black-tailed and white-tailed prairie dogs (*Cynomys ludovicianus* and *C. leucurus*, respectively) are both social species, but black-tails live in larger, denser groups (Hoogland 1981; 1995). Hoogland has used variation in group size and density within and between these two species to explore the adaptive significance of such group living. Three lines of evidence indicate that social living conveys antipredator benefits. First, prairie dogs in larger social groups respond with shorter latencies to simulated predatory attacks by badgers (*Taxidea taxus*), and black-tails detect predators more quickly than white-tails. Second, this advantage of group size and/or density holds even though individuals in larger groups devote proportionately less time to antipredator vigilance than those in smaller groups, and black-tails are less vigilant per capita than white-tails. Third, both breeding synchrony and center-edge differences in individual alertness indicate that social living provides selfish-herd effects. These social differences may have originated for antipredator reasons, as white-tail habitats contain significantly more protective cover than black-tail habitats, and black-tails enhance the visibility of their surroundings by cutting down vegetation near their burrows.

Relational Similarities Generate Systems Convergence

When relations between predator and prey are similar to those among conspecifics, social and antipredator systems are likely to converge. Here we explore two such relational similarities, temporally extended (tonic) proximity between individuals and symmetry in power, and corresponding convergent properties of social and antipredator systems.

Temporally extended (tonic) proximity

Ground squirrels and rattlesnakes share the same vicinity for multiple days (see introductory episode), as conspecifics often do. This is because rattlesnakes are ambush hunters that choose a site and often remain there for several days, awaiting the opportunity to surprise their prey (Hersek 1990; Greene 1997). This can generate tonic states of alertness in the squirrels (Hersek and Owings 1993; Hersek and Owings 1994). Such sustained associations have apparently been a source of selection on ground squirrels, favoring tonic features of antisnake behavior systems similar to the tonic features of behavior systems used in sporadic interactions with conspecifics (e.g., the cumulative effects of repeated interactions, as in Hinde 1974, and the design of signaling systems to produce tonic effects, as in Schleidt 1973). Tonic features of antisnake behavior were evident in tail-flagging, a visual signal that is specific to snake contexts (Hennessy et al. 1981). Although tail flagging is snake specific, 90% of the tail-flagging episodes recorded in the field were emitted outside the context of direct encounters with snakes. Nevertheless, this tonic tail flagging was linked to indirect tonic sources of variation in squirrel vulnerability to snakes, such as the presence of a rattlesnake in the colony that day, or developmental changes in pup vulnerability to snake predation (Hersek and Owings 1993). The tonic

dimension of behavior varies adaptively with temporal extension of predatory threat. Rock squirrels (*Spermophilus variegatus*) persist in their alertness about snakes for longer after encounters with rattlesnakes than with gopher snakes (*Pituophis melanoleucus*), who are less likely than rattlesnakes to remain for extended periods in one area (Hanson 2003). This difference in persistence is independent of the *amount* of danger posed by the snakes' presence, as inferred from squirrel behavior toward the snakes during actual encounters (Hanson, unpublished analysis).

Symmetry in power

As illustrated by our opening squirrel-snake scenario, the relationship between California ground squirrels and their principal predator, northern Pacific rattlesnakes, involves symmetry in power. Rattlesnakes are the major source of pup mortality (Fitch 1948a, 1949), but adult ground squirrels can be formidable opponents of rattlesnakes, protected by both skillful antisnake behavior (Owings and Coss 1977; Hennessy and Owings 1988), and blood serum proteins that neutralize rattlesnake venom (Poran et al. 1987; Poran and Coss 1990; Biardi et al. 2000). While rattlesnakes normally cannot kill adult ground squirrels, the lower serum volume in pups falls short of that needed to prevent death (Poran and Coss 1990). This combination of rattlesnake predation on pups and adult capacity to mount an effective defense has set the stage for extended bouts of conflict between adult squirrels and snakes that can involve injury and even death to both parties (Hennessy and Owings 1988; Coss and Owings, 1989; Hersek 1990; Owings 2002). The repertoire of maneuvers that each party brings to these conflicts is illustrated by our opening episode (Owings and Coss 1977; Hennessy and Owings 1988; Rowe and Owings 1990).

Due to these commonalities, many of the concepts and phenomena of intraspecies conflict also characterize interactions between adult squirrels and snakes (Swaisgood, Owings, and Rowe 1999). In both systems, fatal injuries are rare but can occur for either participant (Fitch 1949; Hersek 1990). Further, ground squirrels use "conventional fighting methods" (Parker 1974) to probe and assess the danger posed by rattlesnakes, and vary their antisnake behavior appropriately. These squirrels evoke rattling by confronting rattlesnakes (Rowe and Owings 1978), and use the resulting acoustic feedback to adjust their assertiveness downward when rattle structure is a cue of high risk (rattling from a large or warm snake) or upward with acoustic cues of low risk (rattling from a small or cold snake; Rowe and Owings 1978; Swaisgood, Owings, and Rowe 1999, Swaisgood, Rowe, and Owings 1999; Swaisgood et al. 2003). Ground squirrel antisnake behavior is also sensitive to the value of the contested resource (the pups; cf. Parker and Rubenstein 1981). Relative to other adults, females with young pups spend more time confronting snakes and are more discriminating with regard to the level of threat posed by the rattlesnake (Swaisgood, Owings, and Rowe 1999; Swaisgood, Rowe, and Owings 1999; Swaisgood et al. 2003).

We have only limited data on the defensive behavior of rattlesnakes while dealing with ground squirrels. But we do know that rattlesnakes also proceed in ways generally characteristic of social conflicts, defending themselves in risk-sensitive ways while dealing with squirrels, experimental squirrel models, and other sources of danger. For example, rattlesnakes defend themselves less readily by rattling and striking when colder body temperature renders them less able to follow through on such a threat (Rowe and Owings 1990; Rowe and Owings 1996; Owings et al. 2002; also unpublished observations). On the offensive end, these rattlesnakes are active exploiters of prey-related cues, choosing hunting sites rich in both prey odors and microhabitat features preferred by their prey (Theodoratus and Chiszar 2000). These snakes also use active probing in an apparent quest for squirrel reactions that leak cues about whether a female has nearby young. Mother squirrels resist such leakage, standing their ground while engaging in little additional activity until forced into it by a persistent rattlesnake that may eventually get dangerously close to the burrow containing the pups (Hennessy and Owings 1988).

The relation between ground-dwelling sciurids and mustelids of about the same size appears to provide another example of relative parity between predator and prey. For example, black-tailed prairie dogs effectively mob and harass black-footed ferrets (*Mustela nigripes*) and plug the burrows they invade (Henderson et al. 1974; Martin et al. 1984). Similarly, female Belding's ground squirrels attack and chase long-tailed weasels (*Mustela frenata*; Sherman 1977).

How Social and Antipredator Systems Acquire Their Functional States, Linkages, and Similarities

Biological systems are adjusted via a variety of processes that act in immediate, tonic, developmental, and evolutionary time frames (Coss and Owings 1985; Owings 1994). These proximate and ultimate processes dovetail in the production of an adaptive organism-environment relationship, typically acting in complementary rather than alternative ways (Cairns et al. 1990; West et al. 1994; Stamps 2003). Proximate and ultimate contributions interact with each other rather than working independently. For example, evolutionary processes shape development through hetero-

chrony, that is, by changing the relative rates of development in different systems (Gould 1977; Mason 1979). In turn, developmental processes are central to generating the heritable variation upon which natural selection acts (Bateson 1988; Cairns et al. 1990; Mateo, chap. 17 this volume). A central feature of this latter point is that parents shape the development of their offspring by providing not only genes but also many reliable environmental contexts for development (West et al. 1994). Such experientially induced effects can either amplify or reduce the behavioral manifestation of genetic variation. For example, allowing mice to have only one social encounter can increase the expression of variation in aggressiveness. This variation can be used in artificial selection experiments to generate very rapid divergence of lines of high and low aggressiveness. Conversely, these selected differences in aggressiveness can be masked if each mouse has multiple social encounters, which generates increased aggressive activity in the low-aggression line and therefore developmental convergence with the high-aggression line (Cairns et al. 1990).

Proximate and ultimate processes also interact with each other when the current structure of behavioral systems creates constraints and opportunities for subsequent evolutionary change. As a consequence, adjustment in biological systems has a cascading quality. Self-grooming in an agonistic context by California ground squirrels provides a case in point (Durant et al. 1988; Bursten et al. 2000). When males of this species engage in a territorial boundary encounter, agonistic contact may be interspersed with breaks during which one or both individuals use forepaws and mouth to self-groom in a highly stereotyped cephalocaudal pattern. The focus of this grooming starts with the muzzle and moves in a caudal progression across the head, down the body, and along the tail to its tip. Such grooming is associated with other noncontact agonistic activities, including tail piloerection, staring at the adversary, and scent-rubbing against stationary objects. Agonistic encounters containing such grooming bouts are less likely to escalate to fighting than those without grooming.

In other words, this cephalocaudal self-grooming has become a social signal, perhaps through the following scenario derived by Spruijt and colleagues (1992) from the literature on grooming. They argue that grooming originated evolutionarily to care for the body surface in part by defending against ectoparasite infestation (see also Hart et al. 1987). This defense requires that individuals not only groom in immediate reaction to irritation from such sources, but also engage in tonic grooming as prophylaxis against infestation (programmed grooming; Hart et al. 1992). However, tonic grooming does not produce the short-term payoff of reduced irritation, and so requires a more endogenous form of payoff to sustain it, such as arousal reduction and endorphin release (Keverne et al. 1989; Spruijt et al. 1992). The resulting capacity of grooming to produce more positive psychological states appears to have led to its use for arousal regulation in agonistic encounters; such use, combined with the fact that tonic grooming is a low-priority activity, has made tonic grooming a reliable perceptual cue that the groomer is reluctant to escalate (Spruijt et al. 1992). Groomers' adversaries may have responded to this cue by reducing escalation, a type of exchange that could have been the source of selection leading to ritualization of an agonistic display (Tinbergen 1952), now seen as highly stereotyped cephalocaudal self-grooming in California ground squirrels (Bursten et al. 2000). A similar de-escalation process, followed by ritualization, may have operated with rodent gaze aversion (Grant and Mackintosh 1963; see also Coss et al. 2002).

Such evolutionary proliferation of the functions of self-grooming illustrates how cascading change in biological systems might be driven by active assessment and management (Owings and Morton 1997, 1998). Assessment is the active quest for useful cues, such as those provided by the grooming activity of an adversary when this activity correlates with a reluctance to escalate. Management is synonymous with regulation; animals regulate themselves and others in their own interest. Self-interested assessment and management give behavioral systems an opportunistic quality that makes different functional systems potential resources for each other. In the case of ground squirrel self-grooming, the expropriation of grooming for the management of arousal in agonistic contexts led to the availability of grooming as an assessment cue that escalation was unlikely. Use of this cue in assessment by adversaries may in turn have favored the specialization of grooming for a management (signal) function. The functions of this self-grooming pattern appear to have proliferated even further, becoming a part of olfactory communication systems. Male meadow voles and prairie voles (*Microtus pennsylvanicus* and *M. ochrogaster,* respectively) use grooming as a means for broadcasting scent to potential sexual partners (Ferkin et al. 1996; Wolff et al. 2002). And some ground-dwelling sciurids use the latter portions of the cephalocaudal grooming pattern to anoint their flanks and tails with snake scent (Kobayashi and Watanabe 1986; Owings et al. 2001), an activity that may serve an antipredator function (Kobayashi and Watanabe 1986).

Sociodevelopmental sources of antipredator inferences

Early in ontogeny, social living can generate experiences that foster development of sophisticated inferences about predators. These can take at least three different forms in which information is connected to affordances via specific perceptual cues: (1) generalizing from conspecific intention cues to predator intention cues; (2) using social cues picked up early from social interactions to detect or locate predators; and (3) recognizing the distinctiveness of nonmammalian predators, such as snakes and raptors. All of these opportunities should be readily available from interactions with littermates and adults during early development.

Generalizing from conspecific to predator-intention cues

Interactions with parents and peers can produce experiences that foster development of skills useful in anticipating the behavior of others, some of which are probably generic enough to be used during interactions with predators. For instance, the schema of two facing eyes is a widespread cue used by animals to detect when another animal is looking at them; this cue is used during interactions with both conspecifics and heterospecifics (Coss 1978; Topal and Csanyi 1994). Use of this cue for assessing intentions may have favored the use of directed gaze as a signal (Coss et al. 2002). In fact, staring is an aggressive act in California ground squirrels (Owings et al. 1977). Aggressive use of staring may subsequently have led to signal-enhancing devices such as the eye rings of sciurids, and even mimicry of the pattern of two facing eyes in other vertebrates (Coss and Goldthwaite 1995). Additional examples that could be derived from social experience include anticipating the actions of predators on the basis of their direction of orientation, patterns of movement and immobility, and postures, as well as discovering a predator's intentions by noting that its behavior is contingent on the prey's behavior (e.g., Levin 1997).

Using social cues to detect or locate predators

As noted earlier, Belding's ground squirrels appear to exploit the gaze direction of alarm-calling conspecifics as a source of information about predator location (Sherman 1977). Such an ability might have its developmental roots in learning about the significance of a littermate's focus of attention on a third littermate, perhaps in the context of the third individual's joining a play bout. Interactions of this sort can provide a context for learning correlations between an individual's current behavior and subsequent meaningful events, such as initiation of an interaction.

Does early social learning enhance the distinctiveness of snakes and raptors?

Norway rats undergo the functional equivalent of filial imprinting, developing an attachment to their own mother through olfactory learning prior to birth, and extension of that learning to tactile cues during the early postnatal period (Polan and Hofer 1999). Such early learning contributes to the development of a same-species preference based on odor cues in a number of rodent species (see Vasilieva et al. 2001 for a review). Little is known about such processes in more visually oriented, diurnal rodents, but ample opportunities exist for extension of such learning to the visual domain once interactions outside the nest begin. Development of a preference for conspecifics provides a schema against which other organisms in a rodent's environment might be compared. Those predators whose form contrasts most with this schema, i.e., snakes and raptors, should be least likely to be misidentified as a conspecific. In fact, laboratory-born, snake-inexperienced California ground squirrels respond differently to snakes and heterospecific mammals (Owings and Coss 1977; Coss 1991a; Coss 1993; Coss and Biardi 1997). This difference in response to novel mammalian and reptilian species may be based in part on differences in their similarity to the learned ground-squirrel schema.

Conversely, associative learning typically involves generalization to similar configurations—for example, to the similar forms of other mammals, such as other rodents and cats. Consistent with this prediction, laboratory-reared California ground squirrel pups and juveniles responded with much greater caution to a gopher snake than to either a guinea pig (*Cavia porcellus*) or a domestic cat (*Felis catus*; Coss 1991a; Coss and Biardi 1997). In contrast, an experimentally presented dog evoked more cautious responses by juveniles than by adults a few weeks postemergence, suggesting that rapid learning of the danger posed by mammalian predators counters this generalization effect (Hanson and Coss 1997). The juveniles in this study were more aroused than the adults (Hanson and Coss 2001a), a condition known to facilitate associative learning (McGaugh 1989). Such learning may have been mediated by the evocativeness of this species' alarm calls, which serve as unconditioned stimuli, evoking antipredator behavior in lab-born, predator-inexperienced members of this species (Tromborg 1999). So encounters with mammalian predators may involve pairing conspecific vocal unconditioned stimuli for antipredator behavior with conditioned stimuli, such as the sight of novel mammalian predators (see also Shriner 1999; Mateo, chap. 17 this volume).

Ontogenetic adjustment of the stress-response system

The hypothalamic-pituitary-adrenal (HPA) stress-response system that supports social and antipredator behavior varies adaptively in ways that are shaped through both proximate and ultimate processes. This is an ancient system that has undergone evolutionary diversification in its structure and function (Stoddart and Bradley 1991; Eilam et al. 1999; DeVries 2002; Romero 2002). And this system varies intraspecifically in the details of its functional organization, depending on the environmental conditions prevailing during each individual's development, as influenced by such factors as the levels of food availability, predation intensity, or social conflict (Francis and Meaney 1999). However, many such developmentally important environmental conditions exist outside the nest, and so are unavailable for direct assessment by the developing neonate.

Work with laboratory rats indicates that pups adjust to such environmental variation through differences in the maternal care they receive (Francis and Meaney 1999). Conditions that curtail contact between mother and infant limit the amount of licking, grooming, and arched-back nursing that the mother provides to her infants. Pups respond to such low levels of maternal behavior by developing more active stress-response systems. Conversely, conditions that foster licking, grooming, and arched-back nursing between mother and infant also facilitate development of a less excitable stress-response system in the offspring. These effects have been discovered through experimenter manipulation of pups, and confirmed through laboratory studies of natural variation in maternal behavior (Liu et al. 1997). Such effects on HPA system development have not been demonstrated in the field, and therefore raise important questions for future field research (see Mateo, chap. 17 this volume).

Two-way connections between proximate and ultimate processes

The work on maternal effects on HPA function illustrates how developmental processes are not only shaped through evolution but also influence evolutionary processes. Mother rats with more active HPA systems engage in less licking, grooming, and arched-back nursing with their pups, and this induces more active HPA systems in their pups (Francis and Meaney 1999). Consequently, such variation in HPA activity is heritable via maternal effects, and so is potential raw material for the action of natural selection.

Patterns of development may also influence the evolutionary persistence of a phenotypic system. Evidence for persistence of systems under relaxed selection was revealed by systematic comparisons of the antisnake behavior of eleven populations of California ground squirrels, six of which had been relatively free of selection from rattlesnakes or gopher snakes for estimated time frames of 70,000 to 300,000 years (Coss 1991a; Coss 1993; Coss 1999). Estimates of these time frames were derived by analyzing genetic variation among thirty-one populations and calibrating this variation to time using geological evidence for the formation of a barrier to gene flow. These results were combined with data on interpopulation variation in rattlesnake venom resistance and current densities of rattlesnakes and gopher snakes. It was assumed that snakes not present during the current warm interglacial period would have been even less likely to be present during past colder glacial periods. The systems subserving the antisnake behavior of California ground squirrels proved to be remarkably persistent, spanning gaps of up to 300,000 years in which selection from snakes had been relaxed. This persistence was not simply the result of occasional encounters with rattlesnakes in snake-rare areas. Not a single population in a snake-rare area exhibited high levels of venom resistance, even though they continued to recognize snakes as dangerous (Coss and Goldthwaite 1995).

Persistence of behavioral antisnake defenses may be a byproduct of selection for reliable expression of this important antipredator system in individuals large enough to withstand envenomation. Such selection may have had the effect of locating the neural substrates for antisnake behavior on the early-developing proximal processes of developing neurons (Coss 1991b; Coss 1999). This pattern of early development protects the neural system from alterations via the less predictable experiences mediating neural growth as the older animal engages with a broader, less-predictable world outside the nest (Coss 1991b). But early "installation" of the neural substrate subserving antisnake behavior produces developmental entrenchment of this system by making it part of the foundation upon which development of the rest of the phenotype depends (as in Schank and Wimsatt 2000). Entrenchment in turn increases the negative developmental consequences of modifying the system and thereby adds to its evolutionary persistence (Coss and Moore 2002).

Installation of the neural bases of antisnake behavior early in development has had a cascading development consequence that may well have modified the social component of antisnake behavior. As we described in our model episode, California ground squirrel pups express adult-like antisnake behavior early, before they have the venom-neutralizing capacity or behavioral skills to handle the risk posed by confronting a rattlesnake (Poran and Coss 1990; Coss 1991a; Owings 2002). Such precocious, risky confrontation of snakes not only places pups in greater danger but also complicates the protective mother's social task of managing her pups while dealing with snakes (see fig. 26.1). Precocious snake confrontation does not appear to involve nepotistic self-sacrifice for siblings; the antisnake behavior of these pups is more self-interested than nepotistic (Hersek and Owings 1994).

Finally, a major theme of this chapter, the sharing of components by different systems, also appears to have had evolutionary consequences. Another important part of California ground squirrel defense against rattlesnakes is the venom-neutralizing system, a system that shows much greater evidence of adaptive intraspecies variation than the behavioral defense system (Poran et al. 1987; Coss 1999).

This indicates that the venom-neutralizing system is more evolutionarily labile than the behavioral system, a difference that may be due in part to differential sharing of components with other systems. The venom-neutralizing system is dedicated to that function (Biardi et al. 2000), whereas the behavior system shares many parts with other functional systems, especially social systems, as discussed in this chapter. Continuing selection associated with those social functions has the potential to maintain a multiplexed system like the behavioral antisnake system even when selection associated with the antisnake functions has been relaxed for thousands of generations (Hodgkin 1998; Coss 1999). This idea—that dedicated systems disintegrate more rapidly than multiplex systems—seems more applicable to these results than the idea that this differential persistence is generated by differential maintenance costs. From the perspective of differential cost, the behavioral system should disintegrate more rapidly because it is far more complex and thus presumably more costly.

Questions for Future Research

The observations and ideas discussed in this chapter raise some important questions for future research. First, we have seen that shared components can cause one system to limit useful change in another. What other kinds of connections between organismic systems can have that effect? Means-end relations are another obvious candidate, but are there additional ones? Next, the kinds of factors that we have discussed also indicate that adaptive modification in organismic systems may be limited by the need for the organism to remain functionally coherent. Given this constraint, how is it that complex organisms manage to adjust individual systems to new conditions? For evolution, learning, or any other form of change to be feasible, organismic systems need to be at least quasi-independent (Lewontin and Levins 1978; Hodgkin 1998). Such independence may be fostered by such processes as gene copying and the establishment of modular systems (Schank and Wimsatt 2000). These considerations help us to identify one of the most challenging current questions in the study of the adaptive adjustment of behavioral systems: how does adaptive adjustment of individual organismic systems coexist with functional coherence of whole organisms (Lewontin and Levins 1978)?

Summary

This chapter has explored connections and similarities between rodent social and antipredator behavior as well as the processes that generate them. We have relied heavily on the literature on ground-dwelling sciurids, emphasizing our own work on ground squirrels, but have also cited additional relevant rodent literature. In the first part of this chapter we explored the variety of avenues whereby social and antipredator systems are connected, including shared components at multiple organizational levels and means-end relations. Components shared by social and antipredator systems include the following: (1) shared motor patterns at the level of basic behavioral units; and at higher organizational levels, (2) the dual social and antipredator roles of the stress-response system, (3) shared perceptual cues with context-dependent meaning, and (4) shared utility of knowledge of the spatial layout and nature of the resources in the animal's surroundings. With regard to means-end relations between social and antipredator systems, we discussed the familiar fact that features of social behavior and organization are means (mechanisms) for dealing with the threat of predation. We then examined how social and antipredator systems can converge in form when their participants share certain properties of their relations, specifically parity in the threat posed by each participant and extended occupation of the same vicinity. These properties are more typical of social relationships, but are also found in certain predator-prey relationships, such as that between ground squirrels and rattlesnakes. At least two advantages can be derived from identifying such connections and similarities between social and antipredator systems. We are reminded that systems that are distinct functionally may share quite a bit of causal common ground. And we can gain a better understanding of each type of system by appealing to insights available from the other.

We completed this chapter by exploring the proximate and ultimate processes that generate connections and similarities between systems. Key themes of this section were that (1) organisms undergo modification in multiple time frames, both proximate (including immediate causation and development) and ultimate (such as current effects of selection and evolutionary persistence of traits), and (2) organisms are active participants in all such changes, rather than simply passive responders to environmental "pressures" (Lewontin 2001a, 2001b). This idea, that organisms participate actively in the changes they undergo, is founded on three kinds of observations: (1) animals capitalize on the resources available from their environments and their own structure by expropriating components from existing systems for use in new functional contexts; (2) the way systems currently work generates constraints and opportunities for future change, in both proximate and ultimate time frames; and (3) organismic systems not only need to meet environmental demands, but also must be compatible with the other systems that comprise the organism.

Chapter 27 The Evolution of Alarm Communication in Rodents: Structure, Function, and the Puzzle of Apparently Altruistic Calling

Daniel T. Blumstein

O N A SUMMER DAY a number of years ago I looked down from an arête high up on the Creststone Needle, a peak in southwestern Colorado, and watched, with awe, two golden eagles (*Aquila chrysaetos*) cooperatively hunting in a meadow filled with yellow-bellied marmots (*Marmota flaviventris*). One eagle circled slowly above the meadow, which then erupted in a cacophony of loud chirps that radiated up the rock face. Marmots scampered to their burrows and reared up to attention. Meanwhile the second eagle flew low, and using the contours of the glacial moraines as cover, tried to attack the marmots, who were focused on its companion. That day the marmots were lucky and, after several sorties, the eagles flew off. Why were these chirps given? Did they repel the eagles? How could such conspicuous signals evolve? What might they mean? This chapter discusses the adaptive significance of alarm signals in rodents. I focus mostly on airborne vocal signals, but also mention substrate-born seismic alarm signals such as foot drumming.

When alarmed by predators, individuals of many species emit loud vocalizations known as *alarm calls* (Klump and Shalter 1984). Calls may be directed to conspecifics to warn them about the presence of a predator (Sherman 1977; Blumstein and Armitage 1997a), or to create pandemonium (Neill and Cullen 1974; Sherman 1985), during which time the caller may escape. Calls that function in these contexts should occur in social species. Calls may also be directed to the predator and may function to discourage pursuit (Hasson 1991), or perhaps to attract other predators —which would create competition, or predation on one predator by another, thus allowing the prey to escape (Högstedt 1983).

Snake-elicited foot-drumming by banner-tailed kangaroo rats (*Dipodomys spectabilis*) is a pursuit deterrent signal that informs the snake that it has been detected (Randall and Matocq 1997). Similarly, California ground squirrels (*Spermophilus beecheyi*; Owings and Coss 1977), black-tailed prairie dogs (*Cynomys ludovicianus*; Loughry 1988; Owings and Loughry 1985), and the Formosan squirrel (*Callosciurus erythraeus thaiwanensis*; Tamura 1989) directly mob snakes, and their mobbing is associated with both vocal and visual displays. In these cases, animals obtain phenotypic (self-preserving) benefits from producing alarms. Such behavior requires no complex explanation. However, when signals are directed toward conspecifics, the very act of signaling may also alert the predator to the caller's presence. Thus, explaining why animals emit potentially costly alarm calls to help others has been a topic of considerable interest for some time (Maynard Smith 1965; Charnov and Krebs 1975; Sherman 1977; Blumstein et al. 1997).

The structure and function of alarm signals are interrelated. For instance, we expect signals that are directed to a predator to be obvious. Marler (1955) argued that mobbing calls of songbirds illustrate this in that they are broadband, rapidly repeated sounds that are easy to localize. In contrast, alarm calls of songbirds that are elicited by aerial predators are difficult to localize because they have a relatively narrow bandwidth and fade in and out (Marler 1955). Thus, a complementary line of research seeks to understand the adaptive significance of alarm signal structure.

Studies of rodents have been influential in developing a better understanding of both aspects of alarm communication.

Which Rodents Produce Alarm Calls?

First, an apology. Because most rodents are nocturnal, solitary, and semifossorial, they are generally difficult to study. To properly categorize a vocalization as an alarm call one must observe an individual interacting with a predator. For most nocturnal species, this has not been done. Some species are described as producing whines or squeals when held (e.g., Watts 1975; Verts and Kirkland 1988), but this by itself is not evidence of an alarm call. Thus any review of alarm calling in rodents is unavoidably biased towards diurnal, terrestrial, and social species.

Functionally, alarm calls would be most valuable to diurnal, social, or colonial species. Alarm calls are long-distance signals. If calling increases predation risk then such signals should be produced only if calls carry relatively long distances, callers can accurately assess their own vulnerability, and callers benefit from communicating alarm to someone. Calling in the dark, when a caller might not be able to evaluate risk or track predators accurately, or calling underground when sounds attenuate quickly, could expose callers to excessive risk with little benefit and might therefore be disfavored.

Of course, there are exceptions. For example, the plains viscacha (*Lagostomus maximus*) is a highly social, nocturnal species that has been referred to as especially 'loquacious' (Hudson 1872). It has a rich repertoire of vocalizations, including two types of alarm calls (Branch 1993). Moreover, naked mole-rats (*Heterocephalus glaber*), which are totally subterranean, have at least six different calls that are associated with predator avoidance or colony defense (Pepper et al. 1991). Although long-distance seismic communication is common in fossorial mammals (Francescoli 2000), these signals are more commonly used to communicate territorial ownership and dominance or submission rather than to signal alarm (Francescoli 2000; Randall 2001).

In fact, alarm calling seems to be most common in diurnal rodents: it has been reported in twenty rodent families, and has probably evolved multiple independent times (table 27.1). Ultrasonic alarm calls have been recently reported in laboratory rats (*Rattus norvegicus;* Brudzynski 2001), but given the lack of comparative data, it is difficult to know how common these signals are in other species. Moreover, because ultrasonic signals attenuate quickly (Bradbury and Vehrencamp 1998), the active space of these alarm calls must be relatively small. Alarm calling has been best studied in the sciurid rodents (particularly in ground squirrels, prairie dogs, and marmots). Tree squirrels in several genera give alarm calls in addition to territorial calls. Some social gerbils, and muroid rodents—voles, bamboo rats, and whistling rats – also give alarm calls. Brush-tailed porcupines reportedly shake their quills and stomp their feet in alarm, cane rats and kangaroo rats, jerboas, and gerbils foot drum in alarm, and beavers slap their tail in alarm, but none of these mammals has been reported to produce alarm vocalizations. Alarm calling has also been reported in many South American hystricognath rodents (Eisenberg 1974), but the details have been little studied. It would be particularly rewarding to do so because these animals represent a radiation of complex sociality (Ebensperger 1998c; Ebensperger and Cofré 2001) that is phylogenetically independent from the better-studied sciurid rodents. Thus further study can evaluate the generality of adaptive hypotheses developed in convergent social systems.

What Are the Costs of Calling or Responding to Calls?

Understanding the adaptive significance of alarm calling has often focused on investigating its costs. Like other behaviors, if calling has no costs, it is not difficult to envision its evolution. Calling may have three types of fitness costs: energy, opportunity, or predation.

No studies have been conducted on the energetic costs of alarm calling in any species of rodent. Because individual alarm calls often are brief (< 5 sec in duration), energetic costs of producing a single utterance are probably trivial. However, animals often engage in tonic bouts of alarm calling whereby calls are repeated over time (Schleidt 1973; Owings and Hennessy 1984). Tonic calls have been reported in several ground squirrels (e.g., Balph and Balph 1966; Leger et al. 1984; Loughry and McDonough 1988), prairie dogs (e.g., Smith et al. 1977), marmots (e.g., Waring 1966; Heard 1977; Barash 1989; Blumstein and Armitage 1997a; Blumstein 1999a), and tree squirrels (Emmons 1978). In some cases, tonic communication persists for long periods after a predator has apparently left the area (Owings and Hennessy 1984; Loughry and McDonough 1988). The energetic costs of these bouts of calling are likely to be slightly greater than emitting a single call.

Opportunity costs—the costs of not engaging in other important behaviors—are experienced by both the signaler and the receiver. From the signaler's perspective, alarm calling seems to preclude foraging and engaging in activities other than vigilance. While analyses of time budgets may be used to contrast the opportunity costs of calling (i.e., the cost of not engaging in an alternative behavior), the link between opportunity costs and fitness is unstudied. From the receiver's perspective, responding to calls modifies current

Table 27.1 Genera in which acoustic alarm communication has been reported

Family	Genus	Common name	Reference	Alarm calls	Other alarm signals[a]
Sciuridae	Aethosciurus	Tree squirrels	Emmons 19781		
	Ammospermophilus	Ground squirrels	Bolles 1988	yes	
	Callosciurus	Tree squirrels	Tamura and Yong 1993	yes	
	Citellus	Ground squirrels	Nikolskii 1979	yes	
	Cynomys	Prairie dogs	Blumstein and Armitage 1997b; Hoogland 1995	yes	yes
	Epixerus	Tree squirrels	Emmons 1978	yes	
	Eutamias	Chipmunks	Smith 1978	yes	
	Euxerus	Ground squirrels	Haltenorth and Diller 1980	yes	
	Funambulus	Palm squirrel	Roberts 1977	yes	
	Funisciurus	Tree squirrels	Emmons 1978	yes	
	Heliosciurus	Tree squirrels	Emmons 1978	yes	
	Marmota	Marmots	Blumstein and Armitage 1997b; Roberts 1977	yes	yes
	Microsciurus	Tree squirrels	Emmons 1997	yes	
	Myosciurus	Tree squirrels	Emmons 1978	yes	
	Paraxerus	Bush squirrels	de Graaff 1981	yes	
	Petinomys	Flying squirrels	Medway 1978	yes	
	Protoxerus	Tree squirrels	Emmons 1978	yes	
	Ratufa	Giant squirrels	Nowak 1991	yes	
	Rhinosciurus	Tree squirrels	Medway 1978	yes	
	Sciurus	Tree squirrels	Lishak 1984; Farentinos 1974	yes	yes
	Spermophilus	Ground squirrels	Blumstein and Armitage 1997b	yes	
	Tamias	Chipmunks	Weary and Kramer 1995	yes	
	Tamiasciurus	Tree squirrels	Greene and Meagher 1998	yes	
	Xerus	Ground squirrels	Haltenorth and Diller 1980	yes	
Castoridae	Castor	Beavers	Hodgdon and Larson 1973		yes
Dipodidae	Jaculus	Jerboas	Randall 1994		yes
Muridae	Clethrionomys	Red-backed voles	Nowak 1991	yes	
	Dicrostonyx	Lemming	Brooks and Banks 1973	yes	
	Gerbillurus	Hairy-footed gerbils	Dempster et al. 1998		yes
	Lemmus	Lemming	Krebs 1984	yes	
	Meriones	Gerbils	Roberts 1977	yes	yes
	Microtus	Voles	Youngman 1975; Wolff 1980b; Nikolskii and Sukhanova 1992	yes	
	Neotoma	Woodrats	Randall 1994		yes
	Onychomys	Grasshopper mice	McCarty 1978	yes	
	Parotomys	Whistling rats	de Graaff 1981	yes	
	Peromyscus	Deer mice	Lackey et al. 1985; Johnson and Armstrong 1987	yes	yes
	Praomys	Multi-mammate mice	de Graaff 1981	yes	
	Rhizomys	Bamboo rats	Medway 1978	yes	
	Rhombomys	Gerbils	Randall et al. 2000; Randall and Rogovin 2002	yes	yes
	Spalax	Lesser mole rats	van der Brink 1968	yes	
Myoxidae	Dryomys	Dormice	Roberts 1977	yes	
Geomyidae	Dipodomys	Kangaroo rats	Randall 1994		yes
	Microdipodops	Kangaroo mice	O'Farrell and Blaustein 1974	yes	
Pedetidae	Pedetes	Spring hare	Haltenorth and Diller 1980	yes	
Ctenodactylidae	Ctenodactylus	Gundis	Haltenorth and Diller 1980	yes	
	Felovia		Nowak 1991	yes	yes
	Massouteria		Nowak 1991	yes	yes
	Pectinator	Pectinators	Nowak 1991	yes	
Hystricidae	Atherurus = Hystrix	Brush-tailed porcupines	Haltenorth and Diller 1980; Roberts 1977	yes	yes
Petromuridae	Petromus	Dassie rats	Nowak 1991	yes	
Thryonomyidae	Thryonomys	Cane rats	Haltenorth and Diller 1980; de Graaff 1981	yes	yes

(continued)

Table 27.1 (continued)

Family	Genus	Common name	Reference	Alarm calls	Other alarm signals[a]
Bathyergidae	*Heterocephalus*	Naked mole rats	Pepper et al. 1991	yes	
Agoutidae	*Agouti*	Pacas	Eisenberg 1974	yes	yes
	Dasyprocta	Agouti	Emmons 1997	yes	yes
	Myoprocta	Acouchy	Emmons 1997	yes	yes
Dinomyidae	*Dinomys*	Pacarana	Eisenberg 1974	yes	yes
Caviidae	*Cavia*	Guinea pigs, cavies	Eisenberg 1974	yes	
	Dolichotis = Pediolagus	Maras	Eisenberg 1974	yes	yes
	Galea	Cuis	Eisenberg 1974	yes	yes
	Microcavia	Cavies	Eisenberg 1974	yes	yes
Hydrochaeridae	*Hydrochaeris*	Capybara	Emmons 1997	yes	
Octodontidae	*Ctenomys*	Tucu-tucus	Eisenberg 1974	yes	
	Octodon	Degus	Eisenberg 1974	yes	yes
	Octodontomys	Long-tailed octodons	Eisenberg 1974	yes	
	Spalacopus	Cururos	Eisenberg 1974	yes	yes
Echimyidae	*Hoplomys*	Armored rat	Emmons 1997	yes	
	Dactylomys	Bamboo rats	Emmons 1997	yes	
	Kannabateomys	Southern bamboo rats	Redford and Eisenberg 1992	yes	
	Proechimys	Spiny rats	Emmons 1997	yes	yes
Capromyidae	*Capromys*	Hutias	Eisenberg 1974	yes	yes
	Geocapromys	Ground hutias	Eisenberg 1974	yes	
	Plagiodontia	Hispaniolan hutias	Eisenberg 1974		yes
Chinchillidae	*Chinchilla*	Chinchillas	Eisenberg 1974	yes	
	Lagidium	Mountain viscachas	Eisenberg 1974	yes	yes
	Lagostomus	Vizcachas	Eisenberg 1974	yes	yes

NOTES: This summary is inevitably incomplete and has a number of intrinsic biases (see text). Nonetheless, it does illustrate that alarm communication has been reported in 20 of the 53 families of rodents.
[a]Other signals include tooth-chattering, quill-shaking, tail-slapping, foot-thumping.

behavior (Baack and Switzer 2000). Typically, individuals immediately increase vigilance but, over time, receivers habituate to tonic signals (i.e., individuals assess that the danger has passed and return to what they were doing previously: Loughry and McDonough 1988; Nikolskii 2000; Hare and Atkins 2001; Blumstein and Daniel 2004). Again, the fitness consequences of this opportunity cost have not been investigated in any species of rodent. However, as long as there is some opportunity cost, there is a selective pressure on receivers to evaluate the reliability of callers.

If a caller calls when no predator is present, it is an unreliable informant. Caller reliability can be evaluated in two ways: receivers could either assess characteristics of reliable and unreliable *classes* of callers, or they could discriminate among *individuals*. For instance, if calls from juveniles were on average less reliable than calls from adults, and if juveniles had acoustically distinctive calls (as has been demonstrated in California ground squirrels; Hanson and Coss 2001b) and steppe marmots (Nesterova 1996), then receivers might "de-value" the calls from juveniles.

Individually distinctive acoustic signals do occur in rodents. For example, banner-tailed kangaroo rats (*Dipodomys spectabilis*) have individually distinctive foot-

drumming signatures that are used as territorial advertisements (Randall 1989a). The postcopulatory chirps of male Belding's ground squirrels are individually distinctive (Leger et al. 1984). However, individually distinctive alarm calls are relatively unstudied. Juvenile Richardson's ground squirrels (*Spermophilus richardsonii*) were the first rodent (Hare 1998b), and the second mammal (the first being vervet monkeys: Cheney and Seyfarth 1980) in which discrimination among individual alarm callers has been inferred. Hare and Atkins (2001) selectively manipulated the reliability of Richardson's ground squirrel callers by either pairing playbacks of calls from an individual with the appearance of a stuffed badger (thus creating "reliable" callers) or broadcasting the calls without a badger present (thus creating "unreliable" callers). When the calls from reliable or unreliable individuals were later played back, reliable calls elicited a higher level of response.

Subsequently, using a habituation-recovery playback design (Evans 1997), yearling and adult yellow-bellied marmots discriminated among individuals as well as some age-sex classes (Blumstein and Daniel 2004). In this study, marmots were first habituated, with repeated playback, to different exemplars of calls from an individual that was not a member of their social group. Subjects were then "probed" with either a novel call from the same individual or a novel call from a different individual nongroup member. Marmots increased vigilance and suppressed foraging in response to the call from the novel individual. Additional playbacks of calls from different age-sex classes demonstrated that these marmots were particularly responsive to calls from young (Blumstein and Daniel 2004), which is interesting because calls from young were initially hypothesized to be less reliable than calls from adults. In contrast, calls from young (which are demonstrably less reliable) were less evocative in both California ground squirrels (Hanson and Coss 1997, 2001b) and steppe marmots (*Marmota bobak*; Nesterova 1996). For yellow-bellied marmots, something other than reliability must favor individual discrimination abilities; an alternative nepotistic explanation is discussed as follows.

If, by calling, individuals exposed themselves to a greater risk of predation than noncallers, then calling behavior would be a phenotypically altruistic behavior (Alexander 1974). How such behaviors are maintained by natural selection is an interesting puzzle.

Is alarm calling in fact a risky behavior? Unfortunately, evidence for predation costs of calling is difficult to obtain. Most people who study alarm communication use these vocalizations to help locate individual callers (e.g., Gurnell 1987; Barash 1989). An obvious inference is that predators can do this as well. However, predation events are rare and hard to observe, and unlike studies in birds (e.g., Klump et al. 1986; Wood et al. 2000), there have been no experimental studies focusing on predator's responses to sciurid alarm calls (Lima 2002).

However, there have been studies of predators' responses to foot thumps by banner-tailed kangaroo rats (Randall and Matcoq 1997). Randall and Matcoq reported that hungry snakes were attracted to territorial foot drumming whereas recently fed snakes were repelled by foot drumming. Since snakes did not differentiate between the antipredator and territorial foot drumming, foot drumming may be costly.

Sherman (1977) found that when a terrestrial predator appeared, Belding's ground squirrels emitting calls were tracked and killed more often than noncallers, whereas calling in response to an aerial predator enhanced an individual's likelihood of escape over that of noncallers (who probably were unaware of the raptor's presence; Sherman 1985). Other researchers also have observed diurnal sciurids being attacked and killed by predators (e.g., Armitage 1982; Barash 1989; Murie 1992), but I am aware of no studies other than Sherman's, which simultaneously compared the fate of callers with noncallers. For instance, in 18 years of fieldwork, Barash observed thirteen cases of predation, but in none of these cases was a calling animal observed to be killed (Barash 1989).

Belding's ground squirrels sit up in place and call when they detect a terrestrial predator, whereas they scurry for cover before or while calling when closely pursued by a rapidly moving (aerial) predator (Sherman 1985). Likewise, Columbian ground squirrels (*Spermophilus columbianus*) modify their calling behavior as a function of predation risk (MacWhirter 1992). When suddenly surprised by a simulated aerial predator attack (a flying disk thrown directly at them), individuals bolted into the nearest burrow. In contrast, upon sighting a distant flying disk, or a taxidermically mounted badger from a distance—individuals gave repeated calls, often while running to their burrow. In some other species individuals that call only do so after they have sought cover. For instance, great gerbils (*Rhombomys opimus*; Randall et al. 2000), black-tailed prairie dogs (Hoogland 1996b), and yellow-bellied marmots (Blumstein and Armitage 1997a) generally call from burrow entrances, while taiga voles (*Microtus xanthognathus*) may stop calling if a predator comes too close and, following release from capture, call only after they have reached safety (Wolff 1980b).

One reason that evidence may be equivocal about the cost of calling is that callers may also be communicating with the predator, and there may be variation between species in the value of such communication. Thus while the alarm calls of some ground squirrels (Sherman 1985) and marmots become more cryptic as risk increases (Blumstein 1995a; Blumstein and Arnold 1995), some other rodents call more, and at greater rates, as risk increases (e.g., Harris et al. 1983; Nikolskii and Nesterova 1989, 1990; Nikol'skii and Pereladova 1994; Nikol'skii et al. 1994; Nikol'skii 2000; Blumstein and Armitage 1997a; Randall and Rogovin 2002). For instance, yellow-bellied marmots produce more calls and calls at a faster rate as a human approaches them (Blumstein and Armitage 1997a). Making one's self more obvious as risk increases is consistent with the hypothesis that calls are directed to the predator. Thus, calls may simultaneously have a pursuit-deterrent function while they also communicate relative risk to conspecifics. Identifying the relative importance of both of these factors is a worthy goal for future research.

What Are the Benefits of Calling?

There are several possible solutions to the problem of phenotypically altruistic acts. I believe that we gain insight into the workings of evolution by decomposing inclusive fitness benefits into direct and indirect components (Brown 1987). Others (e.g., Hauber and Sherman 1998) question the use of "direct" and "indirect" to describe fitness gains through descendents or nondescendents because, once appropriately weighted by relatedness, the Hamiltonian logic (Hamilton 1964) of inclusive fitness is agnostic about its source. Nonetheless, by calling, individuals may warn descendent or nondescendent kin, or both. Paths to obtaining direct fitness include reciprocity and directly increasing the probability of their own survival by calling, or the survival of their descendent kin (Sherman 1977, 1985).

Individuals could conceivably engage in reciprocal calling (Trivers 1971), whereby individual A might call one time and individual B might call another time. Without defectors, such a strategy might explain costly alarm calling. All such reciprocal arrangements rely on individual recognition and memory (Wilkinson 2002). Some ground squirrels and marmots have such abilities (Hare 1998b; Blumstein and Daniel 2004; Blumstein et al. 2004). Using olfactory cues, Belding's ground squirrels can remember individuals for at least 9 months (Mateo and Johnston 2000). However, there is no evidence from any rodent that callers "take turns," or that when surrounded by unreliable callers, other individuals cease calling. Moreover, alarm calls are unlikely to be reciprocal because they are broadcast widely. This means that eavesdropping "cheaters" can hear and benefit from calls but not take their turn at calling. Moreover, there is no way for a caller to select its audience so as to not warn cheaters if calls are loud and have a large active space. Reciprocity only works when there is a direct transfer of benefits from individual A to B and vice versa; if eavesdropping individuals C, D, and E also benefit, reciprocity is destabilized. (I thank Paul Sherman for clearly articulating this important point.)

If callers are in fact communicating to predators, then calling should reduce individuals' predation risk. Differentiating the degree to which callers are communicating to the predator or to conspecifics is difficult. Imagine a coyote or a mountain lion walking through a colony of prairie dogs or plains viscachas. As the predator passes through, multiple individuals may call (e.g., Branch 1993; Hoogland 1995). Calls evoke escape and heightened vigilance in conspecifics, and the predator walks on and leaves the colony. Is each caller calling to encourage the predator to move on? Is this a form of collective defense? Or, because individuals may be in different social groups, could each caller be calling to warn their family members? In this case we would see multiple callers, because many individuals have kin nearby.

Callers may obtain indirect fitness benefits by increasing the survival of collateral kin. There is some controversy over the relative importance of warning descendent versus collateral kin for explaining the adaptive significance of alarm calling. Sherman (1977) and Dunford (1977a) independently reported that by calling, individual Belding's and round-tailed ground squirrels (*Spermophilus tereticaudus*) respectively, were alerting descendent and nondescendent kin. Callers therefore received nepotistic fitness benefits from calling. Calling to increase indirect fitness has subsequently been reported to occur in chipmunks (Smith 1978; Burke da Silva et al. 2002), prairie dogs (Hoogland 1995, 1996a), as well as in several other ground squirrels (e.g., Schwagmeyer 1980; Davis 1984a; MacWhirter 1992). Sherman's (1977) study quantified the frequency of calling when animals were surrounded by different audiences (*sensu* Gyger 1990), but many other studies did not, and evidence for kin-selected benefits from calling often is based on a caller being surrounded by relatives.

There have been several suggestions (Shields 1980; Blumstein et al. 1997; Blumstein and Armitage 1998a) that such evidence of kin-selection *sensu lato* fails to clarify the relative importance of indirect fitness in explaining the evolution of alarm-calling behavior. On one hand, fitness is fitness however it is obtained, and indirect fitness should not be considered a special type of fitness (Dawkins 1979; Sherman 1980b; Hauber and Sherman 1998). On the other hand, viewing calling as a behavior that functions solely to protect descendents (which may have evolved as a form of parental care) is different from hypothesizing that alarm calling behavior functions solely to protect nondescendents. Admittedly, most researchers do not make this strong dichotomy; rather, they point out that calling is nepotistic and then determine which relatives are beneficiaries. Sherman (1977), studying Belding's ground squirrels, and my colleagues and myself, studying yellow-bellied marmots, found that adult females with emergent (and vulnerable) young-of-the-year are the age/sex class most likely to call. In the ground squirrels, females with older offspring (or more collateral relatives around) called at higher frequencies than females with fewer nearby relatives. However, in the marmots, numbers of adult kin did not affect calling frequencies. These differences suggest that nepotism in the form of alarm calling extends to descendent and collateral kin in Belding's ground squirrels, but only to descendents in marmots. Sherman (1980a, 1981a) discussed how demography (dispersal and mortality) affect the limits of nepotism. Demographic differences between marmots and Belding's ground squirrels may affect the different limits of nepotism, as evidenced by alarm calling in these two species.

A recent experiment suggests that both male and female marmots pay attention to vulnerable young (Blumstein and Daniel 2004). Following experimental playbacks of alarm calls from different age/sex classes, yellow-bellied marmots suppressed foraging the most after hearing calls from young. We inferred from this that marmots are particularly attuned to the status of vulnerable young. Note, this finding is inconsistent with the hypothesis that young callers were less reliable, and that calls from young should therefore communicate less risk than calls from adults. This finding is consistent with Sherman's (1980a, 1981a) limits of nepotism framework because adult yellow-bellied marmots are likely to be surrounded primarily by their offspring and by offspring from female relatives. Compared to Belding's ground squirrels, yellow-bellied marmots may limit their nepotistic behavior toward offspring because they evolved in a patchier habitat and live in a matrilineal group structure. In contrast, ground squirrels live in relatively higher-density meadows and many more relatives are likely to be within earshot of an alarm call. These demographic differences may help explain interspecific variation in the evolution and adaptive utility of alarm communication.

How Does the Acoustic Environment Affect Communication?

All signals must be transmitted from the signaler to the receiver, during which time they may degrade (i.e., lose fidelity) and attenuate (i.e., lose amplitude—Bradbury and Vehrencamp 1998). It follows that the environment should favor certain types of vocalizations. Several predictions can be drawn from first principles about the structure of long-distance signals like rodent alarm calls.

First, forest-dwelling species should have lower frequency vocalizations than species in open habitats to maximize transmission distance, because low-frequency sounds travel around objects and attenuate less than high-frequency sounds. The fundamental and dominant frequencies of alarm calls of southern African tree squirrels (*Paraxerus* spp. and *Funisciurus congicus*) are as predicted: forest species have lower frequency calls than do savannah species (Viljoen 1983). Emmons (1978) studied nine species of West African rainforest squirrels and contrasted their vocalizations to temperate *Sciurus* and *Tamiasciurus* species living in more open habitats. She found that certain long-distance calls from rainforest species were lower in frequency than similar calls from temperate tree squirrels. Smith (1978) studied two species of *Tamiasciurus* tree squirrels and found that they also produced relatively low-frequency alarm calls. Perla and Slobodchikoff (2002) found that frequency components of Gunnison prairie dog (*Cynomys gunnisoni*) alarm calls varied seasonally in ways that were consistent with the hypothesis that calls were modified to be transmitted through different microhabitats, which themselves changed seasonally. And Le Roux et al. (2002) found that a forest-dwelling whistling rat (*Parotomys* sp.) had a lower-frequency alarm call than a sibling species living in more open habitat.

Second, low frequency calls are predicted in subterra-nean species, because of the rapid attenuation of high frequency sounds in earthen burrows. Studies of subterranean mammals generally (Francescoli 2000), and naked mole-rats particularly (Pepper et al. 1991; Judd and Sherman 1996), have shown that their alarm calls are indeed very low in frequency.

Third, dense forest habitat should select against rapid frequency modulation because rapidly paced calls would reverberate off trees and thus degrade. In the open we might expect selection to act against long pure tonal calls because they will be degraded by heat waves reflecting off the open ground. The antipredator vocalizations of antelope squirrels (*Ammospermophilus* spp.) vary with habitat, but not as predicted from first principles (Bolles 1988). Specifically, species in open desert habitats where we might expect selection against tonal calls have long-duration pure-toned trills. In contrast, those species in more closed, rocky/prairie habitats have shorter-duration harsh trills. Because habitat complexity and vertical relief might increase reverberation, selection should favor short and potentially redundant calls in such habitats. However, the opposite has been reported in two rodents. In Gunnison's prairie dogs (*Cynomys gunnisoni*), the number of syllables and the total call length are positively associated with habitat complexity (Slobodchikoff and Coast 1980). Populations in areas with more vegetative cover, rocks, and tree stumps emit longer calls and calls with more syllables than populations in more open country. Slobodchikoff and Coast (1980) suggested that these calls are longer and more complex in more structurally complex habitat, where callers might not be able to see other individuals, to ensure that kin are alerted to the presence of a predator. Nikol'skii (Nikol'skii 1974, 1984; Nikol'skii 1994; Nikol'skii et al. 2002) has found that marmot species (and populations) in habitats with greater relief have more rapidly paced alarm calls than species (and populations) in flatter terrain. If rapidly paced calls communicate greater risk (e.g., Blumstein and Armitage 1997a), then it is possible that habitat-specific perceptions of risk influence call structure. The relationship between habitat-specific predation risk and call complexity remains to be tested directly.

Studies on birds provide some support for the hypothesis that evolution has designed long-distance signals to maximize transmission through a species' habitat (e.g., Wiley 1991; Bradbury and Vehrencamp 1998). However, Daniel and Blumstein (1998) found no support for this acoustic adaptation hypothesis in marmots. While there was variation in how well marmot alarm calls were transmitted through habitats, and there was evidence that some habitats generally degraded calls more than other habitats, there was no statistical interaction between habitat and species. Thus, a species' own call was not transmitted best in its native habitat—an essential prediction of the acoustic adaptation hypothesis.

I found no support for the hypothesis that overall repertoire size is constrained by the acoustic transmission fidelity of the habitat. Blumstein (2003) broadcast and rerecorded pure tones through the habitats of eight marmot species, thus generating a metric of habitat transmission fidelity (Blumstein and Daniel 1997). No relationship occurred between habitat transmission fidelity and alarm call repertoire size. However, after removing variation in alarm call repertoire size explained by the acoustic environment, there was a relationship between social complexity and repertoire size (see also Blumstein and Armitage 1997b).

Taken together, we might generally expect the habitat to select for gross frequency characteristics of calls, whereas it might not have a direct effect on temporal characteristics or on microstructural differences. A recent comparative study (Blumstein and Turner 2005) drew similar conclusions for birdsong.

What Might Explain Variation in Call Microstructure?

Unlike birdsong (Catchpole and Slater 1995) or many of the calls of insects or anurans (Gerhardt and Huber 2002), alarm calls are not directly involved in species identification, territorial defense, or mate choice. And the diversity of alarm calls in rodents requires special explanation, especially because of the contrast to alarm calls in songbirds, which often are convergent (Marler 1955). What, other than gross habitat differences, might favor the calls of rodents to diverge?

Character displacement (Schulter 2002) has been suggested to be important among sympatric species. In three species of Townsend chipmunks (*Eutamius* spp.), alarm chirps were most distinctive and less variable in populations at species boundaries (Gannon and Lawlor 1989). Because character displacement results from resource competition, there should be a demonstrable cost to having less distinctive alarm calls in sympatry. While sympatric species may respond to each other's alarm calls (Blumstein and Armitage 1997a; Shriner 1998), it is likely that variation in reliability of heterospecific callers may select for divergence in sympatry. Specifically, if a small species has more predators than a larger species, the smaller species will be more likely to give alarm calls in situations that are not threatening to the larger species than vice versa. Thus, from the perspective of the larger species, calls from the smaller species are not reliable, but not vice versa. Selection within a species living in sympatry with alarm-calling heterospecifics might thus favor divergent calls. No data are currently available to evaluate this.

Genetic drift has been suggested to lead to call variation over time (Daniel and Blumstein 1998). Evidence of heritable genetic variation in call structure comes from studies of hybrids that have been reported to have calls of intermediate structure (Nikol'skiy et al. 1984; Nikol'skii and Starikov 1997), or structures that resemble one parent more than another (Koeppl et al. 1978). Sibling species have calls more similar to each other than more distant relatives (Hoffmann et al. 1979; Bibikow 1996; Nikol'skii 1996; Blumstein 1999a). Divergence of alarm calls may occur relatively rapidly once populations are isolated on islands or by glaciers (Nikol'skii et al. 1999). For instance, Nikolsky (1981) reported divergence in arctic ground squirrel alarm calls after 7,500 years of isolation on islands. Given the potential importance of drift, it is surprising that the alarm calls of geographically isolated populations of yellow-bellied marmots have not diverged (Blumstein and Armitage 1997a).

Finally, in at least one species (the yellow-bellied marmot), variation in temporal characteristics of calls seems to be important in communicating risk, while variation in the frequency structure of calls seems to be important for individual discrimination (Blumstein and Armitage 1997a). Selection for individual recognition systems can act on signalers, receivers, or both (Beecher and Stoddard 1990). For instance, if it is in the best interest of the signaler to indicate its identity, selection should favor signalers to produce distinctive calls. Such selection is likely to be common in territorial and nepotistic signaling systems. By contrast, there may be no particular benefit from producing individually specific variation, but there is a benefit to receivers for discriminating among callers. In this case, calls may not be distinctive, but receivers may nevertheless be able to discriminate among them.

Repertoire Size and the Evolution of Functional Reference

Human language is unquestionably unique relative to the diversity of nonhuman vocalizations (Hockett 1960; Pinker 1994). A comparative perspective allows us to gain novel insights into language evolution (Blumstein 1999b). A complementary line of research on alarm vocalizations in rodents has focused on the evolution of meaningful communication. While much has been written about avian repertoire size (Kroodsma 1982; Irwin 1990; Catchpole and Slater 1995), birdsong is hypothesized to only have one or two functions (mate choice and territory defense—Catchpole and Slater; 1995). Each song (or syllable) that a bird sings is not typically hypothesized to have a particular function per se. In contrast, alarm call variants of birds and mammals may in fact refer to external objects or events. Such functionally referential communication has been reported in fowl (Evans et al. 1993), in some nonhuman primates (e.g., Cheney and Seyfarth 1990; Zuberbühler 2000), and in a social mongoose (Manser 2001; Manser et al. 2001). Predator-specific, functionally referential calls have been reported in one study of Gunnison's prairie dogs (*Cynomys gunnisoni;* Slobodchikoff et al. 1991), but not in another (Fitzgerald and Lechleitner 1974), and in one study of alpine marmots (*M. marmota;* Lenti Boero 1992), but not another (Blumstein and Arnold 1995). Greene and Meagher (1998) reported that red squirrel (*Tamiasciurus hudsonicus*) alarm calls had a high degree of production specificity and were likely to be functionally referential. Functionally referential alarm calls may indicate specific types of predators (e.g., aerial or terrestrial), specific predatory species (e.g., snake, raptor, canid), or commands for recipients to follow (run away, stand alert, climb a tree). To study functional reference, two pieces of complementary evidence are required (Evans 1997).

First, there should be a strong association between a specific external object or event (e.g., the appearance of a coyote) and a particular call. This call should be different from calls elicited when, say, an eagle appears. Satisfying this condition means that calls have a high degree of "production specificity."

Second, calls should elicit unique behavioral responses. Communication can only be understood by studying the behavior of the signaler and the receiver; playback experiments help us understand meaning. Simply documenting variable alarm calls does not necessarily imply that individuals will respond differently to them (Blumstein 1995b). To demonstrate functional reference there must be predator-specific responses. Thus playback of a "coyote alarm call" or an "eagle alarm call" should evoke responses typically observed when the relevant predator is seen. If so, we can infer a high degree of response specificity.

Some support has been provided for the production specificity criterion, but less so for the response specificity criterion in rodent alarm communication. Greene and Meagher (1998) provide experimental evidence that red squirrels produced predator-class specific alarm calls. Slobodchikoff et al. (1991) and Ackers and Slobodchikoff (1999) simulated different predators by walking toward Gunnison's prairie dogs wearing different colored shirts. They reported that the animals modified the structure of their calls to potentially communicate information about the individual predator, as well as aspects of the size and shape of silhouette models of actual predators. Three species of Malaysian tree squirrels (*Callosciurus* spp.) reported to have a high degree of production specificity are also reported to vary their responses as a function of alarm call type (Tamura and Yong 1993). In none of these three cases were playback experiments conducted, so the degree to which calls alone can elicit unique responses is unknown.

Most ground squirrels produce two different types of alarm calls (Blumstein and Armitage 1997b). The first, a short whistle, is often elicited by aerial predators, while the second, a longer trill, is often elicited by terrestrial predators. Ground squirrels also have predator-specific response differences: the sudden appearance of a raptor causes them to run to the nearest burrow, whereas they do not necessarily return to the nearest burrow after discovering a weasel (e.g., Turner 1973; Sherman 1985). However, closer examination typically reveals that "aerial" calls are actually elicited in high-risk situations (e.g., Robinson 1981; Owings and Hennessy 1984; Leger et al. 1984; Sherman 1985) and the production specificity is not high. Thus rather than communicating predator type, calls are likely to communicate degree of risk, which may reflect the time an individual has to escape the predator (e.g., Leger et al. 1979, 1984; Blumstein and Armitage 1997a; Robinson 1981; Sherman 1985) or may encode information about distance to the predator (Burke da Silva et al. 1994; Blumstein 1995a). Even when there is some degree of production specificity, playback experiments typically lead to graded responses, which are more indicative of risk, rather than information about a specific type of predator (e.g., Blumstein and Armitage 1997a; Blumstein 1999b).

Does Lack of Functional Reference Limit Complex Communication?

A reasonable question emerges from the observation that variable repertoires are not necessarily functionally referential: does a limited "vocabulary" prevent meaningful communication? At one level this question reveals an anthropocentric bias. Because humans have language, we classify language-like communication as especially complex. However, if we have learned anything by studying biological diversity, it is that there are multiple ways to solve a problem.

Rodents illustrate some of the ways in which animals can dynamically communicate the degree of risk. First, animals communicate risk by varying the number of calls emitted, or the rate at which they call, as seen in great gerbils (Randall and Rogovin 2002), yellow-bellied (Blumstein and Armitage 1997a), steppe (Nikol'skii 2000), and alpine marmots (Hofer and Ingold 1984; Blumstein and Arnold 1995), tassel-eared squirrels (*Sciurus aberti;* Farentinos 1974), chipmunks (Weary and Kramer 1995), and California (Leger et al. 1979) and Richardson's ground squirrels (Warkentin et al. 2001). Second, individuals can vary how they "package" calls into multi-note units, as seen in golden marmots (Blumstein 1995a). Third, individuals can vary the duration of a nonreferential whistle, as seen with Brant's whistling rats (*Parotomys brantsii*). Whistling rats produce longer whistles in lower-risk situations (a distant human or snake) and shorter whistles in higher-risk situations, which are followed by disappearing into their burrows. Fourth, individuals can use different calls, as seen in yellow-bellied marmots (Blumstein and Armitage 1997a) and in the plains viscacha (Branch 1993). Or, combinations of calls may be used to dynamically communicate risk, as seen in Vancouver Island marmots (*Marmota vancouverensis;* Blumstein 1999a) and in great gerbils (Randall and Rogovin 2002). Fifth, call amplitude can communicate degree of risk, as seen in chipmunks (Weary and Kramer 1995), California (Leger et al. 1979) and Columbian ground squirrels (Harris et al. 1983), and yellow-bellied marmots (Blumstein and Armitage 1997a). Sixth, the existence of multiple callers rather than a single caller can communicate risk, as seen with chipmunks (Weary and Kramer 1995). Seventh, the duration of calling bouts may communicate degree of risk. For instance, snake-elicited antipredator behavior persists for longer periods of time when California ground squirrels are responding to a dangerous snake (large and warm) than to a less dangerous snake (small and cold; Swiasgood et al. 1999a, 1999b). Finally, some rodents have multichannel systems, such as the great gerbil (Randall et al. 2000) and the California ground squirrel (Owings and Hennessy 1984), which may have elements that both communicate risk to conspecifics and discourage attack by predators. Thus, while referential communication is a special case of alarm communication, it need not be viewed as the epitome of complex alarm communication. And, while complex communicative abilities may emerge from being able to produce different calls, they need not necessarily emerge from being able to produce different calls (Blumstein 1999b).

A Model for the Evolution of Alarm Communication in Rodents

Any model for the evolution of alarm communication must address three questions: (1) what factors influence whether a species produces alarm calls? (2) what is the function of alarm calling? and (3) what explains variation in call structure? I summarize the conclusions of this review schematically (fig. 27.1) and discuss them as follows.

Habitat, sociality, and behavior influence the evolution of alarm calling. Social, terrestrial, and diurnal species are those most likely to produce alarm calls, although there are some notable exceptions. Rodents produce alarm calls to increase personal, direct, and indirect fitness. We expect the degree of aggregation (sociality: e.g., Randall 1994; Randall 2001) and demography (Sherman 1980a, 1981a) to constrain the types of fitness benefits. For instance, solitary rodents, such as kangaroo rats (Randall and Matocq 1997), or a female ground squirrel with a snake in its reproduc-

tive burrow (Swaisgood, Owings, and Rowe 1999), produce alarm signals directed toward the predator to discourage it from hunting or to drive it away. By driving off a predator, individuals will save themselves as well as vulnerable offspring.

If solitary adult females are surrounded by mature offspring, or if animals live in more complex social groups formed by delayed dispersal and characterized by overlapping generations (Blumstein and Armitage 1998b, 1999), there exists the opportunity for animals to obtain indirect fitness benefits from calling. Enhancing this indirect fitness may be more important to some species than others, and not all species will have evolved alarm calling behavior the same way (e.g., Holmes 2001).

The evolution of call structure and repertoire size is influenced by a combination of environmental, social, and functional considerations. Available evidence suggests that a call's dominant frequency is influenced by the openness of the habitat; species living in closed, forested habitats have lower dominant frequencies than those in more open habitats. There remains, however, a need for studies to properly control for phylogeny and body size (e.g., Wiley 1991; Blumstein and Turner 2005) when testing for these effects. However, the acoustic environment seemingly has no other consistent influence on call structure. Interspecific variation may result from drift, although character displacement also is a possibility. In some cases, variation could result from advantages of communicating individual identity.

Functional considerations also influence the structure of alarm calls. More socially complex sciurid rodents emit more types of alarm calls (Blumstein and Armitage 1997b; Blumstein 2003) and there are indications of this in other taxa (e.g., naked mole-rats—Pepper et al. 1991). And while functionally referential communication is uncommon in rodents, modulating the number, rate, amplitude, and duration of alarm calls, using different calls or modalities, and manipulating call order are all ways rodents communicate degree of risk. Interestingly, in the species that are reported to have a high degree of production specificity (Gunnison's prairie dogs, red squirrels, and three species of Malaysian tree squirrels), complex and species-specific antipredator behavior is employed. Thus, the Macedonia and Evans' (1993) model for the evolution of functionally referential communication, which suggested that the need to communicate about different mutually exclusive escape strategies may have general, explanatory value for rodent alarm calls.

Diurnal, social rodents will continue to be an outstanding model system to study questions of the adaptive utility of alarm-calling behavior. New studies that test hypotheses generated from studies of sciurid rodents will increase our general understanding of factors responsible for the evolution and maintenance of alarm calling.

The evolution of alarm calling in rodents

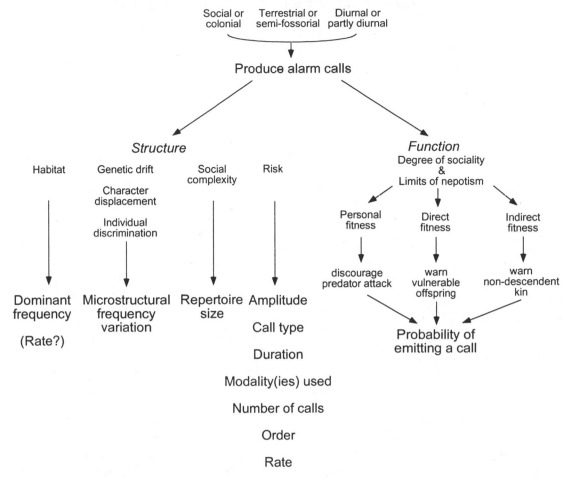

Figure 27.1 Summary of factors influencing the evolution of alarm calling in rodents. Alarm calling is most commonly reported in social or colonial species that are terrestrial or semifossorial, and are at least partly active by day. Various components of the structure of alarm calls are influenced by different factors. The dominant frequency of calls is higher in open habitats than closed habitats, and habitat complexity may influence calling rate via its effect on risk perception. Several factors influence microstructural frequency variation in calls. Social complexity influences repertoire size. Variation in risk influences the amplitude at which a call is emitted, call type, call duration, the modality or modalities used to communicate risk, the number of calls emitted, the rate at which calls are emitted, and the order in which calls are emitted. The probability of emitting an alarm call is a function of the benefits obtained, which are influenced by the degree of sociality. Highly social species, or species in which potential callers are surrounded by kin, may obtain indirect fitness by calling. Other species may obtain personal and direct fitness by calling.

Summary

Alarm calls are signals elicited by predators that may be directed to predators, most likely to discourage attack, or to conspecifics to warn them about the presence of a predator. Social, terrestrial, and diurnal species are those most likely to produce alarm calls, although there are some notable exceptions. The evolution of call structure and repertoire size is influenced by a combination of environmental, social, and functional considerations. More socially complex sciurid rodents emit more types of alarm calls. If solitary adult females are surrounded by mature offspring, or if animals live in more complex social groups formed by delayed dispersal and characterized by overlapping generations, there exists the opportunity for animals to obtain indirect fitness benefits from calling. Overall, rodents may receive both personal, direct, and indirect benefits from calling. Enhancing indirect fitness may be more important to some species than others, and not all species will have evolved alarm-calling behavior the same way. Diurnal social rodents will continue to be an outstanding model system to study questions of the adaptive utility of alarm calling behavior.

Chapter 28 Fear and the Foraging, Breeding, and Sociality of Rodents

Hannu Ylönen and Joel S. Brown

GOOD NIGHT, sleep well." After a good-night kiss, the parent leaves the child's room and turns off the lights. Immediately bogeymen, ghosts, and all manner of fearsome things crawl out of drawers, threaten from underneath the bed, and creep from behind the curtains. The child hides under its blankets and pillows, remains wide-eyed in the darkness, and waits—until merciful sleep saves the night. Actually, the child's responses to these more or less imaginary fears fit effectively within the "ecology of fear" (Brown et al. 1999). Hiding under or behind obstacles, remaining still, and being vigilant are among the most frequent adaptive responses to increased predation risk. Additionally, sociality may influence fear responses and their effectiveness. Hiding in the parents' bed might be an even better strategy for a frightened child. Sociality and social behaviors influence predation risk. And fear responses may in turn influence social behaviors such as breeding. Our objective in this chapter is to describe how the interaction of fear, sociality, breeding, and foraging produces adaptive and ecological feedbacks.

Predators have both lethal and nonlethal effects on their prey. The lethal effects manifest through the predator's functional response when some number or proportion of the prey population succumbs to predators (Taylor 1984). The nonlethal effects emerge as fear responses of the prey to their predators (Lima and Dill 1990). Sometimes the dramatic lethal aspects of predation obscure the drama of the nonlethal effects induced by the mere presence of predators. During the last decade, however, antipredatory responses of prey animals toward direct or indirect cues of a predator's presence have received increasing attention (reviews

by Lima and Dill 1990; Lima 1998; Norrdahl and Korpimäki 2000; Ylönen 2001; Owings and Coss chap. 26, this volume).

The ecology of fear examines the behavioral, population, and community consequences of the nonlethal effects of predators. Predation risk may reduce activity levels, shift activity from risky to safe places, alter mating behaviors and reproduction, increase levels of vigilance and apprehension, and alter intra- and interspecific competitive interactions (Brown et al. 2001). Furthermore, fear in response to predation risk may alter life-history characteristics such as growth rates, dispersal, and age at maturity (see Lima 1998 for review). These fear responses are adaptations that serve to maximize the prey's fitness in the face of predation risk. Predation risk is an activity cost (Brown 1988; Brown and Kotler 2004) that influences the costs and benefits of foraging, breeding, or engaging in other social activities.

In this chapter, our aim is to examine fear and its effects on the foraging, breeding, and social behavior of rodents. Based on our own research biases, we will often refer to boreal voles (*Microtus* and *Clethrionomys*) living under relatively dense vegetation, and desert rodents (e.g., *Gerbillus*) exposed to predation in more open environments (fig. 28.1). All of these rodents share risks from terrestrial predators (mammals and snakes) and avian predators (hawks and owls). Different predators pose different challenges and may present different cues to their rodent prey (Lima and Dill 1990). We begin with a brief description of boreal voles and desert rodent systems. We then examine predation risk as an activity cost of foraging, breeding, and social interac-

Figure 28.1 The field vole (*Microtus agrestis,* top), bank vole (*Clethrionomys glareolus,* middle), and sand gerbil (*Gerbillus pyramidum,* bottom) are model species in many studies on predator-prey interactions and antipredatory behavior in boreal, temperate, and desert rodent communities, respectively. Field vole photo by X. Lambin, bank vole by Ines Klemme, and gerbil by J. Brown.

tions such as territoriality, interference competition, and group vigilance. We discuss the implications of the cost of predation for (1) foraging in terms of space use and habitat selection, (2) timing and/or suppression of breeding, (3) social foraging in terms of vigilance, alarm calls, and various

forms of safety in numbers, and (4) ecological implications in terms of population dynamics and seasonality.

Boreal Voles and Desert Rodents: Why?

Boreal voles and desert rodents have been traditional subjects for studies on foraging ecology, breeding ecology, and predator-prey interactions. The "vole cycle" (the periodic fluctuations of boreal vole populations) has inspired strong debate ever since Elton's (1924) pioneering analysis. Current evidence implies a strong role for predation risk in affecting the nature of the fluctuations (e.g., Hanski and Henttonen 1996; Korpimäki and Krebs 1996; but see also Graham and Lambin 2002). In addition to detecting strong lethal effects of predators on boreal voles, research has identified potentially important effects of predators on foraging behaviors, space use, and reproduction (e.g., Ylönen 1994; L. Oksanen and Lundberg 1995; Norrdahl and Korpimäki 2000).

Deserts and semi-arid environments support unusually diverse assemblages of rodents. An assemblage often includes nocturnal granivores whose members include families with striking adaptations for hot, arid conditions (Heteromyidae and subfamilies of Muridae and Dipodidae), omnivores that consume plant material and insects (such genera as *Peromyscus* and *Onychomys*), nocturnal and diurnal herbivores/granivores (Sciuridae, Dipodidae, Octodontidae), and herbivores that live in groups (*Cynomys,* some *Spermophilus, Octodon, Mastomys*). These rodent communities often provide geographical replicates and natural experiments of coevolution and community structure. Desert rodent communities have taught us aspects of species interactions, competition, and resource use, and especially how these respond to habitat properties and resource renewal (e.g., Rosenzweig and Winakur 1969; Price 1978; Mitchell et al. 1990; Ovadia et al. 2001).

Both the lethal (Webster and Webster 1971) and non-lethal effects (Kotler 1984) of predators have obvious and important impacts on desert rodents—affecting foraging, habitat selection in time and space, and species coexistence (Brown et al. 1988; Longland and Price 1991; Kotler et al. 1992; Abramsky et al. 1996). The openness of desert habitats leaves rodents more exposed than habitats with denser vegetation cover, such as those occupied by voles. Yet research starting in the 1990s suggests major roles for predators and predation risk in voles as well (Jedrzejewski and Jedrzejewska 1990; Dickman 1992; Ylönen et al. 1992; Korpimäki et al. 1994; Bolbroe et al. 2000). During recent years, research approaches to desert rodents and to boreal voles have converged. A synergy emerges from studies of these two model systems.

Fear and Foraging

In deserts, it appears that most, if not all, small rodents end their days at the hands, claws, jaws, or talons of a predator (Brown and Harney 1993). In one study, Merriam's kangaroo rats (*Dipodomys merriami*) suffered one predation loss per 134 nights of activity (Behrends et al. 1986; Daly et al. 1992). Mortality came from snakes, owls, shrikes, and coyotes. The risks faced by rodents vary temporally and spatially. At small spatial scales, proximity to shrub cover seems critical to predation risk. All else being equal, most desert rodents favor the shrub microhabitat over the open (Brown and Lieberman 1973; Price 1978; Thompson 1982; Brown 1989; Hughes and Ward 1993; Brown et al.). Even the Indian crested porcupine (*Hystrix indica*), the largest species of desert rodent, prefers areas with high vegetation cover (Brown and Alkon 1990).

Voles appear to prefer areas of taller and denser vegetation to areas with less dense vegetation (Jacob and Brown 2000; Pusenius and Schmidt 2002). Similarly, perception of predation risk increases as thirteen-lined ground squirrels (*Spermophilus tridecemlineatus*) venture farther from their burrow (Thorson et al. 1998), as tree squirrels (*Sciurus carolinensis* and *S. niger*) forage farther from trees (Lima et al. 1985; Dill and Houtman 1989; Bowers et al. 1993; Brown and Morgan 1995), and as rodents such as white-footed mice (*Peromyscus leucopus*) and some voles (*Clethrionomys sp.*) inhabit margins of woods relative to forest interiors (Morris and Davidson 2000; Wolf and Batzli 2004). Australian house mice (*Mus musculus*) inhabiting an open agricultural habitat consisting of grain fields, pastures, and nonmanaged fence lines, avoided pastures and preferred dense crop fields until harvest. Following the grain harvest and the removal of cover from the cropland, the mice switched preferences and activity to the fence-line habitat, the only remaining habitat with protective cover (Ylönen et al. 2002). Regardless of scale, rodents respond to spatial heterogeneity in predation risk by biasing activity toward safer areas, by demanding a higher harvest rate from risky areas, and by using food patches less thoroughly in risky relative to safe habitats.

Presence of predators: Do rodents learn fear from prior experience?

Temporal patterns of risk influence foraging behavior. Abramsky et al. (2002) flew trained barn owls for short durations over rodent enclosures at a sand-dune habitat in the Negev Desert, Israel. In response, gerbils (*Gerbillus sp.*) greatly reduced foraging activity during the period of flights, but by the end of the night total gerbil activity was the same on grids with and without trained-owl flights. In response to owl flights, the gerbils simply rearranged their nightly activity to avoid periods when owls were active. Aviary experiments have shown how desert rodents can respond nightly to the presence or absence of predators such as owls (Brown et al. 1988; Kotler et al. 1991; Longland and Price 1991). One species of gerbil (*Gerbillus pyramidum*) returns to baseline levels of activity within one night of the removal of owls from the aviary, whereas a smaller, more vulnerable species (*G. andersoni*) requires up to three days to return to baseline activity levels following the "trauma" of a night with owls (Kotler 1997).

Circadian rhythms and predation risk

Besides the irregular arrival and departure of predators from an area, rodents experience three important scales of predation risk: day versus night, changes in moon phase, and seasonal changes in predator abundance. It is interesting that most rodents opt for either complete nocturnality or diurnality (Bartness and Albers 2000). Given that their food resources do not seem to vary strongly between night and day, it seems most likely that rodents are avoiding climatic or predation stresses by adopting diurnal or nocturnal activity. Nocturnal rodents tend to seek cover, avoid being detected by predators, lack complex social structures such as colonies, and do not seem to use alarm calls. In contrast, diurnal rodents (especially sciurids) rely heavily on vision to detect predators, prefer open sight lines, and provide examples of sociality. In colonies, alarm calls and other social interactions assist with detecting and avoiding predators (Blumstein, chap. 27 this volume).

Voles and spiny mice (Elvert et al. 1999) provide two prominent examples of rodents that break the "rule" by having 24-hour patterns of foraging activity. Despite considerable variation, voles appear to be "polyphasic," with a 2- to 3-hour cycle of rest and foraging (Madison 1985; Ylönen 1988; Halle 2000b). Polyphasic activity may be an adaptation to herbivory and a relatively low-quality diet. Such activity patterns are common for large ungulates (e.g., black rhinoceros, *Diceros bicornis;* Kiwia 1989).

Predation risk sculpts the polyphasic activity pattern in a manner that makes boreal vole activity more like that of nocturnal desert rodents (Halle 2000a). Voles become more nocturnal and less synchronous in their activity times in fragmented habitats with microhabitats of cover and open. Furthermore, their use of the open microhabitat becomes almost entirely nocturnal. Open microhabitats may present rodents with intense exposure to predators. Voles integrate habitat structure, predation risk, and the activity of other voles into their own foraging decisions. While this is merely speculation, a polyphasic activity pattern may influence their predator's hunting success by overdispersing the vole population's temporal distribution of activity. A uniform level of activity across time may thwart the predators by precluding predictable periods of very high and very low activity. A further reduction in activity synchrony in fragmented habitats may further overdisperse activity. Alternatively, other factors may primarily influence the circadian rhythm of microtines, with the predators adjusting their behavior around that of the voles (Halle 1993).

Activity and moonphase

Among nocturnal rodents a strong pattern of moonlight avoidance has been documented experimentally (Kotler 1984; Brown et al. 1988) and via field data (Lockard and Owings 1974; Price et al. 1984; Rosenzweig 1974). White-footed mice, in an experiment, adjusted their behavior to both moon phase and the crunchiness of their leaf substrate (Fitzgerald and Wolff 1988). Moonlight seems to favor owls over their rodent prey. The pattern of mortality, however, may be complex. Radio-tagged Merriam's kangaroo rats suffered higher predation mortality rates from typically diurnal predators (e.g., shrikes) during full moon nights but higher predation mortality from typically nocturnal predators during the new moon (Daly et al. 1992). This seemingly odd pattern of predation risk may be a consequence of the foraging game between rodents and their predators (Brown et al. 2001).

With the full moon, kangaroo rats may reduce activity, remain in safer microhabitats, and become more crepuscular and thus more exposed to diurnal predators. Bannertail kangaroo rats (*Dipodomys spectabilis*) show more activity at new than full moon, range farther from their burrows, and show equal or more crepuscular activity on full moons compared to moons (Rosenzweig 1974). In aviary experiments with owls and heteromyid rodents, more rodents were captured on dark nights than nights with artificial illumination (Kotler et al. 1988). This apparently anomalous result can be explained through activity: rodents were much less active on nights with illumination than without. When active, however, rodents suffered much higher mortality rates per unit of foraging activity on nights with illumination. Gerbils at a Negev Desert site exhibit the greatest apprehension toward predators early in the night and near full moon (Kotler et al. 2002). Interestingly, apprehension increases again near dawn, in accord with Daly et al.'s (2000) suggestion that crepuscular activity incurs risk from shrikes.

Experiments with giving-up densities

Giving-up densities in natural or experimental food patches provide an effective tool for measuring spatial and temporal variability in rodents' perceived predation risk (fig. 28.2; see also Brown and Kotler 2004). When presented with a depletable food patch, a rodent should forage less thor-

Figure 28.2 Seed trays such as this are used for experimentally measuring giving-up densities (GUDs) in response to predation risk. The tray and burrow are on semistabilized sand dunes at Bir Asluj, Negev Desert, Israel. One night of foraging at this patch revealed spoor from the Egyptian sand gerbil, *Gerbillus pyramidum*, Allenby's gerbil, *G. a. allenbyi*, and the common jerboa, *Jaculus jaculus*. Photo by Joel Brown.

oughly and hence leave more food behind (the giving-up density, or GUD) in risky food patches than safe food patches (Brown 1988; Brown et al. 1992). All else being equal, a forager should have a higher GUD as predation risk increases, a higher GUD as its energy reserves increase (asset protection principle Clark 1994), and a higher GUD as the marginal value of the resource declines.

Giving-up densities track a forager's optimal behavior in a risky environment. Most desert rodents exhibit higher GUDs in the open than in shrub microhabitat (Kotler et al. 1991); voles tend to exhibit increasing GUDs as vegetation cover declines (Jacob and Brown 2000); GUDs increase with moonlight or artificial illumination (Brown et al. 1988; Kotler et al. 1991); and GUDs increase with the presence of predators (Brown et al. 1988). Giving-up densities show that predation risk imposes an additional foraging cost on rodents, and this cost can be substantial. For desert rodents the cost of predation risk appears to be 3 to 8 times higher than the metabolic cost of foraging (Brown et al. 1994b).

Rodents must balance trade-offs of food and safety among their different activities. Lima and Bednekoff (1999) observed that many foraging experiments use an on/off system of predation risk. They modeled how antipredatory responses should be affected by previous experience with predators, and with the temporal scale of changes in predation risk. According to their "predation risk allocation hypothesis," prey should trade off their feeding effort and vigilance in relation to temporal variation of predation risk. Feeding efforts should be lowest during short periods of high risk that punctuate longer periods of low risk. Conversely, foraging efforts should be highest during short periods of low risk that punctuate longer periods of high risk.

In terms of the model (Lima and Bednekoff 1999), studies with boreal voles and desert rodents have yielded mixed results. Koivisto and Pusenius (2003) found that field voles (*Microtus agrestis*) in the laboratory responded as predicted to temporal variation in the presence of a least weasel (*Mustela nivalis*; fig. 28.3). Sundell et al. (2004) exposed free-ranging bank voles in large enclosures to cages with weasels or weasel odor to mimic short- versus long-term exposures to predation risk. While increased perceived predation risk reduced the voles' use of feeding trays and increased their GUDs, neither prior experience nor the temporal scale of exposure influenced the bank voles' GUDs or foraging efforts.

In aviary studies with heteromyid and gerbilline rodents (Brown et al. 1988; Kotler et al. 1988, 1991; Kotler 1997), animals showed clear and striking changes in GUDs in response to nightly changes in owl activity or artificial illumination. If the owls remained in the aviary for long sequences of nights, the rodents eventually increased their

Figure 28.3 The least weasel (*Mustela nivalis*), the world's smallest, but most voracious, carnivore, is responsible for vole mortality and fear in northern boreal habitats. Additionally, the small weasels provide an alternative food source for larger predators, especially owls during winter. The white winter coat is an important survival adaptation. Photo by Seppo Laakso.

foraging efforts, reduced their GUDs, and suffered higher mortality. This change in activity was most striking when kangaroo rats were offered no shrub cover. When owls were present, kangaroo rats became inactive; all activity occurred on nights without owls. However, during a long sequence of nights with owls, the kangaroo rats shifted abruptly from no activity to uncharacteristically high levels of activity (Kotler et al. 1988).

Do rodents fit the risk allocation hypothesis?

Thus, the clearest results from the predation risk allocation hypothesis (Lima and Bednekoff's 1999) show how energetic demands force individuals to increase foraging despite known risks, provided the risky periods are persistent. A valuable contribution of future research will be to determine how prey develop fear responses from combining prior experiences of predator activity with immediate exposures to predation risk. It may be that the best survival strategy is to adopt a continuous and rather high level of alertness to predators, regardless of past experiences. For instance, a bird like the willow tit (*Parus montanus*) seems to maintain a certain state of alarm at all times (Haftorn 2000).

Alternatively, experimental results that seemingly contradict the predation risk allocation hypothesis may have been too short and missed the timescale over which foragers cultivate an underlying behavioral pattern for balancing energy gain and safety. In the study by Kotler et al. (1988), the experiment lasted over several days with changes in predation risk occurring nightly. Such studies reveal that heteromyid rodents (Brown et al. 1988) and gerbils (Kotler et al. 1991) make appropriate and nightly adjustments to their foraging behavior. In a study on voles exposed to three nights of predators, the voles shifted from their normally polyphasic activity pattern toward a decidedly nocturnal activity pattern (J. Eccard et al. 2001). In this study, the main predator, the least weasel, is primarily diurnal (Sundell et al. 2000). In both studies by Kotler et al. (1988) and by J. Eccard et al. (submitted ms.), the predators were free-ranging, actively hunting, and posing a real and imminent threat to the rodents.

An important issue in experimental tests of predation risk concerns mimicking increased risk. Risk has most often been simulated by either olfactory or visual cues, and more rarely by the introduction of actual predators capable of preying upon the experimental animals. It may be that persistent indirect cues of predation risk such as microhabitats and moon phases have fixed, long-term effects, whereas some short-lived, direct cues of predation risk may simply cause spiked responses because the animal only responds to a perception of imminent peril.

Youth and predation

A common perception among vole ecologists is that young, inexperienced voles and males are more vulnerable to avian predation (e.g., Pearson and Pearson 1947; Lagerström and Häkkinen 1978; Halle 1988; Mappes et al. 1993), whereas female voles may be the preferred target of mammalian predators because of scents associated with estrus (Cushing 1985). Both generalizations are not fully supported by recent prey selection studies. Koivunen et al. (1996) identified a "doomed surplus" of animals in which young field voles, regardless of sex, were more prone to owl predation than older individuals (but see Koivunen et al. 1998). While inexperience may contribute to vulnerability in young voles, their patterns of habitat use and mobility enforced by social circumstances may play a larger role. Increased mobility increases individual vulnerability to avian predation (Norrdahl and Korpimäki 1995, 1998). In response to experimental reductions in small mustelid predators, all voles increased their mobility (Norrdahl and Korpimäki 1998). Furthermore, in accord with Cushing's (1985) prediction, small mustelids preferentially prey upon female voles.

Desert rodents such as the kangaroo rat differ from most voles by having smaller litter sizes, fewer litters per year, and greater investment in young. Whereas young voles may become independent at less than one-half final adult mass, young kangaroo rats do not become independent until they are around two-thirds or three-quarters adult mass. This greater investment in young may be an antipredator strategy that insures higher chances of survival of the young. Patterns of dispersal in kangaroo rats may further reduce

predation risk to the young. For example, bannertail kangaroo rats maintain elaborate multi-entranced burrow systems that appear as mounds (ca. 3 m across and 0.5 m high) across the landscape. Mothers may either disperse themselves, leaving their present mound to an offspring, or protect an unoccupied mound that is then bequeathed to an offspring (Randall 1993). In the more nomadic Merriam's kangaroo rat, adults rather than juveniles frequently disperse to new home ranges. In both species, females may be indirectly bequeathing safety to their offspring.

Juvenile desert rodents purportedly face greater risks of predation than adults for two possible reasons (however, empirical proof of this conjecture is lacking at present). First, juveniles may be less experienced in detecting and evading predators. In fact, juvenile eastern chipmunks (*Tamias striatus*) appear highly susceptible to avian predation (e.g., Cooper's Hawk, *Accipiter cooperii*; Bosakowski and Smith 1992). Second, juveniles, even more than adults, may bias their foraging toward high-food, high-risk microhabitats because of their lack of foraging experience, less efficient foraging, or greater need for energy.

The open landscapes of deserts may be uniquely risky to rodents faced with avian and mammalian predators. Female kangaroo rats may reduce predation risk to their offspring by (1) postponing independence, (2) bequeathing burrows of known location, and (3) bequeathing a feeding range with a known landscape of fear.

Odors and olfactory signals

Odors emitted by voles (and presumably other taxa of rodents) vary with gender, age, and social status. These odors influence social interactions and communication. Ylönen et al. (2003) studied the effect of sex- and age-specific odors on the ability of weasels to hunt reproductive females, mature males, and nonreproductive females and males. They found no demonstrable olfactory biases in the weasels' hunting success—to a hunting weasel, lunch is lunch.

Olfactory signs of a prey individual are but one part in the hunting sequence of a predator (Ylönen et al. 2003). While directing a predator, like a mustelid, to sites with abundant prey, olfactory signatures may play less of a role than acoustic signals in facilitating a successful attack on a vole. Movement by males and young may provide visual and acoustic cues, while females attached to nest sites may consequently leave more olfactory signals for predators (Pusenius and Ostfeld 2000).

Female deer mice (*Peromyscus maniculatus*) frequently carry newborn pups to new nest sites (Sharpe and Millar 1980). This relocation may serve to deter predators by reducing cues that might otherwise accumulate. Also as a means for suppressing the accumulation of scent signs for searching predators, lemmings (*Dicrostonyx groenlandicus*) defecate in special underground latrines during the summer, but not during the winter when snow protects their outdoor toilets (Boonstra et al. 1996).

Mobility and scent marking are interconnected by sexual advertisement. Despite the negative result of Ylönen et al. (2003), several studies demonstrate a connection between rodent scent marking and increased risk of predation. High-marking mice (*Mus musculus*) decreased marking intensity when exposed to ferret (*Mustela putorius*) scent, but not when exposed to scent of naked mole-rats (*Heterocephalus glaber*) in a laboratory experiment (Roberts et al. 2001). In an unpublished study, D. Dudek and H. Ylönen failed to find any difference in urine marking of male bank voles exposed to either weasel scent or moose (*Alces alces*) scent as a control. Similarly, Wolff (2004) found that prairie voles (*Microtus ochrogaster*) did not decrease scent marking in response to odors of mink (*Mustela vison*) or a bullsnake (*Pituophis melanoleucus*) in the field or the laboratory. In the field, however, artificially increasing vole scent marking led to increased predation and lower survival of radio-collared voles (Koivula and Korpimäki 2001).

Acoustical and auditory cues

To their predators, nocturnal desert rodents likely emit acoustic signatures, and perhaps olfactory, visual, and—to pit vipers—heat signatures as well. Barn owls (*Tyto alba*) are thought to use acoustic cues to detect rodents at a distance and then to use visual cues to fine-tune strikes. Canids (*Canis sp., Vulpes sp.*) likely follow the scents of rodents in addition to responding to visual and acoustical cues. The presence of fox scent raised the GUDs of Merriam's kangaroo rats, although this effect was only significant in the winter and in the bush microhabitat (Herman and Valone 2000). Snakes may use olfactory cues to determine areas with higher likelihoods of encountering rodents (Duvall and Chiszar 1990). Sand vipers in Israel are known to stake out the burrow entrances of gerbils. In an experiment with the Australian house mouse (*Mus musculus*), mice did not utilize seed trays closer than 20 m to rabbit nests occupied by a brown snake (genus species). The absence of mice from these trays was due either to direct predation by the snake or, more likely, to the fear caused by the snake's presence (Ylönen et al. 2002).

In response to predators, nocturnal desert rodents may rely heavily on hearing to detect owls. Relative to other rodent taxa, nocturnal rodents often possess inflated auditory bullae, which aid in the detection of low-frequency sounds such as those emitted by owls in flight (Webster and Web-

ster 1971). Kotler (1984, 1985) found that a species' use of the open microhabitat increased with the size of its auditory bullae. Furthermore, in samples collected from owl pellets, the owls' diets favored heteromyid species with smaller auditory bullae.

Variation in noise level is one of the most striking features of rodent-owl experiments in aviaries. Nights without owls provide a cacophony of sounds. Kangaroo rats (Brown et al. 1988) or gerbils (Kotler et al. 1991) can be heard noisily scurrying from brush pile to brush pile. In addition, clearly audible sound signatures reveal the digging and scratching activities of rodents within the food patches. A "deathly" stillness descends over the aviary during nights with owls. The overall silence is punctuated on occasion by the crash of owl into a brush pile and the sudden noisy scurry of an escaping rodent. Then all is silent again. Yet, in the morning, even when owls have been present, footprints, evidence of digging, and seed removal from food patches indicate considerable rodent activity. While these findings have never been quantified, it appears in these aviary experiments that the amount of sound emitted per seed harvested is much lower on nights with owls than on nights without owls.

Auditory cues probably influence predation rates on boreal voles as well. Trapping success of voles typically increases with rainfall in both forest and field habitats (Sidorowicz 1960; Ansorge 1983). Similarly, field and laboratory studies with *Peromyscus* showed that animals are more active on rainy nights or on wet substrate than on dry nights or on leaves that are crunchy (Fitzgerald and Wolff 1988). The conclusions from these studies are that rain dampens auditory cues emitted by foraging rodents, cues that are used by predators to locate their prey.

Fear- versus mortality-driven predator-prey systems

Boreal voles and desert rodents can have similar predators (owls, foxes, and snakes), and they show behavioral response to predation risk that involves microhabitat selection and shifts in activity patterns. But here the similarities may end. Predator-prey interactions are population-size driven (N-driven) when the prey pay the cost of predation by high fecundity and feeding rates. Classical predator-prey models apply to N-driven systems in that doubling the number of predators doubles the prey's predation risk. Predator-prey systems are driven by fear (μ-driven) when prey opt to forgo feeding and breeding opportunities in the face of predators. Increasing the numbers of predators results in more vigilant or less active prey that are harder to catch; doubling the number of predators does not double the prey's predation risk. N-driven systems are likely to see relatively large populations of prey and predators and to display the oscillatory dynamics of classical predator-prey models. Fear-driven systems are likely to see relatively smaller population sizes of predators and to display more stable population dynamics. In μ-driven systems predators have strong negative direct effects on themselves through the fear responses they induce in their prey (Brown et al. 1999).

We suggest that voles and their predators (particularly weasels) represent primarily N-driven systems, whereas most desert rodents represent μ-driven systems. Productivity and diet may explain these differences. Most voles, especially *Microtus,* feed from a food source that is of low quality and occurs at high productivity, in contrast to most desert rodents, for which the opposite is true. Voles exhibit less long-term caching behavior, except for the root vole (*Microtus oeconomus*) and *Clethrionomys* species. The lack of food caching requires voles in general to feed frequently, even in the face of danger, and to "store" food in the form of fecundity. Granivorous desert rodents enjoy a high-quality, highly cacheable food. This food and foraging behavior allows desert rodents to forgo both foraging and reproduction during times or places of high predation risk. While denying itself immediate offspring, the granivorous desert rodent also denies the predator its lunch. Obviously, voles can and do respond to predation risk, and obviously desert rodents suffer predation, but on the continuum between totally N-driven and totally μ-driven systems it is likely that boreal voles are more N-driven and less μ-driven than are most desert rodents.

Fear from Multiple Predators: Birds, Mammals, and Snakes

The expression "[the] fangs of the snakes are driving gerbils into the talons of the owls" (Kotler et al. 1992, p. 155) describes a trade-off in the decision-making process of rodents that are faced with risks from multiple predator species (Lima 1992). *Predator facilitation* (Charnov et al. 1976) is the term used for tactics that reduce mortality from one type of predator while increasing the chances of falling prey to another predator species. This pattern as described for desert habitats is true for more covered habitats as well (Korpimäki et al. 1996). Terrestrial predators such as mammals and snakes often use ambush tactics in hunting. Habitats with cover offer the predator concealment while it awaits the approach of a potential prey. In contrast, complex vegetation cover reduces the effectiveness and lethality of avian predators such as owls and hawks. Thus, rodents face multiple fears and potentially conflicting hazards when choosing activity patterns among spatial (shrub versus open) and/or temporal habitats (day versus night, moon phase, etc.).

An important alternative activity for rodents (or any prey) is to remain inactive in a safe place or burrow. In optimal foraging models, remaining safely inactive becomes attractive when all foraging opportunities have been depleted to the point that the harvest rate, H, from foraging no longer exceeds the metabolic cost, C, predation cost, P, and missed opportunity cost of foraging, MOC; i.e., individuals cease foraging activity when $H = C + P + MOC$ (Brown 1988, 1992). Here the missed opportunity cost is the negative of the sum of the resting metabolic cost and the cost of predation while inactive in a burrow or safe retreat. Presumably, the metabolic and predation costs of resting are lower than those when foraging. If safety in the rodents' burrow is relatively low due to predators, such as snakes and weasels, that are able to hunt in the burrows of rodents (L. Oksanen and Lundberg 1995; Ylönen et al. 2003), then there are fewer advantages for decreasing foraging activity and remaining inactive in one's burrow. This may be another driver for the different activity patterns seen in desert rodents and voles. Desert rodents may experience higher costs of predation while foraging than do voles, and they may experience lower costs of predation while remaining inactive in their burrow. The difference in degree of safety in and out of the burrow may be less for voles than for desert rodents.

Biologists have studied antipredatory responses of rodents by offering them choices of habitats that differ in risk. As expected, under equal feeding opportunities rodents bias their activity toward safer habitats (Brown et al. 1992; Kotler and Blaustein 1995; Jacob and Brown 2000). In the case of predator facilitation from terrestrial predators and avian predators, rodents bias their foraging activity toward the microhabitat with cover in response to owls (Brown et al. 1988; Kotler et al. 1991; Brown et al. 1992; Kotler and Blaustein 1995; Kotler 1984) and toward the open microhabitat in response to snakes (Brown 1989; Bouskila 1995) or terrestrial mammals (Korpimäki et al. 1996). Lima (1992) suggests that the responses of prey to the presence of multiple predators should differ from situations with just one predator species. The addition of the second predator into the system may actually decrease prey vigilance, because the prey aim to minimize their time exposed to predators by decreasing the time they spend foraging.

Experiments with simultaneous risks from two predator species are more difficult to design and control than those with just one predator species. In desert rodents, Kotler and colleagues have manipulated owl and snake predation, illumination, and habitat properties in large aviaries (e.g.,

Kotler et al. 1992, 2001). In boreal voles, large-scale field or aviary studies with multiple predator species have been lacking. A study with combined avian and weasel predation risk (Korpimäki et al. 1996) was conducted in small laboratory arenas. However, J. Eccard and colleagues (2001) managed to document significant increases in risk taking in the foraging behavior of bank voles in an enclosure experiment where voles were exposed to freely hunting weasels and long-eared owls simultaneously. Despite the risk posed by weasels hunting in covered microhabitats, voles increased their use of this microhabitat because of the even stronger threat posed by owls in more open microhabitats. Interestingly, the presence of owls forced the least weasels to be more cautious when moving through open areas, and the presence of owls may have reduced the lethality of weasels to the voles. To the owls, the presence of weasels may have provided a prey in itself and made the voles easier to catch. To the weasels, the owls may have reduced their hunting efficiency, because the weasels now had a heightened foraging cost of predation.

Breeding and Predation Risk

Applying a Russian proverb to female mating strategies, L. Oksanen and Lunberg (1995, p.46) point out that a female "cannot be a little bit of pregnant." A female faces the trade-off between reproduction and survivorship. The breeding suppression hypothesis (Ylönen 1989, 1994b; Ylönen and Ronkainen 1994) predicts that females should favor survivorship over reproduction during the declining phase of a vole cycle and fecundity over survivorship during the increase phase. The hypothesis assumes that female reproductive activities increase the risk of predation, especially when vole-specialist predators such as small weasels are abundant (*sensu* Cushing 1985). In northern cyclic vole populations where predation risk also fluctuates cyclically, females that delay breeding during the high-risk phase enhance their fitness by increasing their survivorship and by depositing their offspring into the low-risk phase of the cycle (see L. Oksanen and Lundberg 1995).

Initial testing of breeding suppression hypothesis in the laboratory reported obvious effects of simulated weasel predation risk on the reproduction and reproductive behavior of voles (e.g., Heikkilä et al. 1993; Ylönen and Ronkainen 1994; Ronkainen and Ylönen 1994; Koskela and Ylönen 1995; Koskela et al. 1995). When high weasel abundance foiled an otherwise optimal breeding season, reproduction by voles in the field was lower than predicted (Korpimäki et al. 1994). However, all field attempts to use mustelid scent to simulate increased predation risk have failed to produce changes in female reproductive behavior consistent with the breeding suppression hypothesis (see Parsons and Bondrup-Nielsen 1996; Wolff and Davis-Born 1997; Ylönen 2001, for review). Despite obvious behavioral responses to predation risk, trading off reproduction for future survival does not seem to maximize individual female reproductive success. For voles that are short-lived and cannot store food for long periods and that must feed frequently, it may be maladaptive to forgo breeding during the short breeding seasons offered by higher latitudes. In fact, voles have adaptations like postpartum estrus and other physiological adaptations for avoiding even short delays in subsequent litters. Maximizing breeding effort and producing as many young as possible and as quickly as possible is probably the best strategy for maximizing lifetime reproductive success (Kokko and Ranta 1996).

The field studies previously mentioned simulated increased predation risk during the best of breeding circumstances: namely, under conditions of moderate vole densities and abundant food resources. The breeding suppression hypothesis may apply to voles under less favorable feeding conditions—for instance, very early in the breeding season or for late breeders. During these less favorable feeding conditions, reductions in activity caused by increased risk of predation may place female voles below a threshold breeding condition. Theoretically this strategy would resemble the "waiter" strategy and represent a compromise between the strict reproducers or strict survivors of L. Oksanen and Lundberg (1995). Carlsen et al. (1999) examined the effects of predator scent on the body mass and survivorship of field voles over winter. They observed the same trend that Dickmann (1992) noted for house mice and Norrdahl and Korpimäki (1998) for field voles. Under low predation risk, survivors were heavier, either because predation that targeted heavier individuals was lacking or because voles fed more under low predation risk.

A study of grey-sided voles (*Clethrionomys rufocanus*) in subalpine tundra in northern Norway supported the predictions of the breeding suppression hypothesis even during a summer breeding season (Fuelling and Halle 2004). A significantly higher proportion of reproductively inactive adult females occurred on plots treated with weasel scent as compared to control plots. Control plots saw a higher number of juveniles recruited into the adult populations. One interpretation of these results is that breeding suppression may occur as a consequence of energy limitations in harsh environments or under less favorable conditions at the boundaries of the breeding season (Norrdahl and Korpimäki 2000). Breeding suppression by voles is probably not an evolved antipredator strategy. Evolved breeding suppression as outlined in the breeding suppression hypothesis applies to μ-driven rather than the N-driven system of voles.

Predation Risk Affecting Social Interactions

Rodents exhibit a broad range of social complexity levels, ranging from primarily solitary taxa to the highly integrated colonies of naked mole-rats (Randall 1994). The family Sciuridae has a Social Complexity Index of 0.27 for the woodchuck (*Marmota monax*) up to 1.12 for the black-tailed prairie dog (*Cynomys ludovicianus*) (as defined and calculated by Blumstein and Armitage 1997). Such a score for the woodchuck indicates a low level of social interaction and structure, one that is relatively typical for many rodent species. In this system, solitary adult females maintain small home ranges; males maintain larger home ranges that often include the ranges of several females; males and females live apart; breeding is promiscuous, with some antagonistic behavior of males; and the young may reside with the mother or within the mother's home range into or close to adulthood. Prairie dogs represent an example of a colonial lifestyle without strict division of labor and with only inconspicuous dominance hierarchies. Coteries of often related individuals maintain a shared system of burrows, share sentinel duties, and cultivate social bonds through ritualized behaviors (Hoogland 1995). Levels of sociality likely influence and have been influenced by numerous life history and environmental factors. For instance, Devillard et al. (2004) showed that among ground-dwelling sciurids male-biased dispersal increases with social complexity. To what extent has predation risk sculpted and in turn been sculpted by social interactions among rodents?

A striking feature of sociality in rodents is that most colonial rodents are diurnal. Diurnal and colonial rodent species exist among species of *Cynomys, Marmota, Degus, Xerus, Rhombomys, Parotomys, Petromys, Spermophilus, Rhabdomys,* and others. The naked mole-rat and Damara mole-rat (*Cryptomys damarensis*) provide conspicuous exceptions to coloniality and diurnality, but these are fossorial rodents. The nature of predation risk likely plays a role. The veil of darkness conceals both prey and predator from each other's view, rendering the night less propitious for the benefits of group wariness and group foraging.

Alarm calls are rare among nocturnal rodents. Sight lines and vision may be requisites for either the recognition of the need for, or the effectiveness of, group vigilance and group defense. Not surprisingly, the complexity of alarm calls increases with the size and complexity of rodent colonies (Blumstein and Armitage 1997b, 1998a). Furthermore, most colonial rodents occupy relatively open landscapes. Prairie dogs will actually destroy vegetation and shrubs that obstruct lines of sight. The black-tailed prairie dog is notable but not unique for a repertoire of alarm calls that varies and adjusts for different species of predators and different approaches of predators (Hoogland 1995; see Boero 1992 for discussion of European marmot, *Marmota marmota*).

The great gerbil (*Rhombomys opimus*) of the Kyzylkum desert, Uzbekistan, is diurnal and colonial (Rogovin et al. 2004). It exhibits a social structure also found in the cape ground squirrel (*Xerus inauris*) of the deserts of southern Africa (Waterman, chap. 3 this volume). Both species tend to have colonies of up to thirty individuals built around several adult females and their offspring. A single adult male may continuously or periodically share the colony. Most males live alone or in pairs, sometimes moving in and out of colonies (Waterman 1995, 1997). In the great gerbil a rhythmic alarm warns of a distant predator, whereas an intense whistle warns of imminent threat from a predator nearby (Randall and Rogovin 2002). Foot drumming also warns of dangers when emitted outside (in response to snakes) or inside (in response to a dog) of the burrow. Furthermore, solitary gerbils do not emit alarm calls, and adult gerbils call more frequently in the presence of young gerbils (Randall et al. 2000).

Predation risk clearly plays a role in the colonial life and alarm calls of diurnal rodents (Blumstein, chap. 27 this volume). Yet a critical research question remains. Did some other benefit to social complexity drive the evolution of coloniality in these rodents? If this is the case, sentinels, alarm calls, and group wariness may have evolved as adaptations useful to coloniality. Or did the antipredator advantages of sociality drive the evolution of coloniality, as beautifully demonstrated by Hoogland (1981) for prairie dogs? We think the latter hypothesis is more likely and believe that in the absence of predation risk most forms of coloniality would not have evolved in rodents.

While lacking coloniality, nocturnal desert rodents do exhibit rudimentary forms of sociality and calls in response to predators. Foot drumming is widespread across diverse families of desert rodents (Randall 2001). Randall (2001) noted that most species foot drum in response to predation risk. Solitary, territorial rodents such as the bannertail kangaroo rat have complex patterns of airborne and seismically transmitted foot drums that communicate territoriality and the presence of predators. When it occurs in response to predators, foot drumming in the bannertail kangaroo rat serves as parental care (warning to a female's current litter) or individual defense (indicates to the predator that it has been detected; Randall and Matocq 1997).

In accord with the hypothesis that these nocturnal prey are communicating directly with the predator, larger (and presumably less vulnerable) kangaroo rat species engage in more frequent behaviors of foot drumming, sand kicking, and approach to predators such as snakes than do smaller kangaroo rat species (Randall et al. 1995; Randall and King 2001; Randall, chap. 31 this volume). Although diurnal and living a solitary life, the eastern chipmunk exhibits highly audible alarm calls that may serve to alert kin (Burke da Silva et al. 2002) or to alert others that, if they escape, reduces the patch quality to the predator.

A series of trained-owl experiments at a site in the Negev Desert, Israel, revealed the tendency to seek safety in numbers in a nonsocial gerbil species (Allenby's gerbil) and also revealed that the negative effect of a dominant competitor (Egyptian sand gerbil) becomes nullified in the face of predation risk (overflights by barn owls). Whereas ordinarily the gerbils would distribute their activity evenly between two halves of an enclosure (in accord with an ideal free distribution), Allenby's gerbil would actually bunch up into one half or the other in response to increased numbers of owl flights even when the owl flights were evenly distributed between halves of the enclosure (Rosenzweig et al. 1997). The larger Egyptian sand gerbil, by virtue of interference competition—and disproportionate to its own numbers—will drive the smaller Allenby's gerbil into the half of the enclosure free from the larger species. However, in the presence of increased predation risk from owl flights, the interference effect of *G. pyramidum* on *G. a. allenbyi* evaporates as responding to fear seems to trump interspecific competition (Abramsky et al. 1998).

Predation Risk and Seasonality

A striking difference between boreal and arid environments emerges from the stronger temperature seasonality of boreal habitats and the greater precipitation seasonality of arid habitats. For boreal and subarctic voles the winter provides long-lasting permanent snow cover that alters predation pressures and risk. Overwinter survival of the bank vole suggests a late autumn period of high mortality before the onset of permanent snow, and another spike in mortality associated with the spring thaw (e.g., Ylönen and Viitala 1985). This pattern does not necessarily demonstrate changing predation rates and may simply represent the mortality consequences of harsh environmental conditions. However, a persistent blanket of snow provides rodents with subnivean spaces that protect them from severe temperatures and from the attacks of most predators. Large owls, like the great grey owl (*Strix uralensis*), can attack vole prey through thick snow cover, and least weasels and possibly small stoats (*Mustela erminea*) are able to enter the subnivean cavities of voles, but in general the snow hinders most predators' hunting. Winter predation risk is further reduced by the absence of migratory raptors. While it offers increased protection from most predators, the snow cover may also protect and facilitate the hunting success of the vole specialists, like the least weasel. T. Oksanen et al. (2001) suggested that this winter tightening of the one-on-one predation between voles and the weasel shifts the pattern of vole population dynamics from one of annual fluctuation to the regular and multi-annual cycles observed between northern and southern Fennoscandia (Hansson and Henttonen 1985, 1988).

Many rodent species of the northern hemisphere may overwinter socially in aggregations (West 1977; Wolff and Lidicker 1981; Ylönen and Viitala 1985), although good empirical evidence for most species is lacking. The main benefit of social aggregation in the winter is probably the saving of energy through communal huddling (Madison 1984; Wolff 1980b). Furthermore, if your nest mate/mates remain in the nest when you go foraging, then you return to a warmer nest compared to the cold nest of a solitary individual (Vickery and Millar 1984; Wolff and Lidicker 1981). Staying aggregated, however, may attract predators and increase predation risk, especially because the snake-like least weasel can search the subnivean space and is likely to detect more easily the odor or heat signatures of aggregated voles. However, during the winter the cost-benefit ratio between huddling and predation risk may favor winter aggregations as the benefits of huddling go up and the predation risks go down relative to huddling during other seasons. Thus, the common pattern is that most voles are territorial or solitary during the breeding season and social during the winter (Madison 1984).

Deserts lie primarily at 30 degrees latitude and close to the west sides of continents, creating diverse seasonal patterns that influence predation risk on desert rodents. Most if not all deserts lie along migratory bird routes, including those of raptors. The GUDs of gerbils increase dramatically at a Negev Desert site during the spring raptor migration (Brown et al. 1994a). Deserts see sharp seasonal shifts in temperature, although these are less dramatic than the shifts that take place in boreal landscapes. At a Sonoran desert site, rattlesnakes drove kangaroo rats (*D. merriami*) to prefer the open microhabitat during the summer, whereas ground squirrels (*Spermophilus tereticaudus*) enjoyed relative safety from diurnal raptors rendered scarce by high temperatures. These same squirrels have increased GUDs due to the presence of migratory and wintering raptors. The winter brings a reprieve for the kangaroo rats as the rattlesnakes go torpid, and the presence of owls scares the rodents into preferring the bush microhabitat (Brown 1989).

In addition to climate-driven shifts in predation risk, deserts see pulses of productivity following winter and/or summer rains. During this period desert rodents almost universally employ one of two strategies: food caching and/or seasonal dormancy. For pocket mice (*Perognathus* and *Chaedopidus*) and ground squirrels dormancy means that the rodents only experience risk during their activity season. Owls may time their breeding to correspond to periods of peak activity and densities of pocket mice, and indeed pocket mice are more vulnerable to owl predation than are kangaroo rats (Kotler 1985). Caching behavior provides desert rodents with a powerful antipredator tool: namely, the opportunity to focus foraging behavior toward periods high in food and/or low in predation risk. Caching behavior probably evolved in response to seasonal food production and the low perishability of seeds. Once evolved, however, it becomes a "bank account" that allows the rodents to forgo any costly foraging periods. The latitude of antipredator behaviors afforded by caching may be what renders desert rodent systems noncyclic and μ-driven.

Conclusions and Future Prospects

If we draw a line from the Negev desert in Israel to Finnish Lappland by Kilpisjärvi and Pallasjärvi, both well studied in terms of empirical and theoretical rodent ecology, we have a productivity and cover gradient over a range of familiar vegetation types. Productivity is very low in the desert, increases strongly in temperate grasslands and forests, remains high in boreal habitats and decreases again toward northern subarctic habitats (T. Oksanen and Henttonen 1996). Cover and shelter from predation follow the same curve but change in northern boreal and subarctic habitats, where strong seasonality and permanent snow during winter provide shelter from cold and predation (Hansson and Henttonen 1985). Rodent abundance is low in the desert, high and fluctuating in temperate regions (Tkadlec and Stenseth 2001), and high but strongly fluctuating (and often cyclic) in boreal and subarctic habitats (e.g., Hanski et al. 2001).

Predator assemblages change less, however, having the greatest diversity and numbers in temperate regions and southern parts of boreal habitats. Stronger seasonality and long-lasting permanent snow cover cause a strong seasonal change in the predator guild in the north. Most raptors depart the areas during winter, leaving a small number of resident mammalian predators and owls (T. Oksanen et al. 2000). Let us now try to combine the impact of productivity, shelter, and prey and predator numbers to assess the joint *per capita predation risk* for gerbils, temperate-region mice, and voles across the gradient (table 28.1).

Desert, temperate, and boreal regions probably offer rodents quite different risk regimes. In deserts, a very high per-unit-foraging-time predation risk likely promotes immediate and density-dependent responses in gerbils to risk. Gerbils respond by spending little time active per night. Temperate regions see high numbers of prey together with high numbers of diverse predators, but due to a dilution effect, the per-unit-foraging-time predation risk is lower than in the desert. Effects of predators on prey populations— and behavior—are directly density-dependent and probably more N-driven than in desert systems and more μ-driven than in boreal systems (T. Oksanen et al. 2000). While risk per unit activity is lower than in deserts, each individual probably spends more time actively foraging. In the northern boreal areas and the subarctic, per capita predation risk even on a per-foraging-time basis is higher than for temperate mice, probably due to the impact of vole-specialist predators such as weasels and stoats. The dynamics of the predators and voles exhibit time lags and delayed density-dependent effects. Furthermore, boreal voles spend a relatively large amount of time foraging and exposed to predation risk, and predation risk may remain relatively high even when the voles are in their burrows.

The gradient of fear, foraging, breeding, and population dynamics proposed here is not yet proven but is rather a series of interconnected hypotheses (fig. 28.4). As a more synthetic conceptual framework develops that is applicable to all rodents, new research avenues will develop that apply similar research tools over similar spatial and time scales. Research tools that can be used more systematically across desert, temperate, and boreal systems include censusing protocols, measurements of populations, social and breeding structures, GUDs, the duration and scale of enclosure and aviary experiments, and the protocols for food enrichments, habitat modifications, and manipulations of actual or perceived predation risk.

The synthesis of conceptual tools will require an understanding of how fear influences foraging, breeding, and social behaviors and of how predation risk interacts with specific environmental circumstances to produce μ-driven versus N-driven predator-prey systems. Fear, the nonlethal effect of predators on their prey, has far-reaching consequences for the life histories of rodents. Rodent feeding, breeding, and social behaviors are all impacted by fear of

Table 28.1 Comparing and contrasting the life-history properties of boreal voles and desert gerbils relevant to their predators and to the rodents' responses to the risk of predation

	Boreal voles	Desert gerbils
Diet	Lower quality: herbs and grasses	Higher quality: seeds
Reproductive ecology	Larger litters; shorter gestation time; shorter weaning period	Smaller litter; longer gestation time; longer weaning period
Activity time	Polyphasic: Longer activity time; roughly 2 hours of activity followed by 2 hours of rest.	Nocturnal: Short activity time; highest activity early in the night.
Food storage	Noncaching	Extensive caching
Burrows	Susceptible to predation.	Relatively immune to predation; snakes may sit and wait at burrow entrances.
Specialist predators	Weasel and stoat	Probably none?
Generalist predators	Fox, hawks, owls	Fox, owls, shrikes, monitor lizard, snakes
Metabolism	High mass-specific metabolic rate (may be adaptation to high activity times)	Low mass specific metabolic rate (may be antipredator adaptation by lowering time spent active).
Direct cues of predation risk	Predator odor, weasel, owls?, hawks?	Owls, foxes, snakes; likely respond to sound and visual spotting of predators; predator odors?
Indirect cues	Microhabitat: cover safer than open; moonlight?	Microhabitat: Cover generally safer than open; moonlight avoidance
Population dynamics	Multiannual fluctuations or cycles with large amplitudes; likely due to time lags in prey and specialist-predator dynamics.	Seasonal fluctuations in response to pulse of seed from spring bloom; year to year fluctuations likely due to year to year variability in rainfall amounts and patterns.
Exposure to predation	Per-unit activity risk is likely lower, but daily exposure to risk is likely higher.	Per-unit activity risk is high, but mitigated by low activity times biased toward particularly profitable periods; lower daily exposure to risk.
Behavioral flexiblity	Need to forage frequently and inability to cache likely reduces latitude for antipredator behaviors.	Ability to cache increases greatly the latitude for anti-predator behaviors, either through apprehension while foraging or time allocation to safer places and times.
Fear system	Population size driven (N-driven): the population response to risk is to have high feeding rates and high fecundity rates to compensate for high predation rates. In effect, voles pay their cost of predation by feeding the predators.	Fear driven (μ-driven): the population response to risk is to forgo fecundity by forgoing feeding during risky times or risky places. In effect, gerbils pay the cost of predation by not feeding the predators through apprehensive foraging and lower fecundity.

NOTES: Although there should be much generality with respect to comparing boreal rodents with desert rodents in general, the specific systems from which these conclusions are drawn emerge mostly from studies in Fennoscandia on *Microtus agrestis* and *Clethrionomys glareolus,* and studies in the Negev Desert with *Gerbillus pyramidum* and *G. andersoni.*

predators, and these behaviors feed back on the predators. Rodents have provided and will continue to provide a crucible for formulating and testing theories of the nonlethal effects of predators.

Summary

The ecology of fear examines the behavioral, population, and community consequences of the nonlethal effects of predators. Predation risk may reduce activity levels, shift activity from risky to safe places, alter mating behaviors and reproduction, increase levels of vigilance and apprehension, and alter intra- and interspecific competitive interactions. Furthermore, fear in response to predation risk may alter life-history characteristics such as growth rates, dispersal, and age at maturity. These fear responses are adaptations that serve to maximize the prey's fitness in the face of predation risk. In this chapter, our aim was to examine fear and its effects on the foraging, breeding, and social behavior of rodents. We refer mostly to boreal voles (*Microtus* and *Clethrionomys*) living under relatively dense vegetation and to desert rodents (e.g., *Gerbillus*) exposed to predation in more open environments. All of these rodents share risks from terrestrial predators (mammals and snakes) and avian predators (raptors and owls). Different predators pose different challenges and may present different cues to their rodent prey. Predation risk is a cost-benefit assessment of foraging, breeding, and social interactions such as territoriality, interference competition, and group vigilance. The costs of predation may have implications for foraging in terms of space use, habitat selection, and timing and/or suppression of breeding; social foraging in terms of vigilance, alarm calls, and various forms of safety in num-

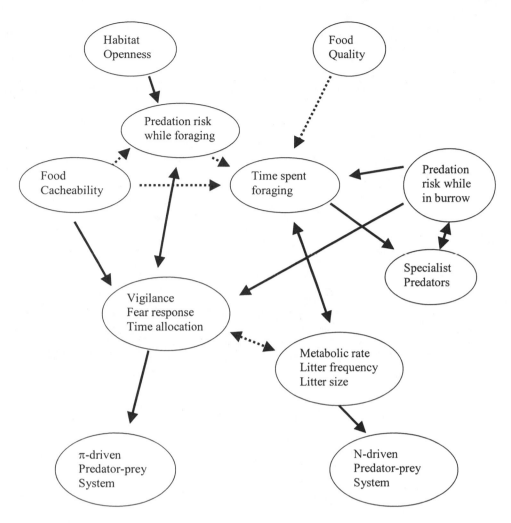

Figure 28.4 A sequence of hypotheses for factors that may lead to π-driven versus N-driven predator-prey dynamics in rodents. Seed-eating desert rodents and boreal voles provide the models for each system, respectively. Solid arrows represent positive relationships between factors, and dotted arrows indicate negative relationships. Bidirectional arrows indicate feedback loops.

bers; and population dynamics in seasonal environments. Desert, temperate, and boreal regions probably offer rodents quite different risk regimes. Accumulation of comparative and experimental data from these regions allow us a synthesis, which suggests that not only the direct effect of predators on mortality, but also the indirect effects are essential to an understanding of predator-prey interactions that affect population growth and population and community structure.

Comparative
Socioecology

Chapter 29 Ecology, Kinship, and Ground Squirrel Sociality: Insights from Comparative Analyses

James F. Hare and Jan O. Murie

E O. WILSON (1975) defines sociality as "the combined properties and processes of social existence." Only a definition this broad could allow for the myriad components that embody sociality proper, including: (1) multiple individuals that overlap in space and time such that a social group is formed, (2) the underlying ecological and evolutionary factors promoting and maintaining group living (e.g., Ebensperger 2001a), (3) interactions among group members that impart both benefits and costs (social behavior), and (4) the emergent properties of groups themselves, and in particular, their success relative to other groups (D. S. Wilson 1975).

Unlike morphological traits, behavior does not leave fossils. Thus inferences regarding the evolutionary origin of behavioral phenomena must be made indirectly, using a combination of functional, phylogenetic, and comparative data (Crespi and Choe 1997a). Alexander (1974) regarded the revival of the comparative approach, along with the more general acceptance that the structural and functional attributes of organisms could be interpreted as products of selection operating on individuals, as factors revolutionizing the study of social behavior. Comparative analyses contrast the expression of behavioral or morphological traits among members of the same or of closely related species, relating apparent differences in those traits to differences in the organisms' environments (e.g., Crook 1970; Felsenstein 1985; Lacey 2000). As such, comparative analyses may reveal convergences in traits, or illuminate transitions by combining presumed phylogenetic relationships with microevolutionary analyses based upon functional design, optimality, and even the measurement of selection (Crespi and Choe 1997b).

The broadest comparative literature pertaining to sociality exists for the social insects, where the application of the comparative approach has provided many important insights into the origins and maintenance of sociality (Hamilton 1964; Wilson 1971; Lin and Michener 1972). Insight into factors promoting reproductive skew (Vehrencamp 1979) within animal societies has also been gleaned from more recent models integrating data from both vertebrate and invertebrate societies (Vehrencamp 1983; Keller and Reeve 1994; Reeve and Keller 1995; Sherman et al. 1995). In this chapter, however, we adopt a taxon-centered approach to the comparative analysis of social behavior, in agreement with Crespi and Choe's (1997b) suggestion that the causal factors underlying the origin and maintenance of sociality, while overlapping, are unlikely to be universal across taxa.

Of the rodent societies examined to date, the most comprehensive data in terms of combined taxonomic breadth, adequate descriptions of spatial, life-history, and behavioral relationships, and consideration of environmental correlates, exist for the ground-dwelling squirrels (marmots, prairie dogs, and ground squirrels proper). Indeed, in reviewing Murie and Michener's (1984) volume "The Biology of Ground-Dwelling Squirrels," Barash (1985) suggested that ground squirrels could be as important to understanding vertebrate sociality as the giant squid motor axon had been to the history of neurophysiology, given the ground squirrels' diurnal habits, wide geographic distribution, relatively large number of species evidencing a diversity of social systems, and relative insensitivity to human observers.

In this chapter, we review comparative schemes examin-

Figure 29.1 An alarm-calling Richardson's ground squirrel (*Spermophilus richardsonii*) warns conspecifics of the presence of a predatory threat. Photo by J. Hare.

ing ground squirrel sociality, in an attempt to: (1) summarize the ecological factors underlying the appearance and maintenance of sociality, (2) clarify the role that kinship plays in defining spatial and behavioral relationships among members of ground squirrel societies, and (3) identify previously underemphasized correlates of increasing sociality. These interrelationships are summarized in figure 29.4. Finally, we identify fruitful avenues for future research.

The Comparative Sociality of Ground Squirrels

David Barash (1973, 1974a) was the first to recognize the full utility of the comparative method in examining the ecological basis of sociality in the genus *Marmota*. Barash contrasted his own detailed observations of the biology and behavior of Olympic marmots (*M. olympus*) with those made by other investigators on yellow-bellied marmots (*M. flaviventris*) and woodchucks (*M. monax*), and related those observations to environmental differences experienced by those North American marmots in nature. Specifically, Barash (1973) reported a decreasing incidence of aggression, increasing incidences of "greeting" and play fighting, and the relaxation of territoriality with increasing social integration as woodchucks, yellow-bellied and Olympic marmots are considered in series. Advancing sociality was noted to be coincident with increasing elevation, which Barash in turn related to the progressively diminishing length of the growing season, and hence foraging time available for young of the year of these relatively large-bodied rodents to amass sufficient fat reserves to survive hibernation (see Armitage, chap. 30 this volume).

Further, Barash noted that whereas woodchucks would typically disperse from the natal area during the summer in which they were born, yellow-bellied and Olympic marmots would typically delay dispersal for 1 or 2 years respectively. From the individual's perspective, delays in dispersal are also accompanied by delayed reproductive maturity. Thus dispersal of young from the natal area was increasingly delayed with increased elevation, and related inversely to the length of the growing season. Taken together, Barash (1973, 1974a) concluded that the reduction in the length of the growing season imposed by increasing elevation selects for reduced aggression and increasing social cohesion, including delayed dispersal as a means of ensuring the survival of young.

Within the prairie dogs (*Cynomys*), Hoogland (1981) contrasted the ecology and behavior of the more highly social black-tailed prairie dog (*C. ludovicianus*) with that of the less social white-tailed prairie dog (*C. leucurus*). As was the case for marmots, delays in both dispersal and reproduction were evident for the more socially complex species, though these differences were not correlated with differences in elevation or length of growing season. Rather, Hoogland (1981) marshaled data suggesting that coloniality in prairie dogs was attributable to antipredator benefits of group living (see fig. 29.1 and Bertram 1978; Alcock 2005 for reviews), and that interspecific differences could be ascribed to habitat differences promoting different antipredator strategies. White-tailed prairie dog habitats offered significantly more protective cover than black-tailed habitats. Thus white-tailed prairie dogs were interpreted by Hoogland (1981) to have a greater reliance on avoiding predation through the use of cover, thereby limiting the size and density of groups to the number of individuals that could effectively exploit the cover available in a given area. By contrast, for black-tailed prairie dogs, without extensive cover, selection has favored a dependence on the rapid detection of predators (Hoogland 1981), which is enhanced via increases in the size and density of a social group (Hoogland 1979b; Kildaw 1995; Hoogland, chap. 37 this volume).

In a more taxonomically comprehensive effort, Armitage (1981) examined correlations between an ordinal "sociality index" and the life-history traits of twenty-four populations of eighteen different ground-dwelling squirrel species. While admitting that his sociality index served only as a first approximation, Armitage categorized the species he included in his analysis as Level: 1 = solitary, 2 = individual females forming an aggregation or "colony" in suitable habitat, 3 = the expression of male territoriality within a colony of individual females, 4 = the aggregation of female colony members into a male-controlled harem, and 5 = existence in multiharem colonies. He then reduced the overall data set to four principal component axes reflecting body size, seasonality, reproductive effort, and age at reproductive maturity, and examined correlations between those axes and his sociality index. Sociality was not correlated directly with body size, seasonality, or reproductive effort, but was highly correlated with age of first reproduction, such that delayed reproduction could be equated with enhanced sociality. While body size emerged as a highly significant predictor of reproductive effort, the lack of any correlation between sociality and reproductive effort or body size suggested to Armitage that reproductive effort was a product of size-dependent energetics, while sociality was a consequence of other factors. Among those factors, Armitage (1981), like Hoogland (1981), suggested antipredator benefits of group living as the necessary precursor to sociality, and allowed that kin selection may have played a role in the evolution of ground squirrel sociality, given evidence of kin-biased patterns of alarm calling (Dunford 1977a; Sherman 1977). In more direct accord with his own analysis, however, Armitage (1981) indicated that a nutritionally poor environment (as would be the case with short growing seasons) could result in delayed maturity, which in turn would delay dispersal and promote sociality via interactions among individuals of different sex/age cohorts. To Armitage, then, sociality is the product of life-history tradeoffs, particularly in the timing of an animal's development toward reproductive independence, which is shaped ultimately by extrinsic environmental factors.

Michener (1983a) also recognized five grades of sociality, and presented a similar classification scheme to that of Armitage (1981), but focused more directly on the role of kinship in defining sociality. Based upon the spatial overlap of sex/age cohorts, and the relatedness of females within groups, Michener delineated grades as: 1 = asocial, 2 = single-family female kin cluster, 3 = female kin cluster with territoriality, 4 = female kin cluster with male dominance, and 5 = egalitarian, polygynous harems. In comparing species, Michener (1983a) noted that while more social species tend to possess large body size, the active seasons of those species were not necessarily short. Delays in dispersal and reproductive maturity were characteristic of the more social species, though the primary correlate of advancing social grade in Michener's analysis (accounting for some 84% of the variance) was the temporal overlap between adults and subadults. Michener recognized that in addition to delayed dispersal, factors postponing the fall immergence of adults, or even year-round activity (as is apparent in black-tailed prairie dogs), could enhance overlap and thereby promote sociality. The driving force, though, was kin selection, as Michener noted that the clusters of females forming the core of social groups were related females comprising matrilines. Thus from her comparative assessment Michener recognized the following emergent trends characterizing the evolution of ground squirrels from a presumptive ancestral asocial state to advanced societies composed of multiple polygynous harems: (1) the retention of daughters in the female's home range, (2) the relaxation of distinctions between litters of adjacent females, and (3) male territoriality extending beyond the breeding season and ultimately expanding in scope to incorporate the ranges of multiple females.

Rayor and Armitage (1991) contrasted patterns of spatial overlap and social behavior among highly social Gunnison's prairie dogs (*Cynomys gunnisoni*), less social Columbian ground squirrels, and relatively asocial thirteen-lined ground squirrels. They found that amicable social interactions between mothers and their offspring and among littermates were more prevalent with complex sociality. Further, spatial overlap was greater and was maintained for a longer period of time in more social species. Such trends were not restricted to littermates: increased spatial overlap and cohesive behavioral interactions with young from other litters were also apparent with advances in sociality (Rayor and Armitage 1991). Thus while sociality in ground-dwelling squirrels is associated with the retention of offspring within the mother's range, it is the extension of cooperative behavior among individuals that is the earmark of increasing social complexity.

The terms *sociality* and *social complexity* are often used in the literature without explicit quantification, though comparative analyses by Blumstein and Armitage (1997b, 1998) present a quantitative and continuous index of social complexity. The index itself is based on information theory (Shannon and Weaver 1949), whereby a higher index expresses the increased amount of information necessary to describe a more complex social construct. Their overall index is, in effect, based on a grand sum of the number of bits of information required to describe the variation in the frequencies with which certain demographic (age/sex) cohorts appear within groups of individuals of each species of interest, multiplied by a factor that increases with delays in dispersal. Complexity, then, is equated with variability in the probability of interaction of individuals in various sex/age cohorts, indirectly capturing factors that have been used historically to describe sociality (e.g., group size, Eisenberg 1981, spatiotemporal overlap of sex/age cohorts, Michener 1983a). The multiplicative factor that scales this information index for each species reflects the period over which young are retained in the natal area, and hence amplifies the overall social complexity score where kin have the potential to overlap in space and time (Blumstein and Armitage 1997b).

While recognizing that their index of social complexity does not reflect the actual nature of behavioral interactions, Blumstein and Armitage used the indices calculated for multiple *Cynomys*, *Marmota*, and *Spermophilus* species to examine correlations between social complexity and communicative complexity in alarm vocalizations (twenty-two species; Blumstein and Armitage 1997b) and between so-cial complexity and ground squirrel life-history tactics, in-cluding: the proportion of females breeding within a group, age of first reproduction, length of gestation, litter size, length of the lactation period, offspring survival, and female body mass (twenty-five species: Blumstein and Armitage 1998). Alarm-call repertoire size was positively correlated with social complexity (Blumstein and Armitage 1997b), supporting Marler's (1977, p. 46) suggestion that the great-est elaboration of communication would be expected where "trends towards increasing interindividual cooperation con-verge with the emergence of social groupings consisting of close kin." This overall relationship remained in phylo-genetically controlled contrasts, though in considering the genera independently, the relationship between social com-plexity and alarm-call repertoire size was upheld only within *Marmota,* not within *Cynomys* or *Spermophilus.* Blumstein and Armitage (1997b) conclude that social complexity has contributed to the evolution of alarm-call repertoire size in the ground squirrels, but that other factors, including morphology, the acoustic properties of the environment, and the specific functional needs dictated by the organism's ecology are also significant in determining communicative complexity. Increased social complexity was also accompa-nied by a decrease in the proportion of females within a group that bred, an increase in the age of first reproduction, a decrease in litter size, and enhanced survival of offspring (Blumstein and Armitage 1998). Neither the duration of gestation or lactation varied consistently with changing so-cial complexity, though it is apparent that sociality imposes a trade-off between per capita reproductive output and off-spring survival (Blumstein and Armitage 1998).

The elucidation of factors underlying sociality in ground squirrels is complicated by variation in the criteria investi-gators have employed to quantify sociality. While this prob-lem is not unique to comparative work on ground squirrels (e.g., Lacey 2000), it is necessary to consider the limitations of previous attempts to make progress in the future. Barash (1973) and Rayor and Armitage (1991) contrasted species in terms of social interactions, identifying life-history and ecological correlates of more frequent and amicable inter-actions among members of each species under considera-tion. The schemes put forth by Armitage (1981) and Mich-ener (1983a), while broader taxonomically, are based upon qualitative criteria that in effect prejudge the importance of delayed dispersal and the spatial overlap of kin, respec-tively. While providing a continuous rather than categori-cal metric of social classification, Blumstein and Armitage's (1997b, 1998) social complexity index also presumes that advancing sociality can be equated with enhanced overlap of sex/age cohorts within the species considered. As such, their identification of life-history traits associated with in-

creased social complexity (Blumstein and Armitage 1998) is circular. Further, it is unclear whether the factors emerg-ing from contrasts of social systems across the Marmotini as a whole can explain the narrower range of variation in sociality expressed within certain genera, such as *Spermo-philus.* Given these problems with broader comparative classifications, it is useful to revisit the ecological and be-havioral underpinnings of sociality and to critically evalu-ate evidence as to how such factors contribute to the evolu-tionary appearance and maintenance of sociality.

Ecological Factors Promoting Sociality

Spatial and temporal overlap among individuals is required for benefits to accrue via sociality (Holmes and Sherman 1983). Thus a comprehensive examination of sociality in the ground squirrels must first identify the factors promot-ing the aggregation of individuals into groups (Ebensper-ger 2001a). Ecological factors figure prominently in setting the stage for sociality (Alexander 1974; Slobodchikoff and Shields 1988).

Contagious distributions of individuals can be the prod-uct of habitat heterogeneity, such that individuals are con-strained to co-occupy suitable habitat patches (e.g., Malizia 1998; Stallman and Holmes 2002). While comparative data are not available to examine the correlation between over-all habitat patchiness and sociality, there is little doubt that the distribution of burrow systems, underlain by suitably drained and arable subsoils, promotes the aggregation of in-dividuals. Subterranean burrows represent critical resources for ground squirrels in that burrows facilitate escape from predators (Blumstein 1998), and provide a stable and rela-tively secure microenvironment during periods of reduced activity and/or increased vulnerability such as resting, rear-ing young, and hibernation (King 1984). Indeed, burrows serve the same functions as expansible and defensible nests, which Alexander et al. (1991) argued to be a primary fac-tor promoting eusociality in naked mole-rats and other social animals. The spatial distribution of burrows is inti-mately linked to soil geomorphology (Moss 1940; Svendsen 1976) and land use (Henderson and Gilbert 1978). Thus in-dividuals may be clustered in areas that contain or at least provide suitable soil architectures for the development of burrow systems.

Several lines of empirical evidence suggest that burrow systems may ultimately influence the expression of social-ity in ground-dwelling squirrels. Holmes (1984a) invoked the wide dispersion of suitable hibernacula as a factor pro-moting monogamy in hoary marmots (*Marmota caligata*). Armitage (1988) turned to burrow systems to explain vari-ation in the expression of yellow-bellied marmot sociality.

Armitage (1988) noted that large patches of suitable habitat support yellow-bellied marmot colonies composed of multiple matrilines, while small habitat patches (satellite sites) support only a single matriline with few individuals. As habitat area increases, more burrow systems are available and the number of matrilines increases. The mean size of matrilines, however, is inversely correlated with habitat area, reflecting the opportunity of individual females to establish their own smaller matrilines where burrow systems are available, and thereby avoid the potential cost of reproductive suppression in a larger group (Armitage 1988). Among yellow-bellied marmots certain burrows also constitute preferred hibernacula that are used year after year, and cooperating individuals may outcompete others for these choice hibernacula (Armitage 1988). Selection may thus favor social relationships in the face of competition for burrows as resources (King 1984).

Basic habitat variables also interact with antipredator benefits. Holmes (1991) has shown that pika (*Ochotona princeps*) will extend their foraging into previously unoccupied areas if rocky outcrops providing refuge from predation are extended into those areas. The ground squirrels are also known to be risk-sensitive foragers, adjusting escape responses relative to their distance from a burrow (Bonenfant and Kramer 1996) and the trade-off between foraging and vigilance according to their proximity to the center versus edge of the group (Armitage 1962; Hoogland 1979b, 1981). Indeed, Hoogland's (1981) study clearly implicates the relative density of cover as a factor influencing the expression of coloniality in prairie dogs. Individuals aggregating together as a group enjoy a variety of antipredator benefits (Sherman 1977; Hoogland 1981; Kildaw 1995). Thus at least some component of antipredator benefit likely served as a primary impetus for group living early in the evolution of ground squirrel sociality.

The spatial distribution and general abundance of food resources do not, in and of themselves, appear to have been instrumental in promoting ground squirrel sociality. Food is widely distributed within patches of suitable habitat, and individuals do not appear to have been selected to cooperate in the defense of food proper. The distribution of food in time, however, is invariably clumped, with a discrete growing season providing an abundance of resources at one time, and a paucity of resources at another. If the climate is such that the growing season is short, as it is for ground squirrels occupying high latitude or high elevation sites, then the attainment of adult body mass, and hence breeding, may be delayed, and selection may favor delayed dispersal (1973, 1974a; Armitage 1981, 1999).

The harsh winter climate imposed by high elevation and high latitude sites may also act directly to promote social tolerance. Arnold (1990b) has demonstrated that social

thermoregulation is critical to successful hibernation in alpine marmots, and thus a harsh environment alone can select for social grouping (Armitage, chap. 30 this volume). Group hibernation in alpine marmots reduces winter mortality of all group members, though the presence of infants, particularly in poor-quality hibernacula, imposes costs on other group members (Arnold 1990b). These costs, however, are not borne equitably by all subordinate subadults: full siblings of infants within the group lose more mass than others, suggesting investment on their part in nonlittermate siblings (Arnold 1990b). Thus in addition to direct fitness benefits of reduced mortality and the increased likelihood of territory inheritance, delayed dispersal may allow individuals to accrue indirect fitness benefits by assisting nondescendant kin.

Kinship and Ground Squirrel Sociality

As summarized previously, clumped burrow resources, a short growing season, a harsh winter environment, and intense predation pressure set the stage for group living in the ground squirrels. Michener (1983a), however, regarded kin selection as the driving force behind the evolution of ground squirrel sociality, in that the retention of daughters in the natal area results in the formation of matrilines composed of related females. Where relatives are available for social interactions, as is commonly the case for ground squirrels (Vestal and McCarley 1984), kin selection may promote the evolution of nepotism. Favoritism toward kin among ground squirrels appears in a variety of contexts. In both Belding's (*Spermophilus beldingi;* Sherman 1977) and round-tailed ground squirrels (*S. tereticaudus;* Dunford 1977a), females with either descendent or nondescendent kin in the local population are more likely to emit alarm vocalizations in response to a terrestrial predator than females without kin present. Further, female Belding's ground squirrels are more likely to participate in chases of territorial intruders with close female kin than with nonkin (Sherman 1981a). In Arctic ground squirrels, neighboring female relatives have been reported to pool recently weaned young into common burrows (McLean 1982). The most universal expression of kin-biased behavior among ground squirrels, however, appears in the context of social tolerance, manifested both in terms of behavioral interactions and the use of space. In *Spermophilus* alone, evidence of kin-differential behavior has been obtained for thirteen-lined (*S. tridecemlineatus;* Holmes 1984b; Vestal and McCarley 1984), Franklin's (*S. franklinii;* Hare 2004), round-tailed (Dunford 1977a), Belding's (Sherman 1980a, 1981a; Holmes and Sherman 1982), Richardson's (*S. richardsonii;* Yeaton 1972; Michener 1973a), Arctic (*S. parryii;* McLean 1982), and

Columbian ground squirrels (*S. columbianus;* King 1989a, 1989b). Similarly, kin-differential behavior and the spatial aggregation of female kin are common to both marmots (Armitage, chap. 30 this volume) and prairie dogs (Hoogland, chap. 37 this volume).

Given the potential importance of kin selection to ground squirrel sociality, the mechanisms allowing beneficent behavior to be directed toward kin have also been the subject of comparative analysis (Holmes and Sherman 1982; Holmes 1984b; Schwagmeyer 1988a; Holmes and Mateo, chap. 19 this volume). Holmes (1984b) noted preliminarily that an increasing reliance on more direct mechanisms of kin recognition (i.e., those in which phenotype matching or perhaps recognition alleles precluded a reliance upon social familiarization) paralleled increasing social complexity in contrasting thirteen-lined and Belding's ground squirrel kin recognition abilities. Consistent with that notion, Schwagmeyer (1988a) reported that greater spatial clustering and more complex social behaviors, such as frequent allogrooming and even joint hibernation, coincided with the appearance of kin recognition mechanisms that were not dependent upon familiarity in her review of the social behavior and kin recognition mechanisms of thirteen-lined, Richardson's, Belding's, and Arctic ground squirrels. Schwagmeyer (1988a) ultimately concluded that more refined kin recognition abilities evolve where opportunities to express nepotism exist. The subsequent findings of Hare and Murie (1996) that social discrimination in highly social Columbian ground squirrels is dependent entirely upon social familiarization, and the more recent demonstration by Mateo (2002) that relatively asocial golden-mantled ground squirrels (*Spermophilus lateralis*) make exacting discriminations on the basis of kinship in the absence of familiarity, suggest, however, that no overall correlation exists between social recognition mechanisms and sociality. Exacting kin discrimination among solitary species may result from selection pressure relating to mate choice (Mateo 2002), and may serve to promote an optimal balance between inbreeding and outbreeding (Bateson 1983). Hence, the nature of recognition mechanisms is unlikely to prove useful in assessing the contribution of kin selection to the evolution of sociality.

Regardless of the mechanism promoting kin-biased behavior, it is undeniable that kin selection can play a substantial role in maintaining and defining the limits of sociality in certain ground squirrel species. What is not clear, however, is the extent to which kin selection enjoys primacy in terms of promoting the evolutionary appearance of sociality, or to what extent kin selection is necessary to account for the wide range of social constructs evident among ground squirrels at present. Comparative analyses of marmot sociality both among (Armitage 1987b) and within species (yellow-bellied marmots; Armitage 1988) call into ques-

tion the extent to which evolutionary explanations of sociality need to invoke kin selection. Armitage (1987b) agrees with Michener (1983a) that matrilines are the fundamental unit of ground squirrel societies. Unlike Michener, however, Armitage argues that the formation of matrilines is an expression of parental investment (a component enhancing direct fitness), and only incidentally allows benefits to accrue via kin selection (promoting indirect fitness). In support of his contention, Armitage (1987b) notes that yellow-bellied marmot matrilines begin as mother/daughter associations, and that sister/sister associations result only from the joint recruitment of daughters to the natal area and the subsequent death of their mother. He also reports that there is no evidence that females associating in a matriline trade off direct for indirect fitness benefits, as matriline size has no effect on the per capita production of descendants. Finally, he notes that the small size of matrilines, owing in part to mortality, but more tellingly through fissioning of larger matrilines, is inconsistent with the operation of a mechanism promoting the ongoing association of kin, as would be expected if sociality were predicated upon kin selection. For yellow-bellied marmots, the average relatedness of members of a matriline was typically 0.5, and when relatedness decreased, the matriline subdivided (Armitage 1988). The fact that matrilines split when burrow systems are available (Armitage 1988), however, again suggests that both relatedness and resources contribute to the spatial aggregation of female kin.

In yellow-bellied marmots, the recruitment of daughters was positively correlated with female reproductive success, and thus cooperation among females would prove selectively advantageous if that cooperation facilitated the exclusion of immigrants from access to the resources of the natal area (Armitage 1988). From the subordinate's perspective, any loss of reproductive opportunities early in life would be more than compensated for by the reproductive payoffs of later becoming the dominant female within a matriline. Further, where individuals have a limited probability of successfully dispersing and establishing themselves as immigrants (Armitage 1999), natal philopatry may simply make the best of a bad situation and allow some fitness benefit to accrue via kin selection (Emlen 1982; Woolfenden and Fitzpatrick 1990). The adult female's selective recruitment of daughters over other potential allies reveals that parental investment also contributes to matriline formation (Armitage 1988). To illustrate the broader applicability of this principle, Armitage (1988) notes that the delayed dispersal of black-tailed relative to white-tailed prairie dog young reported by Hoogland (1981) was most readily interpreted as a product of parental investment increasing offspring survival, and enhancing coterie membership in such a way that facilitates the defense of the resources of the natal area. That one's own reproductive success remains the primary focus in such social endeavors is amply illustrated by the occurrence of infanticide among close kin in black-tailed prairie dogs (Hoogland 1985). In killing nieces, coterie members enhance their own direct fitness by increasing the probability that their daughters will remain in the natal area. Such philopatry may also promote the inheritance of resources of the natal area (Harris and Murie 1984; Lindström 1986; Myles 1988).

It is also evident that kinship is not a definitive correlate of amicable social interactions and advanced sociality in ground-dwelling squirrels. Hoogland (1986) noted nepotism in general among both male and female black-tailed prairie dogs, but found that amicable and aggressive behavior varied with seasonal differences in competition between individuals rather than with their coefficient of relatedness. Thus neither Hamilton's (1964) prediction that the expression of nepotism should vary with the coefficient of relatedness, nor Altmann's (1979) prediction that individuals should direct all beneficent behavior toward their closest available relative were upheld. Indeed, the killing of young of closely related females (Hoogland 1985) clearly indicates that competition to recruit one's own daughters to the natal area takes precedence over any component of indirect fitness in black-tailed prairie dogs. Similarly, yellow-bellied marmots limit amicable behavior to uterine kin (mothers, daughters, and sisters), having a coefficient of relatedness of 0.5, but are agonistic toward all other conspecific females, including their more distant relatives (Armitage 1988). In highly social Olympic marmots, holding individuals away from the group and reintroducing them at a later time results in aggression from colony members that is correlated in its intensity with the duration of the individuals' absence, and not contingent upon its relatedness to those group members (Barash 1973). This finding in itself suggests that reinforcement of group membership, perhaps via olfactory contacts in the context of extensive greeting described by Barash (1973), may in the proximate sense underlie amicable coloniality. A similar diminution in the importance of relatedness among members of the society may also occur in the genus *Spermophilus*. Columbian ground squirrels, which were regarded as the most highly social members of that genus by both Armitage (1981) and Michener (1983a), and which received the highest social complexity index of all the *Spermophilus* species considered by Blumstein and Armitage (1997b, 1998), are amicable to all familiar conspecifics (Hare 1994) and fail to manifest kin-differential behavior (see fig. 29.2 and Hare and Murie 1996). Both Michener (1983a) and Rayor and Armitage (1991) foreshadowed this possibility in noting that relaxed discrimination among members of neighboring litters, both in terms of more cohesive behavioral interactions and increased spatial overlap, accompanied complex sociality in the ground-dwelling squirrels. Even among Bel-

Figure 29.2 Highly social Columbian ground squirrel (*Spermophilus columbianus*) juveniles discriminate group members from other conspecifics but not kin from nonkin. Photo by J. Hare.

ding's ground squirrels, where kin discrimination is highly refined (Holmes 1986b) and nepotism well documented, Sherman (1980a) noted the adoption of recently emerged juveniles by nesting adult females, and indicated that as yearlings the adopted females treated their true littermates as nonrelatives. Where neighbors are kin, benefits of social cooperation will accrue via kin selection. Payoffs for cooperation, however, are not limited to kin (Trivers 1971; Axelrod and Hamilton 1981; Connor 1995), and thus societies may appear and be maintained by social bonds that transcend lines of kinship. Arguments predicated upon kinship are not necessary to explain the formation or maintenance of social groups (Jaisson 1991; Myles 1988; Waser 1988), and high relatedness in isolation of particular ecological factors may be insufficient to account for advanced sociality (Andersson 1984). To understand sociality, one must consider all possible levels at which both benefits and costs may accrue.

Additional Factors Contributing to Sociality

The retention of daughters in the natal area clearly pays direct fitness dividends in terms of offspring survival (Armitage 1988). Further, the direct fitness interests of parents and young alike will be served where offspring inherit maternal resources (Myles 1988), as would be the case where daughters inherit burrow systems in their mother's territory (Harris and Murie 1984).

Explanations extending beyond direct fitness are typically invoked to account for behavior that appears to place the survival and/or reproductive interests of the actor in jeopardy. Among ground squirrels, such apparent altruism is most often cited to occur in the alarm-calling system (Blumstein, chap. 27 this volume), where calling individuals may place themselves at greater risk of predation in warning fellow group members of the presence of a predator (Sherman 1985). While kin selection may ultimately explain such behavior—where potential benefits associated with self-sacrifice are directed toward kin (Holmes and Mateo, chap. 19, this volume; Dunford 1977a; Sherman 1977)—several alternative mechanisms must be considered to explain cooperation among nonkin (Connor 1995), and may also contribute to the benefits accruing via social interactions among relatives.

Individuals interacting cooperatively within a group may simultaneously enhance each other's fitness, as is the case where nonrelatives participate in assemblages of hibernating individuals (Arnold 1990). Such elements of social behavior are best considered mutualisms. Individuals that manifest apparent altruism but are repaid later in kind by their beneficiary are offsetting any costs of their initial sacrifice through reciprocal altruism (Trivers 1971). It is conceivable that an individual placing itself at greater risk by alarm calling in one instant will be repaid by receiving a warning from another individual at a later time. Recent demonstrations that Richardson's ground squirrels (Hare 1998a) and yellow-bellied marmots (Blumstein and Daniel 2004) discriminate among callers as individuals, and that Richardson's ground squirrels selectively attend to callers that have proven reliable in the past (Hare and Atkins 2001), are consistent with the recognition of individuals and the memory of how those individuals behaved in past encounters, as required for the evolution of reciprocal altruism (Trivers 1971; Axelrod and Hamilton 1981). Data are not available for any ground squirrel species, however, to address the extent to which individuals modify the expression of apparent altruism according to that of their social contemporaries. Even outside the context of strict reciprocity, costs associated with alarm signaling could be offset by benefits associated with receiving signals from others where alarm signaling becomes a social convention within a group (i.e., byproduct mutualism; Connor 1995), though no data are available to address this hypothesis. Similarly, no data are available to address the extent to which an individual's apparent personal sacrifice may enhance its social status within the group, thereby improving the apparent altruist's lot by virtue of direct fitness payoffs associated with that enhanced social standing (Zahavi 2003).

Future Directions

Comparative analysis has implicated a variety of ecological factors, and has identified components of both direct and indirect fitness contributing to the expression of ground squirrel sociality. Sociality imposes tangible costs on individuals (Alexander 1974; Wilson 1975; Hoogland 1979a), including reproductive suppression (Blumstein and Armitage 1998). Such costs are offset, however, by benefits derived from (1) increased survival of young, (2) potential breeding opportunities as a subordinate, (3) potential for territory inheritance and reproductive benefits associated with becoming a territory holder, (4) participation in social thermoregulation during hibernation (e.g., Arnold 1990), and (5) alloparental care (Armitage 1999). Further research, however, is necessary.

While both King (1984) and Armitage (1988) highlighted burrow systems as resources critical to survival and reproduction, no systematic study has examined how the dispersion and availability of burrow systems influences social relationships, either within or among ground squirrel species. Similarly, further insight is likely to be gained by applying the comparative analysis of vegetative cover, both to the density and size of ground squirrel colonies, just as Hoogland (1981) did for prairie dogs. Such analyses could make fruitful use of Geographical Information Systems to address the patchiness of available habitat at multiple scales, to determine whether habitat availability at the landscape level correlates with the expression of sociality.

Considerable effort should also be devoted to consider plasticity in the expression of sociality within certain ground squirrel species, so that the factors underlying such variability might become apparent. Woodchucks, for instance, have traditionally been regarded as solitary (Bronson 1964; Ferron and Ouellet 1989a), though Meier (1992) reported considerable overlap in space use and frequent and amicable interactions among female kin and their coincident adult male. Chris Maher (personal communication) has observed similar social interactions among woodchucks, and thus both the behavior of woodchucks themselves, and comparative schemes treating woodchucks as solitary, bear further examination (fig. 29.3). For Columbian ground squirrels, both litter size (Murie et al. 1980; Risch et al. 1995) and age of first reproduction (Zammuto and Millar 1985; Murie 1995) have been shown to vary with elevation. It remains to be seen, however, whether those differences in life-history traits translate into differences in social behavior.

Future comparative analyses require a well-supported phylogeny of the ground-dwelling squirrels. Molecular phylogenies are available for marmots (Steppan et al. 1999), and are now beginning to appear for a wide range of sciurid species (Harrison et al. 2003; Herron et al. 2004). The phylogenies resolved by such analyses, however, are often at odds with traditional phylogenetic schemes based largely upon morphological traits (e.g., Hafner 1984). Once phylogenetic relationships among species are established with confidence, the mapping of specific social traits (e.g., joint

Figure 29.3 Intraspecific variation in sociality requires further examination, as is evident from field studies of woodchucks (*Marmota monax*). Photo by J. Hare.

hibernation, delayed dispersal) as attempted by Blumstein and Armitage (1997b, 1998) onto those phylogenies may ultimately provide the greatest insight into the evolution of the traits in question (Dobson 1985).

Our efforts to understand ground squirrel sociality must also extend beyond the application of comparative methods. Indeed, while comparative methods have proven invaluable in suggesting both proximate and ultimate hypotheses to test, comparative data are correlative and thus do not necessarily reveal underlying causation. This problem is particularly evident in attempts to uncover the underlying basis of ground squirrel sociality, as the retention of young within the natal area may be the product of attraction to the resources of the natal area and direct fitness benefits, but inescapably results in, and is thus correlated with, the clustering of female kin (Michener 1983a; Armitage 1988). To determine whether such clustering is the result of an active attraction to kin proper, as opposed to a mutual attraction of those individuals to a common area, manipulative experimentation is required. A transplant of individuals to new sites using methods similar to those of Wiggett and Boag (1993a) or Michener (1996), so that groups composed exclusively of kin or of nonkin are formed, could begin to tease out the role that kinship itself plays in determining aspects of spatial overlap and social behavior. The removal of close kin within a colony of Richardson's ground squirrels by Davis (1984c) increased interruptions of foraging with bouts of vigilance, increased chasing and fleeing, reduced sharing of core areas by neighbors, and decreased breeding success relative to what was documented on the other half of the same colony, where removals left females with close kin as contiguous neighbors. The absence of replication, particularly in light of extensive

badger predation within the nonkin area, and potential bias introduced into spatial relationships via the attraction of individuals to familiar burrow systems within their natal areas, preclude attributing the behavioral asymmetries documented by Davis to relatedness per se, but suggest that a similar approach holds promise.

Subsequent investigations must also explore the generality and strength of certain patterns of kin-biased behavior and the presumptive costs associated with those. Data suggesting costs associated with ground squirrel alarm calling have been presented only for Belding's ground squirrels (Sherman 1977, 1985), and though cited extensively, even these data must be interpreted cautiously. Statistically signifiicant differences were evident only for pursuits, and predators killed equal numbers of noncallers and callers (Sherman 1985). Similarly, kin bias in cooperative chasing (Sherman 1981a) merits further empirical study. The greater likelihood of displacement in chases of nonkin relative to close kin, as indicated for Richardson's ground squirrels by Davis (1984b), might give the appearance of cooperation between kin when both of them are simply responding to intrusions of nonkin on their territories.

Our efforts to explain sociality in ground squirrels are also hampered by our limited understanding of the functional aspects of social discrimination. Kin discrimination clearly functions to promote nepotism in some species, but may serve other purposes in more solitary species (Mateo 2002). While in the past it was commonplace to equate kin discrimination and kin-differential behavior with kinship-based sociality, such conclusions are not always warranted. We must work toward a greater understanding of the roles that individual (Hare 1998a), neighbor (Hare 1998b), and even group-member discrimination (Hare 1992, 1994) play in promoting the fitness of individuals within ground squirrel societies. Progress toward understanding the function of social discrimination at any level is less likely to be made by studying sociality overall than by observational and manipulative studies focusing on how such discrimination contributes to the expression of specific social traits within and among species (e.g., joint hibernation, allogrooming, play, territory defense, alarm calling).

It is particularly revealing that three of what are commonly regarded as the most social ground squirrel species within their respective genera (Olympic marmots in *Marmota*, black-tailed prairie dogs in *Cynomys*, and Columbian ground squirrels in *Spermophilus*) fail to show kin-biased behavior in certain contexts (Barash 1973; Hoogland 1986; Hare and Murie 1996, respectively). Clearly, advancing sociality cannot be predicated entirely upon kin selection, but must involve social benefits that accrue via other mechanisms. Just as kin selection recognizes the fitness payoffs implicit in the broader inclusion of collateral kin over direct descendents, increasing sociality may involve the relaxation of restrictions on interactions among group members, resulting in a more inclusive and egalitarian society. This is not always the case, as many eusocial species form highly regimented societies characterized by extreme reproductive skew (Sherman et al. 1995). For ground squirrels, however, both Michener (1983a) and Rayor and Armitage (1991) suggested that discrimination among individuals from neighboring litters waned with more complex sociality, consistent with the eventual appearance of more inclusive social constructs. Barash (1973) also noted that with more complex sociality, as seen in Olympic marmots, all parts of the colony become equally available to all colony members. Measures of inclusivity, both in terms of space use and behavioral interactions, may thus prove as useful in cross-taxa comparisons examining the adaptive basis of sociality as measures of reproductive skew.

The acknowledgment that benefits of sociality extend beyond those encompassed under kin selection does not deny the fundamental importance of kin selection to ground squirrel sociality. The formation of kin clusters, regardless of the primary selective force producing such clustering, ensures that kin overlap in space and time. Thus cooperation among individuals will be subject to kin selection. Perhaps the most important question, not only for our understanding of ground squirrel sociality, but for the survival of our own species, is the extent to which the genetic selfishness underlying social evolution can ultimately be directed toward the common good.

Summary

Comparative analyses reveal that multiple ecological factors, including the reduced length of the growing season at higher elevations, nutritionally poor environments that promote delayed dispersal, a harsh winter climate favoring group hibernation, the clumped distribution of burrows, and antipredator benefits of group living contribute to the evolution and maintenance of ground squirrel sociality (fig. 29.4).

Female philopatry, resulting in spatial and temporal overlap of kin, is common among ground squirrels, and has promoted the appearance of nepotism. Kin selection may thus contribute to ground squirrel sociality, though payoffs accruing via indirect fitness are likely to be secondary to those accruing via descendent kin, either through enhanced survivorship or benefits associated with the inheritance of territories. Uncertainty as to the function of social discrimination, variation in the measures used to quantify sociality, and the absence of well-resolved molecular phylogenies for some ground squirrel taxa limit our ability to understand the evolution of ground squirrel sociality. We advocate broader replication and the performance of manipula-

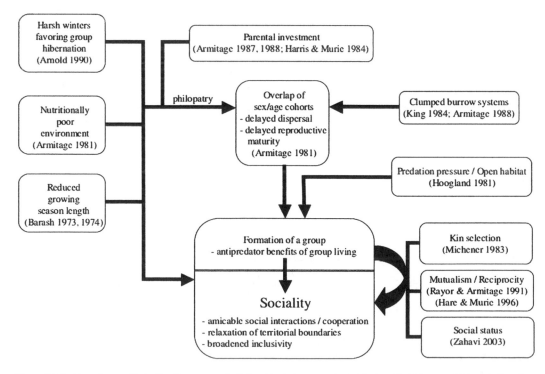

Figure 29.4   Flow diagram illustrating the proposed relationships between ecological factors, life history, and the evolution of sociality in ground-dwelling squirrels.

tive experiments to clarify the role kinship plays in influencing social behavior. We also suggest that insight into the evolution of specific social traits will best be gleaned by mapping those traits onto more robust phylogenies. The examination of intraspecific variation in social expression should also prove informative, and we identify species that will provide useful systems for such studies. Finally, we note that cooperation in ground squirrels is not limited to genetic relatives but expands with more complex sociality to incorporate nonkin. Future investigations must also consider societal benefits that accrue beyond lines dictated by kinship.

Chapter 30 Evolution of Sociality in Marmots: It Begins with Hibernation

Kenneth B. Armitage

I T IS NOW GENERALLY AGREED that sociality, that is, group living, has both costs and benefits. Among the benefits are increased awareness of predators through increased vigilance (Armitage et al. 1996; Blumstein 1996) and alarm-calling (Sherman 1977; Blumstein and Armitage 1997a). Major costs are a loss of reproduction; increased social complexity in ground-dwelling sciurids is associated with a smaller proportion of females breeding, a greater time to age of first reproduction, and a decrease in litter size (Blumstein and Armitage 1998). Many groups are cooperative breeders; these groups are typified by considerable reproductive skew (Keller and Reeve 1994; Sherman et al. 1995), in which only one or few mature females reproduce in a reproductive season (Solomon and French 1997; Blumstein and Armitage 1999). However, one benefit of increased social complexity and cooperative breeding is increased juvenile survival.

In this paper I describe the costs and benefits of group living in marmots (*Marmota*) and present a historical analysis of the probable evolutionary events that led to sociality.

Social Systems of Marmots

Currently, fourteen species of marmots restricted to the northern hemisphere are recognized (Barash 1989). Six species occur in western North America; only the range of the woodchuck, *M. monax,* extends into eastern Canada and the United States. Two species, the alpine marmot (*M. marmota*) and the steppe marmot (*M. bobak*), occur in Europe; the remaining six species occur in Asia. Typically marmots occur in alpine or subalpine environments, with the exception of the steppe marmot, which lives on the Eurasian prairie, and the woodchuck, which lives in low-elevation woodland/meadow habitats. Range distribution and habitat characteristics of the fourteen species are described elsewhere (Armitage 2000).

Marmot species have been placed in either three (Allainé 2000) or four (Armitage 1996b; Blumstein and Armitage 1999; Armitage 2000; Armitage and Blumstein 2002) social groups (table 30.1). The major difference between three- and four-group categories is that in the three-group system, the restricted family and extended family systems are combined into a group of "species with complex level of sociality" (Allainé 2000; p. 23). The basic social unit of this group is the family (Bibikow 1996), but my interpretation of the available literature is that family structure differs between the restricted and extended family groups, in that subordinate adults are not normally present in the restricted family groups. Information is unavailable for assigning the Himalayan marmot (*M. himalayana*) and Menzbier's marmot (*M. menzbieri*) to social groups, but they probably have extended families.

The social systems should be considered "types," as they vary considerably. In *M. monax,* in Ohio, 25% of the female offspring remained with their mother through the first hibernation, and moved to different burrows as yearlings (Meier 1992). Nearly 73% of the yellow-bellied marmot (*M. flaviventris*) matrilines consist of one female; large matrilines occur only on larger habitat patches (Armitage and Schwartz 2000). On smaller habitat patches, only a reproductive pair may be present, and mating is monogamous

Table 30.1 Social systems of *Marmota*

Social system	Species	Characteristics
Solitary	*M. monax*	Mating system polygynous; little overlap of female home ranges; disperse as young; solitary hibernation
Female kin group	*M. flaviventris*	Mother : daughter : sister groups persist through time as matrilines; territorial male defends one or more matrilines; thus mating polygynous; yearling dispersal; solitary or group hibernation
Restricted family	*M. caligata, M. olympus, M. vancouverensis*	Adult male typically with one to three females and yearlings; mating within the family; disperse as 2-year-olds; group hibernation
Extended family	*M. baibacina, M. bobak, M. broweri, M. camtschatica, M. caudata, M. marmota, M. siberica*	Typical family of adult territorial pair, subordinate adults, and yearlings; mating monogamous or polyandrous; disperse at age three or older; group hibernation

SOURCE: Modified from Armitage 2000 and Blumstein et al. 2004.

(Armitage and Downhower 1974; Armitage 1991). However, some males achieve polygymy by expanding their home ranges to include two or more small patches with females (Salsbury and Armitage 1994). Many alpine marmot families consist only of a reproductive pair, with or without infants (Arnold 1990a, 1990b; Perrin et al. 1992; Sala et al. 1996; Lenti Boero 1999).

This variation within social systems is primarily a consequence of demographic processes. In yellow-bellied marmots, a female immigrant to a habitat patch forms a matriline of one (Armitage 1991, 2002). Recruitment of 2-year-old daughters increases matriline size; mortality decreases matriline size, and may reduce a matriline to one adult female (Armitage 1973, 1991, 1996b, 2002, 2003a). Similar processes occur in alpine marmots, with one major exception: solitary individuals are virtually nonexistent. An immigrant becomes resident in a family either by filling a vacancy or by evicting an adult resident (Arnold 1990a; Sala et al. 1992; Lenti Boero 1999; Grimm et al. 2003). The family structure of grey marmots (*M. baibacina*) is simpler on less favorable habitat than on favorable habitat (Mikhailuta 1991).

There may be a fifth social system. In the steppe marmot (*M. bobak*) in central Kazakstan, families consisted of one mature female and male plus old nonbreeding males and subadults. Subordinate adults were not reported, and dispersal apparently occurred at age two (Mashkin 1991). Families of steppe marmots in the Ukraine consisted of an adult female with or without offspring and two or three adult males. Again, no subordinate adult females were reported (Nikol' skii and Savchenko 1999). One group had two adult females, but one soon dispersed. The black-capped marmot (*M. camtschatica*) may be a superspecies, and one or more of the three sub-species may be a separate species (Boyeskorov et al. 1996). Family groups of *M. c. camtschatica* on average consisted of 1.8 adult males,

1.7 adult females, and 4.3 young (less than half of the families produced young in any given year; Mosolov and Tokarsky 1994). Families of *M. c. bungei* usually consisted of parents and progeny; about half of the marmots lived alone or in pairs (Yakovlev and Shadrina 1996). Neither of these studies reports the presence of subordinate adults, a major characteristic of the extended family.

The uncertainty regarding the full variety of marmot social systems derives from the lack of long-term studies of social structure and dynamics. Because marmots are long-lived species, known individuals must be followed for several generations to determine the typical social system and its variability. Except for *M. marmota*, marmot research on the Eurasian marmots concentrated on population changes following exploitation of marmots for food, fat, and fur or reduction and recovery of marmot populations in programs to evaluate the role of marmots as a plague vector (Bibikow 1996). Results are presented as population averages; for example, percent of females of different ages breeding, size of families (e.g., Pole and Bibikov 1991; Shubin 1991). These studies enable one to infer the nature of the social system and its relationship to the age of first reproduction and of dispersal; some of the characteristics of social systems to be discussed later are based on such population data.

Sociality evolved at least twice (Kruckenhauser et al. 1998; Steppan et al. 1999), once in the Eurasian clade and once in the North American clade. The solitary *M. monax* has a basal position in the Eurasian clade and suggests that ancestral marmots were asocial. The female kin group and restricted family social groups occur in the North American or *caligata* clade and the extended family species occur in the Eurasian clade.

Uncertainty of the exact nature of the social system of all fourteen marmot species does not refute that this monophyletic genus (Steppan et al. 1999) has diverse social systems. This diversity raises two questions: (1) why did sociality in this genus evolve, and (2) why are there diverse social systems? The attempt to answer these questions requires understanding the evolutionary history of *Marmota*.

Evolution in a Harsh Environment

Historical record

Both DNA hybridization studies (Giboulet et al. 2002) and the fossil record (Black 1972; Mein 1992) indicate that the first Sciuridae evolved in the Oligocene in North America. The first true ground squirrel, *Miospermophilus,* appeared by late Oligocene and was the ancestor of the spermophiles (*Spermophilus*), prairie dogs (*Cynomys*), and marmots. These ground-dwelling squirrels radiated in North America in the Miocene and Pliocene. Although *M. vetus* appeared in the Miocene in what is now the United States (Mein 1992), the earliest incontrovertible fossils are *M. minor* from the late Miocene, about eight million years ago (Polly 2003). One radiation led to large-bodied forms, such as *M. nevadensis* and *Paenemarmota barbouri,* the largest of the ground-dwelling squirrels (Kurtén and Anderson 1980). These large forms did not survive into the Pleistocene. Marmots reached Eurasia in the late Pliocene or early Pleistocene. All extant species evolved in the Pleistocene (Black 1972; Mein 1992).

Early marmots probably inhabited cool, moist habitats in the basin-and-range country of the North American west (Polly 2003), and occurred as far south as Mexico (Cushing 1945). In Europe, all modern species are known only from the late Pleistocene. Marmots were associated with the tundra-forest-steppe fauna, which occupied the periglacial landscape (Zimina and Gerasimov 1973; Zimina 1996). Marmots were widespread in European lower mountain ranges (Kalthoff 1999a) and are common species of the "loess fauna" of the late Pleistocene (Rumiantsev and Bibikov 1994). The environment was characterized by cold winters and short, warm summers in a grassy landscape (Zimina and Gerasimov 1973).

Marmot distribution changed dramatically with warming and advance of the forests at the end of the last glaciation. Marmot populations from Porcupine Cave, Colorado, about 750,000 to 800,000 years ago, are related to *M. monax* (Polly 2003). By about 500,000 years ago *M. monax* appeared in Pleistocene locations in the eastern United States and was prominent in fossil deposits in Missouri about 15,000 years ago. *Marmota flaviventris* now lives in Colorado; fossils were found in numerous sites in western North America dating from less than 125,000 years ago (Kurtén and Anderson 1980). Some of these sites are south of the present range of *M. flaviventris,* such as the Newberry

Mountains of southern California (Goodwin 1989). The steppe marmot retreated from central Europe and the alpine marmot became restricted to the higher Alps Mountains. *Marmota primigenia,* found in middle and late Pleistocene loess deposits along with some *M. bobak* fossils in the middle Rhine region, became extinct as reforestation occurred north of the Alps (Kalthoff 1999b). In Italy, fossils from the floor level of cave deposits are *M. marmota.* Marmots from other levels (Upper or late Middle Pleistocene) have a more massive skull and a larger mandible (Aimar 1992). Several characters are in concordance with *M. marmota* and indicate a possible shift toward a smaller size as warming occurred.

This brief review of the historical record indicates that marmots have been large-bodied ground-dwelling squirrels for several million years and adapted to an open, cool landscape. Adaptation to a cool environment is evident in current species. Activity of *M. marmota* (Turk and Arnold 1988) and *M. flaviventris* (Webb 1980; Melcher et al. 1990) is much more restricted by heat than by cold during the active season.

The harsh environment

Considerable evidence supports the interpretation that currently species of marmots occupy harsh environments (Armitage and Blumstein 2002). Environmental harshness or severity is not readily defined and can include physical factors such as cold and drought and biological factors such as intense predation and disease. I will emphasize physical factors, such as length of winter and air temperature, which affect growth, reproduction, and survival of marmots. Because climatic records from marmot habitats are generally unavailable, biological features will be presented as evidence for environmental harshness.

Many species of marmots lose mass after emergence from hibernation. Heavy snow causes mass loss in the Olympic marmot (*M. Olympus,* fig. 30.1), the hoary marmot (*M. caligata*), and the steppe marmot. Mass loss typically occurs for up to several weeks in the woodchuck and in the alpine and long-tailed (*M. caudata*) marmots. Mass loss is associated with the use of fat reserves in the grey and black-headed marmots (Armitage and Blumstein 2002 and references therein).

In the long-tailed and grey marmots, up to 48% of the embryos are reabsorbed in bad years (e.g., a cold spring). Because fat accumulation is much lower in reproductive females than in barren females in many marmot species, reproduction is often not possible in successive years unless the litter is small (Bonesi et al. 1996; Armitage and Blumstein 2002). *Marmota bobak, M. marmota, M. olympus,* and *M. sibirica* usually skip 1 year between successful re-

Figure 30.1 The Olympic marmot overlooking a snowfield in mid June. Photo by K. Armitage.

productions, and *M. baibacina, M. caligata, M. camtschatica, M. caudata, M. menzbieri,* and *M. vancouverensis* often skip 2 or more years (Blumstein and Arnold 1998; Armitage 2000; Armitage and Blumstein 2002).

Because the length of the active season is short (mean length is 4.8 months), marmots must initiate reproduction as early as possible in the spring so time will be sufficient for all members of the population, but especially for reproductive females and young, to accumulate fat for hibernation (Armitage 2000). However, if a female emerges from hibernation too early, she may use considerable resources coping with a cold, snowy environment. At least six marmot species reduce early post-hibernation energetic stress by mating before emergence (Armitage and Blumstein 2002). Birth and early development of the young (e.g., *M. broweri;* Rausch and Bridgens 1989), may occur in the burrow before surface activity begins.

These known biological and climatic factors that demonstrate that marmots live in a harsh environment raises the question: what are the major adaptations for coping with environmental harshness?

Hibernation and large body size

Hibernation is a means of conserving energy during a period of food unavailability. Energy savings for animals hibernating singly during the hibernation period (compared to constant euthermy) averaged 83.3% for the yellow-bellied marmot, 43.8% for the alpine marmot (Armitage et al. 2003) and 43.2% for the woodchuck (Armitage et al. 2000). Marmots are the largest true hibernator; large body size in-

creases energetic efficiency (French 1986). Energy (fat) stores scale directly with mass; energy use scales to mass$^{3/4}$ at environmental temperatures within thermoneutrality and mass$^{1/2}$ at colder temperatures. Thus larger marmots accumulate more fat and use it relatively more slowly.

Body mass varies considerably among marmot species; emergence mass of the largest species, *M. bobak,* is 1.76 times that of *M. flaviventris,* the smallest species (Armitage 1999a). Larger species of marmots not only accumulate more mass because of larger size; they may accumulate more mass than expected based on body size. The accumulated fat not only fuels the metabolism of hibernation, it also sustains activity, including reproduction, following termination of hibernation and is the source of energy for coping with unfavorable climatic conditions until vegetation again becomes available. Large marmot species store more fat because they use more fat to cope with their harsh environment (Armitage 1999a).

Large body size has a second major advantage. A larger body size has a larger absolute gut capacity relative to metabolic rate. Larger animals can use more fibrous diets by means of longer retention of digesta in the gut (Hume et al. 1993). The gastrointestinal tract is one of the most intensive energy-use organs of the vertebrate body (Hume et al. 2002) and could exact high costs during hibernation. However, the system is reduced in size at the time of hibernation and increases in size and activity following emergence (Rausch and Rausch 1971; Bassano et al. 1992; Hume et al. 2002). The body-size effects of diet and hibernation form an integrative system; large body size enables marmots to exploit a diet consisting of grasses and forbs (Armitage

2000) that, coupled with a metabolic rate lower than that predicted from body mass (metabolism equations and a tissue growth efficiency about five times greater than that of a typical homeotherm; Kilgore and Armitage 1978; Armitage and Salsbury 1992) allows marmots to accumulate fat reserves and use them efficiently during the hibernation and early post-hibernation seasons.

Consequences of large size

The major consequence of large size is that there is insufficient time during the active season for young to mature before their first hibernation, with the exception of the woodchuck. The active season of the woodchuck, which is about 7.5 months, is 2 to 4 months longer than that of any other marmot (Armitage and Blumstein 2002). As a result of the combination of large size and a short active season, the young of all other species of marmots remain in their natal environment for their first hibernation or longer. This retention of young forms the basic social unit of marmots (Armitage 1981, 1996b, 2000).

The age of maturity may be estimated by calculating a maturity index (MI) by dividing the body mass at a specified age by the body mass of an adult. For this calculation, body mass values must be used from the same time of year to minimize the effects of fattening. Dispersal can occur when MI $\geq$ 0.5, but only *M. monax* achieves that value in its first summer (Blumstein and Armitage 1999) and disperses. The young of all other species of marmots have MI values $\leq$ 0.45 and do not disperse. However, yearlings of all species reach an MI $\geq$ 0.76, but only *M. flaviventris* disperses (Armitage 1999a). Thus, most species of marmots have delayed dispersal; that is, dispersal occurs 1 or more years beyond the year in which the MI for dispersal is reached.

A second consequence of large size is delayed reproduction. Reproduction is possible the year after an MI of about 0.65 is achieved (Blumstein and Armitage 1999). Only *M. monax* reaches that MI as a young and is the only species of marmot to reproduce as a yearling. All marmots have a MI > 0.7 as yearlings, and apparently could reproduce at age 2 years (Armitage 1999a). However, many species do not reproduce until age three. Delayed reproduction refers to those species that reproduce at least 1 year later than they otherwise could as indicated by the MI. However, for many species reproduction is delayed for more than 1 year for many members of the social group. Population means are available for some species from life table analyses; otherwise I made rough estimates based on population structure and the percentages of various age groups that reproduce (table 30.2). In all species for which data are available, the realized age of first reproduction is later than the age of reproductive maturity, even though all species have been reported to reproduce at the recorded age of maturity. Furthermore, the age of dispersal for at least three species, *M. sibirica*, *M. marmota*, and *M. bobak*, is older than the age of first reproduction (Blumstein and Armitage 1999; Armitage 1999a). Delayed dispersal and delayed reproduction characterize the more complex marmot social systems (table 30.1).

Group Living and Social Behavior

As a consequence of retaining offspring in the family or matriline, marmot social groups are based on kinship. Burrow-

Table 30.2 Age of first reproduction for female marmots

| Species[a] | Age (years) | | References |
	Reproductive maturity	Population mean	
M. flaviventris	2	3.02	Schwartz et al. 1998
M. caudata aurea	3	>3	Blumstein and Arnold 1998
M. c. caudata	4		Davydov 1991
M. sibirica	3		Bibikow 1996
M. marmota	2	>3	Arnold 1990a
M. bobak	2, 3	>3	Mashkin 1991; Shubin 1991
M. monax	1	ca. 1.5[b]	Snyder and Christian 1960
M. vancouverensis	3	4.33	Bryant 1996
M. baibacina	2 (3)	>4	Pole and Bibikov 1991; Bibikov 1991
M. caligata	3	>3	Barash 1974d; Holmes 1984
M. olympus	3	$\geq$3.6[b]	Barash 1973

[a]Species are listed in order of increasing body mass at immergence.
[b]Author's calculations.

mates are more closely related than a random sample of the population. Burrow-mates in *M. olympus* on average are related by 0.41 (Barash 1989); those of *M. caligata*, 0.39 (Armitage 1996b); and those of *M. flaviventris*, by 0.5 (Armitage and Johns 1982). The dynamic processes leading to patterns of relatedness have been best analyzed in *M. marmota* and *M. flaviventris*. In the alpine marmot, average relatedness between dominant and subordinate females is 0.33 (Allainé 2000); in the yellow-bellied marmot it is 0.5 (Armitage 1991). The difference in average relatedness is a consequence of the recruitment patterns of the two species.

In the yellow-bellied marmot, daughters are recruited by their mothers into the matriline. Daughters from different year classes or littermate sisters may be recruited. However, turnover in the territorial male may result in maternal half-sibs being recruited, and demographic processes may lead to grandmother–granddaughter or aunt–niece relationships. Under these conditions, average relatedness can temporarily decrease to 0.25. These matrilines divide to form two groups, with each group having an average relatedness of 0.5 (Armitage 1984, 1991). One result of the fission of matrilines is that they are small; mean matriline size measured over eleven habitat sites was 1.38 (*N* = 544) (Armitage and Schwartz 2000). Matrilines are territorial and share resources, such as burrow sites and foraging areas, with other members of the matriline—but not with members of neighboring matrilines. Use of space serves as an indirect measure of resource sharing. Space-use overlap averages about 43% for adult females related by 0.5 and is much less than predicted for all other degrees of relatedness (Armitage 1996a). For adults related as aunt–niece, space-use overlap averages about 4%.

The major factor that reduces average relatedness in alpine marmots is that only 15% (Vanoise, France; Allainé 2000) to 18.3% (Berchtesgaden, Germany; Arnold 1993) of the territorial females inherited the territory of their birth. On average, 17.5% to 21.5% of the females take over a neighboring territory. In an 8-year study of a high altitude colony in northern Italy, no female became reproductive in her natal family, one female was an immigrant from an adjacent family, and three female immigrants from outside the colony successfully reproduced (Lenti Boero 1994, 1999). Similarly, immigrant males replaced resident males; no resident male was known to have two successive female partners (Lenti Boero 1999). Regardless of the degree of relatedness within alpine marmot families, territories are stable and overlap is slight (Perrin et al. 1993; Lenti Boero 1996; Sala et al. 1996). Transfer among groups commonly occurs in golden marmots (*M. caudata aurea*); most transfers are from larger to smaller groups (Blumstein and Arnold 1998). Translocations from one family to another also occur in *M. menzbieri*, *M. c. caudata*, *M. baibacina*, *M. bobak*, *M. sibirica*, and *M. camtschatica* (Mashkin 2003 and refer-

ences therein). Intergroup transfer occurs more frequently in golden marmots than in alpine marmots (Blumstein and Arnold 1998), but the dynamics of intergroup transfer are generally unknown. Long-term studies of the demographic characteristics of the immigrating animals and of the residents and the consequences of such transfers on the lifetime reproductive success of all the participants are needed. Some consequences of the variable kinship in extended families will be discussed later.

In general, rates of amicable and agonistic behavior (Armitage 1977) are related to the social system. Amicable behavior among woodchucks is mostly limited to mother–offspring interactions; agonistic behavior characterizes the social interactions of this solitary species. The rate of greetings is much higher in the species in the restricted family groups than in the yellow-bellied marmot. Agonistic behavior in the restricted family groups is rare and occurs primarily when the adult male repels intruders or chases peripheral males (reviewed in Armitage 2003a). In the extended families, agonistic behavior occurs primarily between members of different groups, and amicable behavior predominates within groups (Armitage 1996b). However, when a territorial takeover occurs, subordinates may be chased and subsequently disappear (Lenti Boero 1994). The territorial pair defends against adult same-sex intruders (Arnold 1993; Lenti Boero 1999) and fighting may occur, with the loser being expelled almost immediately after the fight (Hackländer and Arnold 1999). Sometimes the intruder male does not expel the old territory owner and both males remain in the group. Because of variable relatedness in the extended family groups, the relationship of rates of social behavior and nearest neighbor associations, such as during foraging, should be determined. This study would be especially interesting, because evidence suggests that reproductive suppression (to be discussed subsequently) is kin-biased.

The relationship between kinship and social behavior has been most intensively studied in the yellow-bellied marmot. Mother–daughter and sister–sister amicable behavior greatly exceeds agonistic behavior, whereas aunt–niece agonistic behavior greatly exceeds amicable behavior in three colonies observed for 30 years (Armitage 2002). Mothers, daughters, and sisters typically associate in the same matriline; their aunts and/or nieces live in different matrilines. The same pattern is evident when social behavior of adult females in four colonies is analyzed as a function of the degree of relatedness (*r*). Amicable behavior exceeds agonistic behavior by 5.5 times when *r* = 0.5; at all *r* ≤ 0.25, agonistic behavior exceeds amicable behavior by an average of 5.4 times (Armitage 2002). Behavior is related to recruitment of female yearlings; recruitment is more likely when the amicable/agonistic ratio is high and when amicable behaviors predominate among adult females (Armitage 1986a, 1996b).

Benefits of Group Living

Delayed dispersal in marmots with a family social structure is associated with increased survivorship. After the first year, survivorship of the golden marmot did not differ significantly from that of the Olympic and Vancouver Island marmots, and was significantly greater than that of the yellow-bellied marmot (Blumstein et al. 2002). The life table of the steppe marmot (Shubin 1991) is similar to that of the golden marmot and indicates higher survivorship, similar to that of the other family social groups. Only for the yellow-bellied marmot did males have lower survivorship than females (Schwartz et al. 1998; Blumstein et al. 2002). Lower survivorship in males is associated, in part, with much higher rates of dispersal by male yearlings and extensive movements by older males, either searching for females or defending a very large territory (Van Vuren and Armitage 1994b; Schwartz et al. 1998).

Group size affects fitness, measured as survivorship and net reproductive rate (NRR), in yellow-bellied marmots. Mean matriline size for eleven habitat sites varied from 1.17 to 2.24. Survivorship and NRR increased as mean matriline size increased, decreased slightly in large matrilines, and varied between sites with similar mean matriline sizes. Although there was no direct measure of site quality, survivorship, but not NRR, was closely related to the area of the habitat patch (Spearman rank correlation = 0.45, 0.1 > P > 0.05). The number of females was directly correlated with habitat area; possibly the larger number of females on a patch provides increased detection of predators (Armitage and Schwartz 2000). Marmots usually alarm-call when a predator is detected. Alarm-calling is not altruistic; we have never observed a caller to become a victim. Similarly, callers did not suffer predation in other marmot species (Barash 1975b). Variation in the rate of alarm-calling by yellow-bellied marmots was best explained by direct parental care (Blumstein et al. 1997). More females on a habitat patch is associated with more litters, hence the likelihood that alarm-calling will occur increases. Although a female may call to warn the offspring, other marmots will also be alerted.

When matriline size was considered over all habitat sites and the average per year calculated, both survivorship and NRR were significantly related to matriline size. The relationships were curvilinear; survivorship was only slightly affected by matriline size, but the relationship explained most of the variation ($R^2 = 0.97$). Matriline size explained less of the variation ($R^2 = 0.67$) when sites were treated separately. The difference between the amounts of variation in survivorship explained by the two analyses suggests that there is a major effect of habitat quality on survivorship. NRR was strongly affected by matriline size; NRR increased as matriline size increased to a maximum at three, then decreased in matrilines of four and five (Armitage and Schwartz 2000).

Although the optimal matriline size is three, matrilines of this size occur only 4.7% of the time. The low frequency of the optimal matriline size raises the question of why it does not occur more often. The three major reasons seem to be habitat size, demography, and reproductive competition. Small habitat patches never develop large matrilines; 55% of the large matrilines ($\geq$ 3) occurred at three localities. Although these are among the larger localities, size is a necessary but not a sufficient condition for the development of large matrilines. These localities have numerous burrow sites and large meadows with abundant forage (Svendsen 1974; Kilgore and Armitage 1978; Frase and Armitage 1989); larger areas with widely spaced, small resource patches have smaller matrilines. Demography affects the formation of large matrilines as follows. Assume that a female first reproduces at age 3 years (the population mean). She would be five years old when her first daughter would become an adult at age 2, and 6 years old when she could recruit her second daughter. Only 31% of the 3-year-old females reach age 6 years (Schwartz et al. 1998), and a female's first 2-year-old has a 66% probability of reaching age three. Thus there is a 0.22 probability that both these demographic events will occur within a matriline. However, even if favorable demographic events occur in a favorable habitat, matriline size may be limited by reproductive competition (Armitage 2003d), which will be discussed as a cost of group living.

In marmots with a family social system, there is no indication that group size affects NRR. Regardless of group size, only the dominant territorial female reproduces (Bibikow 1996; Blumstein and Armitage 1999; Armitage 1999a). However, group size may affect survivorship through social thermoregulation during hibernation (Arnold 1988, 1993). In the alpine marmot, mortality may be high when parents hibernate with juveniles; the entire group may die (Arnold 1990b; Lenti Boero 1999). Mortality results from a greater loss of mass when juveniles are present, and is especially severe in hibernacula of low quality (Arnold 1990b).

Increased survivorship occurs under two conditions: (1) the absence of juveniles and (2) the presence of subordinate adults with the territorial pair and juveniles. When juveniles were absent, loss of mass was lower and survivorship greater the larger the number hibernating together. When juveniles were present, there was no influence of group size on winter mortality (Arnold 1993), but the presence of older, subordinate animals significantly reduced mortality. Only potential full sibs engaged in alloparental care by warming infants; mass loss of the subordinates was greater but survivorship of infants was greater. Overall, there was almost no mortality when territorials, subordinates, and infants were present (Arnold 1990b). Interestingly, less-related subordinates lost less mass when hibernating in groups with infants. Because the other animals increase burrow temperatures, these subordinates profit from the increased heat but apparently do not increase metabolism to contribute to burrow warming and alloparental care. Two essential points should be emphasized: social thermoregulation and alloparental care require the presence of subordinate adults, and alloparental care is strongly biased to favor close kin.

Although golden marmots live in groups containing up to seven adults and hibernate in groups, total group size does not influence overwinter survival of juveniles. For nonjuveniles, total group size, including juveniles, does not influence the probability of survival. There is a tendency for nonjuvenile survival to decrease as the number of nonjuveniles increased. For juveniles, when only parents and littermate sibs were present, there is less mortality than in groups containing other animals. When juveniles are excluded from the count, overwinter survival of juveniles decreased when hibernating in groups with more nonjuveniles (Blumstein and Arnold 1998). Thus, I find no compelling evidence for social themoregulation, and some indication that group hibernation may be costly.

Social thermoregulation

Social thermoregulation refers to the active contribution of heat to other individuals during hibernation at a cost in increased mass loss in the donor. It requires synchrony in the arousal-torpor patterns, which reduces mass loss (Ruf and Arnold 2000). The presence of juveniles can decrease overall synchrony and offset the reduction in mass loss provided by the increased insulation of group huddling. Decreased synchrony probably explains high mass loss and mortality when only juveniles and the two territorial adults hibernate together, and the mass loss of related subordinate adults who provide the alloparental care. Juvenile yellow-bellied marmots hibernating in groups had higher rates of mass loss than animals hibernating singly; torpor cycles in the grouped young were asynchronous (Armitage and Woods 2003). Furthermore, there was no difference in overwinter survivorship of yellow-bellied marmot young hibernating in groups ranging from one to eight individuals. Although these young probably hibernate with their mother (Lenihan and Van Vuren 1996), there is no evidence for benefits from joint hibernation.

Probably all species of marmots, except *M. monax*, hibernate in groups (I use the term *group hibernation* to indicate that members of a social group share a hibernaculum and have the potential to jointly benefit from huddling). Group hibernation raises the question of its evolution to social hibernation and of the importance of social hibernation

in the evolution of marmot sociality. Social hibernation apparently requires the presence of subordinate adults; this social group is present only in species that have extended families (table 30.1). Although delayed dispersal was interpreted to occur so that subordinate adults could provide alloparental care (Arnold 1990a, 1990b), this interpretation seems unlikely. Alloparental care would have to occur before delayed dispersal; in other words, the nondispersers would have to "anticipate" future social thermoregulation. It seems more likely that delayed dispersal occurred to increase the probability that subordinates would become more competitive for a territorial position. By remaining in the family reproduction is delayed and a direct fitness cost is incurred. However, by hibernating with other family members, the subordinates have an opportunity to recoup some of the fitness loss by gaining indirect fitness through alloparental care directed to close kin. The territorials benefit from delayed dispersal. The fitness of the territorials is zero unless their offspring reproduce. By accepting delayed dispersal, the territorial dominants increase the probability of having grand-offspring while increasing their fitness through their increased survival and that of the next generation of their offspring through social hibernation. In effect, delayed dispersal incorporates subordinate adults into the social group and makes possible the evolution of social thermoregulation from group hibernation.

Group hibernation probably evolved as a consequence of retaining infants at home for their first hibernation. A good hibernaculum may be a limited resource (Holmes 1984a), and may be used over many generations (Lenti Boero 2001; Armitage 2003b). Thus survival of young was likely greater if they hibernated in a proven hibernaculum independent of any possible benefits from huddling. Group or social hibernation likely is more important in those species that are relatively small and have a long hibernation period. For example, the alpine marmot at the onset of hibernation is relatively small and hibernates for about 6.5 months; thus social thermoregulation characterizes this species. By contrast, the larger woodchuck hibernates for only about 4.5 months and relies solely on its fat stores during solitary hibernation (Armitage et al. 2000). Although woodchucks at more northern latitudes hibernate about a month longer under more harsh winter conditions, they survive by using a larger percentage of their mass (Ferron 1996). Two of the largest marmots, the Olympic and hoary, lose the largest amount of mass during a hibernation of about 7.5 months (Armitage and Blumstein 2002). This large mass loss may be a consequence of a long hibernation in a harsh environment; possibly the adults provide heat to their young and thus incur a mass loss (Armitage 1999a). Virtually nothing is known about hibernation energetics outside the woodchuck and the alpine and yellow-bellied marmots. Considerable research is needed to evaluate the role and evolution of social thermoregulation in marmots.

Costs of Group Living

One potential cost of group living is increased disease transmission. A group of four Vancouver Island marmots hibernated together and all died from a bacterial infection (Bryant et al. 2002). Female yellow-bellied marmots that failed to reproduce had more fleas than those that reproduced, and marmots that died during hibernation had more fleas than the survivors. Yearlings with greater flea infestations grew more slowly (Van Vuren 1996). Infant winter mortality of alpine marmots increased with the ectoparasite load of the family, but there was no relationship between ectoparasite load and group size or marmot density (Arnold and Lichtenstein 1993). Both of these reports suggest that ectoparasites affected individual fitness, but neither found evidence that the parasitism was a cost of sociality. Plague foci—concentrations of infected marmots—occur regularly in marmots such as *M. baibacina* and *M. sibirica*. Plague apparently is most detrimental when associated with drought or food shortage. Mortality during plague epizootics is small, and may reach 10 to 15%; populations quickly recover (Bibikow 1996). Some populations of *M. sibirica* develop genetic resistance to plague, which is reflected in a difference in gene frequencies and the level of heterozygosity between plague-free and infected populations (Batbold 2002). There is no evidence that plague infections are a cost of sociality; plague is present in many other nonsocial species in the same geographic area (Bibikow 1996). Thus available evidence does not support disease as a cost of sociality; in fact, some of the severe viral infections seem to be restricted to the asocial *M. monax* (Armitage 2003a).

The major cost of sociality is decreased reproduction. In yellow-bellied marmots, reproductive competition occurs between adjacent matrilines. A numerically dominant matriline reduces reproductive frequency in the numerically inferior matriline. Within a matriline, adults may chase yearlings, even when the yearlings are full sibs, from the colony. This activity results in resources becoming available to the offspring of the adult rather than to other individuals (Armitage 1986b, 2003c).

The primary cause of decreased reproduction is reproductive suppression. Evidence indicates that reproductive suppression occurs in every marmot species (Armitage 1996b; Blumstein and Armitage 1999). For example, in the biennial-breeding hoary marmot a female skips an additional year if her coresident breeds. When subordinate females breed, they produce half as many young as dominant females (Wasser and Barash 1983).

In the alpine marmot, *M. marmota,* the dominant territorial female suppresses the reproduction of the subordinate females, even in those years when the dominant female skips reproduction. Although subordinate females copulate, they fail to wean a litter (Arnold 1990a; Perrin et al. 1993; Lenti Boero 1999). Reproductive suppression in alpine marmots apparently is mediated through social behavior. Social interactions were most frequent during gestation, and the dominant females initiated most of the agonistic behavior (Hackländer et al. 2003). Subordinate females were not denied access to males; some copulated and developed nipples, which indicates pregnancy (Armitage and Wynne-Edwards 2002). Social status was unrelated to estradiol concentrations during mating; glucocorticoids increased in all adult females, with higher levels in subordinates (Hackländer et al. 2003). The authors suggest that suppression occurs by the negative effects of stress acting on the hypothalamic-pituitary-gonadal axis. Dominant females apparently pay a cost for reproductive suppression; reproductive success decreased as the number of adult subordinate females in the family increased.

But what of males? Extended family groups often contain several adult males, but only one is a dominant territorial. Among adult subordinate male alpine marmots at least 3 years old, sons had androgen levels as high as the territorial male, whereas non-sons had significantly lower levels, similar to those of 2-year-olds (Arnold and Dittami 1997). Corticosteroid and androgen levels were positively correlated in territorial males and negatively correlated in subordinate males. Corticosteroids were negatively correlated with body mass in sons and were high regardless of body mass in non-sons. High corticosteroid levels are associated with less mass gain; only those individuals with sufficient summer mass gain are capable of both surviving hibernation and sustaining the energetic costs of reproduction. Thus, suppression of competitive males may act through direct conflict during the mating season and through impairment of energetic processes needed to sustain reproduction.

Male suppression is kin-biased. Non-sons received more injuries from fighting, and large sons are tolerated. Large sons are critical for social thermoregulation and may also be more competitive for taking over an adjacent territory (Arnold and Dittami 1997). Although there may be fitness gains from the favorable treatment of sons, there is a possible fitness cost. Sons compete with their fathers for reproductive success, such that the territorial male does not father all the offspring (Arnold et al. 1994).

In alpine marmots, males may cause reproductive failure of females. Territory takeover by males impaired reproduction of dominant females if the takeover occurred after the mating period, despite clear evidence of pregnancy in the form of enlarged nipples and late molt (Hackländer and

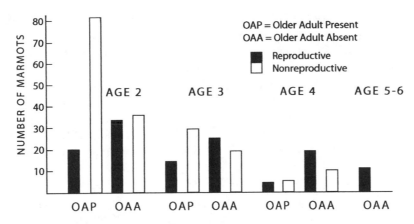

Figure 30.2 Age and reproductive suppression in yellow-bellied marmots.

Arnold 1999). Progesterone decreased, and there was only slight evidence of infanticide. However, male takeover may result in infanticide (Perrin et al. 1994). The females that failed to wean young gained sufficient mass to reproduce the following year, rather than skipping reproduction. Thus the males father young at the first opportunity rather than having to wait an additional year. The females regained some of the lost fitness by producing a larger litter in the subsequent mating with the new male (Hackländer and Arnold 1999).

In yellow-bellied marmots there is strong directional selection for early maturity (Oli and Armitage 2003). Despite females reproducing earlier having a higher fitness than those that delayed maturity, only 53 of 171 (31.0%) of 2-year-old females (age of first reproduction) successfully weaned litters. However, only 19.6% reproduced when an older female was present in the matriline, compared to 47.8% when older adults were absent (Armitage 2003d). Reproductive suppression acted to delay the age of first reproduction and was independent of the number of females in the matriline. As age of first reproduction increased, a smaller percentage of females was suppressed (fig. 30.2). Reproductive suppression occurred when the mother was the older female present; however, reproduction by a 2-year-old was more likely if the mother was the older female rather than some other adult (Armitage 2003d).

Clearly, reproduction was suppressed more in 2-year-olds than in older animals (fig. 30.2). Possibly the 2-year-olds failed to reproduce because they were immature rather than suppressed. However, when 2-year-olds were matched for condition (Armitage 2003d) or received supplemental food (Woods 2001), reproduction was significantly less likely if older females were present.

The effect of the presence of older adult females on the reproduction of younger adult females raises the question of whether the decline in fitness (NRR) in large matrilines (4 or 5 adult females) could be related to their age distribu-

tion and reproductive suppression. In matrilines in which one or no female reproduced (low success), half of the females aged five or older reproduced, whereas only 12% of females aged two to four reproduced (fig. 30.3). Two females reproduced in matrilines of moderate success; a greater percentage of older females than of younger females reproduced (fig. 30.3). In the matrilines in which three or four females (high success) reproduced, both old and young females reproduced at high rates (fig. 30.3). The combined low- and moderate-success matrilines are characterized by significant differences in the reproductive rates of old and young females, whereas reproductive rates of young and old females in the high success matrilines do not differ (Armitage 2003d). Thus reproductive suppression is the major factor decreasing mean NRR in large matrilines; however, reproductive suppression is not total, which results in high NRR in some large matrilines. Reproductive suppression

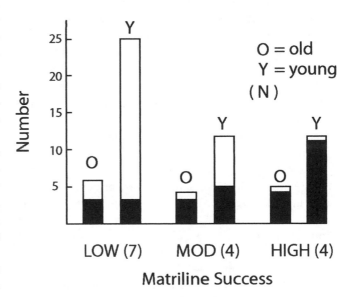

Figure 30.3 Reproductive success in large matrilines by young and old adult females.

through its effect on the age of first reproduction is a major demographic mechanism of population dynamics. Retrospective analyses of a life table response experiment revealed that age of first reproduction made a major contribution to annual changes in the projected population growth rate (Oli and Armitage 2004).

Why Remain in the Group, Given Suppression?

The major reason for remaining in the group is the high cost of dispersal. In the alpine marmot, dispersers and evicted, former territorials form the floating part of the population. As many as 50% of the floaters may be killed by predators; survival of those floaters that do not establish territorial residency is virtually nil, whereas annual mortality of subordinates living with resident territorials is about 4% (Arnold 1993; Grimm et al. 2003). By contrast, mortality of dispersing yearling yellow-bellied marmots is only 14% more than that of philopatric yearlings, and overwinter survival does not differ between dispersing and philopatric yearlings (overall mean = 89.3%).

Why the difference between the two species? Alpine marmots live in open habitat in forests; all available space is filled with family territories of about 2.5 ha (see fig. 13 in Arnold 1999). Similarly, steppe marmot families occurred over millions of hectares until about the 11th century (Tokarski et al. 1991). In effect, the only available habitat was occupied, and a disperser had almost no probability of surviving unless it found a vacant territorial position or was sufficiently large and strong to displace a resident territorial. Presumably older and larger marmots had a much higher probability of success. By contrast, radio-tracking of yellow-bellied marmots revealed that unoccupied patches of habitat were widely available (Downhower 1968; Van Vuren 1990). Although many of these sites did not have resources sufficient for long-term residency, they did provide shelter and hibernation sites during dispersal. Marmots that delayed dispersal until age two did not have increased survival, presumably because increased size was not an advantage for utilizing the widely available habitat patches. Similarly, burrow sites are readily available to dispersing young woodchucks, and young individuals readily dig new burrows (Armitage 2003b). Thus delayed dispersal does not occur in these species, as there is no known competitive advantage that can be gained by delay. By contrast, alpine marmots cannot dig a hibernaculum in 1 year (Arnold 1993), and delay dispersal until they are competitive for obtaining a territorial position in a saturated environment (Armitage 1996b). I believe that differences in the probability of successful dispersal, which results in delayed dispersal in many

species of marmots, is the major factor leading to the diversity of social systems.

Furthermore, marmots may gain direct fitness in a variety of ways. Subordinate male alpine marmots may father some of the offspring (Arnold et al. 1994; Goossens et al. 1998). In addition, males gain indirect fitness benefits from alloparental care, which leads to a male-biased population sex ratio (Allainé et al. 2000) and polyandry. A subordinate female may reproduce in the family group (Mikhailuta 1991; Goossens et al. 1998) or may bud off to form a new group or matriline, or ascend to the territorial position in her natal group (Arnold 1990a; Armitage and Schwartz 2000; Oli and Armitage 2003). She may also mate in one group and escape suppression by the territorial dominant by joining a neighboring group (Goossens et al. 1996). Finally, by being a member of a matriline of two or three adult females, her reproductive success is higher than that of living alone (Armitage and Schwartz 2000).

Summary

Marmots evolved as large-bodied, ground-dwelling squirrels that occupied cool, moist, open landscapes. Large size allowed marmots to use more fibrous diets and to efficiently store and use fat during hibernation. Warming and the advance of the forests at the end of the last glaciation restricted the distribution of most species of marmots to high elevations, harsh environments that are characterized by a short growing season. Thus all species of marmots except the woodchuck require at least one additional growing season for young to reach maturity. The retention of young in their natal environment for their first hibernation or longer forms the basic social unit of marmots. Thus the evolution of adaptations for hibernation led to the evolution of sociality.

Marmot sociality evolved twice in two subgenera to form four social groups: solitary and extended family groups in the subgenus *Marmota*, and female kin groups and restricted family groups in the subgenus *Petromarmota*. Social behaviors are generally amicable within social groups and agonistic between social groups. Behaviors are kin-biased, with closely related kin treated more favorably than more distantly related kin. Most social groups are characterized by delayed dispersal and reproduction.

Delayed dispersal occurs because habitat saturation provides almost no opportunity for a small animal to survive by becoming resident as an immigrant in a social group. Only large individuals have a high probability of successfully competing for the dominant position within a family. Delayed reproduction is the result of reproductive competition; the dominant female (or male in extended families) suppresses the reproduction of subordinates. This loss of reproduction is the major cost of sociality. A major benefit is increased survivorship. In yellow-bellied marmots, net reproductive rate increases as the number of matrilineal members increases, but decreases in the largest matrilines. Reproductively suppressed subordinates apparently choose the best alternative available to them by remaining in the social group until they can escape suppression, bud off to form a new group, or take over the dominant position within their group or an adjoining group.

Chapter 31 Environmental Constraints and the Evolution of Sociality in Semifossorial Desert Rodents

Jan A. Randall

DESERTS ARE DIFFICULT PLACES in which to live. Animals must survive and reproduce under extreme and unpredictable environmental conditions that can include extended droughts lasting for months, even years. Dramatic seasonal fluctuations in environmental conditions are common, with temperatures extremely high in the summer and below freezing in the winter. High temperatures and low humidity during the day are followed by cold nights. Strong winds and unstable sandy soils hinder the growth of vegetation (Cloudsley-Thompson 1975; Shenbrot et al. 1999). Deserts are widely distributed on all continents except Antarctica, and despite the high temperatures and aridity, desert habitats can be quite diverse, leading to considerable taxonomic diversity in desert-dwelling rodents (Mares 1993; Degen 1996; Shenbrot et al. 1999).

To survive above ground in desert environments rodents often must withstand periods of limited plant growth, resulting in low primary production of vegetation, a sparse and patchy distribution of food, and unpredictable food abundance (Shenbrot et al. 1999; Randall et al., 2005). Desert habitats usually have limited cover, and rodents are preyed on by numerous predators (Randall et al. 2000; Rogovin et al. 2004). Finally, desert rodents are faced with the physiological problem of maintaining a constant body temperature while minimizing water loss in arid conditions.

Although the physiology and ecology of desert rodents are well known (e.g., Genoways and Brown 1993; Shenbrot et al. 1999), there is still much to be learned about behavioral adaptations to the challenges of arid conditions. My goal in this chapter is to examine how environmental conditions in deserts limit or promote the social behavior of semifossorial desert rodents that are unable to escape extreme temperatures and aridity by living totally underground. For a better understanding, I sometimes contrast the behavior of semifossorial rodents with the behavior of subterranean rodents. See Nevo chap. 25, Lacey and Sherman chap. 21, Lacey and Ebensperger chap. 34, and Faulkes and Bennett chap. 36 this volume. I emphasize social adaptations in what is fundamentally a solitary group of animals (Randall 1994). Although my analysis is somewhat constrained by available information from studies under natural field conditions, enough is known to develop these themes: (1) physiological adaptations to water scarcity and high temperatures place broad constraints on the behavior of semifossorial desert rodents, (2) conditions of limited abundance and patchy distribution of food inhibit the formation of stable social groups and promote a solitary life style, (3) desert rodents are often flexible and opportunistic in their social adaptations to changing and unpredictable conditions of arid environments, (4) when social groups do emerge, dispersal is delayed, leading to the retention of young in the natal area, social tolerance, and the sharing of space by relatives, and (5) because of unpredictable conditions of food abundance, social groups in deserts are often unstable and density dependent. I will emphasize comparative studies of semifossorial kangaroo rats (*Dipodomys*, family Heteromyidae) and gerbils (family Muridae, subfamily Gerbillinae), with pertinent research on other rodents to provide insights into the evolution of social behavior in arid environments.

Factors Influencing Social Behavior of Desert Rodents

Physiological constraints

High temperatures and limited water are perhaps the most unrelenting environmental pressures on the characteristics of desert rodents. The major adaptations to these conditions include low metabolic rates, efficient kidneys for water conservation, and water produced through oxidative metabolism (Schmidt-Nielsen 1975; French 1993). There are also behavioral adaptations that have physiological benefits. A majority of semifossorial rodents confine their activity to the cooler temperatures of night time (Randall 1994). The rodents also avoid temperature extremes by retreating into subterranean burrows, where cool temperatures and high humidity provide a favorable microclimate for thermoregulation and water conservation, as well as protection against predators and a place to store food (Kinlaw 1999).

Because burrows are so important to survival and reproduction, communal digging of burrows has been proposed as a cause for the evolution of social groups of bathyergid and hystricognath rodents (Faulkes et al. 1997; Ebensperger and Cofré 2001; Ebensperger and Bozinovic 2000a). Cooperative digging of burrows to locate food is especially adaptive for subterranean mole-rats in arid habitats (Faulkes et al. 1997). The role of communal digging of burrows is less clear in other desert rodents (Kinlaw 1999). Under arid conditions, burrow systems used for shelter can remain intact for years or decades, and are passed from one generation to the next. In addition, solitary species seem perfectly capable of digging sufficient burrow space for survival and reproduction under most conditions; some semifossorial desert rodents use existing rocky crevices and niches as shelters rather than digging burrows (Shenbrot et al. 1999).

Stress is another physiological factor that could impact the survival and reproduction of desert rodents. Although semifossorial desert rodents are usually solitary, under some conditions densities become high, forcing individuals into group situations. These high-density conditions could increase stress, leading to group instability, dispersal, and increased concentrations of glucocorticoids (Sapolsky 1983). The effects of group living on stress in a social desert rodent across dramatic changes in food availability and population densities have been tested in breeding males of the great gerbil, *Rhombomys opimus* (Rogovin, Randall et al. 2003). Concentrations of fecal testosterone were lower and corticosterone higher during periods of high population densities when social groups were large compared with higher testosterone and lower corticosterone concentrations during lower densities when social groups were small (Rogovin, Randall et al. 2003; Rogovin, Moshkin et al. 2003, and

unpublished data). Regression analysis of the relative role of social factors showed a positive and significant relationship between concentrations of male fecal corticosterone and the number of adult females in a social group. These findings suggest that social stress adversely influences survival of male gerbils and may be greater in higher population densities because of increased social contact with multiple females in larger groups (Sapolsky 1983; Wingfield et al. 1990).

Social interaction may also be stressful to solitary rodents. Ganem and Nevo (1996) proposed that social tolerance is an adaptation to conserve energy and water as well as to minimize stress. Ganen and Nevo compared genetically distinct populations of solitary mole-rats (*Spalax ehrenbergi*) from different environments and found that the arid-adapted rodents were more tolerant of conspecfics than more aggressive individuals from a mesic habitat. The desert rodents had lower concentrations of corticosterone and higher urine osmolarity than the more aggressive mesic rodents. (See Nevo chap. 25 for more information on behavior of fossorial *S. ehrenbergi*.)

Constraints of food abundance and distribution

The costs and benefits of obtaining and defending food can be significant factors influencing the formation of social groups in arid habitats (Jarvis et al. 1994). Arid-adapted plants range from those that produce underground storage structures in roots and tubers to permanent succulent plants and widely distributed shrubs and grasses of low quality and quantity. Because of the range in type, productivity, and distribution of plants, desert rodents vary in their adaptations for obtaining food (Shenbrot et al. 1999). Besides feeding on underground storage organs, desert rodents feed on seeds (granivores), plant leaves and stems (folivores), insects (insectivores), and a combination of food types (omnivores; table 31.1).

The dispersed, patchy food and unpredictable rainfall in arid habitats have been proposed as primary causes of group formation in the subterranean bathyergid mole-rats (Jarvis et al. 1994). According to the "aridity-food-distribution hypothesis," eusociality evolved in *Heterocephalus glaber* and *Cryptomys damarensis* because cooperation was necessary to dig burrows during short, unpredictable periods of rainfall to maximize the harvest of the patchily distributed roots and tubers. Social mole-rats gain from cooperation because the high quality and quantity of the resource feed many animals once the food is found, which presumably compensates for the high costs of burrowing (Jarvis et al. 1994; Faulkes et al. 1997; Faulkes and Bennett chap. 36 this volume). Individual mole-rats in habitats with higher rainfall

Table 31.1 A comparison of sociality, vegetative cover, feeding adaptations, pattern of activity, mode of locomotion and body mass of genera in three families of semifossorial desert rodents, Heteromyidae, Muridae and Dipodidae

Genera	Sociality	Habitat openness	Feeding adaptation	Activity	Loco-motion	Body mass (g)	Reference
Heteromyidae							
Microdipodops	Sol.	O, I	G	N	B	10–12	O'Farrell and Blaustein 1974a,b; Genoways and Brown 1993
Dipodomys	Sol.	O, I	G, GF	N	B	27–178	Eisenberg 1963a, 1967; Genoways and Brown 1993; Randall 1993
Chaetodipus	Sol.	O, I	G	N	B	12–40	Eisenberg 1963a, 1967; Hoffmeister, 1986; Genoways and Brown 1993; Nowak 1999; Shenbrot et al. 1999
Perognathus	Sol.	O, I	G	N	B	8–18	Eisenberg 1963a, 1967; Genoways and Brown 1993; Nowak 1999; Shenbrot et al. 1999
Muridae, Gerbillinae							
Gerbillus	Sol., I	O, I	G, GF, GI	N	Q	10–50	Roberts 1977; Osborn and Hemly, 1980; Griffin 1990; Skinner and Smither 1990; Pavlinov et al. 1990; Harrison and Bate, 1991; Nowak 1999; Shenbrot 1999; Gromov 2000
Gerbillurus[a]	Sol., I	O, I	G, GF	N	Q	22–48	Skinner and Smithers 1990; Pavlinov et al. 1990; Shenbrot et al. 1999
Meriones[b]	Sol., I, Soc.	O, I, C	G, GF, F	N, ND, D	Q	40–250	Naumov and Lobachev 1975; Roberts 1977; Sludsky et al. 1978; Osborn and Hemly 1980; Eigelis 1980; Prakash 1981; Harrison and Bates 1991; Pavlinov et al. 1990; Gromov et al. 1996; Gromov 2000; Shenbrot et al. 1999
Tatera[c]	I, Soc.	I, C	O	N	Q	90–150	Prakash 1975; Roberts 1977; Pavlinov et al. 1990
Psammomys	Sol.	O, I, C	F	D	Q	200	Daly and Daly 1975a; Osborn and Helmy 1980; Gromov 2000; Tchabovsky et al. 2001
Rhombomys	Soc.	O, I	F	D	Q	200	Kutcheruk et al. 1972; Naumov and Lobachev 1975; Goltzman et al. 1977; Dubrovsky 1978; Pavlinov et al. 1990; Gromov 2000
Dipodidae							
Euchoreutes	Sol.	O	I	N	B	33	Ma et al. 1987; Sokolov et al. 1996; Rogovin et al. 1987
Cardiocraniu	Sol.	O	G	N	B	12	Fokin 1978; Shenbrot et al. 1995; Rogovin et al. 1987; Sludsky et al. 1977; Sokolov et al. 1996; Rogovin 1999
Salpingotus	Sol.	O, I, C	G, GI	N	B	12–20	Sludsky et al. 1977; Fokin 1978; Shenbrot et al. 1995 ; Sokolov et al. 1996; Rogovin 1999
Paradipus	Sol.	O	F	N	B	120	Sludsky et al. 1977; Fokin 1978; Shenbrot et al. 1995; Rogovin 1999
Stylodipus	Sol.	O, I	GF	N	B	60–80	Sludsky et al. 1977; Shenbrot et al. 1995; Sokolov et al. 1998; Rogovin 1999
Dipus	Sol.	O	G	N	B	70	Sludsky et al. 1977; Fokin 1978; Shenbrot et al. 1995; Sokolov et al. 1996; Rogovin 1999
Eremodipus	Sol.	O, I	GF	N	B	50	Sludsky et al. 1977; Fokin 1978; Shenbrot et al. 1995; Rogovin 1999
Jaculus	Sol.	O	GF	N	B	65–130	Fokin 1978; Shenbrot et al. 1999; Rogovin 1999
Allactaga	Sol.	O	GF, O	N	B	50–50	Sludsky et al. 1977; Fokin 1978; Shenbrot et al. 1995; Rogovin 1999
Allactodipus	Sol.	O	F	N	B	70	Fokin 1978; Shenbrot et al. 1995; Rogovin 1999
Pygeretmus	Sol.	O	F	N	B	40–60	Sludsky et al. 1977; Fokin 1978; Shenbrot et al. 1995; Rogovin 1999

NOTES: Genera with known social species are in **bold** type. Sol. = solitary; I (in Sociality column) = intermediate tolerance; Soc. = social; O = open; I (in Habitat openness column) = intermediate (patchy, coarse-grained); C = closed (fine-grained, with cover in a lower layer > 30%); G = granivore; GF = granivore-folivore; F = primarily folivore; I (in Feeding adaptation column) = insectivore; GI = granivore-insectivore; O = omnivore; N = nocturnal; D = diurnal; ND = all day active or seasonal change; B = bipedal; Q = quadrupedal.

[a] *Gerbillurus* according to Skinner and Smithers (1990) and Pavlinov et al. (1990) is mainly a granivore and nocturnal. *G. vallinus* was considered by Shenbrot et al. (1999) as social. In reality it probably forms groups in high population densities.

[b] *Meriones unguiculatus* is social, diurnal, and primarily eats green vegetation; *M. vinogradovi* is social, diurnal, and feeds on seeds and vegetation; *M. hurrianae* is intermediate tolerant or social. *M. libycus,* in deserts of Turkestan, has season-dependent daily activity and is intermediate tolerant.

[c] Communal use of burrows is typical for *Tatera indica*. According to characteristics given by different authors this species can be classified as social or intermediate tolerant (Pavlinov et al. 1990).

are able to excavate burrows to locate the smaller, more evenly dispersed underground vegetation without the cooperation of conspecifics (Jarvis et al. 1994).

In contrast to the fossorial mole-rats, the costs of obtaining food seem to impose significant restrictions on group formation in many species of semifossorial desert rodents (Daly and Daly 1975a, 1975b; Randall 1993, 1994; Shenbrot et al 1999). For seed-eating rodents, heterogeneous topography and scattered vegetation cause patchiness in characteristics of the soil surface and the subsequent distribution of food (Reichman and Price 1993). This patchy dispersion of food is especially true for seeds that are distributed by wind. Green vegetation in desert environments is relatively sparse, of low quality, and patchily distributed, which is probably why folivory is limited in desert rodents (Shenbrot et al. 1999). Furthermore, temporal availability of food can have major effects on efficiency of food harvesting (Price and Joyner 1997). In most desert conditions, therefore, food may be too dispersed, low in quality and temporally unpredictable to support more than a single animal, resulting in a solitary social structure (Daly and Daly 1975a, 1975b; Eisenberg 1975; Shenbrot et al. 1999).

Comparison of arid-adapted species in different habitats illustrates how constraints of food abundance and distribution may affect group formation. The great gerbil (R. opimus) and the fat sand rat (Psammomys obesus) are ecologically similar. The two species have similar body mass, feed primarily on green vegetation, and are diurnal, but P. obesus is solitary and R. opimus is social (table 31.1). The main difference between the two species is that P. obesus inhabits very arid areas of the Middle East and North Africa where vegetation can be quite limited and competition for food is high while R. opimus inhabits more productive habitats in Central Asia that can support social groups (Daly and Daly 1974; Rogovin 1996; Tchabovsky et al. 2001).

Group sizes and structure in social desert rodents can be constrained by food availability (Ågren et al. 1989a, 1989b; Randall 1994; Rogovin, Moshkin et al. 2003; table 31.1). The Mongolian gerbil (Meriones unguiculutus) forms larger groups and attains higher population densities where vegetation is more abundant and predictable compared with smaller groups in more arid regions with little herbaceous growth (Xia et al. 1982; Orlenev 1987; Ågren et al. 1989a, 1989b). Social groups of the Indian gerbil (Tatera indica) are much larger in urban areas near human habitation where resources are plentiful and consistent compared with solitary or male-female pairs in arid grasslands (Idris and Prakash 1985). In R. opimus, large social groups form when herbaceous vegetation becomes plentiful and population densities are high. When food becomes limited and densities are low, social groups are small and many females are solitary (Tchabovsky et al. 2001; Randall et al. 2005). So-

cial structure is also based on food type in the southern African striped mouse (Rhabdomys pumilio; Schradin and Pillay 2004; Schradin and Pillay 2005b). Social groups form in the desert habitat because of a stable, year-round food source of succulent plants. In grasslands, the mice feed on a more ephemeral food source of seeds and are solitary (Schradin and Pillay 2005b).

Predation

Individuals in social groups often benefit from an increased ability to detect and escape from predators (Ebensperger 2001a). Individual risk of predation decreases with an increase in group size because of increased alternative targets and early detection of predators (Betram 1978; Inman and Krebs 1987; Lima and Dill 1990). Ground squirrels under predatory risk in open, riskier habitats exhibit larger group sizes and more cooperative defenses compared with ground squirrels in more vegetated, safer habitats (Dunbar 1989; Blumstein and Armitage 1997b). However, this relationship does not hold for desert rodents. Few species have group predator defenses, possibly because a majority are nocturnal (70–90% in a local fauna; Shenbrot et al. 1999). Nocturnal activity may eliminate the group benefit in predator detection because long-distance visual identification of predators is ineffective in the dark.

Both diurnal great gerbils, R. opimus, and subterranean naked mole-rats, H. glaber, however, have evolved group defenses against predators (Lacey and Sherman 1991; Randall et al. 2000). Alexander (1991) proposed that group living in the naked mole-rats evolved as a response to predation, and the well-developed defenses against predators suggest that predation was a factor in the evolution of sociality in R. opimus. The gerbils benefit from mutual vigilance and the use of alarm signals to alert group members of approaching danger (Randall and Rogovin 2002). Rogovin et al. (2004) found that a common predator of R. opimus, the monitor lizard (Varanus greseus caspius), visited dispersed groups of gerbils more frequently than areas where family groups were close together, but there was no difference by group size and composition. The alarm-call system of gerbils may be more effective when family groups live in close proximity and neighbors can hear each other. Rhombomys opimus has an elaborate system of alarm communication, consisting of three different vocalizations and seismic footdrumming signals (Randall et al. 2000; Randall and Rogovin 2002). The calls communicate response urgency to gerbils outside the burrow, and the footdrumming probably communicates to family members inside. Meriones unguiculatus footdrums, but does not call, when disturbed, as a possible alarm signal (Ågren et al. 1989a).

Most species of gerbil footdrum when they are disturbed (Randall 2001). Whether they drum in response to predators is unknown.

Solitary desert rodents have individual predator defenses that depend on auditory and olfactory information for detection of predators at night (Randall 1993, 1994; Randall et al. 1995; DeAngelo 2002; Busch 2003). In comparison with semifossorial rodents in mesic environments, arid-adapted rodents have a more specialized ear, which includes expanded auditory bullae and specialized cochlea for hearing the low-frequency sounds made by predator movements (Webster and Webster 1984; Webster and Plassmann 1992). Low-frequency hearing for predator detection seems to be a highly valuable adaptation, because it has evolved independently in at least four different lineages of semifossorial desert rodents: kangaroo rats in North America, (Webster and Webster 1984), gerbils in Africa and Asia (Plassmann et al. 1987; Pavlinov 1988), jerboas (Dipodidae) in Asia (Pavlinov and Rogovin 2000), and vizcacha rats (*Tympanoctomys barrerae*) in South America (Ojeda et al. 1999).

Desert rodents also exhibit fast and efficient locomotion to escape predators (Shenbrot et al. 1999). Bipedal locomotion, which provides an advantage in speed to escape predators in the initial stages of locomotion in open habitats, has evolved independently in the deserts of Asia (Dipodidae), Africa (Pedetidae), Australia (Muridae), and North America (Heteromyidae; Mares 1993; Randall 1994; Shenbrot et al. 1999). Species of larger body mass usually show the most specialization for a fast, bipedal gate, whereas medium- or smaller-sized rodents usually hide in shelter burrows or under vegetation (Rogovin 1999; Shenbrot et al. 1999).

Flexible strategies of reproduction

Mammalian behavior often reflects opportunistic responses to changes in the characteristics of the environment (Austad 1984; Wrangham and Rubenstein 1986; Lott 1991). In particular, mixed strategies allow animals to deal with uncertainty (Haccou and Iwasa 1995; Flaxman 2000). For instance, although marmots are usually polygynous, sparse resources and spacing of females can lead to strictly monogamous mating (Holmes 1984a). Social behavior is also related to the distribution and predictability of resources (Bradbury and Vehrencamp 1976; Waser 1981; Wrangham and Rubenstein 1986); temporal and spatial patterns of resource availability can affect group sizes (Pulliam and Caraco 1984; Johnson et al. 2002).

Behavior of desert rodents seems especially opportunistic and flexible. Instead of a strictly seasonal pattern of reproduction, many species reproduce at any time of the year, during variable periods of rainfall and peak food availability (Schmidt-Nielsen 1975; Randall 1993, 1994). Most desert rodents are sexually mature before their first year of life and thus are able to take advantage of favorable conditions for reproduction. These adaptations are common to different lineages of desert rodents; fast maturation and irregular reproductive schedules are reported in rodents from deserts in North and South America, Africa, Asia, and Australia (Happold 1975; Prakash 1975; Taylor and Green 1976; Breed 1990; Jones 1993; Randall 1993, 1994).

Flexible behavior also appears during mating. In kangaroo rats, mating ranges from being exclusive between male and female neighbors to competition between two or more males for access to a female (Kenagy 1976; Randall 1987a, 1987b, 1989b, 1989c, 1991a; Randall et al. 2002). Exclusive mating usually occurs at low densities as a result of spatial isolation when a single male overlaps the home range of a single female. Multiple males are much more likely to converge at the home area of an estrous female and compete for access to her when population densities are higher and the operational sex ratio favors males (Kenagy 1976; Randall 1987b, 1989b, 1991a; Randall et al. 2002). Exclusive mating can also occur at the beginning of the breeding season, when a female enters estrus outside of regular breeding activity, or when females occupy territories that are spaced far apart (Randall 1993; Randall et al. 2002).

Because the timing of female receptivity is variable and short, males must continuously monitor females for signs of a mating opportunity (Randall 1989b, 1991a). Although vaginal swelling can last for several days, behavioral estrus is brief, lasting only 4–6 hours per cycle (Eisenberg 1963a; Daly et al. 1984; Randall 1991a; Yoerg and Shier 2000; Randall et al. 2002). Estrus is not evident in the appearance of the vaginal area (Wilson et al. 1985), but may be detected by olfactory cues (Roberts, chap. 22 this volume). During my studies of mating behavior, I found that females mated from two to eight nights after the vagina became swollen and open, and judging by visits of males to estrous females, neither the males nor I could predict behavioral estrus until females accepted advances by the males (Randall 1989b; Randall 1991a; Randall et. al 2002). Moreover, females of some species, such as *D. merriami,* do not seem to signal their reproductive readiness ahead of time by scent from either their dorsal gland or their urine (Randall 1993); *D. merriami* and *D. agilis* females, but not males, prefer conspecific odors when in reproductive condition (Daly et al. 1980). These results suggest that males may obtain limited information in advance about the reproductive status of females in these species.

Male kangaroo rats may gain information about whether an estrous female is ready to mate only at close proximity, when he engages in nasogenital circling and smells the vaginal opening or when he is allowed to mount. These behaviors may give females more control over their choice of mates and minimize harassment by males. Future studies will have to determine whether this flexibility in receptivity provides females with more opportunity for mate choice and timing of pregnancies or whether the behavior is a result of some other factor, such as a reluctance to allow contact in solitary animals that compete for resources.

Dispersal and philopatry

Delayed dispersal is proposed as a key to understanding the evolution of sociality in many birds and mammals (Emlen 1997). Long-lasting, stable societies of related individuals allow for the development of complex social systems, and philopatry (the retention of young in the natal area) is usually a prerequisite for the formation of social groups (Greenwood 1980; Armitage 1981; Solomon and Getz 1997). Offspring often remain in the natal area in response to ecological constraints of limited territories and other resources (Emlen 1997; Hatchwell and Komedeur 2000). Young that remain in the natal area in cooperative breeding species become helpers and forgo their own reproduction but gain inclusive fitness benefits (Emlen 1991).

Although philopatry is commonly associated with highly social animals, there are many solitary species in which dispersal is delayed and relatives continue to share space. Young benefit by gaining access to territories important for survival and reproduction, and the dangers of leaving the natal area are eliminated. Adults may lose little in space or food, and may gain in inclusive fitness because of the higher fitness of their offspring (Waser and Jones 1983; Waser 1988; Wolff 1994b).

Philopatry has been observed in both social and solitary desert rodents. There is direct evidence for philopatry in two species of kangaroo rat, *D. spectabilis* and *D. merriami*, and indirect evidence in several other species (Fitch 1948b; Rogovin 1981; Jones 1984, 1989; Randall 1993, 1994; Winters and Waser 2003). Despite their solitary lifestyle as adults, juvenile male and female *D. spectabilis* may remain in the maternal territory for several months after weaning. When juveniles do disperse, they settle within 20 meters of their natal site, which can result in dispersed kin clusters (Jones et al. 1988; Winters and Waser 2003). Once established in a burrow, the majority of males and females remain in that same territory throughout their lives (Jones 1993; Randall 1993, 1994). Philopatry occurs because mothers share territories with offspring, donate a secondary territory, or abdicate their territory and move to a new burrow nearby as young mature (Waser 1988).

Philopatry leads to the continued association of kin, which facilitates cooperation and social interactions in social gerbils in many of the same ways as in other social mammals (Emlen 1997). Females of the social gerbil, *R. opimus,* are philopatric, and are related to each other (Randall et al. 2005). This relationship may facilitate the high degree of cooperation in these female groups.

The advantage of philopatry might not seem as obvious in solitary rodents that compete with neighbors for food and space. Recognition of neighbors as kin, however, could minimize agonistic interactions between solitary animals. A recent study using microsatellite genotypes over five successive breeding seasons confirmed that *D. spectabilis* neighbors in adjacent territories are often close relatives (Winter and Waser 2003). Because kin recognition is a prerequisite for the formation of more gregarious groups (Sherman et al. 1997), future research should examine whether kangaroo rats differentiate kin from nonkin and are more tolerant of related neighbors than unrelated neighbors.

Kangaroo Rats as a Model of Social Evolution in Solitary Species of Desert Rodent

How social are kangaroo rats?

A comparison of closely related species that share similar ecological conditions is a useful way to establish evolutionary relationships (Sherman et al. 1995). In rodents, social continua have been successfully constructed in terrestrial sciurids (Armitage 1981, 1999a; Michener 1983a; Blumstein and Armitage 1997b), fossorial mole-rats (Jarvis and Bennett 1991), and voles (*Microtus;* Wolff 1985a). Comparisons of evolutionary relationships within these groups have been especially revealing because of a wide range of social structures and clear correlations between sociality, environmental variables, and life-history traits (Armitage 1981; Michener 1983a).

Why, then, compare sociality in a taxonomic group of desert rodents that is considered intolerant and asocial, such as kangaroo rats? As expressed in the title of a recent study by Yoerg (1999, p. 317), "solitary is not asocial." Sexes meet for mating, juveniles associate with their mothers, and adults interact with neighboring conspecifics with the general objective of maintaining individual spacing (Eisenberg 1966; Randall 1993, 1994; Lacey 2000; Shier 2003).

Kangaroo rats have evident, if rudimentary, social structures (table 31.2). They are philopatric (Jones 1984, 1989, 1993), show a range of social tolerance, and have well developed means of communication (Randall 1993). Extension and development of any of these aspects of sociality

Table 31.2 Traits of *Dipodomys* on a social continuum from the least to the most social

	D. deserti (1)	D. spectabilis (2)	D. ingens (3)	D. heermanni (4)	D. ordii (5)	D. merriami (6)
Body size	Large, 83–148 g	Large, 98–132 g	Large, 93–195 g	Medium, 70–80 g	Small, 50–96 g	Smallest, 33–54 g
Spacing						
Territorial						
Males	Yes	Yes	Yes	No	No	No
Females	Yes	Yes	Yes	Yes	No	No
Food hoarding	Larder	Larder	Larder	Larder?	Scatter	Scatter
Tolerance of conspecifics	Low	Moderate	Moderate	Moderate	Moderate	High
Males visit	?	Yes	Yes	Yes	Yes	Yes, females to mate
Philopatric	?	Both sexes	Probably	Maybe	Probably	Yes
Neighbors related	?	Yes	Probably	?	?	Probably
Neighbor recognition						
Olfactory	?	Yes	Yes	?	Yes	Yes
In encounters	No	Yes	Yes	No	Yes	Yes
Scent gland	No	No	No	No	Seasonal	Yes, dimorphic
Footdrum						
To conspecifics	Yes	Yes	Yes	Yes	No	No
Individual signatures	No	Yes	Maybe	No	No	No
Drum back	Seldom	Yes	Yes	Yes	No	No
Predators	Yes	Yes	Yes	No	No	No defense

SOURCE: See Randall 1993, 1994 for references. For column (1): Randall 1997, 2001; Sullivan 2000; Randall and Boltas King 2001. For column (2): Jones 1993; Randall 1997; 2001; Randall and Matocq 1997; Winters and Waser 2003. For column (3): Randall 1997, 2001; Murdock and Randall 2001; Randall et al. 2002; Cooper 2002; Busch 2003. For column (4): Fitch 1948b; Yoerg 1999; Yoerg and Shier 1997; Shier and Yoerg 1999; Randall 2001; Shier 2003; Shier and Randall 2004. For column (5): Perri and Randall 1999. For column (6): Randall 1989b; Perri and Randall 1999. For body size: Kays and Wilson 2002.

could lead to a more complex social system (Linn 1984). Thus a comparison of behavior of closely related solitary species could provide new insights into the processes that lead to the evolution of more gregarious species (Brockman 1984; Waser 1988).

Spacing of kangaroo rats

Because individuals within a distribution are influenced by the proximity or absence of other individuals, spatial organization can provide clues to the social structure of a population (Lott 1991). In kangaroo rats, spacing varies with body size and methods of storing food and ranges from individual, nonsexually dimorphic territories to overlapping home ranges (tables 31.2 and 31.3). Species with larger body masses defend individual, dispersed territories against members of both sexes, in which they store larders of seeds, their main food source. Smaller-sized species inhabit overlapping home ranges in which they scatter-hoard seeds. Males have home ranges that overlap those of one or more females and each other, whereas females maintain area exclusive of other females. Although a large sex difference in home range size might be expected in kangaroo rats, sex differences in spacing behavior are usually small, and home range sizes of males are similar to those of females in most species outside the breeding season (table 31.3). During breeding, in all species investigated thus far, male home

ranges expand to overlap those of females, because males visit female territories or core areas for mating (Maza et al. 1973; Behrends et al. 1986; Randall 1984, 1989b, 1991a; Perri and Randall 1999; Randall et al. 2002; Cooper 2002; Shier and Randall 2004).

Social behavior and food

The more social species of kangaroo rat are smaller in body size, have overlapping home ranges, and hoard seeds in widely spaced caches throughout their home ranges (table 31.2). The chances of an individual recovering seeds are enhanced by being able to remember where the seeds are cached (Jacobs 1992). Although the kangaroo rats meet conspecifics in the process of gathering and caching seeds, they do not defend their caches, and *D. merriami* only responds to another rat's presence if they observe caching behavior (Preston and Jacobs 2001). Because seeds are usually abundant enough to support the solitary kangaroo rats gathering them, and pilfering of each other's seed caches is reciprocal (Daly et al. 1992; Vander Wall and Jenkins 2003), it is probably adaptive for the kangaroo rats with overlapping home ranges to tolerate the presence of conspecifics that are also gathering and caching seeds. Aggression would be costly in time, energy, and water conservation (Ganem and Nevo 1996). Because *D. merriami* is relatively long-lived and individuals remain in the same area for most

Table 31.3 Comparison of size and percentage overlap of homes ranges for six species of *Dipodomy*, ranging from the largest body mass to the smallest

Species		Home range (ha)		Home range overlap (%)		
				Opposite sex (male on female)	Same sex	
		Male	Female		Male	Female
D. ingens	July–Aug.	0.02	0.02	12.1	0.3	48.6[a]
	Feb.–March[b]	0.10	0.02	63.3	19.3	0
D. spectabilis		0.05[c]	0.05	<2	<2	<2
D. deserti	March–April[b]	0.18	0.12	3.5	<2	<1
	Nov.–Dec.[d]	0.08	0.13	6.69	16.62	8.86
D. heermanni[b]		0.11	0.04	16.8	10.7	1.3
D. ordii[b,e]		0.07	0.07	25.4	9.9	3.14
D. merriami	California[b,f]	0.34	0.31	22.2	16.2	10.4
	Arizona[b,e]	0.17	0.07	50.7	20.7	8.12

[a]Overlapping cluster of five females in nonbreeding season.
[b]Breeding season.
[c]Estimate without regard to sex.
[d]Beginning of breeding; density increased from 4.4/ha to 6.4/ha.
[e]High density.
[f]Data from three periods combined.
SOURCE: For *D. ingens:* Cooper 2002. For *D. spectabilis:* Schroder 1979. For *D. deserti:* Sullivan 2000. For *D. heermanni:* Shier and Randall 2004. For *D. merriami:* in California, Behrends et al. 1986; in Arizona, Perri and Randall 1999.

of their lives, the same neighbors may share overlapping areas for extended periods (Zeng and Brown 1987; Jones 1989; Randall 1989b). *Dipodomys merriami* are philopatric; individuals living near each other may be related, so that most pilferers are close kin. For all these reasons, the benefit of chasing conspecifics does not seem worth the costs in time, energy, and water loss that would result from social intolerance.

In contrast, the larger-sized species of kangaroo rats are highly territorial within and between sexes. Although territoriality often evolves for reasons other than defense of food, such as to prevent infanticide (Wolff and Peterson 1998), these larger-sized kangaroo rats store food in defensible caches. Both males and females store seeds in a larder in a defined burrow area, and they actively defend the area from other kangaroo rats by chasing and footdrumming (Randall 1993). The larger body size allows the kangaroo rats to dig and manage burrow areas for their seed caches that might be less economically feasible for the smaller-sized species.

Dipodomys deserti is the most territorial and intolerant species of kangaroo rat studied thus far. The lack of social tolerance is probably related to the harshness of the environment and competition for scarce resources in a sand dune habitat that has limited space to construct burrows, and limited seed production (Sullivan 2000). Competition is intense—the kangaroo rats often visit neighboring territories in attempts to enter burrows to pilfer seeds (Randall 1997; Sullivan 2000; Randall and Boltas King 2001). Ago-

nistic interactions are frequent; the territorial owner usually responds by rushing the visitor and chasing aggressively, often until contact is made and a rollover fight ensues.

Role of familiarity and social tolerance in kangaroo rats

Familiarity and neighbor recognition play important roles in the maintenance of spacing and social order in several species of kangaroo rat (*D. merriami;* Randall 1989b; *D. ordii;* Perri and Randall 1999; *D. ingens;* Murdock and Randall 2001; *D. spectabilis;* Randall 1989c). Familiarity is independent of the type of spacing, and species such as *D. merriami* in overlapping home ranges and *D. spectabilis* and *D. ingens* in individual territories recognize and tolerate familiar neighbors more than unfamiliar strangers.

Social experience and familiarity apparently are important in coordinating the mating interactions of kangaroo rats. Female *D. heermanni* housed in social isolation exhibited longer estrous cycles that were disrupted for months compared with females housed under conditions in which they could see and smell conspecifics (Yoerg 1999). Female *D. merriami, D. ordii, D. spectabilis,* and *D. ingens* prefer the scent of familiar males (Randall 1993; Murdock and Randall 2001), and females tolerate and interact with familiar males more than with unfamiliar males in paired encounters and during mating (Randall 1989b, 1989c, 1991a; Perri and Randall 1999; Randall et al. 2002). Before mating, males spend considerable time in the home area of estrous females, probably to establish familiarity.

Although behavioral observations provided reliable evidence that female *D. spectabilis* mated with neighbors (Randall 1991a), DNA analysis revealed that many offspring were fathered by nonneighboring males (Winters and Waser 2003). Because neighbors can be related, females may mate with more distant males to avoid inbreeding. Males travel considerable distances to visit the territories of estrous females, and females mate more than once during behavioral estrus (Randall 1991a). In addition to promoting tolerance among neighbors, it seems that recognition of related neighbors could facilitate mate choice by females to prevent inbreeding.

Recognition of social status may be another means of promoting social order in kangaroo rats (Shier and Yoerg 1999; Newmark and Jenkins 2000; Shier 2003). Shier (2003) found that *D. herrmanni* formed a nearly linear dominance hierarchy almost immediately upon meeting; familiarization with dyad partners enhanced the relationships. For *D. heermanni*, socialization (exposure to conspecifics in general) rather than familiarization with particular individuals seems to be the main factor mediating aggression (Yoerg 1999). Subordinate individuals avoid contact with more aggressive individuals by retreating into the burrow and footdrumming (Shier and Yoerg 1999).

Not all species of kangaroo rat, however, are more tolerant of familiar neighbors than unfamiliar strangers. *Dipodomys deserti* did not direct less aggression toward neighbors than nonneighbors in paired encounters in the field (Sullivan 2000). During playback experiments of footdrumming, *D. deserti* approached the speaker, presumably to chase away the intruder, rather than stand and footdrum in response to playbacks of conspecific footdrums, as seen in *D. spectabilis* and *D. ingens* (Randall 1997).

Communication by scent and footdrumming

Kangaroo rats have evolved complex systems to communicate identity. Familiarity is communicated by scent at sandbathing sites and through footdrumming exchanges (table 31.2; Randall 1993; Randall 1997; Perri and Randall 1999; Murdock and Randall 2001). Territorial kangaroo rats, in which both sexes defend territories, have sexually monomorphic dorsal glands that are used for territorial marking independent of age, sex, season, and gonadal hormones (table 31.2; Randall 1993). Sebum from the gland is deposited during sandbathing, and both males and females discriminate familiar from unfamiliar scent at these sites. Species with overlapping home ranges, however, have a sexually dimorphic scent gland, larger in males than in females, controlled by androgens. Scent from the gland is probably important for identification of males by females during mating; females, but not males, discriminate between dorsal gland scent of familiar and unfamiliar males (Randall 1993).

Kangaroo rats use both seismic and airborne signals generated by drumming their feet to communicate to conspecifics (Randall and Lewis 1997; Randall 2001). Footdrumming patterns are species-specific and range from single thumps in *D. deserti* to individually specific footdrumming signatures in *D. spectabilis* and possibly *D. ingens* (Randall 1993, 1994, 1995, 1997, 2001). The footdrumming effectively communicates ownership of a territory and usually is sufficient to deter a visitor without chases or fights, except in the case of *D. deserti*, mentioned earlier.

Footdrumming occurs in contexts other than territorial advertisement. Male kangaroo rats footdrum during competitive interactions when they compete for estrous females (Randall 1987b; table 31.2). Footdrumming in *D. heermanni* is done inside the burrow by the kangaroo rat, avoiding contact with a more dominant individual outside the burrow (Shier and Yoerg 1999) and outside the burrow after mating (Shier, personal communication).

Kangaroo rats also footdrum to communicate to predators. *Dipodomys spectabilis*, *D. ingens*, and *D. deserti* actively approach snakes, jump back, and footdrum to communicate directly to the snake that the kangaroo rat is alert, is not easy prey, and to go away (Randall and Matocq 1997; Randall and Boltas King 2001). *Dipodomys ingens* also footdrums on hearing the approach of a kit fox (*Vulpes macrotus*), its major predator. In a field experiment, *D. ingens* discriminated playbacks of footfalls of an approaching kit fox and the movement of snakes from controls of a rabbit hopping and tape noise (Busch 2003). Kangaroo rats also discriminated the odor of snakes and kit foxes from controls in laboratory experiments (Randall et al. 1995; DeAngelo 2002).

Social Gerbils

In contrast to the kangaroo rats that are strictly solitary, gerbils show a social continuum ranging from solitary to communal (Eisenberg 1967; Daly and Daly 1975a, 1975b; Dempster and Perrin 1989a; Pavlinov et al. 1990; Randall 1994; Shenbrot et al. 1999; Gromov 2000; table 31.1). Russian scientists provide the best information about a social continuum in the genus *Meriones*. From comparative studies the following species have been classified from the least to the most social: *Meriones tamariscinus*, *M. meridianus*, *M. libycus*, and *M. unguiculatus* (Shilova et al. 1983; Chabovsky et al 1990; Goltsman and Borisova 1993; Chabovsky and Popov 1994; Goltsman et al. 1994; Gromov et al. 1996). Social designation was based on the sharing of a burrow by a male-female pair and the stability of male-female relationships over time. Social behavior, however, varied with density and season.

Two species of gerbils have been studied in enough detail to understand some of the dynamics of group living in desert habitats. These two species are the great gerbil, *R. opimus*, in Central Asia (e.g., references in table 31.1 and Randall et al. 2000; Tchabovsky et al. 2001; Randall and Rogovin 2002; Rogovin, Randall et al. 2003) and the Mongolian gerbil, *M. unguiculatus*, in Russia and China (Ågren 1984a, 1984b, 1984c; Ågren et al. 1989a, 1989b; Orlenev 1987; Gromov 2000). Both species are arid-adapted, diurnal rodents that live in family groups in complex burrow systems constructed in open habitats. The gerbils cooperatively defend nonoverlapping, multipurpose territories from encroachment by neighboring groups. Animals are active all year, and members of the family group cooperatively gather and store vegetation in common stores for survival during times of the years when food is scarce (Ågren et al. 1989b; Tchabovsky et al. 2001). In *R. opimus*, the adult male may store a larger proportion of the food than other members of the group. Males of both species actively scent mark the territory by building heaps of soil and then rubbing their ventrums on the heaps to deposit a secretion from a large ventral scent gland (Roper and Polioudakis 1977; Popov and Tchabovsky 1996; Gromov 2000).

Although there are many similarities, the two species of social gerbils exhibit some interesting differences in group composition and behavior. Family groups of *M. unguiculatus* seem to consist of a monogamous male-female pair and their offspring, except at high densities, when males associate with several females (Ågren et al. 1989a, 1989b; Gromov 2000). Mating is not entirely monogamous, because estrous females enter adjacent territories to mate with strangers. Mongolian gerbils have a dominance structure in which males dominate females and larger animals dominate smaller ones. Ågren et al. (1989a) suggest a reproductive skew and suppression of reproduction via suppression of maturation. In contrast, family groups of *R. opimus* may be larger and more complex than those of *M. unguiculatus*. Families consist of (at least at higher densities; see the following) one adult male, from one to six adult females, juveniles, and subadults of previous litters. *Rhombomy opimus* seems to have an egalitarian social structure. Males do not dominate females, and both males and females store food and give alarm calls. Suppression of reproduction does not seem to occur with any regularity, and the majority of females in family groups exhibit signs of pregnancy and lactation. All females in a family group interact socially with pups of all litters of different ages (Randall et al. 2005; Randall and Rogovin unpublished data). *Rhombomy opimus* has a well-developed system of alarm communication consisting of both vocal and seismic signals (Randall et al. 2000; Randall and Rogovin 2002).

An 8-year field study of *R. opimus* revealed that family groups of *R. opimus* are inherently unstable and form and expand based on the availability of vegetation and subsequent changes in population density (Randall et al. 2005). Reproduction of individual females, as measured by the number of pups per female emerging from the burrow and survival for 1 season depended on favorable environmental conditions in the form of rainfall, mild temperatures, and availability of green vegetation, the main food source. Reproduction is associated with the production of green vegetation in the spring, and gerbils will continue to produce litters while conditions are favorable. In years with good production of green vegetation (1996–1999) the majority of females lived in multifemale family groups with a resident male (fig. 31.1). During a drought extending from late spring 1999 to spring 2002 the abundance and temporal availability of green vegetation were limited. There was high mortality and increased predation, causing population densities to decrease and the social structure to shift to solitary females living alone with their offspring (fig. 31.1). As dispersion of females changed, adult males altered their behavior. Instead of remaining with family groups and providing parental care by storing food and marking territories, males traveled the study area visiting solitary females (Tchabovsky et al. 2001; Rogovin et al. 2003; Randall et al., in press). Thus far, we have found few differences in reproductive success in the different-sized groups. Although solitary females behave differently from females in groups and spend more time in the burrow and are more vigilant when foraging (Tchabovsky et al. 2001), they still seem to have about the same reproductive success as individual females in groups. Great gerbils are facultatively social. Females live solitarily under conditions of limited food and high mortality, which disrupt social behavior and group formation; they share territories with female kin under favorable conditions for survival and reproduction when kin groups can be maintained. Males adjust to the distribution of females.

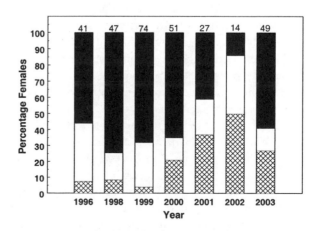

Figure 31.1 Percentage of *Rhombomys opimus* females living in groups of two or more with a resident male (cross-hatched), in a male-female pair (white) and solitary (black) in years of high (1996–2000), low (2001–2002), and recovering population densities. *N* females in the population at top of bars.

Conclusions

Water scarcity, extreme temperatures, and unpredictable and limited productivity of food place broad constraints on the behavior of semifossorial desert rodents. As a result, the majority are "solitary specialists." The rodents included in this review are primarily solitary and nocturnal. Feeding adaptations vary and include granivores, folivores, omnivores, and insectivores. The social semifossorial rodents, on the other hand, are diurnal folivores (table 31.1).

High productivity of green vegetation is an important factor necessary to support group living in social rodents. Under most desert conditions food may be too dispersed, low in quality, and temporally unpredictable to support more than a single animal and a solitary social structure. When social groups do form, group sizes tend to be unstable and expand and contract with conditions of food abundance; group living may increase social stress enough to influence survival (Rogovin, Randall et al. 2003). For these reasons, although social desert rodents should be subjected to the same rules predicated for the evolution of sociality in other animals (Alexander 1974; Pulliam and Caraco 1984; Wrangham and Rubenstein 1986), the link between sociality and traditional costs and benefits of sociality are not as compelling for the desert rodents examined in this paper as they are in other rodent groups (see reviews by Ebensperger [2001a] and Lacey and Sherman, chap. 21 this volume). Both group formation and reproductive strategies seem to be opportunistic responses to harsh environments with highly variable and limited food productivity.

Although semifossorial desert rodents occupy open habitats in which group vigilance and detection of predators would be predicted, a complex system of alarm signals has not evolved in them, with the exception of the great gerbil. Constraints on group predator defenses may occur because of nocturnal activity that has evolved to facilitate water conservation. Visual detection of approaching predators and subsequent alarm calls to communicate predation risk would not be effective at night. Instead, some desert rodents have evolved acute low-frequency hearing to detect approaching predators, effective bipedal locomotion to escape from predators in open habitats, and footdrumming behavior for individual defense.

Social tolerance is important for desert rodents. Besides saving time and energy, avoidance of aggressive interactions minimizes the effects of physiological stress and water loss. Social tolerance is promoted by philopatry in both social and solitary species. In social species, philopatry is important for group formation; delayed dispersal leads to cooperative groups of related females in *R. opimus*. In solitary species, philopatry might lead to recognition of related neighbors to increase tolerance and decrease inbreeding.

Kangaroo rats provide a model system for understanding the social behavior of solitary rodents. Behavioral differences among species are related to body size, spacing, food storage, social tolerance, communication, and response to predators. Social tolerance varies on a social continuum from the most socially tolerant species, *D. merriami*, to the least tolerant, *D. deserti*. If future research shows that solitary kangaroo rats are more socially tolerant of related neighbors than unrelated neighbors, a behavioral link in the evolution of mammalian sociality will be made. In the future, there would be a much better understanding of social structures in general if solitary mammals were not considered asocial and dismissed as having no clear social relationships.

Social flexibility may be more common in desert rodents than previously assumed. Rodents seen together in high numbers may reflect high population densities rather than a stable social structure (Randall 1994). Whether group living is an evolved strategy to increase inclusive fitness, a consequence of local food abundance, or delayed emigration resulting from habitat saturation has not been demonstrated for most species of desert rodents (Lacey 2000). Further research that experimentally tests the relative influence of food abundance, habitat saturation, and density-dependent factors leading to delayed dispersal and shared space by female kin will help to elucidate the proximate causation and ultimate benefits of group living in deserts (Wolff 1993b, 1994b). However, because even the most social desert rodents (e.g., the Mongolian gerbil) seem to exhibit flexible social behavior, researchers should exercise caution when generalizing about the social structure of desert rodents.

Summary

Desert rodents must often survive and reproduce in harsh conditions of scarce water, extreme temperatures, and limited plant growth. These environmental constraints, especially the limited abundance and patchy distribution of food, impose limitations on the formation of social groups in semifossorial desert rodents. Both social and solitary desert rodents exhibit behavioral flexibility, probably to facilitate responses to the changing and unpredictable conditions of arid environments. Social tolerance is important to save time and energy and to minimize the effects of aggression on physiological stress and water loss. When social groups do form, group sizes tend to be unstable; they expand and contract with conditions of food abundance. Gerbils are semifossorial desert rodents that range from solitary to communal. The great gerbil (*Rhombomys opimus*) is one of a few social gerbils in a group of primarily solitary species. An apparent advantage to group living in the great gerbil is the detection of predators and warning of group members via a complex system of alarm signals. Group size and composition are flexible and vary with availability of vegetation and subsequent changes in population density. Another group of semifossorial rodents, kangaroo rats (*Dipodomys*), provides a model system for understanding social behavior of solitary rodents. Kangaroo rats are philopatric, show a range of social tolerance, and have well-developed means of communication by scent marking and footdrumming. Social tolerance varies on a social continuum from the most to the least tolerant species. Because behavioral flexibility seems important to survival and reproduction in both solitary and social desert rodents, social flexibility may be common in desert rodents. Caution, therefore, should be exercised when generalizing about their social behavior.

Chapter 32 Comparative Social Organization and Life History of *Rattus* and *Mus*

Manuel Berdoy and Lee C. Drickamer

RATS AND MICE provide the best examples of commensal living between rodent and human societies. First, as notorious pests, rats and mice have a significant impact on the world economy. As the archetype of commensal rodents (fig. 32.1), they have contributed to famines and continue to be responsible for the loss, through feeding or destruction, of about one fifth of the world's harvest. While their impact is most crippling in developing countries (Singleton et al. 1999), rats and mice continue to have an impact even in the world's most industrial cities (Sullivan 2004). In this chapter we will refer to populations living in close association with humans as *urban* or *commensal* as distinct from *field* or *feral* populations. As successful invaders, rats and mice can also have a significant effect on endemic species from invertebrates to other mammals and birds through direct predation, competition, and by modifying the physical environment (Atkinson 1985, 1996; Jones et al. 1997; Dickman 1999).

The very aspects of their biology that make rats and mice such successful species (and indeed, competitors) have also contributed to their use as laboratory animals. The Norway rat (*Rattus norvegicus*) was the first species to be domesticated for mammalian research (in the early nineteenth century) and the first rat and mouse inbred strains were established at the turn of the twentieth century (Lindsey 1979; Hedrich 2000). One century later, hundreds of inbred strains (see www.informatics.jax.org) have revolutionized experimental biomedical science (Krinke 2000; Festing et al. 2002); rats and mice now constitute about 80% of the species used in the laboratory (European Commission 2003). Although the transgenics revolution is starting to blur this distinction, mice have traditionally been the model of choice

Figure 32.1 Rats, like mice, are commensal, literally meaning "eating at the same table." Photo by M. Berdoy.

for geneticists, whereas rats have been used as a model for a wide range of biomedical endeavors, from psychology to biochemistry. Moreover, since 2004, *R. norvegicus* has joined *Mus domesticus* and *Homo sapiens* as the first three mammals to have their genomes sequenced (International Human Genome Sequencing Consortium 2001; Mouse Genome Sequencing Consortium 2002; Rat Genome Sequencing Project Consortium 2004). While the motivation behind these large-scale sequencing efforts is obviously medical, some of the current and most probably future results stemming from such genomic comparisons should be heeded by whole animal and field biologists.

Although rats and mice are two mammal species that have had profound effects on humans, both positive and negative, no attempt has been made to evaluate and/or compare various aspects of their social evolution or life history that may have preadapted these species for this outcome. Therefore, the objective of this chapter is to present a brief summary of what is known about the social organization and life history of two of the most commonly studied rodents, rats (genus *Rattus*) and house mice (genus *Mus*). Given the overall paucity of information on the various species within each genus, our focus here will be on *Rattus norvegicus* and *Mus musculus*. In this chapter we present the basic social and behavioral ecology of these two species.

Evolutionary Origins

Molecular data suggest that humans and rodents shared a common ancestor some 80 million years ago (mya), and that rats and mice diverged about 12 to 24 mya (Adkins et al. 2001; Springer et al. 2003; Rat Genome Sequencing Project Consortium 2004). With between fifty to fifty-six species, the genus *Rattus* is, in terms of number of species, among the largest of mammalian genera (Wilson and Reeder 1993; Nowak 1999). The genus is sorted into five groups: the xanthurus group (species occurring in Sulawesi and nearby islands), the native New Guinea species, the native Australian species, the *rattus* group, and the *norvegicus* group (Wilson and Reeder 1993).

The Norway rat (*Rattus norvegicus* Berkenhout 1769), also called Brown Norway, is not always brown (Watson 1944; Smith 1958; Figala 1963), does not come from Norway and was not always called a rat. The Norway rat was initially called *Mus*, later replaced by *Epimys* before its formal Linnean nomenclatural description by Berkenhout (1769), who mistakenly believed that the species came from Norway (Barnett 2002; see also Yosida 1980). Rather, fossil evidence suggests that the rat (perhaps suitably the first sign of the Chinese zodiac) likely originated from the vast Asian plains of northern China and Mongolia (Meng et al. 1994). Although it is unclear when the rat became com-

mensal, its later spread to other parts of the world is directly attributed to its relationship with humans (Southern 1964; Robinson 1965). While the black rat (*R. rattus*), the species responsible for the bubonic plague (but see also Ruiz 2001 for *R. norvegicus*) via the flea vector (*Xenopsylla cheopis*) is thought to have been part of the European landscape at least since the middle ages, reliable records of *R. norvegicus* in European countries date back only to the eighteenth century. Rats may have colonized Europe after crossing the Volga River following an earthquake in 1727 (Grzimek 1968; and see Barnett 2002), and later came to America on the ships of the new settlers in the latter part of the eighteenth century (Lantz 1909; Grzimek 1968; see also Twigg 1975).

In contrast, archeological evidence suggests that house mice likely exhibited social and behavioral flexibility enabling them to coexist with humans at least 10,000 years before the present (Auffray et al. 1988). Fossil and molecular data suggest that the early representative of the mouse genus probably first appeared in the Miocene, about 5 mya, and underwent several radiations before the appearance of *M. musculus* by the early Pleistocene, about 0.5 mya (Brothwell 1981; Boursot et al. 1993; Galtier et al. 2004). Fossil and subfossil finds indicate that these species occurred at various locations in Asia, Europe, and Africa (Dixon 1972; Corbet 1974; Tchernov 1975; Storch and Uerpmann 1976).

One recent classification of species of the genus *Mus*, using morphometric and biochemical techniques, divides the group into as many as thirty-seven distinct species (Wilson and Reeder 1993). There are disagreements as to whether *M. domesticus* is a distinct species or is, instead, a subspecies of *M. musculus* (Boursot et al. 1993), although we are not concerned with that taxonomic discussion in this chapter. *Mus musculus* (Linnaeus 1758) is recognized as the taxon that has become worldwide in its distribution. The work that we summarize was all done with this species. *Mus musculus* is also the taxon from which most of the laboratory strains of house mice are derived.

Our primary focus in this review is on field studies of rats and mice, which, by their very nature, are both feral and commensal, and thus it is often not possible to completely distinguish confined conditions from truly feral or wild conditions.

Social Organization

Radio-tracking and trapping studies highlight the flexibility of rat population dynamics and social organization. Data from poor environments with sparsely distributed food, such as rural environment and field settings, suggest that feral rat societies are characterized by low population density, slow sexual maturity, and single male territoriality, with exclusive male ranges overlapping those of several females. In rich environments, where food is abundant and clumped (such as urban areas, grain stores, and rubbish tips) commensal rat societies are generally characterized by high densities, small home ranges, rapid sexual maturity, and polygynandry (Calhoun 1962a; Fenn 1989; Macdonald and Fenn 1995; also in Macdonald et al. 1999). In stable, rich environments, rat colonies are likely to be organized in multimale/multifemale "clan" territories (see also Calhoun 1962a), within which males compete for estrous females. Within home ranges, rats retreat to secure sites (burrow systems in feral environments) that can act as "information centers" where rats can passively transmit information about the environment to each other.

House mice also have a flexible and variable social system and spatial dispersion pattern, depending on where they live. Male territoriality and a polygynous mating system characterize most house mice that are truly commensal with humans. Feral house mice living in field settings and away from human habitation exhibit a flexible pattern of overlapping home ranges, both within each sex and across sexes (Brown 1962; Berry 1968, 1981; Bronson 1979; Mikesic and Drickamer 1992), and their mating system, though not fully understood, involves both constrained promiscuity and multiple paternity. Multiple paternity occurs in about 35% of the litters in a field setting (Drickamer, unpublished data). Home range overlap within and between sexes constrains promiscuity by limiting mating with those individuals that share breeding space (Drickamer and Robinson, unpublished).

Density and home ranges

Plasticity in rats and mice social organization is matched by flexibility in home ranges, and consequently the ability to live in highly variable densities. Estimates of rat abundance in urban areas show 1,000-fold differences in rat density, from 1 rat per 100 people to 1,000 rats per 100 people. Because rat densities and movements are heavily influenced by the spatial and temporal distribution of food resources such estimates, even when reliable (e.g., see Davis and Fales 1949, 1950; Davis 1950), are of specific rather than general value. The reported killing of 2,650 rats in a single night and 16,000 in one month in a horse slaughterhouse in France in 1840 provides a relative estimate of rat abundance in a small area (Shipley 1908 in Twigg 1975). While current control programs would prevent the buildup of such densities, these accounts demonstrate the flexibility of the rat population dynamics. Rat colonies have typically been maintained at a density of several rats per m² in outdoor enclosures and, while social processes eventually limit population size (Davis and Christian 1956, 1958), rats show

little constraint by self-regulation when food is unlimited (Calhoun 1962). In laboratory cages, current guidelines recommend a maximum density of adult rats equivalent to about 25 rats per m² (or 250,000/ha), while mice can be stocked in densities equivalent to about 100 adults per m² (equivalent to 1,000,000/ha, see the following).

Even in relatively rugged environments, rat populations can be viable away from human habitations (e.g., South Georgia Island; Pye and Bonner 1980). Kozlov (1979) reported that rat populations could be found 10 km away from villages in northern Kazakhstan, and within a year had moved up to 70 km away (in Pye and Bonner 1980). Following Elton and Chitty's pioneer work (Chitty 1954a—see also Errington 1935; Aisenstadt 1945; Davis et al. 1948; Davis 1949; Barnett 1955), most studies of rats in the wild have been carried out under the framework of their role as agricultural pests. In farmland, up to three-quarters of rat populations may occupy stable home ranges (Hartley and Bishop 1979), while the rest are transient animals. Studies of farm rats in temperate regions (Czechoslovakia; Zapletal 1964; Moravia; Homolka 1983; United Kingdom; Brodie 1981; Hartley and Bishop 1979; Huson and Rennison 1981; Middleton 1954; Cowan et al. 2003; United States; Errington 1935; USSR; Aisenstadt 1945) suggest that hedges and fields constitute only temporary habitats, although they can provide an important reservoir for disease (see the following) as well as a source for the reinfestation of buildings.

Sizes of home ranges are equally flexible and dependent upon food availability, including seasonal variation in harvest. Linear home range sizes vary from less than 100 m in habitat where food is plentiful (e.g., urban areas, grain stores) to almost a kilometer in poorer rural environments (Davis et al. 1948; Taylor 1978; Taylor and Quy 1978; Hardy and Taylor 1979; Hartley and Bishop 1979; Stroud 1982; Fenn et al. 1987; Macdonald and Fenn 1995). Studies based on radio-tracking typically show greater movements than those based on capture-mark-recapture and, in addition, rats can also travel long distances on foraging sorties. The longest radio-tracked movement in a single night was a 3.3 km round trip made by a male rat (Taylor and Quy 1978; Hardy and Taylor 1979). Steiniger also reported that rats would travel an estimated 6 km on an overnight round trip (including swimming across a 50-m river) to take eels from the nets at ebb tide in North Germany (in Twigg 1975).

House mice living in commensal situations in and around human-made structures can attain densities of more than 14 mice/m² or 10,000/ha (Laurie 1946; Rowe et al. 1963), although the area involved is often only a fraction of this size, as the mice are highly concentrated. When outbreaks occur (fig. 32.2), densities of commensal house mice can exceed the equivalent of 100,000/ha (Hall 1927), but

Figure 32.2 Mice seem to have been associated with humans for over 10,000 years, and are also well known for their impressive population outbreaks. Photo by G. Singleton.

are more often at densities approaching 1,000/ha (Pearson 1963). Outbreaks of field populations of mice are common in the main cereal-growing regions of Australia (Singleton et al. 2005) and in the beech forests of New Zealand following mast years (Ruscoe et al. 2001). House mice living under feral conditions are generally found at densities ranging from 10–100 mice/ha (Elton 1942), even when living in field enclosures (Drickamer and Gillie 1998). Changes in densities of mouse populations across different habitats in an agricultural landscape and the processes likely to be driving these dynamics have been detailed in the Australian wheatlands (Newsome 1969; Singleton 1989; Singleton and Redhead 1989).

When resources are scarce or widely dispersed, as in natural or field settings, house mice are often not territorial, but exhibit overlapping home ranges whose size can be highly variable (e.g., 0.0002–8 ha; Chambers et al. 2000). Studies in the agricultural landscape of southern Queensland (Australia) reported that most individuals were site attached during the breeding season, with an extensive overlap of home range of both sexes and with breeding males occupying home ranges about twice as large as breeding females. After breeding, home ranges increased by about tenfold as most mice became nomadic. Similarly, Chambers et al. (2000) reported from studies in an agricultural site of northwestern Victoria (Australia) that nonbreeding mice seemed to be nomadic when densities were low, as well evidence for exclusive home range by females (at all densities), low to moderate home range overlap for males when densities were low, and a switch to a more gregarious phase after breeding and when densities were high. The flexible social system and related changes in breeding structure and care of young have likely contributed to this species success throughout its worldwide distribution (Berry 1981).

Although there is considerable variation among methodologies, results, and interpretations of the numerous studies of rats and mice, we can make several generalizations. Mice appear to have a more flexible social system than rats, and both species are opportunistic with respect to reproduction and rapid population growth when food availability is high.

Dispersal

Dispersal has not been studied directly in the wild, but laboratory studies suggest that it is socially as well as food driven. Calhoun (1962a), in his pioneering work on rats housed in a large outdoor enclosure, observed that subordinate males were excluded from the more favorable sites. Farmland data suggest that colonists are usually young males approaching sexual maturity (Zapletal 1964; Telle 1966; Bishop and Hartley 1976; Farhang-Azad and Southwick 1979), with females more likely to stay in stable breeding environments (Leslie et al. 1952; Kendal 1984; see also Calhoun 1962a for results in large outdoor enclosures). It appears thus that dispersal is male biased, and it is likely that around reliable food supplies resident males exclude transient immigrants (see Nunes, chap. 13 this volume for examples and theory of dispersal).

A similar pattern of dispersal appears to occur in mice. Young females have a greater tendency to remain philopatric than males; however, both sexes do disperse to some extent. Males emigrate from their maternal site as early as 40 days of age and females at about 70 days of age (Brown 1953; Berry and Bronson 1992; Drickamer unpublished data). As in most species, males may disperse farther than females, but no clear sex differences have been noted in the tendency to disperse (Rowe et al. 1963; Drickamer unpublished data). Aggressive behavior by adult males toward juvenile males has been reported as a precipitating factor in juvenile dispersal (Bronson 1979); however, this has not been well documented for mice or other species of mammals (Wolff 1993a, 1994a).

Foraging and feeding

Rats and mice are mostly nocturnal, presumably to avoid predation, although both are able to modify activity patterns. In a long-term monitoring of feeding activity, Berdoy (1994) reported how wild rats in an outdoor enclosure (fig. 32.3) adjusted feeding intensity to maintain nocturnality. As night length decreased during the summer months, rats compensated by increasing feeding activity, mostly in the last quarter of the night. This nonuniform compensation may point to time-budget conflicts between feeding and social interactions, which took place in the first part of the night. It may also highlight the fact that rats feed in a way

Figure 32.3 Because rats are small, brown, nocturnal, and difficult to mark, most detailed behavioral data of wild rats have been recorded in large outdoor enclosures, such as this one near Oxford (UK) where rat colonies can be presented with problems that they would normally encounter in the wild, such as a diverse and multigenerational environment and a dispersed food supply. Photo by M. Berdoy.

that anticipates further energy needs; heavy predawn feeding presumably enabled the rats to last through the correspondingly longer daylight hours during the summer. Rats, however, can abandon nocturnality altogether when subject to heavy predation. Fenn and Macdonald (1995) found that a population of rats in a rubbish tip that was regularly visited by foxes at night reverted to a diurnal activity pattern.

Foraging behavior in rats is typically described as the outcome of conflicting motivations between curiosity (*neophilia*) and caution (*neophobia*, meaning "fear of the new"). Being omnivorous is an undeniable asset when colonizing new environments; rats are typically faced with a wide variety of foods, some of which are palatable and others poisonous. This feeding behavior results in conflicting selective pressures, known as "the omnivore's paradox" (Rozin 1976) i.e., the benefits of exploiting new food sources ver-

sus the cost of being harmed by them. Rats have evolved several mechanisms that help them in diet selection. (1) Rats have developed the capacity to make the association between what they eat and how they feel several hours afterward ("long delay learning"; Garcia et al. 1966) thus enabling them to assess the metabolic consequences of the food that they ingest, therefore avoiding poisonous items. (2) Rats may be physiologically adapted to survive the occasional mistake: recent findings concerning the rapid evolution of enzymes responsible for metabolic detoxification in rats (Rat Genome Sequencing Project Consortium 2004), cytochrome P450 (Danielson 2002) may highlight a hitherto underestimated weapon in its armory to survive the omnivore's paradox. (3) Rats can learn from other individuals in the colony. A large body of evidence (mainly from the laboratory) indicates that feeding *preferences* are socially transmitted (reviewed in Galef, chap. 18 this volume). Young juveniles follow their mother during her foraging sorties, but food preferences can also be influenced at an earlier age by their mother's milk, as well as when adults by the smell emanating from other individuals in the colony. Rats are more likely to try new foods if they smell them on a healthy individual—the "Demonstrator Effect"—(see Berdoy 1994 for evidence in a wild rat colony). While there is no evidence that food *aversions* can be socially transmitted, a rat is more likely to avoid a novel food if, following ingestion, it finds itself in the presence of an individual that is ill—"The Poisoned Partner Effect." (4) Rats are cautious (neophobic). In contrast to mice, which are generally regarded as neophilic, rats are often described as among the most neophobic of mammals (fig. 32.4). Neophobia varies between rat species (Cowan 1977), between wild and laboratory strains where it has been partially bred out (Mitchell 1976; see also Berdoy 2002), and in the wild between populations (Brunton and Macdonald 1996; Brunton, Macdonald, and Buckle, 1993; Quy et al 1992;). Neophobia also varies between individuals (Cavigelli and McClintock 2003), and can be modulated by social factors (see Galef, chap. 18 this volume).

House mice are also mainly nocturnal, though they often feed in daylight hours (Rowe 1981). House mice also shift their activity rhythms and times of feeding depending on season and climate. Much of the work on wild house mice concerns their effects as commensals in terms of consumption of and damage to grain stores (Rowe et al. 1963; Jackson 1977; Stoddard 1979; Stenseth et al. 2003). The issue of transmission of food preferences, so thoroughly explored in rats by Galef and colleagues (Galef, chap. 18 this volume), has not been examined as comprehensively in house mice, although they also show social transfer of information (Valsecchi et al. 1996; Choleris et al. 1997). Unlike rats, house mice are typically described as neophilic

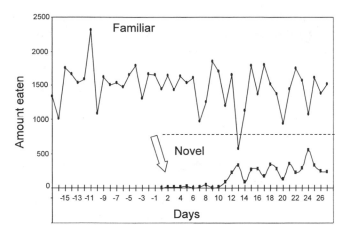

Figure 32.4 Neophobia: grain consumption (in g) of a "stable" enclosed colony of wild rats at familiar and novel containers (but same food) introduced at day 0 (X axis). The numbers of rats were constant throughout the observations and the containers were accessible in equal measure to all colony rats (thus controlling for possible spatial factors). The dotted line indicates the average amount expected to be consumed if both types of containers were used in equal measure. Note the lack of consumption for the first 10 days, the relatively rapid change over the next three days (days 11–13) and the still very reduced (although slowly increasing) consumption after about 1 month (Berdoy, unpublished data).

with regard to their feeding habits, and are generally willing to try new foods (Southern 1954; Meehan 1984); however, under certain conditions they appear to avoid novel foods (Misslin 1982; Humphries et al. 2000). Neophilia, like neophobia, incurs costs and benefits. Benefits could include the ability to find new foods readily, which would enhance the capacity to exploit new environments. Nevertheless the extent of the difference with rats remains somewhat of a puzzle.

Aggression and social status

Rats and mice differ in the ways that agonistic interactions are used to establish and maintain social relations, partly reflecting the greater range of the rats' behavioral repertoire. Overall, both male rats and mice display more aggression than females, although rats appear to be able to establish dominance hierarchies within their living groups that perpetuate themselves despite changes in relative physical prowess of males in that group. In contrast, male mice constantly compete for dominance among each other. Comparatively few data exist for aggression in female rats, although aggression in female mice is well documented in a variety of contexts.

Most studies on aggression have been conducted in the laboratory, although observations of wild rats confirm that aggression, while not very common, is highest among males (see descriptive drawings, including "play fighting," in Pellis and Pellis 1987; and Berdoy 2002 for film footage of ag-

onistic interactions). Agonistic interactions between sexes and between females are relatively infrequent; a dominance hierarchy among females is less readily apparent (Calhoun 1962a; Robitaille and Bovet 1976; Adams and Boice 1983, 1989; Berdoy et al. 1995; for laboratory rats see also Blanchard and Blanchard 1977, Rampaud 1984; and Blanchard et al. 1988).

Predictably, larger rats are more likely to win contests upon first encounter with smaller individuals. However, observations of wild rats in *stable groups* also show stable linear dominance hierarchies among males, where social status can become "settled" as the outcome of previous encounters and determines the outcome of future ones (Berdoy et al. 1995; see also Smith et al. 1994 for laboratory rats). In these conditions, age becomes a better predictor of social status than weight, as dominant individuals maintain their social status even though initial body weight asymmetries with younger (and at the time lighter) individuals have disappeared or been reversed. It is noteworthy that the existence of such "settled dominance" (Berdoy et al. 1995) may mitigate against an easy demonstration of the role of competitive abilities in maintaining social status because, once dominance has been established, changes in competitive abilities may not be immediately reflected by a change in social status. This relationship may explain why, even in laboratory settings, the evidence that dominance is correlated with body weight has been conflicting, with only about half the studies reporting a correlation (reviewed in Berdoy et al. 1995). It is not known to what extent such settled dominance is also found in rats living at low densities (and therefore with less frequent contacts), but its existence among wild rats in stable groups raises the question as to why young individuals with greater competitive abilities should accept the status quo. First, the benefits of living in groups may outweigh the cost of being relegated to a subordinate position, particularly if the costs of escalated aggression are great relative to the value of the contested resource. Second, as a consequence of some of the features of rat mating behavior (see the following), dominant males often cannot monopolize access to estrous females or to food resources (Calhoun 1962a; Berdoy and Macdonald 1991; Berdoy 1994; Berdoy et al. 1995). Finally, dominance hierarchies, while settled, are not static, and can change substantially between six monthly periods, thus allowing competitive subordinates to rise to prominent social status eventually (M. Berdoy, unpublished data). Clearly, the relative tolerance of such social dynamics allows rats to live at high densities.

Aggression has also been studied in a variety of strains of house mice, including wild house mice (Lagerspetz 1964; Brain et al. 1989). Similar to rat aggression tendencies, male-male aggression is important for the establishment and maintenance of territories, in some instances for dominance hierarchies, and for inducing dispersal of males (Crowcroft 1955; Anderson and Hill 1965). In feral house mice, male-male dominance likely is significant for access to resources such as food or mates, and influences dispersal via male-male contact. To our knowledge, these male-male effects have not been observed directly in feral house mice, but are inferred from trapping data, radio-tracking, and spatial dispersion patterns.

As in rats, aggression by female house mice occurs at lower frequencies and intensities than in males (Mackintosh 1981), although it has been investigated in more detail. The ecological functions of aggression by females involve competition for resources, possible influences on dispersal, and maternal aggression, exhibited by postpartum females toward almost all intruders (Svare 1977, 1981). Both ecology and evolution have likely influenced aggression in female house mice. In their ancestral habitat, with a relatively even distribution of resources such as food, there would be little need for females to compete with other females, but they would need to protect their young through maternal aggression if breeding space became limited (Wolff and Peterson 1998). In the commensal system, with males holding territories and each territory potentially containing more than one female, female-female tolerance is important to maintain cohesive breeding within the deme, especially among related females (Solomon and Keane, chap. 4 this volume).

Social odors

As social nocturnal animals (Calhoun 1962a; Taylor et al. 1991; Berdoy 1994) it is not surprising that olfaction plays an important role in rat societies (there are also suggestions that rats use ultrasound for short range echolocation; Rosenzweig et al. 1955; Riley and Rosenzweig 1957; Kaltwasser and Schnitzler 1981; and for social communication; Sales and Pye 1974; Barfield and Geyer 1975; Sales and Smith 1978; White et al. 1990; Blanchard at al. 1991). Olfactory signals emanating from the animals' body, breath, urine, and feces convey an enormous level of information, ranging from individual recognition, sexual status, and dominance status to stress levels in both rats and mice as well as rodents in general (see Roberts, chap. 22 this volume). Olfactory signals might be encoded through the presence of single components, but also through the ratio of several chemicals or through the calibration of a fluctuating signal with another stable, odorous one. The scent from sebaceous glands, which are located on a rat's hindquarters, is often sniffed by individuals of both sexes, and has at least 22 volatile components that can vary between individuals (Natynczuck and Macdonald 1994a, 1994b). Olfaction is also useful in predator avoidance. Rats, including labora-

tory rats that have never been in contact with predators for generations, remain innately aversive to predator odors (Vernet-Maury et al. 1968, 1984; Berdoy and Macdonald 1991; Tanapat et al. 2001; see the following for how this aversion can be manipulated by parasites). Social odors are relevant to virtually all aspects of rats' lives, including mating behavior, aggression, parental behavior, and food selection, and have been reviewed extensively for rats (Brown and Macdonald 1985) and other rodents (Roberts, chap. 22 this volume). The finding that rats have an estimated 2,070 genes and pseudogenes involved with olfaction (37% more than mice) suggests that we are a long way from unraveling its complexity (Rat Genome Sequencing Project Consortium 2004; table 32.1).

Although rats and mice of both sexes are known to scent mark by depositing urine, this has been generally investigated more in mice, where scent marking clearly plays a significant role in social behavior (Bronson 1976; Hurst 1987, 1989, 1990a, 1990b, 1990c, and see extensive review for rodents in general by Roberts, chap. 22 this volume). Dominant males typically mark territorial boundaries with urine, and intensely counter-mark both resident and nonresident males. Urine marking can also be used to predict social dominance in male house mice in that males destined to become dominant leave more marks on a greater area than those that will likely become subordinate (Drickamer 2001). Female house mice respond to urine by investigation and counter-marking toward urine from resident and neighboring females, and females mark at a higher frequency to indicate higher social and breeding status. Between the sexes, females are more attracted to marks from their resident dominant males than other males. In addition to the importance of urine cues for social behavior, mice of all ages and sexes appear to use urine marks as a means of orientation (Hurst 1990b, 1990c, 1990d). For a complete discussion of urine marking in mice, rats, and other rodents see Roberts, chap. 22 this volume.

Mate choice

Rats and mice clearly exert mate choice. Mating behavior is more complex in rats, but the proximate mechanisms of discrimination, such as the possible role of the Major Histocompatibililty Complex (MHC) and Major Urinary Proteins (MUPs; Penn and Potts 1999; Hurst et al. 2001; Roberts and Gosling 2003), as well as the fitness consequences of free mate choice have been more investigated in mice.

The dynamics of mating in rats appears to be strongly dependent upon density, but their mating patterns show multimale-multifemale mating, with a potentially complex (and largely not understood) pattern of intromissions (e.g., see McClintock, Anisko, and Adler 1982 for observations

Table 32.1 Comparative social systems, behavior, and life-history traits for *Mus* and *Rattus*

	Rattus norvegicus	*Mus domesticus*
No. of species in genus	51–57	37
Distribution	Worldwide except in polar regions	Worldwide except in polar regions
Common names	Common, barn, brown, gray Norway, water, wharf, sewer	Common house mouse, domestic house mouse
Weight (as adults)		
Males	250–500g	17–23g
Females	250–300g	15–21g
Food intake (per day)	5–10g/100g body weight	15g/100g body weight
Heart rate (beat/min)	250–450	310–840
Size of genome	2.75 Gb	2.6 Gb (Human = 2.9)
Genes and pseudogenes for olfactory reception	c. 2,070	c. 1,510
Sexual maturity	65–110 days	55–80 days
Estrous cycle	4–5 days if not mated	4–5 days if not mated
Vaginal plug	Yes	Yes
Post-partum estrus	Yes	Yes
Gestation	20–23 days	19–21 days
Litter size	6–10 in wild (variable), up to 20 in laboratory	2–7, up to 15 in laboratory
Lifespan	About 12–18 months in the wild (3–4 years in laboratory)	Up to 18 months in the wild (1.5–3 years in laboratory)
Mating system	Flexible: from polygynous to polygynandrous	Flexible: generally polygynous, but also promiscuous
Spatial dispersion pattern	Overlapping home ranges (feral) to group territories (commensal)	Home ranges (feral); territories (commensal)
Reaction to novelty	Neophobic	Neophillic
Use of urinary cues	Moderate (limited data?)	Extensive
Dispersal	Male biased	Both sexes
Involvement in spread of disease	Extensive	Limited data

NOTE: The values of many of the reproductive parameters strongly depend upon environmental factors such as densities and ambient conditions. Gb = billion base pairs.

in the laboratory). While some observations may represent an artifact of laboratory conditions, studies of wild and laboratory rats in large outdoor enclosures (fig. 32.3) show that females, which are typically in estrus for only one night, both solicit/recruit and flee from neighboring males in a series of mating chases. In high-density environments females can be followed assiduously by a string of up to 15 males whenever they move outside their burrows (Calhoun 1962a; Robitaille and Bovet 1976; Berdoy 1994 for wild rats; see Berdoy et al. 1995 and also film footage in Berdoy 2002 for lab rats). Chases are punctuated by mating in a type of scramble competition (following lordosis by the female) and/or disappearance of the female inside a burrow until her next sortie. As mating reduces the probability of infanticide by the male (Brown 1986; Wolff and Macdonald 2004) the female may benefit from such multiple matings. By encouraging male-male competition, mating chases may help the female to assess male fitness independently of male dominance and exert mate choice by controlling mating opportunities (e.g., delaying lordosis or escaping into a burrow). Since males have little time to interact with each other during the pursuit (as they would lose a mating opportunity), even the most dominant males cannot prevent other males from attempting to participate in mating chases. Thus

although dominant males are more likely to mate than lower subordinates (confirmed by paternity data; Berdoy, Stanley, Macdonald, unpublished data for wild rats, see also Berdoy et al. 1995), they cannot always monopolize access to the female, particularly as colony size increases (fig. 32.5). Although anecdotal evidence suggests that wild male rats can also exert mate choice in some situations (Berdoy, unpublished data), to our knowledge this has not been formally investigated.

Mate choice in feral house mice involves some degree of selection between partners of the opposite sex that share overlapping areas of the home range. DNA-based parental assignments for house mice living in outdoor enclosures (fig. 32.6) suggest mate selectivity exhibited by members of both sexes (Robinson 2000; see also Rolland et al. 2004). A number of factors are associated with mate choice in wild house mice studied in the laboratory, including odors (Mainardi, Marson, and Pasquali, 1965; Drickamer et al. 2001), odors associated with the T locus genotypes (Lenington and Egid 1985; Lenington et al. 1988; Egid and Brown 1989; Lenington and Heisler 1991), odors associated with presence or absence of parasites (Kavaliers and Colwell 1995b), early learning experiences (Mainardi, Marson, and Pasquali, 1965, Mainardi, Scudo, and Barbieri, 1965), so-

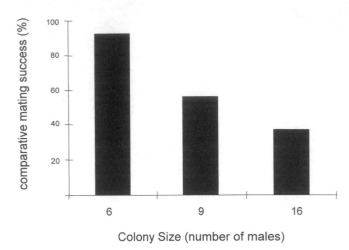

Figure 32.5 Relative mating success of the top four males (to control for a dilution effect) in four colonies of wild rats. When colony size was small, the dominant male exhibited the greater mating success. But, in larger colony sizes, its mating success was more variable and sometimes lower than the next two highest ranking males (Berdoy and Macdonald, unpublished data; see also Macdonald et al. 1999).

cial dominance and aggression (Oakeshott 1974), and chance (Berry and Jakobson 1974). Thus numerous experimental studies have demonstrated that various proximate mechanisms, which have fitness consequences, are used by mice in assessing mates. Evidence from field and laboratory studies indicates fitness consequences for males and females choosing a preferred mating partner (Drickamer et al.

2000; Gowaty et al. 2003). The progeny of females mated with a preferred male showed greater pre- and post-weaning survival and more reproduction in field enclosures than the progeny from females mated with a nonpreferred male. Males mated with nonpreferred females produced fewer offspring with lower viability than males mated with preferred females. Progeny of both sexes when males mated with preferred females built better nests, and males were more aggressive than progeny from matings between males and nonpreferred females.

Reproduction and maturation

Rats and mice are well known for their high reproductive rate. Both species show postpartum estrus and are theoretically capable of producing a new litter every month or so (table 32.1). Reproductive output is, however, strongly influenced by environmental factors, and life-history parameters in the wild are better documented in mice.

Rats can breed throughout the year (Barnett 2002); reproduction is affected by a seasonal fluctuation in food availability. Pregnant females have been captured from early spring until late summer (Davis and Hall 1951; Farhang-Azad and Southwick 1979; Fenn 1989). Cold winters were associated with no reproduction (Leslie et al. 1952; Andrews et al. 1972; Lattanzio and Chapman 1980; Pye and Bonner 1980), but Fenn (1989) found that winter breeding in the UK was stimulated by provision of additional food in

Figure 32.6 A series of eight 0.1-ha enclosures were used for extensive studies of the social system, reproduction, effects of olfactory cues, and mate choice in feral house mice. Photo by L. Drickamer.

winter. Rats show delayed onset of sexual maturity and facultative cessation of spermatogenesis when food is scarce (Fenn 1989; Macdonald et al. 1999).

In mice, breeding intensity and the proportion of sexually mature females that are pregnant or lactating at any one time also varies with climate, resource availability, and whether the mice are commensal or feral (Pelikan 1981). In feral populations, the average number of females breeding at any one time is around 40%, but this can peak at more than 80% during summer months (Pelikan 1981; Drickamer unpublished data; Singleton et al. 2001). In commensal locations, it is possible that some females may have eight to ten litters or more in a lifetime, though this would be exceptional. In feral situations, most females can be expected to have two to four litters during a breeding season (Pelikan 1981; Drickamer, unpublished data).

Litter size exhibits wide variation; in Europe, it varies seasonally and with habitat. Average litter size for commensal mice in winter ranges from 5.0–5.5 pups, rising to 6.5–6.7 pups in summer months (Pelikan 1981). Feral mice in Europe have summer litter sizes of about 8.0 pups. In laboratory colonies of wild house mice and field populations of feral house mice in seminatural enclosures in the United States, summer litter sizes average 5.4–5.6 pups (Drickamer 1977b, unpublished data). Feral mice in agricultural areas in Australia show average litter sizes ranging from five to nine pups, depending on seasons and years (Singleton et al. 2001).

Females attain sexual maturity at 50–60 days of age, as evidenced by first vaginal estrus (Drickamer 1979); its timing can be influenced by a combination of factors. Factors that stimulate early estrus include high nutrition followed by rapid weight gain and heavier body mass (Monteiro and Falconer 1966; Vandenbergh et al. 1972), and social factors, particularly urinary chemosignals from adult males (Vandenbergh and Coppola 1986; Drickamer 1986a, 1999; see also Drickamer, chap. 9 this volume).

House mice can breed throughout the year and generally do so in commensal situations if sufficient food resources are available (Southern 1964; Berry 1970; Pelikan 1981). In natural or feral settings house mice exhibit seasonal breeding, which involves either cessation of reproduction for winter months, or in some instances, a summer seasonal peak with diminished reproduction in winter (Berry 1970; Bronson 1979; Singleton et al. 2001). Bronson and Perrigo (1987) argue that virtually all of the variance in reproduction for house mice can be explained by energetics, and that readily available food is the key to variation in reproduction in different habitats and climatic conditions. From a dataset on 18 years of breeding, Singleton et al. (2001) suggest that reproductive changes may be driven by food quality, particularly protein.

Perhaps the only life table published for house mice is that of Berry and Jakobson (1974) for Skokholm Island (see also Krebs et al. 1995). Life expectancy at birth varies seasonally with time of birth, from a high of over 15 weeks for mice born in late spring to a low of 11 weeks for mice born in fall. Longevity was up to 18 months for a very few individuals born in late spring and was somewhat less for mice born later in the breeding season. As would be expected for an opportunistic species, about 50% of all young die before attaining puberty. Data for feral house mice living in field enclosures indicate that only 22.3% of young survive to recruitment (Rossinni et al. unpublished). Analyses of these same data revealed that several factors influence survival in wild house mice, including genetics, weather, density, body mass, and season of birth. The only published study with a life table indicated that these factors influence survival for house mice; consequently, additional experimental data are needed to confirm the relevance of ecological factors to life-history parameters.

Parental care and infanticide

As altricial species, both rats and mice need prolonged and intense parental care, which is provided almost entirely by the female (Lonstein and De Vries 2000c), although laboratory evidence hints at a possible influence of males in mice (Wright and Brown 2000).

While the proximate mechanisms of parental behavior (Rosenblatt et al. 1979; Brown 1993; Lonstein and De Vries 2000b) and infanticide (Brown 1986) have been investigated extensively in laboratory rats, virtually nothing is known about parental care and infanticide in wild rats and it is unclear to what extent some of the laboratory findings on these behaviors (e.g., communal nesting; Mennella et al. 1990; Schultz and Lore 1993) occur in wild rats. Parental care is provided by the female although, perhaps surprisingly, laboratory evidence indicates that, among preweaning juveniles, males show more parental *responsiveness* toward pups than do females. These gender differences are reversed after weaning (see reviews in Stern 1996; Lonstein and De Vries 2000c). Infanticide in wild rats is likely to be common, if not frequent (Calhoun 1962a; Robitaille and Bovet 1976; Mennella and Moltz 1988). M. Berdoy (unpublished data) observed infanticidal events by both males and females in an enclosed colony of wild rats (see also film footage in Berdoy 2002 for lab rats). In one instance, the dominant male in the colony was observed to wait outside the burrow of a female at dusk and kill a pup about every 30 minutes whenever she was not there. However, by the following day, about half of the litter remained unharmed and subsequently survived to sexual maturity (M. Berdoy, unpublished data). Given the evidence of olfaction-based kin recognition and the multiple paternity of litters, could it be that the male was able to kill selectively the pups that were not his within a litter? See Ebensperger and Blumstein, chap. 23 this volume for further discussion on infanticide.

In their feral environment, house mice are not readily observable, so conclusions about their parental behavior in these conditions are limited. Females are often observed with their litters under both feral and commensal situations, but males are rarely present (Rowe 1981). Communal nesting has been recorded in which at least two litters were being cared for in the same nest cavity by females related as sisters or as mother and daughter (L. Drickamer, unpublished data based on the DNA-parental assignment used by Robinson 2000). Konig (1994) has investigated aspects of communal nesting in wild house mice in a laboratory setting. She reported that communally nesting females enjoy greater lifetime reproductive success than those nesting alone. This effect is pronounced when the communally nesting mice are related females as opposed to unrelated females. It is noteworthy that there may be some connections between the lower levels of female-female aggression, sharing nests, and reproductive success. This notion would require additional testing for confirmation.

Considerably more is known about parental care in commensal than feral house mice. A system of chemical and auditory communication cues mediates the relationship between mother and pups (reviewed by Smith 1981). Communal nesting with nursing of own and other's pups is recorded by a number of investigators studying commensal house mice (e.g., Crowcroft 1955; Brown 1962; Drickamer personal observations). There are clear reproductive success advantages from communal nursing relative to monogamously mated females and those that rear litters alone, and there may be some overall effects due to related versus unrelated mothers (Konig 1993, 1994). The cost-benefit analysis of communal nursing likely revolves around several factors, which could include food abundance and distribution, external threats such as predators or other house mice, and the availability of suitable nesting habitat. Further study would be required to test these factors. As in many polygynous mammals, males have a lesser role with regard to rearing of young. Males are found with mothers and litters in nests but nothing really is known of their potential paternal behavior toward pups or young.

Infanticide has been studied extensively in the laboratory with mice, but virtually nothing is known from wild populations. As for rats, the general results from laboratory studies show that nonsire males will kill unfamiliar pups, the act of copulation inhibits males from committing infanticide, and that females will come into estrus after losing a litter thus providing the perpetrator with a mating opportunity (e.g., Labov 1980; vom Saal and Howard 1982;

Parmigiani and vom Saal 1994). This "sexual selection" infanticide also occurs in rats (Mennella and Moltz 1988) and is common among rodents, and may be an adaptive reproductive strategy for males (see Ebensperger and Blumstein, chap. 23 this volume).

Diseases

Rat and diseases are synonymous in popular culture; rat-borne diseases are thought to have contributed to more deaths in the last 10 centuries than all wars and revolutions ever fought (Nowak 1999). Recent comprehensive surveys of rats in rural and urban areas have revealed more than 10 zoonotic agents (i.e., animal pathogens transmissible to humans), including Leptospira, Toxoplasma, Yersinia, and Hantavirus (Webster and Macdonald 1995a, 1995b; Battersby et al. 2002). While this may be of particular relevance to public health, it should also be of interest to behavioral ecologists, because of their impact to other wild species generally (which is essentially unknown), but also on behavior more specifically. *Toxoplasma goondii*, the causative agent of toxoplasmosis, has been shown to increase general activity levels in infected rats (Webster 1994) and to reduce their neophobia (Berdoy et al. 1995; Webster et al. 1994) as well as their innate aversion to cat odors (Berdoy et al 2000; fig. 32.7). Such host manipulation is likely to increase rat predation by cats and therefore will benefit the parasite that, although capable of infecting all mammals (including humans), needs to find its way into a cat host to complete its indirect life cycle. It is noteworthy that while humans clearly represent a dead-end host for the parasite, reports ranging from altered personality to higher incidence of car crashes among *T. gondii*-infected patients (Flegr et al. 2002) may represent the intriguing side effects of a parasite evolved to manipulate the behavior of another host (Berdoy et al 2000; Webster 2001). Mice also exhibit behavioral changes, although the effect of *T. gondii* infection is more acute and sometime lethal; the response of infected mice to the risk of cat predation has not been tested so far (see Hay et al. 1983; Hrda et al. 2000; Jackson et al. 1986; and review in Webster 2001).

House mice can also act as both reservoirs of and vectors for a variety of disease organisms (Blackwell 1981; Moro et al. 1999; Rowe 1981; Singleton et al. 1993; 2000; Smith et al. 1993), although there are far fewer recorded instances of house mice acting as reservoirs or vectors of human-related diseases than for rats. House mice are associated with food poisoning (*Salmonella*; Rowe 1981; *Streptobacillus*; Taylor et al. 1994) in connections with both humans and livestock, and are reservoirs for lymphocytic choriomeningitis virus (Smith et al. 1993), cryptosporidium (Morgan et al. 1999), and possibly rickettsial diseases, but do

Figure 32.7 Rats have evolved an innate, and sensible, aversion to cat odors, which is present even among individuals (e.g., laboratory rats) that have never seen a cat for generations. However, this can be manipulated by *Toxoplasma gondii,* a parasite that can infect rats and that will benefit from finding its way back to its definitive host, the cat. Photo by M. Berdoy.

not appear to play any significant role in diseases such as plague (*Yersinia pestis*). Mice transmit *Salmonella* but in general appear to be genetically resistant to this bacterium, as measured by the failure of inoculations to cause illness or death. Further study of the disease resistance of rats and mice could prove quite significant with regard to the evolutionary history and socioecology of these widespread and commensal species, because disease/parasite resistance conveys advantages in terms of inhabiting areas where there is regular exposure to particular pathogens.

Summary and Unanswered Questions

Rats and mice are similar with respect to their reproductive biology, timing of sexual maturation, gestation, and litter size when adjusting for body size (table 32.1). Both species also exhibit flexible spatial distributions and mating systems that are generally polygynous or promiscuous, can live either commensally or in feral populations, and tolerate high densities. These traits have undoubtedly contributed to their worldwide distribution, as well as preadapted them to laboratory breeding. Their comparison also highlights differences however—some stemming from their biology, others likely stemming from differences in research efforts. For example, rat density is usually described in terms of animals per inhabitant rather than (as for mice) surface area,

and detailed behavioral data on social interaction in natural or seminatural circumstances are more available for rats (partly reflecting their greater behavioral repertoire—see www.ratbehavior.org) while life-history data in the wild is more abundant for mice. Nevertheless, given the broad similarity in their reproductive biology, an obvious question is: why are population explosions more associated with mice than rats? Research efforts in ecology have so far not resolved this deceptively simple question.

An important factor in the ecological success of rats and mice is the way that they are both able to colonize new environments; a noticeable difference, however, is that mice are reportedly attracted to novelty (neophilic), whereas rats tend to avoid novel objects or foods (neophobic). Although the rats' extreme caution is often viewed as a feature of diet selection, in that it may assist in the development of conditioned aversion, it is worth considering that it may be linked instead to a general antipredation response. Rat colonies, for example, can also react strongly to the presence of a new food *container* (fig. 32.4), and Inglis et al. (1996) and Beck et al. (1988) showed that the pattern of feeding was not consistent with general diet assessment (see also Galef, chap. 18 this volume). Moreover, neophobia has a cost, that of delaying the exploitation of a resource—although this also seems to be subject to environmental plasticity: rat populations can afford to be most neophobic in stable environments where food is plentiful (e.g., as in fig. 32.4), and this is often where control efforts are most intensive (Brunton, Macdonald, and Buckle 1993; 1996; Quy et al. 1992). The merit of neophobia as a general feature of diet selection is therefore generally assumed rather than proven, but the rat's paranoid tendencies are of undeniable value in a poison-rich environment. Yet, considering these marked differences in attraction versus avoidance of novel (often human-produced) foods, both species are extremely successful in their association with humans. Not much is known about dispersal patterns or distances in wild populations, but they must be pronounced, considering the widespread distribution of both species.

It is also not clear why diseases are more readily associated with rats. Perhaps their more predatory nature compared to mice and their greater predilection for water may make them more likely to harbor zoonoses and waterborne diseases, respectively. It is more likely, however, that the rat's reputation may be a consequence of its greater visibility, its involvement in some very prominent diseases such as the bubonic plague (transmitted by rat fleas) that decimated over a quarter of the human population in Europe in the Middle Ages, and more comprehensive disease surveys on wild rats compared to wild mice.

Mammals in general are olfactory oriented; however, the extensive research on scent marking and MHC discrimination in mice may appear to suggest that olfactory cues play a greater role in the social system of house mice than rats. This difference is puzzling, and may reflect a bias in research efforts. Indeed, recent genomic evidence shows that rats have in fact 37% more genes and pseudogenes involved in olfaction than mice, thus suggesting further avenues for behavioral and physiological work.

This apparent discrepancy between behavioral and genetic data also highlights a more general phenomenon; perhaps more than any other mammalian species, our knowledge of rats and mice reveals a tale of (at least) two cultures in the biological sciences: broadly speaking, those who are interested in animals for their own sake and those who are interested in them as models for research related to human concerns. These two approaches inevitably involve different techniques and are currently separated by a barrier of jargon and concepts. As a result, their relevance to each other is underexploited: on the one hand, work in whole animal biology can help those that are interested in causal explanations (mostly biomedical scientists). Although the language of sociobiology, for example, has already proved a

useful metaphor for investigating intragenomic conflicts, it is only recently that zoologists have been able to convince biomedical scientists that an understanding of why animals do what they do can assist in the design and interpretation of experimental work on these animals (e.g., appropriate husbandry, objective interpretation of results, more external validity, increased statistical power). Because of their ubiquitous use as laboratory animals, this is an example where fundamental knowledge of basic evolutionary history and biology can have applied benefits.

But the converse is equally true: biomedically driven laboratory discoveries ranging from psychology to genetics should be used by field biologists to understand better the socioecology of a species. For example, although the rats' ability to avoid ingesting poisons had been experienced by rat catchers worldwide as an impediment to the success of control programs (bait shyness), it was laboratory psychologists who demonstrated the existence of long delay learning, and the conditions under which it can be achieved (Garcia et al. 1966; Garcia and Koelling 1966; Kalat and Rozin 1973). Similarly, the two main types of feeding patterns observed by nutritional biologists in the laboratory (pre- and postprandial correlations) provide insight into the physiological mechanisms underlying feeding in mammals (Le Magnen 1985) but can also be used to infer how animals might feed in the wild, and provide testable predictions about the constraints of social and environmental factors on feeding (Berdoy 1994; Slater 1981). Finally, the recent completion of the rat and mouse genomes will bring important insights into some aspects of rodent evolution of interest to behavioral ecologists. As rats and mice tend, all too often, to be tarred by the same brush, the genetic evidence (contrary to initial expectation) actually emphasizes the fact that, at least in terms of number of genes, there is as much difference between mice and rats as there is between rats and humans (see table 32.1). More specific comparisons between rats and mice also highlight that the genomic regions that appear to have evolved fastest (those involved in chemosensation, olfactory receptors, pheromones, cytochromes P450, proteases, and protease inhibitors) are those associated with special characteristics of the rats feeding behavior, social interaction, and reproduction. And these are the very traits that fascinate us. The time is therefore ripe for those interested in genetic societies (sensu Haig 1997) and those interested in rodent societies to take a further step toward each other.

Chapter 33 Social Organization and Resource Use in Capybaras and Maras

David W. Macdonald, Emilio A. Herrera, Andrew B. Taber, and José Roberto Moreira

PATTERNS IN THE AVAILABILITY and utilization of resources are important constraints on the social organization of mammals, as established by early studies of primates (Crook and Gartlan 1966), antelopes (Jarman 1974), and carnivores (Macdonald 1983). In particular, the temporal and spatial dispersion of food and shelter, along with other factors such as predation pressure, affect the costs and benefits of forming groups. In turn, the social behavior and grouping patterns of individuals of any species have a strong influence on the structure and dynamics of their populations. Although studies of rodents have explored relationships between resource dispersion, spatial systems, and demography (e.g., Ims 1988; Ostfeld 1990; Wolff 1993b), the cascade of links between resources and sociality have been a lesser focus in studies of this Order than those mentioned previously.

Here we review data from studies of two species of hystricognath rodents, the mara (*Dolichotis patagonum*) and the capybara (*Hydrochaerus hydrochaeris*). The suborder Hystricognathi includes rodents from the Old World, such as porcupines (Old World Hystricidae) and the mole-rats (family Bathyergidae). Caviomorphs is the term commonly used for South American Hystricognaths. Having been isolated from other rodents some 50 million years ago, caviomorphs evolved separately from those on other continents. They occupy a diversity of niches and, as a result, many contemporary species of South American histricognaths are comparable in ecological and morphological terms to other terrestrial herbivores such as ungulates, macropods, and lagomorphs (Moreira and Macdonald 1996). The mara is a member of the family Caviidae, while capybaras are in a family of their own: Hydrochoeridae (Mones and Ojasti 1986). We compare the two species in terms of their social organization, behavior, and resource use in the habitats they occupy, and highlight ecological factors influencing their sociality (summarized in table 33.1). We conclude that although maras and capybaras live in conspicuously different landscapes, their societies can be interpreted as the products of remarkably similar resource dispersion features. Furthermore, our studies lay a foundation that reveals these species to be models for future studies of the interaction between resource availability and social organization.

The Species and Their Habitats

The capybara (fig. 33.1) is a 50 kg semiaquatic grazer whose range spans much of South America. Capybaras have been studied particularly in Venezuela (e.g., Ojasti 1973; Macdonald 1981a, 1981b), but also Columbia (Jorgenson 1986), Brazilian Pantanal Matogrossense (Schaller and Crawshaw 1981; Alho et al. 1987a, 1987b) and Brazilian Amazonia (Moreira and Macdonald 1993). Capybaras are abundant in the low Llanos region of Venezuela, where the habitat consists of tropical savannas—vast, flat plains dominated by grasses—and where the climate is characterized by marked wet/dry seasonality, producing a severe water shortage at the height of the dry season and extensive flooding in the wettest months. The region comprises three physiographic units: (1) *esteros* (ponds), most of which dry up

Table 33.1 Comparative behavior, ecology, and life history of capybaras and maras

	Capybara	Mara
Order	Rodentia	Rodentia
Family	Hydrochaeridae	Caviidae
Weight	35–66 kg; average 50 kg	7–9 kg; average 8 kg
Range	South America: Venezuela, Brazil, Colombia, Argentina	South America: Central and southern Argentina
Habitat	Flooded savannah or grassland next to waterholes; also along ponds and rivers in tropical forests	Dry scrubland deserts
Habitat Use	Changes seasonally to follow the availability of water, grasses and dry land	Changes seasonally to follow food availability
Home range size	2–200 ha; average 10–20 ha	Up to 200 ha; average 98 ha
Territoriality	Minimal overlap between groups. Territories are defended by all adult members of a group against conspecific intruders.	Drifting territories, with a high degree of overlap between pairs.
Social structure	Mixed sex groups-dominant male, females and young and subordinate males	Monogamous pairs, although settlements of several pairs may form in resource rich areas
Group size	10–30	Male-female pair
Diet	Grasses	Grasses and herbs
Scent marking	Males have snout gland used to signal their dominance. Anal glands in both males and females used to mark territories	Males mark daily ranges using anal dragging. Enurination and anal dragging by both sexes is greatest during breeding season
Breeding season	Year round with seasonal peaks that varies according to area	Twice a year, June–July and September–October
Reproduction	Group living, allosuckling,	Communal denning, allosuckling uncommon
Gestation period	150 days	90 days
Litter size	1–8, average 4	1–3, average 2
Resource Dispersion Hypothesis	Group size is determined by size, heterogeneity and richness of habitat patch	Monogamous pairs form when resources are sparse, groups of pairs may form when resources are clumped and plentiful

in the dry season, (2) *bajios*, areas covered with short, highly palatable grasses which flood in the wet season, and (3) *bancos*, areas vegetated with tall grasses or bushes, which rarely flood.

The mara, also known as the Patagonian cavy, is a 7 to 9 kg cursorial member of the Caviidae. Its habitat contrasts with that of the capybara, being the dry scrub deserts and grasslands of south and central Argentina. Maras have been

Figure 33.1 Capybara male. The small size of the capybara's eyes and ears may be adaptations to their semi-aquatic lifestyle, whereas the prominence of the male's morrillo gland indicates the importance of social odors. Jacanas have a commensal relationship with capybaras, which act as "beaters," disturbing insects that the birds catch (Macdonald 1981b). Photo by D. W. Macdonald.

most intensively studied in the wild on sheep ranches on the Valdés Peninsula in Argentina (e.g., Taber and Macdonald 1992a, 1992b; Parkhurst 2002 provides parallel observations in captivity). These areas consist of semiarid thorn scrub, broken by ephemeral lagoons of a few ha in size. Bushes predominate, growing to a height of around 1 m, and sparse grasses and forbs grow in the spaces between them. There is a clear seasonality, with winter (June–August) being the wet season, and summer (December–February) the dry season. Summer temperatures average 17°C, while winter temperatures rarely drop below freezing.

Habitat Use, Home Ranges, and Territoriality

Herrera and Macdonald (1987, 1989) marked and observed groups of capybaras in the Venezuelan Llanos, where social group sizes varied between four and sixteen adult animals (mean = ten, but see the following). Capybaras use a progression of habitats, with animals following the availability of water, grasses, and dry land as the season progresses. The areas of highest primary productivity, the grassy ponds and flooded bajios (Escobar and Gonzalez-Jimenez 1976), typically support *Leersia hexandra* and the *Hymenachne* species. These species are nutritious for grazers and predominate in the capybaras' diet (Ramia 1974; Escobar and Gonzalez-Jimenez 1976). Consequently, capybaras feed on bajios throughout the year, but particularly in the dry and early-wet seasons; animals spend less time in this habitat as it becomes more deeply flooded late in the wet season. Bancos are virtually unused until the peak of the rains, when deep flooding and displacement of the best grasses by reed and water hyacinth force the animals up to these higher, bushier areas, which become the only dry land available. These seasonal shifts in habitat use do not involve migratory movements. On the contrary, in their search for dry land in the wet season, capybaras rarely move their center of activity more than 300 m from their dry season wallow. Because capybaras are confined to the vicinity of water their distribution is restricted, and the nutritious grazing, together with an apparent paucity of competitors, facilitates high local population densities. Home ranges are therefore relatively small (10 ha for a ten-adult strong social group) compared to those of other similarly sized tropical herbivores (c.f. Eisenberg 1981). For instance, the density and biomass of capybaras within an average home range is roughly ten times greater than those of the Bohor reedbuck (*Redunca redunca,* Hendrichs 1975). This species is perhaps the closest equivalent of capybaras in terms of size (30–50 kg) and habit (grazing herbivore of tropical marshes in Africa), but the reedbuck shares its habitat with as many as ten sympatric species of grazing antelope.

Capybaras did not substantially alter the size or juxtaposition of their home ranges between years (Herrera and Macdonald 1989), in contrast to the large-scale movements of some other tropical, mammalian grazers living in seasonal regions (e.g., Leuthold 1977; Maddock 1979). The dispersion of group ranges showed minimal overlap between neighboring home ranges and an effectively contiguous blanket of home ranges throughout the study area. Herrera and Macdonald (1989) found that home range size was significantly correlated with both the number of adults in the group, and the total group size. They argued that this trend was related to the habitat composition of the home ranges, with larger home ranges also having greater areas of bushy banco and bajio. Furthermore, the total area of bushy banco in each home range was also highly correlated with the number of surviving young per female.

The conclusion that groups are territorially organized is supported by Herrera and Macdonald's (1987) observations of the expulsion of intruders—principally males, excluded by males. Analyses of territory economics have shown that for territoriality to evolve, there must be some critical resource that makes a territory economically defendable; that is, the costs of defense are lower than the benefits thereby obtained. The territoriality of capybaras may be explained by their dependence on water. Water sustains their food plants, and capybaras also thermoregulate in water, mate in water, and seek refuge from predators in water (Ojasti 1973; Azcarate et al. 1980; Macdonald 1981a; Herrera 1986). As the dry season progresses, lakes that form the flooded landscape contract, becoming little more than puddles. These remaining pools and the associated food become limiting and highly patchy resources. It seems likely that it is the need to guarantee access to a permanent dry-season lagoon that favors territoriality; Herrera and Macdonald (1989) found that every territory did indeed contain a water hole next to a grazing patch, making the piece of land viable and defensible.

In their study of maras in Argentina, Taber and Macdonald (1992b) found that radio-collared animals also differed seasonally in their use of habitat, predominantly staying in thorn scrub in autumn and winter (the wet season), while grazing on dry lagoon beds in summer. The daily activities of maras on the Valdés Peninsula involved almost continuous movement; excluding travel and search time, maras devoted at least 36% of daylight hours to grazing on grasses and forbs, probably due to the paucity of available food. Mara pairs had unstable (drifting) home ranges, in that they were continually utilizing new areas and vacating old ones. Over a year, pairs ranged through areas of around 200 ha. While ranges overlapped substantially, utilization differed within areas that overlapped, and pairs generally avoided each other. The segregation of pairs seemed to arise through avoidance, and active exclusion by males defending a 20 m space around their mate. Daily ranges were scent marked by males using anal dragging.

It seems likely that the functional explanation of this spatial avoidance is that the grazing of an area by one pair renders it useless to other pairs for at least several weeks. The grass and forbs on which maras graze are swiftly depleted, slow to replenish, and, on the fine scale of foraging movements, patchily dispersed. The need for a period of fallowing before an area is reusable necessitates a shifting pattern of range utilization; it is also advantageous for foraging animals to avoid areas recently depleted by others. Although these arguments would favor solitary territoriality, there are countermanding pressures (see the following) that result in a unique social compromise: maras forage in pairs that occupy drifting territories whose members convene in the vicinity of a shared breeding den.

The argument that a system of fallowing and depletion explains drifting territoriality is based on the deduction that a fixed territory accommodating sufficient areas of recuperating vegetation would be too large to defend against other maras, and against sheep, from whom, unlike capybara, maras face considerable competition. Therefore pairs defend a prevailing range, vacating exhausted patches and drifting into the next fruitful one they encounter. In contrast, the similarly sized antelopes of steppe, savannah, and open forest, klipspringer (*Oreotragus oreotragus*), blue duiker

(*Cephalophus monticola*), and Kirk's dikdik (*Madoqua kirkii*) all pair bond and live in much smaller ranges (< 10 ha) compared to maras (as reviewed in Taber and Macdonald 1992b). None of these species has to cope with abundant sheep or the need for scarce shared pupping dens. The maras' pupping dens are large, generally with multiple entrances; their scarcity suggests they may be a limiting resource, but we have no evidence on how difficult they are to dig.

Social Structure: Monogamy versus Group Living

Taber and Macdonald (1992b) observed and radio-collared wild maras over 3 seasons on the Valdés Peninsula. They found that maras' social units could be easily distinguished by the close proximity of their members (within 15 m of each other) and the wide spacing between them (usually > 100 m). Of 342 independent sightings of such units, 65% consisted of a male-female pair, and, apart from aggregations, groups of more than three pairs were never seen. In short, most maras in the studied population lived in monogamous pairs.

Monogamy, having one social mate at a time, is in itself a relatively rare mating system among mammals (Crook 1977; Kleiman 1977; Clutton-Brock 1989b; Waterman, chap. 3 this volume). A number of hypotheses have been put forward concerning the adaptive significance of monogamy in mammals (e.g., sequestering females, indivisible paternal care; Kleiman 1977, 1981; Dunbar 1984; Zeveloff and Boyce 1980; Wittenberger and Tilson 1980). Monogamy may be favored by patterns of resource dispersion that determine that the carrying capacity of a territory is insufficient to support additional breeding females (e.g., gibbons, *Hyoblates klossi*; Tenaza 1975; see also Kleiman 1977). This finding is supported by Komers and Brotherton's (1997) model, which showed that female dispersion was the main factor favoring monogamy in mammals. This assertion is exemplified by Hendrichs and Hendrichs (1971), who concluded that defense of rich, widely dispersed food patches precluded larger groups of dik diks. Taber and Macdonald (1992b) suggest that, because the grasses and forbs on which maras feed are swiftly depleted, slow to replenish, and patchily dispersed, interference competition forces maras to avoid each other; the paucity of the resources and their dispersion militates against social units larger than a pair. Further, synchrony in births suggests that most females come into estrus (which lasts for only a few hours) in relative synchrony during the wetter season, when more forage is available throughout the thorn scrub and maras are at their most dispersed. Rather than solitary territoriality, though, the spatial organization and movement patterns imposed by the mara's diet, and the brevity of the females' sexual receptivity, means that males probably reproduce

most successfully by monopolizing one female. Males do not care for the young directly, but they do save time for the foraging female by being vigilant. Furthermore, with few opportunities for promiscuous mating, males may maximize their reproductive success by guarding their young and their mate from predation.

The mara's social system is strikingly different from that of other large caviomorphs such as the capybara. The only other caviomorph rodent (out of 188 species; Macdonald 2001) that is reported to live in pairs is the agouti (Genus *Dasyprocta*; Smythe 1978). Agoutis live in territorial pairs in dense tropical rainforests, and differ markedly from maras in that the paired animals are rarely together and live an essentially solitary existence (Kleiman's [1977] facultative monogamy). Interestingly, the mara demonstrates a morphological convergence with some ungulates, with its long legs and body and its reliance on speed and crypsis to escape from predators (some of these, such as dik diks and duikers, are also pair-living). Most strikingly, stotting, characteristic ungulate escape behavior otherwise unknown in other orders (Caro 1986) is part of their predator aversion repertoire. Although polygyny is common in ungulates, five species are known to be monogamous, and probably represent convergence with the circumstances of maras. (These fall into a category between small solitary ruminants of dense forests and generally larger, group-living ungulates of open savannahs). However, the mara's drifting range distinguishes it from the monogamous ruminants, which maintain stable, exclusive territories. This difference may arise as a compromise necessitated by the mara's communal denning behavior, which forces pairs into closer contact than foraging considerations alone would dictate.

By contrast, capybaras live in mixed-sex groups that typically comprise a dominant male, one or more females (which are probably kin), several infants and young, and one or more subordinate males. At El Frio ranch in Venezuela, group sizes ranged from four to sixteen individuals (Herrera and Macdonald 1987, 1989), while on a ranch just 50 km away in the same general ecosystem (El Cedral ranch) groups ranged from eight to twenty-five individuals (Salas 1999). The differences between the Salas (1999) and Herrera and Macdonald (1987, 1989) studies seem to be related to resources: El Cedral had a more homogeneously distributed, more abundant, and less seasonally variable resource base than El Frio. In both study areas, there is a well-defined dominance hierarchy among the males, maintained by aggressive interactions that usually take the form of simple chases. Dominant males are heavier and obtain a greater proportion of matings than do subordinates (Herrera and Macdonald 1993; Salas 1999). Females are much more tolerant of each other, although the precise details of their social relationships, hierarchical or otherwise, are unknown. The membership of these groups appears to be quite stable, with few changes occurring, even across years (Herrera and Macdonald 1987). This stability is comparable to that of several territorial species of primates and carnivores (e.g., Hrdy 1977b; Rasa 1985).

Why do capybaras live in groups? The niche occupied by the capybara has much in common with those of many species of tropical ungulates. Like them, the capybara is a relatively large grazer and is subject to predation on open plains. Capybaras are, or have been in the recent past, prey to puma, jaguar, feral dogs, caiman, and man. Macdonald (1981a) described how a group of capybaras, under attack by feral dogs, bunched together with youngsters within a protective cordon. There are three reasons why capybaras might form groups in response to predation: (1) increased vigilance to forestall attack, (2) increased possibility of intimidating the predator through greater strength of numbers, and (3) the *selfish herd* effect (Hamilton 1971), in which individuals lower their risk of capture by hiding in the group. Yáber and Herrera (1994) found that total vigilance (number of heads up per hour) increased with group size whereas individual vigilance (number of times an animal would lift its head per hour) decreased. Also, subordinate males made the greatest contribution to vigilance and paid the higher cost in terms of frequency of head-up behavior. Indeed, Ebensperger and Cofré (2001) suggest that capybara sociality may be related to predation, whereas in other caviomorphs it is more related to cooperation in burrowing and limited burrowing sites. Apart from defense against predators, it is also possible that aggregation by herbivores may help maintain a high quality and quantity of forage (McNaughton 1984).

Despite the differences in social structure, both capybaras and maras are, at times, gregarious feeders. In the dry season, capybara groups coalesce around dwindling pools, forming temporary aggregations of 100 or more animals. When the wet season resumes, these large aggregations split up into the original groups that formed them (Herrera and Macdonald 1989). Similarly, despite their monogamy, as many as twenty-five pairs of maras often form aggregations, known as *settlements*. Taber and Macdonald (1992b) found pairs in a settlement maintained proximity to clearings and lagoon beds. Settlement members also invariably chose sites adjoining a shepherd's outstation with a water trough, corral, and a shepherd (the latter perhaps having an impact on predator distribution and behavior). The authors suggest that, although the lagoons are dry during the dry season, the vegetation around them would have had the most recent access to water and be closer to the water table. Furthermore, in the study area, sheep congregated around the dry lagoons; their nitrogenous waste in the form of dung may have resulted in faster regeneration and/or higher nutritive value of the maras' forage. Taber and Macdonald (1992b) also suggested that maras might den near outsta-

tions so that their young can benefit from the reduced number or increased timorousness of predators there. This benefit could be enjoyed where food resources (possibly also affected by the shepherd's activities) allowed maras to cluster around dwellings. In this context, Sunquist and Sunquist (1989) noted that, perhaps in a similar way, the aggregations of chital deer (*Axis axis*) around park dwellings at dusk afforded the deer protection against predators.

Importance of Olfactory Communication

A conspicuous feature of the sociality of both capybaras and maras is scent marking, but it is far more frequent in capybaras (Herrera and Macdonald 1994). Capybaras have two kinds of scent gland (Macdonald et al. 1984). The first, highly developed in males but almost nonexistent in females, is located on top of the snout and is a dark, oval shaped, naked protrusion known as the *morrillo* (fig. 33.2). The second is found in both sexes and takes the form of two glandular pockets located on either side of the anus. Male anal glands are filled with easily detachable hairs abundantly coated with layers of hard, crystalline calcium salts (Macdonald et al. 1984). Female anal pockets also have hairs, but theirs are not detachable and are coated in a greasy secretion rather than with crystalline layers. The proportions of each chemical present in the secretions of individual capybaras are different, providing a potential for individual recognition via personal olfactory fingerprints.

Capybara males scent mark more and have more aggressive interactions than females (Herrera and Macdonald 1987). It appears that the snout scent gland is mainly involved in signaling dominance status, while the anal gland may be important in group membership recognition and perhaps territoriality. Larger groups had lower scent marking rates than did smaller groups, not only as a unit, but

Figure 33.2 Scent marking. A male deposits a white sticky secretion from its morrillo. Photo by D. W. Macdonald.

also individually. Large groups also had more subordinates that have lower marking rates. Since larger groups defended larger territories (Herrera and Macdonald 1989) it is likely that the home ranges of large groups were less thoroughly covered with scent than were those of smaller groups. Male members of large groups tended to be more loosely associated with their social units than were those in smaller groups (Herrera and Macdonald 1987), and this difference in cohesiveness of membership may also have had bearing on the reduced marking activity of members of large groups.

The testes mass of the capybara is closely associated with the size of androgen-dependent scent glands (Herrera 1992) and with the proportion of nonspermatogenic tissue in the testes. This suggests that testes size in capybaras is more associated with androgen-dependent chemical signalling than with sperm production (Moreira, Clarke, and Macdonald 1997, Moreira, Macdonald, and Clarke 1997). Although capybaras do not burrow, their partly nocturnal habits may increase the importance of olfactory signals. Overall, the pattern appears to be of female-defense polygyny, with dominant males marking ranges that they strive to make at least partly exclusive, and with males within the group competing for sexual access to females.

In comparison to the capybara, scent marking in the mara is less frequent (Taber and Macdonald 1984). Olfactory communication among maras includes anal dragging, enurination (the spraying of urine over a conspecific), and the formation of dung piles. Both anal dragging and enurination are more frequent during the breeding season and are most commonly done by males. The function of both behavior patterns is unknown. Scent marking in maras is associated with intense social interaction in diverse contexts, and thus generally in circumstances where individual recognition might be important. It seems probable that at least some functions of enurination are different for the two sexes, with males typically marking estrous females, whereas female urine spraying seems to be associated with repelling the advances of males and of unrelated pups trying to steal suckles. Although anal marking and enurination are widespread amongst hystricomorph rodents (reviewed in Macdonald 1985), proof of their function is elusive. While it seems plausible that male enurination on females could be olfactory mate guarding, the seemingly diverse contexts of this behavior leave plenty of room for hypothesizing alternative functions.

Reproductive Biology

The hystricognath rodents share several reproductive characteristics that are unusual among rodents. Compared to myomorph rodents, the hystricognath group are character-ized by long lifespans, long and variable estrous cycles, protracted gestations, precocious young, and moderate to small litter sizes (Weir 1974; Kleiman et al. 1979), resulting in the majority of hystricognaths having low reproductive potential. However, the capybara is an outlier to the negative correlation between litter size and body weight found among the hystricognaths (Kleiman et al. 1979), having one of the largest litter sizes (mean = close to four), and being the heaviest species within the suborder. It is these traits that make the capybara particularly suitable for sustainable harvest, with the result that they have been hunted throughout their range for many years (Ojasti 1991). While the impact of hunting is difficult to estimate, it is believed to be selective, thereby affecting the social and age structure of populations (Moreira and Macdonald 1993). A reduction in capybara numbers in Venezuela indicates the need for further evaluation of hunting pressures (Ojasti 1991); models that predict birth probabilities from hunting data (e.g., Moreira and Macdonald, unpublished) may offer an opportunity for exploring ways of maintaining populations while allowing cropping by impoverished local communities.

Capybaras den communally when breeding, and females appear to suckle infants indiscriminately within their group (Macdonald 1981a). An average of 4.2 capybara young are born per litter, each weighing approximately 1.5 kg (Ojasti 1973). The gestation period averages 150 days. On Marajó Island, Moreira et al. (2001) found that around 18% of implanted capybara embryos are lost during gestation. Further analyses of the 275 embryos of 50 females in the last 4 months of gestation (Macdonald and Moreira unpublished) revealed that male embryos were heavier than female ones, and appeared to demand more nutrients due to their higher rate of development. Following sex allocation theory (Trivers and Willard 1973), we suggested that, if one sex gains more than the other from extra parental investment, then parents with relatively more resources should bias their investment toward the sex with the greater rate of reproductive return. Since a dominant male capybara secures most of the copulations within his multimale group, his reproductive success is potentially much higher than that of any female or subordinate male. Females in better condition, therefore, might be expected to produce a relatively higher proportion of sons compared with weaker females. Moreira et al. (2001) and Macdonald and Moreira (in press) did indeed find that females in better condition had more dead fetuses in their uteri, and that their litters were more likely to be male biased. Indeed, they found that, irrespective of its sex, a fetus positioned between two male fetuses weighed significantly less than one positioned between two female fetuses. Further, female fetuses were found more often, and male fetuses were found less often, between male fetuses than expected by a binomial distribution. Male fetuses were found adjoining one female neighbor at the head of the uterine horn significantly more than expected. These results led us to conclude that female capybara overproduce embryos and subsequently adjust their litter size and sex ratio downward. In the different circumstances of the Venezuelan llanos, Herrera (1998) suspected a different system, whereby females were more likely to resorb male embryos likely to be born in the dry season.

Female capybaras are fertile throughout the year, but there is strong evidence for seasonal peaks in reproductive effort. This seasonal component is not typical of most tropical mammalian species, which give birth at any time of year, showing only slight peaks in reproduction (Bronson 1989). The departure of capybaras from this generalization might be explained by the seasonality of flooding and therefore the availability of food and shelter, and the risk of predation. Moreira and Macdonald (unpublished) described how the timing of these seasonal peaks in reproduction varies according to area. For example, in Venezuela (Ojasti 1973) and the Brazilian Pantanal Matogrossense (Schaller and Crawshaw 1981) the peak is at the end of the rainy season, whereas on Marajó Island it coincides with the onset of the rainy season. The explanation probably lies in the pattern and severity of the floods, which are less extreme in Marajó than in the llanos (where Azcarate et al. [1980] found a high rate of newborn mortality due to drowning in the wet season). Indeed, in the Llanos, dry shelters (offered by banco habitat) were essential for the survival of newborn, and the availability of dry land during the peak wet season was related to capybara group sizes (Herrera and Macdonald 1989). In this habitat, the risk of drowning diminishes only as the floods subside, in turn making the flush of vegetation accessible in many areas. A further step along this continuum is illustrated by the natural savannahs of central Brazil, where there is no flooding at all (Moreira et al. 2002).

These are adaptive explanations for seasonal breeding, sensu Rutberg (1987) in tropical ruminants. The onset of the rainy season is swiftly followed by the appearance of green foliage, especially grasses (Talbot and Talbot 1963). Grasses account for approximately 70% of capybara diet, depending on the period of the year (Escobar and González-Jiménez 1976). Therefore, coincidence between the birth of capybaras and a season when forage is available is obviously advantageous for the survival and growth of the young and the survival and future reproductive success of the female (fig. 33.3). Furthermore, reproductive synchrony may reduce predation in a number of ways (Estes 1976; Bertram 1978). Capybaras do not burrow but benefit from aspects of sociality for predator defense (see Ebensperger and Cofre 2001). Members of larger groups may also benefit from dilution (Hamilton 1971), confusion, and enhanced vigilance

Figure 33.3 Capybara young. Capybaras are born after a long gestation of over 5 months, and attain sexual maturity after more than 1 year. Photo by D. W. Macdonald.

(Yáber and Herrera 1994). Lastly, Macdonald (1981a) reported that allosuckling is common within capybara groups. This may increase efficiency through cooperation, since females share the costs of nursing the young. In turn, this provides another selective force for birth synchrony in capybaras.

The mara's social system embraces not only social monogamy (there are no genetic data) but, often, communal denning as well (figs. 33.4, 33.5), a combination otherwise unknown among mammals (Taber and Macdonald 1992a). Communal denning is taxonomically widespread among mammals and is often associated with cooperative breeding, defined as situations where individuals other than the parents provide care in rearing the young. It has been reported in both suborders of the Rodentia; for example, Sciurognathi: prairie dogs, *Cynomys* spp. (Hoogland 1979a) and *Alticola strelzowi* (Eshelkin 1976); Hystricognathi: naked mole-rats, *Heterocephalus glaber* (Jarvis 1981), and most of the cavies (e.g., *Cavia aperea, Galea musteloides,* and *Microcavia australis;* Rood 1972; Weir 1974: Asher

Figure 33.4 Adult mara grazing. Photo by D. W. Macdonald.

Figure 33.5 Occasionally two pairs of maras try to approach the communal den simultaneously, with resulting hostility. Here, the males of two rival pairs engage in a standoff to decide precedence in the queue to retrieve their young from the crèche. Photo by D. W. Macdonald.

Figure 33.6 A crèche of young maras at a communal den. Photo by D. W. Macdonald.

2004). Generally, communal denning species that behave alloparentally are polygynous, or reproduction is confined to one female and one male in the group. The mara's social system is in marked contrast and, so far, appears to be unique among mammals.

Maras are socially monogamous due to the effect of sparse, patchily dispersed grazing and the risk of interference competition; nevertheless, it is advantageous for pairs to become gregarious during the breeding season. Communal denning in this way may have a range of advantages (Taber and Macdonald (1992a). First, there is cooperation in terms of group defense and vigilance. At warrens with larger membership, more adults were present for more of the time, and therefore total vigilance was greater at larger crèches, even though each pair spent less time there (fig. 33.6). Adults therefore had more time to forage over wider areas away from the den. Older pups were also active in increasing the corporate vigilance. Second, female maras were seen to nurse young other than their own. However, it is not clear whether this represented communal suckling or milk theft, since, in contrast to capybaras, they seemed anxious to repel usurpers. Third, coloniality may play an important role in thermoregulation. Taber and Macdonald (1992a) found that mortality of pups was inversely related to crèche size. Patagonia is at the southern end of the species' range, and nightly minimum temperatures during the pupping season fall to less than −5 °C; the shared body heat of larger crèches may be critical to the survival of newborn pups. Warrens housing fewer than ten pups over the season might, at any one time, have insufficient corporate warmth to survive a colder-than-average night. Consistent with this, in the warmer northern part of the species' range, large denning settlements have not been reported. The unusual breeding system of the mara may thus be a compromise, conferring

on the pups the benefits of coloniality in an environment wherein association between pairs is otherwise apparently disadvantageous (fig. 33.5).

The mara's blend of monogamy and communal denning has loose analogues among certain birds. Up to four pairs of insectivorous groove-billed anis (*Crotophaga sulcirostris*) share the tasks of building a communal nest, incubation, and rearing young, and may thereby reduce adult mortality during incubation (Vehrencamp 1977, 1978). The clearest example of stable bonds between different pairs within freely mixing groups occurs in the black vulture (*Coragyps atratus*). Coalitions of several nuclear families occur together consistently at roosts and carcasses, and apparently cooperate to compete over food (Rabenold 1985), while pairs remain strictly monogamous (Decker et al. 1993). The glossy black-cockatoo (*Calyptorhynchus lathami halmaturinus*), combines group living with long-term monogamy, with large groups nesting together at night, then dispersing as mated pairs or trios of the mated pair and a juvenile to feed (Pepper 1996). Crook (1964) described six insectivorous species of socially monogamous weaverbirds (Ploceinae) that nested colonially but were constrained to forage in pairs due to the disadvantages of interference competition in larger flocks. Colonial nesting may arise (1) without advantages from sociality per se, due to aggregation at limited nest sites or safe places (*Ploceus pelzelni* normally nests solitarily but forms colonies in isolated mangrove bushes), or (2) because of the advantages of sociality (*Bubalornis albirostris* builds thorny fortress nests). By analogy, sociality is incidental to the advantages to maras of aggregating near resources at sheep stations where, for example, predation may be low. Similarly, communal denning may have arisen without the benefit of sociality if den sites were limiting (as may occur in rabbits [Cowan and Garson

1985]). However, since the soil was easily dug and dens were widespread in the study area, it seems unlikely that a limitation on den sites forced maras into communal denning.

Some maras den solitarily rather than communally; it seems likely that the interaction of resource dispersion, protection from predators, and the advantages of sociality, including vigilance, communal suckling, and thermoregulation, determine denning behavior. The relative importance of each variable could be investigated by comparing mara behavior in different parts of their range. Such comparisons may shed light on the two possible evolutionary routes to the mara's social system: (1) from a monogamous ancestral system, modified to accommodate the advantages of a creche, or (2) from a communal polygynous ancestor on which monogamy has been imposed by, for example, habitat becoming more patchy, sparse, and unpredictable. The fact that coloniality and polygyny are widespread in the mara's caviomorph relatives (Kleiman 1977; Lacher 1981; Ebensperger and Cofré 2001) suggests that the latter route is the more probable.

Group Size and the Resource Dispersion Hypothesis

Arguments about the determinants of social group size often involve the richness of resource patches; for example, the Resource Dispersion Hypothesis (see Macdonald and Carr 1989; Johnson et al. 2002). In Herrera and Macdonald's (1987, 1989) capybara study area, each home range included patches of each major habitat type, including a body of water. However, in the low Llanos of Venezuela there are vast expanses of bajio without permanent surface water, and these are devoid of capybaras. Around waterholes big enough to retain water well into the dry season, capybara territories contiguously cover the land available. As all other habitat types can exist in the absence of waterholes, it appears that the waterhole is the key resource that makes a territory viable and defendable for year-round survival. Components of a territory necessary to sustain capybaras throughout the changing seasons were (1) bushy banco (higher-lying land) to provide food and harborage during the height of the wet season, (2) ponds for drinking water, thermoregulation, mating, and sanctuary from predators in the dry season, and (3) bajio for grazing during the peak of the dry season. The size and shape of an economically defendable territory with these minimal requirements will be determined by the dispersion of these three resources. In such a minimum territory (sensu Carr and Macdonald 1986), that is, one that supports the requirements of a minimum social unit, the resource patches may be sufficiently rich to support a larger group. Insofar as group size is determined by resource availability (as opposed to, for example, predation), it will be limited by whichever was the most limited of the three critical resources: banco, bajio, or pond. In the groups, studied over 3 years, the close correlation between adult group size and area of banco indicates that this was the limiting resource.

In these circumstances, territory and group size might be determined independently and respectively by the dispersion and abundance (richness) of available resources, a possibility described by the Resource Dispersion Hypothesis. In minimum territories, group and territory size could be explained solely by a correlation between patch dispersion and richness; for example, if territories sufficiently large to encompass widely spaced ponds and/or bajios in the dry season automatically encompassed additional banco for the wet season. In this case, the dispersion of one or two critical resources in one season would be correlated with the richness of the third critical resource in the other season. An observation by Herrera and Macdonald (1989) that larger territories were apparently configured to embrace two ponds, and thereby acquire additional intervening banco (and bajio), is compatible with this suggestion.

In the case of maras, the Resource Dispersion Hypothesis may help explain why settlements develop in certain locations. The vicinity of the lagoon and outstation represents a single rich patch: a carpet that can be utilised simultaneously by all the maras in a settlement during a bottleneck in food availability in the adjoining scrub, and possibly a safer place for the pups. This rich patch is effectively indefensible from the other maras or sheep which flock to it. The result is that year-round local carrying capacity for maras is elevated, by either or both of water or sheep dung, to a level where one or more pairs can be supported within daily travelling distance of a warren.

We have already seen that group size in capybaras is determined by the size of particular habitat patches. According to the Resource Dispersion Hypothesis, the development of larger social groups is more likely where resources are spatiotemporally heterogeneous in availability and patch richness is high. However, the diverse habitats occupied by capybaras make the species a good model for exploring intraspecific variation in the consequences of resource dispersion. For example, Salas (1999) found larger groups in a location with both greater habitat homogeneity and more abundant (and less seasonal) resources, whereas rather small groups are more likely where resources are homogeneously dispersed and/or patch richness is low. Maras, in contrast, appear to be confronted with conditions that lie at either extreme of the resource dispersion continuum. In the dry season maras depend on a single, rich patch, which can be viewed as either shareable or undefendable but which is rapidly exhausted. Rain brings regrowth of grasses and forbs in the scrub, where the dispersion of sparse grazing forces maras towards isolation. While their ranges drift, maras are constrained for other reasons (the benefits of communality and reduced predation risk, discussed earlier), to remain in the vicinity of the clearing and the outstation. As the next summer advances, regeneration time of fallowed grazing patches increases as vegetation becomes parched, at which point the pairs of maras follow the sheep (and their dung) to the lagoons/corrals, where the flush of grass is more attractive. According to this model, where resource dispersion alternates between favoring separation and congregation, the size of settlement memberships will be determined by whichever is the narrower seasonal bottleneck in resource availability. Due to the apparent uniformity of scrub and the variability in features of the outstations, and the fact that mara settlements develop only near lagoons/outstations, it seems likely that qualities of the dry season resources, in terms of lagoon-outstation patch richness, are a good determinant of settlement size.

Summary and Conclusions

Two large, unusual, South American rodents, the capybara and mara, have complicated societies that reveal both inter- and intraspecific variation that appears to be related to the extremes of resource dispersion that they face. We propose that these species provide fertile models for testing relationships between patterns of resource availability and sociality.

Capybaras live in groups, often in seasonally flooded habitats. Changes in availability of favorable grassland, caused by the alternation of drought and flooding, result in shifting habitat use, but their home ranges remain stable and territorially organized. Territory size is positively correlated with group size and, within their home range, areas of key habitat types are also correlated with group size. Capybaras depend upon access to permanent surface water; where there is such water, it appears that their territories are configured to encompass sufficient resources to ensure survival under widely differing seasonal conditions.

Maras face extremes of resource dispersion between the wet and dry seasons, with associated pressures that lead to a different social system. In the wet season, sparsely dis-

persed grazing and interference competition favor spacing out, leading to the formation of monogamous pairs. In the dry season, clumping of resources facilitates pairs congregating in settlements around outstations and dry lagoons. Superimposed upon the ecological factors favoring spacing out during the wet season are the sociological factors that cause maras to den communally—for example, protection from predators through increased vigilance, and better thermoregulation. The resulting compromise is a social system so far unique among mammals.

The Resource Dispersion Hypothesis proposes that the spatiotemporal dispersion and richness of resources can lead to the passive formation and maintenance of groups, even in the absence of any direct benefits of group living per se. Put simply, where resources are patchily distributed, territory owners must defend an area that guarantees minimum food security. However, much of the time this minimum food security is likely to be exceeded, in which case additional animals can join the territory with low competition costs, forming a group. Additional factors, notably predator defense, also clearly affect the functional significance of group living. In both species, competition arising from group living would seem to be lower among females than males, and resource dispersion may particularly favor female sociality. Other issues, such as an inability to chase other males away, kin selection, and sharing in the defense of the territory may be playing a role in male tolerance of each other (concepts such as *peace incentives* and *staying incentives*, as modeled by Kokko and Johnstone (1999), may prove relevant). Capybaras offer an interesting subject on which to test such ideas. Indeed, within groups of capybaras there is a dominance hierarchy, which seems to function much as modeled by Kokko and Johnstone (1999), while females are grouped because they can share resources and protect themselves against predators. Although doubtless this is only one factor among many that affects the function of their societies, despite their contrasting life histories, we suggest that resource dispersion provides a framework for explaining why capybaras live in groups and why monogamous maras form settlements. Indeed, as Johnson et al. (2002) argue, the Resource Dispersion Hypothesis lays a foundation for considering the sociological factors that shape mammalian societies and the underlying influence of ecological conditions on social organization.

Chapter 34 Social Structure in Octodontid and Ctenomyid Rodents

Eileen A. Lacey and Luis A. Ebensperger

THE OCTODONTIDAE AND CTENOMYIDAE offer numerous intriguing opportunities to explore the evolution of rodent societies. Although closely related, these hystricognath families vary markedly with regard to patterns of phyletic, geographic, and ecological diversification (Mares and Ojeda 1981; Contreras et al. 1987; Reig et al. 1990; Redford and Eisenberg 1992). Correspondingly, the animals are behaviorally diverse, with each family including solitary as well as social species (Weir 1974; Redford and Eisenberg 1992; Muñoz-Pedreros 1992). Collectively, these attributes suggest that octodontids and ctenomyids provide a rich array of possibilities for comparative studies of rodent social structure.

Given their potential to yield new insights into the evolution of rodent social systems, it is disappointing that so little behavioral research has been conducted on octodontids and ctenomyids. To date, information on social structure has been obtained for fewer than 10% of the species in these families, making quantitative analyses of the correlates of social structure difficult. Of those species that have been the subjects of behavioral research, none has been studied as extensively as "model" taxa such as voles (Tamarin 1985; Tamarin et al. 1990), African mole-rats (Sherman et al. 1991; Faulkes and Bennett, chap. 36 this volume), or ground squirrels (Murie and Michener 1984; Barash 1989; Hoogland 1995; Yensen and Sherman 2003; Hare and Murie, chap. 29 this volume), and much remains to be learned regarding the behavior of even the best-known octodontids and ctenomyids.

In this chapter, we review the behavioral biology of the Octodontidae and Ctenomyidae, with emphasis on the vari-ation in social structure found among these animals. We begin by comparing patterns of taxonomic, ecological, and morphological diversification in these families, which, despite their close phylogenetic affinity, have followed quite different evolutionary trajectories. To illustrate how these animals can contribute to our understanding of rodent societies, we present detailed information regarding the social systems of three members of these families, much of it drawn from our own research programs in rodent behavioral biology. Using this framework, we review the adaptive hypotheses that have been proposed to explain social structure in these animals. We also consider how several of the distinctive life history characteristics of these families may influence their social behavior. Based upon the available data, we suggest that comparative studies of these taxa will substantially improve our understanding of the evolutionary bases for sociality and communal breeding in rodents. We conclude by identifying several directions for future research in rodent social behavior, for which octodontids and ctenomyids represent particularly appropriate study subjects.

Patterns of Evolutionary Diversification

The Octodontidae (degus, cururos, viscacha rats) and Ctenomyidae (tuco-tucos) are caviomorph rodents in the Suborder Hystricognathi (Woods 1993). Together with the Abrocomidae (chinchilla rats), Capromyidae (huitas), Echimyidae (spiny rats) and Myocastoridae (coipu), these animals comprise the superfamily Octodontoidea (Nedbal et al. 1994; Huchon and Douzery 2001; Honeycutt et al. 2003). The octodontoids are thought to have originated 13 to 18 million years ago (Houchon and Douzery 2001), perhaps in response to major vegetation changes associated with the uplifting of the Andes and concomitant changes in rainfall regimes (Honeycutt et al. 2003).

Within the Octodontoidea, the exact relationship between octodontids and ctenomyids is disputed. Some authors consider the tuco-tucos to be a subfamily (Ctenomyinae) within the Octodontidae (Reig et al. 1990) whereas others recognize them as a distinct family, the Ctenomyidae (Woods 1993). Recent molecular analyses (Mascheretti et al. 2000; Gallardo and Kirsch 2001; Honeycutt et al. 2003) support the latter interpretation, with divergence between the octodontids and ctenomyids occurring ca. 10 million years before present, during the late Miocene (Lessa and Cook 1998). For the purposes of this review, we concur with Wilson and Reeder (1993) in treating the Octodontidae and Ctenomyidae as distinct families and apparent sister taxa to one another.

Although closely allied phylogenetically, octodontids and ctenomyids exhibit strikingly different patterns of evolutionary diversification. The Ctenomyidae consist of a single genus (*Ctenomys*) containing fifty to sixty named forms (Reig et al. 1990; Woods 1993; Mascheretti et al. 2000). In contrast, only ten species of octodontids are recognized (Woods 1993), but these taxa are divided among six genera (*Aconaemys, Octodon, Octodontomys, Octomys, Spalacopus, Tympanoctomys;* Woods 1993, although see Hutterer 1994). Thus, while ctenomyids show extensive species-level diversification but relatively little other intrafamilial phylogenetic structure, octodontids are characterized by multiple, deeper (generic) phylogenetic divisions but little species-level diversification within each genus. The reasons for these differences in diversity are unclear. Although early ctenomyids are thought to have undergone a burst of speciation (Lessa and Cook 1998), rates of diversification within this lineage are not significantly greater than those for octodontids (Cook and Lessa 1998), suggesting that the different patterns of phyletic diversity in these families are not due solely to more rapid evolutionary divergence among ctenomyids.

Ecological diversity

In addition to their distinct patterns of phyletic diversity, the Octodontidae and Ctenomyidae display markedly different patterns of ecological specialization. Both families are endemic to subAmazonian South America. The geographic distribution of ctenomyids, however, is more extensive (fig. 34.1), suggesting that members of this family may occupy a greater diversity of habitats. Although the ar-

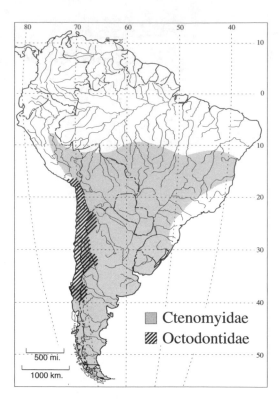

Figure 34.1 Geographic distributions of the *Octodontidae* and *Ctenomyidae.* Range data were compiled from geographic data provided in Woods (1984) and Redford and Eisenberg (1992).

eas in which ctenomyids are found range from arid, coastal sand dunes to wet, montane meadows (Redford and Eisenberg 1992), the actual variety of environmental conditions experienced may be less than is implied by this description; all ctenomyids are subterranean (Reig et al. 1990) and, hence, the environments in which they live are generally assumed to be similar due to the buffering of surface conditions by subterranean burrows (Reichman and Smith 1987; Buffenstein 1996; but see Lacey et al. 2000). While some aspects of ctenomyid ecology clearly differ among species (e.g., type of soils inhabited, nature and distribution of plants consumed; Reig et al. 1990; Busch et al. 2000), how these differences have interacted with the shared challenges of subterranean life to shape ctenomyid evolution remains largely unknown.

In contrast, although geographically more restricted, the octodontids are ecologically more diverse, and include surface-dwelling and semisubterranean as well as truly subterranean forms. Collectively, these animals occur in a wide array of habitats, including open grasslands, dense forests, and extremely arid salt flats (Mares and Ojeda 1981; Redford and Eisenberg 1992). Although most octodontids use subterranean burrows as nest sites, members of many species are thought to spend a substantial proportion of their time above ground (e.g., Vásquez 1997). As a result, these animals may be more affected by variation in surface habitats than are ctenomyids, suggesting that habitat structure may have played a more conspicuous role in the evolutionary diversification of octodontids.

Morphological diversity

Associated with the general ecological differences outlined in the previous section are differences in the range of morphotypes evident within each family. In keeping with their shared specialization for subterranean habitats, all ctenomyids exhibit similar morphological features associated with life in underground burrows, including reduced eyes and ear pinnae, shortened limbs, and generally squat, tubular bodies (Stein 2000). Among octodontids, the fully subterranean *Spalacopus* shares a number of the general morphological traits identified for ctenomyids. In contrast, members of the semisubterranean genera (e.g., *Octodon, Octodontomys,* and *Tympanoctomys*) exhibit a markedly different body structure, characterized by large eyes and ear pinnae and, often, relatively enlarged hind feet. While all ctenomyids possess comparatively short, unornamented tails, tail structure among octodontids varies; although the tails of the more subterranean genera (*Spalacopus, Aconaemys*) resemble those of ctenomyids, the tails of the more surface-active genera (e.g., *Octodon, Octomys*) are often characterized by pronounced brushes or plumes at the distal end. Additionally, octodontids display marked diversity in the structure of the masticatory apparatus, which may be linked to the variety of habitats and diets used by these animals (Olivares et al. 2004). In short, while all ctenomyids are characterized by the same basic body plan, the octodontids can be divided into two distinct morphological types: the subterranean specialists (*Spalacopus* and *Aconaemys*) and the burrow-dwelling but surface-active genera (*Octodon, Octodontomys, Tympanoctomys, Octomys*).

Behavioral diversity

Few octodontids or ctenomyids have been the subjects of detailed behavioral research. At present, quantitative data regarding social organization are available for only two species (20%) of octodontids and four species (ca. 8%) of ctenomyids. Anecdotal accounts provide intriguing hints as to the social structures of a number of other species in each family, although these reports require verification through field studies of marked individuals. In addition, both families contain multiple species for which no information regarding social behavior is available (tables 34.1 and 34.2).

Despite the paucity of behavioral studies of these rodents, data obtained to date suggest that patterns of social structure differ markedly between octodontids and ctenomyids. In particular, while sociality (defined as group living) appears to be quite common among octodontids, it is thought to be rare among ctenomyids. Although data for many octodontids are largely anecdotal, comparisons of the behavior of poorly known species with that of species which have been studied in some detail (e.g., *Octodon degus, Spalacopus cyanus*) suggest that some form of group living likely occurs in at least half the members of this family (table 34.1). While even anecdotal data are lacking for several species (e.g., *Aconaemys sagei, Octomys mimax*), at present the only species specifically identified as solitary is the red viscacha rat, *Tympanoctomys barrerae*.

In contrast, even anecdotal suggestions of sociality are limited to less than 10% of ctenomyid species (Reig et al. 1990; Lacey 2000). The only species for which quantitative evidence of sociality has been obtained is the colonial tuco-tuco, *Ctenomys sociabilis* (Lacey et al. 1997; Lacey 2004). In comparison, solitary burrow use has been demonstrated for at least eleven species (table 34.1). These data suggest that the ratio of social to solitary species is considerably smaller for ctenomyids than it is for octodontids. While we suspect that these values will change as studies of the behavioral biology of these animals increase, it seems unlikely that the prevalence of sociality among ctenomyids will come to equal, let alone exceed, that among octodontids.

Case Studies of Social Structure

Given the apparently pronounced difference in the prevalence of group living in octodontids versus ctenomyids, an obvious question that must be addressed when comparing the social biology of these taxa is: why is sociality so much more common among octodontids? As a first step toward answering this question, we review the behavioral and ecological data available for those octodontids and ctenomyids whose social systems have been studied in some detail: degus, cururos, and colonial tuco-tucos. As this list suggests, behavioral studies of these animals have focused on social species, with few comparative data available for solitary taxa. Consequently, it is not yet possible to exploit the full range of behavioral variation in these families to explore the adaptive bases for differences in social structure. The information provided here, much of which is drawn from our own ongoing studies of these species, reveals intriguing parallels as well as striking differences between the social systems of octodontid and ctenomyid rodents. These contrasts form the basis for the remaining sections of this chapter, which explore the evolution of variation in social structure among these animals.

Table 34.1 Summary of natural and life history attributes of octodontid rodents

Species (common name)	Body mass (g)	Principal habitat	Pattern of burrow use	Social structure	Reference
Aconaemys fuscus (tunduco)	100–143	Forests, bunchgrass	Partially subterranean, frequent digger	Social (?)	Mann 1978; Pearson 1984; Reise and Gallardo 1989; Redford and Eisenberg 1992
Aconaemys porteri (great tunduco)	?	Forests	Partially subterranean, frequent digger (?)	Social (?)	Muñoz-Pedreros 2000
Aconaemys sagei (tunduco)	83–110	Forests, bunchgrass	Partially subterranean, frequent digger	?	Pearson 1984
Octodon bridgesi (Bridge's degu)	142	Forests	Nests underground, rarely digs	Social (?)	Ipinza et al. 1971; Mann 1978; Muñoz and Murúa 1987
Octodon degus (common degu)	170–260	Open thorn-bush savannah	Nests underground, frequent digger	Social	Fulk 1976; Mann 1978; Redford and Eisenberg 1992; Ebensperger et al. 2004
Octodon lunatus (coastal degu)	233	Dense thorny scrub	Nests underground, rarely digs (?)	?	Ipinza et al. 1971; Reford and Eisenberg 1992
Octodontomys gliroides (soco)	115–176	Open areas with rocks	Nests underground, climbs rocks, rarely digs	?	Mann 1978; Redford and Eisenberg 1992
Octomys mimax (hairy-tailed rat)	121–144	Arid scrub, rock outcrops	Nests undergound, climbs rocks	?	Mares and Ojeda 1981
Spalacopus cyanus (cururo)	75–190	Open to moderately dense scrub, montane meadows	Subterranean	Social	Reig 1970; Torres-Mura and Contreras 1998
Tympanoctomys barrerae (red viscacha rat)	52–91	Open arid basins and dunes	Nests underground frequent digger,	Solitary	Mares et al. 1997; Díaz et al. 2000

NOTE: Measures of body mass are from Redford and Eisenberg (1992).

Table 34.2 Summary of natural and life history attributes of ctenomyid rodents

Species	Body mass (g)	Principal habitat	Social structure	References
C. australis	248–500	Coastal sand dunes	Solitary	Contreras and Reig 1965; Comparatore et al. 1991; Redford and Eisenberg 1992
C. flamarioni*	194–260[a]	Sandy coastal areas	Solitary	El Jundi and de Freitas 2004
C. fulvus	300–400	Sandy desert flats	Solitary	Mann 1978, Miller et al. 1983
C. haigi	110–180[b]	Arid steppe grasslands	Solitary	Pearson and Christie 1985; Lacey et al. 1998; Lacey and Wieczorek 2003
C. maulinus	220–300	Mesic meadows, *Notofagus* forest	Solitary	Gallardo and Anrique 1991; E. Lacey, personal observations
C. mendocinus	145–180[c]	Arid shrubs and grasslands	Solitary	Puig et al. 1992; Rosi et al. 2002
C. opimus	230–530	Altiplano	Solitary	Pearson 1951, 1959; Pine et al. 1979; Mares et al. 1981
C. pearsoni	165–300	Coastal dunes and grasslands	Solitary	Altuna et al. 1999; G. Francescoli and G. Izquierdo, personal communication
C. peruanus	340–663[d]	Mesic meadows	Social	Pearson 1959
C. rionegrensis*	110–230[5]	Stable sand dunes	Solitary	Lessa et al. in 2005
C. sociabilis	180–350[2]	Mesic meadows ("mallines")	Social	Pearson and Christie 1985; Lacey et al. 1997; Lacey and Wieczorek 2004
C. talarum	92–193	Coastal grasslands	Solitary	Pearson et al. 1969; Busch et al. 2000
C. torquatus	156–304	Open savannas	Solitary	Redford and Eisenberg 1992

NOTES: The taxonomy follows Woods (1993) but includes several more recently described species (indicated with an asterisk [*]). Unless otherwise noted, measures of body mass are from Redford and Eisenberg (1992). No data regarding social structure were available for an additional ca. 28 species of ctenomyids included in Woods (1993) or described subsequently.
[a]Data on body mass from El Jundi and de Freitas (2004).
[b]Data on body mass from E. A. Lacey, unpublished data.
[c]Data on body mass from Rosi et al. (2002).
[d]Data on body mass from Pearson (1959).
[e]Data on body mass from E. Lessa and B. Tassino (personal communication).

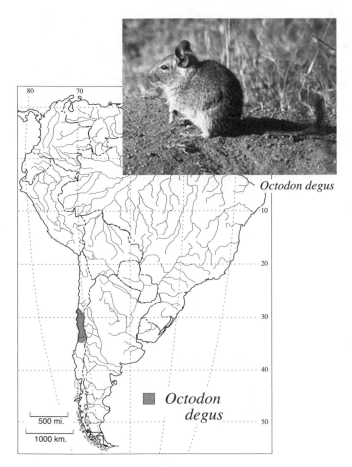

Octodon degus

Octodon degus

500 mi.
1000 km.

Figure 34.2 Geographic distribution of *Octodon degus*. Range data were compiled from Redford and Eisenberg (1992). Inset: photo of *O. degus*, taken by Luis A. Ebensperger.

Example 1: Social structure in degus

Degus (*Octodon degus*) are medium-sized (ca. 180 g), diurnal octodontids that occur in semiarid and Mediterranean habitats in north-central Chile (fig. 34.2; Woods and Boraker 1975). Although degus construct underground burrows and nest in subterranean chambers, the animals spend a considerable portion of their time above ground, during which they may travel up to several meters from the nearest burrow entrance (Fulk 1976; Yáñez 1976; Vásquez et al. 2002; Soto-Gamboa 2004). The animals forage exclusively on surface-growing vegetation, including the leaves of a variety of grasses, forbs, and, to a lesser extent, shrubs (Fulk 1976; Mann 1978; Meserve et al. 1983, 1984). Young leaves are preferred (Simonetti and Montenegro 1981), which may reflect selection of food items of high nutritional quality (e.g., high water, nitrogen content) and low fiber or secondary plant metabolite content (Bozinovic 1995, 1997; Torres-Contreras and Bozinovic 1997; Gutiérrez and Bozinovic 1998; Bozinovic and Torres-Contreras 1998).

Because the animals occur in habitats characterized by seasonal precipitation, the availability of green vegetation varies throughout the year; this variation appears to have significant implications for other aspects of degu biology, including annual patterns of reproduction. Degus typically mate during the early winter (June–July), with parturition occurring during September–October (Contreras and Bustos Obregón 1977; Rojas et al. 1977; Meserve et al. 1984; Kenagy et al. 1999, Ebensperger and Hurtado 2005b). Females in at least some populations are thought to undergo a postpartum estrus (Rojas et al. 1977), resulting in the production of a second litter of young during December–January. Mean litter size is six pups (Veloso 1997). The young are precocial; newborn degu pups have pelage and their eyes are open (Reynolds and Wright 1979; Ebensperger, unpublished data). Although the young quickly become ambulatory and begin eating vegetation at about 6 to 10 days of age, they continue to nurse until they are approximately 4 weeks old (Reynolds and Wright 1979; Veloso 1997). Based on the weights of field-caught animals, juveniles first become active above ground about 3–4 weeks after birth, after which they can frequently be observed interacting with conspecifics while outside of their natal burrow system.

Degus are social, with groups consisting of two to five adult females and their dependent young (Fulk 1976; Yáñez 1976). The number of adult males per group is unclear; while some accounts indicate only a single adult male per group (Soto-Gamboa 2004), others suggest that groups contain multiple adult males (Fulk 1976; Yáñez 1976). Members of a social group share the same burrow system, including a communal nest site (Soto-Gamboa 2004; Ebensperger et al. 2004). Genetic estimates of kinship indicate that female nestmates are significantly more related to one another than to randomly selected females in the population, with a mean coefficient of relatedness among conesting females of ca. 0.25 (Ebensperger et al. 2004). Evidence of kin discrimination, however, is inconsistent. For example, while preliminary observations of interactions among lab-reared juveniles indicate that unfamiliar individuals display less aggression toward siblings than nonsiblings (Márquez et al. 2002), lab-reared pups do not associate more with their mothers than with unrelated lactating females (White et al. 1982), and captive females retrieve their own and unfamiliar pups indiscriminately (Ebensperger, unpublished results). Ongoing field and laboratory studies of known individuals should serve to clarify how kin discrimination contributes to observed patterns of kin structure in this species.

Typically, all adult female degus in a group produce offspring (Soto-Gamboa 2004; Ebensperger and Hurtado, unpublished data). How reproductive success is partitioned among female nestmates or among the males in multi-male groups, however, is unknown. Within social groups, individuals appear to cooperate in several ways. For example, multiple animals contribute to the excavation of tunnels; although tunnel construction is not as conspicuously coordinated as it is in some social bathyergids (Jarvis and Sale 1971), participation by multiple individuals may reduce the per capita energetic cost of tunnel construction or may allow the animals to excavate more extensive tunnel systems than would be possible for lone individuals (Ebensperger and Bozinovic 2000a). When above ground, the animals alarm-call in response to predators and other apparent threats (Fulk 1976; Yáñez 1976). Preliminary evidence suggests that alarm calls differ with predator type (Cecchi et al. 2003) and these calls may serve to alert group-mates to perceived dangers, thereby facilitating the survival of individuals with whom the caller lives and interacts. Finally, allonursing of young has been observed among captive degus (Ebensperger et al. 2002); field studies are currently underway to document this form of cooperative care of young among free-living animals.

Example 2: Social structure in cururos

Cururos (*Spalacopus cyanus*) are subterranean octodontids that are patchily distributed throughout central Chile (fig. 34.3; Saavedra and Simonetti 2003). Perhaps more so than the other species reviewed here, cururos exhibit considerable intraspecific variation in ecology. Populations are found in arid, coastal scrub habitat (where they may co-occur with degus) as well as in mesic, montane meadows (Contreras et al. 1987; Torres-Mura and Contreras 1998). Associated with this range of habitat types are pronounced differences in body mass, with adult males ranging from ca. 80 g in arid areas to ca. 180 g in more mesic regions (Contreras 1986; Lacey, Ebensperger, and Wieczorek, unpublished data). Given the conspicuous differences in the floral communities characteristic of these habitat types, diet composition also presumably varies substantially among conspecifics, as does the proportion of the diet consisting of surface-growing versus subterranean plant parts (Reig 1970; Torres-Mura 1990). As a result of this intraspecific variation in habitats, studies of cururos provide a particularly important opportunity to examine the effects of specific ecological variables on patterns of social structure.

In general, the natural history of cururos has been less thoroughly documented than that of degus. With regard to their annual reproductive cycle, cururos are thought to breed between June and March in coastal areas, with the duration of reproductive activity expected to be shorter in Andean populations (Unda et al. 1980). Whether individual females rear more than one litter of young per year remains unknown, and this and other aspects of the annual repro-

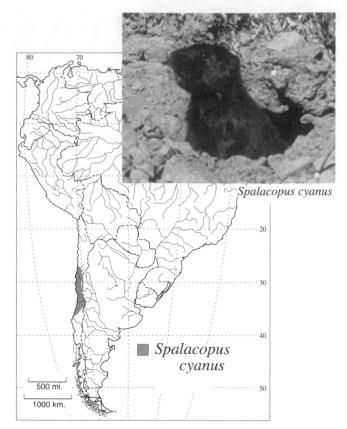

Figure 34.3 Geographic distribution of *Spalacopus cyanus*. Range data were compiled from Redford and Eisenberg (1992). Inset: photo of *S. cyanus*, taken by Luis A. Ebensperger.

ductive cycle may vary among populations and habitats. Based on observations of mating and parturition in captive animals, gestation lasts for approximately 77 days (Begall et al. 1999). Like degus, neonatal cururos are precocial, but the timing of several critical ontogenetic events appears to be somewhat delayed relative to degus. For example, although the young are born with pelage, their eyes do not open until 2 to 8 days after birth (Begall et al. 1999; but see Torres-Mura and Contreras 1998). Pups do not begin consuming solid food until ca. 18 days after birth and, in captivity, weaning does not occur until pups are 60 days old (Begall et al. 1999). Thus, while cururos follow the general octodontid pattern of producing precocial young, postparturition development of pups is not as rapid in this species as in degus.

All populations of cururos surveyed to date are social, meaning that multiple adults share the same burrow system (Reig 1970; Begall et al. 1999; Lacey, Ebensperger and Wieczorek, unpublished data). Up to ten adults, including multiple animals of both sexes, have been captured in the same burrow system (Reig 1970; Begall et al. 1999). Animals resident in the same burrow system share a single nest site, which may remain in the same location for multiple years (Lacey, Ebensperger, and Wieczorek, unpublished data). Within a group, all adult females show evidence of reproduction, including pregnancy and lactation (Begall et al. 1999; Lacey, Ebensperger, and Wieczorek, unpublished data). At present, however, the relationship between group size and female direct fitness is unknown, as is the distribution of direct fitness among adult groupmates of the same sex. Intergroup transfer by males has been detected, as has natal philopatry by members of both sexes (Lacey, Ebensperger, and Wieczorek, unpublished data). These observations suggest that groups are likely to contain a mixture of close kin and unrelated, immigrant individuals. Long-term field studies of cururos currently in progress at Parque Nacional Fray Jorge and Santuario de la Naturaleza Yerba Loca, Chile, promise to yield critical information regarding philopatry, kinship, and social structure in this species.

Like degus, cururos appear to cooperate to excavate burrows and, frequently, multiple members of a social group can be observed simultaneously expelling loose soil from adjacent burrow entrances. Radiotelemetry studies indicate that cururos are strictly diurnal (Urrejola et al. 2005); when active during the daytime, the animals frequently alarm-call in response to predators and other disturbances, which may function to warn groupmates, including kin, of potential threats. Although allonursing has not been reported for captive cururos, the tendency for groupmates, including multiple reproductive females, to share a single nest suggests that this form of cooperation is possible for free-living members of this species (Hayes 2000).

Example 3: Social structure in colonial tuco-tucos

Colonial tuco-tucos (*C. sociabilis*) are medium-sized (females: 200 g; males: 300 g), diurnal ctenomyids that are endemic to an approximately 1500 km² area of Neuquén Province, Argentina (fig. 34.4). Like cururos, colonial tuco-tucos are almost exclusively subterranean. Although both species feed on surface vegetation, they rarely emerge more than one body length from their burrows when foraging or expelling loose soil from their tunnels. Unlike cururos, however, colonial tuco-tucos are habitat specialists; the latter species occurs primarily in mesic seeps and meadows at elevations ranging from 800 to 1600 m. Dietary analyses based on fecal samples collected during the spring and summer breeding season indicate that the animals consume primarily grasses and sedges (Lacey and Wieczorek 2003); anecdotal observations during the autumn and winter suggest that the animals forage on these same types of vegetation throughout the year.

Colonial tuco-tucos breed only once per year, with litters

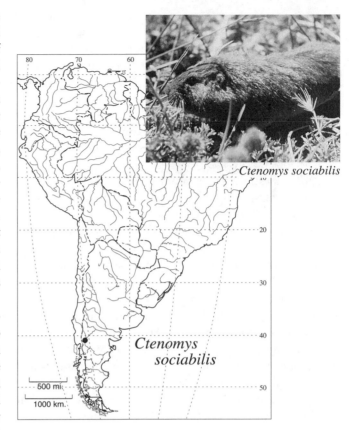

Figure 34.4 Geographic distribution of *Ctenomys sociabilis*. Range data were compiled from Pearson and Christie (1985) and from unpublished data collected by E. A. Lacey. Inset: photo of *C. sociabilis*, taken by Eileen A. Lacey.

of one to seven young born during September–November. The length of gestation is not known for this species; estimates for other ctenomyids range from 90–110 days (Weir 1974; Busch et al. 2000), implying that *C. sociabilis* mate during June–August. Like degus and cururos, young colonial tuco-tucos are precocial; neonates have pelage, their eyes open within several days of birth, and they begin eating solid food at ca. 7–10 days of age. The young first begin foraging for themselves on surface-growing vegetation ca. 4 weeks after birth and, although they are weaned shortly thereafter, they remain in the natal burrow for at least an additional 1–2 months.

As their common name implies, colonial tuco-tucos are social. Burrow systems are inhabited by up to six adult females, with ca. 45% of burrow systems also containing a single adult male during the period when young are present (Lacey and Wieczorek 2004). All residents of a burrow system share a single nest site (Lacey et al. 1997). Groups form due to natal philopatry by females; in contrast, all males disperse from their natal burrow at the end of their juvenile summer (Lacey and Wieczorek 2004). As a result, social groups are composed of several closely related adult females, their dependent young, and, in some cases, an immigrant adult male. Most dispersal occurs within the local population; simulation analyses based on observed demographic patterns indicate that, on average, the male and female(s) in a burrow system share a common maternal ancestor within the last five to seven generations (Lacey and Wieczorek 2004), suggesting that even putative reproductive partners may be relatively close kin. To date, efforts to determine parentage within social groups have been unsuccessful due to very low levels of genetic variation among conspecifics (Lacey 2000).

All females in a group produce offspring, although the per capita number of young reared to weaning declines significantly as group size increases (Lacey 2004). At present, it is not known if this decrease in direct fitness is equitably distributed among the females resident in a burrow system. Adults of both sexes contribute to the care of young (Soares 2004). Studies of free-living animals indicate that the percentage of time that the nest is unattended (i.e., no adult present) is significantly less for multiadult groups than for lone females (Izquierdo 2005). Studies of captive animals indicate that presence of an adult significantly increases the probability that juveniles will remain in the nest (Soares 2004). Collectively, these data suggest that one benefit of group living may be the increased retention of young in the relative safety of the shared nest. As with degus, allonursing by female colonial tuco-tucos has been observed for captive animals (Lacey, unpublished data), although this form of cooperation has not yet been documented in the field. Other potential forms of cooperation among colonial tuco-tucos include shared participation in burrow excavation and alarm-calling to alert groupmates to the presence of predators. All adults excavate burrows and call in response to raptors or other threats (Lacey, unpublished data); as with degus and cururos, sharing of these activities may represent an adaptive benefit to group living.

In summary, the social systems of the octodontids and ctenomyids considered here are similar in that all are group living and communally nesting. While the degree of specialization for subterranean life varies among species, all use underground burrows and, hence, the excavation of new tunnels represents a significant component of their behavior. In each species, groups contain multiple adult females, all of whom are thought to produce offspring. In contrast, one of the most striking differences between the social systems of these animals is the number of males per burrow system; while groups of colonial tuco-tucos never contain more than one adult male, multimale groups occur regularly in cururos, and, apparently, in some populations of degus. Accordingly, patterns of natal philopatry vary between species. Specifically, while philopatry by both sexes has been reported for cururos, only female colonial tuco-

tucos remain in their natal burrows. The habitats in which the animals occur also vary markedly, with cururos spanning much of the range of environmental conditions occupied by degus and colonial tuco-tucos. Thus, these species provide ample opportunity to explore the adaptive bases for variation in several fundamental aspects of rodent social structure.

Adaptive Bases for Sociality

Given that the octodontids and ctenomyids examined to date are social, it is not surprising that behavioral studies of these families have focused on the adaptive bases for group living. The conceptual and methodological approaches that have been employed, however, are varied, with surprisingly little overlap among the adaptive hypotheses examined for each species. To help forge a more comprehensive picture of social structure in these animals, we briefly review the variety of research strategies applied to these species and the resulting insights into the ecological causes and fitness consequences of group living. We then consider several more general attributes of octodontid and ctenomyid rodents that may contribute to observed patterns of sociality in these animals. A primary objective of this discussion is to draw attention to adaptive patterns not evident from studies of individual species in these families.

Conceptual approaches

Numerous adaptive benefits have been proposed to explain the occurrence of group living and communal nesting in rodents (Hayes 2000; Ebensperger 2001a). While the selective factors identified by these explanations represent multiple adaptive hypotheses, they are frequently divided into the following two categories: (1) benefits of philopatry that are intrinsic to social groups, and (2) constraints on natal dispersal that are extrinsic to social groups (Solomon and Getz 1997; Hatchwell and Komdeur 2000; Hayes 2000). These categories are not mutually exclusive; both types of factors may contribute to sociality in a given species (Lacey and Sherman, chap. 21 this volume), and the same variable may simultaneously represent a benefit of philopatry and a constraint on dispersal. Although there is increasing awareness that these arguments represent two sides of the same coin (Mumme 1997; Lacey and Sherman 1997), the distinction between benefits of philopatry and constraints on dispersal remains entrenched in the literature and is evident in the conceptual and methodological approaches applied to studies of the adaptive bases of group living in octodontid and ctenomyid rodents.

Intrinsic benefits of group living in degus

Studies of degus have emphasized the intrinsic benefits of group living and communal nesting. For example, it has been hypothesized that increased protection from predators represents a primary benefit of sociality in this species (Ebensperger and Wallem 2002). Above ground, degus are preyed on by numerous species, including foxes and several types of raptors (Jaksic et al. 1981; 1993) and individuals appear to adjust their behavior in response to predator risk (Lagos et al. 1995; Vásquez et al. 2002; Ebensperger and Hurtado 2005a). In particular, animals that are active in sparsely vegetated areas form larger groups and spend more time vigilant than do animals in areas with more substantive vegetative cover (Ebensperger and Wallem 2002; Vásquez et al. 2002). In open habitats, degus in larger groups are able to detect predators at a greater distance than are degus in small groups (Vásquez 1997; Ebensperger and Wallem 2002), suggesting that group living reduces predation risk in this type of environment.

It has also been proposed that group living may confer energetic benefits on individuals that nest together (e.g., Ebensperger and Wallem 2002). Captive degus have been shown to reduce their energetic expenditure by huddling with conspecifics (Canals et al. 1989); this benefit may extend to natural habitats in which individuals are expected to experience greater extremes in ambient temperature. Group living may also reduce the energetic costs of burrow construction (Ebensperger and Bozinovic 2000a). Laboratory studies indicate that while degus housed in groups do not spend less time burrowing than solitary animals, group-living individuals coordinate their digging activities and, per capita, move more soil than do solitary diggers (Ebensperger and Bozinovic 2000a). As a result, cooperative burrowing may lead to a reduction in the energy required per individual to move a given amount of soil. Burrow excavation, however, represents a relatively limited portion (ca. 1–2%; Ebensperger, unpublished data) of an animal's activity budget, leading to questions concerning the ability of this apparent energetic savings to maintain group living in this species (Ebensperger and Wallem 2002).

Other adaptive benefits proposed to explain sociality in this species include the defense of critical food resources. For at least some degu populations, the abundance and quality of food appear to vary both spatially and temporally (Jaksic and Fuentes 1980; Holmgren et al. 2000). If, as has been argued for group-living bathyergid mole-rats (Jarvis et al. 1994; Bennett and Faulkes 2000), cooperative foraging increases the animals' ability to locate or to defend patchily distributed resources, then this may represent another benefit to group living in degus. Relationships between food resource distributions, social structure, and individual fitness have not been quantified for this species, however, and thus this hypothesis remains largely untested for degus.

Extrinsic constraints on dispersal in tuco-tucos

In contrast to studies of degus, studies of colonial tuco-tucos have focused on ecological factors that constrain natal dispersal, thereby leading to the formation of social groups. A primary objective of these studies has been to determine if group living in *C. sociabilis* is associated with the same ecological factors thought to favor sociality in African mole-rats (family Bathyergidae). Sociality among bathyergid rodents is thought to reflect the combined effects of hard-to-excavate soils, patchily distributed food resources, and limited and unpredictable rainfall (Jarvis et al. 1994; Lacey and Sherman 1997; Lacey and Sherman, chap. 21 this volume). If these variables represent general ecological correlates of sociality among burrow-dwelling rodents, then the same factors should be associated with group living in other, phylogenetically and biogeographically independent taxa.

To test the generality of this hypothesis, Lacey and Wieczorek (2003) compared patterns of resource use by the colonial tuco-tuco and its solitary congener, the Patagonian tuco-tuco (*C. haigi*; Lacey et al. 1997, 1998). Both species occur in the Limay Valley of southwestern Argentina, where populations of heterospecifics are separated only by the width of the Río Limay (ca. 500 m). The habitat in the valley consists primarily of arid, steppe grassland that is punctuated at irregular intervals by mesic patches known as mallines. Comparative analyses of habitat use have revealed that while the solitary *C. haigi* occurs in steppe and mallín portions of the habitat, the group-living *C. sociabilis* is restricted to the borders of mallín patches. Soil humidity and the distribution of critical food resources do not differ between habitat types, but steppe soils are significantly more difficult to penetrate than are mallín soils (Lacey and Wieczorek 2003). Although direct comparisons of the energetic costs of burrow excavation in the two habitat types have not been completed, the data on soil penetrability suggest that burrowing is more difficult in habitats occupied by the solitary *C. haigi*, leading Lacey and Wieczorek (2003) to conclude that adaptive hypotheses proposed to explain group living in African mole-rats do not account for sociality in colonial tuco-tucos. Instead, they propose that greater habitat specialization by *C. sociabilis* increases the difficulty of dispersing from one habitat patch to another, thereby favoring natal philopatry and group living. Thus, ecological constraints on dispersal rather than intrinsic benefits of group living appear to favor sociality in this species and, accordingly, the direct fitness of lone females is greater than that of group-living animals (Lacey 2004).

As this synopsis suggests, the approaches used to explore the adaptive bases for group living have differed between studies of octodontids and ctenomyids, with the former making greater use of experimental manipulation of natural and laboratory environments and the latter emphasizing long-term monitoring of free-living populations. Clearly, both methodological approaches are important, and both yield critical information regarding potential causes of sociality in these animals. Future studies of these and other rodent species will benefit from an integrated research strategy that combines observational studies of behavior, ecology, and demography with experimental manipulations of relevant environmental variables. Analyses of the adaptive bases for sociality in cururos are just beginning and, hence, this species provides an ideal opportunity to pursue a comprehensive research program aimed at identifying intrinsic as well as extrinsic factors that favor group living.

Potential Life-History Correlates

Although the Octodontidae and Ctenomyidae are sister families (Honeycutt chap. 2 this volume), as discussed previously in this chapter, they appear to have followed quite different evolutionary trajectories in terms of phyletic, ecological, and morphological differentiation. This marked variation among otherwise relatively closely related species provides an unusual opportunity to explore how variability in the basic life-history traits of octodontids and ctenomy-

ids may have contributed to their differential tendency to live in groups. Salient life-history attributes of these animals that may influence sociality include the following:

Production of precocial young

Like other hystricognath rodents, octodontids and ctenomyids give birth to relatively precocial young (Weir 1974; Künkele and Trillmich 1997). This is in marked contrast to sciurognath rodents, most of which produce altricial offspring (Ferron 1984). Although the social consequences of this life history variation have not been explored in detail, it has been suggested that group living may be favored in species with more precocial young. Specifically, group defense or babysitting of neonates may be particularly beneficial in species in which pups quickly become mobile yet remain vulnerable to predators (Kleiman 1974).

Predicted relationships between precociality and sociality
The predator protection hypothesis suggests that within the Octodontidae and Ctenomyidae there should be a positive relationship between precociality and degree of sociality. By extension, between families, precociality should be more pronounced among octodontids, in which sociality appears to be more common.

Available evidence
Within the Ctenomyidae, the degree of precociality appears to vary among the few species for which appropriate data

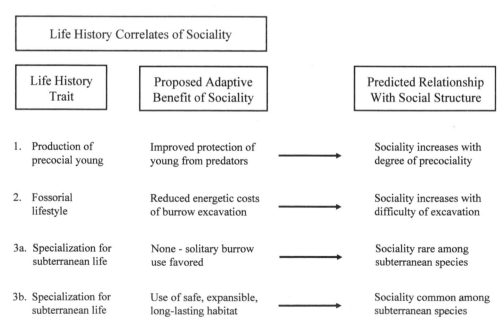

Figure 34.5 Summary of proposed life history correlates of sociality in octodontid and ctenomyid rodents. For each hypothesis, the relevant life-history trait is indicated, as is the primary adaptive benefit of sociality and the predicted relationship with social structure.

are available. Descriptions of juvenile *C. talarum* indicate that at birth the eyes are closed and young lack a full pelage (Weir 1974; Cutrera, pers. comm.). In contrast, neonatal *C. peruanus* have open eyes, are fully furred, and are capable of leaving the nest and feeding on green vegetation almost immediately after birth. Juvenile *C. sociabilis* appear to fall somewhere in between; the young are born with a short but complete pelage, the eyes open within a few days of birth, and young begin leaving the nest 48–72 hours after birth (Soares 2004). While this sample is minimal, it is intriguing to note that the two species with more precocial young are reportedly social (Lacey et al. 1997; Pearson 1959; Lacey 2000), while the species with the least precocial young is solitary (Pearson et al. 1968).

Comparable data for octodontids are even more limited and thus, at present, analyses of precociality and social structure within this family are not possible. Given that sociality, including communal nesting, occurs in numerous nonhystricognath species (Solomon and Getz 1997; Hayes 2000; chapters 24, 30, and 31 this volume), we expect that production of precocial young is not a primary selective factor favoring group living in rodents. Future studies of hystricognath rodents, however, should help to clarify relationships between precociality and group living by providing additional data regarding neonate ontogeny and degree of sociality.

Fossorial lifestyle

All ctenomyids and the majority of octodontids are fossorial, meaning that individuals exhibit morphological or other specializations for digging (table 34.1; Nevo 1979; Reig et al. 1990; Lacey et al. 2000). Energetic costs of burrow excavation have been implicated in the evolution of sociality in at least one other hystricognath lineage, the bathyergid mole-rats. Specifically, for animals living in difficult-to-excavate soils (i.e., high energetic costs of excavation), individuals may live in groups and dig cooperatively in order to obtain access to patchily distributed but locally abundant food resources (Lovegrove and Wissell 1988; Lovegrove 1991; Jarvis et al. 1994; Faulkes and Bennett, chap. 36 this volume). Under this scenario, the costs of burrow excavation may vary as a function of the physical properties of soil (Lovegrove 1989; Ebensperger and Bozinovic 2000b), the unpredictable distribution of food or other critical resources (Jarvis et al. 1994), or both.

Predicted relationships between burrowing costs and sociality

Among species with similar degrees of specialization for subterranean life (i.e., similar reliance on burrowing to accrue resources), a positive relationship should be evident between sociality and the energetic costs of burrow excavation. Thus within the Octodontidae and Ctenomyidae, sociality should occur in species that occupy particularly difficult (e.g., hard, dry) soils or in species that must burrow particularly long distances to reach critical resources. Comparisons between families are potentially confounded by differences in degree to which animals are specialized for digging. Given, however, that total burrowing effort should be greater for taxa that are more specialized for excavation, we would expect the effects of soil conditions and resource distributions to be more pronounced among ctenomyids.

Available evidence

Relationships between soil conditions, resource distributions, and group living have not been examined for octodontids. Within ctenomyids, however, group living does not appear to reflect more difficult-to-excavate soils or more sparsely distributed food resources. Comparative studies of the colonial tuco-tuco and its syntopic congener, the Patagonian tuco-tuco (*C. haigi*), have revealed no differences in food resource distributions for populations of these species in the Limay Valley of southwestern Argentina (Lacey and Wieczorek 2003). Further, these analyses indicate that it is the solitary *C. haigi* that occurs in harder, presumably more costly-to-excavate soils (Lacey and Wieczorek 2003). Although the energetics of tunnel excavation have not been examined directly, the available data suggest that burrowing costs due to soil conditions or resource distributions are unlikely to explain sociality in ctenomyids.

More generally, sociality may be influenced by the relative extent to which a species engages in digging and burrow construction. A phylogenetically controlled analysis of twenty-six New World hystricognath species revealed that apparent energetic costs of burrow excavation provide a better predictor of social group size than do measures of predation risk or parental investment (Ebensperger and Cofré 2001). Because this analysis included burrowing as well as nonburrowing species, it is likely that this outcome reflects the distinction between species that routinely dig tunnels or nests and those that rarely, if ever, excavate subterranean structures. Applying this interpretation to octodontids and ctenomyids, we would expect sociality to be more prevalent among the latter, since these are the animals that should be most affected by soil conditions or other factors that increase burrowing costs. Again, the quantitative data needed to test this hypothesis within octodontids are lacking. The rarity of social ctenomyids, however, suggests that simply engaging more frequently or more extensively in burrow excavation is not sufficient to favor group living. Thus, while we cannot exclude the possibility that energetic costs of burrow excavation contribute to sociality in at least some of the species considered in this review, these costs do not provide a comprehensive explanation for patterns of sociality in octodontids and ctenomyids.

Specialization for subterranean life

The different degrees to which octodontids and ctenomyids use subterranean burrows may influence social structure independently of the energetic costs of burrowing. Among truly subterranean rodents, group living is rare (Nevo 1979; Lacey 2000). It has been suggested that life underground favors a solitary existence because the spatial limits of a burrow system facilitate territory defense and, hence, exclusion of conspecifics from an individual's area of activity (Nevo 1979). In contrast, Alexander et al. (1991) have proposed that the use of safe, expansible habitats such as subterranean burrows facilitates sociality. Specifically, because burrow systems can be modified and may be used for multiple generations, it may be advantageous for individuals to be philopatric and to form groups rather than to leave the safety and known resource availability of their natal burrow (Alexander et al. 1991).

Predictions regarding subterranean life and sociality

Although these hypotheses represent different levels of analysis (Sherman 1988), they yield strikingly different predictions, which can be explored using available data for octodontids and ctenomyids. Based on the territory defense hypothesis, we would expect that, within the Octodontidae, sociality should be negatively related to the degree of specialization for life underground. Although all members of the Ctenomyidae are considered truly subterranean (Lacey et al. 2000), we would expect any interspecific variation in the tendency for animals to be active above ground to show a similar relationship with social structure. Between families, the consistently greater specialization for subterranean life displayed by ctenomyids leads us to predict that sociality should be less prevalent in this family. Conversely, the expansible burrow hypothesis predicts that sociality should be positively related to the degree of specialization for life underground. Accordingly, this hypothesis suggests that sociality should be more prevalent among ctenomyids than among octodontids.

Available evidence

Across families, the apparent rarity of social ctenomyids appears to be consistent with the predictions of the territory defense hypothesis. Data on behavioral variation with each family are more difficult to interpret. The extent to which different ctenomyid species are active above versus below ground has not been quantified. Anecdotal accounts, however, suggest that the two reportedly social species, *C. sociabilis* and *C. peruanus*, spend more time at burrow entrances than do members of at least some solitary species in this family (Pearson 1959; Lacey et al. 1997). While intriguing, these observations are inconclusive; because members of these species do not typically leave their burrows or interact above ground with residents of other burrow systems, it is not clear whether greater time spent at the surface would affect the proposed benefits of the territory defense hypothesis. Data from octodontids are more problematic; the cururo is the octodontid that is most specialized for subterranean life, yet burrow systems of this species are routinely inhabited by multiple adults (Reig 1970; Begall et al. 1999; Lacey, Ebensperger, and Wieczorek, unpublished data).

The rarity of social ctenomyids appears to contradict the predictions of the expansible burrow hypothesis. Although use of a protected nest or den site is associated with group living in a number of vertebrate lineages (Andersson 1984; Alexander et al. 1991), this relationship is not apparent among octodontids or ctenomyids. More generally, the absence of social species in several lineages of subterranean rodents (e.g., geomyids, mysospalacines, rhizomyines; Nevo 1979; Lacey 2000) suggests that this hypothesis does not provide a comprehensive explanation for patterns of social structure among semi- and fully subterranean rodents. Although more detailed, quantitative studies are required to determine whether social structure varies with the degree of specialization for subterranean life in octodontids and ctenomyids, it is clear that other factors must contribute to patterns of sociality among semisubterranean versus truly subterranean species of rodents.

Summary and Future Directions

As evident from this review, comparative studies of octodontids and ctenomyids provide important opportunities to explore the adaptive bases for variation in rodent societies. In particular, the distinctive patterns of phyletic, ecological, and behavioral diversity evident in these families offer multiple opportunities to explore relationships between current environmental conditions and social structure. To date, analyses of sociality in octodontids have emphasized intrinsic benefits to group living, while studies of ctenomyids have focused on extrinsic constraints on natal dispersal that lead to the formation of social groups. Although the number of species for which quantitative data are available is small, neither soil conditions nor the spatial distribution of food resources appear to explain the occurrence of sociality in these animals. Similarly, no relationship is evident between social structure and key life-history attributes of these families.

The absence of a single, consistent ecological or life-history predictor of group living suggests that multiple factors contribute to the social systems of octodontid and ctenomyid rodents. While resource distributions, burrowing costs, predation pressures, and the production of precocial young may all influence the behavior of these animals, the specific blend of selective pressures and adaptive consequences that shape social structure seems likely to vary among species. Acknowledging the multivariate nature of interactions between environment and social structure does not preclude the search for general correlations between environmental conditions and sociality, although it may render the identification of those relationships more challenging. Thus while we suspect that many of the ecological and other potential selective forces identified here are important, we expect that their contributions to social structure vary not only between octodontids and ctenomyids but also within each family.

To exploit fully the comparative opportunities afforded by these animals, we suggest that future studies of these animals should address the following objectives:

1. *Characterization of social systems.* At present, the behavior of most species in these families remains unknown, and thus a primary goal of future research should be to generate comparative information regarding basic aspects of social structure such as the number of adults of each sex that live together, the kin structure of social units, and the social determinants of reproductive success.

2. *Analyses of intraspecific variation in ecology and behavior.* Comparative studies of conspecifics living in different habitats provide a powerful means of assessing the role of specific environmental variables in shaping social behavior. Because such comparisons effectively control for differences in evolutionary history that may confound cross-taxon analyses, identifying causal relationships between environmental conditions and behavior is facilitated.

3. *Experimental manipulation of causal factors.* Controlled manipulation of environmental variables is a compelling approach to testing proposed causal relationships between ecology, life history, and variation in social behavior. Experiments conducted in field settings are often challenging, but even relatively simple manipulations may yield important information regarding the effects of specific factors on social structure.

Although much work is required to generate a comprehensive picture of the social biology of octodontid and ctenomyid rodents, the growing number of studies of these animals suggests that such data will be forthcoming. As our knowledge of these families increases, we expect that octodontids and ctenomyids will come to play an increasingly prominent role in our understanding of rodent societies.

Chapter 35 Socioecology of Rock-Dwelling Rodents

Karen J. Nutt

ROCKY HABITAT is found throughout the world, and almost everywhere that it exists, one or more rock-dwelling species of mammal has made a home for itself among the crevices. In Australia, 11% of all rodents and marsupials are considered to be rock-specialists (Freeland et al. 1988). This list includes species of antechinus (*Parantechinus bilarni* and *Pseudantechinus macdonnellensis*), rock wallabies (*Petrogale* sp.), wallaroos (*Marcopus* sp.), and rock-rats (*Zyzomys* sp.; Freeland et al. 1988). The African continent is home to another group of rock-dwelling mammals, which includes red rabbits (*Pronolagus* sp.), hyrax (*Procavia* sp. and *Heterohyrax* sp.), and dassie rats (*Petromus typicus;* George and Crowther 1981; Hoeck 1989; Nowak 1991). Sixty percent of Namibia's endemic mammals are in fact rock-dwellers (Griffin 1998). In Asia, rocky habitat is home to pikas (*Ochotona* sp.), high mountain voles (*Alticola* sp.), and the recently rediscovered woolly flying squirrel (*Eupetaurus cinereus*), whereas in South America, chinchillas (*Chinchilla* sp.), rock cavy (*Kerodon rupestris*), and punaré (*Thrichomys apereoides*) are the local inhabitants of rock outcrops (Lacher 1981; Nowak 1991; Zahler 1996; Nadachowski and Mead 1999a; Nadachowski and Mead 1999b; dos Reis and Pessoa 2004; Spotorno et al. 2004). Even on the remote island of East Plana Cay in the Bahamas, an endangered rock-dwelling mammal, the Bahaman hutia (*Geocapromys ingrahami*), has managed to find a place to live (Clough 1972).

Although comparative studies have discussed similarities among mammals that live in other types of habitat (subterranean [Nevo 1979; Lacey and Sherman, chap. 21 this volume]; desert [Prakash and Ghosh 1975; Randall 1994]; aquatic [Wolff and Guthrie 1985]), few studies have sought to elucidate similarities among taxa that live in rocky habitat. One notable exception is the comparative study of Mares and Lacher (1987), which contrasted numerous aspects of the ecology, morphology, and behavior of sixteen species of rock-dwelling mammals against similar attributes of non-rock-dwelling taxa. Mares and Lacher's (1987) study revealed convergent evolution among rock-dwelling mammals in several traits. Morphologically, rock-dwellers tend to have padded feet, reduced claws, and (in approximately a third of the genera) a specialized grooming claw. Ecologically, approximately half of the rock-dwelling mammals are arboreal; most species utilize lookout posts for predator detection, and all species live in habitat that is patchily distributed. Behaviorally, most species have warning calls and the majority of taxa use the disjunct rockpiles in their habitat as a defensible resource. Most species also use communal defecation sites called middens, which are increasingly being used for phylogeographic studies because they contain valuable historical information on regional flora (Kuch et al. 2002). Information on social structure was compiled for ten of the rock-dwelling mammals. Remarkably, 90% could be classified as social; seven species live in harems that contain multiple females per male, and two species live in family groups (table 35.1). The only species that was not classified as social was the solitary American pika (*Ochotona princeps*).

The high degree of social behavior exhibited by rock-dwelling mammals in Mares and Lacher's (1987) study is of particular interest to mammalian sociobiology and warrants further investigation. This chapter extends the findings of Mares and Lacher (1987) by addressing how living within rock outcrops is likely to affect the social behavior

Table 35.1 Sociobiology of obligate rock-specialists and additional rock-dwelling mammals discussed in Mares and Lacher (1987)

Family	Species	Common name	Distribution	Social structure
Sciuridae	*Tamias dorsalis*	Cliff chipmunk	Western U.S. south into Mexico	Dens solitarily?; Forages communally
Sciuridae	*Eupetaurus cinereus*	Woolly flying squirrel	Northern Pakistan	Solitary
Muridae	*Alticola strelzovi*	Flat-headed vole	Asia	Family groups
Muridae	*Chionomys nivalis*	Snow vole	SW Europe to Iran	Females do not have overlapping home ranges
Muridae	*Acomys russatus*	Golden spiny mouse	Eastern Egypt through Sinai peninsula to Arabian peninsula	Females have overlapping home ranges; Social
Muridae	*Aethomys namaquensis*	Namaqua rock rat	Southern Africa	Females do not have overlapping home ranges; Family groups
Muridae	*Zyzomys palatalis*	Carpentarian rock-rat	Gulf of Carpentaria region of Australia	Females have overlapping home ranges
Muridae	*Neotoma cinerea**	Bushy-tailed woodrat	Western North America	Loose social groups
Muridae	*Neotoma magister*	Allegheny woodrat	Eastern U.S.	Colonial
Caviidae	*Kerodon rupestris**	Rock cavy	NE Brazil	Multiple females per male
Chinchillidae	*Chinchilla laniger**	Chilean chinchilla	North-central Chile	Colonial; Multi-family groups
Chinchillidae	*Lagidium peruanum**	Mountain viscacha	Western South America	Colonial; Family groups
Capromyidae	*Geocapromys brownii*	Jamaican hutia	Jamaica	Family groups
Capromyidae	*Geocapromys ingrahami*	Bahaman hutia	Bahamas	Family groups
Abrocomidae	*Abrocoma cinerea* species complex	Chinchilla rats	West-central South America	Colonial
Petromuridae	*Petromus typicus*	Dassie rat	Angola, Namibia, South Africa	Social
Ctenodactylidae	*Ctenodactylus gundi*	Common gundi	Morocco to Libya	Family groups; multiple females per male; multifamily groups; cooperative breeding units
Ctenodactylidae	*Ctenodactylus vali*	Val's gundi	Algeria and Libya	Family groups
Ctenodactylidae	*Felovia vae*	Felou gundi	Mali	Family groups
Ctenodactylidae	*Massoutiera mzabi*	Mzab gundi	Algeria, Niger, Chad	Family groups
Ctenodactylidae	*Pectinator spekei*	Speke's gundi	Ethiopia, Eritrea, Somalia, Djibouti	Extended family groups
Additional mammals studied by Mares and Lacher (1987)				
Sciuridae	*Marmota flaviventris*	Yellow-bellied marmot	Western U.S. and southern Canada	Multiple females per male
Sciuridae	*Marmota caligata*	Hoary marmot	Alaska to Washington and Montana	Multiple females per male
Sciuridae	*Marmota olympus*	Olympic marmot	Olympic peninsula of western Washington	Multiple females per male
Hyracoidea: Procaviidae	*Procavia johnstoni =* (*P. capensis*)	Rock hyrax	Africa, Arabia, Israel, Jordan, Lebanon, Syria	Multiple females per male
Hyracoidea: Procaviidae	*Heterohyrax brucei*	Bush hyrax	SE Egypt to Angola and NE South Africa	Multiple females per male
Lagomorpha: Ochotonidae	*Ochotona princeps*	American pika	Western North America	Solitary

Sources: Pearson (1948); Howe and Clough (1971); Clough (1972); George (1974); Howe (1974); Rowlands (1974); George (1978); Frase and Hoffmann (1980); George (1981); George and Crowther (1981); Lacher (1981); Hoeck (1982); Olds and Shoshani (1982); Anderson et al. (1983); Mares and Lacher (1987); Gouat (1988a); Rice (1988); Smith and Weston (1990); Nowak (1991); Hart (1992); Moses and Millar (1992); Kronfeld et al. (1994); Braun and Mares (1996); Churchill (1996); Zahler (1996); Smith (1997); Nadachowski and Mead (1999b); Shargal et al. (2000); Castleberry et al. (2001); Braun and Mares (2002); Nutt (2003); Fleming and Nicolson (2004); Luque-Larena et al. (2004); Puckey et al. (2004); Spotorno et al. (2004); Nutt (2005).
* Denotes species whose social structure is also described in Mares and Lacher (1987).

of rock-specialist rodents (rodents that are confined to living in crevices within rocks). The study carried out by Mares and Lacher (1987) was less inclusive than this study, because it did not assess the degree of social behavior in smaller-bodied rock-dwelling rodents, yet it was more inclusive than the current study because it discussed the social behavior of mammals other than rodents. The first part of this chapter will discuss rocky habitat—how many rodents utilize rocky substrate, and potential benefits they may gain from doing so. The second part of the chapter will delineate the types of social behavior observed in obligate rock-

specialists and how various hypotheses for group living may explain observed levels of social affiliation.

How Many Rodents Utilize Rocky Habitat?

A vast number of rodents utilize rocky substrate. According to the rodent generic accounts in Nowak (1991), at least 114 genera of rodents (over one-fourth of all rodent genera!) from eighteen different families (over one-half of all rodent families!) contain at least one species that uses rocks

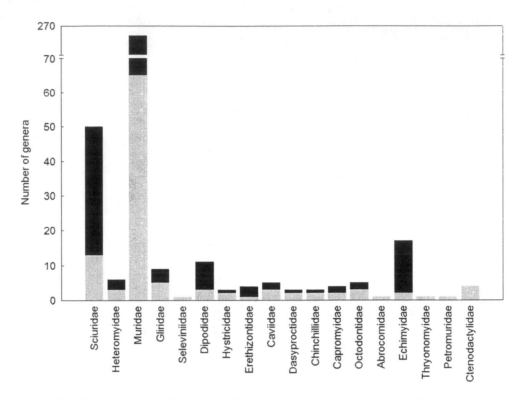

Figure 35.1 Number of genera within each rodent family that contain at least one species that utilizes rocky substrate (gray bars) or contain only species that do not associate with rocks (black bars). Data were compiled from generic accounts in Nowak (1991). Not shown are the eleven rodent families that do not contain any genera that associate with rocks.

or rocky substrate for one reason or another (fig. 35.1). This figure includes genera that sometimes live in rocky habitat (*Ototylomys;* big-eared climbing rats and *Cardiocranius;* five-toed dwarf jerboas); den in caves or crevices (*Petaurista;* giant flying squirrels and *Erethizon;* North American porcupine); dig burrows under rocks or shelter among rocks (*Ichthyomys;* fish-eating rats and *Selevinia;* desert dormice); have been collected among rocks (*Vernaya;* Vernay's climbing mice and *Diomys;* Crump's mouse), and others. Although expansive, the list in figure 35.1 is certainly not exhaustive, as some genera like *Marmota* contain species that are clearly associated with rocks and boulders (Tyser 1980), yet the generic account in Nowak (1991) did not mention this association. Regardless, figure 35.1 illustrates the extensive diversity of rodents that utilize rocky substrate. What is it about rocks that make them such an essential component of the habitat for so many rodents?

Why Live in Rocky Habitat?

Complex nature of the habitat

In its broadest sense, rocky habitat may be described as any locality that contains boulders, rocks, scree, pebbles, outcrops, cliffs, or caves. Rocky substrate may extend without interruption across the landscape, as exemplified by the continuous mountain ranges of south-central Tunisia inhabited by common gundis (*Ctenodactylus gundi*) and elephant shrews (*Elephantulus rozeti;* K. Nutt, pers. obs.), or may be clumped and isolated, such as the disjunct Serengeti kopjes of bush and rock hyrax (*Heterohyrax brucei* and *Procavia johnstoni,* respectively; Hoeck 1982; Gerlach and Hoeck 2001). One of the reasons why so many rodents live in rocky habitat is because it is topographically very complex, containing a multitude of crevices and cavities in which animals can nest, seek shelter, and find shade (George 1986; Trainor et al. 2000). Studies of the ecology and behavior of rock-dwelling mammals have provided valuable insight into how such topographically complex habitat is partitioned (George and Crowther 1981; Channing 1984; Etheredge et al. 1989; Trainor et al. 2000; Jones et al. 2001). For example, in South Africa the namtap (*Graphiurus ocularis*) is able to share its rocky habitat with both Namaqua rock rats (*Aethomys namaquensis*) and elephant shrews (*Elephantulus edwardii*) because the namtap preferentially makes use of vertical rock crevices, whereas the other two species use horizontal crevices (Channing 1984). In the rock formations of Israeli deserts, a difference in diel cycle allows two spiny mice (*Acomys* sp.) to coexist (Jones et al. 2001). In many rocky habitats, smaller-bodied rock-dwelling rodents coexist with larger-bodied mammals

(Bolivian big-eared mice; *Auliscomys boliviensis* and mountain viscachas; *Lagidium peruanum*; Pearson 1948; dassie rats and rock hyrax; George and Crowther 1981). In the latter case, crevices were divided between the two species according to height, the maximum crevice size used by the smaller-bodied dassie rat equaling the skull height of the rock hyrax. The highly complex nature of rocky substrate, with its abundance of crevices and fissures, is clearly one reason why so many rodents make use of rocky habitat.

Predator avoidance

Mammals also make use of rocky substrate because it can enhance their ability to protect themselves against predators. Some species use rock promontories and ledges as lookout posts for predator detection (Tyser 1980; Mares and Lacher 1987). In gundis, this behavior appears to be cooperative, with group members taking turns reposing on ledges and watching out for intruders (Gouat and Gouat 1989). Other species prefer rocky habitat because rocks can be used to hide a burrow or nest. For example, burrows of both the golden-mantled ground squirrel (*Spermophilus lateralis*) and the least chipmunk (*Tamias minimus*) are located under large rocks more often than would be predicted from the distribution of boulders in their habitat (Bihr and Smith 1998). Species that live in more complex rocky environs conceal themselves from predators by foraging beneath rocky overhangs, or evade capture by darting into nearby cracks and crevices. For the common spiny mouse (*Acomys cahirinus*), the pygmy rock mouse (*Petromyscus collinus*), and collared pikas (*Ochotona collaris*), the choice of foraging location depends greatly on the presence of such nearby cover (Holmes 1991; Brown et al. 1998; Mandelik et al. 2003). In all of these instances, rocks are used in an attempt to minimize predation pressure.

Rock-dwelling rodents are also able to evade predation through camouflage. In at least three rock-dwelling species, the canyon mouse (*Peromyscus crinitus*), a spiny mouse (*Acomys* sp.), and the rock pocket mouse (*Chaetodipus intermedius*), alternative color morphs exist for individuals inhabiting different colored substrate (Johnson and Armstrong 1987; Sicard and Tranier 1996; Hoekstra and Nachman 2003). For the *Acomys* sp., three color morphs (brown, red, and gray) can be found in Burkina Faso, each of which appears to be adapted to the coloration of the soil in its particular habitat (Sicard and Tranier 1996). Populations of the rock pocket mouse contain either light or dark-colored morphs, although this is a deceptively simple categorization, since alternative genes are responsible for the adaptive melanism in different dark-colored populations (Hoekstra and Nachman 2003). Habitat-dependent selection maintains the positive association between coat color and habitat color observed in rock pocket mice (Hoekstra et al. 2004). This ability of rodents to camouflage themselves on rocky substrate to avoid predation may partially explain why there are such a large number of rodents that associate with rocks.

Thermoregulation

Rock-dwelling rodents occur in almost every ecosystem, ranging from extremely hot, dry deserts to very cold alpine regions. The geographic ranges of several species, including golden spiny mice (*Acomys russatus*), common gundis, and cliff chipmunk (*Tamias dorsalis*) extend from one environmental extreme to the other, making it necessary for these species to be able to adapt to a wide range of climatic conditions (Haim and Borut 1975; Gouat and Gouat 1984; Hart 1992). Rock-dwelling rodents are able to survive in such extreme environments in part because their rocky habitat can help them to properly thermoregulate.

Many rock-dwelling rodents are able to survive in harsh environments because of the insulating ability of rocks to moderate extreme fluctuations in ambient temperature. Rocks cool down and heat up slowly, so temperatures inside crevices are generally cooler than ambient during the heat of the day and warmer than ambient at night (Clough 1972; George 1986). Temperatures inside crevices are also generally much more stable than ambient temperature (Sale 1966). For example, the temperature of bushy-tailed woodrat (*Neotoma cinerea*) burrows constructed in rock crevices fluctuates by only a few degrees Celsius daily, whereas ambient temperature may fluctuate by up to 20°C (Brown 1968, as cited in Smith 1997). This thermal stability of rocky environs may be enhanced by snow cover insulation, perhaps explaining why some alpine rodents such as the snow vole (*Chionomys nivalis*) and the bank vole (*Clethrionomys glareolus*) use rocks as their preferred over-wintering microhabitat (Karlsson 1988; Luque-Larena et al. 2002b). The moderated temperatures found within rocky crevices help to provide a warm home for species living in cold climates and a cool refuge for inhabitants of warm environments.

Rodents also use the thermal properties of rock surfaces to help with thermoregulation. In hot climates, rock surface temperature is generally higher than that of ambient temperature (George 1986; Fredericksen et al. 2003). Many diurnal species make use of the higher temperatures on rock surfaces to warm up on cold mornings. Social rock-dwelling rodents, including mountain viscachas and gundis, often huddle together in piles during this early morning bask (Pearson 1948; Gouat 1991). The heat-retaining and insulating properties of rock help many rock-dwelling species to properly thermoregulate, thereby enabling them to survive in extreme environments.

Rocky substrate as a water basin

Rocky substrate is important for species living in xeric regions of the world because depressions in rocks form small water catchments during periods of rain (Streilein 1982). These rock pools can increase the humidity of the region around the rocks, leading to a more mesic habitat and providing the opportunity for species to live where they otherwise could not (Streilein 1982; Morris 2000). Red-back voles (*Clethrionomys gapperi*), for example, live in forests in the northern part of their range, but in the southern part of their range they are confined to living in rock outcrops, presumably because outcrops are the only habitat capable of providing a sufficiently high humidity to avoid dehydration (Wolff and Dueser 1986). The Australian pebble-mound mouse (*Pseudomys chapmani*) uses resourcefulness in combination with rocky substrate to create its own mesic microhabitat. The pebble-mound mouse scatters a mound of pebbles around its burrows. Each morning, the air temperature around the pebbles warms up faster than the pebbles themselves, causing small droplets of dew to form by condensation. The dew collected on pebbles within each 1-meter-diameter mound can then be ingested by the pebble-mound mouse (Nowak 1991). Through the use of ingenuity or merely a strategic choice of habitat, rodents ensure their survival in what would otherwise be harsh, inhospitable environments by using rocky substrate to obtain water.

Unique microenvironments and floral diversity

The complex topography of rocky habitat, including the presence of catchment sites for water, can lead to the existence of unique microenvironments throughout rocky substrate and increased plant species richness (Freeland et al. 1988; Trainor et al. 2000; Fredericksen et al. 2003). A survey of plants surrounding a rock escarpment in western Arnhemland, Australia, revealed a larger number of plant species, a larger number of tree species, and greater tree species diversity with increasing proximity to the escarpment (Freeland et al. 1988). Higher levels of floral diversity near the escarpment are thought to result from increased water retention by the rock surface during seasonal rains (either due to water absorption by the rock, water catchment sites within the rock face, or water runoff from the escarpment). This particular escarpment in Arnhemland is home to six species of mammal that are restricted to rock outcrops. Because of the high diversity of rock-specialists on this outcrop and the extensive floral diversity surrounding the outcrop, Freeland et al. (1988) proposed that the uniqueness of microflora environments surrounding rocky habitat might lead to speciation within rock-dwelling taxa. If correct, this supposition could help to explain the large diversity of

mammals that associate with rocky habitat throughout the world. Further studies are needed to test this hypothesis, however, against the more traditional explanation of speciation through geographic isolation (see Braun and Mares 2002, for example).

Effect of fire and over-grazing

Two of the major factors threatening faunal diversity within habitats of Australian rock-dwelling rodents are fire and the overgrazing of food plants by large herbivores (Begg 1981; Begg et al. 1981; Morris 2000; Trainor et al. 2000; Brook et al. 2002). Populations of both the critically endangered Carpentarian rock-rat (*Zyzomys palatalis*) and the Arnhem Land rock-rat (*Zyzomys maini*) are thought to be highly susceptible to these concerns (Begg et al. 1981; Churchill 1996; Trainor et al. 2000; Brook et al. 2002). Both fire and overgrazing, however, appear to be less of a problem the rockier the habitat (Trainor et al. 2000). For the Arnhem Land rock-rat, a fire greatly reduced population numbers and led to decreased reproduction the following breeding season (Begg et al. 1981). Although this species was initially most abundant in closed forest, that habitat was severely impacted by the fire, so the Arnhem Land rock-rat relocated after the fire to less disturbed scree habitat (Begg et al. 1981). This study suggests that many Australian mammals may choose to live in rocky habitat so as to reduce their exposure to fire and problems of a similar nature caused by overgrazing.

More stable microclimate

Because rocky substrate can provide abundant refuges, enhance thermoregulation, lead to increased floral diversity, improve water availability, and reduce the extent of damage from fires and overgrazing, a more stable microclimate often exists in and around rocky habitat compared to surrounding regions (Streilein 1982; Trainor et al. 2000). In some cases, such favorable conditions can lead to the availability of year-round resources, and may result in smaller home range sizes for rock-dwelling mammals in comparison to related taxa that do not live in rocks (Pavey et al. 2003). The overall stability of rocky habitat also makes it suitable for use as a temporary refuge by species that normally reside in other habitats (Withers and Edward 1997). For example, a large-scale temporal analysis of the distribution of mammals in Australia indicates that rocky habitat is increasingly occupied during times of reduced groundwater levels (Braithwaite and Muller 1997). Since rocky habitat provides a stable microclimate that is exploitable by rock-dwelling species all year round and non-rock-dwelling taxa during critical times, it is not surprising that over one-fourth of all rodent genera are known to associate with rocky habitat (fig. 35.1).

Defining an Obligate Rock-Specialist

Obligate rock-specialists—those rodents that live exclusively within rocks or rocky crevices—can be called saxatile (meaning that they live among rocks) or petrophilic (rock-loving). There are approximately seventy-seven species of obligate rock-specialists in forty-five genera and eleven different families (table 35.2). Included in this list are two families of rodents that contain only saxatile species (the monotypic petromurid dassie rat and the ctenodactylid gundis), several genera of rodents with adaptive radiations of rock-specialist species, including Australian rock-rats (*Zyzomys*), high mountain voles (*Alticola*), rock rats (*Aethomys*), chinchilla rats (*Abrocoma*), snow voles (*Chionomys*), hutias (*Geocapromys*), and mountain viscachas (*Lagidium*), and a few species that have atypically specialized on rocky habitat, including the punaré (the only echimyid or spiny rat to be predominantly petrophilic; Streilein 1982; note, however, that one population of the punaré occurs in an area with sandy soil but no rocks; Lacher and Alho 1989) and the montane guinea pig (*Cavia tschudii*,

which appears to be the only guinea pig to have been classified as an outcrop specialist; Fredericksen et al. 2003).

Sociality in Obligate Rock-Specialists

Despite the fact that there are so many saxatile rodents found throughout the world, the extent of social behavior exhibited by most species is virtually unknown. There are perhaps four reasons why so little information is known about the natural history, behavior, and ecology of most rock-dwelling rodents: (1) many saxatile rodents inhabit extreme environments in which it is often very taxing to work, (2) the rocky terrain inhabited by petrophilic rodents often makes it extremely difficult to observe their behavior, track them, or trap them, (3) the excellent climbing abilities and shy nature of many rock-specialists makes capturing them an extremely complicated process (Rowlands 1974; Lacher 1981; Streilein 1982; Nutt 2003; Zahler and Khan 2003; Spotorno et al. 2004), and (4) some rock-dwelling species are rare, threatened, or on the verge of extinction (Oliver 1977; Balcom and Yahner 1996; Jimenez 1996; Zahler 1996).

In the following I describe the social behavior of those rock-specialist species for which any information on social

Table 35.2 Rodents that are obligate rock-specialists

Family	Species
Sciuridae	*Ammospermophilus insularis?, Atlantoxerus getulus?, Eupetaurus cinereus, Rheithrosciurus macrotis?, Sciurotamias davidianus?, S. forresti?, Spermophilus atricapillus?, Tamias dorsalis, Trogopterus xanthipes?*
Heteromyidae	*Chaetodipus intermedius*
Muridae	
Cricetinae	*Calomyscus bailwardi?*
Gerbillinae	*Gerbillus campestris?, G. dasyurus?. G. lowei?, G. rupicola, Sekeetamys calurus*
Microtinae	*Alticola argentatus, A. roylei?, A. strelzovi, Chionomys gud, C. nivalis, C. roberti, Microtus chrotorrhinus?*
Murinae	*Acomys russatus, Aethomys chrysophilus?, A. namaquensis, A. nyikae?, Cremnomys cutchicus, Niveventer hinpoon, Rattus neilli, Vernaya fulva?, Zyzomys argurus, Z. maini, Z. palatalis, Z. pedunculatus, Z. woodwardi*
Petromyscinae	*Petromyscus collinus, P. monticularis*
Sigmodontinae	*Auliscomys boliviensis, A. sublimis, Chinchillula sahamae?, Euneomys chinchilloides, Galenomys garleppi?, Nelsonia neotomodon?, Neotoma cinerea, N. magister, N. mexicana, Peromyscus attwateri?, P. crinitus?, P. difficilis?, Phyllotis xanthopygus, Other Phyllotis sp?, Punomys lemminus*
Gliridae	*Dryomys laniger, Graphiurus ocularis, G. platyops?*
Caviidae	*Cavia tschudii, Kerodon rupestris*
Chinchillidae	*Chinchilla brevicaudata, C. laniger, Lagidium boxi, L. peruanum, L. viscacia, L. wolffsohni*
Capromyidae	*Geocapromys brownii, G. ingrahami*
Abrocomidae	*Abrocoma budini, A. famatina, A. schistacea, A. uspallata, A. vaccarum*
Echimyidae	*Thrichomys apereoides?*
Petromuridae	*Petromus typicus*
Ctenodactylidae	*Ctenodactylus gundi, C. vali, Felovia vae, Massoutiera mzabi, Pectinator spekei*

SOURCES: Hershkovitz (1962); George (1974); Schmidly (1974); Haim and Borut (1975); Lekagul and McNeely (1977); Pine et al. (1979); George and Crowther (1981); Withers (1983); Cornely and Baker (1986); Johnson and Armstrong (1987); Roberts et al. (1988); Dempster and Perrin (1989b); Nowak (1991); Hart (1992); Alvarez-Castaneda et al. (1996); Braun and Mares (1996); Withers and Edward (1997); Brown et al. (1998); Galindo-Leal and Krebs (1998); Kramer et al. (1999); Shenbrot et al. (1999); Cole and Woinarski (2000); Shargal et al. (2000); Trainor et al. (2000); Woinarski (2000); Castleberry et al. (2001); Braun and Mares (2002); dos Reis and Pessoa (2004); Puckey et al. (2004).

structure could be obtained from the literature (these data are compiled in table 35.1). Where possible, literature reviews such as mammalian species accounts were used for citations. The following accounts are intended to summarize the types of social behavior observed in petrophilic rodents while providing background information on several rodent taxa not described elsewhere in this volume.

Sciuridae

The cliff chipmunk inhabits the Rocky Mountains of the western United States and Mexico (Hart 1992). The social structure of the cliff chipmunk is somewhat unclear from the species account; individuals exhibit territorial behavior toward conspecifics near the home den (a crevice in rocks or cliffs that contains a food cache), yet join feeding aggregations away from the den (Hart 1992). Up to ten individuals, mostly female (males tend not to associate with cliff den areas in the summer months), have been observed in such feeding aggregations (Hart 1992).

The woolly flying squirrel was recently rediscovered inhabiting the high, cold desert regions of northern Pakistan (Zahler 1996). Local inhabitants of this region claim that the woolly flying squirrel lives solitarily in remote caves on vertical cliff walls and that it feeds on pine needles obtained by climbing into conifers within nearby open forests (Zahler 1996; Zahler and Khan 2003).

Muridae: Microtinae

The flat-headed vole (*Alticola strelzovi*) inhabits rocky steppes in Asia (Nadachowski and Mead 1999b). Although this species usually builds its nest in fissures and crevices in rocks, it does occasionally dig its own burrow in soft soil. The flat-headed vole lives in family groups comprised of a pair of adults and between five and ten offspring. In Kazakhstan, each social group collects plant material, dries it under a pile of pebbles, and stores it within rocky fissures until it is needed during the winter (Nadachowski and Mead 1999b).

The snow vole lives in patchy rock habitat distributed between Europe and Iran. Female home ranges do not overlap during the breeding season, whereas male home ranges overlap extensively with those of both females and males, leading to what appears to be a promiscuous mating system (Luque-Larena et al. 2004). Interestingly, although many vole species tend to aggregate for over-winter survival, the snow vole is solitary and nomadic during this time (Le Louarn and Janeau 1975; Luque-Larena et al. 2002d; Luque-Larena et al. 2002b). It has been hypothesized that the snow vole does not need to socially thermoregulate because it has better insulating properties than other voles

and because its rocky home helps to modulate winter temperatures, making social thermoregulation unnecessary (Luque-Larena et al. 2002b).

Muridae: Murinae

The golden spiny mouse is distributed from Egypt to the Arabian Peninsula. In the hot rocky deserts of southern Israel, it is sympatric with the common spiny mouse, although the common spiny mouse is much less of a specialist on rocky habitat (Shargal et al. 2000). Both species appear to be social; in the wild, females and males have overlapping home ranges (Kronfeld et al. 1994; Shargal et al. 2000) and in captivity both species preferentially aggregate in nest boxes (Shargal et al. 2000).

The Namaqua rock rat inhabits rock outcrops in southern Africa. Females have nonoverlapping territories, whereas male territories overlap considerably with those of several females and with other males (Fleming and Nicolson 2004). The Namaqua rock rat is thought to exhibit scramble-competition polygyny during the breeding season because nonresident males enter the population during this time and vie for mating opportunities (Fleming and Nicolson 2004). Previous field and captive studies of the Namaqua rock rat, however, suggested that this species is social and that it lives in family groups (Nowak 1991). The red rock rat (*Aethomys chrysophilus*) is also reported to live in paired social groups (Nowak 1991).

The Carpentarian rock-rat is the only one of the petrophilic Australian rock-rats (table 35.2) for which any information on territorial overlap among the sexes is known. This critically endangered species is restricted to the scree slopes of monsoon rainforests in the Gulf of Carpentaria region of Australia (Churchill 1996). Telemetry data have revealed that home ranges of females, males, and juveniles are nonexclusive (averaging 41% overlap), with the greatest degree of overlap occurring between males and juveniles (Puckey et al. 2004).

Muridae: Sigmodontinae

Although most woodrats construct stick houses, at least three species appear to be obligate rock-specialists that build nests within rocky crevices (table 35.2). The social behavior of two species, the bushy-tailed woodrat and the Allegheny woodrat (*Neotoma magister*) have been discussed in the literature. The bushy-tailed woodrat inhabits rock outcrops throughout western North America. From one to twelve breeding females and one to seven breeding males occupy each outcrop (Moses and Millar 1992). Telemetry studies have revealed extensive intrasexual and intersexual home range overlap among individuals (Topping and Millar 1996c). Although this spacing arrangement would suggest a promiscuous mating system, only one male sires all offspring in a given litter (Topping and Millar 1998). The Allegheny woodrat of the Appalachian Mountains in the eastern United States apparently lives in colonies (Castleberry et al. 2001), although little information could be found on proximity of individuals to one another or on crevice usage within an outcrop. Both of these woodrats cache food (Smith 1997; Castleberry et al. 2001).

Caviidae

The rock cavy inhabits rock outcrops in the semiarid Caatinga of northeastern Brazil (Lacher 1981). Field observations suggest that rock cavy males defend rockpiles inhabited by more than one female in a resource defense polygyny type of mating strategy (Lacher 1981). Such a hypothesis was supported by the observation of a skewed operational sex ratio in the field (three males : fifteen females; Lacher 1981). Additional evidence on wild and captive animals supports the notion that the rock cavy is social; in the field, groups of up to five individuals have been observed in one small area (Streilein 1982) and in captivity males provide parental care (Tasse 1986).

Chinchillidae

The Chilean chinchilla (*Chinchilla laniger*) inhabits the rugged, barren mountain ranges of north-central Chile (Spotorno et al. 2004). The Chilean chinchilla lives in colonies of up to 500 individuals (Spotorno et al. 2004) with several pairs reportedly coinhabiting the same den (Rice 1988). The distribution of the Chilean chinchilla is reliant on the availability of suitable rocky habitat (Spotorno et al. 2004). The Chilean chinchilla was severely hunted in the past for its fur and is now considered endangered, while its congener the short-tailed chinchilla (*C. brevicaudata*) is already thought to be extinct in the wild (Jimenez 1996).

The northern mountain viscacha (*Lagidium peruanum*) lives in rock outcrops and cliffs in the altiplano of southwestern South America (Rowlands 1974). Northern mountain viscachas are highly gregarious, living in colonies of up to seventy-five individuals where rock formations are able to support such numbers (Pearson 1948). Each colony is composed of family groups that contain between two and five individuals each (usually an adult male, a parous female, and offspring of varying ages). These family groups are located in close proximity to one another, although there is very little aggression between them most of the year. During the breeding season, however, aggression levels within the colony increase substantially; females apparently evict adult breeding males from their family group

and aggressively attack any who approach them, while males exhibit promiscuous behavior and attempt to guard females from other males. Males are allowed to reassociate with females once again after the females have become pregnant (Pearson 1948). The social group structure of the other mountain viscachas (*L. wolffsohni* and *L. boxi*) have also been described as colonial, although *L. boxi* is thought to be more irascible than *L. peruanum* (Rowlands 1974; Walker 2001, as cited in Walker et al. 2003).

Capromyidae

The Jamaican hutia (*Geocapromys brownii*) and the Bahaman hutia (*G. ingrahami*) live in limestone crevices on their respective islands (Clough 1974; Nadachowski and Mead 1999b). Observations in the wild are scant for both species. Observations of captive Bahaman hutias before and after their release into the wild suggest that this species lives in small family groups of between two and six individuals (sometimes up to ten individuals; Nadachowski and Mead 1999b). Captive studies of the Jamaican hutia indicate that this species is also likely to be gregarious (Clough 1972; Howe 1974). Captive studies also suggest that both hutia species may live in paired social groups that contain only one breeding female (Howe and Clough 1971; Oliver et al. 1986).

Abrocomidae

The genus *Abrocoma* (chinchilla rats) contains six species that are very closely allied within the *Abrocoma cinerea* species complex (Braun and Mares 2002). These six species are distributed throughout arid, rocky areas in west-central South America and have specific morphological adaptations for living in rocky habitat, including padded feet and reduced claws (Braun and Mares 2002). Five of the six members of this species complex (all except *A. cinerea*) appear to be obligate rock-specialists (table 35.2; Braun and Mares 2002). *Abrocoma cinerea* is exceptional because, although it usually associates with rocks, it sometimes lives in crevices, but at other times digs its own burrow (Braun and Mares 2002). *Abrocoma cinerea* is reported to live in small colonies. The social structure of the five other species is unknown, although the presence of extremely large latrine piles in fissures inhabited by *Abrocoma vaccarum* suggest that this species may also live in colonies (Braun and Mares 1996).

Petromuridae

The dassie rat lives in rock outcrops in Angola, Namibia, and South Africa (Nowak 1991). The dassie rat is a diurnal species that has been described as social (George and Crowther 1981). Although no information is available on group size, the population size at a six-hectare study site in the Namib desert was sixteen individuals (Withers 1983).

Ctenodactylidae

The ctenodactylid family of rodents contains five species, all of which are obligate rock-specialists (table 35.1). Ctenodactylid rodents inhabit the semidesert and desert mountainous regions of northern Africa (George 1974). All five species live in social groups, although densities range from 0.3 gundis/hectare in the widely dispersed Mzab gundi (*Massoutiera mzabi*) to 237 gundis/hectare in Speke's gundi (*Pectinator spekei*; George 1981). The common gundi has been the most extensively studied of all five ctenodactylid rodents. This species is thought to be a cooperative breeder that lives in groups of up to twenty individuals (Gouat and Gouat 1983; Nutt 2005). Each social group contains between one and three breeding males and females and their offspring, the number of breeding individuals in a social group increasing with group size (Nutt 2003). Preliminary field observations of Val's gundi (*C. vali*) and the Mzab gundi suggest that these species live in paired social groups (Gouat 1988a). The diversity of social structure exhibited by ctenodactylid rodents provides an excellent opportunity for comparative studies on the effect of density and patchiness of rocky habitat on degree of sociality (Gouat 1988a; Nutt 2005).

Hypotheses for Observed Levels of Social Behavior in Rock-Dwelling Rodents

Of the twenty-one saxatile rodents listed in table 35.1, at least fourteen (67%) have been classified as group living (this includes species classified as social, as well as those that have been described as living in family groups, extended family groups, loose social groups, multifamily groups, or groups that contain multiple females per male). Note that the Namaqua rock mouse is not included within these fourteen species, even though one study claimed that this species lives in family groups. Furthermore, some "colonial" species may also be social. Regardless, over half of all rock-specialist rodents appear to live in groups. Lacey and Sherman (chapter 21 this volume) suggest that there are at least seventy species of social rodents. According to this estimate, the number of group-living petrophilic rodents represents approximately 20% of all social rodents.

Several hypotheses have been put forward to account for group living in mammals: (1) the phylogenetic constraints hypothesis (Rowe and Honeycutt 2002), (2) the need for extended parental care (Armitage 1999), (3) high predation

levels (Jarman 1974), and (4) the clumped distribution of a limited resource (Emlen and Oring 1977). There is some evidence to suggest that to some extent, each of these four constraints may influence the degree of social behavior observed in rock-dwelling rodents.

The phylogenetic constraints hypothesis

It has been suggested that sociality within some rodent lineages is constrained by phylogeny (Rowe and Honeycutt 2002, but see Trillmich et al. 2004 for rebuttal). Is it possible that shared ancestry leads to the observed levels of social behavior in rock-dwelling rodents? Considering the highly diverse taxonomic distribution of petrophilic rodents (table 35.2), it is unlikely that similarities in social behavior among taxa are governed by shared ancestry. Phylogeny may, however, influence social behavior indirectly in one of two ways: (1) phylogeny may influence choice of habitat, which may directly influence degree of sociality (shared ancestry for habitat preference is exemplified by the adaptive radiations of rock-specialist species in several rodent genera; table 35.2), and (2) phylogeny may limit rather than define the type of social behavior that is observed. For example, the restricted availability of crevices in rock outcrops forces bushy-tailed woodrat and Allegheny woodrat individuals into close proximity. However, unlike many other petrophilic rodents, the rock-dwelling woodrats do not form cohesive social groups. Instead, loose social groups or colonies are formed, possibly because woodrats (*Neotoma* sp.) in general tend to be solitary (Nowak 1991) and are not prone to exhibiting high degrees of social behavior. Although phylogeny may minimally influence the degree of social behavior observed in some petrophilic rodents, ecological constraints are likely to have a greater effect on observed levels of social affiliation.

Need for extended parental care in extreme environments

For some rodents, group living seems to have evolved because of a need for extended parental care in harsh environments (Barash 1974a; Armitage 1999). Since many petrophilic rodents also live in extreme environments, it is possible that they too form social groups as a result of delayed dispersal and the need for extended parental care. There are two lines of evidence to suggest that this may be the case. First, many rock-specialist rodents are K-selected, meaning that they have only a few young to which they give a large amount of parental care. As examples, the rock cavy has the longest gestation length, the smallest mean litter size, and the lowest mean litter weight of all the caviinae (cavies; Roberts et al. 1984), the punaré has the longest gestation length by 50% and the lowest annual and lifetime reproductive rates of any echimyid (spiny mouse; Roberts et al. 1988), and the canyon mouse has a relatively long weaning time and a very high relative birth mass compared to other species of deer mice (*Peromyscus;* Modi 1984). There is also some evidence to suggest that the extreme environments inhabited by some rock-specialist rodents influences their survival and tendency to remain philopatric. For example, bushy-tailed woodrat yearling females experience increased over-winter survivorship when they spatially aggregate near their mothers on an outcrop (Moses and Millar 1992; Moses and Millar 1994). Similarly, although it is thought that male gundis normally disperse out of their natal group, many remained philopatric during the course of a drought (Gouat 1988a; Nutt 2003; Nutt 2005). These observations suggest that the degree of sociality observed in at least some rock-dwelling rodents may be influenced by extreme environmental conditions and the need for extended parental care.

High predation levels

Group living is thought to have evolved in some mammals to alleviate high predation pressure on individuals (Jarman 1974; Waterman 1997; Hill and Lee 1998). It is possible that rocky habitat tends to have a high abundance of predators, and that this is why some rock-dwelling mammals are social. I have stated previously that topographically complex rocky habitat is beneficial to prey because it helps them to hide from predators. Such complex habitat may also work against prey by concealing predators until it is too late. It is perhaps because of high predation pressures that so many rock-dwelling mammals have alarm calls and utilize lookout posts (Mares and Lacher 1987). The ctenodactylid rodents are notable in this respect, as all five species have characteristic alarm calls (George 1981).

The degree of predation pressure on each rock-dwelling rodent will vary considerably depending on the predators in the environment and the particular habitat. Not only is it difficult to discern the exact level of predation pressure experienced by each rock-dwelling mammal, predators are also likely to affect the sociality of each species in different ways. For example, the American pika dens solitarily, yet exhibits high levels of philopatry (Peacock 1997). By living in close proximity to relatives, American pikas can increase the nepotistic effect of alarm calling without having to form large social groups (Sherman 1977; Smith 2001). The cliff chipmunk also seems to den solitarily, yet this species presumably minimizes predation pressure through the use of communal foraging (Hart 1992). Finally, rock and bush hyrax form heterospecific aggregations while basking, apparently to increase the survivorship of young offspring (Barry and Mundy 2002). Clearly, predation pressure has a large influence on the type of aggregate behavior observed in at least some rock-dwelling mammals.

The clumped distribution of a limiting resource

Some evidence suggests that many mammals live in groups because of the clumped distribution of a limiting resource such as food or shelter (Emlen and Oring 1977). For rock-dwelling species, the distribution and abundance of rocks, crevices, and outcrops is likely to be one of the most important resources influencing social behavior in this regard. The size and distribution of rocky substrate is likely to affect the sociobiology of rock-dwelling mammals in three distinct ways: (1) by influencing whether social groups form, (2) by influencing the size of social groups and density of animals that live on an outcrop, and (3) by influencing the larger geographic spatial distribution of animals.

How distribution of rocky substrate may influence social group formation

Living in crevices may lead to high levels of sociality and group living because crevices are a nonconsumable resource that may easily be shared. Waser (1988) asserted that species requiring nonconsumable resources, such as burrows or elaborate refuges, tend to be philopatric. Philopatry is expected to evolve in such a situation because the cost to parents of sharing a refuge will be relatively low, whereas the cost to offspring of dispersing and finding a new refuge that is limited in number will be fairly high (Waser 1988). Once natal philopatry occurs, sociality may evolve.

There is much evidence within Rodentia to support the hypothesis that the use of elaborate refuges may lead to group living. In New World hystricognath rodents, living in groups appears to be more associated with the need to reduce burrowing costs than the need to reduce predation pressure or the need for extended parental care (Ebensperger and Cofré 2001). Similarly, group living in the bathyergid mole-rats is thought to have evolved in environments where burrowing costs are extremely high (Faulkes et al. 1997; Faulkes and Bennett, chap. 36 this volume). These examples illustrate that sociality tends to evolve when burrowing is costly. If burrowing is not an option at all (as is the case for petrophilic rodents), then individuals will be completely reliant on the availability of suitable crevices and will be forced to clump together when these resources are limited.

How distribution of rocky substrate may influence population density and group size

The distribution and abundance of crevices within an outcrop is likely to influence the number of animals that can live in a region and the size of social groups that may form. The population densities of several rock-specialist rodents, including the mountain viscacha, the cutch rock rat (*Cremnomys cutchicus*), the namtap, the bushy-tailed woodrat, and the dassie rat, are all thought to be limited by the availability of suitable crevices (Pearson 1948; George and Crowther 1981; Channing 1984; Moses and Millar 1992; Prakash and Singh 2001). This effect of crevice number on population density has been confirmed in the bushy-tailed woodrat by experimental manipulation of den and nest sites (Hickling 1987, as cited in Moses and Millar 1992). In the case of the dassie rat, competitive exclusion from shelters by the rock hyrax led to a 97% difference in population densities between the two species (George and Crowther 1981). Both intraspecific and interspecific competition for crevices limits the number of rock-dwellers that can occupy any one site.

Social structure is also influenced by the size and distribution of crevices. Small groups of petrophilic rodents are found where crevices are small or few in number, whereas large groups tend to occupy outcrops that have many large crevices (Anderson et al. 1983; Hickling 1987, as cited in Smith 1997; K. Nutt, pers. obs). In studies of the common gundi, territory size was comparable among all social groups, yet the smallest group (of three individuals) occupied a territory that contained only two small primary shelters, whereas the largest group (with over twenty gundis) occupied an isolated territory that contained a vast number of sizeable shelters (K. Nutt, pers. obs.). Interspecific competition for shelters may also influence group size in the ctenodactylid rodents. Field observations indicate that Val's gundi may live in smaller-sized social groups because the common gundi prevents it from occupying better outcrops with larger crevices (Gouat 1988b; Gouat 1988a). More detailed quantitative studies are needed to verify this effect of crevice size and distribution on group size in rock-dwelling rodents.

How distribution of rocky substrate may influence a species' geographic distribution

Many rock-dwelling mammals inhabit rock outcrops that are disjunct and patchily distributed (Smith 1980; Hoeck 1982; Christian and Daniels 1985; Kilpatrick and Crowell 1985; Churchill 1996; Kim et al. 1998; Braun and Mares 2002; Walker et al. 2003; Spotorno et al. 2004). These rock islands vary greatly in size and proximity to one another, and together form what is known as a metapopulation. The occupancy of each rock outcrop within a metapopulation varies over time as some populations go extinct and others are recolonized. Models have shown that patch occupancy

of mountain viscacha outcrops is primarily determined by the abundance and depth of crevices at a given site (Walker et al. 2003). Alternatively, metapopulation studies of the American pika have shown that patch occupancy depends on size of rocky habitat (smaller habitat patches going extinct more often than larger patches) and the distance between patches (more isolated islands of rocky habitat having slower rates of recolonization; Smith 1980). Additional studies have shown that low rates of recolonization in some large patches may lead to long-term population declines (Moilanen et al. 1998; Clinchy et al. 2002). The knowledge of such a process should prove invaluable to conservation efforts of rare, endangered, or threatened rock-dwelling rodents (Oliver 1977; Balcom and Yahner 1996; Jimenez 1996; Zahler 1996).

Summary

Over one-fourth of all rodent genera contain at least one species that utilizes rocky substrate. These species do so because of the highly complex nature of rocky habitat, to avoid or detect predators, for the thermoregulatory and water-retaining properties of rocks, to avoid habitat damage by fires and grazers, to take advantage of unique microenvironments within rocky habitat, and to obtain year-round resources.

Approximately seventy-seven species of rodents can be considered to be obligate rock-specialists. Although very little is known about the sociobiology of most saxatile rodents, documented social behavior ranges from highly colonial to solitary, with the majority of species tending to live in social groups. Several factors undoubtedly influence the type of social behavior exhibited by each petrophilic rodent including shared ancestry, the need for extended parental care in harsh environments, predation pressure, and the limited distribution of favorable rocky habitat. The special features of rocky habitat provide a unique opportunity to study the selective pressures influencing the social behavior of rock-dwelling mammals. Hopefully the very limited but suggestive information currently known about saxatile rodents will serve to motivate future studies of the relationship between ecological constraints and social evolution in this select group of mammals.

Chapter 36 African Mole-Rats: Social and Ecological Diversity

Chris G. Faulkes and Nigel C. Bennett

THE FAMILY BATHYERGIDAE (African mole-rats) are subterranean rodents that range in sociality from species that are strictly solitary to the naked mole-rat, arguably the most social of all rodents. This chapter will review our current understanding of social evolution in the Bathyergidae within the framework of phylogeny, biogeography, and speciation. Beginning first with a brief review of their palaeontology, we will use this in combination with molecular phylogenetic data to build a picture of the adaptive radiation of the family. With reference to both inter- and intraspecific comparisons, we will explore the evidence for a relationship between sociality, and ecological constraints, focusing in particular on the possible role of habitat aridity and food distribution.

Species Diversity in the Bathyergidae

Species descriptions and distributions

Despite their cryptic nature, many bathyergid holotypes were named and described by naturalists in the early days of African exploration. Summarizing these studies, Ellerman (1940) listed a total of sixty-two species in five genera, as follows: *Heterocephalus* ($n = 4$); *Heliophobius* ($n = 8$); *Georychus* ($n = 3$); *Bathyergus* ($n = 3$); *Cryptomys* ($n = 49$). Recent molecular phylogenetic studies of both nuclear and mitochondrial genes have produced congruent trees that support the high species diversity of these early descriptions (e.g., Allard and Honeycutt 1992; Faulkes, Bennett et al. 1997, Faulkes et al. 2004; Huchon et al. 1999; Walton et al.

2000; Ingram et al. 2004). Figure 36.1 gives approximate ranges for the five extant genera of the Bathyergidae.

The solitary mole-rats (*Heliophobius, Bathyergus,* and *Georychus*) are generally restricted to regions of higher precipitation (greater than 400 mm per annum). Of these, *Heliophobius* has the widest distributional range; it occurs in the sandy soils of savannas and woodlands of southern Kenya, throughout Tanzania and parts of southeastern Zaire, through Malawi to central Mozambique (fig. 36.1). These areas are characterized by a higher annual rainfall, which on average exceeds 900 mm. The other two genera are much more restricted in their ranges. *Bathyergus* is endemic to some coastal areas of South Africa and southern Namibia, whereas *Georychus* occurs in disjunct populations within South Africa (fig. 36.1).

In contrast, the two social genera (*Cryptomys* and *Heterocephalus*) are found in both mesic and xeric regions. Extant populations of *Heterocephalus* are found in the arid regions of East Africa from the Rift Valley of Ethiopia eastward into the north of Somalia, and from east of Lake Turkana in Kenya eastward to the coast of Somalia, and south as far as Tsavo National Park in Kenya. These areas are characterized by low (less than 400 mm per annum) and unpredictable rainfall, with on average only 4 months per year having more than 25 mm of rain. This rainfall is approximately the quantity required to soften the soil at the depth of foraging tunnels and facilitate burrowing (Jarvis et al. 1994). The *Cryptomys* genus is the most speciose and widely distributed of all the extant bathyergids. Molecular and morphological data strongly suggest that the genus should be considered as two distinct lineages, the *Crypto-*

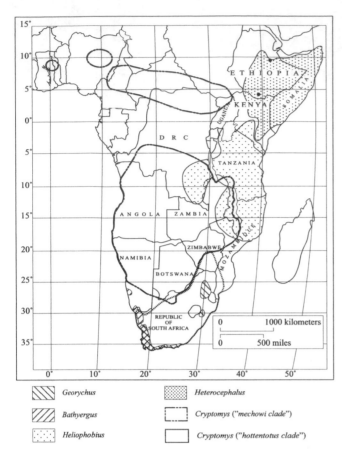

Georychus

Bathyergus

Heliophobius

Heterocephalus

Cryptomys ("mechowi clade")

Cryptomys ("hottentotus clade")

Figure 36.1 Map showing the approximate ranges of extant genera, separating the genus *Cryptomys* into the two subclades suggested by molecular phylogenies (adapted from Bennett and Faulkes 2000).

mys mechowi clade and the *Cryptomys hottentotus* clade (fig. 36.2) or even two genera (*Coetomys* and *Cryptomys*; Ingram et al. 2004). While the latter is distributed throughout South Africa, extending into part of Mozambique and Zimbabwe, species in the *Cryptomys mechowi* clade are found in southern, central, and western Africa, but are absent from the horn of Africa, tropical rainforests of Central and West Africa, and the Sahara (fig. 36.1). As with *Heterocephalus,* the ranges of some of these social cryptomids extend into areas of very low rainfall, which is sporadic and unpredictable (sometimes < 200 mm per annum). However, some species also occur in mesic areas, like the common mole-rat, *Cryptomys hottentotus hottentotus.* The relationships between rainfall patterns, food distribution, and sociality have received much discussion (e.g., Jarvis et al. 1994; Faulkes et al. 1997; Bennett and Faulkes 2000; Burda et al. 2000), and will be reviewed in the following.

The bathyergid fossil record

Using a molecular clock approach inferred from DNA sequence analysis, an ancient (Eocene) origin for the Bathyer-

gidae has been consistently suggested (see the following). Unfortunately, fossil-bearing strata of the Eocene/Oligocene are extremely rare in Africa (Lavocat 1978), so if these molecular timings are correct, from a palaeontological point of view, this earliest stage of bathyergid evolution may remain shrouded in mystery. Fossil bathyergids first appear in the more common early Miocene deposits, and three extinct genera have been identified from this period (around 20 million years ago) in both East Africa and Namibia (Lavocat 1973, 1978). Of the extant genera, *Heterocephalus* is the first to appear in the fossil record. The earliest (but unnamed) fossil resembling *H. glaber* was discovered in Early Miocene deposits at Napak in Uganda, together with fossil Bathyergoidea, a sister taxon to the Bathergidae (Bishop 1962). Bishop et al. (1969) subsequently dated these strata to a minimum age of 17.8 million years (myr). The temporal association of *Heterocephalus* fossils with extinct bathyergid ancestors supports the early divergence of *Heterocephalus* within the family that is suggested by molecular phylogenies.

Phylogeography and patterns of speciation

The availability of robust molecular phylogenies and the application of molecular clocks to estimate divergence times now enable us to infer the potential phylogeographic influences that underlie the adaptive radiation of the Bathyergidae and patterns of social evolution. Molecular phylogenies of the Bathyergidae are firmly rooted in East Africa (*Heterocephalus* and *Heliophobius* forming the basal lineages; fig. 36.2). It has been suggested (Honeycutt et al. 1991) that a possible route for the spread of the Bathyergidae was via a corridor of fluctuating aridity linking east and south West Africa (Bakker 1967). This arid corridor has been implicated in the distribution of both modern and Early Miocene fossil faunas (Van Couvering and Van Couvering 1976). Thus arid- or mesic-adapted ancestral bathyergids could have exploited this corridor, according to the prevailing climatic conditions. At the same time, large-scale physical and climatic changes were also occurring in this part of Africa as a result of continental drift, leading to rifting and increased volcanism. Apart from its possible role as a physical barrier, in the form of volcanic uplands and deep valleys (some eventually forming the great lakes of Africa), it is also likely that the climatic and vegetative changes that indirectly resulted from the rifting process have been of importance in the distribution and social evolution of the Bathyergidae.

Molecular clock-based timing of the major divergences within the family suggest that the initial cladogenesis of the Bathyergidae was sufficiently early to be independent of rifting, and that a general radiation occurred from East Africa into southern and south-central Africa. Divergence of

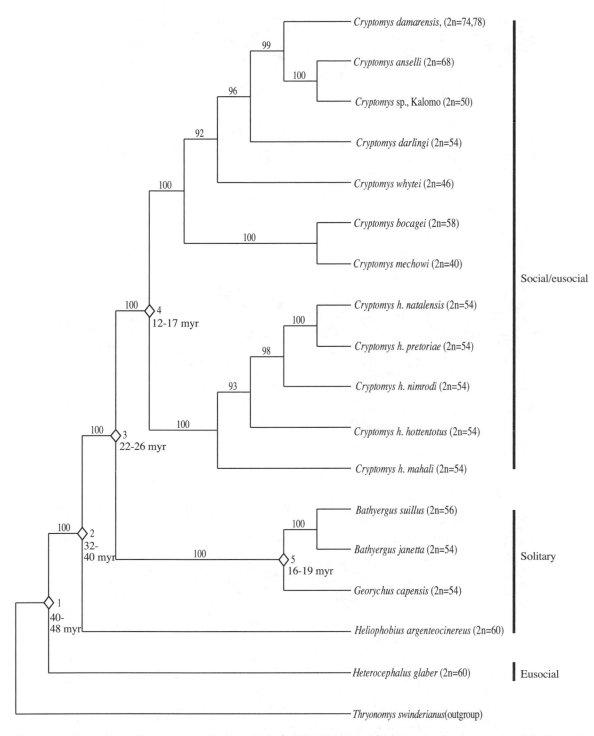

Figure 36-2 Phylogram based on maximum parsimony analysis of 18 bathyergid mtDNA haplotypes (combined 12S rRNA and cytochrome *b* gene sequences) and outgroup species *Thryonomys swinderianus* (cane rat). Numbers above each branch refer to the percent bootstrap values following 100 replications, after weighting sites with the rescaled consistency index. Divergence times of selected internal nodes (numbered 1–5 for reference in text) are in million years before present (myr; data adapted from Faulkes et al. 1997a, 2004).

the basal lineages in the Bathyergidae, *Heterocephalus* (40–48 myr), and *Heliophobius* (32–40 myr), may have been constrained northward by the Ethiopian highlands but was relatively unimpeded south and west (fig. 36.2, nodes 1 and 2; fig. 36.3a to 36.3c; Faulkes et al. 2004). Estimates of the divergence of *Georychus/Bathyergus* from their common

ancestor with *Cryptomys* (20–26 myr; fig. 36.2, node 3; fig. 36.3c) coincide with the beginning of volcanism in the Kenya rift, and possibly favored the expansion into southern Africa rather than to the north and west. Aridification of the Saharo-Arabian belt was also beginning at this time and would have further restricted the Bathyergidae to sub-

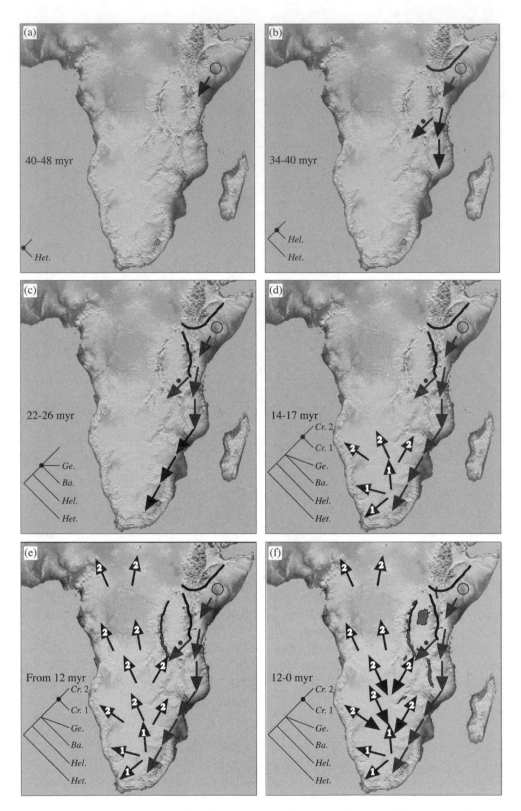

Figure 36.3 Phylogeographic trends in the Bathyergidae inferred from analysis of mitochondrial 12S rRNA and cytochrome *b* sequence differences. (a) Initial divergence of the *Heterocephalus* (*Het.*) lineage from the common ancestor of the family in East Africa; (b) Radiation of *Heliophobius* (*Hel.*) and movement from East Africa into Southern Africa, with some populations crossing the Rift Valley (*); (c) *Bathyergus* (*Ba.*) and *Georychus* (*Ge.*) lineages diverge in South Africa; (d) *Cryptomys* (*Cr.*) diverges into two clades, one radiating predominantly in South Africa (1), and the other spreading north into Southern and Central Africa (2); (e) Formation of Lake Tanganyika (f) Later diversification of *Cryptomys* clade 2 in south-central Africa. (Data adapted from Faulkes et al. 2004).

Saharan Africa. The first fossil bathyergids are known from this period (Early Miocene) in both East Africa and Namibia, confirming that southern Africa was also being colonized at this time. During the later Early and Middle Miocene the *Cryptomys* genus diverged into its two subclades (at 12–17 myr; fig. 36.2, node 4; fig. 36.3d, 36.3e). Again, this coincides with a critical period when rifting was progressing in the Kenya rift and also just beginning in the Western rift. While the *C. hottentotus* subclade appears to have speciated almost exclusively in South Africa, the *C. mechowi* subclade underwent a more extensive radiation, particularly in Zambia and Central Africa, resulting in a wide diversity of genetically divergent chromosomal forms. All extant species in these clades are social, cooperative breeders. Interestingly, the initial radiation of *Cryptomys* is coincident with the onset of volcanism in the adjacent Western rift, which presumably resulted in considerable environmental challenges as climate and vegetation changed. It is possible that in these circumstances, social elaboration could have been adaptive. This period has previously been reported as a time of faunal turnover (Van Couvering and Van Couvering 1976). The increasing volcanism and formation of the East Africa rift during the Miocene appears to have almost completely isolated the populations of *Heterocephalus* and *Heliophobius* to the east, and restricted *Cryptomys* to the west of the Rift. Exceptions to this are a few populations of *Cryptomys whytei* in western Tanzania, and *Heliophobius* found in Malawi and Zambia. The latter appear to have diverged before local rifting restricted their movement (fig. 36.2, node 5; fig. 36.3f; Faulkes et al. 2004).

Social Diversity in the Bathyergidae

Despite many early species descriptions, current interest in the Bathyergidae began to increase in the 1980s, after the eusocial system of the naked mole-rat was first reported (Jarvis 1981). While it is now clear that there is wide social diversity between genera across the family (solitary-dwelling genera, *Heliophobius*, *Georychus*, and *Bathyergus*, versus social genera, *Heterocephalus* and *Cryptomys*), there may also be considerable variation within the social genus *Cryptomys*. Robust molecular phylogenies now make comparative analyses possible, enabling the evolution of social traits to be correlated with environmental factors (e.g., Faulkes et al. 1997). However, fundamental difficulties remain in quantifying both social behavior and ecological parameters.

Definitions of sociality

Ironically, a major problem with the quantitative study of social evolution is defining what is meant by sociality.

The term "eusocial" was first used to describe groups of insects living in close-knit communities but which have a reproductive division of labor. By definition, only a small number of individuals are actually involved in direct reproduction (reproductive skew). The remainder of the social group is composed of overlapping generations of functionally or irreversibly sterile nonbreeding helpers that cooperate in the rearing of offspring (Michener 1969; Wilson 1971).

Rigid, categorical definitions, such as those of Michener (1969) and Wilson (1971), which have been used to describe eusocial species, are less useful for quantitative comparative studies than a continuous measure of social complexity. In comparative studies of mole-rats, Faulkes et al. (1997) used maximum group size as an indication of sociality and constraint on dispersal. The use of group size as a criterion was criticized by Burda et al. 2000, who argued that eusocial mole-rats should be defined by "permanent philopatry." However, while all social species are philopatric to some degree, not even the most social of the mole-rats are permanently so, e.g., *H. glaber* (Braude 2000) and *C. damarensis* (Hazell et al. 2000; Burland et al. 2002, 2004). It is generally accepted that some kind of continuous measure of the type proposed by Sherman et al. (1995) and Keller and Perrin (1995) will prove to be the most useful estimate of the degree of sociality. These indexes of eusociality and reproductive skew are difficult to quantify empirically because estimates of relatedness and lifetime reproductive success are needed. However, studies of group kin-structure, parentage, and dispersal using molecular genetic markers are now beginning to gather these kinds of data, and show great promise for the future (e.g., Bishop et al. 2004; Burland et al. 2002, 2004; Griffin et al. 2003).

The common ancestor of mole-rats: Was it solitary or social?

The question of the social status of the ancestral bathyergid is an interesting and important one, but difficult to answer. Because few other subterranean mammals are social (Nevo 1979), Jarvis and Bennett (1990) have suggested that the first bathyergids were probably solitary. The earliest known fossil Bathyergoidea (a sister taxon to the Bathyergidae) were large animals (Lavocat 1978) and the largest living bathyergids are solitary (Jarvis and Bennett 1990). By inference this might suggest that if these large fossil forms were completely subterranean they were also solitary. Recent studies suggest an arid, savanna-type habitat at the fossil sites for naked mole-rats, e.g., Acacia savanna at Laetoli (Denys 1987), and assemblages similar to Rushinga and Songhor (i.e., savanna) at Napak I (Bishop 1962). Conversely, the presence of *Heterocephalus jaegeri* at Olduvai is used to infer aridity of the habitat (Denys 1989). However, there is no a priori reason to suppose the ancestor was solitary, as the social traits of a common ancestor could equally have been lost. This so-called secondary solitarity has been reported in some species of bees (Wcislo and Danforth 1997).

It is impossible to verify the status of the common ancestor directly. In the absence of such data, clues may be sought in the close hystricomorph relatives of the Bathyergidae. There is evidence that some of the South American caviomorphs, like tuco-tucos (family Ctenomyidae), exhibit some form of social grouping (Lacey and Sherman, chap. 21 this volume). However, these New World hystricomorphs are divergent from the Old World families, and according to the molecular phylogeny of Nedbal et al. (1994) the closest relatives to the Bathyergidae are Old World porcupines (family Hystricidae), cane rats (family Thryonomyidae), and dassie rats (family Petromuridae). At least one species in the Hystricidae, the Cape porcupine (*Hystrix africaeaustralis*), has colonial habits. Cape porcupines live in colonies of six to eight individuals, consisting of an adult pair and consecutive litters of offspring, which are normally singletons, occasionally twins, or rarely triplets. Adult males protect the young, are aggressive towards foreign males and females, and accompany the young on foraging trips until 6–7 months of age, after which they tend to become solitary feeders (Van Aarde 1987). However, in the cane rats, both the greater cane rat (*Thryonomys swinderianus*) and the lesser cane rat (*T. gregorianus*) are generally reported to be solitary, although individuals may live in close proximity in reed beds (Skinner and Smithers 1990). Among the dassie rats, *Petromus typicus* is reported to live in pairs or families in the crevices that occur in their rocky habitat, although information is limited (Skinner and Smithers 1990). However, none of the species in these three families are subterranean, so it remains difficult to extrapolate to make any definite inferences about the ancestral bathyergid.

If we assume that the ancestral condition of the Bathyergidae was social (node 1 in fig. 36.2), Burda et al. (2000) suggest that the appropriate questions about the evolution of eusociality in mole-rats should be: why did certain species of the bathyergid family become solitary, and why have *Cryptomys* and *Heterocephalus* not abandoned their social way of life? While this approach might be of interest from the point of view of phylogeny, it has little or no effect on the hypotheses that address convergence of lifestyles within the family. All phylogenies published to date (Allard and Honeycutt 1992; Faulkes et al. 1997; Walton et al. 2000; Faulkes et al. 2004; Ingram et al. 2004) show that naked mole-rats and the *Cryptomys* genus are separated by a number of common ancestors leading to lineages of solitary species, indicating repeated losses and gains of

varying degrees of social elaboration (e.g., nodes 2 and 3 in fig. 36.2).

Social evolution: Ultimate factors

There are two principal ways in which societies may form. The parasocial route involves nonoverlapping reproductive generations, "shared nests" composed of groups of related, or related and unrelated individuals. The second principal way in which societies may form, and the one generally agreed as the most likely precursor of cooperatively breeding/eusocial vertebrates, is the so-called subsocial route. In this case, groups of overlapping generations arise as a result of natal philopatry, where offspring delay dispersal and remain to form a family group. Thus a family, as defined by Emlen (1995, 1997) is a social group where offspring continue to interact with their parents into adulthood; that is, beyond the age of sexual maturity. Emlen (1995, 1997) further differentiates families into simple or extended. In simple families only one pair breeds, a situation akin to the social bathyergids, although with time immigration of nonfamily members into the group may also occur through dispersal (Jarvis and Bennett 1993; Jarvis et al. 1994; O'Riain et al. 1996; Bishop et al. 2004; Burland et al. 2004). In extended families, more than one individual of either one or both sexes breed.

There has been much debate in the literature regarding the ultimate causes of natal philopatry and sociality in the Bathyergidae (for review see Bennett and Faulkes 2000; Burda et al. 2000). Irrespective of where, or even if, the line between eusociality and other forms of sociality is eventually drawn, most authors concur that the evolutionary origins of eusociality should no longer be considered to lie in intrinsic or genetic factors (e.g., haplodiploidy). Thus while haplodiploidy might predispose Hymenopteran insects toward eusociality, similar social systems in diploid termites and mammals discount it as a necessary prerequisite for the evolution of eusociality. The ultimate evolutionary factors that have thus far been suggested as important in the evolution of sociality in mammals are (1) predator vigilance and protection, e.g., meerkats (Clutton-Brock et al. 1998), dwarf mongoose (Rasa 1977), naked mole-rats (Alexander et al. 1991), and (2) increased efficiency in procuring food, e.g., naked mole-rats (Jarvis and Sale 1971), Damaraland mole-rats (Jarvis and Bennett 1993), wild dog (Frame et al. 1979), and wolves (Zimen 1976). Several authors have proposed that the distribution, size, and digestibility of the geophytes (underground roots and tubers) upon which mole-rats feed, as well as the variation and predictability of rainfall, have played a pivotal role in the evolution of sociality in the Bathyergidae (Jarvis 1978; Bennett 1988; Lovegrove and Wissel 1988; Lovegrove 1991; Jarvis et al. 1994; Faulkes et al. 1997). This model has become known as the *aridity-food-distribution hypothesis* (AFDH), and contrasts markedly with the proposition put forward by Randall (chap. 31 this volume), in which scattered food resources appear to restrict group formation in semifossorial desert rodents.

The AFDH proposes that cooperative breeding and eusociality in African mole-rats evolved as an adaptive response to a combination of the patterns of rainfall and the distribution of food and the subsequent costs and risks of foraging and dispersal/independent reproduction. Mole-rats generally burrow when the substrate is softened by rain, and in *H. glaber* and *C. damarensis* heavy rainfall triggers a frenzy of digging activity (Brett 1991; Jarvis et al. 1998). Thus when rainfall is both low and unpredictable, the opportunities for extending the burrow to search for food and/or to disperse and form new colonies are limited. Furthermore, in such habitats the plants are arid adapted. The consequence of this is that they tend to produce swollen roots and tubers to store their reserves, and this makes an excellent food and moisture resource for mole-rats. Such plants reproduce vegetatively and therefore tend to occur in patchily distributed and/or widely dispersed clumps compared to those in mesic regions. This increased dispersion of food increases the risk of unsuccessful foraging when individuals are blindly excavating energetically costly foraging burrows. It is suggested that this cost can be offset by the cooperative foraging seen in social mole-rats. Clearly these high costs, but potentially large benefits, are very different from the challenges facing a surface-foraging desert rodent. Many such semifossorial rodents are folivores and/or granivores and are thus exploiting food resources that are also often scattered but of low quality, and hence can only support solitary foragers. Physiological constraints imposed by surface activity are also different and more variable than the highly stable environment of the subterranean niche of mole-rats. The need to meet these different physiological demands may also lead to divergence in social behavior in semifossorial versus subterranean rodents (see Randall, chap. 31 this volume). One consequence of high diurnal surface temperatures in deserts is that many mammals forage at night, resulting in a reduction in visual cues of food, although other sensory modalities remain unimpaired. Again, this contrasts with the subterranean niche, where sensory cues to food resources are presumably severely limited or nonexistent, leading to largely blind foraging. In a study of Damaraland mole-rats, Jarvis et al. (1998) have shown that initial foraging is indeed done blindly, as tunnels often missed geophyte-rich areas, burrowing past them as little as one meter away. Conversely, Heth et al. (2002) argue that

chemosignals from plants (known as kairomones) may attract burrowing mole-rats. They have shown that in T-maze choice tests performed in captivity, *Cryptomys anselli* and *Heterocephalus glaber* preferred to dig in soil in which food plants had previously been growing. Such a preference suggests that the animals were attracted by kairomones from the plants that were contained within the soil. In the wild it seems likely that any such effects act over a short range and possibly after rains have washed chemosignals from the plants into the surrounding soil. In accordance with the hypothesis that social species are able to forage more efficiently than solitary species, Le Comber et al. (2002) measured the fractal dimension of burrows to quantify the complexity of their architecture and found evidence that the burrows of social mole-rats explore their surrounding area more thoroughly.

Further evidence in support of the AFDH comes from comparative studies across the family, which showed convergence in sociality among unrelated taxa in similar habitats. Specifically, three ecological variables were significantly correlated with social group size: geophyte density, the mean number of months per year that rainfall was greater than 25 mm (the quantity that is estimated as sufficient to penetrate the ground enough to stimulate burrowing), and the coefficient of rainfall variation (Faulkes et al. 1997). Comparative analysis of this kind makes independent contrasts with no a priori assumption of the character state of the common ancestor. Burda et al. (2000) suggest that if the AFDH is supported we should see evidence of divergence in social structure among related taxa in different habitats, as well as convergence in behavior among unrelated, divergent taxa. Spinks et al. (1998, 2000) have indeed shown that within a single species (*Cryptomys h. hottentotus*) the social structure varied along an aridity gradient according to the predictions of the AFDH. Colonies of this species that occurred in more arid regions had a social structure more similar to that of the eusocial species: there was a greater degree of philopatry, and increased reproductive skew and overlap in generations. In a different study of *C. damarensis*, a link between the environment and social traits in keeping with the AFDH was also apparent. In this case, the average value for within-colony relatedness between individuals was found to be greater in more arid regions when compared to a region of higher rainfall (Burland et al. 2002). Such an observation fits with the hypothesis that philopatry is increased when ecological constraints on dispersal are higher. When the costs of dispersal are constantly high, offspring will benefit from staying at home until ecological conditions improve and the benefit to cost ratio is favorable for dispersal (see Nunes, chap. 13 this volume). It follows that conditions that allow for reproduction but limit dispersal will result in a trend toward larger group size, provided other variables (e.g., predation) are fairly constant. Further, in areas where the food density is low or of low energetic value the average body size within a given species will be reduced. Evidence for the latter has been shown in the naked mole-rat (Jarvis 1985), where the mean size of nonbreeders was smaller compared to those for non-breeders in colonies living in areas where food resources were more abundant. Spinks (1998) reports a similar trend in *Cryptomys h. hottentotus,* where individuals in arid regions were smaller than those from mesic parts of their distribution. These observations support the utility of a combination of group size, reproductive skew, and a measure of philopatry in defining the degree of sociality in African mole-rats.

The occurrence of solitary dwelling mole-rats in arid regions would argue against the AFDH, which is the case for *Bathyergus janetta,* a species that ranges into areas of very low rainfall in the northwestern Cape region of South Africa and parts of southwestern Namibia. However, this apparent contradiction to the AFDH is an oversimplification, as the AFDH depends on a combination of aridity and food distribution. Not only is *B. janetta* often found in association with subterranean moisture seepage areas, which mitigate against the low rainfall, but due to the unique flora of Namaqualand, with many plants with swollen roots, subterranean stems and tubers, food resources are abundant in these otherwise very arid regions (Herbst 2002). Recently, a study of *H. argenteocinereus* in Malawi has suggested that this solitary species is also successfully exploiting a region of low geophyte density (Sumbera et al. 2003). While the reported mean geophyte density was very low (2.5 geophytes/m²) and comparable with areas inhabited by naked mole-rats, the mean geophyte biomass was relatively high at 192 g/m², and falls within the range of other solitary mole-rats (e.g., *G. capensis:* 120-600 geophytes/m²; Du Toit et al. 1985; Bennett 1988). Following on, it has also been argued that social or eusocial species of mole-rats should not occur in mesic areas (e.g., *C. mechowi;* Burda and Kawalika 1993). However, the AFDH does not propose that social species cannot occur in mesic areas, and the limited data available suggest that in such areas social group dynamics fit the predictions of the AFDH (e.g., *C. h. hottentotus* in mesic versus arid habitats; Spinks et al. 1998, 2000). Nevertheless, unambiguous data on group sizes, kin structure, and patterns of dispersal/philopatry are needed, especially on the diverse populations of social *Cryptomys* in Zambia that inhabit regions of high rainfall.

Burda et al. (2000) argued against the AFDH, and suggested that the ultimate evolutionary reason for eusociality in mole-rats is that it is a phylogenetically constrained phenomenon resulting from a social common ancestor rather than a response to ecological conditions, and that a solitary lifestyle is the derived trait in some species of Bathyergidae. While this is an interesting point to consider, this proposal merely sidesteps the issue of what selective factors ultimately may have given rise to natal philopatry and cooperation in ancestral bathyergid species in the first place. It also ignores the comparative approach, whereby phylogeny can be used to make valid independent contrasts between sociality and ecological factors irrespective of the status of the common ancestor (e.g., Faulkes et al. 1997).

Other factors that have been suggested as causative in the social evolution of some mole-rats include their apparent inability to store fat and their increased postnatal development rates. It is suggested that these lead to an inability of the breeding female to rear young on her own, hence the retention of helpers to provision food during the long periods of gestation and lactation (Burda 1990; Burda et al. 2000). In the absence of a comprehensive study on fat stores in mole-rats, the relevance of fat storage ability to the evolution of sociality remains unclear. However, field observations of autopsied, freshly captured breeding and nonbreeding females of *C. damarensis, C. darlingi, C. h. hottentotus,* and *H. glaber* show clear evidence of stored fat (Bennett and Jarvis, unpublished data). Lack of an ability to store fat might predispose such species to the retention of offspring, but it does not explain why such a social system evolved in the first place. Females that cannot store fat but that can nevertheless maintain a positive energy balance throughout reproduction because of ample food supplies need not retain either their mates or their offspring. In contrast, females that live, for example, in an arid region where food resources are difficult to obtain may be under selection to retain their partners and their offspring to ensure continued mating and to lower the costs and risks with obtaining such food. An ability to store fat might be a proximate factor favoring the evolution of sociality, but seems unlikely as an ultimate evolutionary factor.

Burda et al. (2000) also argue that postnatal developmental rates are a cause rather than a consequence of eusociality in naked and Damaraland mole-rats, although there is no consistent evidence for this being a causative factor. Developmental length differs dramatically in naked and Damaraland mole-rats, both of which are considered to be eusocial. Naked mole-rats are comparatively altricial at birth and progress to solid foods at approximately 2 weeks of age, and begin to leave the nest voluntarily at 3 weeks. In contrast, Damaraland mole-rats are relatively precocious, and may eat solids within a few days after birth and explore the burrow system as early as days one to five. Developmental length may well be phylogenetically constrained in the Hystricognaths, but the variance within the family Bathyergidae does not appear to correlate with the degree of sociality (see Faulkes et al. 1997).

Social evolution: Proximate factors

The maintenance of highly social behavior in the Bathyergidae has also received much attention, and was again initiated in studies of the naked mole-rat. An increasing body of work is now revealing that the proximate mechanisms underlying the characteristic reproductive division of labor (reproductive skew) differ among mole-rat species. In particular, there is debate about the exact role the breeding female plays in maintaining optimal levels of reproductive skew. Theoretical models of incomplete control by dominants (Clutton-Brock 1998a) and concession theory, involving peace and staying incentives (Keller and Reeve 1994; Reeve et al. 1998) have been proposed to explain both inter- and intraspecific differences in reproductive skew among animals. However, implicit in both models is the notion that the dominant breeder exerts, or attempts to exert, some kind of reproductive control over subordinate nonbreeders in the group. This control could manifest itself in different ways, ranging from infanticide of subordinates' offspring or interference by dominants with subordinates' mating attempts, to actual suppression of subordinates' reproductive physiology (Faulkes and Abbott 1997). If most mole-rats live in extended family groups, incest avoidance alone could be sufficient to maintain reproductive skew, as the only unrelated individuals in the group would be the breeding pair (Jarvis and Bennett 1993; Bennett et al. 1994, 1997; Burda 1995; Cooney and Bennett 2000; Faulkes and Bennett 2001). Almost all animals have evolved mechanisms that prevent them from breeding with close relatives because of the fitness costs of harmful recessive traits and decreased heterozygosity that might be manifest in offspring arising from such mating (Pusey and Wolf 1996).

All cooperatively breeding mole-rats studied to date have unequal reproduction, with a single female normally breeding with one, but on occasion two or three breeding males (Faulkes et al. 1997; Bishop et al. 2004; Burland et al. 2004). However, skew in terms of lifetime reproductive success may differ considerably between species. The social mole-rats studied so far in the genus *Cryptomys* all have a mating system that involves obligate outbreeding, whereas naked mole-rats are unusual within the family and among mammals in general because they may inbreed to a high degree. These incestuous tendencies were originally proposed as an important factor in explaining their eusociality, and that inbreeding produced a within-kin group genetic structure analogous to haplodiploidy in the Hymenoptera, with

intracolony relatedness estimated in some groups at 0.8 (Reeve et al. 1990). This estimate of relatedness is actually greater than the average three-quarter relatedness in haplodiploid organisms where the queen is singly mated (Hamilton 1964).

Although the inclusive fitness benefits of high relatedness to nonbreeding colony members are readily apparent, the puzzle remains that field data from other cooperatively breeding mammals, including other social mole-rat species, follow the typical pattern of incest avoidance. In the Damaraland mole-rat, mean within-colony relatedness has been estimated at 0.46 (Burland et al. 2002), indicating that normal familial levels of relatedness are sufficient for highly social behavior to evolve for a given cost/benefit ratio for such behavior (average relatedness for first order relatives is 0.50). It appears that inbreeding might be a derived trait peculiar to the naked mole-rat, which may have evolved as an adaptive response to the high costs of dispersal. Despite the potential costs of inbreeding, once deleterious recessive traits were purged from a population, inbreeders would have had an advantage over obligate outbreeders, who could potentially suffer fitness costs if opportunities for finding unrelated mates were rare. Without incestuous mating, colonies in which a breeder dies could face extinction before sufficient (unpredictable) rainfall facilitates emigration. The lack of inbreeding avoidance, together with genetic studies of wild-caught colonies, indicates that successful dispersal in naked mole-rats is infrequent (Reeve et al. 1990; Faulkes et al. 1997). More recently, mark-recapture data from field studies have supported the contention that dispersal is highly risky. Twenty out of twenty-one nascent colonies that contained one to four immigrants from nearby colonies went extinct within a year (O'Riain and Braude 2001), despite the fact that other field and laboratory data suggest outbreeding may be the preferred mating system if the opportunity arises (Clarke and Faulkes 1999; Braude 2000; Cisek 2000).

Given that naked mole-rats will quite readily inbreed, the reproductive monopoly of the queen must be maintained by active means, and nonbreeders of both sexes within colonies are physiologically suppressed by the queen. In females, ovarian cyclicity and ovulation are blocked, whereas in males most nonbreeders have spermatozoa within the reproductive tract that are both reduced in number and lack normal levels of motility. In both sexes, reduced secretion of the pituitary gonadotrophin luteinizing hormone is evident. These extreme (but reversible) reproductive blocks are suggested to be brought about by physical threats and aggression from the dominant breeding queen (Faulkes and Abbott 1993; 1997). Unlike the facultative inbreeding of the naked mole-rat, the eusocial Damaraland mole-rat is an obligate outbreeder. In this species, a clear physiological block to reproduction is observed only in nonbreeding females (Bennett et al. 1996; Molteno and Bennett 2000). As with the naked mole-rat, ovulation is blocked, although possibly through a different physiological mechanism, at the level of the ovary (Bennett et al. 1999). Nonbreeding males are not physiologically suppressed in the same way as naked mole-rats; however, they do possess increased proportions of immature sperm (Maswanganye et al. 1999). Although the significance of these sperm abnormalities for fertility is unclear, nonbreeding males make no attempt to mate with their female colony mates, presumably as a result of an incest avoidance mechanism, as they are usually close kin. Both wild and captive colonies in which the breeding female has died will remain reproductively quiescent (sometimes for years) until a foreign, unrelated, individual becomes available or dispersal/fragmentation of the colony occurs (Jarvis and Bennett 1993, Rickard and Bennett, 1997). However, new genetic studies indicate that in the wild, conditions exist in which nonreproductive females may come into contact with unrelated males, even when they do not disperse from their natal colony. Multiple and unidentified paternity was found to be widespread, and immigrants of both sexes were identified. In addition, unrelated opposite-sex nonbreeders were identified in two colonies out of the eighteen studied, yet all the Damaraland mole-rat colonies studied to date contained a single breeding female (Burland et al. 2004). Incest avoidance alone is therefore insufficient to maintain the high levels of reproductive skew identified in this species, supporting the observation of a physiological block among females.

Other social species in the genus *Cryptomys* in which there is a single breeding female appear to lack a physiological block to reproduction in either sex; e.g., the Mashona mole-rat, (*Cryptomys darlingi;* Bennett et al. 1997), the Giant mole-rat (*Cryptomys mechowi;* Bennett et al. 2000), Ansell's mole-rat, (*Cryptomys anselli;* Burda 1995), and the common mole-rat (*Cryptomys h. hottentotus;* Spinks et al. 2000). The kin structure of colonies in these species is assumed to be predominantly the breeding pair and their offspring. In such cases, maintenance of reproductive skew could be achieved by incest avoidance alone. However, detailed genetic data are lacking in all but the common mole-rat, in which parentage analysis has revealed that while only a single breeding female was found per colony, both extra-pair and extra-colony paternity were common (Bishop et al. 2004). The presence of both adult and subadult foreign conspecifics within colonies means that inbreeding avoidance may not be sufficient to maintain reproductive skew, and that in the absence of physiological suppression, other factors now need to be considered.

Clearly, the interplay between many factors has potentially influenced both phenotype and social evolution in the Bathyergidae. Figure 36.4 displays these in a hierarchical flow chart, which attempts to link cause and effect among these factors. While sociality can be seen as an adaptation to arid environments in subterranean bathyergids, the way in which ecological constraints may influence social group formation may vary markedly among different niches in a given environment. For example, aridity has been shown to favor a solitary lifestyle in surface-active desert rodents like kangaroo rats (Randall, chap. 31 this volume).

Seasonal aspects of reproduction

In some species of mole-rat, environmental cues as well as social factors play a role in regulating reproductive and life-history traits. Many of the cues available to terrestrial animals normally used by seasonally breeding organisms (e.g., changing annual photoperiod) may be precluded, as African mole-rats rarely come to the surface. Thus the onset of breeding may be triggered by other physical cues, such as changes in temperature, changes in moisture content, or the associated sudden flush of vegetation associated with abundant precipitation (Bennett et al. 1988).

All of the solitary-dwelling mole-rats of southern African breed seasonally. While many of the social species of bathyergids reproduce throughout the year (Bennett et al. 1991), two exceptions to this are known to occur. The common mole-rat inhabits a winter rainfall region and rears young during the southern hemisphere summer (late November to January; Spinks et al. 1997, 1999). In contrast, the highveld mole-rat (*Cryptomys hottentotus pretoriae*) occurs in a summer rainfall zone and rears the young in the winter (early June to August; Janse van Rensburg et al. 2002, 2003). One of the challenges that now remains is to link both seasonal and social effects and examine in detail the variation in mating patterns, philopatry, and lifetime reproductive success across the family.

Summary

There is extensive social diversity within the taxonomic diversity of the Bathyergidae, with convergent gains and/or losses of sociality. Inter- and intraspecific studies generally support an ecological constraints model for the ultimate cause of sociality, whereby foraging and dispersal risks, in part resulting from rainfall patterns, lead to natal philopatry, cooperation, and skew in reproduction. Sociality may be maintained at a proximate level by kin selection, but the high relatedness observed in naked mole-rats is not

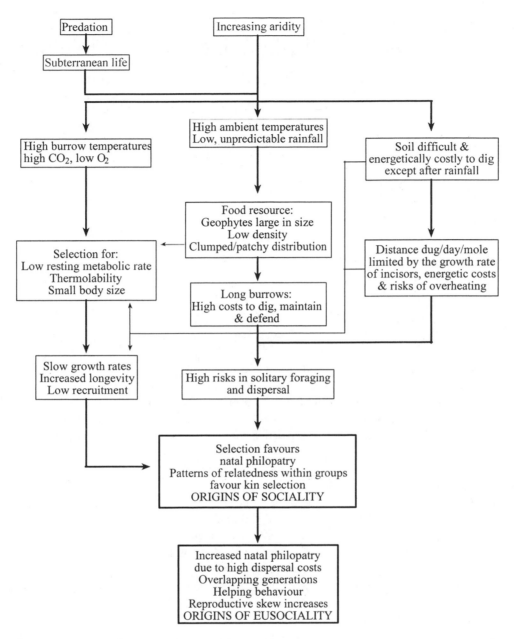

Figure 36.4 Summary of the principal factors, and their interrelationships, thought to be important in the evolution of cooperative breeding and eusociality in the Bathyergidae. (Modified from J. U. M. Jarvis, unpubl.; Brett 1986 and Spinks 1998).

a prerequisite and may be a derived trait. Phylogenetically constrained factors such as long gestation and developmental time may also play a role. Comparative endocrine studies of African mole-rats suggest that the frequency and mode of dispersal and nascent colony formation, which in turn are influenced by ecological factors, may be fundamental in shaping sociality and the proximate mechanisms controlling the reproductive skew so characteristic of mole-rats. New data on group kin structure and parentage high-light the potential diversity in reproductive success among males, and suggest that combinations of dominant control, self-restraint (arising from incest avoidance), and possibly other models of reproductive suppression operate within colonies. The recent characterization of several new social species and the accumulating molecular genetic data offer the opportunity for further, more extensive and detailed comparative studies within the family.

Chapter 37 Alarm Calling, Multiple Mating, and Infanticide among Black-Tailed, Gunnison's, and Utah Prairie Dogs

John L. Hoogland

Aᴠᴛᴇʀ ᴅᴇᴛᴇᴄᴛɪɴɢ ᴀ ᴘʀᴇᴅᴀᴛᴏʀ, individuals of numerous avian and mammalian species give alarm calls that warn conspecifics. But why should an individual incur the cost of attracting a predator's attention to itself by vocalizing (table 37.1)? This question has puzzled evolutionary and behavioral ecologists for > 50 years (see reviews in Sherman 1977, 1985; Blumstein, chap. 27 this volume). Breakthroughs came when Christopher Dunford (1977a) and Paul Sherman (1977, 1980a) discovered that alarm calls of round-tailed (*Spermophilus tereticaudus*) and Belding's ground squirrels (*S. beldingi*), respectively, function primarily to warn kin, especially juvenile offspring. Similar results soon appeared for numerous other ground-dwelling squirrels (Smith 1978; Yahner 1978b; Noyes and Holmes 1979; Schwagmeyer 1980; Davis 1984a; Owings et al. 1986; MacWhirter 1992). To explain alarm calling, some behavioral ecologists (e.g., Shields 1980; Blumstein et al. 1997) have focused on the importance of direct selection versus indirect selection (Brown 1987)—i.e., calling to warn offspring and grandoffspring versus calling to warn nondescendent kin. Other behavioral ecologists (e.g., Sherman 1980b; Hauber and Sherman 1998; see also Dawkins 1979) have de-emphasized this dichotomy. Either way, we can all agree that observing marked individuals of known genealogies is pivotal for a better understanding of the adaptive significance of alarm calling. In this chapter I summarize information on alarm calling for three species of prairie dogs. As expected, kinship is once again a big part of the story, but I discuss two other factors that also affect alarm calling: demography and obesity. Demography is an ultimate factor that also has been important in the evolution

of alarm calling among Belding's ground squirrels (Sherman 1977, 1980a). Obesity is a proximate factor, whose relevance to alarm calling has not been previously considered.

Male reproductive success in most species varies directly with the number of inseminations (Trivers 1972; Emlen and Oring 1977), and multiple mating by males is widespread. Sperm numbers from a single insemination are usually sufficient to fertilize all a female's ova (Bateman 1948; Petrie et al. 1992; Birkhead 2000), however, so the observation that females of so many species routinely mate with several males is puzzling (Birkhead and Moller 1992; Kano 1992; Solomon and Keane, chap. 4 this volume). Multiple mating by females entails both costs and benefits (table 37.1). Evaluating these costs and benefits is difficult because it requires information on both the copulatory and rearing success for the same females. In this chapter I discuss two factors that affect the frequency of multiple mating by female prairie dogs: the ability of males to monopolize females, and the effect of multiple copulations on litter size.

Nonparental infanticide—i.e., the killing of a conspecific's juvenile offspring—is one of the most intriguing, controversial, and misunderstood issues in behavioral ecology and population biology (Hrdy 1979; Sherman 1981b; Hausfater and Hrdy 1984; Ebensberger 1998d; Ebensberger and Blumstein, chap. 23 this volume). Infanticide involves both costs and benefits (table 37.1), and seems to be especially common and observable among rodents such as gray-tailed voles (*Microtus canicaudus*; Wolff et al. 2002), ground-dwelling squirrels (Sherman 1981b; Hoogland 1985), Mongolian gerbils (*Meriones unguiculatus*; Elwood and Ostermeyer 1984b), and house mice (*Mus musculus*; vom Saal

Table 37.1 Possible costs and benefits of alarm calling, multiple mating by females, and nonparental infanticide

Behavior	Possible costs	Possible benefits
Alarm calling	Increased susceptibility to predation	Warning kin; warning others likely to reciprocate with later alarm calls; diversion of predator's attention to other prey; discouragement of pursuit by predator; reduced probability of later attacks by same predator
Multiple mating by females	Increased susceptibility to predation; increased probability of physical harm; increased exposure to diseases and parasites; reduced paternal care, if males do not help to rear offspring of mate that also copulates with other males	Increased sustenance from courtship feeding or nutritious spermatophores; higher probability of conception; reduced harassment from courting males; reduced probability of losing offspring to infanticide; fresh sperm for fertilization; promotion of sperm competition; increased survivorship of offspring resulting from higher genetic diversity within litters via multiple paternity; increased paternal care, if male helps to rear offspring of all females with whom he copulated
Nonparental infanticide	Retaliation by victimized parent(s); killing of nondescendant kin	Removal of future competitors for both killer and killer's offspring; sustenance via cannibalism of victim(s); lower probability of misdirecting parental care; higher reproductive success for male killers if infanticide induces victimized mother to become sexually receptive more quickly

SOURCE: References for alarm calling: Sherman 1977, 1985; Blumstein, chapter 27, this volume. References for multiple mating: Schwagmeyer 1984; Westneat et al. 1990; Eberhard 1996; Newcomer et al. 1999; Wolff and Macdonald 2004. References for infanticide: Hrdy 1979; Sherman 1981b; Hausfater and Hrdy 1984; Hoogland 1985; Ebensberger and Blumstein, chap. 23, this volume.

1984). In this chapter, I discuss interspecific differences among three species of prairie dogs regarding the frequency of infanticide and the sex of killers.

The Study Animals

Prairie dogs (*Cynomys*) are diurnal, colonial, herbivorous, harem-polygynous, burrowing rodents of the squirrel family (Sciuridae), and thus are akin to marmots (*Marmota*), tree squirrels (*Sciurus* and *Tamiasciurus*), flying squirrels (*Glaucomys*), chipmunks (*Tamias*), and ground squirrels (*Spermophilus* and *Ammospermophilus*; Hafner 1984; Harrison et al. 2003). On a typical day, prairie dogs emerge from their burrows at dawn and forage aboveground until dusk. Reasons to submerge temporarily during daylight hours include nest-building, copulation, infanticide, nursing of offspring, and retreat from either predators or severe weather.

Mammalogists currently recognize five species of prairie dogs, which comprise three distinct groups (Hollister 1916; Kelson 1949; Pizzimenti 1975; Hoogland 1995, chap. 40 this volume): black-tailed and Mexican (*C. ludovicianus* and *C. mexicanus*), Gunnison's (*C. gunnisoni*), and Utah and white-tailed (*C. parvidens* and *C. leucurus*). In this chapter I summarize information from one species of all three groups.

Colonies of prairie dogs are subdivided into territorial family groups that typically contain one breeding male and several breeding females. For Gunnison's and Utah prairie dogs, these groups are called *clans* (Fitzgerald and Lechleitner 1974; Wright-Smith 1978; Travis et al. 1995, 1996). For black-tailed prairie dogs, family groups are called *coteries* (King 1955).

Methods

Periods of research

I studied black-tailed prairie dogs at Wind Cave National Park, South Dakota, from February through June of 1975–1988 (Hoogland 1995, 1996a); Gunnison's prairie dogs at Petrified Forest National Park, Arizona, from March through June of 1989–1995 (Hoogland 1998a, 1999); and Utah prairie dogs at Bryce Canyon National Park, Utah, from March through June of 1996–2005 (Hoogland 2001, 2003b).

Aging, livetrapping, eartagging, marking

To capture adult prairie dogs, I use double-door livetraps (15 cm × 15 cm × 60 cm), and for juveniles I use single-door livetraps (13 cm × 13 cm × 40 cm; Tomahawk Livetrap Company, Tomahawk, Wisconsin). For permanent identification of individuals, I insert one fingerling eartag in each ear (National Band and Tag Company, Newport, Kentucky). To identify prairie dogs from a distance, I apply unique markers to the pelage of each individual with Nyanzol-D black dye (Greenville Colorants, Clifton, New

Jersey). Using binoculars and a 60-power telescope, students and I observe marked individuals from 4-meter-high towers.

Because female prairie dogs rear their offspring in separate nursery burrows, maternity is usually easy to establish (Rayor 1988; Hoogland 1997, 1999, 2001). By surrounding nursery burrows with livetraps on the day after juveniles first appear aboveground, when they are about 5.5 weeks old, I capture, eartag, and mark all littermate siblings before they mix with juveniles from other litters.

Juvenile prairie dogs are (1) individuals that have not yet emerged from the natal burrow, or (2) individuals that first emerged from the natal burrow < 8 months ago (Hoogland 1995). Adults (including yearlings) are individuals that first emerged from the natal burrow > 8 months ago.

For each species of prairie dog that I have studied, almost all my information has come from a single colony. Consequently, I have been able to collect detailed, long-term data from individuals of known age, kinship, and reproductive success. I must assume, however, that information from a single colony is representative of the entire species.

Experiments with stuffed badgers

In response to predators, black-tailed, Gunnison's, and Utah prairie dogs run to burrow mounds and then sometimes give an alarm call (King 1955; Waring 1970; Fitzgerald and Lechleitner 1974; Wright-Smith 1978). Predators strike quickly at prairie dog colonies, however. A prairie falcon (*Falco mexicanus*), for example, can capture a young prairie dog and fly away in < 15 seconds, and a coyote (*Canis latrans*) can run away with an adult or juvenile victim < 30 seconds after charging into a colony. Accurate identification of alarm callers during such attacks is almost impossible. To investigate alarm calling, I use stuffed specimens ($N = 4$) of a natural predator, the American badger (*Taxidea taxus*), mounted in a lifelike pose and attached to a Plexiglas sled (Hoogland 1995). Using fishing wire wound around a garden hose reel, one person pulls the badger across a family's home territory, and another person records whether or not each aboveground member of the family gives an alarm call. We perform all experimental runs in the 4 weeks after juveniles first emerge from their natal burrows. The assumption is that prairie dogs respond to the stuffed badger as if it were real. Attacks by live badgers indicate that this assumption is valid for all three species. Specifically, as in our experiments, some individuals call in response to live badgers, others remain silent, and calling is more likely for individuals with nearby kin (Hoogland 1995, unpublished data).

Alpine and hoary marmots (*M. marmota* and *M. cali-*

gata) and several species of ground squirrels have at least two alarm calls for different types of predators (e.g., aerial versus terrestrial), or for different levels of danger from the same predator (Balph and Balph 1966; Melchior 1971; Taulman 1977; Davis 1984a; Hofer and Ingold 1984; Sherman 1985). Black-tailed, Gunnison's, and Utah prairie dogs, on the other hand, resemble Olympic marmots (*M. olympus*; Barash 1973, 1989) and Columbian and thirteen-lined ground squirrels (*S. columbianus* and *S. tridecemlineatus*; Betts 1976; Matocha 1977; Schwagmeyer 1980) by having only one distinct alarm call. Subtle variation in prairie dog alarm calls, however, might indicate different types of predators or different levels of urgency (Slobodchikoff et al. 1986, 1991). For all three species, for example, individuals seem to call at a faster rate (i.e., more barks per minute) when danger is imminent (King 1955; Waring 1970), but I have not quantitatively investigated this possibility. Nor have I investigated why some individuals give short alarm calls (< 5 seconds) whereas other individuals give longer alarm calls (> 10 minutes).

The stuffed badger allows me to expose each prairie dog to the same level of (simulated) danger, and also enables me to obtain large sample sizes. For black-tailed prairie dogs, results are from > 4,000 responses of 323 individuals (Hoogland 1995); for Gunnison's prairie dogs, results are from 287 responses from 88 individuals (Hoogland 1996b); and for Utah prairie dogs, results are from > 1,500 responses from 242 individuals. After five to ten experimental runs, I calculate the percentage of times that each individual gave an alarm call, and then use a single data point (i.e., the percentage) for each individual for all statistical analyses. Prairie dogs call as often for late experimental runs as for earlier runs (Hoogland 1983, 1995), and therefore do not seem to habituate to the stuffed badger.

For all three species, calling by juveniles is rarer than calling by adults, and for this chapter I only consider calling by adults. Calling among adults does not vary systematically with age for black-tailed prairie dogs (Hoogland 1995), but I have not investigated calling versus age for Gunnison's and Utah prairie dogs.

Observations of mating behaviors

Like females of other ground dwelling squirrels (Hanken and Sherman 1981; Sherman 1989; Boellstorff et al. 1994; Murie 1996), females of black-tailed, Gunnison's, and Utah prairie dogs are sexually receptive for several hours on only 1 day of each year, usually in the afternoon. To detect copulations, students and I watch marked individuals from dawn until dusk during the period of 3–5 weeks when matings occur (Hoogland 1995, 1998a, 1998b, 2001). Most copulations occur underground, but aboveground courtship and guarding behaviors indicate which male(s) copulates with each estrous female. Courting male black-tailed prairie dogs, for example, sometimes take nest material into a burrow before copulating there, and males of all three species commonly give a unique mating call just before or just after copulation. Three independent lines of evidence indicate that my inferences of estrus and copulation are accurate (Hoogland 1995, 1998b; Haynie et al. 2003). Black-tailed and Gunnison's prairie dogs avoid extreme inbreeding, but at least 10% of matings among Utah prairie dogs involve parent/offspring or sister/brother (Hoogland 1982, 1992, 1999, unpublished data).

Several of the costs and benefits that result from copulating with three different males also might accrue to females that copulate three times with the same male (Dewsbury 1982b; Birkhead 2000). In this chapter I only consider the number of different male sexual partners versus litter size, but I recognize the importance of trying to separate the effects of total number of copulations versus number of different sexual partners.

Alarm Calling

Females

For all three species, females with kin in the home territory are more likely to call than females with no nearby kin (fig. 37.1a). For Gunnison's and Utah prairie dogs, females with *juvenile offspring* in the home territory are more likely to call than females with only nondescendent adult or juvenile kin such as siblings, nieces, or cousins nearby (fig. 37.2a). By contrast, female black-tailed prairie dogs call as often when nondescendent adult or juvenile kin are in the home territory as when nearby adult or juvenile offspring are present (fig. 37.2a).

Why do female black-tailed prairie dogs differ from female Gunnison's and Utah prairie dogs by calling so often for nondescendent kin? Demography, I speculate, has played a key role in the evolution of these interspecific differences (Sherman 1980a). For Gunnison's and Utah prairie dogs, > 90% of adult females give birth each year (Hoogland 1999, 2001). Natural selection thus frequently can favor calling by mothers, but selection for calling by (rarer) females with only nondescendent kin within earshot can only occur infrequently. For black-tailed prairie dogs, by contrast, < 50% of adult females give birth each year—mainly because females usually do not copulate as yearlings (Hoogland 1995). Adult female black-tailed prairie dogs with only nondescendent kin in the home territory are

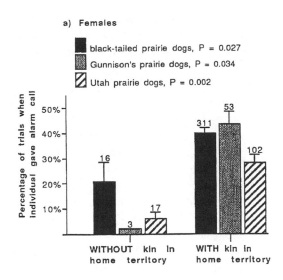

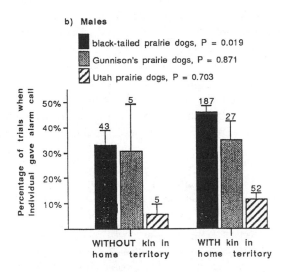

Figure 37.1 Alarm calling by prairie dogs with and without kin in the home territory for (a) females and (b) males. Shown here are means + 1 SE; the number above each SE line indicates the number of individuals observed (each approximately seven times per year). *P*-values are from Mann-Whitney U test. Data are from Hoogland 1995, 1996, and unpublished data.

therefore common. Indeed, in some years such females are as common as females with nearby offspring. Natural selection thus frequently can favor calling by females with only nondescendent kin nearby.

Males

Breeding male black-tailed, Gunnison's, and Utah prairie dogs typically copulate with most or all females in the home territory, and copulating males usually remain in the same home territory until after the emergences of juveniles from their natal burrows. Further, 1-year-old males that are sexually immature usually reside in the natal territory. Consequently, most adult male prairie dogs are surrounded by juvenile kin when they are exposed to badgers in June: breeding males have nearby juvenile offspring, and nonbreeding 1-year-old males have nearby parents, adult and juvenile siblings, juvenile nieces and nephews, adult and juvenile cousins, and so forth. Male prairie dogs with kin in the home territory are more likely to call than are males without nearby kin—but only for black-tailed prairie dogs is this difference significant (fig. 37.1b). Sample sizes are larger for black-tailed prairie dogs than for the other two species, and perhaps additional data for Gunnison's and Utah prairie dogs would lead to significance as well. Notice, however, that 33% of male black-tailed prairie dogs and 31% of male Gunnison's prairie dogs call when they have *no kin* within the home territory (fig. 37.1b). These high frequencies indicate that some factor in addition to kinship is important for alarm calling among male prairie dogs. Perhaps calling to warn potential *mates,* which are usually unrelated or only distantly related, is important for males. Or perhaps a male's alarm call announces his competitive status, good health, freedom from parasites, and willingness to call in the future—and therefore is important in both male-male competition and female choice.

Figure 37.2b shows alarm calling by adult male prairie dogs that have one or more of the following in the home territory: (1) juvenile offspring, (2) parent or adult littermate sibling, or (3) more distant adult and juvenile nondescendent kin. For black-tailed and Gunnison's prairie dogs, calling frequencies are relatively high (i.e., > 25% for all three classes), and males with nondescendent kin in the home territory call as often as males with nearby offspring. The reason for this latter equivalence, I speculate, pertains to demography once again. The probability of copulating in the first year is only 6% for male black-tailed prairie dogs and only 24% for male Gunnison's prairie dogs (Hoogland 2001). Further, nonbreeding yearling males of both species usually remain in the natal territory, and thus commonly have nondescendent kin nearby. Natural selection thus easily can favor calling by male black-tailed and Gunnison's prairie dogs with only nondescendent kin nearby, as argued earlier for females.

But why do male Utah prairie dogs call so rarely in response to the badger (figs. 37.1b, 37.2b)? The males least likely to call are fathers (at a frequency of 7%), and this paternal reticence, coupled with the calling frequency of 25% for males with nearby parents or siblings, is primarily responsible for the significant difference for male Utah prairie dogs in figure 37.2b. The silence of male Utah prairie dogs has baffled me for years. Recently I formulated a testable hypothesis that pertains to male body mass. Male and

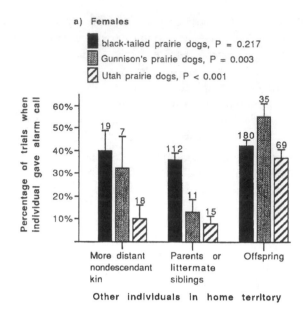

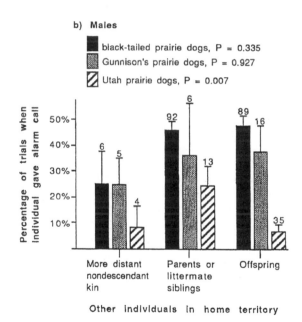

Figure 37.2 Alarm calling by prairie dogs with different categories of nearby kin in the home territory for (a) females and (b) males. Individuals with parents or littermate siblings in the home territory did not also have offspring there, and individuals with more distant nondescendent kin in the home territory did not also have either parents, siblings, or offspring there. Shown here are means + 1 SE; the number above each SE line indicates the number of individuals observed (each approximately seven times per year). *P*-values are from Kruskal-Wallis analysis of variance. Data are from Hoogland 1995, 1996, and unpublished data.

female black-tailed prairie dogs, which do not hibernate, attain maximal body mass in late fall (Hoogland 2003a)—several months after completion of our experiments with the badger. Male and female Gunnison's prairie dogs, which do hibernate, usually attain maximal body mass and initiate torpor in early fall (Fitzgerald and Lechleitner 1974; Rayor et al. 1987; Hoogland 1999)—once again months after completion of our experiments. Utah prairie dogs also hibernate. Adult females usually initiate torpor in August or September, but adult males attain maximal body mass and initiate hibernation as early as July (Wright-Smith 1978; Hoogland 2003b, unpublished data)—soon after experiments with the stuffed badgers. Body mass of adult male Utah prairie dogs when juveniles are emerging from their natal burrows averages 147% of body mass during the breeding season; for adult male black-tailed and Gunnison's prairie dogs, by contrast, body mass at first juvenile emergences averages only 108% and 129%, respectively, of body mass during the breeding season (fig. 37.3). Like obese Belding's ground squirrels (Trombulak 1989), obese male Utah prairie dogs that are ready to initiate hibernation cannot run at full midseason speed, and they sometimes trip and fall when racing toward a burrow.

I hypothesize that male Utah prairie dogs remain silent in response to badgers in June because obesity makes calling too risky. Specifically, males do not call shortly after first juvenile emergences because they cannot run quickly, and therefore might be unable to reach safety after an alarm call would make them more vulnerable to detection by a predator. In terms of Hamilton's (1964) rule, the cost of calling (i.e., increased vulnerability) is greater than the recipient's benefit, which is devalued by the coefficient of genetic relatedness (i.e., $c > rb$). Data from 2004 support this hypothesis. As in previous years, students and I introduced badgers in June 2004, when males were obese. But we also introduced badgers in April 2004, soon after the breeding season, when males were still lean and fast. The mean calling frequency of obese males in June ($N = 38$) was only 6%, but the mean calling frequency by lean males in April ($N = 28$) was 19% ($P = 0.043$, Mann-Whitney U test). These results indicate that obesity is a proximate factor that

influences alarm calling of male Utah prairie dogs. In ultimate terms, the silence in June of males that are surrounded by kin remains puzzling.

Increased vulnerability associated with obesity might explain why a male Utah prairie dog away from a burrow entrance does not utter an alarm call to warn nearby kin. But why don't males call after reaching the safety of a burrow entrance, perhaps while partially submerged in the entrance? I have no answer for this pertinent question.

Body mass is one easy-to-quantify factor that might affect a prairie dog's vulnerability to predation, and thus its probability of giving an alarm call. Perhaps other factors that are more difficult to quantify also are important. Minor injuries, weak eyesight, and poor health, for example, might reduce the probability that an individual will give an alarm call. Might these other factors that affect vulnerability help to explain the extreme variation in alarm calling that I have observed among individual prairie dogs, and that other researchers have observed among individuals of other species?

As noted previously, breeding male prairie dogs consistently have *some* offspring in the home territory in June for our experiments with the badgers. But litters of all three species sometimes show multiple paternity, and the resident breeding male is cuckolded whenever another male sires some of the juveniles in his territory. Might interspecific differences in the frequency of multiple paternity help to explain interspecific differences in the frequency of alarm calling by breeding males? Specifically, is natural selection more likely to favor calling by breeding males when 75% of the juveniles in the home territory consistently are offspring than when only $= 25\%$ are offspring? If multiple paternity and cuckoldry are rampant, so that juvenile beneficiaries of alarm calls by breeding males frequently will include more non-offspring than offspring, then perhaps natural selection will favor mechanisms less dangerous than alarm calling by which breeding males can increase reproductive success. Consistent with this hypothesis, the calling frequency by breeding males is highest among black-tailed prairie dogs, for which the frequency of multiple paternity is lowest, and the calling frequency is lowest among Utah prairie dogs, for which the frequency of multiple paternity is highest (see below). Note, however, that the estimated frequencies of multiple paternity for Gunnison's and Utah prairie dogs are almost identical (77% versus 82%), but calling frequencies for breeding male Gunnison's and Utah prairie dogs differ substantially (37% versus 7%).

In summary, data in this chapter show that kinship is an important predictor of alarm calling for three species of prairie dogs, as it is for many other animals. I have identified two other factors—demography and body mass—that also affect the frequency of alarm calling. Data from prairie

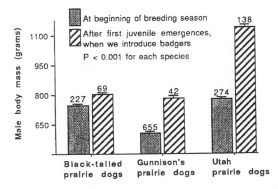

Figure 37.3 Body mass of adult male prairie dogs at the beginning of the breeding season versus body mass of adult males after first juvenile emergences. Shown here are means + 1 SE; the number above each SE line indicates the number of males weighed. P-values are from Mann-Whitney U test.

dogs and other rodents show that alarm calling is a complicated behavior that defies simple explanation.

Multiple Mating by Females

Like females of other species, female prairie dogs commonly copulate with more than one male. Behavioral observations indicate that 33% of female black-tailed prairie dogs copulate with two or more males each year, and that 65% of female Gunnison's and Utah prairie dogs copulate with two or more males each year (fig. 37.4). Consistent with these interspecific differences, the minimal frequency of litters with multiple paternity is only 5% for black-tailed prairie dogs, but is 77% and 82%, respectively, for Gunnison's and Utah prairie dogs. Some of this disparity might result because detection of multiple paternity for black-tailed prairie dogs results from use of starch-gel electrophoresis (Hoogland 1995), whereas detection for Gunnison's and Utah prairie dogs results from use of microsatellites (Haynie et al. 2003).

If 65% of female Gunnison's and Utah prairie dogs copulate with two or more males, how can frequencies of multiple paternity for both species be > 65% (i.e., 77% and 82%, respectively)? Two possible explanations come to mind. First, for some of the litters that we sampled for multiple paternity, we had no information on the mother's number of sexual partners (Haynie et al. 2003). Second, despite our long hours and best efforts, students and I probably failed to detect certain copulations. Failure was probably more likely when a copulation was especially short or secretive, or when numerous females from the same and different clans were sexually receptive on the same day.

Why the interspecific differences in the frequency of multiple mating by females? Figure 37.5 provides one possible

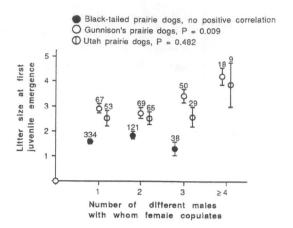

Figure 37.5 Litter size for prairie dogs versus mother's number of sexual partners. Shown here are means + 1 SE; the number above each SE line indicates the number of females for which I have information on both number of copulations and litter size at first juvenile emergence. *P*-values are from Spearman rank correlation test.

explanation. Female Gunnison's prairie dogs that copulate with several males rear larger litters than do females that copulate with only one male; multiple regression shows that this effect is real, and does not result simply because heavier, older, more fecund females are more likely to copulate with several males (Hoogland 1998b). Female Utah prairie dogs also might rear larger litters by copulating with several males, but sample sizes to this point are too small for acceptance or rejection of this hypothesis (fig. 37.5). Female black-tailed prairie dogs, by contrast, evidently do not enhance litter size by copulating with several males (fig. 37.5). The trends in figure 37.5 thus might explain why female black-tailed prairie dogs are less likely than female Gunnison's and Utah prairie dogs to copulate with second and third males. Why incur the risks of copulation if additional matings do not enhance reproductive success?

How might copulating with several males increase litter size for prairie dogs? As noted earlier, multiple mating commonly leads to multiple paternity. Multiple paternity promotes genetic diversity among littermates and thus maximizes the advantages of sexual reproduction (Williams 1975; Hamilton and Zuk 1982; Seger and Hamilton 1988; Birkhead and Moller 1992). In an unpredictable environment that includes diseases and parasites, genetic diversity might be especially important to fetal and neonatal survivorship. Multiple copulations also might enhance the quality of offspring via intrauterine sperm competition (Parker 1984b; Madsen et al. 1992; Birkhead et al. 1993; Keller and Reeve 1995)—so that mortality of fetuses and neonates is less likely and larger litters at first emergence are more likely.

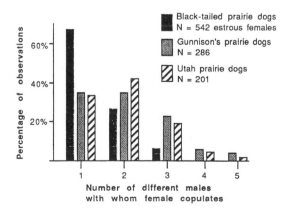

Figure 37.4 Number of different males with whom female prairie dogs copulate.

Female black-tailed prairie dogs that copulated with exactly two males reared (insignificantly) larger litters than females that copulated with only one male (fig. 37.5). When I examine the number of juvenile offspring that survived for at least 1 year, then the difference between doubly- and singly-copulating females is almost significant (mean ± SD = 1.00 ± 1.20 versus 0.815 ± 0.123; P = 0.069, Mann-Whitney U test). Thus, females of all three prairie dog species that I have studied might rear more offspring by copulating with more than one male. If so, then why is the frequency of multiple copulations for Gunnison's and Utah prairie dogs twice the frequency for black-tailed prairie dogs? The answer, I speculate, relates to interspecific variation in the ability of males to monopolize sexual access to females (Orians 1969; Emlen and Oring 1977).

A female might gain from copulating with second and third males, but the first male that copulates seldom gains by allowing the female to copulate with additional males. The first male therefore should try to prevent copulations with additional males, and first males of all three species vigorously engage in such mate-guarding. Why are male black-tailed prairie dogs more successful than male Gunnison's and Utah prairie dogs at stopping a female from copulating with a second male? As explained below, the answer might relate to visibility within the home territory, number of entrances per mating burrow, and reproductive synchrony among females.

Visibility within the home territory

Mate-guarding will be easier and more effective when males can see females. One of the most conspicuous features of a colony of black-tailed prairie dogs is the paucity and shortness of vegetation (Scheffer 1947; King 1955). During the breeding season in February–March, for example, a typical home territory is devoid of all vegetation >15 cm tall (fig. 37.6, top). Consequently, a male black-tailed prairie dog easily can see an estrous female that is aboveground, where she predictably is during daylight hours of the mating season unless she is actually copulating. In colonies of Gunnison's and Utah prairie dogs, by contrast, tall vegetation (> 30 cm) is conspicuous throughout the year (fig. 37.6, bottom). Estrous females of the latter two species can conceal themselves behind sagebrush (*Artemisia*), rabbitbrush (*Chrysothamnus*), and other bushes/shrubs, so that watching and guarding of estrous females is probably more difficult for male Gunnison's and Utah prairie dogs than for male black-tailed prairie dogs. One thing is certain: visual tracking of estrous Gunnison's and Utah prairie dogs is more difficult for students and me than is tracking of estrous black-tailed prairie dogs.

Number of entrances per mating burrow

One way for the first copulating male to deter copulations with additional males is to sequester the estrous female in a burrow so that she cannot exit and so that other males cannot enter (fig. 37.7). Like male Idaho ground squirrels (*S. brunneus*; Sherman 1989), male black-tailed, Gunnison's, and Utah prairie dogs all use this tactic to enhance mate-guarding. Such incarceration is easy when a burrow has only one entrance, but is more difficult when a burrow has numerous entrances.

Because she submerges before the male (Hoogland 1995, 1998a, 1998b, unpublished data), the female is primarily responsible for choosing the burrow where copulation occurs (the *mating burrow*) for all three species. Mating burrows for black-tailed prairie dogs usually have two entrances, and occasionally have three or four (Hoogland 1995, unpublished data; see also King 1955; Sheets et al. 1971). Mating burrows of Gunnison's prairie dogs, by contrast, commonly have as many as five to six entrances (Hoogland 1998b, unpublished data; see also Rayor 1988). Mating burrows of Utah prairie dogs almost always have a minimum of five to six entrances, and sometimes have > twenty entrances. Incarceration of estrous females thus is probably easier for male black-tailed prairie dogs than for male Gunnison's and Utah prairie dogs.

I have argued that male ability to monopolize sexual access to females probably varies *directly* with visibility within the home territory and *inversely* with the number of entrances per mating burrow. A key assumption here is that the incentive for females to avoid monopolization is equal for all three species. But, as noted previously (fig. 37.5), perhaps the payoff for avoiding monopolization and for copulating with additional males is higher for female Gunnison's and Utah prairie dogs than for female black-tailed prairie dogs; if so, this factor might explain the interspecific differences of figure 37.4. Opposite of male ability to monopolize, female ability to escape from mate-guarding and to seek additional copulations probably varies *inversely* with visibility within the home territory and *directly* with the number of entrances per mating burrow.

Reproductive synchrony among females

The typical clan or coterie of prairie dogs usually has one reproductive male and several reproductive females; further, each female is sexually receptive for only several hours on a single day each year. If only one female per coterie or clan comes into estrus per day, then the resident reproductive male can devote all his attention to courting and guarding that female—and thereby can increase the probability

Figure 37.6 Vegetation within home territory during the breeding season for (top) black-tailed prairie dogs, and (bottom) Utah prairie dogs. Because of these differences, male black-tailed prairie dogs can guard estrous females more easily than can male Utah prairie dogs. Top photo courtesy of Wind Cave National Park; bottom photo by John L. Hoogland.

that he will be the only male to copulate with her. If several females come into estrus on the same day, however, then monopolization of each female by the resident breeding male inevitably will be more difficult. Reproductive synchrony of females thus affects mate-guarding, and consequently the frequency at which females copulate with a second or third male (Emlen and Oring 1977; Sherman 1989).

Because they are seasonal breeders, a certain level of re-

productive synchrony occurs for all three species of prairie dogs—probably because natural selection has favored females who wean offspring when new vegetation is available in late spring. At my study colonies, for example, black-tailed prairie dogs always copulate in February and March, and Gunnison's and Utah prairie dogs always copulate in March and April. But do factors other than seasonality affect reproductive synchrony, and are all three species

Figure 37.7 After copulating with a female, black-tailed prairie dog male-05 tries to preclude copulations with other males by sequestering the female in burrow 40. Photo by John L. Hoogland.

equally synchronous? In particular, are female black-tailed prairie dogs less synchronous than female Gunnison's and Utah prairie dogs, and might this difference be another reason why mate-guarding is easier for male black-tailed prairie dogs?

Two lines of evidence indicate that female black-tailed prairie dogs breed less synchronously than female Gunnison's and Utah prairie dogs (Hoogland 1995, 1999, unpublished data). First, even though students and I observe approximately the same number of estrous females each year for all three species, the mean interval between the first and last copulation is consistently about 5 weeks for black-tailed prairie dogs, but is only 3–4 weeks for both Gunnison's and Utah prairie dogs ($P < 0.001$ for black-tailed versus Gunnison's, $P < 0.001$ for black-tailed versus Utah, and $P > 0.100$ for Gunnison's versus Utah, Mann-Whitney U test). Second, and more important, females of the same coterie of black-tailed prairie dogs are less likely to come into estrus on the same day than are females of the same clan of Gunnison's and Utah prairie dogs. For sixty-nine female black-tailed prairie dogs in 1981, for example, only 12% copulated on a day when another female of the home coterie also copulated. For seventy-eight female Gunnison's prairie dogs in 1994 and fifty-three female Utah prairie dogs in 2004, by contrast, 27% and 36%, respectively, synchronized estrus on the same day with at least one other female of the home clan ($P = 0.020$ for black-tailed versus Gunnison's, $P = 0.001$ for black-tailed versus Utah, and $P = 0.276$ for Gunnison's versus Utah, 2 × 2 chi-square test; for all three species, I used data from the year when students and I observed the highest number of copulations).

As for yellow-bellied marmots (*M. flaviventris*; Armitage 1986), male and female prairie dogs probably have a con-

flict of interest regarding the optimal number of different sexual partners per female: the copulating male gains by precluding matings with other males, but the female probably benefits from additional matings. Figure 37.4 shows that neither sex for any of my study species is completely winning the conflict of interest.

In summary, female black-tailed prairie dogs probably are less likely than female Gunnison's and Utah prairie dogs to copulate with more than one male for two reasons. First, the enhancement of litter size from second and third copulations is probably lower for female black-tailed prairie dogs. Second, because of shorter vegetation, fewer entrances per mating burrow, and reduced reproductive synchrony, male black-tailed prairie dogs probably more easily can monopolize estrous females and thereby deter copulations by additional males.

Infanticide

For black-tailed prairie dogs, nonparental infanticide affects 39% of litters and is the major cause of juvenile mortality ($N = 130$ observed cases). Females are the more common killers ($N =$ thirty-eight different killers), but males also kill ($N =$ twenty different killers; fig. 37.8). By contrast, during my 7-year study of Gunnison's prairie dogs, I never observed a single unequivocal case of infanticide. For Utah prairie dogs, infanticide affects 15% of litters, but only males kill ($N =$ forty-eight cases, involving twenty-five different killers; fig. 37.8). For both species that show killing of juveniles, cannibalism usually follows infanticide (table 37.2). What might account for the striking interspecific differences depicted in figure 37.8? I first will investigate this question for females, then for males.

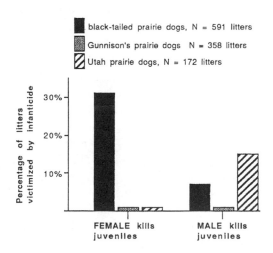

Figure 37.8 Infanticide by male and female prairie dogs.

Table 37.2 Cannibalism for black-tailed, Gunnison's, and Utah prairie dogs

Species	Sex of killer	Does killer consume victim in >90% of cases?
Black-tailed	Male	Yes?
	Female	Yes
Gunnison's	Male	Males do not kill
	Female	Females do not kill
Utah	Male	Yes
	Female	Females do not kill

SOURCE: Hoogland 1985, 1995, 1999, 2001.
NOTES: Infanticide by male Utah prairie dogs usually occurs aboveground, so verification of frequent cannibalism is easy. Infanticide by female black-tailed prairie dogs occurs underground, but a bloody face and other diagnostic signs indicate that cannibalism consistently follows infanticide. When infanticide by male black-tailed prairie dogs occurs aboveground, cannibalism is sporadic; when males kill underground, diagnostic signs of cannibalism are common.

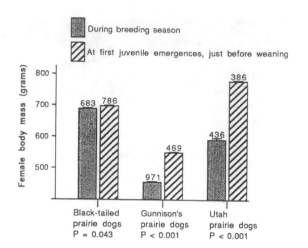

Figure 37.9 Body mass of female prairie dogs during the breeding season versus female body mass at first juvenile emergences. Shown here are means + 1 SE; the number above each SE line indicates the number of females weighed. *P*-values are from Mann-Whitney U test.

Females

Female black-tailed prairie dogs reap at least five benefits from infanticide (Hoogland 1995): (1) removal of future competitors from themselves and their offspring, (2) increased sustenance, because infanticide almost always includes cannibalism, (3) larger area in which to forage, because victimized females no longer defend a territory around the home nursery burrow, (4) more help with alarm calling, defense of the coterie territory, and excavation and maintenance of burrows, because victimized mothers are emancipated from maternal duties, and (5) increased safety of offspring, because victimized mothers are unlikely to kill. Increased sustenance from cannibalism is probably the primary benefit of infanticide, but I have not rigorously evaluated the relative importance of different payoffs.

As for mammals in general (Clutton-Brock et al. 1989) and for ground-dwelling squirrels in particular (Michener 1989), lactation probably is physiologically stressful for prairie dog mothers. In late winter and early spring, when new vegetation is still sparse, mothers not only must feed themselves, but also must produce milk for their offspring. Perhaps some mothers can only wean a litter if they can obtain crucial additional sustenance—e.g., protein, rare minerals—via infanticide and cannibalism sometime during lactation. If so, then perhaps infanticide by females should be more common in species in which lactation is more stressful. To investigate this possibility, I have compared female body mass during the breeding season versus female body mass at first juvenile emergences (i.e., just before weaning). I predicted that lactation would depress body mass for all three species, and that the reduction would be most pronounced for black-tailed prairie dogs (the only species in which females engage in nonparental infanticide). Figure 37.9 shows an unexpected result: females of all three

species are *heavier* at first juvenile emergences than during the breeding season. Female Gunnison's and Utah prairie dogs are heavier by 21% and 31%, respectively, however, but female black-tailed prairie dogs are heavier by only 1%. These results support the notion that lactation is more stressful for black-tailed prairie dogs than for Gunnison's and Utah prairie dogs. If so, the reason might be that vegetation in early spring is less abundant for black-tailed prairie dogs than for the other two species (fig. 37.6).

Males

Infanticide and cannibalism by male black-tailed prairie dogs occur after a male invades a new territory with either preemergent (unweaned) juveniles or emergent juveniles that have been coming aboveground for < 4 weeks; invaders most commonly are yearlings. The invading male is usually unrelated, or only distantly related, to his victims (29/39 = 74%). Infanticides of unweaned juveniles occur underground, but killings of emergent juveniles frequently occur aboveground. Killers usually eliminate all juveniles of a victimized litter (29/39 = 74%), and such infanticide affects 7% of litters (Hoogland 1995).

For Utah prairie dogs, infanticidal males are residents of the territory in which killing occurs (48/48 = 100%), and killing and cannibalism usually occur aboveground after a male emerges from a nursery burrow with a live unweaned juvenile (45/48 = 94%). Marauders are of three types: (a) male that copulated with the victimized mother (N = 9), (b) breeding male that did not copulate with the victimized mother (N =19), and (c) nonbreeding yearling male (N = 20). Marauders usually kill only one juvenile within a litter (N = 22 different killers), but occasionally eliminate an entire litter via serial infanticides over several days ($N = 3$ killers).

Infanticide after invasion by a new male is common for many harem-polygynous mammals, and the payoff is that victimized mothers come into estrus and conceive more quickly than do mothers that continue to lactate (Hrdy 1977a, 1979; Fossey 1984; Packer and Pusey 1984). For prairie dogs, however, females come into estrus only once each year—so that infanticide by an invading male does not reduce the time until the next estrus. Further, benefits from the removal and cannibalism of future competitors are probably minimal, because infanticidal male black-tailed and Utah prairie dogs do not seem to experience either improved survivorship or increased reproductive success in the year following infanticide (Hoogland 1995, unpublished data). Finally, males of neither species seem to gain via improved condition of victimized mothers in the next breeding season, because victimized females do not survive better than females that wean a litter, are not more likely to rear offspring in the following year, and do not produce larger litters in the following year (Hoogland 1995, unpublished data). The adaptive significance of infanticide by male black-tailed and Utah prairie dogs thus remains unclear.

Infanticidal male black-tailed prairie dogs usually kill unrelated juveniles, but $10/39 = 26\%$ of infanticides involved males killing half-siblings. In every case, a yearling, nonbreeding male ($N = 6$ different killers) invaded a territory on which his father was the resident breeding male, and then killed his father's offspring (i.e., the killer's half-siblings). Why has natural selection favored the killing of half-siblings by invading yearling males? Perhaps the loss of offspring to invading sons is a secondary consequence of helping sons to establish residency in a breeding territory. If so, however, why hasn't natural selection worked against such a detrimental secondary consequence?

For $9/48 = 19\%$ of infanticides among Utah prairie dogs, a male killed the offspring of a female with whom he had copulated—that is, the male might have killed his own offspring. Why? I do not have an answer, but offer the following relevant observations/thoughts. (1) As noted in the previous section, female Utah prairie dogs usually copulate with more than one male, and multiple paternity is common. When males kill offspring of females with whom they copulated, perhaps they target litters for which the mother also copulated with more than one other male—so that multiple paternity is likely and so that the male has a reduced probability of killing his own offspring. In support of this hypothesis, in $8/9 = 89\%$ of cases in which a male killed offspring of a female with whom he copulated, the female also copulated with at least one other male. (2) When a male kills offspring of a female that copulated with him and other males, perhaps he somehow can discriminate between

his own offspring and juveniles sired by the other male(s), and then can victimize only the other male's offspring. I am investigating this hypothesis via blood/tissue samples from victims (when anything remains aboveground after cannibalism) and killers, but the low level of variation among microsatellites (Haynie et al. 2003) is hindering my analysis.

Multiple mating by females sometimes might reduce the probability of later infanticide by males. Specifically, because multiple mating might make it difficult to determine which male sires offspring, a male might refrain from killing juveniles that could be his own offspring (Hrdy 1979; Wolff and Macdonald 2004; but see Buchan et al. 2003 and Sherman and Neff 2004). As a group, the three species of prairie dogs that I have studied offer little support for the notion that multiple mating by females functions to deter infanticide. Gunnison's prairie dogs show the lowest frequency of infanticide by males, for example, but Gunnison's prairie dogs also have the same high frequency of multiple mating as do Utah prairie dogs, which show the highest frequency of infanticide by males.

In summary, interspecific differences in the frequency and details of infanticide among three species of prairie dogs are salient and bewildering. Data for a better understanding will be elusive. Even though detectable frequencies of killing among black-tailed and Utah prairie dogs are high, for example, students and I observe infanticide only about once every 300 person-hours of observations during the period when juveniles are vulnerable.

Summary

From observations of marked individuals, students and I studied alarm calling, multiple mating by females, and infanticide among black-tailed, Gunnison's, and Utah prairie dogs. For Gunnison's and Utah prairie dogs, females with juvenile offspring in the home territory are more likely to call than females with only nondescendent adult or juvenile kin such as siblings, nieces, and cousins nearby. Female black-tailed prairie dogs, by contrast, call as often when only nondescendent adult or juvenile kin are in the home territory as when nearby adult or juvenile offspring are present. Male prairie dogs with kin in the home territory are more likely to call than are males without nearby kin, but only for black-tailed prairie dogs is this difference significant. In addition to kinship, demography and body mass also have been important in the evolution of alarm calling among prairie dogs.

The percentage of females that copulate with two or more males is 33% for black-tailed prairie dogs, but is > 65% for Gunnison's and Utah prairie dogs. One possible reason for these differences is that female Gunnison's prairie dogs, and perhaps female Utah prairie dogs as well, rear larger litters after copulating with several males. Another reason is that monopolization of estrous females by males is probably easier for black-tailed prairie dogs than for Gunnison's and Utah prairie dogs. Factors that affect the monopolization of estrous females include visibility within the home territory, number of entrances per mating burrow, and reproductive synchrony among females.

Nonparental infanticide among black-tailed prairie dogs involves killing by adults of both sexes, and affects 39% of litters. Infanticide among Gunnison's prairie dogs, by contrast, is exceedingly rare. Among Utah prairie dogs, infanticide involves killing by adult males only, and affects 15% of litters. Reasons for these striking interspecific differences are elusive.

Conservation
& Disease

Chapter 38 Issues in Rodent Conservation

William Z. Lidicker Jr.

RODENTS ARE A HARD SELL when it comes to conservation. In most human cultures, rodents are generally viewed as vermin. Only in one country, Canada, is a rodent considered a national symbol; namely, the beaver (*Castor canadensis*). At least two rodents have positive images as mascots of American university football teams (beaver, gopher [= ground squirrel]). And in some cases, rodents supplement human diets (several species in parts of Southeast Asia and the capybara, *Hydrochaeris hydrochaeris* [Macdonald et al., chap. 33 this volume] and nutria [*Myocaster coypus*] in South America). Admittedly, rodents are in fact often pests, being guilty of damaging crops, pilfering stored grains, invading households, and spreading zoonoses. However, these faults, from a human perspective, are perpetrated by a very small percentage of rodent species. A recent compilation of species (Wilson and Reeder 1993; see also Honeycutt, Frabotta, and Rowe, chap. 2 this volume) lists 2,021 species of rodents. These species are arranged in 438 genera and twenty-nine families, and they occur in all the continents except Antarctica. Thus rodents comprise the largest and most diverse Order of living mammals.

Aside from their numbers and diversity, there are many reasons why rodents should be major concerns in conservation planning. Perhaps of least importance, there are actually some quite charismatic species. These include beavers (*Castor*), dormice (twenty-six species in the Family Myoxidae), hamsters (especially *Mesocricetus auratus*), flying squirrels (forty-four species), chinchillas (species in the family Chinchillidae), and the huge (up to 79 kg) capybara. A few species have been harvested commercially for their fur:

muskrat (*Ondatra zibethicus*; fig. 38.1), beavers (*C. canadensis, C. fiber*), chinchillas (*Chinchilla brevicaudata, C. lanigera*), and nutria. Several species have been domesticated for use as pets, and particularly for biomedical research: house mice (*Mus musculus*), rats (*Rattus norvegicus, R. rattus*), hamsters (*Mesocricetus auratus*), and guinea pigs (*Cavia porcellus*). The contributions of these species of rodents to human welfare, while often under-appreciated, is so tremendous that it defies quantitative estimation. Moreover, there is a huge potential for many more species to serve human needs in this way.

Of course the important reasons for conserving rodent species are their taxonomic and ecological diversity, and the fact that they are major players in ecological communities worldwide. Rodents occur in all of the terrestrial biomes; because of their generally small size and mostly herbivorous food habits they often provide the foundation for all higher trophic levels in a community. Life forms run the gamut, from subterranean and semiaquatic to arboreal and gliding. Some species are carnivores (e.g., *Onychomys*, *Hydromis Mayermys*) and many are omnivores. Among herbivores, there are grazers, browsers, granivores, bark-feeders, frugivores, and root and/or bulb specialists. Behaviorally, rodents exhibit the full range of social systems found among mammals, and this includes the only eusocial (social insect-like) species known for mammals (the naked mole-rat, *Heterocephalus glaber*; Sherman et al.1991; Faulkes and Bennett, chap. 36 this volume). Rodents, therefore, are not only essential components in the functioning of communities everywhere, they provide innumerable opportunities for evolutionary, ecological, and behavioral investigations.

Figure 38.1 Muskrat (*Ondatra zibethicus*), a popular and widespread rodent furbearer. Ogden Bay, Utah; 20 July, 1950; photo by author.

Table 38.1 Numbers and percentages of North American species and subspecies of rodents in various categories of conservation concern

Status	Species No.	Species %	Subspecies No.	Subspecies %
Extinct	0	—	7	0.5
Endangered	3	1.5	19	1.3
Vulnerable	6	2.9	25	1.8
Low risk	69	33.5	37	2.6
Insufficient data	5	2.4	36	2.5
Totals	83	40.3	124	8.7

SOURCE: Data from Hafner et al. (1998).
NOTE: $N = 206$ species and 1,426 subspecies.

Given the overwhelming importance of rodents to human welfare, both directly and indirectly, it is pertinent to ask how we are doing in our efforts to keep the Earth's rodent fauna intact. The answer seems to be "Not well," although it is difficult to be very precise about this, as the rodent fauna in much of the world is poorly known. Basic information on life-history traits, ecological relations, and demographic patterns of most species remains unknown or poorly understood. A particularly grievous example of our conservation failures is that of the Vancouver Island marmot (*Marmota vancouverensis*). There are only two species of mammals endemic to Canada (both rodents); this is one of them, and it is on the brink of extinction (Johnson 1989).

There have been two reviews of the conservation status of rodents on a continental scale. The first of these (Lidicker 1989) attempted to survey rodents worldwide. However, Central America was not covered, and South America and most of Asia received only superficial coverage. This report was a product of the Rodent Specialist Group, Species Survival Commission, International Union for the Conservation of Nature and Natural Resources (IUCN), and served mainly to call attention to the conservation needs of rodents and to set a baseline for future research. In 1998, the North American contingent of the Rodent Specialist Group published a detailed survey of the conservation status of North American rodents, including action plans where appropriate (Hafner et al. 1998). Table 38.1 is compiled from the summarizing statistics in that report. In addition, The Nature Conservancy lists two extinct subspecies that are not included in table 38.1; adding in these would raise the percentage of recently extinct subspecies to 0.6. These data clearly illustrate that even in North America, which has a relatively well-known rodent fauna and supports a relatively strong conservation program, extinctions have oc-

curred, many more species and subspecies are in a precarious state, and data are inadequate for determining the status of many more. The only other continent-sized area with a comparable level of knowledge and concern is Australia (Calaby and Lee 1989; MacPhee and Flemming 1999). For this continent, at least eight species of rodents are known to be extinct since European colonization (14.3%), nine more are considered vulnerable (16.1%), and at least five are insufficiently known, for a total of twenty-two species of conservation concern (39.3%). MacPhee and Flemming (1999) reported that over the last 500 years, forty-three species of rodents have disappeared from Australia; and this represents 52.3% of all mammalian losses known for that period.

In this review, I will summarize the known factors that influence the conservation status of the Earth's rodent fauna. This will be addressed in the context of life-history variables, as they may impact demographic behavior, spatial structure, movements, and interspecific relations. Special emphasis will be given to the role of social behavior as a contributing variable. As a group, rodents face the same suite of threats to their welfare and existence as do other organisms. Paramount are habitat destruction and introduction of exotic species. Other well-known factors are pollution, over-exploitation by humans, and disease. A new threat of unknown but potentially major influence is rapid global warming. Additionally, rodent conservation is hampered by indifference and complacency. The former is the result of a conservation focus on larger, more charismatic and generally familiar species, as well as residual negative attitudes toward rodents. The latter stems from the view that rodents are abundant, have high reproductive rates, widespread distributions, and are adaptable. Therefore, they can take care of themselves, or so it is thought.

Rodent Demographic Patterns

Rodents have been the subjects of innumerable investigations on population dynamics, and so relatively a great deal is known about their demographic patterns. We can say with confidence that rodents exhibit the full known range of demographic behaviors. Effective conservation action requires that we know the demographic details for each species of concern.

Common species

Some species are indeed common. They may be widespread as well as abundant. Such species inhabit readily available habitats, are habitat generalists, or utilize agricultural crops. Mostly they are herbivores, have high reproductive rates, and short life spans. They are thus low on the food chain and are adapted for rapid population growth. Such common species are generally not of conservation concern, although commonness does have its hazards. Common species attract predators and parasites, may experience intense intraspecific competition, and invite persecution by humans. Natural selection may push common species toward poor dispersal capabilities, and large gene pools experience genetic inertia. We know from our experience with the Passenger Pigeon (*Ectopistes migratorius*) that even superabundant species can become extinct in only a few decades.

Common species are the stereotypes of the rodent world. Examples include species of *Microtus, Peromyscus, Apodemus, Mus,* and *Rattus.* Because of their commonness, such species are often major players (keystone species) in their communities. As such, they may have significant top-down effects on vegetation, as well as bottom-up impacts as substantial contributors to higher trophic levels. Lastly, they can have significant nontrophic interactions with other species that share their habitat. An example is the relation between California voles (*Microtus californicus*; fig. 38.2) and western harvest mice (*Reithrodontomys megalotis*) in central coastal California (Heske et al. 1984). The harvest mouse is favored at low vole densities, presumably because it profits from use of vole runways (commensalism), whereas at high vole densities harvest mice are completely excluded from the habitat (amensalism).

Rare species

Rarity comes in various flavors, and the conservation implications are not the same for all types. Rabinowitz (1981) classifies uncommon species into seven types, based on three criteria: geographic range large or small, habitat requirements broad or narrow, and local population size. Spe-

Figure 38.2 California vole (*Microtus californicus*), an example of a common and intensively studied arvicoline rodent that has contributed considerably to our understanding of grassland communities. Richmond Field Station, Richmond, California; 3 June, 1994; photo by author.

cies most prone to extinction would have small geographic ranges, narrow habitat requirements, and small local population sizes. Most resilient of the rare forms would be those that occur at low densities, but have large geographic distributions and can live in a broad range of habitat types.

Rodents illustrate all of these forms of rarity. For example, the southern bog lemming (*Synaptomys cooperi*) is nowhere common, but has a large geographic range and occurs widely in grasslands across eastern North America. Similarly, the North American beaver (fig. 38.3) is very widespread, but is a habitat specialist and reaches densities of only a very few individuals per square kilometer (see also Busher, chap. 24 this volume). More vulnerable would be the Point Reyes jumping mouse (*Zapus trinotatus orarius*),

Figure 38.3 Beaver dam and pond (*Castor canadensis*). This rodent is a national symbol of Canada, a mascot of Oregon State University, a model for industriousness, and a major player in the European colonization of North America. Danby, New York; 7 October, 1950; photo by author.

which occurs only in mesic grassland along the western edge of Marin County, California, and is isolated by over 100 km from the nearest adjacent subspecies to the north (Lidicker 1998). In the last 3.5 decades, it has been found only within Point Reyes National Seashore. Highly vulnerable examples would include the Vancouver Island marmot, which occurs only in a narrow band of subalpine habitat in the southern half of Vancouver Island (Johnson 1989), an area heavily impacted by timber harvesting and recreational facilities. Another is the saltmarsh harvest mouse (*Reithrodontomys raviventris*), which is restricted to the saltmarshes of San Francisco Bay (California), especially to areas dominated by pickleweed (*Salicornia*) and that are adjacent to upland areas suitable as refuges during very high tides (Shellhammer 1998).

Ephemeral patterns

In this category, I include species whose role in the community is temporally strongly variable. Hibernators, for example, largely withdraw from active participation for significant parts of the annual cycle. Among rodents, these include marmots, ground squirrels, dormice, and zapodids. Seasonally migratory species similarly come and go from communities. This is quite common among rodents. In cold climates, house mice tend to move into buildings in the winter and crop fields in warmer periods. The root vole (*Microtus oeconomus*) in Finland moves between upland meadows and marsh borders (Tast 1966), Norwegian lemmings (*Lemmus lemmus*) breed in alpine tundra but winter in subalpine scrub (Kalela 1961; Kalela et al. 1971), various African rodents move in and out of a Zambian flood plain on a seasonal basis (Sheppe 1972), and where California voles live in grasslands adjacent to saltmarshes, they will mostly move into the marshes during the dry summers and then migrate back into the uplands when winter rains commence (Harding 2000). Many other examples are known for such interhabitat migrations. From a conservation perspective, it is essential that protected areas include all the required habitat types, not only the one associated with the breeding season.

An especially important and widespread ephemeral demographic pattern is that of multiannual cycles in numbers. These are best known among arvicoline rodents, where the phenomenon has been intensively studied (Taitt and Krebs 1985; Lidicker 1988, 1991; Stenseth 1999). Cyclic behavior raises several conservation concerns. First of all, populations must be large enough to survive periods of low density. Related to this, the effective population size (N_e) will often be closer to the low densities experienced by a population than to the more conspicuous peaks. Low N_e risks the genetic consequences of small population sizes (Allen-

dorf 1986; Ralls et al. 1988; Lacy 1997). These dangers include inbreeding, genetic drift, skewed sex ratios, and possibly the loss of adaptive polymorphisms. The second area of concern is that of the habitat mosaic available to such cyclic populations. Often when cyclic species crash in numbers they survive only in especially favorable patches or refugia. As numbers grow, they progressively occupy the less favorable matrix around these refugia, and at peak numbers will inhabit all habitat types in which survival is possible, even for the short term (Wolff 1980a; Lidicker 1985). This scenario has several critical implications. The refugia must be large enough and numerous enough to assure survival through the low-density episodes. They also must be close enough together that recolonization can readily occur into patches that do become extinct. It may also be the case that secondary habitats are essential for such species to persist. This is because without dispersal sinks (Lidicker 1975) refugia may sustain high densities for long intervals, such that significant vegetation damage is perpetrated, and/or predators become abundant, and/or parasites may invade and spread easily. Evans (1942) has documented a case where bank voles (*Clethrionomys glareolus*) were able to survive population crashes only because some individuals persisted in secondary habitat.

The final class of ephemeral patterns is that of irruptive species. Such populations occur at low densities most of the time, but occasionally irrupt to very high numbers. Irruptions are usually attributed to an unusual combination of favorable conditions that permit a large increase in numbers. Irruptive species generally have high potential reproductive rates, so they can opportunistically take advantage of favorable circumstances. In South Australian wheat-growing areas, house mice irrupt to "plague" proportions when rare summer rains permit the construction of burrows needed for successful reproduction at the same time that food is abundant (Newsome 1969). Rodents in temperate forests increase greatly during so-called mast years (Pucek et al. 1993; Ostfeld et al. 1996). In Japan (Tanaka 1957) and southern South America, rodents have been observed to reach plague proportions following the massive flowering of bamboo, a phenomenon which seems to occur only rarely. Jaksic and Lima (2003) summarize data for rodent outbreaks (ratadas) in South America that include twenty-eight associated with bamboo flowering and twenty-seven with above-average rainfall.

Persistence of irruptive species will depend on conditions being suitable for long-term survival at low numbers. However, it may also be critical for these species to occasionally experience irruptions. A brief episode of high numbers will allow recolonization of empty or new habitat patches, and most importantly, a thorough mixing of local gene pools. Such periodic genetic revolutions will serve to reverse any deleterious effects that may accumulate over years at low densities, and occasionally will produce genetic novelties that will be given a chance to test their fitness potentials. A particularly instructive example of this life style is that of *Rattus villosissimus*, which lives in semiarid and arid Australia (Newsome and Corbett 1975). This species persists in refugia that result from unpredictable local rainfall events. The locations of such refugia are therefore unpredictable. Occasionally, the desert experiences a more general precipitation episode, which allows the rat to increase greatly in numbers, disperse widely, followed by restriction back to survival in a new set of small refugia.

Spatial Structure and Movements

The spatial structure (dispersion) of a species is closely related to its demographic behavior, which, as we have seen, signals its propensity to suffer extinction and also gives us guidelines for appropriate conservation actions. Spatial structure also strongly influences genetic structure, which can feed back to demographic processes (Chepko-Sade and Halpin 1987; Lidicker and Patton 1987). Clearly this spatial structure/demography connection is something we need to understand for all species of conservation concern. A species' distribution in space depends on its geographic distribution, fragmentation of habitat, dispersal capabilities, and social system. Two of these factors sustain strong anthropogenic influences. Geographic range is affected by habitat destruction and global climate changes, the latter including temperature, the amount, seasonality, and regularity of precipitation, changes in sea level, and frequency of severe storms. Habitat destruction clearly causes the fragmentation of destroyed habitats and the coalescence and availability of new habitats. The other two variables are less affected by humans as they are inherent species properties. The social system can interact with genetic structure in various ways (Chepko-Sade and Halpin 1987; Lidicker and Patton 1987). The proportion and sex ratio of young that remain philopatric or that disperse to various distances can be strongly influenced by social factors (Anderson 1989; Wolff 1999). At high densities, dispersal can be inhibited by a tight mosaic of hostile territorial defenders, producing the so-called "social fence effect" (Lidicker 1976; Hestbeck 1982; Wolff 1999).

Increasing fragmentation of habitats means that more and more species are living in a metapopulation structure, that is, as an array of variously isolated populations (demes), variously connected by dispersal (McCullough 1996; Hanski and Gilpin 1997; Krohne 1997; Hanski 1999). The success of a metapopulation depends on the balance between the chances of extinction in the habitat patches (demic mor-

tality rate) and the chances of colonization of empty patches by a sufficient number of individuals for establishing a viable population (demic reproductive rate). In general, large and high-quality habitat patches suffer less demic extinction than small and low-quality patches. But this is not invariably the case, because large patches may permit the residency of predators that would be absent or transient in small patches, and hence result in reduced numbers or even demic extinctions (Oksanen and Schneider 1995; Lidicker 1999). Moreover, large and dense demes risk decimation by parasites or other pathogens.

The rate of colonization is a function of successful dispersal among patches. This in turn is influenced by a species' inherent vagility, the timing and magnitude of emigration from patches, the nature of the matrix habitat, the presence of dispersal barriers, a species' response to habitat edges and corridors, mortality rates during dispersal, distances among patches, and the social system (Wolff 2003b). With all of these variables at play, it is notoriously difficult to predict successful colonization rates for any particular metapopulation. Moreover, empirical measures of colonization rates are often difficult and may require long-term studies.

One feature of organisms that can impact a number of these relevant variables is the presence or absence of *presaturation dispersal* (Lidicker 1975; Stenseth and Lidicker 1992). This type of dispersal behavior generates emigrants even during the growth phase of populations. Moreover, these emigrants are in relatively good condition and generally travel during favorable times of the year. Both the quality and quantity of emigrants are thus enhanced by presaturation dispersal, and species exhibiting this behavior are likely to have a relatively high colonization rate. A second critical variable is the species' response to habitat edges (Lidicker and Koenig 1996; Lidicker 1999; Lidicker and Peterson 1999). Species that avoid edges or that are reluctant to cross into matrix habitats will be unlikely to locate corridors between patches and even more unlikely to search for distant habitat patches. In his study of forest/clearcut edges in Oregon, Mills (1995, 1996) illustrates a variety of edge responses by rodents. Chipmunks (*Tamias townsendii*) avoided clearcuts but not the forest edge; red-backed voles (*Clethrionomys californicus*) avoided both edges (to a distance of 45 m into the forest) and clearcuts; and deermice (*Peromyscus maniculatus*) preferred clearcuts but extended into the forest in diminishing numbers to a distance of 45 m. A nonrodent (*Sorex trowbridgii*) did not seem to recognize the boundary as an edge. In patches of grassland maintained by mowing the matrix, *Microtus pennsylvanicus* favored patch edges, presumably because it could feed on freshly sprouted vegetation in the adjacent matrix (Bowers and Dooley 1999), whereas *Microtus canicaudus* females preferred the interior of such patches (Lidicker and Peterson 1999).

A third intrinsic variable affecting colonization potential is the nature of a species' social system. Although often underappreciated as a conservation issue, social behavior has the potential for making a real difference (Wolff 1999). Strong social bonds provide a centripetal force, inhibiting dispersal (Laurance 1990; 1991), and they also may make it difficult for potential immigrants to establish in a new habitat patch, by resisting recruitment of strangers to the social group. Such resistance may furthermore be biased as to sex or age of the would-be immigrant. Dispersers searching for a new home will also be preferentially drawn to inhabited patches as opposed to uninhabited ones. Successful colonization of empty patches may only be possible if a group of individuals either travel together (as in humans) or nearly simultaneously arrive at an empty patch. Sex-biased dispersal may make it less likely that a suitable social group can be assembled in this way. Taken together, there is likely a negative, albeit noisy, relationship between colonizing ability and increasing levels of social complexity. By contrast, the beneficial effects of social living should improve success for groups moving into uninhabited or marginally productive habitat fragments.

An extreme form of spatial structure is for species or populations to inhabit islands or island-like habitat patches. For rodents this means actual islands (surrounded by water) or terrestrial "islands" such as isolated mountaintops (Lomolino et al. 1989). The essence of island life is that there is complete or almost complete isolation from immigration. Therefore, to the extent that this is true, persistence depends only on the probability of local extinction, there being almost no chance of the population being reestablished (rescued) from the outside. It is not therefore surprising that the extinction rate for species restricted to islands is much higher than for those on similar-sized mainland fragments, which maintain at least some level of connectivity with other patches (MacPhee et al. 1989; Steadman 1995; Brown and Lomolino 1998; Lomolino and Perault 2001). MacPhee and Flemming (1999) reported that of forty-three well-documented rodent species extinctions in the last 500 years, 78.3% have been from islands. In the West Indies, fifty-two species of rodents are known to have been present at the close of the Pleistocene, but only thirteen (25%) remain today, and almost all of these are in a precarious state (Woods 1989).

Life and Death of Small Populations

When species become of conservation concern, they generally are persisting as small populations or are rapidly moving in that direction. Hence conservation biologists must pay attention to the behavior of small and persistently declining populations. In an ultimate sense, small populations are at a high risk of extinction simply because their numbers are inherently close to zero, and any negative perturbations are likely to send them to their demise. Therefore, if populations are small, conservationists must work to maintain as many of them as possible so that inevitable demic extinctions can be reversed by immigration from surviving demes. Given this first principle, there remain many other properties of small population size that will be useful to know if management procedures are to be as effective as possible.

Factors that affect population numbers are of two sorts: deterministic and stochastic (Lidicker 1978, 2002). The first are influences that are either predictably related to population density or are consistent or regular in their impacts while being unrelated to population density. Such factors may vary temporally and/or spatially, but once understood, the effects are predictable. Secondly, there are stochastic (chance) influences. These factors occur inconsistently, irregularly, and independently of the status of the focal population. Hence their effects are unpredictable, both in their frequency and magnitude. Both kinds of factors influence all populations, but small populations are especially vulnerable to the latter, and hence they are of special concern in the present context. Stochastic influences are of three types: (1) environmental—factors extrinsic to the subject population that unpredictably impact its size and welfare, that is, weather, food supplies, arrival of novel predators, competitors, or parasites, and other catastrophes, (2) demographic—chance variations in population age structure, sex ratio, or reproductive output, any or all of which could affect demographic behavior, and (3) genetic—random changes (drift) in the genetic constitution of the deme, including loss of advantageous alleles or other polymorphisms, disruption of coadapted genetic complexes, and arrival of new mutations, most of which are likely to be deleterious. Of special note is that these various stochastic influences are not mutually exclusive; various combinations are likely to be important in small populations, and their effects can be multiplicative (synergistic rather than additive).

Small populations are also influenced by deterministic factors. On the positive side, density-regulating factors (Lidicker 1978, 2002) exert their strongest growth-stimulating influence at low densities and hence encourage population growth. On the other hand, density antiregulating forces may also be present, and these have the effect of making small populations decline more rapidly and dense populations increase faster (the Allee Effect, inverse density dependence). They are inherently destabilizing, and may drive small populations to extinction. When antiregulating forces combine with regulating ones, they produce an un-

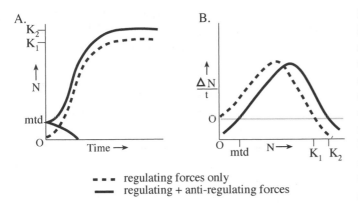

Figure 38.4 Social benefits may generate anti-regulating (inverse density dependent) forces that will contribute to a minimum threshold density (mtd); K_1 and K_2 are equilibrium densities. A. Deterministic changes in population size (N) over time when regulating forces only are operating and when anti-regulating influences are added. B. Population growth rate as a function of population size, with and without anti-regulating forces.

stable equilibrium called the *minimum threshold density* (fig. 38.4). Densities falling below this threshold are headed for extinction. Factors that act in this negative way may be genetic, demographic, or social. In dioecious species like rodents, at least one male and one female with reproductive capacities are minimally required for demic survival. Beyond this, a more substantial group may be needed to maintain sufficient genetic variation for avoiding inbreeding depression, or for preserving an essential polymorphic complex such as is needed for effective immune responses. More than a few individuals may be needed to provide adequate care of young, achieve a viable age structure, or to survive an onslaught of predation. The latter may involve *predator swamping*, the synchronization of breeding so that vulnerable young are available to predators in large numbers but only for a brief period, assuring that an adequate number survive (Darling 1938). Predation may also be antiregulating in multiannual cycling populations in which prey declines precede that of their predators. This generates a situation in which the predator/prey ratio increases as prey numbers drop, thus accelerating the intensity of predation as prey decline (Hansson 1984; Pearson 1985; Lidicker 1988, 1991, 1994). Finally, social behavior may play a major role in influencing the minimum threshold. Species that live in social groups presumably profit from their group behavior. Therefore, at least up to some point, larger numbers means more benefits. This is the classic Allee Effect (Allee 1931, 1951). Conversely, as numbers drop there is a group size that is too small for the social benefits to be maintained. Antiregulating effects at low densities are appropriately called the Darling Effect (Darling 1938), although they are sometimes labeled "the Allee Effect" (Courchamp et al. 1999; Lidicker 2002). Whatever it is called, the more complex a species' social system is the higher will be

its minimum threshold density, and therefore the greater its risk of local extinction when numbers are small. For rodents, this implies that highly social species, such as naked mole-rats (Sherman et al. 1991; Faulkes and Bennett, chap. 36 this volume), prairie dogs (*Cynomys*; Hoogland, chaps. 37 and 39 this volume), various ground squirrels (*Spermophilus*; Murie and Michener 1984, Hare and Murie, chap. 29 this volume; Van Horne, chap. 39 this volume), beaver (Busher, chap. 24 this volume), and flying squirrels (*Glaucomys*; Koprowski, chap. 7 this volume) will carry this added risk factor. Even species with much simpler social behaviors overall, but which require group behavior at certain times (e.g., thermoregulatory advantage for overwintering, such as in *Microtus xanthognathus*; Wolff and Lidicker 1981), will have high minimum thresholds.

Social activities engaged in by rodents that could be lost at low densities include construction of burrows and dams, reproductive stimulation, cooperative breeding, predator swamping when vulnerable young are present, alarm calling, food storage, group thermoregulation, and management of appropriate vegetation height. Lastly, we need to be reminded that deterministic influences at low densities can interact synergistically or additively with various stochastic factors.

Community and Landscape Contexts

It has become increasingly obvious over the past several decades that the welfare of target species of interest cannot be fully understood unless it is placed in the context of the community and landscape in which they live (Lidicker 1995; Barrett and Peles 1999). All species interact (coact) with other species in their communities. Coactions can be trophic (generally exploitative, $+ / -$), competitive ($- / -$), cooperative (mutualistic, $+ / +$), commensalistic ($+ / \circ$), or amensalistic ($- / \circ$), and often they are dynamic in nature (Lidicker 1979b). As has been abundantly documented, it is difficult to predict the effect on species when exotic species become new members of their community. The fact that such exotic introductions are widely acknowledged as second only to habitat destruction in causing species extinctions emphasizes that often the impacts of introductions are negative for resident species (Holdaway 1999). Disruption also occurs when community members are removed. The magnitude of change occurring when this happens directly measures the dominance relations of the removed species.

When species with high dominance are removed, massive changes in the community occur. Imagine the impact if redwood trees (*Sequoia sempervirens*) were removed from a redwood-dominated forest. Very few species of that forest community would persist on the site. Plotnick and McKinney (1993) model the relationship between the number and intensity of connections among species (which they call *dependencies*) on the impact of species removal. As the level of dependencies among species goes up, the number of species expected to go extinct when a species is removed rises in an S-shaped curve. Their model does not distinguish positive and negative coactions, nor does it consider indirect effects. They nevertheless come to the cautious conclusion that communities that rarely experience species removals will evolve to increasing levels of dependencies. This in turn will make them extremely vulnerable to mass extinctions if a major disturbance does occur. By contrast, frequently perturbed communities will evolve toward lower levels of dependency, and hence acquire more extinction resistance. For conservationists, this implies that for a species of conservation concern we need to know about its dependencies with other species to judge both (1) its community dominance, and therefore its importance to the integrity of the community as a whole, and (2) its vulnerability to nonspecific disturbances.

Because rodents are often abundant in communities, they will be strongly connected to other species and hence relatively dominant. Rare species may be connected to common ones through food competition and predator sharing. A few examples will illustrate the importance of rodents in diverse communities. Ostfeld and colleagues (Jones et al. 1998) have assembled a remarkable account of how white-footed mice (*Peromyscus leucopus*) of eastern North American deciduous forests are complexly connected to white-tailed deer (*Odocoileus virginiana*), gypsy moth (*Lymantria dispar*), Lyme disease (*Borrelia burgdorferi*), and oak trees (*Quercus*). Keesing (1998b) has shown how the small mammal community in African savannahs interacts strongly with the large herbivores. In an experimental investigation of the effects of sheep grazing in Norwegian mountain pastures, Steen et al. (2005) demonstrate that sheep strongly depressed populations of *Microtus agrestis*, but had no apparent effect on *Clethrionomys glareolus*. Rodents also have been implicated as major dispersers of mycorrhizal spores in coniferous forests (Maser et al. 1978). Many other examples of rodents being complexly connected within their communities will undoubtedly be reported as investigators focus more on this level of ecological analysis.

A more recent, but extremely important, insight has been the realization that not only is the community context critical for conservation action, but landscape variables may also be important determinants of a species' well-being (Forman and Godron 1986; Forman 1995; Lidicker 1995; Barrett and Peles 1999). Recently (Lidicker 2000), I argued that the interaction between the ratio of optimal to marginal habitat (ROMPA) in a landscape combined with the presence or absence of generalist and/or specialist predators can determine the kind of demographic pattern shown by

species of *Microtus*. In general, the presence of various predators will depend on the sizes of the habitat patches and their proximity to other community types in which the predators may reside. Moreover, *spillover predation* may impact species living in habitats insufficiently productive or too small to support resident predators (Oksanen and Schneider 1995; Lidicker 1999). Different combinations of contiguous community types can lead to quite different population dynamics in species resident in only one of the patches (Danielson 1991; Lidicker 1995, 1999). For example, a habitat patch suitable for some rodent species that is surrounded by a matrix that absorbs emigrants from the patch but does not supply immigrants in return (dispersal sink; Lidicker 1975), will result in a chronically depressed rodent population in the isolated patch (demographic sink; Pulliam 1988, 1996). Conservation planning must also acknowledge that migratory species will require preservation of both winter and summer habitats. Even nonmigratory species may benefit from or even require the juxtaposition of two or more habitat types (Lidicker et al. 1992; Shellhammer 1998).

Another landscape feature receiving increased attention is the nature of the ecotone (edge) that develops between two community types (fig. 38.5), and the responses of organisms to this zone (Wiens 1992; Lidicker and Koenig 1996; Lidicker 1999; Lidicker and Peterson 1999; Wolff 1999, 2003). Sometimes ecotones have positive impacts on biodiversity. In other situations they may serve as conduits for the invasion of predators, competitors, and parasites that are deleterious to the original community members. Ecotones also can strongly decrease the effective size of the habitat patches if edge effects penetrate deeply into the adjacent communities, as they often do (Laurance, Bierregard et al. 1997, Laurance, Laurance et al. 1997). As pointed out earlier, it is important to know how target species respond to ecotones, as this may significantly affect their success in the landscape. Some species are ecotonalphobic, and hence are unlikely to cross gaps in their habitat to find new suitable patches. They also will not readily be able to find corridors that could allow them to emigrate from their home habitat fragment (Crome et al. 1994; Lidicker and Koenig 1996). Consequently, both colonization of empty patches and genetic exchange among patches will be rare.

Corridors are also landscape features that can facilitate connectedness among habitat fragments (Forman and Godron 1986; Lidicker and Koenig 1996; Lidicker 1999), and therefore may be critical assets in metapopulation or metacommunity persistence (fig. 38.6). Negative effects can also be associated with corridors (Simberloff and Cox 1987). Corridors can simply increase the amount of edge habitat available to exotic invaders and thereby negatively impact the habitat patches (Crooks and Soulé 1999; Hawkins et al. 1999; Panetta and Hopkins 1991). Or, corridors can facilitate movement of predators, parasites, and disease organisms across the landscape (Hess 1994; Watson 1991). A more subtle effect is that corridors, like matrix, may act as dispersal sinks. Individuals might be attracted to use corridors as residences, but then fail to reproduce enough to replace themselves. Such corridors would then act as a continual drain on the populations living in more productive patches (sources). Baranga (1991) describes an example of this from east Africa. Corridors are also likely to be community filters in that only some members of the community will be able to use them successfully (Downes et al. 1997; Forman 1995; Laurance 1995). This bias will likely lead to distortions in the community compositions on either end of the corridor, with all of the potential negative effects, discussed earlier, that this entails. Corridors can in principle be designed to be efficient transporters of individuals and not suboptimal habitat space, but generally the specifications for this will be very species specific, and therefore not conducive to community connnectivity. Lastly, it needs to be mentioned that corridors can be expensive to incorporate into regional conservation efforts, and planning efforts must weigh the potential benefits and disadvantages of landscape corridors. My guess is that most of the time the benefits will prevail.

Conclusions

1. Rodents are ubiquitous and extremely diverse. They have suffered a large number of anthropogenic extinctions, 78.3% of which have been on islands.
2. In spite of the great need for conservation attention, the public remains largely indifferent and/or complacent

Figure 38.5 View of Østerdalen at Evenstad, Hedmark, Norway, showing numerous edges (ecotones) generated by natural landscape features (riparian zone, altitudinal zonation of communities) and anthropogenic modifications (timber harvesting, agriculture, experimental enclosures for investigating landscape influences on root voles (*Microtus oeconomus*). 12 June, 1990; photo by author.

Figure 38.6 Experimental landscapes at Oregon State University used for studying population dynamics and social behavior of gray-tailed voles (*Microtus canicaudus*). Corridors can be seen connecting some of the habitat patches. June 1997; photo by J. O. Wolff.

regarding rodent conservation. These attitudes emphasize the need to focus conservation efforts on entire
communities rather than on focal species.

3. We need to greatly expand our knowledge of rodents in
most parts of the world.

4. Rodents are subject to the same principles of conservation biology as other organisms, and in fact because of their extraordinary diversity (2,021 species in
twenty-nine families) they comprise a representative
subsample of mammals, except that there are few large-
bodied species. Observed demographic patterns cover
the full range of commonness, all forms of rarity, and
ephemerality (migration, hibernation, multiannual
cycles, irruptions).

5. Social behavior is a significant factor in rodent conservation, and collectively rodents exhibit a wide range of
social complexity.

6. Social factors affect: (1) spatial structure, (2) genetic
structure, (3) the chances of small-population extinction by modifying the minimum threshold density,
(4) colonization ability, (5) dispersal behavior, (6) resistance to immigration, (7) the success of metapopulations, and (8) responses to landscape features such as
edges and corridors.

7. Our understanding of what constitutes effective conservation action has expanded greatly as we have moved
from a focus on simply measuring population vital
rates to an appreciation for the complex array of factors that generally influences these rates, and that can
vary over time and space. Moreover, we have begun
to incorporate community context and landscape elements into our conceptual framework.

8. Research on rodents has been an essential component
in this intellectual development, and hopefully they will
reap the benefits of this continuing enterprise, along
with the rest of the Earth's biota, including most particularly our own species.

Summary

Rodent conservation is disadvantaged by widespread indifference and complacency. Yet, rodents comprise a large proportion of the mammalian fauna on six continents. Rodents
are extremely diverse (438 genera in twenty-nine families),
and often play key roles in the communities in which they
live. Rodents impact human welfare as agricultural pests,
carriers of zoonoses, pilferers of stored food, and as adept
and damaging alien invaders of natural communities. However, they also are incredibly valuable in biomedical research, as components of healthy ecosystems, as pets, furbearers, and even food. Most importantly, rodents provide
innumerable opportunities for evolutionary, ecological, and
behavioral research. In spite of their overwhelming importance, rodents are poorly known on most continents. Even
in relatively well-known areas such as North America and
Australia, 40.3 and 39.3 % of extant species respectively are
currently of conservation concern. Rodents show all of the
known patterns of demographic behavior, the various types
posing different modalities of conservation concern. Distribution in space depends on geographic range, habitat fragmentation, dispersal capabilities, and social system. Small
ranges, fragmented habitat, poor dispersal ability, and complex social systems all increase the risk of extinctions. Small
populations carry additional risks from stochastic impacts
as well as deterministic antiregulating forces, which generate minimum threshold densities often well above zero. Social behavior often contributes to such minimum thresholds. Effective conservation requires an understanding of
the often complex community context of focal populations,
and this may vary spatially and temporally. This context
includes community dominance relations as well as the nature and magnitude of interspecific interactions. Finally, it
is critical to understand how species respond to landscape
factors such as habitat patch size, quality, and distribution,
and edge effects, corridors, and matrix features.

Chapter 39 Conservation of Ground Squirrels

Beatrice Van Horne

SPECIES OF GROUND SQUIRRELS (*Spermophilus*) and prairie dogs (*Cynomys*) comprise a monophyletic group of fossorial species (Black 1963; Hafner 1984; Harrison et al. 2003; Nadler et al. 1971). *Cynomys* represents a specialized form of highly social species (Armitage 1981; Michener 1983a) that probably arose in the late Pliocene (Black 1963) in association with an adaptive radiation of sciurids resulting from mountain uplift and drying trends in western North America (Hafner 1984). Ground squirrels have long been considered by many to be pests and varmints that compete with cattle for food, dig holes that are treacherous to domestic ungulates, provide shelter for poisonous snakes, carry plague, and create unsightly disturbances in otherwise tidy landscapes. Such problems are primarily identified by farmers, ranchers, and residents of urban edges, and have provided a rationale for localized and/or widespread shooting and poisoning, primarily of prairie dogs. Others, especially some urban dwellers as well as ecologists interested in maintaining ecosystems in which burrowing mammals influence a wide variety of other species, wish to see them protected. Prairie dogs, with their high level of sociality, larger size, and appealing visage may be especially targeted by the public for conservation. Hoogland (chap. 40 this volume) describes problems in conservation of prairie dogs. This chapter provides a more general overview of ground squirrel conservation.

My intent here is to explore the biology of these species to better understand the underlying causes of broad population declines. It may seem counterintuitive that a group composed of a number of widespread and seemingly robust species that are well adapted to habitat disturbance may yet contain several species and subspecies at risk of near-term extinction. I will review the conservation status of these species and summarize characteristics of species that are of concern. This review will largely focus on North American species, for which the most relevant information is available. I will then explore how the combination of their body size, fossorial habits, and largely herbivorous diets greatly restricts the range of adaptations that can be exhibited by this group, and allows for some commonalities in patterns of genetic variation across the landscape and in the regulation of population size. From these commonalities, I will suggest some generalizations regarding conservation.

Distribution and Conservation Status

For a visual overview of ranges of north American ground squirrels, I have separated species of immediate conservation concern (small to large, fig. 39.1) from smaller (fig. 39.2) and larger (figs. 39.3, 39.4) species of less concern, and have shown the ranges of prairie dogs separately (fig. 39.5). Conservation status of these species varies (table 39.1). Of the thirty-two species of ground squirrels and prairie dogs in North America, one, the Utah prairie dog, is considered by NatureServe (NatureServe 2004) to be "critically imperiled" and is listed by the US federal government as threatened. Four species (the Idaho, Washington, and Mojave ground squirrels, and the Mexican prairie dog are considered "imperiled" on the same basis, and the Washington ground squirrel is a candidate for US federal listing (scientific names are presented in tables 39.1 and 39.2). One (the

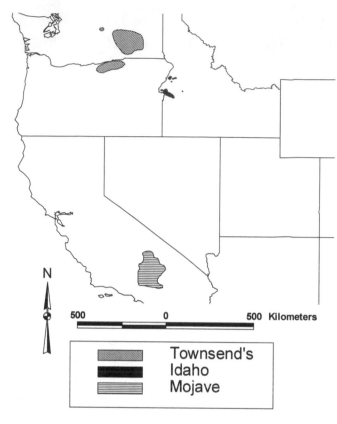

Figure 39.1 Ranges of ground squirrel species of conservation concern (Patterson et al. 2003). These include species designated "critically imperiled" or "imperiled" by NatureServe. Range sizes are distorted by map projections.

black-tailed prairie dog) is "vulnerable." Merriam's and Townsend's ground squirrels are "apparently secure," that is, uncommon but not rare, but with possible cause for long-term concern. The remaining eighteen species are considered "secure," that is, common, widespread, and abundant, although Franklin's ground squirrel has been declining rapidly because of loss of its tall grass habitat resulting from agricultural and urban expansion (Johnson and Choromanski-Norris 1992), and is considered "vulnerable" by the IUCN.

In Europe and Asia there is uneven knowledge about the status of the thirteen *Spermophilus* species. The European ground squirrel has declined in Europe because of habitat destruction by modern agricultural techniques and urbanization (Smit and Van Wijngaarden 1981). The European and speckled ground squirrels are considered "vulnerable" by the IUCN (World Conservation Union: IUCN 2003), while the Russet ground squirrel is "near threatened."

The causes of imperiled status are varied. Control programs (Pizzimenti and Collier 1975), along with a restricted range following the last major episode of climate change (see Goodwin 1995), may account for the status of the Utah prairie dog (fig. 39.5). Remaining imperiled Idaho ground squirrels consist of two subpopulations, each having different factors that have led to habitat loss. Northern Idaho ground squirrels are threatened by thick growth of young trees into their preferred open stands of conifers with herbaceous understory, a condition attributable to fire

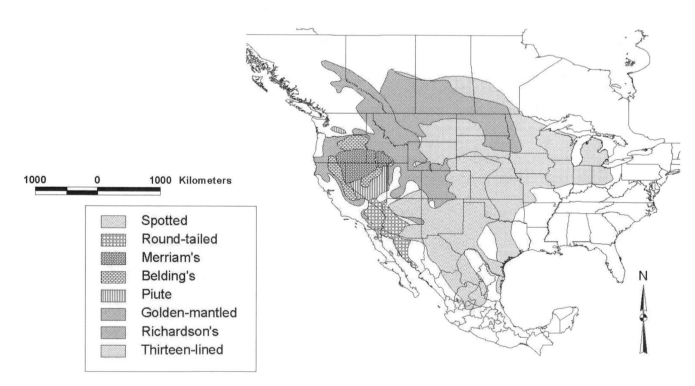

Figure 39.2 Ranges of "smaller" ground squirrels (average body length 13.9–21.9 cm; Patterson et al. 2003) Range sizes are distorted by map projections.

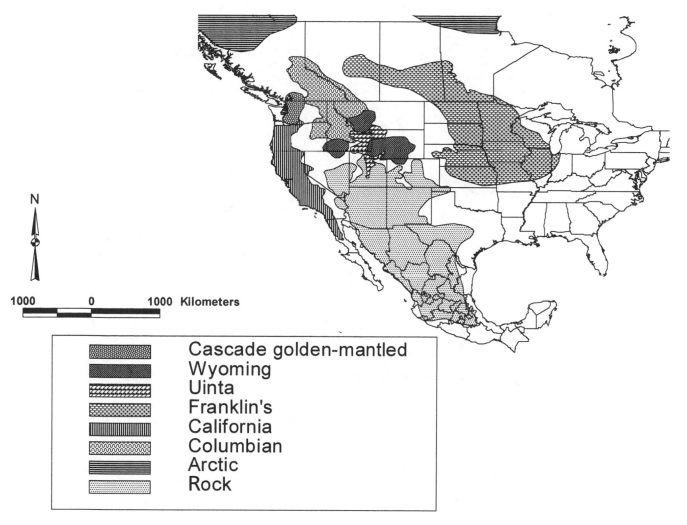

Figure 39.3   Ranges of "larger" ground squirrels (average body length 22.5–30.5 cm; Patterson et al. 2003). Range sizes are distorted by map projections.

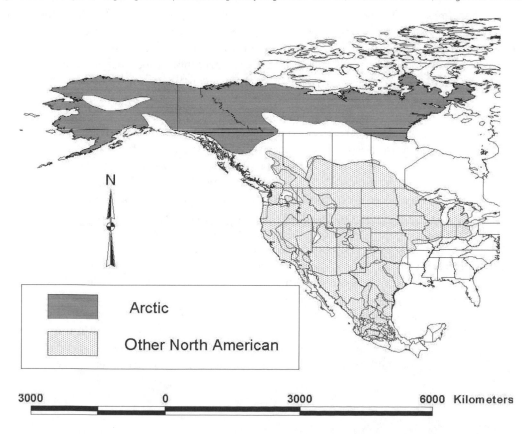

Figure 39.4   Range of the arctic ground squirrel compared with a composite range of other North American ground squirrels (Patterson et al. 2003). Range sizes are distorted by map projections.

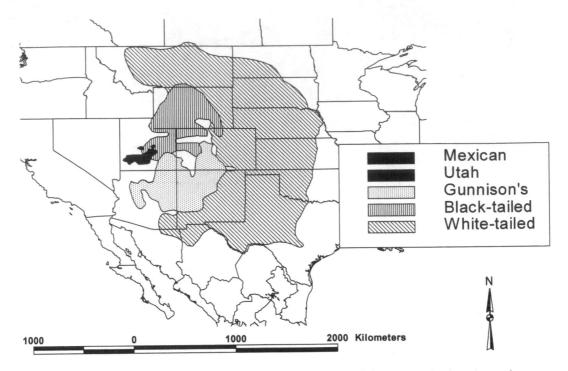

Figure 39.5 Ranges of North American prairie dogs (Patterson et al. 2003). Black-tailed, Mexican, and Utah species are of conservation concern. These include species designated "vulnerable," "critically imperiled," or "imperiled" by NatureServe. Range sizes are distorted by map projections.

suppression, logging, and grazing (Crane and Fischer 1986; Steele et al. 1986; Sherman and Runge 2002), and by invasion of exotic grasses that lack adequate seed heads to provide overwinter food (Sherman and Runge 2002). Southern Idaho ground squirrels have a small range in west-central Idaho. The native shrub and bunchgrass habitat favored by this subspecies has been mostly converted to a fire-dependent grassland dominated by exotic annuals, or to intensive agriculture; rodent control negatively affects the remaining small populations (Whisenant 1990, Eric Yensen, pers. comm). The Washington ground squirrel has been affected by agricultural conversion of grasslands, invasion of exotic plants, intensive grazing, and control measures (Betts 1999, Rickart and Yensen 1991). Habitat conversion to agricultural, mining, and urban uses, as well as desertification associated with climate change affect the Mojave ground squirrel near the Los Angeles urban area (Best 1995).

The imperiled Mohave, Idaho, and Washington ground squirrels are below the median body length of North American ground squirrels and are found in relatively moist habitat islands within the drier areas of ground squirrel ranges in North America (fig. 39.1). These ground squirrels may have reached an adaptive limit for tolerating arid environments and have therefore declined as their relatitvely mesic habitat islands have shrunk following the most recent Pleistocene glacial period. They are then vulnerable to further climatic drying, increasing irregularlity of rainfall associ-

ated with climate change, invasion by exotic annual species that make quick use of surface moisture before drying out (as compared to the deeper-rooted native bunchgrasses), and shrinkage and fragmentation of remaining habitat.

Where populations are disjunct, population genetics demonstrate spatial structuring (c.f. Antolin et al. 2001). Maintaining the genetic variability represented by such structuring may be important in both localized adaptation and longer-term evolution, especially in response to climatic change. At a minimum, conservation of recognized subspecies should be important to maintaining a portion of this genetic variability. There are nineteen subspecies of ground squirrels and prairie dogs recognized in North America; sixteen of these are of particular conservation concern or of unknown status (table 39.2). Most of these subspecies are associated with globally common species: the northern and southern Idaho ground squirrels are an exception, however, as subspecies of an imperiled species. The Coachella round-tailed ground squirrel is considered "critically imperiled" and is a candidate for US federal listing. The northern and southern subspecies of the Idaho ground squirrel are considered "imperiled" (NatureServe 2004); the northern subspecies is federally listed in the US as "threatened," while the southern is a candidate for listing. Seven subspecies are "vulnerable." Three subspecies are "apparently secure." The remaining five are too poorly known to rank. All of the imperiled subspecies have his-

Table 39.1 North American species of *Cynomys* and *Spermophilus,* and *Spermophilus* spp. elsewhere known to be endangered or vulnerable

Common name	Scientific name	IUCN status	Global conservation status rank
Gunnison's prairie dog	C. gunnisoni		
White-tailed prairie dog	C. leucurus		
Black-tailed prairie dog	C. ludovicianus		Vulnerable
Mexican prairie dog	C. mexicanus		Imperiled
Utah prairie dog	C. parvidens		Critically imperiled
Uinta ground squirrel	S. armatus		
California ground squirrel	S. beecheyi		
Belding's ground squirrel	S. beldingi		
Idaho ground squirrel	S. brunneus	Endangered	Imperiled
Merriam's ground squirrel	S. canus		Apparently secure
Columbian ground squirrel	S. columbianus		
Wyoming ground squirrel	S. elegans		
Franklin's ground squirrel	S. franklinii	Vulnerable	
Golden-mantled ground squirrel	S. lateralis		
Mexican ground squirrel	S. mexicanus		
Mohave ground squirrel	S. mohavensis	Vulnerable	Imperiled or vulnerable
Piute ground squirrel	S. mollis		
Arctic ground squirrel	S. parryii		
Richardson's ground squirrel	S. richardsonii		
Cascade golden-mantled ground squirrel	S. saturatus		
Spotted ground squirrel	S. spilosoma		
Round-tailed ground squirrel	S. tereticaudus		
Townsend's ground squirrel	S. townsendii	Data deficient	Apparently secure
Thirteen-lined ground squirrel	S. tridecemlineatus		
Rock squirrel	S. variegatus		
Washington ground squirrel	S. washingtoni	Vulnerable	Imperiled

SOURCE: IUCN (2003); NatureServe (2004).
NOTES: C. = *Cynomys*; S. = *Spermophilus*. IUCN red list status: Vulnerable = facing a high risk of extinction in the medium-term future. Data deficient = inadequate information to assess risk of extinction based on distribution and/or population status. Global conservation status rank (NatureServe G1–G3): Critically imperiled = at very high risk of extinction due to extreme rarity (often 5 or fewer populations), very steep declines, or other factors. Imperiled = at high risk of extinction due to very restricted range, very few populations (often 20 or fewer), steep declines, or other factors. Vulnerable = at moderate risk of extinction due to a restricted range, relatively few populations (often 80 or fewer), recent and widespread declines, or other factors. Apparently secure = Uncommon but not rare; some cause for long-term concern due to declines or other factors. Unranked = not yet assessed. If no rank is given, the species is considered to be Secure (common, widespread, abundant).

torically limited ranges. The round-tailed squirrel has lost about half of its historic range to urban land conversion.

It seems that island endemism and/or habitat loss is associated with most imperiled status classifications for species and subspecies. Habitat loss is most apparent in the forest encroachment at the dry northwestern edge of the range of intermountain prairie dogs, desertification at the southwestern edge of the range, and anthropogenic conversion of these dry northern and southern grasslands. Continental ground squirrels and prairie dogs that are less dependent on dry grasslands, including those found in mountainous areas, populations in the far north, west of the Coast range, in shrubsteppe (or high desert), in tallgrass prairie, or adapted to a wide variety of habitats generally are less threatened. Habitat conversion by replacement of deep-rooted, perennial bunchgrasses with invasive annual monocots and dicots is a concern throughout, however, but especially in north-central shrubsteppe habitats.

Understanding how this broad picture of intertwined habitat loss and population fragmentation affects ground squirrels requires that we consider how species attributes related to subterranean burrow construction and energy budgeting place limits on adaptation to habitat change. Within this context, predation, disease, and genetic structuring also influence conservation status and long-term prospects for these species.

Energetic Adaptations

Given that diurnal ground squirrels are an important food source for avian and mammalian predators, one might expect "top-down" predation to limit population growth.

Table 39.2 Conservation status of all North American ground squirrel subspecies known to be endangered, imperiled, or vulnerable

Common name	Scientific name	Taxon conservation status rank
Northern Idaho ground squirrel	S. brunneus brunneus	Imperiled
Southern Idaho ground squirrel	S. brunneus endemicus	Imperiled
Wyoming ground squirrel	S. elegans nevadensis	Apparently secure
Wind river golden-mantled ground squirrel	S. lateralis lateralis	Apparently secure
Golden-mantled ground squirrel	S. lateralis wortmani	Apparently secure
Kodiak island arctic ground squirrel	S. parryii kodiacensis	Vulnerable
St. Lawrence Island ground squirrel	S. parryii lyratus	Vulnerable
Shumagin Islands Arctic ground squirrel	S. parryii nebulicola	Vulnerable
Odgood's arctic ground squirrel	S. parryii osgoodi	Vulnerable (uncertain)
Spotted ground squirrel	S. spilosoma marginatus	Unranked
	S. spilosoma obsoletus	Unranked
Coachella round-tailed ground squirrel	S. tereticaudus chlorus	Critically imperiled or Imperiled
Allen's thirteen-lined ground squirrel	S. tridecemlineatus alleni	Possibly extinct, Questionable taxonomic status
Ground squirrel	S. tridecemlineatus arenicola	Unranked
Ground squirrel	S. tridecemlineatus blanca	Vulnerable
White mountains ground squirrel	S. tridecemlineatus monticola	Vulnerable
Ground squirrel	S. tridecemlineatus pallidus	Unranked
Ground squirrel	S. tridecemlineatus parvus	Unranked
Rock squirrel	S. variegatus tularosae	Critically imperiled, Imperiled, or Vulnerable

SOURCE: NatureServe (2004).
NOTES: S. = *Spermophilus*. For infraspecific taxon rank (NatureServe), see table 41.1.

However, burrow use and diets place strong energetic constraints on these animals, and it appears that these "bottom-up" factors are often limiting (except in cases of episodic plague and anthropogenic control efforts).

Burrows are essential for both active and torpid ground squirrels for thermoregulation and predator protection, and anchor much of the above-ground activity. Burrow systems are not completely safe, and may be invaded by small-bodied snakes as well as larger-bodied digging predators such as badgers (*Taxidea taxus*). Badgers take ground squirrels regardless of their size or torpor status (Michener 2000; Michener and Iwaniuk 2001). Ground squirrels are thus spatially restricted, but their largely herbaceous diets (e.g., Van Horne et al. 1998) require a high volume of consumption. Ranges in adult male weights are 135–625 g for *Spermophilus* and 600–750 g for *Cynomys*. A 400-g Richardson's ground squirrel requires 30.6 g of fresh gypsyflower (*Cynoglossum officinale*) leaves per day, or 7.65% of its body mass, to meet its metabolic requirements (Blintz 1984). At this rate, ground squirrels and the larger-bodied prairie dogs can consume a large fraction of the preferred foods near their burrows, which, when risk of predation is high and food productivity is low, can create a spatially restricted food limitation.

For hibernators, caloric intake must also support a steep increase in body mass through the active season. Body mass of Piute ground squirrels (*S. mollis*, formerly *S. townsendii*) generally doubled during their 4–5-month active season (Van Horne et al. 1997). The types of foods consumed by ground squirrels generally cannot be stored for long periods in moist burrow systems. Indeed, caching behavior has not been reported; caching is found in ground-dwelling mammals in very dry environments, or in moist environments where caches can be kept relatively dry and/or seed coats are very hard.

Small mammals generally use favorable microclimates to avoid exposure to inclement weather. For ground squirrels, these favorable microclimates often are available only in the burrow systems, where food is unavailable. When weather conditions are such that foraging requires more energy than can be gained, animals must use stored energy. Under these conditions, food availability can easily become limiting. Adaptations to such conditions, particularly in very hot or very cold climates, involve combinations of maximizing food storage (mostly as fat) and minimizing metabolic energy expenditure through the use of seasonal or temporary heterothermy (Lehmer et al. 2001).

So it seems that energy budgeting could be challenging for these species. What evidence do we have that energy availability actually limits populations? Surely the pervasiveness of heterothermy among these species and populations indicates the need to conserve energy. Metabolic adaptations to the problem of obtaining energy in ground squirrels vary from obligate hibernation to opportunistic use of torpor bouts. Heterothermy occurs in response to lack of moisture and/or low temperatures. Facultative torpor in prairie dogs is influenced by body condition, as it occurs at warmer temperatures in late winter when animals are near their minimum annual body mass, as compared to late fall/early winter, when larger temperature drops are required to stimulate torpor (Lehmer et al. 2001, Lehmer and Van Horne 2001, Lehmer et al. 2003).

From heterothermy we may infer that energy conservation is important, but there is also abundant evidence that energy availability influences population change. For example, Van Horne et al. (1997) described near-zero survival of the juvenile cohort; and reduced survival of adults, especially females, when a drought caused primary foods to dessicate during what is normally the active season. Low survival was associated with low body mass. This lower survival of females during periods of low food supply is likely associated with the high energetic cost of reproduction. A dramatic population decline in Northern Idaho ground squirrels was attributed to inadequate food resources, especially seeds, resulting from drying of the habitat and changes to plant species composition associated with incursion of conifers and with grazing (Sherman and Runge 2002).

Columbian ground squirrels have been the subject of exacting physiological and demographic studies that demonstrate the effects of energy limitation on survival and reproduction. Bennett (1999) found that females that did not produce litters had higher chances of surviving, and that females that lost their young during lactation showed higher survival rates than those completing the lactation period. Juvenile Columbian ground squirrels with larger body masses survived better over winter. Body mass was, in turn, related to the availability of digestible energy, with forbs being a better source than grasses. Similarly, another study showed that the proportion of Columbian ground squirrels breeding as yearlings increased as seasonal energy intake increased (Ritchie and Belovsky 1990). A snowstorm early in the above-ground season was associated with mass loss in females, and those losing more mass produced fewer young. (Neuhas et al. 1999). The importance of food limitation and body condition during reproduction for female Columbian ground squirrels is further supported by the work of Neuhaus (2000). Heavier females were more likely to produce litters (the same pattern is evident in King et al. 1991), and females that did not produce litters were heavier going into hibernation and emerging the following spring, and had higher survival. When young were added to litters, the females seemed to adjust the litter size back down, according to their ability to wean the young. Females that did not raise litters had a shorter active and longer hibernation season (Neuhaus 2000). Columbian ground squirrel females that gained the most mass also weaned the largest litters (Dobson et al. 1999).

This pattern of apparent energetic limitation is evident in other ground squirrels. As in Columbian ground squirrels, Morton and Sherman (1978) found that a spring snow-

storm substantially reduced survival of Belding's ground squirrels, and the same circumstance was observed in Piute (formerly Townsend's) ground squirrels (Van Horne et al. 1997). Because Belding's ground squirrels have a short time to store sufficient fat to survive over winter, maintaining habitat quality in meadows (abundant grasses) is essential. Where habitat quality declines, long winters may be associated with low survival rates (Morhardt and Gates 1974). Golden-mantled ground squirrels subjected to a shorter than normal above-ground season reproduced at a lower rate (Phillips 1984). Female European ground squirrels with high reproductive output and investment exhibited lower overwinter survival or delayed estrus the following season (Millesi et al. 1999).

Drought can affect the seasonal energy balance by reducing the length of the above-ground season. Many ground-dwelling sciurids satisfy their water needs through consumption of succulent food. Lack of water stimulates animals in dry environments to immerge into seasonal torpor (Blintz 1984; Harlow 1997; Van Horne et al. 1997). Piute ground squirrels in Idaho entered torpor earlier in drought than in nondrought years (Smith and Johnson 1985; Van Horne et al. 1997).

Experiments involving adding individuals, and thus putting pressure on resources, or adding or improving the quality of food also indicate the importance of bottom-up processes. Arctic ground squirrel populations to which individuals were added declined to match unmanipulated low-density populations as a result of declines in weaning rates and overwinter survival (Karels and Boonstra 2000). Because overwinter survival was not a function of predation, it appears that bottom-up processes involving nutritional status were responsible for the density-dependent population declines. Forage improved in quality (water, nitrogen) by point fertilization with urine was consumed preferentially by Columbian ground squirrels. Females with access to such forage gained more mass and produced larger litters than those without high-quality forage (Boag and Wiggett 1994a, 1994b). Dobson and Kjelgaard (1985) observed a similar effect from enriched forage, including larger litters, earlier female maturation, higher juvenile survival, and greater female body mass.

Food supplementation experiments generally support the linkage between food availability and survival/fecundity. Dobson and Oli (2001) found that food supplementation produced both earlier age at maturity and increased fertility rates in Columbian ground squirrels. Supplementation (Dobson and Kjelgaard 1985) also led to increased reproduction and a higher survival rate for resident young animals.

Broad-scale experiments conducted by Byrom et al. (2000) on populations of arctic ground squirrels in the northern boreal forest demonstrated a two-fold increase in population density associated with removal of predators, but a four-fold increase associated with food supplementation. Predation may have been particularly intense in their system, as the experiment was conducted during a decline and low phase in the snowshoe hare cycle, when predators were high and alternative prey low in density. The added food resulted in improved body condition, earlier emergence of juveniles, and increased reproductive rates (Karels et al. 2000). Although more than 50% of juvenile Columbian ground squirrels were lost to badgers one year (Murie 1992), higher survivorship of remaining juveniles may have compensated.

Given that production of food plants is critically important to maintaining viable populations of ground squirrels, competition with invasive plants may be the greatest cause of decline (aside from outright loss of habitat to agricultural or other development.) Exotic annuals compete with native perennials for scarce water in dry environments, but provide only an ephemeral food supply to herbivores. Fire, overgrazing, and other anthropogenic disturbances can encourage these invasives and the invasives, in turn, encourage the spread of fire. Reduced or more irregular rainfall associated with climate change will exacerbate this problem.

Habitat Fragmentation

Habitat fragmentation for ground squirrels is produced by conversion of land for agriculture, roads, and urban development. Prairie dogs live in dense colonies, and so are already unevenly distributed across the landscape. Perhaps their extreme sociality makes prairie dogs more susceptible to both control efforts and plague. Indeed, sylvatic plague (*Yersinia pestis;* Barnes 1982) has caused local extinction of many colonies. Disease is not restricted to prairie dogs; a widespread decline, possibly resulting from plague, led to apparent extinction across much of the range for Townsend ground squirrels in Utah beginning in 1936 and persisting at least 18 years (Hansen 1956).

For prairie dogs, areas that appear on the basis of vegetation to be uniform habitat may contain soils unsuitable for colonization, adding to population fragmentation. Variation in topography, soils, and land-use history contribute to fragmentation (Koford 1958). The need to be able to dig burrows that will remain sufficiently dry without excessive energy expenditure restricts these species to reasonably uniform, colloidial, and arid to somewhat mesic soils that are lacking in aggregate rock. Sandy soils are particularly unsuitable for animals with larger body sizes, so that we may find kangaroo rats (*Dipodomys* spp.), for instance, in soils too sandy for burrows of larger ground squirrels or prairie

dogs. Prairie dog colonies in the Pawnee National Grasslands of Colorado were located less often than expected on Olney fine, sandy loam, based on its availability, possibly because of avoidance of soils with deep, sandy horizons.

In a fragmented population, animals that disperse to genetically dissimilar colonies may produce more successful and robust offspring than those that remain and breed with more genetically similar animals. Long-distance dispersal has been more closely associated with inbreeding avoidance (Dobson et al. 1997) than with lack of food resources or other factors. Prairie dogs occasionally disperse long distances, sometimes more than 5 km (Garret et al. 1982; Garrett and Franklin 1988). Dispersal is risky, however, exposing animals to predators in unfamiliar territory. Dispersing juvenile arctic ground squirrels suffered higher mortality than philopatric squirrels, and this effect increased with dispersal distance (Byrom and Krebs 1999).

Dependence on burrows for hibernation, thermoregulation, escape from predators, and reproduction raises the costs of dispersal away from the home burrow system. Constructing burrow systems *de novo* is challenging and risky, as digging animals are exposed to predators and weather, so that animals most often settle in areas with existing burrow systems (Boag and Murie 1981; Murie and Harris 1984; Knowles 1985; Weddell 1989), even if the entrances to such systems have been effectively covered over with loose soil. Finding existing burrows can be difficult for highly social species that must cover wide areas of unoccupied habitat to disperse between social groups. Hence, ground squirrels show pronounced genetic population structuring, in which similarity decreases directly with distance or with distance along dispersal routes such as drainages (Gavin et al. 1999; Roach et al. 2001; Dobson, chap. 14 this volume). Because of the energetic costs and risks of burrow construction, and the corresponding energetic savings and safety they offer, burrow systems are valuable resources (King 1984).

Conservation and Management Implications

Ground squirrels can strongly affect ecosystem processes as a rich diurnal prey item (Marti et al. 1993) by foraging and defecating and through burrow construction and soil mixing (Yensen and Sherman 2003). In nature, ground squirrels are largely regulated by environmental "crunches" in bottom-up nutrient availability. There is abundant evidence of the sensitivity of these species to reduction in nutrient availability through prolonged winter, periodic drought, and invasion of exotic annuals. Climate change, especially where it affects cycles of precipitation and their predictability, could push extensive areas of the ranges of these species below a critical production threshold, particularly in arid parts of the ranges. Two or three years of unsuccessful reproduction on a level with what has been observed in single years (Van Horne et al. 1997) would be devastating. Because of their dependence on existing burrow systems, range expansion following localized extinction can be slow and is sensitive to physiographic barriers. Although conservation concern is reduced for species found across a broad area, range extent is no assurance of population persistence.

Several ground squirrels are subject to some of the same threats to habitat as have been experienced by the more imperiled species. For instance, Piute ground squirrels have been reduced by invasion of exotic annuals (Van Horne et al. 1997). Species and subspecies with very small ranges are also threatened by broadscale climate and habitat change. Like the Utah prairie dog, the Mexican prairie dog is found in habitats potentially affected by climate change, and has a limited range amid the dry and hot southern limits of prairie dog extent.

Understanding and predicting these threats is essential to the long-term persistence of these species. Recording general habitat associations provides insufficient data to predict population behavior. It is essential, for both common and rare taxa, to understand the processes that affect the populations, particularly food availability. Once understood, these linkages must be placed in the context of past and predicted climatic trends and fluctuations, current and expected habitat changes caused by invasive species and land use changes, and current and expected disease epidemiology. Successful conservation of populations requires scientists and managers to identify the important regulating processes and predict how they will change. The effects of these changes on population persistence will need to be further deduced. It is important to concentrate on the bottom-up understanding of food habits and factors influencing energy and nutrient flow through the primary producers to ground squirrels.

Habitat fragmentation can be viewed as a secondary phenomenon that interacts with habitat change to produce both genetic and population effects. Fragmentation of habitat reduces interconnectedness of populations. Such fragmentation can interact with disease, such as plague, by reducing the likelihood of recolonization following localized extinction. Understanding and predicting the effects of habitat fragmentation requires an understanding of population dynamics across a broad scale. For socially structured species, such as prairie dogs, metapopulation structure may be pronounced. In such a situation, synchronous change in populations (such as produced by climate) can have a much greater effect on extinction than asynchronous change (as might be produced by episodic plague; see den Boer 1979). The degree of population interconnectedness can be ascertained through genetic studies. Effects of genetic discontinuities in space on the evolution of adaptations, such as plague resistance, should not be discounted. For this reason, the transport of animals for recolonization should be considered only as a last resort for preventing extinction of the larger population or metapopulation.

The bottom line is that maintaining these species requires that we first understand broadscale habitat change, as from invasives or climate change, and the interaction of habitat change with food habits of target species (Sherman and Runge 2002). Second, we need to understand the genetic structuring and dynamics of populations across the landscape and the role of dispersal in influencing genetic change and recolonization. This is particularly important for the more highly social species that exhibit a metapopulation structure. Low mobility of those species that are tied to burrow systems and terrestrial navigation means that recolonization and range expansion is slow and difficult, as in Idaho ground squirrels (Gavin et al. 1999; Sherman and Runge 2002). Hence, for both habitat change and fragmentation considerations conservation efforts must focus on long-term climate and population change, not just the short-term fixes to population density.

Conserving ground squirrels will require some change in human perception and values. The historic classification of these species as varmints seems to imply an ability of the populations to rebound and recolonize, if not overrun, areas of occupancy. Yet ground squirrels are far less mobile than avian species, large-bodied mammals, or most small rodents because of their dependence on burrow systems. Populations are easily diminished by disease or lack of food. Hence, as a group, they are exceptionally sensitive to habitat change associated with factors such as climate change, invasives, fragmentation, and land-use changes. Avian species receive far more attention relative to such threats; indeed, much of the ground squirrel research has been stimulated by the role of these species as avian prey. A comprehensive, multinational evaluation of ground squirrel habitat, and current and projected threats to that habitat, could provide the foundation for educational and political changes that would help preserve not only these species but also the ecosystems on which they depend. This would be an important first step in avoiding multiple near-term extinctions.

Chapter 40 Conservation of Prairie Dogs

John L. Hoogland

THE BLACK-TAILED PRAIRIE DOG (*Cynomys ludovicianus,* hereafter "prairie dog") currently inhabits < 2% of the area that it occupied about 200 years ago (Manes 2006; Proctor et al. 2006). In response to this precipitous decline, the United States Fish and Wildlife Service (USFWS) determined in 2000 that the prairie dog should be added to the Federal List of Endangered and Threatened Wildlife and Plants (FLETWP; USFWS 2000; see also Manes 2006). In 2004, USFWS reversed its earlier designation by concluding that the prairie dog is no longer a candidate species (USFWS 2004; Manes 2006). Regardless of designation, the inescapable conclusion is that prairie dog populations have decreased sharply over the last 200 years, are still declining today throughout much of the geographic range, and need better conservation.

In this chapter I examine factors that have contributed to the decline of the prairie dog, its ecological importance, and ways to promote its longterm survival. I focus my discussion on the black-tailed prairie dog, the species of *Cynomys* for which we have the most information. Almost every argument also applies to the other four species of prairie dogs. The Mexican prairie dog (*C. mexicanus*) is on FLETWP (USFWS 1970), as is the rarest species of all, the Utah prairie dog (*C. parvidens;* USFWS 1984; Roberts et al. 2000). White-tailed and Gunnison's prairie dogs (*C. leucurus* and *C. gunnisoni*) are currently under consideration for FLETWP as well (Center for Native Ecosystems et al. 2002; Rosmarino 2004).

At least four aspects regarding the conservation of prairie dogs are noteworthy. First, we usually know little about the natural history of species that are under consideration for FLETWP. Prairie dogs, by contrast, have been the focus of several longterm studies (King 1955; Tileston and Lechleitner 1966; Halpin 1983, 1987; Knowles 1985, 1987; Garrett and Franklin 1988; Hoogland 1995, 2003, chap. 37 this volume). The prospects thus are higher than usual for using information on ecology, demography, and population dynamics to formulate realistic, promising plans for conservation. Second, many endangered animals affect only a small geographic area. The northern Idaho ground squirrel (*Spermophilus brunneus*), for example, is on FLETWP and inhabits only two counties in west-central Idaho (Clark 2000; Sherman and Runge 2002; Yensen and Sherman 2003). The prairie dog, by contrast, inhabits ten states (Hollister 1916; Pizzimenti 1975), and its perceived impact on ranching and farming in these states is gargantuan. Third, people rarely see individuals of most endangered species, and, except for accounts in the media, are unaware of their existence. Prairie dogs, by contrast, are highly conspicuous, because they are diurnal and live in large colonies, and because they frequently fight, chase, kiss, and allogroom aboveground (King 1955). Fourth, the rarity of most endangered species has resulted, incidentally rather than deliberately, from anthropocentric activities such as conversion of habitat for agriculture, suppression of fire, and construction of factories and houses (Yensen and Sherman 1997, 2003). These incidental activities also have contributed to the decline of prairie dogs, but in addition has been a war specifically against prairie dogs over the last

200 years: toxic baits have killed billions of prairie dogs, and recreational shooting has eliminated millions more.

Where Have All the Prairie Dogs Gone?

In the 1800s, the geographic range of prairie dogs encompassed more than 160 million hectares, and extended from southern Canada to northern Mexico and from eastern Nebraska to the foothills of the Rocky Mountains (Hollister 1916; Hall 1981; Proctor et al. 2005). Prairie dogs did not live everywhere in this huge range, but rather inhabited about 30 million of the 160 million hectares. Today's prairie dogs, however, inhabit only 0.5–0.8 million hectares, and over two-thirds live in small, isolated colonies (USFWS 2000; Lomolino and Smith 2003; Proctor et al. 2006). As explained in the following, at least five factors have contributed to the decline of prairie dog numbers, and these factors are sometimes synergistic (Wuerthner 1997; Manes 2006). Before converting a colony site to cropland, for example, the landowner might first poison all the prairie dogs. Several of these factors have contributed to the rarity of other ground-dwelling squirrels (e.g., Yensen and Sherman 1997, 2003).

Destruction of suitable habitat

Approximately one-third of the suitable habitat within the former geographic range of prairie dogs has disappeared over the last 200 years (Mac et al. 1998; Manes 2005; Proctor et al. 2006). Most of the destruction of habitat has involved conversion to cropland, but conversion to accommodate industry, urbanization, and livestock also has occurred.

Poisoning

The first organized campaign to poison prairie dogs began in Texas in the late 1890s (Palmer 1899; Knowles 1986a). By the 1920s, federal programs to kill prairie dogs were operational in every state where they lived. The numbers of victimized prairie dogs are staggering. In 1923 alone, for example, 1.5 million hectares inhabited by prairie dogs were poisoned (Forrest and Luchsinger 2006). And from 1912 through 1923 in Colorado, approximately 30,000,000 prairie dogs were killed by 595,926 liters of strychnine-treated grain (Clark 1989).

Recreational shooting

Shooting of prairie dogs in western states has occurred for > 100 years (Smith 1967). People sometimes shoot prairie dogs for food, but most shooting has been, and continues to be, primarily for recreation/sport. By itself, recreational shooting probably has not seriously depressed populations of prairie dogs over the last century. The combination of poisoning with follow-up shooting, however, sometimes eradicates small colonies (Reeve and Vosburgh 2006).

In the last decade or so, recreational shooting of prairie dogs has begun to attract marksmen with high-technology, long-range rifles, some of which enable accuracy from > 1,350 meters. In North Dakota, for example, the number of out-of-state recreational shooters increased from 163 in 1989 to 1,326 in 2001 (Reeve and Vosburgh 2006). With this greater interest and better weaponry, the toll from recreational shooting can be substantial. In South Dakota in 2000, for example, recreational shooting eliminated 1.2 million prairie dogs (Reeve and Vosburgh 2006).

Prairie dogs as pets

If captured when young, prairie dogs make excellent pets (Ferrara 1985). Prairie dogs do not readily breed under laboratory or pet-store conditions, however, and catching juveniles in the wild is arduous. Further, prairie dogs sometimes transmit diseases such as plague and monkeypox to humans (Long et al. 2006). Consequently, the commerce of pet prairie dogs has been dormant for most of the last 200 years. In the late 1990s, however, Gary Balfour designed a truck that can vacuum young prairie dogs from their natal burrows (Long et al. 2006). Capturing large numbers of juveniles from the wild is now relatively easy, and national and international interest in prairie dogs as pets has skyrocketed. The pet trade thus poses another obstacle to the long-term survival of prairie dogs. Pets increase public awareness of the charm and importance of prairie dogs, however. Pet owners thus might be more likely to support efforts to save prairie dogs from extinction. The overall effect of the pet industry on the conservation of prairie dogs is unclear (Miller and Reading 2006).

Plague

Plague (called *sylvatic plague* when it occurs among wild rodents) is another factor responsible for recent population crashes among prairie dogs. Plague is caused by a bacterium (*Yersinia pestis*), and fleas (Siphonaptera) are its most common vectors (Barnes 1993; Hoogland et al. 2004). Plague probably arrived into the United States about 100 years ago via flea-infested rats from Asian ships; the first incidence of plague among prairie dogs occurred in the 1940s (Pollitzer 1951; Biggins and Kosoy 2001; Cully et al. 2006). Perhaps because of this short exposure to an introduced disease, prairie dogs have not evolved a good defense against plague, and consequently remain highly susceptible—so that mortality within an infected colony usually is $> 95\%$ (Barnes 1993; Cully et al. 2006).

Over the last 20 years or so, plague has been especially problematic for prairie dogs (Cully et al. 2005). For unknown reasons, plague currently is absent throughout most of the eastern one-third of the prairie dog's geographic range (i.e., North and South Dakota, Nebraska, Kansas, Oklahoma, and Texas; Cully et al. 2006). In Montana, for example, plague has been primarily responsible for reducing the cumulative area inhabited by prairie dogs by 50% since 1986 (Luce et al. 2006). Because of the combination of catastrophic mortality with unpredictability regarding the place and timing of epidemics, plague poses a formidable obstacle for the long-term survival of prairie dogs (Cully and Williams 2001; see the following, also).

The Prairie Dog as a Keystone Species

An organism that dramatically and uniquely affects the composition and functioning of an ecosystem is called a *keystone species* (Paine 1969; Power et al 1996; Kotliar 2000). As summarized in the following, the prairie dog qualifies as a keystone species because (1) so many other organisms associate with it, and (2) its grazing and burrowing radically alter the landscape and affect ecosystem processes. These two effects frequently overlap; the altered landscape at colony sites, for example, is one reason that colonies attract certain plants and animals.

Organisms that associate with prairie dogs

Because >200 species of plants and animals frequently associate with them, prairie dog colonies increase biological diversity and species richness (King 1955; Clark et al. 1982; Miller et al. 1994, 2000; Kotliar et al. 1999, 2006). Prairie dogs are prey, for example, not only for mammalian predators such as American badgers (*Taxidea taxus*), bobcats (*Lynx rufus*), coyotes (*Canis latrans*), and red foxes (*Vulpes vulpes*), but also for avian predators such as golden eagles (*Aquila chrysaetos*), prairie falcons (*Falco mexicanus*), Cooper's hawks (*Accipiter cooperi*), ferruginous hawks (*Buteo regalis*), and red-tailed hawks (*B. jamaicensis*). Prairie dog burrows provide shelter and nesting sites for many kinds of animals, including amphibians (e.g., tiger salamanders; *Ambystoma tigrinum*), reptiles (e.g., prairie rattlesnakes; *Crotalus viridis*), birds (e.g., burrowing owls; *Athene cunicularia*), and myriad insects (Agnew et al. 1986; Butts and Lewis 1982; Desmond et al. 2000). Grazing and clipping of vegetation by prairie dogs create open habitats preferred by

animals such as horned larks (*Eremophila alpestris*), mountain plovers (*Charadrius montanus*), American bison (*Bison bison*), and pronghorn (*Antilocapra americana*; Knowles et al. 1982; Olson 1985; Kotliar et al. 1999).

At one extreme, many of the species that associate with prairie dogs also occur in grassland habitats with no prairie dogs (Kotliar et al. 1999). Examples that show such casual association with prairie dogs include American robins (*Turdus migratorius*), slate-colored juncos (*Junco hyemalis*), eastern cottontails (*Sylvilagus floridanus*), and northern pocket gophers (*Thomomys talpoides*). At the other extreme, many plants and animals depend heavily on prairie dogs. At least four species of plants, for example, survive and reproduce better at colony sites: black nightshade (*Solanum nigrum*), pigweed (*Amaranthus retroflexus*), scarlet globemallow (*Sphaeralcea coccinea*), and fetid marigold (*Dyssodia papposa*; King 1955; Detling 2006). The species most dependent on prairie dogs is probably the black-footed ferret (*Mustela nigripes*). Prairie dogs comprise about 90% of the ferret's diet, and their burrows provide shelter and nesting chambers for ferrets (Clark 1989; Seal et al. 1989; Miller et al. 1996). The drastic decline of prairie dogs over the last 200 years has been calamitous for the ferret, which is on FLETWP and is at the brink of extinction. No known natural populations of ferrets exist, and survival of the species will depend on the success of reintroducing laboratory-reared ferrets into the wild (Chadwick 1993; Miller et al. 1996; Vargas et al. 1998; Dobson and Lyles 2000; Kotliar et al. 2006).

Altered landscape and ecosystem processes

Like some other ground-dwelling squirrels (e.g., Yensen and Sherman 1997, 2003), prairie dogs markedly alter the grassland ecosystems they inhabit (Kotliar et al. 1999, 2006). Their excavations promote mixing of topsoil and subsoil, for example. They consume grasses, forbs, and other herbaceous plants (Detling 2006). In addition, prairie dogs sometimes clip tall vegetation but do not consume it—presumably to enhance detection of predators (King 1955; Hoogland 1995). The combination of consumption and clipping decreases the height of vegetation, and also alters floral species composition (King 1955; Koford 1958; Uresk and Bjugstad 1983).

Even though prairie dogs reduce vegetational biomass by their clipping and foraging, the quality of grasses and forbs at colony sites is sometimes better (i.e., has a higher concentration of proteins) because of (1) fertilization of the topsoil via the feces and urine of prairie dogs, (2) redistribution of minerals and nutrients in soil via burrowing, and (3) better penetration and retention of water within soil via burrows (Munn 1993; Detling 2006). Enhanced quality of forage partially offsets the lower plant biomass at colony sites, and the higher quality might help to explain the preferential grazing at colony sites by American bison and pronghorn (O'Meilia et al. 1982; Coppock et al. 1983; Whicker and Detling 1988; Detling 1998).

Because it is a keystone species, conservation is important not only for the prairie dog, but also for its grassland ecosystem. The two are not necessarily synonymous. A conservation plan that specifies thousands of small colonies scattered throughout the former geographic range, for example, might support the longterm survival of prairie dogs. But such a plan probably would not ensure the longterm survival of either those species that depend on prairie dogs or the ecosystem processes typical of the western grasslands. Black-footed ferrets, for example, need *large* prairie dog colonies (Clark 1989; Miller et al. 1996). Similarly, mountain plovers and burrowing owls are especially attracted to large prairie dog colonies (Knowles et al. 1982; Knowles and Knowles 1984; Olson 1985; Griebel 2000).

To promote the longterm survival of prairie dogs and their grassland ecosystem, conservation biologists should formulate plans that identify the minimal area that will allow not only the coexistence of prairie dogs and other organisms that associate with them, but also the ecological processes discussed previously. But what is that minimal area? Bigger is always better, but 4,000 hectares for a colony site (or complex of nearby colony sites) is probably the minimal area necessary for a fully functional grassland/prairie dog ecosystem (Proctor et al. 2006). Density varies among colonies (Biggins et al. 2006), but a colony that occupies 4,000 hectares usually contains > 100,000 adult and yearling prairie dogs (Hoogland 1995; Proctor et al. 2006).

How Can We Save Prairie Dogs from Extinction?

In theory, the solution for saving prairie dogs is simple, and it involves the following steps: (1) restore all the suitable habitat that has been converted for farming and urban/commercial development, (2) kill fleas at all colonies, and thereby deter transmission of plague, (3) prohibit all poisoning and recreational shooting, (4) abolish the capture and sale of prairie dogs as pets, and (5) change negative attitudes toward prairie dogs. In practice, of course, these steps are impractical and unrealistic. In this section I propose several steps that are more reasonable.

Identify focal areas

Restoring prairie dogs over their entire former geographic range is impossible, but finding representative areas where the grassland ecosystem can be restored is feasible (Proctor

et al. 2006; see also Miller et al. 1994; Wuerthner 1997). A *focal area* is a locale where we can allow prairie dogs to occupy at least 4,000 hectares, with minimal impact on other areas—so that species that associate with prairie dogs can thrive, and so that ecological processes such as changes in species composition and recycling of nutrients can occur. Four criteria are important for the evaluation of potential focal areas (Proctor et al. 2006): (1) suitability of habitat, (2) opportunities for management, (3) geography, and (4) probability of plague. (1) Evaluation of the suitability of habitat includes an examination of soil type, slope, and height and type of vegetation. The best proof that habitat is suitable for prairie dogs is to find colonies already there. Consequently, the most promising regions for restoration and conservation are those where prairie dogs currently live. (2) Opportunities for management improve the suitability of focal areas. Public lands—e.g., those belonging to the Bureau of Land Management and the United States Forest Service—offer tremendous potential for restoration and management because they often occur in large, contiguous blocks; they are not cultivated for crops; conservation of biodiversity is a priority there; and large colonies of prairie dogs already live there in many cases. Similar potential exists on certain private lands owned by groups interested in the conservation of prairie dogs. Some of these groups are Turner Enterprises (with lands in South Dakota, Kansas, and New Mexico), The Nature Conservancy (Montana, Wyoming, and Colorado), the Southern Plains Land Trust (Colorado), and the Gray Ranch (New Mexico). (3) In order to preserve possible genetic differences associated with geography, the restoration and conservation of prairie dogs should include focal areas distributed throughout the former geographic range, including areas at the periphery (Lomolino and Smith 2001, 2003). (4) Finally, focal areas should be sufficiently numerous and widespread to maximize the probability of long-term survival against plague.

Proctor et al. (2006) have identified eighty-four focal areas, ranging in size from 4,300 hectares to 2 million hectares. Of highest priority are the five focal areas that are already inhabited by > 4,000 hectares of prairie dogs; three of these are on the Cheyenne River, Pine Ridge, and Rosebud Native American reservations in South Dakota. Second in priority are the three focal areas where restoration of a 4,000-hectare prairie dog complex is currently underway: Vermejo Park, New Mexico; Bad River, South Dakota; and the Northern Cheyenne Reservation, Montana (USFWS 2002; Long et al. 2006). Most of the other focal areas are on federal lands and already have prairie dogs, but the area inhabited is usually < 4,000 hectares.

For focal areas that currently contain no prairie dogs, translocations from other areas will be necessary. For years I have been telling people that < 5% of translocated prairie dogs will remain at the new site for > 1 week; most of the others will perish as they disperse in search of their genetic relatives back home (e.g., Radcliffe 1992; Long et al. 2006). In the past 10 years or so, however, wildlife managers have developed ploys that induce as many as 71% of translocated prairie dogs to remain at their new colony site for > 2 months, and most of the sedentary survivors rear offspring in the year after translocation (Long et al. 2005; Shier 2006). One ploy is to mow tall vegetation at new colony sites; another is to provide underground nest chambers; a third is to move entire families together; and a fourth involves temporary, large aboveground enclosures that deter dispersal in the first week or so after translocation. With these new techniques, translocation is now an excellent mechanism by which we can initiate colonies in areas of suitable habitat.

Reduce poisoning and recreational shooting

Because so many people disdain prairie dogs, regulations are unlikely to halt all poisoning and shooting. If we emphasize the importance of focal areas and improve enforcement of regulations there, however, then killing of prairie dogs in these areas, at least, should cease.

Landowners sometimes profit by charging a fee for recreational shooting of prairie dogs. Paradoxically, colonies where marksmen pay to shoot might persist longer than colonies with no shooting—because income from recreational shooting gives landowners an incentive to manage colonies for long-term survival (Reeve and Vosburgh 2006; Hoogland 2006).

Stop plague

Plague is an introduced disease, and questions abound regarding transmission within and among colonies (see preceding and Barnes 1993; Cully et al. 2006). Consequently, we cannot easily predict how plague will impede our efforts to conserve prairie dogs. Three measures should help to counter plague. (1) For now, we should give higher priority to conservation of focal areas that are in the plague-free eastern one-third of the prairie dog's geographic range (Cully 2006; Proctor et al. 2006). (2) To kill fleas and thereby halt transmission of plague, we should infuse burrows with a biodegradable insecticide such as Deltadust (Bayer Environmental Science, Montvale, New Jersey) at the first sign that plague might be present (Seery et al. 2003; Hoogland et al. 2004). Infusion also is an advisable prophylactic for those colonies that are especially valuable for some reason (e.g., research, habitat for black-footed ferrets). (3) Because of uncertainties with plague, the "precau-tionary principle" should apply to the conservation of prairie dogs—that is, as insurance against the unpredictable loss of colonies to plague, we should designate more sanctuaries than otherwise might seem necessary (Miller and Reading 2005).

Consider politics, legislation

Addition of the prairie dog to FLETWP would necessitate federal regulations and restrictions for state governments and private landowners. To avoid such entanglements, states have responded in several ways. For example, the ten states with prairie dogs (plus an eleventh state, Arizona, where the prairie dog has been eradicated) have formed the Interstate Prairie Dog Conservation Team, from which a cooperative plan for the conservation of prairie dogs has emerged (Luce et al. 2006). Cooperation also is underway between Native American tribes and states of Montana, North Dakota, and South Dakota. Many county governments and extension agents, however, still oppose conservation of prairie dogs, and continue to promote eradication. Will saving the prairie dog require its addition to FLETWP? The answer to this pivotal question is not yet clear (Hoogland 2006; Manes 2006; Miller and Reading 2006).

The Endangered Species Act (ESA) became effective in 1973—before the full development of concepts such as viable populations, keystone species, and ecological function. In the last 30 years we have learned much about these issues, but current ESA policy does not incorporate much of this new information (Pyare and Berger 2003; Soulé et al. 2003). A strict legal interpretation of the current version of ESA must focus on single species. USFWS (2000) has argued, for example, that interspecific interactions are irrelevant for evaluating the status of prairie dogs under ESA—even though black-footed ferrets, burrowing owls, mountain plovers, and other organisms depend on prairie dogs for survival (Kotliar et al. 2006). Strict legal interpretation of ESA does not *require* USFWS to consider the ecological role of prairie dogs in grassland ecosystems, but ESA does *allow* such consideration (Miller and Reading 2006). The USFWS website lists more than seventy-five recovery plans that include multiple species or subspecies. For prairie dogs and the grassland ecosystem, perhaps conservation biologists can write a multispecies recovery plan similar to the South Florida Multispecies Recovery Plan for the Everglades, which considers sixty-eight species (USFWS 1998). A single multispecies recovery plan will be more economical and more effective than separate plans for prairie dogs and the many species that depend on them for survival (Miller and Reading 2006).

Mollify negative attitudes

With the possible exception of plague, the most serious threat to the long-term survival of prairie dogs is our own species (Wuerthner 1997; Vermeire et al. 2004; Lamb et al. 2006). Farmers complain that prairie dogs eat crops. Ranchers argue that prairie dogs compete with livestock for forage, and that horses and cows break legs when they step into burrows. Other negative attitudes stem from concerns about diseases such as plague and monkeypox, limits on urban development, losing local control over management, and the restrictive nature of ESA. These complaints are all legitimate, but commonly are exaggerated. Prairie dogs do compete with livestock for some types of forage, for example—but the presence of prairie dogs sometimes improves the quality of certain plants, so that livestock often prefer to forage at colony sites (see the preceding and Detling 2006). Further, prairie dogs are especially likely to colonize areas that livestock already have overgrazed, and thus commonly are the effect, rather than the cause, of overgrazing (Koford 1958; Knowles 1986b; Detling 2006). A horse or cow might break a leg after stepping into a prairie dog burrow, but the probability of such a fracture is trivial (Hoogland 1995). And a person might contract plague from either a prairie dog or its fleas, but such transmission is exceedingly rare (Levy and Gage 1999). And so on. For successful conservation, we must try to dispel myths and inaccuracies about prairie dogs—an onerous task, because these misconceptions have persisted for generations.

In 2001, the Interstate Prairie Dog Conservation Team submitted a proposal to the United States Department of the Interior that would provide financial incentives—and thereby presumably would mollify negative attitudes—for farmers and ranchers who agree to conserve prairie dogs. The proposal contained six components that are important to landowners (Luce et al. 2006): (1) provisions for flexibility in land management, including grazing by livestock that is compatible with the needs of prairie dogs, (2) protection from future regulatory actions regarding prairie dogs, (3) financial benefits, especially direct payments ($4.05 per hectare per year, plus a one-time payment of $40.50 per hectare), (4) assistance with management of colonies that expand into areas not covered in the incentive program, (5) freedom to use enrolled lands to generate income by means that will not harm prairie dogs (e.g., ecotourism), and (6) other features, such as allowances for wildlife-based recreational income (e.g., off-site recreational shooting). The proposal was not funded, but will serve as a blueprint for future efforts.

Summary

The outlook for many endangered species is dim. For prairie dogs, however, the potential for conservation is enormous. Researchers have identified eighty-four focal areas that are suitable for large sanctuaries, for example. With the recent realization that the prairie dog is a keystone species, many people now appreciate that prairie dogs are crucial for the grassland ecosystems of western North America. And by surviving 200 years of shooting, poisoning, and destruction of habitat, prairie dogs have demonstrated that they can overcome formidable obstacles. With prudent conservation, the battle to save prairie dogs is one that we can win (Hoogland 2006; Manes 2006; Miller and Reading 2006).

Chapter 41 Social Behavior, Demography, and Rodent-Borne Pathogens

Richard S. Ostfeld and James N. Mills

POPULATION AND COMMUNITY ECOLOGISTS tend to focus on the obvious. Unlike many other branches of biology, the focal entities (populations of whole organisms) typically are macroscopic and the key processes (or their consequences) often can be observed directly, without highly specialized equipment. In the case of rodents, interactions between the focal population and its resources, predators, and competitors tend to be relatively easily observable and the consequences discernable with simple, nontechnical tools (e.g., live traps, field enclosure/exclosure designs, vegetation sampling). As a consequence, much is known about the two-way interactions between rodents and their food resources, predators, and competitors. In contrast, beyond lists of species involved, the interactions between rodents and their pathogens and parasites are not well understood. Perhaps this relative neglect exists because these pathogens and parasites tend to be microscopic, are often hard to detect and monitor without specialized laboratory procedures, and have until recently been left out of the ecological/behavioral mainstream.

Clearly, though, such neglect is not justified on the basis of the strength of the interspecific interactions. Pathogens and parasites of rodents are likely at least as important in influencing population dynamics as are macroscopic predators, resources, and competitors, and they are much more numerous, both in terms of species and individuals. Similarly, demographic and behavioral traits of rodents probably are at least as important to the dynamics of their pathogens as to those of predators, resources, and competitors. Unlike most predators, biotic resources, and competitors, patho-

gens are often highly specialized on, and therefore tightly coupled to, their hosts. This chapter focuses on what we know and need to know about social and demographic factors that influence the *maintenance* and *transmission* of pathogens (broadly defined to include viruses, *Rickettsia*, and bacteria, as well as eukaryotic parasites) in rodent populations and between rodents and tangential hosts, particularly humans.

Prior Impediments to an Understanding of Rodent-Pathogen Interactions

Rodent-pathogen interactions are bidirectional. Early studies of these interactions focused on pathogens and disease as factors regulating rodent populations (Elton 1931; Elton et al. 1935; Chitty 1954b). As advocated in the scientific philosophy of Dennis Chitty (1996) and perpetuated by his academic descendents (e.g., Lambin et al. 2002), these studies tended to limit their inquiries to the question of whether disease is both necessary and sufficient to cause cycles in host population density. They (Elton 1942; Chitty 1996) have concluded that, despite being pervasive in rodents, disease has not been demonstrated as both necessary and sufficient to cause cycles. Therefore, like many other factors deemed not necessary and sufficient, disease should be dropped from consideration as a factor influencing population dynamics. This position appears to have been an influential one in that relatively little research has been conducted regarding the role of disease in rodent population dynamics other than a few well-studied cases, such as plague (etiological agent, *Yersinia pestis*) decimating populations of prairie dogs (*Cynomys* spp.; Cully et al. 1997).

In contrast, epidemiologists have long been interested in rodent-borne zoonoses, including plague, tularemia, leptospirosis, leishmaniasis, trypanosomiasis, and various viral hemorrhagic fevers. But, the focus has been largely on determining the primary reservoirs of zoonotic pathogens. Documenting the important reservoir species is helpful but inadequate for assessing how rodent population dynamics influence changes in risk of human exposure to zoonoses. Until recently, little effort has been devoted to understanding the determinants of transmission rates within rodent populations or from rodents to people.

Detection of Pathogens within Rodent Populations

Rodent pathogens and parasites of epidemiological importance include viruses, bacteria, and protists, all of which are microscopic. Three major types of methods have been used to assess their presence, and in some cases to quantify abundance. First, the presence of pathogens can be detected directly by one of several means. Preparations of tissues invaded by the pathogen can be examined microscopically, sometimes using immunohistochemical staining, to visualize the microbes. Pathogens can sometimes be isolated from host tissue by growth in culture for later identification. An increasingly popular method is the use of polymerase chain reaction (PCR) to detect nucleic acid (usually DNA, occasionally RNA) specific to pathogens, which often can be recovered from host tissues without death or injury to the host. Typically, both isolation and PCR are qualitative methods (pathogens present or absent), although quantitative PCR methods have been developed. Finally, specific antigen may be detected in blood or tissue using immunologic assays that employ antibody that binds specifically with the pathogen when present. A second kind of immunological assay employs antigen to a specific pathogen to detect recent or remote infection. Typically, host blood is drawn from free-ranging or laboratory-held animals, and specific antibody is assayed from serum. In the case of antibody detection via serology, results can be considered qualitative (hosts are categorized as infected or not), or semiquantitative, based on antibody titers. Seroprevalence is the proportion of a population with detectable antibody to a pathogen; seroconversion describes the change in antibody status of an individual host, usually from negative to positive or from low titer to higher titer (usually a four-fold or greater rise in antibody titer), and indicates recent infection. An important difference between antibody assays and di-

rect detection of pathogens is that the presence of antibody does not necessarily indicate current infection, but demonstrates that the host was infected at some unknown time in the past. Detection of different immunoglobulin fractions may reveal more specific information about infection. The presence of IgG antibody only indicates that the host has been infected in the past (weeks, months, or years previously). However, the current status of the host could be infected, previously infected but recovered, or even immune to further infection. The presence of IgM antibody indicates a very recent infection, and the pathogen may still be present in host blood or tissues. In the special cases of hantaviruses and arenaviruses, infected hosts typically develop a chronic infection that involves persistent shedding of virus into the environment (in urine, feces, and saliva) for extended periods, perhaps the lifetime of the rodent. In this case, hosts with IgG antibody are often assumed to be currently infected and infectious.

A third method for detecting infection in rodent hosts, termed xenodiagnosis, is limited to vector-borne pathogens. For these types of pathogens, naïve (uninfected) arthropod vectors are allowed to take a blood meal from a host and the vector is then subjected to an assay (e.g., PCR, microscopy, and others) for the pathogen. If the vector tests positive, the host must have been infected; however, if the vector tests negative, the host might still have been infected but did not transmit the pathogen to the vector.

Factors Influencing Pathogen Transmission and Maintenance within Rodent Populations

Background

Pathogens disperse from one individual host to another via several different modes, including direct transmission, blood-feeding arthropod vectors, consumption of pathogens in water or food, or sexual contact. The direct transmission category typically includes both the deposition of pathogens via bites and scratches, and deposition into urine and feces of pathogens that enter other individuals through mucous membranes (e.g., inhalation) or the digestive system (i.e., consumption). Most pathogens probably use only one mode as the exclusive means of dispersing from host to host, although some use more than one method. An example is the bacterium (*Francisella tularensis*) that causes tularemia, which can be transmitted by tick vectors or by consumption of contaminated materials (Reintjes et al. 2002).

To understand disease dynamics within populations, it is useful to categorize individuals by their status with respect to the pathogen, that is, whether they are infected (I) or uninfected and susceptible (S). In some cases, when infection is followed by immunity, a third category is added to represent uninfected and recovered (not susceptible) individuals (R; Kermack and McKendrick 1927; Anderson and May 1978; May and Anderson 1978). For an infection that is transmitted directly between individuals, the spread of disease is thought to depend largely on the rate of contact between (S) and (I) individuals. If the population has no spatial structuring and (S) and (I) individuals show no bias in their probability of associating with other individuals, their rate of contact should be a function of the combined density of (S) and (I). Under these conditions, disease spread is expected to be density dependent.

The assumption that pathogen transmission rates are density dependent arises from epidemiological models of the basic reproductive rate of a pathogen, R_0, which is usually defined as the average number of new (secondary) infections generated by a single infectious host entering a naïve (susceptible) host population. R_0 is a positive function of the population abundance of the host species (S), the rate of transmission between individual hosts (T), and the length of time infected individuals remain infectious (L; e.g., Anderson and May 1978), or

$$R_0 = (S \times T \times L)$$

Greater population abundance provides more opportunities for transmission; the transmission rate defines the proportion of those opportunities that are realized; and length of time hosts are infectious defines how long those opportunities will persist. If $R_0 > 1$, then the disease spreads; if $R_0 < 1$, then the disease declines to extinction.

In reality, the basic reproductive rate of an infection under ideal conditions is probably overemphasized in epidemiology, as is the threshold value described previously. Rodent ecologists are well aware that populations tend to fluctuate through time, sometimes dramatically. Therefore, the idealized reproductive rate of an infection might rarely be reached or be transient. This is particularly important for diseases with R_0 values that are near unity; small fluctuations in host density can cause R_0 to oscillate around the critical threshold separating disease spread from disease extinction. Perhaps more important to predicting disease spread than R_0 is R_E, or the effective reproductive ratio, which can be defined as the number of secondary cases produced in a host population that is not entirely naïve, that is, one consisting of a mixture of susceptible, infected, and recovered individuals. If the pathogen reduces survival or fecundity of the host, and therefore population growth rate, then the pathogen should tend to stabilize host density (Anderson and May 1978; May and Anderson 1978). As a consequence, R_E, which increases with increasing host density, should also be stabilized. Therefore, these types of epidemiological models predict more or less constant rates of infection and the coexistence of pathogen and host. Measuring the rate of disease spread across a continuous range of host population densities, particularly in taxa such as rodents, would be useful for predicting both the impact of host population dynamics on pathogens and the effects of pathogens on host population dynamics. Such studies are rare (see the following discussion).

In contrast, some pathogens are not transmitted directly among individuals that interact randomly in the absence of spatial structuring. These pathogens include those associated with vector-borne diseases and those that are transmitted during sexual or aggressive encounters. For the latter types of transmission, disease spread is more likely to depend on the *proportion* of individuals that are infected than on their absolute abundance or density; therefore, in these situations disease spread is thought to be frequency dependent (May and Anderson 1978; Getz and Pickering 1983). For vector-borne diseases, frequency dependence arises because an individual arthropod vector is limited in the number of hosts it can bite, and therefore, the number of bites per vector (a surrogate for disease transmission) will be largely independent of host density. Instead, vector bites resulting in pathogen transmission will be more closely tied to the probability that any given bite results in acquisition or transmission of a pathogen, and this value should vary with the frequency of infected individuals in the population. Similarly, because the number of sexual or aggressive encounters (but see the following for possible exceptions) should tend to be independent of population density, pathogen transmission will more likely vary with the probability that the fixed number of sexual or aggressive encounters per individual involve an infectious individual.

Pathogens with frequency-dependent transmission do not incorporate the stabilizing effect of density-dependent processes, but instead are expected to cause highly unstable dynamics of both pathogen and host (Getz and Pickering 1983). When transmission rates increase with the proportion (frequency) of individuals infected, a positive feedback loop ensues, such that low frequencies foster the extinction of the pathogen and high frequencies lead to increasingly rapid spread. Frequency dependence therefore results in the existence of a threshold proportion infected, below which the infection rapidly ceases and above which the infection, if lethal, causes the demise of hosts and consequently of the pathogen. Declining host density does not, in this case, rescue the host from extinction. New epidemics would be expected to arise following recolonization events or dispersal events that establish new populations temporarily free of infection.

Clearly, the consequences of density-dependent versus frequency-dependent transmission for both hosts and pathogens are profound. Rodents host many arthropod vectors and their associated pathogens, and rodent populations are often highly spatially and behaviorally structured. Because both of these features are associated with frequency dependence, one might expect pathogens with frequency-dependent transmission to predominate. Moreover, rodents are notorious for their dispersal and colonizing abilities, which would be able to promote the global persistence of pathogens with exclusively frequency-dependent transmission. Unfortunately, tests that attempt to measure density and frequency dependence and distinguish between them are rare, although interest in this issue appears to be increasing.

Population density

The primary means of assessing whether transmission rates or infection prevalence within rodent populations increase with density is to monitor rates of seroprevalence or seroconversion in natural populations over sufficiently long periods that some ability to detect a trend exists. Although such correlative studies can be criticized as not addressing cause-and-effect relationships, the potential for significant correlations to be spurious seems low. Clear mechanisms exist that would explain how increasing host density can increase disease prevalence, but mechanisms that would account for high disease prevalence causing high density (i.e., the reverse causal direction) do not seem plausible. Nevertheless, we suspect that experimental manipulations of host density would contribute importantly to the assessment of density dependence in rodent-disease interactions.

Correlative studies of the relationship between rodent population density and prevalence of infection have demonstrated positive associations for several types of pathogens and rodent hosts (table 41.1). In some cases, unusually high prevalence has been detected in low-density rodent populations, but only after a recent decline from high density (Smith et al. 1993; Niklasson et al. 1995; Abbott et al. 1999). These would seem to represent cases of delayed density dependence rather than a lack of density dependence. The only studies we are aware of that reject density-dependent and support frequency dependent transmission of rodent pathogens involve cowpox virus in bank voles (*Clethrionomys glareolus*) and wood mice (*Apodemus sylvaticus*; Begon et al. 1999, 2003; Hazel et al. 2000). However, the relative lack of published studies that support frequency dependence might simply reflect a lack of exploration rather than rarity of frequency-dependent transmission.

Finally, we suspect that the *pattern* of population dynamics might be as important to the maintenance and transmission of pathogens in rodents as is density per se. Populations that fluctuate strongly are characterized by prolonged periods of new recruitment, which would represent a rapid influx of new, susceptible individuals—a situation that should promote epizootics. Moreover, we would expect crashes in rodent populations to pose a strong risk of extinction to pathogens, which might then require immigration events by infected rodent hosts for reinvasion (see below). Consequently, we expect that disease dynamics might be linked to features such as population growth rates and the length and severity of crashes, possibilities that so far have not received attention (fig. 41.1).

Demographic biases

Embedded within an apparent trend for infection prevalence to increase with increasing population density of rodents is the frequent overrepresentation of some demographic categories in the infected fraction (fig. 41.1). For hantaviruses and some arenaviruses, males and older indi-

Table 41.1 Studies finding a link between rodent population density and some measure of transmission or maintenance of pathogens in rodent host populations

Disease	Disease measure	Type of pathogen	Rodent host	Source
Hemorrhagic fever with renal syndrome	Antibody prevalence in host population	Puumala hantavirus	Bank vole (*Clethrionomys glareolus*)	Escutenaire et al. 1997; Niklasson et al. 1995; Olsson et al.2002
Hantavirus pulmonary syndrome	Antibody prevalence in host population	Sin Nombre hantavirus	Deer mouse (*Peromyscus maniculatus*)	Kuenzi et al. 1999; Mills et al. 1999a; Biggs et al. 2000
	Abundance of infected mice	Sin Nombre hantavirus	Deer mouse (*Peromyscus maniculatus*)	Yates et al. 2002
Argentine hemorrhagic fever	Abundance of infected mice	Junin arenavirus	Drylands vesper mouse (*Calomys musculinus*)	Mills et al. 1992
Lyme disease	Population density of infected ticks	spirochete bacterium	White-footed mouse (*Peromyscus leucopus*)	Ostfeld et al. 2001

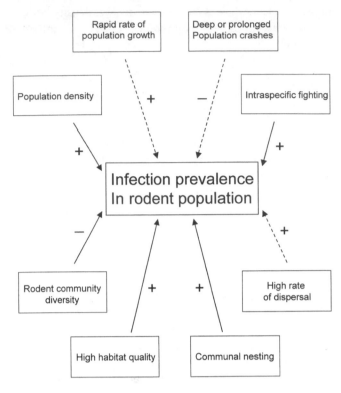

Figure 41.1 Selected factors known or suspected to affect intraspecific rates of transmission or prevalence of infection with a zoonotic pathogen. Plus signs near arrows indicate a positive effect on infection prevalence, and minus signs indicate a negative effect. Dashed arrows indicate relationships suspected to occur but without strong empirical support, whereas solid arrows represent established relationships.

viduals are more likely to be infected than are females and younger individuals (Childs et al. 1987; Glass et al. 1988; Niklasson et al. 1995; Mills et al. 1997; Mills and Childs 1998; Douglass et al. 2001; Yahnke et al. 2001). This pattern indicates that transmission of these agents within host populations is predominantly horizontal (from adult to adult) and by a specific mechanism that favors males. On the other hand, there seems to be no age or sex bias in multimammate rats (*Mastomys* spp.) infected with Lassa arenavirus (Demby et al. 2001), suggesting vertical transmission of virus from dam to pups. Lymphocytic choriomeningitis virus, another Old World arenavirus, may also be transmitted vertically, probably in utero, in populations of its host, the house mouse (*Mus musculus;* Mims 1966).

The degree of male bias in antibody prevalence varies among the various hantavirus-host pairings. For example, the ratio of antibody prevalence in males to antibody prevalence in females ranges from 1:1 for Norway rats (*Rattus norvegicus*), reservoir of Seoul virus, to 2:1 for deer mice, host of Sin Nombre virus, and 7:1 for brush mice (*Peromyscus boylii*), host of Limestone Canyon virus (Glass et al. 1988; Mills et al. 1997). These differences presumably result from differences in mechanisms of transmission (e.g.,

fighting versus venereal versus communal nesting), or differences in the relative frequency of such behaviors between genders in different species.

Habitat biases

Studies of rodent populations that incorporate specific habitat types are beginning to reveal sometimes dramatic differences among subpopulations in seroprevalence. For instance, working in the Paraguayan Chaco, Yahnke et al. (2001) found that hantavirus antibody prevalence in populations of small vesper mice (*Calomys laucha*) inhabiting croplands was higher than those inhabiting either pastures or native thorn scrub. Similar among-habitat variation in prevalence of infection with Junin arenavirus has been found in drylands vesper mice (*Calomys musculinus;* Mills et al. 1992, 1994). Kuenzi et al. (2001) found higher prevalence of antibody to Sin Nombre virus in deer mice from peridomestic habitats in Montana than in nearby sylvan habitats, and Mills et al. (1997) found substantial variation in prevalence of antibody to hantaviruses among natural habitat types. Correlation of antibody prevalence with habitat on the scale of a single trapping grid has also been described (Abbott et al. 1999; Mills, Ksiazek et al. 1999). Prevalence of antibody to Limestone Canyon virus in brush mice was associated with islands of apparently preferred microhabitat. Nevertheless, these pockets of virus activity became blurred during periods of high population density, indicating an interaction between habitat selection and population density. The mechanisms that underlie the observed patterns of spatial variation are not well understood, although local population density of rodent hosts and abiotic conditions conducive to survival of pathogens in the environment have been implicated.

Social behavior

Social behavior—fighting—has repeatedly been implicated as increasing the probability of pathogen transmission between individuals (fig. 41.1). However, the evidence to support the association between fighting and exposure is indirect; individuals with wounds or scars are more likely to be seropositive for hantaviruses (Glass et al. 1988; Douglass et al. 2001) and some arenaviruses (Mills and Childs 1998). In addition, demographic categories (i.e., older males) most likely to fight tend to demonstrate the highest antibody prevalence. Lower prevalence in females than conspecific males suggests that sexual transmission of zoonotic pathogens is relatively unimportant for hantaviruses and at least one arenavirus that has been well studied (Junin virus; Mills et al. 1994). The possibility exists that greater seroprevalence in males results from a biased sexual transmission from females to males, compared to male-to-female transmission. To our knowledge, this possibility has not been assessed. With the possible exceptions of the arenaviruses, Machupo virus (Johnson 1985), Lassa virus, and lymphocytic choriomeningitis virus (Childs and Peters 1993), no evidence exists to suggest that parent-offspring social interactions represent a common pathway for pathogen transmission. In fact, in some host-virus systems, vertical transfer of maternal antibody during gestation and lactation appears to protect dependent young against hantavirus infection for at least 3 months postpartum (Bernshtein et al. 1999). In other systems (e.g., Lassa virus; Demby et al. 2001) such antibody may not be protective.

A study in Colorado (Calisher et al. 1999) showed that there were two seasonal peaks in seroconversions in a population of deer mice infected with Sin Nombre virus. A peak in seroconversions during the breeding season affected mostly males, while a second over-winter peak affected males and females equally, suggesting different mechanisms of virus transmission during the two periods. Transmission during the breeding season may result from agonistic encounters (primarily between males), while winter transmission may occur during communal nesting (Mills, Yates et al. 1999).

Preliminary evidence suggests that the male bias in antibody prevalence does not occur for vesper mice (*Calomys* sp.) infected with Machupo arenavirus (D. Carroll and J. Mills, unpublished data). There is some laboratory evidence that Machupo virus may be maintained by venereal transmission in its rodent host, and that chronically infected females are rendered effectively sterile. A model has been proposed whereby Machupo virus causes cyclic epizootics and subsequent crashes in host populations; this rodent cycle, in turn, controls the incidence of Bolivian hemorrhagic fever in humans (Johnson 1985). Field studies are needed to test this hypothesis.

Another behavioral phenomenon, dispersal, is likely to be profoundly important to the dynamics of disease in rodents, but is poorly studied in this context (fig. 41.1). Given the unstable dynamics expected under frequency-dependent transmission, dispersal between populations or demes, or dispersal events leading to colonization, would be critical in maintaining both host and pathogen populations. Dispersal following population crashes of prairie dogs (*Cynomys* spp.) afflicted with plague is thought to be important in reestablishing extinct or nearly extinct populations of prairie dogs (Anderson and Williams 1997; Roach et al. 2001). In many cases, hantaviruses appear to become locally extinct when rodent populations decrease to very low densities (Kuenzi et al. 1999; Calisher et al. 2005). Under such situations, the virus may be locally absent for a few months to a few years, but always seems to reappear, presumably

via reintroduction by dispersing individuals from adjacent populations. The persistence of pathogens during unfavorable conditions is currently an area of active investigation. It has been hypothesized that higher host densities (and consequently hantavirus transmission) are maintained in refugia of ideal habitat, and that less favorable habitats are repopulated via dispersal by infected individuals from these refugia during periods of more favorable environmental conditions.

Establishment and defense of territories is another behavior that may be associated with the transmission of pathogens. Males defending territories may be more likely to be involved in aggressive encounters than are females or males without territories. Several studies have shown a positive correlation between hantavirus antibody prevalence and longevity on trapping sites (when corrected for age; Mills et al. 1998; L. Ruedas and others, unpublished data), and long-lived residents have been hypothesized to be important in the trans-seasonal maintenance of hantaviruses (Abbott et al. 1999; Calisher et al. 2001).

Community composition and diversity

Evidence is mounting that pathogen transmission within host populations is inhibited by high species diversity within the rodent community (fig. 41.1). Lyme disease is a zoonosis caused by the bacterium *Borrelia burgdorferi*, which is transmitted by *Ixodes* ticks. The presence of a high diversity of small mammals results in reduced abundance of *Borrelia*-infected ticks (Ostfeld and Keesing 2000a, 2000c), which in turn reduces the inoculation rate of both competent disease reservoirs (e.g., the white-footed mouse, *Peromyscus leucopus*) and incompetent reservoirs (Schauber and Ostfeld 2002). In this case, high diversity of small mammals and other vertebrates, most of which are incompetent at transmitting infection to feeding ticks, dilutes the impact of white-footed mice and reduces pathogen transmission rates (LoGiudice et al. 2003; Ostfeld and LoGiudice 2003).

A similar pattern appears to exist for directly transmitted viral and bacterial diseases, including hantaviruses, arenaviruses, and possibly *Bartonella*. Yahnke et al. (2001) found that the percent of vesper mice seropositive for Laguna Negra hantavirus decreased with increasing community diversity of small mammals. Mills (in press) found a similar pattern with hantaviruses in the US Southwest. A reanalysis of data in Kosoy et al. (1997) similarly demonstrated a pattern of reduced prevalence of antibody to the bacterium *Bartonella* in rodent communities of high species diversity (Ostfeld and Keesing 2000c). Mills (2005) suggests that the primary mechanism by which high species diversity reduces the transmission of these pathogens within their principal host is the increase in interspecific encounters at the expense of intraspecific ones. Because interactions between the principal host and heterospecifics typically result in a "dead-end" infection (the pathogen is not passed on to other hosts), the presence of high species diversity results in "wasted" encounters (from the perspective of the pathogen). An additional mechanism, proposed for Lyme disease by Schmidt and Ostfeld (2001), is the potential for the absolute density of the primary host species to be reduced in communities of high diversity, owing to stronger regulation by competitors and predators. A recent review of effects of predators on rodent-borne pathogen transmission found support for the hypothesis that predators can suppress disease transmission in rodent reservoirs, although some exceptions exist (Ostfeld and Holt 2004).

Factors Influencing Pathogen Transmission from Rodents to Humans

Background

Zoonotic pathogens often use the same mode of transmission between individuals within rodent populations as they do in cross-species transmission, including from rodents to humans. Some of the most epidemiologically important rodent-borne pathogens are most frequently transmitted either via inhalation of viral aerosols or virus-contaminated dust (e.g., the hantaviruses and arenaviruses) or via the bites of haematophagous arthropods (e.g., Lyme disease, ehrlichiosis, leishmaniasis, Rocky Mountain spotted fever). For both these modes, the force of transmission potentially could vary positively with: (1) the population density (or size) of the rodent reservoir; (2) the frequency of infection (infection prevalence or seroprevalence) in the rodent reservoir; and (3) the density of infected individuals in the reservoir population (fig. 41.2).

Although prevalence of infection within reservoir populations often has been used as a determinant of disease risk to humans (Mills and Childs 1998), we suspect that prevalence by itself is unlikely to be informative in human risk assessment. Consider two populations of a rodent reservoir species, one consisting of 100 individuals ha^{-1} with 25% prevalence of infection and the other at ten individuals ha^{-1} with 50% prevalence. We suggest that twenty-five infected individuals ha^{-1} would pose a higher risk to nearby humans than five infected individuals ha^{-1}, despite the lower prevalence in the former. Instead, we expect that total population density of rodent reservoirs, or the density of infectious individuals, will better predict disease risk to people.

Rodent population density and dynamics

Studies from several different parts of Europe have recently demonstrated temporal correlations between superannual

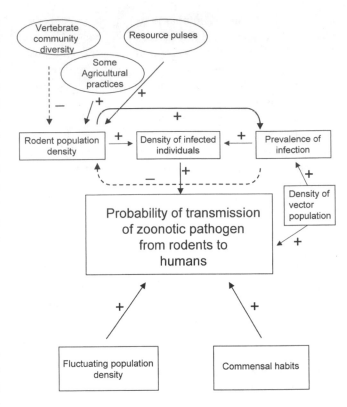

Figure 41.2 Selected factors known or suspected to affect the probability of transmission of a zoonotic pathogen from rodent hosts to humans. Plus signs near arrows indicate a positive effect on infection prevalence, and minus signs indicate a negative effect. Dashed arrows indicate relationships suspected to occur but without strong empirical support, whereas solid arrows represent established relationships.

peaks in fluctuating populations of bank voles and outbreaks of nephropathia epidemica in humans caused by Puumala hantavirus (Niklasson et al. 1995; Escutenaire et al. 1997; Brummer-Korvenkontio et al. 1999). Population outbreaks of deer mice in the US Southwest are sometimes, but not always, associated with epidemics of hantavirus pulmonary syndrome (Yates et al. 2002; Brown and Earnest 2002). An abrupt increase in the population density of corn mice was followed by a similar increase in cases of Argentine hemorrhagic fever in central Argentina (Mills et al. 1992). For Lyme disease in the northeastern US, annual risk of human exposure, as measured by the density of infected nymphal ticks, is a positive linear function of the prior year's population density of white-footed mice (Ostfeld et al. 2001). Risk of human exposure to Lyme disease also has been shown to increase with decreasing size of forest fragments, ostensibly as a result of the loss of vertebrate diversity and/or increases in abundance of white-footed mice (Allan et al. 2003).

In stark contrast to these examples of positive associations between density of rodent reservoirs and human disease risk or incidence, the culling of Norway rats (*Rattus norvegicus*) could result in the initiation or exacerbation of human outbreaks of Bubonic plague. Using a metapopulation model, Keeling and Gilligan (2000) found that strong reductions in abundance of rats, which serve as the zoonotic reservoir for *Yersinia pestis* and as the primary host for the flea vectors, can cause fleas to switch from rats to humans. The consequence of rat population crashes, counterintuitively, can therefore be increased rates of contacts between fleas and people and therefore human epidemics; this scenario is thought to have played a role in the Bubonic plague epidemics of Europe in the previous millennium (Keeling and Gilligan 2000).

Ultimate causes

The emerging pattern of increased risk or incidence of human disease with increased density of rodent reservoirs begs the question of what controls rodent abundance. Several chapters in this volume address this question. For herbivorous rodents such as voles, evidence is mounting that top-down effects of predators, often combined with bottom-up impacts of food supply, play a major role (Berryman 2002). However, for the most epidemiologically important rodent-borne diseases, the rodent hosts tend to be granivorous or omnivorous. In this category we include the sigmodontine rodents that serve as reservoirs for New World hantaviruses, arenaviruses, and bacteria (*Borrelia, Anaplasma* [= *Ehrlichia*]), the murine reservoirs for *Borrelia* and Old World hemorrhagic fever viruses, and the murine and gerbilline reservoirs for the agents of visceral and cutaneous leishmaniasis in Africa, Asia, and southern Europe. For these granivorous rodents, it appears that bottom-up effects of food supply predominate in determining abundance (fig. 41.2).

In some arid parts of South America and North America, El Niño events produce heavy rains followed by dramatically increased primary production. El Niño-induced seed production by annual plants, and masting in oak- or beech-dominated forests, constitute resource pulses that drive population increases in many rodent species (Ostfeld and Keesing 2000b). Epidemics of hantavirus pulmonary syndrome (HPS) have been associated with El Niño years in both North and South America (Yates et al. 2002; Toro et al. 1998). Similarly, high densities of ticks infected with Lyme disease bacteria have been detected following heavy mast years (Ostfeld et al. 2001).

El Niño events and oak/beech masting are natural events (although some evidence suggests that El Niño years will become more frequent and more intense with human-caused global warming [Herbert and Dixon 2003]). Human-induced changes to the environment also can induce local increases in rodent reservoir populations, or decreases in species diversity, both of which can increase disease incidence in people. Clearing of forests or agricultural practices in Central and South America have been associated with localized irruptions of rodent hosts or more generalized

changes in rodent community composition and associated risk of transmission of arenaviruses (Enría et al. 1999) and hantaviruses (Ruedas et al. 2004, Carroll et al. 2005). Habitat fragmentation in the northeastern US is associated with increased risk of Lyme disease in humans (Allan et al. 2003). Irrigation for local agriculture promotes populations of both fat sand rat (*Psammomys obesus*) reservoirs and sandfly (*Phlebotomus papatasi*) vectors of the etiologic agent of cutaneous leishmaniasis in Israel (Wasserberg et al. 2003).

Behavior

A behavioral trait with critical consequences for human disease is dispersal by rodents from sylvan to peridomestic environments (fig. 41.2). The presence of deer mice in or around human dwellings is a clear risk factor for HPS in the southwestern US (Zeitz et al. 1995) and probably elsewhere. Unfortunately, little is known about the factors that affect either rodent dispersal to human dwellings or those that regulate rodent populations in a commensal setting. Commensal populations of deer mice appear to have less stable composition (i.e., greater turnover of individuals) than do nearby sylvan populations (Douglass et al. 2003), but the generality of this difference is not known. We suggest that the behavioral and demographic causes and consequences of rodent commensalism are critical areas for future research involving behaviorists and epidemiologists.

Concluding Thoughts

To illustrate the patterns we have described, we relied heavily on examples from a few reasonably well-studied systems, such as the rodent-borne hemorrhagic fever viruses and Lyme disease. Clearly, many more studies are needed before we can conclude that the patterns observed in these systems can be generalized. One reason for the dearth of studies on the ecology of host-pathogen interactions may be that few investigators are schooled in ecology *and* microbiology *and* immunology/infectious diseases. Furthermore, some questions (e.g., the possibility of venereal transmission and the protectiveness of maternal antibody) must be addressed using laboratory studies. Nevertheless, because laboratory results do not always reflect what happens in nature, conclusions from laboratory studies should be tested in the field (Mills and Childs 1998; Wolff 2003c). Additionally, ecologists and epidemiologists have historically conducted their studies with little interdisciplinary consultation, and have published them in their own separate literature. These facts underscore the need for multidisciplinary studies involving ecologists, microbiologists, and public health researchers.

Ecologists generally prefer to work in pristine, sylvan ecosystems. However, diseases in rodent populations are probably most prevalent in perturbed ecosystems where biodiversity, community composition, and population dynamics have been altered. Studies are also quite rare in peridomestic habitats. These are precisely the environments where most transmission of zoonotic pathogens to humans takes place. Not only are disturbed and peridomestic habitats some of the most interesting to study in terms of host-pathogen dynamics, but the potential rewards in terms of understanding and prevention of zoonotic diseases in humans can be great.

Many of the reservoirs for serious human pathogens are highly opportunistic taxa (*Mastomys, Mus, Rattus, Calomys, Peromyscus, Sigmodon, Zygodontomys, Apodemus*). It is unclear whether this relationship is artefactual—zoonotic diseases carried by nonopportunistic species may remain unknown to us because we rarely encounter them —or if the ability to reproduce quickly and reach high population densities in temporarily ideal habitats is conducive to the evolution and maintenance of pathogens. If the apparent association between deadly zoonotic pathogens and opportunistic rodent taxa is real, then anthropogenic environmental changes (e.g., habitat fragmentation, conversion to agriculture, climate change) might be expected to increase the burden of human disease in the future.

Summary

As a group, rodents are probably the predominant natural reservoirs for pathogens that cause disease in humans. Nev-ertheless, beyond documenting the associations between specific rodents and their pathogens, little is known about how behavioral and population dynamics of rodents influence transmission either between individuals within rodent populations or between rodents and humans or other mammals. In this chapter, we provided an overview of how pathogens are detected within rodent hosts and what factors contribute to pathogen transmission. High population density, old age structure, fighting, and occupation of commensal or other disturbed (e.g., agricultural) habitats are associated with high rates of pathogen transmission. Although disperal has a high potential to influence pathogen transmission and disease dynamics, the role of host dispersal is poorly studied. Pathogen transmission and maintenance tend to be maximized in low rodent-diversity communities and in habitats where predators have been reduced or removed. Neither the generality of nor the precise mechanisms that underlie these apparent patterns are well understood, and therefore linkages between rodent demographic and behavioral dynamics and disease dynamics comprise a major research frontier. Lastly, we note an apparent pattern whereby rodent species of great importance to human disease tend to be widespread, opportunistic, and resilient species that are favored by anthropogenic environmental change. Whether this pattern is simply an artifact of heavy research focus on these species, or represents a true correlation between rodent population and life-history traits and disease dynamics, is unknown. If the pattern is real, the implication is that further anthropogenic environmental changes will result in further health risks to humans via their impacts on rodents.

Conclusions

Chapter 42 Conclusions and Future Directions

Paul W. Sherman and Jerry O. Wolff

A S YOU ARE UNDOUBTEDLY AWARE BY NOW, *Rodent Societies* is unabashedly a work of enlightened natural history—that is, authors have presented details of the natural behavior, physiology, and social ecology of various rodents and have then interpreted the facts in light of modern evolutionary theory. Every chapter has proudly and enthusiastically highlighted new results and synthesized existing information about the sociobiology of one or more rodent species (much of which was scattered hitherto), and has then analyzed the results through developing and testing alternative hypotheses at one or more of the four levels of analysis: mechanisms, ontogenies, fitness consequences, or evolutionary histories.

Every chapter in *Rodent Societies* concludes with a brief summary. Perusal of these summaries indicates how much we presently know about that topic. On the one hand, the amount of information is impressive, especially about certain subjects and taxa. On the other hand, it is obvious that a tremendous amount of information remains to be discovered. This is hardly surprising, since long-term studies of behavior and ecology have been conducted on less than 10% of the more than 2,000 species of rodents. At this rate, many species probably will go extinct without ever having been studied in detail! Even in well-studied species, we know very little about intraspecific (interpopulation) variations—for example, in annual cycles, digestive and hibernation physiology, demography, food preferences, mating systems, and social behavior. In no rodent species has any social or reproductive behavior been completely elucidated at all four levels of analysis.

Various authors infer from observing the behavior of their study animals that rodents are making complex cost/benefit decisions about how to behave so as to maximize fitness. However, actual measurement of fitness differences among behavioral variants is rare, and no one understands the physiological, genetic, or ontogenetic mechanisms underlying adaptive decision making. Similarly, theory about the functions of facultative sex ratio adjustment is mature, but the extent of the predicted adjustments at conception and their underlying physiological and ontogenetic mechanisms are unknown. Neophobia and neophilia are well documented in some rodents, but their mechanisms, ontogenies, and evolutionary histories are essentially unstudied. And despite the current revolution in molecular biology, we know very little about the genetic and ontogenetic bases, or the evolutionary histories of variations in life histories, sex ratios, and developmental programs, nor social behaviors such as kin recognition, alloparental care, dispersal, or alarm calling.

Even regarding fitness consequences—the level of analysis at which we and most other behavioral ecologists work—many intriguing and important questions about rodent sociobiology remain. For example, no one has quantified the fitness consequences of recognizing or not recognizing kin, giving or not giving an alarm call, or committing or not committing infanticide. Alternative reproductive tactics, infanticide, and natal dispersal are widespread in rodents, and multiple hypotheses have been proposed to explain the adaptive significance of each phenomenon. However, few of these hypotheses have been adequately tested in nature for even a single species—that is, using the strong-inference approach and manipulative experimentation. A great deal is now known about the neurohormonal mechanisms of aggression and pairbonding in some rodents (e.g., voles), but we know almost nothing about how selection favors these behaviors in various species with different mating systems. Likewise, nearly ubiquitous sexual dimorphisms in social rodents, such as body size, timing of emergence from hibernation, dispersal, and life spans have yet to be adequately analyzed. We still do not understand the function of dispersal, nor why natal dispersal occurs in some species, postbreeding dispersal occurs in others, and both types of dispersal (or neither type) occur in a few species. And when dispersal occurs, why is it sometimes male biased, sometimes female biased, and sometimes neither? Of course, part of the problem is that some social behaviors are nearly impossible to study in the field, including parental behavior, scent marking, facultative sex-ratio adjustment, reproductive suppression and activation, mate choice, and social learning, whereas other social behaviors are nearly impossible to study in the laboratory, including social group formation, alarm calling, dispersal, and life histories. Whereas it often is difficult to apply results of lab studies to field situations, field studies often are unable to control variables known from lab studies to be important.

Much also remains to be learned about rodent socioecology. In particular, there is the perennial question of whether group living (sociality) is a nonadaptive result of patchy distributions of critical resources, or is an adaptive response due to advantages of foraging socially for widely scattered and unpredictable food sources, or rapidly detecting and effectively avoiding predators. The answer to this question leads to different predictions about how the resulting group or society will be structured and the types of social and reproductive interactions that will occur (e.g., complex cooperation and reproductive skew may occur in the adaptive case, but is not predicted in the nonadaptive case). Understanding the ecological causes of sociality are also important for conserving and managing populations of rodents, especially those that are threatened and endangered. For example, rodents that group adaptively are more likely to carry and transmit parasites and pathogens than are those species that live in groups only when suitable habitats are limited or patchy.

When rodent societies form because of advantages of living near conspecifics, they are usually composed of close kin. A question worth addressing is whether postweaning young remain near home because of demographic benefits of philopatry or costs of dispersal (e.g., safety, food), or because there are inclusive fitness benefits to be derived from assisting parents in raising siblings. Another interesting question is the degree to which rodent societies are convergent with those of other taxa. Conceptual analyses of social behavior remain sharply divided between schema that

attempt to differentiate cooperative breeding from eusociality and those that view eusociality as part of a spectrum of cooperative social systems. Although this distinction is not explicitly taxonomic, proponents of the more restrictive view typically regard eusociality as unique to insects, while supporters of the more expansive view contend that patterns of social structure transcend taxonomic boundaries, and therefore apply to mammals and particularly to rodents (e.g., African mole-rats, prairie dogs, beavers).

Chapters in this book reveal that, in general, ecologists emphasize studies of rodent populations, communities, and population-level outcomes of behaviors, but sometimes they lose sight of the individual. Psychologists are particularly good at teasing apart cognitive processes and developmental ontogenies of individuals, but sometimes they are unaware of fitness outcomes, or use of the comparative method to infer evolutionary histories. Physiologists have taught us much about rodent behavioral mechanisms, and geneticists have elucidated the details of genetic mechanisms of rodents, but sometimes these biologists are not completely aware of the environmental influences, fitness consequences, or evolutionary histories of the mechanisms they have discovered. Conservation biologists investigate population numbers and distributions, habitat choices and resource needs, but they sometimes ignore the importance of social behaviors when developing management plans. And behavioral ecologists specialize in studying rodent sociobiology in nature, but they sometimes are oblivious to the underlying mechanistic and developmental constraints on those behaviors. Bridging these gaps and preconceptions, and expansion of our knowledge, will require new programs of collaborative research. There are exciting possibilities here, because scientific fields often grow most rapidly at their intersections. We suggest that a modest and worthy goal over the next decade would be a collaborative effort to thoroughly investigate and achieve a complete understanding of just one social behavior (e.g., alarm calling, infanticide, alloparental care, sex-ratio adjustment) in one social rodent species at all four levels of analysis.

In our introduction to this anthology we explained its conception, scope, and intended audience. We expressed the hope that this book would contribute not only to a better understanding of the social biology of rodents but also to evolutionary and ecological theory in general. The extent to which the editors and authors have been successful in meeting those goals are up to you and future generations to decide. At this point, we wish to express three final hopes: that you have enjoyed reading *Rodent Societies,* that at least one chapter has touched you personally and professionally, and that as a result you, your colleagues and students, and your institutional libraries will find it to be a valuable reference for many years to come.

References

Aars, J., and R. A. Ims. 1999. The effect of habitat corridors on rates of transfer and interbreeding between vole demes. *Ecology* 80:1648–55.

———. 2000. Population dynamic and genetic consequences of spatial density-dependent dispersal in patchy populations. *American Naturalist* 155:252–65.

———. 2002. Intrinsic and climatic determinants of population demography: The winter dynamics of tundra voles. *Ecology* 83:3449–56.

Aars, J., E. Johannesen, and R. A. Ims. 1999. Demographic consequences of movements in subdivided root vole populations. *Oikos* 85:204–16.

Aars, J., X. Lambin, R. Denny, and A. C. Griffin. 2001. Water vole in the Scottish uplands: Distribution patterns of disturbed and pristine populations ahead and behind the American mink invasion front. *Animal Conservation* 4:187–94.

Abbott, K. D., T. G. Ksiazek, and J. N. Mills. 1999. Long-term hantavirus persistence in rodent populations in central Arizona. *Emerging Infectious Diseases* 5:102–12.

Abercrombie, E. D., K. A. Keefe, D. S. DiFrischia, and M. J. Zigmond. 1989. Differential effect of stress on in vivo dopamine release in striatum, nucleus accumbens, and medial frontal cortex. *Journal of Neurochemistry* 52:1655–58.

Abramsky, Z., M. L. Rosenzweig, and A. Subach. 1998. Do gerbils care more about competition or predation? *Oikos* 83:75–84.

Abramsky, Z., M. L. Rosenzweig, and A. Subach. 2002. The costs of apprehensive foraging. *Ecology* 83:1330–40.

Abramsky, Z., E. Strauss, A. Subach, B. P. Kotler, and A. Riechman. 1996. The effects of barn owls (*Tyto alba*) on the activity and microhabitat selection of *Gerbillus allenbyi* and *G. pyramidium*. *Oecologia* 105:313–19.

Ackers, S. H., and C. N. Slobodchikoff. 1999. Communication of stimulus size and shape in alarm calls of Gunnison's prairie dogs, *Cynomys gunnisoni*. *Ethology* 105:149–62.

Adams, C. E. 1976. Measurements and characteristics of fox squirrel, *Sciurus niger rufiventer*, home ranges. *American Midland Naturalist* 95:211–15.

Adams, E. J., and P. Parham. 2001. Species-specific evolution of MHC class I genes in the higher primates. *Immunological Reviews* 183:41–64.

Adams, N., and R. Boice. 1983. A longitudinal study of dominance in an outdoor colony of domestic rats. *Journal of Comparative Psychology* 97:24–33.

———. 1989. Development of dominance in domestic rats in laboratory and semi-natural environments. *Behaviour Proceedings* 19:127–42.

Ader, R., and L. J. Grota. 1970. Rhythmicity in the maternal behaviour of *Rattus norvegicus*. *Animal Behaviour* 18:144–50.

Adkins, R. M., E. L. Gelke, D. Rowe, and R. L. Honeycutt. 2001. Molecular phylogeny and divergence time estimates for major rodent groups: Evidence from multiple genes. *Molecular Biology and Evolution* 18:777–91.

Adkins, R. M., A. H. Walton, and R. L. Honeycutt. 2003. Higher-level systematics of rodents and divergence time estimates based on two congruent nuclear genes. *Molecular Phylogenetics and Evolution* 26:409–20.

Adler, G. H., M. L. Wilson, and M. J. Derosa. 1987. Effects of adults on survival and recruitment of *Peromyscus leucopus*. *Canadian Journal of Zoology* 65:2519–44.

Adler, N. T. 1969. Effects of the male's copulatory behavior in successful pregnancy of the female rat. *Journal of Comparative and Physiological Psychology* 69:613–22.

Adler, N. T., and S. R. Zoloth. 1970. Copulatory behavior can inhibit pregnancy in female rats. *Science* 168:1480–82.

Agnew, W., D. W. Uresk, and R. M. Hansen. 1986. Flora and fauna associated with prairie dog colonies and adjacent ungrazed mixed grass prairie in western South Dakota. *Journal of Range Management* 39:135–139.

Agrell, J. 1995. A shift in female social organization independent of relatedness: An experimental study on the field vole (*Microtus agrestis*). *Behavioral Ecology* 6:182–91.

Agrell, J., S. Erlinge, J. Nelson, and M. Sandell. 1996. Shifting spacing behaviour of male field voles (*Microtus agrestis*) over the reproductive season. *Annales Zoologici Fennici* 33:243–48.

Agrell, J., J. O. Wolff, and H. Ylönen. 1998. Counter-strategies to infanticide in mammals: Costs and consequences. *Oikos* 83:507–17.

Ågren, G. 1981. Two laboratory experiments on inbreeding avoidance in the Mongolian gerbil. *Behavioural Processes* 6:291–97.

———. 1984a. Alternative mating strategies in the Mongolian gerbil. *Behaviour* 91:229–44.

———. 1984b. Incest avoidance and bonding between siblings in gerbils. *Behavioral Ecology and Sociobiology* 14:161–69.

———. 1984c. Pair formation in the Mongolian gerbil. *Animal Behaviour* 32:528–35.

Ågren, G. 1990. Sperm competition, pregnancy initiation and litter size: Influence of the amount of copulatory behaviour in Mongolian gerbils, *Meriones unguiculatus*. *Animal Behaviour* 40:417–27.

Ågren, G., Q. Zhou, and W. Zhong. 1989a. Ecology and social behaviour of Mongolian gerbils, *Meriones unguiculatus*, at Xilinhot, Inner Mongolia, China. *Animal Behaviour* 37:11–27.

———. 1989b. Territoriality, cooperation and resource priority: Hoarding in the Mongolian gerbil, *Meriones unguiculatus*. *Animal Behaviour* 37:28–32.

Aguilera, G. 1998. Corticotropin releasing hormone, receptor regulation and the stress response. *Trends in Endocrinology and Metabolism* 9:329–36.

Ahnesjo, I., A. Vincent, R. Alatalo, T. Halliday, and W. J. Sutherland. 1993. The role of females in influencing mating patterns. *Behavioral Ecology* 4:187–89.

Aimar, A. 1992. A morphometric analysis of Pleistocene marmots. In *First international symposium of alpine marmot* (Marmota marmota) *and on genus* Marmota, ed. B. Bassano, P. Durio, U. Gallo Orsi, and E. Macchi, 179–84. Torino.

Aisenstadt, O. S. 1945. Habitat distribution and ecology of the brown rat (*Rattus norvegicus,* Berkenh.) in a northwestern district of the European part of the USSR. *Zoologicheskii zhurnal* 24:182–89.

Aisner, R., and J. Terkel. 1992. Ontogeny of pine-cone opening behaviour in the black rat (*Rattus rattus*). *Animal Behaviour* 44:327–36.

Albers, H. E., and M. Bamshad. 1998. Role of vasopressin and oxytocin in the control of social behavior in Syrian hamsters (*Mesocricetus auratus*). *Progress in Brain Research* 119: 395–408.

Alberts, A. C. 1992. Constraints on the design of chemical communication systems in terrestrial vertebrates. *American Naturalist* 139 (suppl.):S62–S89.

Alberts, J. R. 1978. Huddling by rat pups: Group behavioral mechanisms of temperature regulation and energy conservation. *Journal of Comparative and Physiological Psychology* 92:231–45.

———. 1984. Sensory-perceptual development in the Norway rat: A view toward comparative studies. In *Comparative perspective on the development of memory,* ed. R. V. Kail and N. E. Spear, 65–101. Hillsdale, NJ: Erlbaum.

———. 1987. Early learning and ontogenetic adaptation. In *Perinatal development: A psychobiological perspective,* ed. N. A. Krasnegor, E. M. Blass, M. A. Hofer, and W. P. Smotherman, 11–37. Orlando, FL: Academic Press.

Alberts, J. R., and D. J. Gubernick. 1983. Reciprocity and resource exchange. A symbiotic model of parent-offspring interactions. In *Symbiosis in parent-offspring interactions,* ed. L. A. Rosenblum and H. Moltz, 7–44. New York: Plenum.

Alberts, S. C. 1999. Paternal kin discrimination in wild baboons. *Proceedings of the Royal Society of London, Series B* 266: 1501–6.

Alberts, S. C., and C. Ober. 1993. Genetic variability in the major histocompatibility complex: A review of non-pathogen-meditated selective mechanisms. *Yearbook of Physical Anthropology* 36:71–89.

Alcock, J. 2005. *Animal behavior: An evolutionary approach.* Sunderland, MA: Sinauer.

Alcorn, J. R. 1940. Life history notes on the piute ground squirrel. *Journal of Mammalogy* 21:160–70.

Aleksiuk, M. 1968. Scent-mound communication, territoriality, and population regulation in beaver (*Castor canadensis* Kuhl). *Journal of Mammalogy* 49:759–62.

Alexander, R. D. 1974. The evolution of social behavior. *Annual Review of Ecology and Systematics* 5:325–83.

———. 1990. Epigenetic rules and Darwinian algorithms. *Ethology and Sociobiology* 11:241–303.

———. 1991. Some unanswered questions about naked mole-rats. In *The biology of the naked mole-rat,* ed. P. W. Sherman, J. U. M. Jarvis, and R. D. Alexander, 446–65. Princeton, NJ: Princeton University Press.

Alexander, R. D., J. L. Hoogland, R. D. Howard, K. M. Noonan, and P. W. Sherman. 1979. Sexual dimorphism and breeding systems in pinnipeds, ungulates, primates and humans. In *Evolutionary biology and human social behavior,* ed.

N. A. Chagnon and W. Irons, 402–35. North Scituate, MA: Duxbury.

Alexander, R. D., K. M. Noonan, and B. J. Crespi. 1991. The evolution of eusociality. In *The biology of the naked mole-rat,* ed. P. W. Sherman, J. U. M. Jarvis, and R. D. Alexander, 3–44. Princeton, NJ: Princeton University Press.

Alho, C. J. R., M. S. C. Campos, and H. C. Gonçalves. 1987a. Ecologia de capivara (*Hydrochoerus hydrochaeris,* Rodentia) do Pantanal: I. Habitats, densidades e tamanho de grupo. *Revisita Brasileira de Biologica* 47:87–97.

———. 1987b. Ecologia de capivara (*Hydrochoerus hydrochaeris,* Rodentia) do Pantanal: II. Atividade, sazonalidade, uso do espaço e manejo. *Revisita Brasileira de Biologica* 47:99–110.

Al-kahtani, M. A., C. Zuleta, E. Caviedes-Vidal, and T. Garland, Jr. 2004. Kidney mass and relative medullary thickness of rodents in relation to habitat, body size, and phylogeny. *Physiological and Biochemical Zoology* 77:346–65.

Allainé, D. 2000. Sociality, mating system, and reproductive skew in marmots: Evidence and hypotheses. *Behavioural Processes* 51:21–34.

Allainé, D., F. Brondex, L. Graziani, J. Coulon, and I. Till-Battraud. 2000. Male-biased sex ratio in litters of alpine marmots supports the helper repayment hypothesis. *Behavioral Ecology* 11:507–14.

Allan, B. F., F. Keesing, and R. S. Ostfeld. 2003. Effects of habitat fragmentation on Lyme disease risk. *Conservation Biology* 17:267–72.

Allard, M. W., and R. L. Honeycutt. 1992. Nucleotide sequence variation in the mitochondrial 12S rRNA gene and the phylogeny of African mole–rats (Rodentia: Bathyergidae). *Molecular Biology and Evolution* 9:27–40.

Allee, W. C. 1931. *Animal aggregations: A study in general sociology.* Chicago: University of Chicago Press.

———. 1951. *Cooperation among animals, with human implications.* New York: Henry Schuman.

Allen, D. S., and W. P. Aspey. 1986. Determinants of social dominance in eastern gray squirrels (*Sciurus carolinensis*): A quantitative assessment. *Animal Behaviour* 34:81–89.

Allendorf, F. W. 1986. Genetic drift and the loss of alleles versus heterozygosity. *Zoo Biology* 5:181–90.

Alleva, E., A. Caprioli, and G. Laviola. 1989. Litter gender composition affects maternal behavior of the primiparous mouse dam (*Mus musculus*). *Journal of Comparative Psychology* 103:83–87.

Alonzo, S. H., and R. R. Warner. 2000. Female choice, conflict between the sexes and the evolution of male alternative reproductive behaviours. *Evolutionary Ecology Research* 2:149–70.

Altbacker, V., R. Hudson, and A. Bilko. 1995. Rabbit-mothers' diet influences pups' later food choice. *Ethology* 99:107–16.

Altmann, S. A. 1979. Altruistic behaviour: The fallacy of kin deployment. *Animal Behaviour* 27:958–62.

Alvarez-Buylla, A., and J. M. Garcia-Verdugo. 2002. Neurogenesis in adult subventricular zone. *Journal of Neuroscience* 22: 629–34.

Alvarez-Castaneda, S. T., G. Arnaud, and E. Yensen. 1996. *Spermophilus atricapillus. Mammalian Species* 521:1–3.

Amrein, I., L. Slomianka, I. I. Poletaeva, N. V. Bologova, and H.-P. Lipp. 2004. Marked species and age-dependent differences in cell proliferation and neurogenesis in the hippocampus of wild-living rodents. *Hippocampus* 14:1000–1010.

Anderson, P. K. 1989. *Dispersal in rodents: A resident fitness hypothesis.* Special publication no. 9. Provo, UT: American Society of Mammalogists.

Anderson, P. K., and J. L. Hill. 1965. *Mus musculus*: Experimental induction of territory formation. *Science* 148:1753–55.

Anderson, R. M., and R. M. May. 1978. Regulation and stability of host-parasite population interactions. I. Regulatory processes. *Journal of Animal Ecology* 47:219–47.

Anderson, S. 1985. Taxonomy and systematics. In *Biology of New World Microtus,* Special publication no. 8, ed. R. H. Tamarin, 52–83. Manhattan, KS: The American Society of Mammalogists.

Anderson, S., C. A. Woods, G. S. Morgan, and W. L. R. Oliver. 1983. *Geocapromys brownii. Mammalian Species* 201: 1–5.

Anderson, S. H., and E. S. Williams. 1997. Plague in a complex of white-tailed prairie dogs and associated small mammals in Wyoming. *Journal of Wildlife Diseases* 33:720–32.

Andersson, M. 1984. The evolution of eusociality. *Annual Review of Ecology and Systematics* 15:165–89.

———. 1994. *Sexual selection.* Princeton, NJ: Princeton University Press.

Andervont, H. B. 1944. Influence of environment on mammary cancer in mice. *Journal of the National Cancer Institute* 4:579–81.

Andreassen, H. P., and R. A. Ims. 2001. Dispersal in patchy vole populations: Role of patch configuration, density dependence, and demography. *Ecology* 82:2911–26.

Andreassen, H. P., K. Hertzberg, and R. A. Ims. 1998. Space-use responses to habitat fragmentation and connectivity in the root vole, *Microtus oeconomus. Ecology* 79:1223–35.

Andrews, R. V., R. W. Belknap, S. Southard, M. Lorincz, and S. Hess. 1972. Physiological, demographic and pathological changes in wild Norway rat populations over an annual cycle. *Comparative Biochemistry and Physiology* 41: 149–65.

Ansorge, H. 1983. Zur Wertung der Quadratmethode beim Kleinsäugerfang. *Säugetierkundliche Informationen* 2:13–18.

Antolin, M. F., B. Van Horne, M. D Berger, Jr., A. K. Holloway, J. L. Roach, and R. D. Weeks, Jr. 2001. Effective population size and genetic structure of a Piute ground squirrel (*Spermophilus mollis*) population. *Canadian Journal of Zoology* 79: 26–34.

Aragona, B. J., Y. Liu, J. T. Curtis, F. K. Stephan, and Z. Wang. 2003a. A critical role for nucleus accumbens dopamine in partner-preference formation in male prairie voles. *Journal of Neuroscience* 23:3483–90.

Aragona, B. J., Y. Liu, and Z. X. Wang. 2003b. Nucleus accumbens dopamine is important for social choice and the maintenance of social bonds. Program No. 757.7. *Society for Neuroscience.*

Arbetan, P. T. 1993. The mating system of the red squirrel, (*Tamiasciurus hudsonicus*). Master's thesis, University of Kansas, Lawrence.

Arcaro, K. F., and A. Eklund. 1998. A review of MHC-based mate preferences and fostering experiments in two congenic strains of mice. *Genetica* 104:241–44.

Archibald, J. D. 2003. Timing and biogeography of the eutherian radiation: Fossils and molecules compared. *Molecular Phylogenetics and Evolution* 28:350–59.

Ardlie, K. G., and L. M. Silver. 1998. Low frequency of *t* haplotypes in natural populations of house mice (*Mus musculus domesticus*). *Evolution* 52:1185–96.

Arellano, E., D. S. Rogers, and F. A. Cervantes. 2003. Genic differentiation and phylogenetic relationships among tropical harvest mice (*Reithrodontomys*: Subgenus *Aporodon*). *Journal of Mammalogy* 84:129–43.

Arkin, A., T. R. Saito, K. Takahashi, H. Amao, S. Aoki-Komori, and K. W. Takahashi. 2003. Age-related changes on marking, marking-like behavior and the scent gland in adult Mongolian gerbils (*Meriones unguiculatus*). *Experimental Animals* 52: 17–24.

Armitage, K. B. 1962. Social behaviour of a colony of the yellow-bellied marmot (*Marmota flaviventris*). *Animal Behaviour* 10: 319–31.

———. 1973. Population changes and social behavior following colonization by the yellow-bellied marmot. *Journal of Mammalogy* 58:842–54.

———. 1977. Social variety in the yellow-bellied marmot: A population-behavioural system. *Animal Behaviour* 25: 585–93.

———. 1981. Sociality as a life-history tactic of ground squirrels. *Oecologia* 48:36–49.

———. 1982. Marmots and coyotes: Behavior of prey and predator. *Journal of Mammalogy* 63:503–5.

———. 1984. Recruitment in yellow-bellied marmot populations: Kinship, philopatry, and individual variability. In *Biology of ground-dwelling squirrels,* ed. J. O. Murie and G. R. Michener, 377–403. Lincoln: University of Nebraska Press.

———. 1986a. Individuality, social behavior, and reproductive success in yellow-bellied marmots. *Ecology* 67:1186–93.

———. 1986b. Marmot polygyny revisited: Determinants of male and female reproductive strategies. In *Ecological aspects of social evolution,* ed. D. I. Rubenstein and R. W. Wrangham, 303–31. Princeton, NJ: Princeton University Press.

———. 1987a. Do female yellow-bellied marmots adjust the sex ratios of their young? *American Naturalist* 129:501–19.

———. 1987b. Social dynamics of mammals: Reproductive success, kinship and individual fitness. *Trends in Ecology and Evolution* 2:279–84.

———. 1988. Resources and social organization of ground-dwelling squirrels. In *The ecology of social behavior,* ed. C. N. Slobodchikoff, 131–55. San Diego, CA: Academic Press.

———. 1991. Social and population-dynamics of yellow-bellied marmots—results from long-term research. *Annual Review of Ecology and Systematics* 22:379–407.

———. 1996a. Resource sharing and kinship in yellow-bellied marmots. In *Biodiversity in marmots,* ed. M. Le Berre, R. Ramousse, and L. Le Guelte, 129–34. Lyon: International Network on Marmots.

———. 1996b. Social dynamics, kinship, and population dynamics of marmots. In *Biodiversity in marmots,* ed. M. Le Berre, R. Ramousse, and L. Le Guelte, 113–28. Lyon: International Network on Marmots, Lyon.

———. 1998. Reproductive strategies of yellow-bellied marmots: Energy conservation and differences between the sexes. *Journal of Mammalogy* 79:385–93.

———. 1999a. Evolution of sociality in marmots. *Journal of Mammalogy* 80:1–10.

———. 1999b. Social and population dynamics of yellow-bellied marmots: Results from long-term research. *Annual Review of Ecology and Systematics* 22:379–407.

———. 2000. The evolution, ecology, and systematics of marmots. *Oecologia Montana* 9:1–18.

———. 2002. Social dynamics of yellow-bellied marmots: Strategies for evolutionary success. In *Holarctic marmots as a factor of biodiversity,* ed. K. B. Armitage and V. Yu. Rumiantsev, 10–16. Moscow: ABF.

———. 2003a. Marmots, *Marmota monax* and allies. In *Wild mammals of North America: Biology, management, and conservation,* 2nd ed., ed. G. Feldhamer, B. Thompson, and J. Chapman, 188–210. Baltimore: Johns Hopkins University Press.

———. 2003b. Nesting activities of yellow-bellied marmots. In *Adaptive strategies and diversity in marmots,* ed. R. Ramousse, D. Allainé, and M. Le Berre, 27–32. Lyon: International Network on Marmots.

———. 2003c. Recovery of a yellow-bellied marmot population following a weather-induced decline. In *Adaptive strategies and diversity in marmots,* ed. R. Ramousse, D. Allainé, and M. Le Berre, 217–24. Lyon: International Network on Marmots.

———. 2003d. Reproductive competition in female yellow-bellied marmots. In *Adaptive strategies and diversity in marmots,* ed. R. Ramousse, D. Allainé, and M. Le Berre, 133–42. Lyon: International Network on Marmots.

Armitage, K. B., and D. T. Blumstein. 2002. Body-mass diversity in marmots. In *Holarctic marmots as a factor of biodiversity,* ed. K. B. Armitage and V. Yu. Rumiantsev, 22–32. Moscow: ABF.

Armitage, K. B., D. T. Blumstein, and B. C. Woods. 2003. Energetics of hibernating yellow-bellied marmots (*Marmota flaviventris*). *Comparative Biochemistry and Physiology A* 134:101–14.

Armitage, K. B., and J. F. Downhower. 1970. Interment behavior in the yellow-bellied marmot (*Marmota flaviventris*). *Journal of Mammalogy* 51:177–78.

———. 1974. Demography of yellow-bellied marmot populations. *Ecology* 55:1233–45.

Armitage, K. B., J. F. Downhower, and G. E. Svendsen. 1976. Seasonal changes in weights of marmots. *American Midland Naturalist* 96:36–51.

Armitage, K. B., and G. E. Gurri-Glass. 1994. Communal nesting in yellow-bellied marmots. In *Actual problems of marmots investigation,* ed. V. Y. Rumiantsev, 14–26. Moscow, RUS: ABF.

Armitage, K. B., and D. W. Johns. 1982. Kinship, reproductive strategies and social dynamics of yellow-bellied marmots. *Behavioral Ecology and Sociobiology* 11:55–63.

Armitage, K. B., D. Johns, and D. C. Andersen. 1979. Cannibalism among yellow-bellied marmots. *Journal of Mammalogy* 60:205–7.

Armitage, K. B., and C. M. Salsbury. 1992. Factors affecting oxygen consumption in wild-caught yellow-bellied marmots (*Marmota flaviventris*). *Comparative Biochemistry and Physiology A* 103:729–37.

Armitage, K. B., C. M. Salsbury, E. L. Barthelmess, R. C. Gray, and A. Kovach. 1996. Population time budget for the yellow-bellied marmot. *Ethology Ecology and Evolution* 8:67–95.

Armitage, K. B., and O. A. Schwartz. 2000. Social enhancement of fitness in yellow-bellied marmots. *Proceedings of the National Academy of Sciences* 97:12149–52.

Armitage, K. B., and B. C. Woods. 2003. Group hibernation does not reduce energetic costs of young yellow-bellied marmots. *Physiological and Biochemical Zoology* 76:888–98.

Armitage, K. B., B. C. Woods, and C. M. Salsbury. 2000. Energetics of hibernation in woodchucks (*Marmota monax*). In *Life in the cold,* ed. G. Heldmaier and M. Klingenspor, 73–80. Berlin: Springer-Verlag.

Armitage, K. B., and K. E. Wynne-Edwards. 2002. Progesterone concentrations in wild-caught yellow-bellied marmots. In *Holarctic marmots as a factor of biodiversity,* ed. K. B. Armitage and V. Yu. Rumiantsev, 41–47. Moscow: ABF.

Arnason, U., J. A. Adegoke, K. Bodin, E. W. Born, Y. B. Esa, A. Gullberg, M. Nilsson, R. Short, X. Xu, and A. Janke. 2002. Mammalian mitogenomic relationships and the root of the eutherian tree. *Proceedings of the National Academy of Sciences (USA)* 99:8151–56.

Arnason, U., A. Gullberg, and A. Janke. 1997. Phylogenetic analyses of mitochondrial DNA suggest a sister group relationship between Xenarthra (Edentata) and ferungulates. *Molecular Biology and Evolution* 14:762–68.

Arnold, K. E., and I. P. F. Owens. 1998. Cooperative breeding in birds: A comparative test of the life history hypothesis. *Proceedings of the Royal Society of London, Series B* 265:739–45.

Arnold, S. J., and D. Duvall. 1994. Animal mating systems: A synthesis based on selection theory. *American Naturalist* 143:317–48.

Arnold, W. 1988. Social thermoregulation during hibernation in alpine marmots (*Marmota marmota*). *Journal of Comparative Physiology B* 158:151–56.

———. 1990a. The evolution of marmot sociality: I. Why disperse late? *Behavioral Ecology and Sociobiology* 27:229–37.

———. 1990b. The evolution of marmot sociality: II. Costs and benefits of joint hibernation. *Behavioral Ecology and Sociobiology* 27:239–46.

———. 1993. Social evolution in marmots and the adaptive value of joint hibernations. *Verhandlungen der Deutschen Zoologischen Gesellschaft* 86:79–93.

———. 1999. Allgemeine Biologie und Lebensweise des Alpenmurmeltieres (*Marmota marmota*). *Stapfia* 63:1–20.

Arnold, W., and J. Dittami. 1997. Reproductive suppression in male alpine marmots. *Animal Behaviour* 53:53–66.

Arnold, W., M. Klinkicht, K. Rabmann, and D. Tautz. 1994. Molecular analysis of the mating system of alpine marmots (*Marmota marmota*). *Verhandlungen der Deutschen Zoologischen Gesellschaft* 86:27.

Arnold, W., and A. V. Lichtenstein. 1993. Ectoparasite loads decrease the fitness of alpine marmots (*Marmota marmota*) but are not a cost of sociality. *Behavioral Ecology* 4:36–39.

Artzt, K., P. McCormick, and D. Bennett. 1982. Gene mapping within the *T/t* complex of the mouse. I. *t*-lethal genes are nonallelic. *Cell* 28:463–70.

Arvola, A., M. Ilmén, and T. Koponen. 1962. On the aggressive behaviour of the Norwegian lemming (*Lemmus lemmus*) with special reference to the sounds produced. *Archivum Societatis Zoologicae Botanicae Fennicae "Vanamo"* 17:80–101.

Asher, M. 2004. Social system and spatial organization of wild guinea pigs (*Cavia aperea*) in a natural low density population. *Journal of Mammalogy* 85:788–96.

Asher, M. R., E. S. De Oliveira, and N. Sachser. 2004. Social system and spatial organization of wild guinea pigs (*Cavia aperea*) in a natural population. *Journal of Mammalogy* 85:788–96.

Ashton, K. G., M. C. Tracy, and A. de Queiroz. 2000. Is Bergmann's rule valid for mammals? *American Naturalist* 156: 390–415.

Atchley, W. R., C. T. Gaskins, and D. Anderson. 1976. Statistical properties of ratios: Empirical results. *Systematic Zoology* 27:137–48.

Atkinson, I. A. E. 1985. The spread of commensal species of *Rattus* to oceanic islands and their effects on island avifaunas. In *Conservation of island birds*, ed. O. J. Moors, 35–81. Cambridge, UK: International Council for Bird Preservation.

———. 1996. Introduction of wildlife as a cause of species extinctions. *Wildlife Biology* 2:135–41.

Auffray, J.-C., E. Tchernov, and E. Nevo. 1988. Origin of the commensalism of the house mouse (*Mus musculus domesticus*) in relation to man. *Comptes Rendus de l'Académie des Sciences. Série III-Vie* 307:517–22.

Austad, S. N. 1984. A classification of alternative reproductive behaviors and methods of field testing ESS models. *American Zoologist* 24:309–19.

———. 1993. Retarded senescence in an insular population of opossums. *Journal of Zoology* 229:695–708.

Austin, D., and D. A. Dewsbury. 1986. Reproductive capacity of male laboratory rats. *Physiology and Behavior* 37:627–32.

Axelrod, R., and W. D. Hamilton. 1981. The evolution of cooperation. *Science* 211:1390–96.

Axelson, J. F., M. Smith, and M. Duarte. 1999. Prenatal flutamide treatment eliminates the adult male rat's dependency upon vasopressin when forming social-olfactory memories. *Hormones and Behavior* 36:109–18.

Ayer, M. L., and J. M. Whitsett. 1980. Aggressive behaviour of female prairie deer mice in laboratory populations. *Animal Behaviour* 28:763–71.

Azcarate, T., F. Alvarez, and F. Braza. 1980. Tendencias gregarias del capibara (*Hydrochoerus hydrochaeris*) en los Llanos de Venezueal. *Reunion Iberoamericana de Zoologia Vertebrata* 1:285–92.

Azzara, A. V., R. J. Bodnar, A. R. Delamater, and A. Sclafani. 2001. D1 but not D2 dopamine receptor antagonism blocks the acquisition of a flavor preference conditioned by intragastric carbohydrate infusions. *Pharmacology Biochemistry and Behavior* 68:709–20.

Baack, J. K., and P. V. Switzer. 2000. Alarm calls affect foraging behavior in Eastern chipmunks (*Tamias striatus*, Rodentia: Sciuridae). *Ethology* 106:1057–66.

Bacchini, A., E. Gaetani, and A. Cavaggioni. 1992. Pheromone binding proteins of the mouse, *Mus musculus*. *Experientia* 48:419–21.

Baddaloo, E. G. Y., and F. V. Clulow. 1981. Effects of the male on growth, sexual maturation, and ovulation of young female meadow voles, *Microtus pennsylvanicus*. *Canadian Journal of Zoology* 59:415–21.

Badyaev, A. V., G. E. Hill, A. M. Stoehr, P. M. Nolan, and K. J. McGraw. 2000. The evolution of sexual dimorphism in the house finch: II. Population divergence in relation to local selection. *Evolution* 54:2134–44.

Bailey, R. C., and J. Byrnes. 1990. A new, old method for assessing measurement error in both univariate and multivariate morphometric studies. *Systematic Zoology* 39:124–30.

Baker, A. E. M. 1981. Gene flow in house mice: Introduction of a new allele into free-living populations. *Evolution* 35:243–58.

Baker, R. H. 1983. *Michigan mammals*. Detroit: Wayne State University.

Baker, R. J., K. D. Makova, and R. K. Chesser. 1999. Microsatellites indicate a high frequency of multiple paternity in *Apodemus* (Rodentia). *Molecular Ecology* 8:107–11.

Bakken, A. 1959. Behavior of gray squirrels. *Proceedings of the Southeastern Association of Game and Fish Commissioners* 13:393–406.

Bakker, E. M. van Zinderen. 1967. The "arid corridor" between southwest Africa and the Horn of Africa. In *Paleoecology of Africa*, vol. 2, ed. E. M. van Zinderen Bakker, 76–79. Amsterdam: A. A. Balkema.

Balcom, B. J., and R. H. Yahner. 1996. Microhabitat and landscape characteristics associated with the threatened Allegheny woodrat. *Conservation Biology* 10:515–25.

Balcombe, J. P. 1990. Vocal recognition of pups by mother Mexican free-tailed bats, *Tadarida brasiliensis mexicana*. *Animal Behaviour* 39:960–66.

Balfour, D. 1983. Infanticide in the Columbian ground squirrel, *Spermophilus columbianus*. *Animal Behaviour* 31: 949–50.

Balph, D. F. 1961. Underground concealment as a method of predation. *Journal of Mammalogy* 42:423–24.

———. 1984. Spatial and social behavior in a population of Uinta ground squirrels: Interrelations with climate and annual cycle. In *The biology of ground-dwelling squirrels*, ed. J. O. Murie and G. R. Michener, 336–52. Lincoln: University of Nebraska Press.

Balph, D. M., and D. F. Balph. 1966. Sound communication of Uinta ground squirrels. *Journal of Mammalogy* 47: 440–50.

Balph, D. R., and A. W. Stokes. 1963. On the ethology of a population of Uinta ground squirrels. *The American Midland Naturalist* 69:106–26.

Bamshad, M., M. A. Novak, and G. J. De Vries. 1993. Sex and species differences in the vasopressin innervation of sexually naive and parental prairie voles, *Microtus ochrogaster* and meadow voles, *Microtus pennsylvanicus*. *Journal of Neuroendocrinology* 5:247–55.

Baranga, J. 1991. Kibale forest game corridor: Man or wildlife? In *Nature conservation 2: The role of corridors*, ed. D. A. Saunders and R. J. Hobbs, 371–76. Norton, Australia: Surrey Beatty and Sons.

Barash, D. P. 1973. The social biology of the Olympic marmot. *Animal Behaviour Monographs* 6:173–245.

———. 1974a. The evolution of marmot societies: A general theory. *Science* 185:415–20.

———. 1974b. Mother-infant relations in captive woodchucks (*Marmota monax*). *Animal Behaviour* 22:446–48.

———. 1974c. The social behaviour of the hoary marmot (*Marmota caligata*). *Animal Behaviour* 22:256–61.

———. 1975a. Ecology of paternal behavior in the hoary marmot (*Marmota caligata*): An evolutionary investigation. *Journal of Mammalogy* 56:613–18.

———. 1975b. Marmot alarm-calling and the question of altruistic behavior. *American Midland Natrualist* 94: 468–70.

———. 1981. Mate-guarding and gallivanting in male hoary marmots (*Marmota caligula*). *Behavioral Ecology and Sociobiology* 9:187–93.

———. 1989. *Marmots: Social behavior and ecology*. Stanford, CA: Stanford University Press.

Barash, D. P. 1985. Review of *The biology of ground-dwelling squirrels,* ed. J. O. Murie and G. R. Michener. *Zeitschrift für Tierpsychologie* 70:342–43.

Barbazanges, A., P. V. Piazza, M. Le Moal, and S. Maccari. 1996. Maternal glucocorticoid secretion mediates long-term effects of prenatal stress. *Journal of Neuroscience* 16: 3943–49.

Bard, J., K. Yamazaki, M. Curran, E. A. Boyse, and G. K. Beauchamp. 2000. Effect of B2m gene disruption on MHC-determined odortypes. *Immunogenetics* 51:514–18.

Barfield, R. J., and L. A. Geyer. 1975. The ultrasonic postejaculatory vocalization and the postejaculatory refractory period of the male rat. *Journal of Comparative and Physiological Psychology* 88:723–34.

Barkalow, F. S., Jr., and R. F. Soots, Jr. 1975. Lifespan and reproductive longevity of the gray squirrel, *Sciurus c. carolinensis* Gmelin. *Journal of Mammalogy* 56:522–24.

Barker, J.M., J. M. Wojtowicz, and R. Boonstra. 2005. Where's my dinner? Adult neurogenesis in free-living food-storing rodents. *Genes, Brain and Behavior* 4:89–98.

Barlow, G. 1988. Monogamy in relation to resources. In *The ecology of social behavior,* ed. C. N. Slobodchikoff, 55–79. San Diego: Academic Press.

Barnard, C. J. 1990. Kin recognition: Problems, prospects, and the evolution of discrimination systems. *Advances in the Study of Behavior* 19:29–81.

Barnard, C. J., and J. Fitzsimons. 1988. Kin recognition and mate choice in mice: The effects of kinship, familiarity and social interference on intersexual interaction. *Animal Behaviour* 36:1078–90.

———. 1989. Kin recognition and mate choice in mice: Fitness consequences of mating with kin. *Animal Behaviour* 38: 35–40.

Barnes, A. M. 1982. Surveillance and control of bubonic plague in the United States. *Symposium of the Zoological Society of London* 50:237–70.

———. 1993. A review of plague and its relevance to prairie dog populations and the black-footed ferret. In *Management of prairie dog complexes for the reintroduction of the black-footed ferret,* ed. J. L. Oldemeyer, D. E. Biggins, and B. J. Miller, 28–37. *United States Fish and Wildlife Service Biological Report* 13:1–96.

Barnett, S. A. 1955. Competition among wild rats. *Nature* 175: 126–27.

———. 2002. *The story of rats: Their impact on us, and our impact on them.* Crows Nest, Australia: Allen and Unwin.

Barrett, G. W., and J. D. Peles, eds. 1999. *Landscape ecology of small mammals.* New York: Springer-Verlag.

Barry, R. E., and P. J. Mundy. 2002. Seasonal variation in the degree of heterospecific association of two syntopic hyraxes (*Heterohyrax brucei* and *Procavia capensis*) exhibiting synchronous parturition. *Behavioral Ecology and Sociobiology* 52:177–81.

Bartmann, S., and G. Gerlach. 2001. Multiple paternity and similar variance in reproductive success of male and female wood mice (*Apodemus sylvaticus*) housed in an enclosure. *Ethology* 107:889–99.

Bartness, T. J., and H. E. Albers. 2000. Activity patterns and the biological clock in mammals. In *Activity patterns in small mammals,* Ecological Studies no. 141, ed. S. Halle and N. C. Stenseth, 23–47. Berlin: Springer.

Bassano, B., B. Sabatier, L. Rossi, and E. Macchi. 1992. Parasitic fauna of the digestive tract of *Marmota marmota* in the western Alps. In *First international symposium of alpine marmot* (Marmota marmota) *and on genus* Marmota, ed. B. Bassano, P. Durio, U. Gallo Orsi, and E. Macchi, 13–24. Torino.

Bassareo, V., M. A. De Luca, and G. Di Chiara. 2002. Differential expression of motivational stimulus properties by dopamine in nucleus accumbens shell versus core and prefrontal cortex. *Journal of Neuroscience* 22:4709–19.

Basset, P., F. Balloux, and N. Perrin. 2001. Testing demographic models of effective population size. *Proceedings of the Royal Society of London, Series B* 268:311–17.

Batbold, J. 2002. A study on population genetic structure of the Mongolian marmot and some problems of plague. In *Holarctic marmots as a factor of biodiversity,* ed K. B. Armitage and V. Yu. Rumiantsev, 54–61. Moscow: ABF.

Bateman, A. J. 1948. Intrasexual selection in *Drosophila. Heredity* 2:349–68.

Bateman, G. C. 1967. Home range studies of a desert nocturnal rodent fauna. PhD diss., University of Arizona, Tucson.

Bateson, P. P. G. 1982. Behavioural development and evolutionary processes. In *Current problems in sociobiology,* ed. King's College Sociobiology Group, 133–51. New York: Cambridge University Press.

———. 1983. Optimal outbreeding. In *Mate choice,* ed. P. Bateson, 257–77. Cambridge: Cambridge University Press.

———. 1988. The active role of behaviour in evolution. In *Evolutionary processes and metaphors,* ed . M.-W. Ho and S. W. Fox, 191–207. Chichester, UK: Wiley.

Battersby, S. A., R. Parson, and J. P. Webster. 2002. Urban rat infestation and the risk to public health. *Journal of Environmental Health Research* 1:57–65.

Batzli, G., L. Getz, and S. Huxley. 1977. Suppression of growth and reproduction in Microtine rodents by social factors. *Journal of Mammalogy* 58:583–91.

Baumgardner, D. J., T. G. Hartung, D. K. Sawrey, D. G. Webster, and D. A. Dewsbury. 1982. Muroid copulatory plugs and female reproductive tracts: A comparative investigation. *Journal of Mammalogy* 63:110–17.

Beacham, T. D. 1980. Dispersal during population fluctuations of the vole, *Microtus townsendii. Journal of Animal Ecology* 49:867–77.

Beauchamp, G. K., and J. L. Wellington. 1984. Habituation to individual odors occurs in brief, widely-spaced presentations. *Physiology and Behavior* 32:511–14.

Beauchamp, G. K., K. Yamazaki, J. Bard, and E. A. Boyse. 1988. Preweaning experience in the control of mating preferences by genes in the major histocompatibility complex of the mouse. *Behavior Genetics* 18:537–47.

Beauchamp, G. K., K. Yamazaki, and E. A. Boyse. 1985. The chemosensory recognition of genetic individuality. *Scientific American* 253:86–92.

Beauchamp, G. K., K. Yamazaki, C. J. Wysocki, B. M. Slotnick, L. Thomas, and E. A. Boyse. 1985. Chemosensory recognition of mouse major histocompatibility types by another species. *Proceedings of the National Academy of Sciences* (USA) 82:4186–88.

Beck, M., and B. G. Galef, Jr. 1989. Social influences on the selection of protein-sufficient diet by Norway rats. *Journal of Comparative Psychology* 103:132–39.

Beck, M., C. L. Hitchcock, and B. G. Galef, Jr. 1988. Diet sampling by wild Norway rats offered several unfamiliar foods. *Animal Learning and Behaviour* 16:224–30.

Becker, J. B., C. N. Rudick, and W. J. Jenkins. 2001. The role of dopamine in the nucleus accumbens and striatum during sexual behavior in the female rat. *Journal of Neuroscience* 21: 3236–41.

Bediz, G. M., and J. M. Whitsett. 1979. Social inhibition of sexual maturation in male prairie deer mice. *Journal of Comparative and Physiological Psychology* 93:493–500.

Beecher, M. D. 1982. Signature systems and kin recognition. *American Zoologist* 22:477–90.

———. 1991. Successes and failures of parent-offspring recognition in animals. In *Kin recognition*, ed. P. G. Hepper, 94–124. Cambridge: Cambridge University Press.

Beecher, M. D., P. Loesche, P. K. Stoddard, and M. B. Medvin. 1989. Individual recognition by voice in swallows: Signal or perceptual adaptation? In *Comparative psychology of audition: Perceiving complex sounds,* ed. R. J. Dooling and S. H. Hulse, 227–94. Hillsdale, NJ: Erlbaum.

Beecher, M. D., and P. K. Stoddard. 1990. The role of bird song and calls in individual recognition: Contrasting field and laboratory perspectives. In *Comparative perception—vol. II: Complex sounds,* ed. M. Berkley and W. C. Stebbins, 375–408. New York: Wiley.

Begall, S., H. Burda, and M. H. Gallardo. 1999. Reproduction, postnatal development, and growth of social coruros, *Spalacopus cyanus* (Rodentia: Octodontidae), from Chile. *Journal of Mammalogy* 80:210–17.

Begg, R. J. 1981. The small mammals of Little Nourlangie Rock, N.T. IV. Ecology of *Zyzomys woodwardi,* the large rock-rat, and *Zyzomys argurus,* the common rock-rat, (Rodentia: Muridae). *Australian Wildlife Research* 8:307–20.

Begg, R. J., K. C. Martin, and N. F. Price. 1981. The small mammals of Little Nourlangie Rock, N. T. V. The effects of fire. *Australian Wildlife Research* 8:515–27.

Begon, M., S. Hazel, D. Baxby, K. Bown, R. Cavanagh, J. Chantrey, T. Jones, and M. Bennett. 1999. Transmission dynamics of a zoonotic pathogen within and between wildlife host species. *Proceedings of the Royal Society of London, Series B* 266:1939–45.

Begon, M., S. M. Hazel, S. Telfer, K. Bown, D. Carslake, R. Cavanagh, J. Cnantrey, T. Jones, and M. Bennett. 2003. Rodents, cowpox virus and islands: Densities, numbers and thresholds. *Journal of Animal Ecology* 72:343-55.

Behrends, P. M., M. Daly, and M. I. Wilson. 1986. Range use patterns and spatial relationships of Merriam's kangaroo rats (*Dipodomys merriami*). *Behaviour* 96:187–209.

Bekoff, M. 1977. Mammalian dispersal and the ontogeny of individual behavioral phenotypes. *American Naturalist* 111: 715–32.

———. 1981. Mammalian sibling interactions: Genes, facilitative environments, and the coefficient of familiarity. In *Parental care in mammals,* ed. D. J. Gubernick and P. H. Klopfer, 307–46. New York: Plenum.

Bel, M. C., C. Porteret, and J. Coulon. 1995. Scent deposition by cheek rubbing in the alpine marmot (*Marmota marmota*) in the French Alps. *Canadian Journal of Zoology* 73:2065–71.

Belk, M. C., and H. D. Smith. 1991. *Ammospermophilus leucurus. Mammalian Species* 368:1–8.

Belluscio, L., C. Lodovichi, P. Feinstein, P. Mombaerts, and

L. C. Katz. 2002. Odorant receptors instruct functional circuitry in the mouse olfactory bulb. *Nature* 419:296–300.

Bennett, N. C. 1988. The trend towards sociality in three species of southern African mole–rats (Bathyergidae): Causes and consequences. PhD diss., University of Cape Town, South Africa.

Bennett, N. C. and C. G. Faulkes. 2000. *African mole-rats: Ecology and eusociality.* Cambridge: Cambridge University Press.

Bennett, N. C., C. G. Faulkes, and J. U. M. Jarvis. 1999. Reproduction in subterranean rodents. In *The Biology of subterranean rodents: Evolutionary challenges and opportunities,* ed. G. Cameron, E. A. Lacey, and J. Patton, 145–77. Chicago: University of Chicago Press.

Bennett, N. C., C. G. Faulkes, and A. J. Molteno. 1996. Reproductive suppression in subordinate, non–breeding female Damaraland mole–rats: Two components to a lifetime of socially–induced infertility. *Proceedings of the Royal Society of London, Series B* 263:1599–1603.

———. 2000. Reproduction in subterranean rodents. In *Life underground: The biology of subterranean rodents,* ed. E. A Lacey, J. L. Patton, and G. N. Cameron, 145–77. Chicago: University of Chicago Press.

Bennett, N. C., C. G. Faulkes, and A. C. Spinks. 1997. LH responses to single doses of exogenous GnRH by social Mashona mole–rats: A continuum of socially–induced infertility in the family Bathyergidae. *Proceedings of the Royal Society of London, Series B* 264:1001–6.

Bennett, N. C., J. U. M. Jarvis, G. H. Aguilar, and E. McDaid. 1991. Growth rates and development in six species of African mole–rats (Family; Bathyergidae). *Journal of Zoology* 225: 13–26.

Bennett, N. C., J. U. M. Jarvis, and F. P. D. Cotterill. 1994. The colony structure and reproductive biology of the afrotropical Mashona mole–rat, *Cryptomys darlingi. Journal of Zoology* 234:477–87.

Bennett, N. C., J. U. M. Jarvis, and K. C. Davies. 1988. Daily and seasonal temperatures in the burrows of African rodent moles. *South African Journal of Zoology* 23:189–95.

Bennett, N. C., A. J. Molteno, and A. C. Spinks. 2000. Pituitary sensitivity to exogenous GnRH in giant Zambian mole–rats, *Cryptomys mechowi* (Rodentia: Bathyergidae): Support for the 'socially induced infertility continuum.' *Journal of Zoology* 252:447–52.

Bennett, R. I., A. G. Motulsky, A. Bittles, L. Hudgins, S. Uhrich, D. L. Doyle, K. Silvey, et al. 2002. Genetic counseling and screening of consanguineous couples and their offspring: Recommendations of the national society of genetic counselors. *Journal of Genetic Counseling* 11:97–119.

Bennett, R. P. 1999. Effects of food quality on growth and survival of juvenile Columbian ground squirrels (*Spermophilus columbianus*). *Canadian Journal of Zoology* 77: 1555–61.

Benson, B. N. 1980. Dominance relationships, mating behaviour, and scent marking in fox squirrels (*Sciurus niger*). *Mammalia* 44:143–60.

Berdoy, M. 1994. Making decisions in the wild: Constraints, conflicts and communication in foraging rats. In *Behavioural aspects of feeding,* ed. B. G. Galef, Jr. and P. Valsechhi, 289–313. Ettore Majorana International Life Science Series, vol 12. Chur, Switzerland: Harwood Academic.

———. 2002. *The laboratory rat: A natural history.* (27 min. film). Accessed at www.ratlife.org.

Berdoy, M., and D. W. Macdonald. 1991. Factors affecting feeding in wild rats. *Acta Oecologica* 12: 261–79.

Berdoy, M., P. Smith, and D. W. Macdonald. 1995. Stability of social status in wild rats: Age and the role of settled dominance. *Behaviour* 132:193–212.

Berdoy, M., J. P. Webster, and D. W. Macdonald. 1995. Parasite-altered behaviour: Is the effect of *Toxoplasma gondii* on *Rattus norvegicus* specific? *Parasitology* 111:403–9.

———. 2000. Fatal attractions in rats infected with *Toxoplasma gondii. Proceedings of the Royal Society of London, Series B* 267:1591–94.

Berger, P. J., N. C. Negus, and M. Day. 1997. Recognition of kin and avoidance of inbreeding in the montane vole, *Microtus montanus. Journal of Mammalogy* 78:1182–86.

Bergerud, A. T., and D. R. Miller. 1977. Population dynamics of Newfoundland beaver. *Canadian Journal of Zoology* 55: 1480–92.

Berkenhout, J. 1769. *Outlines of the natural history of Great Britain and Ireland.* London.

Bernshtein, A. D., N. S. Apekina, T. V. Mikhailova, Y. A. Myasnikov, L. A. Khlyap, and Y. S. Korotkov. 1999. Dynamics of Puumala hantavirus infection in naturally infected bank voles (*Clethrionomys glareolus*). *Archives of Virology* 144: 2415–28.

Berry, R. J. 1968. The ecology of an island population of the house mouse. *Journal of Animal Ecology* 37:445–70.

———. 1970. The natural history of the house mouse. *Field Studies* 3:219–62.

———. 1981. Town mouse, country mouse: Adaptation and adaptability in *Mus domesticus* (*M. musculus domesticus*). *Mammal Review* 11:91–136.

Berry, R. J., and F. H. Bronson. 1992. Life history and bioeconomy of the house mouse. *Biological Reviews* 67:519–50.

Berry, R. J., and M. E. Jakobson. 1974. Vagility in an island population of the house mouse. *Journal of Zoology, London* 173:341–54.

Berry, R. J., G. S. Triggs, P. King, H. R. Nash, and L. R. Noble. 1991. Hybridization and gene flow in house mice introduced into an existing population on an island. *Journal of Zoology* 225:615–32.

Berryman, A. A. 2002. Population cycles: Causes and analysis. In *Population cycles: The case for trophic interactions,* ed. A. Berryman, 3–28. Oxford: Oxford University Press.

Berteaux, D., and S. Boutin. 2000. Breeding dispersal in female North American red squirrels. *Ecology* 81:1311–26.

Berteaux, D., J. Bety, E. Rengifo, and J. Bergeron. 1999. Multiple paternity in meadow voles (*Microtus pennsylvanicus*): Investigating the role of the female. *Behavioral Ecology and Sociobiology* 45:283–91.

Bertram, B. C. R. 1978. Living in groups: Predators and prey. In *Behavioural ecology: An evolutionary approach,* ed. J. R. Krebs and N. B. Davies, 64–96. Oxford: Blackwell Scientific.

Best, T. L. 1993. Patterns of morphological and morphometric variation in heteromyid rodents. In *Biology of the heteromyidae.* Special publication no. 10, ed. H. H. Genoways and J. H. Brown, 197–235. American Society of Mammalogists.

———. 1995 *Spermophilus mohavensis. Mammalian Species* 509:1–7.

Best, T. L., and S. Riedel. 1995. *Sciurus arizonensis. Mammalian Species* 496:1–5.

Bester-Meredith, J. K., and C. A. Marler. 2001. Vasopressin and aggression in cross-fostered California mice (*Peromyscus californicus*) and white-footed mice (*Peromyscus leucopus*). *Hormones and Behavior* 40:51–64.

Bester-Meredith, J. K., and C. A. Marler. 2003. Vasopressin and the transmission of paternal behavior across generations in mated, cross-fostered *Peromyscus* mice. *Behavioral Neuroscience* 117:455–63.

Bester-Meredith, J. K., L. J. Young, and C. A. Marler. 1999. Species differences in paternal behavior and aggression in *Peromyscus* and their associations with vasopressin immunoreactivity and receptors. *Hormones and Behavior* 36:25–38.

Betts, B. J. 1976. Behavior in a population of Columbian ground squirrels, *Spermophilus columbianus columbianus. Animal Behaviour* 24:652–80.

———. 1999. Current status of Washington ground squirrels in Oregon and Washington. *Northwestern Naturalist* 80: 35–38.

Beynon, R. J., and J. L. Hurst. 2003. Multiple roles of major urinary proteins in the house mouse, *Mus domesticus. Biochemical Society Transactions* 31:142–46.

Beynon, R. J., J. L. Hurst, S. J. Gaskell, S. J. Hubbard, R. E. Humphries, N. Malone, A. D. Marie, et al. 2001. Mice, MUPs and myths: Structure-function relationships of the major urinary proteins. In *Chemical signals in vertebrates IX,* ed. A. Marchlewska-Koj, J. Lepri, and D. Müller-Schwarze, 149–56. New York: Kluwer Academic/ Plenum.

Biardi, J., R. G. Coss, and D. G. Smith. 2000. California ground squirrel (*Spermophilus beecheyi*) blood sera inhibits crotalid venom proteolytic activity. *Toxicon* 38:713–21.

Bibikov, D. I. 1991. Population structure and strategy of reproduction in marmots. In *Population structure of the marmot,* ed. D. I. Bibikov, A. A. Nikolski, V. Ju. Rumiantsev, and T. A. Seredneva, 6–31. Moscow: USSR Theriological Society.

———. 1996. *Die murmeltiere der welt.* Bd. 388. Heidelberg, Germany: Spektrum Akademischer.

Biggins, D. E., and M. Kosoy. 2001. Influence of introduced plague on North American mammals: Implications from ecology of plague in Asia. *Journal of Mammalogy* 82:906–16.

Biggins, D. E., J. G. Sidle, D. B. Seery, and A. E. Ernst. 2006. Estimating the abundance of prairie dogs. 2005. In *Conservation of the black-tailed prairie dog,* ed. J. L. Hoogland, 94–107. Washington, DC: Island Press.

Biggs, J. R., K. D. Bennett, M. A. Mullen, T. K. Haarmann, M. Salisbury, R. J. Robinson, D. Keller, N. Torrez-Martinez, and B. Hjelle. 2000. Relationship of ecological variables to Sin Nombre virus antibody seroprevalence in populations of deer mice. *Journal of Mammalogy* 81:676–82.

Bihr, K. J., and R. J. Smith. 1998. Location, structure, and contents of burrows of *Spermophilus lateralis* and *Tamias minimus,* two ground-dwelling sciurids. *Southwestern Naturalist* 43:352–62.

Birdsall, D. A., and D. Nash. 1973. Occurrence of successful multiple insemination of females in natural populations of deer mice (*Peromyscus maniculatus*). *Evolution* 27:106–10.

Birkenholz, D. E. 1963. A study of the life history and ecology of the round-tailed muskrat (*Neofiber alleni* True) in north-central Florida. *Ecological Monographs* 33:255–80.

Birkhead, T. 2000. *Promiscuity.* Cambridge, MA: Harvard University Press.

Birkhead, T. R., and A. P. Møller. 1993. Female control of paternity. *Trends in Ecology and Evolution* 8:100–104.

———. 1998. *Sperm competition and sexual selection*. London: Academic Press.

Birkhead, T. R., A. P. Moller, and W. J. Sutherland. 1993. Why do females make it so difficult for males to fertilize their eggs? *Journal of Theoretical Biology* 161:51–60.

Birney, E. C. 1973. Systematics of three species of woodrats (Genus *Neotoma*) in central North America. *University of Kansas, Museum of Natural History Miscellaneous Publications* 58:1–173.

Bishop, J. A., and D. J. Hartley. 1976. The size and age structure of rural populations of *Rattus norvegicus* containing individuals resistant to the anticoagulant poison Warfarin. *Journal of Animal Ecology* 45:623–46.

Bishop, J. M., J. U. M. Jarvis, A. C. Spinks, N. C. Bennett, and C. O'Ryan. 2004. Molecular insight into patterns of colony composition and paternity in the common mole–rat *Cryptomys hottentotus hottentotus*. *Molecular Ecology* 13:1217–29.

Bishop, M. J., and P. F. D. Chevins. 1987. Urine odours and marking patterns in territorial laboratory mice (*Mus musculus*). *Behavioral Processes* 15:233–48.

Bishop, W. W. 1962. The mammalian fauna and geomorphological relations of the Napak volcanics, Karamoja. *Uganda Geological Survey,* 1957–1958:1–18.

Bishop, W. W., J. A. Miller, and F. J. Fitch. 1969. New potassium–argon age determinations relevant to the Miocene fossil mammal sequence in East Africa. *American Journal of Science* 267:669–99.

Black, C. C. 1963. A review of the North American Tertiary Sciuridae. *Bulletin of the Museum of Comparative Zoology, Harvard* 130:109–248.

———. 1972. Holarctic evolution and dispersal of squirrels (Rodentia: Sciuridae). *Evolutionary Biology* 6:305–22.

Blackwell, B. F., P. D. Doerr, J. M. Reed, and J. R. Walters. 1995. Inbreeding rate and effective population size: A comparison of estimates from pedigree analysis and a demographic model. *Biological Conservation* 71:299–304.

Blackwell, J. M. 1981. The role of the house mouse in disease and zoonoses. *Symposia of the Zoological Society of London* 47:591–616.

Blair, W. F. 1941. Observations on the life history of *Baiomys taylori subater*. *Journal of Mammalogy* 22:378–82.

———. 1943. Populations of the deermouse and associated small mammals in the mesquite association of southern New Mexico. *Contributions from the Laboratory of Vertebrate Biology* 21:1–40.

———. 1951. Population structure, social behavior, and environmental relations in a natural population of the beach mouse (*Peromyscus polionotus leucocephalus*). *Contributions from the Laboratory of Vertebrate Biology* 48:1–47.

Blake, B. H., and K. E. Gillet. 1988. Estrous cycle and related aspects of reproduction in captive Asian chipmunks, *Tamias sibricus*. *Journal of Mammalogy* 69:598–603.

Blanchard, R. J., and D. C. Blanchard. 1977. Aggressive behavior in the rat. *Behavioural Biology* 21:197–224.

Blanchard, R. J., D. C. Blanchard, R. Agullana, and S. M. Weiss. 1991. Twenty-two kHz alarm cries to presentation of a predator by laboratory rats living in visible burrow systems. *Physiology and Behavior* 50:967–72.

Blanchard, R. J., K. J. Flannelly, and D. C. Blanchard. 1988. Lifespan studies of dominance and aggression in established colonies of laboratory rats. *Physiology and Behavior* 43:1–7.

Blanchard, R. J., J. N. Nikulina, R. R. Sakai, C. McKittrick, B. McEwen, and D. C. Blanchard. 1998. Behavioral and endocrine change following chronic predatory stress. *Physiology and Behavior* 63:561–69.

Blanckenhorn, W. U. 2000. The evolution of body size: What keeps organisms small? *Quarterly Review of Biology* 75: 385–407.

Blaustein, A. R., M. Bekoff, J. A. Byers, and T. J. Daniels. 1991. Kin recognition in vertebrates: What do we really know about adaptive value? *Animal Behaviour* 41:1079–83.

Blaustein, A. R., M. Bekoff, and T. J. Daniels. 1987. Kin recognition in vertebrates (excluding primates): Mechanisms, functions and future research. In *Kin recognition in animals,* ed. D. J. C. Fletcher and C. D. Michener, 333–57. New York: Wiley.

Blaustein, A. R., and S. I. Rothstein. 1978. Multiple captures of *Reithrodontomys megalotis* in Santa Barbara, California. *American Midland Naturalist* 100:376–83.

Blaustein, A. R., and B. Waldman. 1992. Kin recognition in anuran amphibians. *Animal Behaviour* 44:207–21.

Bleich, V., and O. Schwartz. 1975. Observations on the home range of the desert woodrat, *Neotoma lepida intermedia*. *Journal of Mammalogy* 56:518–19.

Blintz, G. L. 1984. Water balance, water stress, and seasonal torpor in ground-dwelling sciurids. In *The biology of ground-dwelling squirrels,* ed. J. O. Murie and G. R. Michener, 142–65. Lincoln: University of Nebraska Press.

Blondel, J., P. Perret, M.-C. Anstett, and C. Thébaud. 2002. Evolution of sexual size dimorphism in birds: Test of hypotheses using blue tits in contrasted Mediterranean habitats. *Journal of Evolutionary Biology* 15:440–50.

Blouin, S. F., and M. Blouin. 1988. Inbreeding avoidance behaviors. *Trends in Ecology and Evolution* 3:230–33.

Blumstein, D. T. 1993. Infanticide: a search for an adaptive explanation. *Natura—World Wide Fund for Nature—Pakistan* 19:14–15.

———. 1995a. Golden-marmot alarm calls: I. The production of situationally specific vocalizations. *Ethology* 100:113–25.

———. 1995b. Golden-marmot alarm calls: II. Asymmetrical production and perception of situationally specific vocalizations? *Ethology* 101:25–32.

———. 1996. How much does social group size influence golden marmot vigilance? *Behaviour* 133:1133–51.

———. 1997. Infanticide among golden marmots (*Marmota caudata aurea*). *Ethology Ecology and Evolution* 9:169–73.

———. 1998. Quantifying predation risk for refuging animals: A case study with golden marmots. *Ethology* 104:501–16.

———. 1999a. Alarm calling in three species of marmots. *Behaviour* 136:731–57.

———. 1999b. The evolution of functionally referential alarm communication: Multiple adaptations; multiple constraints. *Evolution of Communication* 3:135–47.

———. 2000. The evolution of infanticide in rodents: A comparative analysis. In *Infanticide by males and its implications,* ed. C. P. Van Schaick and C. H. Jason, 178–97. Cambridge: Cambridge University Press.

———. 2003. Social complexity but not the acoustic environment is responsible for the evolution of complex alarm communication. In *Adaptative strategies and diversity in marmots,* ed. R. Ramousse, D. Allainé, and M. Le Berre, 41–49. Lyon: International Marmot Network.

Blumstein, D. T., and K. B. Armitage. 1997a. Alarm calling in yellow-bellied marmots: I. The meaning of situationally variable alarm calls. *Animal Behaviour* 53:143–71.

———. 1997b. Does sociality drive the evolution of communicative complexity? A comparative test with ground-dwelling sciurid alarm calls. *American Naturalist* 150:179–200.

———. 1998a. Life history consequences of social complexity: A comparative study of ground-dwelling sciurids. *Behavioral Ecology* 9:8–19.

———. 1998b. Why do yellow-bellied marmots call? *Animal Behaviour* 56:1053–55.

———. 1999. Cooperative breeding in marmots. *Oikos* 84:369–82.

Blumstein, D. T., and W. Arnold. 1995. Situational specificity in alpine-marmot alarm communication. *Ethology* 100:1–13.

———. 1998. Ecology and social behavior of golden marmots (*Marmota caudata aurea*). *Journal of Mammalogy* 79:873–86.

Blumstein, D. T., and J. C. Daniel. 1997. Inter- and intraspecific variation in the acoustic habitats of three marmot species. *Ethology* 103:325–38.

———. 2004. Yellow-bellied marmots discriminate among the alarm calls of individuals and are more responsive to the calls from pups. *Animal Behaviour* 68:1257–65.

Blumstein, D. T., J. C. Daniel, and W. Arnold. 2002. Survivorship of golden marmots (*Marmota caudata aurea*) in Pakistan. In *Holarctic marmots as a factor of biodiversity,* ed. K. B. Armitage and V. Yu. Rumiantsev, 82–85. Moscow: ABF.

Blumstein, D. T., and S. J. Henderson. 1996. Cheek-rubbing in golden marmots (*Marmota caudata aurea*). *Journal of Zoology* 238:113–23.

Blumstein, D. T., S. Im, A. Nicodemus, and C. Zugmeyer. 2004. Yellow-bellied marmots (*Marmota flaviventris*) hibernate socially. *Journal of Mammalogy* 85:25–29.

Blumstein, D. T., J. Steinmetz, K. B. Armitage, and J. C. Daniel. 1997. Alarm calling in yellow-bellied marmots: II. The importance of direct fitness. *Animal Behaviour* 53:173–84.

Blumstein, D. T., and A. C. Turner. 2005. Can the acoustic adaptation hypothesis predict the structure of Australian birdsong? *Acta Ethologica* 15:35–44.

Blumstein, D. T., L. Verenyre, and J. C. Daniel. 2004. Reliability and the adaptive utility of discrimination among alarm callers. *Proceedings of the Royal Society of London, Series B* 271:1851–57.

Blundell, G. M., M. Ben-David, and R. T. Bowyer. 2002. Sociality in river otters: cooperative foraging or reproductive strategies? *Behavioral Ecology* 13:134–41.

Bluthe, R. M., and R. Dantzer. 1993. Role of the vomeronasal system in vasopressinergic modulation of social recognition in rats. *Brain Research* 604:205–10.

Boag, D. A., and J. O. Murie. 1981. Population ecology of Columbian ground squirrels in southwestern Alberta. *Canadian Journal of Zoology* 59:2230–40.

Boag, D. A., and D. R. Wiggett. 1994a. Food and space: Resources defended by territorial parous female Columbian ground squirrels. *Canadian Journal of Zoology* 72:1908–14.

———. 1994b. The impact of foraging by Columbian ground squirrels, *Spermophilus columbianus,* on vegetation growing on patches fertilized with urine. *Canadian Field-Naturalist* 108:282–87.

Boellstorff, D. E., and D. H. Owings. 1995. Home range, population structure, and spatial organization of California ground squirrels. *Journal of Mammalogy* 76:551–61.

Boellstorff, D. E., D. H. Owings, M. C. T. Penedo, and M. J. Hersek. 1994. Reproductive behaviour and multiple paternity of California ground squirrels. *Animal Behaviour* 47:1057–64.

Boer, P. J. den. 1979. The significance of dispersal power for the survival of species, with special reference to the carabid beetles in a cultivated countryside. *Fortschrift Zoology* 25:79–94.

Boero, D. L. 1992. Alarm calling in alpine marmot (*Marmota marmota* L): Evidence for semantic communication. *Ethology, Ecology and Evolution* 4:125–38.

Boesch, C. 1996. The emergence of cultures among wild chimpanzees. In *Reaching into thought: The minds of great apes,* ed. A. E. Russon, K. A. Bard, and S. T. Parker, 251–68. Cambridge: Cambridge University Press.

Boggs, J. R. 1974. Social ecology of the white throated wood rat (*Neotoma albigula*) in Arizona. PhD diss., Arizona State University, Tempe.

Bolbroe, T., L. L. Jeppsen, and H. Leirs. 2000. Behavioural response of field voles under mustelid predation risk in the laboratory: More than neophobia. *Annales Zoologici Fennici* 37:169–78.

Bolhuis, J. J., and L.-A. Giraldeau. (eds.). 2005. *The behavior of animals.* Malden, MA: Blackwell.

Bolhuis, J. J., A. M. Strijkstra, E. Moor, and K. Vanderlende. 1988. Preferences for odours of conspecific non-siblings in the common vole, *Microtus arvalis. Animal Behaviour* 36:1551–53.

Bolles, K. 1988. Evolution and variation of antipredator vocalisations of Antelope squirrels, *Ammospermophilus* (Rodentia: Sciuridae). *Zeitschrift für Säugetierkunde* 53:129–47.

Bolles, R. C. 1970. Species-specific defense reactions and avoidance learning. *Psychological Reviews* 77:32–48.

Bollinger, E. K., S. J. Harper, and G. W. Barrett. 1993. Inbreeding avoidance increases dispersal movements of the meadow vole. *Ecology* 74:1153–56.

Bond, M. B., and J. O. Wolff. 1999. Does access to females or competition among males limit home-range size in a promiscuous rodent? *Journal of Mammalogy* 80:1243–50.

Bond, M. L., J. O. Wolff, and S. Krackow. 2002. Recruitment sex ratios in gray-tailed voles (*Microtus canicaudatus*) in response to density, sex ratio, and season. *Canadian Journal of Zoology* 81:1306–11.

Bondrup-Nielsen, S. 1993. Early malnutrition increases emigration of adult female meadow voles, *Microtus pennsylvanicus. Oikos* 67:317–20.

Bondrup-Nielsen, S., and R. A. Ims. 1990. Reversed sexual size dimorphism in microtines: Are females larger than males or are males smaller than females? *Evolutionary Ecology* 4:261–72.

Bonenfant, M., and D. L. Kramer. 1996. The influence of distance to burrow on flight initiation distance in the woodchuck, *Marmota monax. Behavioral Ecology* 7:299–303.

Bonesi, L., L. Lepini, and G. Gregori. 1996. Temporal analysis of activities in alpine marmot (*Marmota marmota* L.). In *Biodiversity in marmots,* ed. M. Le Berre, R. Ramousse, and L. Le Guelte, 157–68. Lyon: International Network on Marmots.

Boonstra, R. 1978. Effect of adult Townsend voles (*Microtus townsendii*) on survival of young. *Ecology* 59:242–48.

———. 1989. Life-history variation in maturation in fluctuating meadow vole populations. (*Microtus pennsylvanicus*). *Oikos* 54:265–74.

———. 1994. Population cycles in microtines: The senescence hypothesis. *Evolutionary Ecology* 8:196–219.

———. 2005. Equipped for life: The adaptive role of the stress axis in male mammals. *Journal of Mammalogy* 86:236–47.

Boonstra, R., and P. T. Boag. 1992. Spring declines in *Microtus pennsylvanicus* and the role of steroid hormones. *Journal of Animal Ecology* 61:339–52.

Boonstra, R., L. Galea, S. Matthews, and J. M. Wojtowicz. 2001a. Adult neurogenesis in natural populations. *Canadian Journal of Physiological Pharmacology* 79:297–302.

Boonstra, R., B. S. Gilbert, and C. J. Krebs. 1993. Mating systems and sexual dimorphism in mass in microtines. *Journal of Mammalogy* 74:224–29.

Boonstra, R., D. Hik, G. R. Singleton, and A. Tinnikov. 1998. The impact of predator-induced stress on the snowshoe hare cycle. *Ecological Monographs* 68:371–94.

Boonstra, R., and I. Hogg. 1988. Friends and strangers: A test of the Charnov-Finerty hypothesis. *Oecologia* 77:95–100.

Boonstra, R., A. H. Hubbs, E. A. Lacey, and C. J. McColl. 2001b. Seasonal changes in glucocorticoids and testosterone in free-living arctic ground squirrels from the boreal forest of the Yukon. *Canadian Journal of Zoology* 79:49–58.

Boonstra, R., and C. J. Krebs. 1977. A fencing experiment on a high-density population of *Microtus townsendii*. *Canadian Journal of Zoology* 55:1166–75.

Boonstra, R., C. J. Krebs, M. S. Gaines, M. L. Johnson, and I. M. T. Craine. 1987. Natal philopatry and breeding systems in *Microtus*. *Journal of Animal Ecology* 56:655–73.

Boonstra, R., C. J. Krebs, and A. Kenney. 1996. Why lemmings have indoor plumbing in summer. *Canadian Journal of Zoology* 74:1947–49.

Boonstra, R., and C. J. McColl. 2000. Contrasting stress response of male arctic ground squirrels and red squirrels. *Journal of Experimental Zoology* 286:390–404.

Boonstra, R., C. J. McColl, and T. J. Karels. 2001c. Reproduction at all costs: The adaptive stress response of male arctic ground squirrels. *Ecology* 82:1930–46.

Boonstra, R., and F. H. Rodd. 1983. Regulation of breeding density in *Microtus pennsylvanicus*. *Journal of Animal Ecology* 52:757–80.

Boonstra, R., X. Xia, and L. Pavone. 1993. Mating system of the meadow vole, *Microtus pennsylvanicus*. *Behavioral Ecology* 4:83–89.

Bosakowski, T., and D. G. Smith. 1992. Comparative diets of sympatric nesting raptors in the eastern deciduous forest biome. *Canadian Journal of Zoology* 70:984–92.

Boswell, T., S. C. Woods, and G. J. Kenagy, 1994. Seasonal changes in body mass, insulin, and glucocorticoids of free-living golden-mantled ground squirrels. *General and Comparative Endocrinology* 96:339–46.

Boursot, P., J.-C. Auffray, J. Britton-Davidson, and F. Bonhomme. 1993. The evolution of house mice. *Annual Review of Ecology and Systematics* 24:119–52.

Bouskila, A. 1995. Interactions between predation risk and competition: A field study of kangaroo rats and snakes. *Ecology* 76:165–78.

Boutin, S. 1990. Food supplementation experiments with terrestrial vertebrates: patterns, problems, and the future. *Canadian Journal of Zoology* 68:203–20.

Boutin, S., and K. W. Larsen. 1993. Does food availability affect growth and survival of males and females differently in a promiscuous small mammal, *Tamiasciurus hudsonicus*. *Journal of Animal Ecology* 62:364–70.

Bowers, J. M., and B. K. Alexander. 1967. Mice: Individual recognition by olfactory cues. *Science* 158:1208–10.

Bowers, M. A., and J. L. Dooley, Jr. 1999. EMS studies at the individual, patch, and landscape scale: Designing landscapes to measure scale-specific responses to habitat fragmentation. In *Landscape ecology of small mammals*, ed. G. W. Barrett and J. D. Peles, 147–74. New York: Springer-Verlag.

Bowers, M. A., J. L. Jefferson, and M. G. Kuebler. 1993. Variation in giving-up densities of foraging chipmunks (*Tamias striatus*) and squirrels (*Sciurus carolinensis*). *Oikos* 66:229–36.

Bowler, C. M., B. S. Cushing, and C. S. Carter. 2002. Social factors regulate female-female aggression and affiliation in prairie voles. *Physiology and Behavior* 76:559–66.

Bowman, J. C., M. Edwards, L. S. Sheppard, and G. J. Forbes. 1999. Record distance for a non-homing movement by a deer mouse, *Peromyscus maniculatus*. *Canadian Field-Naturalist* 113:292–93.

Boyd, S. K., and A. R. Blaustein. 1985. Familiarity and inbreeding avoidance in the gray-tailed vole (*Microtus canicaudus*). *Journal of Mammalogy* 66:348–52.

Boyeskorov, G. G., M. V. Shchelchkova, and V. N. Vasiliev. 1996. Divergence in *Marmota camtschatica*. In *Biodiversity in marmots*, ed. M. Le Berre, R. Ramousse, and L. Le Guelte, 229–30. Lyon: International Network on Marmots.

Boyse, E. A., G. K. Beauchamp, K. Yamazaki, and J. Bard. 1991. Genetic components of kin recognition in mammals. In *Kin recognition*, ed. P. G. Hepper, 148–61. Cambridge: Cambridge University Press.

Bozinovic, F. 1995. Nutritional energetics and digestive responses of an herbivorous rodent (*Octodon degus*) to different levels of dietary fiber. *Journal of Mammalogy* 76:627–37.

———. 1997. Diet selection in rodents: An experimental test of the effect of dietary fiber and tannins on feeding behavior. *Revista Chilena de Historia Natural* 70:67–71.

Bozinovic, F., and H. Torres-Contreras. 1998. Does digestion rate affect diet selection? A study in *Octodon degus*, a generalist herbivorous rodent. *Acta Theriologica* 43:205–12.

Bradbury, J. W., and S. L. Vehrencamp. 1976. Social organization and foraging in emballonurid bats. II. A model for the determination of group size. *Behavioral Ecology and Sociobiology* 2:383–404.

———. 1998. *Principles of animal communication.* Sunderland, MA: Sinauer.

Bradley, A. J. 2003. Stress, hormones, and mortality in small carnivorous marsupials. In *Predators with pouches: The biology of carnivorous marsupials*, ed. M. Jones, C. Dickman, and M. Archer, 250–63. Sydney: CSIRO Press.

Bradley, R. D., C. W. Edwards, D. S. Carroll, and C. W. Kilpatrick. 2004. Phylogenetic relationships of neotomine-peromyscine rodents: Based on DNA sequences from the mitochondrial cytochrome-*b* gene. *Journal of Mammalogy* 85:389–95.

Bradt, G. W. 1938. A study of beaver colonies in Michigan. *Journal of Mammalogy* 19:139–62.

Brady, C. A., and G. E. Svendsen. 1981. Social behaviour in a family of beaver, *Castor canadensis*. *Biological Behaviour* 65:99–114.

Brady, K. M., and K. B. Armitage. 1999. Scent-marking in the yellow-bellied marmot (*Marmota flaviventris*). *Ethology, Ecology and Evolution* 11:35–47.

Brain, P. F., M. H. Homady, and M. Mainardi. 1983. Preputial glands, dominance and aggressiveness, in mice. *Bollettino di Zoologia* 50:173–87.

Brain, P. F., D. Mainardi, and S. Parmagiani (eds.). 1989. *House mouse aggression.* London: Harwood Academic.

Braithwaite, R. W., and A. K. Lee. 1979 A mammalian example of semelparity. *American Naturalist* 113:151–56.

Braithwaite, R. W., and W. J. Muller. 1997. Rainfall, groundwater and refuges: Predicting extinctions of Australian tropical mammal species. *Australian Journal of Ecology* 22:57–67.

Brambell, F. W. R. 1970. The transmission of passive immunity from mother to young. In *Frontiers of Biology*, vol. 18, ed. A. Neuberger and E. L. Tatum, 218. Amsterdam: North-Holland.

Branch, L. C. 1993. Social organization and mating system of the plains viscacha (*Lagostomus maximus*). *Journal of Zoology, London* 229:473–91.

Brandt, J. K. 1855. Beiträge zur nahern Kenntnissder Säugethiere Russlands. *Mémoires de l'Academie Imperiale des Sciences de St. Petersbourg* 69:1–375.

Braude, S. H. 1991. Behavior and demographics of the naked mole-rat *Heterocephalus glaber.* PhD diss., University of Michigan, Ann Arbor.

———. 2000. Dispersal and new colony formation in wild naked mole-rats: Evidence against inbreeding as the system of mating. *Behavioral Ecology* 11:7–12.

Braun, J. K. and M. A. Mares. 1989. *Neotoma micropus. Mammalian Species* 330:1–9.

———. 1996. Unusual morphological and behavioral traits in *Abrocoma* (Rodentia: Abrocomidae) from Argentina. *Journal of Mammalogy* 77:891–97.

———. 2002. Systematics of the *Abrocoma cinerea* species complex (Rodentia : Abrocomidae), with a description of a new species of *Abrocoma. Journal of Mammalogy* 83:1–19.

Braun, K., P. Kremz, W. Wetzel, and T. Wagner. 2003. Influence of parental deprivation on the behavioral development in *Octodon degus*: Modulation by maternal vocalizations. *Developmental Psychobiology* 42:237–45.

Breck, S. W., K. R. Wilson, and D. C. Andersen. 2001. The demographic response of bank-dwelling beavers to flow regulation: A comparison of the Green and Yampa rivers. *Canadian Journal of Zoology* 79:1957–64.

Breed, W. G. 1983. Sexual dimorphism in the Australian hopping mouse, *Notomys alexis. Journal of Mammalogy* 64: 536–39.

———. 1990. Comparative studies on the timing of reproduction and foetal number in six species of Australian conilurine rodents (Muridae: Hydromyinae). *Journal of Zoology* 221:1–10.

Brennan, P. 2001. How mice make their mark. *Nature* 414: 590–91.

Brennan, P. A., H. M. Schellinck, E. B. and Keverne. 1999. Patterns of expression of the immediate-early gene egr-1 in the accessory olfactory bulb of female mice exposed to pheromonal constituents of male urine. *Neuroscience* 90: 1463–70.

Brett, R. A. 1991. The ecology of naked mole-rat colonies: Burrowing, food and limiting factors. In *The biology of the naked mole–rat*, ed. P. W. Sherman, J. U. M. Jarvis, and R. D. Alexander, 137–84. Princeton, NJ: Princeton University Press.

Brockman, H. J. 1984. The evolution of social behaviour in insects. In *Behavioral ecology: An evolutionary approach,* ed.

J. R. Krebs and J. B. Davies, 340–61. Sunderland, MA: Sinauer.

———. 2001. The evolution of alternative strategies and tactics. *Advances in the Study of Behavior* 30:1–51.

Brodie, J. 1981. Norway rats (*Rattus norvegicus*) on cereal stubble. *Journal of Zoology, London* 195:542–46.

Brody, A. K., and K. B. Armitage. 1985. The effects of adult removal on dispersal of yearling yellow-bellied marmots. *Canadian Journal of Zoology* 63:2560–64.

Brody, A. K., and J. Melcher. 1985. Infanticide in yellow-bellied marmots. *Animal Behaviour* 33:673–74.

Bromham, L., D. Penny, A. Rambaut, and M. D. Hendy. 2000. The power of relative rates tests depends on the data. *Journal of Molecular Evolution* 50:296–301.

Bronson, F. H. 1964. Agonistic behaviour in woodchucks. *Animal Behaviour* 12:470–78.

———. 1975. Male-induced precocial puberty in female mice: Confirmation of the role of estrogen. *Endocrinology* 96: 511–14.

———. 1976. Urine marking in mice: Causes and effects. In *Mammalian olfaction, reproductive processes and behavior,* ed. R. L. Doty, 119–41. New York: Academic Press.

———. 1979a. The reproductive ecology of the house mouse. *The Quarterly Review of Biology* 54:265–99.

———. 1979b. The role of priming pheromones in mammalian reproductive strategies. In *Chemical ecology: Odour communication in animals*, ed. F. J. Ritter, 97–104. Amsterdam: Elsevier/North Holland.

———. 1983. Chemical communication in house mice and deer mice: Functional roles in reproduction of wild populations. In *Advances in the study of mammalian behavior.* Special publication no. 7, ed. J. G. Eisenberg and D. G. Kleiman, 198–238. American Society of Mammalogists.

———. 1985. Mammalian reproduction: An ecological perspective. *Biology of Reproduction* 32:1–26.

———. 1989. *Mammalian reproductive biology.* Chicago: University of Chicago Press.

Bronson, F. H., and A. Coquelin. 1980. The modulation of reproduction by priming pheromones in house mice: Speculation on adaptive function. In *Chemical signals in vertebrates,* ed. D. Muller-Schwarze and R. M. Silverstein, 243–65. New York: Plenum.

Bronson, F. H., and H. E. Dezell. 1968. Studies on the estrus-inducing (pheromonal) action of male deermouse urine. *General and Comparative Endocrinology* 10:339–43.

Bronson, F. H., and B. E. Eleftheriou. 1963. Influence of strange males on implantation in the deermouse. *General and Comparative Endocrinology* 3:515–18.

Bronson, F. H., B. A. Eleftheriou, and H. E. Dezell. 1969. Strange male pregnancy block in deermice: Prolactin and adrenocortical hormones. *Biology of Reproduction* 1: 302–6.

Bronson, F. H., B. E. Eleftheriou, and E. I. Garrick. 1964. Effects of intra- and inter-specific social stimulation on implantation in deermice. *Journal of Reproduction and Fertility* 8:23–27.

Bronson, F. H., and B. Macmillan. 1983. Hormonal responses to primer pheromones. In *Pheromones and reproduction in mammals*, ed. J. G. Vandenbergh, 175–97. New York: Academic Press.

Bronson, F. H., and H. M. Marsden. 1964. Male-induced synchrony of estrus in deermice. *General and Comparative Endocrinology* 4:634–37.

———. 1973. The preputial gland as an indicator of social dominance in male mice. *Behavioral Biology* 9:625–28.

Bronson, F. H., and J. A. Maruniak. 1976. Differential effects of male stimuli on follicle-stimulating hormone, luteinizing hormone, and prolactin secretion in prepubertal female mice. *Endocrinology* 98:1101–08.

Bronson, F. H., and G. Perrigo. 1987. Seasonal regulation of reproduction in muroid rodents. *American Zoologist* 27:929–40.

Bronson, F. H., and M. H. Stetson. 1973. Gonadotropin release in prepubertal female mice following male exposure: A comparison with the adult cycle. *Biology of Reproduction* 9:449–59.

Bronson, F. H., and W. K. Whitten. 1968. Oestrus-accelerating pheromone of mice: Assay, androgen-dependency and presence in bladder urine. *Journal of Reproduction and Fertility* 15:131–34.

Bronson, M. T. 1979. Altitudinal variation in the life history of the golden-mantled ground squirrels (*Spermophilus lateralis*). *Ecology* 60:272–79.

Bronstein, P. M., M. J. Levine, and M. Marcus. 1975. A rat's first bite: The non-genetic, cross-generational transfer of information. *Journal of Comparative and Physiological Psychology* 89:295–98.

Brook, B. W., A. D. Griffiths, and H. L. Puckey. 2002. Modelling strategies for the management of the critically endangered Carpentarian rock-rat (*Zyzomys palatalis*) of northern Australia. *Journal of Environmental Management* 65:355–68.

Brooks, R. J. 1984. Causes and consequences of infanticide in populations of rodents. In *Infanticide: Comparative and evolutionary perspectives*, ed. G. Hausfater and S. B. Hrdy, 331–48. New York: Aldine.

Brooks, R. J., and E. M. Banks. 1973. Behavioural biology of the collared lemming [*Dicrostonyx groenlandians* (Trail)]: An analysis of acoustic communication. *Animal Behaviour Monographs* 6:1–83.

Brooks, R. J., and L. Schwarzkopf. 1983. Factors affecting incidence and discrimination of related and unrelated neonates in male *Mus musculus*. *Behavioral and Neural Biology* 37:149–61.

Brooks, R. P., M. W. Fleming, and J. J. Kennelly. 1980. Beaver colony response to fertility control: Evaluating a concept. *Journal of Wildlife Management* 44:568–75.

Brotherton, P. N. M., and P. E. Komers. 2003. Mate guarding and the evolution of social monogamy in mammals. In *Monogamy: Mating strategies and partnerships in birds, humans and other mammals*, ed. U. H. Reichard and C. Boesch, 42–58. Cambridge: Cambridge University Press.

Brotherton, P. N. M., and M. B. Manser. 1997. Female dispersion and the evolution of monogamy in the dik-dik. *Animal Behaviour* 54:1413–24.

Brotherton, P. N. M., and A. Rhodes. 1996. Monogamy without biparental care in a dwarf antelope. *Proceedings of the Royal Society of London, Series B* 263:23–29.

Brothwell, D. R. 1981. The Pleistocene and Holocene archeology of the house mouse and related species. *Symposia of the Zoological Society of London* 47:1–13.

Brouette-Lahlou, I., F. Godinot, and E. Vernet-Maury. 1999. The mother rat's vomeronasal organ is involved in detection of

dodecyl propionate, the pup's preputial gland pheromone. *Physiology and Behavior* 66:427–36.

Broussard, D. R., T. S. Risch, F. S. Dobson, and J. O. Murie. 2003. Senescence and age-related reproduction of Columbian ground squirrels. *Journal of Animal Ecology* 72:212–19.

Brown, C. R., and M. B. Brown. 1996. *Coloniality in the cliff swallow*. Chicago: University of Chicago Press.

Brown, G. R., and J. B. Silk. 2002. Reconsidering the null hypothesis: Is maternal rank associated with birth sex ratios in primate groups? *Proceedings National Academy of Science* 99:11252–55.

Brown, J. A. 1984. Parental care and the ontogeny of predator-avoidance in two species of centrarchid fish. *Animal Behaviour* 32:113–19.

Brown, J. H. 1968. Adaptation to environmental temperature in two species of woodrats, *Neotoma cinerea* and *N. albigula*. *Miscellaneous Publications of the Museum of Zoology, University of Michigan* 135:1–48.

———. 1995. *Macroecology*. Chicago: University of Chicago Press.

Brown, J. H., and S. K. M. Ernest. 2002. Rain and rodents: Complex dynamics of desert consumers. *BioScience* 52:979–87.

Brown, J. H., and B. A. Harney. 1993. Population and community ecology of Heteromyid rodents in temperate habitats. In *Biology of Heteromyidae*, Special publication no. 10, ed. H. H. Genoways and J. H. Brown, 618–51. Lawrence, KS: American Society of Mammalogists.

Brown, J. H., and G. A. Lieberman. 1973. Resource utilization and coexistence of seed-eating rodents in sand dune habitats. *Ecology* 54:788–97.

Brown, J. H., and M. V. Lomolino. 1998. *Biogeography*, 2nd ed. Sunderland, MA: Sinauer.

Brown, J. L. 1974. Alternate routes to sociality in jays—with a theory for the evolution of altruism and communal breeding. *American Zoologist* 14:63–80.

———. 1987. *Helping and communal breeding in birds: Ecology and evolution*. Princeton, NJ: Princeton University Press.

———. 1998. The new heterozygosity theory of mate choice and the MHC. *Genetica* 104:215–21.

Brown, J. L., and A. Eklund. 1994. Kin recognition and the major histocompatibility complex: An integrative review. *American Naturalist* 143:435–61.

Brown, J. L., and G. H. Orians. 1970. Spacing patterns in mobile animals. *Annual Review of Ecological Systems* 1:239–62.

Brown, J. S. 1988. Patch use as indicator of habitat preference, predation risk and competition. *Behavioral Ecology and Sociobiology* 22:37–47.

———. 1989. Desert rodent community structure: A test of four mechanisms of coexistence. *Ecological Monographs* 59:1–20.

———. 1992. Patch use under predation risk: I. Models and predictions. *Annales Zoologici Fenneci* 29:301–9.

Brown, J. S., and P. U. Alkon. 1990. Testing values of crested porcupine habitats by experimental food patches. *Oecologia* 83:512–18.

Brown, J. S., and B. P. Kotler. 2004. Hazardous duty pay and the foraging cost of predation. *Ecology Letters* 7:999–1014.

Brown, J. S., B. P. Kotler, and A. Bouskila. 2001. Ecology of fear: Foraging games between predators and prey with pulsed resources. *Annales Zoologici Fennici* 38:71–87.

Brown, J. S., B. P. Kotler, and M. H. Knight. 1998. Patch use in the pygmy rock mouse (*Petromyscus collinus*). *Mammalia* 62:108–12.

Brown, J. S., B. P. Kotler, and W. A. Mitchell. 1994a. Foraging theory, patch use, and the structure of a Negev Desert granivore community. *Ecology* 75:2286–2300.

Brown, J. S., B. P. Kotler, R. J. Smith, and W. O. Wirtz II. 1988. The effects of owl predation on the foraging behaviour of heteromyid rodents. *Oecologia* 76:408–15.

Brown, J. S., B. P. Kotler, and T. J. Valone. 1994b. Foraging under predation: A comparison of energetic and predation costs in rodent communities of the Negev and Sonoran Deserts. *Australian Journal of Zoology* 42:435–48.

Brown, J. S., J. W. Laundre, and M. Gurung. 1999. The ecology of fear: Optimal foraging, game theory, and trophic interactions. *Journal of Mammalogy* 80:385–99.

Brown, J. S., and R. A. Morgan. 1995. Effects of foraging behavior and spatial scale on diet selectivity: A test with fox squirrels. *Oikos* 74:122–36.

Brown, J. S., R. A. Morgan, and B. D. Dow. 1992. Patch use under predation risk II: A test with fox squirrels, *Sciurus niger*. *Annales Zoologici Fennici* 29:311–18.

Brown, L. E. 1962. Home range in small mammal communities. *Survey of Biological Processes* 4:131–79.

Brown, L. N. 1964. Ecology of three species of *Peromyscus* from southern Missouri. *Journal of Mammalogy* 45:189–202.

———. 1986. Sex ratio bias among gray squirrels foraging at a single attractive seasonal food source. *Journal of Mammalogy* 67:582–83.

Brown, R. E. 1985a. The rodents I: Effects of odours on reproductive physiology (primer effects). In *Social odours in mammals*, ed. R. E. Brown, 245–344. Oxford: Oxford University Press.

———. 1985b. The rodents II: Suborder Myomorpha. In *Social odours in mammals*, ed. R. E. Brown, 345–457. Oxford: Oxford University Press.

———. 1986a. Paternal behavior in the male Long-Evans rat (*Rattus norvegicus*). *Journal of Comparative Psychology* 100: 162–72.

———. 1986b. Social and hormonal factors influencing infanticide and its suppression in adult male Long-Evans rats. *Journal of Comparative Psychology* 100:155–61.

———. 1993. Hormonal and experiential factors influencing parental behaviour in male rodents: An integrative approach. *Behavioural Processes* 30:1–28.

Brown, R. E., and D. W. Macdonald. 1985. *Social odours in mammals*, vol. 1. New York: Oxford University Press.

Brown, R. E., B. Roser, and P. B. Singh. 1990. The MHC and individual odours in rats. In *Chemical signals in vertebrates V*, ed. D. W. McDonald and S. Natynczuk, 228–43. Oxford: Oxford University Press.

Brown, R. E., P. B. Singh, and B. Roser. 1987. The major histocompatibility complex and the chemosensory recognition of individuality in rats. *Physiology and Behavior* 40:65–73.

Brown, R. Z. 1953. Social behavior, reproduction, and population changes in the house mouse (*Mus musculus* L.). *Ecological Monographs* 23:217–40.

Bruce, H. M. 1959. An exteroceptive block to pregnancy in the mouse. *Nature, London* 184:105.

———. 1960. A block to pregnancy in the mouse caused by proximity of strange males. *Journal of Reproduction and Fertility* 1:96–103.

———. 1961. Time relations in the pregnancy-block induced in mice by strange males. *Journal of Reproduction and Fertility* 2:138–42.

Bruce, H. M., and D. M. V. Parrott. 1960. Role of olfactory sense in pregnancy block by strange males. *Science* 131:1526.

Bruck, B. 1957. Male segregation ratio advantage as a factor in maintaining lethal alleles in wild populations of house mice. *Genetics* 43:152–58.

Brudzynski, S. M. 2001. Pharmacological and behavioral characteristics of 22 kHz alarm calls in rats. *Neuroscience and Biobehavioral Reviews* 25:611–17.

Bruemmer, F. 1994. Rough rookeries. *Natural History* 103: 26–33.

Brummer-Korvenkontio, M., O. Vapalahti, H. Henttonen, P. Koskela, P. Kuusisto, and A. Vaheri. 1999. Epidemiological study of nephropathia epidemica in Finland 1989–96. *Scandinavian Journal of Infectious Diseases* 31:427–35.

Brunjes, P. C., and J. R. Alberts. 1979. Olfactory stimulation induces filial preferences for huddling in rat pups. *Journal of Comparative and Physiological Psychology* 93:548–55.

Bruns, U., M. Muller, W. Hofer, G. Heth, and E. Nevo. 1988. Inner ear structure and electrophysiological and audiograms of the subterranean mole rat, *Spalax ehrenbergi*. *Hearing Research* 33:1–10.

Brunton, C. F. A., and D. W. Macdonald. 1996. Measuring the neophobia of individuals in different populations of wild brown rats. *Journal of Wildlife Research* 1:7–14.

Brunton, C. F. A., D. W. Macdonald, and A. P. Buckle. 1993. Behavioural resistance towards poison baits in brown rats *Rattus norvegicus*. *Applied Animal Behaviour Science* 38:159–174.

Bryant, A. A. 1996. Reproduction and persistence of Vancouver Island marmots (*Marmota vancouverensis*) in natural and logged habitats. *Canadian Journal of Zoology* 74:678–87.

Bryant, A. A., H. M. Schwantje, and N. I. de With. 2002. Disease and unsuccessful reintroduction of Vancouver Island marmots (*Marmota vancouverensis*). In *Holarctic marmots as a factor of biodiversity*, ed. K. B. Armitage and V. Yu. Rumiantsev, 101–7. Moscow: ABF.

Buchan, J. C., S. C. Alberts, J. B. Silk, and J. Altmann. 2003. True paternal care in a multi-male primate society. *Nature* 425:179–81.

Buck, C. L., and B. M. Barnes. 1999. Annual cycle of body composition and hibernation in free-living arctic ground squirrels. *Journal of Mammalogy* 80:430–42.

Buech, R. R. 1987. Environmental relations and adaptive strategies of beavers (*Castor canadensis*) in a near-boreal region. PhD diss., University of Minnesota, Minneapolis-St.Paul.

———. 1995. Sex difference in behavior of beavers living in near-boreal lake habitat. *Canadian Journal of Zoology* 73:2133–43.

Buffenstein, R. 1996. Ecophysiological responses to a subterranean habitat; a bathyergid perspective. *Mammalia* 60: 591–605.

———. 2000. Ecophysiological responses of subterranean rodents to underground habitats. In *Life underground: The biology of subterranean rodents*, ed. E. A. Lacey, J. L. Patton, and G. N. Cameron, 62–110. Chicago: University of Chicago Press.

Bugge, J. 1985. Systematic value of the carotid arterial pattern in rodents. In *Evolutionary relationships among rodents: A mul-*

tidisciplinary analysis, ed. W. P. Luckett and J. L. Hartenberger, 381–402. New York: Plenum.

Bujalska, G. 1970. Reproductive stabilizing elements in an island population of *Clethrionomys glareolus* (Schreber 1780). *Acta Theriologica* 15:381–412.

———. 1973. The role of spacing behaviour among females in the regulation of reproduction in the bank vole. *Journal of Reproduction and Fertility* 19:465–74.

Bull, J. J., and E. L. Charnov. 1988. How fundamental are Fisherian sex ratios? In *Oxford surveys in evolutionary biology,* ed. P. H. Harvey and L. Partridge, 96–135. Oxford: Oxford University Press.

Bullock, J. M., R. E. Kenward, and R. S. Hails (eds). 2002. *Dispersal ecology.* Oxford: Blackwell.

Bulmer, M. 1988. Sex ratio evolution in lemmings. *Heredity* 61:231–33.

Burda, H. 1989. Relationships among rodent taxa, as indicated by reproductive biology. *Zeitschrift für Zoologische Systematik und Evolutionsforschung* 27:49–50.

———. 1990. Constraints of pregnancy and evolution of sociality in mole-rats—with special reference to reproductive and social patterns in *Cryptomys hottentotus* (Bathyergidae, Rodentia). *Zeitschrift Fur Zoologische Systematik Und Evolutionsforschung* 28:26–39.

———. 1995. Individual recognition and incest avoidance in eusocial common mole–rats rather than reproductive suppression by parents. *Experientia* 51:411–13.

———. 1999. Syndrome of eusociality in African subterranean mole rats (Bathyergidae, Rodentia): Its diagnosis, and aetiology. In *Evolutionary theory and processes: Modern perspectives: Papers in honour of Eviatar Nevo,* ed. S. P. Wasser, 385–418. Dordrecht: Kluwer Academic.

Burda, H., R. H. Honeycutt, S. Begall, O. Löcker–Grutjen, and A. Scharff. 2000. Are naked and common mole–rats eusocial and if so, why? *Behavioral Ecology and Sociobiology* 47:293–303.

Burda, H., and M. Kawalika. 1993. Evolution of eusociality in the Bathyergidae: The case of the giant mole rats (*Cryptomys mechowi*). *Naturwissenschaften* 80:235–37.

Burda, H., V. Bruns, and M. Muller. 1990. Sensory adaptations in subterranean mammals. In *Evolution of subterranean mammals at the organismal and molecular levels,* ed. E. Nevo and O. A. Reig, 269–94. New York: Wiley-Liss.

Burda, H., V. Bruns, and E. Nevo. 1989. Middle ear and cochlear receptors in the subterranean mole rat, *Spalax ehrenbergi. Hearing Research* 39:225–30.

Burke da Silva, K., D. L. Kramer, and D. M. Weary. 1994. Context-specific alarm calls of the eastern chipmunk, *Tamias striatus. Canadian Journal of Zoology* 72:1087–92.

Burke da Silva, K., C. Mahan, and J. da Silva. 2002. The trill of the chase: Eastern chipmunks call to warn kin. *Journal of Mammalogy* 83:546–52.

Burland, T. M., N. C. Bennett, J. U. M. Jarvis, and C. G. Faulkes. 2004. Colony structure and parentage in wild colonies of co-operatively breeding Damaraland mole-rats suggests incest avoidance alone may not maintain reproductive skew. *Molecular Ecology* 13:2371–79.

———. 2002. Eusociality in African mole–rats: New insights from patterns of genetic relatedness in the Damaraland mole-rat (*Cryptomys damarensis*). *Proceedings of the Royal Society of London, Series B* 269:1025–30.

———. 2004. Colony structure and parentage in wild colonies of co-operatively breeding Damaraland mole-rats suggests a role for reproductive suppression. *Molecular Ecology* 13:2371–79.

Burley, N. 1981. Mate choice by multiple criteria in a monogamous species. *The American Naturalist* 117:515–28.

Burley, N., and P. J. Bartels. 1990. Phenotypic similarities of sibling zebra finches. *Animal Behaviour* 39:174–80.

Bursten, S. N., K. C. Berridge, and D. H. Owings. 2000. Do California ground squirrels (*Spermophilus beecheyi*) use ritualized syntactic cephalocaudal grooming as an agonistic signal? *Journal of Comparative Psychology* 114:281–90.

Burt, W. H. 1943. Territoriality and home range concepts as applied to mammals. *Journal of Mammalogy* 24:346–52.

Busch, A. 2003. Predator defense from auditory cues. Master's thesis, San Francisco State University.

Busch, C., C. D. Antinuchi, J. C. del Valle, M. J. Kittlein, A. I. Malizia, A. I. Vassallo, and R. R. Zenuto. 2000. Population ecology of subterranean rodents. In *Life underground: The biology of subterranean rodents,* ed. E. A. Lacey, J. L. Patton, and G. N. Cameron, 183–226. Chicago: University of Chicago Press.

Busher, P. E. 1975. Movements and activities of beavers, *Castor canadensis,* on Sagehen Creek, California. Master's thesis, San Francisco State University.

———. 1980. The population dynamics and behavior of beavers in the Sierra Nevada. PhD diss., University of Nevada, Reno.

———. 1983a. Interactions between beavers in a montane population in California, *Acta Zoologica Fennica* 174:109–10.

———. 1983b. Interrelationships between behaviors in a beaver, *Castor canadensis,* population. *Journal of the Tennessee Academy of Science.* 58:50–53.

———. 1987. Population parameters and family composition of beaver in California. *Journal of Mammalogy* 68:860–64.

Busher, P. E., and G. Hartman. 2001. Beavers. In *New encyclopedia of mammals,* ed. D. W. Macdonald, 590–95. Oxford: Oxford University Press.

Busher, P. E., and S. H. Jenkins 1985. Behavioral patterns of a beaver family in California. *Biology of Behaviour* 10:41–54.

Busher, P. E., and P. J. Lyons. 1999. Long-term population dynamics of the North American beaver, *Castor canadensis,* on Quabbin Reservation Massachusetts, and Sagehen Creek, California. In *Beaver protection, management, and utilization in Europe and North America,* ed. P. E. Busher and R. M. Dzieciolowski, 147–60. New York: Kluwer Academic/Plenum.

Busher, P. E., R. J. Warner, and S. H. Jenkins. 1983. Population density, colony composition, and local movements in two Sierra Nevadan beaver populations. *Journal of Mammalogy* 64:314–18.

Butler, P. M., E. Nevo, A. Beiles, and S. Simson. 1993. Variations of molar morphology in the *Spalax ehrenbergi* superspecies: Adaptive and phylogenetic significance. *Journal of Zoology, London* 229:191–216.

Butts, K. O., and J. C. Lewis. 1982. The importance of prairie dog towns to burrowing owls in Oklahoma. *Proceedings of Oklahoma Academy of Science* 62:46–52.

Butynski, T. M. 1982. Harem-male replacement and infanticide in the blue monkey (*Cercopithecus mitis stuhlmanni*) in the Kibale forest, Uganda. *American Journal of Primatology* 3:1–22.

Byers, J. A., and M. Bekoff. 1991. Development, the conveniently forgotten variable in true kin recognition. *Animal Behavior* 41:1088–90.

Byrom, A. E., T. J. Karels, C. J. Krebs, and R. Boonstra. 2000. Experimental manipulation of predation and food supply of arctic ground squirrels in the boreal forest. *Canadian Journal of Zoology* 78:1309–19.

Byrom, A. E., and C. J. Krebs. 1999. Natal dispersal of juvenile arctic ground squirrels in the boreal forest. *Canadian Journal of Zoology* 77:1048–59.

Cade, W. H. 1980. Alternative male reproductive behaviors. *Florida Entomologist* 63:30–45.

Caffe, A. R., F. W. van Leeuwen, and P. G. Luiten. 1987. Vasopressin cells in the medial amygdala of the rat project to the lateral septum and ventral hippocampus. *Journal of Comparative Neurology* 261:237–52.

Cairns, R. B., J.-L. Gariepy, and K. E. Hood. 1990. Development, microevolution, and social behavior. *Psychological Review* 97:49–65.

Calaby, J. H., and A. K. Lee. 1989. The rare and endangered rodents of the Australian region. In *Rodents, a world survey of species of conservation concern*, ed. W. Z. Lidicker, Jr., 53–57. Gland, Switzerland: Occasional Papers, IUCN Species Survival Commission, Rodent Specialist Group.

Calder, W. A., III. 1984. *Size, function and life history.* Cambridge, MA: Harvard University Press.

Caley, J., and S. Boutin. 1985. Infanticide in wild populations of *Ondatra zibethicus* and *Microtus pennsylvanicus*. *Animal Behaviour* 33:1036–37.

Caley, M. J., and T. D. Nudds. 1987. Sex ratio adjustment in *Odocoileus*: Does local resource competition play a role? *American Naturalist* 129:452–57.

Calhoun, J. B. 1962a. *The ecology and sociology of the Norway rat.* Public Health Service Publication no. 1008. Bethesda, MD: US Department of Health, Education, and Welfare.

———. 1962b. Population density and social pathology. *Scientific American* 206:139–48.

Calisher, C. H., J. N. Mills, W. P. Sweeney, J. R. Choate, D. E. Sharp, K. M. Canestorp, and B. J. Beaty. 2001. Do unusual site-specific population dynamics of rodent reservoirs provide clues to the natural history of hantaviruses? *Journal of Wildlife Diseases* 37:280–88.

Calisher, C. H., J. J. Root, J. N. Mills, J. E. Rowe, S. A. Reeder, E. S. Jentes, and B. J. Beaty. 2005. Epizootiology of Sin Nombre and El Moro Canyon hantaviruses, southeastern Colorado, 1995–2000. *Journal of Wildlife Diseases* 41:1–11.

Calisher, C. H., W. Sweeney, J. N. Mills, and B. J. Beaty. 1999. Natural history of Sin Nombre virus in western Colorado. *Emerging Infectious Diseases* 5:126–34.

Callahan, J. R. 1981. Vocal solicitation and parental investment in female *Eutamias*. *American Naturalist* 118:872–75.

Calsbeek, R., and B. Sinervo. 2002. Uncoupling direct and indirect components of female choice in the wild. *Proceedings of the National Academy of Sciences (USA)* 99:14897–902.

Cameron, E. Z. 2004. Facultative adjustment of mammalian sex ratios in support of the Trivers-Willard hypothesis: Evidence for a mechanism. *Proceedings of the Royal Society of London, Series B* 271:1723–28.

Cameron, G. C., and W. B. Kincaid. 1982. Species removal effects on movements of *Sigmodon hispidus* and *Reithrodontomys fulvescens*. *American Midland Naturalist* 108:60–67.

Cameron, G. N. 1973. Effects of litter size on postnatal growth and survival in the desert woodrat. *Journal of Mammalogy* 54:489–93.

Cameron, G. N, and S. R. Spencer. 1981. *Sigmodon hispidus*. *Mammalian Species* 158:1–9.

———. 1985. Assessment of space-use patterns in the hispid cotton rat (*Sigmodon hispidus*). *Oecologia* 68:133–39.

Cameron, H. A., and E. Gould. 1994. Adult neurogenesis is regulated by adrenal steroids in the dentate gyrus. *Neuroscience* 61:203–9.

Campagna, C., B. J. Le Boeuf, and H. L. Cappozzo. 1988. Pup abduction and infanticide in southern sea lions. *Behaviour* 107:44–60.

Campbell, F. M. and C. M. Heyes. 2002. Rats smell: Odour-mediated local enhnacement in a vertical movement two-action test. *Animal Behaviour* 63:1055–63.

Canals, M., M. Rosenmann, and F. Bozinovic. 1989. Energetics and geometry of huddling in small mammals. *Journal of Theoretical Biology* 141:181–89.

Cant, M. A. 2000. Social control of reproduction in banded mongooses. *Animal Behaviour* 59:147–58.

Cantoni, D., and R. Brown. 1997. Paternal investment and reproductive success in the California mouse, *Peromyscus californicus*. *Animal Behaviour* 54:377–86.

Cao, Y., J. Adachi, T. Yano, and M. Hasegawa. 1994. Phylogenetic place of guinea pigs: No support of the rodent polyphyly hypothesis from maximum-likelihood analyses of multiple protein sequences. *Molecular Biology and Evolution* 11:593–604.

Capek, K., and J. Jelinek. 1956. The development of the control of water metabolism I. The excretion of urine by young rats. *Physiologia Bohemoslovenca* 5:91–96.

Carleton, A., L. T. Petreanu, R. Lansford, A. Alvarez-Buylla, and P. M. Lledo. 2003. Becoming a new neuron in the adult olfactory bulb. *Nature Neuroscience* 6:507–18.

Carleton, M. D. 1989. Systematics and evolution. In *Advances in the study of* Peromyscus *(Rodentia)*, ed. G. Kirkland Jr. and J. N. Layne, 7–141. Lubbock: Texas Tech University Press.

Carleton, M. D., and G. G. Musser. 1984. Muroid rodents. In *Orders and families of recent mammals of the world*, ed. S. Anderson and J. K. Jones, Jr., 289–379. New York: Wiley.

Carlsen, M., J. Lodal, H. Leirs, and T. S. Jensen. 1999. The effect of predation risk on body weight in the field vole, *Microtus agrestis*. *Oikos* 87:277–85.

Caro, T. M. 1986. The functions of stotting: A review of hypotheses. *Animal Behaviour* 34:649–62.

Caro, T. M., and P. Bateson. 1986. Organization and ontogeny of alternative tactics. *Animal Behaviour* 34:1483–99.

Carr, D. B., and S. R. Sesack. 2000. Projections from the rat prefrontal cortex to the ventral tegmental area: Target specificity in the synaptic associations with mesoaccumbens and mesocortical neurons. *Journal of Neuroscience* 20:3864–73.

Carr, G. M., and D. W. Macdonald. 1986. The sociality of solitary foragers: A model based on resource dispersion. *Animal Behaviour* 34:1540–49.

Carr, W. J., K. R. Kimmel, S. L. Anthony, and D. E. Schlocker. 1982. Female rats prefer to mate with dominant rather than subordinate males. *Bulletin of Psychonomic Society* 20:89–91.

Carr, W. J., L. Krames, and D. J. Constanzo. 1970a. Previous sexual experience and olfactory preference for novel versus original sex partners in rats. *Journal of Comparative Physiology and Psychology* 71:216–22.

Carr, W. J., R. D. Martorano, and L. Krames. 1970b. Responses of mice to odors associated with stress. *Journal of Comparative Physiology and Psychology* 71:223–28.

Carraway, L. N., and B. J. Verts. 1991. *Neotoma fuscipes*. *Mammalian Species* 386:1–10.

Carrington, M., G. W. Nelson, M. P. Martin, T. Kissner, D. Vlahov, J. J. Goedert, R. Kaslow, et al. 1999. HLA and HIV-1: heterozygote advantage and B*35-Cw*04 disadvantage. *Science* 283:1748–52.

Carroll, D. S., J. N. Mills, J. M. Montgomery, D. G. Bausch, P. J. Blair, J. P. Burans, V. Felices et al. 2005. Hantavirus pulmonary syndrome in central Bolivia: Relationships between reservoir hosts, habitats, and viral genotypes. *American Journal of Tropical Medicine and Hygiene* 72:42–46.

Carroll, L. E., and H. H. Genoways. 1980. *Lagurus curtatus*. *Mammalian Species* 124:1–6.

Carroll, L. S. 2001. Major histocompatibility genes and selfish genes: A chemosensory and ecological analysis of sexual selection. PhD diss., University of Utah, Salt Lake City.

Carroll, L. S., S. Meagher, L. Morrison, D. J. Penn, and W. K. Potts. 2004. Fitness effects of a selfish gene (the *Must* complex) are revealed in an ecological context. *Evolution* 58:1318–28.

Carroll, L. S., D. J. Penn, and W. K. Potts. 2002. Discrimination of MHC-derived odors by untrained mice is consistent with divergence in peptide-binding region residues. *Proceedings of the National Academy of Sciences* (USA) 99:2187–92.

Carter, C. S. 1973. Stimuli contributing to the decrement in sexual receptivity of female golden hamsters (*Mesocricetus auratus*). *Animal Behaviour* 21:827–34.

———. 1998. Neuroendocrine perspectives on social attachment and love. *Psychoneuroendocrinology* 23:779–818.

Carter, C. S., and L. L. Getz. 1993. Monogamy and the prairie vole. *Scientific American* 268:100–10.

Carter, C. S., L. L. Getz, and M. Cohen-Parsons. 1986. Relationships between social organization and behavioral endocrinology in a monogamous mammal. *Advances in the Study of Behavior* 16:109–45.

Carter, C. S., L. L. Getz, L. Gavish, J. L. McDermott, and P. Arnold. 1980. Male-related pheromones and the activation of female reproduction in the prairie vole (*Microtus ochrogaster*). *Biology of Reproduction* 23:1038–45.

Carter, C. S., and R. L. Roberts. 1997. The psychobiological basis of cooperative breeding in rodents. In *Cooperative breeding in mammals*, ed. N. G. Solomon and J. A. French, 231–66. Cambridge: Cambridge University Press.

Carter, C. S., D. M. Witt, J. Schneider, Z. L. Harris, and D. Volkening. 1987. Male stimuli are necessary for female sexual behavior and uterine growth in prairie voles (*Microtus ochrogaster*). *Hormones and Behavior* 21:74–82.

Castilho, V., C. Macedo, and M. Brandão. 2002. Role of benzodiazepine and serotonergic mechanisms in conditioned freezing and antinociception using electrical stimulation of the dorsal periaqueductal gray as unconditioned stimulus in rats. *Psychopharmacology* 165:77–85.

Castleberry, S. B., W. M. Ford, P. B. Wood, N. L. Castleberry, and M. T. Mengak. 2001. Movements of Allegheny woodrats in relation to timber harvesting. *Journal of Wildlife Management* 65:148–56.

Caswell, H. 2001. *Matrix population models: Construction, analysis, and interpretation*. Sunderland, MA: Sinauer.

Catalani, A., P. Casolini, S. Scaccianoce, F. R. Patacchioli, P. Spinozzi, and L. Angelucci. 2000. Maternal corticosterone during lactation permanently affects brain corticosteroid receptors, stress response and behaviour in rat progeny. *Neuroscience* 100:319–25.

Catchpole, C., and P. J. B. Slater. 1995. *Bird song: Biological themes and variations*. Cambridge: Cambridge University Press.

Catzeflis, F. M., E. Nevo, J. E. Ahlquist, and C. G. Sibley. 1989. Relationships of the chromosomal species in the Eurasian mole rats of the *Spalax ehrenbergi* group as determined by DNA-DNA hybridization, and an estimate of the spalacid-murid divergence time. *Journal of Molecular Evolution* 29:223–32.

Cavigelli, S. A., and M. K. McClintock. 2003. Fear of novelty in infant rats predicts adult glucocorticoid dynamics and an early death. *Proceedings of the National Academy of Sciences (USA)* 100:16131–36.

Cawthorn, J. M., and R. K. Rose. 1989. The population ecology of the eastern harvest mouse (*Reithrodontomys humulis*) in southeastern Virginia. *American Midland Naturalist* 122: 1–10.

Cecchi, C., M. Penna, and R. A. Vásquez. 2003. Alarm calls in the social caviomorph rodent *Octodon degus*. Paper presented at the XXVIII International Ethological Conference, Florianópolis, Brasil. *Revista de Etología* (Suplemento) 5:150.

Center for Native Ecosystems, Biodiversity Conservation Alliance, Southern Utah Wilderness Alliance, American Lands Alliance, Forest Guardians, T. T. Williams, Ecology Center, and Sinapu. 2002. Petition to list the white-tailed prairie dog (*Cynomys leucurus*) as a threatened or endangered species and to designate critical habitat under the Endangered Species Act of 1973, as amended, presented to United States Fish and Wildlife Service, Washington, DC.

Chabovsky, A. V., V. A. Lapin, and S. V. Popov. 1990. Seasonal dynamics of social relations in the group of *Meriones libycus*. [In Russian.] *Zoologicheskii Zhurnal* 69:111–26.

Chabovsky, A. V., and S. V. Popov. 1994. Interspecific relations of three species of gerbils in southeast Turkmenia. [In Russian.] *Bulletin of Moscow Naturalists Society* 99:30–38.

Chadwick, D. H. 1993. The American prairie. *National Geographic* 184:90–119.

Chaline, J., P. Brunet-Lecomte, S. Montuire, L. Viroit, and F. Courant. 1999. Anatomy of the arvicoline radiation (Rodentia): Palaeogeographical, palaeoecological history and evolutionary data. *Annual Zoological Fennica* 36:239–67.

Chambers, L. K., G. R. Singleton, and C. J. Krebs. 2000. Movements and social organization of wild house mice (*Mus domesticus*) in the wheatlands of northwestern Victoria, Australia. *Journal of Mammalogy* 81:59–69.

Chandler, M. C. 1984. Life history aspects of *Reithrodontomys humulis* in southeastern Virginia. Master's thesis, Old Dominion University, Norfolk, VA.

Channing, A. 1984. Ecology of the namtap *Graphiurus ocularis* (Rodentia: Gliridae) in the Cedarberg, South Africa. *South African Journal of Zoology* 19:144–49.

Chapman, M., and G. Hausfater. 1979. The reproductive consequences of infanticide in langurs: A mathematical model. *Behavioral Ecology and Sociobiology* 5:227–40.

Chapman, V. M., and W. K. Whitten. 1969. The occurrence and inheritance of pregnancy block in inbred mice. *Genetics* 61:59.

Charlesworth, D., and B. Charlesworth. 1987. Inbreeding depression and its evolutionary consequences. *Annual Review of Ecology and Systematics* 18:237–68.

Charnov, E. L. 1982. *The theory of sex allocation.* Princeton, NJ: Princeton University Press.

Charnov, E. L., and J. P. Finerty. 1980. Vole population cycles: A case for kin-selection? *Oecologia* 45:1–2.

Charnov, E. L., and J. R. Krebs. 1975. The evolution of alarm calls: Altruism or manipulation? *American Naturalist* 109:107–12.

Charnov, E. L., G. H. Orians, and K. Hyatt. 1976. Ecological implications of resource depression. *American Naturalist* 110:247–59.

Cheney, D. L., and R. M. Seyfarth. 1980. Vocal recognition in free-ranging vervet monkeys. *Animal Behaviour* 28:362–67.

———. 1990. *How monkeys see the world.* Chicago: University of Chicago Press.

———. 1999. Recognition of other individuals' social relationships by female baboons. *Animal Behaviour* 58:67–75.

Chepko-Sade, B. D., and Z. T. Halpin, eds. 1987. *Mammalian dispersal patterns: The effects of social structure on population genetics.* Chicago: University of Chicago Press.

Cherry, B. A., B. Cadigan, N. Mansourian, C. Nelson, and E. L. Bradley. 2002. Adrenal gland differences associated with puberty and reproductive inhibition in *Peromyscus maniculatus. General and Comparative Endocrinology* 129:104–13.

Chesser, R. K. 1983. Genetic variability within and among populations of the black-tailed prairie dog. *Evolution* 37:320–31.

———. 1991a. Gene diversity and female philopatry. *Genetics* 127:437–447.

———. 1991b. Influence of gene flow and breeding tactics on gene diversity within populations. *Genetics* 129:573–83.

———. 1998. Relativity of behavioral interactions in socially structured populations. *Journal of Mammalogy* 79:713–24.

Chesser, R. K., O. E. Rhodes, Jr., D. W. Sugg, and A. Schnabel. 1993. Effective sizes for subdivided populations. *Genetics* 135:1221–32.

Chesser, R. K., D. W. Sugg, O. E. Rhodes, Jr., and M. H. Smith. 1996. Conservation. In *Population processes in ecological space and time,* ed. O. E. Rhodes, R. K. Chesser, and M. H. Smith, 237–52. Chicago: University of Chicago Press.

Chester, R. V., and I. Zucker. 1970. Influence of male copulatory behavior on sperm transport, pregnancy and pseudopregnancy in female rats. *Physiology and Behavior* 5:35–43.

Chew, R. M., and A. E. Chew. 1970. Energy relationships of the mammals of a desert shrub (*Larrea tridentata*) community. *Ecological Monographs* 40:1–21.

Childs, J. E., G. W. Korch, G. E. Glass, J. W. LeDuc, and K. V. Shah. 1987. Epizootiology of hantavirus infections in Baltimore—Isolation of a virus from Norway rats, and characteristics of infected rat populations. *American Journal of Epidemiology* 126:55–68.

Childs, J. E., and C. J. Peters. 1993. Ecology and epidemiology of arenaviruses and their hosts. In *The Arenaviridae,* ed. M. S. Salvato, 331–84. New York: Plenum.

Chitty, D. 1954a. *The control of rats and mice.* Oxford: Clarendon.

———. 1954b. Tuberculosis among wild voles: With a discussion of other pathological conditions among certain mammals and birds. *Ecology* 35:227–37.

———. 1996. *Do lemmings commit suicide? Beautiful hypotheses and ugly facts.* Oxford: Oxford University Press.

Cho, M. M., A. C. DeVries, J. R. Williams, and C. S. Carter. 1999. The effects of oxytocin and vasopressin on partner preferences in male and female prairie voles (*Microtus ochrogaster*). *Behavioral Neuroscience* 113:1071–79.

Choleris, E., C. Guo, H. Liu, M. Mainardi, and P. Valsecchi. 1997. The effect of demonstrator age and number on duration of socially-induced food preferences in house mouse (*Mus domesticus*). *Behavioural Processes* 41:69–77.

Chou, L.-S., R. E. Marsh, and P. J. Richerson. 2002. Constraints on social transmission of food selection by roof rats, *Rattus rattus. Acta Zoologica Taiwanica* 11:95–109.

Chovnick, A., N. J. Yasukawa, H. Monder, and J. J. Christian. 1987. Female behavior in populations of mice in the presence and absence of male hierarchy. *Aggressive Behavior* 13:367–75.

Christian, D. P., and J. M. Daniels. 1985. Distributional records of rock voles, *Microtus chrotorrhinus,* in northeastern Minnesota. *Canadian Field-Naturalist* 99:356–59.

Churchill, S. K. 1996. Distribution, habitat and status of the Carpentarian rock-rat, *Zyzomys palatalis. Wildlife Research* 23:77–91.

Ciani, A. C. 1984. A case of infanticide in a free-ranging group of rhesus monkeys (*Macaca mulatta*) in the Jackoo forest, Simla, India. *Primates* 25:372–77.

Ciszek, D. 2000. New colony formation in the "highly inbred" eusocial naked mole-rat: Outbreeding is preferred. *Behavioral Ecology* 11:1–6.

Cittadino, E. A., D. N. Bilenca, M. Busch, and F. O. Kravetz. 2002. Characteristics of dispersing pampean grassland mice (*Akodon azarae*) in agroecosystems of central Argentina. *Studies on Neotropical Fauna and Environment* 37:1–7.

Cittadino, E. A., M. Busch, and F. O. Kravetz. 1998. Population abundance and dispersal in *Akodon azarae* (pampean grassland mouse) in Argentina. *Studies on Neotropical Fauna and Environment* 37:1–7.

Clark, A. B. 1978. Sex ratio and local resource competition in a prosimian primate. *Science* 201:163–65.

Clark, C. W. 1994. Antipredator behavior and the asset-protection principle. *Behavioral Ecology* 5:159–70.

Clark, J. R. 2000. Endangered and threatened wildlife and plants; determination of threatened status for the northern Idaho ground squirrel. *Federal Register* 65 (66):17779–86.

Clark, M. M., S. Bone, and B. G. Galef, Jr. 1990. Evidence of sex-biased postnatal maternal investment by Mongolian gerbils. *Animal Behaviour* 39:735–44.

Clark, M. M., and B. G. Galef, Jr. 1995. Prenatal influences on reproductive life history strategies. *Trends in Ecology and Evolution* 10:151–53.

Clark, M. M., S. A. Malenfant, D. A. Winter, and B. G. Galef. 1990. Fetal uterine position affects copulation and scent marking by adult male gerbils. *Physiology and Behavior* 47:301–5.

Clark, M. M., L. Tucker, and B. G. Galef. 1992a. Stud males and dud males: Intra-uterine position effects on the reproductive success of male gerbils. *Animal Behaviour* 43:215–21.

Clark, M. M., F. S. Vom Saal, and B. G. Galef, Jr. 1992b. Intra-uterine positions and testosterone levels of adult male gerbils are correlated. *Physiology and Behavior* 51:957–60.

Clark, M. M., C. L. Waddingham, and B. G. Galef, Jr. 1991.

Further evidence of sex-biased maternal investment by Mongolian gerbil dams. *Animal Behaviour* 42:161–62.

Clark, T. W. 1970. Early growth, development, and behavior of the Richardson ground squirrel (*Spermophilus richardsoni elegans*). *American Midland Naturalist* 83:197–205.

———. 1989. *Conservation biology of the black-footed ferret*, Mustela nigripes. Special Scientific Report, no. 3. Phildaelphia: Wildlife Preservation Trust International.

Clark, T. W., T. M. Campbell, D. G. Socha, and D. E. Casey. 1982. Prairie dog colony attributes and associated vertebrate species. *Great Basin Naturalist* 42:572–82.

Clarke, F. M., and C. G. Faulkes. 1999. Kin discrimination and female mate choice in the naked mole-rat *Heterocephalus glaber*. *Proceedings of the Royal Society of London, Series B* 266:1995–2002.

Clarke, J. R., and F. V. Clulow. 1973. The effect of successive matings upon bank vole (*Clethrionomys glareolus*) and field vole (*Microtus agrestis*) ovaries. In *The development and maturation of the ovary and its functions*, ed. H. Peters, 160–70. Amsterdam: Excerpta Medica.

Clayton, D. H. 1991. The influence of parasites on host sexual selection. *Parasitology Today* 7:329–34.

Clinchy, M., D. T. Haydon, and A. T. Smith. 2002. Pattern does not equal process: What does patch occupancy really tell us about metapopulation dynamics? *American Naturalist* 159:351–62.

Clinchy, M., C. J. Krebs, and P. J. Jarman. 2001. Dispersal sinks and handling effects: Interpreting the role of immigration in common brushtail possum populations. *Journal of Animal Ecology* 70:515–526.

Clobert, J., E. Danchin, A. A. Dhondt, and J. D. Nichols (eds). 2001. *Dispersal*. Oxford: Oxford University Press.

Cloudsley-Thompson, J. L. 1975. The desert as habitat. In *Rodents in desert environments*, ed. I. Prakash and P. K. Ghosh, 1–13. The Hague: Junk.

Clough, G. C. 1972. Biology of the Bahaman hutia, *Geocapromys ingrahami*. *Journal of Mammalogy* 53:807–23.

———. 1974. Additional notes on the biology of the Bahamian hutia, *Geocapromys ingrahami*. *Journal of Mammalogy* 55:670–72.

Clulow, F. V., and J. R. Clarke. 1968. Pregnancy block in *Microtus agrestis* an induced ovulator. *Nature, London* 219:511.

Clulow, F. V., and P. E. Langford. 1971. Pregnancy-block in the meadow vole, *Microtus pennsylvanicus*. *Journal of Reproduction and Fertility* 24:275–77.

Clulow, F. V., and F. F. Mallory. 1974. Ovaries of meadow voles, *Microtus pennsylvanicus*, after copulation with a series of males. *Canadian Journal of Zoology* 52:265–67.

Clutton-Brock, T. H. 1989a. Female transfer and inbreeding avoidance in social mammals. *Nature* 337:70–72.

———. 1989b. Mammalian mating systems. *Proceedings of the Royal Society of London, Series B* 236:339–72.

———. 1991. *The evolution of parental care*. Princeton, NJ: Princeton University Press.

———. 1998a. Reproductive skew, concessions and limited control. *Trends in Ecology and Evolution* 13:288–92.

———. 1998b. Reproductive skew: Disentangling concessions from control—reply. *Trends in Ecology and Evolution* 13:459.

Clutton-Brock, T. H., and S. D. Albon. 1979. The roaring of red deer and the evolution of honest advertisement. *Behaviour* 69:145–70.

Clutton-Brock, T. H., S. D. Albon, R. M. Gibson, and F. E. Guinness. 1979. The logical stag: Adaptive aspects of fighting in red deer (*Cervus elaphus* L.). *Animal Behaviour* 27:211–25.

Clutton-Brock, T. H., S. D. Albon, and F. E. Guiness. 1982. Competition between female relatives in a matrilocal mammal. *Nature* 300:178–80.

———. 1985. Parental investment and sex differences in juvenile mortality in birds and mammals. *Nature* 313:131–33.

———. 1986. Great expectations: Dominance, breeding success and offspring sex ratios in red deer. *Animal Behaviour* 34:460–71.

———. 1988. Reproductive success in male and female red deer. In *Reproductive success,* ed. T. H. Clutton-Brock, 325–43. Chicago: University of Chicago Press.

———. 1989. Fitness costs of gestation and lactation in wild mammals. *Nature* 337:260–62.

Clutton-Brock, T. H., P. N. M. Brotherton, R. Smith, G. M. McIlrath, R. Kansky, D. Gaynor, M. J. O'Riain, and J. D. Skinner. 1998. Infanticide and expulsion of females in a cooperative mammal. *Proceedings of the Royal Society of London, Series B* 265:2291–95.

Clutton-Brock, T. H., D. Gaynor, R. Kansky, A. D. C. MacColl, G. McIlrath, P. Chadwick, P. N. M. Brotherton, J. M. O'Riain, M. Manser, and J. D. Skinner. 1998. Costs of cooperative behaviour in suricates (*Suricata suricatta*). *Proceedings of the Royal Society of London, Series B* 265:185–90.

Clutton-Brock, T. H., and G. R. Iason. 1986. Sex ratio variation in mammals. *Quarterly Review of Biology* 61:339–74.

Clutton-Brock, T. H., and G. A. Parker. 1992. Potential reproductive rates and the operation of sexual selection. *The Quarterly Review of Biology* 67:437–56.

Cockburn, A. 1988. *Social behaviour in fluctuating populations*. London: Croom Helm.

Cockburn, A., S. Legge, and M. C. Double. 2002. Sex ratios in birds and mammals: Can the hypotheses be disentangled? In *Sex ratios: Concepts and research methods*, ed. I. C. W. Hardy, 266–86. Cambridge: Cambridge University Press.

Cockburn, A., M. P. Scott, and C. R. Dickman. 1985. Sex ratio and intrasexual kin competition in mammals. *Oecologia* 66:427–29.

Cockerham, C. C. 1967. Group inbreeding and coancestry. *Genetics* 56:89–104.

———. 1969. Variance of gene frequencies. *Evolution* 23:72–84.

———. 1973. Analyses of gene frequencies. *Genetics* 74:679–700.

Cohen-Salmon, C. 1987. Differences in patterns of pup care in *Mus musculus domesticus*. VIII. Effects of previous experience and parity in XLII inbred mice. *Physiology and Behavior* 40:177–80.

Cole, J. R., and J. C. Z. Woinarski. 2000. Rodents of the arid Northern Territory: Conservation status and distribution. *Wildlife Research* 27:437–49.

Collett, T. S., B. A. Cartwright, and B. A. Smith. 1986. Landmark learning and visuo-spatial memories in gerbils *Meriones unguiculatus*. *Journal of Comparative Physiology A* 158:835–51.

Collins, R. L. 1988. Observational learning of a left-right behavioral asymmetry in mice (*Mus musculus*). *Journal of Comparative Psychology* 102: 222–24.

Collins, S. A., L. M. Gosling, J. Hudson, and D. Cowan. 1997. Does behaviour after weaning affect the dominance status of adult male mice (*Mus musculus*)? *Behaviour* 134: 989–1002.

Coltman, D.W, J. A. Smith, D. R. Bancroft, J. Pilkington, A. D. C. MacColl, T. H. Clutton-Brock, and J. M. Pemberton. 1999. Density-dependent variation in lifetime breeding success and natural and sexual selection in Soay rams. *American Naturalist* 154:730–46.

Compaan, J. C., R. M. Buijs, C. W. Pool, A. J. De Ruiter, and J. M. Koolhaas. 1993. Differential lateral septal vasopressin innervation in aggressive and nonaggressive male mice. *Brain Research Bulletin* 30:1–6.

Comparatore, V. M., N. Maceira, and C. Busch. 1991. Habitat relations in *Ctenomys talarum* (Caviomorpha: Octodontidae) in a natural grassland. *Zeitschrift für Saugetierkunde* 56: 112–18.

Conditt, S. A., and D. O. Ribble. 1997. Social organization of *Neotoma micropus*, the southern plains woodrat. *American Midland Naturalist* 137:290–97.

Connor, J. L., and M. J. Bellucci. 1979. Natural selection resisting inbreeding depression in captive wild house mice (*Mus musculus*). *Evolution* 33:929–40.

Connor, R. C. 1995. Altruism among non-relatives: Alternatives to the 'Prisoner's dilemma.' *Trends in Ecology and Evolution* 10:84–86.

Conroy, C. J., and J. A. Cook. 2000. Molecular systematics of a Holarctic rodent (*Microtus*: Muridae). *Journal of Mammalogy* 81:344–59.

Contreras, J. R., and O. A. Reig. 1965. Datos sobre la distribución del género *Ctenomys* (Rodentia, Octodontidae) en la zona costera de la provincia de Buenos Aires comprendida entre Necochea y Bahía Blanca. *Physis* 25:169–86.

Contreras, L. C. 1986. Bioenergetics and distribution of fossorial *Spalacopus cyanus* (Rodentia): Thermal stress, or cost of burrowing? *Physiological Zoology* 59:20–28.

Contreras, L. C., and E. Bustos-Obregón. 1977. Ciclo reproductivo anual en *Octodon degus* (Molina) macho. *Medio Ambiente (Chile)* 3:83–90.

Contreras, L. C., J. C. Torres-Mura, and J. L. Yáñez. 1987. Biogeography of octodontid rodents: An eco-evolutionary hypothesis. *Fieldiana Zoology* 39:401–11.

Cook, J. A., and E. P. Lessa. 1998. Are rates of diversification in subterranean South American tuco-tucos (genus *Ctenomys*, Rodentia: Octodontidae) unusually high? *Evolution* 52: 1521–27.

Cooke, B. M., W. Chowanadisai, and S. M. Breedlove. 2000. Post-weaning social isolation of male rats reduces the volume of the medial amygdala and leads to deficits in adult sexual behavior. *Behavioural Brain Research* 117: 107–13.

Cooney, R., and N. C. Bennett. 2000. Inbreeding avoidance and reproductive skew in a cooperative mammal. *Proceedings of the Royal Society of London, Series B* 267:801–6.

Cooper, A., and R. Fortey. 1998. Evolutionary explosions and the phylogenetic fuse. *Trends in Ecology and Evolution* 13: 151–56.

Cooper, H. M., M. Herbin, and E. Nevo. 1993a. Ocular regression conceals adaptive progression of the visual system in a blind subterranean mammal. *Nature* 361:156–59.

Cooper, H. M., M. Herbin, and E. Nevo. 1993b. Visual system of a naturally microphthalmic mammal: The blind mole rat, *Spalax ehrenbergi*. *Journal of Comparative Neurology* 328: 313–50.

Cooper, L. D. 2002. Spatial dynamics and social structure of the giant kangaroo rat (*Dipodomys ingens*). Master's thesis, San Francisco State University.

Coopersmith, C. B., and E. M. Banks. 1983. Effects of olfactory cues on sexual behavior in the brown lemming, *Lemmus trimucronatus*. *Journal of Comparative Psychology* 97: 120–26.

Coopersmith, C. B., and S. Lenington. 1990. Preferences of female mice for males whose *t*-haplotype differs from their own. *Animal Behaviour* 40:1179–81.

———. 1996. The relationship between pregnancy block and infanticide in house mice (*Mus musculus domesticus*) during lactational pregnancy. *Behaviour* 133:1023–50.

Coppock, D. L., J. K. Detling, J. E. Ellis, and M. I. Dyer. 1983. Plant-herbivore interactions in a North American mixed-grass prairie. I. Effects of black-tailed prairie dogs on intraseasonal aboveground plant biomass and nutrient dynamics and plant species diversity. *Oecologia* 56:1–9.

Coppola, D. M., and J. G. Vandenbergh. 1985. Effect of density, duration of grouping and age of urine stimulus on the puberty delay pheromone in female mice. *Journal of Reproduction and Fertility* 73:517–22.

———. 1987. Induction of a puberty-regulating chemosignal in wild mouse populations. *Journal of Mammalogy* 68:86–91.

Corbet, G. B. 1974. The distribution of mammals in historic times. In *The changing fauna and flora of Britain*, ed. D. L. Hawksworth, 179–202. London: Academic Press.

———. 1978. *The mammals of the palearctic region: A taxonomic review.* London: British Museum of Natural History.

Corbet, N. U., and R. J. van Aarde. 1996. Social organization and space use in the Cape porcupine in a Southern African savanna. *African Journal of Ecology* 34:1–14.

Cornely, J. E., and R. J. Baker. 1986. *Neotoma mexicana*. *Mammalian Species* 262:1–7.

Coss, R. G. 1978. Perceptual determinants of gaze aversion by the lesser mouse lemur (*Microcebus murinus*): The role of two facing eyes. *Behaviour* 64:248–70.

———. 1991a. Context and animal behavior III: The relationship between early development and evolutionary persistence of ground squirrel antisnake behavior. *Ecological Psychology* 3:277–315.

———. 1991b. Evolutionary persistence of memory-like processes. *Concepts in Neuroscience* 2:129–68.

———. 1993. Evolutionary persistence of ground squirrel antisnake behavior: Reflections on Burton's commentary. *Ecological Psychology* 5:171–94.

———. 1999. Effects of relaxed natural selection on the evolution of behavior. In *Geographic variation in behavior: Perspectives on evolutionary mechanisms,* ed. S. A. Foster and J. A. Endler, 180–208. Oxford: Oxford University Press.

Coss, R. G., and J. Biardi. 1997. Individual variation in the antisnake behavior of California ground squirrels (*Spermophilus beecheyi*). *Journal of Mammalogy* 73:294–310.

Coss, R. G., and R. O. Goldthwaite. 1995. The persistence of old designs for perception, In *Perspectives in ethology, vol. 11: Behavioral design,* ed. N. S. Thompson, 83–148. New York: Plenum.

Coss, R. G., K. L. Guse, N. S. Poran, and D. G. Smith. 1993.

Development of antisnake defenses in California ground squirrels (*Spermophilus beecheyi*): II. Microevolutionary effects of relaxed selection from rattlesnakes. *Behaviour* 124:137–64.

Coss, R. G., S. Marks, and U. Ramakrishnan. 2002. Early environment shapes the development of gaze aversion by wild bonnet Macaques (*Macaca radiata*). *Primates* 43:217–22.

Coss, R. G., and M. Moore. 2002. Precocious knowledge of trees as antipredator refuge in preschool children: An examination of aesthetics, attributive judgments, and relic sexual dinichism. *Ecological Psychology* 14:181–222.

Coss, R. G., and D. H. Owings. 1978. Snake-directed behavior by snake naive and experienced California ground squirrels in a simulated burrow. *Zeitschrift für Tierpsychologie* 48:421–35.

———. 1985. Restraints on ground squirrel antipredator behavior: Adjustments over multiple time scales. In *Issues in the ecological study of learning*, ed. T. D. Johnston and A. T. Pietrewicz, 167–200. Hillsdale, NJ: Lawrence Erlbaum.

———. 1989. Rattler battlers. *Natural History* 5:30–35.

Cottam, C. 1948. Aquatic habits of the Norway rat. *Journal of Mammalogy* 29:299.

Coulon, J., L. Graziani, D. Allainé, M. C. Bel, and S. Pouderoux. 1995. Infanticide in the Alpine marmot (*Marmota marmota*). *Ethology Ecology and Evolution* 7:191–94.

Courchamp, R., T. Clutton-Brock, and B. Grenfell. 1999. Inverse density dependence and the Allee effect. *Trends in Ecology and Evolution* 14:405–10.

Cowan, D. P., and P. J. Garson. 1985. Variations in the social structure of rabbit populations: Causes and demographic consequences. In *Behavioural ecology*, ed. R. M. Sibley and R. H. Smith, 537–55. Oxford: Blackwell Scientific.

Cowan, D. P., R. J. Quy, and M. S. Lambert. 2003. Ecological perspectives on the management of commensal rodents. In *Rats, mice and people: Rodent biology and management*, ed. G. R. Singleton, L. A. Hinds, C. J. Krebs, and D. M. Spratt, 433–39. Canberra: Australian Centre for International Agricultural Research.

Cowan, P. E. 1977. Neophobia and neophilia: New object and new place reactions of three *Rattus* species. *Journal of Comparative and Physiological Psychology* 91:63–71.

Crane, M. F., and W. C. Fischer. 1986. Fire ecology of the forest habitat types of central Idaho. USDA Forest Service, Intermountain Research Station, General Technical Report INT-218:1–86.

Creel, S. 2001. Social dominance and stress hormones. *Trends in Ecology and Evolution* 16:491–97.

Crespi, B. J. 1996. Comparative analysis of the origins and losses of eusociality: Casual mosaics and historical uniqueness. In *Phylogenies and the comparative method in animal behavior*, ed. E. P. Martins, 253–87. Oxford: Oxford University Press.

Crespi, B. J., and J. C. Choe. 1997a. Explanation and evolution of social systems. In *The evolution of social behaviour in insects and arachnids*, ed. J. C. Choe and B. J. Crespi, 499–524. Cambridge: Cambridge University Press.

———. 1997b. Introduction. In *The evolution of social behaviour in insects and arachnids*, ed. J. C. Choe and B. J. Crespi, 1–7. Cambridge: Cambridge University Press.

Crome, F., J. Isaacs, and L. Moore 1994. The utility to birds and mammals of remnant riparian vegetation and associated windbreaks in the tropical Queensland uplands. *Pacific Conservation Biology* 1:328–43.

Cronin, H. 1991. *The ant and the peacock: Altruism and sexual selection from Darwin to today.* Cambridge: Cambridge University Press.

Crook, J. H. 1964. The evolution of social organisation and communication in the weaver birds (Ploceinae). *Behaviour* (supplement) 10:1–178.

———. 1970. Social organization and the environment: Aspects of contemporary social ethology. *Animal Behaviour* 18:197–209.

———. 1977. On the integration of gender strategies in mammalian social systems. In *Reproductive behaviour and evolution,* ed. J. S. Rosenblatt and B. R. Komisaruk, 17–38. New York: Plenum.

Crook, J. H., and J. S. Gartlan. 1966. Evolution of primate societies. *Nature* 210:1200–3.

Crooks, K. R., and M. E. Soulé. 1999. Mesopredator release and avifaunal extinctions in a fragmented system. *Nature* 400:563–66.

Cross, S. P. 1969. Behavioral aspects of western gray squirrel ecology. PhD diss., University of Arizona, Tucson.

Crow, J. F., and D. Denniston. 1988. Inbreeding and variance effective population numbers. *Evolution* 42:482–95.

Crowcroft, P. 1955. Territoriality in wild house mice, *Mus musculus* L. *Journal of Mammalogy* 36:299–301.

Cully, J. F., Jr. 1989. Plague in prairie dog ecosystems: Importance for black-footed ferret management. In *The prairie dog ecosystem: Managing for biological diversity*, 47–55. Montana BLM Wildlife Technical Bulletin no. 2. Billings, MT: Bureau of Land Management.

Cully, J. F., Jr., A. M. Barnes, T. J. Quan, and G. Maupin. 1997. Dynamics of plague in a Gunnison's prairie dog colony complex from New Mexico. *Journal of Wildlife Diseases* 33:706–19.

Cully, J. F., Jr., D. E. Biggins, and D. B. Seery. 2006. Conservation of prairie dogs in areas with sylvatic plague. In *Conservation of the black-tailed prairie dog,* ed. J. L. Hoogland, 157–68. Washington, DC: Island Press.

Cully, J. F., Jr., and J. D. Ligon. 1976. Comparative mobbing behavior of scrub and Mexican jays. *Auk* 93:116–25.

Cully, J. F., Jr., and E. S. Williams. 2001. Interspecific comparisons of sylvatic plague in prairie dogs. *Journal of Mammalogy* 82:894–905.

Curio, E. 1993. Proximate and developmental aspects of antipredator behavior. In *Advances in the study of behavior,* vol. 22, ed. P. J. B. Slater, J. S. Rosenblatt, C. T. Snowdon, and M. Milinski, 135–238. New York: Academic Press.

Curio, E., U. Ernst, and W. Vieth. 1978. The adaptive significance of avian mobbing. II. Cultural transmission of enemy recognition in blackbirds: Effectiveness and some constraints. *Zeitschrift fur Tierpsychologie* 48:184–202.

Curtin, R., and P. Dolhinow. 1978. Primate social behavior in a changing world. *American Scientist* 66:468–75.

Curtis, J. T., Y. Liu, and Z. X. Wang. 2001. Lesions of the vomeronasal organ disrupt mating-induced pair bonding in female prairie voles (*Microtus ochrogaster*). *Brain Research* 901:167–74.

Curtis, J. T., and Z. X. Wang. 2003. Forebrain *c*-fos expression under conditions conducive to pair bonding in female prairie voles (*Microtus ochrogaster*). *Physiology and Behavior* 80:95–101.

Cushing, B. S. 1985. Estrous mice and vulnerability to weasel predation. *Ecology* 66:1976–78.

Cushing, B. S., J. O. Martin, L. J. Young, and C. S. Carter. 2001. The effects of peptides on partner preference formation are predicted by habitat in prairie voles. *Hormones and Behavior* 39:48–58.

Cushing, B. S., N. Mogekwu, W. W. Le, G. E. Hoffman, and C. S. Carter. 2003. Cohabitation induced Fos immunoreactivity in the monogamous prairie vole. *Brain Research* 965:203–11.

Cushing, J. E., Jr. 1945. Quaternary rodents and lagomorphs of San Josecita Cave, Nueva Leon, Mexico. *Journal of Mammalogy* 26:182–85.

Czéh, B., T. Welt, A. K. Fischer, A. Erhardt, W. Schmitt, M. B. Müller, N. Toschi, E. Fuchs, and M. E. Keck. 2002. Chronic psychosocial stress and concomitant repetitive transcranial magnetic stimulation: Effects on stress hormone levels and adult hippocampal neurogenesis. *Biological Psychiatry* 52:1057–65.

Da Cunha-Nogueira, S. S., S. L. G. Nogueira-Filho, E. Otta, C. T. dos Santos-Dias, and A. de Carvalho. 1999. Determination of the causes of infanticide in capybara (*Hydrochaeris hydrochaeris*) groups in captivity. *Applied Animal Behavior* 62:351–57.

Daenen, E. W., G. Wolterink, M. A. Gerrits, and J. M. Van Ree. 2002. The effects of neonatal lesions in the amygdala or ventral hippocampus on social behaviour later in life. *Behavioural Brain Research* 136:571–82.

Dagg, A. I., and D. E. Windsor. 1971. Olfactory discrimination limits in gerbils. *Canadian Journal of Zoology* 49:284–85.

Dalby, P. 1975. Biology of pampa rodents. *Michigan State University Museum Publications Biological Series* 5:149–272.

Daley, J. G. 1992. Population reductions and genetic variability in black-tailed prairie dogs. *Journal of Wildlife Management* 56:212–20.

Dalton, C. L. 2000. Effects of female kin groups on reproduction and demography in the gray-sided vole, *Microtus canicaudus*. *Oikos* 90:153–59.

Daly, J. C., and J. L. Patton. 1990. Dispersal, gene flow, and allelic diversity between local populations of *Thomomys bottae* pocket gophers in the coastal ranges of California. *Evolution* 44:1283–94.

Daly, M., and S. Daly. 1974. Spatial distribution of a leaf-eating Saharan gerbil (*Psammomys obesus*) in relation to its food. *Mammalia* 38:591–603.

———. 1975a. Behavior of *Psammomys obesus* (Rodentia: Gerbillinae) in the Algerian Sahara. *Zeitschrift für Tierpsychologie* 37:298–321.

———. 1975b. Socio-ecology of Saharan gerbils, especially *Meriones libycus*. *Mammalia* 39:289–311.

Daly, M., L. F. Jacobs, M. I. Wilson, and P. R. Behrends. 1992. Scatter-hoarding by kangaroo rats (*Dipodomys merriami*) and pilferage from their caches. *Behavioral Ecology* 3:102–11.

Daly, M., M. I. Wilson, and P. R. Behrends. 1980. Factors affecting rodents' responses to odours of strangers encountered in the field: Experiments with odour-baited traps. *Behavioral Ecology and Sociobiology* 6:323–29.

———. 1984. Breeding of captive kangaroo rats, *Dipodomys merriami* and *D. microps*. *Journal of Mammalogy* 65:338–41.

Daly, M., M. I. Wilson, P. R. Behrends, and L. F. Jacobs. 1992. Sexually differentiated effects of radio-transmitters on predation risk and behaviour in kangaroo rats, *Dipodomys merriami*. *Canadian Journal of Zoology* 70:1851–55.

Daly, M., P. R. Behrends, and M. I. Wilson. 2000. Activity patterns of kangaroo rats: Granivores in a desert habitat. In *Activity patterns in small mammals*, Ecological Studies no. 141, ed. S. Halle and N. C. Stenseth, 145–58. Berlin: Springer.

Damsma, G., J. G. Pfaus, D. Wenkstern, A. G. Phillips, and H. C. Fibiger. 1992. Sexual behavior increases dopamine transmission in the nucleus accumbens and striatum of male rats: Comparison with novelty and locomotion. *Behavioral Neuroscience* 106:181–91.

Danchin, E., and R. H. Wagner. 1997. The evolution of coloniality: The emergence of new perspectives. *Trends in Ecology and Evolution* 12:342–47.

Daniel, J. C., and D. T. Blumstein. 1998. A test of the acoustic adaptation hypothesis in four species of marmots. *Animal Behaviour* 56:1517–28.

Danielson, B. J. 1991. Communities in a landscape: The influence of habitat heterogeneity on the interactions between species. *American Naturalist* 138:1105–20.

Danielson, P. B. 2002. The cytochrome P450 superfamily: Biochemistry, evolution and drug metabolism in humans. *Current Drug Metabolism* 3:561–97.

Danilov, P. I. 1995. Canadian and European beavers in Russian Northwest (distribution, number, comparative ecology). In *Proceedings of the third Nordic beaver symposium*, ed. A. Ermala and S. Lahti, 10–16. Helsinki: Finnish Game and Fisheries Research Institute.

Danilov, P. I., and V. Ya. Kan'shiev. 1983. The state of populations and ecological characteristics of European (*Castor fiber* L.) and Canadian (*Castor canadensis* Kuhl.) beavers in the northwestern USSR. *Acta Zoologica Fennica* 174:95–97.

Dantzer, R., and R. M. Bluthe. 1992. Vasopressin involvement in antipyresis, social communication, and social recognition: A synthesis. *Critical Reviews in Neurobiology* 6:243–55.

Dantzer, R., G. F. Koob, R. M. Bluthe, and M. Le Moal. 1988. Septal vasopressin modulates social memory in male rats. *Brain Research* 457:143–47.

Darling, F. F. 1938. *Bird flocks and the breeding cycle: A contribution to the study of avian sociality*. Cambridge: Cambridge University Press.

Darwin, C. 1859. *On the origin of species*. London: John Murray.

———. 1871. *The descent of man and selection in relation to sex*. London: John Murray.

David-Gray, Z. K., J. Gurnell, and D. M. Hunt. 1998. DNA fingerprinting reveals high levels of genetic diversity within British populations of the introduced non-native grey squirrel (*Sciurus carolinensis*). *Journal of Zoology* 246:443–45.

———. 1999. Estimating the relatedness in a population of grey squirrels *Sciurus carolinensis*, using DNA fingerprinting. *Acta Theriologica* 44:243–51.

Davies, N. B., B. J. Hatchwell, B. J. Robson, and T. Burke. 1992. Paternity and parental effort in dunnocks *Prunella modularis*: How good are male chick-feeding rules? *Animal Behaviour* 43:729 45.

Davis, D. E. 1949. The weight of wild brown rats at sexual maturity. *Journal of Mammalogy* 39:125–30.

———. 1950. The rat population of New York, 1949. *American Journal of Hygiene* 52:147–52.

———. 1958. The role of density in aggressive behaviour in house mice. *Animal Behaviour* 6:207–10.

Davis, D. E., and J. J. Christian. 1956. Changes in Norway rat populations induced by the introduction of rats. *Journal of Wildlife Management* 20:378–83.

———. 1958. Population consequences of a sustained yield program for Norway rats. *Ecology* 39:217–22.

Davis, D. E., J. T. Emlen, and A. W. Stroke. 1948. Studies on home range in the brown rat. *Journal of Mammalogy* 29: 207–25.

Davis, D. E., and W. T. Fales. 1949. The distribution of rats in Baltimore, Maryland. *American Journal of Hygiene* 52:247–54.

———. 1950. The rat population of Baltimore, 1949. *American Journal of Hygiene* 52:143–46.

Davis, H. N., G. D. Gray, M. Zerylnick, and D. A. Dewsbury. 1974. Ovulation and implantation in montane voles (*Microtus montanus*) as a function of varying amounts of copulatory stimulation. *Hormones and Behavior* 5:383–88.

Davis, D. E., and O. Hall. 1951. The seasonal reproductive condition of the female Norway (brown) rat in Baltimore, Maryland. *Physiological Zoology* 24:9–20.

Davis, L. S. 1982a. Copulatory behaviour of Richardson's ground squirrels (*Spermophilus richardsonii*) in the wild. *Canadian Journal of Zoology* 60:2953–55.

———. 1982b. Sibling recognition in Richardson's ground squirrels (*Spermophilus richardsonii*). *Behavioral Ecology and Sociobiology* 11:65–70.

Davis, L. S. 1984a. Alarm calling in Richardson's ground squirrels (*Spermophilus richardsonii*). *Zeitschrift für Tierpsychologie* 66:152–64.

Davis, L. S. 1984b. Behavioral interactions of Richardson's ground squirrels. In *The biology of ground-dwelling squirrels*, ed. J. O. Murie and G. R. Mitchell, 424–444. Lincoln: University of Nebraska Press.

Davis, L. S. 1984c. Kin selection and adult female Richardson's ground squirrels: A test. *Canadian Journal of Zoology* 62: 2344–48.

Davis, L. S., and J. O. Murie. 1985. Male territoriality and the mating system of Richardson's ground squirrels (*Spermophilus richardsonii*). *Journal of Mammalogy* 66:268–79.

Davis, W. B. 1966. The mammals of Texas. *Texas Parks and Wildlife Department Bulletin* 41:1–267.

Davis-Born, R., and J. O. Wolff. 2000. Age- and sex-specific responses of the gray-tailed vole, *Microtus canicaudus*, to connected and unconnected habitat patches. *Canadian Journal of Zoology* 78:864–70.

Davydov, G. S. 1991. Some characters of two populations of the long-tailed marmot. In *Population structure of the marmot*, ed. D. I. Bibikov, A. A. Nikolski, V. Yu. Rumiantsev, and T. A. Seredneva, 188–216. Moscow: USSR Theriological Society.

Dawkins, R. 1976. *The selfish gene.* New York: Oxford University Press.

———. 1979. Twelve misunderstandings of kin selection. *Zeitschrift für Tierpsychologie* 51:184–200.

Dayan, T., and D. Simberloff. 1994. Character displacement, sexual dimorphism, and morphological variation among British and Irish mustelids. *Ecology* 75:1063–73.

DeAngelo, M. 2002. Response of a kangaroo rat (*Dipodomys ingens*) to a mammalian predator (*Vulpes macrotus*). Master's thesis, San Francisco State University.

Dearborn, D. C., A. D. Anders, and P. G. Parker. 2001. Sexual dimorphism, extrapair fertilizations, and operational sex ratio in great frigatebirds (*Fregata minor*). *Behavioral Ecology* 12: 746–52.

De Boer, S. F., and J. M. Koolhaas. 2003. Defensive burying in rodents: Ethology, neurobiology and psychopharmacology. *European Journal of Pharmacology* 463:145–61.

DeBry, R. W. 2003. Identifying conflicting signal in a multi-gene analysis reveals a highly resolved tree: The phylog-

eny of Rodentia (Mammalia). *Systematic Biology* 52: 604–17.

DeBry, R. W., and R. M. Sagel. 2001. Phylogeny of Rodentia (Mammalia) inferred from the nuclear-encoded gene IRBP. *Molecular Phylogenetics and Evolution* 19:290–301.

De Catanzaro, D., and E. Macniven. 1992. Psychogenic pregnancy disruptions in mammals. *Neuroscience and Biobehavioral Reviews* 16:43–53.

De Catanzaro, D., P. Wyngaarden, J. Griffiths, H. Ham, J. Hancox, and D. Brain. 1995. Interactions of contact, odor cues, and androgens in strange-male-induced early pregnancy disruptions in mice (*Mus musculus*). *Journal of Comparative Psychology* 109:115–22.

Decker, M. D., P. G. Parker, D. J. Minchella, and K. N. Rabenold. 1993. Monogamy in black vultures: Genetic evidence from DNA fingerprinting. *Behavioral Ecology* 4:29–35.

Degen, A. A. 1996. *Ecophysiology of small desert mammals.* Berlin: Springer.

De Graaff, G. 1981. *The rodents of Southern Africa: Notes on their identification, distribution, ecology and taxonomy.* Durban, UK: Butterworths.

De Jong, W. W., W. Hendriks, S. Sanyal, and E. Nevo. 1990. The eye of the blind mole rats (*Spalax ehrenbergi*): Regressive evolution at the molecular level. In *Evolution of subterranean mammals at the organismal and molecular levels*, ed. E. Nevo and A. O. Reig, 383–95. New York: Alan R. Liss.

De Kloet, E. R., M. S. Oitzl, and M. Joels. 1993. Functional implications of brain corticosteroid receptor diversity. *Cellular and Molecular Neurobiology* 13:433–55.

———. 1999. Stress and cognition: Are corticosteroids good or bad guys? *Trends in Neuroscience* 22:422–26.

De Kock, L. L., and I. Rohn. 1972. Intra-specific behaviour during the up-swing of groups of Norway lemmings, kept under semi-natural conditions. *Zeitschrift für Tierpsychologie* 30: 405–15.

De la Maza, H. M., J. O. Wolff, and A. Lindsey. 1999. Exposure to strange adults does not cause pregnancy disruption or infanticide in the gray-tailed vole. *Behavioral Ecology and Sociobiology* 45:107–13.

Delarbre, C., T. T. Kashi, P. Boursot, J. S. Beckmann, F. Kourilsky, F. Bonhomme, and G. Gachelin. 1988. Phylogenetic distribution in the genus *Mus* of *t*-complex-specific DNA and protein markers: Inferences on the origin of *t*-haplotypes. *Molecular and Biological Evolution* 5:120–33.

DelBarco-Trillo, J., and M. H. Ferkin. 2004. Male mammals respond to a risk of sperm competition conveyed by odours of conspecific males. *Nature* 431:446–49.

Dell'omo, G., M. Fiore, and E. Alleva. 1994. Strain differences in mouse response to odours of predators. *Behavioural Processes* 32:105–15.

DeLong, K. T. 1966. Population ecology of feral house mice: interference by *Microtus*. *Ecology* 47:481–84.

Delsuc, F., M. Scally, O. Madsen, M. J. Stanhope, W. W. de Jong, F. M. Catzeflis, M. S. Springer, and E. J. P. Douzery. 2002. Molecular phylogeny of living xenarthrans and the impact of character and taxon sampling on the placental tree rooting. *Molecular Biology and Evolution* 19:1656–71.

Delville, Y., K. M. Mansour, and C. F. Ferris. 1996. Serotonin blocks vasopressin-facilitated offensive aggression: Interactions within the ventrolateral hypothalamus of golden hamsters. *Physiology and Behavior* 59:813–16.

Demas, G. E., and R. J. Nelson. 1998. Social, but not photoperiodic, influences on reproductive function in male *Peromyscus aztecus*. *Biology of Reproduction* 58:385–89.

Demas, G. E., J. M. Williams, and R. J. Nelson. 1997. Amygdala but not hippocampal lesions impair olfactory memory for mate in prairie voles (*Microtus ochrogaster*). *American Journal of Physiology* 273:R1683–1689.

Demby, A. H., A. Inapogui, K. Kargbo, J. Koninga, K. Kourouma, J. Kanu, M. Coulibaly et al. 2001. Lassa fever in Guinea: II. Distribution and prevalence of Lassa virus infection in small mammals. *Vector Borne and Zoonotic Diseases* 1:283–97.

Dempster, E. R., and M. R. Perrin. 1989a. A comparative study of agonistic behaviour in hairy-footed gerbils (genus *Gerbillurus*). *Ethology* 83:43–59.

Dempster, E. R., and M. R. Perrin. 1989b. Maternal behaviour and neonatal development in three species of Namib Desert rodents. *Journal of Zoology* 218:407–19.

Dempster, E. R., M. R. Perrin, C. T. Downs, and M. Griffin. 1998. *Gerbillurus setzeri*. *Mammalian Species* 598:1–4.

Denenberg, V. H., R. H. Taylor, and M. X. Zarrow. 1969. Maternal behavior in the rat: An investigation and quantification of nest building. *Behaviour* 34:1–16.

Dennell, R. E., and B. H. Judd. 1961. Segregation distorter in *D. melanogaster* males: An effect of female genotype on recovery. *Molecular Genetics and Genomics* 105:262–74.

Denniston, R. H. 1957. Notes on breeding and size of young in the Richardson's ground squirrel. *Journal of Mammalogy* 38:414–16.

Denny, M. R., C. F. Clos, and R. C. Bell 1988. Learning in the rat of a choice response by observation of S-S contingencies. In *Social learning: Psychological and biological perspectives,* ed. T. R. Zentall and B. G. Galef, Jr., 207–38. Hillsdale, NJ: Lawrence Erlbaum.

Denys, C. 1987. Rodentia and Lagomorpha. 6.1: Fossil rodents (other than Pedetidae) from Laetoli. In *Laetoli a Pliocene site in Tanzania,* ed. M. D. Leakey and J. M. Harris, 118–70. London: Oxford University Press.

———. 1989. A new species of bathyergid rodent from Olduvai Bed I (Tanzania, Lower Pleistocene). *Neue Jahrbuecher Geologica Palaeontologica* 5:257–64.

D'Erchia, A. M., C. Gissi, G. Pesole, C. Saccone, and U. Arnason. 1996. The guinea-pig is not a rodent. *Nature* 381:597–600.

Derting, T. L., J. H. Kruper, J. L. Wiles, M. L. Carter, and H. M. Furlong. 1999. Physiological bases of male olfactory cues and mate preferences in prairie voles. In *Advances in chemical signals in vertebrates,* ed. R. E. Johnston, D. Müller-Schwarze, and P. W. Sorensen, 463–73. New York: Plenum.

Desjardins, C., J. A. Maruniak, and F. H. Bronson. 1973. Social rank in house mice: Differentiation revealed by ultraviolet visualization of urinary marking patterns. *Science* 182:939–41.

Desmond, M. J., J. A. Savidge, and K. M. Eskridge. 2000. Correlations between burrowing owl and black-tailed prairie dog declines: A 7-year analysis. *Journal of Wildlife Management* 64:1067–75.

Desy, E. A., G. O. Batzli, and J. Liu. 1990. Effects of food and predation on behaviour of prairie voles: A field experiment. *Oikos* 58:159–68.

Detling, J. K. 1998. Mammalian herbivores: Ecosystem-level effects in two grassland national parks. *Wildlife Society Bulletin* 26:438–48.

———. 2006. Do prairie dogs compete with livestock? In *Conservation of the black-tailed prairie dog,* ed. J. L. Hoogland, 65–88. Washington, DC: Island Press.

Devenport, J. A., and L. D. Devenport. 1994. Spatial navigation in natural habitats by ground-dwelling sciurids. *Animal Behaviour* 47:727–29.

Devenport, J. A., L. D. Luna, and L. D. Devenport. 2000. Placement, retrieval, and memory of caches by thirteen-lined ground squirrels. *Ethology* 106:171–83.

Devillard, S., D. Allaine, J. Gaillard, and D. Pontier. 2004. Does social complexity lead to sex-biased dispersal in polygynous mammals? A test on ground-dwelling sciurids. *Behavioral Ecology* 15:83–87.

De Villiers, D. J. 1986. Infanticide in the tree squirrel, *Paraxerus cepapi*. *South African Journal of Zoology* 21:183–84.

DeVries, A. C. 2002. Interaction among social environment, the hypothalamic–pituitary–adrenal axis, and behavior. *Hormones and Behavior* 41:405–13.

DeVries, A. C., and C. S. Carter. 1999. Sex differences in temporal parameters of partner preference in prairie voles (*Microtus ochrogaster*). *Canadian Journal of Zoology* 77:885–89.

DeVries, A. C., M. B. DeVries, S. E. Taymans, and C. S. Carter. 1995. Modulation of pair bonding in female prairie voles (*Microtus ochrogaster*) by corticosterone. *Proceedings of the National Academy of Science (USA)* 92:7744–48.

———. 1996. The effects of stress on social preferences are sexually dimorphic in prairie voles. *Proceedings of the National Academy of Science (USA)* 93:11980–84.

DeVries, A. C., S. E. Taymans, and C. S. Carter. 1997. Social modulation of corticosteroid responses in male prairie voles. *Annals New York Academy of Science* 807:494–97.

De Vries, G. J., and M. A. Miller. 1998. Anatomy and function of extrahypothalamic vasopressin systems in the brain. *Progress in Brain Research* 119:3–20.

Dewar, G. 2004. Social and asocial cues about food: Cue reliability influences intake in rats. *Learning and Behavior* 32:82–89.

Dewsbury, D. A. 1972. Patterns of copulatory behavior in male mammals. *Quarterly Review of Biology* 47:1–33.

———. 1975. Diversity and adaptation in rodent copulatory behavior. *Science* 190:947–54.

———. 1981. An exercise in the prediction of monogamy in the field from laboratory data on 42 species of muroid rodents. *The Biologist* 63:138–62.

———. 1982a. Dominance rank, copulatory behavior, and differential reproduction. *The Quarterly Review of Biology* 57:135–59.

———. 1982b. Ejaculate cost and male choice. *The American Naturalist* 119:601–10.

———. 1982c. Pregnancy blockage following multiple-male copulation or exposure at the time of mating in deer mice, *Peromyscus maniculatus*. *Behavioral Ecology and Sociobiology* 11:37–42.

———. 1984. Sperm competition in muroid rodents. In *Sperm competition and the evolution of animal mating systems,* ed. R. L. Smith, 547–71. New York: Academic Press.

———. 1985. Paternal behavior in rodents. *American Zoologist* 25:841–52.

———. 1987. The comparative psychology of monogamy. *Nebraska Symposium on Motivation* 35:1–50.

———. 1988. Kin discrimination and reproductive behavior in muroid rodents. *Behavior Genetics* 18:525–36.

———. 1990. Tests of preferences of adult deer mice (*Peromyscus maniculatus bairdi*) for siblings versus nonsiblings. *Journal of Comparative Psychology* 104:177–82.

Dewsbury, D. A., S. M. Bauer, J. D. Pierce, Jr., L. E. Shapiro, and

S. A. Taylor. 1992. Ejaculate disruption in two species of voles (*Microtus*): On the PEI matching law. *Journal of Comparative Psychology* 106:383–87.

Dewsbury, D. A., and D. J. Baumgardner. 1981. Studies of sperm competition in two species of muroid rodents. *Behavioral Ecology and Sociobiology* 9:121–33.

Dewsbury, D. A., and D. Q. Estep. 1975. Pregnancy in cactus mice: Effects of prolonged copulation. *Science* 187:552–53.

Dewsbury, D. A., and T. G. Hartung. 1980. Copulatory behaviour and differential reproduction of laboratory rats in a two-male, one-female competitive situation. *Animal Behaviour* 28:95–102.

Dhabhar, F. S., and B. S. McEwen. 1999. Enhancing versus suppressive effects of stress hormones on skin immune function. *Proceedings of the National Academy of Science* 96:1059–64.

Diamond, M. 1970. Intromission pattern and species vaginal code in relation to the induction of pseudopregnancy. *Science* 169:995–97.

Diaz, G. B., R. A. Ojeda, M. H. Gallardo, and S. M. Giannoni. 2000. *Tympanoctomys barrerae. Mammalian Species* 646:1–4.

Dickman, C. R. 1992. Predation and habitat shift in the house mouse *Mus domesticus. Ecology* 73:313–22.

———. 1999. Rodent-ecosystem relationships: A review. In *Ecologically-based rodent management*, ed. G. R. Singleton, L. Hinds, H. Leirs, and Z. Zhang, 113–33. Canberra: Australian Centre for International Agricultural Research.

Dielenberg, R. A., and I. S. McGregor. 2001. Defensive behavior in rats towards predatory odors: A review. *Neuroscience and Biobehavioral Reviews* 25:597–609.

Dieterlen, V. F. 1962. Geburt und Geburtshilfe bei der Stachelmaux, *Acomys cahirinus. Zeitschrift für Tierpsychologie* 19:191–222.

Diffendorfer, J., and N. Slade. 2002. Long-distance movements in cotton rats (*Sigmodon hispidus*) and prairie voles (*Microtus ochrogaster*) in northeastern Kansas. *American Midland Naturalist* 148:309–19.

Dill, L. M., and R. Houtman. 1989. The influence of distance of refuge on flight initiation distance in the gray squirrel (*Sciurus carolinensis*). *Canadian Journal of Zoology* 67:233–35.

Dixon, D. M. 1972. Population, pollution and death in ancient Egypt. In *Population and pollution*, ed. P. R. Cox and J. Peel, 29–36. London: Academic Press.

Djeridane, Y. 2002. Integumentary odoriferous glands in *Meriones libycus*: A histological study and related behavior. *Folia Zoologica* 51:37–58.

Djoshkin, W. W., and V. G. Safonov. 1972. *Die biber der alten und neuen welt. A.* Wittenberg: Ziemsen Verlag.

Dluzen, D. E., S. Muraoka, and R. Landgraf. 1998. Olfactory bulb norepinephrine depletion abolishes vasopressin and oxytocin preservation of social recognition responses in rats. *Neuroscience Letters* 254:161–64.

Doboszynska, T., and W. Zurowski. 1983. Reproduction of the European beaver. *Acta Zoologica Fennica* 174:123–26.

Dobson, A., and A. Lyles. 2000. Ecology—black-footed ferret recovery. *Science* 288:985.

Dobson, F. S. 1979. An experimental study of dispersal in the California ground squirrel. *Ecology* 60:1103–9.

———. 1981. An experimental examination of an artificial dispersal sink. *Journal of Mammalogy* 62:74–81.

———. 1982. Competition for mates and predominant juvenile male dispersal in mammals. *Animal Behaviour* 30:1183–92.

———. 1983. Agonism and territoriality in the California ground squirrel. *Journal of Mammalogy* 64:218–25.

———. 1984. Environmental influences on sciurid mating systems. In *The biology of ground-dwelling squirrels,* ed. J. O. Murie, and G. R. Michener, 229–49. Lincoln: University of Nebraska Press.

———. 1985. The use of phylogeny in behavior and ecology. *Evolution* 39:1384–88.

———. 1990. Environmental influences on infanticide in Columbian ground squirrels. *Ethology* 84:3–14.

———. 1992. Body mass, structural size, and life-history patterns of the Columbian ground squirrel. *American Naturalist* 140:109–25.

———. 1998. Social structure and gene dynamics in mammals. *Journal of Mammalogy* 79:667–70.

Dobson, F. S., and C. Baudoin. 2002. Experimental tests of spatial association and kinship in monogamous mice (*Mus spicilegus*) and polygynous mice (*Mus musculus domesticus*). *Canadian Journal of Zoology* 80:980–86.

Dobson, F. S., R. K. Chesser, J. L. Hoogland, D. W. Sugg, and D. W. Foltz. 1997. Do black-tailed prairie dogs minimize inbreeding? *Evolution* 51:970–78.

———. 1998. Breeding groups and gene dynamics in a socially-structured population of prairie dogs. *Journal of Mammalogy* 79:671–80.

———. 2004. The influence of social breeding groups on effective population size in black-tailed prairie dogs. *Journal of Mammalogy* 85:146–54.

Dobson, F. S., R. K. Chesser, and B. Zinner. 2000. The evolution of infanticide: Genetic benefits of extreme nepotism and spite. *Ethology, Ecology & Evolution* 12:131–48.

Dobson, F. S., and D. E. Davis. 1986. Hibernation and sociality in the California ground squirrel. *Journal of Mammalogy* 67:416–21.

Dobson, F. S., C. Jacquot, and C. Baudoin. 2000. An experimental test of kin association in the house mouse. *Canadian Journal of Zoology* 78:1806–12.

Dobson, F. S., and W. T. Jones. 1985. Multiple causes of dispersal. *American Naturalist* 126:855–58.

Dobson, F. S., and J. D. Kjelgaard. 1985. The influence of food resources on life history in Columbian ground squirrels. *Canadian Journal of Zoology* 63:2105–9.

Dobson, F. S., and G. R. Michener. 1995. Maternal traits and reproduction in Richardson's ground squirrels. *Ecology* 76:851–62.

Dobson, F. S., and M. K. Oli. 2001. The demographic basis of population regulation in Columbian ground squirrels. *The American Naturalist* 158:236–47.

Dobson, F. S., T. S. Risch, and J. O. Murie. 1999. Increasing returns in the life history of Columbian ground squirrels. *Journal of Animal Ecology* 68:73–86.

Dobson, F. S., A. T. Smith, and X. G. Wang. 1998. Social and ecological influences on dispersal and philopatry in the plateau pika (*Ochotona curzoniae*). *Behavioral Ecology* 9:622–35.

———. 2000. The mating system and gene dynamics of plateau pikas. *Behavioural Processes* 51:101–10.

Dobson, F. S., and J. D. Wigginton. 1996. Environmental influences on the sexual dimorphism in body size of western bobcats. *Oecologia* 108:610–16.

Dobson, F. S., and B. Zinner. 2003. Social groups, genetic structure, and conservation. In *Animal behavior and wildlife conservation*, ed. M. Festa-Bianchet and M. Apollonio, 211–28. Washington, DC: Island Press.

Dod, B., C. Litel, P. Makoundou, A. Orth, and P. Boursot, P. 2003. Identification and characterization of *t* haplotypes in wild mice populations using molecular markers. *Genetic Research* 81:103–14.

Doherty, P. C., and R. M. Zinkernagel. 1975. Enhanced immunological surveillance in mice heterozygous at the H-2 gene complex. *Nature* 256:50–52.

Dolan, P. G., and D. C. Carter. 1977. *Glaucomys volans*. *Mammalian Species* 78:1–6.

Dominey, W. J. 1984. Alternative mating tactics and evolutionarily stable strategies. *American Zoologist* 24:385–96.

Dominguez, J., J. V. Riolo, Z. Xu, and E. M. Hull. 2001. Regulation by the medial amygdala of copulation and medial preoptic dopamine release. *Journal of Neuroscience* 21:349–55.

Dominic, C. J. 1969. Pheromonal mechanisms regulating mammalian reproduction. *General and Comparative Endocrinology Supplement* 2:260–67.

Doonan, T. J., and N. Slade. 1995. Effects of supplemental food on population dynamics of cotton rats, *Sigmodon hispidus*. *Ecology* 76:814–26.

Dos Reis, S. F., and L. M. Pessoa. 2004. *Thrichomys apereoides*. *Mammalian Species* 741:1–5.

Douglass, R. J., A. J. Kuenzi, C. Y. Williams, S. J. Douglass, and J. N. Mills. 2003. Removing deer mice from buildings and the risk for human exposure to Sin Nombre Virus. *Emerging Infectious Diseases* 9:390–92.

Douglass, R. J., T. Wilson, W. J. Semmens, S. N. Zanto, C. W. Bond, R. C. Van Horn, and J. N. Mills. 2001. Longitudinal studies of Sin Nombre virus in deer mouse-dominated ecosystems of Montana. *American Journal of Tropical Medicine and Hygiene* 65:33–41.

Douzery, E. J. P., and D. Huchon. 2004. Rabbits, if anything, are likely Glires. *Molecular Phylogenetics and Evolution* 33:922–35.

Downes, S. J., K. A. Handasyde, and M. A. Edgar. 1997. The use of corridors by mammals in fragmented Australian eucalypt forests. *Conservation Biology* 11:718–26.

Downhower, J. F. 1968. Factors affecting the dispersal of yearling yellow-bellied marmots (*Marmota flaviventris*). PhD diss., University of Kansas, Lawrence.

Downhower, J. F., and K. B. Armitage. 1981. Dispersal of yearling yellow-bellied marmots (*Marmota flaviventris*). *Animal Behaviour* 29:1064–69.

Drago, F., J. D. Caldwell, C. A. Pedersen, G. Continella, U. Scapagnini, and A. J. Prange, Jr. 1986. Dopamine neurotransmission in the nucleus accumbens may be involved in oxytocin-enhanced grooming behavior of the rat. *Pharmacology Biochemistry and Behavior* 24:1185–88.

Drago, F., M. Stanciu, S. Salehi, and U. Scapagnini. 1997. The block of central vasopressin V1 but not V2 receptors suppresses grooming behavior and hypothermia induced by intracerebroventricular vasopressin in male rats. *Peptides* 18:1389–92.

Drewett, R. F. 1983. Sucking, milk synthesis, and milk ejection in the Norway rat. In *Parental behaviour of rodents*, ed. R. W. Elwood, 181–203. New York: Wiley.

Drickamer, L. C. 1974a. Contact stimulation, androgenized females and accelerated sexual maturation in female mice. *Behavioral Biology* 12:101–10.

———. 1974b. Sexual maturation of female mice: Social inhibition. *Developmental Psychobiology* 7:257–65.

———. 1977a. Delay of sexual maturation in female house mice by exposure to grouped females or urine from grouped females. *Journal of Reproduction and Fertility* 51:77–81.

———. 1977b. Seasonal variation in litter size, body weight and sexual maturation in juvenile female house mice (*Mus musculus*). *Laboratory Animal* 11:159–62.

———. 1979. Acceleration and delay of first estrus in wild *Mus musculus*. *Journal of Mammalogy* 60:215–16.

———. 1982. Acceleration and delay of first vaginal oestrus in female mice by urinary chemosignals: Dose levels and mixing urine sources. *Animal Behaviour* 30:456–60.

———. 1983. Chemosignal effects on puberty in young female mice: Urine from pregnant and lactating females. *Developmental Psychobiology* 16:207–17.

———. 1984a. Effects of very small doses of urine on acceleration and delay of sexual maturation in female house mice. *Journal of Reproduction and Fertility* 71:475–77.

———. 1984b. Pregnancy in female house mice exposed to urinary chemosignals from other females. *Journal of Reproduction and Fertility* 115:233–41.

———. 1984c. Seasonal variation in acceleration and delay of sexual maturation in female mice by urinary chemosignals. *Journal of Reproduction and Fertility* 72:55–58.

———. 1984d. Urinary chemosignals and puberty in female house mice: Effects of photoperiod and food deprivation. *Physiology and Behavior* 33:907–11.

———. 1984e. Urinary chemosignals from mice (*Mus musculus*): Acceleration and delay of puberty in related and unrelated young females. *Journal of Comparative Psychology* 89:414–20.

———. 1986a. Effects of urine from females in oestrus on puberty in female mice. *Journal of Reproduction and Fertility* 77:613–22.

———. 1986b. Peripheral anosmia affects puberty-influencing chemosignals in mice: Donors and recipients. *Physiology and Behavior* 37:741–46.

———. 1986c. Puberty-influencing chemosignals in mice: Ecological and evolutionary considerations. In *Chemical signals in vertebrates IV,* ed. D. Duvall, D. Muller-Schwarze, and R. M. Silverstein, 441–45. New York: Plenum.

———. 1987. Seasonal variations in the effectiveness of urinary chemosignals influencing puberty in female house mice. *Journal of Reproduction and Fertility* 80:295–300.

———. 1988. Acceleration and delay of sexual maturation in female house mice (*Mus domesticus*) by urinary chemosignals: Mixing urine sources in unequal proportions. *Journal of Comparative Psychology* 102:215–21.

———. 1989. Pregnancy block in wild house mice, *Mus domesticus*: Olfactory preferences of females during gestation. *Animal Behaviour* 37:690–98.

———. 1992a. Oestrous female house mice discriminate dominant from subordinate males and sons of dominant from sons of subordinate males by odour cues. *Animal Behaviour* 43:868–70.

———. 1992b. Urinary chemosignals affect reproduction of adult female mice. In *Chemical signals in vertebrates VI,* ed. R. Doty, D. Muller-Schwarze, and R. M. Silverstein, 245–51. New York: Plenum.

———. 1999. Pregnancy in female house mice exposed to urinary chemosignals from other females. *Journal of Reproduction and Fertility* 115:233–41.

————. 2001. Urine marking and social dominance in male house mice (*Mus musculus domesticus*). *Behavioural Processes* 53:113–20.

Drickamer, L. C., and S. Assmann. 1981. Acceleration and delay of puberty in female house mice: Methods of delivery of the urinary stimulus. *Developmental Psychobiology* 15:433–42.

Drickamer, L. C., and L. L. Gillie. 1998. Integrating proximate and ultimate causation in the study of vertebrate behavior: Methods considerations. *American Zoologist* 38:43–58.

Drickamer, L. C., P. A. Gowaty, and C. M. Holmes. 2000. Free female mate choice in house mice affects reproductive success and offspring viability and performance. *Animal Behaviour* 59:371–78.

Drickamer, L. C., P. A. Gowaty, and D. M. Wagner. 2003. Free mutual mate preferences in house mice affect reproductive success and offspring performance. *Animal Behaviour* 65:106–14.

Drickamer, L. C., and D. G. Mikesic. 1990. Urinary chemosignals, reproduction, puberty, and population size in wild house mice in enclosures. *Journal of Chemical Ecology* 16:2955–68.

Drickamer, L. C., and R. X. Murphy. 1978. Female mouse maturation: Effects of excreted and bladder urine from juvenile and adult males. *Developmental Psychobiology* 11:63–72.

Drickamer, L. C., A. S. Robinson, and C. A. Mossman. 2001. Differential responses to same and opposite sex odors by adult house mice are associated with anogenital distance. *Ethology* 107:509–19.

Driessen, M. M., and R. K. Rose. 1999. *Pseudomys higginsi*. *Mammalian Species* 623:1–5.

Dubois, J.-Y., D. Rakotondravony, C. Hänni, P. Sourrouille, and F. F. Catzeflis. 1996. Molecular evolutionary relationships of three genera of Nesomyinae, endemic rodent taxa from Madagascar. *Journal of Mammalian Evolution* 3:239–60.

Dubrovsky, Yu. A. 1978. *Jirds and natural nidi of the cutaneous leishmaniasis.* [In Russian.] Moscow: Nauka.

Dudley, D. 1974. Contributions of paternal care to the growth and development of the young in *Peromyscus californicus*. *Behavioral Biology* 11:155–66.

Dugatkin, A. A. 2003. *Principles of animal behavior.* New York: W. W. Norton.

Dukel'skaya, N. M. 1935. The mole rat *Spalax* and mole rat trapping. In *The thin toed ground squirrel, the fat doormouse, the mole rat and the chipmunk,* ed. S. P. Naumov, N. P. Lavarov, E. P. Spangenberg, N. M. Dukel'skaya, I. M. Zaleskii, and M. D. Zverev, 71–79. Moscow: All-Union Coop. Union Publications House. (In Russian. English translation 142 F 395 D Bureau of Animal Populations, Oxford University).

Dunaway, P. B. 1968. Life history and populational aspects of the eastern harvest mouse. *American Midland Naturalist* 79:48–67.

Dunbar, R. I. M. 1982. Intraspecific variations in mating strategy. *Perspectives in Ethology* 5:385–431.

————. 1984. The ecology of monogamy. *New Scientist* 1419:12–15.

————. 1989. Social systems as optimal strategy sets: The costs and benefits of sociality. In *Comparative socioecology: The behavioral ecology of humans and other mammals,* ed. V. Standen and R. A. Foley, 131–49. Oxford: Blackwell Scientific.

Dunford, C. 1977a. Kin selection for ground squirrel alarm calls. *American Naturalist* 111:782–85.

————. 1977b. Social system of round-tailed ground squirrels. *Animal Behaviour* 25:885–906.

Dunn, A. J. 1988. Stress-related changes in cerebral catecholamine and indoleamine metabolism: Lack of effect of adrenalectomy and corticosterone. *Journal of Neurochemistry* 51:406–12.

Dunn, L. C., A. B. Beasley, and H. Tinker. 1958. Relative fitness of wild house mice heterozygous for a lethal allele. *American Naturalist* 92:215–20.

Dunn, L. C., and H. Levene. 1961. Population dynamics of a variant *t* allele in a confined population of wild house mice. *Evolution* 15:385–93.

Dunn, L. C., and J. A. Suckling. 1955. A preliminary comparison of the fertilities of wild house mice with and without a mutant at locus T. *American Naturalist* 89:231–33.

Duquette, L. S., and J. S. Millar. 1995a. The effect of supplemental food on life-history traits and demography of a tropical mouse *Peromyscus mexicanus*. *Journal of Animal Ecology* 64:348–60.

————. 1995b. Reproductive response of a tropical mouse, *Peromyscus mexicanus*, to changes in food availability. *Journal of Mammalogy* 76:596–602.

Duquette, L. S., and J. S. Millar. 1998. Litter sex ratios in a food-supplemented population of *Peromyscus mexicanus*. *Canadian Journal of Zoology* 76:623–29.

Durand, D., K. Ardlie, L. Buttel, S. A. Levin, and L. M. Silver. 1997. Impact of migration and fitness on the stability of lethal t-haplotype polymorphism in *Mus musculus*: A computer study. *Genetics* 145:1093–1108.

Durant, P., J. W. Dole, and G. F. Fisler. 1988. Agonistic behavior of the California ground squirrel, *Spermophilus beecheyi,* at an artificial food source. *Great Basin Naturalist* 48:19–24.

Du Toit, J. T., J. U. M. Jarvis, and G. N. Louw. 1985. Nutrition and burrowing energetics of the Cape mole–rat, *Georychus capensis*. *Oecologia* 66:81–87.

Duvall, D., and D. Chiszar. 1990. Behavior and chemical ecology of vernal migration and pre- and post-strike predatory activity in prairie rattlesnakes. In *Chemical Signals in Vertebrates 5,* ed. D. W. MacDonald, D. Mueller-Scwarze, and S. E. Natynczuk, 539–54. Oxford: Oxford University Press.

Ebensperger, L. A. 1998a. Do female rodents use promiscuity to prevent male infanticide? *Ethology Ecology & Evolution* 10:129–41.

————. 1998b. The potential effects of protected nests and cage complexity on maternal aggression in house mice. *Aggressive Behavior* 24:385–96.

————. 1998c. Sociality in rodents: The New World fossorial hystricognaths as a study model. *Revista Chilena Historia Natural* 71:65–77.

————. 1998d. Strategies and counterstrategies to infanticide in mammals. *Biological Reviews* 73:321–46.

————. 2001a. A review on the evolutionary causes of rodent group living. *Acta Theriologica* 46:115–44.

————. 2001b. No infanticide in the hystricognath rodent, *Octodon degus*: Does ecology play a role? *Acta Ethologica* 3:89–93.

Ebensperger, L. A., C. Botto-Mahan, and R. H. Tamarin. 2000. Nonparental infanticide in meadow voles, *Microtus pennsyl-*

vanicus: The influence of nutritional benefits. *Ethology Ecology and Evolution* 12:149–60.

Ebensperger, L. A., and F. Bozinovic. 2000a. Communal burrowing in the hystricognath rodent, *Octodon degus*: A benefit of sociality? *Behavioral Ecology and Sociobiology* 47: 365–69.

———. 2000b. Energetics and burrowing behaviour in the semifossorial degu, *Octodon degus* (Rodentia: Octodontidae). *Journal of Zoology, London* 252:179–86.

Ebensperger, L. A., and H. Cofré. 2001. On the evolution of group-living in the New World cursorial hystricognath rodents. *Behavioral Ecology* 12:227–36.

Ebensperger, L. A., and M. J. Hurtado. 2005a. On the relationship between herbaceous cover and vigilance activity of degus (*Octodon degus*). *Ethology* 111:593–608.

———. 2005b. Seasonal changes in the time budget of degus, *Octodon degus*. *Behaviour* 142:91–112.

Ebensperger, L. A., M. J. Hurtado, M. Soto-Gamboa, E. A. Lacey, and A. T. Chang. 2004. Communal nesting and kinship in degus (*Octodon degus*). *Naturwissenschaften* 91: 391–95.

Ebensperger, L. A., C. Veloso, and P. K. Wallem. 2002. Do female degus communally nest and nurse their pups? *Journal of Ethology* 20:143–46.

Ebensperger, L. A., and P. K. Wallem. 2002. Grouping increases the ability of the social rodent, *Octodon degus,* to detect predators when using exposed microhabitats. *Oikos* 98: 491–97.

Eberhard, W. G. 1996. *Female control: Sexual selection by cryptic female choice.* Princeton, NJ: Princeton University Press.

Eccard, J. A., J. Pusenius, J. Sundell, R. Tiilikainen, H. Ylönen, and S. Halle. 2001. Activity and foraging behaviour of bank voles under avian and mamalian predation risk. *Mammalian Biology* 66 (Suppl:11).

Ecke, D. H., and C. W. Johnson. 1952. Plague in Colorado and Texas. Part I. Plague in Colorado. *Public Health Monograph* 6:1–38.

Edelman, A. J. 2004. The ecology of an introduced population of Abert's squirrels in a mixed-conifer forest. MS thesis. University of Arizona, Tucson.

Edwards, C. W., and R. D. Bradley. 2002. Molecular systematics of the genus *Neotoma. Molecular Phylogenetics and Evolution* 25:489–500.

Edwards, M., J. M. Serrao, J. P. Gent, and C. S. Goodchild. 1990. On the mechanism by which midazolam causes spinally mediated analgesia. *Anesthesiology* 73:273–77.

Edwards, R. L. 1946. Some notes on the life history of the Mexican ground squirrel. *Journal of Mammalogy* 27: 105–21.

Egid, K., and J. L. Brown. 1989. The major histocompatability complex and female mating preferences in mice. *Animal Behaviour* 38:548–50.

Egid, K., and S. Lenington. 1985. Responses of male mice to odors of females: Effects of T- and H-2-locus genotype. *Behavior Genetics* 15:287–95.

Ehman, K. D., and M. E. Scott. 2001. Urinary odour preferences of MHC congenic female mice, *Mus domesticus*: Implications for kin recognition and detection of parasitized males. *Animal Behaviour* 62:781–89.

Ehninger, D., and G. Kempermann. 2003. Regional effects of wheel running and environmental enrichment on cell genesis and microglia proliferation in the adult murine neocortex. *Cerebral Cortex* 13:845–851.

Ehrich, D., P. E. Jorde, C. J. Krebs, A. J. Kenney, J. E. Stacy, and N. C. Stenseth. 2001a. Spatial structure of lemming populations (*Dicrostonyx groenlandicus*) fluctuating in density. *Molecular Ecology* 10:481–95.

Ehrich, D., C. J. Krebs, A. J. Kenney, and N. C. Stenseth. 2001b. Comparing the genetic population structure of two species of arctic lemmings: More local differentiation in *Lemmus trimucronatus* than in *Dicrostonyx groenlandicus. Oikos* 94: 143–50.

Ehrich, D., and N. C. Stenseth. 2001. Genetic structure of Siberian lemmings (*Lemmus sibiricus*) in a continuous habitat: Large patches rather than isolation by distance. *Heredity* 86:716–30.

Eichenbaum, H. 1997. How does the brain organize memories? *Science* 277:330–32.

Eigelis, Yu. K. 1980. *Rodents of the Eastern Caucasus and the problems of sanitation of the local plague nidi.* [In Russian.] Saratov: Saratov State University Press.

Eilam, D., T. Dayan, S. Ben-Eliyahu, I. Schulman, G. Shefer, and C. A. Hendrie. 1999. Differential behavioural and hormonal responses of voles and spiny mice to owl calls. *Animal Behaviour* 58:1085–93.

Eisenberg, J. F. 1963a. The behavior of heteromyid rodents. *University of California Publications in Zoology* 69:1–100.

———. 1963b. The intrapecific social behavior of some cricetine rodents of the genus *Peromyscus. American Midland Naturalist* 69:240–46.

———. 1966. The social organization of mammals. *Handbook of Zoology* 10:1–92.

———. 1967. A comparative study in rodent ethology with emphasis on evolution of social behavior. *Proceedings United States National Museum* 122:1–51.

———. 1968. Behavior patterns. In *Biology of* Peromyscus, ed. J. A. King, 451–95. Stillwater, OK: American Society of Mammalogists.

———. 1974. The function and motivational basis of Hystricomorph vocalizations. *Symposium Zoological Society of London* 34:211–47.

———. 1975. The behavior patterns of desert rodents. In *Rodents in desert environments*, ed. I. Prakash and C. K. Ghosh, 189–224. The Hague: Junk.

———. 1981. *The mammalian radiations.* Chicago: University of Chicago Press.

Eizirik, E., W. J. Murphy, and S. J. O'Brien. 2001. Molecular dating and biogeography of the early placental mammal radiation. *Journal of Heredity* 92:212–19.

Eklund, A. 1997. The effect of early experience on MHC-based mate preferences in two B10.W strains of mice *(Mus domesticus). Behavior Genetics* 27:223–29.

Eklund, A. C. 1997. The major histocompatibility complex and mating preferences in wild house mice (*Mus domesticus*). *Behavioral Ecology* 8:630–34.

———. 1999. Use of the MHC for mate choice in wild house mice (*Mus domesticus*). *Genetica* 104:245–48.

El Jundi, T. A. R. J., and T. R. O. de Freitas. 2004. Genetic and demographic structure in a population of *Ctenomys lami* (Rodentia-Ctenomyidae). *Hereditas* 140:18–23.

Elgar, M. A. 1989. Predator vigilance and group size in mammals and birds: A critical review of the empirical evidence. *Biological Review* 64:13–33.

Elgar, M. A., and B. J. Crespi, eds. 1992. *Cannibalism: Ecology and evolution among diverse taxa*. Oxford: Oxford University Press.

El-Haddad, M., J. S. Millar, and X. Xia. 1988. Offspring recognition by male *Peromyscus maniculatus*. *Journal of Mammalogy* 69:811–13.

Ellerman, J. R. 1940. *The families and genera of living rodents*, vol 1. London: Trustees of the British Musuem (Natural History).

Ellis, L. 1995. Dominance and reproductive success among nonhuman animals: A cross-species comparison. *Ethology and Sociobiology* 16:257–333.

Elton, C. 1931. The study of epidemic diseases among wild animals. *Journal of Hygiene (Cambridge)* 31:435–56.

———. 1942. *Voles, mice and lemmings*. Oxford: Clarendon.

Elton, C., D. H. S. Davis, and G. M. Findlay. 1935. An epidemic among voles (*Microtus agrestis*) on the Scottish border in the spring of 1934. *Journal of Animal Ecology* 4:277–88.

Elton, C. S. 1924. Periodic fluctuations in numbers of animals: Their causes and effects. *British Journal of Experimental Biology* 2:119–63.

Elvert, R., N. Kronfeld, T. Dayan, A. Haim, N. Zisapel, and G. Heldmeier. 1999. Telemetric field studies of body temperature and activity rhythms of *Acomys russatus* and *A. cahirinus* in the Judean Desert of Israel. *Oecologia* 199:484–92.

Elwood, R. W. 1975. Paternal and maternal behaviour in the Mongolian gerbil. *Animal Behaviour* 23:766–72.

———. 1977. Changes in the responses of male and female gerbils (*Meriones unguiculatus*) towards test pups during the pregnancy of the female. *Animal Behaviour* 25:46–51.

———. 1980. The development, inhibition and disinhibition of pup-cannibalism in the Mongolian gerbil. *Animal Behaviour* 28:1188–94.

———. 1983. Paternal care in rodents. In *Parental behaviour of rodents*, ed. R. W. Elwood, 235–57. New York: Wiley.

———. 1985. Inhibition of infanticide and onset of paternal care in male mice (*Mus musculus*). *Journal of Comparative Psychology* 99:457–67.

———. 1986. What makes male mice paternal? *Behavioral and Neural Biology* 46:54–63.

———. 1991. Ethical implications of studies on infanticide and maternal aggression in rodents. *Animal Behaviour* 42:841–49.

———. 1992. Pup-cannibalism in rodents: Cause and consequences. In *Cannibalism: Ecology and evolution among diverse taxa*, ed. M. A. Elgar and B. J. Crespi, 299–322. Oxford: Oxford University Press.

Elwood, R. W., and J. M. Broom. 1978. The influence of litter size and parental behaviour on the development of Mongolian gerbil pups. *Animal Behaviour* 26:438–54.

Elwood, R. W., and H. F. Kennedy. 1990. The relationship between infanticide and pregnancy block in mice. *Behavioral and Neural Biology* 53:277–83.

———. 1991. Selectivity in paternal and infanticidal responses by male mice: Effects of relatedness, location, and previous sexual partners. *Behavioral and Neural Biology* 56:129–47.

Elwood, R. W., A. A. Nesbitt, and H. F. Kennedy. 1990. Mater-nal aggression in response to the risk of infanticide by male mice, *Mus domesticus*. *Animal Behaviour* 40:1080–86.

Elwood, R. W., and M. C. Ostermeyer. 1984a. The effects of food deprivation, aggression, and isolation on infanticide in the male Mongolian gerbil. *Aggressive Behavior* 10:293–301.

———. 1984b. Infanticide by male and female Mongolian gerbils: Ontogeny, causation, and function. In *Infanticide: Comparative and evolutionary perspectives*, ed. G. Hausfater and S. Hrdy, 367–86. New York: Aldine.

———. 1986. Discrimination between conspecific and allospecific infants by male gerbils and mice before and after experience of their own young. *Developmental Psychobiology* 19:327–34.

Emlen, S. T. 1982. The evolution of helping. I. An ecological constraints model. *The American Naturalist* 119:29–39.

———. 1984. Cooperative breeding in birds and mammals. In *Behavioural ecology: An evolutionary approach*, ed. J. R. Krebs and N. B. Davies, 305–39. Oxford: Blackwell.

———. 1991. Evolution of cooperative breeding in birds and mammals. In *Behavioural ecology: An evolutionary approach*, 3rd ed., ed. J. R. Krebs and N. B. Davies, 301–37. Oxford: Blackwell Scientific.

———. 1994. Benefits, constraints and the evolution of the family. *Trends in Ecology and Evolution* 9:282–84.

———. 1995. An evolutionary theory of the family. *Proceedings of the National Academy of Sciences (USA)* 92:8092–99

———. 1996. Reproductive sharing in different types of kin associations. *American Naturalist* 148:756–63.

———. 1997. Predicting family dynamics in social vertebrates. In *Behavioural ecology: An evolutionary approach*, 4th ed., ed. J. R. Krebs and N. B. Davies, 228–53. Oxford: Blackwell.

Emlen, S. T., and L. W. Oring. 1977. Ecology, sexual selection and the evolution of mating systems. *Science* 197:215–23.

Emmons, L. H. 1978. Sound communication among African rainforest squirrels. *Zeitschrift für Tierpsychologie* 47:1–47.

———. 1997. *Neotropical rainforest mammals: A field guide*, 2nd ed. Chicago: University of Chicago Press.

Emry, R. A., and R. W. Thorington, Jr. 1982. Descriptive and comparative osteology of the oldest fossil, *Protosciurus* (Rodentia: Sciuridae). *Smithsonian Contributions in Paleobiology* 47:1–35.

Endler, J. A. 1986. *Natural selection in the wild*. Princeon, NJ: Princeton University Press.

Engel, S. R., K. M. Hogan, J. F. Taylor, and S. K. Davis. 1998. Molecular systematics and paleobiogeography of the South American sigmodontine rodents. *Molecular Biology and Evolution* 15:35–49.

Engeman, R. M., and D. L. Campbell. 1999. Pocket gopher reoccupation of burrow systems following population reduction. *Crop Protection* 18:523–25.

Enría, D. A., M. D. Bowen, J. N. Mills, G. J. Shieh, D. Bausch, and C. J. Peters. 1999. Arenaviruses. In *Tropical infectious diseases, principles, pathogens, and practice*, ed. R. L. Guerrant, D. H. Walker, and P. F. Weller, 1191–1212. New York: W. B. Saunders.

Enrico, P., M. Bouma, J. B. de Vries, and B. H. Westerink. 1998. The role of afferents to the ventral tegmental area in the handling of stress-induced increase in the release of dopamine in the medial prefrontal cortex: A dual-probe microdialysis study in the rat brain. *Brain Research* 779:205–13.

Ericsson, G., K. Wallin, J. P. Ball, and M. Broberg. 2001. Age-related reproductive effort and senescence in free-ranging moose, *Alces alces*. *Ecology* 82:1613–20.

Erpino, M. J. 1968. Copulatory behavior in the white-tailed prairie dog. *American Midland Naturalist* 79:250–51.

Errington, P. L. 1935. Wintering of field-living Norway rats in south-central Wisconsin. *Ecology* 16:122–23.

———. 1963. *Muskrat populations*. Aimes: The Iowa State University Press.

Erskine, M. S., and M. J. Baum. 1982. Effects of paced coital stimulation on termination of estrus and brain indoleamine levels in female rats. *Pharmacology Biochemistry and Behavior* 17:857–61.

Erskine, M. S., V. H. Denenberg, and B. D. Goldman. 1978. Aggression in the lactating rat: Effects of intruder age and test arena. *Behavioral Biology* 23:52–66.

Escobar, A., and E. Gonzalez-Jimenez. 1976. Estudio de la competencia alimenticia de los herbivoros mayores del llano inundable con especial al chiguire (*Hydrochoerus hydrochaeris*). *Agronomia Tropical* 23:215–27.

Escutenaire, S., I. Thomas, J. Clément, R. Verhagen, P. Chalon, and P.-P. Pastoret. 1997. Epidémiologie de l'hantavirose chez le campagnol roussâtre (*Clethrionomys glareolus*). *Annales de Médecine Vétérinaire* 141:471–76.

Eshelkin, I. I. 1976. On the reproduction of *Alticola strelzowi* in the Southeast Altai. [In Russian.] *Zoologicheskii Zhurnal* 55:437–42.

Eshelman, B. D., and G. N. Cameron. 1987. *Baiomys taylori*. *Mammalian Species* 285:1–7.

Eshelman, B. D., and C. S. Sonnemann. 2000. *Spermophilus armatus*. *Mammalian Species* 637:1–6.

Eskelinen, O. 1997. On the population fluctuations and structure of the wood lemming *Myopus schisticolor*. *Zeitschrift für Saugetierkunde* 62:293–302.

Espinosa, M. B., and A. D. Vitullo. 1996. Offspring sex ratio and reproductive performance in heterogametic females of the South American field mouse *Akodon azarae*. *Hereditas* 124:57–62.

Estes, R. D. 1976. The significance of breeding synchrony in the wildebeest. *East African Wildlife Journal* 14:135–52.

Estes, W. K. 1994. *Classification and cognition*. New York: Oxford University Press.

Etheredge, D. R., M. D. Engstrom, and R. C. Stone. 1989. Habitat discrimination between sympatric populations of *Peromyscus attwateri* and *Peromyscus pectoralis* in west-central Texas. *Journal of Mammalogy* 70:300–7.

European Commission. 2003. *Third report from the Commission to the Council and the European Parliament on the statistics on the number of animals used for experimental and other scientific purposes in the member states of the European Union*. Brussels: European Commission.

Evans, C. S. 1997. Referential communication. *Perspectives in Ethology* 12:99–143.

Evans, C. S., L. Evans, and P. Marler. 1993. On the meaning of alarm calls: Functional reference in an avian vocal system. *Animal Behaviour* 46:23–38.

Evans, F. C. 1942. Studies of a small mammal population in Bagley Wood, Berkshire. *Journal of Animal Ecology* 11:182–97.

Everts, H. G., and J. M. Koolhaas. 1999. Differential modulation of lateral septal vasopressin receptor blockade in spatial learning, social recognition, and anxiety-related behaviors in rats. *Behavioural Brain Research* 99:7–16.

Everts, H. G., A. J. De Ruiter, and J. M. Koolhaas. 1997. Differential lateral septal vasopressin in wild-type rats: Correlation with aggression. *Hormones and Behavior* 31:136–44.

Evsikov, V. I., G. G. Nazarova, and M. A. Potapov. 1994. Female odor choice, male social rank, and sex ratio in the water vole. *Advances in Biosciences* 93:303–7.

Ewer, R. P. 1971. The biology and behaviour of a free-living population of black rats (*Rattus rattus*). *Animal Behaviour Monographs* 4:127–74.

Fadao, T., S. Ruyong, and W. Tingzheng. 2002. Does low fecundity reflect kin recognition and inbreeding avoidance in the mandarin vole (*Microtus mandarinus*)? *Canadian Journal of Zoology* 80:2151–55.

Fadao, T., W. Tingzheng, and Z. Yajun. 2000. Inbreeding avoidance and mate choice in the mandarin vole (*Microtus mandarinus*). *Canadian Journal of Zoology* 78:2119–25.

Farentinos, R. C. 1972. Social dominance and mating activity in the tassel-eared squirrel (*Sciurus aberti ferreus*). *Animal Behaviour* 20:316–26.

———. 1974. Social communication of the tassel-eared squirrel (*Sciurus aberti*): A descriptive analysis. *Zeitschrift für Tierpsychologie* 34:441–58.

———. 1980. Sexual solicitation of subordinate males by female tassel-eared squirrels (*Sciurus aberti*). *Journal of Mammalogy* 61:337–41.

Farhang-Azad, A., and C. H. Southwick. 1979. Population ecology of Norway rats in the Baltimore Zoo and Druid Hill Park, Baltimore, Maryland. *Annals of Zoology* 15:1–42.

Faulkes, C. G. 1999. Social transmission of information in a eusocial rodent, the naked mole-rat (*Heterocephalus glaber*). In *Mammalian social learning: Comparative and ecological perspectives*, ed. H. O. Box and K. R. Gibson, 205–20. Cambridge: Cambridge University Press.

Faulkes, C. G., and D. H. Abbott. 1993. Evidence that primer pheromones do not cause social suppression of reproduction in male and female naked mole-rats, *Heterocephalus glaber*. *Journal of Reproduction and Fertility* 99:225–30.

———. 1997. Proximate mechanisms regulating a reproductive dictatorship: A single dominant female controls male and female reproduction in colonies of naked mole–rats. In *Cooperative breeding in mammals*, ed. N. G. Solomon and J. A. French, 302–34. New York: Cambridge University Press.

Faulkes, C. G., D. H. Abbott, and J. U. M. Jarvis. 1990. Social suppression of ovarian cyclicity in captive and wild colonies of naked mole-rats, *Heterocephalus glaber*. *Journal of Reproduction and Fertility* 88:559–68.

———. 1991. Social suppression of reproduction in male naked mole-rats, *Heterocephalus glaber*. *Journal of Reproduction and Fertility* 91:593–604.

Faulkes, C. G., D. H. Abbott, H. P. O'Brien, L. Lau, M. R. Roy, R. K. Wayne, and M. W. Bruford. 1997. Micro- and macrogeographic genetic structure of colonies of naked mole–rats, *Heterocephalus glaber*. *Molecular Ecology* 6:615–28.

Faulkes, C. G., and N. C. Bennett. 2001. Family values: Group dynamics and social control of reproduction in African mole-rats. *Trends in Ecology and Evolution* 16:184–90.

Faulkes, C. G., N. C. Bennett, M. W. Bruford, H. P. O'Brien, G. H. Aguilar, and J. U. M. Jarvis. 1997. Ecological constraints drive social evolution in the African mole-rats. *Proceedings of the Royal Society of London, Series B* 264:1619–27.

Faulkes, C. G., E. Verheyen, W. Verheyen, J. U. M. Jarvis, and N. C. Bennett. 2004. Phylogeographic patterns of speciation and genetic divergence in African mole–rats (Family Bathyergidae). *Molecular Ecology* 13:613–29.

Fedorka, K. M., and T. A. Mousseau. 2002. Material and genetic benefits of female multiple mating and polyandry. *Animal Behaviour* 64:361–67.

Feldhammer, G., B. Thompson, and J. Chapman. (eds.). 2003. *Wild mammals of North America: Biology, management, and conservation.* 2nd edition. Baltimore: Johns Hopkins University Press.

Felsenstein, J. 1985. Phylogenies and the comparative method. *American Naturalist* 125:1–15.

Fenn, M. G. P. 1989. The ecology of the Norway rat (*Rattus Norvegicus*) on lowland farms in England. PhD diss., University of Oxford.

Fenn, M. G. P., and D. W. Macdonald. 1995. Use of middens by red foxes: Risk reverses rhythms of rats. *Journal of Mammalogy* 76:130–36.

Fenn, M. G. P., T. E. Tew, and D. W. Macdonald. 1987. Rat movements and control on an Oxfordshire farm. *Journal of Zoology, London* 213:745–49.

Ferkin, M. H. 1987. Parental care and social interactions in captive plateau mice, *Peromyscus melanophrys. Journal of Mammalogy* 68:266–74.

———. 1999. Meadow voles (*Microtus pennsylvanicus,* Arvicolidae) over-mark and adjacent-mark the scent marks of same-sex conspecifics. *Ethology* 105:825–37.

———. 2001. Patterns of sexually distinct scents in *Microtus* spp. *Canadian Journal of Zoology* 79:1621–25.

Ferkin, M. H., J. Dunsavage, and R. E. Johnston. 1999. What kind of information do meadow voles (*Microtus pennsylvanicus*) use to distinguish between the top and bottom scent of an over-mark? *Journal of Comparative Psychology* 113:43–51.

Ferkin, M. H., S. T. Leonard, K. Bartos, and M. K. Schmick. 2001. Meadow voles and prairie voles differ in the length of time they prefer the top-scent donor of an over-mark. *Ethology* 107:1099–1114.

Ferkin, M. H., S. T. Leonard, L. A. Heath, and G. Paz-y-Mino C. 2001. Self-grooming as a tactic used by prairie voles *Microtus ochrogaster* to enhance sexual communication. *Ethology* 107:939–49.

Ferkin, M. H., and R. E. Johnston. 1995. Meadow voles, *Microtus pennsylvanicus,* use multiple sources of scent for sex recognition. *Animal Behaviour* 49:37–44.

Ferkin, M. H., and T. F. Rutka. 1990. Mechanisms of sibling recognition in meadow voles. *Canadian Journal of Zoology* 68:609–13.

Ferkin, M. H., E. S. Sorokin, and R. E. Johnston. 1996. Self-grooming as a sexually dimorphic communicative behaviour in meadow voles, *Microtus pennsylvanicus. Animal Behaviour* 51:801–10.

Ferkin, M. H., E. S. Sorokin, R. E. Johnston, and C. J. Lee. 1997. Attractiveness of scents varies with protein content of the diet in meadow voles. *Animal Behaviour* 53:133–41.

Ferrara, J. 1985. Prairie home companions. *National Wildlife* 23:48–53.

Ferris, C. F., and Y. Delville. 1994. Vasopressin and serotonin interactions in the control of agonistic behavior. *Psychoneuroendocrinology* 19:593–601.

Ferris, C. F., Y. Delville, S. Bonigut, and M. A. Miller. 1999. Galanin antagonizes vasopressin-stimulated flank marking in male golden hamsters. *Brain Research* 832:1–6.

Ferris, C. F., R. H. Melloni, Jr., G. Koppel, K. W. Perry, R. W.

Fuller, and Y. Delville. 1997. Vasopressin/serotonin interactions in the anterior hypothalamus control aggressive behavior in golden hamsters. *Journal of Neuroscience* 17:4331–40.

Ferron, J. 1984. Behavioral ontogeny analysis of sciurid rodents, with emphasis on the social behavior of ground squirrels. In *The biology of ground-dwelling squirrels,* ed. J. O. Murie and G. R. Michener, 24–42. Lincoln: University of Nebraska Press.

———. 1996. How do woodchucks (*Marmota monax*) cope with harsh winter conditions? *Journal of Mammalogy* 77:412–16.

Ferron, J., and J.-P. Ouellet. 1989a. Behavioural context and possible function of scent marking by cheek rubbing in the red squirrel (*Tamiasciurus hudsonicus*). *Canadian Journal of Zoology* 67:1650–53.

———. 1989b. Temporal and intersexual variation in the use of space with regard to social organization in the woodchuck (*Marmota monax*). *Canadian Journal of Zoology* 67:1642–49.

Festa-Bianchet, M. 1996. Offspring sex ratio studies of mammals: Does publication depend upon the quality of the research or the direction of the results? *Ecoscience* 3:42–44.

Festing, M, P. Overend, R. Gaine Das, M. Cortina Borja, and M. Berdoy. 2002. *The design of animal experiments: Reducing the use of animals in research through better experimental design.* London: Royal Society of Medicine Press.

Figala, J. 1963. Yellow mutant in the brown rat, *Rattus Norvegicus* (Berk.). *Acta Societatis Zoological Bohemoslovenicae* 28:209–10.

Figueroa, F., E. Neufeld, U. Ritte, and J. Klein. 1988. *t*-specific DNA polymorphisms among wild mice from Israel and Spain. *Genetics* 119:157–60.

Finch, C. E. 1990. *Longevity, senescence, and the genome.* Chicago: University of Chicago Press.

Fischer, R. B., and P. S. Brown. 1993. Vaginal secretions increase the likelihood of intermale aggression in Syrian hamsters. *Physiology and Behavior* 54:213–14.

Fischer, R. B., and J. McQuiston. 1991. A possible role for Syrian hamster, *Mesocricetus auratus,* vaginal secretion in inter-female competition. *Animal Behaviour* 42:949–54.

Fisher, R. A. 1930. *The genetical theory of natural selection.* Oxford: Oxford University Press.

Fisler, G. F. 1963. Adaptations and speciation in harvest mice of the San Francisco Bay region. *Journal of Mammalogy* 48:549–56.

Fitch, H. S. 1948a. Ecology of the California ground squirrel on grazing lands. *American Midland Naturalist* 39:513–96.

———. 1948b. Habits and economic relationships of the Tulare kangaroo rat. *Journal of Mammalogy* 29:5–35.

———. 1949. Study of snake populations in central California. *American Midland Naturalist* 41:513–79.

———. 1958. Home ranges, territories, and seasonal movements of vertebrates of the natural history reservation. *University of Kansas Publications, Museum of Natural History* 11:63–326.

Fitch, H. S., and D. G. Rainey. 1956. Ecological observations on the woodrat, *Neotoma floridana. University of Kansas Publications, Museum of Natural History* 8:499–533.

Fitzgerald, J. P., and R. R. Lechleitner. 1974. Observations on the biology of Gunnison's prairie dog in central Colorado. *American Midland Naturalist* 92:146–63.

Fitzgerald, R. W., and D. M. Madison. 1983. Social organization of a free-ranging population of pine voles, *Microtus pinetorum. Behavioral Ecology and Sociobiology* 13:183–87.

Fitzgerald, V. J., and J. O. Wolff. 1988. Behavioral responses of escaping *Peromyscus* to wet and dry substrates. *Journal of Mammalogy* 69:825–28.

Flandera, V., and V. Novakova. 1974. Effect of the mother on the development of aggressive behavior in rats. *Developmental Psychobiology* 8:49–54.

Flannelly, K. J., and L. Flannelly. 1985. Opponents' size influences maternal aggression. *Psychological Reports* 57:883–86.

Flaxman, S. M. 2000. The evolutionary stability of mixed strategies. *Trends in Ecology and Evolution* 15:482–84.

Flegr, J., J. Havlíček, P. Kodym, M. Mal'y, and Z. Šmahel. 2002. Increased risk of traffic accidents in subjects with latent toxoplasmosis: A retrospective case-control study. *BMC Infectious Diseases* 2:11:1–13.

Fleming, A. S. 1979. Maternal nest defense in the desert woodrat *Neotoma lepida lepida*. *Behavioral and Neural Biology* 26:41–63.

Fleming, A. S., C. Kuchera, A. Lee, and G. Winocur. 1994. Olfactory-based social learning varies as a function of parity in female rats. *Psychobiology* 22:37–43.

Fleming, P. A., and S. W. Nicolson. 2004. Sex differences in space use, body condition and survivorship during the breeding season in the Namaqua rock mouse, *Aethomys namaquensis*. *African Zoology* 39:123–32.

Flower, D. R. 1996. The lipocalin protein family: Structure and function. *Biochemical Journal* 318:1–14.

Flynn, L. J., L. L. Jacobs, and E. H. Lindsay. 1985. Problems in muroid phylogeny: Relationship to other rodents and origin of major groups. In *Evolutionary relationships among rodents: A multidisciplinary analysis,* ed. W. P. Luckett and J. L. Hartenberger, 589–616. New York: Plenum.

Flynn, L. J., E. Nevo, and G. Heth. 1987. Incisor enamel microstructure in subterranean mole rats (*Spalax ehrenbergi* superspecies): Adaptive and phylogenetic significance. *Journal of Mammalogy* 68:500–507.

Fokin, I. M. 1978. *Locomotion and morphology of locomotor apparatus of jerboas.* [In Russian.] Leningrad: Nauka.

Foltz, D. W. 1981. Genetic evidence for long-term monogamy in a small rodent, *Peromyscus polionotus*. *The American Naturalist* 117:665–75.

Foltz, D. W., and J. L. Hoogland. 1981. Analysis of the mating system in the black-tailed prairie dog (*Cynomys ludovicianus*). *Journal of Mammalogy* 64:706–12.

———. 1983. Genetic evidence of outbreeding in the black-tailed prairie dog (*Cynomys ludovicianus*). *Evolution* 37:273–81.

Foltz, D. W., and P. L. Schwagmeyer. 1989. Sperm competition in the thirteen-lined ground squirrel: Differential fertilization success under field conditions. *The American Naturalist* 133:257–65.

Foote, M., J. P. Hunter, C. M. Janis, and J. J. Sepkoski, Jr. 1999. Evolutionary and preservational constraints on origins of biologic groups: Divergence times of eutherian mammals. *Science* 283:1310–14.

Forchhammer, M. C. 2000. Timing of foetal growth spurts can explain sex ratio variation in polygynous mammals. *Ecology Letters* 3:1–4.

Forest Guardians. 2004. Petition to the U.S. Fish and Wildlife Service to list the Gunnison's prairie dog as an endangered or threatened species under the Endangered Species Act, 16 U. S. C & 1531 et Seq. (1973 as amended), and to designate critical habitat. Accessed from www.fguardians.org.

Forkman, B. 1991. Social facilitation is shown by gerbils when presented with novel but not with familiar food. *Animal Behaviour* 42:860–61.

Forman, R. T. T. 1995. *Land mosaics: The ecology of landscapes and regions.* Cambridge: University of Cambridge Press.

Forman, R. T. T., and M. Godron. 1986. *Landscape ecology.* New York: Wiley.

Forrest, S. C., and J. C. Luchsinger. 2006. Past and current chemical control of prairie dogs. In *Conservation of the black-tailed prairie dog,* ed. J. L. Hoogland, 115–28. Washington, DC: Island Press.

Fortier, G. M., and R. H. Tamarin. 1998. Movement of meadow voles in response to food and density manipulations: A test of the food-defense and pup-defense hypotheses. *Journal of Mammalogy* 79:337–45.

Fossey, D. 1984. Infanticide in mountain gorillas (*Gorilla gorilla beringei*) with comparative notes on chimpanzees. In *Infanticide: Comparative and evolutionary perspectives,* ed. G. Hausfater and S. B. Hrdy, 217–35. New York: Aldine.

Fowler, C. D., Y. Liu, C. Ouimet, and Z. X. Wang. 2002. The effects of social environment on adult neurogenesis in the female prairie vole. *Journal of Neurobiology* 51:115–28.

Fox, M. W., ed. 1968. *Abnormal behavior in animals.* Philadelphia, PA: Saunders.

Frame, L., J. R. Malcolm, G. W. Frame, and H. van Lawick. 1979. Social organization of African wild dogs (*Lycaon pictus*) on the Serengeti Plains, Tanzania, 1967–78. *Zeitschrift für Tierpsychologie* 50:225–49.

Francescoli, G. 2000. Sensory capabilities and communication in subterranean rodents. In *Life underground: The biology of subterranean rodents,* ed. E. A. Lacey, J. L. Patton, and G. N. Cameron, 111–44. Chicago: University of Chicago Press.

Francis, D. D., and M. J. Meaney. 1999. Maternal care and the development of stress responses. *Current Opinion in Neurobiology* 9:128–34.

Frank, D. H. 1989. Spatial organization, social behavior, and mating strategies of the southern grasshopper mouse (*Onychomys torridus*) in southeastern Arizona. PhD diss., Cornell University, Ithaca, NY.

Frank, D. H., and E. J. Heske. 1992. Seasonal changes in space use patterns in the southern grasshopper mouse *Onychomys torridus torridus*. *Journal of Mammalogy* 73:292–98.

Frank, S. A. 1990. Sex allocation theory for birds and mammals. *Annual Review of Ecology and Systematics* 21:13–55.

Frankham, R. 1995. Effective population size/adult population size ratios in wildlife: A review. *Genetical Research* 66:95–107.

Franks, P., and S. Lenington. 1986. Dominance and reproductive behavior of wild house mice in a seminatural environment correlated with T-locus genotype. *Behavioral Ecology and Sociobiology* 18:395–404.

Frase, B. A., and K. B. Armitage. 1989. Yellow-bellied marmots are generalist herbivores. *Ethology, Ecology and Evolution* 1:353–66.

Frase, B. A., and R. S. Hoffmann. 1980. *Marmota flaviventris*. *Mammalian Species* 135:1–8.

Fredericksen, N. J., T. S. Fredericksen, B. Flores, E. McDonald, and D. Rumiz. 2003. Importance of granitic rock outcrops to vertebrate species in a Bolivian tropical forest. *Tropical Ecology* 44:183–94.

Fredga, K. 1988. Aberrant chromosomal sex-determining mechanism in mammals, with special reference to species with XY females. *Proceedings of the Royal Society of London, Series B* 322:83–95.

Freeland, W. J., J. W. Winter, and S. Raskin. 1988. Australian rock-mammals: A phenomenon of the seasonally dry tropics. *Biotropica* 20:70–79.

French, A. R. 1986. Patterns of thermoregulation during hibernation. In *Living in the cold: Physiological and biochemical adaptations,* ed. H. C. Heller, X. J. Musacchia, and L. C. H. Wang, 393–402. New York: Elsevier.

———. 1993. Physiological ecology of the heteromyidae: Economics of energy and water utilization. In *Biology of the heteromyidae,* ed. H. H. Genoways and J. H. Brown, 509–38. Shippensburg, PA: American Society of Mammalogists.

French, J. A. 1994. Alloparents in the Mongolian gerbil: Impact on long-term reproductive performance of breeders and opportunities for independent reproduction. *Behavioral Ecology* 5:273–79.

Fried, J. J. 1987. The role of juvenile pine voles (*Microtus pinetorum*) in the caretaking of their younger siblings. Master's thesis, North Carolina State University, Raleigh.

Fuchs, S. 1982. Optimality of parental investment: The influence of nursing on reproductive success of mother and female young house mice. *Behavioral Ecology and Sociobiology* 10:39–51.

Fuelling, O., and S. Halle. 2004. Breeding suppression in free-ranging grey-sided voles under the influence of predator odour. *Oecologia* 138:151–59.

Fuemm, H., and P. Driscoll. 1981. Litter size manipulations do not alter maternal behaviour traits in selected lines of rats. *Animal Behaviour* 29:1267–69.

Fulk, G. W. 1976. Notes on the activity, reproduction, and social behavior of *Octodon degus. Journal of Mammalogy* 57:495–505.

Gabathuler, U., N. C. Bennett, and J. U. M. Jarvis. 1996. The social structure and dominance hierarchy of the Mashona mole-rat, *Cryptomys darlingi* (Rodentia: Bathyergidae) from Zimbabwe. *Journal of Zoology, London* 240:221–31.

Gaillard, J.-M., M. Festa-Bianchet, N. G. Yoccoz, A. Loison, and C. Toïgo. 2000. Temporal variation in fitness components and population dynamics of large herbivores. *Annual Review of Ecology and Systematics* 31:367–93.

Gaillard, J.-M., D. Pontier, D. Allainé, J. D. Lebreton, J. Trouvilliez, and J. Clobert. 1989. An analysis of demographic tactics in birds and mammals. *Oikos* 56:59–76.

Gaines, M. S., and L. R. McClenaghan, Jr. 1980. Dispersal in small mammals. *Annual Review of Ecology and Systematics* 11:163–96.

Galea, L. A. M., and B. S. McEwen. 1999. Sex and seasonal differences in the rate of cell proliferation in the dentate gyrus of adult wild meadow voles. *Neuroscience* 89:955–64.

Galef, B. G., Jr. 1977. Mechanisms for the social transmission of food preferences from adult to weanling rats. In *Learning mechanisms in food selection,* ed. L. M. Barker, M. Best, and M. Domjan, 123–48. Waco, TX: Baylor University Press.

———. 1980. Diving for food: Analysis of a possible case of social learning in wild rats (*Rattus norvegicus*). *Journal of Comparative and Physiological Psychology* 94:416–25.

———. 1981a. Development of olfactory control of feeding-site selection in rat pups. *Journal of Comparative and Physiological Psychology* 95:615–22.

———. 1981b. The ecology of weaning: Parasitism and the achievement of independence by altricial mammals. In *Paren-

tal care in mammals,* ed. D. J. Gubernick and P. H. Klopfer, 211–41. New York: Plenum.

———. 1984. Reciprocal heuristics: A discussion of the study of learned behavior in laboratory and field. *Learning and Motivation* 15:479–93.

———. 1986. Social interaction modifies learned aversions, sodium appetite, and both palatability and handling-time induced dietary preference in rats (*Rattus norvegicus*). *Journal of Comparative Psychology* 100:432–39.

———. 1988a. Communication of information concerning distant diets in a social, central-place foraging species (*Rattus norvegicus*). In *Social learning: Psychological and biological perspectives,* ed. T. R. Zentall and B. G. Galef, Jr., 119–40. Hillsdale, NJ: Lawrence Erlbaum.

———. 1988b. Imitation in animals: History, definition and interpretation of data from the psychological laboratory. In *Social learning: Psychological and biological perspectives,* ed. T. R. Zentall and B. G. Galef, Jr., 3–28. Hillsdale, NJ: Lawrence Erlbaum.

———. 1989. Enduring enhancement of rats' preferences for the palatable and the piquant. *Appetite* 13:81–92

———. 1991. A contrarian view of the wisdom of the body as it relates to food selection. *Psychological Review* 98:218–24.

———. 1996a. Social enhancement of food preferences in Norway rats. In *Social learning and imitation: The roots of culture,* ed. C. M. Heyes and B. G. Galef, Jr., 49–64. New York: Academic Press.

———. 1996b. Social influences on food preferences and feeding behaviors of vertebrates. In *Why we eat what we eat,* ed. E. Capaldi, 207–32. Washington, DC: American Psychological Association.

———. 1996c. Tradition in animals: Field observations and laboratory analyses. In *Readings in animal cognition,* ed. M. Bekoff and D. Jamieson, 91–106. Cambridge: MIT Press.

———. 2004. Methods of studying traditional behaviors of free-living animals. *Learning and Behavior* 32:53–61.

Galef, B. G., Jr., and C. Allen. 1995. A new model system for studying animal traditions. *Animal Behaviour* 50:705–17.

Galef, B. G., Jr., K. S. Attenborough, and E. E. Whiskin. 1990. Responses of observer rats to complex, diet-related signals emitted by demonstrator rats. *Journal of Comparative Psychology* 104:11–19.

Galef, B. G., Jr., and M. Beck. 1985. Aversive and attractive marking of toxic and safe foods by Norway rats. *Behavioral and Neural Biology* 43:298–310.

Galef, B. G., Jr., M. Beck, and E. E. Whiskin. 1991. Effects of protein deficiency on susceptibility of Norway rats to social influences on food choice. *Journal of Comparative Psychology* 105:55–59.

Galef, B. G., Jr., and L. L. Buckley. 1996. Use of foraging trails by Norway rats. *Animal Behaviour* 51:765–71.

Galef, B. G., Jr., and M. M. Clark. 1971a. Parent-offspring interactions determine time and place of first ingestion of solid food by wild rat pups. *Psychonomic Science* 25:15–16.

———. 1971b. Social factors in the poison avoidance and feeding behavior of wild and domesticated rat pups. *Journal of Comparative and Physiological Psychology* 25:341–57.

Galef, B. G., Jr., and Giraldeau, L.-A. 2002. Social influences on foraging in vertebrates: Causal mechanisms and adaptive functions. *Animal Behaviour* 61:3–15.

Galef, B. G., Jr., and L. Heiber. 1976. The role of residual olfactory cues in the determination of feeding site selection and exploration patterns of domestic rats. *Journal of Comparative and Physiological Psychology* 90:727–39.

Galef, B. G., Jr., and P. W. Henderson. 1972. Mother's milk: A determinant of food preferences of weaning rats. *Journal of Comparative and Physiological Psychology* 78:213–19.

Galef, B. G., Jr., and C. P. Iliffe.1994. Social enhancement of food odors in rats: Is there something special about odors associated with foods. *Journal of Comparative Psychology* 108:266–73.

Galef, B. G., Jr., D. J. Kennett, and M. Stein. 1985. Demonstrator influence on observer diet preference: Effects of simple exposure and presence of a demonstrator. *Animal Learning and Behavior* 13:25-30.

Galef, B. G., Jr., D. J. Kennett, and S. W. Wigmore. 1984. Transfer of information concerning distant foods in rats: A robust phenomenon. *Animal Learning and Behavior* 12:292–96.

Galef, B. G., Jr., J. R. Mason, G. Preti, and N. J. Bean. 1988. Carbon disulfide: A semiochemical mediating socially-induced diet choice in rats. *Physiology and Behaviour* 42:119–24.

Galef, B. G., Jr., L. M. McQuoid, and E. E. Whiskin. 1990. Further evidence that Norway rats do not socially transmit learned aversions to toxic baits. *Animal Learning and Behavior* 18:199–205.

Galef, B. G., Jr., B. Rudolf, E. E. Whiskin, E. Choleris, M. Mainardi, and P. Valsecchi. 1998. Familiarity and relatedness: effects on social learning about foods by Norway rats and Mongolian gerbils. *Animal Learning and Behavior* 26:448–54.

Galef, B. G., Jr., and D. F. Sherry. 1973. Mother's milk: A medium for transmission of information about mother's diet. *Journal of Comparative and Physiological Psychology* 83:374–78.

Galef, B. G., Jr., and M. Stein. 1985. Demonstrator influence on observer diet preference: Analyses of critical social interactions and olfactory signals. *Animal Learning and Behavior* 13:31–38.

Galef, B. G., Jr., and E. E. Whiskin. 1997. Effects of asocial learning on longevity of food-preference traditions. *Animal Behaviour* 53:1313–22.

———. 1998. Determinants of the longevity of socially learned food preferences of Norway rats. *Animal Behaviour* 55:967–75.

———. 2003. Socially transmitted food preferences can be used to study long-term memory in rats. *Learning and Behavior* 31:160–64.

Galef, B. G., Jr., and S. W. Wigmore. 1983. Transfer of information concerning distant foods: A laboratory investigation of the "information-centre" hypothesis. *Animal Behaviour* 31:748–58.

Galef, B. G., Jr., S. W. Wigmore, and D. J. Kennett. 1983. A failure to find socially mediated taste-aversion learning in Norway rats (*Rattus norvegicus*). *Journal of Comparative Psychology* 97:358–63.

Galindo, C., and C. J. Krebs. 1987. Population regulation in deer mice: The role of females. *Journal of Animal Ecology* 56:11–23.

Galindo-Leal, C., and C. J. Krebs. 1998. Effects of food abundance on individuals and populations of the rock mouse (*Peromyscus difficilis*). *Journal of Mammalogy* 79:1131–42.

Gallardo, M. H., and J. A. Anrique. 1991. Population parame-

ters and burrow systems in *Ctenomys maulinus brunneus* (Rodentia, Ctenomyidae). *Medio Ambiente* 11:48–53.

Gallardo, M. H., and J. A. W. Kirsch. 2001. Molecular relationships among Octodontidae (Mammalia: Rodentia: Caviomorpha). *Journal of Mammalian Evolution* 8:73–89.

Galtier, N., F. Bonhomme, C. Moulia, K. Belkhir, P. Caminade, E. Desmarais, J. J. Duquesne, A. Orth, B. Dod, and P. Boursot. 2004. Mouse biodiversity in the genomic era. *Cytogenetics and Genome Research* 105:385–94.

Gamboa, G. J., H. K. Reeve, and W. G. Holmes. 1991. Conceptual issues and methodology in kin-recognition research: A critical discussion. *Ethology* 88:109–27.

Gamboa, G. J., H. K. Reeve, and D. W. Pfennig. 1986. The evolution and ontogeny of nestmate recognition in social wasps. *Annual Review of Entomology* 31:431–54.

Gammie, S. C., and R. J. Nelson. 2000. Maternal and mating-induced aggression is associated with elevated citrulline immunoreactivity in the paraventricular nucleus in prairie voles. *Journal of Comparative Neurology* 418:182–92.

———. 2001. cFOS and pCREB activation and maternal aggression in mice. *Brain Research* 898:232–41.

Gandelman, R. 1972. Induction of pup killing in female mice by androgenization. *Physiology and Behavior* 9:101–2.

Gandolfi, G., and V. Parisi. 1972. Predazione su Unio pictorum L. da parte del ratto, *Rattus norvegicus* (Berkenhout). *Acta Naturalia* 8:1–27.

———. 1973. Ethological aspects of predation by rats, *Rattus norvegicus* (Berkenhout) on bivalves, *Unio pictorum*, L. and *Cerastoderma lamarcki* (Reeve). *Bullettino di Zoologia* 40:69–74.

Ganem, G., and E. Nevo. 1996. Ecophysiological constraints associated with aggression and evolution toward pacifism in *Spalax ehrenbergi*. *Behavioral Ecology and Sociobiology* 38:245–52.

Gannon, W. L., and T. E. Lawlor. 1989. Variation of the chip vocalization of three species of Townsend chipmunks (Genus *Eutamias*). *Journal of Mammalogy* 70:740–53.

Garcia, J., F. R. Ervin, and R. A. Koelling. 1966. Learning with prolonged delay of reinforcement. *Psychonomic Science* 5:121–22.

Garcia, J., W. G. Hankins, and K. W. Rusiniak. 1974. Behavioral regulation of the internal milieu in man and rat. *Science* 185:824–31.

Garcia, J., and R. A. Koelling. 1966. Relation of cue to consequence in avoidance learning. *Psychonomic Science* 4:123–24.

Garrett, M. G., and W. L. Franklin. 1988. Behavioral ecology of dispersal in the black-tailed prairie dog. *Journal of Mammalogy* 69:236–50.

Garrett, M. G., J. L. Hoogland, and W. L. Franklin. 1982. Demographic differences between an old and a new colony of black-tailed prairie dogs (*Cynomys ludovicianus*). *American Midland Naturalist* 108:51–59.

Gaulin, S. J. C., and R. W. Fitzgerald. 1989. Sexual selection for spatial-learning ability. *Animal Behaviour* 37:322–31.

Gavin, T. A., P. W. Sherman, E. Yensen, and B. May. 1999. Population genetic structure of the northern Idaho ground squirrel (*Spermophilus brunneus brunneus*). *Journal of Mammalogy* 80:156–68.

Gavish, L., C. S. Carter, and L. Getz. 1983. Male-female interactions in prairie voles. *Animal Behaviour* 31:511–17.

Gavish, L., J. E. Hofmann, and L. L. Getz. 1984. Sibling recognition in the prairie vole, *Microtus ochrogaster*. *Animal Behaviour* 32:362–66.

Gaylard, A., Y. Harrison, and N. C. Bennett. 1998. Temporal changes in the social structure of a captive colony of the Damaraland mole-rat, *Cryptomys damarensis*: The relationship of sex and age to dominance and burrow-maintenance activity. *Journal of Zoology, London* 244:313–21.

Geist, V. 1964. On the rutting behaviour of the mountain goat. *Journal of Mammalogy* 45:551–68.

Genelly, R. E. 1965. Ecology of the common mole-rat (*Cryptomys hottentotus*) in Rhodesia. *Journal of Mammalogy* 46:647–65.

Genoways, H. H. 1973. *Systematics and evolutionary relationships of spiny pocket mice, Genus* Liomys. Lubbock: Texas Tech University Press.

Genoways, H. H., and J. H. Brown 1993. *Biology of the heteromyidae*. Shippensburg, PA: American Society of Mammalogists.

Gentile, R., P. S. D. A. Andrea, and R. Cerqueira. 1997. Home ranges of *Philander frenata* and *Akodon cursor* in a Brazilian restinga (coastal shrubland). *Mastozoologia Neotropical* 4:105–12.

Gentry, J., and M. Smith. 1968. Food habits and burrow associates of *Peromyscus polionotus*. *Journal of Mammalogy* 49:562–65.

George, W. 1974. Notes on the ecology of gundis (F. Ctenodactylidae). *Symposium of the Zoological Society of London* 34:143–60.

———. 1978. Reproduction in female gundis (Rodentia: Ctenodactylidae). *Journal of Zoology, London* 185:57–71.

———. 1981. Species-typical calls in the Ctenodactylidae (Rodentia). *Journal of Zoology* 195:39–52.

———. 1986. The thermal niche: Desert sand and desert rock. *Journal of Arid Environments* 10:213–24.

George, W., and G. Crowther. 1981. Space partitioning between two small mammals in a rocky desert. *Biological Journal of the Linnean Society* 15:195–200.

Georges, F., and G. Aston-Jones. 2002. Activation of ventral tegmental area cells by the bed nucleus of the stria terminalis: A novel excitatory amino acid input to midbrain dopamine neurons. *Journal of Neuroscience* 22:5173–87.

Gerhardt, H. C., and F. Huber. 2002. *Acoustic communication in insects and anurans*. Chicago: University of Chicago Press.

Gerlach, G., and H. N. Hoeck. 2001. Islands on the plains: Metapopulation dynamics and female biased dispersal in hyraxes (Hyracoidea) in the Serengeti National Park. *Molecular Ecology* 10:2307–17.

Getz, L. L., C. M. Larson, and K. A. Lindstrom. 1992. *Blarina brevicauda* as a predator on nestling voles. *Journal of Mammalogy* 73:591–96.

Getz, L. L., and B. McGuire. 1993. A comparison of living singly and in male-female pairs in the prairie vole, *Microtus ochrogaster*. *Ethology* 94:265–78.

Getz, L. L., B. McGuire, and T. Pizzuto. 2004. Does mate choice take place in free-living prairie voles, *Microtus ochrogaster*? Evidence from field data. *Acta Zoologica Sinica* 50:527–34.

Getz, L. L., B. McGuire, T. Pizzuto, J. E. Hofmann, and B. Frase, B. 1993. Social organization of the prairie vole (*Microtus ochrogaster*). *Journal of Mammalogy* 74:44–58.

Getz, W. M. 1981. Genetically based kin recognition systems. *Journal of Theoretical Biology* 92:209–26.

Getz, W. M., and J. Pickering. 1983. Epidemic models: Thresholds and population regulation. *American Naturalist* 121:892–98.

Gibber, J. R., Y. Piontkewitz, and J. Terkel. 1984. Response of male and female Siberian hamsters towards pups. *Behavioral and Neural Biology* 42:177–82.

Giboulet, O., R. Ramousse, and M. Le Berre. 2002. Evolution of life history traits and molecular phylogenies: Sociality in ground dwelling squirrels as an example. In *Holarctic marmots as a factor of biodiversity*, ed. K. B. Armitage and V. Yu. Rumiantsev, 171–75. Moscow: ABF.

Gibson, J. J. 1979 [1986]. *The ecological approach to visual perception*. Hillsdale, NJ: Lawrence Erlbaum.

Gilbert, A. N. 1984. Postpartum and lactational estrus: A comparative analysis in Rodentia. *Journal of Comparative Psychology* 98:232–45.

———. 1986. Mammary number and litter size in Rodentia: The "one-half" rule. *Proceedings of the National Academy of Sciences* 83:4828–30.

Gilbert, A. N., D. A. Burgoon, K. A. Sullivan, and N. T. Adler. 1983. Mother-weanling interactions in Norway rats in the presence of a successive litter produced by postpartum mating. *Physiology and Behavior* 30:267–71.

Gilbert, A. N., R. J. Pelchat, and N. T. Adler. 1984. Sexual and maternal behaviour at the postpartum oestrus: The role of experience in time-sharing. *Animal Behaviour* 32:1045–53.

Gilbert, B. S., C. J. Krebs, D. Talarico, and D. B. Cichowski. 1986. Do *Clethrionomys rutilus* females suppress maturation of juvenile females? *Journal of Animal Ecology* 55:543–52.

Gilbert, S. F. 2001. Ecological developmental biology: Developmental biology meets the real world. *Developmental Biology* 233:1–12.

Gilder, P. M., and P. J. B. Slater. 1978. Interest of mice in conspecific male odours is influenced by degree of kinship. *Nature* 274:364–65.

Gileva, E. A. 1998. Inbreeding and sex ratio in two captive colonies of *Dicrostonyx torquatus* Pall., 1779: A reply to Jarrell, G. H., *Hereditas* 128:185–88.

Gileva, E. A., I. E. Benenson, L. A. Konopistseva, V. F. Puchkov, and I. A. Makarenets. 1982. XO females in the varying wood lemming, *Dicrostonyx torquatus*: Reproductive performance and its evolutionary significance. *Evolution* 36:601–9.

Gileva, E. A., and V. B. Federov. 1991. Sex ratio, XY females and absence of inbreeding in a population of the wood lemming *Myopus schisticolor* Lilleborg, 1844. *Heredity* 66:351–55.

Gillis E. A. 2003. Breeding dispersal, male mating tactics, and population dynamics of arctic ground squirrels. PhD diss., University of British Columbia, Vancouver.

Gillis, E. A., and C. J. Krebs. 1999. Natal dispersal of snowshoe hares during a cyclic population increase. *Journal of Mammalogy* 80:933–39.

Gingrich, B., Y. Liu, C. Cascio, Z. X. Wang, and T. R. Insel. 2000. Dopamine D2 receptors in the nucleus accumbens are important for social attachment in female prairie voles (*Microtus ochrogaster*). *Behavioral Neuroscience* 114:173–83.

Giordano, A. L., H. I. Siegel, and J. S. Rosenblatt. 1984. Effects of mother-litter separation and reunion on maternal aggression and pup mortality in lactating hamsters. *Physiology and Behavior* 33:903–6.

Gipps, J. H. W., M. J. Taitt, C. J. Krebs, and Z. Dundjerski. 1981. Male aggression and the population dynamics of the vole, *Microtus townsendii*. *Canadian Journal of Zoology* 59:147–57.

Gittleman, J., and P. Harvey. 1982. Carnivore home-range size needs and ecology. *Behavioral Ecology and Sociobiology* 10:57–63.

Gittleman, J. L., and S. D. Thompson. 1988. Energy allocation in mammalian reproduction. *American Zoologist* 28:863–75.

Glass, G. E., J. E. Childs, G. W. Korch, and J. W. LeDuc. 1988. Association of intraspecific wounding with hantaviral infection in wild rats (*Rattus norvegicus*). *Epidemiology and Infection* 101:459–72.

Glazier, D. S. 1980. Ecological shifts and the evolution of geographically restricted species of North American *Peromyscus* (mice). *Journal of Biogeography* 7:63–83.

Glendinning, J., and L. Brower. 1990. Feeding and breeding responses of five mice species to overwintering aggregations of the monarch butterfly. *Journal of Animal Ecology* 59:1091–112.

Godfrey, G., and P. Crowcroft. 1960. *Life of the Mole* (Talpa europaea L.). 1–152. London: Museum Press.

Goel, N., and T. M. Lee. 1996. Relationship of circadian activity and social behaviors in reentrainment rates in diurnal Octodon degus (Rodentia). *Physiology and Behavior* 59:817–26.

Goldman, L., and H. Swanson. 1975. Population control in confined colonies of golden hamsters (*Mesocricetus auratus* Waterhouse). *Zeitschrift für Tierpsychologie* 37:225–36.

Goltsman, M. E., and N. G. Borisova. 1993. Comparative analysis of interaction duration in three species of gerbils (Genus *Meriones*). *Ethology* 94:177–86.

Goltsman, M. E., N. P. Naumov, A. A. Nikolsky, N. G. Ovsianikov, N. M. Paskhina, and V. M. Smirin. 1977. Social behavior of a great gerbil (*Rhombomys opimus*). [In Russian.] In *Mammalian behavior*, ed. V. E. Sokolov, 5–69. Moscow: Nauka.

Goltsman, M. E., S. V. Popov, A. V. Tchabovsky, and N. G. Borisova. 1994. Syndrome of sociality: Comparative study of gerbils' behaviour. [In Russian.] *Zhurnal Obschei Biologii* 55:49–69.

Gomendio, M., and E. R. S. Roldan. 1991. Sperm competition influences sperm size in mammals. *Proceedings of the Royal Society of London, Series B,* 243:181–85.

Goodrum, P. D. 1961. *The gray squirrel in Texas.* Austin: Texas Wildlife and Parks.

Goodwin, H. T. 1989. *Marmota flaviventris* from the Central Mohave Desert of California: Biogeographic implications. *Southwestern Naturalist* 34:284–87.

———. 1995. Pliocene-pleistocene biogeographic history of prairie dogs genus *Cynomys* (Sciuridae). *Journal of Mammalogy* 76:100–122.

Goossens, B., J. Coulon, D. Allainé, L. Graziani, M.-C. Bel, and P. Taberlet. 1996. Immigration of a pregnant female in an alpine marmot family group: behavioural and genetic data. *Comptes Rendus de l'Académie des Sciences Paris, Sciences de la vie* 319:241–46.

Goossens, B., L. Graziani, L. P. Waits, E. Farand, S. Magnolon, J. Coulon, M. Bel, P. Taberlet, and D. Allaine. 1998. Extrapair paternity in the monogamous Alpine marmot revealed by nuclear DNA microsatellite analysis. *Behavioral Ecology and Sociobiology* 43:281–88.

Gosling, L. M. 1980. The duration of lactation in feral coypus (*Myocastor coypus*). Journal of Zoology (London) 191:461–74.

———. 1981. Demarkation in a gerenuk territory: An economic approach. *Zeitschrift für Tierpsychologie* 56:305–22.

———. 1982. A reassessment of the function of scent marking in territories. *Zeitschrift für Tierpsychologie* 60:89–118.

———. 1986. Selective abortion of entire litters in the coypu: Adaptive control of offspring production in relation to quality and sex. *American Naturalist* 127:772–95.

———. 1990. Scent marking by resource holders: Alternative mechanisms for advertising the costs of competition. In *Chemical signals in vertebrates* V, ed. D. W. Macdonald, S. Natynczuk, and D. Müller-Schwarze, 315–28. Oxford: Oxford University Press.

Gosling, L. M., N. W. Atkinson, S. A. Collins, R. J. Roberts, and R. L. Walters. 1996a. Avoidance of scent-marked areas depends on the intruder's body size. *Behaviour* 133:491–502.

Gosling, L. M., N. W. Atkinson, S. Dunn, and S. A. Collins. 1996b. The response of subordinate male mice to scent marks varies in relation to their own competitive ability. *Animal Behaviour* 52:1185–91.

Gosling, L. M., S. J. Baker, and K. M. H. Wright. 1984. Differential investment by female coypus (*Myocastor coypus*) during lactation. *Symposium of the Zoological Society of London* 51:273–300.

Gosling, L. M., and H. V. McKay. 1990. Competitor assessment by scent matching: An experimental test. *Behavioral Ecology and Sociobiology* 26:415–20.

Gosling, L. M., and S. C. Roberts. 2001a. Scent-marking by male mammals: Cheat-proof signals to competitors and mates. *Advances in the Study of Behavior* 30:169–217.

———. 2001b. Testing ideas about the function of scent marks in territories from spatial patterns. *Animal Behaviour* 62:F7–F10.

Gosling, L. M., S. C. Roberts, E. A. Thornton, and M. J. Andrew. 2000. Life history costs of olfactory status signalling in mice. *Behavioral Ecology and Sociobiology* 48:328–32.

Gosling, L. M., and K. H. M. Wright. 1994. Scent marking and resource defence by male coypus (*Myocastor coypus*). *Journal of Zoology* 234:423–36.

Gosling, L. M., K. M. H. Wright, and G. D. Few. 1988. Facultative variation in the timing of parturition by female coypus (*Myocastor coypus*) and the cost of delay. *Journal of Zoology (London)* 214:407–15.

Gottlieb, G. 1976. The roles of experience in the development of behavior and the nervous system. In *Studies on the Development of Behavior and the Nervous System,* vol. 3, *Neural and Behavioral Specificity,* ed. G. Gottlieb, 25–53. New York: Academic Press.

———. 1981. Roles of early experience in species-specific perceptual development. In *Development of perception,* vol. 1, ed. R. N. Aslin, J. R. Alberts, and M. R. Petersen, 5–44. New York: Academic Press.

Gouat, J., and P. Gouat. 1984. Distribution and habitat of the gundis in Algeria (Rodentia: Ctenodactylidae). *Mammalia* 48:227–38.

Gouat, P. 1988a. Etude socioecologique de trois espaces de rongeurs ctenodactylides d'Algerie. PhD diss., Claude Bernard, Lyon.

———. 1988b. Interspecific competition and space utilization

study of two ctenodactylid rodents in Algeria. *Sciences et Techniques de l'Animal de Laboratorie* 13:123–27.

———. 1991. Adaptation comportementale à la température chez trois espèces de Cténodactylidés sahariens, In *Le Rongeur et l'Espace*, ed. M. Le Berre and L. Le Guelte, 79–89. Paris: R. Chabaud.

Gouat, P., and J. Gouat. 1983. L'habitat du goundi (*Ctenodactylus gundi*) dans le massif de l'Aurès (Algérie). *Mammalia* 47:507–18.

———. 1989. Replacement sequences in *Ctenodactylus gundi*: Competition or cooperation? *Behavioural Processes* 18:107–18.

Gouat, P., K. Katona, and C. Poteaux. 2003. Is the socio-spatial distribution of mound-building mice, *Mus spicilegus*, compatible with a monogamous mating system? *Mammalia* 67: 15–24.

Goudet, J. 1995. FSTAT (Version 1.2): A computer program to calculate *F*-statistics. *Journal of Heredity* 86:485–86.

Gould, E., and C. G. Gross. 2002. Neurogenesis in adult mammals: Some progress and problems. *Journal of Neuroscience* 22:619–23.

Gould, S. J. 1977. *Ontogeny and phylogeny.* Cambridge, MA: Belknap Press of Harvard University Press.

Goundie, T. R., and S. R. Vessey. 1986. Survival and dispersal of young white-footed mice born in nest boxes. *Journal of Mammalogy* 67:53–60.

Gowaty, P. A., L. C. Drickamer, and S. Schmid-Holmes. 2003. Male house mice produce fewer offspring with lower viability and poorer performance when mated with females they do not prefer. *Animal Behaviour* 65:95–03.

Grafen, A. 1990. Do animals really recognize kin? *Animal Behaviour* 39:42–54.

Graham, I. M., and X. Lambin. 2002. The impact of weasel predation on cyclic field-vole survival: The specialist predator hypothesis contradicted. *Journal of Animal Ecology* 71:946–56.

Grant, E. C., and J. H. Mackintosh. 1963. A comparison of the social postures of some common laboratory rodents. *Behaviour* 21:246–59.

Grant, P. R. 1972. Interspecific competition among rodents. *Annual Review of Ecology and Systematics* 3:79–106.

Grau, H. J. 1982. Kin recognition in white-footed deermice (*Peromyscus leucopus*). *Animal Behaviour* 30:497–505.

Graur, D., L. Duret, and M. Guoy. 1996. Phylogenetic position of the order Lagomorpha (rabbits, hares and allies). *Nature* 379:333–35.

Graur, D., W. A. Hide, and W.–H. Li. 1991. Is the guinea-pig a rodent? *Nature* 351:649–52.

———. 1992. The biochemical phylogeny of guinea pigs and gundies and the paraphyly of the order Rodentia. *Comparative Biochemical Physiology* B 101:495–98.

Gray, G. D., M. Zerylnick, H. N. Davis, and D. A. Dewsbury. 1974. Effects of variations in male copulatory behavior on ovulation and implantation in prairie voles, *Microtus ochrogaster*. *Hormones and Behavior* 5:389–96.

Gray, S., and J. L. Hurst. 1995. The effects of cage cleaning on aggression within groups of male laboratory mice. *Animal Behaviour* 49:821–26.

Green, A. J. 2001. Mass/length residuals: measures of body condition or generators of spurious results? *Ecology* 82:1473–83.

Greene, E., and T. Meagher. 1998. Red squirrels, *Tamiasciurus hudsonicus,* produce predator-class specific alarm calls. *Animal Behaviour* 55:511–18.

Greene, H. W. 1997. *Snakes : The evolution of mystery in nature.* Berkeley: University of California Press.

Greenwood, P. J. 1980. Mating systems, philopatry and dispersal in birds and mammals. *Animal Behaviour* 28:1140–62.

———. 1983. Mating systems and the evolutionary consequences of dispersal. In *The ecology of animal movement,* ed. I. R. Swingland and P. J. Greenwood, 116–31. Oxford: Clarendon.

Greenwood, P. J., and J. Adams. 1987. Sexual selection, size dimorphism and a fallacy. *Oikos* 48:106–8.

Gregory, W. K. 1910. The orders of mammals. *Bulletin of the American Museum of Natural History* 27:1–524.

Griebel, R. L. 2000. Ecological and physiological factors affecting nesting success of burrowing owls in Buffalo Gap National Grassland. Master's thesis, University of Nebraska, Lincoln.

Griffin, A. S., J. M. Pemberton, P. N. M. Brotherton, G. McIlrath, D. Gaynor, R. Kansky, J. O'Riain, and T. H. Clutton-Brock. 2003. A genetic analysis of breeding success in the cooperative meerkat (*Suricata suricatta*). *Behavioral Ecology* 14:472–80.

Griffin, A. S., and S. A. West. 2003. Kin discrimination and the benefit of helping in cooperatively breeding vertebrates. *Science* 302:634–36.

Griffin, M. 1990. A review of taxomomy and ecology of gerbilline rodents of the central Namib desert with keys to the species (Rodentia: Muridae). In *Namib ecology: 25 years of Namib research,* ed. M. K. Seely, 83–98. Transvaal Museum Monograph. Pretoria: Transvaal Museum.

———. 1998. The species diversity, distribution and conservation of Namibian mammals. *Biodiversity and Conservation* 7:483–94.

Grimm, V., N. Dorndorf, F. Frey-Roos, C. Wissel, T. Wyszomirski, and W. Arnold. 2003. Modelling the role of social behavior in the persistence of the alpine marmot, *Marmota marmota*. *Oikos* 102:124–36.

Gromov, V. S. 2000. *Ethological mechanisms of population homeostasis in gerbils* (Mammalia, Rodentia). [In Russian.] Moscow: A. N. Severtzov Institute of Ecology and Evolution.

Gromov, V. S., A. V. Tchabovsky, D. V. Paramonov, and A. P. Pavlov. 1996. The seasonal dynamics of demographic and spatial structure in a population of tamarisk gerbils (*Meriones tamariscinus*) in southern Kalmykia. [In Russian.] *Zoologicheskii Zhurnal* 75:413–28.

Gross, C. G. 2000. Neurogenesis in the adult brain: Death of a dogma. *Nature Reviews: Neuroscience* 1:67–73.

Gross, M.R. 1996. Alternative reproductive strategies and tactics: Diversity within sexes. *Trends in Ecology and Evolution* 11:92–98.

Grota, L. J. 1973. Effects of litter size, age of young, and parity on foster mother behaviour in *Rattus norvegicus*. *Animal Behaviour* 21:78–82.

Grota, L. J., and R. Ader. 1969. Continuous recording of maternal behaviour in *Rattus norvegicus*. *Animal Behaviour* 17:722–29.

Grzimek, B., ed. 1968. *Enzyklopädie des Tierreiches*, vol. 11. Zurich: Kindler.

Gubernick, D. J. 1994. Biparental care and male-female relations in mammals. In *Infanticide and parental care,* ed. S. Parmigiani and F. S. Vom Saal, 427–63. Chur, Switzerland: Harwood.

Gubernick, D. J., and R. L. Addington. 1994. The stability of female social and mating preferences in the monogamous California mouse, *Peromyscus californicus*. *Animal Behaviour* 47:559–67.

Gubernick, D. J., and J. R. Alberts. 1987. The biparental care system of the California mouse, *Peromyscus californicus*. *Journal of Comparative Psychology* 101:169–77.

Gubernick, D. J., and J. C. Nordby. 1992. Parental influences on female puberty in the monogamous California mouse, *Peromyscus californicus*. *Animal Behaviour* 44:259–67.

Gubernick, D. J., and T. Teferi. 2000. Adaptive significance of male parental care in a monogamous mammal. *Proceedings of the Royal Society of London, Series B*, 267:147–50.

Gubernick, D. J., S. L. Wright, and R. E. Brown. 1993. The significance of the father's presence for offspring survival in the monogamous California mouse, *Peromyscus californicus*. *Animal Behaviour* 46:539–46.

Guerra, R. F., and C. R. Nunes. 2001. Effects of litter size on maternal care, body weight and infant development in golden hamsters (*Mesocricetus auratus*). *Behavioural Processes* 55:127–42.

Guichon, M. L., M. Borgnia, C. F. Righi, G. H. Cassini, and M. H. Cassini. 2003. Social behavior and group formation in the coypu (*Myocastor coypus*) in the Argentinean pampas. *Journal of Mammalogy* 84:254–62.

Gundersen, G., and H. P. Andreassen. 1998. Causes and consequences of natal dispersal in root voles, *Microtus oeconomus*. *Animal Behaviour* 56:1355–66.

Gundersen, G., J. A. Moe, H. P. Andreassen, R. G. Carlsen, and H. Gundersen. 1999. Intersexual attraction in natal dispersing root voles. *Acta Theriologica* 44:283–90.

Gunson, J. R. 1970. Dynamics of the beaver of Saskatchewan's northern forest. Master's thesis. University of Alberta, Edmonton.

Gurnell, J. 1987. *The natural history of squirrels*. New York: Facts on File.

Gustafsson, T., B. Andersson, and P. Muerling. 1980. Effect of social rank on the growth of the preputial glands in male bank voles, *Clethrionomys glareolus*. *Physiology and Behavior* 24:689–92.

Gustin, M. K., and G. F. McCracken. 1987. Scent recognition between females and pups in the bat *Tadarida brasiliensis mexicana*. *Animal Behaviour* 35:13–19.

Gutiérrez, J. R., and F. Bozinovic. 1998. Diet selection in captivity by a generalist herbivorous rodent (*Octodon degus*) from the Chilean coastal desert. *Journal of Arid Environments* 39:601–7.

Gutierrez, P. J., J. S. Meyer, and M. Novak. 1989. Comparison of postnatal brain development in meadow voles (*Microtus pennsylvanicus*) and pine voles (*M. pinetorum*). *Journal of Mammalogy* 70:292–99.

Guttman R., G. Naftali, and E. Nevo. 1975. Aggression patterns in three chromosome forms of the mole rat, *Spalax ehrenbergi*. *Animal Behaviour* 23:485–93.

Gyger, M. 1990. Audience effects on alarm calling. *Ethology, Ecology and Evolution* 2:227–32.

Haccou, P., and Y. Iwasa. 1995. Optimal mixed strategies in stochastic environments. *Theoretical Population Biology* 47:212–43.

Hackländer, K., and W. Arnold. 1999. Male-caused failure in female reproduction and its adaptive value in alpine marmots (*Marmota marmota*). *Behavioral Ecology* 10:592–97.

Hackländer, K., E. Möstl, and W. Arnold. 2003. Reproductive suppression in female Alpine marmots, *Marmota marmota*. *Animal Behaviour* 65:1133–40.

Hafner, D. J. 1984. Evolutionary relationships of the nearctic Sciuridae. In *The biology of ground-dwelling squirrels*, ed. J. O. Murie and G. R. Michener, 3–23. Lincoln: University of Nebraska Press.

Hafner, D. J., E. Yensen, and G. L. Kirkland, Jr., eds. 1998. *North American rodents, status survey and conservation action plan*. Cambridge: Rodent Specialist Group, Species Survival Commission, IUCN.

Haftorn, S. 2000. Contexts and possible functions of alarm calling in the willow tit, *Parus montanus:* The principle of "better safe than sorry." *Behaviour* 137:437–49.

Haig, D. 1997. The social gene. In *Behavioural ecology: An evolutionary approach*, ed. J. R. Krebs and N. B. Davies, 284–304. Oxford: Blackwell Science.

Haigh, G. R. 1987. Reproductive inhibition of female *Peromyscus leucopus*: Female competition and behavioral regulation. *American Zoologist* 27:867–78.

Haigh, G. R., D. S. Cushing, and F. H. Bronson. 1988. A novel postcopulatory block of reproduction in white-footed mice. *Biology of Reproduction* 38:623–26.

Haigh, G. R., D. M. Lounsbury, and T. A. Gordon. 1985. Pheromone-induced reproductive inhibition in young female *Peromyscus leucopus*. *Biology of Reproduction* 33:271–76.

Haim, A., and A. Borut. 1975. Size and activity of a cold resistant population of the golden spiny mouse (*Acomys russatus:* Muridae). *Mammalia* 39:605–12.

Halanych, K. M. 1998. Lagomorphs misplaced by more characters and fewer taxa. *Systematic Biology* 47:138–46.

Hall, E. R. 1927. An outbreak of house mice in Kern County, California. *University of California Publications in Zoology* 30:189–203.

———. 1981. *The mammals of North America*. New York: Wiley.

Hall, J. G. 1960. Willow and aspen in the ecology of beaver on Sagchen Creek, California. *Ecology* 41:484–94.

Hall, L., and M. Morrison. 1997. Den and relocation site characteristics and home ranges of *Peromyscus truei* in the white mountains of California. *Great Basin Naturalist* 57:124–30.

Halle, S. 1988. Avian predation upon a mixed community of common voles (Microtus arvalis) and wood mice (Apodemus sylvaticus). *Oecologia* 75:451–55.

———. 1993. Diel pattern of predation risk in microtine rodents. *Oikos* 68:510–18.

———. 2000a. Ecological relevance of daily activity patterns. In *Activity patterns in small mammals*, Ecological Studies no. 141, ed. S. Halle and N. C. Stenseth, 67–90. Berlin: Springer.

———. 2000b. Voles: Small graminivores and polyphasic activity. In *Activity patterns in small mammals*, Ecological Studies no. 141, ed. S. Halle and N. C. Stenseth, 191–215. Berlin: Springer.

Halpin, Z. T. 1976. The role of individual recognition by odours in the social interactions of the Mongolian gerbil (*Meriones unguiculatus*). *Behaviour* 58:117–30.

———. 1983. Naturally occurring encounters between black-tailed prairie dogs (*Cynomys ludovicianus*) and snakes. *American Midland Naturalist* 109:50–54.

———. 1985. The rodents. III. Suborder Sciuromorpha. In *Social odours in mammals*, vol. 1, ed. R. E. Brown and D. W. Macdonald, 458–79. Oxford: Oxford Science Publications.

———. 1986. Individual odors among mammals: Origins and functions. *Advances in the Study of Behavior* 16:39–70.

———. 1987. Natal dispersal and the formation of new social groups in a newly established town of black-tailed prairie dogs (*Cynomys ludovicianus*). In *Mammalian dispersal patterns: The effects of social structure on population genetics,* ed. B. D. Chepko-Sade and Z. T. Halpin, 104–18. Chicago: University of Chicago Press.

———. 1991. Kin recognition cues of vertebrates. In *Kin recognition,* ed. P. G. Hepper, 220–58. Cambridge: Cambridge University Press.

Haltenorth, T., and H. Diller. 1980. *A field guide to the mammals of African including Madagascar.* London: Collins.

Hambuch, T. M., and E. A. Lacey. 2002. Enhanced selection for MHC diversity in social tuco-tucos. *Evolution* 56:841–45.

Hamilton, W. D. 1964. The genetical evolution of social behavior. I and II. *Journal of Theoretical Biology* 7:1–52.

———. 1967. Extraordinary sex ratios. *Science* 1565:477–78.

———. 1971. Geometry for the selfish herd. *Journal of Theoretical Biology* 31:295–311.

Hamilton, W. D., and M Zuk. 1982. Heritable true fitness and bright birds: A role for parasites. *Science* 218:384–87.

Hammer, M. F., J. Schimenti, and L. M. Silver. 1989. Evolution of mouse chromosome 17 and the origin of inversions associated with *t* haplotypes. *Proceedings of the National Academy of Sciences* (USA) 86:3261–65.

Hampton, J. A. 1995. Testing the prototype theory of concepts. *Journal of Memory and Language* 34:686–708.

Handa, R. J., L. H. Burgess, J. E. Kerr, and J. A. O'keefe. 1994. Gonadal steroid hormone receptors and sex differences in the hypothalamo-pituitary-adrenal axis. *Hormones and Behavior* 28:464–76.

Hanken, J., and P. W. Sherman. 1981. Multiple paternity in Belding's ground squirrels. *Science* 212:351–53.

Hansen, R. M. 1956. Decline in Townsend ground squirrels in Utah. *Journal of Mammalogy* 37:123–24.

———. 1957. Communal litters in *Peromyscus maniculatus*. *Journal of Mammalogy* 38:523.

Hansen, R. M., and I. K. Gold. 1977. Blacktail prairie dogs, desert cottontails and cattle trophic relations on shortgrass range. *Journal of Range Management* 30:210–14.

Hanski, I. A. 1999. *Metapopulation ecology.* Oxford: Oxford University Press.

Hanski, I. A., and M. E. Gilpin, eds. 1997. *Metapopulation biology: Ecology, genetics and evolution.* San Diego, CA: Academic Press.

Hanski, I., and H. Henttonen. 1996. Predation on competing rodent species: A simple explanation of complex patterns. *Journal of Animal Ecology* 65:220–32.

Hanski, I., H. Henttonen, E. Korpimäki, L. Oksanen, and P. Turchin. 2001. Small-rodent dynamics and predation. *Ecology* 82:1505–20.

Hanson, A. F. 2003. Plasticity and tonic processes in the antipredator behavior of rock squirrels (*Spermophilus variegatus*). PhD diss., University of California, Davis.

Hanson, M. T., and R. G. Coss. 1997. Age differences in the response of California ground squirrels (*Spermophilus beecheyi*) to avian and mammalian predators. *Journal of Comparative Psychology* 111:174–84.

———. 2001a. Age differences in arousal and vigilance in California ground squirrels (*Spermophilus beecheyi*). *Developmental Psychobiology* 39:199–206.

———. 2001b. Age differences in the response of California ground squirrels (*Spermophilus beecheyi*) to conspecific alarm calls. *Ethology* 107:259–75.

Hansson, L. 1984. Predation as the factor causing extended low densities in microtine cycles. *Oikos* 43:255–56.

Hansson, L., and H. Henttonen. 1985. Gradients in density variations of small rodents: The importance of latitude and snow cover. *Oecologia* 67:394–402.

———. 1988. Rodent dynamics as community process. *Trends in Ecology and Evolution* 3:195–200.

Hanwell, A., and M. Peaker. 1977. Physiological effects of lactation on the mother. *Symposium of the Zoological Society of London* 41:297–312.

Happold, D. C. D. 1975. The ecology of rodents in the Northern Sudan. In *Rodents in desert environments,* ed. I. Prakash and P. K. Ghosh, 14–45. The Hague: Junk.

Harding, E. K. 2000. Landscape heterogeneity and its importance for community dynamics and conservation of a marsh-grassland system. PhD diss., University of California, Santa Cruz.

Hardy, A. R., and K. D. Taylor. 1979. Radiotracking of *Rattus norvegicus* on farms. In *A handbook on biotelemetry and radiotracking,* ed. C. J. Amlaner and D. W. Macdonald, 657–65. Oxford: Pergamon.

Hardy, D. F., and J. F. DeBold. 1972. Effects of coital stimulation upon behavior of the female rat. *Journal of Comparative and Physiological Psychology* 78:400–408.

Hardy, I. C. W. 2002. *Sex ratios: Concepts and research methods.* Cambridge: Cambridge University Press.

Hare, J. F. 1991. Intraspecific killing of preweaned young in the Columbian ground squirrel, *Spermophilus columbianus.* *Canadian Journal of Zoology* 69:797–800.

———. 1992. Colony member discrimination by juvenile Columbian ground squirrels (*Spermophilus columbianus*). *Ethology* 92:301–15.

———. 1994. Group member discrimination by Columbian ground squirrels via familiarity with substrate-borne chemical cues. *Animal Behaviour* 47:803–13.

———. 1998a. Juvenile Richardson's ground squirrels (*Spermophilus richardsonii*), discriminate among individual alarm callers. *Animal Behaviour* 55:451–60.

———. 1998b. Juvenile Richardson's ground squirrels (*Spermophilus richardsonii*) manifest both littermate and neighbour/stranger discrimination. *Ethology* 104:991–1002.

———. 2004. Kin discrimination by asocial Franklin's ground squirrels (*Spermophilus franklinii*): Is there a relationship between kin discrimination and ground squirrel sociality? *Ethology Ecology and Evolution* 16:157–69.

Hare, J. F., and B. A. Atkins 2001. The squirrel that cried wolf: Reliability detection by juvenile Richardson's ground squirrels (*Spermophilus richardsonii*). *Behavioral Ecology and Sociobiology* 51:108–12.

Hare, J. F., and J. O. Murie 1996. Ground squirrel sociality and the quest for the "holy grail": Does kinship influence behavioural discrimination by juvenile Columbian ground squirrels? *Behavioral Ecology* 7:76–81.

Harlow, H. J. 1997. Winter body fat, food consumption and nonshivering thermogenesis of representative spontaneous and facultative hibernators: The white-tailed prairie dog and black-tailed prairie dog. *Journal of Thermal Biology* 22:21–30.

Harper, S. J., and G. O. Batzli. 1996. Effects of predators on structure of the burrows of voles. *Journal of Mammalogy* 77:1114–21.

Harrington, F. H. 1981. Urine-marking and caching behaviour in the wolf. *Behaviour* 76:280–88.

Harris, M. A., and J. O. Murie. 1982. Responses to oral gland scents from different males in Columbian ground squirrels. *Animal Behaviour* 30:140–48.

———. 1984. Inheritance of nest sites in female Columbian ground squirrels. *Behavioral Ecology and Sociobiology* 15:97–102.

Harris, M. A., J. O. Murie, and J. A. Duncan. 1983. Responses of Columbian ground squirrels to playback of recorded calls. *Zeitschrift für Tierpsychologie* 63:318–30.

Harrison, D. L., and P. J. J. Bates. 1991. *The mammals of Arabia*, 2nd ed. Kent, England: Zoological Museum Publication.

Harrison, R. G., S. M. Bogdanowicz, R. S. Hoffmann, E. Yensen, and P. W. Sherman. 2003. Phylogeny and evolutionary history of the ground squirrels (Rodentia: Marmotinae). *Journal of Mammalian Evolution* 10:249–76.

Hart, B. L., L. A. Hart, M. S. Mooring, and R. Olubayo. 1992. Biological basis of grooming behaviour in antelope: The body-size, vigilance, and habitat principles. *Animal Behaviour* 44:615–31.

Hart, B. L., E. Korinek, and P. Brennan. 1987. Postcopulatory genital grooming in male rats: Prevention of sexually transmitted infections. *Physiology and Behavior* 41:321–25.

Hart, E. B. 1992. *Tamias dorsalis. Mammalian Species* 399:1–6.

Hartenberger, J.-L. 1985. The order Rodentia: Major questions on their evolutionary origin, relationships and suprafamily systematics. In *Evolutionary relationships among rodents: A multidisciplinary analysis,* ed. W. P. Luckett and J. L. Hartenberger, 1–33. New York: Plenum.

———. 1998. Description of the radiation of the Rodentia (Mammalia) from the Late Paleocene to the Miocene: Phylogenetic consequences. *Comptes Rendus de l'Académie Des Sciences* 326:439–44.

Hartl, D. L., and A. G. Clark. 1997. *Principles of population genetics*, 3rd ed. Sunderland, MA: Sinauer.

Hartley, D. J., and J. A. Bishop. 1979. Home range and movement in populations of *Rattus norvegicus* polymorphic for Warfarin resistance. *Biological Journal of the Linnean Society* 12:19–43.

Hartman, G. 1994. Long-term population development of a reintroduced beaver (*Castor fiber*) population in Sweden. *Conservation Biology* 8:713–17.

———. 1995. Patterns of spread of a reintroduced beaver *Castor fiber* population in Sweden. *Wildlife Biology* 1:97–103.

———. 1999. Beaver management and utilization in Scandinavia. In *Beaver protection, management, and utilization in Europe and North America,* ed. P. E. Busher and R. M. Dzieciolowski, 1–6. New York: Kluwer Academic/Plenum.

———. 2003. Irruptive population development of European beaver (*Castor fiber*) in southwest Sweden. *Lutra* 46:103–8.

Hartung, T. G., and D. A. Dewsbury. 1978. A comparative analysis of copulatory plugs in muroid rodents and their relationship to copulatory behavior. *Journal of Mammalogy* 59:717–23.

———. 1979. Paternal behaviour in six species of muroid rodents. *Behavioral and Neural Biology* 26:466–78.

Harvey, P. H., and M. D. Pagel. 1991. *The comparative method in evolutionary biology.* Oxford: Oxford University Press.

Harvey, P. H., and A. Purvis. 1999. Understanding the ecological and evolutionary reasons for life history variation: Mammals as a case study. In *Advanced ecological theory: Principles and applications,* ed. J. McGlade, 232–48. Malden, MA: Blackwell Science.

Harvey, S., B. Jemiolo, and M. Novotny. 1989. Pattern of volatile compounds in dominant and subordinate male mouse urine. *Journal of Chemical Ecology* 15:2061–72.

Hasler, M. J., and A. V. Nalbandov. 1974. The effect of weanling and adult males on sexual maturation in female voles (*Microtus ochrogaster*). *General and Comparative Endocrinology* 23:237–38.

Hasson, O. 1991. Pursuit-deterrent signals: Communication between prey and predators. *Trends in Ecology and Evolution* 6:325–29.

Hastings, N. B., M. Orchinik, M. V. Aubourg, and B. S. McEwen. 1999. Pharmacological characterization of central and peripheral type I and type II adrenal steroid receptors in the prairie vole, a glucocorticoid-resistant rodent. *Endocrinology* 140:4459–69.

Hatchwell, B. J., and J. Komdeur. 2000. Ecological constraints, life history traits and the evolution of cooperative breeding. *Animal Behaviour* 59:1079–86.

Hatton, D., and M. Myer. 1973. Paternal behavior in cactus mice (*Peromyscus eremicus*). *Southwestern Naturalist* 17:85–93.

Hauber, M E., and P. W. Sherman. 1998. Nepotism and marmot alarm calling. *Animal Behaviour* 56:1049–52.

———. 2001. Self-referent phenotype matching: Theoretical considerations and empirical evidence. *Trends in Neurosciences* 24:609–16.

Hausfater, G. 1984. Infanticide in langurs: Strategies, counterstrategies, and parameter values. In *Infanticide: Comparative and evolutionary perspectives,* ed. G. Hausfater and S. B. Hrdy, 257–81. New York: Aldine.

Hausfater, G., and S. B. Hrdy, eds. 1984. *Infanticide: Comparative and evolutionary perspectives.* New York: Aldine.

Havelka, M. A., and J. S. Millar. 1997. Sex ratio of offspring in *Peromyscus maniculatus borealis. Journal of Mammalogy* 78:623–37.

Havera, S. P., and C. M. Nixon. 1978. Interaction among female fox squirrels during their winter breeding season. *Transactions of the Illinois Academy of Science* 71:24–38.

Hawkins, C. C., W. E. Grant, and M. T. Longnecker. 1999. Effect of subsidized house cats on California birds and rodents. *Transactions Western Section Wildlife Society* 35:29–33.

Hawkins, L. K., and J. A. Cranford. 1992. Long-term effects of intraspecific and interspecific cross-fostering on two species of *Peromyscus. Journal of Mammalogy* 73:802–7.

Hay J., W. M. Hutchison, P. P. Aitken, D. I. Graham. 1983. The effect of congenital and adult-acquired Toxoplasma infections on the motor performance of mice. *Annals Tropical Medical Parasitology* 77:261–77.

Hay, K. G. 1958. Beaver census methods in the Rocky Mountain region. *Journal of Wildlife Management* 22:395–402.

Hayashi, S. 1986. Effects of a cohabitant on preputial gland weight of male mice. *Physiology and Behavior* 38:299–300.

———. 1990. Social condition influences sexual attractiveness of dominant male mice. *Zoological Science* 7:889–94.

Hayashi, S., and T. Kimura. 1978. Effects of exposure to males on sexual preference in female mice. *Animal Behaviour* 26:290–95.

———. 1983. Degree of kinship as a factor regulating preferences among conspecifics in mice. *Animal Behaviour* 31: 81–85.

Hayes, L. D. 2000. To nest communally or not to nest communally: A review of rodent communal nesting and nursing. *Animal Behaviour* 59:677–88.

Hayes, L. D., and N. G. Solomon. 2004. Costs and benefits of communal rearing to female prairie voles (*Microtus ochrogaster*). *Behavioral Ecology and Sociobiology* 56:585–93.

Hayflick, L. 2000. The future of ageing. *Nature* 408:267–69.

Haynie, M. L., R. A. van den Bussche, J. L. Hoogland, and D. A. Gilbert. 2003. Parentage, multiple paternity, and reproductive success in Gunnison's and Utah prairie dogs. *Journal of Mammalogy* 84:1244–53.

Hayssen, V. R., A. van Tienhoven, and A. van Tienhoven. 1993. *Asdell's paterns of mammalian reproduction. A compendium of species-specific data.* Ithaca, NY: Cornell University Press.

Hazel, S. M., M. Bennett, J. Changrey, K. Bown, R. Cavanagh, T. R. Jones, D. Baxby, and M. Begon. 2000. A longitudinal study of an endemic disease in its wildlife reservoir: Cowpox and wild rodents. *Epidemiology and Infection* 124:551–62.

Hazell, R. W. A., N. C. Bennett, J. U. M. Jarvis, and M. Griffin. 2000. Adult dispersal in the co-operatively breeding Damaraland mole-rat (*Cryptomys damarensis*): A case study from the Waterberg region of Namibia. *Journal of Zoology, London* 252:19–25.

Heaney, L. R. 1984. Climatic influences on life-history tactics and behavior of North American tree squirrels. In *The biology of ground-dwelling squirrels: Annual cycles, behavioral ecology, and sociality,* ed. J. O. Murie and G. R. Michener, 43–78. Lincoln: University of Nebraska Press.

Heard, D. C. 1977. The behaviour of Vancouver Island marmots, *Marmota vancouverensis.* Master's thesis, University of British Columbia.

Hediger, H. 1949. Säugetier-Territorien und ihre Markierung. *Bijdr. Dierkd.* 28:172–84.

Hedrich, H. J. 2000. History, strains and models. In *The laboratory rat,* ed. G. J. Krinke, 3–16. San Diego, CA: Academic Press.

Hedrick, A. V., and E. J. Temeles. 1989. The evolution of sexual dimorphism in animals: Hypotheses and tests. *Trends in Ecology and Evolution* 4:136–38.

Hedrick, P. W. 1992. Female choice and variation in the major histocompatibility complex. *Genetics* 132:575–81.

Heikkilä, J., K. Kaarsalo, O. Mustonen, and P. Pekkarinen. 1993. Influence of predation risk on early development and maturation in three species of *Clethrionomys* voles. *Annales Zoologici Fennici* 30:153–61.

Heise, S., and J. L. Hurst. 1994. Territorial experience causes a shift in the responsiveness of female house mice to odours from dominant males. *Advances in the Biosciences* 93: 291–96.

Heise, S., and J. Lippke. 1997. Role of female aggression in prevention of infanticidal behavior in male common voles, *Microtus arvalis* (Pallas, 1779). *Aggressive Behavior* 23: 293–98.

Heise, S. R., and F. M. Rozenfeld. 1999. Reproduction and urine marking in laboratory groups of female common voles, *Microtus arvalis. Journal of Chemical Ecology* 25:1671–85.

———. 2002. Effect of odour cues on the exploratory behaviour of female common voles living in matriarchal groups. *Behaviour* 139:897–911.

Henderson, F. R., P. F. Springer, and R. Adrian. 1974. The black-footed ferret in South Dakota. *Technical Bulletin 4: South Dakota Department of Fish, Game and Parks*:1–32.

Henderson, J. A., and F. F. Gilbert. 1978. Distribution and density of woodchuck burrow systems in relation to land-use practices. *Canadian Field-Naturalist* 92:128–36.

Hendrichs, H. 1975. Observations on a population of Bohor reedbuck (*Redunca redunca;* Pallas, 1767). *Zeitschrift fur Tierpsycologie* 38:44–54.

Hendrichs, H., and U. Hendrichs. 1971. *Dikdik und Elefanten Okologie und Soziologie zwei afrikanischer Huftiere.* Munich: R. Piper.

Hendrie, C. A. 1991. The calls of murine predators activate endogenous analgesia mechanisms in laboratory mice. *Physiology and Behavior* 49:569–73.

Hendriks, W., J. Luenisson, E. Nevo, H. Bloemendal, and W. W. De Jong. 1987. The lens protein aA crystalline in the blind mole rat *Spalax ehrenbergi*: Evolutionary change and functional constraints. *Proceedings of the National Academy of Science, (USA)* 84:5320–24.

Hennessey, A. C., D. C. Whitman, and H. E. Albers. 1992. Microinjection of arginine-vasopressin into the periaqueductal gray stimulates flank marking in Syrian hamsters (*Mesocricetus auratus*). *Brain Research* 569:136–40.

Hennessy, D. F., and D. H. Owings. 1988. Rattlesnakes create a context for localizing their search for potential prey. *Ethology* 77:317–29.

Hennessy, D. F., D. H. Owings, M. P. Rowe, R. G. Coss, and D. W. Leger. 1981. The information afforded by a variable signal: Constraints on snake-elicited tail flagging by California ground squirrels. *Behaviour* 78:188–226.

Hennessy, M. B. 2003. Enduring maternal influences in a precocial rodent. *Developmental Psychobiology* 42:225–36.

Henry, D. B., and T. A. Bookhout.1969. Productivity of beavers in northeastern Ohio. *Journal of Wildlife Management* 33:927–32.

Henry, J. D. 1976. The use of urine marking in the scavenging behavior of the red fox (*Vulpes vulpes*). *Behaviour* 61: 82–105.

Hensley, M. M., and J. B. Cope. 1951. Further data on removal and repopulation of the breeding birds in a spruce-fir forest community. *Auk* 68:483–93.

Hepper, P. G. 1986. Kin recognition: Functions and mechanisms. A review. *Biological Reviews of the Cambridge Philosophical Society* 61:63–93.

———. 1987. The amniotic fluid: An important priming role in kin recognition. *Animal Behaviour* 35:1343–46.

———. 1988. Adaptive fetal learning: Prenatal exposure to garlic affects postnatal preference. *Animal Behaviour* 36:935–36.

Herbert, J. M., and R. W. Dixon. 2002. Is the ENSO phenomenon changing as a result of global warming? *Physical Geography* 23:196–211.

Herbst, M. 2002. The biology and population ecology of the Namaqua dune mole-rat, *Bathyergus janetta,* from the Northern Cape Province, South Africa. Master's thesis, University of Pretoria, South Africa.

Herman, C. S., and T. J. Valone. 2000. The effect of mammalian predator scent on the foraging behaviour of *Dipodomys merriami. Oikos* 91:139–45.

Herr, J., and F. Rosell. 2004. Use of space and movement patterns in monogamous adult Eurasian beavers (*Castor fiber*). *Journal of Zoology* 262:257–64.

Herrera, E. A. 1986. *The behavioural ecology of the capybara* Hydrochoerus hydrochaeris. PhD diss., University of Oxford.

———. 1992. Size of testes and scent glands in capybaras, *Hydrochaeris hydrochaeris* (Rodentia, Caviomorpha). *Journal of Mammalogy* 73:871–75.

———. 1998. Reproductive strategies of female capybaras: Dry season gestations. *Symposium of the Zoological Society of London* 71:281–92.

Herrera, E. A., and D. W. Macdonald. 1987. Group stability and the structure of a capybara population. *Symposium of the Zoological Society of London* 58:1–350.

———. 1989. Resource utilization and territoriality in group-living Capybaras *Hydrochoerus hydrochaeris. Journal of Animal Ecology* 58:667–79.

———. 1993. Aggression, dominance and mating success among capybara males (*Hydrochoerus hydrochaeris*). *Behavioural Ecology* 4:114–19.

———. 1994. Social significance of scent marking in capybaras. *Journal of Mammalogy* 75:410–15.

Herrmann, B., M. Bucan, P. E. Mains, A. M. Frischauf, L. M. Silver, and H. Lehrach. 1986. Genetic analysis of the proximal portion of the mouse *t* complex: Evidence for a second inversion within *t* haplotypes. *Cell* 44:469–76.

Herrmann, B. G., B. Koschorz, K. Wertz, K. J. McLaughlin, and A. Kispert. 1999. A protein kinase encoded by the *t* complex responder gene causes non-mendelian inheritance. *Nature* 402:141–46.

Herron, M. D., T. A. Castoe, and C. L. Parkinson. 2004. Sciurid phylogeny and the paraphyly of Holarctic ground squirrels (*Spermophilus*). *Molecular Phylogenetics and Evolution* 31:1015–30.

Hersek, M. J. 1990. Behavior of predator and prey in a highly coevolved system: Northern Pacific rattlesnakes and California ground squirrels. PhD diss., University of California, Davis.

Hersek, M. J., and D. H. Owings. 1993. Tail flagging by adult California ground squirrels: A tonic signal that serves different functions for males and females. *Animal Behaviour* 46:129–38.

———. 1994. Tail flagging by young California ground squirrels, *Spermophilus beecheyi*: Age-specific participation in a tonic communicative system. *Animal Behaviour* 48:803–11.

Hershkovitz, P. 1962. Evolution of neotropical cricetine rodents (Muridae) with special reference to the Phyllotine group. *Fieldiana Zoology* 46:1–524.

———. 1966. Mice, land bridges and Latin American faunal interchange. In *Ectoparasites of Panama*, ed. R. L. Wenzel and V. J. Tipton, 725–51. Chicago: Field Museum of Natural History.

Herzog, H. A., and G. M. Burghardt. 1986. Development of antipredator responses in snakes: I. Defensive and open-field behaviors in newborns and adults of three species of garter snakes (*Thamnophis melanogaster, T. sirtalis, T. butleri*). *Journal of Comparative Psychology* 100:372–79.

Heske, E. J. 1987. Pregnancy interruption by strange males in the California vole. *Journal of Mammalogy* 68:406–10.

Heske, E. J., and R. S. Ostfeld. 1990. Sexual dimorphism in size, relative size of testes, and mating systems in North American voles. *Journal of Mammalogy* 71:510–19.

Heske, E. J., R. S. Ostfeld, and W. Z. Lidicker, Jr. 1984. Competitive interactions between *Microtus californicus* and *Reithrodontomys megalotis* during two peaks of *Microtus* abundance. *Journal of Mammalogy* 65:271–80.

Hess, G. R. 1994. Conservation corridors and contagious disease: A cautionary note. *Conservation Biology* 8:256–62.

Hestbeck, J. B. 1982. Population regulation of cyclic mammals: The social fence hypothesis. *Oikos* 39:157–63.

Heth, G. 1989. Burrow patterns of the mole rat *Spalax ehrenbergi* in two soil types (terra rossa and rendzina) in Mount Carmel. *Israel Journal of Zoology* 7:39–56.

———. 1991. The environmental impact of subterranean mole rats (*Spalax ehrenbergi*) and their burrows. *Symposia of the Zoological Society, London* 63:265–80.

Heth, G., G. Beauchamp, E. Nevo, and K. Yamazaki. 1996. Species, population and individual specific odors in urine of mole rats (*Spalax ehrenbergi*) detected by laboratory rats. *Chemoecology* 7:107–11.

Heth, G., A. Beiles, and E. Nevo. 1988. Adaptive variation of pelage color within and between species of the subterranean mole rat (*Spalax ehrenbergi*) in Israel. *Oecologia* 74: 617–22.

Heth, G., E. Frankenberg, and E. Nevo. 1986. Adaptive optimal sound for vocal communication in tunnels of a subterranean mammal (*Spalax ehrenbergi*). *Experientia* 42:1287–89.

———. 1988. "Courtship" call of subterranean mole rats (*Spalax ehrenbergi*): Physical analysis. *Journal of Mammalogy* 69:121–25.

Heth, G., E. Frankenberg, H. Pratt, and E. Nevo. 1991. Seismic communication in the blind subterranean mole rats: Patterns of head thumping and of their detection in the *Spalax ehrenbergi* superspecies in Israel. *Journal of Zoology, London* 224:633–38.

Heth, G., E. Frankenberg, A. Raz, and E. Nevo. 1987. Vibrational communication in subterranean mole rats (*Spalax ehrenbergi*). *Behaviorial Ecology and Sociobiology* 21:31–33.

Heth, G., E. Nevo, R. Ikan, V. Weinstein, U. Ravid, and H. Duncan. 1992. Differential olfactory perception of enantiomeric compounds by blind subterranean mole rats (*Spalax ehrenbergi*). *Experientia* 48:897–902.

Heth, G., and J. Todrank. 1995. Assessing chemosensory perception in subterranean mole rats: Different responses to smelling versus touching odorous stimuli. *Animal Behaviour* 49:1009–15.

———. 2000. Individual odour similarities across species parallel phylogenetic relationships in the *S. ehrenbergi* superspecies of mole-rats. *Animal Behaviour* 2000:789–95.

Heth, G., J. Todrank, S. Begall, R. Koch, Y. Zilbiger, E. Nevo, S. H. Braude, and H. Burda. 2002. Odours underground: Subterranean rodents may not forage "blindly." *Behavioral Ecology and Sociobiology* 52:53–58.

Heth, G., J. Todrank, and H. Burda. 2002. Individual odor similarities within colonies and across species of *Cryptomys* mole rats. *Journal of Mammalogy* 83:569–75.

Heyes, C. M. 1993. Imitation, culture and cognition. *Animal Behaviour* 46:999–1010.

———. 1996. Genuine imitation? In *Social learning and imitation: The roots of culture*, ed. C. M. Heyes and B. G. Galef, Jr., 371–90. New York: Academic Press.

Heyes, C. M., and B. G. Galef, Jr. 1996. *Social learning in animals: The roots of culture*. San Diego, CA: Academic Press.

Heyes, C. M., and G. R. Dawson. 1990. A demonstration of observational learning using a bidirectional control. *Quarterly Journal of Experimental Psychology* B 42:59–71.

Heyes, C. M., G. R. Dawson, and T. Nokes. 1992. Imitation in rats: Initial responding and transfer evidence. *Quarterly Journal of Experimental Psychology* B 45:229–40.

Heyes, C. M., and P. J. Durlach. 1990. "Social blockade" of taste-aversion learning in Norway rats (R. norvegicus): Is it a social phenomenon? *Journal of Comparative Psychology* 104:82–87.

Hickling, G. J. 1987. Seasonal reproduction and group dynamics of bushy-tailed woodrats (*Neotoma cinerea* L.). PhD diss., University of Western Ontario, London, Ontario.

Hight, M. E., M. Goodman, and W. Prychodke, W. 1974. Immunological studies of the Sciuridae. *Systematic Zoology* 23:12–25.

Hik, D. 1995. Does risk of predation influence population dynamics? Evidence from the cyclic decline of snowshoe hares. *Wildlife Research* 22:115–29.

Hik, D., C. J. McColl, and R. Boonstra. 2001. Why are arctic ground squirrels more stressed in the boreal forest than in alpine meadows? *Ecoscience* 8:275–88.

Hill, E. P. 1982. Beaver (*Castor canadensis*). In *Wild mammals of North America: Biology, management, and economics*, ed. J. A. Chapman and G. A. Feldhamer, 256–81. Baltimore: John Hopkins University Press.

Hill, R. A., and P. C. Lee. 1998. Predation risk as an influence on group size in cercopithecoid primates: Implications for social structure. *Journal of Zoology* 245:447–56.

Hill, W. C. O., A. Porter, R. T. Bloom, J. Seago, and M. D. Southwick. 1957. Field and laboratory studies on the naked mole rat, *Heterocephalus glaber*. *Proceedings of the Zoological Society London* 128:455–514.

Hinde, R. A. 1974. *Biological bases of human social behaviour*. New York: McGraw-Hill.

Hine, E., S. Lachish, M. Higgie, and M. W. Blows. 2002. Positive genetic correlation between female preference and offspring fitness. *Proceedings of the Royal Society of London, Series B* 269:2215–19.

Hinton, M. A. C. 1926. *Monograph of voles and lemmings, living and extinct*, vol. 1. London: British Museum (Natural History).

Hiraiwa-Hasegawa, M. 1988. Adaptive significance of infanticide in primates. *Trends in Ecology and Evolution* 3:102–5.

Hockett, C. F. 1960. The origin of speech. *Scientific American* 203:89–96.

Hodgdon, H. E. 1978. Social dynamics and behavior within an unexploited beaver (*Castor canadensis*) population. PhD diss., University of Massachusetts, Amherst.

Hodgdon, H. E., and R. A. Lancia. 1983. Behavior of the North American beaver, *Castor canadensis*. *Acta Zoologica Fennica* 174:99–103.

Hodgdon, H. E., and J. S. Larson. 1973. Some sexual differences in behaviour within a colony of marked beavers (*Castor canadensis*). *Animal Behaviour* 21:147–52.

Hodges, K. E., S. Mech, and J. O. Wolff. 2002. Sex and the single vole: Effects of social grouping on prairie vole reproductive success. *Ethology* 108:871–84.

Hodgkin, J. A. 1998. Seven types of pleiotropy. *International Journal of Developmental Biology* 42:501–5.

Hoeck, H. N. 1982. Population dynamics, dispersal and genetic isolation in two species of hyrax (*Heterohyrax brucei* and *Procavia johnstoni*) on habitat islands in the Serengeti. *Zeitschrift Fur Tierpsychologie-Journal of Comparative Ethology* 59:177–210.

———. 1989. Demography and competition in hyrax—a 17 year study. *Oecologia* 79:353–60.

Hoekstra, H. E., K. E. Drumm, and M. W. Nachman. 2004. Ecological genetics of adaptive color polymorphism in pocket mice: Geographic variation in selected and neutral genes. *Evolution* 58:1329–41.

Hoekstra, H. E., and J. M. Hoekstra. 2001. An unusual sex-determination system in South American field mice (genus *Akodon*): The role of mutation, selection, and meiotic drive in maintaining XY females. *Evolution* 55:190–97.

Hoekstra, H. E., J. M. Hoekstra, D. Berrigan, S. N. Vignieri, A. Hoang, C. E. Hill, P. Beerli, et al. 2001. Strength and tempo of directional selection in the wild. *Proceedings of the National Academy of Sciences (USA)* 98:9157–60.

Hoekstra, H. E., and M. W. Nachman. 2003. Different genes underlie adaptive melanism in different populations of rock pocket mice. *Molecular Ecology* 12:1185–94.

Hofer, S., and P. Ingold. 1984. The whistles of the alpine marmot (*Marmota marmota*): Their structure and occurrence in the antipredator context. *Revue Suisse de Zoologia* 91:861–65.

Hoffmann, R. S., C. G. Anderson, R. W. Thorington, Jr., and L. R. Heaney. 1993. Family Sciuridae. In *Mammal species of the world*, ed. D. E. Wilson and D. M. Reeder, 419–65. Washington, DC: Smithsonian Institution Press.

Hoffmann, R. S., J. W. Koeppl, and C. F. Nadler. 1979. The relationships of the amphiberingian marmots (Mammalia: Sciuridae). *University of Kansas: Occasional Papers of the Museum of Natural History* 83:1–56.

Hoffmeister, D. F. 1981. *Peromyscus truei. Mammalian Species* 161:1–5.

———. 1986. *Mammals of Arizona*. Tucson: University of Arizona Press.

Hoffmeyer, I. 1982. Responses of female bank voles (*Clethrionomys glareolus*) to dominant vs. subordinate conspecific males and to urine odors from dominant vs. subordinate males. *Behavioral and Neural Biology* 36:178–88.

Hofmann, J., B. McGuire, and T. Pizzuto. 1989. Parental care in the sagebrush vole (*Lemmiscus curtatus*). *Journal of Mammalogy* 70:162–65.

Hofmeijer, G. K., and H. De Bruijn. 1985. The mammals from the lower Miocene of Aliveri (Island of Evia, Greece) Part 4: The Spalacidae and Anomalomyidae. *Proceedings Koninklijke Nederlanse Akadamie van Weteschappen B* 88:185–98.

Högstedt, G. 1983. Adaptation unto death: Function of fear screams. *American Naturalist* 121:562–70.

Hohoff, C., K. Franzen, and N. Sachser. 2003. Female choice in a promiscuous wild guinea pig, the yellow-toothed cavy (*Galea musteloides*). *Behavioral Ecology and Sociobiology* 53:341–49.

Holdaway, R. N. 1999. Introduced predators and avifaunal extinction in New Zealand. In *Extinctions in near time: Causes, contexts, and consequences*, ed. R. D. E. MacPhee, 189–238. New York: Plenum.

Holekamp, K. E. 1984. Natal dispersal in Belding's ground squirrels (*Spermophilus beldingi*). *Behavioral Ecology and Sociobiology* 16:21–30.

———. 1986. Proximal causes of natal dispersal in Belding's ground squirrels (*Spermophilus beldingi*). *Ecological Monographs* 56:365–91.

Holekamp, K. E., L. Smale, H. B. Simpson, and N. A. Holekamp. 1984. Hormonal influences on natal dispersal in free-living Belding's ground squirrels (*Spermophilus beldingi*). *Hormones and Behavior* 18:465–83.

Hollister, N. 1916. A systematic account of the prairie dogs. *North American Fauna* 40:1–37.

Holm, R. F. 1976. Observations on a cannibalistic gray squirrel. *Natural History, Miscellanea* 197:1–2.

Holmes, W. G. 1977. Cannibalism in the arctic ground squirrel (*Spermophilus parryii*). *Journal of Mammalogy* 58:437–38.

———. 1984a. The ecological basis of monogamy in Alaskan hoary marmots. In *The biology of ground-dwelling squirrels*, ed. J. O. Murie and G. R. Michener, 250–74. Lincoln: University of Nebraska Press.

———. 1984b. Sibling recognition in thirteen-lined ground squirrels: Effects of genetic relatedness, rearing association, and olfaction. *Behavioral Ecology and Sociobiology* 14:225–33.

———. 1986a. Identification of paternal half-siblings by captive Belding's ground squirrels. *Animal Behaviour* 34:321–27.

———. 1986b. Kin recognition by phenotype matching in female Belding's ground squirrels. *Animal Behaviour* 34:38–47.

———. 1990. Parent-offspring recognition in mammals: A proximate and ultimate perspective. In *Mammalian parenting*, ed. N. A. Krasnegor and R. S. Bridges, 441–60. New York: Oxford University Press.

———. 1991. Predator risk affects foraging behaviour of pikas: Observational and experimental evidence. *Animal Behaviour* 42:111–19.

———. 1994. The development of littermate preferences in juvenile Belding's ground squirrels. *Animal Behaviour* 48:1071–84.

———. 1995. The ontogeny of littermate preferences in juvenile golden-mantled ground squirrels: Effects of rearing and relatedness. *Animal Behaviour* 50:309–22.

———. 2001. The development and function of nepotism: Why kinship matters in social relationships. In *Developmental Psychobiology: Developmental Neurobiology and Behavioral Ecology*. Vol. 13, *Handbook of Behavioral Neurobiology*, ed. E. Blass, 281–316. New York: Kluwer Academic/Plenum.

———. 2004. The early history of Hamiltonian-based research on kin recognition. *Annales Zoologici Fennici* 41:691–711.

Holmes, W. G., and P. W. Sherman. 1982. The ontogeny of kin recognition in two species of ground squirrels. *American Zoologist* 22:491–517.

———. 1983. Kin recognition in animals. *American Scientist* 71:46–55.

Holmgren, M., R. Avilés, L. Sierralta, A. M. Segura, and E. R. Fuentes. 2000. Why have European herbs so successfully invaded the Chilean matorral? Effects of herbivory, soil nutrients, and fire. *Journal of Arid Environments* 44:197–211.

Holst, D. V., and F. Eichmann. 1998. Sex-specific regulation of marking behavior by sex hormones and conspecifics scent in tree shrews (*Tupaia belangeri*). *Physiology and Behavior* 63:157–64.

Homolka, M. 1983. On the problem of the exanthropic occurrence of *Rattus norvegicus*. *Folia Zoologica* 32:203–11.

Honey, P. L., and B. G. Galef, Jr. 2004. Long-lasting effects of rearing by an ethanol-consuming dam on ethanol consumption by rats. *Appetite* 43:261–68.

Honeycutt, R. L., and R. M. Adkins. 1993. Higher level systematics of eutherian mammals: An assessment of molecular characters and phylogenetic hypotheses. *Annual Review of Ecology and Evolution* 24:279–305.

Honeycutt, R. L., M. W. Allard, S. V. Edwards, and D. A. Schlitter. 1991. Systematics and evolution of the family Bathyergidae. In *The biology of the naked mole–rat*, ed. P. W. Sherman, J. U. M. Jarvis, and R. D. Alexander, 45–65. Princeton, NJ: Princeton University Press.

Honeycutt, R. L., D. L. Rowe, and M. H. Gallardo. 2003. Molecular systematics of the South American caviomorph rodents: Relationships among species and genera in the family Octodontidae. *Molecular Phylogenetics and Evolution* 26:476–89.

Hoogland, J. L. 1979a. Aggression, ectoparasitism, and other possible costs of prairie dog (Sciuridae: *Cynomys* spp.) coloniality. *Behaviour* 69:1–35.

———. 1979b. The effect of colony size on individual alertness of prairie dogs (Sciuridae; *Cynomys* spp.). *Animal Behaviour* 27:394–407.

———. 1981. The evolution of coloniality in white-tailed prairie dog and black-tailed prairie dogs (Sciuridae: *Cynomys leucurus* and *Cynomys ludovicianus*). *Ecology* 62:252–72.

———. 1982. Prairie dogs avoid extreme inbreeding. *Science* 215:1639–41.

———. 1983. Nepotism and alarm calling in the black-tailed prairie dog (*Cynomys ludovicianus*). *Animal Behaviour* 31:472–79.

———. 1985. Infanticide in prairie dogs: Lactating females kill offspring of close kin. *Science* 230:1037–40.

———. 1986. Nepotism in prairie dogs (*Cynomys ludovicianus*) varies with competition but not with kinship. *Animal Behaviour* 34:263–70.

———. 1992. Levels of inbreeding among prairie dogs. *American Naturalist* 139:591–602.

———. 1995. *The black-tailed prairie dog: Social life of a burrowing mammal.* Chicago: University of Chicago Press.

———. 1996a. The black-tailed prairie dog. *Mammalian Species* 535:1–10.

———. 1996b. Why do Gunnison's prairie dogs give antipredator calls? *Animal Behaviour* 51:871–80.

———. 1997. Duration of gestation and lactation for Gunnison's prairie dogs. *Journal of Mammalogy* 78:173–80.

———. 1998a. Estrus and copulation among Gunnison's prairie dogs. *Journal of Mammalogy* 79:887–97.

———. 1998b. Why do female Gunnison's prairie dogs copulate with more than one male? *Animal Behaviour* 55:351–59.

———. 1999. Philopatry, dispersal, and social organization of Gunnison's prairie dogs. *Journal of Mammalogy* 80:243–51.

———. 2001. Black-tailed, Gunnison's, and Utah prairie dogs all reproduce slowly. *Journal of Mammalogy* 82:917–27.

Hoogland, J. L. 2003a. Black-tailed prairie dog. In *Wild mammals of North America*, ed. G. A. Feldhamer, B. C. Thompson, and J. A. Chapman, 232–47. Baltimore: Johns Hopkins University Press.

Hoogland, J. L. 2003b. Sexual dimorphism in five species of prairie dogs. *Journal of Mammalogy* 84:1254–66.

Hoogland, J. L. 2006. Saving prairie dogs: Can we? Should we? In *Conservation of the black-tailed prairie dog*, ed. J. L. Hoogland, 261–66. Washington, DC: Island Press.

Hoogland, J. L., M. A. Hoogland, S. Davis, D. Kaulfus, and D. LaBruna. 2004. Pyraperm halts plague among Utah prairie dogs. *Southwestern Naturalist* 49:376–83.

Hoogland, J. L., and P. W. Sherman. 1976. Advantages and disadvantages of bank swallow (*Riparia riparia*) coloniality. *Ecological Monographs* 46:33–58.

Hoogland, J. L., R. H. Tamarin, and C. K. Levy. 1989. Communal nursing in prairie dogs. *Behavioral Ecology and Sociobiology* 24:91–95.

Hooper, E. T., and G. G. Musser. 1964. The glans penis in Neotropical cricetines (family Muridae), with comments on classification of muroid rodents. *Miscellaneous Publications of the Museum of Zoology of the University of Michigan* 123:1–57.

Hoppe, P. C. 1975. Genetic and endocrine studies of the pregnancy-blocking pheromone of mice. *Journal of Reproduction and Fertility* 45:109–15.

Horne, T. J., and H. Ylönen. 1996. Female bank voles (*Clethrionomys glareolus*) prefer dominant males; but what if there is no choice? *Behavioral Ecology and Sociobiology* 38:401-5.

———. 1998. Heritabilities of dominance-related traits in male bank voles (*Clethrionomys glareolus*). *Evolution* 52:894–99.

Horner, B. E. 1947. Paternal care of young mice in the genus *Peromyscus*. *Journal of Mammalogy* 28:31–36

———. Paternal care of the young and convulsive seizures in the grasshopper mouse. *American Zoologist* 1:360.

Horner, B. E., and J. M. Taylor. 1968. Growth and reproduction in the southern grasshopper mouse. *Journal of Mammalogy* 49:644–60.

Hornig, L. E., and M. K. McClintock. 1994. Unmasking sex-ratio biasing through targeted analysis. *Animal Behaviour* 47:1224–26.

———. 1996. Male sexual rest affects sex ratio of newborn Norway rats. *Animal Behaviour* 51:991–1005.

Howard, H. E. 1920. *Territory in bird life*. London: Collins.

Howard, W. E. 1949. Dispersal, amount of inbreeding, and longevity in a local population of prairie deermice on the George Reserve, Michigan. *Contributions from the Laboratory of Vertebrate Biology* 43:1–50.

Howard, W. E., and H. Childs. 1959. Ecology of pocket gophers with emphasis on *Thomomys Bottae mewa*. *Hilgardia* 29:277–358.

Howe, R. J. 1974. Marking behaviour of the Bahaman hutia (*Geocapromys ingrahami*). *Animal Behaviour* 22:645–49.

Howe, R. J., and G. C. Clough. 1971. The Bahaman hutia (*Geocapromys ingrahami*) in captivity. *International Zoo Yearbook* 11:89–93.

Howland, J. G., P. Taepavarapruk, and A. G. Phillips. 2002. Glutamate receptor-dependent modulation of dopamine efflux in the nucleus accumbens by basolateral, but not central, nucleus of the amygdala in rats. *Journal of Neuroscience* 22:1137–45.

Hoyt, S. Y., and S. F. Hoyt. 1950. Gestation period of the woodchuck, *Marmota monax*. *Journal of Mammalogy* 31:454.

Hrda S., J. Votypka, P. Kodym, and J. Flegr. 2000. Transient nature of *Toxoplasma gondii*-induced behavioural changes in mice. *Journal of Parasitology* 86:657–63.

Hrdy, S. B. 1974. Male-male competition and infanticide among the langurs (*Presbytis entellus*) of Abu, Rajasthan. *Folia Primatologica* 22:19–58.

———. 1977a. Infanticide as a primate reproductive strategy. *American Scientist* 65:40–49.

———. 1977b. *The langurs of Abu: Female and male strategies of reproduction*. Cambridge, MA: Harvard University Press.

———. 1979. Infanticide among animals: A review, classification, and examination of the implications for the reproductive strategies of females. *Ethology and Sociobiology* 1:13–40.

———. 1981. *The woman that never evolved*. Cambridge, MA: Harvard University Press.

———. 1986. Empathy, polyandry, and the myth of the coy female. In *Feminist approaches to science*, ed. R. Bleier, 119–46. New York: Pergamon.

Huang, L. Y., G. J. De Vries, and E. L. Bittman. 1998. Photoperiod regulates neuronal bromodeoxyuridine labeling in the brain of a seasonally breeding mammal. *Journal of Neurobiology* 36:410–20.

Huang, S.-W., K. G. Ardlie, and H.-T. Yu. 2001. Frequency and distribution of *t*-haplotypes in the Southeast Asian house mouse (*Mus musculus castaneus*) in Taiwan. *Molecular Ecology* 10:2349–54.

Hubbs, A. H., and R. Boonstra. 1997. Population limitation in arctic ground squirrels: Effects of food and predation. *Journal of Animal Ecology* 66:527–41.

Hubbs, A. H., J. S. Millar, and J. P. Wiebe. 2000. Effect of brief exposure to a potential predator on cortisol concentrations in female Columbian ground squirrels (*Spermophilus columbianus*). *Canadian Journal of Zoology* 78:578–87.

Huber, S., I. E. Hoffmann, E. Millesi, J. Dittami, and W. Arnold. 2001. Explaining the seasonal decline in litter size in European ground squirrels. *Ecography* 24:205–11.

Huber, S., E. Millesi, and J. Dittami. 2002. Paternal effort and its relation to mating success in the European ground squirrel. *Animal Behaviour* 63:157–64.

Huchon, D., F. M. Catzeflis, and E. J. P. Douzery. 1999. Molecular evolution of the nuclear von Willebrand factor gene in mammals and the phylogeny of rodents. *Molecular Biology and Evolution* 16:577–89.

———. 2000. Variance of molecular datings, evolution of rodents, and the phylogenetic affinities between Ctenodactylidae and Hystricognathi. *Proceedings of the Royal Society of London, Series B* 276:393–402.

Huchon, D., and E. J. P. Douzery. 2001. From the Old World to the New World: A molecular chronicle of the phylogeny and biogeography of Hystricognath rodents. *Molecular Phylogenetics and Evolution* 20:238–51.

Huchon, D., O. Madsen, M. J. J. B. Sibbald, K. Ament, M. J. Stanhope, F. Catzeflis, W. W. de Jong, and E. J. P. Douzery. 2002. Rodent phylogeny and a timescale for the evolution of Glires: Evidence from an extensive taxon sampling using three nuclear genes. *Molecular Biology and Evolution* 19:1053–65.

Huck, U. W. 1982. Pregnancy block in laboratory mice as a function of male social-status. *Journal of Reproduction and Fertility* 66:181–84.

———. 1984. Infanticide and the evolution of pregnancy block in rodents. In *Infanticide: Comparative and evolutionary perspectives*, ed. G. Hausfater and S. B. Hrdy, 349–65. New York: Aldine.

Huck, U. W., and E. M. Banks. 1979. Behavioral components of individual recognition in the collared lemming (*Dicrostonyx groenlandicus*). *Behavioral Ecology and Sociobiology* 6:85–90.

———. 1982. Male dominance status, female choice and mating success in the brown lemming, *Lemmus trimucronatus*. *Animal Behaviour* 30:665–75.

Huck, U. W., A. C. Bracken, and R. D. Lisk. 1983. Female-induced pregnancy block in the golden hamster. *Behavioral and Neural Biology* 38:190–93.

Huck, U. W., and R. D. Lisk. 1985. Determinants of mating success in the golden hamster (*Mesocricetus auratus*): I. Male capacity. *Journal of Comparative Psychology* 99:98–107.

———. 1986. Mating-induced inhibition of receptivity in the female golden hamster. *Behavioral and Neural Biology* 45:107–19.

Huck, U. W., R. D. Lisk, J. C. Allison, and C. G. Van Dongen. 1986. Determinants of mating success in the golden hamster (*Mesocricetus auratus*): Social dominance and mating tactics under seminatural conditions. *Animal Behaviour* 34:971–89.

Huck, U. W., R. L. Soltis, and C. B. Coopersmith. 1982. Infanticide in male laboratory mice: Effects of social status, prior sexual experience, and basis for discrimination between related and unrelated young. *Animal Behaviour* 30:1158–65.

Hudson, J. W. 1974. The estrous cycle, reproduction, growth, and development of temperature regulation in the pygmy mouse, *Baiomys taylori*, in Texas. *Journal of Mammalogy* 55:572–88.

Hudson, W. H. 1872. On the habits of the vizcacha (*Lagostomus trichodactylus*). *Proceedings of the Zoological Society of London* 1872:822–33.

Hughes, J. J., and D. Ward. 1993. Predation risk and distance to cover affect foraging behavior in Namib desert gerbils. *Animal Behaviour* 46:1243–45.

Hume, I. D., C. Beiglböck, T. Ruf, F. Frey-Roos, U. Bruns, and W. Arnold. 2002. Seasonal changes in morphology and function of the gastrointestinal tract of free-living alpine marmots (*Marmota marmota*). *Journal of Comparative Physiology B* 172:197–207.

Hume, I. D., K. R. Morgan, and G. J. Kenagy. 1993. Digesta retention and digestive performance in sciurid and microtine rodents: Effects of hindgut morphology and body size. *Physiological Zoology* 66:396–411.

Humle, T., and T. Matsuzawa. 2002. Ant-dipping among the chimpanzees of Boussou, Guinea, and some comparisons with other sites. *American Journal of Primatology* 58:133–48.

Humphries, R. E., D. H. L. Robertson, R. J. Beynon, and J. L. Hurst. 1999. Unravelling the chemical basis of competitive scent marking in house mice. *Animal Behaviour* 58:1177–90.

Humphries, R. E., R. M. Silby, and A. P. Meehan. 2000. Cereal aversion in behaviourally resistant house mice in Birmingham, England. *Applied Animal Behaviour Science* 66:323–33.

Hunt, P. S., J. L. Holloway, and E. M. Scordalakes. 2001. Social interaction with an intoxicated sibling can result in increased intake of ethanol by periadolescent rats. *Developmental Psychobiology* 38:101–9.

Hunt, P. S., G. M. Lant, and C. A. Carroll. 2000. Enhanced intake of ethanol in preweanling rats following interaction with intoxicated siblings. *Developmental Psychobiology* 37:90–99.

Hunter, R. H. F. 1975. Transport, migration and survival of spermatozoa in the female genital tract: Species with intra-uterine deposition of sperm. In *The biology of spermatozoa,* ed. E. S. E. Hafez and C. G. Thibault, 145–55. New York: Basel.

Hurley, K. M., H. Herbert, M. M. Moga, and C. B. Saper. 1991. Efferent projections of the infralimbic cortex of the rat. *Journal of Comparative Neurology* 308:249–76.

Hurst, J. L. 1987. The functions of urine marking in a free-living population of house mice, *Mus domesticus* Rutty. *Animal Behaviour* 35:1433–42.

———. 1989. The complex network of olfactory communication in populations of wild house mice *Mus domesticus* Rutty: Urine marking and investigation within family groups. *Animal Behaviour* 37:705–25.

———. 1990a. The network of olfactory communications operating in populations of wild house mice. In *Chemical signals in vertebrates* V, ed. D. W. Macdonald, D. Müller-Scharze, and S. E. Natynczuk, 401–14. New York: Plenum.

———. 1990b. Urine marking in populations of wild house mice *Mus domesticus* Rutty. I. Communication between males. *Animal Behaviour* 40:209–22.

———. 1990c. Urine marking in populations of wild house mice *Mus domesticus* Rutty. II. Communication between females. *Animal Behaviour* 40:223–32.

———. 1990d. Urine marking in populations of wild house mice *Mus domesticus* Rutty. III. Communication between the sexes. *Animal Behaviour* 40:233–43.

Hurst, J. L., and C. J. Barnard. 1995. Kinship and social tolerance among female and juvenile wild house mice: Kin bias but not kin discrimination. *Behavioral Ecology and Sociobiology* 36:333–42.

Hurst, J. L., R. J. Beynon, R. E. Humphries, N. Malone, C. M. Nevison, C. E. Payne, D. H. L. Robertson, and C. Veggerby. 2001. Information in scent signals of competitive social status: The interface between behaviour and chemistry. In *Chemical signals in vertebrates,* 9, ed. A. Marchlewska-Koj, J. L. Lepri, and D. Muller-Schwarze, 43–52. New York: Plenum.

Hurst, J. L., L. Hayden, M. Kingston, R. Luck, and K. Sorensen. 1994. Response of the aboriginal house mouse *Mus spretus* Lataste to tunnels bearing the odours of conspecifics. *Animal Behaviour* 48:1219–29.

Hurst, J. L., C. E. Payne, C. M. Nevison, A. D. Marie, R. E. Humphries, D. H. L. Robertson, A. Cavaggioni et al. 2001. Individual recognition in mice mediated by major urinary proteins. *Nature* 414:631–34.

Hurst, J. L., D. H. L. Robertson, U. Tolladay, and R. J. Beynon. 1999. Proteins in urine scent marks of male house mice extend the longevity of olfactory signals. *Animal Behaviour* 55:1289–97.

Huson, L. W., and B. D. Rennison. 1981. Seasonal variability of Norway rat (*Rattus norvegicus*) infestation of agricultural premises. *Journal of Zoology, London* 194:257–60.

Hutterer, R. 1994. Island rodents: A new species of *Octodon* from Isla Mocha, Chile (Mammalia: Octodontidae). *Zeitschrift für Säugetierkunde* 59:27–41.

Idris, M., and I. Prakash. 1985. Social and scent marking behaviour in Indian gerbil. *Tatera indica. Biology of Behavior* 10:31–39.

Ims, R. A. 1987a. Determinants of competitive success in *Clethrionomys rufocanus. Ecology* 68:1812–18.

———. 1987b. Responses in spatial-organization and behavior to manipulations of the food resource in the vole *Clethrionomys rufocanus. Journal of Animal Ecology* 56:585–96.

———. 1988. Spatial clumping of sexually receptive females induces space sharing among male voles. *Nature* 335:541–43.

———. 1989. Kinship and origin effects on dispersal and space sharing in *Clethrionomys rufocanus. Ecology* 70:607–16.

———. 1990. Determinants of natal dispersal and space use in grey-sided voles, *Clethrionomys rufocanus*: A combined field and laboratory experiment. *Oikos* 57:106–13.

Ims, R. A., and H. P. Andreassen. 1999. Effects of experimental habitat fragmentation and connectivity on root vole demography. *Journal of Animal Ecology* 68:839–52.

Ingles, L. G. 1947. Ecology and life history of the California gray squirrel. *California Fish and Game* 33:138–58.

Inglis, I. R., D. S. Shepherd, P. Smith, P. J. Haynes, D. S. Bull, D. P. Cowan, and D. Whitehead. 1996. Foraging behaviour of wild rats (*Rattus norvegicus*) towards new foods and bait containers. *Applied Animal Behaviour Science* 47:175–90.

Ingram, C. M., H. Burda, and R. L. Honeycutt. 2004. Molecular phylogenetics and taxonomy of the African mole-rats, genus *Cryptomys* and the new genus *Coetomys* Gray, 1864. *Molecular Phylogenetics and Evolution* 31:997–1014.

Inman, A. J., and J. R. Krebs. 1987. Predation and group living. *Trends in Ecology and Evolution* 2:31–32.

Inouye, D. W., B. Barr, K. B. Armitage, and B. D. Inouye. 2000. Climate change is affecting altitudinal migrants and hibernating species. *Proceedings of the National Academy of Sciences (USA)* 97:1630–33.

Insel, T. R. 2003. Is social attachment an addictive disorder? *Physiology and Behavior* 79:351–57.

Insel, T. R., R. Gelhard, and L. E. Shapiro. 1991. The comparative distribution of forebrain receptors for neurohypophyseal peptides in monogamous and polygamous mice. *Neuroscience* 43:623–30.

Insel, T. R., S. Preston, and J. T. Winslow. 1995. Mating in the monogamous male: Behavioral consequences. *Physiology and Behavior* 57:615–27.

Insel, T. R., and L. E. Shapiro. 1992. Oxytocin receptor distribution reflects social organization in monogamous and polygamous voles. *Proceedings of the National Academy of Science (USA)* 89:5981–85.

Insel, T. R., Z. X. Wang, and C. F. Ferris. 1994. Patterns of brain vasopressin receptor distribution associated with social organization in microtine rodents. *Journal of Neuroscience* 14:5381–92.

Insley, S. J. 2000. Long-term vocal recognition in the northern fur seal. *Nature* 406:404–5.

———. 2001. Mother-offspring vocal recognition in northern fur seals is mutual but asymmetrical. *Animal Behaviour* 61:129–37.

International Human Genome Sequencing Consortium. 2001. Initial sequencing and analysis of the human genome. *Nature* 409:860–921.

Ipinza, J., M. Tamayo, and J. Rottmann. 1971. Octodontidae en Chile. *Noticiario Mensual del Museo Nacional de Historia Natural, (Chile)* 183:3–10.

Irvin, R. W., P. Szot, D. M. Dorsa, M. Potegal, and C. F. Ferris. 1990. Vasopressin in the septal area of the golden hamster controls scent marking and grooming. *Physiology and Behavior* 48:693–99.

Irwin, R. E. 1990. Directional sexual selection cannot explain variation in song repertoire size in the New World blackbirds (Icterinae). *Ethology* 85:212–24.

———. 1996. The phylogenetic content of avian courtship display and song evolution. In *Phylogenies and the comparative method in animal behavior*, ed. E. P. Martins, 234–52. Oxford: Oxford University Press.

Ishizawa, H., B. Tabakoff, I. N. Mefford, and P. L. Hoffman. 1990. Reduction of arginine vasopressin binding sites in mouse lateral septum by treatment with 6-hydroxydopamine. *Brain Research* 507:189–94.

Isles, A. R., M. J. Baum, D. Ma, E. B. Keverne, and N. D. Allen. 2001. Urinary odour preferences in mice. *Nature* 409:783–84.

IUCN. 2003. *2003 IUCN red list of threatened species*. Accessed 4 February, 2004, at www.redlist.org.

Ivanitskaya, E., and E. Nevo. 1998. Cytogenetics of mole rats of the *Spalax ehrenberg* superspecies from Jordan (Spalacidae, Rodentia.) *Zeit Saugetierkunde* 63:336–46.

Izquierdo, G. 2005. La cría cooperativa en el tucu-tucu colonial *Ctenomys sociabilis* (Rodentia, Ctenomyidae): Efecto del tamaño de grupo. Master's thesis, Universidad de la República, Montevideo,Uruguay.

Jaarola, M., N. Martínkova, I. Gunduz, C. Brunhoff, J. Zima, A. Nadachowski, G. Amori, et al. 2004. Molecular phylogeny of the speciose vole genus *Microtus* (Arvicolinae, Rodentia) inferred from mitochondrial DNA sequences. *Molecular Phylogenetics and Evolution* 33:647–63.

Jackel, M., and F. Trillmich. 2003. Olfactory individual recognition of mothers by young guinea-pigs (*Cavia porcellus*). *Ethology* 109:197–208.

Jackson, J. E., L. C. Branch, and D. Villareal. 1996. *Lagostomus maximus*. *Mammalian Species* 543:1–6.

Jackson, M. H., W. M. Hutchison, and C. J. Siim. 1986. Toxoplasmosis in a wild rodent population of central Scotland and a possible explanation of the mode of transmission. *Journal of Zoology, London* 209:549–57.

Jackson, T. P. 1999. The social organization and breeding system of Brants' whistling rat (*Parotomys brantsii*). *Journal of Zoology, London* 247:323–31.

Jackson, W. B. 1977. Evaluation of rodent depredations to crops and stored products. *European and Mediterranean Plant Protection Organization Bulletin* 7:439–58.

Jacob, J., and J. S. Brown. 2000. Microhabitat use, giving-up densities and temporal activity as short- and long-term antipredator behaviors in common voles. *Oikos* 91:131–38.

Jacobs, D. S., and S. Kuiper. 2000. Individual recognition in the Damaraland mole-rat, *Cryptomys damarensis* (Rodentia: Bathyergidae). *Journal of Zoology* 251:411–15.

Jacobs, L. F. 1992. Memory for cache locations in Merriam's kangaroo rats. *Animal Behaviour* 43:585–93.

Jacobs, L. L. 1984. Rodentia. In *Mammals: Notes for a short course, Paleontological Society*, ed. P. D. Gingerich and C. E. Badgley, 155–66. Knoxville: University of Tennessee Studies in Geology.

Jacobs, L. L., L. J. Flynn, and W. R. Downs. 1989. Neogene rodents of southern Asia. In *Papers on fossil rodents in honor of Albert Elmer Wood*, ed. C. C. Black and M. R. Dawson, 157–77. Science Series No. 33. Los Angeles: Natural History Museum of Los Angeles County.

Jacobs, L. L., and D. Pilbeam, 1980. Of mice and men: Fossil-based divergence dates and molecular "clocks." *Journal of Human Evolution* 9:551–55.

Jacquot, J. J., and S. H. Vessey. 1994. Non-offspring nursing in the white-footed mouse, *Peromyscus leucopus*. *Animal Behaviour* 48:1238–40.

———. 1995. Influence of the natal environment of dispersal of white-footed mice. *Behavioral Ecology and Sociobiology* 37:407–12.

Jaeger, J.-J. 1988. Rodent phylogeny: New data and old problems. In *The phylogeny and classification of the tetrapods*, ed. M. J. Benton, 177–99. Oxford: Clarendon.

Jaisson, P. 1991. Kinship and fellowship in ants and social wasps. In *Kin recognition*, ed. P.G. Hepper, 60–93. Cambridge: Cambridge University Press.

Jaksic, F. M., and E. R. Fuentes. 1980. Why are native herbs in the Chilean matorral more abundant beneath bushes: Microclimate or grazing? *Journal of Ecology* 68:665–69.

Jaksic, F. M., H. W. Greene, and J. L. Yáñez. 1981. The guild structure of a community of predatory vertebrates in central Chile. *Oecologia* 49:21–28.

Jaksic, F. M., and M. Lima. 2003. Myths and facts on ratadas: Bamboo blooms, rainfall peaks and rodent outbreaks in South America. *Austral Ecology* 28:237–51.

Jaksic, F. M., P. L. Meserve, J. R. Gutiérrez, and E. L. Tabilo. 1993. The components of predation on small mammals in semiarid Chile: Preliminary results. *Revista Chilena de Historia Natural* 66:305–21.

Jakubowski, M., and J. Terkel. 1982. Infanticide and caretaking in non-lactating *Mus musculus*: Influence of genotype, family group and sex. *Animal Behaviour* 30:1029–35.

———. 1985a. Incidence of pup killing and parental behavior in virgin female and male rats (*Rattus norvegicus*): Differences between Wistar and Sprague-Dawley stocks. *Journal of Comparative Psychology* 99:93–97.

———. 1985b. Transition from pup killing to parental behavior in male and virgin female albino rats. *Physiology and Behavior* 34:683–86.

James, W. H. 1996. Evidence that mammalian sex ratios at birth are partially controlled by parental hormone levels at the time of conception. *Journal of Theoretical Biology* 180:271–86.

Jannett, F. J. 1978. Density-dependent formation of extended maternal families of montane vole, *Microtus montanus nanus*. *Behavioral Ecology and Sociobiology* 3:245–63.

Jannett, F. J., Jr. 1980. Social dynamics of the montane vole, *Microtus montanus*, as a paradigm. *The Biologist* 62:3–19.

Jans, J. E., and M. Leon. 1983. Determinants of mother-young contact in Norway rats. *Physiology and Behavior* 30:919–35.

Jansa, S. A., S. M. Goodman, and P. K. Tucker. 1999. Molecular phylogenetics of the native rodents of Madagascar: A test of the single origin hypothesis. *Cladistics* 15:253–70.

Jansa, S. A., and M. Weksler. 2004. Phylogeny of muroid rodents: Relationships within and among major lineages as determined by IRBP gene sequences. *Molecular Phylogenetics and Evolution* 31:256–76.

Janse van Rensburg, L., N. C. Bennett, M. van der Merwe, and A. S. Schoeman. 2002. Seasonal reproduction in the highveld mole–rat, *Cryptomys hottentotus pretoriae*. *Canadian Journal of Zoology* 80:810–20.

Janse van Rensburg, L., N. C. Bennett, M. van der Merwe, A. S. Schoeman, and J. Brinders. 2003. Are non–reproductive male highveld mole–rats, *Cryptomys hottentotus pretoriae* physiologically suppressed while in the confines of the natal colony? *Journal of Zoology* 260:73–78.

Jarman, P. J. 1974. The social organization of antelope in relation to their ecology. *Behaviour* 48:215–67.

Jarrell, G. H. 1995. A male-biased natal sex ratio in inbred collared lemmings, *Dicrostonyx groenlandicus*. *Hereditas* 123:31–37.

Jarvis, J. U. M. 1978. Energetics of survival in *Heterocephalus glaber* (Rüppell), the naked mole–rat (Rodentia: Bathyergidae). *Bulletin of the Carnegie Museum of Natural History* 6:81–87.

———. 1981. Eusociality in a mammal: Co-operative breeding in naked mole-rat *Heterocephalus glaber* colonies. *Science* 212:571–73.

———. 1985. Ecological studies of *Heterocephalus glaber*, the naked mole–rat, in Kenya. *National Geographic Society Research Reports* 20:429–37.

Jarvis, J. U. M., and N. C. Bennett. 1990. The evolution history, population biology, and social structure of African mole rats: Family Bathyergidae. In *Evolution of subterranean mammals at the organismal and molecular levels*, ed. E. Nevo and O. A. Rieg, 97–128. New York: Wiley-Liss.

———. 1991. Ecology and behavior of the family Bathyergidae. In *The biology of the naked mole-rat*, ed. P. W. Sherman, J. U. M. Jarvis, and R. D. Alexander, 66–96. Princeton, NJ: Princeton University Press.

———. 1993. Eusociality has evolved independently in two genera of bathyergid mole–rats—but occurs in no other subterranean mammal. *Behavioral Ecology and Sociobiology* 33:253–60.

Jarvis, J. U. M., N. C. Bennett, and A. C. Spinks. 1998. Food availability and foraging by wild colonies of Damaraland mole–rats (*Cryptomys damarensis*): Implications for sociality. *Oecologia* 113:290–98.

Jarvis, J. U. M., M. J. O'Riain, N. C. Bennett, and P. W. Sherman. 1994. Mammalian eusociality: A family affair. *Trends in Ecology and Evolution* 9:47–51.

Jarvis, J. U. M., and J. B. Sale. 1971. Burrowing and burrow patterns of East African mole-rats *Tachyoryctes, Heliophobius* and *Heterocephalus*. *Journal of Zoology, London* 163:451–79.

Jarvis, J. U. M., and P. W. Sherman. 2002. *Heterocephalus glaber*. *Mammalian Species* 706:1–9.

Jedrzejewska, B., and W. Jedrzejewski. 1990. Antipredatory behaviour of bank voles and prey choice of weasels: Enclosure experiments. *Annales Zoologici Fennici* 27:321–28.

Jekel, P. A., C. Ciabatti, C. Schuller, J. J. Beintema, and E. Nevo. 1990. Ribonuclease in different chromosomal species of the mole rats, superspecies *Spalax ehrenbergi*: Concentration in the pancreas and primary structure. In *Evolution of subterranean mammals at the organismal and molecular levels*, ed. E. Nevo and A. O. Reig, 367–81. New York: Alan R. Liss.

Jemiolo, B., J. Alberts, S. Sochinski-Wiggins, S. Harvey, and M. Novotny. 1985. Behavioural and endocrine responses of female mice to synthetic analogues of volatile compounds in male urine. *Animal Behaviour* 33:1114–18.

Jemiolo, B., F. Andreolini, T. Xie, D. Wiesler, and M. Novotny. 1989. Puberty- affecting synthetic analogs of urinary chemosignals in the house mouse, *Mus domesticus*. *Physiology and Behavior* 46:293–98.

Jenkins, S. H., and P. E. Busher. 1979. *Castor Canadensis*. *Mammalian Species* 120:1–8.

Jennions, M. D., and D. W. Macdonald. 1994. Cooperative breeding in mammals. *Trends in Ecology and Evolution* 9:89–93.

Jennions, M. D., and M. Petrie. 2000. Why do females mate multiply? A review of the genetic benefits. *Biological Review* 75:21–64.

Jeppsson, B. 1986. Mating by pregnant water voles (*Arvicola terrestris*): A strategy to counter infanticide by males? *Behavioral Ecology and Sociobiology* 19:293–96.

Jesseau, S. A. 2004. Kin discrimination and social behavior in communally-nesting degus (*Octodon degus*). PhD diss., University of Michigan, Ann Arbor.

Jiménez, J. A., K. A. Hughes, G. Alaks, L. Graham, and R. C. Lacy. 1994. An experimental study of inbreeding depression in a natural habitat. *Science* 266:271–73.

Jimenez, J. E. 1996. The extirpation and current status of wild chinchillas *Chinchilla lanigera* and *C. brevicaudata*. *Biological Conservation* 77:1–6.

Johnson, C. N. 1988. Dispersal and the sex ratio at birth in primates. *Nature* 332:726–28.

Johnson, D. D., P. R. Kays, P. G. Blackwell, and D. W. Mcdonald. 2002. Does the resource dispersion hypothesis explain group living? *Trends in Ecology and Evolution* 17:563–70.

Johnson, D. W., and D. P. Armstrong. 1987. *Peromyscus crinitus*. *Mammalian Species* 287:1–8.

Johnson, K. 1981. Social organization in a colony of rock squirrels (*Spermophilus variegates*, Sciuridae). *The Southwestern Naturalist* 26:237–42.

Johnson, K. M. 1985. Arenaviruses. In *Virology*, ed. B. N. Fields and K. N. Knipe, 1033–53. New York: Raven Press.

Johnson, M. L. 1989. Threatened rodents in the northwestern North American region. In *Rodents, a world survey of species of conservation concern*, ed. W. Z. Lidicker, Jr., 1–3. Occasional Papers no. 4. Gland, Switzerland: IUCN Rodent Specialist Group, Species Survival Commission.

Johnson, R. P. 1973. Scent marking in mammals. *Animal Behaviour* 21:521–35.

———. 1975. Scent marking with urine in two races of the bank vole (*Clethrionomys glareolus*). *Behaviour* 55: 81–93.

Johnson, S. A., and J. Choromanski-Norris. 1992. Reduction in the eastern limit of the range of the Franklin's ground squirrel (*Spermophilus franklinii*). *American Midland Naturalist* 128: 325–31.

Johnston, P. G., and G. H. Brown. 1969. A comparison of the relative fitness of genotypes segregating for the $tw2$ allele in laboratory stock and its possible effect on gene frequency in mouse populations. *American Naturalist* 103:5–21.

Johnston, R. E. 1977. The causation of two scent marking behavior patterns in female hamsters (*Mesocricetus auratus*). *Animal Behaviour* 25:317–27.

———. 1981. Attraction to odors in hamsters: An evaluation of methods. *Journal of Comparative and Physiological Psychology* 95:951–60.

———. 1992. Vomeronasal and/or olfactory mediation of ultrasonic calling and scent marking by female golden hamsters. *Physiology and Behavior* 51:437–48.

———. 1993. Memory for individual scent in hamsters (*Mesocricetus auratus*) as assessed by habituation methods. *Journal of Comparative Psychology* 107:201–7.

———. 2003. Chemical communication in rodents: From pheromones to individual recognition. *Journal of Mammalogy* 84: 1141–62.

Johnston, R. E., and A. Bhorade. 1998. Perception of scent over-marks by golden hamsters (*Mesocricetus auratus*): Novel mechanisms for determining which individual's mark is on top. *Journal of Comparative Psychology* 112:230–43.

Johnston, R. E., and T. A. Bullock. 2001. Individual recognition by use of odours in golden hamsters: the nature of individual representations. *Animal Behaviour* 61:545–57.

Johnston, R. E., G. Chiang, and C. Tung. 1994. The information in scent over-marks of golden hamsters. *Animal Behaviour* 48:323–30.

Johnston, R. E., R. Munver, and C. Tung. 1995. Scent counter marks: Selective memory for the top scent by golden hamsters. *Animal Behaviour* 49:1435–42.

Johnston, R. E., and M. Peng. 2000. The vomeronasal organ is involved in discrimination of individual odors by males but not by females in golden hamsters. *Physiology and Behavior* 70:537–49.

Johnston, R. E., and K. Rasmussen. 1983. Individual recognition of female hamsters by males: roles of chemical cues on the olfactory and vomeronasal systems. *Physiology and Behavior* 33:95–104.

Johnston, R. E., and T. A. Robinson. 1993. Cross-species discrimination of individual odors by hamsters (Muridae, *Mesocricetus auratus, Phodopus campbelli*). *Ethology* 94: 317–25.

Johnston, R. E., E. S. Sorokin, and M. H. Ferkin. 1997a. Female

voles discriminate males' over-marks and prefer top-scent males. *Animal Behaviour* 54:679–90.

Johnston, R. E., E. S. Sorokin, and M. H. Ferkin. 1997b. Scent counter-marking by male meadow voles: Females prefer the top-scent male. *Ethology* 103:443–53.

Johnston, T. D. 1982. Selective costs and benefits in the evolution of learning. In *Advances in the study of behavior*, vol. 12, ed. J. S. Rosenblatt, R. A. Hinde, B. C., and M.-C. Busnel, 65–106. New York: Academic Press.

———. 1987. The persistence of dichotomies in the study of behavioral development. *Developmental Review* 7:149–82.

Johnstone, R. A., and M. A. Cant. 1999. Reproductive skew and indiscriminate infanticide. *Animal Behaviour* 57:243–49.

Jones, C., and N. Hildreth. 1989. *Neotoma stephensi. Mammalian Species* 328:1–3.

Jones, C. G., J. H. Lawton, and M. Schachack. 1997. Positive and negative effects of organisms as physical ecosystem engineers. *Ecology* 78:1946–57.

Jones, C. G., R. S. Ostfeld, M. P. Richard, E. M. Schauber, and J. O. Wolff. 1998. Chain reactions linking acorns to gypsy moth outbreaks and Lyme disease risk. *Science* 279: 1023–26.

Jones, J. K., and H. H. Genoways. 1978. *Neotoma phenax. Mammalian Species* 108:1–3.

Jones, J. S., and K. E. Wynne-Edwards. 2000. Paternal hamsters mechanically assist the delivery, consume amniotic fluid and placenta, remove fetal membranes, and provide parental care during the birth process. *Hormones and Behavior* 37:116–25.

Jones, M., Y. Mandelik, and T. Dayan. 2001. Coexistence of temporally partitioned spiny mice: Roles of habitat structure and foraging behavior. *Ecology* 82:2164–76.

Jones, R. B., and N. W. Nowell. 1973. The effect of urine on the investigatory behaviour of male albino mice. *Physiology and Behavior* 11:35–38.

———. 1989. Aversive potency of urine from dominant and subordinate male laboratory mice (*Mus musculus*): Resolution of a conflict. *Aggressive Behavior* 15:291–96.

Jones, T. 1993. Social systems of Heteromyid rodents. In *Biology of the Heteromyidae*, Special publication no. 10, ed. H. H. Genoways and J. H. Brown, 575–95. Provo, UT: American Society of Mammalogists.

Jones, W. T. 1984. Natal philopatry in bannertailed kangaroo rats. *Behavioral Ecology and Sociobiology* 15:151–55.

———. 1986. Survivorship in philopatric and dispersing kangaroo rats (*Dipodomys spectabilis*). *Ecology* 67:202–7.

———. 1989. Dispersal distance and the range of nightly movements in Merriam's kangaroo rats. *Journal of Mammalogy* 70:27–34.

———. T. 1993. The social systems of heteromyid rodents. In *Biology of the heteromyidae*, ed. H. H. Genoways and J. H. Brown, 575–93. Shippensburg, PA: American Society of Mammalogists.

Jones, W. T., P. M. Waser, L. F. Elliott, N. E. Link, and B. B. Bush. 1988. Philopatry, dispersal, and habitat saturation in the banner-tailed kangaroo rats, *Dipodomys spectabilis*. *Ecology* 69:1466–73.

Jonsson, P., J. Agrell, E. Koskela, and T. Mappes. 2002. Effects of litter size on pup defence and weaning success of neighbouring bank vole females. *Canadian Journal of Zoology* 80:1–5.

Jordan, W. C., and M. W. Bruford. 1998. New perspectives on mate choice and the MHC. *Heredity* 81:127–33.

Jorgenson, J. P. 1986. Notes on the ecology and behaviour of capybaras in Northeastern Columbia. *Vida Silvestre Neotropical* 1:31–40.

Joy, J. E. 1984. Population differences in circannual cycles of thirteen-lined ground squirrels. In *The biology of ground-dwelling squirrels: Annual cycles, behavioral ecology, and sociality,* ed. J. O. Murie, and G. R. Michener, 125–41. Lincoln: University of Nebraska Press.

Judd, T. M. and P. W. Sherman. 1996. Naked mole-rats recruit colony mates to food sources. *Animal Behaviour* 52:957–69.

Kaczmarski, F. 1966. Bioenergetics of pregnancy and lactation in the bank vole. *Acta Theriologica* 11:409–17.

Kalat, J. W., and P. Rozin. 1973 "Learned Safety" as a mechanism in long-delay taste-aversion learning in rats. *Journal of Comparative and Physiological Psychology* 83:198–207.

Kalcounis-Rüppell, M. C. 2000. Breeding systems, habitat overlap and activity patterns of monogamous and promiscuous mating in *Peromyscus californicus* and *P. boylii.* PhD diss., University of Western Ontario, London, Ontario.

Kalcounis-Rüppell, M. C., and J. S. Millar. 2002. Partitioning of space, food, and time by syntopic *Peromyscus boylii* and *P. californicus. Journal of Mammalogy* 82:614–25.

Kalcounis-Rüppell, M. C., J. S. Millar, and E. J. Herdman. 2002. Beating the odds: Weather effects on a short season population of mice. *Canadian Journal of Zoology* 80: 1594–1601.

Kalcounis-Rüppell, M. C., J. S. Millar, and J. R. Speakman. 2002. The energetic cost of reproduction for monogamous and promiscuous small mammals. Abstract published in the proceedings of the 9th International Society for Behavioral Ecology Congress. Montreal.

Kalcounis-Rüppell, M. C., and T. R. Spoon. In revision. *Peromyscus boylii. Mammalian Species.*

Kalela, O. 1961. Seasonal change of habitat in the Norwegian lemming, *Lemmus lemmus* (L.). *Annales Academiae Scientiarum Fennicae Ser. A, IV* 55:1–72.

Kalela, O., L. Kilpeläinen, T. Kopenen, and J. Tast. 1971. Seasonal differences in habitats of the Norwegian lemming, *Lemmus lemmus* (L.), in 1959 and 1960 at Kilpisjärvi. *Annales Academiae Scientiarum Fennicae, Ser. A, IV* 178:1–22.

Kalthoff, D. C. 1999a. Jungpleistozäne Murmeltiere (Rodentia, Sciuridae) vom Mittelrhein (Deutschland) und ihre vervandtschaftlichen Beziehungen Zu den beiden rezenten europäischen Arten. *Staphia* 63:119–28.

Kalthoff, D. C. 1999b. Ist *Marmota primigenia* (Kaup) eine eigenständige Art? Osteologische Variabilität pleistozäner *Marmota*-Populationen (Rodentia:Sciuridae) in Neuwieder Becken (Rheinland-Pfalz, Deutschland) und benachbarter Gebiete. *Kaupia* 9:127–86.

Kaltwasser, M. T., and H. U. Schnitzler. 1981. Echolocation signals confirmed in rats. *Zietschrift für Saugetierkunde* 46: 394–95.

Kano, T. 1992. *The last ape: Pygmy chimpanzee behavior and ecology.* Trans. E. O. Vineberg. Stanford, CA: Stanford University Press.

Kareem, A. M., and C. J. Barnard. 1982. The importance of kinship and familiarity in social interactions between mice. *Animal Behaviour* 30:594–601.

Karels, T. J., and R. Boonstra. 2000. Concurrent density dependence and independence in populations of arctic ground squirrels. *Nature* 408:460–63.

Karels, T. J., E. Byrom, R. Boonstra, and C. J. Krebs. 2000. The interactive effects of food and predators on reproduction and overwinter survival of arctic ground squirrels. *Journal of Animal Ecology* 69:235–47.

Karlsson, A. F. 1988. Over-winter survival in a boreal population of the bank vole, *Clethrionomys glareolus. Canadian Journal of Zoology* 66:1835–40.

Kaufman, D. W., and G. A. Kaufman. 1989. Population biology. In *Advances in the study of* Peromyscus, ed. G. L. Kirkland, Jr. and J. N. Layne, 233–70. Lubbock: Texas Tech University Press.

Kavaliers, M. 1988. Brief exposure to a natural predator, the short-tailed weasel, induces benzodiazepine-sensitive analgesia in white-footed mice. *Physiology and Behavior* 43:187–93.

Kavaliers, M., and D. D. Colwell. 1993. Aversive responses of female mice to the odors of parasitized males: Neuromodulatory mechanisms and implications for mate choice. *Ethology* 95:202–12.

Kavaliers, M., and D. D. Colwell. 1995a. Discrimination by female mice between the odours of parasitized and non-parasitized males. *Proceedings of the Royal Society of London, Series B* 261:31–35.

———. 1995b. Odours of parasitized males induce aversive responses in female mice. *Animal Behaviour* 50:1161–69.

Kavaliers, M., M. A. Fudge, D. D. Colwell, and E. Choleris. 2003. Aversive and avoidance responses of female mice to the odors of males infected with an ectoparasite and the effects of prior familiarity. *Behavioral Ecology and Sociobiology* 54: 423–30.

Kavaliers, M., and K. P. Ossenkopp. 2001. Corticosterone rapidly reduces male odor preferences in female mice. *Neuroreport* 12:2999–3002.

Kawata, M. 1987. Pregnancy failure and suppression by female-female interaction in enclosed populations of the red-backed vole *Clethrionomys rufocanus bedfordiae. Behavioural Ecology and Sociobiology* 20:89–97.

———. 1988. Mating succcess, spatial organization, and male characteristics in experimental field populations of the red-backed vole, *Clethrionomys rufocanus bedfordiae. Journal of Animal Ecology* 57:217–35.

Kaye, S. V. 1961. Movements of harvest mice tagged with gold-198. *Journal of Mammalogy* 42:323–37.

Kays, R. W., and D. E. Wilson. 2002. *Mammals of North America.* Princeton, NJ: Princeton University Press.

Keane, B. 1990a. Dispersal and inbreeding avoidance in the white-footed mouse, *Peromyscus leucopus. Animal Behaviour* 40:143–52.

———. 1990b. The effect of relatedness on reproductive success and mate choice in the white-footed mouse, *Peromyscus leucopus. Animal Behaviour* 39:264–73.

Keeling, M., and C. A. Gilligan. 2000. Metapopulation dynamics of Bubonic Plague. *Nature* 407:903–6.

Keer, S. E., and J. M. Stern. 1999. Dopamine receptor blockade in the nucleus accumbens inhibits maternal retrieval and licking, but enhances nursing behavior in lactating rats. *Physiology and Behavior* 67:659–69.

Keesing, F. 1998a. Ecology and behavior of the pouched mouse, *Saccostomus mearnsi,* in central Kenya. *Journal of Mammalogy* 79:919–31.

———. 1998b. Impacts of ungulates on the demography and diversity of small mammals in central Kenya. *Oecologia* 116: 381–89.

Keevin, T. M., Z. T. Halpin, and N. McCurdy. 1981. Individual and sex-specific odors in male and female eastern chipmunks (*Tamias striatus*). *Behavioral Biology* 6:329–38.

Keightley, M-C., and P. J. Fuller. 1996. Anomalies in the endocrine axes of the guinea pig: Relevance to human physiology and disease. *Endocrine Review* 17:30–44.

Keil, A., J. T. Epplen, and N. Sachser. 1999. Reproductive success of males in the promiscuous-mating yellow-toothed cavy (*Galea musteloides*). *Journal of Mammalogy* 80:1257–63.

Keil, A., and N. Sacher. 1998. Reproductive benefits from female promiscuous mating in a small mammal. *Ethology* 104:897–903.

Keller, L. 1997. Indiscriminate altruism: Unduly nice parents and siblings. *Trends in Ecology and Evolution* 12:99–103.

Keller, L., and N. Perrin. 1995. Quantifying the level of eusociality. *Proceedings of the Royal Society of London, Series B* 260:311–15.

Keller, L., and H. K. Reeve. 1994. Partitioning of reproduction in animal societies. *Trends in Ecology and Evolution* 9:98–102.

———. 1995. Why do females mate with multiple males? The sexually selected sperm hypothesis. In *Advances in the study of behavior*, vol. 24, ed. P. J. B. Slater, J. S. Rosenblatt, C. T. Snowden, and M. Milinski, 291–315. New York: Academic Press.

Kelley, M. J. 1985. Species-typical taxic behavior and event-reinforcer interactions in conditioning. *Learning and Motivation* 16:301–14.

Kelly, P. 1989. Population ecology and social organization of dusky-footed woodrats, *Neotoma fuscipes*. PhD diss., University of California Berkeley.

Kelly, S. J., and T. D. Tran. 1997. Alcohol exposure during development alters social recognition and social communication in rats. *Neurotoxicology and Teratology* 19:383–89.

Kelson, K. R. 1949. Speciation of rodents in the Colorado River drainage of eastern Utah. PhD diss., University of Utah.

Kemble, E. D. 1984. Effects of preweaning predatory or consummatory experience and litter size on cricket predation in northern grasshopper mice (*Onychomys leucogaster*). *Aggressive Behavior* 10:55–58.

Kenagy, G. J. 1976. Field observation of male fighting, drumming and copulation in the Great Basin kangaroo rat, *Dipodomys microps*. *Journal of Mammalogy* 57:781–85.

Kenagy, G. J., and N. J. Place. 2000. Seasonal changes in plasma glucocorticoids of free-living yellow-pine chipmunks: Effects of reproduction and capture and handling. *General and Comparative Endocrinology* 117:189–99.

Kenagy, G. J., N. J. Place, and C. Veloso. 1999. Relation of glucocorticosteroids and testosterone to the annual cycle of free-living degus in semiarid central Chile. *General Comparative Endocrinology* 115:236–43.

Kenagy, G. J., and S. C. Trombulak. 1986. Size and function of mammalian testes in relation to body size. *Journal of Mammalogy* 67:1–22.

Kendall, P. B. 1984. Seasonal changes of sex ratio in Norway rat (*Rattus norvegicus*) populations in Wales. *Journal of Zoology, London* 203:208–11.

Kendrick, K. M., A. P. da Costa, A. E. Leigh, M. R. Hinton, and J. W. Peirce. 2001. Sheep don't forget a face. *Nature* 414:165–66.

Kenney, C. McM., R. L. Evans, and D. A. Dewsbury. 1977. Postimplantation pregnancy disruption in *Microtus ochrogaster*, *M. pennsylvanicus* and *Peromyscus maniculatus*. *Journal of Reproduction and Fertility* 49:365–67.

Kenward, R. E. 1985. Ranging behaviour and population dynamics in grey squirrels. *Symposium of the British Ecological Society* 25:319–30.

Kermack, W. O., and A. G. McKendrick. 1927. A contribution to the mathematical theory of epidemics. *Proceedings of the Royal Society of London, Series A* 115:700–21.

Keverne, E. B. 2002. Mammalian pheromones: From genes to behaviour. *Current Biology* 12:R807–9.

Keverne, E. B., N. D. Martensz, and B. Tuite. 1989. Beta-endorphin concentrations in cerebrospinal fluid of monkeys are influenced by grooming relationships. *Psychoneuroendocrinology* 14:155–61.

Khokhlova, I., B. R. Krasnov, G. I. Shenbrot, and A. Degen. 2001. Body mass and environment: A study in Negev rodents. *Israel Journal of Zoology* 47:1–13.

Kildaw, S. D. 1995. The effect of group size manipulations on the foraging behavior of black-tailed prairie dogs. *Behavioral Ecology* 6:353–58.

Kilgore, D. L., Jr., and K. B. Armitage. 1978. Energetics of yellow-bellied marmot populations. *Ecology* 59:78–88.

Kilpatrick, C. W., and K. L. Crowell. 1985. Genic variation of the rock vole, *Microtus chrotorrhinus*. *Journal of Mammalogy* 66:94–101.

Kim, I., C. J. Phillips, J. A. Monjeau, E. C. Birney, K. Noack, D. E. Pumo, R. S. Sikes et al. 1998. Habitat islands, genetic diversity, and gene flow in a Patagonian rodent. *Molecular Ecology* 7:667–78.

King, J. A. 1955. Social behavior, social organization, and population dynamics in a black-tailed prairie dog town in the Black Hills of South Dakota. *Contributions of the Laboratory of Vertebrate Biology, University of Michigan* 67:1–123.

———. 1984. Historical ventilations on a prairie dog town. *The biology of ground-dwelling squirrels*, ed. J. O. Murie and G. R. Michener, 447–56. Lincoln: University of Nebraska Press.

King, W. J. 1989a. Kin-differential behaviour of adult female Columbian ground squirrels. *Animal Behaviour* 38:354–56.

———. 1989b. Spacing of female kin in Columbian ground squirrels (*Spermophilus columbianus*). *Canadian Journal of Zoology* 67:91–95.

King, W. J., and D. Allainé. 2002. Social, maternal, and environmental influences on reproductive success in female Alpine marmots (*Marmota marmota*). *Canadian Journal of Zoology* 80:2137–43.

King, W. J., M. Festa-Bianchet, and S. E. Hatfield. 1991. Determinants of reproductive success in female Columbian ground squirrels. *Oecologia* 86:528–34.

Kinlaw, A. 1999. A review of burrowing by semi-fossorial vertebrates in arid environments. *Journal of Arid Environments* 41:127–45.

Kinsley, C. H. 1994. Developmental psychobiological influences on rodent parental behavior. *Neuroscience and Biobehavioral Reviews* 18:269–80.

Kirkpatrick, B., C. S. Carter, S. W. Newman, and T. R. Insel. 1994. Axon-sparing lesions of the medial nucleus of the amygdala decrease affiliative behaviors in the prairie vole (*Microtus ochrogaster*): Behavioral and anatomical specificity. *Behavioral Neuroscience* 108:501–13.

Kivanç, E. 1988. Geographic variations of Turkish *Spalax* species (Spalacidae, Rodentia, Mammalia) PhD diss., Ankara University, Turkey (in Turkish with English summary).

Kivastik, T., K. Vuorikallas, T. P. Piepponen, A. Zharkovsky, and L. Ahtee. 1996. Morphine- and cocaine-induced conditioned place preference: Effects of quinpirole and preclamol. *Pharmacology Biochemistry and Behavior* 54:371–75.

Kivett, V. K., J. O. Murie, and A. L. Steiner. 1976. A comparative study of scent-gland location and related behavior in some northwestern neararctic ground squirrel species (*Sciuridae*): An evolutionary approach. *Canadian Journal of Zoology* 54:1294–1306.

Kiwia, H.Y.D. 1989. Ranging patterns of the black rhinoceros (*Diceros bicornis L.*) in Ngorongoro Crater, Tanzania. *African Journal of Ecology* 27:305–12.

Klauer, G., H. Burda, and E. Nevo. 1997. Adaptive differentiations of the skin of the head in a subterranean rodent, *Spalax ehrenbergi. Journal of Morphology* 233:53–66.

Kleiman, D. 1966. Scent marking in the Canidae. *Symposium of the Zoological Society of London* 18:167–77.

Kleiman, D. G. 1970. Reproduction in the female green acouchi, *Myoprocta pratti* Pocock. *Journal of Reproduction and Fertility* 23:55–65.

———. 1972. Maternal behaviour of the green acouchi (*Myoprocta pratti* Pocock), a South American caviomorph rodent. *Behaviour* 43:48–84.

———. 1974. Patterns of behaviour in hystricomorph rodents. In *The biology of the hystricomorph rodents,* ed. I. W. Rowlands and B. J. Weir, 171–209. Symposium of the Zoological Society of London 34. London: Academic Press.

———. 1977. Monogamy in mammals. *The Quarterly Review of Biology* 52:39–69.

———. 1981. Correlations among life history characteristics of mammalian species exhibiting two extreme forms of monogamy. In *Natural selection and social behavior,* ed. R. D. Alexander and D. W. Tinkle, 332–44. Oxford: Blackwell Scientific.

Kleiman, D. G., J. F. Eisenberg, and E. Maliniak. 1979. Reproductive parameters and productivity of caviomorph rodents. In *Vertebrate ecology in the northern neotropics,* ed. J. F. Eisenberg, 173–83. Washington, DC: Smithsonian Institution Press.

Kleiman, D. G., and J. R. Malcolm. 1981. The evolution of male parental investment in mammals. In *Parental care in mammals,* ed. D. J. Gubernick and P. H. Klopfer, 347–87. New York: Plenum.

Klein, J. 1986. *Natural history of the histocompatibility complex.* New York: Wiley.

Klein, S. L., H. R. Gamble, and R. J. Nelson. 1999. *Trichinella spiralis* infection in voles alters female odor preference but not partner preference. *Behavioral Ecology and Sociobiology* 45:323–29.

Kleinschmidt, T., E. Nevo, M. Goodman, and G. Braunitzer. 1985. Mole rat hemoglobin: Primary structure and evolutionary aspects in a second karyotype of *Spalax ehrenbergi, Rodentia, (2n = 52). Biological Chemistry* 366:679–85.

Kling, A. S., and L. A. Brothers. 1992. The amygdala and social behavior. In *The amygdala: Neurobiological aspects of emotion, memory, and mental dysfunction,* ed. J. P. Aggleton, 353–77. New York: Wiley-Liss.

Klump, G. M., E. Kretzschmar, and E. Curio. 1986. The hearing of an avian predator and its avian prey. *Behavioral Ecology and Sociobiology* 18:317–23.

Klump, G. M., and M. D. Shalter. 1984. Acoustic behaviour of birds and mammals in the predator context. I. Factors affecting the structure of alarm signals. II. The functional significance and evolution of alarm signals. *Zeitschrift für Tierpsychologie* 66:189–226.

Knopf, F. L., and D. F. Balph. 1969. Badgers plug burrows to confine prey. *Journal of Mammalogy* 50:635–36.

Knopf, J. L., J. F. Gallagher, and W. A. Held. 1983. Differential multihormonal regulation of the mouse major urinary protein gene family in the liver. *Molecular and Cell Biology* 3:2232–40.

Knowles, C. J. 1985. Observations on prairie dog dispersal in Montana. *The Prairie Naturalist* 17:33–40.

———. 1986a. Population recovery of black-tailed prairie dogs following control with zinc phosphide. *Journal of Range Management* 39:249–50.

———. 1986b. Some relationships of black-tailed prairie dogs to livestock grazing. *Great Basin Naturalist* 46:198–203.

———. 1987. Reproductive ecology of black-tailed prairie dogs in Montana. *Great Basin Naturalist* 47:202–6.

Knowles, C. J., and P. R. Knowles. 1984. Additional records of mountain plovers using prairie dog towns in Montana. *Prairie Naturalist* 16:183–86.

Knowles, C. J, C. J. Stoner, and S. P. Gieb. 1982. Selective use of black-tailed prairie dog towns by mountain plovers. *Condor* 84:71–74.

Kobayashi, T., and M. Watanabe. 1986. An analysis of snake-scent application behaviour in Siberian chipmunks (*Eutamias sibiricus asiaticus*). *Ethology* 72:40–52.

Koenig, W. D., and J. L. Dickinson, eds. 2004. *Ecology and evolution of cooperative breeding in birds.* Cambridge: Cambridge University Press.

Koenig, W. D., and F. A. Pitelka. 1981. Ecological factors and kin selection in the evolution of cooperative breeding in birds. In *Natural selection and social behavior,* ed. R. D. Alexander and D. W. Tinkle, 261–80. New York: Chiron.

Koenig, W. D., F. A. Pitelka, W. J. Carmen, R. L. Mumme, and M. T. Stanback. 1992. The evolution of delayed dispersal in cooperative breeders. *Quarterly Review of Biology* 67:111–50.

Koeppl, J. W., and R. S. Hoffmann. 1981. Comparative postnatal growth of four ground squirrel species. *Journal of Mammalogy* 62:41–57.

Koeppl, J. W., R. S. Hoffmann, and C. F. Nadler. 1978. Pattern analysis of acoustical behavior in four species of ground squirrels. *Journal of Mammalogy* 59:677–96.

Koford, C. B. 1958. Prairie dogs, whitefaces, and blue grama. *Wildlife Monographs* 3:1–78.

Koford, R. R. 1982. Mating system of a territorial tree squirrel (*Tamiasciurus douglasii*) in California. *Journal of Mammalogy* 63:274–83.

Kogan, J. H., P. W. Frankland, J. A. Blendy, J. Coblentz, Z. Marowitz, G. Shutz, and A. J. Silva. 1997. Spaced training induces normal long-term memory in CREB muutant mice. *Current Biology* 7:1–16.

Kohli, K. L., and M. H. Ferkin. 1999. Over-marking and adjacent marking are influenced by sibship in male prairie voles, *Microtus ochrogaster. Ethology* 105:1–11.

Koivisto, E., and J. Pusenius. 2003. Effects of temporal variation in the risk of predation by least weasel (*Mustela nivalis*) on feeding behaviour of field vole (*Microtus agrestis*). *Evolutionary Ecology* 17:477–89.

Koivula, M., and E. Korpimäki. 2001. Do scent marks increase predation risk of microtine rodents? *Oikos* 95:275–81.

Koivula, M., E. Koskela, and J. Viitala. 1999. Sex and age-specific differences in ultraviolet reflectance of scent marks of bank voles (*Clethrionomys glareolus*). *Journal of Comparative Physiology* 185:561–64.

Koivula, M., and J. Viitala. 1999. Rough-legged buzzards use vole scent marks to assess hunting areas. *Journal of Avian Biology* 30:329–32.

Koivula, M., J. Viitala, and E. Korpimäki. 1999. Kestrels prefer scent marks according to species and reproductive status of voles. *Ecoscience* 6:415–20.

Koivunen, V., E. Korpimäki, and H. Hakkarainen. 1996. Differential avian predation on sex and size classes of small mammals: Doomed surplus or dominant individuals? *Annales Zoologici Fennici* 33:293–301.

———. 1998. Refuge sites of voles under owl predation risk: Priority of dominant individuals? *Behavioral Ecology* 9:261–66.

Kokko, H., and E. Ranta. 1996. Evolutionary optimality of delayed breeding in voles. *Oikos* 77:173–75.

Kokko, H., and R. A. Johnstone. 1999. Social queuing in animal societies: A dynamic model of reproductive skew. *Proceedings of the Royal Society of London, Series B* 266:571–78.

Kokko, H., R. A. Johnstone, and T. H. Clutton-Brock. 2001. The evolution of cooperative breeding through group augmentation. *Proceedings of the Royal Society of London, Series B* 268:187–96.

Kokko, H., and P. Lundberg. 2001. Dispersal, migration, and offspring retention in saturated habitats. *American Naturalist* 157:188–202.

Kollack-Walker, S., and S. W. Newman. 1995. Mating and agonistic behavior produce different patterns of Fos immunolabeling in the male Syrian hamster brain. *Neuroscience* 66:721–36.

Komdeur, J., and B. J. Hatchwell. 1999. Kin recognition: Function and mechanism in avian societies. *Trends in Ecology and Evolution* 14:237–41.

Komers, P. E., and P. N. M Brotherton. 1997. Female space use is the best predictor of monogamy in mammals. *Proceedings of the Royal Society of London, Series B* 264:1261–70.

König, B. 1989a. Behavioural ecology of kin recognition in house mice. *Ethology Ecology and Evolution* 1:99–110.

———. 1989b. Kin recognition and maternal care under restricted feeding in house mice (*Mus domesticus*). *Ethology* 82:328–43.

———. 1993. Maternal investment of communally nursing female house mice (*Mus musculus domesticus*). *Behavioural Processes* 30:61–74.

———. 1994a. Communal nursing in mammals. *Verhandlungen der Deutschen Zoologische Gesellschaft* 87:115–27.

———. 1994b. Components of lifetime reproductive success in communally and solitarily nursing house mice: A laboratory study. *Behavioral Ecology and Sociobiology* 34:275–83.

———. 1997. Cooperative care of young in mammals. *Naturwissenschaften* 84:95–104.

König, B., and H. Markl. 1987. Maternal care in house mice. I. The weaning strategy as a means for parental manipulation of offspring quality. *Behavioral Ecology and Sociobiology* 20:1–9.

König, B., J. Riester, and H. Markl. 1988. Maternal care in house mice (*Mus musculus*): II. The energy cost of lactation as a function of litter size. *Journal of Zoology, London* 216:195–210.

Koolhaas, J. M., S. M. Korte, S. F. De Boer, B. J. Van Der Veg, C. G. Van Reenen, H. Hopster, I. C. De Jong et al. 1999. Coping styles in animals: Current status in behavior and stress-physiology. *Neuroscience and Biobehavioral Reviews* 23:925–35.

Koprowski, J. L. 1991. Mixed-species mating chases of fox squirrels and gray squirrels. *Canadian Field-Naturalist* 105:117–18.

———. 1992. Removal of copulatory plugs by female tree squirrels. *Journal of Mammalogy* 73:572–76.

———. 1993a. Alternative reproductive tactics in male eastern gray squirrels: "Making the best of a bad job". *Behavioral Ecology* 4:165–71.

———. 1993b. Behavioral tactics, dominance, and copulatory success among male fox squirrels. *Ethology Ecology and Evolution* 5:169–76.

———. 1993c. Do estrous female gray squirrels, *Sciurus carolinensis*, advertise their receptivity? *Canadian Field-Naturalist* 106:392–94.

———. 1993d. Sex and species biases in scent marking by fox squirrels and eastern gray squirrels. *Journal of Zoology* 230:319–23.

———. 1994a. *Sciurus carolinensis*. *Mammalian Species* 480:1–9.

———. 1994b. *Sciurus niger*. *Mammalian Species* 479:1–9.

———. 1996. Natal philopatry, communal nesting, and kinship in fox squirrels and eastern gray squirrels. *Journal of Mammalogy* 77:1006–16.

———. 1998. Conflict between the sexes: A review of social and mating systems of the tree squirrels. In *Ecology and evolutionary biology of tree squirrels*, ed. M. A. Steele, J. F. Merritt, and D. A. Zegers, 33–41. Martinsville: Virginia Museum of Natural History.

Koprowski, J. L., J. L. Roseberry, and W D. Klimstra. 1988. Longevity records for the fox squirrel. *Journal of Mammalogy* 69:383–84.

Korpimäki, E., and C. J. Krebs. 1996. Predation and population cycles of small mammals. *BioScience* 46:754–64.

Korpimäki, E., K. Norrdahl, and J. Valkama. 1994. Reproductive investment under fluctuating predation risk: Microtine rodents and small mustelids. *Evolutionary Ecology* 8:357–68.

Korpimäki, E., V. Koivunen, and H. Hakkarainen. 1996. Microhabitat use and behaviour of voles under weasel and raptor predation risk: Predator facilitation? *Behavioral Ecology* 7:30–34.

Koskela, E., T. Horne, T. Mappes, and H. Ylönen. 1995. Does risk of mustelid predation affect oestrus cycle in the bank vole *Clethrionomys glareolus*? *Animal Behaviour* 51:1159–63.

Koskela, E., T. Mappes, and H. Ylönen. 1997. Territorial behaviour and reproductive success of bank vole *Clethrionomys glareolus* females. *Journal of Animal Ecology* 66:341–49.

Koskela, E., and H. Ylönen. 1995. Suppressed breeding in the field vole (*Microtus agrestis*): An adaptation to cyclically fluctuating predation risk. *Behavioral Ecology* 6:311–15.

Kosoy, M. L., R. L. Regnery, T. Tzianabos, E. L. Marsont, D. C. Jones, D. Green, G. O. Maupin, J. A. Olson, and J. E. Childs. 1997. Distribution, diversity, and host specificity of *Bartonella* in rodents from the southeastern United States. *American Journal of Tropical Medicine and Hygiene* 57:578–88.

Kotler, B. P. 1984. Predation risk and the structure of desert rodent communities. *Ecology* 65:689–701.

Kotler, B. P. 1985. Owl predation on desert rodents that differ in morphology and behavior. *Journal of Mammalogy* 66:824–28.

Kotler, B. P. 1997. Patch use by gerbils in a risky environment: Manipulating food and safety to test four models. *Oikos* 78:274–82.

Kotler, B. P., and L. Blaustein. 1995. Titrating food and safety in a heterogeneous environment: When are the risky and safe patches of equal value? *Oikos* 74:251–58.

Kotler, B. P., L. Blaustein, and J. S. Brown. 1992. Predator facilitation: The combined effect of snakes and owls on the foraging behaviour of gerbils. *Annales Zoologici Fennici* 29:199–206.

Kotler, B. P., J. S. Brown, S. R. X. Dall, S. Gresser, D. Ganey, and A. Bouskila. 2002. Foraging games between gerbils and their predators: Temporal dynamics of resource depletion and apprehension in gerbils. *Evolutionary Ecology Research* 4:495–518.

Kotler, B. P., J. S. Brown, and O. Hasson. 1991. Owl predation on gerbils: The role of body size, illumination and habitat structure on rates of predation. *Ecology* 72:2249–60.

Kotler, B. P., J. S. Brown, A. Oldfield, J. Thorson, and D. Cohen. 2001. Foraging substrate and escape substrate: Patch use by three species of gerbils. *Ecology* 82:1781–90.

Kotler, B. P., J. S. Brown, R. H. Slotow, W. L. Goodfriend, and M. I. Strauss. 1993. The influence of snakes on the foraging behavior of gerbils. *Oikos* 67:309–18.

Kotler, B. P., J. S. Brown, R. J. Smith, and W. O. Wirtz II. 1988. The effects of morphology and body size on rates of owl predation on desert rodents. *Oikos* 53:145–52.

Kotliar, N. B. 2000. Application of the new keystone-concept to prairie dogs: How well does it work? *Conservation Biology* 14:1715–21.

Kotliar, N. B., B. W. Baker, A. D. Whicker, and G. Plumb. 1999. A critical review of assumptions about the prairie dog as a keystone species. *Environmental Management* 24:177–92.

Kotliar, N. B., B. Miller, R. P. Reading, and T. W. Clark. 2006. The prairie dog as a keystone species. In *Conservation of the black-tailed prairie dog*, ed. J. L. Hoogland, 53–64. Washington, DC: Island Press.

Kozlov, A. N. 1979. Population of northern Kazakhstan by the brown rat. *The Soviet Journal of Ecology* 10:572–75.

Krackow, S. 1992. Sex ratio manipulation in wild house mice: The effect of fetal resorption in relation to the mode of reproduction. *Biology of Reproduction* 47:541–48.

Krackow, S. 1995. The developmental asynchrony hypothesis for sex ratio manipulation. *Journal of Theoretical Biology* 176:273–80.

Krackow, S. 1997. Maternal investment, sex differential prospects, and the sex ratio in wild house mice. *Behavioral Ecology and Sociobiology* 41:435–43.

Krackow, S. 2002. Why parental sex ratio manipulation is rare in higher vertebrates. *Ethology* 108:1041–56.

Krackow, S., and P. S. Burgoyne. 1998. Timing of mating, developmental asynchrony and the sex ratio in mice. *Physiology and Behavior* 63:81–84.

Krackow, S., and H. N. Hoeck. 1989. Sex ratio manipulation, maternal investment and behavior during concurrent pregnancy and lactation in house mice. *Animal Behaviour* 37:177–86.

Krackow, S., and B. Matuschak. 1991. Mate choice for non-siblings in wild house mice: Evidence from a choice test and a reproductive test. *Ethology* 88:99–108.

Kramer, K. M., B. S. Cushing, and C. S. Carter. 2003. Developmental effects of oxytocin on stress response: Single versus repeated exposure. *Physiology and Behavior* 79:775–82.

Kramer, K. M., J. A. Monjeau, E. C. Birney, and R. S. Sikes. 1999. *Phyllotis xanthopygus*. *Mammalian Species* 617:1–7.

Krames, L., and L. A. Mastromatteo. 1973. Role of olfactory stimuli during copulation in male and female rats. *Journal of Comparative and Physiological Psychology* 85:528–35.

Kraus, C., J. Kenkele, and F. Trillmich. 2003. Spacing behaviour and its implications for the mating system of a precocial small mammal: An almost asocial cavy *Cavia magna*? *Animal Behaviour* 66:225–38.

Krause, J., and G. D. Ruxton. 2002. *Living in groups*. Oxford: Oxford University Press.

Krebs, C. J. 1966. Demographic changes in fluctuating populations of *Microtus californicus*. *Ecological Monographs* 36:239–73.

———. 1978. A review of Chitty's hypothesis of population regulation. *Canadian Journal of Zoology* 56:2463–80.

———. 1984. Voles and lemmings. In *The encyclopedia of mammals*, ed. D. W. Macdonald, 651–55. New York: Facts on File.

———. 1985. Do changes in spacing behaviour drive population cycles in small mammals? In *Behavioural ecology: Ecological consequences of adaptive behaviour*, ed. R. M. Sibly and R. H. Smith, 295–312. Oxford: Blackwell.

———. 1996. Population cycles revisited. *Journal of Mammalogy* 77:8–24.

———. 2003. How does rodent behaviour impact on population dynamics? In *Rats, mice and people: Rodent biology and management*, ed. G. R. Singleton, L. A. Hinds, C. J. Krebs, and D. M. Spratt, 117–23. Canberra: Australian Centre for International Agricultural Research.

Krebs, C. J., B. L. Keller, and R. H. Tamarin. 1969. *Microtus* population biology: Demographic changes in fluctuating populations of *M. ochrogaster* and *M. pennsylvanicus* in southern Indiana. *Ecology* 50:587–607.

Krebs, C. J., A. J. Kenney, and G. R. Singleton. 1995. Movements of feral house mice in agricultural landscapes. *Australian Journal of Zoology* 43:293–302.

Krebs, C. J., J. A. Redfield, and M. J. Taitt. 1978. A pulsed removal experiment on the vole *Microtus townsendii*. *Canadian Journal of Zoology* 56:2253–62.

Krebs, C. J., G. R. Singleton, and A. J. Kenney. 1994. Six reasons why feral house mouse populations might have low recapture rates. *Wildlife Research* 21:559–67.

Krinke, G. H., ed. 2000. *The laboratory rat*. San Diego, CA: Academic Press.

Krohne, D. T. 1997. Dynamics of metapopulations of small mammals. *Journal of Mammalogy* 78:1014–26.

Kronfeld, N., T. Dayan, N. Zisapel, and A. Haim. 1994. Coexisting populations of *Acomys cahirinus* and *A. russatus*: A preliminary report. *Israel Journal of Zoology* 40:177–83.

Kroodsma, D. E. 1982. Song repertoires: Problems in their definition and use. In *Acoustic communication in birds*, ed. D. E. Kroodsma and E. H. Miller, 125–46. New York: Academic Press.

Kruckenhauser, L., W. Pinsker, E. Haring, and W. Arnold. 1999. Marmot phylogeny revisited: Molecular evidence for a diphyletic origin of sociality. *Journal of Zoological Systematics and Evolutionary Research* 37:49–56.

Kruczek, M. 1997. Male rank and female choice in the bank vole, *Clethrionomys glareolus*. *Behavioural Processes* 40:171–76.

Kruczek, M., and A. Golas. 2003. Behavioural development of conspecific odour preferences in bank voles, *Clethrionomys glareolus*. *Behavioural Processes* 64:31–39.

Kruczek, M., and E. Pochron. 1997. Chemical signals from conspecifics modify the activity of female bank voles *Clethrionomys glareolus*. *Acta Theriologica* 42:71–78.

Kuch, M., N. Rohland, J. L. Betancourt, C. Latorre, S. Steppan, and H. N. Poinar. 2002. Molecular analysis of an 11,700-year-old rodent midden from the Atacama Desert, Chile. *Molecular Ecology* 11:913–24.

Kuenzi, A. J., R. J. Douglass, D. White, C. W. Bond, and J. N. Mills. 2001. Antibody to Sin Nombre virus in rodents associated with peridomestic habitats in west central Montana. *American Journal of Tropical Medicine and Hygiene* 64:137–46.

Kuenzi, A. J., M. L. Morrison, D. E. Swann, P. C. Hardy, and G. T. Downard. 1999. A longitudinal study of Sin Nombre virus activity in rodents of southeastern Arizona. *Emerging Infectious Diseases* 5:113–17.

Kumar, S., and B. Hedges. 1998. A molecular tree for vertebrate evolution. *Nature* 392:917–19.

Kunkel, P., and I. Kunkel. 1964. Report on the ethology of guinea pigs *Cavia apera f. porcellus* (L.). *Zeitschrift für Tierpsychologie* 21:602–41.

Künkele, J., and H. N. Hoeck. 1989. Age-dependent discrimination of unfamiliar pups in *Galea musteloides* (Mammalia, Caviidae). *Ethology* 83:316–19.

———. 1995. Communal suckling in the cavy *Galea musteloides*. *Behavioral Ecology and Sociobiology* 37:385–91.

Künkele, J., and F. Trillmich. 1997. Are precocial young cheaper? Lactation energetics in the guinea pig. *Physiological Zoology* 70:589–96.

Künzl, C., S. Kaiser, E. Meier, and N. Sachser. 2003. Is a wild mammal kept and reared in captivity still a wild animal? *Hormones and Behavior* 43:187–96.

Kurtén, B., and E. Anderson. 1980. *Pleistocene mammals of North America*. New York: Columbia University Press.

Kushnirov, D., F. Beolchini, F. Lombardini, and E. Nevo. 1998. Radiotracking studies in the blind mole rat, *Spalax ehrenbergi*. (Abstract 381). Euro-American Mammal Congress, 19–24 July 1998, Santiago de Compostela, Spain.

Kutcheruk, V. V., I. L. Kulik, and I. A. Dubrovsky. 1972. Great gerbil as a life form of a desert rodent. [In Russian.] In *Fauna and ecology of rodents*. vol. 11, ed. V. V. Kutcheruk, 5–70. Moscow: Moscow University Press.

Kvarnemo, C., and I. Ahnesjo. 1996. The dynamics of operational sex ratios and competition for mates. *Trends in Ecology, Evolution and Ecology* 11:404–8.

Labov, J. B. 1980. Factors influencing infanticidal behavior in wild male house mice (*Mus musculus*). *Behavioral Ecology and Sociobiology* 6:297–303.

Labov, J. B. 1981a. Male social status, physiology, and ability to block pregnancies in female house mice (*Mus musculus*). *Behavioral Ecology and Sociobiology* 8:287–91.

———. 1981b. Pregnancy blocking in rodents: Adaptive advantages for females. *American Naturalist* 118:361–71.

Labov, J. B., U. W. Huck, R. W. Elwood, and R. J. Brooks. 1985. Current problems in the study of infanticidal behavior of rodents. *Quarterly Review of Biology* 60:1–20.

Labov, J. B., U. W. Huck, P. Vaswani, and R. D. Lisk. 1986. Sex ratio manipulation and decreased growth of male offspring in undernourished golden hamsters (*Mesocricetus auratus*). *Behavioral Ecology and Sociobiology* 18:241–49.

Lacey, E. A. 1991. Reproductive and dispersal strategies of male arctic ground squirrels (*Spermophilus parryii plesius*). PhD diss., University of Michigan.

———. 1992. Infanticide in Arctic ground squirrels. *American Zoologist* 32:169A.

———. 2000. Spatial and social systems of subterranean rodents. In *Life underground: The biology of subterranean*

rodents, ed. E. A. Lacey, J. L. Patton, and G. N. Cameron, 257–96. Chicago: University of Chicago Press.

———. 2004. Sociality reduces individual direct fitness in a communally breeding rodent, the colonial tuco-tuco (*Ctenomys sociabilis*). *Behavioral Ecology and Sociobiology* 56:449–57.

Lacey, E. A., S. H. Braude, and J. R. Wieczorek. 1997. Burrow sharing by colonial tuco-tucos (*Ctenomys sociabilis*). *Journal of Mammalogy* 78:556–62.

———. 1998. Solitary burrow use by adult Patagonian tuco-tucos (*Ctenomys haigi*). *Journal of Mammalogy* 79:986–91.

Lacey, E. A., J. L. Patton, and G. N. Cameron, eds. 2000. *Life underground: The biology of subterranean rodents*. Chicago: University of Chicago Press.

Lacey, E. A., and P. W. Sherman. 1991. Social organization of naked mole-rat colonies: Evidence for division of labor. In *The biology of the naked mole-rat*, ed. P. W. Sherman, J. U. M. Jarvis, and R. D. Alexander, 275–336. Princeton, NJ: Princeton University Press.

———. 1997. Cooperative breeding in naked mole rats: Implications for vertebrate and invertebrate sociality. In *Cooperative breeding in mammals*, ed. N. G. Solomon and J. A. French, 267–301. Cambridge: Cambridge University Press.

———. 2005. Redefining eusociality: Goals, concepts, and levels of analysis. *Annales Zoologici Fenneci* 42:573–77.

Lacey, E. A., and J. R. Wieczorek. 2001. Territoriality and male reproductive success in arctic ground squirrels. *Behavioral Ecology* 12:626–32.

———. 2003. Ecology of sociality in rodents: A ctenomyid perspective. *Journal of Mammalogy* 84:1198–1211.

———. 2004. Kinship in colonial tuco-tucos: Evidence from group composition and population structure. *Behavioral Ecology* 15:988–96.

Lacey, E. A., J. R. Wieczorek, and P. K. Tucker. 1997. Male mating behaviour and patterns of sperm precedence in Arctic ground squirrels. *Animal Behaviour* 53:767–79.

Lacher, T. E., Jr. 1981. The comparative social behavior of *Kerodon rupestris* and *Galea spixii* and the evolution of behavior in the Caviidae. In *Bulletin of the Carnegie Museum of Natural History*, ed. H. H. Genoways, D. A. Schlitter, and S. L. Williams, 1–71. No. 17. Pittsburgh: Trustees of Carnegie Institute.

Lacher, T. E., Jr., and C. J. R. Alho. 1989. Microhabitat use among small mammals in the Brazilian Pantanal. *Journal of Mammalogy* 70:396–401.

Lack, D. 1954. *The natural regulation of animal numbers*. Oxford: Clarendon.

Lackey, J. A., D. G. Huckaby, and B. G. Ormiston. 1985. *Peromyscus leucopus*. *Mammalian Species* 247:1–10.

Lacy, R. C. 1997. Importance of genetic variation to the viability of mammalian populations. *Journal of Mammalogy* 78:320–35.

Lacy, R. C., A. Petric, and M. Warneke. 1993. Inbreeding and outbreeding in captive populations of wild animal species. In *The natural history of inbreeding and outbreeding*, ed. N. W. Thornhill, 352–74. Chicago: University of Chicago Press.

Lacy, R. C., and P. W. Sherman. 1983. Kin recognition by phenotype matching. *American Naturalist* 121:489–512.

Lagerspetz, K. 1964. Studies on the aggressive behaviour of mice. *Annales Academiae Scientiarum Fennicae* 131:1–131.

Lagerström, M., and I. Häkkinen. 1978. Uneven sex ratio of

voles in the food of *Aogolius funereus*. *Ornis Fennica* 55: 149–53.

Lagos, V. O., L. C. Contreras, P. L. Meserve, J. R. Gutiérrez, and F. M. Jaksic. 1995. Effects of predation risk on space use by small mammals: A field experiment with a Neotropical rodent. *Oikos* 74:259–64.

Lai, S.-C., N. Y. Vasilieva, and R. E. Johnston. 1996. Odors providing sexual information in Djungarian hamsters: evidence for an across-odor code. *Hormones and Behavior* 30:26–36.

Lai, W. S., and R. E. Johnston. 2002. Individual recognition after fighting by golden hamsters: A new method. *Physiology and Behavior* 76:225–39.

Laland, K. N., P. J. Richerson, and R. Boyd. 1996. Developing a theory of animal social learning. In *Social learning in animals: The roots of culture,* ed. C. M. Heyes and B. G. Galef, Jr., 129–54. San Diego: Academic Press.

Laland, K. N., and H. C. Plotkin. 1990. Social learning and social transmission of foraging information in Norway rats (*Rattus norvegicus*). *Animal Learning and Behavior* 18:246–51.

———. 1991. Excretory deposits surrounding food sites facilitate social learning of food preferences in Norway rats. *Animal Behaviour* 41:997–1005.

———. 1992. Further experimental analysis of the social learning and transmission of foraging information amongst Norway rats. *Behavioural Processes* 27:53–64.

Lamb, B. L., R. P. Reading, and W. F. Andelt. 2006. Attitudes and perceptions about prairie dogs. In *Conservation of the black-tailed prairie dog,* ed. J. L. Hoogland, 108–14. Washington, DC: Island Press.

Lamb, C. E., and R. J. Van Aarde. 2001. Maternal dietary protein intake and sex-specific investment in *Mastomys coucha* (Rodentia: Muridae). *Journal of Zoology, London* 253:505–12.

Lambin, X. 1994a. Litter sex ratio does not determine natal dispersal tendency in female Townsend's voles. *Oikos* 69:353–56.

———. 1994b. Natal philopatry, competition for resources and inbreeding avoidance in Townsend's voles (*Microtus townsendii*). *Ecology* 75:224–35.

———. 1994c. Sex ratio variation in relation to female philopatry in Townsend's voles. *Journal of Animal Ecology* 63:945–53.

———. 1997. Home range shifts by breeding female Townsend's voles (*Microtus townsendii*): A test of the territory bequeathal hypothesis. *Behavioral Ecology and Sociobiology* 40:363–72.

Lambin, X., J. Aars, and S. B. Piertney. 2001. Dispersal, intraspecific competition, kin competition and kin facilitation: A review of the empirical evidence. In *Dispersal,* ed. J. Clobert, E. Danchin, A. A. Dhondt, and J. D. Nichols, 110–22. Oxford: Oxford University Press.

Lambin, X., and C. J. Krebs. 1991a. Can changes in female relatedness influence microtine population dynamics? *Oikos* 61:126–32.

———. 1991b. Spatial organization and mating system of *Microtus townsendii*. *Behavioral Ecology and Sociobiology* 28:353–63.

———. 1993. Influence of female relatedness on the demography of female Townsend's vole populations in the spring. *Journal of Animal Ecology* 62:536–50.

Lambin, X., C. J. Krebs, R. Moss, and N. G. Yoccoz. 2002. Population cycles: Inferences from experimental, modeling, and time series approaches. In *Population cycles: The case for trophic interactions,* ed. A. Berryman, 155–76. Oxford: Oxford University Press.

Lambin, X., and N. G. Yoccoz. 1998. The impact of population kin-structure on nestling survival in Townsend's voles, *Microtus townsendii*. *Journal of Animal Ecology* 67:1–16.

Lammers, A. R., H. A. Dziech, and R. Z. German. 2001. Ontogeny of sexual dimorphism in *Chinchilla lanigera* (Rodentia: Chinchillidae). *Journal of Mammalogy* 82:179–89.

Lancia, R. A. 1979. Year-long activity patterns of radio-marked beaver (*Castor canadensis*). PhD diss., University of Massachusetts, Amherst.

Land, R. B., and T. E. McGill. 1967. The effects of the mating pattern of the mouse on the formation of corpora lutea. *Journal of Reproduction and Fertility* 13:121–25.

Lande, R. 1980. Sexual dimorphism, sexual selection, and adaptation in polygenic characters. *Evolution* 34:292–305.

Landry, C., D. Garant, P. Duchesne, and L. Bernatchez. 2001. 'Good genes as heterozygosity': The major histocompatibility complex and mate choice in Atlantic salmon (*Salmo salar*). *Proceedings of the Royal Society of London, Series B* 268:1279–85.

Landry, P.-A., and F.-J. Lapointe. 2001. Within population craniometric variability in insular populations of deer mice, *Peromyscus maniculatus,* elucidated by landscape configuration. *Oikos* 95:136–46.

Landry, S. O. 1999. A proposal for a new classification and nomenclature for the Glires (Lagomorpha and Rodentia). *Mitteilungen aus dem Zoologischen Museum in Berlin* 75:283–316.

Langtimm, C. A., and D. A. Dewsbury. 1991. Phylogeny and evolution of rodent copulatory behaviour. *Animal Behaviour* 41:217–26.

Lanier, D. L., D. Q. Estep, and D. A. Dewsbury. 1979. Role of prolonged copulatory behavior in facilitating reproductive success in a competitive mating situation in laboratory rats. *Journal of Comparative and Physiological Psychology* 93:781–92.

Lantz, D. E. 1909. *The brown rat in the United States.* Biological Survey Bulletin 33. Washington, DC: United States Department of Agriculture.

Larsen, K. W., and S. Boutin. 1998. Sex-unbiased philopatry in the North American red squirrel: (*Tamiasciurus hudsonicus*). In *Ecology and evolutionary biology of tree squirrels,* ed. M. A. Steele, J. F. Merritt, and D. A. Zegers, 21–32. Martinsville: Virginia Museum of Natural History.

Lattanzio, R. M., and J. A. Chapman. 1980. Reproduction and physiological cycles in an island population of Norway rats. *Bulletin of the Chicago Academy of Science* 12:1–68.

Latter, B. D. H. 1998. Mutant alleles of small effect are primarily responsible for the loss of fitness with slow inbreeding in *Drosophila melanogaster*. *Genetics* 148:1143–58.

Laurance, W. F. 1990. Comparative responses of five arboreal marsupials to tropical forest fragmentation. *Journal of Mammalogy* 71:641–53.

———. 1991. Ecological correlates of extinction proneness in Australian tropical rainforest mammals. *Conservation Biology* 5:79–89.

———. 1995. Rainforest mammals in a fragmented landscape. In *Landscape approaches in mammalian ecology and conservation,* ed. W. Z. Lidicker, Jr., 46–63. Minneapolis: University of Minnesota Press.

Laurance, W. F., R. O. Bierregaard, Jr., C. Gascon, R. K. Didham, A. P. Smith, A. T. Lynam, V. M. Viana, et al. 1997. Tropical forest fragmentation: Synthesis of a diverse and dynamic discipline. In *Tropical forest remnants,* ed. W. F. Laurance and R. O. Bierregaard, Jr., 502–14. Chicago: University of Chicago Press.

Laurance, W. F., S. G. Laurance, L. V. Ferreira, J. M. Rankin-de-Merona, C. Gascon, T. E. Lovejoy, et al. 1997. Biomass collapse in Amazonian forest fragments. *Science* 278:1117–18.

Laurie, E. M. O. 1946. The reproduction of the house mouse (*Mus musculus*) living in different environments. *Proceedings of the Royal Society of London, Series B* 133:248–81.

Laurien-Kehnen, C., and F. Trillmich. 2003. Lactation performance of guinea pigs (*Cavia porcellus*) does not respond to experimental manipulation of pup demands. *Behavioral Ecology and Sociobiology* 53:145–52.

Lavenex, P., M. A. Steele, and L. F. Jacobs. 2000. The seasonal pattern of cell proliferation and neuron number in the dentate gyrus of wild adult eastern grey squirrels. *European Journal of Neuroscience* 12:643–48.

Lavocat, R. 1973. Les rongeurs du Miocene d'Afrique Orientale. *Mémoires et Travaux de l'Institut de Montpellier de l'École Pratique des Hautes Études, Institut de Montpellier* 1:1–284.

———. 1978. Rodentia and Lagomorpha. In *Evolution of African mammals,* ed. V. J. Maglio and H. B. S. Cooke, 69–89. Cambridge, MA: Harvard University Press.

Lavocat, R., and J. P. Parent. 1985. Phylogenetic analysis of middle ear features in fossils and living rodents. In *Evolutionary relationships among rodents: A multidisciplinary analysis,* ed. W. P. Luckett and J. L. Hartenberger, 333–54. New York: Plenum.

Lavrov, L. S. 1983. Evolutionary development of the genus *Castor* and taxonomy of the contemporary beavers of Eurasia. *Acta Zoologica Fennica* 174:87–90.

Lavrov, L. S., and V. N. Orlov. 1973. Karyotypes and taxonomy of modern beavers (*Castor,* Castoridae, Mammalia). *Zoologicheskii-Zhurnal* 52:734–42.

Lay, D. M., and C. F. Nadler. 1972. Cytogenetics and origin of North African *Spalax* (Rodentia: Spalacidae). *Cytogenetics* 11:279–85.

Layne, J. N. 1954. The biology of the red squirrel *Tamiasciurus hudsonicus loquax* (Bangs) in central New York. *Ecological Monographs* 24:227–68.

———. 1968. Ontogeny. In *Biology of* Peromyscus *(Rodentia),* Special publication no. 2, ed. J. A. King, 148–253. Shippensburg, PA: American Society of Mammalogists.

———. 1998. Nest box use and reproduction in the gray squirrel (*Sciurus carolinensis*) in Florida. In *Ecology and evolutionary biology of tree squirrels,* ed. M. A. Steele, J. F. Merritt, and D. A. Zegers, 61–70. Martinsville: Virginia Museum of Natural History.

Layne, J. N., and M. A. V. Raymond. 1994. Communal nesting of southern flying squirrels in Florida. *Journal of Mammalogy* 75:110–20.

Le Comber, S., A. C. Spinks, N. C. Bennett, J. U. M. Jarvis, and C. G. Faulkes. 2002. Fractal dimension of African mole-rat burrows. *Canadian Journal of Zoology* 80:436–41.

Le Louarn, H., and G. Janeau. 1975. Répartition et biologie du campagnol des neiges *Microtus nivalis* Martins dans la région de Briançon. *Mammalia* 39:589–604.

Le Magnen, J. 1985. *Hunger.* Cambridge: Cambridge University Press.

Le Roux, A., T. P. Jackson, and M. I. Cherry. 2001. Does Brants' whistling rat (*Parotomys brantsii*) use an urgency-based alarm system in reaction to aerial and terrestrial predators? *Behaviour* 138:757–73.

———. 2002. Differences in alarm vocalizations of sympatric populations of the whistling rats, *Parotomys brantsii* and *P. littledalei* (Rodentia: Muridae). *Journal of Zoology, London* 257:189–94.

Leamy, L. J., S. Meagher, S. Taylor, L. Carroll, and W. K. Potts. 2001. Size and fluctuating asymmetry of morphometric characters in mice: Their associations with inbreeding and *t*-haplotype. *Evolution* 55:2333–41.

Leckie, P. A., J. G. Watson, and S. Chaykin. 1973. An improved method for the artificial insemination of the mouse (*Mus musculus*). *Biology of Reproduction* 9:420–25.

Lee, A. W., and R. E. Brown. 2002. The presence of the male facilitates parturition in California mice (*Peromyscus californicus*). *Canadian Journal of Zoology* 80:926–33.

Lee, D. W., L. E. Miyasoto, and N. S. Clayton. 1998. Neurobiological bases of spatial learning in the natural environment: Neurogenesis and growth in the avian and mammalian hippocampus. *NeuroReport* 9:R15–R27.

Lee, P. C. 1994. Social structure and evolution. In *Behaviour and evolution,* ed. P. J. B. Slater and T. R. Halliday, 266–303. Cambridge: Cambridge University Press.

Lee, S., and L. M. van der Boot. 1955. Spontaneous pseudopregnancy in mice. *Acta Physiologica Pharmacologia Neerlandische* 4:442–44.

———. 1956. Spontaneous pseudopregnancy in mice II. *Acta Physiologica Pharmacologia Neerlandische* 5:213–15.

Lee, T. M., and M. Gorman. 2000. Environmental control of seasonal reproduction: Photoperiod, maternal history and diet. In *Reproduction in context,* ed. K. Whalen and J. Schneider, 191–218. Cambridge: MIT Press.

Leege, T. A., and R. M. Williams. 1967. Beaver productivity in Idaho. *Journal of Wildlife Management* 31:326–32.

Leger, D. W., S. D. Berney-Key, and P. W. Sherman. 1984. Vocalizations of Belding's ground squirrels (*Spermophilus beldingi*). *Animal Behaviour* 32:753–64.

Leger, D. W., and D. H. Owings. 1978. Responses to alarm calls by California ground squirrels: Effects of call structure and maternal status. *Behavioral Ecology and Sociobiology* 3:177–86.

Leger, D. W., D. H. Owings, and L. M. Boal. 1979. Contextual information and differential responses to alarm whistles in California ground squirrels. *Zeitschrift für Tierpsychologie* 49:142–55.

Leger, D. W., D. H. Owings, and R. G. Coss. 1983. Behavioral ecology of time allocation in California ground squirrels (*Spermophilus beecheyi*): Microhabitat effects. *Journal of Comparative Psychology* 97:283–91.

Leger, D. W., D. H. Owings, and D. L. Gelfand. 1980. Single-note vocalizations of California ground squirrels: Graded signals and situation-specificity of predator and socially evoked calls. *Zeitschrift für Tierpsychologie* 52:227–46.

Lehmann, J., and J. Feldon. 2000. Long-term biobehavioral effects of maternal separation in the rat: Consistent or confusing? *Reviews in the Neurosciences* 11:383–408.

Lehmann, L., and N. Perrin. 2002. Altruism, dispersal, and phenotype-matching kin recognition. *American Naturalist* 159:451–68.

———. 2003. Inbreeding avoidance through kin recognition:

Choosy females boost male dispersal. *American Naturalist* 162:638–52.

Lehmer, E. M., J. M. Bossenbroek, and B. Van Horne. 2003. The influence of environment, sex, and innate timing mechanisms on body temperature patterns of free-ranging black-tailed prairie dogs (*Cynomys ludovicianus*). *Physiological and Biochemical Zoology* 76:72–83.

Lehmer, E. M., and B. Van Horne. 2001. Seasonal changes in lipids, diet, and body composition of free-ranging black-tailed prairie dogs (*Cynomys ludovicianus*). *Canadian Journal of Zoology* 79:955–65.

Lehmer, E. M., B. Van Horne, B. Kulbartz, and G. L. Florant. 2001. Facultative torpor in free-ranging black-tailed prairie dogs (*Cynomys ludovicianus*). *Journal of Mammalogy* 82:551–57.

Lehrman, D. S. 1970. Semantic and conceptual issues in the nature-nurture problem. In *Development and evolution of behavior,* ed. L. R. Aronson, E. Tobach, D. S. Lehrman, and J. S. Rosenblatt, 17–52. San Francisco: W. H. Freeman.

Leigh, H., and M. Hofer. 1973. Behavioral and physiologic effects of littermate removal on the remaining single pup and mother during the pre-weaning period in rats. *Psychosomatic Medicine* 35:497–508.

Leinders-Zufall, T., P. Brennan, P. Widmayer, P. Chandramani, A. Maul-Pavicic, M. Jäger, X. Li, H. Breer, F. Zufall, and T. Boehm. 2004. MHC class I peptides as chemosensory signals in the vomeronasal organ. *Science* 306:1033–37.

Lekagul, B., and J. A. McNeely. 1977. *Mammals of Thailand.* Bangkok: Sahakarnbhat.

Lenihan, C., and D. Van Vuren. 1996. Growth and survival of juvenile yellow-bellied marmots (*Marmota flaviventris*). *Canadian Journal of Zoology* 74:297–302.

Lenington, S. 1983. Social preferences for partners carrying "good genes" in wild house mice. *Animal Behaviour* 31:325–33.

———. 1991. The *t* complex: A story of genes, behavior, and populations. *Advances in the Study of Behavior* 20:51–86.

Lenington, S., C. B. Coopersmith, and M. Erhart, M. 1994. Female preference and variability among *t*-haplotypes in wild house mice. *American Naturalist* 143:766–84.

Lenington, S., L. C. Drickamer, A. S. Robinson, and M. Erhart, M. 1996. Genetic basis for male aggression and survivorship in wild house mice (*Mus domesticus*). *Aggressive Behavior* 22:135–45.

Lenington, S., and K. Egid. 1985. Female discrimination of male odors correlated with male genotype at the T locus: A response to T-locus and H-2-locus variability? *Behavior Genetics* 15:53–67.

Lenington, S., K. Egid, and J. Williams. 1988. Analysis of a genetic recognition system in wild house mice. *Behavior Genetics* 18:549–64.

Lenington, S., P. Franks, and J. Williams. 1988. Distribution of *t*-haplotypes in natural populations of wild house mice. *Journal of Mammalogy* 69:489–99.

Lenington, S., and I. L. Heisler. 1991. Behavioral reduction in the transmission of deleterious *t* haplotypes by wild house mice. *The American Naturalist* 137:366–78.

Lenti Boero, D. 1992. Alarm calling in Alpine marmot (*Marmota marmota* L.): Evidence for semantic communication. *Ethology, Ecology and Evolution* 4:125–38.

———. 1994. Survivorship among young alpine marmots and their permanence in their natal territory in a high altitude colony. *IBEX Journal of Mountain Ecology* 2:9–16.

———. 1996. Space and resource use in alpine marmots (*Marmota marmota* L.). In *Biodiversity in marmots,* ed. M. Le Berre, R. Ramousse, and L. Le Guelte, 175–80. Lyon: International Network on Marmots.

———. 1999. Population dynamics, mating system and philopatry in a high altitude colony of alpine marmots (*Marmota marmota* L.). *Ethology Ecology and Evolution* 11:105–22.

———. 2001. Occupation of hibernacula, seasonal activity, and body size in a high altitude colony of Alpine marmots (*Marmota marmota*). *Ethology Ecology and Evolution* 13:209–23.

Leonard, S. T., M. H. Ferkin, and M. M. Johnson. 2001. The response of meadow voles to an over-mark in which the two donors differ in gonadal hormone status. *Animal Behaviour* 62:1171–77.

Lepri, J., and J. G. Vandenbergh. 1986. Puberty in pine voles, *Microtus pinetorum,* and the influence of chemosignals on female reproduction. *Biology of Reproduction* 3:370–77.

Lepri, J. J., and C. J. Wysocki. 1987. Removal of the vomeronasal organ disrupts the activation of reproduction in female voles. *Physiology and Behavior* 40:349–55.

Leslie, P. H., U. M. Venables, and L. S. V. Venables. 1952. The fertility and population structure of the brown rat (*Rattus norvegicus*) in corn ricks and some other habitats. *Proceedings of the Zoological Society, London* 122:187–238.

Lessa, E. P., and J. A. Cook. 1998. The molecular phylogenetics of tuco-tucos (genus *Ctenomys,* Rodentia: Octodontidae) suggests an early burst of speciation. *Molecular Phylogenetics and Evolution* 9:88–99.

Lessa, E. P., G. Wlasiuk, and C. Garza. 2005. Dynamics of genetic differentiation in the Río Negro tuco-tuco (*Ctenomys rionegrensis*) at the local and geographical levels. In *Mammalian diversification: From chromosomes to phylogeography,* ed. E. A. Lacey and P. Myers, 155–74. University of California Press, Berkeley, CA.

Lester, L. S., and M. S. Fanselow. 1985. Exposure to a cat produces opioid analgesia in rats. *Behavioral Neuroscience* 99:756–59.

Leuthold, W. 1977. *African ungulates: A comparative review of their ethology and behavioural ecology.* Berlin: Springer-Verlag.

Levenson, H. 1990. Sexual size dimorphism in chipmunks. *Journal of Mammalogy* 71:161–70.

Levin, B. R., M. L. Petras, and D. I. Rasmussen. 1969. The effect of migration on the maintenance of a lethal polymorphism in the house mouse. *American Naturalist* 103:647–61.

Levin, L. E. 1997. Kinetic dialogs in predator-prey recognition. *Behavioural Processes* 40:113–20.

Levine, L., R. F. Rockwell, and J. Grossfield. 1980. Sexual selection in mice. V. Reproductive competition between +/+ and +/tw5 males. *American Naturalist* 116:150–56.

Levine, S. 1994. The ontogeny of the hypothalamic-pituitary-adrenal axis: The influence of maternal factors. *Annals of the New York Academy of Sciences* 746:275–88.

———. 2001. Primary social relationships influence the development of the hypothalamic-pituitary-adrenal axis in the rat. *Physiology and Behavior* 73:255–60.

Levy, C. E., and K. L. Gage. 1999. Plague in the United States, 1995–1997. *Infections in Medicine* 16:54–64.

Levy, F., A. Melo, B. G. Galef, Jr., M. Madden, and A. Fleming. 2003. Complete maternal deprivation affects social but not non-social learning in adult rats. *Learning and Cognition* 43:177–91.

Levy, N. 1977. Sound communication in the California ground squirrel. Master's thesis, California State University at Northridge.

Lewis, A. W. 1973. Seasonal population changes in the cactus mouse, *Peromyscus eremicus*. *Southwestern Naturalist* 18:85–93.

Lewis, S. E., and A. E. Pusey. 1997. Factors influencing the occurrence of communal care in plural breeding mammals. In *Cooperative breeding in mammals*, ed. N. G. Solomon and J. A. French, 335–63. Cambridge: Cambridge University Press.

Lewis, S. M., and S. N. Austad. 1994. Sexual selection in flour beetles: The relationship between sperm precedence and male olfactory attractiveness. *Behavioral Ecology* 5:219–24.

Lewison, R. 1998. Infanticide in the hippopotamus: Evidence for polygynous ungulates. *Ethology Ecology and Evolution* 10:277–86.

Lewontin, R. C. 1968. The effect of differential viability on the population dynamics of *t* alleles in the house mouse. *Evolution* 22:262–73.

Lewontin, R. C. 2001a. Gene, organism and environment. In *Cycles of contingency: Developmental systems and evolution*, ed. S. Oyama, P. E. Griffiths, and R. D. Gray, 59–66. Cambridge: The MIT Press.

———. 2001b. Gene, organism and environment: A new introduction. In *Cycles of contingency: Developmental systems and evolution*, ed. S. Oyama, P. E. Griffiths, and R. D. Gray, 55–57. Cambridge: The MIT Press.

Lewontin, R., and L. Dunn. 1960. The evolutionary dynamics of a polymorphism in the house mouse. *Genetics* 45:705–22.

Lewontin, R. C., and R. Levins. 1978. Evoluzione. *Enciclopedia Einaudi*, vol. 5, 995–1051.

Li, W. H., W. A. Hide, and D. Graur. 1992. Origin of rodents and guinea-pigs. *Nature* 359:277–78.

Libhaber, N., and D. Eilam. 2002. Social vole parents force their mates to baby-sit. *Developmental Psychobiology* 41:236–40.

———. 2004. Parental investment in social voles varies and is relatively independent of litter size. *Journal of Mammalogy* 85:748–55.

Lidicker, W. Z., Jr. 1975. The role of dispersal in the demography of small mammals. In *Small mammals: Their production and population dynamics*, ed. F. B. Golley, K. Petrusewicz, and L. Ryszkowski, 103–28. Cambridge: Cambridge University Press.

———. 1976. Social behaviour and density regulation in house mice living in large enclosures. *Journal of Animal Ecology* 45:677–97.

———. 1978. Regulation of numbers in small mammal populations—historical reflections and a synthesis. In *Populations of small mammals under natural conditions*, Special publication vol. 5, ed. D. P. Snyder, 122–41. Pittsburgh: Pymatuning Laboratory of Ecology, University of Pittsburgh.

———. 1979a. Analysis of two freely-growing enclosed populations of the California vole. *Journal of Mammalogy* 60:447–66.

———. 1979b. A clarification of interactions in ecological systems. *BioScience* 29:475–77.

———. 1985. Population structuring as a factor in understanding microtine cycles. *Acta Zoologica Fennica* 173:23–27.

———. 1988. Solving the enigma of microtine "cycles." *Journal of Mammalogy* 69:225–35.

———, ed. 1989. *Rodents, a world survey of species of conservation concern*. Occasional Papers no. 4. Gland, Switzerland: IUCN Rodent Specialist Group, Species Survival Commission.

———. 1991. In defense of a multifactor perspective in population ecology. *Journal of Mammalogy* 72:631–35.

———. 1994. Population ecology. In *Seventy-five years of mammalogy (1919–1994)*, Special publication no. 11, ed. E. C. Birney and J. R. Choate, 323–47. American Society of Mammalogists.

———. 1995. The landscape concept: Something old, something new. In *Landscape approaches in mammalian ecology and conservation*, ed. W. Z. Lidicker, Jr., 3–19. Minneapolis: University of Minnesota Press.

———. 1998. *Zapus trinotatus* Rhoads 1895, Pacific jumping mouse. In *North American rodents, status survey and conservation action plan*, ed. D. J. Hafner, E. Yensen, and G. L. Kirkland, Jr. 123–24. Gland, Switzerland: Rodent Specialist Group, Species Survival Commission, IUCN.

———. 1999. Responses of mammals to habitat edges: An overview. *Landscape Ecology* 14:333–43.

———. 2000. A food web/landscape interaction model for microtine rodent density cycles. *Oikos* 91:435–45.

———. 2002. From dispersal to landscapes: Progress in our understanding of population dynamics. *Acta Theriologica* Suppl. no. 1. 47:23–37.

Lidicker, W. Z., Jr., and W. D. Koenig. 1996. Responses of terrestrial vertebrates to habitat edges and corridors. In *Metapopulations and wildlife conservation*, ed. D. R. McCullough, 85–109. Covelo, CA: Island Press.

Lidicker, W. Z., Jr., and J. L. Patton. 1987. Patterns of dispersal and genetic structure in populations of small rodents. In *Mammalian dispersal patterns: The effects of social structure on population genetics*, ed. B. D. Chepko-Sade and Z. T. Halpin, 144–61. Chicago: University of Chicago Press.

Lidicker, W. Z., Jr., and J. A. Peterson. 1999. Responses of small mammals to habitat edges. In *Landscape ecology of small mammals*, ed. G. W. Barrett and J. D. Peles, 211–27. New York: Springer-Verlag.

Lidicker, W. Z., Jr., J. O. Wolff, L. N. Lidicker, and M. H. Smith. 1992. Utilization of a habitat mosaic by cotton rats during a population decline. *Landscape Ecology* 6:259–68.

Liebert, A. E., and P. T. Starks. (2004). The action component of recognition systems: A focus on the response. *Annales Zoologici Fennici* 41:747–64.

Lifjeld, J. T., P. O. Dunn, and D. F. Westneat. 1994. Sexual selection by sperm competition in birds: Male-male competition or female choice? *Journal of Avian Biology* 25:244–50.

Lima, S. L. 1992. Life in a multi-predator environment: Some considerations for anti-predatory vigilance. *Annales Zoologici Fennici* 29:217–26.

Lima, S. L. 1998. Nonlethal effects in the ecology of predator-prey interactions. *Bioscience* 48:25–34.

———. 2002. Putting predators back into behavioral predator-prey interactions. *Trends in Ecology and Evolution* 17:70–75.

Lima, S. L., and P. A. Bednekoff. 1999. Temporal variation in danger drives antipredatory behaviour: The predation risk allocation hypothesis. *American Naturalist* 153:649–59.

Lima, S. L., and L. M. Dill. 1990. Behavioral decisions made under the risk of predation: A review and prospectus. *Canadian Journal of Zoology* 68:619–40.

Lima, S. L., T. J. Valone, and T. Caraco. 1985. Foraging-efficiency-predation-risk trade-off in the grey squirrel. *Animal Behaviour* 33:155–65.

Lin, N., and C. D. Michener. 1972. Evolution of sociality in insects. *Quarterly Review of Biology* 47:131–59.

Lin, Y.-H., P. J. Waddell, and D. Penny. 2002. Pika and vole mitochondrial genomes increase support for both rodent monophyly and glires. *Gene* 294:119–29.

Lindsey, J. R. 1979. Historical foundations. In *The laboratory rat,* ed. H. J. Baker, J. R. Lindsey, and S. H. Weisbroth, 1–36. New York: Academic Press.

Lindström, E. 1986. Territory inheritance and the evolution of group-living in carnivores. *Animal Behaviour* 34:1825–35.

Linn, I. 1984. Home ranges and social systems in solitary mammals. *Acta Zoologica Fennica* 171:245–49.

Linnaeus, C. 1758. *Systema Naturae per regna tria naturae, secundum classis, ordines, genera, species cum characteribus, differentiis, synonymis, locis.* 10th ed., vol. 1. Stockholm: Laurentii Salvii.

Linsdale, J. M., and L. P. Tevis. 1951. *The dusky-footed woodrat.* Berkeley: University of California Press.

Lis, E. J., P. S. Brown, and R. B. Fischer. 1990. A role for the hamster's flank gland in mate selection. *Biology of Behaviour* 15:205–12.

Lishak, R. S. 1982. Gray squirrel mating calls: A spectrographic and ontogenic analysis. *Journal of Mammalogy* 63:661–63.

———. 1984. Alarm vocalizations of adult gray squirrels. *Journal of Mammalogy* 65:681–84.

Lisk, R. D, U. W. Huck, A. C. Gore, and M. X. Armstrong. 1989. Mate choice, mate guarding and other mating tactics in golden hamsters maintained under seminatural conditions. *Behaviour* 69:58–75.

Litvin, V. Y., B. E. Karulin, Y. V. Okhotsky, and Y. S. Pavlovsky. 1977. An experimental study of cannibalism of the common voles in straw stacks. *Zoologicheskii Zhurnal* 56:1693–99.

Liu, D., B. Tannenbaum, C. Caldji, D. D. Francis, A. Freedman, S. Sharma, D. Pearson, P. M. Plotsky, and M. J. Meaney. 1997. Maternal care, hippocampal glucocorticoid receptor gene expression and hypothalamic-pituitary-adrenal responses to stress. *Science* 277:1659–62.

Liu, Y., J. T. Curtis, and Z. X. Wang. 2001a. Pair bond formation in male prairie voles is regulated by vasopressin in the lateral septum. *Hormones and Behavior* 39:311.

———. 2001b. Vasopressin in the lateral septum regulates pair bond formation in male prairie voles (*Microtus ochrogaster*). *Behavioral Neuroscience* 115:910–19.

Liu, Y., C. D. Fowler, and Z. X. Wang. 2001. Ontogeny of brain-derived neurotrophic factor gene expression in the forebrain of prairie and montane voles. *Developmental Brain Research* 127:51–61.

Liu, Y., and Z. X. Wang. 2003. Nucleus accumbens oxytocin and dopamine interact to regulate pair bond formation in female prairie voles. *Neuroscience* 121:537–44.

Lizarralde, M. S. 1993. Current status of the introduced beaver (*Castor canadensis*) population in Tierra del Fuego, Argentina. *Ambio* 22:351–58.

Lockard, R. B., and D. H. Owings. 1974. Seasonal variation in moonlight avoidance by bannertail kangaroo rats. *Journal of Mammalogy* 55:189–93.

LoGiudice, K., R. S. Ostfeld, K. A. Schmidt, and F. Keesing. 2003. The ecology of infectious disease: Effects of host diversity and community composition on Lyme disease risk. *Proceedings of the National Academy of Sciences* 100:567–71.

Loison, A., M. Festa-Bianchet, J.-M. Gaillard, J. T. Jorgenson, and J. M. Jullien. 1999. Age-specific survival in five populations of ungulates: Evidence of senescence. *Ecology* 80:2539–54.

Loison, A., J-M. Gaillard, C. Pélabon, and N. G. Yoccoz. 1999. What factors shape sexual size dimorphism in ungulates? *Evolutionary Ecology Research* 1:611–33.

Lombardi, J. R., and J. M. Whitsett. 1980. Effects of urine from conspecifics on sexual maturation in female prairie deermice, *Peromyscus maniculatus bairdii. Journal of Mammalogy* 61:766–68.

Lomolino, M. V., J. H. Brown, and R. Davis. 1989. Island biogeography of montane forest mammals in the American Southwest. *Ecology* 70:180–94.

Lomolino, M. V., and D. R. Perault. 2001. Island biogeography and landscape ecology of mammals inhabiting fragmented, temperate rain forests. *Global Ecology and Biogeography* 10:113–32.

Lomolino, M. V., and G. A. Smith. 2001. Dynamic biogeography of prairie dog (*Cynomys ludovicianus*) towns near the edge of their range. *Journal of Mammalogy* 82:937–45.

———. 2003. Prairie dog towns as islands: Applications of island biogeography and landscape ecology for conserving nonvolant terrestrial vertebrates. *Global Ecology and Biogeography* 12:275–86.

Long, D., K. Bly-Honness, J. Truett, and D. B. Seery. 2006. Establishment of new prairie dog colonies by translocation. In *Conservation of the black-tailed prairie dog,* ed. J. L. Hoogland, 188–209. Washington, DC: Island Press.

Longland, W. S., and M. V. Price. 1991. Direct observations of owls and heteromyid rodents: Can predation risk explain microhabitat use? *Ecology* 72:2261–73.

Lonstein, J. S., and G. J. De Vries. 1999. Comparison of the parental behavior of pair-bonded female and male prairie voles (*Microtus ochrogaster*). *Physiology and Behavior* 66:33–40.

Lonstein, J. S., and G. J. De Vries. 2000a. Influence of gonadal hormones on the development of parental behavior in adult virgin prairie voles (*Microtus ochrogaster*). *Behavioural Brain Research* 114:79–87.

Lonstein, J. S., and G. J. De Vries. 2000b. Maternal behaviour in lactating rats stimulates c-fos in glutamate decarboxylase-synthesizing neurons of the medial preoptic area, ventral bed nucleus of the stria terminalis, and ventrocaudal periaqueductal gray. *Neuroscience* 100:557–68.

———. 2000c. Sex differences in the parental behavior of rodents. *Neuroscience and Biobehavioral Reviews* 24:669–86.

Lore, R., A. Blanc, and P. Suedfeld. 1971. Empathic learning of a passive-avoidance response in domesticated *Rattus norvegicus. Animal Behaviour* 19:112–14.

Lorrain, D. S., J. V. Riolo, L. Matuszewich, and E. M. Hull. 1999. Lateral hypothalamic serotonin inhibits nucleus accumbens dopamine: Implications for sexual satiety. *Journal of Neuroscience* 19:7648–52.

Lott, D. F. 1991. *Intraspecific variation in the social systems of wild vertebrates.* Cambridge: Cambridge University Press.

Louch, C. D. 1956. Adrenocortical activity in relation to the density and dynamics of three confined populations of *Microtus pennsylvanicus. Ecology* 37:701–13.

Lougheed, S. C., T. W. Arnold, and R. C. Bailey. 1991. Measurement error of external and skeletal variables in birds and its effects on principal components. *Auk* 108:432–36.

Loughry, W. J. 1988. Population differences in how black-tailed prairie dogs deal with snakes. *Behavioral Ecology and Sociobiology* 22:61–67.

Loughry, W. J., and C. M. McDonough. 1988. Calling and vigilance in California ground squirrels: A test of the tonic communication hypothesis. *Animal Behaviour* 36:1533–40.

Lovecky, D. V., D. Q. Estep, and D. A. Dewsbury. 1979. Copulatory behaviour of cotton mice (*Peromyscus gossypinus*) and their reciprocal hybrids with white-footed mice (*P. leucopus*). *Animal Behaviour* 27:371–75.

Lovegrove, B. G. 1989. The cost of burrowing by the social mole rats (Bathyergidae) *Cryptomys damarensis* and *Heterocephalus glaber*: The role of soil moisture. *Physiological Zoology* 62:449–69.

———. 1991. The evolution of eusociality in molerats (Bathyergidae): A question of risks, numbers, and costs. *Behavioral Ecology and Sociobiology* 28:37–45.

Lovegrove, B. G., and C. Wissel. 1988. Sociality in mole–rats: Metabolic scaling and the role of risk sensitivity. *Oecologia* 74:600–606.

Lovich, J. E., and J. W. Gibbons. 1992. A review of techniques for quantifying sexual size dimorphism. *Growth, Aging and Development* 56:269–81.

Lozada, M., J. A. Monjeau, K. M. Heinemann, N. Guthmann, and E. C. Birney. 1996. *Abrothrix xanthorhinus*. *Mammalian Species* 540:1–6.

Lu, L., G. Bao, H. Chen, P. Xia, X. Fan, J. Zhang, G. Pei, and L. Ma. 2003. Modification of hippocampal neurogenesis and neuroplasticity by social environments. *Experimental Neurology* 183:600–609.

Luce, R. J., R. Manes, and B. van Pelt. 2006. A multi-state approach to conserve and manage black-tailed prairie dogs. In *Conservation of the black-tailed prairie dog*, ed. J. L. Hoogland, 210–17. Washington, DC: Island Press.

Luckett, W. P. 1985. Superordinal and intraordinal affinities of rodents: Developmental evidence from the dentition and placentation. In *Evolutionary relationships among rodents: A multidisciplinary analysis*, ed. W. P. Luckett and J. L. Hartenberger, 227–76. New York: Plenum.

Luckett, W. P., and J.-L. Hartenberger. 1985. Evolutionary relationships among rodents: Comments and conclusions. In *Evolutionary relationships among rodents: A multidisciplinary analysis*, ed. W. P. Luckett and J.-L. Hartenberger, 685–712. New York: Plenum.

———. 1993. Monophyly or polyphyly of the order Rodentia: Possible conflict between morphological and molecular interpretations. *Journal of Mammalian Evolution* 1:127–47.

Ludwig, D. R. 1984. *Microtus richardsoni*. *Mammalian Species* 223:1–6.

Luis, J., A. Carmona, J. Delgado, F. A. Cervantes, and R. Cardenas. 2000. Parental behavior of the volcano mouse, *Neotomodon alstoni* (Rodentia: Muridae), in captivity. *Journal of Mammalogy* 81:600–605.

Luis, J., F. A. Cervantes, M. Martínez, R. Cardenas, J. Delgado, and A. Carmona. 2004. Male influence in maternal behavior and offspring of captive volcano mice (*Neotomodon alstoni*) from Mexico. *Journal of Mammalogy* 85:268–72.

Lumley, L. A., M. L. Sipos, R. C. Charles, R. F. Charles, and J. L. Meyerhoff. 1999. Social stress effects on territorial marking and ultrasonic vocalizations in mice. *Physiology and Behavior* 67:769–75.

Luo, M., M. S. Fee, and L. C. Katz. 2003. Encoding pheromonal signals in accessory olfactory bulb of behaving mice. *Science* 299:1196–1201.

Lupfer, G., J. Friedman, and D. Coonfield. 2003. Social transmission of flavor preferences in two species of hamsters. *Journal of Comparative Psychology* 117:449–55.

Lupien, S. J., and B. S. McEwen. 1997. The acute effects of corticosteroids on cognition: Integration of animal and human model studies. *Brain Research Reviews* 24:1–27.

Luque-Larena, J. J., P. López, and J. Gosálbez. 2001. Scent matching modulates space use and agonistic behaviour between male snow voles, *Chionomys nivalis*. *Animal Behaviour* 62:1089–95.

———. 2002a. Levels of social tolerance between snow voles *Chionomys nivalis* during over-wintering periods. *Acta Theriologica* 47:163–73.

———. 2002b. Microhabitat use by the snow vole *Chionomys nivalis* in alpine environments reflects rock-dwelling preferences. *Canadian Journal of Zoology-Revue Canadienne De Zoologie* 80:36–41.

———. 2002c. Relative dominance affects use of scent-marked areas in male snow voles *Chionomys nivalis*. *Ethology* 108:273–85.

———. 2002d. Responses of snow voles, *Chionomys nivalis*, towards conspecific cues reflect social organization during overwintering periods. *Ethology* 108:947–59.

———. 2003. Male dominance and female chemosensory preferences in the rock-dwelling snow vole. *Behaviour* 140:665–81.

———. 2004. Spacing behavior and morphology predict promiscuous mating strategies in the rock-dwelling snow vole, *Chionomys nivalis*. *Canadian Journal of Zoology-Revue Canadienne De Zoologie* 82:1051–60.

Lurz, P. W. W., P. J. Garson, and L. A. Wauters. 1997. Effects of temporal and spatial variation in habitat quality on red squirrel dispersal behavior. *Animal Behaviour* 54:427–35.

Lynch, C. B. 1977. Inbreeding effects upon animals derived from a wild population of *Mus musculus*. *Evolution* 31:526–37.

Lyon, M. F. 1984. Transmission ratio distortion in mouse *t*-haplotypes is due to multiple distorter genes acting on a responder locus. *Cell* 37:621–28.

Ma, Y., F. Wang, S. Jin, and S. Li. 1987. *Glires (rodents and lagomorphs) of Northern Xinjiang and their zoogeographical distribution*. [In Chinese]. Beijing: Scientific Press.

Macêdo, R., and M. A. Mares. 1988. *Neotoma albigula*. *Mammalian Species* 310:1–7.

Mac, M. J., P. A. Opler, E. E. Puckett Haecker, and P. D. Doran. 1998. In *Status and trends of the nation's biological resources*, vol. 2, 437–964. Reston, VA: United States Department of the Interior, United States Geological Survey.

MacArthur, R. H., and E. O. Wilson. 1967. *The theory of island biogeography*. Princeton, NJ: Princeton University Press.

Maccari, S., P. V. Piazza, A. Barbazanges, H. Simon, and M. Le Moal. 1995. Adoption reverses the long-term impairment in glucocorticoid feedback induced by prenatal stress. *Journal of Neuroscience* 15:110–16.

Macdonald, D. W. 1981a. Dwindling resources and the social behaviour of Capybaras (*Hydrochoerus hydrochaeris*) (Mammalia). *Journal of Zoology, London* 194:371–91.

———. 1981b. Feeding associations between capybaras *Hydrochoerus hydrochaeris* and some bird species. *Ibis* 123:364–66.

———. 1983. The ecology of carnivore social behaviour. *Nature* 301:379–84.

———. 1985. The rodents. IV. Suborder Hystricomorpha. In *Social odours in mammals*, vol. 1, ed. R. E. Brown and D. W. Macdonald, 480–506. Oxford: Oxford Scientific Publications.

———. 2001. *The new encyclopedia of mammals.* Oxford: Oxford University Press.

Macdonald, D. W., and R. E. Brown. 1985. Introduction: the pheromone concept in mammalian chemical communication. In *Social odours in mammals,* vol. 1, eds. R. E. Brown and D. W. Macdonald, 1–18. Oxford: Clarendon Press.

Macdonald, D. W., and G. M. Carr. 1989. Food security and the rewards of tolerance. In *Comparative socioecology: The behavioural ecology of humans and other mammals,* ed. V. Standen and R. A. Folley, 75–99. Oxford: Blackwell Scientific.

Macdonald, D. W., and M. P. G. Fenn. 1995. Rat ranges in arable areas. *Journal of Zoology, London* 236:349–53.

Macdonald, D. W., K. Krantz, and R. T. Aplin. 1984. Behavioural, anatomical and chemical aspects of scent marking amongst Capybaras (*Hydrochoerus hydrochaeris*) (Rodentia: Caviomorpha). *Journal of Zoology, London* 202:341–60.

Macdonald, D. W., F. Mathews, and M. Berdoy. 1999. The behaviour and ecology of *Rattus norvegicus*: From opportunism to kamikaze tendencies. In *Ecologically-based rodent management,* ed. G. R. Singleton, L. Hinds, H. Leirs, and Z. Zhang, 49–80. Canberra: Australian Centre for International Agricultural Research.

Macdonald, D. W., and J. R. Moreira. In press. Foetal positioning, selective death and sex allocation in capybaras. *Journal of Mammalogy.*

Macedo, R. H., and M. A. Mares. 1988. *Neotoma albigula. Mammalian Species* 310:1–7.

Macedonia, J. M., and C. S. Evans. 1993. Variation among mammalian alarm call systems and the problem of meaning in animal signals. *Ethology* 93:177–97.

Mackintosh, J. H. 1981. Behaviour of the house mouse. *Symposia of the Zoological Society of London* 47:337–65.

Mackintosh, J. H., and E. C. Grant. 1966. The effect of olfactory stimuli in the agonistic behaviour of laboratory mice. *Zeitschrift für Tierpsychologie* 23:584–87.

MacMillen, R. 1964. Population ecology, water relations, and social behavior of a southern California semidesert rodent fauna. *University of California Publications in Zoology* 71:1–59.

MacPhee, R. D. E., and C. Flemming. 1999. *Requiem aeternum,* the last five hundred years of mammalian species extinctions. In *Extinctions in near time,* ed. R. D. E. MacPhee, 333–71. New York: Kluwer Academic/Plenum.

MacPhee, R. D. E., D. C. Ford, and D. A. McFarlane. 1989. Pre-Wisconsinian mammals from Jamaica and models of late Quaternary extinction in the Greater Antilles. *Quaternary Research* 31:94–106.

MacWhirter, R. B. 1992. Vocal and escape responses of Colombian ground squirrels to simulated terrestrial and aerial predator attacks. *Ethology* 91:311–25.

Madden, J. R. 1974. Female territoriality in a Suffolk County, Long Island, population of *Glaucomys volans. Journal of Mammalogy* 55:647–652.

Maddison, W. P. 2000. Testing character correlation using pairwise comparisons on a phylogeny. *Journal of Theoretical Biology* 202:195–204.

Maddison, W. P., and D. R. Maddison. 1992. *MacClade: Analysis of phylogeny and character evolution.* Sunderland, Massachusetts: Sinauer.

———. 2004. Mesquite: A modular system for evolutionary analysis. Version 1.02.

Maddock, L. 1979. The migration and grazing succession. In *Serengeti: Dynamics of an ecosystem,* ed. A. R. E. Sinclair and M. Norton-Griffiths, 104–29. Chicago: University of Chicago Press.

Madison, D. M. 1980a. An integrated view of the social biology of *Microtus pennsylvanicus. The Biologist* 62:20–33.

———. 1980b. Space use and social structure in meadow voles, *Microtus pennsylvanicus. Behavioral Ecology and Sociobiology* 7:65–71.

———. 1984. Group nesting and its ecological and evolutionary significance in overwintering microtine rodents. *Special Publications of the Carnegie Museum of Natural History* 10:267–74.

———. 1985. Activity rhythm and spacing. In *Biology of New World Microtus,* Special publication no. 8, ed. R. H. Tamarin, 373–419. Shippensburg, PA: American Society of Mammalogists.

Madison, D. M., R. W. Fitzgerald, and W. J. McShea. 1984. Dynamics of social nesting in overwintering meadow voles (*Microtus pennsylvanicus*): Possible consequences for population cycling. *Behavioral Ecology and Sociobiology* 15:9–17.

Madsen, O., M. Scally, C. J. Douady, D. J. Kao, R. W. DeBryk, R. Adkins, H. M. Amrine, M. J. Stanhope, W. W. de Jong, and M. S. Springer. 2001. Parallel adaptive radiations in two major clades of placental mammals. *Nature* 409, 610–14.

Madsen, T., Shine, R., Loman, J. and Hakansson, T. 1992. Why do female adders copulate so frequently? *Nature* 355:440–41.

Maestripieri, D. 1992. Functional aspects of maternal aggression in mammals. *Canadian Journal of Zoology* 70:1069–77.

Maestripieri, D., and E. Alleva, E. 1990. Maternal aggression and litter size in the female house mouse. *Ethology* 84:27–34.

Maestripieri, D., and C. Rossi-Arnaud. 1991. Kinship does not affect litter defense in pairs of communally nesting female house mice. *Aggressive Behavior* 17:223–28.

Magurran, A. E. 1990. The inheritance and development of minnow anti-predator behaviour. *Animal Behaviour* 39:834–42.

Mahady, S. J., and J. O. Wolff. 2002. A field test of the Bruce effect in the monogamous prairie vole (*Microtus ochrogaster*). *Behavioral Ecology and Sociobiology* 52:31–37.

Maier, T. J. 2002. Long-distance movements by female white-footed mice, *Peromyscus leucopus,* in extensive mixed-wood forest. *Canadian Field-Naturalist* 116:108–11.

Mainardi, D., M. Marsan, and A. Pasquali. 1965. Causation of sexual preferences of the house mouse. The behaviour of mice reared by parents whose odour was artificially altered. *Atti della Societa Italiana di Scienze Naturali e del Museo Civico di Storia Naturale di Milano* 54:325–38.

Mainardi, D., F. M. Scudo, and D. Barbieri. 1965. Assortative mating based on early learning: Population genetics. *Acta Bio-Medica* 36:583–605.

Makin, J. W., and R. H. Porter. 1984. Paternal behavior in the spiny mouse (*Acomys cahirinus*). *Behavioral and Neural Biology* 41:135–51.

Malizia, A. I. 1998. Population dynamics of the fossorial rodent *Ctenomys talarum* (Rodentia: Octodontidae). *Journal of Zoology, London* 244:545–51.

Mallory, F. F., and R. J. Brooks. 1978. Infanticide and other reproductive strategies in the collared lemming, *Dicrostonyx groenlandicus. Nature* 273:144–46.

———. 1980. Infanticide and pregnancy failure: Reproductive strategies in the female collared lemming (*Dicrostonyx groenlandicus*). *Biology of Reproduction* 22:192–96.

Mallory, F. F., and F. V. Clulow. 1977. Evidence of pregnancy failure in the wild meadow vole, *Microtus pennsylvanicus*. *Canadian Journal of Zoology* 55:1–17.

Mandelik, Y., M. Jones, and T. Dayan. 2003. Structurally complex habitat and sensory adaptations mediate the behavioural responses of a desert rodent to an indirect cue for increased predation risk. *Evolutionary Ecology Research* 5:501–15.

Manes, R. 2006. Does the prairie dog merit protection via the Endangered Species Act? In *Conservation of the black-tailed prairie dog*, ed. J. L. Hoogland, 169–84. Washington, DC: Island Press.

Mann, G. 1978. Los pequeños mamíferos de Chile: Marsupiales, quirópteros, edentados y roedores. *Gallana Zoología (Chile)* 40:1–342.

Mann, T., and C. Lutwak-Mann. 1981. *Male reproductive function and semen*. Berlin: Springer-Verlag.

Manning, C. J., D. A. Dewsbury, E. K. Wakeland, and W. K. Potts. 1995. Communal nesting and communal nursing in house mice, *Mus musculus domesticus*. *Animal Behavior* 50:741–51.

Manning, C. J., W. K. Potts, E. K. Wakeland, and D. A. Dewsbury. 1992. What's wrong with MHC mate choice experiments? In *Chemical signals in vertebrates VI*, ed. R. L. Doty and D. Müller-Schwarze, 229–35. New York: Plenum.

Manning, C. J., E. K. Wakeland, and W. K. Potts. 1992. Communal nesting patterns in mice implicate MHC genes in kin recognition. *Nature* 360:581–83.

Manser, M. 2001. The acoustic structure of suricates' alarm calls varies with predator type and the level of response urgency. *Proceedings of the Royal Society of London, Series B* 268:2315–24.

Manser, M. B., M. B. Bell, and L. B. Fletcher. 2001. The information receivers extract from alarm calls in suricates. *Proceedings of the Royal Society of London, Series B* 268:2485–491.

Mantalenakis, S. J., and M. M. Ketchel. 1966. Frequency and extent of delayed implantation in lactating rats and mice. *Journal of Reproduction and Fertility* 12:391–94.

Mappes, T., M. Halonen, J. Suhonen, and H. Ylönen. 1993. Selective avian predation on declining population of the field vole *Microtus agrestis*: Age and sex preferences. *Ethology, Ecology and Evolution* 5:519–27.

Mappes, T., E. Koskela, and H. Ylönen. 1998. Breeding suppression in voles under predation risk of small mustelids: Laboratory or methodological artifact? *Oikos* 82:365–69.

Mappes, T., H. Ylonen, and J. Viitala. 1995. Higher reproductive success among kin groups of bank voles (*Clethrionomys glareolus*). *Ecology* 76:1276–82.

Marchlewska-Koj, A. 1983. Pregnancy blocking by pheromones. In *Pheromones and reproduction in mammals*, ed. J. G. Vandenbergh, 151–74. New York: Academic Press.

Marchlewska-Koj, A., M. Kruczek, J. Kapusta, and E. Pochron. 2003. Prenatal stress affects the rate of sexual maturation and attractiveness in bank voles. *Physiology and Behavior* 79:305–10.

Marchlewska-Koj, A., M. Kruczek, and P. Olejniczak, P. 2003. Mating behaviour of bank voles (*Clethrionomys glareolus*) modified by hormonal and social factors. *Mammalian Biology* 68:144–52.

Mares, M. A., J. K. Braun, and R. Chennell. 1997. Ecological observations on the octodontid rodent, *Tympanoctomys barrerae*, in Argentina. *Southwestern Naturalist* 42:488–504.

Mares, M. A., and T. E. Lacher, Jr. 1987. Ecological, morpho-logical, and behavioral convergence in rock-dwelling mammals. *Current Mammalogy* 1:307–48.

Mares, M. A., and R. A. Ojeda. 1981. Patterns of diversity and adaptation in South American histricognath rodents. In *Mammalian biology in South America*, ed. M. A. Mares and H. H. Genoways, 393–432. Pittsburgh, PA: Special publication of the Pymatuning Laboratory of Ecology.

Mares, M. A., R. A. Ojeda, and M. P. Kosco. 1982. Observations on the distribution and ecology of the mammals of Salta Province, Argentina. *Annals of the Carnegie Museum* 50:151–206.

Marin, G., and A. Pilastro. 1994. Communally breeding dormice, *Glis glis*, are close kin. *Animal Behaviour* 47:1485–87.

Marinelli, L., and F. Messier. 1995. Parental-care strategies among muskrats in a female-biased population. *Canadian Journal of Zoology* 73:1503–10.

Marinelli, L., F. Messier, and Y. Plante. 1997. Consequences of following a mixed reproductive strategy in muskrats. *Journal of Mammalogy* 78:163–72.

Marivaux, L., M. Vianey-Liaud, and J. J. Jaeger. 2004. Higher-level phylogeny of early Tertiary rodents: Dental evidence. *Zoological Journal of the Linnean Society* 142:105–34.

Marler, P. 1955. Characteristics of some animal calls. *Nature* 176:6–8.

———. 1977. The evolution of communication. In *How animals communicate*, ed. T. A. Sebeok, 45–70. Bloomington: Indiana University Press.

Marques, D. M., and E. S. Valenstein. 1976. Another hamster paradox: More males carry pups and fewer kill and cannibalize young than do females. *Journal of Comparative and Physiological Psychology* 90:653–57.

Márquez, N., C. Villavicencio, M. C. Cecchi, and R. A. Vásquez. 2002. Reconocimiento de parentesco en el roedor caviomorfo *Octodon degus*: Relaciones hermano-hermano y padre-hijo. Sextas Jornadas de Etología, Santiago, Chile, *Libro de Resúmenes*: 20.

Marsden, H. M., and F. H. Bronson. 1964. Estrus synchrony in mice: Alteration by exposure to male urine. *Science, New York* 144:1469.

———. 1965a. Strange male block to pregnancy: Its absence in inbred mouse strains. *Nature, London* 207:878.

———. 1965b. The synchrony of oestrus in mice: Relative roles of the male and female environments. *Journal of Endocrinology* 32:313–19.

Martan, J., C. S. Adams, and B. L. Perkins. 1970. Epididymal spermatozoa of two species of squirrels. *Journal of Mammalogy* 51:376–78.

Martan, J., and Z. Hruban. 1970. Unusual spermatozoan formations in epididymis of flying squirrel (*Glaucomys volans*). *Journal of Reproduction and Fertility* 21:167–70.

Marti, C. D., K. S. Steenhof, M. N. Kochert, and J. S. Marks. 1993. Community trophic structure: The roles of diet, body size, and activity time in vertebrate predators. *Oikos* 67:6–18.

Martin, I. G., and G. K. Beauchamp. 1982. Olfactory recognition of individuals by male cavies (*Cavia aperea*). *Journal of Chemical Ecology* 8:1241–49.

Martin, L. T., and J. R. Alberts. 1979. Taste aversions to mother's milk: The age-related role of nursing in the acquisition and expression of a learned association. *Journal of Comparative and Physiological Psychology* 93:430–45.

Martin, S. J., M. H. Schroeder, and H. Tietjen. 1984. Burrow plugging by prairie dogs *Cynomys ssp*. in response to Siberian polecats *Mustela-eversmanni*. *Great Basin Naturalist* 44: 447–49.

Martin, T. 1993. Early rodent incisor enamel evolution: Phylogenetic implications. *Journal of Mammalian Evolution* 1:227–54.

Martin, Y., G. Gerlach, C. Schlötterer, and A. Meyer. 2000. Molecular phylogeny of European muroid rodents based on complete cytochrome b sequences. *Molecular Phylogenetics and Evolution* 16:37–47.

Martínez, J. C., F. Cardenas, M. Lamprea, and S. Morato. 2002. The role of vision and proprioception in the aversion of rats to the open arms of an elevated plus-maze. *Behavioural Processes* 60:15–26.

Martins, E. P., and T. F. Hansen. 1997. Phylogenies and the comparative method: A general approach to incorporating phylogenetic information in to the analysis of interspecific data. *American Naturalist* 149:646–67.

Maruniak, J. A., K. Owen, F. H. Bronson, and C. Desjardins. 1975. Urinary marking in female house mice: Effects of ovarian steroids, sex experience, and type of stimulus. *Behavioral Biology* 13:211–17.

Mascheretti, S., P. M. Mirol, M. D. Giménez, C. J. Bidau, J. R. Contreras, and J. B. Searle. 2000. Phylogentics of the speciose and chromosomally variable rodent genus *Ctenomys* (Ctenomyidae, Octodontoidea), based on mitochondrial cytochrome b sequences. *Biological Journal of the Linnean Society* 70: 361–76.

Maser, C., J. M. Trappe, and R. A. Nussbaum. 1978. Fungal–small mammal interrelationships with emphasis on Oregon coniferous forests. *Ecology* 59:799–809.

Mashkin, V. I. 1991. Hunting press and bobac population structure. In *Population structure of the marmot,* ed. D. I. Bibikov, A. A. Nikolski, V. Yu. Rumiantsev, and T. A. Seredneva, 119–47. Moscow: USSR Theriological Society.

———. 2003. Interfamily regrouping of Eurasian marmots. In *Adaptive strategies and diversity in marmots,* ed. R. Ramousse, D. Allainé, and M. Le Berre, 183–88. Lyon: International Network on Marmots.

Mason, W. A. 1979. Ontogeny of social behavior. In *Handbook of behavioral neurobiology,* vol. 3, ed. P. Marler and J. G. Vandenbergh, 1–28. New York: Plenum.

Massey, A., and J. G. Vandenbergh. 1980. Puberty delay by a urinary cue from female house mice in feral populations. *Science, New York* 209:821–22.

———. 1981. Puberty acceleration by a urinary cue from male mice in feral populations. *Biology of Reproduction* 24:523–27.

Maswanganye, K. A., N. C. Bennett, J. Brinders, and R. Cooney. 1999. Oligospermia and azoospermia in non–reproductive male Damaraland mole–rats *Cryptomys damarensis* (Rodentia : Bathyergidae). *Journal of Zoology* 248:411–18.

Mateo, J. M. 1995. The development of alarm-call responses in free-living and captive Belding's ground squirrels, *Spermophilus beldingi*. PhD diss., University of Michigan, Ann Arbor.

———. 1996a. The development of alarm-call response behaviour in free-living juvenile Belding's ground squirrels. *Animal Behaviour* 52:489–505.

———. 1996b. Early auditory experience and the ontogeny of alarm-call discrimination in Belding's ground squirrels (*Spermophilus beldingi*). *Journal of Comparative Psychology* 110:115–24.

———. 2002. Kin-recognition abilities and nepotism as a function of sociality. *Proceedings of the Royal Society of London, Series B* 269:721–27.

———. 2003. Kin recognition in ground squirrels and other rodents. *Journal of Mammalogy* 84:1163–81.

———. 2004. Recognition systems and biological organization: The perception component of social recognition. *Annales Zoologici Fennici* 41:729–45.

Mateo, J. M., and W. G. Holmes. 1997. Development of alarm-call responses in Belding's ground squirrels: The role of dams. *Animal Behaviour* 54:509–24.

———. 1999a. How rearing history affects alarm-call responses of Belding's ground squirrels (*Spermophilus beldingi*, Sciuridae). *Ethology* 105:207–22.

———. 1999b. Plasticity of alarm-call response development in Belding's ground squirrels (*Spermophilus beldingi*, Sciuridae). *Ethology* 105:193–206.

———. 2004. Design and interpretation of cross-fostering studies. *Animal Behaviour* 68: 1451–59.

Mateo, J. M., and R. E. Johnston. 2000a. Kin recognition and the 'armpit effect': Evidence of self-referent phenotype matching. *Proceedings of the Royal Society of London, Series B* 267:695–700.

———. 2000b. Retention of social recognition after hibernation in Belding's ground squirrels. *Animal Behaviour* 59: 491–99.

Matocha, K. 1977. The vocal repertoire of *Spermophilus tridecemlineatus*. *American Midland Naturalist* 98:482–87.

Matochik, J. A., and R. J. Barfield. 1991. Hormonal control of precopulatory sebaceous scent marking and ultrasonic mating vocalizations in male rats. *Hormones and Behavior* 25:445–60.

Matochik, J. A., N. R. White, and R. J. Barfield. 1992. Variations in scent marking and ultrasonic vocalizations by Long-Evans rats across the estrous cycle. *Physiology and Behavior* 51: 783–86.

Matocq, M. D. 2002. Morphological and molecular analysis of a contact zone in the *Neotoma fuscipes* species complex. *Journal of Mammalogy* 83:866–83.

———. 2004. Reproductive success and effective population size in woodrats (*Neotoma macrotis*). *Molecular Ecology* 13: 1635–42.

Matocq, M. D., and E. A. Lacey. 2004. Philopatry, kin clusters, and genetic relatedness in a population of woodrats (*Neotoma macrotis*). *Behavioral Ecology* 15:647–53.

Matthews, M., and N. T. Adler. 1977. Facilitative and inhibitory influences of reproductive behavior on sperm transport in rats. *Journal of Comparative and Physiological Psychology* 91:727–41.

———. 1979. Relative efficiency of sperm transport and number of sperm ejaculated in the female rat. *Biology of Reproduction* 20:540–44.

Matthews, S. G. 2002. Early programming of the hypothalamo-pituitary-adrenal axis. *Trends in Endocrinology and Metabolism* 13:373–80.

Maxwell, C. S., and M. L. Morton. 1975. Comparative thermoregulatory capabilities of neonatal ground squirrels. *Journal of Mammalogy* 56:821–28.

May, R. M., and R. M. Anderson. 1978. Regulation and stability of host-parasite population interactions. II. Destabilizing processes. *Journal of Animal Ecology* 47:249–67.

Maynard-Smith, J. 1965. The evolution of alarm calls. *American Naturalist* 99:59–63.

————. 1982. *Evolution and the theory of games*. Cambridge: Cambridge University Press.

————. 1996. The games lizards play. *Nature* 380:198–99.

Maynard-Smith, J., and N. C. Stenseth. 1978. On the evolutionary stability of female biased sex ratio in the wood lemming. *Heredity* 41:205–14.

Maynard-Smith, J., and Price, G. R. 1973. The logic of animal conflict. *Nature (London)* 246:15–18.

Mayr, E. 1963. *Animal species and evolution*. Cambridge, MA: Harvard University Press.

Maza, B. G., N. R. French, and A. P. Aschwanden. 1973. Home range dynamics in a population of heteromyid rodents. *Journal of Mammalogy* 54:405–25.

McAdam, A. G., and J. S. Millar. 1998. Breeding by young-of-the-year female deer mice: Why weight? *Ecoscience* 6:400–405.

McCarthy, M. M., and F. S. Vom Saal. 1985. The influence of reproductive state on infanticide by wild female house mice (*Mus musculus*). *Physiology and Behavior* 35:843–49.

————. 1986a. Infanticide by virgin CF-1 and wild house mice (*Mus musculus*): Effects of age, prolonged isolation, and testing procedure. *Developmental Psychobiology* 19:279–90.

————. 1986b. Inhibition of infanticide after mating by wild male house mice. *Physiology and Behavior* 36:203–9.

McCarty, R. 1975. *Onychomys torridus*. *Mammalian Species* 59:1–5.

————. 1978. *Onychomys leucogaster*. *Mammalian Species* 87:1–6.

McCarty, R., and C. H. Southwick. 1977a. Paternal care and the development of behavior in the southern grasshopper mouse, *Onychomys torridus*. *Behavioral Biology* 19:476–90.

————. 1977b. Patterns of parental care in two cricetid rodents, *Onychomys torridus* and *Peromyscus leucopus*. *Animal Behaviour* 25:945–48.

McClelland, E. E., D. J. Penn, and W. K. Potts. 2003. Major histocompatibility complex heterozygote superiority during coinfection. *Infection and Immunity* 71:2079–86.

McClintock, M. K. 1983. Pheromonal regulation of the ovarian cycle: Enhancement, suppression, and synchrony. In *Pheromones and reproduction in mammals*, ed. J. G. Vandenbergh, 113–49. New York: Academic Press.

————. 1984. Group mating in the domestic rat as a context for sexual selection: Consequences for the analysis of sexual behavior and neuroendocrine responses. *Advances in the Study of Behavior* 14:1–50.

McClintock, M. K., J. J. Anisko, and N. T. Adler. 1982. Group mating among Norway rats II. The social dynamics of copulation: Competition, cooperation, and mate choice. *Animal Behaviour* 30:410–25.

McClintock, M. K., J. P. Toner, N. T. Adler, and J. J. Anisko. 1982. Post-ejaculatory quiescence in female and male rats: Consequences for sperm transport during group mating. *Journal of Comparative and Physiological Psychology* 96:268–77.

McCloskey, R. J., and K. C. Shaw. 1977. Copulatory behavior of fox squirrel. *Journal of Mammalogy* 58:633–65.

McClure, P. A. 1981. Sex-biased litter reduction in food restricted woodrats (*Neotoma floridana*). *Science* 211:1058–60.

McCullough, D. R., ed. 1996. *Metapopulations and wildlife conservation*. Covelo, CA: Island Press.

McDonald, D. B., J. W. Fitzpatrick, and G. E. Woolfenden. 1996. Actuarial senescence and demographic heterogeneity in the Florida Scrub Jay. *Ecology* 77:2373–81.

McEwen, B. S., ed. 2001. *Coping with the environment: Neural and endocrine mechanisms*. New York: Oxford University Press.

McEwen, B. S., and R. M. Sapolsky. 1995. Stress and cognitive function. *Current Opinion in Neurobiology* 5:205–16.

McEwen, B. S., and J. C. Wingfield. 2003. The concept of allostasis in biology and biomedicine. *Hormones and Behavior* 43:2–15.

McFadyen-Ketchum, S. A., and R. H. Porter. 1989. Transmission of food preferences in spiny mice (*Acomys cahirinus*) via nose-mouth interaction. *Behavioral Ecology and Sociobiology* 24:59–62.

McGaugh, J. L. 1989. Involvement of hormonal and neuromodulatory systems in the regulation of memory storage. *Annual Review of Neuroscience* 12:255–87.

McGill, T. E., 1962. Sexual behavior in three inbred strains of mice. *Behaviour* 19:341–50.

McGill, T. E., D. M. Corwin, and D. T. Harrison. 1968. Copulatory plug does not induce luteal activity in the mouse *Mus musculus*. *Journal of Reproduction and Fertility* 15:149–51.

McGrew, W. C. 1974. Tool use by wild chimpanzees in feeding upon driver ants. *Journal of Human Evolution* 3:501–8.

McGuire, B. 1988. The effects of cross-fostering on the parental behavior of the meadow vole (*Microtus pennsylvanicus*). *Journal of Mammalogy* 69:332–41.

————. 1997. Influence of father and pregnancy on maternal care in red-backed voles. *Journal of Mammalogy* 78:839–49.

————. 1998. Suckling behavior of prairie voles (*Microtus ochrogaster*). *Journal of Mammalogy* 79:1184–90.

McGuire, B., and L. L. Getz. 1991. Response of young female prairie voles (*Microtus ochrogaster*) to nonresident males: Implications for population regulation. *Canadian Journal of Zoology* 69:1348–55.

————. 1998. The nature and frequency of social interactions among free-living prairie voles (*Microtus ochrogaster*). *Behavioral Ecology and Sociobiology* 43:271–79.

McGuire, B., L. L. Getz, J. E. Hofmann, T. Pizzuto, and B. Frase. 1993. Natal dispersal and philopatry in prairie voles (*Microtus ochrogaster*) in relation to population density, season, and natal social environment. *Behavioral Ecology and Sociobiology* 32:293–302.

McGuire, B., E. Henyey, E. McCue, and W. E. Bemis. 2003. Parental behavior at parturition in prairie voles (*Microtus ochrogaster*). *Journal of Mammalogy* 84:513–23.

McGuire, B., and M. Novak. 1984. A comparison of maternal behaviour in the meadow vole (*Microtus pennsylvanicus*), prairie vole (*M. ochrogaster*) and pine vole (*M. pinetorum*). *Animal Behaviour* 32:1132–41.

————. 1986. Parental care and its relationship to social organization in the montane vole (*Microtus montanus*). *Journal of Mammalogy* 67:305–11.

————. 1987. The effects of cross-fostering on the development of social preferences in meadow voles (*Microtus pennsylvanicus*). *Behavioral and Neural Biology* 47:167–72.

McGuire, B., K. D. Russell, T. Mahoney, and M. Novak. 1992. The effects of mate removal on pregnancy success in prairie voles (*Microtus ochrogaster*) and meadow voles (*M. pennsylvanicus*). *Biology of Reproduction* 47:37–42.

McGuire, B., and S. Sullivan. 2001. Suckling behavior of pine voles (*Microtus pinetorum*). *Journal of Mammalogy* 82:690–99.

McGuire, M. R., and L. L. Getz. 1981. Incest taboo between sibling *Microtus ochrogaster*. *Journal of Mammalogy* 62:213–15.

McGuire, M. R., L. L. Getz, J. E. Hofmann, T. Pizzuto, and B. Frase. 1993. Natal dispersal and philiopatry in prairie voles (*Microtus ochrogaster*) in relation to population density, season, and anatal social environment. *Behavioral Ecology and Sociobiology* 32:293–302.

McKenna, M. C. 1975. Toward a phylogenetic classification of the Mammalia. In *Phylogeny of primates,* ed. W. P. Luckett and F. S. Szalay, 21–46. New York: Plenum.

McKinstry, M. C., and S. H. Anderson. 2002. Survival, fates, and success of transplanted beavers, *Castor canadensis,* in Wyoming. *Canadian Field-Naturalist* 116:60–68.

McLean, I. G. 1982. The association of female kin in the arctic ground squirrel (*Spermophilus parryii*). *Behavioral Ecology and Sociobiology* 10:91–99.

———. 1983. Paternal behaviour and killing of young in Arctic ground squirrels. *Animal Behaviour* 31:32–44.

McMillan, R. E., and T. J. Garland, Jr. 1989. Adaptive physiology. In *Advances in the study of* Peromyscus *(Rodentia),* ed. G. Kirkland Jr. and J. N. Layne, 143–68. Lubbock: Texas Tech University Press.

McNab, B. K. 2002. *The physiological ecology of vertebrates: A view from energetics.* Ithaca, NY: Cornell University Press.

McNaughton, S. J. 1984. Animals in herds, plant form and co-evolution. *American Naturalist* 124:863–86.

McShea, W. J., and D. M. Madison. 1984. Communal nursing between reproductively active females in a spring population of *Microtus pennsylvanicus*. *Canadian Journal of Zoology* 62:344–46.

———. 1986. Sex ratio shifts within litters of meadow voles (*Microtus pennsylvanicus*): Possible consequences for population cycling. *Behavioral Ecology and Sociobiology* 18:431–36.

———. 1989. Measurements of reproductive traits in a field population of meadow voles. *Journal of Mammalogy* 70:132–41.

Meagher, S., D. Penn, and W. K. Potts. 2000. Male-male competition magnifies inbreeding depression in wild house mice. *Proceedings of the National Academy of Sciences (USA)* 97:3324–29.

Meaney, M. J. 2001. Maternal care, gene expression, and the transmission of individual differences in stress reactivity across generations. *Annual Review of Neuroscience* 24:1161–92.

Mech, S. G., A. S. Dunlap, and J. O. Wolff. 2003. Female prairie voles do not choose males based on their frequency of scent marking. *Behavioral Processes* 61:101–8.

Medway, L. 1978. *The wild mammals of Malaya (Peninsular Malaysia) and Singapore,* 2nd ed. Kuala Lumpur, Malaysia: Oxford University Press.

Meehan, A. P. 1984. *Rats and mice: Their biology and control.* Sussex, England: Rentokill.

Meier, P. T. 1992. Social organization of woodchucks (*Marmota monax*). *Behavioral Ecology and Sociobiology* 31:393–400.

Meier, P. T., and G. Svendsen. 1983. Alternative mating strategies of male woodchucks (*Marmota monax*). *American Zoologist* 23:932–932.

Meikle, D. B., and L. C. Drickamer. 1986. Food availability and secondary sex ratios variation in wild and laboratory house mice (*Mus musculus*). *Journal of Reproduction and Fertility* 78:587–91.

Meikle, D. B., and M. W. Thornton. 1995. Premating and gestation effects of maternal nutrition on secondary sex ratio in house mice. *Journal of Reproduction and Fertility* 105:193–96.

Mein, P. 1992. Taxonomy. In *First international symposium of alpine marmot* (Marmota marmota) *and on genus* Marmota, ed. B. Bassano, P. Durio, U. Gallo Orsi, and E. Macchi, 6–12. Torino.

Meites, J., and J. K. H. Lu. 1994. Reproductive ageing and neuroendocrine function. *Reviews of Reproductive Biology* 16:215–47.

Melcher, J. C., K. B. Armitage, and W. P. Porter. 1990. Thermal influences on the activity and energetics of yellow-bellied marmots (*Marmota flaviventris*). *Physiological Zoology* 63:803–20.

Melchior, H. R. 1971. Characteristics of arctic ground squirrel alarm calls. *Oecologia* 7:184–90.

Mendl, M. 1988. The effects of litter size variation on mother-offspring relationships and behavioural and physical development in several mammalian species (principally rodents). *Journal of Zoology, London* 215:15–34.

Meng, J. 1990. The auditory region of *Reithroparamys delicatissimus* (Mammalia, Rodentia) and its systematic implications. *American Museum Novitates* 2972:1–35.

Meng, J., A. R. Wyss, M. R. Dawson, and R. Zhai. 1994. Primitive fossil rodent from Inner Mongolia and its implications for mammalian phylogeny. *Nature* 370:134–36.

Mennella, J. A., M. S. Blumberg, M. K. McClintock, and H. Moltz. 1990. Inter-litter competition and communal nursing among Norway rats—advantages of birth synchrony. *Behavioral Ecology and Sociobiology* 27:183–90.

Mennella, J. A., and H. Moltz. 1988. Infanticide in rats: Male strategy and female counter-strategy. *Physiology and Behavior* 42:19–28.

Menzies, R. A., G. Heth, R. Ikan, V. Weinstein, and E. Nevo. 1992. Sexual pheromones in lipids and other fractions from urine of the male mole rat, *Spalax ehrenbergi*. *Physiology and Behavior* 52:741–47.

Mercer, J. M., and V. L. Roth. 2003. The effects of Cenozoic global change on squirrel phylogeny. *Science* 299:1568–72.

Merilä, J., B. C. Sheldon, and H. Ellegren. 1998. Quantitative genetics of sexual size dimorphism in the collared flycatcher, *Ficedula albicollis*. *Evolution* 52:870–76.

Mermelstein, P. G., and J. B. Becker. 1995. Increased extracellular dopamine in the nucleus accumbens and striatum of the female rat during paced copulatory behavior. *Behavioral Neuroscience* 109:354–65.

Merriam, C. H. 1901. The prairie dog of the great plains *Yearbook of the United States Department of Agriculture,* 257–70. Washington, DC: U.S. Government Printing Office.

Merritt, J. F. 1978. *Peromyscus californicus*. *Mammalian Species* 85:1–6.

Merritt, R. B., and B. J. Wu. 1975. Quantification of promiscuity (or "*Promyscus*" *maniculatus*?). *Evolution* 29:575–78.

Meserve, P. L., R. E. Martin, and J. Rodríguez. 1983. Feeding ecology of two Chilean caviomorphs in a central mediterranean savanna. *Journal of Mammalogy* 64:322–25.

———. 1984. Comparative ecology of the caviomorph rodent

Octodon degus in two Chilean Mediterranean-type communities. *Revista Chilena de Historia Natural* 57:79–89.

Metzgar, L. M. 1967. An experimental comparison of owl predation on resident and transient white-footed mice (*Peromyscus leucopus*). *Journal of Mammalogy* 48:387–91.

Michaux, J., and F. Catzeflis. 2000. The bushlike radiation of muroid rodents is exemplified by the molecular phylogeny of the LCAT nuclear gene. *Molecular Phylogenetics and Evolution* 17:280–93.

Michaux, J., A. Reyes, and F. Catzeflis. 2001. Evolutionary history of the most speciose mammals: Molecular phylogeny of muroid rodents. *Molecular Biology and Evolution* 18:2017–31.

Michener, C. D. 1969. Comparative social behaviour of bees. *Annual Review of Entomology* 14:299–342.

Michener, G. R. 1971. Maternal behaviour in Richardson's ground squirrel, *Spermophilus richardsonii richardsoniiI*: Retrieval of young by lactating females. *Animal Behaviour* 19:653–56.

———. 1973a. Field observations on the social relationships between adult female and juvenile Richardson's ground squirrels. *Canadian Journal of Zoology* 51:33–38.

———. 1973b. Maternal behaviour in Richardson's ground squirrel (*Spermophilus richardsonii richardsonii*): Retrieval of young by non-lactating females. *Animal Behaviour* 21:157–59.

———. 1974. Development of adult-young identification in Richardson's ground squirrel. *Developmental Psychobiology* 7:375–84.

———. 1978. Effect of age and parity on weight gain and entry into hibernation in Richardson's ground squirrels. *Canadian Journal of Zoology* 56:2573–77.

———. 1981. Ontogeny of spatial relationships and social behavior in juvenile Richardson's ground squirrels. *Canadian Journal of Zoology* 59:1666–76.

———. 1983a. Kin identification, matriarchies and the evolution of sociality in ground-dwelling sciurids. In *Recent advances in the study of mammalian behavior*, Special publication no. 7, ed. J. F. Eisenberg and D. G. Kleiman, 528–72. American Society of Mammalogists.

———. 1983b. Spring emergence schedules and vernal behavior of Richardson's ground squirrels: Why do males emerge from hibernation before females? *Behavioral Ecology and Sociobiology* 14:29–38.

———. 1984. Sexual differences in body weight patterns of Richardson's ground squirrels during the breeding season. *Journal of Mammalogy* 65:59–66.

———. 1989. Reproductive effort during gestation and lactation by Richardson's ground squirrels. *Oecologia* 78:77–86.

———. 1996. Establishment of a colony of Richardson's ground squirrels in southern Alberta. In *Proceedings of the fourth prairie conservation and endangered species workshop*, ed. W. D. Willms and J. F. Dormaar, 303–8. Provincial Museum of Alberta Natural History Occasional Paper 23:1–337.

———. 2000. Caching of Richardson's ground squirrels by North American badgers. *Journal of Mammalogy* 81:1106–17.

Michener, G. R., and A. N. Iwaniuk. 2001. Killing technique of North American badgers preying on Richardson's ground squirrels. *Canadian Journal of Zoology* 79:2109–13.

Michener, G. R., and D. Michener. 1977. Population structure and dispersal in Richardson's ground squirrels. *Ecology* 58:359–68.

Michener, G. R., and I. G. McLean. 1996. Reproductive behaviour and operational sex ratio in Richardson's ground squirrels. *Animal Behaviour* 52:743–58.

Middleton, A. D. 1954. Rural rat control. In *Control of rats and mice*, ed. D. Chitty, 414–48. Oxford: Clarendon.

Mihalcin, J. A. 2002. Dominance rank, parasite infection, and mate choice in house mice. Master's thesis. Miami University, Oxford, Ohio.

Mihok, S. 1979. Behavioral structure and demography of subarctic *Clethrionomys gapperi* and *Peromyscus maniculatus*. *Canadian Journal of Zoology* 57:1520–35.

Mihok, S., and R. Boonstra. 1992. Breeding performance in captivity of meadow voles (*Microtus pennsylvanicus*) from decline- and increase-phase populations. *Canadian Journal of Zoology* 70:1561–66.

Mikesic, D. G., and L. C. Drickamer. 1992. Factors affecting home-range size in house mice (*Mus musculus domesticus*) living in outdoor enclosures. *American Midland Naturalist* 127:31–40.

Mikhailuta, A. A. 1991. Family structure in grey marmots. In Pages *Population structure of the marmot*, ed. D. I. Bibikov, A. A. Nikolski, V. Yu. Rumiantsev, and T. A. Seredneva, 172–87. Moscow: USSR Theriological Society.

Miles, D. B., and A. E. Dunham. 1992. Comparative analyses of phylogenetic effects in the life-history patterns of iguanid reptiles. *American Naturalist* 139:848–69.

Millar, J. S. 1970. The breeding season and reproductive cycle of the western red squirrel. *Canadian Journal of Zoology* 48:471–73.

———. 1979. Energetics of lactation in *Peromyscus maniculatus*. *Canadian Journal of Zoology* 57:1015–19.

———. 1987. Energy reserves in breeding small rodents. *Symposium of the Zoological Society of London* 57:231–40.

———. 1989. Reproduction and development. In *Advances in the study of Peromyscus (Rodentia)*, ed. G. Kirkland Jr. and J. N. Layne, 169–231. Lubbock: Texas Tech University Press.

Millar, J. S., and E. M. Derrickson. 1992. Group nesting in *Peromyscus maniculatus*. *Journal of Mammalogy* 73:403–7.

Miller, B., G. Ceballos, and R. P. Reading. 1994. The prairie dog and biotic diversity. *Conservation Biology* 8:677–81.

Miller, B., and R. P. Reading. 2006. A proposal for more effective conservation of prairie dogs. In *Conservation of the black-tailed prairie dog*, ed. J. L. Hoogland, 248–60. Washington, DC: Island Press.

Miller, B., R. P. Reading, and S. C. Forrest. 1996. *Prairie night: Black-footed ferrets and the recovery of endangered species*. Washington, DC: Smithsonian Institution Press.

Miller, B., R. Reading, J. Hoogland, T. Clark, G. Ceballos, R. List, S. Forrest, et al. 2000. The role of prairie dogs as keystone species: A response to Stapp. *Conservation Biology* 14:318–21.

Miller, D. B. 1981. Conceptual strategies in behavioral development: Normal development and plasticity. In *Behavioral development*, ed. K. Immelmann, G. W. Barlow, L. Petrinovich, and M. Main, 58–85. New York: Cambridge University Press.

Miller, R. S. 1964. Ecology and distribution of pocket gophers (Geomyidae) in Colorado. *Ecology* 45:256–72.

Miller, S. D., J. Rottman, K. J. Raedeke, and R. D. Taber. 1983. Endangered mammals of Chile: Status and conservation. *Biological Conservation* 25:335–52.

Millesi, E., I. E. Hoffmann, S. Steurer, M. Metvaly, and J. P. Dittami. 2002. Vernal changes in the behavioral and endocrine responses to GnRH application in male European ground squirrels. *Hormones and Behavior* 41:51–58.

Millesi, E., S. Huber, J. Dittami, I. Hoffman, and S. Daan. 1998. Parameters of mating effort and success in male European ground squirrels, *Spermophilus citellus*. *Ethology* 104: 298–313.

Millesi, E., S. Huber, L. G. Everts, and J. P. Dittami. 1999. Reproductive decisions in female European ground squirrels: Factors affecting reproductive output and maternal investment. *Ethology* 105:163–75.

Milligan, S. R. 1980. Pheromones and rodent reproductive physiology. *Symposia of the Zoological Society of London* 45:251–75.

———. 1982. Induced ovulation in mammals. *Oxford Reviews of Reproductive Biology* 4:1–46.

Mills, J. N., and J. E. Childs. 1998. Ecologic studies of rodent reservoirs: Their relevance for human health. *Emerging Infectious Diseases* 4:529–37.

Mills, J. N., B. A. Ellis, J. E. Childs, K. T. McKee, Jr., J. I. Maiztegui, C. J. Peters, T. G. Ksiazek, and P. B. Jahrling. 1994. Prevalence of infection with Junín virus in rodent populations in the epidemic area of Argentine hemorrhagic fever. *American Journal of Tropical Medicine and Hygiene* 51:554–62.

Mills, J. N., B. A. Ellis, K. T. McKee, G. E. Calderón, J. I. Maiztegui, G. O. Nelson, T. G. Ksiazek, C. J. Peters, and J. E. Childs. 1992. A longitudinal study of Junín virus activity in the rodent reservoir of Argentine hemorrhagic fever. *American Journal of Tropical Medicine and Hygiene* 47:749–63.

Mills, J. N., T. G. Ksiazek, B. A. Ellis, P. E. Rollin, S. T. Nichol, T. L. Yates, W. L. Gannon et al. 1997. Patterns of association with host and habitat: Antibody reactive with Sin Nombre virus in small mammals in the major biotic communities of the southwestern United States. *American Journal of Tropical Medicine and Hygiene* 56:273–84.

Mills, J. N., T. G. Ksiazek, C. J. Peters, and J. E. Childs. 1999. Long-term studies of hantavirus reservoir populations in the southwestern United States: A synthesis. *Emerging Infectious Diseases* 5:135–42.

Mills, J. N., K. Schmidt, B. E. Ellis, and T. G. Ksiazek. 1998. Epizootiology of hantaviruses in sigmodontine rodents on the pampa of central Argentina. Paper presented at the Euro-American Mammal Congress, Santiago de Compostela, Spain, July 19–24, 1998.

Mills, J. N., T. L. Yates, T. G. Ksiazek, C. J. Peters, and J. E. Childs. 1999. Long-term studies of hantavirus reservoir populations in the southwestern United States: Rationale, potential, and methods. *Emerging Infectious Diseases* 5:95–101.

Mills, J. N. 2005. Regulation of rodent-borne viruses in the natural host: implications for human disease. *Archives of Virology* 19:45–57.

Mills, L. S. 1995. Edge effects and isolation: Red-backed voles on forest remnants. *Conservation Biology* 9:395–403.

———. 1996. Fragmentation of a natural area: Dynamics of isolation for small mammals on forest remnants. In *National parks and protected areas: Their role in environmental protection,* ed. R. G. Wright, 199–219. Cambridge, MA: Blackwell Science.

Mims, C. A. 1966. Immunofluorescence study of the carrier state and mechanism of vertical transmission in lymphocytic chori-

omeningitis virus infection in mice. *Journal of Pathology and Bacteriology* 91:395–402.

Mineau, P., and D. Madison. 1977. Radio-Tracking of *Peromyscus leucopus*. *Canadian Journal of Zoology* 55:465–68.

Misawa, K., and A. Janke. 2003. Revisiting the Glires concept—phylogenetic analysis of nuclear sequences. *Molecular Phylogenetics and Evolution* 28:320–27.

Misslin, R. 1982. Some determinants of the new object reaction of the mouse. *Biology of Behavior* 3:209–14.

Mitchell, C. J., C. M. Heyes, M. R. Gardner, and G. R. Dawson. 1999. Limitations of a bidirectional control procedure for the investigation in rats: Odour cues on the manipulandum. *Quarterly Journal of Experimental Psychology* B 52:193–202.

Mitchell, D. 1976. Experiments on Neophobia in wild and laboratory rats: A re-evaluation. *Journal of Comparative and Physiological Psychology* 90:190–97.

Mitchell, J. B., and A. Gratton. 1992. Mesolimbic dopamine release elicited by activation of the accessory olfactory system: A high speed chronoamperometric study. *Neuroscience Letters* 140:81–84.

Mitchell, W. A., Z. Abramsky, B. P. Kotler, B. P. Pinshow, and J. S. Brown. 1990. The effect of competition on foraging activity in desert rodents: Theory and experiments. *Ecology* 71:844–54.

Modi, W. S. 1984. Reproductive tactics among deer mice of the genus *Peromyscus*. *Canadian Journal of Zoology* 62:2576–81.

Moffatt, C. A., G. F. Ball, and R. J. Nelson. 1995. The effects of photoperiod on olfactory c-fos expression in prairie voles, *Microtus ochrogaster*. *Brain Research* 677:82–88.

Moilanen, A., A. T. Smith, and I. Hanski. 1998. Long-term dynamics in a metapopulation of the American pika. *American Naturalist* 152:530–42.

Møller, A. P., and R. V. Alatalo. 1999. Good-genes effects in sexual selection. *Proceedings of the Royal Society of London, Series B* 266:85–91.

Møller, A. P., and T. R. Birkhead. 1989. Copulation behaviour in mammals: Evidence that sperm competition is widespread. *Biological Journal of the Linnean Society* 38:119–31.

Møller, A.P., and F. De Lope. 1999. Senescence in a short-lived migratory bird: Age-dependent morphology, migration, reproduction and parasitism. *Journal of Animal Ecology* 68:163–71.

Møller, A. P., and J. P. L. Swaddle. 1997. *Asymmetry, developmental stability and evolution.* Oxford: Oxford University Press.

Molteno, A. J., and N. C. Bennett. 2000. Anovulation in non-reproductive female Damaraland mole–rats (*Cryptomys damarensis*). *Journal of Reproduction and Fertility* 119: 35–41.

———. 2002. Rainfall, dispersal and reproductive inhibition in eusocial Damaraland mole-rats (*Cryptomys damarensis*). *Journal of Zoology, London* 256:445–48.

Moltz, H., and D. Robbins. 1965. Maternal behavior of primiparous and multiparous rats. *Journal of Comparative and Physiological Psychology* 60:417–21.

Monard, A. M., P. Duncan, H. Fritz, and C. Feh. 1997. Variations in the birth sex ratio and neonatal mortality in a natural herd of horses. *Behavioral Ecology and Sociobiology* 41: 243–49.

Mones, A., and J. Ojasti. 1986. *Hydrochoerus hydrochaeris*. *Mammalian species* 264:1–7.

Monteiro, L. S., and D. S. Falconer. 1966. Compensatory growth and sexual maturity in mice. *Animal Production* 8: 179–85.

Montgelard, C., S. Bentz, C. Tirard, O. Verneau, and F. M. Catzeflis. 2002. Molecular systematics of Sciurognathi (Rodentia): The mitochondrial cytochrome *b* and 12S rRNA genes support the Anomaluroidea (Pedetidae and Anomaluridae). *Molecular Phylogenetics and Evolution* 22:220–33.

Moore, A. J., N. L. Reagan, and K. F. Haynes. 1995. Conditional signalling strategies: Effects of ontogeny, social experience and social status on the pheromonal signal of male cockroaches. *Animal Behaviour* 50: 191–202.

Moore, B. 1996. The evolution of imitative learning. In *Social learning and imitation: The roots of culture,* ed. C. M. Heyes and B. G. Galef, Jr., 245–65. New York: Academic Press.

Moore, C. L. 1981. An olfactory basis for maternal discrimination of sex of offspring in rats (*Rattus norvegicus*). *Animal Behaviour* 29:383–86.

Moore, C. L., and G. A. Morelli. 1979. Mother rats interact differently with male and female offspring. *Journal of Comparative and Physiological Psychology* 93:677–84.

Moore, C. L., and B. R. Samonte. 1986. Preputial glands of infant rats (*Rattus norvegicus*) provide chemosignals for maternal discrimination of sex. *Journal of Comparative Psychology* 100:76–80.

Moore, J., and R. Ali. 1984. Are dispersal and inbreeding avoidance related? *Animal Behaviour* 32:94–112.

Moreira, J. R. 1995. The reproduction, demography and management of capybaras (*Hydrochaeris hydrochaeris*) on Marajó Island—Brazil. PhD diss., University of Oxford.

Moreira, J. R., J. R. Clarke, and D. W. Macdonald. 1997. The testis of capybaras (*Hydrochoerus hydrochaeris*). *Journal of Mammalogy* 78:1096–1100.

Moreira, J. R., H. J. Cunha, P. R. S. Pinha, J. P. Carvalho, and A. P. Hercos. 2002. Estação de nascimentos de capivaras (*Hydrochoerus hydrochaeris*) no cerrado. Simpósio Ecologia e Biodiversidade do cerrado: Perspectives e desafios para o Século 21:p34. Brasília: Emprapa Sede Brasília.

Moreira, J. R., and D. W. Macdonald. 1993. The population ecology of capybaras (*Hydrochoerus hydrochaeris*) and their management for conservation in Brazilian Amazonia. In *Biodiversity and environment: Brazilian themes for the future,* ed. S. J. Mayo and D. C. Zappi, 26–27. Kew, London: The Linnean Society of London/the Royal Botanic Gardens.

———. 1996. Capybara use and conservation in South America. In *The exploitation of mammal populations,* ed. V. J. Taylor and N. Dunstone, 88–101. London: Chapman and Hall.

Moreira, J. R., D. W. Macdonald, and J. R. Clarke. 1997. Correlates of testes mass in capybaras (*Hydrochoerus hydrochaeris*): Dominance assurance or sperm production? *Journal of Zoology, London* 241:457–63.

———. 2001. Alguns aspectos comportamentais da reprodução da capivara. *Revista Brasileira de Reprodução Animal* 25: 120–22.

Morgan, L. H. 1868. *The American beaver and his works.* Philadelphia: J. B. Lippincott.

Morgan, U. M., A. P. Sturdee, G. R. Singleton, M. S. Gomez, M. Gracenea, J. Torres, S. G. Hamilton, D. P. Woodside, and R. C. A. Thompson. 1999. The *Cryptosporidium* "Mouse" genotype is conserved across geographic areas. *Journal of Clinical Microbiology* 37:1302–5.

Morhardt, S. S., and D. M. Gates. 1974. Energy-exchange analysis of the Belding ground squirrel and its habitat. *Ecological Monographs* 44:17–44.

Morin, P. A., J. J. Moore, R. Chakraborty, L. Jin, J. Goodall, and D. S. Woodruff. 1994. Kin selection, social structure, gene flow and the evolution of chimpanzees. *Science* 265: 1193–1201.

Moro, D., K. L. Lloyd, A. L. Smith, G. R. Shellam, and M. A. Lawson. 1999. Murine viruses in an island population of introduced house mice and endemic short-tailed mice in Western Australia. *Journal of Wildlife Diseases* 35:301–10.

Morris, D. W., and D. L. Davidson. 2000. Optimally foraging mice match patch use with habitat differences in fitness. *Ecology* 81:2061–66.

Morris, K. D. 2000. The value of granite outcrops for mammal conservation in Western Australia. *Journal of the Royal Society of Western Australia* 83:169–72.

Morrison, P., R. Dietrich, and D. Preston. 1977. Body growth in sixteen rodent species and subspecies maintained in laboratory colonies. *Physiological Zoology* 50:294–310.

Morton, M. L., and H. L. Tung. 1971. Growth and development in the Belding ground squirrel (*Spermophilus beldingi beldingi*). *Journal of Mammalogy* 52:611–16.

Morton, M. L., and P. W. Sherman. 1978. Effects of a spring snowstorm on behavior, reproduction, and survival of Belding's ground squirrels. *Canadian Journal of Zoology* 56: 2578–90.

Moses, R. A., S. Boutin, and T. Teferi. 1998. Sex-biased mortality in woodrats occurs in the absence of parental intervention. *Animal Behaviour* 55:563–71.

Moses, R. A., G. J. Hickling, and J. S. Millar. 1995. Variation in sex ratios of offspring in wild bushy-tailed woodrats. *Journal of Mammalogy* 76:1047–55.

Moses, R. A., and J. S. Millar. 1992. Behavioural asymmetries and cohesive mother-offspring sociality in bushy-tailed wood rats. *Canadian Journal of Zoology* 70:597–604.

———. 1994. Philopatry and mother-daughter associations in bushy-tailed woodrats: Space use and reproductive success. *Behavioral Ecology and Sociobiology* 35:131–40.

Mosolov, V. I., and V. A. Tokarsky. 1994. The black-capped marmot (*Marmota camtschatica* Pall.) in the Kronotsky Reserve. In *Actual problems of marmots investigation,* ed. V. Yu. Rumiantsev, 98–110. Moscow: ABF.

Moss, A. E. 1940. The woodchuck as a soil expert. *Journal of Wildlife Management* 4:441–43.

Moss, R., R. Parr, and X. Lambin. 1994. Effects of testosterone on breeding density, breeding success and survival of red grouse. *Proceedings of the Royal Society of London, Series B* 258:175–80.

Mossman, H. W., R. S. Hoffmann, and C. M. Kirkpatrick. 1955. The accessory genital glands of male gray and fox squirrels correlated with age and reproductive cycles. *American Journal of Anatomy* 97:257–301.

Mougeot, F., S. M. Redpath, F. M. Leckie, and P. J. Hudson. 2003. The effect of aggressiveness on the population dynamics of a territorial bird. *Nature* 421:737–39.

Mougeot, F., S. M. Redpath, R. Moss, J. Matthiopoulos, and P. J. Hudson. 2003. Territorial behaviour and population dynamics in red grouse *Lagopus lagopus scoticus.* I. Population experiments. *Journal of Animal Ecology* 72:1073–82.

Mourão, G., and L. Carvalho. 2001. Cannibalism among giant otters (*Pteronura brasiliensis*). *Mammalia* 65:225–27.

Mouse Genome Sequencing Consortium. 2002. Initial sequencing and comparative analysis of the mouse genome. *Nature* 420:520–62.

Mucignat-Caretta, C., A. Caretta, and A. Caviggioni. 1995. Acceleration of puberty onset in female mice by male urinary proteins. *Journal of Physiology* 486:517–22.

Müller-Schwarze, D. 1992. Castoreum of beaver (*Castor canadensis*): Function, chemistry and biological activity of its components. In *Chemical signals in vertebrates 6,* ed. R. L. Doty and D. Müller-Schwarze, 457–64. New York: Plenum.

Müller-Schwarze, D., and B. A. Schulte. 1999. Behavioral and ecological characteristics of a "climax" population of beaver (*Castor canadensis*). In *Beaver protection, management, and utilization in Europe and North America,* ed. P. E. Busher and R. M. Dzieciolowski, 161–77. New York: Kluwer Academic/Plenum.

Müller-Schwarze, D., and L. Sun. 2003. *The beaver: Natural history of a wetlands engineer.* Ithaca, NY: Cornell University Press.

Mumme, R. L. 1997. A bird's eye view of mammalian cooperative breeding. In *Cooperative breeding in mammals,* ed. N. G. Solomon and J. A. French, 364–88. Cambridge: Cambridge University Press.

Munn, L. C. 1993. Effects of prairie dogs on physical and chemical properties of soils. In *Management of Prairie Dog Complexes for the Reintroduction of the Black-footed Ferret,* ed. J. L. Oldemeyer, D. E. Biggins, and B. J. Miller, 11–17. Washington, DC: United States Department of the Interior.

Muñoz, A., and R. Murúa. 1987. Biología de *Octodon bridgesi bridgesi* (Rodentia, Octodontidae) en la zona costera de Chile central. *Boletín de la Sociedad de Biología de Concepción (Chile)* 58:107–17.

Muñoz-Pedreros, A. 1992. Ecología del ensamble de micro-mamíferos en un agroecosistema forestal de Chile central: Una comparación latitudinal. *Revista Chilena de Historia Natural* 65:417–28.

Murdock, H. G., and J. A. Randall. 2001. Olfactory communication and neighbor recognition in giant kangaroo rats. *Ethology* 107:149–60.

Murie, J. O. 1992. Predation by badgers on Columbian ground squirrels. *Journal of Mammalogy* 73:385–94.

———. 1995. Mating behavior of Columbian ground squirrels. I. Multiple mating by females and multiple paternity. *Canadian Journal of Zoology* 73:1819–26.

Murie, J. O., and D. A. Boag. 1984. The relationship of body weight to overwinter survival in Columbian ground squirrels. *Journal of Mammalogy* 65:688–90.

Murie, J. O., D. A. Boag, and K. Kivett. 1980. Litter size in Columbian ground squirrels (*Spermophilus columbianus*). *Journal of Mammalogy* 61:237–44.

Murie, J. O., and M. A. Harris. 1984. The history of individuals in a population of Columbian ground squirrels: Source, settlement, and site attachment. In *The biology of ground dwelling squirrels,* ed. J. O. Murie and G. R. Michener, 353–74. Lincoln: University of Nebraska Press.

———. 1994. Social interactions and dominance relationships between female and male Columbian ground squirrels. *Canadian Journal of Zoology* 66:1414–20.

Murie, J. O., and I. G. McLean. 1980. Copulatory plugs in ground squirrels. *Journal of Mammalogy* 61:355–56.

Murie, J. O., and G. R. Michener, eds. 1984. *The biology of ground-dwelling squirrels: Annual cycles, behavioral ecology, and sociality.* Lincoln: University of Nebraska Press.

Murphy, G. 1968. Pattern in life history and the environment. *American Naturalist* 102:390–404.

Murphy, W. J., E. Eizirik, W. E. Johnson, Y. P. Zhang, O. A. Ryder, and S. J. O'Brien. 2001a. Molecular phylogenetics and the origins of placental mammals. *Nature* 409:614–18.

Murphy, W. J., E. Eizirik, S. J. O'Brien, O. Madsen, M. Scally, C. J. Douady, E. Teeling, O. A. Ryder, M. J. Stanhope, W. W. de Jong., and M. S. Springer. 2001b. Resolution of the early placental mammal radiation using Bayesian phylogenetics. *Science* 294:2348–51.

Musser, G. G., and M. D. Carleton. 1993. Family Muridae. In *Mammal species of the world,* ed. D. E. Wilson and D. M. Reeder, 501–755. Washington, DC: Smithsonian Institution Press.

Myers, J. 1978. Sex ratio adjustment under food stress: Maximization of quality or numbers of offspring? *American Naturalist* 112:381–88.

Myers, J. H. 1973. The absence of *t* alleles in feral populations of house mice. *Evolution* 27:702–4.

Myers, J. H., and C. J. Krebs. 1971. Genetic, behavioral, and reproductive attributes of dispersing field voles, *Microtus pennsylvanicus* and *Microtus ochrogaster*. *Ecological Monographs* 41:53–78.

Myers, P., and L. L. Masters. 1983. Reproduction by *Peromyscus maniculatus*: Size and compromise. *Journal of Mammalogy* 64:1–18.

Myers, P., L. L. Master, and A. Garrett. 1985. Ambient temperature and rainfall: An effect on sex ratio and litter size in deer mice. *Journal of Mammalogy* 66:289–98.

Myles, T. G. 1988. Resource inheritance in social evolution from termites to man. In *The ecology of social behavior,* ed. C. N. Slobodchikoff, 379–423. San Diego, CA: Academic Press.

Myllymaki, A. 1977. Intraspecific competition and home range dynamics in the field vole *Microtus agrestis*. *Oikos* 29:553–69.

Nadachowski, A., and J. I. Mead. 1999a. *Alticola argentatus*. *Mammalian Species* 625:1–4.

———. 1999b. *Alticola strelzovi*. *Mammalian Species* 626:1–3.

Nadeau, J. H. 1985. Ontogeny. In *Biology of New World Microtus.* Special publication no. 8, ed. R. H. Tamarin, 254–85. Shippensburg, PA: American Society of Mammalogists.

Nadler, C. F., R. S. Hoffmann, and J. J. Pizzimenti. 1971. Chromosomes and serum proteins of prairie dogs and a model of *Cynomys* evolution. *Journal of Mammalogy* 52:545–55.

Nagorsen, D. W. 1987. *Marmota vancouverensis*. *Mammalian Species* 270:1–5.

Nathan, R. 2001. The challenges of studying dispersal. *Trends in Ecology and Evolution* 16:481–83.

National Oceanic and Atmospheric Association (NOAA). 1994. US divisional and station climatic data and normals vol. 1. United States Department of Commerce. Retrieved from http://www.ncdc.noaa.gov/oa/ncdc.html.

NatureServe. 2004. NatureServe Explorer: An online encyclopedia of life (Web application). Version 4.0. Available http://www.natureserve.org/explorer. Arlington, VA: NatureServe.

Natynczuk, S. E., and D. W. Macdonald. 1994a. Scent, sex, and the self-calibrating rat. *Journal of Chemical Ecology* 20:1843–57.

———. 1994b. Analysis of rat haunch odour using the dynamic solvent effect and principal component analysis. *Journal of Chemical Ecology* 20:1859–66.

Naumov, N. P., and V. S. Lobachev. 1975. Ecology of desert

rodents of the U.S.S.R. (jerboas and gerbils). In *Rodents in desert environments*, ed. I. Prakash and C. K. Ghosh, 465–598. The Hague: Junk.

Nedbal, M. A., M. W. Allard, and R. L. Honeycutt. 1994. Molecular systematics of hystricognath rodents: Evidence from the mitochondrial 12S rRNA gene. *Molecular Phylogenetics and Evolution* 3:206–20.

Nedbal, M. A., R. L. Honeycutt, and D. A. Schlitter. 1996. Higher-level systematics of rodents (Mammalia, Rodentia): Evidence from the mitochondrial 12S rRNA gene. *Journal of Mammalian Evolution* 3:201–37.

Neill, S. R. S. J., and J. M. Cullen. 1974. Experiments on whether schooling by their prey affects the hunting behaviour of cephalopods and fish predators. *Journal of Zoology, London* 172:549–69.

Nelson, J. 1995. Determinants of male spacing behavior in microtines: An experimental manipulation of female spatial distribution and density. *Behavioral Ecology and Sociobiology* 37:217–32.

Nelson, R. J. 1991. Maternal diet influences reproductive development in male prairie vole offspring. *Physiology and Behavior* 50:1063–66.

Nesterova, N. L. 1996. Age-dependent alarm behaviour and response to alarm call in bobac marmots (*Marmota bobac* Müll.). In *Biodiversity in marmots*, ed. M. Le Berre, R. Ramousse, and L. Le Guelte, 181–86. Moscow-Lyon: International Network on Marmots.

Neuhaus, P. 2000. Timing of hibernation and molt in female Columbian ground squirrels. *Journal of Mammalogy* 81:571–77.

Neuhaus, P., R. Bennett, and A. Hubbs. 1999. Effect of a late snow-storm and rain on survival and reproductive success in Columbian ground squirrels. *Canadian Journal of Zoology* 77:879–84.

Neuhaus, P., and N. Pelletier. 2001. Mortality in relation to season, age, sex, and reproduction in Columbian ground squirrels (*Spermophilus columbianus*). *Canadian Journal of Zoology* 79:465–70.

Neumann, I. D., H. A. Johnstone, M. Hatzinger, et al. 1998. Attenuated neuroendocrine responses to emotional and physical stressors in pregnant rats involve adenohypophysial changes. *Journal of Physiology* 508:289–300.

Nevison, C. M., S. Armstrong, R. J. Beynon, R. E. Humphries, and J. L. Hurst. 2003. The ownership signature in mouse scent marks is involatile. *Proceedings of the Royal Society of London, Series B* 270:1957–63.

Nevison, C. M., C. J. Barnard, J. L. Hurst, and R. J. Beynon. 2000. The consequences of inbreeding for recognising competitors. *Proceedings of the Royal Society of London, Series B* 267:687–94.

Nevo, E. 1961. Observations on Israeli populations of the mole rat, *Spalax ehrenbergi* Nehring 1898. *Mammalia* 25:127–44.

———. 1969. Mole Rat *Spalax ehrenbergi*: Mating behavior and its evolutionary significance. *Science* 163:484–86.

———. 1979. Adaptive convergence and divergence of subterranean mammals. *Annual Review of Ecology and Systematics* 10:269–308.

———. 1985. Speciation in action and adaptation in subterranean mole rats: Patterns and theory. *Bolletino Zoologico* 52:65–95.

———. 1989. Modes of speciation: The nature and role of peripheral isolates in the origin of species. In *Genetics, speciation and the founder principle*, ed. L. V. Giddings, K. Y.

Kaneshiro, and W. W. Anderson, 205–36. Oxford: Oxford University Press.

———. 1990. Evolution of nonvisual communication and photoperiodic perception in speciation and adaptation of blind subterranean mole rats. *Behaviour* 114:249–73.

———. 1991. Evolutionary theory and processes of active speciation and adaptive radiation in subterranean mole rats, *Spalax ehrenbergi* superspecies, in Israel. *Evolutionary Biology* 25:1–125.

———. 1998. Evolution of a visual system for life without light: Optimization via tinkering in blind mole rats. In *Principles of animal design. The optimization and symmorphosis debate*, ed. E. R. Weibel, C. R. Taylor, and C. Bolis, 288–98. Cambridge: Cambridge University Press.

———. 1999. *Mosaic evolution of subterranean mammals: Regression, progression and global convergence*. Oxford: Oxford University Press.

Nevo, E., M. G. Filippucci, C. Redi, A. Korol, and A. Beiles. 1994. Chromosomal speciation and adaptive radiation of mole rats in Asia Minor correlated with increased ecological stress. *Proceedings of the National Academy of Science (USA)* 91:8160–64.

Nevo, E., M. G. Filippucci, C. Redi, S. Simson, G. Heth, and A. Beiles. 1995. Karyotype and genetic evolution in speciation of subterranean mole rats of the genus *Spalax* in Turkey. *Biological Journal of the Linnean Society* 54:203–29.

Nevo, E., and H. Bar-El. 1976. Hybridization and speciation in fossorial mole rats. *Evolution* 30:831–40.

Nevo, E., and G. Heth. 1976. Assortative mating between chromosome forms of the mole rat, *Spalax ehrenbergi*. *Experientia* 32:1509–11.

Nevo, E., G. Heth, J.C. Auffray, and A. Beiles. 1987. Preliminary studies of sex pheromones in the urine of actively speciating subterranean mole rats of the *Spalax ehrenbergi* superspecies in Israel. *Phytoparasitica* 15:163–64.

Nevo, E., G. Heth, and A. Beiles. 1982. Population structure and evolution in subterranean mole rats. *Evolution* 36:1283–89.

———. 1986. Aggression patterns in adaptation and speciation of subterranean mole rats. *Journal of Genetics* 65:65–78.

Nevo, E., G. Heth, A. Beiles, and E. Frankenberg. 1987. Geographic dialects in blind mole rats: Role of vocal communication in active speciation. *Proceedings of the National Academy of Science (USA)* 84:3312–15.

Nevo, E., G. Heth, and H. Pratt. 1991. Seismic communication in a blind subterranean mammal: A major somatosensory mechanism in adaptive evolution underground. *Proceedings of the National Academy of Science (USA)* 88:1256–60.

Nevo, E., G. Heth, and S. Simson. 1991. Evolution of adaptive pacifistic behavior in Sahara Desert ecology originating from aggressive subterranean Israeli mole rats: Prologue to sociality? (Abstract). International Society Research on Aggression. VI European Conference at Hebrew University, Jerusalem, Israel. June 23–28, 1991.

Nevo, E., E. Ivanitskaya, and A. Beiles. 2001. *Adaptive radiation of blind subterranean mole rats: Naming and revisiting the four sibling species of the* Spalax ehrenbergi *superspecies in Israel:* Spalax galili ($2n = 52$), S. golani ($2n = 54$), S. carmeli ($2n = 58$), S. judaei ($2n = 60$). Leiden: Bachkuys.

Nevo, E., E. Ivanitskaya, M. G. Filippucci, and A. Beiles. 2000. Speciation and adaptive radiation of subterranean mole rats,

Spalax ehrenbergi superspecies, in Jordan. *Biological Journal of the Linnean Society* 69:263–81.

Nevo, E., V. Kirzhner, A. Beiles, and A. Korol. 1997. Selection versus random drift: Long-term polymorphism persistence in small populations (evidence and modeling). *Philosophical Transactions Royal Society of London, Series B* 352:381–89.

Nevo, E., G. Naftali, and R. Buttman. 1975. Aggression patterns and speciation. *Proceedings of the National Academy of Science (USA)* 72:3250–54.

Nevo, E., and A. Shkolnik 1974. Adaptive metabolic variation of chromosome forms in mole rats, *Spalax*. *Experientia* 30:724–26.

Nevo, E., S. Simson, G. Heth, and A. Beiles. 1992. Adaptive pacifistic behaviour in subterranean mole rats in the Sahara Desert, contrasting to and originating from polymorphic aggression in Israeli species. *Behaviour* 123:70–76.

Nevo, E., S. Simson, G. Heth, C. Redi, and G. Filippucci. 1991. Recent speciation of subterranean mole rats of the *Spalax ehrenbergi* superspecies in the El-Hamam isolate, northern Egypt. (Abstract) 6th International colloquium on the ecology and taxonomy of small African mammals. August 11–16, 1991. Mitzpe Ramon, Israel.

Newcomer, S. D., J. A. Zeh, and D. W. Zeh. 1999. Genetic benefits enhance the reproductive success of polyandrous females. *Proceedings of the National Academy of Sciences (USA)* 96:10236–41.

Newman, K. S., and Z. T. Halpin. 1988. Individual odours and mate discrimination in the prairie vole (*Microtus ochrogaster*). *Animal Behaviour* 36:1779–87.

Newmark, J. E., and S. H. Jenkins. 2000. Sex differences in agonistic behavior of Merriam's kangaroo rats *(Diopodomys merriami)*. *American Midland Naturalist* 143:377–88.

Newsome, A. E. 1969. A population study of house mice temporarily inhabiting a South Australian wheatfield. *Journal of Animal Ecology* 38:341–59.

Newsome, A. E., and L. K. Corbett. 1975. Outbreaks of rodents in semi-arid and arid Australia: Causes, preventions, and evolutionary considerations. In *Rodents in desert environments,* ed. I. Prakash and P. K. Ghosh, 117–53. The Hague: Junk.

Newton, I., and P. Rothery. 1997. Senescence and reproductive value in sparrowhawks. *Ecology* 78:1000–1008.

Nice, M. M. 1937. Studies in the life History of the song sparrow. I. *Transactions of the Linnean Society of New York* 4:1–247.

Nichols, N. R., M. Zieba, and N. Bye. 2001. Do glucocorticoids contribute to brain aging? *Brain Research Reviews* 37:273–86.

Nieder, L., M. Cagnin, and V. Parisi.1982. Burrowing and feeding behaviour in the rat. *Animal Behaviour* 30:837–844.

Niklasson, B., B. Hornfeldt, A. Lundkvist, S. Bjorsten, and J. Leduc. 1995. Temporal dynamics of Puumala virus antibody prevalence in voles and of nephropathia epidemica incidence in humans. *American Journal of Tropical Medicine and Hygiene* 53:134–40.

Nikol'skii, A. A. 1974. Geographical variability of sound call rhythmic organization in marmots of the group Bobac (Rodentia, Sciuridae). *Zoologicheskii Zhurnal* 53:436–44.

———. 1979. Species specificity of the alarm call in suslics (*Citellus*: Sciuridae) of Eurasia. *Zoologicheskii Zhurnal* 58:1183–94.

———. 1984. *Sound signals of the mammals in the evolutionary process.* Moscow: Nauka.

Nikol'skii, A. A. 1994. Geographical variability of the alarm call rhythmical structure in *Marmota baibacina*. In *Actual problems of marmots investigation,* ed. V. Y. Rumiantsev, 111–26. Moscow: ABF.

———. 1996. Species specificity and interspecies parallelisms of alarm call in eurasian marmots. In *Biodiversity in marmots,* ed. M. Le Berre, R. Ramousse, and L. Le Guelte, 187–92. Moscow-Lyon: International Network on Marmots.

———. 2000. Influence of spatial and temporal factors on vocal activity of the steppe marmot, *Marmota bobak*. *Zoologicheskii Zhurnal* 79:338–47.

Nikol'skii, A. A., V. P. Denisov, T. G. Stojko, and N. A. Formosov. 1984. The alarm call in F1 hybrids *Citellus pygmaeus* x *C. suslica* (Sciuridae, Rodentia). *Zoologicheskii Zhurnal* 63:1216–25.

Nikol'skii, A. A., V. M. Kotlyakov, and D. T. Blumstein. 1999. Glaciation as a factor of geographic variation in the long-tailed marmot (bioacoustical analysis). *Doklady Biological Sciences* 368:509–13.

Nikol'skii, A. A., and N. L. Nesterova. 1989. Responses of *Marmota baibacina* to con- and heterospecific vocalizations. *Vest. Moscow Univ.* 16:50–55.

———. 1990. Perception of acoustic signals by *Marmota caudata* (Sciuridae, Rodentia). *Vest. Moscow Univ.* 16:53–59.

Nikol'skii, A. A., N. L. Nesterova, and M. V. Suchanova. 1994. Situational variations of spectral structure in *Marmota bobac* Müll. alarm signal. In *Actual problems of marmots investigation,* ed. V. Y. Rumiantsev, 127–48. Moscow: ABF Publishing House.

Nikol'skii, A. A., and O. B. Pereladova. 1994. An alarm call of Menzbier's marmot (*Marmota menzbieri* Kaschk., 1925). In *Actual problems of marmots investigation,* ed. V. Y. Rumiantsev, 149–68. Moscow: ABF Publishing House.

Nikol'skii, A. A., E. E. Roschina, and O. V. Soroka. 2002. New conformation of correlation between the relief's desmemberment and the alarm call's temporal organisation (on instance of *Marmota bobac*). In *Holarctic marmots as a factor of biodiversity—Proceedings of the 3d International Conference on Marmots,* ed. K. B. Armitage, and V. Y. Rumiantsev. Cheboksary, Russia, 25–30 August 1997. Moscow: ABF.

Nikol'skii, A. A., and G. A. Savchenko. 1999. Structure of family groups and space use by steppe marmots (*Marmota bobac*): Preliminary results. *Vestnik Zoologii* 33:67–72.

Nikol'skii, A. A., and V. P. Starikov. 1997. Variation in the alarm call in ground squirrels *Spermophilus major* and *S. erythrogenys* (Rodentia, Sciuridae) in the contact zone in the Kurgan Oblast. *Russian Journal of Zoology* 1:340–51.

Nikol'skii, A. A., and M. V. Sukhanova. 1992. Situation dependent variations of a call emitted by great gerbil, *Rhombomys optimus* and by Brandt's vole, *Microtus brandti,* retreating to burrows. *Zoologicheskii Zhurnal* 71:125–32.

Nikolsky, A. A. 1981. Initial stage of ethologic divergence of isolated populations of arctic ground squirrel (*Citellus parrgi* Rich.). *Journal of General Biology* 42:193–98.

Nishimura, K., K. Utsumi, M. Yuhara, Y. Fujitani, and A. Iritani. 1989. Identification of puberty-accelerating pheromones in male mouse urine. *The Journal of Experimental Zoology* 251:300–305.

Nixon, C. M., L. P. Hansen, and S. P. Havera, S. P. 1991. Growth patterns of fox squirrels in east-central Illinois. *American Midland Naturalist* 125:168–72.

Noble, J., P. M. Todd, and E. Tuci. 2001. Explaining social learning of food preferences without aversions: An evolutionary simulation model of Norway rats. *Proceedings of the Royal Society of London, Series B* 268:141–49.

Noguchi, T., Y. Yoshida, and S. Chiba. 2001. Effects of psychological stress on monoamine systems in subregions of the frontal cortex and nucleus accumbens of the rat. *Brain Research* 916:91–100.

Nolet, B. A., and F. Rosell. 1994. Territoriality and time budgets in beavers during sequential settlement. *Canadian Journal of Zoology* 72:1227–37.

———. 1998. Comeback of the beaver *Castor fiber*: An overview of old and new conservation problems. *Biological Conservation* 83:165–73.

Norrdahl, K., and E. Korpimäki. 1995. Mortality factors in a cyclic vole population. *Proceedings of the Royal Society of London B* 261:49–53.

———. 1998. Does mobility or sex of voles affect risk of predation by mammalian predators? *Ecology* 79:226–32.

———. 2000. The impact of predation risk from small mustelids on prey populations. *Mammal Review* 30:147–56.

Norris, R. W., K. Zhou, C. Zhou, G. Yang, C. W. Kilpatrick, and R. L. Honeycutt. 2004. The phylogenetic position of the zokors (Myospalacinae) and comments on the families of muroids (Rodentia). *Molecular Phylogenetics and Evolution* 31:972–78.

Novacek, M. J. 1985. Cranial evidence for rodent affinities. In *Evolutionary relationships among rodents: A multidisciplinary analysis*, ed. W. P. Luckett and J. L. Hartenberger, 59–81. New York: Plenum.

———. 1992. Mammal phylogenies: Shaking the tree. *Nature* 356:121–25.

Novakowski, N. S. 1967. The winter bioenergetics of a beaver population in northern latitudes. *Canadian Journal of Zoology* 45:1107–18.

Novotny, M., S. Harvey, and B. Jemiolo. 1990. Chemistry of male dominance in the house mouse, *Mus domesticus*. *Experientia* 46:109–13.

Novotny, M. V., B. Jemiolo, and S. Harvey. 1990. Chemistry of rodent pheromones: Molecular insights into chemical signaling in mammals. In *Chemical signals in vertebrates, 5*, ed. D. W. Macdonald, D. Muller-Schwarze, and S. E. Natynczuk, 1–22. Oxford: Oxford University Press.

Novotny, M., W. Ma, D. Wiesler, and L. Zidek, L. 1999a. Positive identification of the puberty-accelerating pheromone of the house mouse: The volatile ligands associating with the major urinary protein. *Proceedings of the Royal Society of London, Series B* 266:2017–22.

Novotny, M. V., W. Ma, L. Zidek, and E. Daev. 1999b. Recent biochemical insights into puberty acceleration, estrus induction and puberty delay in the house mouse. In *Advances in chemical signals in vertebrates*, ed. R. E. Johnston, D. Muller-Schwarze, and P. W. Sorenson, 99–116. New York: Kluwer Academic.

Novotny, M., F. J. Schwende, D. Wiesler, J. W. Jorgenson, and M. Carmack. 1984. Identification of a testosterone-dependent unique volatile constituent of male mouse urine: 7-exo-ethyl-5-methyl-6,8-dioxabicyclo[3.2.1]-3-octene. *Experientia* 40:217–19.

Nowak, R. M. 1991. *Walker's mammals of the world*. 5th ed. Baltimore: Johns Hopkins University Press.

———, ed. 1999. *Walker's mammals of the world*, 6th ed. Baltimore: Johns Hopkins University Press. (Also available online at: www.press.jhu.edu/books/walkers_mammals_of_the _world/w-contents.html)

Noyes, D. H., and W. G. Holmes. 1979. Behavioral responses of free-living hoary marmots to a model golden eagle. *Journal of Mammalogy* 60:408–11.

Nunes, S., T. R. Duniec, S. A. Schweppe, and K. E. Holekamp. 1999. Energetic and endocrine mediation of dispersal behavior in Belding's ground squirrels. *Hormones and Behavior* 35:113–24.

Nunes, S., C.-D. T. Ha, P. J. Garrett, E.-M. Muecke, L. Smale, and K. E. Holekamp. 1998. Body fat and time of year interact to mediate dispersal behaviour in ground squirrels. *Animal Behaviour* 55:605–14.

Nunes, S., and K. E. Holekamp. 1996. Mass and fat influence the timing of natal dispersal in Belding's ground squirrels. *Journal of Mammalogy* 77:807–17.

Nunes, S., E.-M. Muecke, L. T. Lancaster, N. A. Miller, M. A. Mueller, J. Muelhaus, and L. Castro. 2004. Functions and consequences of play behaviour in juvenile Belding's ground squirrels. *Animal Behaviour* 68:27–37.

Nunes, S., P. A. Zugger, A. L. Engh, K. O. Reinhart, and K. E. Holekamp. 1997. Why do female Belding's ground squirrels disperse away from food resources. *Behavioral Ecology and Sociobiology* 40:199–207.

Nunney, L., 1993. The influence of mating system and overlapping generations on effective population size. *Evolution* 47:1329–41.

———. 1999. The effective size of a hierarchically structured population. *Evolution* 53:1–10.

Nunney, L., and A. Campbell. 1993. Assessing viable population size: Demography meets population genetics. *Trends in Ecology and Evolution* 8:234–39.

Nunney, L., and D. R. Elam. 1994. Estimating the effective population size of conserved populations. *Conservation Biology* 8:175–84.

Nutt, K. J. 2003. The social group structure, mating system, and dispersal patterns of *Ctenodactylus gundi* (Rodentia: Ctenodactylidae): Understanding multi-male groups using field research and molecular data. PhD diss., University of California, Berkeley.

———. 2005. Philopatry of both sexes leads to the formation of multi-male, multi-female groups in *Ctenodactylus gundi* (Rodentia: Ctenodactylidae). *Journal of Mammalogy* 86:961–68.

O'Brien, S. J., M. Menotti-Raymond, W. J. Murphy, W. G. Nash, J. Wienberg, R. Stanyon, N. G. Copeland, N. A. Jenkins, J. E. Womack, and J. A. Marshall-Graves. 1999. The promise of comparative genomics in mammals. *Science* 286:458–81.

O'Farrell, M. J., and A. R. Blaustein. 1974. *Microdipodops pallidus*. *Mammalian Species* 47:1–2.

———. 1974a. *Microdipodops megacephalus*. *Mammalian Species* 46:1–3.

———. 1974b. *Microdipodops pallidus*. *Mammalian Species* 47:1–2.

O'Riain, M. J., and S. Braude. 2001. Inbreeding versus outbreeding in captive and wild populations of naked mole-rats. In *Dispersal*, ed. J. Clobert, E. Danchin, A. A. Dhont, and J. D. Nichols, 143–54. New York: Oxford University Press.

O'Riain, M. J., J. U. M. Jarvis, and C. G. Faulkes. 1996. A dispersive morph in the naked mole-rat. *Nature* 380:619–21.

Oakeshott, J. G. 1974. Social dominance, aggressiveness and mating success among male house mice (*Mus musculus*). *Oecologia* 15:143–58.

Obbard, M. E. 1987. Red squirrel. In *Wild furbearers management and conservation in North America,* ed. M. Novak, M. E. Obbard, and B. Malloch, 265–81. Toronto: Ontario Ministry of Natural Resources.

Ober, C., L. R. Weitkamp, N, Cox, H. Dytch, D. Kostyu, and S. Elias. 1997. HLA and mate choice in humans. *American Journal of Human Genetics* 61:497–504.

Oglesby, J. M., D. L. Lanier, and D. A. Dewsbury. 1981. The role of prolonged copulatory behavior in facilitating reproductive success in male Syrian golden hamsters (*Mesocricetus auratus*) in a competitive mating situation. *Behavioral Ecology and Sociobiology* 8:47–54.

Ognev, S. I. 1940. *Animals of the USSR and adjacent countries. IV. Rodents.* Jerusalem: Israel Programme for Scientific Translations.

———. 1947 [1963]. *Mammals of USSR and adjacent countries, rodents.* Jerusalem: Israel Program Scientific Translation, 5.

Ohnishi, N., T. Saitoh, and Y. Ishibashi. 2000. Spatial genetic relationships in a population of the Japanese wood mouse *Apodemus argenteus*. *Ecological Research* 15:285–92.

Ojasti, J. 1973. *Estudio biologico del chiguire o capibara.* Caracas: FONAIAP.

———. 1991. Human exploitation of capybara. In *Neotropical wildlife use and conservation,* ed. J. G. Robinson and K. H. Redford, 236–52. Chicago: University of Chicago Press.

Ojeda, R. A., C. E. Borghi, G. B. Diaz, S. M. Giannoni, M. Mares, and J. K. Braun. 1999. Evolutionary convergences of the highly adapted desert rodent *Tympanoctomys barrarae* (Octodontidae). *Journal Arid Environments* 41:443–52.

Okasanen, T. A., R. V. Alatalo, T. J. Horne, E. Koskela, J. Mappes, and T. Mappes. 1999. Maternal effort and male quality in the bank vole, *Clethrionomys glareolus*. *Proceedings of the Royal Society of London, Series B* 266:1495–99.

Oksanen, L., and P. Lundberg. 1995. Optimization of reproductive effort and foraging time in mammals: The influence of resource level and predation risk. *Evolutionary Ecology* 9:45–56.

Oksanen, T., and H. Henttonen. 1996. Dynamics of voles and mustelids in the taiga landscape of northern Fennoscandia in relation to habitat quality. *Ecography* 19:432–43.

Oksanen, T., L. Oksanen, W. Jedrzejewski, B. Jedrzejewska, E. Korpimäki, and K. Norrdahl. 2000. Predation and the dynamics of the bank vole, *Clethrionomys glareolus*. *Polish Journal of Ecology* 48 (suppl.):197–217.

Oksanen, T., L. Oksanen, M. Schneider, and M. Aunapuu. 2001. Regulation, cycles and stability in northern carnivore-herbivore systems: Back to the principles. *Oikos* 94:101–17.

Oksanen, T., and M. Schneider. 1995. Predator-prey dynamics as influenced by habitat heterogeneity. In *Landscape approaches in mammalian ecology and conservation,* ed. W. Z. Lidicker, Jr., 122–50. Minneapolis: University of Minnesota Press.

Olds, N., and J. Shoshani. 1982. *Procavia capensis. Mammalian Species* 171:1–7.

Oli, M. K. 2004. The fast-slow continuum and mammalian life-history patterns: An empirical evaluation. *Basic and Applied Ecology* 5:449–63.

Oli, M. K., and K. B. Armitage. 2003. Sociality and individual fitness in yellow-bellied marmots: Insights from a long-term study (1962–2001). *Oecologia* 136:543–50.

———. 2004. Yellow-bellied marmot population dynamics: Demographic mechanisms of growth and decline. *Ecology* 85:2446–55.

Oli, M. K., and F. S. Dobson. 1999. Population cycles in small mammals: The role of age at sexual maturity. *Oikos* 86:557–65.

———. 2001. Population cycles in small mammals: The a-hypothesis. *Journal of Mammalogy* 82:573–81.

———. 2003. The relative importance of life-history variables to population growth rate in mammals: Cole's prediction revisited. *American Naturalist* 161:422–40.

———. 2005. Generation time, elasticity patterns, and mammalian life histories: A reply to Gaillard et al. *American Naturalist* 166:124–28.

Oli, M. K., and B. Zinner. 2001. Partial life-cycle analysis: A model for birth-pulse populations. *Ecology* 82:1180–90.

Olivares, A. I., D. H. Verzi, and A. I. Vasallo. 2004. Masticatory morphological diversity and chewing modes in South American caviomorph rodents (family Octodontidae). *Journal of Zoology, London* 263:167–77.

Oliver, W. L. R. 1977. The hutias (Capromyidae) of the West Indies. *International Zoo Yearbook* 17:14–20.

Oliver, W. L. R., L. Wilkins, R. H. Kerr, and D. L. Kelly. 1986. The Jamaican hutia (*Geocapromys brownii*) captive breeding and reintroduction programme—history and progress. *Dodo* 23:32–58.

Oliveras, D., and M. Novak. 1986. A comparison of paternal behaviour in the meadow vole *Microtus pennsylvanicus,* the pine vole *M. pinetorum,* and the prairie vole *M. ochrogaster*. *Animal Behaviour* 34:519–26.

Olson, S. L. 1985. Mountain plover food items on and adjacent to a prairie dog town. *Prairie Naturalist* 17:83–90.

Olsson, G. E., N. White, C. Ahlm, E. Fredrik, A.-C. Verlemyr, P. Juto, and R. T. Palo. 2002. Demographic factors associated with hantavirus infection in bank voles (*Clethrionomys glareolus*). *Emerging Infectious Diseases* 8:924–29.

Olsson, I. A. S., C. M. Nevison, E. G. Patterson-Kane, C. M. Sherwin, H. A. van de Weerd, and H. Wurbel. 2003. Understanding behaviour: the relevance of ethological approaches in laboratory animal science. *Applied Animal Behaviour Science* 81:245–64.

O'Meilia, M. E., F. L. Knopf, and J. C. Lewis. 1982. Some consequences of competition between prairie dogs and beef cattle. *Journal of Range Management* 35:580–85.

Onaka, T., and K. Yagi. 1998. Role of noradrenergic projections to the bed nucleus of the stria terminalis in neuroendocrine and behavioral responses to fear-related stimuli in rats. *Brain Research* 788:287–93.

Orians, G. 1969. On the evolution of mating systems in birds and mammals. *American Naturalist* 103:589–603.

Orlenev, D. P. 1987. Spatial and ethological structure of mongolian gerbils' population in natural conditions and in conditions of experimental modulation of density. [In Russian.] Published summary of Candidate of Biological Sciences diss., Moscow.

Osborn, D. J., and I. Hemly. 1980. *The contemporary land mammals of Egypt (including Sinai).* Fieldiana Zoology. Publications Field Museum of Natural History New Series N5.

Ostermeyer, M. 1983. Maternal aggression. In *Parental behaviour in rodents,* ed. R. W. Elwood, 151–79. Chichester, NY: Wiley.

Ostermeyer, M. C., and R. W. Elwood. 1983. Pup recognition in *Mus musculus*: Parental discrimination between own and alien young. *Developmental Psychobiology* 16:75–82.

Ostfeld, R. S. 1985. Limiting resources and territoriality in microtine rodents. *American Naturalist* 126:1–15.

———. 1990. The ecology of territoriality in small mammals. *Trends in Ecology and Evolution* 5:411–15.

———. 1994. The fence effect reconsidered. *Oikos* 70:340–48.

Ostfeld, R. S., and E. J. Heske. 1993. Sexual dimorphism and mating systems in voles. *Journal of Mammalogy* 74:230–33.

Ostfeld, R. S., and R. D. Holt. 2004. Are predators good for your health? Evaluating evidence for top-down regulation of zoonotic disease reservoirs. *Frontiers in Ecology and the Environment* 2:13–20.

Ostfeld, R. S., C. G. Jones, and J. O. Wolff. 1996. Of mice and mast: Ecological connections in eastern deciduous forests. *BioScience* 46:323–30.

Ostfeld, R. S., and F. Keesing. 2000a. Biodiversity and disease risk: The case of Lyme disease. *Conservation Biology* 14:722–28.

———. 2000b. Pulsed resources and community dynamics of consumers in terrestrial ecosystems. *Trends in Ecology and Evolution* 15:232–37.

———. 2000c. The role of biodiversity in the ecology of vector-borne zoonotic diseases. *Canadian Journal of Zoology* 78:2061–78.

Ostfeld, R. S., and K. LoGiudice. 2003. Community disassembly, biodiversity loss, and the erosion of an ecosystem service. *Ecology* 84:1421–27.

Ostfeld, R. S., E. M. Schauber, C. D. Canham, F. Keesing, C. G. Jones, and J. O. Wolff. 2001. Effects of acorn production and mouse abundance on abundance and *Borrelia burgdorferi*-infection prevalence of nymphal *Ixodes scapularis*. *Vector-borne and Zoonotic Diseases* 1:55–63.

Oswald, C., and P. A. McClure. 1987. Energy allocation during concurrent pregnancy and lactation in Norway rats with delayed and undelayed implantation. *Journal of Experimental Zoology* 241:343–57.

Otte, D. 1974. Effects and functions in the evolution of signaling systems. In *Annual review of ecology and systematics,* vol. 5, ed. R. F. Johnston, 385–417. Palo Alto, CA: Annual Reviews.

Ovadia, O., Y. Ziv, Z. Abramsky, B. Pinshow, and B. P. Kotler. 2001. Harvest rates and foraging strategies in Negev desert gerbils. *Behavioral Ecology* 12:219–26.

Ovaska, K., and T. B. Herman. 1988. Life history characteristics and movements of the woodland jumping mouse, *Napaeozapus insignis*, in Nova Scotia. *Canadian Journal of Zoology* 66:1752–62.

Owings, D. H. 1994. How monkeys feel about the world: A review of *How monkeys see the world*. *Language and Communication* 14:15–30.

———. 2002. The cognitive defender: How ground squirrels assess their predators. In *The cognitive animal,* ed. M. Bekoff, C. Allen, and G. Burghardt, 19–25. Cambridge: MIT Press.

Owings, D. H., M. Borchert, and R. Virginia. 1977. The behaviour of California ground squirrels. *Animal Behaviour* 25:221–30.

Owings, D. H., and R. G. Coss. 1977. Snake mobbing by California ground squirrels: Adaptive variation and ontogeny. *Behaviour* 62:50–69.

Owings, D. H., R. G. Coss, D. Mckernon, M. P. Rowe, and P. C. Arrowood. 2001. Snake-directed antipredator behavior of rock squirrels (*Spermophilus variegatus*): Population differences and snake-species discrimination. *Behaviour* 138:575–95.

Owings, D. H., and D. F. Hennessy. 1984. The importance of variation in sciurid visual and vocal communication. In *The biology of ground-dwelling squirrels: Annual cycles, behavioral ecology, and sociality,* ed. J. O. Murie and G. R. Michener, 169–200. Lincoln: University of Nebraska Press.

Owings, D. H., D. F. Hennessy, D. W. Leger, and A. B. Gladney. 1986. Different functions of "alarm" calling for different time scales: A preliminary report on ground squirrels. *Behaviour* 99:101–16.

Owings, D. H., and D. W. Leger. 1980. Chatter vocalizations of California ground squirrels: Predator- and social-role specificity. *Zeitschrift für Tierpsychologie* 54:163–84.

Owings, D. H., and W. J. Loughry. 1985. Variation in snake-elicited jump-yipping by black-tailed prairie dogs: Ontogeny and snake-specificity. *Zeitschrift für Tierpsychologie* 70:177–200.

Owings, D. H., and E. S. Morton. 1997. The role of information in communication: An assessment/management approach. In *Perspectives in ethology, vol. 12: Communication,* ed. D. H. Owings, M. D. Beecher, and N. S. Thompson, 359–90. New York: Plenum.

———. 1998, *Animal vocal communication: A new approach.* Cambridge: Cambridge University Press.

Owings, D. H., M. P. Rowe, and A. S. Rundus. 2002. The rattling sound of rattlesnakes (*Crotalus viridis*) as a communicative resource for ground squirrels (*Spermophilus beecheyi*) and burrowing owls (*Athene cunicularia*). *Journal of Comparative Psychology* 116:197–205.

Owings, D. H., and R. A. Virginia. 1978. Alarm calls of California ground squirrels (*Spermophilus beecheyi*). *Zeitschrift für Tierpsychologie* 46:58–70.

Owings, D. H., R. A. Virginia, and D. Paussa. 1979. Time budgets of California ground squirrels during reproduction. *Southwestern Naturalist* 24:191–95.

Pack, J. C., H. S. Mosby, and P. B. Siegel. 1967. Influence of social hierarchy on gray squirrel behavior. *Journal of Wildlife Management* 31:720–28.

Packard, R. L. 1956. The tree squirrels of Kansas: Ecology and economic importance. *Miscellaneous Publications, Museum of Natural History, University of Kansas* 11:1–67.

———. 1960. Speciation and evolution of the pygmy mice, genus *Baiomys*. *University of Kansas Publications, Museum of Natural History* 9:579–670.

———. 1968. An ecological study of the fulvous harvest mouse in eastern Texas. *American Midland Naturalist* 79:68–88.

Packer, C., S. Lewis, and A. Pusey. 1992. A comparative analysis of non-offspring nursing. *Animal Behaviour* 43:265–81.

Packer, C., and A. E. Pusey. 1984. Infanticide in carnivores. In *Infanticide: Comparative and evolutionary perspectives,* ed. G. Hausfater and S. B. Hrdy, 31–42. New York: Aldine.

Packer, C., M. Tatar, and A. Collins. 1998. Reproductive cessation in female mammals. *Nature* 392:807–11.

Paine, R. 1969. A note on trophic complexity and community stability. *American Naturalist* 103:91–93.

Palanza, P., F. Morellini, S. Parmigiani, and F. S. vom Saal. 1999. Prenatal exposure to endocrine disrupting chemicals: Effects on behavioral development. *Neuroscience and Biobehavioral Reviews* 23:1011–27.

Palanza, P., and S. Parmigiani. 1991. Inhibition of infanticide in male Swiss mice: Behavioral polymorphism in response to multiple mediating factors. *Physiology and Behavior* 49:797–802.

———. 1994. Functional analysis of maternal aggression in the house mouse (*Mus Musculus*). *Behavioural Processes* 32:1–16.

Palanza, P., S. Parmigiani, F. S. Vom Saal, and F. Farabollini. 1994. Maternal aggression toward infanticidal males of different social status in wild house mice (*Mus musculus domesticus*). *Aggressive Behavior* 20:267–74.

Palanza, P., L. Re, D. Mainardi, P. F. Brain, and S. Parmigiani. 1996. Male and female competitive strategies of wild house mice pairs (*Mus musculus domesticus*) confronted with intruders of different sex and age in artificial territories. *Behaviour* 133:863–82.

Palmer, T. S. 1899. Report of Acting Chief of Division of Biological Survey. In *Annual Reports for the fiscal year ended June 30, 1899, 5967*. Washington, DC: United States Department of Agriculture.

Panetta, F. D., and A. J. M. Hopkins 1991. Weeds in corridors: Invasion and management. In *Nature conservation 2: The role of corridors*, ed. D. A. Saunders and R. J. Hobbs, 341–52. Chipping Norton, Australia: Surrey Beatty and Sons.

Pankhurst, S. J. 2002. The social organisation of the mara (*Dolichotis patagonum*) at Whipsnade Wild Animal Park, UK. *Advances in Ethology* 37: 61.

Parisi, V., and G. Gandolfi. 1974. Further aspects of the predation by rats on various mollusk species. *Bollettino di Zoologia* 41:87–106.

Parker, G. A. 1970. Sperm competition and its evolutionary consequences in the insects. *Biological Reviews* 45:525–67.

———. 1974. Assessment strategy and the evolution of fighting behaviour. *Journal of Theoretical Biology* 47:223–43.

Parker, G. A. 1984a. Evolutionarily stable strategies. In *Behavioral ecology: An evolutionary approach*, ed. J.R. Krebs and N.B. Davies, 30–61. Oxford: Blackwell Scientific.

———. 1984b. Sperm competition and the evolution of animal mating strategies. In *Sperm competition and the evolution of animal mating systems*, ed. R. L. Smith, 1–60. London: Academic Press.

Parker, G. A., and D. I. Rubenstein. 1981. Role assessment, reserve strategy, and acquisition of information in asymmetric animal conflicts. *Animal Behaviour* 29:221–40.

Parker, K. J., and T. M. Lee. 2001. Social and environmental factors influence the suppression of pup-directed aggression and development of paternal behavior in captive meadow voles (*Microtus pennsylvanicus*). *Journal of Comparative Psychology* 115:331–36.

Parkes, A. S., and H. M. Bruce. 1961. Olfactory stimuli in mammalian reproduction. *Science, New York* 134:1049–54.

———. 1962. Pregnancy-block in female mice placed in boxes soiled by males. *Journal of Reproduction and Fertility* 4:303–8.

Parmigiani, S. 1986. Rank order in pairs of communally nursing female mice (*Mus musculus domesticus*) and maternal aggression towards conspecific intruders of differing sex. *Aggressive Behavior* 12:377–86.

———. 1989. Inhibition of infanticide in male mice (*Mus domesticus*): Is kin recognition involved? *Ethology Ecology and Evolution* 1:93–98.

Parmigiani, S., P. F. Brain, D. Mainardi, and V. Brunoni. 1988. Different patterns of biting attack employed by lactating female mice (*Mus domesticus*) in encounters with male and female conspecific intruders. *Journal of Comparative Psychology* 102:287–93.

Parmigiani, S., V. Brunoni, and A. Pasquali. 1982. Behavioural influences of dominant, isolated and subordinated male mice on female socio-sexual preferences. *Bollettino di Zoologia* 49:31–35.

Parmigiani, S., P. Palanza, and P. F. Brain. 1989. Intraspecific maternal aggression in the house mouse (*Mus domesticus*): A

counterstrategy to infanticide by male? *Ethology Ecology and Evolution* 1:341–52.

Parmigiani, S., P. Palanza, D. Mainardi, and P. F. Brain. 1994. Infanticide and protection of young in house mice (*Mus domesticus*): Female and male strategies. In *Infanticide and parental care*, ed. S. Parmigiani and F. S. Vom Saal, 341–63. Chur, Switzerland: Harwood.

Parmigiani, S., A. Sgoifo, and D. Mainardi 1988. Parental aggression displayed by female mice in relation to the sex, reproductive status and infanticidal potential of conspecific intruders. *Monitore Zoologico Italiano* (Nuova Serie) 22:193–201.

Parmigiani, S., and F. S. Vom Saal, eds. 1994. *Infanticide and parental care*. Chur, Switzerland: Harwood Academic.

Parr, L. A., and F. B. M. de Waal. 1999. Visual kin recognition in chimpanzees. *Nature* 399:647–48.

Parsons, G. J., and S. Bondrup-Nielsen. 1996. Experimental analysis of behaviour of meadow voles (*Microtus pennsylvanicus*) to odours of the short-tailed weasel (*Mustela erminea*). *Ecoscience* 3:63–69.

Pascale, E., E. Valle, and A. V. Furano. 1990. Amplification of an ancestral mammalian L1 family of long interspersed repeated DNA occurred just before the murine radiation. *Proceedings of the National Academy of Sciences (USA)* 87:9481–85.

Pasley, J. N., and T. D. McKinney. 1973. Grouping and ovulation in *Microtus pennsylvanicus*. *Journal of Reproduction and Fertility* 34:527–30.

Patenaude, F. 1983. Care of the young in a family of wild beavers, *Castor canadensis*. *Acta Zoologica Fennica* 174: 121–22.

Patenaude, F., and J. Bovet. 1983. Parturition and related behavior in wild American beavers (*Castor canadensis*). *Zeitschrift für Saeugetierkundee-International Journal of Mammalian Biology* 48:136–45.

———. 1984. Self-grooming and social grooming in the North American beaver, *Castor canadensis*. *Canadian Journal of Zoology* 62:1872–78.

Patris, B., and C. Baudoin. 2000. A comparative study of parental care between two rodent species: Implications for the mating system of the mound-building mouse *Mus spicilegus*. *Behavioural Processes* 51:35–43.

Patterson, B. D., G. Ceballos, W. Sechrest, M. F. Tognelli, T. Brooks, L. Luna, P. Ortega, I. Salazar, and B. E. Young. 2003. *Digital distribution maps of the mammals of the western hemisphere, version 1.0*. Arlington, VA: NatureServe.

Patterson, B., and A. E. Wood. 1982. Rodents from the Deseadan Oligocene of Bolivia and the relationships of Caviomorpha. *Bulletin of the Museum of Comparative Zoology* 149: 371–543.

Patterson, I. A. P., R. J. Reid, B. Wilson, K. Greillier, H. M. Ross, and P. M. Thompson. 1998. Evidence for infanticide in bottlenose dolphins: An explanation for violent interactions with harbour porpoises? *Proceedings of the Royal Society of London, Series B* 265:1167–70.

Patton, J. L., and J. H. Feder. 1981. Microspatial genetic heterogeneity in pocket gophers: Non-random breeding and drift. *Evolution* 35:912–20.

Patton, J. L., and M. F. Smith. 1990. The evolutionary dynamics of the pocket gopher *Thomomys bottae*, with emphasis on California populations. *University of California Publications in Zoology* 123:1–161.

Paul, L. 1986. Infanticide and maternal aggression: Synchrony of male and female reproductive strategies in mice. *Aggressive Behavior* 12:1–11.

Paul, L., and J. Kupferschmidt. 1975. Killing of conspecific and mouse young by male rats. *Journal of Comparative and Physiological Psychology* 88:755–63.

Pavey, C. R., N. Goodship, and F. Geiser. 2003. Home range and spatial organisation of rock-dwelling carnivorous marsupial, *Pseudantechinus macdonnellensis*. *Wildlife Research* 30:135–42.

Pavlinov, I. Y. 1988. Evolution of mastoid part of the bulla tympani in specialized arid rodents. [In Russian.] *Zoologicheskii Zhurnal*, 67:739–50.

Pavlinov, I. Y., Yu. A. Dubrovsky, O. L. Rossolimo, and E. G. Potapova. 1990. *Gerbils of the world.* [In Russian.] Moscow: Nauka.

Pavlinov, I. Y., and K. A. Rogovin. 2000. Relation between size of pinna and auditory bulla in specialized desert rodents. [In Russian.] *Zhurnal Obschey Biologii* 61:87–101.

Payne, N. F. 1982. Colony size, age, and sex structure of New foundland beaver. *Journal of Wildlife Management* 46:655–61.

———. 1984. Mortality rates of beaver in Newfoundland. *Journal of Wildlife Management* 48:912–17.

Paz y Miño C., G., and Z. Tang-Martínez. 1999a. Effects of exposures to siblings or sibling odors on sibling recognition in prairie voles (*Microtus ochrogaster*). *Canadian Journal of Zoology* 77:118–23.

Paz y Miño C., G., and Z. Tang-Martínez. 1999b. Effects of isolation on sibling recognition in prairie voles, *Microtus ochrogaster*. *Animal Behaviour* 57:1091–98.

———. 1999c. Social interactions, cross-fostering, and sibling recognition in prairie voles, *Microtus ochrogaster*. *Canadian Journal of Zoology* 77:1631–36.

Peacock, M. M. 1997. Determining natal dispersal patterns in a population of North American pikas (*Ochotona princeps*) using direct mark-resight and indirect genetic methods. *Behavioral Ecology* 8:340–50.

Peacock, M. M., and S. H. Jenkins. 1988. Development of food preferences: Social learning by Belding's ground squirrels *Spermophilus beldingi*. *Behavioural Ecology and Sociobiology* 22:393–99.

Pearson, C. R., and A. K. Pearson. 1947. Owl predation in Pennsylvania, with notes of small mammals of Delaware County. *Journal of Mammalogy* 28:137–47.

Pearson, D., R. Shine, and A. Williams. 2002. Geographic variation in sexual size dimorphism within a single snake species (*Morelia spilota*, Pythonidae). *Oecologia* 131:418–26.

Pearson, O. P. 1948. Life history of mountain viscachas in Peru. *Journal of Mammalogy* 29:345–74.

———. 1951. Mammals in the highlands of southern Peru. *Bulletin of the Museum of Comparative Zoology* 106:117–74.

———. 1959. Biology of the subterranean rodents, *Ctenomys*, in Peru. *Memorias del Museo de Historia Natural "Javier Prado"* 9:1–56.

———. 1963. History of two local outbreaks of feral house mice. *Ecology* 44:540–49.

———. 1983. Characteristics of a mammalian fauna from forests in Patagonia, southern Argentina. *Journal of Mammalogy* 64:476–92.

———. 1984. Taxonomy and natural history of some fossorial rodents of Patagonia, southern Argentina. *Journal of Zoology, London* 202:225–37.

———. 1985. Predation. In *Biology of New World* Microtus, Special publication no. 8, ed. R. H. Tamarin, 535–66. American Society of Mammalogists.

Pearson, O. P., N. Binsztein, L. Boiry, C. Bush, M. Di Pace, G. Gallopin, P. Penchaszadeh, and M. Piantanida. 1968. Estructura social, distribución espacial y composición por edades de una población de tuco-tucos (*Ctenomys talarum*). *Investigaciones Zoológicas Chilenas* 13:47–80.

Pearson, O. P., and M. I. Christie. 1985. Los tuco-tucos (genéro *Ctenomys*) de los Parques Nacionales Lanin y Nahuel Huapi, Argentina. *Historia Natural* 5:337–44.

Pearson, P. G. 1953. A field study of *Peromyscus* populations in Gulf hammock, Florida. *Ecology* 34:199–207.

Pedersen, C. A., J. A. Ascher, Y. L. Monroe, and A. J. Prange, Jr. 1982. Oxytocin induces maternal behavior in virgin female rats. *Science* 216:648–50.

Pedersen, W. A., R. Wan, and M. P. Mattson. 2001. Impact of aging on stress-responsive neuroendocrine systems. *Mechanisms of Ageing and Development* 122:963–83.

Pelikan, J. 1981. Patterns of reproduction in the house mouse. *Symposia of the Zoological Society of London* 47:205–29.

Pellis, S. M., and A. N. Iwaniuk. 1999. The roles of phylogeny and sociality in the evolution of social play in muroid rodents. *Animal Behaviour* 58:361–73.

Pellis, S. M., and V. C. Pellis. 1987. Play-fighting differs from serious fighting in both target of attack and tactics of fighting in the laboratory rat *Rattus norvegicus*. *Aggressive Behavior* 13:227–42.

Pellis, S. M., V. C. Pellis, and D. A. Dewsbury. 1989. Different levels of complexity in the play-fighting by Muroid rodents appear to result from different levels of intensity of attack and defense. *Aggressive Behavior* 15:297–310.

Pen, I., and F. J. Weissing. 2000. Towards a unified theory of cooperative breeding: The role of ecology and life history re-examined. *Proceedings of the Royal Society of London, Series B* 267:2411–18.

———. 2002. Optimal sex allocation: Steps towards a mechanistic theory. In *Sex ratios: Concepts and research methods*, ed. I. C. W. Hardy, 26–45. Cambridge: Cambridge University Press.

Penn, D. 2002. The scent of genetic compatibility: Sexual selection and the major histocompatibility complex. *Ethology* 108:1–21.

Penn, D. J., K. Damjanovich, and W. K. Potts. 2002. MHC heterozygosity confers a selective advantage against multiple-strain infections. *Proceedings of the National Academy of Sciences (USA)* 99:11260–64.

Penn, D. J., and W. K. Potts. 1998a. Chemical signals and parasite-mediated sexual selection. *Trends in Ecology and Evolution* 13:391–6.

———. 1998b. MHC-disassortative mating preferences reversed by cross-fostering. *Proceedings of the Royal Society of London, Series B* 265:1299–1306.

———. 1999. The evolution of mating preferences and major histocompatibility complex genes. *American Naturalist* 153:145–64.

Penn, D., G. Schneider, K. White, P. Slev, and W. Potts. 1998. Influenza infection neutralizes the attractiveness of male odour to female mice (*Mus musculus*). *Ethology* 104:685–94.

Pepper, J. W. 1996. The behavioral ecology of the glossy black-cockatoo *Calyptorhynchus lathami halmaturinus*. PhD diss., University of Michigan, Ann Arbor.

Pepper, J. W., S. H. Braude, E. A. Lacey, and P. W. Sherman. 1991. Vocalizations of the naked mole-rat. In *The biology of the naked mole-rat*, ed. P. W. Sherman, J. U. M. Jarvis, and R. D. Alexander, 243–74. Princeton, NJ: Princeton University Press.

Perla, B. S., and C. N. Slobodchikoff. 2002. Habitat structure and alarm call dialects in Gunnison's prairie dog (*Cynomys gunnisoni*). *Behavioral Ecology* 13:844–50.

Perri, L. M., and J. A. Randall. 1999. Behavioral mechanisms of coexistence in sympatric species of desert rodents, *Dipodomys ordii* and *D. merriami*. *Journal of Mammalogy* 80:1297–1310.

Perrigo, G., L. Belvin, P. Quindry, T. Kadir, J. Becker, C. Van Look, J. Niewoehner, and F. S. Vom Saal. 1993. Genetic mediation of infanticide and parental behavior in male and female domestic and wild stock house mice. *Behavior Genetics* 23:525–31.

Perrin, C., D. Allainé, and M. Le Berre. 1993. Socio-spatial organization and activity distribution of the Alpine Marmot *Marmota marmota*: Preliminary results. *Ethology* 93:21–30.

———. 1994. Intrusion de males et possibilité d' infanticide chez la marmotte alpine. *Mammalia* 58:150–53.

Perrin, C., L. Le Guelte, and M. Le Berre. 1992. Temporal and spatial distribution of activities during summer in the alpine marmot. In *First international symposium of alpine marmot (Marmota marmota) and on genus Marmota*, ed. B. Bassano, P. Durio, U. Gallo Orsi, and E. Macchi, 101–10. Torino.

Perrin, N., and L. Lehmann. 2001. Is sociality driven by the costs of dispersal or the benefits of philopatry? A role for kin-discrimination mechanisms. *American Naturalist* 158:471–83.

Perrot-Sinal, T. S., K.-P. Ossenkopp, and M. Kavaliers. 1999. Brief predator odour exposure activates the HPA axis independent of locomotor changes. *Neuroreport: An International Journal for the Rapid Communication of Research in Neuroscience* 10:775–80.

Peters, L. C., and M. B. Kristal. 1983. Suppression of infanticide in mother rats. *Journal of Comparative Psychology* 97:167–77.

Peters, R. H. 1983. *The ecological implications of body size*. Cambridge: Cambridge University Press.

Peters, R. P., and L. D. Mech. 1975. Scent-marking in wolves. *American Scientist* 63:628–37.

Peterson, R. P., and N. F. Payne. 1986. Productivity, size, age, and sex structure of nuisance beaver colonies in Wisconsin. *Journal of Wildlife Management* 50:265–68.

Petras, M., and J. C. Topping. 1983. The maintenance of polymorphism at two loci in house mouse (*Mus musculus*) populations. *Canadian Journal of Genetics and Cytology* 25:190–201.

Petrie, M. 1994. Improved growth and survival of offspring of peacocks with more elaborate trains. *Nature* 371:598–99.

Petrie, M., M. Hall, T. Halliday, H. Budgey, and C. Pierpont. 1992. Multiple mating in a lekking bird: Why do peahens mate with more than one male and with the same male more than once? *Behavioral Ecology and Sociobiology* 31:349–58.

Petrie, M., T. Halliday, and C. Sanders. 1991. Peahens prefer peacocks with elaborate trains. *Animal Behaviour* 41:323–31.

Petrulis, A., and R. E. Johnston. 1999. Lesions centered on the medial amygdala impair scent-marking and sex-odor recognition but spare discrimination of individual odors in female golden hamsters. *Behavioral Neuroscience* 113:345–57.

Petrulis, A., M. Peng, and R. E. Johnston. 2000. The role of the hippocampal system in social odor discrimination and scent-marking in female golden hamsters (*Mesocricetus auratus*). *Behavioral Neuroscience* 114:184–95.

Pfaus, J. G., G. Damsma, G. G. Nomikos, D. G. Wenkstern, C. D. Blaha, A. G. Phillips, and H. C. Fibiger. 1990. Sexual behavior enhances central dopamine transmission in the male rat. *Brain Research* 530:345–48.

Pflanz, T. 2002. Age and brood defence in male CRL:NMRI BR laboratory mice, *Mus musculus domesticus. Animal Behaviour* 63:613–16.

Philippe, H. 1997. Rodent monophyly: Pitfalls of molecular phylogenies. *Journal of Molecular Evolution* 45:712–15.

Phillips, J. A. 1981. Growth and its relationship to the initial annual cycle of the golden-mantled ground squirrel, *Spermophilus lateralis. Canadian Journal of Zoology* 59:865–71.

———. 1984. Environmental influences on reproduction in the golden-mantled ground squirrel. In *The biology of ground-dwelling squirrels*, ed. J. O. Murie and G. R. Michener, 108–24. Lincoln: University of Nebraska Press.

Pierce, J. D., Jr., and D. A. Dewsbury. 1991. Female preferences for unmated versus mated males in two species of voles (*Microtus ochrogaster* and *Microtus montanus*). *Journal of Comparative Psychology* 105:165–71.

Pierce, J. D., Jr., V. C. Pellis, D. A. Dewsbury, and S. M. Pellis. 1991. Targets and tactics of agonistic and precopulatory behavior in montane and prairie voles: Their relationship to juvenile play-fighting. *Aggressive Behavior* 17:337–49.

Pierotti, R. 1991. Infanticide versus adoption: An intergenerational conflict. *American Naturalist* 138:1140–58.

Pilastro, A., T. Gomiero, and G. Marin. 1994. Factors affecting body-mass of young fat dormice (*Glis glis*) at weaning and by hibernation. *Journal of Zoology* 234:13–23.

Pilastro, A., E. Missiaglia, and G. Marin. 1996. Age-related reproductive success in solitarily and communally nesting female dormice (*Glis glis*). *Journal of Zoology, London* 239:601–8.

Pine, R. H., S. D. Miller, and M. L. Schamberger. 1979. Contributions to the mammalogy of Chile. *Mammalia* 43:339–76.

Pinker, S. 1994. *The language instinct*. New York: William Morrow.

Pitkow, L. J., C. A. Sharer, X. Ren, T. R. Insel, E. F. Terwilliger, and L. J. Young. 2001. Facilitation of affiliation and pair-bond formation by vasopressin receptor gene transfer into the ventral forebrain of a monogamous vole. *Journal of Neuroscience* 21:7392–96.

Pizzimenti, J. J. 1975. Evolution of the prairie dog genus *Cynomys. Occasional Papers, Museum of Natural History, University of Kansas* 39:1–73.

Pizzimenti, J. J., and G. D. Collier. 1975. *Cynomys parvidens. Mammalian Species* 52:1–3.

Pizzimenti, J. J., and L. R. McClenaghan. 1974. Reproduction, growth and development, and behavior in the Mexican prairie dog, *Cynomys mexicanus* (Merriam). *American Midland Naturalist* 92:130–45.

Pizzuto, T., and L. L. Getz. 1998. Female prairie voles (*Microtus ochrogaster*) fail to form a new pair after loss of mate. *Behavioral Processes* 43:79–86.

Plassmann, W., W. Peetz, and M. Schmidt. 1987. The cochlea in Gerbilline rodents. *Brain Behavior and Evolution* 30:82–101.

Platt, J. R. 1964. Strong inference. *Science* 146:347–53.

Plotnick, R. E., and M. L. McKinney. 1993. Ecosystem organization and extinction dynamics. *Palaios* 8:202–12.

Pluháček, J., and L. Bartoš. 2000. Male infanticide in captive plains zebra, *Equus burchelli*. *Animal Behaviour* 59:689–94.

Polan, H. J., and M. A. Hofer. 1998. Olfactory preference for mother over home nest shavings by newborn rats. *Developmental Psychobiology* 33:5–20.

———. 1999. Psychobiological origins of infant attachment and separation responses. In *Handbook of attachment: Theory, research, and clinical applications,* ed. J. Cassidy and P. R. Shaver, 162–80. New York: Guilford.

Pole, S. B., and D. I. Bibikov. 1991. Dynamics of population structure and mechanisms of maintaining optimal population density in grey marmots. In *Population structure of the marmot,* ed. D. I. Bibikov, A. A. Nikolski, V. Yu. Rumiantsev, and T. A. Seredneva, 148–71. Moscow: USSR Theriological Society.

Pollitzer, R. 1951. Plague studies. 1. A summary of the history and a survey of the present distribution of the disease. *World Health Organization Bulletin* 4:475–533.

Polly, P. D. 2003. Paleophylogeography: The tempo of geographic differentiation in marmots (*Marmota*). *Journal of Mammalogy* 84:369–84.

Pope, T. R. 1992. The influence of dispersal patterns and mating system on genetic differentiation within and between populations of the red howler monkey (*Alouatta seniculus*). *Evolution* 46:1112–28.

———. 1998. Effects of demographic change on group kin structure and gene dynamics of red howling monkey populations. *Journal of Mammalogy* 79:692–712.

Popov, S .V., and A. V. Tchabovsky. 1996. Factors affecting body mass and ventral gland size in great gerbil (*Rhombomys opimus*) in south-eastern Karakum desert. [In Russian.] *Zoologicheskii Zhurnal* 75:1404–11.

Poran, N. S., and R. G. Coss. 1990. Development of antisnake defenses in California ground squirrels (*Spermophilus beecheyi*): I. behavioral and immunological relationships. *Behaviour* 112:222–45.

Poran, N. S., R. G. Coss, and E. Benjamini. 1987. Resistance of California ground squirrels (*Spermophilus beecheyi*) to the venom of the northern Pacific rattlesnake (*Crotalus viridis oreganus*): A study of adaptive variation. *Toxicon* 25:767–77.

Porter, R. H., S. A. Cavallaro, and J. D. Moore. 1980. Developmental parameters of mother-offspring interactions in *Acomys cahirinus*. *Zeitschrift für Tierpsychologie* 53:153–70.

Porter, R. H., and H. M. Doane. 1978. Studies of maternal behavior in spiny mice (*Acomys cahirinus*). *Zeitschrift für Tierpsychologie* 47:225–35.

Porter, R. H., J. A. Matochik, and J. W. Makin. 1983. Evidence for phenotype matching in spiny mice (*Acomys cahirinus*). *Animal Behaviour* 31:978–84.

———. 1984. The role of familiarity in the development of social preferences in spiny mice. *Behavioural Processes* 9:241–54.

Porter, R. H., J. D. Moore, and D. M. White. 1980. Food sharing by sibling vs nonsibling spiny mice (*Acomys cahirinus*). *Behavioral Ecology and Sociobiology* 8:207–12.

Porter, R. H., S. A. McFadyen-Ketchum, and A. P. King. 1989. Underlying bases of recognition signatures in spiny mice, *Acomys cahirinus*. *Animal Behaviour* 37:638–44.

Porter, R. H., and M. Wyrick. 1979. Sibling recognition in spiny mice (*Acomys cahirinus*): Influence of age and isolation. *Animal Behaviour* 27:761–66.

Porteus, I. S., and Pankhurst, S. J. 1998. Social structure of the mara (*Dolichotis patagonum*) as a determinant of gastrointestinal parasitism. *Parasitology* 116:269–75.

Posadas-Andrews, A., and T. J. Roper. 1983. Social transmission of food preferences in adult rats. *Animal Behaviour* 31:265–71.

Post, E., M. C. Forchhammer, N. C. Stenseth, and R. Langvatn. 1999. Extrinsic modification of vertebrate sex ratios by climatic variation. *American Naturalist* 154:194–204.

Post, E., R. Langvatn, M. C. Forchhammer, and N. C. Stenseth. 1999. Environmental variation shapes sexual dimorphism in red deer. *Proceedings of the National Academy of Science (USA)* 96:4467–71.

Potts, W. K., and E. K. Wakeland. 1993. Evolution of MHC genetic diversity: A tale of incest, pestilence and sexual preference. *Trends in Genetics* 9:408–12.

Potts, W. K., C. J. Manning, and E. K. Wakeland. 1991. Mating patterns in seminatural populations of mice influenced by MHC genotype. *Nature* 352:619–21.

———. 1992. MHC-based mating preferences in *Mus* operate through both settlement patterns and female controlled extra-territorial matings. In *Chemical signals in vertebrates VI,* ed. R. L. Doty and D. Müller-Schwarze, 183–87. New York: Plenum.

Pournelle, G. H. 1952. Reproduction and early postnatal development of the cotton mouse, *Peromyscus gossypinus gossypinus*. *Journal of Mammalogy* 33:1–20.

Power, M. E., D. Tilman, J. A. Estes, B. A. Menge, W. J. Bond, L. S. Mills, G. Daily, J. C. Castilla, J. Lubchenco, and R. T. Paine. 1996. Challenges in the quest for keystones. *BioScience* 46:609–20.

Prakash, I. 1975. The population ecology of the rodents of the Rajasthan desert, India. In *Rodents in desert environments,* ed. I. Prakash and C. K. Ghosh, 75–116. The Hague: Junk.

———. 1981. Ecology of the Indian desert gerbil *Meriones hurrianae*. Jodhpur, India: Central Arid Zone Research Institute Press.

Prakash, I., and C. K. Ghosh. 1975. *Rodents in desert environments*. The Hague: Junk.

Prakash, I., and H. Singh. 2001. Composition and species diversity of small mammals in the hilly tracts of southeastern Rajasthan. *Tropical Ecology* 42:25–33.

Preston, S. D., and L. F. Jacobs. 2001. Conspecific pilferage but not presence affects Merriam's kangaroo rat cache strategy. *Behavioral Ecology* 12:517–23.

Previde, E. P., and M. Poli. 1996. Social learning in the golden hamster (*Mesocricetus auratus*). *Journal of Comparative Psychology* 110:203–8.

Preziosi, R. F., and D. J. Fairbairn. 2000. Lifetime selection on adult body size and components of body size in a water-strider: Opposing selection and maintenance of sexual size dimorphism. *Evolution* 54:558–66.

Price, K., and S. Boutin. 1993. Territorial bequeathal by red squirrel mothers. *Behavioral Ecology* 4:144–50.

Price, M. V. 1978. The role of microhabitat specialization in structuring desert rodent communities. *Ecology* 58:1393–99.

Price, M. V., and J. W. Joyner. 1997. What resources are available to desert granivores: Seed rain or soil seed banks? *Ecology* 78:764–73.

Price, M.V., N. W. Waser, and T. A. Bass. 1984. Effects of moonlight on microhabitat use by desert rodents. *Journal of Mammalogy* 65:353–56.

Price, T. D. 1984. The evolution of sexual size dimorphism in Darwin's finches. *American Naturalist* 123:500–518.

Priestnall, R. 1972. Effects of litter size on the behavior of lactating female mice (*Mus musculus*). *Animal Behaviour* 20:386–94.

Priestnall, R., and S. Young. 1978. An observational study of caretaking behavior of male and female mice housed together. *Developmental Psychobiology* 11:23–30.

Probst, R., M. Pavlicev, and J. Viitala. 2002. UV reflecting vole scent marks attract a passerine, the great grey shrike *Lanius excubitor*. *Journal of Avian Biology* 33:437–40.

Proctor, J., B. Haskins, and S. C. Forrest. 2006. Focal areas for restoration and conservation of prairie dogs and the grassland ecosystem of North America's Great Plains. In *Conservation of the black-tailed prairie dog*, ed. J. L. Hoogland, 232–47. Washington, DC: Island Press.

Prohazka, D., M. A. Novak, and J. S. Meyer. 1986. Divergent effects of early hydrocortisone treatment on behavioral and brain development in meadow and pine voles. *Developmenatal Psychobiology* 19:521–35.

Promislow, D. E. L., and P. H. Harvey. 1990. Living fast and dying young: A comparative analysis of life-history variation among mammals. *Journal of Zoology, London* 220:417–37.

———. 1991. Mortality rates and the evolution of mammal life histories. *Acta Oecologica* 12:119–37.

Przybylo, R., B. C. Sheldon, and J. Merilä. 2000. Patterns of natural selection on morphology of male and female collared flycatchers (*Ficedula albicollis*). *Biological Journal of the Linnean Society* 69:213–32.

Pucek, Z., W. Jędrzejewski, B. Jędrzejewska, and M. Pucek. 1993. Rodent population dynamics in a primeval deciduous forest (Białowieża National Park) in relation to weather, seed crop, and predation. *Acta Theriologica* 38:199–232.

Puckey, H., M. Lewis, D. Hooper, and C. Michell. 2004. Home range, movement and habitat utilisation of the Carpentarian rock-rat (*Zyzomys palatalis*) in an isolated habitat patch. *Wildlife Research* 31:327–37.

Puig, S., M. I. Rosi, M. I. Cona, F. Videla, and V. G. Roig. 1992. Estudio ecologico del roedor subterraneo *Ctenomys mendocinus* en la precordillera de Mendoza, Argentina: Densidad poblacional y uso del espacio. *Revista Chilena de Historia Natural* 65:247–54.

Pulliam, H. R. 1988. Sources, sinks, and population regulation. *American Naturalist* 132:652–61.

Pulliam, H. R. 1996. Sources and sinks: Empirical evidence and population consequences. In *Population dynamics in ecological space and time*, ed. O. E. Rhodes, R. K. Chesser, and M. H. Smith, 45–69. Chicago: University of Chicago Press.

Pulliam, R., and T. Caraco. 1984. Optimal group size: Is there an optimal group size? In *Behavioral ecology: An evolutionary approach*, ed. J. R. Krebs and N. B. Davies, 122–47. Sunderland, MA: Sinauer.

Purohit, K. G., L. R. Kametkar, and I. Prakash. 1966. Reproduction biology and post-natal development in the northern palm squirrel, *Funambulus pennanti* Wroughton. *Mammalia* 30:538–46.

Pusenius, J., and R. S. Ostfeld. 2000. Effects of stoat's presence and auditory cues indicating its presence on tree seedling predation by meadow voles. *Oikos* 91:123–30.

Pusenius, J., and K. A. Schmidt. 2002. The effects of habitat manipulation on population distribution and foraging behavior in meadow voles. *Oikos* 98:251–62.

Pusenius, J., J. Viitala, T. Marienberg, and S. Ritvanen. 1998. Matrilineal kin clusters and their effect on reproductive success in the field vole *Microtus agrestis*. *Behavioral Ecology* 9:85–92.

Pusey, A. E. 1987. Sex-biased dispersal and inbreeding avoidance in birds and mammals. *Trends in Ecology and Evolution* 2:295–99.

Pusey, A. E., and C. Packer. 1994. Non-offspring nursing in social carnivores: Minimizing the costs. *Behavioral Ecology* 5:362–74.

Pusey, A. F., and M. Wolf. 1996. Inbreeding avoidance in animals. *Trends in Ecology and Evolution* 11:201–6.

Pyare, S., and J. Berger. 2003. Beyond demography and delisting: Recovery for Yellowstone's wolves and grizzly bears. *Biological Conservation* 113:63–73.

Pye, T., and W. N. Bonner. 1980. Feral brown rats, *Rattus norvegicus*, in South Georgia (South Atlantic Ocean). *Journal of Zoology, London* 192:237–55.

Quy, R. J, D. S. Shepherd, and I. R. Inglis. 1992. Bait avoidance and effectiveness of anticoagulant rodenticides against warfarin-resistant and difenacoum-resistant populations of Norway rats (*Rattus norvegicus*). *Crop Protection* 11:14–20.

Rabenold, K. N. 1985. Cooperation in breeding by nonreproductive wrens: Kinship, reciprocity and demography. *Behavioral Ecology and Sociobiology* 17:1–17.

Rabinowitz, D. 1981. Seven forms of rarity. In *The biological aspects of rare plant conservation*, ed. H. Synge, 205–17. Chichester, UK: Wiley.

Rabinowitz, P. D., M. F. Coffin, and D. Falvey, D. 1983. The separation of Madagascar and Africa. *Science* 220:67–69.

Radcliffe, M. C. 1992. Repopulation of black-tailed prairie dog (*Cynomys ludovicianus*) colonies after artificial reduction. Master's thesis, Frostburg State University, Frostburg, Maryland.

Rado, R., N. Levi, H. Hauser, J. Witcher, N. Adler, N. Intrator, Z. S. Wolberg, and H. Terkel. 1987. Seismic signaling as a means of communication in a subterranean mammal. *Animal Behavior* 35:1249–66.

Rainey, D. G. 1956. Eastern woodrat, *Neotoma floridana*: Life history and ecology. *Museum of Natural History* 8:535–646.

Ralls, K. 1971. Mammalian scent marking. *Science* 171:443–49.

———. 1976. Mammals in which females are larger than males. *Quarterly Review of Biology* 51:245–76.

———. 1977. Sexual dimorphism in mammals: Avian models and unanswered questions. *American Naturalist* 111:917–38.

Ralls, K., J. D. Ballou, and A. Templeton. 1988. Estimates of lethal equivalents and the cost of inbreeding in mammals. *Conservation Biology* 2:185–93.

Ralls, K., P. H. Harvey, and A. M. Lyles. 1986. Inbreeding in natural populations of birds and mammals. In *Conservation biology: Science of scarcity and diversity*, ed. M. E. Soule, 35–56. Sunderland, MA: Sinauer.

Ramia, M. 1974. Estudio ecologico del modulo experimental de Mantecal (Alto Apure). *Boletin de la Sociedad Venezolana de Ciencias Naturales* 31:117–42.

Rampaud, M. 1984. Typology of wild rat populations. *Acta Zoologica Fennica* 171:233–36.

Randall, J. A. 1981. Comparison of sandbathing and grooming in 2 species of kangaroo rat. *Animal Behaviour* 29:1213–19.

———. 1984. Territorial defense and advertisement by foot-drumming in bannertail kangaroo rats (*Dipodomys spectabilis*) at high and low population densities. *Behavioral Ecology and Sociobiology* 16:11–20.

———. 1987a. Field observations of male competition and mating in Merriam's and bannertail kangaroo rats. *American Midland Naturalist* 117:211–13.

———. 1987b. Sandbathing as a territorial scent-mark in the

bannertail kangaroo rat, *Dipodomys spectabilis. Animal Behaviour* 35:426–34.

———. 1989a. Individual footdrumming signatures in bannertailed kangaroo rats *Dipodomys spectabilis. Animal Behaviour* 38:620–30.

———. 1989b. Neighbor recognition in a solitary desert rodent (*Dipodomys merriami*). *Ethology* 81:123–33.

———. 1989c. Territorial-defense interactions with neighbors and strangers in banner-tailed kangaroo rats. *Journal of Mammalogy* 70:308–15.

———. 1991a. Mating strategies of a nocturnal desert rodent (*Dipodomys spectabilis*). *Behavioral Ecology and Sociobiology* 28:215–20.

———. 1991b. Sandbathing to establish familiarity in the Merriam's kangaroo rat, *Dipodomys merriami. Animal Behaviour* 41:267–75.

———. 1993. Behavioural adaptations of desert rodents (Heteromyidae). *Animal Behaviour* 45:263–87.

———. 1994. Convergences and divergences in communication and social organisation of desert rodents. *Australian Journal of Zoology* 42:405–33.

———. 1995. Modification of footdrumming signatures by kangaroo rats, *Dipodomys spectabilis. Animal Behaviour* 49:1227–37.

———. 1997. Species specific footdrumming in kangaroo rats: *Dipodomys ingens, D. deserti, D. spectabilis. Animal Behaviour* 54:1167–75.

———. 2001. Evolution and function of drumming as communication in mammals. *American Zoologist* 41:1143–56.

Randall, J. A., and D. K. Boltas King. 2001. Assessment and defence of solitary kangaroo rats under risk of predation by snakes. *Animal Behaviour* 61:579–87.

Randall, J. A., S. M. Hatch, and E. R. Hekkala. 1995. Interspecific variation in antipredator behavior in sympatric species of kangaroo rat. *Behavioral Ecology and Sociobiology* 36:243–50.

Randall, J. A., E. R. Hekkala, L. D. Cooper, and J. Barfield, J. 2002. Familiarity and flexible mating strategies of a solitary rodent, *Dipodomys ingens. Animal Behaviour* 64:11–21.

Randall, J. A., and E. R. Lewis. 1997. Seismic communication between the burrows of kangaroo rats, *Dipodomys spectabilis. Journal of Comparative Physiology* 181:525–31.

Randall, J. A., and M. D. Matocq. 1997. Why do kangaroo rats (*Dipodomys spectabilis*) footdrum at snakes? *Behavioral Ecology* 8:404–13.

Randall, J. A., and K. A. Rogovin. 2002. Variation in and meaning of alarm calls in a social desert rodent *Rhombomys opimus. Ethology* 108:513–27.

Randall, J. A., K. A. Rogovin, P. G. Parker, and J. A. Eimes. 2005. Flexible social structure of a desert rodent: Philopatry, kinship and ecological constraints. *Behavioral Ecology* 16:961–73.

Randall, J. A., K. A. Rogovin, and D. M. Shier. 2000. Antipredator behavior of a social desert rodent: Footdrumming and alarm calling in the great gerbil, *Rhombomys opiums. Behavioral Ecology and Sociobiology* 48:110–18.

Ranta, E., A. Laurila, and J. Elmberg. 1994. Reinventing the wheel: Analysis of sexual dimorphism in body size. *Oikos* 70:313–21.

Rasa, O. A. E. 1977. The ethology and sociobiology of the dwarf mongoose (*Helogale undulata parvula*). *Zeitschrift für Tierpsychologie* 43:337–406.

———. 1985. *Mongoose watch*. London: John Murray.

Rasoloharijaona, S., B. Rakotosamimanama, and E. Zimmer-

mann. 2000. Infanticide by male Milne-Edwards' sportive lemur (*Lepilemur edwarsi*) in Ampijoroa, NW-Madagascar. *International Journal of Primatology* 21:41–45.

Raspopov, M. P., and Y. A. Isakov. 1980. *Biology of the squirrel*. Springfield, VA: Amerind.

Rat Genome Sequencing Project Consortium. 2004. Genome sequence of the brown Norway rat yields insights into mammalian evolution. *Nature* 428:493–516.

Ratkiewicz, M., and A. Borkowska. 2000. Multiple paternity in the bank vole (*Clethrionomys glareolus*): Field and experimental data. *Zeitschrift für Saugetierkunde* 65:6–14.

Raun, G. G. 1966. A population of woodrats (*Neotoma micropus*) in southern Texas. *Bulletin of Texas Memorial Museum* 11:1-62.

Raun, G. G., and B. J. Wilks. 1964. Natural history of *Baiomys taylori* in southern Texas and competition with *Sigmodon hispidus* in a mixed population. *Texas Journal of Science* 16:28–49.

Rausch, R. L., and J. G. Bridgens. 1989. Structure and function of sudoriferous facial glands in Nearctic marmots, *Marmota* spp. (Rodentia:Sciuridae). *Zoologische Anzeiger* 223:265–82.

Rausch, R. L., and V. R. Rausch. 1971. The somatic chromosomes of some North American marmots (Sciuridae), with remarks on the relationship of *Marmota broweri* Hall and Gilmore. *Mammalia* 35:85–101.

Rayor, L. S. 1985. Effects of habitat quality on growth, age of first reproduction, and dispersal in Gunnison's prairie dogs (*Cynomys gunnisoni*). *Canadian Journal of Zoology* 63:2835–40.

———. 1988. Social organization and space-use in Gunnison's prairie dog. *Behavioral Ecology and Sociobiology* 22:69–78.

Rayor, L. S., and K. B. Armitage. 1991. Social behavior and space-use of young of ground-dwelling squirrel species with different levels of sociality. *Ethology, Ecology and Evolution* 3:185–205.

Rayor, L. S., A. K. Brody, and C. Gilbert. 1987. Hibernation in the Gunnison's prairie dog. *Journal of Mammalogy* 68:147–50.

Read, A. F., and P. H. Harvey. 1989. Life history differences among the eutherian radiations. *Journal of Zoology, London* 219:329–53.

Redfield, J. A., M. J. Taitt, and C. J. Krebs. 1978. Experimental alteration of sex ratios in population of *Microtus townsendii*, a field vole. *Canadian Journal of Zoology* 56:17–27.

Redford, K. H., and J. F. Eisenberg. 1992, *Mammals of the neotropics, vol. 2: The southern cone*. Chicago: University of Chicago Press.

Reeder, S. A., D. S. Carroll, C. W. Edwards, C. W. Kilpatrick, and R. D. Bradley. 2006. Neotomine-peromyscine rodent systematics based on combined analyses of nuclear and mitochondrial DNA sequences. *Molecular Phylogenetics and Evolution* 40:251–58.

Reeve, A. F., and T. C. Vosburgh. 2006. Recreational shooting of prairie dogs. In *Conservation of the black-tailed prairie dog*, ed. J. L. Hoogland, 139–58. Washington, DC: Island Press.

Reeve, H. K. 1989. The evolution of conspecific acceptance thresholds. *American Naturalist* 133:407–35.

———. 1998. Game theory, reproductive skew, and nepotism. In *Game theory and animal behavior*, ed. L. Dugatkin and H. K. Reeve, 118–45. Oxford: Oxford University Press.

Reeve, H. K., S. T. Emlen, and L. Keller. 1998. Reproductive sharing in animal societies: Reproductive incentives or incomplete control by dominant breeders? *Behavioral Ecology* 9:267–78.

Reeve, H. K., and L. Keller. 1995. Partitioning of reproduction in mother-daughter versus sibling associations: A test of optimal skew theory. *American Naturalist* 145:119–32.

Reeve, H. K., and P. W. Sherman. 2001. Optimality and phylogeny: A critique of current thought. In *Adaptationism and optimality*, ed. S. H. Orzack and E. Sober, 64–113. New York: Cambridge University Press.

Reeve, H. K., D. F. Westneat, W. A. Noon, P. W. Sherman, and C. F. Aquadro. 1990. DNA "fingerprinting" reveals high levels of inbreeding in colonies of the eusocial naked mole-rat. *Proceedings of the National Academy of Sciences (USA)* 87: 2496–2500.

Reichard, U. H. 2003. Monogamy: Past and present. In *Monogamy: Mating strategies and partnerships in birds, humans and other mammals*, ed. U. H. Reichard and C. Boesch, 3–25. Cambridge: Cambridge University Press.

Reichard, U. H., and C. Boesch, eds. 2003. *Monogamy: Mating strategies and partnerships in birds, humans and other mammals*. Cambridge: Cambridge University Press.

Reichman, O. J., and M. V. Price. 1993. Ecological aspects of heteromyid foraging. In *Biology of the heteromyidae*, ed. H. H. Genoways and J. H. Brown, 539–95. Shippensburg, PA: American Society of Mammalogists.

Reichman, O. J., and S. C. Smith. 1987. Burrows and burrowing behavior by mammals. *Current Mammalogy* 2:197–244.

Reig, O. A. 1970. Ecological notes on the fossorial octodont rodent *Spalacopus cyanus* (Molina). *Journal of Mammalogy* 51: 592–601.

———. 1986. Diversity patterns and differentiation of high Andean rodents. In *High altitude tropical biogeography*, ed. F. Vuilleumier and M. Monasterio, 404–39. Oxford: Oxford University Press.

Reig, O. A., C. Busch, M. O. Ortells, and J. R. Contreras. 1990. An overview of evolution, systematics, population biology, cytogenetics, molecular biology and speciation in *Ctenomys*. In *Evolution of subterranean mammals at the organismal and molecular levels*, ed. E. Nevo and O. A. Reig, 71–96. New York: Alan R. Liss.

Reig, O. A., A. O. Spotorno, and D. R. Fernandez. 1972. A preliminary survey of chromosomes in populations of the Chilean burrowing octodont rodent *Spalacopus cyanus* Molina (Caviomorpha, Octodontidae). *Biological Journal of the Linnean Society London* 4:29–38.

Reintjes, R., I. Dedushaj, A. Gjini, T. R. Jorgensen, B. Cotter, A. Lieftucht, F. D'Ancona et al. 2002. Tularemia outbreak investigation in Kosovo: Case control and environmental studies. *Emerging Infectious Diseases* 8:69–73.

Reise, D., and M. H. Gallardo. 1989. An extraordinary occurrence of the tunduco *Aconaemys fuscus* (Waterhouse, 1841) (Rodentia, Octodontidae) in the central valley, Chillán, Chile. *Medio Ambiente (Chile)* 10:67–69.

Reiter, J., K. J. Panken, and B. J. Le Boeuf. 1981. Female competition and reproductive success in northern elephant seals. *Animal Behaviour* 29:670–87.

Restle, F. 1957. Discrimination of cues in mazes: A resolution of the "place-vs.-response" question. *Psychological Review* 64: 217–28.

Reusch, T. B. H., M. A. Haeberli, P. B. Aeschlimann, and M. Milinski. 2001. Female sticklebacks count alleles in a strategy of sexual selection explaining MHC polymorphism. *Nature* 414:300–302.

Reyes, A., C. Gissi, G. Pesole, F. M. Catzeflis, and C. Saccone. 2000. Where do rodents fit? Evidence from the complete mitochondrial genome of *Sciurus vulgaris*. *Molecular Biology and Evolution* 17:979–83.

Reyes, A., G. Pesole, and C. Saccone. 1998. Complete mitochondrial DNA sequence of the fat dormouse, *Glis glis*: Further evidence of rodent paraphyly. *Molecular Biology and Evolution* 15:499–505.

Reynolds, J. D. 1996. Animal breeding systems. *Trends in Ecology and Evolution* 11:68–72.

Reynolds, T. J., and J. W. Wright. 1979. Early postnatal physical and behavioural development of degus (*Octodon degus*). *Laboratory Animals* 13:93–99.

Ribble, D. O. 1991. The monogamous mating system of *Peromyscus californicus* as revealed by DNA Fingerprinting. *Behavioral Ecology and Sociobiology* 29:161–66.

Ribble, D. O. 1992. Dispersal in a monogamous rodent, *Peromyscus californicus*. *Ecology* 73:859–66.

———. 2003. The evolution of social and reproductive monogamy in *Peromyscus*, evidence from *Peromyscus californicus* (the California Mouse). In *Monogamy: Mating strategies and partnerships in birds, humans, and other mammals*, ed. U. Reichard and C. Boesh, 81–92. Cambridge, MA: Cambridge University Press.

Ribble, D. O., and J. S. Millar. 1996. The mating system of northern populations of *Peromyscus maniculatus* as revealed by radiotelemetry and DNA fingerprinting. *Ecoscience* 3: 423–28.

Ribble, D. O., and M. Salvioni. 1990. Social organization and nest co-occupancy in *Peromyscus californicus*, a monogamous rodent. *Behavioral Ecology and Sociobiology* 26:9–15.

Ribble, D. O., and S. Stanley. 1998. Home ranges and social organization of syntopic *Peromyscus boylii* and *P. truei*. *Journal of Mammalogy* 79:932–41.

Rice, E. K. 1988. The chinchilla: Endangered whistler of the Andes. *Animal Kingdom* 91:6–7.

Rich, T. J., and J. L. Hurst. 1998. Scent marks as reliable signals of the competitive ability of mates. *Animal Behaviour* 56: 727–35.

———. 1999. The competing countermarks hypothesis: Reliable assessment of competitive ability by potential mates. *Animal Behaviour* 58:1027–37.

Richard, M. M., C. A.Grover, and S. F. Davis. 1987. Galef's transfer-of-information effect occurs in a free-foraging situation. *Psychological Record* 37:79–87.

Richards, M. P. M. 1966. Maternal behaviour in the golden hamster: Responsiveness to young in virgin, pregnant, and lactating females. *Animal Behaviour* 14:310–13.

Richman, A. D., L. G. Herrera, and D. Nash. 2001. MHC class II beta sequence diversity in the deer mouse (*Peromyscus maniculatus*): Implications for models of balancing selection. *Molecular Ecology* 10:2765–73.

Richmond, G., and B. J. Sachs. 1984. Maternal discrimination of pup sex in rats. *Developmental Psychobiology* 17:87–89.

Rickard, C. A., and N. C. Bennett. 1997. Recrudescense of sexual activity in a reproductively quiescent colony of the Damaraland mole-rat, by the introduction of a genetically unrelated male—a case of incest avoidance in "queenless" colonies. *Journal of Zoology* 241:185–202.

Rickart, E. A. 1977. Reproduction, growth and development in two species of cloud forest *Peromyscus* from southern Mexico. *Occasional Papers of the Museum of Natural History, University of Kansas* 67:1–22.

———. 1982. Annual cycles of activity and body composition in *Spermophilus townsendii mollis. Canadian Journal of Zoology* 60: 3298–3306.

———. 1986. Postnatal growth of the Piute ground squirrel (*Spermophilus mollis*). *Journal of Mammalogy* 67:412–16.

Rickart, E. A., and P. B. Robertson. 1985. *Peromyscus melanocarpus. Mammalian Species* 241:1–3.

Rickart, E. A., and E. Yensen. 1991. *Spermophilus washingtoni. Mammalian Species* 371:1–5.

Ricklefs, R., and M. Wikelski. 2002. The physiology/life-history nexus. *Trends in Ecology and Evolution* 17:462–68.

Rijksen, H. D. 1981. Infant killing: A possible consequence of a disputed leader role. *Behaviour* 78:138–68.

Riley, D. A., and M. R. Rosenzweig. 1957. Echolocation in rats. *Journal of Comparative and Physiological Psychology* 50:323–28.

Risch, T. S., F. S. Dobson, and J. O. Murie. 1995. Is mean litter size the most productive? A test in Columbian ground squirrels. *Ecology* 76:1643–54.

Rissman, E. F., and R. E. Johnston. 1985. Female reproductive development is not activated by male California voles exposed to family cues. *Biology of Reproduction* 32:352–60.

Ritchie, M. E., and G. E. Belovsky. 1990. Sociality of Columbian ground squirrels in relation to their seasonal energy intake. *Oecologia* 83:495–503.

Rivers, J. P. W., and M. A. Crawford. 1974. Maternal nutrition and the sex ratio at birth. *Nature* 252:297–98.

Roach, J. S., P. Stapp, B. Van Horne, and M. F. Antolin. 2001. Genetic structure of a metapopulation of black-tailed prairie dogs. *Journal of Mammalogy* 82:946–59.

Roberts, A. M. 1978. The origins of fluctuations in the human secondary sex ratio. *Journal for Biosocial Science* 10:169–82.

Roberts, M., E. Maliniak, and M. Deal. 1984. The reproductive biology of the rock cavy, *Kerodon rupestris*, in captivity: A study of reproductive adaptation in a trophic specialist. *Mammalia* 48:253–66.

Roberts, M. S., K. V. Thompson, and J. A. Cranford. 1988. Reproduction and growth in captive punare (*Thrichomys apereoides* Rodentia:Echimyidae) of the Brazilian Caatinga with reference to the reproductive strategies of the Echimyidae. *Journal of Mammalogy* 69:542–51.

Roberts, R. L., B. S. Cushing, and C. S. Carter. 1998. Intraspecific variation in the induction of female sexual receptivity in prairie voles. *Physiology and Behavior* 64:209–12.

Roberts, R. L., J. R. Williams, A. K. Wang, and C. S. Carter. 1998. Cooperative breeding and monogamy in prairie voles: Influence of the sire and geographical variation. *Animal Behaviour* 55:1131–40.

Roberts, S. C. 1997. Selection of scent-marking sites by klipspringers (*Oreotragus oreotragus*). *Journal of Zoology* 243:555–64.

Roberts, S. C., and R. I. M. Dunbar. 2000. Female territoriality and the function of scent-marking in a monogamous antelope (*Oreotragus oreotragus*). *Behavioral Ecology and Sociobiology* 47:417–23.

Roberts, S. C., and L. M. Gosling. 2003. Genetic quality and similarity interact in mate choice decisions by female mice. *Nature Genetics* 35:103–6.

———. 2004. Manipulation of olfactory signalling and mate choice for conservation breeding: a case study of the harvest mouse. *Conservation Biology* 18:548–56.

Roberts, S. C., L. M. Gosling, E. A. Thornton, and J. McClung.

———. 2001. Scent-marking by male mice under the risk of predation. *Behavioral Ecology* 12:698–705.

Roberts, S. C., and C. Lowen. 1997. Optimal patterns of scent marks in klipspringer (*Oreotragus oreotragus*) territories. *Journal of Zoology* 243:565–78.

Roberts, T. J. 1977. *The mammals of Pakistan.* London: Ernest Benn.

Roberts, W. M., J. P. Rodriguez, T. C. Good, and A. P. Dobson. 2000. *Population viability analysis of the Utah prairie dog.* Washington, DC: Environmental Defense Report.

Robins, J. H., J. S. Scheibe, and K. Laves. 2000. Sexual size dimorphism and allometry in southern flying squirrels, *Glaucomys volans.* In *The biology of gliding mammals,* ed. R. Goldingay and J. S. Scheibe, 229–48. Erlangen, Germany: Filander.

Robinson, A. S. 2000. Factors affecting reproductive success in male house mice (*Mus domesticus*). PhD diss., Southern Illinois University.

Robinson, R. 1965. *Genetics of the Norway rat.* Oxford: Pergamon.

Robinson, S. R. 1981. Alarm communication in Belding's ground squirrels. *Zeitschrift für Tierpsychologie* 56:150–68.

Robinson, S. R., and W. P. Smotherman. 1991. Fetal learning: Implications for the development of kin recognition. In *Kin recognition,* ed. P. G. Hepper, 308–34. Cambridge: Cambridge University Press.

Robitaille, J. A., and J. Bovet. 1976. Field observations on the social behavior of the Norway rat, *Rattus norvegicus* (Berkenhout). *Biology of Behavior* 1:289–308.

Rodd, F. H., and R. Boonstra. 1988. Effects of adult meadow voles, *Microtus pennsylvanicus*, on young conspecifics in field populations. *Journal of Animal Ecology* 57:755–70.

Rodgers, R. J., and J. I. Randall. 1986. Resident's scent: A critical factor in acute analgesic reaction to defeat experience in male mice. *Physiology and Behavior* 37:317–22.

———. 1987. Defensive analgesia in rats and mice. *Psychological Record* 37:335–47.

Roff, D. A. 1992. *The evolution of life histories: Theory and analysis.* New York: Chapman and Hall.

Roff, D. A. 2002. Inbreeding depression: Tests of the overdominance and partial dominance hypotheses. *Evolution* 56:768–75.

Rogovin, K. A. 1981. Utilization of territory by little earth hares (*Alactagulus acontion*) and structure of their groups by materials on marking. [In Russian.] *Zoologicheskii Zhurnal* 60: 568–77.

———. 1996. Social behavior of *Psammomys obesus* (Rodentia, Gerbillidae) under seminatural conditions: A settlement of high density. [In Russian.] *Zoologicheskii Zhurnal* 75:399–412.

———. 1999. On the origin of bipedal locomotion in rodents: Ecological correlates of jerboas bipedal hopping. [In Russian.] *Zoologicheskii Zhurnal* 78:228–39.

Rogovin, K. A., V. F. Kulikov, A. V. Surov, and N. Yu. Vasilieva. 1987. Organization of the desert rodent communities in the Transaltai Gobi in the territory of Mongolia. [In Russian.] *Zoologicheskii Zhurnal* 66:417–29.

Rogovin, K. A., M. P. Moshkin, J. A. Randall, I. E. Kolosova, and Yu. A. Chikin. 2003. Availability of resources, demography and physiological stress in population of great gerbil (*Rhombomys opimus* Licht) in conditions of peak and decline of density. [In Russian.] *Zoologicheskii Zhurnal* 82:497–507.

Rogovin, K. A., J. A. Randall, I. Kolosova, and M. Moshkin. 2003. Social correlates of stress in adult males of the great gerbil, *Rhombomys opimus,* in years of high and low population densities. *Hormones and Behavior* 43:132–39.

———. 2004. Predation on a social desert rodent, *Rhombomys opimus*: Effects of group size, composition and location. *Journal of Mammalogy* 85:723–30.

Rojas, M., O. Rivera, G. Montenegro, and C. Barros. 1977. Algunas observaciones en la reproducción de la hembra silvestre de *Octodon degus,* Molina y su posible relación con la fenología de la vegetación. *Medio Ambiente (Chile)* 3:78–82.

Roldan, E. R. S., M. Gomendio, and A. D. Vitullo. 1992. The evolution of eutherian spermatozoa and underlying selective forces: Female selection and sperm competition. *Biological Reviews* 67:551–93.

Rolland, C., D. W. MacDonald, M. de Fraipont, and M. Berdoy. 2004. Free female choice in house mice: Leaving best for last. *Behaviour* 140:1371–88.

Romero, L. M. 2002. Seasonal changes in plasma glucocorticoid concentrations in free-living vertebrates. *General and Comparative Endocrinology* 128:1–24.

Ronca, A. E., and J. R. Alberts. 1995. Maternal contributions to fetal experience and the transition from prenatal to postnatal life. In *Fetal development: A psychobiological perspective,* ed. J.-P. Lecanuet, W. P. Fifer, N. A. Krasnegor, and W. P. Smotherman, 331–50. Hillsdale, NJ: Lawrence Erlbaum.

Ronkainen, H., and H. Ylönen. 1994. Behaviour of cyclic bank voles under risk of mustelid predation: Do females avoid copulations? *Oecologia* 97:377–81.

Rood, J. P. 1970. Ecology and social behavior of the desert cavy (*Microvia australis*). *American Midland Naturalist* 83:415–54.

———. 1972. Ecological and behavioural comparisons of three genera of Argentine cavies. *Animal Behaviour Monographs* 5:1–83.

Roper, T. J., and E. Polioudakis. 1977. The behaviour of Mongolian gerbils in a semi-natural environment, with special reference to ventral marking, dominance and sociability. *Behaviour* 60:207–37.

Rorie, R. W. 1999. Effects of timing of artificial insemination on sex ratio. *Theriogenology* 52:1273–80.

Rose, M. R. 1991. *Evolutionary biology of aging.* New York: Oxford University Press.

Rosell, F., F. Bergan, and H. Parker. 1998. Scent-marking in the Eurasian beaver (*Castor fiber*) as a means of territory defense. *Journal of Chemical Ecology* 24:207–19.

Rosell, F., and T. Bjorkoyli. 2002. A test of the dear enemy phenomenon in the Eurasian beaver. *Animal Behaviour* 63:1073–78.

Rosell, F., G. Johansen, and H. Parker. 2000. Eurasian beavers (*Castor fiber*) behavioral response to simulated territorial intruders. *Canadian Journal of Zoology* 78:931–35.

Rosell, F., and B. A. Nolet. 1997. Factors affecting scent-marking behavior in the Eurasian beaver (*Castor fiber*). *Journal of Chemical Ecology* 23:673–89.

Rosell, F., and K. V. Pedersen. 1999. *Beaver.* Norway: Landbruksforlaget.

Rosell, F., and L. Sun. 1999. Use of the anal gland secretion to distinguish the two beaver species. *Wildlife Biology* 5:119–23.

Rosenblatt, G. S., and D. S. Lehrman. 1963. Maternal behaviour of the laboratory rat. In *Maternal behaviour in mammals,* ed. H. L. Rheingold, 8–57. London: Wiley.

Rosenblatt, J. S., H. I. Siegel, and A. D. Mayer. 1979. Progress in the study of maternal behavior in the rat: Hormonal, nonhormonal, sensory, and developmental aspects. In *Advances in the study of behavior,* vol. 10, ed. J. S. Rosenblatt, R. A. Hinde, E. Shaw, and C. Beer, 225–311. New York: Academic Press.

Rosenqvist, G., and A. Berglund. 1992. Is female sexual behaviour a neglected topic? *Trends in Ecology and Evolution* 7:174–76.

Rosenzweig, M. L. 1974. On the evolution of habitat selection. *Proceedings of the First International Congress of Ecology*: 401–4. The Hague: Centre for Agricultural Publishing and Documentation.

Rosenzweig, M. L., Z. Abramsky, and A. Subach. 1997. Safety in numbers: Sophisticated vigilance by Allenby's gerbil. *Proceedings of the National Academy of Sciences (USA)* 94:5713–15.

Rosenzweig, M. L., and J. Winakur. 1969. Population ecology of desert rodent communities: Habitat and environmental complexity. *Ecology* 50:558–72.

Rosenzweig, M. R., D. A. Riley, and D. Krech. 1955. Evidence for echolocation in the rat. *Science* 121:600.

Rosi, M. I., M. I. Cona, and V. G. Roig. 2002. Estado actual del conocimiento del roedor fosorial *Ctenomys mendocinus* Philippi 1869 (Rodentia: Ctenomyidae). *Mastozoologia Neotropical* 9:277–95.

Rosmarino, N. J. 2004. Petition to list the Gunnison's prairie dog under the Endangered Species Act. Submitted to United States Fish and Wildlife Service on behalf of Forest Guardians and 73 co-petitioners on February 23, 2004.

Rosner, W. 1990. The functions of corticosteroid-binding globulin and sex hormone-binding globulin: Recent advances. *Endocrine Reviews* 11:80–91.

Ross, P. D. 1995. *Phodopus campbelli. Mammalian Species* 503:1–7.

Roulin, A. 2002. Why do lactating females nurse alien offspring? A review of hypotheses and empirical evidence. *Animal Behaviour* 63:201–8.

Rowe, D. L. 2002. Molecular phylogenetics and evolution of hystricognath rodents. PhD diss., Texas A&M University, College Station.

Rowe, D. L., and R. L. Honeycutt. 2002. Phylogenetic relationships, ecological correlates, and molecular evolution within the Cavioidea (Mammalia, Rodentia). *Molecular Biology and Evolution* 19:263–77.

Rowe, F. P. 1981. Wild house mouse biology and control. *Symposia of the Zoological Society of London* 47:575–89.

Rowe, F. P., E. J. Taylor, and A. H. J. Chudley. 1963. The numbers and movements of house mice (*Mus musculus* L.) in the vicinity of four corn-ricks. *Journal of Animal Ecology* 32:87–97.

Rowe, M. P., R. G. Coss, and D. H. Owings. 1986. Rattlesnake rattles and burrowing owl hisses: A case of acoustic Batesian mimicry. *Ethology* 72:53–71.

Rowe, M. P., and D. H. Owings. 1978. The meaning of the sound of rattling by rattlesnakes to California ground squirrels. *Behaviour* 66:252–67.

———. 1990. Probing, assessment, and management during interactions between ground squirrels and rattlesnakes. Part 1: Risks related to rattlesnake size and body temperature. *Ethology* 86:237–49.

———. 1996. Probing, assessment, and management during interactions between ground squirrels (Rodentia: Sciuridae) and rattlesnakes (Squamata: Viperidae). Part 2: Cues afforded by rattlesnake rattling. *Ethology* 102:856–74.

Rowland, D. L. 1981. Effects of pregnancy on the maintenance of maternal behavior in the rat. *Behavioral and Neural Biology* 31:225–35.

Rowlands, I. W. 1974. Mountain viscacha. *Symposium of the Zoological Society of London* 34:131–41.

Rozenfeld, F. M., E. Le Boulangé, and R. Rasmont. 1987. Urine marking in male bank voles (*Clethrionomys glareolus* Schreber, 1780; Microtidae, Rodentia) in relation to their social rank. *Canadian Journal of Zoology* 65:2594–2601.

Rozenfeld, F. M., and A. Denoel. 1994. Chemical signals involved in spacing behaviour of breeding female bank voles (*Clethrionomys glareolus*, Schreber 1780, Microtidae, Rodentia). *Journal of Chemical Ecology* 20:803–13.

Rozenfeld, F. M., R. Rasmont, and A. Haim. 1994. Home site scent marking with urine and an oral secretion in the golden spiny mouse (*Acomys russatus*). *Israel Journal of Zoology* 40:161–72.

Rozin, P. 1976. The selection of food by rats, humans and other animals. *Advances in the Study of Behaviour* 6:21–76.

Rudran, R. 1973. Adult male replacement in one-male troops of purple-faced langurs (*Presbytis senex senex*) and its effect on population structure. *Folia Primatologica* 19:166–92.

Ruedas, L. A., J. Salazar-Bravo, D. S. Tinnin, B. Armien, L. Caceres, A. Garcia, M. A. Diaz et al. 2004. Community ecology of small mammal populations in Panama following an outbreak of Hantavirus pulmonary syndrome. *Journal of Vector Ecology* 29:177–91.

Ruf, T., and W. Arnold. 2000. Mechanisms of social thermoregulation in hibernating alpine marmots (*Marmota marmota*). In *Life in the cold*, ed. G. Heldmaier and M. Klingenspor, 81–94. Berlin: Springer-Verlag.

Ruiz, A. 2001. Plagues in the Americas. *Emerging Infectious Diseases* 7(3) Supplement:539–40.

Rumiantsev, V. Yu, and D. I. Bibikov. 1994. Marmots in Europe: History and prospects. In *Actual problems of marmots investigation*, ed. V. Yu. Rumiantsev, 193–214. Moscow: ABF.

Ruscoe, W. A., R. Goldsmith, and D. Choquenot. 2001. A comparison of population estimates and abundance indices for house mice inhabiting beech forests in New Zealand. *Wildlife Research* 28:173–78.

Rutberg, A. T. 1987. Adaptive hypotheses of birth synchrony in ruminants: An interspecific test. *American Naturalist* 130:692–710.

Rutherford, W. H. 1964. The beaver in Colorado. *Colorado Game, Fish and Parks Department*. Technical publication 17:1–49.

Ruusila, V., A. Ermala, and H. Hyvarinen. 2000. Costs of reproduction in introduced female Canadian beavers (*Castor canadensis*). *Journal of Zoology* 252:79–82.

Ryan, B. C., and J. G. Vandenbergh. 2002. Intrauterine position effects. *Neuroscience and Biobehavioral Reviews* 26:665–78.

Saal, D., Y. Dong, A. Bonci, and R. C. Malenka. 2003. Drugs of abuse and stress trigger a common synaptic adaptation in dopamine neurons. *Neuron* 37:577–82.

Saavedra, B., and J. A. Simonetti. 2003. Holocene distribution of octodontid rodents in central Chile. *Revista Chilena de Historia Natural* 76:383–89.

Sachser, N., E. Schwarz-Weig, A. Keil, and J. Epplen. 1999.

Behavioural strategies, testis size, and reproductive success in two caviomorph rodents with different mating systems. *Behaviour* 136:1203–17.

Safran, R. J., P. W. Sherman, V. J. Doerr, E. Doerr, and D. W. Winkler. In press. From individual behavior to group breeding: Process and pattern. *Behavioral Ecology*.

Saitoh, T. 1989. Effects of added food on some attributes of an enclosed vole population. *Journal of Mammalogy* 70:772–82.

Sala, L., C. Sola, A. Spampanato, and P. Tongiorgi. 1992. The marmot population of the Tuscon-Emilian Apennine ridge. In *First international symposium of alpine marmot* (Marmota marmota) *and on genus* Marmota, ed. B. Bassano, P. Durio, U. Gallo Orsi, and E. Macchi, 143–49. Torino.

Sala, L., C. Sola, A. Spampanato, M. Magnanini, and P. Tongiorgi. 1996. Space and time use in a population of *Marmota marmota* of the northern Apennines. In *Biodiversity in marmots*, ed. M. Le Berre, R. Ramousse, and L. Le Guelte, 209–16. Lyon: International Network on Marmots.

Salas, V. 1999. Social organisation of capybaras in the Venezuelan Llanos. PhD diss., Cambridge University, Cambridge, UK.

Sale, J. B. 1966. The habitat of the rock hyrax. *Journal of the East African Natural History Society* 25:205–14.

Sales, G. D., and D. Pye. 1974. *Ultrasonic communication by animals*. London: Chapman and Hall.

Sales, G. D., and J. C. Smith. 1978. Comparative studies of the calls of infant murid rodents. *Developmental Psychobiology* 11:595–619.

Salo, A. L., and D. A. Dewsbury. 1995. Three experiments on mate choice in meadow voles (*Microtus pennsylvanicus*). *Journal of Comparative Psychology* 109:42–46.

Salo, A. L., and J. A. French. 1989. Early experience, reproductive success, and development of parental behaviour in Mongolian gerbils. *Animal Behaviour* 38:693–702.

Salsbury, C. M., and K. B. Armitage. 1994. Home-range size and exploratory excursions of adult, male yellow-bellied marmots. *Journal of Mammalogy* 75:648–56.

Sandell, M., J. Agrell, S. Erlinge, and J. Nelson. 1991. Adult philopatry and dispersal in the field vole *Microtus agrestis*. *Oecologia* 86:153–58.

Santos, E. M. dos, H. P. Andreassen, and R. A. Ims. 1995. Differential inbreeding tolerance in two geographically distinct strains of root voles *Microtus oeconomus*. *Ecography* 18:238–47.

Sanyal, S., H. G. Jansen, W. J. de Grip, E. Nevo, and W. W. De Jong. 1990. The eye of the blind mole rat, *Spalax ehrenbergi*: Rudiment with hidden function? *Investigative Ophtalmology Visual Science* 31:1398–1404.

Sapolsky, R. M. 1992. Neuroendocrinology of the stress-response. In *Behavioral endocrinology*, ed. J. B. Becker, S. M. Breedlove, and D. Crews, 287–324. Cambridge: MIT Press.

———. 1983. Endocrine aspects of social instability in the olive baboon (*Papio anubis*). *American Journal of Primatology* 5:365–79.

———. 2002. Neuroendocrinology of the stress-response. In *Behavioral endocrinology*, ed. J. B. Becker, S. M. Breedlove, D. Crews, and M. M. McCarthy, 409–50. Cambridge: MIT Press.

Sapolsky, R. M., L. C. Krey, and B. S. McEwen. 1986. The neuroendocrinology of stress and aging: The glucocorticoid cascade hypothesis. *Endrocrine Reviews* 7:284–301.

Sapolsky, R. M., L. M. Romero, and A. U. Munck. 2000. How do glucocorticoids influence stress responses? Integrating permissive, suppressive, stimulatory, and preparative actions. *Endocrine Reviews* 21:55–89.

Sarich, R. 1985. Rodent macromolecular systematics. In *Evolutionary relationships among rodents: A multidisciplinary analysis,* ed. W. P. Lucket and J. L. Hartenberger, 423–52. New York: Plenum.

Sarich, V. M., and J. E. Cronin. 1980. South American mammal molecular systematics, evolutionary clocks and continental drift. In *Evolutionary biology of the New World monkeys and continental drift,* ed. R. L. Ciochon and A. B. Chiarelli, 399–421. New York: Plenum.

Sarnyai, Z., C. R. McKittrick, B. S. McEwen, and M. J. Kreek. 1998. Selective regulation of dopamine transporter binding in the shell of the nucleus accumbens by adrenalectomy and corticosterone-replacement. *Synapse* 30:334–37.

Sauer, J. R., and N. A. Slade. 1988. Body size as a demographic categorical variable: Ramifications for life-history analysis of mammals. In *Evolution of life-histories of mammals,* ed. M. S. Boyce, 107–21. New Haven, CT: Yale University Press.

Savi'c, I. R. 1973. Ecology of the species *Spalax leucodon* Nordm. *Yugoslavia National Science.* 44:5–70. (In Serbo-Croatian with English summary.)

———. 1982. Familie Spalacidae Gray, 1821: Blindmause. In *Handbuch der Saugetiere Europas,* ed. J. Hiethammer and F. Krapp, 1–649. Wiesbaden: Akadem. Verlagsgesell.

Savi'c, I. R., and E. Nevo. 1990. The Spalacidae: Evolutionary history, speciation, and population biology. In *Evolution of subterranean mammals at the organismal and molecular levels,* ed. E. Nevo and O. A. Reig, 129–54. New York: Wiley-Liss.

Savi'c, I. R., and B. Soldatovi'c. 1979. Distribution range and evolution of chromosomal forms in the Spalacidae of the Balkan Peninsula and bordering regions. *Journal of Biogeography* 6:363–74.

Sawyer, T. F. 1980. Androgen effects on responsiveness to aggression and stress-related odors of male mice. *Physiology and Behaviour* 25:183–87.

Sayler, A., and M. Salmon. 1971. An ethological analysis of communal nursing by the house mouse (*Mus musculus*). *Behaviour* 40:60–85.

Scally, M., O. Madsen, C. J. Douady, W. W. de Jong, J. Stanhope, and M. S. Springer. 2001. Molecular evidence for the major clades of placental mammals. *Journal of Mammalian Evolution* 8:239–77.

Schaller, G. B., and P. G. Crawshaw. 1981. Social dynamics in a capybara population. *Saugetierkundliche Mitteilungen* 29:3–16.

Schank, J. C., and W. C. Wimsatt. 2000. Evolvability, adaptation and modularity. In *Thinking about evolution,* ed. R. S. Singh, 322–35. Cambridge: Cambridge University Press.

Scharff, A., O. Locker-Grütjen, M. Kawalika, and H. Burda. 2001. Natural history of the giant mole-rat, *Cryptomys mechowi* (Rodentia: Bathyergidae), from Zambia. *Journal of Mammalogy* 82:1003–1105.

Schauber, E. M., and R. S. Ostfeld. 2002. Modeling the effects of reservoir competence decay and demographic turnover in Lyme-disease ecology. *Ecological Applications* 12:1142–62.

Schauber, E. M., and J. O. Wolff. 1996. Space use and juvenile recruitment in gray-tailed voles respond more to intruder pressure than to food availability. *Acta Theriologica* 41:35–43.

Scheffer, T. H. 1947. Ecological comparisons of the plains prairie dog and the Zuni species. *Transactions of the Kansas Academy of Science* 49:401–6.

Schellinck, H. M., B. M. Slotnick, and R. E. Brown. 1997. Odors of individuality originating from the major histocompatibility complex are masked by diet cues in the urine of rats. *Animal Learning and Behavior* 25:193–99.

Scherrer, J. A., and G. S. Wilkinson. 1993. Evening bat isolation calls provide evidence for heritable signatures. *Animal Behaviour* 46:847–60.

Schieck, J. O., and J. S. Millar. 1987. Can removal areas be used to assess dispersal of red-backed voles? *Canadian Journal of Zoology* 65:2575–78.

Schlaepfer, M. A., M. C. Runge, and P. W. Sherman. 2002. Ecological and evolutionary traps. *Trends in Ecology and Evolution* 17:474–80.

Schlechte, J. A., and D. Hamilton. 1987. The effect of glucocorticoids on corticosteroid binding globulin. *Clinical Endocrinology* 27:197–203.

Schleidt, W. M. 1973. Tonic communication: Continual effects of discrete signs in animal communication systems. *Journal of Theoretical Biology* 42:359–86.

Schluter, D. 2002. Character displacement. In *Encyclopedia of evolution,* ed. M. Pagel, 149–50. New York: Oxford University Press.

Schmidly, D. J. 1974. *Peromyscus attwateri. Mammalian Species* 48:1–3.

Schmidt, K. A., and R. S. Ostfeld. 2001. Biodiversity and the dilution effect in disease ecology. *Ecology* 82:609–19.

Schmidt-Nielsen, K. 1964. *Desert animals: Physiological problems of heat and water.* Oxford: Clarendon.

———. 1975. Desert rodents: Physiological problems of desert life. In *Rodents in desert environments,* ed. I. Prakash and P. K. Ghosh, 379–96. The Hague: Junk.

———. 1983. *Scaling: Why is animal size so important?* Cambridge: Cambridge University Press.

Schmitz, C., M. E. Rhodes, M. Bludau, S. Kaplan, P. Ong, I. Ueffing, J. Vehoff, H. Korr, and C. A. Frye. 2002. Depression: Reduced number of granule cells in the hippocampus of female, but not male, rats due to prenatal restraint stress. *Molecular Psychiatry* 7:810–13.

Schoffelmeer, A. N., T. J. De Vries, L. J. Vanderschuren, G. H. Tjon, P. Nestby, G. Wardeh, and A. H. Mulder. 1995. Glucocorticoid receptor activation potentiates the morphine-induced adaptive increase in dopamine D-1 receptor efficacy in gamma-aminobutyric acid neurons of rat striatum/nucleus accumbens. *Journal of Pharmacology and Experimental Therapeutics* 274:1154–60.

Schradin, C., and N. Pillay. 2003. Paternal care in the social and diurnal striped mouse (*Rhabdomys pumilio*): Laboratory and field evidence. *Journal of Comparative Psychology* 117:317–24.

———. 2004. The striped mouse (*Rhobdomys pumilio*) from the succulent karoo of South Africa: A territorial group living solitary forager with communal breeding and helpers at the nest. *Journal of Comparative Psychology* 118:37–47.

Schradin, C., and N. Pillay. 2005a. The influence of the father on offspring development in the striped mouse. *Behavioral Ecology* 16:450–55.

Schradin, C., and N. Pillay. 2005b. Intraspecific vatriation in the spatial and social organization of the African striped mouse. *Journal of Mammalogy* 86:99–107.

Schramm, D. L. 1968. A field study of beaver behavior in East Barnard, Vermont. Master's thesis, Dartmouth College, Hanover, MA.

Schroder, G. D. 1979. Foraging behavior and home range utilization of the bannertail kangaroo rat (*Dipodomys spectabilis*). *Ecology* 60:657–65.

Schug, M. D., S. H. Vessey, and E. M. Underwood. 1992. Paternal behavior in a natural population of white-footed mice (*Peromyscus leucopus*). *American Midland Naturalist* 127:373–80.

Schulte, B. A. 1993. Chemical communication and ecology of the North American beaver (*Castor canadensis*). PhD diss., State University of New York, College of Environmental Science and Forestry, Syracuse.

———. 1998. Scent marking and responses to male castor fluid by beavers. *Journal of Mammalogy* 79:191–203.

Schulte, B. A., and D. Müller-Schwarze. 1999. Understanding North American beaver behavior as an aid to management. In *Beaver protection, management, and utilization in Europe and North America*, ed. P. E. Busher and R. M. Dzieciolowski, 109–28. New York: Kluwer Academic/Plenum.

Schulte, B. A., D. Müller-Schwarze, and L. Sun. 1995. Using anal gland secretion to determine sex in beaver. *Journal of Wildlife Management* 59:614–18.

Schulte-Hostedde, A. I. 2004. Sexual selection and mating patterns in a mammal with female-biased sexual size dimorphism. *Behavioral Ecology* 15:351–56.

Schulte-Hostedde, A. I., and J. S. Millar. 2002. 'Little chipmunk' syndrome? Male body size and dominance in captive yellow-pine chipmunks (*Tamias amoenus*). *Ethology* 108:127–37.

———. 2004. Intraspecific variation of testis size and sperm length in the yellow-pine chipmunk (*Tamias amoenus*): Implications for sperm competition and reproductive success. *Behavioral Ecology and Sociobiology* 55:272–77.

Schulte-Hostedde, A. I., J. S. Millar, and H. L. Gibbs. 2002. Female-biased sexual size dimorphism in the yellow-pine chipmunk (*Tamias amoenus*): Sex specific patterns of annual reproductive success and survival. *Evolution* 56:2519–29.

Schulte-Hostedde, A. I., J. S. Millar, and G. J. Hickling. 2001. Sexual dimorphism in body composition of small mammals. *Canadian Journal of Zoology* 79:1016–20.

———. 2003. Intraspecific variation in testis size of small mammals: Implications for muscle mass. *Canadian Journal of Zoology* 81:591–95.

Schultz L. A., and R. K. Lore. 1993. Communal reproductive success in rats (*Rattus norvegicus*)—effects of group composition and prior social experience. *Journal of Comparative Psychology* 107:216–22.

Schultze-Westrum, T. G. 1969. Social communication by chemical signals in flying phalangers *Petaurus breviceps papuanus*. In *Olfaction and taste III*, ed. C. Pfaffman, 268–77. New York: Rockefeller University Press.

Schwagmeyer, P. L. 1979. The Bruce effect: An evaluation of male/female advantages. *The American Naturalist* 114:932–38.

———. 1980. Alarm calling behavior of the thirteen-lined ground squirrel, *Spermophilus tridecemlineatus*. *Behavioral Ecology and Sociobiology* 7:195–200.

———. 1984. Multiple mating and intersexual selection in thirteen-lined ground squirrels. In *The biology of ground-dwelling squirrels*, ed. J. O. Murie and G. R. Michener, 275–93. Lincoln: University of Nebraska Press.

———. 1986. Effects of multiple mating on reproduction in female thirteen-lined ground squirrels. *Animal Behaviour* 34:297–98.

———. 1988a. Ground squirrel kin recognition abilities: Are there social and life-history correlates? *Behavior Genetics* 18:495–510.

———. 1988b. Scramble-competition polygyny in an asocial mammal: Male mobility and mating success. *American Naturalist* 131:885–92.

———. 1990. Ground squirrel reproductive behavior and mating competition: A comparative perspective. In *Contemporary issues in comparative psychology*, ed. D. A. Dewsbury, 175–96. Sunderland, MA: Sinauer.

———. 1994. Competitive mate searching in thirteen-lined ground squirrels (Mammalia, Sciuridac): Potential roles of spatial memory. *Ethology* 98:265–76.

Schwagmeyer, P. L., and C. H. Brown. 1983. Factors affecting male-male competition in thirteen-lined ground squirrels. *Behavioral Ecology and Sociobiology* 13:1–6.

Schwagmeyer, P. L., and D. W. Foltz. 1990. Factors affecting the outcome of sperm competition in thirteen-lined ground squirrels. *Animal Behaviour* 39:156–62.

Schwagmeyer, P. L., and G. A. Parker. 1987. Queuing for mates in thirteen-lined ground squirrels. *Animal Behaviour* 35:1015–25.

———. 1990. Male mate choice as predicted by sperm competition in thirteen-lined ground squirrels. *Nature* 348:62–64.

Schwagmeyer, P. L., G. A. Parker, and D. W. Mock. 1998. Information asymmetries among males: Implications for fertilization success in the thirteen-lined ground squirrel. *Proceedings of the Royal Society of London, Series B* 265:1861–65.

Schwagmeyer, P. L., and S. J. Wootner. 1985. Mating competition in an asocial ground squirrel, *Spermophilus tridemlineatus*. *Behavioral Ecology and Sociobiology* 17:291–96.

———. 1986. Scramble competition polygyny in thirteen-lined ground squirrels: The relative contributions of overt conflict and competitive mate searching. *Behavioral Ecology and Sociobiology* 19:359–64.

Schwartz, O. A., and K. B. Armitage. 1980. Genetic variation in social mammals: The marmot model. *Science* 207:665–67.

Schwartz, O. A., K. B. Armitage, and D. Van Vuren. 1998. A 32-year demography of yellow-bellied marmots (*Marmota flaviventris*). *Journal of Zoology, London* 246:337–46.

Schwartz-Weig, E., and N. Sacher. 1996. Social behaviour, mating system and testes size in cuis (*Galea musteloides*). *Zeitschrift für Saugetierkunde* 61:25–38.

Scott, J. L. 1970. Maternal and pup behaviour in the golden hamster. PhD diss., Cambridge, UK: University of Cambridge.

Scott, J. W., and D. W. Pfaff. 1970. Behavioral and electrophysiological responses of female mice to male urine odors. *Physiology and Behaviour* 5:407–11.

Seal, U. S., E. T. Thorne, M. A. Bogan, and S. H. Anderson, eds. 1989. *Conservation biology and the black-footed ferret*. New Haven, CT: Yale University Press.

Seddon, J. M., and P. R. Baverstock. 1999. Variation on islands: Major histocompatibility complex (Mhc) polymorphism in populations of the Australian bush rat. *Molecular Ecology* 8:2071–79.

Seery, D. B., D. E. Biggins, J. A. Montenieri, R. E. Enscore, D. T. Tanda, and K. L. Gage. 2003. Treatment of black-tailed prairie dog burrows with Deltamethrin to control fleas (Insecta: Siphonaptera) and plague. *Journal of Medical Entomology* 40:718–22.

Seger, J. and W. D. Hamilton. 1988. Parasites and sex. In *The evolution of sex: An examination of current ideas*, ed. R. E. Michod and B. R. Levins, 176–93. Sunderland, MA: Sinauer.

Seitz, P. F. D. 1958. The maternal instinct in animal subjects: I. *Psychomatic Medicine* 20:215–26.

Selander, R. K. 1970. Behavior and genetic variation in natural populations. *American Zoologist* 10:53–66.

Seller, M. J., and K. J. Perkins-Cole. 1987. Sex differences in mouse embryonic development at neurulation. *Journal of Reproductive Fertility* 79:159–61.

Selye, H. 1937. The significance of the adrenals for adaptation. *Science* 85:247–48.

Semb-Johansson, A., R. Wiger, and C. E. Engh. 1979. Dynamics of freely growing, confined populations of the Norwegian lemming *Lemmus lemmus*. *Oikos* 33:246–60.

Sera, W. E., and M. S. Gaines. 1994. The effect of relatedness on spacing behavior and fitness of female prairie voles. *Ecology* 75:1560–66.

Shannon, C. E., and W. Weaver. 1949. *The mathematical theory of communication.* Urbana, IL: University of Illinois Press.

Shapiro, L. E., D. Austin, S. Ward, and D. A. Dewsbury. 1986. Familiarity and female mate choice in two species of voles (*Microtus ochrogaster* and *Microtus montanus*). *Animal Behaviour* 34:90–97.

Shapiro, L. E., and D. A. Dewsbury. 1986. Male dominance, female choice and male copulatory behavior in two species of voles (*Microtus ochrogaster* and *Microtus montanus*). *Behavioral Ecology and Sociobiology* 18:267–74.

Shargal, E., N. Kronfeld-Schor, and T. Dayan. 2000. Population biology and spatial relationships of coexisting spiny mice (*Acomys*) in Israel. *Journal of Mammalogy* 81:1046–52.

Sharpe, F., and F. Rosell. 2003. Time budgets and sex differences in the Eurasian beaver. *Animal Behaviour* 66:1059–67.

Sharpe, S. T., and J. S. Millar. 1980. Relocation of nest sites by female deer mice, *Peromyscus maniculatus* borealis. *Canadian Journal of Zoology* 68:2364–67.

Shaw, K. 2001. *Royal Babylon: The alarming history of European royalty.* New York: Broadway Books.

Shaw, W. T. 1936. Moisture and its relation to the cone-storing habit of the western pine squirrel. *Journal of Mammalogy* 17:337–49.

Sheets, R. G., R. L. Linder, and R. B. Dahlgren. 1971. Burrow systems of prairie dogs in South Dakota. *Journal of Mammalogy* 52:451–53.

Shellhammer, H. S. 1998. *Reithrodontomys raviventris* Dixon 1908, saltmarsh harvest mouse. In *North American rodents, status survey and conservation action plan*, ed. D. J. Hafner, E. Yensen, and G. L. Kirkland, Jr., 114–16. Gland, Switzerland: Rodent Specialist Group, Species Survival Commission, IUCN.

Shenbrot, G. I., B. R. Krasnov, and I. S. Khokhlova. 1999. Notes on the biology of the bushy-tailed jird, *Sekeetamys calurus*, in the Central Negev, Israel. *Mammalia* 63:374–77.

Shenbrot, G. I., B. R. Krasnov, and K. A. Rogovin. 1999. *Spatial ecology of desert rodent communities.* Berlin: Springer-Verlag.

Shenbrot. G. I., V. E. Sokolov, V. G. Heptner, and Yu. M. Kovalskaya. 1995. *Mammals of Russia and adjacent regions. Dipodid rodents.* [In Russian.] Moscow: Nauka.

Sheppard, D. H., and S. M. Yoshida. 1971. Social behavior in captive Richardson's ground squirrels. *Journal of Mammalogy* 52:793–99.

Sheppe, W. 1972. The annual cycle of small mammal populations on a Zambian floodplain. *Journal of Mammalogy* 53:445–60.

Sheridan, M., and R. H. Tamarin. 1986. Kinships in a natural meadow vole population. *Behavioral Ecology and Sociobiology* 19:207–11.

———. 1988. Space use, longevity, and reproductive success in meadow voles. *Behavioral Ecology and Sociobiology* 22:85–90.

Sherman, P. W. 1976. Natural selection among some group-living organisms. PhD diss., University of Michigan, Ann Arbor.

———. 1977. Nepotism and the evolution of alarm calls. *Science* 197:1246–53.

———. 1980a. The limits of ground squirrel nepotism. In *Sociobiology: Beyond nature/nurture?* ed. G. W. Barlow and J. Silverberg, 505–44. Boulder, CO: Westview.

———. 1980b. The meaning of nepotism. *American Naturalist* 116:604–6.

———. 1981a. Kinship, demography, and Belding's ground squirrel nepotism. *Behavioral Ecology and Sociobiology* 8: 251–59.

———. 1981b. Reproductive competition and infanticide in Belding's ground squirrels and other animals. In *Natural selection and social behavior: Recent research and new theory*, ed. R. D. Alexander and R. W. Tinkle, 311–31. New York: Chiron.

———. 1985. Alarm calls of Belding's ground squirrels to aerial predators: Nepotism or self-preservation? *Behavioral Ecology and Sociobiology* 17:313–23.

———. 1988. The levels of analysis. *Animal Behaviour* 36:616–18.

———. 1989. Mate guarding as paternity insurance in Idaho ground squirrels. *Nature* 338:418–20.

Sherman, P. W., S. Braude, and J. U. M. Jarvis. 1999. Litter sizes and mammary numbers of naked mole-rats: Breaking the one-half rule. *Journal of Mammalogy* 80:720–33.

Sherman, P. W., and W. G. Holmes. 1985. Kin recognition: Issues and evidence. In *Experimental behavioral ecology and sociobiology*, ed. B. Holldobler and M. Lindauer, 437–60. Stuttgart: G. Fischer Verlag.

Sherman, P. W., and J. U. M. Jarvis. 2002. Extraordinary life spans of naked mole-rats (*Heterocephalus glaber*). *Journal of Zoology* 258:307–11.

Sherman, P. W., J. U. M. Jarvis, and R. D. Alexander, eds. 1991. *The biology of the naked mole-rat.* Princeton, NJ: Princeton University Press.

Sherman, P. W., E. A. Lacey, H. K. Reeve, and L. Keller. 1995. The eusociality continuum. *Behavioral Ecology* 6:102–8.

Sherman, P. W., and M. L. Morton. 1979. Four months of the ground squirrel. *Natural History* 88:50–57.

———. 1984. Demography of Belding's ground squirrels. *Ecology* 65:1617–28.

Sherman, P. W., and B. D. Neff. 2004. Behavioural ecology: Father knows best. *Nature* 425:136–37.

Sherman, P. W., H. K. Reeve, and D. W. Pfennig. 1997. Recognition systems. In *Behavioural ecology: An evolutionary approach*, ed. J. R. Krebs and N. B. Davies, 69–96. Oxford, UK: Blackwell Science.

Sherman, P. W., and M. C. Runge. 2002. Demography of a population collapse: The northern Idaho ground squirrel (*Spermophilus brunneus brunneus*). *Ecology* 83:2816–31.

Shettleworth, S. 1998. *Cognition, evolution, and behavior.* New York: Oxford University Press.

Shi, H., and T. J. Bartness. 2000. Catecholaminergic enzymes, vasopressin and oxytocin distribution in Siberian hamster brain. *Brain Research Bulletin* 53:833–43.

Shields, W. M. 1980. Ground squirrel alarm calls: Nepotism or parental care? *American Naturalist* 116:599–603.

———. 1982. *Philopatry, inbreeding, and the evolution of sex.* Albany: State University of New York Press.

Shier, D. M. 2003. Sociality and communication in Heerman's kangaroo rats (*Dipodomys heermanni*). Master's thesis, San Francisco State University.

———. 2006. Translocations are more successful when prairie dogs are moved as families. In *Conservation of the black-tailed prairie dog,* ed. J. L. Hoogland, 189–90. Washington, DC: Island Press.

Shier, D. M., and J. A. Randall. 2004. Spacing as a predictor of social organization in kangaroo rats (*Dipodomys heermanni arenae*). *Journal of Mammalogy* 85:1002–8.

Shier, D. M., and S. I. Yoerg, S. I. 1999. What footdrumming signals in kangaroo rats (*Dipodomys heermanni*). *Journal of Comparative Psychology* 113:66–73.

Shilova, S. A., H. V. Derviz, A. I. Shilov, N. A. Shipanov, I. P. Marova, and D. V. Pozjarsky. 1983. Some features of territorial distribution and behavior of noon gerbils (*Meriones meridianus,* Rodentia, Cricetidae). [In Russian.] *Zoologicheskii Zhurnal* 62:916–20.

Shilton, C. M., and R. J. Brooks. 1989. Paternal care in captive collared lemmings (*Dicrostonyx richardsoni*) and its effect on development of the offspring. *Canadian Journal of Zoology* 67:2740–45.

Shriner, W. M. 1998. Yellow-bellied marmot and golden-mantled ground squirrel responses to heterospecific alarm calls. *Animal Behaviour* 55:529–36.

———. 1999. Antipredator responses to a previously neutral sound by free-living adult golden-mantled ground squirrels, *Spermophilus lateralis* (Sciuridae). *Ethology* 105:747–57.

Shubin, V. I. 1991. Population structure and bobac reproduction in the northern part of Kazakh Melkosopotchnik (= low hill area). In *Population structure of the marmot,* ed. D. I. Bibikov, A. A. Nikolski, V. Yu. Rumiantsev, and T. A. Seredneva, 98–118. Moscow: USSR Theriological Society.

Shurtliff, Q. R., D. E. Pearse, and D. S. Rogers. 2005. Parentage analysis of the canyon mouse (*Peromyscus crinitus*): Genetic evidence for multiple paternity. *Journal of Mammalogy* 86:531–40.

Sicard, B., and M. Tranier. 1996. Caractères et répartition de trois phénotypes d'*Acomys* (Rodentia, Muridae) au Burkina Faso. *Mammalia* 60:53–68.

Sidorowicz, J. 1960. Influence of the weather on capture of Micromammalia I: Rodents (Rodentia). *Acta Theriologica* 9:139–58.

Sieber, J., F. Suchentrunk, and G. B. Hartl. 1999. A biochemical-genetic discrimination method for the two beaver species, *Castor fiber* and *Castor canadensis,* as a tool for conservation. In *Beaver protection, management, and utilization in Europe and North America,* ed. P. E. Busher and R.M. Dzieciolowski, 61–65. New York: Kluwer Academic/Plenum.

Sih, A., A. M. Bell, J. C. Johnson, and R. E. Ziemba. 2004. Behavioral syndromes: An integrative overview. *Quarterly Review of Biology* 79:241–77.

Sikes, R. S. 1995. Maternal response to resource limitation in eastern woodrats. *Animal Behaviour* 49:1551–58.

———. 1996a. Effects of maternal nutrition on post-weaning growth in two North American rodents. *Behavioral Ecology and Sociobiology* 38:303–10.

———. 1996b. Tactics of maternal investment of northern grasshopper mice in response to postnatal restriction of food. *Journal of Mammalogy* 77:1092–1101.

Silk, J. B. 1983. Local resource competition and facultative adjustment of sex ratios in relation to competitive abilities. *American Naturalist* 121:56–66.

Silva, M., and J. A. Downing. 1995. *CRC handbook of mammalian body masses.* Boca Raton, FL: CRC Press.

Silver, L. 1993. The peculiar journey of a selfish chromosome: Mouse *t* haplotypes and meiotic drive. *Trends in Genetics* 9:250–54.

Silver, L. M., and P. Olds-Clarke. 1984. Transmission ratio distortion of mouse *t* haplotypes is not a consequence of wild-type sperm degeneration. *Developmental Biology* 105:250–52.

Silver, L. M., and D. Remis. 1987. Five of the nine genetically defined regions of mouse *t* haplotypes are involved in transmission ratio distortion. *Genetical Research* 49:51–56.

Simberloff, D., and J. Cox. 1987. Consequences and costs of conservation corridors. *Conservation Biology* 1:63–71.

Simonetti, J. A., and G. Montenegro. 1981. Food preferences by *Octodon degus* (Rodentia Caviomorpha): Their role in the Chilean matorral composition. *Oecologia* 51:189–90.

Simpson, G. G. 1945. The principles of classification and a classification of mammals. *American Museum of Natural History Bulletin* 85:1–350.

———. 1950. History of the fauna of Latin America. *American Scientist* 38:361–89.

———. 1952. Probabilities of dispersal in geologic time. *Bulletin of the American Museum of Natural History* 99:163–76.

Simson, S., B. Lavie, and E. Nevo. 1993. Penial differentiation in speciation of subterranean mole rats *Spalax ehrenbergi* in Israel. *Journal of Zoology, London* 229:493–503.

Sinclair, A. R. E., D. Chitty, C. I. Stefan, and C. J. Krebs. 2003. Mammal population cycles: Evidence for intrinsic differences during snowshoe hare cycles. *Canadian Journal of Zoology* 81:216–20.

Singer, A. G., G. K. Beauchamp, and K. Yamazaki. 1997. Volatile signals of the major histocompatibility complex in male mouse urine. *Proceedings of the National Academy of Sciences (USA)* 94:2210–14.

Singleton, G. R. 1989. Population dynamics of an outbreak of house mice (*Mus domesticus*) in the mallee wheatlands of Australia—hypothesis of plague formation. *Journal of Zoology, London* 219:495–515.

Singleton, G. R., P. R. Brown, R. P. Pech, J. Jacob, G. J. Mutze, and C. J. Krebs. 2005. One hundred years of eruptions of house mice in Australia—a natural biological curio. *Biological Journal of the Linnean Society* 84:617–27.

Singleton, G. R., C. J. Krebs, S. Davis, L. Chambers, and P. Brown. 2001. Reproductive changes in fluctuating house mouse populations in Southeastern Australia. *Proceedings of the Royal Society of London, Series B* 268:1741–48.

Singleton, G. R., and T. D. Redhead. 1989. House mouse plagues. In *Mediterranean landscapes in Australia: Mallee ecosystems and their management,* ed. J. C. Noble and R. A. Bradstock, 418–33. Melbourne: CSIRO.

Singleton, G. R., A. L. Smith, and C. J. Krebs. 2000. The prevalence of viral antibodies during large fluctuations of house mouse populations in Australia. *Epidemiology and Infection* 125:719–27.

Singleton, G. R., A. L. Smith, G. R. Shellam, N. Fitzgerald, and W. J. Müller. 1993. Prevalence of viral antibodies and helminths in field populations of house mice (*Mus domesticus*) in southeastern Australia. *Epidemiology and Infection* 110: 399–417.

Singleton, I., G. L. Hinds, H. Leirs, and Z. Zhang, eds. 1999. *Ecologically-based rodent management.* Canberra: Australian Centre for International Agricultural Research.

Sivalingam, S. 2002. Neurogenesis in the arctic ground squirrel. MS. thesis. University of Toronto, Toronto.

Skinner, J. D., and H. N. Smithers. 1990. *The mammals of the Southern African subregion,* 2nd ed. Pretoria: University of Pretoria.

Slade, N. A. 1995. Failure to detect senescence in persistence of some grassland rodents. *Ecology* 76:863–70.

Slade, N. A., S. T. McMurray, and R. L. Lochmiller. 1996. Habitat differences in mass-specific litter sizes of hispid cotton rats. *Journal of Mammalogy* 77:346–50.

Slater, P. J. B. 1981. Individual differences in animal behaviour. *Perspectives in Ethology* 4:35–39.

Slatkin, M. 1987. Gene flow and the geographic structure of natural populations. *Science* 236:787–92.

Slattery, S. M., and R. T. Alisauskas. 1995. Egg characteristics and body reserves of neonate Ross' and lesser snow geese. *Condor* 97:970–84.

Slobodchikoff, C. N. 1984. Resources and the evolution of social behavior. In *A new ecology: Novel approaches to interactive systems,* ed. P. W. Price, C. N. Slobodchikoff, and W. S. Gaud, 227–51. New York: John Wiley.

Slobodchikoff, C. N., and R. Coast. 1980. Dialects in the alarm calls of prairie dogs. *Behavioral Ecology and Sociobiology* 7:49–53.

Slobodchikoff, C. N., Fischer, C., and Shapiro, J. 1986. Predator-specific alarm calls of prairie dogs. *American Zoologist* 26:557.

Slobodchikoff, C. N., J. Kiriazis, C. Fischer, and E. Creef. 1991. Semantic information distinguishing individual predators in the alarm calls of Gunnison's prairie dogs. *Animal Behaviour* 42:713–19.

Slobodchikoff, C. N., and W. M. Shields. 1988. Ecological trade-offs and social behaviour. In *The ecology of social behaviour,* ed. C. N. Slobodchikoff, 3–10. New York: Academic Press.

Sludsky, A. A., A. B. Bekenov, B. A. Borisenko, Yu. A. Grachev, and M. I. Ismagilov. 1977. *Mammals of Kazakhstan vol. 1 rodents part 2.* [In Russian.] Alma-Ata, Kazakhstand: Nauka Kazakh SSR Press.

Sludsky, A. A., B. A. Borisenko, B. I. Kapitonov, S. Mahmutov., and N. Yu. Mokrousov. 1978. *Mammals of Kazakhstan. vol. 1 rodents part 3.* [In Russian.] Nauka Kazakh SSR Press, Alma-Ata.

Smale, L., S. Nunes, and K. E. Holekamp. 1997. Sexually dimorphic dispersal in mammals: Patterns, causes, and consequences. *Advances in the Study of Behavior* 26:181–250.

Smale, L., J. M. Pedersen, M. L. Block, and I. Zucker. 1990. Investigation of conspecific male odours by female prairie voles. *Animal Behavior* 39:768–74.

Smit, C. J., and A. Van Wijngaarden. 1981. *Threatened mammals in Europe.* Wiesbaden: Akademische Verlagsgesellschaft.

Smith, A. L., G. R. Singleton, G. M. Hansen, and G. Shellam. 1993. A serologic survey for viruses and *Mycoplasma pulmonis* among wild house mice (*Mus domesticus*) in southeastern Australia. *Journal of Wildlife Diseases* 29:219–29.

Smith, A. T. 1980. Temporal changes in insular populations of the pika (*Ochotona princeps*). *Ecology* 61:8–13.

———. 1993. The natural history of inbreeding and outbreeding in small mammals. In *The natural history of inbreeding and outbreeding: Theoretical and empirical perspectives,* ed. N. W. Thornhill, 329–51. Chicago: University of Chicago Press.

———. 2001. Securing a vacancy: The social organization of the North American pika. In *The encyclopedia of mammals,* ed. D. W. Macdonald, 714–15. New York: Facts on File.

Smith, A. T., and X. G. Wang. 1991. Social relationships of adult black-lipped pikas (*Ochotona curzoniae*). *Journal of Mammalogy* 72:231–47.

Smith, A. T., and M. L. Weston. 1990. *Ochotona princeps. Mammalian Species* 352:1–8.

Smith, C. C. 1968. The adaptive nature of social organization in the genus of tree squirrel *Tamiasciurus. Ecological Monographs* 38:30–63.

———. 1978. Structure and function of the vocalizations of tree squirrels (*Tamiasciurus*). *Journal of Mammalogy* 59:793–808.

Smith, D. W. 1997. Dispersal strategies and cooperative breeding in beavers. PhD diss., University of Nevada, Reno.

Smith, E. E., and D. L. Medin. 1981. *Categories and concepts.* Cambridge, MA: Harvard University Press.

Smith, F. A. 1997. *Neotoma cinerea. Mammalian Species* 564:1–8.

Smith, F. A., H. Browning, and U. L. Shepherd. 1998. The influence of climate change on the body mass of woodrats *Neotoma* in an arid region of New Mexico, USA. *Ecography* 21:140–48.

Smith, G. W., and D. R. Johnson. 1985. Demography of a Townsend ground squirrel population in southwestern Idaho. *Ecology* 66:171–78.

Smith, J. C. 1981. Senses and communication. *Symposium of the Zoological Society of London* 47:367–93.

Smith, K., S. C. Alberts, and J. Altmann. 2003. Wild female baboons bias their social behaviour towards paternal half-sisters. *Proceedings of the Royal Society of London, Series B* 270:503–10.

Smith, M. F., and J. L. Patton. 1999. Phylogenetic relationships and the radiation of sigmodontine rodents in South America: Evidence from cytochrome *b. Journal of Mammalian Evolution* 6:89–128.

Smith, M. H. 1966. Mating behavior in the old-field mouse. *American Zoologist* 6:535.

Smith, N. B., and F. S. Barkalow, Jr. 1967. Precocious breeding in the gray squirrel. *Journal of Mammalogy* 48:328–29.

Smith, P., M. Berdoy, and R. H. Smith. 1994. Body-weight and social dominance in anticoagulant-resistant rats. *Crop Protection* 13:311–16.

Smith, P., M. Berdoy, R. H. Smith, and D. W. Macdonald. 1993. A new aspect of resistance in wild rats: Benefits in the absence of poison. *Functional Ecology* 7:190–94.

Smith, R. E. 1967. Natural history of the prairie dog in Kansas. *Miscellaneous Publications of the Museum of Natural History, University of Kansas* 49:1–39.

Smith, R. J. 1999. Statistics of sexual size dimorphism. *Journal of Human Evolution* 36:423–59.

Smith, S. F. 1978. Alarm calls, their origin and use in *Eutamias sonomae*. *Journal of Mammalogy* 59:888–93.

Smith, W. J., S. L. Smith, J. G. DeVilla, and E. C. Oppenheimer. 1976. The jump-yip display of the black-tailed prairie dog, *Cynomys ludovicianus*. *Animal Behaviour* 24:609–21.

Smith, W. J., S. L. Smith, E. C. Oppenheimer, and J. G. Devilla. 1977. Vocalizations of the black-tailed prairie dog, *Cynomys ludovicianus*. *Animal Behaviour* 25:152–64.

Smith, W. W. 1958. Melanistic *Rattus norvegicus* in southwestern Georgia. *Journal of Mammalogy* 39:304–6.

Smolen, M. J. 1981. *Microtus pinetorum*. *Mammalian Species* 147:1–7.

Smolen, M. J., and B. L. Keller. 1987. *Microtus longicaudus*. *Mammalian Species* 271:1–7.

Smotherman, W. P. 1982. Odor aversion learning by the rat fetus. *Physiology and Behavior* 29:769–71.

Smythe, N. 1978. The natural history of the Central American agouti (*Dasyprocta punctata*). *Smithsonian Contributions to Zoology* 257:1–52.

Snyder, R. L., and J. J. Christian. 1960. Reproductive cycle and litter size of the woodchuck. *Ecology* 41:647–56.

Soares, M. M. 2004. Social structure in two species of communally nesting rodents. PhD diss., University of California, Berkeley.

Soede, N. M., A. K. Nissen, and B. Kemp. 2000. Timing of insemination relative to ovulation in pigs: Effects on sex ratio of offspring. *Theriogenology* 53:1003–11.

Sokolov, V. E., V. S. Lobachev, and B. N. Orlov. 1996. *Mammals of Mongolia. Jerboas: Euchoreutinae, Cardiocraniinae, Dipodinae.* [In Russian.] Moscow: Nauka.

———. 1998. *Mammals of Mongolia. Jerboas: Dipodinae, Allactaginae.* [In Russian.] Moscow: Nauka.

Solomon, N. G. 1991. Current indirect fitness benefits associated with philopatry in juvenile prairie voles. *Behavioral Ecology and Sociobiology* 29:277–82.

———. 1993a. Body size and social preferences of male and female prairie voles, *Microtus ochrogaster*. *Animal Behaviour* 45:1031–33.

———. 1993b. Comparison of parental behavior in male and female prairie voles (*Microtus ochrogaster*). *Canadian Journal of Zoology* 71:434–37.

———. 1994. Effect of the pre-weaning environment on subsequent reproduction in prairie voles, *Microtus ochrogaster*. *Animal Behaviour* 48:331–41.

———. 2003. A reexamination of factors influencing philopatry in rodents. *Journal of Mammalogy* 84:1182–97.

Solomon, N. G., and J. A. French. 1997. The study of mammalian cooperative breeding. In *Cooperative breeding in mammals,* ed. N. G. Solomon and J. A. French, 1–10. Cambridge: Cambridge University Press.

Solomon, N. G., and L. L. Getz. 1997. Examination of alternative hypotheses for cooperative breeding in rodents. In *Cooperative breeding in mammals,* ed. N. G. Solomon and J. A. French, 199–230. Cambridge: Cambridge University Press.

Solomon, N. G., and L. D. Hayes. In press. Biological aspects of alloparenting. In *Alloparenting in human societies,* ed. G. Bentley and R. Mace. Oxford: Berghahn Books.

Solomon, N. G., and J. J. Jacquot. 2002. Characteristics of resident and wandering prairie voles, *Microtus ochrogaster*. *Canadian Journal of Zoology* 80:951–55.

Solomon, N. G., B. Keane, L. R. Knoch, and P. J. Hogan. 2004. Multiple paternity in socially monogamous prairie voles (*Microtus ochrogaster*). *Canadian Journal of Zoology* 82:1667–71.

Solomon, N. G., J. G. Vandenbergh, and W. T. Sullivan, Jr. 1998. Social influences on intergroup transfer by pine voles (*Microtus pinetorium*). *Canadian Journal of Zoology* 76:2131–36.

Solomon, N. G., J. G. Vandenbergh, K. Wekesa, and L. Barghusen. 1996. Chemical cues are necessary but insufficient for reproductive activation of female pine voles (*Microtus pinetorum*). *Biology of Reproduction* 54:1038–45.

Solomon, N. G., C. S. Yaeger, and L. A. Beeler. 2002. Social transmission and memory of food preferences in pine voles (*Microtus pinetorum*). *Journal of Comparative Psychology* 116:35–38.

Sommer, S. 2000. Sex-specific predation on a monogamous rat, *Hypogeomys antimena* (Muridae: Nesomyinae). *Animal Behaviour* 59:1087–94.

———. 2003. Effects of habitat fragmentation and changes of dispersal behaviour after a recent population decline on the genetic variability of noncoding and coding DNA of a monogamous Malagasy rodent. *Molecular Ecology* 12:2845–51.

Sommer, S., D. Schwab, and J. U. Ganzhorn. 2002. MHC diversity of endemic Malagasy rodents in relation to geographic range and social system. *Behavioral Ecology and Sociobiology* 51:214–21.

Sommer, S., and H. Tichy. 1999. Major histocompatibility complex (MHC) class II polymorphism and paternity in the monogamous *Hypogeomys antimena*, the endangered, largest endemic Malagasy rodent. *Molecular Ecology* 8:1259–72.

Sommer, V. 1987. Infanticide among free-ranging langurs (*Presbytis entellus*) at Jodhpur (Rajasthan/India): Recent observations and a reconsideration of hypotheses. *Primates* 28:163–97.

———. 1994. Infanticide among the langurs of Jodhpur: Testing the sexual selection hypothesis with a long-term record. In *Infanticide and parental care,* ed. S. Parmigiani and F. S. vom Saal, 155–98. Chur, Switzerland: Harwood.

Soroker, V., and J. Terkel. 1988. Changes in incidence of infanticidal and parental responses during the reproductive cycle in male and female wild mice *Mus musculus*. *Animal Behaviour* 36:1275–81.

Soto-Gamboa, M. 2004. Formación y estabilidad de estructuras sociales en micromamíferos, su regulación hormonal y la importancia de las interacciones entre machos. PhD diss., P. Universidad Catolica de Chile, Santiago, Chile.

Soulé, M. E., J. A. Estes, J. Berger, and C. M. del Río. 2003. Ecological effectiveness: Conservation goals for interactive species. *Conservation Biology* 17:1238–50.

Southern, H. N. 1954. *Control of rats and mice,* vol. 3. Oxford: Clarendon.

———. 1964. *The handbook of British mammals.* London: Blackwell Scientific.

Southwick, C. H. 1955. Regulatory mechanisms of house mouse populations: Social behavior affecting litter survival. *Ecology* 36:627–34.

Southwood, T. R. E. 1977. Habitat, the templet for ecological strategies? *Journal of Animal Ecology* 46:337–65.

Sözen, M., and E. Kivané. 1998. Two new karyotypic forms of

Spalax leucodon (Nordmann, 1840) (Mammalia: Rodentia) from Turkey. *Z. Saugetierkunde.* 63:307–10.

Speakman, J. R. 1997. *Doubly-labelled water: Theory and practice.* Dortrecht, The Netherlands: Kluwer Academic.

———. 2000. The cost of living: Field metabolic rates of in small mammals. *Advances in Ecological Research* 30:177–297.

Spencer, S. R., and G. Cameron. 1982. *Reithrodontomys fulvescens. Mammalian Species* 174:1–7.

Spinks, A. C. 1998. Sociality in the common mole–rat, *Cryptomys hottentotus hottentotus,* Lesson 1826: The effects of aridity. PhD diss., University of Cape Town, South Africa.

Spinks, A. C., N. C. Bennett, and J. U. M. Jarvis. 1999. Regulation of reproduction in female common mole–rats, *Cryptomys hottentotus hottentotus*: The effects of breeding season and reproductive status. *Journal of Zoology* 248:161–68.

———. 2000. A comparison of the ecology of two populations of the common mole-rat, *Cryptomys hottentotus hottentotus*: The effect of aridity on food, foraging and body mass. *Oecologia* 125:341–49.

Spinks, A. C., J. U. M. Jarvis, and N. C. Bennett. 2000. Comparative patterns of philopatry and dispersal in two common mole-rat populations: Implications for the evolution of mole-rat sociality. *Journal of Animal Ecology* 69:224–34.

Spinks, A. C., M. J. O'Riain, and D. A. Polakow. 1998. Intercolonial encounters and xenophobia in the common mole–rat, of the common mole–rat, *Cryptomys hottentotus hottentotus* (Bathyergidae): The effects of aridity, sex and reproductive status. *Behavioral Ecology* 9:354–59.

Spinks, A. C., G. van der Horst, and N. C. Bennett. 1997. Influence of breeding season and reproductive status on male reproductive characteristics in the common mole–rat, *Cryptomys hottentotus hottentotus. Journal of Reproduction and Fertility* 109:78–86.

Spotorno, A. E., C. A. Zuleta, J. P. Valladares, A. L. Deane, and J. E. Jimenez. 2004. *Chinchilla laniger. Mammalian Species* 758:1–9.

Springer, M. S. 1997. Molecular clocks and the timing of the placental and marsupial radiations in relation to the Cretaceous-Tertiary boundary. *Journal of Mammalian Evolution* 4:285–302.

Springer, M. S., W. J. Murphy, E. Eizirik, and S. J. O'Brien. 2003. Placental mammal diversification and the Cretaceous-Tertiary boundary. *Proceedings of the National Academy of Science (USA)* 100:1056–61.

Spritzer, M. D., D. B. Meikle, and N. G. Solomon. 2005. Female choice based on male spatial ability and aggressiveness among meadow voles. *Animal Behaviour* 69:1121–30.

Spruijt, B. M., J. A. R. A. M. Van Hooff, and W. H. Gispen. 1992. Ethology and neurobiology of grooming behavior. *Physiological Reviews* 72:825–52.

Stacey, P. B., and J. D. Ligon. 1987. Territory quality and dispersal options in the acorn woodpecker, and a challenge to the habitat-saturation model of cooperative breeding. *The American Naturalist* 130:654–76.

———. 1991. The benefits-of-philopatry hypothesis for the evolution of cooperative breeding: Variation in territory quality and group size effects. *The American Naturalist* 137:831–46.

Stalling, D. 1997. *Reithrodontomys humulis. Mammalian Species* 565:1–6.

Stallman, E. L., and W. G. Holmes. 2002. Selective foraging and food distribution of high-elevation yellow-bellied marmots (*Marmota flaviventris*). *Journal of Mammalogy* 83:576–84.

Stamps, J. A. 1995. Motor learning and the value of familiar space. *American Naturalist* 146:41–58.

———. 2003. Behavioural processes affecting development: Tinbergen's fourth question comes of age. *Animal Behaviour* 66:1–13.

Stangl, F. B., Jr., and R. J. Baker. 1984. Evolutionary relationships in *Peromyscus*: Congruence in chromosomal, genic, and classical data sets. *Journal of Mammalogy* 65:643–54.

Stanhope, M. J., V. G. Waddell, O. Madsen, W. W. de Jong, S. B. Hedges, G. C. Cleven, D. Kao, and M. S. Springer. 1998. Molecular evidence for multiple origins of Insectivora and for a new order of endemic African insectivore mammals. *Proceedings of the National Academy of Sciences (USA)* 95: 9967–72.

Stankowich, T., and P. W. Sherman. 2002. Pup shoving by adult naked mole-rats. *Ethology* 108:975–92.

Stapp, P. 1999. Size and habitat characteristics of home ranges of the Northern Grasshopper Mouse (*Onychomys leucogaster*). *Southwestern Naturalist* 44:101–5.

Stapp, P., P. J. Pekins, and W. W. Mautz. 1991. Winter energy expenditure and the distribution of southern flying squirrels. *Canadian Journal of Zoology* 69:2548–55.

Starkey, N. J., and C. A. Hendrie. 1998. Disruption of pairs produces pair-bond disruption in male but not female Mongolian gerbils. *Physiology and Behavior* 65:497–503.

Steadman, D. W. 1995. Prehistoric extinctions of Pacific island birds: Biodiversity meets zooarcheology. *Science* 267: 1123–31.

Stearns, S. C. 1976. Life history tactics: A review of the ideas. *Quarterly Review of Biology* 51:3–47.

———. 1983. The influence of size and phylogeny on patterns of covariation among life history traits in mammals. *Oikos* 41: 173–87.

———. 1992. *The evolution of life histories.* New York: Oxford University Press.

Steel, E. 1984. Effect of the odour of vaginal secretion on non-copulatory behaviour of male hamsters (*Mesocricetus auratus*). *Animal Behaviour* 32:597–608.

Steele, M. A. 1998. *Tamiasciurus hudsonicus. Mammalian Species* 586:1–9.

Steele, M. A., and J. L. Koprowski. 2001. *North American tree squirrels.* Washington, DC: Smithsonian Institution Press.

Steele, R., S. F. Arno, and K. Geier-Hayes. 1986. Wildfire patterns change in Idaho's ponderosa pine–Douglas-fir forest. *Western Journal of Applied Forestry* 1:16–18.

Steen, H., A. Mysterud, and G. Austrheim. 2005. Sheep grazing and rodent populations: Evidence of negative interactions from a landscape scale experiment. *Oecologia* 143:357–64.

Stehn, R. A., and F. J. Jannett, Jr. 1981. Male-induced abortion in various microtine rodents. *Journal of Mammalogy* 62: 369–72.

Stehn, R. A., and M. Richmond. 1975. Male-induced pregnancy termination in the prairie vole, *Microtus ochrogaster. Science* 75:1211–13.

Stein, B. R. 2000. Morphology of subterranean rodents. In *Life underground: The biology of subterranean rodents,* ed. E. A. Lacey, J. L. Patton, and G. N. Cameron, 19–61. Chicago: University of Chicago Press.

Steinberg, E. K., and J. L. Patton. 2000. Genetic structure and the geography of speciation in subterranean rodents: Opportunities and constraints for evolutionary diversification. In *Life underground: The biology of subterranean rodents,* ed. E. A. Lacey, J. L. Patton, and G. N. Cameron, 301–31. Chicago: University of Chicago Press.

Steiner, A. L. 1970. Etude descriptive de quelques activites et comportements de base de *Spermophilus columbianus columbianus*. *Revue Comparative de l'Animal* 4:23–42.

———. 1972. Mortality resulting from intraspecific fighting in some ground squirrel populations. *Journal of Mammalogy* 53:601–3.

Steiniger, F. 1950. Beitrage zur Soziologie und sonstigen Biologie der Wanderratte. *Zeitschrift fur Tierpsychologie* 7:356–79.

Steinmann, A., J. Priotto, M. Provensal, and J. Polop. 1997. Odor incidence in the capture of wild rodents in Argentina. *Mastozoologia Neotropical* 4:17–24.

Stenseth, N. C. 1999. Population cycles in voles and lemmings: Density dependence and phase dependence in a stochastic world. *Oikos* 87:427–61.

Stenseth, N. C., H. Leirs, A. Skonhoft, S. A. Davis, R. P. Pech, H. P. Andreassen, G. R. Singleton, et al. 2003. Mice, rats, and people: The bio-economics of agricultural rodent pests. *Frontiers in Ecology and the Environment* 1:367–75.

Stenseth, N. C., and W. Z. Lidicker, Jr., eds. 1992. *Animal dispersal: Small mammals as a model.* London: Chapman Hall.

———. 1992. Presaturation and saturation dispersal fifteen years later: Some theoretical considerations. In *Animal dispersal: Small mammals as a model,* ed. N. C. Stenseth and W. Z. Lidicker, Jr., 201–23. London: Chapman and Hall.

Steppan, S. J., M. R. Akhverdyan, E. A. Lyapunova, D. G. Fraser, N. N. Vorontsov, R. S. Hoffmann, and M. J. Braun. 1999. Molecular phylogeny of the marmots (Rodentia:Sciuridae): Tests of evolutionary and biogeographic hypotheses. *Systematic Biology* 48:715–34.

Steppan, S. J., B. L. Storz, and R. S. Hoffmann. 2004. Nuclear DNA phylogeny of the squirrels (Mammalia: Rodentia) and the evolution of arboreality from c-myc and RAG1. *Molecular Phylogenetics and Evolution* 30:703–19.

Stern, J. J., and G. Broner. 1970. Effects of litter size on nursing time and weight of the young in guinea pigs. *Psychonomic Science* 21:171–72.

Stern, J. M. 1996. Somatosensation and maternal care in Norway rats. In *Parental care: Evolution, mechanisms, and adaptive significance. Advances in the Study of Behavior,* vol. 25, ed. J. S. Rosenblatt and C. T. Snowdon, 243–94. New York: Academic Press.

Stetter, K. R., L. I. McCann, M. A. Leafgren, and M. T. Segar. 1995. Diet preference in rats (*Rattus norvegicus*) as a function of odor exposure, odor concentration and conspecific presence. *Journal of Comparative Psychology* 109:384–89.

Stevens, S. D. 1998. High incidence of infanticide by lactating females in a population of Columbian ground squirrels (*Spermophilus columbianus*). *Canadian Journal of Zoology* 76:1183–87.

Stockley, P. 1997. Sexual conflict resulting from adaptations to sperm competition. *Trends in Ecology and Evolution* 12:154–59.

———. 2003. Female multiple mating behaviour, early reproductive failure and litter size variation in mammals. *Proceedings of the Royal Society of London, Series B* 270:271–78.

Stockley, P., and B. T. Preston. 2004. Sperm competition and diversity in rodent copulatory behaviour. *Journal of Evolutionary Biology* 17:1048–57.

Stoddard, D. M. 1979. *Ecology of small mammals.* London: Chapman and Hall.

Stoddard, D. M., and A. J. Bradley. 1991. Measurement of short-term changes in heart rate and in plasma concentrations of cortisol and catecholamine in a small marsupial. *Journal of Chemical Ecology* 17:1333–42.

Stoddard, P. K. 1996. Local recognition of neighbors by territorial passerines. In *Ecology and evolution of acoustic communication in birds,* ed. D. E. Kroodsma and H. E. Miller, 356–76. Ithaca, NY: Comstock.

Storch, G., and H. P. Uerpmann. 1976. Die Kleinsaugerknochen vom Castro do Zambujal. *Studien uber fruhe Tiernochenfunde um der Iberischen Halbinsel* 5:130–38.

Storey, A. E. 1986a. Adaptive significance of male-induced pregnancy disruptions in rodents. *Journal of Mammalogy* 62:369–72.

———. 1986b. Influence of sires on male-induced pregnancy disruptions in meadow voles (*Microtus pennsylvanicus*) differs with stage of pregnancy. *Journal of Comparative Psychology* 100:15–20.

———. 1990. Pregnancy disruption by unfamilar males in meadow voles: A comparison of chemical and behavioral cues. In *Chemical signals in vertebrates, 5,* ed. D. W. Macdonald, D. Muller-Schwarze, and S. E. Natynczuk, 199–208. Oxford: Oxford University Press.

———. 1994. Pre-implantation pregnancy disruption in female meadow voles, *Microtus pennsylvanicus* (Rodentia: Muridae): Male competition or female choice. *Ethology* 98:89–100.

Storey, A. E., C. G. Bradbury, and T. L. Joyce. 1994. Nest attendance in male meadow voles: The role of the female in regulating male interactions with pups. *Animal Behaviour* 47:1037–46.

Storey, A. E., and D. T. Snow. 1987. Male identity and enclosure size affect paternal attendance of meadow voles, *Microtus pennsylvanicus. Animal Behaviour* 35:411–19.

———. 1990. Post-implantation pregnancy disruptions in meadow voles: Relationship to variation in male sexual and aggressive behaviour. *Physiology and Behaviour* 47:19–25.

Storz, J. F. 1999. Genetic consequences of mammalian population structure. *Journal of Mammalogy* 80:553–69.

Stowers, L., T. E. Holy, M. Meister, C. Dulac, and G. Koentges. 2002. Loss of sex discrimination and male-male aggression in mice deficient for TRP2. *Science* 295:1493–1500.

Streilein, K. E. 1982. Ecology of small mammals in the semiarid Brazilian Caatinga. I. Climate and faunal composition. *Annals of Carnegie Museum* 51:79–107.

Stribley, J. M., and C. S. Carter. 1999. Developmental exposure to vasopressin increases aggression in adult prairie voles. *Proceedings of the National Academy of Science (USA)* 96:12601–4.

Stroud, D. C. 1982. Population dynamics of *Rattus rattus* and *Rattus norvegicus* in riparian habitat. *Journal of Mammalogy* 63:151–54.

Strupp, B. J., and D. A. Levitsky. 1984. Social transmission of food preferences in adult hooded rats (*Rattus norvegicus*). *Journal of Comparative Psychology* 98:257–66.

Stuart, R. J. 1991. Kin recognition as a functional concept. *Animal Behavior* 41:1093–94.

Stubbe, A., and S. Janke. 1994. Some aspects of social behaviour in the vole *Microtus brandti* (Rade, 1861). *Polish Ecological Studies* 20:449–57.

Suarez, O. V., and F. O. Kravetz. 2001. Male-female interactions during breeding and non-breeding seasons in *Akodon azarae* (Rodentia, Muridae). *Iheringia Serie Zoologia* 91:171–76.

Sugg, D. W., and R. K. Chesser. 1994. Effective population sizes with multiple paternity. *Genetics* 137:1147–55.

Sugg, D. W., R. K. Chesser, F. S. Dobson, and J. L. Hoogland. 1996. Population genetics meets behavioral ecology. *Trends in Ecology and Evolution* 11:338–42.

Sullivan, C. M. 2000. Social biology of the desert kangaroo rat (*Dipodomys deserti*)(Rodentia: Heteromyidae). Master's thesis, San Francisco State University.

Sullivan, J., and D. L. Swofford. 1997. Are guinea pigs rodents? The importance of adequate models in molecular phylogenetics. *Journal of Mammalian Evolution* 4:77–86.

Sullivan, R. 2004. *Rats: Observations on the history and habitat of the city's most unwanted inhabitants.* London: Bloomsbury.

Sullivan, R. M., and T. L. Best. 1997. Effects of environment on phenotypic variation and sexual dimorphism in *Dipodomys simulans* (Rodentia: Heteromyidae). *Journal of Mammalogy* 78:798–810.

Sullivan, T. P., D. S. Sullivan., and E. J. Hogue. 2001. Reinvasion dynamics of northern pocket gopher (*Thomomys talpoides*) population in removal areas. *Crop Protection* 20:189–98.

Sumbera, R., H. Burda, and W. N. Chitaukali. 2003. Reproductive biology of a solitary subterranean bathyergid rodent, the silvery mole-rat (*Heliophobius argenteocinereus*). *Journal of Mammalogy* 84:278–87.

Sumbera, R., H. Burda, W. N. Chitaukali, and J. Kubová. 2003. Silvery mole–rats (*Heliophobius argenteocincereus,* Bathyergidae) change their burrow architecture seasonally. *Naturwissenschaften* 90:370–73.

Summerlin, C. T., and J. L. Wolfe. 1972. Social influences on trap response of the cotton rat, *Sigmodon hispidus. Ecology* 54:1156–59.

Sun, L. 2003. Monogamy correlates, socioecological factors, and mating systems in beavers. In *Monogamy: Mating strategies and partnerships in birds, humans and other mammals,* ed. U. H. Reichard and C. Boesch, 138–46. Cambridge: Cambridge University Press.

Sun, L. X., and D. Müller-Schwarze. 1997. Sibling recognition in the beaver: A field test for phenotype matching. *Animal Behaviour* 54:493–502.

———. 1998a. Anal gland secretion codes for relatedness in the beaver, *Castor canadensis. Ethology* 104:917–27.

———. 1998b. Beaver response to recurrent alien scents: Scent fence or scent match? *Animal Behaviour* 55:1529–36.

Sun, L., D. Müller-Schwarze, and B. A. Schulte. 2000. Dispersal pattern and effective population size of the beaver. *Canadian Journal of Zoology* 78:393–98.

Sundell, J., D. Dudek, I. Klemme, E. Koivisto, J. Pusenius, and H. Ylönen. 2004. Temporal variation of fear and vole behaviour: An experimental field test of the predation risk allocation hypothesis. *Oecologia* 139:157–62.

Sundell, J., K. Norrdahl, E. Korpimäki, and I. Hanski. 2000. Functional response of the least weasel (*Mustela nivalis nivalis*). *Oikos* 90:501–8.

Sunquist, M. E., and F. C. Sunquist. 1989. Ecological constraints on predation by large felids. In *Carnivore behaviour, ecology and evolution,* ed. J. L. Gittleman, 283–301. Ithaca, NY: Cornell University Press.

Sutherland, W. J., and L. M. Gosling. 2000. Advances in the study of behaviour and their role in conservation. In *Behaviour and conservation,* ed. L. M. Gosling and W. J. Sutherland, 3–9. Cambridge: Cambridge University Press.

Svare, B. B. 1977. Maternal aggression in mice: Influence of the young. *Biobehavioral Reviews* 1:151–64.

———. 1981. Maternal aggression in mammals. In *Parental care in mammals,* ed. D. J. Gubernick and P. Klopfer, 179–210. New York: Plenum.

Svare, B. B., and A. Bartke. 1978. Food deprivation induces conspecific pup-killing in mice. *Aggressive Behavior* 4:253–61.

Svare, B. B., and R. Gandelman. 1976. Suckling stimulation induces aggression in virgin female mice. *Nature* 260:606–8.

Svare, B. B., and M. Mann. 1981. Infanticide: Genetic, developmental and hormonal influences in mice. *Physiology and Behavior* 27:921–27.

Svendsen, G. E. 1974. Behavioral and environmental factors in the spatial distribution and population dynamics of a yellow-bellied marmot population. *Ecology* 55:760–71.

Svendsen, G. E. 1976. Structure and location of burrows of yellow-bellied marmot. *Southwest Naturalist* 20:487–94.

———. 1978. Castor and anal glands of the beaver (*Castor canadensis*). *Journal of Mammalogy* 59:618–20.

———. 1980a. Patterns of scent-mounding in a population of beaver (*Castor canadensis*). *Journal of Chemical Ecology* 6:133–47.

———. 1980b. Population parameters and colony composition of beaver (*Castor canadensis*) in southeast Ohio. *American Midland Naturalist* 104:47–56.

———. 1989. Pair formation, duration of pair-bonds, and mate replacement in a population of beavers (*Castor canadensis*). *Canadian Journal of Zoology* 67:336–40.

Svendsen, G. E., and W. D. Huntsman. 1988. A field bioassay of beaver castoreum and some of its components. *American Midland Naturalist* 120:144–49.

Swaisgood, R. R., D. G. Lindburg, X. Zhou, and M. A. Owen. 2000. The effects of sex, reproductive condition and context on discrimination of conspecific odours by giant pandas. *Animal Behaviour* 60:227–37.

Swaisgood, R. R., D. H. Owings, and M. P. Rowe. 1999. Conflict and assessment in a predator-prey system: Ground squirrels versus rattlesnakes. *Animal Behaviour* 57:1033–44.

Swaisgood, R. R., M. P. Rowe, and D. H. Owings. 1999. Assessment of rattlesnake dangerousness by California ground squirrels: Exploitation of cues from rattling sounds. *Animal Behaviour* 57:1301–10.

———. 2003. Antipredator responses of California ground squirrels to rattlesnakes and rattling sounds: The roles of sex, reproductive parity, and offspring age in assessment and decision-making rules. *Behavioral Ecology and Sociobiology* 55:22–31.

Swanson, L. J., and C. S. Campbell. 1979. Maternal behavior in the primiparous and multiparous golden hamster. *Zeitschrift für Tierpsychologie* 50:96–104.

Sweitzer, R. A., and J. Berger. 1997. Sexual dimorphism and evidence for intrasexual selection from quill impalements, injuries, and mate guarding in porcupines (*Erethizon dorsatum*). *Canadian Journal of Zoology* 75:847–54.

———. 1998. Evidence for female-biased dispersal in North American porcupines (*Erethizon dorsatum*). *Journal of Zoology* 244:159–66.

Swilling, W. R, and M. C. Wooten. 2002. Subadult dispersal in a monogamous species: The Alabama beach mouse (*Peromyscus polionotus ammobates*). *Journal of Mammalogy* 83:252–59.

Swofford, D. L. 2003. *PAUP*: Phylogenetic analysis using parsi-*

mony (* and other methods), version 4.ob 10. Sunderland, MA: Sinauer.

Szalay, F. S., M. J. Novacek, and M. C. McKenna, eds. 1993. *Mammal phylogeny. Placental.* Berlin: Springer-Verlag.

Taber, A. B., and D. W. Macdonald. 1984. Scent dispensing papillae and associated behaviour of the Mara, *Dolichotis patagonum*, (Rodentia: Caviomorpha). *Journal of Zoology, London* 203:298–301.

Taber, A. B., and D. W. Macdonald. 1992a. Communal breeding in the mara, *Dolichotis patagonum. Journal of Zoology, London* 227:439–52.

———. 1992b. Spatial organization and monogamy in the mara *Dolichotis patagonum. Journal of Zoology, London* 227: 417–38.

Tachibana, T. 1974. Social facilitation of eating behavior in a novel situation by albino rats. *Japanese Psychological Research* 16:157–61.

Taitt, M. J., and C. J. Krebs. 1981. The effect of supplementary food on small rodent populations. 2. Voles (*Microtus townsendii*). *Journal of Animal Ecology* 50:125–37.

———. 1982. Manipulation of female behaviour in field populations of *Microtus townsendii. Journal of Animal Ecology* 51:681–90.

———. 1985. Population dynamics and cycles. In *Biology of New World* Microtus. Special publication no. 8, ed. R. H. Tamarin, 567–620. Manhattan, KS: American Society of Mammalogists.

Takahata, R., and B. Moghaddam. 1998. Glutamatergic regulation of basal and stimulus-activated dopamine release in the prefrontal cortex. *Journal of Neurochemistry* 71:1443–49.

Takushi, R. Y., K. J. Flannelly, D. C. Blanchard, and R. J. Blanchard. 1983. Maternal aggression in two strains of laboratory rats. *Aggressive Behavior* 9:120.

Talbot, L., and M. Talbot. 1963. The wildebeest in western Masailand, East Africa. *Wildlife Monographs* 12:8–88.

Tamarin, R. H., ed. 1985. *Biology of New World* Microtus. Special publication no. 8. Manhattan, KS: The American Society of Mammalogists.

Tamarin, R. H., R. S. Ostfeld, S. R. Pugh, and G. Bujalska, eds. 1990. *Social systems and population cycles in voles.* Basel, Switzerland: Birkhäuser.

Tamura, N. 1989. Snake-directed mobbing by the Formosan squirrel *Callosciurus erythraeus thaiwanensis. Behavioral Ecology and Sociobiology* 24:175–80.

———. 1993. Role of sound communication in mating of Malaysian *Callosciurus* (Sciuridae). *Journal of Mammalogy* 74:468–76.

———. 1995. Postcopulatory mate guarding by vocalization in the Formosan squirrel. *Behavioral Ecology and Sociobiology* 36:377–86.

Tamura, N., F. Hayashi, and K. Miyashita. 1988. Dominance hierarchy and mating behavior of the Formosan squirrel, *Callosciurus erythraeus taiwanensis. Journal of Mammalogy* 69:320–31.

———. 1989. Spacing and kinship in the Formosan squirrel living in different habitats. *Oecologia* 79:344–52.

Tamura, N., and H.-S. Yong. 1993. Vocalizations in response to predators in three species of Malaysian *Callosciurus* (Sciuridae). *Journal of Mammalogy* 74:703–14.

Tanaka, R. 1957. An ecological review of small-mammal outbreaks with special reference to their association with the flowering of bamboo grasses. *Bulletin of Kochi Women's College* 5:20–30.

Tanapat, P., L. A. M. Galea, and E. Gould. 1998. Stress inhibits the proliferation of granule cell precursors in the developing dentate gyrus. *International Journal of Developmental Neuroscience* 16:235–39.

Tanapat, P., N. B. Hastings, T. A. Rydel, L. A. M. Galea, and E. Gould. 2001. Exposure to fox odor inhibits cell proliferation in the hippocampus of adult rats via an adrenal hormone-dependent mechanism. *Journal of Comparative Neurology* 437:496–504.

Tang, R., F. X. Webster, and D. Müller-Schwarze. 1993. Phenolic compounds from male castoreum of the North American beaver (*Castor canadensis*). *Journal of Chemical Ecology* 19:1491–1500.

———. 1995. Neutral compounds from male castoreum of North American beaver, *Castor canadensis. Journal of Chemical Ecology* 21:1745–62.

Tang-Martínez, Z. 2001. The mechanisms of kin discrimination and the evolution of kin recognition in vertebrates: A critical re-evaluation. *Behavioural Processes* 53:21–40.

Tang-Martínez, Z., L. L. Mueller, and G. T. Taylor. 1993. Individual odours and mating success in the golden hamster *Mesocricetus auratus. Animal Behaviour* 45:1141–51.

Tasse, J. 1986. Maternal and paternal care in the rock cavy, *Kerodon rupestris*, a South American hystricomorph rodent. *Zoo Biology* 5:27–43.

Tast, J. 1966. The root vole, *Microtus oeconomus* (Pallas), as an inhabitant of seasonally flooded land. *Annales Zoologici Fennici* 3:127–71.

Taulman, J. F. 1977. Vocalizations of the hoary marmot, *Marmota caligata. Journal of Mammalogy* 58:681–83.

Taylor, D. 1970. Growth, decline, and equilibrium in a beaver population at Sagehen Creek, California. PhD diss., University of California, Berkeley.

Taylor, J. D., C. P. Stephens, R. G. Duncan, and G. R. Singleton. 1994. Polyarthritis in wild mice (*Mus musculus*) caused by *Streptobacillus moniliformis. Australian Veterinary Journal* 71:143–45.

Taylor, J. M., and J. H. Calaby. 1988. *Rattus lutreolus. Mammalian Species* 299:1–7.

Taylor, K. D. 1978. Range of movement and activity of common rats (*Rattus norvegicus*) on agricultural land. *Journal of Applied Ecology* 15:663–77.

Taylor, K. D., M. P. G. Fenn, and D. W. Macdonald. 1991. Common rat. In *Handbook of British mammals,* ed. G. B. Corbett and S. H. Harris, 248–55. Oxford: Blackwell Scientific.

Taylor, K. D., and M. G. Green. 1976. The influence of rainfall on diet and reproduction in four African rodent species. *Journal of Zoology, London* 180:367–89.

Taylor, K. D., and R. J. Quy. 1978. Long distance movements of a common rat (*Rattus norvegicus*) revealed by radio-tracking. *Mammalia* 42:63–71.

Taylor, R. J. 1984. *Predation.* New York: Chapman and Hall.

Taymans, S. E., A. C. DeVries, M. B. DeVries, R. J. Nelson, T. C. Friedman, M. Castro, S. Detera-Wadleigh, C. S. Carter, and G. P. Chrousos. 1997. The hypothalamic-pituitary-adrenal axis of prairie voles (*Microtus ochrogaster*): Evidence for target tissue glucocorticoid resistance. *General and Comparative Endocrinology* 106:48–61.

Tchabovsky, A. V., S. V. Popov, and B. R. Krasnov. 2001. Intra- and interspecific variation in vigilance and foraging of two gerbillid rodents, *Rhombomys opimus* and *Psammomys obesus*: The effect of social environment. *Animal Behaviour* 62:965–72.

Tchernov, E. 1975. Rodent faunas and environmental changes in the Pleistocene of Israel. In *Rodents in desert environments*, ed. I. Prakash and P. K. Ghosh, 331–62. The Hague: Junk.

Teague, L. G., and E. L. Bradley. 1978. The existence of a puberty accelerating pheromone in the urine of male prairie deermice (*Peromyscus maniculatus bairdii*). *Biology of Reproduction* 19:314–17.

Telfer, S., J. F. Dallas, J. Aars, S. B. Piertney, W. A. Stewart, and X. Lambin. 2003. Demographic and genetic structure of fossorial water voles (*Arvicola terrestris*) on Scottish islands. *Zoology, London* 259:23–29.

Telfer, S., S. B. Piertney, J. F. Dallas, W. A. Stewart, F. Marshall, J. Gow, and X. Lambin. 2003. Parentage assignment reveals widespread and large-scale dispersal in water voles. *Molecular Ecology* 12:1939–51.

Telle, H. J. 1966. Beitrag zur Kenntis der Verhaltensweise von Ratten, vergleichen dargestelt bei *Rattus norvegicus* und *Rattus rattus*. [In German.] *Angewandte Zoologie* 53:129–196 (Translation: MAFF, Tolworth, Rodent Research reprint 4489).

Tenaza, R. R. 1975. Territory and monogamy among Kloss' gibbons (*Hylobates klossi*) in Siberut Island, Indonesia. *Folia primatologica* 24:60–80.

Terkel, J. 1996. Cultural transmission of feeding behavior in the black rat (*Rattus rattus*). In *Social learning in animals: The roots of culture*, ed. C. M. Heyes and B. G. Galef, Jr., 17–49. San Diego: Academic Press.

Terlecki, L. J., J. P. Pinel, and D. Treit. 1979. Conditioned and unconditioned defensive burying in the rat. *Learning and Motivation* 10:337–50.

Terman, C. R. 1965. A study of population growth and control exhibited in the laboratory by deermice. *Ecology* 46:890–95.

———. 1968a. Inhibition of reproductive maturation and function in laboratory populations of prairie deermice: A test of pheromone influence. *Ecology* 49:1169–72.

———. 1968b. Population dynamics. In *Biology of* Peromyscus, ed. J. A. King, 412–50. Special publication no. 2. American Society of Mammalogists.

———. 1984. Sexual maturation of male and female white-footed mice (*Peromyscus leucopus noveboracensis*): Influence of physical or urine contact with adults. *Journal of Mammalogy* 65:97–102.

———. 1992. Reproductive inhibition in female white-footed mice from Virginia. *Journal of Mammalogy* 73:443–48.

———. 1993. Studies of natural populations of white-footed mice: Reduction of reproduction at varying densities. *Journal of Mammalogy* 74:678–87.

Terman, C. R., and E. L. Bradley. 1981. The influence of urine from asymptotic laboratory populations on sexual maturation in prairie deermice. Researches on population. *Ecology* 23:168–76.

Terrazas, A., R. Nowak, N. Serafin, G. Ferreira, F. Levy, and P. Poindron. 2002. Twenty-four-hour-old lambs rely more on maternal behavior than on the learning of individual characteristics to discriminate between their own and an alien mother. *Developmental Psychobiology* 40:408–18.

Terry, L. M., and I. B. Johanson. 1996. Effects of altered olfactory experiences on the development of infant rats' responses to odors. *Developmental Psychobiology* 29:353–77.

Tevis, L., Jr. 1950. Summer behavior of a family of beavers in New York State. *Journal of Mammalogy* 31:40–65.

Theodoratus, D. H., and D. Chiszar. 2000. Habitat selection and prey odor in the foraging behavior of western rattlesnakes (*Crotalus viridis*). *Behaviour* 137:119–35.

Thom, M. D., and J. L. Hurst. (2004). Individual recognition by scent. *Annales Zoologici Fennici* 41:765–87.

Thomas, S. A. 2002. Scent marking and mate choice in the prairie vole, *Microtus ochrogaster*. *Animal Behaviour* 63:1121–27.

Thomas, S. A., and B. K. Kaczmarek. 2002. Scent-marking behaviour by male prairie voles, *Microtus ochrogaster*, in response to the scent of opposite- and same-sex conspecifics. *Behavioral Processes* 60:27–33.

Thomas, S. A., and J. O. Wolff. 2002. Scent marking in voles: A reassessment of over marking, counter marking, and self-advertisement. *Ethology* 108:51–62.

———. 2003. Scent marking in rodents: A reappraisal, problems, and future directions. In *Rats, mice and people: Rodent biology and management*, ed. G. R. Singleton, L. A. Hinds, C. J. Krebs, and D. M. Spratt, 143–47. ACIAR Monograph no. 96. Canberra, Australia: ACIAR

Thompson, D. C. 1976. Accidental mortality and cannibalization of a nestling grey squirrel. *Canadian Field-Naturalist* 90:52–53.

———. 1977. Reproductive behavior of the grey squirrel. *Canadian Journal of Zoology* 55:1176–84.

Thompson, K. V. 1998. Self assessment in juvenile play. In *Animal play: Evolutionary, comparative, and ecological perspectives*, ed. M. Bekoff and J. A. Byers, 183–204. Cambridge: Cambridge University Press.

Thompson, S. D. 1982. Microhabitat utilization and foraging behavior of bipedal and quadrupedal heteromyid rodents. *Ecology* 63:1303–12.

———. 1992. Gestation and lactation in small mammals: Basal metabolic rate and the limits of energy use. In *Mammalian energetics: Interdisciplinary views of metabolism and reproduction*, ed. T. Tomasi and T. Horton, 213–59. Ithaca, NY: Comstock.

Thorington, R. W. 1984. Flying squirrels are monophyletic. *Science* 225:1048–49.

Thorington, R. W., K. Darrow, and A. D. K. Betts. 1997. Comparative myology of the forelimb of squirrels (Sciuridae). *Journal of Morphology* 234:155–82.

Thorington, R. W., D. Pitassy, and S. Jansa. 2002. Phylogenies of flying squirrels (Pteromyinae). *Journal of Mammalian Evolution* 9:99–135.

Thorndike, E. L. 1898. Animal intelligence: An experimental study of the associative process in animals. *Psychological Review Monographs* 2, Whole number 8.

Thorpe, W. H. 1956. *Learning and instinct in animals*. London: Methuen.

Thorson, J. M., R. A. Morgan, J. S. Brown, and J. E. Norman. 1998. Direct and indirect cues of predatory risk and patch use by the fox squirrel and thirteen-lined ground squirrel. *Behavioral Ecology* 9:151–57.

Thursz, M. R., H. C. Thomas, B. M. Greenwood, and A. V. Hill. 1997. Heterozygote advantage for HLA class-II type in hepatitis B virus infection. *Nature Genetics* 17:11–12.

Tidey, J. W., and K. A. Miczek. 1996. Social defeat stress selectively alters mesocorticolimbic dopamine release: An in vivo microdialysis study. *Brain Research* 721:140–49.

Tileston, J. V., and R. R. Lechleitner. 1966. Some comparisons of the black-tailed and white-tailed prairie dogs in north-central Colorado. *American Midland Naturalist* 75:292–316.

Tinbergen, N. 1952. Derived activities: Their causation, biological significance, origin and emancipation during evolution. *Quarterly Review of Biology* 27:1–32.

———. 1953. *Social behaviour in animals.* New York: Wiley.

———. 1963. On the aims and methods of ethology. *Zeitschrift für Tierpsychologie* 20:410–33.

Tkadlec, E., and N. C. Stenseth. 2001. A new geographical gradient in vole population dynamics. *Proceedings of the Royal Society of London, Series B* 268:1547–52.

Todrank, J., and G. Heth. 1996. Individual odors in two chromosomal species of blind, subterranean mole rat (*Spalax ehrenbergi*): Conspecific and cross-species discrimination. *Ethology* 102:806–11.

———. 2001. Rethinking cross-fostering designs for studying kin recognition mechanisms. *Animal Behaviour* 61:503–5.

———. 2003. Odor-genes covariance and genetic relatedness assessments: Rethinking odor-based "recognition" mechanisms in rodents. *Advances in the Study of Behavior* 32:77–130.

Todrank, J., G. Heth, and R. E. Johnston. 1998. Kin recognition in golden hamsters: Evidence for kinship odours. *Animal Behaviour* 55:377–86.

Tognelli, M. F., C. M. Campos, and R. A. Ojeda. 2001. *Microcavia australis. Mammalian Species* 648:1–4.

Tokarski, V. A., O. V. Brandler, and A. V. Zavgorudko. 1991. Spatial structure of the bobac population in the Ukraine. In *Population structure of the marmot,* ed. D. I. Bibikov, A. A. Nikolski, V. Yu. Rumiantsev, and T. A. Seredneva, 45–70. Moscow: USSR Theriological Society.

Tomich, P. Q. 1962. The annual cycle of the California ground squirrel *Citellus beecheyi. University of California Publications in Zoology* 65:213–82.

Topachevskii, V. A. 1969. *The fauna of the USSR: Mammals. Mole rats, spalacidae III.* Leningrad: Nauka.

Topal, J., and V. Csanyi. 1994. The effect of eye-like schema on shuttling activity of wild house mice (*Mus musculus domesticus*): Context-dependent threatening aspects of the eyespot patterns. *Animal Learning and Behavior* 22:96–102.

Topping, M. G., and J. S. Millar. 1996a. Foraging movements of female bushy-tailed wood rats (*Neotoma cinerea*). *Canadian Journal of Zoology* 74:798–801.

———. 1996b. Home range size of bushy-tailed woodrats, *Neotoma cinerea,* in Southwestern Alberta. *The Canadian Field Naturalist* 110:351–53.

———. 1996c. Spatial distribution in the bushy-tailed wood rat (*Neotoma cinerea*) and its implications for the mating system. *Canadian Journal of Zoology* 74:565–69.

———. 1998. Mating patterns and reproductive success in the bushy-tailed woodrat (*Neotoma cinerea*) as revealed by DNA fingerprinting. *Behavioral Ecology and Sociobiology* 43:115–24.

———. 1999. Mating success of male bushy-tailed woodrats: When bigger is not always better. *Behavioral Ecology* 10:161–68.

Toro, J., J. D. Vega, A. S. Khan, J. N. Mills, P. Padula, W. Terry, Z. Yadon et al. 1998. An outbreak of hantavirus pulmonary syndrome, Chile, 1997. *Emerging Infectious Diseases* 4:687–94.

Torres-Contreras, H., and F. Bozinovic. 1997. Food selection in an herbivorous rodent: Balancing nutrition with thermoregulation. *Ecology* 78:2230–37.

Torres-Mura, J. C. 1990. Uso del espacio en el roedor fosorial *Spalacopus cyanus* (Octodontidae). Master's thesis, Universidad de Chile, Santiago, Chile.

Torres-Mura, J. C., and L. C. Contreras. 1998. *Spalacopus cyanus. Mammalian Species* 594:1–5.

Trainor, C., A. Fisher, J. Woinarski, and S. Churchill. 2000. Multiscale patterns of habitat use by the Carpentarian rock-rat (*Zyzomys palatalis*) and the common rock-rat (*Z. argurus*). *Wildlife Research* 27:319–32.

Travis, S. E., and C. N. Slobodchikoff. 1993. Effects of food resource distribution on the social system of Gunnison's prairie dog (*Cynomys gunnisoni*). *Canadian Journal of Zoology* 71:1186–92.

Travis, S. E., C. N. Slobodchikoff, and P. Keim. 1995. Ecological and demographic effects on intraspecific variation in the social system of prairie dogs. *Ecology* 76:1794–1803.

———. 1996. Social assemblages and mating relationships in prairie dogs: A DNA fingerprint analysis. *Behavioral Ecology* 7:95–100.

Trillmich, F. 1986. Are endotherms emancipated? Some considerations on the cost of reproduction. *Oecologia* 69:631–33.

Trillmich, F., C. Kraus, J. Kunkele, M. Asher, M. Clara, G. Dekomien, J. T. Epplen et al. 2004. Species-level differentiation of two cryptic species pairs of wild cavies, genera *Cavia* and *Galea,* with a discussion of the relationship between social systems and phylogeny in the Caviinae. *Canadian Journal of Zoology* 82:516–24.

Trivers, R. L. 1971. The evolution of reciprocal altruism. *Quarterly Review of Biology* 46:35–57.

———. 1972. Parental investment and sexual selection. In *Sexual selection and the descent of man, 1871–1971,* ed. B. Campbell, 136–79. Chicago: Aldine.

———. 1974. Parent-offspring conflict. *American Zoologist* 14:249–64.

Trivers, R. L., and D. E. Willard. 1973. Natural selection of parental ability to vary the sex ratio of offspring. *Science* 179:90–92.

Tromborg, C. T. 1999. Figure and ground squirrels. PhD diss., University of California, Davis.

Trombulak, S. C. 1989. Running speed and body mass in Belding's ground-squirrels. *Journal of Mammalogy* 70:194–97.

———. 1991. Maternal influence on juvenile growth rates in Belding's ground squirrel. *Canadian Journal of Zoology* 69:2140–45.

Trulio, L. A. 1996. The functional significance of infanticide in a population of California ground squirrels (*Spermophilus beecheyi*). *Behavioral Ecology and Sociobiology* 38:97–103.

Trulio, L. A., W. J. Loughry, D. F. Hennessy, and D. H. Owings. 1986. Infanticide in California ground squirrels. *Animal Behaviour* 34:291–94.

Tsutsui, N. D. 2004. Scents of self: The expression component of self/nonself recognition systems. *Annales Zoologici Fennici* 41:713–27.

Tubbiola, M. L., and C. J. Wysocki. 1997. FOS immunoreactivity after exposure to conspecific or heterospecific urine: Where are chemosensory cues sorted. *Physiology and Behavior* 62:867–70.

Tullberg, T. 1899. Ueber das System der Nagethiere: Eine phylogenetische Studie. *Nova Acta Regiae Societatis Scientarium Upsaliensis* 18:1–514.

Turchin, P. 2003. *Complex population dynamics: A theoretical/empirical synthesis.* Princeton, NJ: Princeton University Press.

Türk, A., and W. Arnold. 1988. Thermoregulation as a limit to habitat use in alpine marmots (*Marmota marmota*). *Oecologia* 76:544–48.

Turner, L. W. 1973. Vocal and escape responses of *Spermophilus beldingi* to predators. *Journal of Mammalogy* 54:990–93.

Twigg, G. 1975. *The brown rat.* Newton Abbot, Devon, UK: David and Charles.

Tyser, R. W. 1980. Use of substrate for surveillance behaviors in a community of talus slope mammals. *American Midland Naturalist* 104:32–38.

Ulevicius, A., and L. Balciauskas. 2000. Scent marking intensity of beaver (*Castor fiber*) along rivers of different sizes. *Zeitschrift für Säugetierkunde* 65:286–92.

Unda, C., M. A. Rojas and J. Yánez. 1980. Estudio preliminar del ciclo reproductivo y efecto medioambiental en dos poblaciones de *Spalacopus cyanus. Archivos de Biología y Medicina Experimentales (Chile)* 13:115.

United States Fish and Wildlife Service. 1970. *Conservation of endangered species and other fish or wildlife* (first list of endangered foreign fish and wildlife as appendix A). 35 FR 8491–98.

———. 1984. *Endangered and threatened wildlife and plants: Final rule to reclassify the Utah prairie dog as threatened, with special rule to allow regulated taking.* 49 FR 22330–34.

———. 1998. *Multi-species recovery plan for the threatened and endangered species of south Florida.* vols. I and II: Technical/Agency draft. Vero Beach, Florida: USFWS.

———. 2000. *Endangered and threatened wildlife and plants: 12-month finding for a petition to list the black-tailed prairie dog as threatened.* 65 FR 5476–5488.

———. 2002. *Northern Cheyenne Reservation: Tongue River Enhancement Project.* Lewistown, Montana: USFWS.

———. 2004. *Species assessment and listing priority assignment form.* 69 FR (159):51217–51226.

Upchurch, M., and J. M. Wehner. 1988. Differences between inbred strains of mice in Morris water maze performance. *Behavior Genetics* 18:55–68.

———. 1989. Inheritance of spatial learning ability in inbred mice: A classical genetic analysis. *Behavioral Neuroscience* 103:1251–58.

Uresk, D. W., and A. J. Bjugstad. 1983. Prairie dogs as ecosystem regulators on the northern high plains. In *Proceedings of the Seventh North American Prairie Conference, August 4–6, 1980,* ed. C. L. Kucera, 91–94. Springfield: Southwest Missouri State.

Urrejola, D., E. A. Lacey, J. R. Wieczorek, and L. A. Ebensperger. 2005. Daily activity patterns of free-living coruros (*Spalacopus cyanus*). *Journal of Mammalogy* 86:302–8.

Vaché, M., J. Ferron, and P. Gouat. 2001. The ability of red squirrels (*Tamiasciurus hudsonicus*) to discriminate conspecific olfactory signatures. *Canadian Journal of Zoology* 79:1296–1300.

Valsecchi, P., E. Choleris, A. Moles, C. Guo, and M. Mainardi. 1996. Kinship and familiarity as factors affecting social transfer of food preferences in adult Mongolian gerbils. *Journal of Comparative Psychology* 110:243–51.

Valsecchi, P., and B. G. Galef, Jr. 1989. Social influences on the food preferences of house mice (*Mus musculus*). *International Journal of Comparative Psychology* 2:245–56.

Valsecchi, P., A. Moles, and M. Mainardi. 1993. Does mother's diet affect food selection of weanling wild mice. *Animal Behaviour* 46:827–28.

Valsecchi, P., G. R. Singleton, and W. J. Price, W. J. 1996. Can social behaviour influence food preference of wild mice, *Mus*

domesticus, in confined field populations? *Australian Journal of Zoology* 44:493–501.

Van Aarde, R. 1987. Pre- and postnatal growth of the the Cape porcupine *Hystrix africaeaustralis. Journal of Zoology* 211:25–33.

Van Couvering, J. A. H., and J. A. Van Couvering. 1976. Early Miocene mammal fossils from East Africa: Aspects of geology, faunistics and paleoecology. In *Human origins: Louis Leakey and the East African evidence,* ed. G. L. Isaac and E. R. McCown, 155–207. Menlo Park, CA: W. A. Benjamin.

Van den Berg, C. L., T. Hol, J. M. Van Ree, B. M. Spruijt, H. Everts, and J. M. Koolhaas. 1999. Play is indispensable for an adequate development of coping with social challenges in the rat. Developmental Psychobiology 34:129–38.

Van Den Brink, F. H. 1968. *A field guide to the mammals of Britain and Europe.* Boston: Houghton Mifflin.

Van Horne, B., G. S. Olson, R. L. Schooley, J. G. Corn, and K. P. Burnham. 1997. The effects of drought and prolonged winter on Townsend's ground squirrels in shrubsteppe habitats. *Ecological Monographs* 67:295–315.

Van Horne, B., R. L. Schooley, and P. B. Sharpe. 1998. Influence of age, habitat, and drought on the diet of Townsend's ground squirrels. *Journal of Mammalogy* 79:521–37.

Van Loo, P. L. P., C. L. J. J. Kruitwagen, L. F. M. van Zutphen, J. M. Koolhaas, and V. Baumans. 2000. Modulation of aggression in male mice: Influence of cage cleaning regime and scent marks. *Animal Welfare* 9:281–95.

Van Loo, P. L. P., E. van der Meer, C. L. J. J. Kruitwagen, J. M. Koolhaas, L. F. M. V. Zutphen, and V. Baumans. 2003. Strain-specific aggressive behavior of male mice submitted to different husbandry procedures. *Aggressive Behavior* 29:69–80.

Van Schaik, C. P., M. Ancrenaz, G. Borgen, B. Galdikas, C. Knott, I. Singleton, A. Suzuki, S. S. Utami, and M. Merrill, 2003. Orangutan cultures and their implications. *Science* 299:102–5.

Van Schaik, C. P., and C. H. Janson, eds. 2000. *Infanticide by males and its implications.* Cambridge, UK: Cambridge University Press.

Van Vuren, D. 1990. Dispersal of yellow-bellied marmots. PhD diss., University of Kansas, Lawrence.

———. 1996. Ectoparasites, fitness, and social behaviour of yellow-bellied marmots. *Ethology* 102:686–94.

Van Vuren, D., and K. B. Armitage. 1994a. Reproductive success of colonial and noncolonial female yellow-bellied marmots (*Marmota flaviventris*). *Journal of Mammalogy* 75:950–55.

———. 1994b. Survival of dispersing and philopatric yellow-bellied marmots: What is the cost of dispersal? *Oikos* 69:179–81.

Vandenbergh, J. G. 1967. Effect of the presence of a male on the sexual maturation of female mice. *Endocrinology* 81:345–49.

———. 1969. Male odor accelerates female sexual maturation in mice. *Endocrinology* 84:658–60.

———. 1983. Pheromonal regulation of puberty. In *Pheromones and reproduction in mammals,* ed. J. G. Vandenbergh, 95–112. New York: Academic Press.

Vandenbergh, J. G., and D. M. Coppola. 1986. The physiology and ecology of puberty modulation by primer pheromones. *Advances in the Study of Behavior* 16:71–107.

Vandenbergh, J. G., L. C. Drickamer, and D. R. Colby. 1972. Social and dietary factors in the sexual maturation of female mice. *Journal of Reproduction and Fertility* 28:397–405.

Vandenbergh, J. G., J. S. Finlayson, W. J. Dobrogosz, S. S. Dills, and T. A. Cost. 1976. Chromatographic separation of puberty accelerating pheromone from male mouse urine. *Biology of Reproduction* 15:260–65.

Vandenbergh, J. G., and C. L. Huggett. 1995. The anogenital distance index, a predictor of the intrauterine position effects on reproduction in female house mice. *Laboratory Animal Science* 45:567–73.

Vandenbergh, J. G., J. M. Whitsett, and J. R. Lombardi. 1975. Partial isolation of a pheromone accelerating puberty in female mice. *Journal of Reproduction and Fertility* 43:515–23.

Vander Wall, S. G., and S. H. Jenkins. 2003. Reciprocal pilferage and the evolution of food-hoarding behavior. *Behavioral Ecology* 14:656–67.

VanStaaden, M. J., R. K. Chesser, and G. R. Michener. 1994. Genetic correlations and matrilineal structure in a population of *Spermophilus richardsonii. Journal of Mammalogy* 75: 573–82.

Vargas, A., M. Lockart, P. Marinari, and P. Gober. 1998. Preparing captive-raised black-footed ferrets (*Mustela nigripes*) for survival after release. *Dodo* 34:76–83.

Vasilieva, N. Y., S.-C. Lai, E. V. Petrova, and R. E. Johnston. 2001. Development of species preferences in two hamsters, *Phodopus campbelli* and *Phodopus sungorus*: Effects of cross-fostering. *Ethology* 107:217–36.

Vásquez, R. A. 1997. Vigilance and social foraging in *Octodon degus* (Rodentia: Octodontidae) in central Chile. *Revista Chilena de Historia Natural* 70:557–63.

Vásquez, R. A., L. A. Ebensperger, and F. Bozinovic. 2002. The effect of microhabitat on running velocity, intermittent locomotion, and vigilance in a diurnal rodent. *Behavioral Ecology* 13:182–87.

Vaughan, T. A., and R. M. Hansen. 1964. Experiments on interspecific competition between two species of pocket gophers. *American Midland Naturalist* 72:444–52.

Vaughan, T. A., and S. T. Schwartz. 1980. Behavioral ecology of an insular woodrat (*Neotoma lepida*). *Journal of Mammalogy* 61:205–18.

Veal, R., and W. Caire. 1979. *Peromyscus eremicus. Mammalian Species* 118:1–6.

Vehrencamp, S. L. 1977. Relative fecundity and parental effort in groove-billed anis *Crotophaga sulcirostris. Science* 197: 403–5.

———. 1978. The adaptive significance of communal nesting in groove-billed anis *Crotophaga sulcirostris. Behavioural Ecology and Sociobiology* 4:1–33.

———. 1979. The roles of individual, kin and group selection in the evolution of sociality. In *Handbook of behavioral neurobiology, vol. 3; social behavior and communication,* ed. P. Marler and J. G. Vandenbergh, 351–94. New York: Plenum.

———. 1982. Testicular regression in relation to incubation effort in a tropical cuckoo. *Hormones and Behavior* 16:113–20.

———. 1983. A model for the evolution of despotic versus egalitarian societies. *Animal Behaviour* 31:667–82.

Veloso, C. P. 1997. Energética reproductiva del roedor precocial herbívoro *Octodon degus* (Rodentia: Octodontidae). PhD diss., Universidad de Chile, Santiago, Chile.

Vermeire, L. T., R. K. Heitschmidt, P. S. Johnson, and B. F. Sowell. 2004. The prairie dog story: Do we have it right? *Bioscience* 54: 689–95.

Verner, J. 1964. Evolution of polygamy in the long-billed marsh wren. *Evolution* 18:252–61.

Vernet-Maury, E., J. Le Magnen, and J. Chanel. 1968. Comporte-ment émotif chez le rat: Influence de l'odeur d'un prédateur et d'un nonprédateur. *Paris, Comptes rendus de l'Académie des sciences* 267:331–34.

Vernet-Maury, E., E. H. Polak, and A. Demael. 1984. Structure-activity relationship of stress inducing odourants in the rat. *Journal of Chemical Ecology* 10:1007–17.

Verts, B. J., and G. L. Kirkland, Jr. 1988. *Perognathus parvus. Mammalian Species* 318:1–8.

Vestal, B. M. 1991. Infanticide and cannibalism by male thirteen-lined ground squirrels. *Animal Behaviour* 41:1103–04.

Vestal, B. M., and H. McCarley. 1984. Spatial and social relations of kin in thirteen-lined and other ground squirrels. In *The biology of ground-dwelling squirrels,* ed. J. O. Murie and G. R. Michener, 404–23. Lincoln: University of Nebraska Press.

Vianey-Liaud, M. 1985. Possible evolutionary relationships among Eocene and lower Oligocene rodents of Asia, Europe, and North America. In *Evolutionary relationships among rodents: A multidisciplinary analysis,* ed. W. P. Luckett and J. L. Hartenberger, 277–309. New York: Plenum.

Viau, V., and M. J. Meaney. 1991. Variations in the hypothalamic-pituitary-adrenal response to stress during the estrous cycle in the rat. *Endocrinology* 129:2503–11.

Vickery, W., and J. S. Millar. 1984. The energetics of huddling in endotherms. *Oikos* 43:88–93.

Vieira, M. L., and R. E. Brown. 2003. Effects of the presence of the father on pup development in California mice (*Peromyscus californicus*). *Developmental Psychobiology* 42:246–51.

Viitala, J., E. Korpimäki, P. Palokangas, and M. Koivula. 1995. Attraction of kestrels to vole scent marks visible in ultraviolet light. *Nature* 373:425–27.

Viljoen, S. 1977. Behavior of bush squirrel, *Paraxerus cepapi cepapi* Smith, A, 1836. *Mammalia* 41:119–66.

———. 1983. Communicatory behaviour of southern African tree squirrels, *Paraxerus palliatus ornatus, P. p. tongensis, P. c. cepapi* and *Funisciurus congicus. Mammalia* 47:441–61.

Virgl, J. A., and F. Messier. 1992. Seasonal variation in body composition and morphology of adult muskrats in central Saskatchewan, Canada. *Journal of Zoology, London* 228: 461–77.

Vitale, A. F. 1989. Changes in anti-predator responses of wild rabbits, *Oryctolagus cuniculus* (L.), with age and experience. *Behaviour* 110:47–61.

Vom Saal, F. S. 1984. Proximate and ultimate causes of infanticide and parental behavior in male house mice. In *Infanticide: Comparative and evolutionary perspectives,* ed. G. Hausfater and S. B. Hrdy, 401–24. New York: Aldine.

Vom Saal, F. S., and F. H. Bronson. 1978. In utero proximity of female mouse fetuses to males: Effect on reproductive performance. *Biology of Reproduction* 19:842–853.

———. 1980. Sexual characteristic of adult mice are correlated with their blood testosterone levels during prenatal life. *Science* 208:597–99.

Vom Saal, F. S., P. Franks, M. Boechler, P. Palanza, and S. Parmigiani. 1995. Nest defense and survival of offspring in highly aggressive wild Canadian female house mice. *Physiology and Behavior* 58:669–78.

Vom Saal, F. S., and L. S. Howard. 1982. The regulation of infanticide and parental behavior: Implications for reproductive success in male mice. *Science* 215:1270–72.

Von-Schantz, T., H. Wittzell, G. Göransson, M. Grahn, and K. Persson. 1996. MHC genotype and male ornamentation:

Genetic evidence for the Hamilton-Zuk model. *Proceedings of the Royal Society of London, Series B* 263:265–71.

Voss, R. S. 1979. Male accessory glands and the evolution of copulatory plugs in rodents. *Occasional Papers of the Museum of Zoology* (University of Michigan) 689:1–27.

Voznessenskaya, V. V., V. M. Parfyonova, and E. Zinkevich. 1992. Individual odortypes. In *Chemical signals in vertebrates VI*, ed. R. L. Doty and D. Müller-Schwarze, 503–8. New York: Plenum.

Waddell, P. J., Y. Cao, M. Hasegawa, and D. P. Mindell. 1999. Assessing the Cretaceous superordinal divergence times within birds and placental mammals by using whole mitochondrial protein sequences and an extended statistical framework. *Systematic Biology* 48:119–37.

Waddell, P. J., H. Kishino, and R. Ota. 2001. A phylogenetic foundation for comparative mammalian genomics. *Genome Informatics* 12:141–54.

Waddell, P. J., N. Okada, and M. Hasegawa. 1999. Towards resolving the interordinal relationships of placental mammals. *Systematic Biology* 48:1–5.

Waddell, P. J., and S. Shelley. 2003. Evaluating placental interordinal phylogenies with novel sequences including RAG1, ß-fibrinogen, ND6, and mt-tRNA, pluse MCMC-driven nucleotide, amino acid, and codon models. *Molecular Phylogenetics and Evolution* 28:197–224.

Wahrman, J., R. Goitein, and E. Nevo. 1969a. Geographic variation of chromosome forms in *Spalax*, a subterranean mammal of restricted mobility. In *Comparative mammalian cytogenetics*, ed. K. Benirschke, 30–48. New York: Springer Verlag.

Wahrman, J., R. Goitein, and E. Nevo. 1969b. Mole rat *Spalax*: Evolutionary significance of chromosome variation. *Science* 164:82–84.

Wahrman, J., C. Richler, R. Gamperl, and E. Nevo. 1985. Revisiting *Spalax*: Mitotic and meiotic chromosome variability. *Israel Journal of Zoology* 33:15–38.

Wainwright, P. 1980. Relative effects of maternal and pup heredity on postnatal mouse development. *Developmental Psychobiology* 13:493–98.

Waldman, B. 1982. Sibling association among schooling toad tadpoles: Field evidence and implications. *Animal Behaviour* 30:700–13.

———. 1987. Mechanisms of kin recognition. *Journal of Theoretical Biology* 128:159–85.

Waldman, B., P. C. Frumhoff, and P. W. Sherman. 1988. Problems of kin recognition. *Trends in Ecology and Evolution* 3:8–13.

Walker, R. S. 2001. Effects of landscape structure on the distribution of mountain vizcacha (*Lagidium viscacia*) in the Patagonian steppe. PhD diss., University of Florida, Gainesville.

Walker, R. S., A. J. Novaro, and L. C. Branch. 2003. Effects of patch attributes, barriers, and distance between patches on the distribution of a rock dwelling rodent (*Lagidium viscacia*). *Landscape Ecology* 18:185–92.

Walther, F. R. 1978. Mapping the structure and the marking system of a territory of the Thomson's gazelle. *East African Wildlife Journal* 16:167–76.

Walton, A. H., M. A. Nedbal, and R. L. Honeycutt. 2000. Evidence from Intron 1 of the nuclear transthyretin (prealbumin) gene for the phylogeny of African mole–rats (Bathyergidae). *Molecular Phylogenetics and Evolution* 16:467–74.

Waltz, E. C. 1982. Resource characteristics and the evolution of information centers. *American Naturalist* 119:73–90.

Wang, Z. X. 1991. Effects of social environment and experience on parental care, behavioral development, and reproductive success of prairie voles (*Microtus ochrogaster*). PhD diss., University of Massachusetts Amherst.

———. 1995. Species differences in the vasopressin-immunoreactive pathways in the bed nucleus of the stria terminalis and medial amygdaloid nucleus in prairie voles (*Microtus ochrogaster*) and meadow voles (*Microtus pennsylvanicus*). *Behavioral Neuroscience* 109:305–11.

Wang, Z. X., N. A. Bullock, and G. J. De Vries. 1993. Sexual differentiation of vasopressin projections of the bed nucleus of the stria terminals and medial amygdaloid nucleus in rats. *Endocrinology* 132:2299–2306.

Wang, Z. X., C. F. Ferris, and G. J. De Vries. 1994. Role of septal vasopressin innervation in paternal behavior in prairie voles (*Microtus ochrogaster*). *Proceedings of the National Academy of Science (USA)* 91:400–4.

Wang, Z. X., T. J. Hulihan, and T. R. Insel. 1997. Sexual and social experience is associated with different patterns of behavior and neural activation in male prairie voles. *Brain Research* 767:321–32.

Wang, Z. X., and T. R. Insel. 1996. Parental behavior in voles. In *Advances in the study of behavior*, vol. 25, ed. J. S. Rosenblatt, and C. T. Snowdon, 361–84. New York: Academic Press.

Wang, Z. X., Y. Liu, L. J. Young, and T. R. Insel. 2000. Hypothalamic vasopressin gene expression increases in both males and females postpartum in a biparental rodent. *Journal of Neuroendocrinology* 12:111–20.

Wang, Z. X., and M. A. Novak. 1992. Influence of the social environment on parental behavior and pup development of meadow voles (*Microtus pennsylvanicus*) and prairie voles (*M. ochrogaster*). *Journal of Comparative Psychology* 106:163–71.

———. 1994. Parental care and litter development in primiparous and multiparous prairie voles (*Microtus ochrogaster*). *Journal of Mammalogy* 75:18–23.

Wang, Z. X., W. Smith, D. E. Major, and G. J. De Vries. 1994. Sex and species differences in the effects of cohabitation on vasopressin messenger RNA expression in the bed nucleus of the stria terminalis in prairie voles (*Microtus ochrogaster*) and meadow voles (*Microtus pennsylvanicus*). *Brain Research* 650:212–18.

Wang, Z. X., and L. J. Young. 1997. Ontogeny of oxytocin and vasopressin receptor binding in the lateral septum in prairie and montane voles. *Developmental Brain Research* 104:191–95.

Wang, Z. X., L. J. Young, G. J. De Vries, and T. R. Insel. 1998. Voles and vasopressin: A review of molecular, cellular, and behavioral studies of pair bonding and paternal behaviors. *Progress in Brain Research* 119:483–99.

Wang, Z. X., L. J. Young, Y. Liu, and T. R. Insel. 1997. Species differences in vasopressin receptor binding are evident early in development: Comparative anatomic studies in prairie and montane voles. *Journal of Comparative Neurology* 378:535–46.

Wang, Z. X., G. Yu, C. Cascio, Y. Liu, B. Gingrich, and T. R. Insel. 1999. Dopamine D2 receptor-mediated regulation of partner preferences in female prairie voles (*Microtus ochrogaster*): A mechanism for pair bonding? *Behavioral Neuroscience* 113:602–11.

Ward, S. 1984. An examination of the mating system of Stephen's (sic) woodrat. MS thesis. Northern Arizona University, Flagstaff, AZ.

Waring, G. H. 1966. Sounds and communications of the yellow-bellied marmot (*Marmota flaviventris*). *Animal Behaviour* 14:177–83.

———. 1970. Sound communications of black-tailed, white-tailed, and Gunnison's prairie dogs. *American Midland Naturalist* 83:167–85.

Warkentin, K. J., A. T. H. Keeley, and J. F. Hare. 2001. Repetitive calls of juvenile Richardson's ground squirrels (*Spermophilus richardsonii*) communicate response urgency. *Canadian Journal of Zoology* 79:569–73.

Wartella, J., E. Amory, A. Macbeth, I. McNamara, L. Stevens, K. G. Lambert, and C. H. Kinsley. 2003. Single or multiple reproductive experiences attenuate neurobehavioral stress and fear responses in the female rat. *Physiology and Behavior* 79:373–81.

Waser, P. M. 1981. Sociality or territorial defense? The influence of resource renewal. *Behavioral Ecology and Sociobiology* 8:231–37.

———. 1988. Resources, philopatry, and social interactions among mammals. In *The ecology of social behaviour*, ed. C. N. Slobodchikoff, 109–30. San Diego, CA: Academic Press.

Waser, P. M., S. N. Austad, and B. Keane. 1986. When should animals tolerate inbreeding? *American Naturalist* 128:529–37.

Waser, P. M., and W. T. Jones. 1983. Natal philopatry among solitary mammals. *The Quarterly Review of Biology* 58:355–88.

Wasser, S. K., and D. P. Barash. 1983. Reproductive suppression among female mammals: Implications for biomedicine and sexual selection theory. *Quarterly Review of Biology* 58:513–38.

Wasserberg, G., Z. Abramsky, B. Kotler, R. S. Ostfeld, I. Yarom, and A. Warburg. 2003. Anthropogenic disturbances enhance occurrence of cutaneous leishmaniasis in Israel deserts: Patterns and mechanisms. *Ecological Applications* 13:868–81.

Waterman, J. M. 1984. Infanticide in the Columbian ground squirrel, *Spermophilus columbianus*. *Journal of Mammalogy* 65:137–38.

———. 1986. Behaviour and use of space by juvenile Columbian ground squirrels. *Canadian Journal of Zoology* 64:1121–27.

———. 1988. Social play in free-ranging Columbian ground squirrels, *Spermophilus columbianus*. *Ethology* 77:225–36.

———. 1995. The social organization of the Cape ground squirrel (*Xerus inauris*; Rodentia: Sciuridae). *Ethology* 101:130–47.

———. 1997. Why do male Cape ground squirrels live in groups? *Animal Behaviour* 53:809–17.

———. 1998. Mating tactics of male Cape ground squirrels, *Xerus inauris*: Consequences of year-round breeding. *Animal Behaviour* 56:459–66.

Waterman, J. M., and M. B. Fenton. 2000. The effect of drought on the social structure and use of space in Cape ground squirrels, *Xerus inauris*. *Ecoscience* 7:131–36.

Watson, A., and D. Jenkins. 1968. Experiments on population control by territorial behaviour in red grouse. *Journal of Animal Ecology* 37:595–614.

Watson, A., and R. Moss. 1970. Dominance, spacing behavior and aggression in relation to population limitation in vertebrates. In *Animal populations in relation to their food resources*, ed. A. Watson, 167–220. Oxford: Blackwell.

Watson, J. R. 1991. The identification of river foreshore corridors for nature conservation in the south coast region of Western Australia. In *Nature conservation 2: The role of corridors*, ed. D. A. Saunders and R. J. Hobbs, 63–68. Chipping Norton, Australia: Surrey Beatty and Sons.

Watson, J. S. 1944. The melanic form of *Rattus norvegicus* in London. *Nature* 154:334–35.

Watters, J. J., C. W. Wilkinson, and D. M. Dorsa. 1996. Glucocorticoid regulation of vasopressin V1a receptors in rat forebrain. *Molecular Brain Research* 38:276–84.

Watts, C. H. S. 1975. Vocalizations of Australian hopping mice (Rodentia, Notomys). *Journal of Zoology* 177:247–63.

Watts, C. H. S., and H. J. Aslin. 1981. *The rodents of Australia*. Sydney: Angus and Robertson.

Watts, C. H. S., and C. M. Kemper. 1989. 47. Muridae. In *Fauna of Australia, Volume 1B, Mammalia*, ed. D. W. Walton and B. J. Richardson, 1–35. Fauna of Australia Series, Australian Government Publishing Service.

Wauters, L. A., and A. A. Dhondt. 1989. Body weight, longevity and reproductive success in red squirrels (*Sciurus vulgaris*). *Journal of Animal Ecology* 58:637–51.

———. 1995. Lifetime reproductive success and its correlates in female Eurasian red squirrels. *Oikos* 72:402–10.

Wauters, L., A. A. Dhondt, and R. De Vos. 1990. Factors affecting male mating success in red squirrels (*Sciurus vulgaris*). *Ethology, Ecology, and Evolution* 2:195–204.

Wcislo, W. T., and B. N. Danforth. 1997. Secondarily solitary: The evolutionary loss of social behavior. *Trends in Ecology and Evolution* 12:468–74.

Weary, D. M., and D. L. Kramer. 1995. Response of eastern chipmunks to conspecific alarm calls. *Animal Behaviour* 49:81–93.

Weatherhead, P. J., and R. J. Robertson. 1979. Offspring quality and the polygyny threshold: "The sexy son hypothesis." *American Naturalist* 113:201–8.

Webb, D. R. 1980. Environmental harshness, heat stress, and *Marmota flaviventris*. *Oecologia* 44:390–95.

Webb, D. R., J. L. Fullenwider, P. A. McClure, L. Profeta, and J. Long. 1990. Geometry of maternal-offspring contact in two rodents. *Physiological Zoology* 63:821–44.

Webster, A. B., and R. J. Brooks. 1981. Social behavior of *Microtus pennsylvanicus* in relation to seasonal changes in demography. *Journal of Mammalogy* 62:738–51.

Webster, A. B., R. G. Gartshore, and R. J. Brooks. 1981. Infanticide in the meadow vole, *Microtus pennsylvanicus*: Significance in relation to social system and population cycling. *Behavioral and Neural Biology* 31:342–47.

Webster, D. B., and W. Plassmann. 1992. Parallel evolution of low-frequency sensitivity in old world and new world desert rodents. In *The evolutionary biology of hearing*, ed. R. R. Fry and A. N. Popper, 633–66. New York: Springer.

Webster, D. B., and M. Webster. 1971. Adaptive value of hearing and vision in kangaroo rat predator avoidance. *Brain Behavior and Evolution* 4:310–22.

———. 1984. The specialized auditory system of kangaroo rats. *Sensory Physiology* 8:161–96.

Webster, J. P. 1994. The effects of *Toxoplasma gondii* and other parasites on activity levels in wild and hybrid *Rattus norvegicus*. *Parasitology* 109:583–89.

———. 2001. Rats, cats, people and parasites: The impact of latent toxoplasmosis on behaviour. *Microbes and infection* 3:1037–45.

Webster, J. P., C. F. A. Brunton, and D. W. Macdonald. 1994. Effect of *Toxoplasma gondii* on neophobic behaviour in wild brown rats, *Rattus norvegicus*. *Parasitology* 109:37–43.

Webster, J. P., and D. W. Macdonald. 1995a. Cryptosporidiosis reservoir in wild brown rats (*Rattus norvegicus*): First report in the UK. *Epidemiology and Infection* 115:207–9.

———. 1995b. Parasites of wild brown rats (*Rattus norvegicus*) on UK farms. *Parasitology* 111:247–55.

Webster, W. D., and J. K. Jones, Jr. 1982. *Reithrodontomys megalotis*. *Mammalian Species* 167:1–5.

Weckerly, F. W. 1998. Sexual-size dimorphism: Influence of mass and mating systems in the most dimorphic mammals. *Journal of Mammalogy* 79:33–52.

Weddell, B. J. 1989. Dispersion of Columbian ground squirrels (*Spermophilus columbianus*) in meadow steppe and coniferous forest. *Journal of Mammalogy* 70:842–45.

Wedekind, C., T. Seebeck, F. Bettens, and A. Paepke. 1995. MHC-dependent mate preferences in humans. *Proceedings of the Royal Society of London, Series B* 260:245–49.

Weigl, P. D., and E. V. Hanson. 1980. Observational learning and the feeding behavior of the red squirrel *Tamiasciurus hudsonicus*: the ontogeny of optimization. *Ecology* 61: 213–18.

Weir, B. J. 1974. Reproductive characteristics of hystricognath rodents. *Symposium of the Zoological Society of London* 34:265–301.

Weissenbacher, B. K. H. 1987. Infanticide in tree squirrels—a male reproductive strategy? *South African Journal of Zoology* 22:115–18.

Weksler, M. 2003. Phylogeny of Neotropical oryzomyine rodents (Muridae: Sigmodontinae) based on the nuclear IRBP exon. *Molecular Phylogenetics and Evolution* 29:331–49.

Welberg, L. A. M., and J. R. Seckl. 2001. Prenatal stress, glucocorticoids and the programming of the brain. *Journal of Neuroendocrinology* 13:113–28.

Wells-Gosling, N. M., and L. R. Heaney. 1984. *Glaucomys sabrinus*. *Mammalian species* 229:1–8.

Werren, J. H., and E. L. Charnov. 1978. Facultative sex ratios and population dynamics. *Nature* 272:349–50.

Werren, J. H., and M. J. Hatcher. 2000. Maternal-zygotic gene conflict over sex determination: Effects of inbreeding. *Genetics* 155:1469–79.

West, M. J., A. P. King, and T. M. Freeberg. 1994. The nature and nurture of neo-phenotypes: A case history. In *Behavioral mechanisms in evolutionary ecology*, ed. L. A. Real, 238–57. Chicago: University of Chicago Press.

West, M. J., A. P. King, and D. J. White. 2003. The case for developmental ecology. *Animal Behaviour* 66:617–22.

West, S. A., S. E. Reece, and B. C. Sheldon. 2002. Sex ratios. *Heredity* 88:117–24.

West, S. D. 1977. Mid-winter aggregation in the northern red backed vole, *Clethrionomys rutilus*. *Canadian Journal of Zoology* 55:1404–9.

Western, D. 1979. Size, life history and ecology in mammals. *African Journal of Ecology* 17:185–204.

Western, D., and J. Ssemakula. 1982. Life history patterns in birds and mammals and their evolutionary interpretation. *Oecologia* 54:281–90.

Westneat, D. F., and P. W. Sherman. 1993. Parentage and the evolution of parental behavior. *Behavioral Ecology* 4:66–77.

———. 1997. Density and extra-pair fertilizations in birds. *Behavioral Ecology and Sociobiology* 41:205–15.

Westneat, D. F., P. W. Sherman, and M. L. Morton. 1990. The ecology and evolution of extra-pair copulations in birds. In *Current ornithology*, vol. 7, ed. D. M. Power, 331–69. New York: Plenum.

Wheatley, M. 1993. Report of two pregnant beavers, *Castor canadensis*, at one beaver lodge. *Canadian Field-Naturalist* 107:103.

———. 1994. Boreal beavers (*Castor canadensis*): Home range, territoriality, food habits and genetics of a mid-continent population. PhD diss., University of Manitoba, Winnipeg.

———. 1997. Beaver, *Castor canadensis*, home range size and patterns of use in the taiga of southeastern Manitoba. II. Sex, age and family status. *Canadian Field-Naturalist* 111:211–16.

Whicker, A., and J. K. Detling. 1988. Ecological consequences of prairie dog disturbances. *Bioscience* 38:778–85.

Whisenant, S. G. 1990. Changing fire frequencies on Idaho's Snake River Plains: Ecological and management implications. In *Proceedings—symposium on cheatgrass invasion, shrub die-off, and other aspects of shrub biology and management*, compiled by E. D. McArthur, E. M. Romney, S. D. Smith, and P. T. Tueller, 4–10. General Technical Report INT-276. 5–7 April, 1989, Las Vegas, NV. Ogden, UT: USDA Forest Service, Intermountain Research Station.

Whitaker, J. O., Jr. 1963. A study of the meadow jumping mouse, *Zapus hudsonius* (Zimmerman), in central New York. *Ecological Monographs* 33:215–54.

White, D. J., and B. G. Galef, Jr. 1998. Social influence on avoidance of dangerous stimuli by rats. *Animal Learning and Behavior* 26:433–38.

White, N. R., R. Cagiano, and R. J. Barfield. 1990. Receptivity of the female rat (*Rattus norvegicus*) after male devocalization: A ventral perspective. *Journal of Comparative Psychology* 104:147–51.

White, P., R. Fischer, and G. Meunier. 1982. The lack of recognition of lactating females by infant *Octodon degus*. *Physiology and Behavior* 28:623–25.

Whiten, A., J. Goodall, W. C. McGrew, T. Nishida, V. Reynolds, Y. Sugiyama, C. E. G. Tutin, R. W. Wrangham, and C. Boesch. 1999. Cultures in chimpanzees. *Nature* 399: 682–85.

Whiten, A., and R. Ham. 1992. On the nature and evolution of imitation in the animal kingdom: Reappraisal of a century of research. *Advances in the Study of Behaviour* 21:239–83.

Whitman, D. C., A. C. Hennessey, and H. E. Albers. 1992. Norepinephrine inhibits vasopressin-stimulated flank marking in the Syrian hamster by acting within the medial preoptic-anterior hypothalamus. *Journal of Neuroendocrinology* 4:541–46.

Whitten, W. K. 1958. Modification of the oestrous cycle of the mouse by external stimuli associated with the male: Changes in the oestrous cycle determined by vaginal smears. *Journal of Endocrinology* 17:307–13.

———. 1959. Occurrence of anoestrus in mice caged in groups. *Journal of Endocrinology* 18:102–7.

Whitten, W. K., F. H. Bronson, and J. A. Greenstein. 1968. Estrus-inducing pheromone of male mice: Transport by movement of air. *Science* 161:584–85.

Widdig, A., P. Nurnberg, M. Krawczak, W. J. Streich, and F. B. Bercovitch. 2001. Paternal relatedness and age proximity regulate social relationships among adult female rhesus macaques. *Proceedings of the National Academy of Sciences (USA)* 98:13769–73.

Wiens, J. A. 1992. Ecological flows across landscape boundaries: A conceptual view. In *Landscape boundaries, consequences for biodiversity and ecological flows. Ecological Studies,* vol. 92, ed. A. J. Hansen and F. di Castri, 217–35. New York: Springer-Verlag.

Wiggett, D. R., and D. A. Boag. 1989. Intercolony natal dispersal in the Columbian ground squirrel. *Canadian Journal of Zoology* 67:42–50.

———. 1992. The resident fitness hypothesis and dispersal by yearling female Columbian ground squirrels. *Canadian Journal of Zoology* 70:1984–94.

———. 1993a. Annual reproductive success in three cohorts of Columbian ground squirrels: Founding immigrants, subsequent immigrants, and natal residents. *Canadian Journal of Zoology* 71:1577–84.

———. 1993b. The proximate causes of male-biased emigration in Columbian ground squirrels. *Canadian Journal of Zoology* 71:204–18.

Wiggett, D. R., D. A. Boag, and A. D. R. Wiggett. 1989. Movements of intercolony natal dispersers in the Columbian ground squirrel. *Canadian Journal of Zoology* 67:1447–52.

Wikelski, M., and F. Trillmich. 1997. Body size and sexual size dimorphism in marine iguanas fluctuate as a result of opposing natural and sexual selection: An island comparison. *Evolution* 51:922–36.

Wilcox, R. M., and R. E. Johnston. 1995. Scent counter marks: specialised mechanisms of perception and response to individual odors in golden hamsters, *Mesocricetus auratus*. *Journal of Comparative Psychology* 109:349–56.

Wiley, R. H. 1991. Associations of song properties with habitats for territorial oscine birds of eastern North America. *American Naturalist* 138:973–93.

Wiley, R. W. 1972. Reproduction, postnatal development and growth of the southern plains woodrat (*Neotoma micropus*) in western Texas. PhD diss., Texas Tech University, Lubbock.

———. 1980. *Neotoma floridana*. *Mammalian Species* 139:1–7.

Wilkinson, G. S. 2002. Reciprocal altruism. In *Encyclopedia of evolution*, ed. M. Pagel, 985–90. New York: Oxford University Press.

Wilkinson, G. S., and A. E. M. Baker. 1988. Communal nesting among genetically similar house mice. *Ethology* 77:103–14.

Wilks, B. 1963. Some aspects of the ecology and population dynamics of the pocket gophers (Geomyidae: Rodentia). *American Nature* 96:303–16.

Williams, G. C. 1957. Pleiotropy, natural selection and the evolution of senescence. *Evolution* 11:398–411.

———. 1966. *Adaptation and natural selection*. Princeton, NJ: Princeton University Press.

———. 1975. *Sex and evolution*. Princeton, NJ: Princeton University Press.

———. 1979. The question of adaptive sex ratio in outcrossed vertebrates. *Proceedings of the Royal Society of London, Series B* 205:567–80.

Williams, J. R., K. C. Catania, and C. S. Carter. 1992. Development of partner preferences in female prairie voles (*Microtus ochrogaster*): The role of social and sexual experience. *Hormones and Behavior* 26:339–49.

Williams, J. R., T. R. Insel, C. R. Harbaugh, and C. S. Carter. 1994. Oxytocin administered centrally facilitates formation of a partner preference in female prairie voles (*Microtus ochrogaster*). *Journal of Neuroendocrinology* 6:247–50.

Williams, S. L. 1982. *Geomys personatus*. *Mammalian Species* 170:1–5.

Williams, S. L., and R. J. Baker. 1973. *Geomys arenarius*. *Mammalian Species* 36:1–3.

Williams, S. L., J. Ramirez-Pulido, and R. J. Baker. 1985. *Peromyscus alstoni*. *Mammalian Species* 242:1–4.

Willis, C., and R. Poulin. 2000. Preference of female rats for the odours of non-parasitized males: The smell of good genes? *Folia Parasitologica* 47:6–10.

Wilson, D. E., and D. M. Reeder, eds. 1993. *Mammal species of the world: A taxonomic and geographic reference*, 2nd ed. Washington, DC: Smithsonian Institution Press.

———. 2005. *Mammal species of the world: A taxonomic and geographic reference*, 3rd ed. Baltimore: Johns Hopkins University Press.

Wilson, D. E., and S. Ruff. 1999. *North American mammals*. Washington, DC: Smithsonian Institution Press.

Wilson, D. S. 1975. A theory of group selection. *Proceedings of the National Academy of Science (USA)* 72:143–46.

Wilson, E. O. 1971. *The insect societies*. Cambridge, MA: Belknap Press/Harvard University Press.

———. 1975. *Sociobiology: The new synthesis*. Cambridge, MA: Belknap Press of Harvard University.

Wilson, M. I., M. Daly, and P. Behrends. 1985. The estrous cycle of two species of kangaroo rats (*Dipodomys microps* and *D. merriami*). *Journal of Mammalogy* 66:726–32.

Wilson, S. C. 1982a. The development of social behaviour between siblings and non-siblings of the voles *Microtus ochrogaster* and *Microtus pennsylvanicus*. *Animal Behaviour* 30:426–37.

———. 1982b. Parent-young contact in prairie and meadow voles. *Journal of Mammalogy* 63:300–5.

Wilson, W. L., R. W. Elwood, W. I. and Montgomery. 1993. Infanticide and maternal defense in the wood mouse *Apodemus sylvaticus*. *Ethology Ecology and Evolution* 5:365–70.

Wilsson, L. 1971. Observations and experiments on the ethology of the European Beaver (*Castor fiber* L.). *Viltrevy* 8:115–66.

Wingfield, J. C., R. E. Hegner, A. M. Dufty, and G. F. Ball. 1990. The "challenge hypothesis": Theoretical implications for patterns of testosterone secretion, mating systems, and breeding strategies. *American Naturalist* 136:829–46.

Wingfield, J. C., and L. M. Romero. 2001. Adrenocortical responses to stress and their modulation in free-living vertebrates. In *Coping with the environment: Neural and endocrine mechanisms*, ed. B. S. McEwen, 211–36. New York: Oxford University Press.

Wingfield, J. C., and R. M. Sapolsky. 2003. Reproduction and resistance to stress: When and how. *Journal of Neuroendocrinology* 15:711–24.

Winn, B. E., and B. M. Vestal. 1986. Kin recognition and choice of males by wild female house mice (*Mus musculus*). *Journal of Comparative Psychology* 100:72–75.

Winslow, J. T., N. Hastings, C. S. Carter, C. R. Harbaugh, and T. R. Insel. 1993. A role for central vasopressin in pair bonding in monogamous prairie voles. *Nature* 365:545–48.

Winters, J. B., and P. M. Waser. 2003. Gene dispersal and outbreeding in a philopatric mammal. *Molecular Ecology* 12:2251–59.

Wishner, L. A. 1982. *Eastern chipmunks*. Washington, DC: Smithsonian Institution Press.

Withers, P. C. 1983. Seasonal reproduction by small mammals of the Namib desert. *Mammalia* 47:195–204.

Withers, P. C., and D. H. Edward. 1997. Terrestrial fauna of granite outcrops in Western Australia. *Journal of the Royal Society of Western Australia* 80:159–66.

Wittenberger, J. F. 1979. The evolution of mating systems in birds and mammals. In *Handbook of behavioral neurobiology,* ed. P. Marler and J. Vandenbergh, 271–349. New York: Plenum.

Wittenberger, J. F., and R. L. Tilson. 1980. The evolution of monogamy: Hypotheses and evidence. *Annual Review of Ecology and Systematics* 11:197–232.

Woinarski, J. C. Z. 2000. The conservation status of rodents in the monsoonal tropics of the Northern Territory. *Wildlife Research* 27:421–35.

Wolf, M., and G. O. Batzli. 2004. Forest edge-high or low quality habitat for white-footed mice (*Peromyscus leucopus*)? *Ecology* 85:756–69.

Wolfe, J. L., and A. V. Linzey. 1977. *Peromyscus gossypinus. Mammalian Species* 70:1–5.

Wolff, J. O. 1980a. The role of habitat patchiness in the population dynamics of snowshoe hares. *Ecological Monographs* 50:111–30.

———. 1980b. Social organization of the taiga vole (*Microtus xanthognathus*). *The Biologist* 62:34–45.

———. 1985a. Behavior. In *Biology of New World* Microtus, Special publication no. 8, ed. R. H. Tamarin, 340–72. Shippensburg, PA: American Society of Mammalogists.

———. 1985b. The effects of density, food, and interference on home range size in *Peromyscus leucopus* and *Peromyscus maniculatus. Canadian Journal of Zoology* 63:2657–62.

——— 1985c. Maternal aggression as a deterrent to infanticide in *Peromyscus leucopus* and *P. maniculatus. Animal Behaviour* 33:117–23.

———. 1986. Infanticide in white-footed mice, *Peromyscus leucopus. Animal Behaviour* 34:1568.

———. 1989. Social behavior. In *Advances in the study of* Peromyscus *(Rodentia),* ed. G. L. Kirkland, Jr., and J. N. Layne, 271–91. Lubbock: Texas Tech University Press.

———. 1992. Parents suppress reproduction and stimulate dispersal in opposite-sex juvenile white-footed mice. *Nature* 359:409–10.

———. 1993a. What is the role of adults in mammalian juvenile dispersal? *Oikos* 68:173–76.

———. 1993b. Why are female small mammals territorial? *Oikos* 68:364–70.

———. 1994a. More on juvenile dispersal in mammals. *Oikos* 71:349–52.

———. 1994b. Reproductive success of solitarily and communally nesting white-footed mice and deer mice. *Behavioral Ecology* 5:206–9.

———. 1995. Friends and strangers in vole population cycles. *Oikos* 73:411–13.

———. 1997. Population regulation in mammals: an evolutionary perspective. *Journal of Animal Ecology* 66:1–13.

———. 1999. Behavioral model systems. In *Landscape ecology of small mammals,* ed. G. W. Barrett and J. D. Peles, 11–40. New York: Springer-Verlag.

——— . 2003a. Density-dependence and the socioecology of space use in rodents. In *Rats, mice and people: Rodent biology and management,* ed. G. R. Singleton, L. A. Hinds, C. J. Krebs, and D. M. Spratt, 124–30. Canberra: Australian Centre for International Agricultural Research.

———. 2003b. An evolutionary and behavioral perspective on dispersal and colonization of mammals in fragmented landscapes. In *Mammalian community dynamics: Management and conservation in the coniferous forests of western North America,* ed. C. J. Zabel and R. G. Anthony, 614–30. Cambridge: Cambridge University Press.

———. 2003c. Laboratory studies with rodents: Facts or artifacts. *BioScience* 53:421–27.

———. 2004. Scent marking by voles in response to predation risk: A field-laboratory validation. *Behavioral Ecology* 15:286–89.

———. In press. Alternative reproductive tactics in nonprimate mammals. In *Alternative reproductive tactics: An integrative approach,* ed. R. Oliveira, M. Taborsky, and H. J. Brockmann. Cambridge: Cambridge University Press.

Wolff, J. O., and D. M. Cicirello. 1989. Field evidence for sexual selection and resource competition infanticide in white-footed mice. *Animal Behaviour* 38:637–42.

———. 1990. Mobility versus territoriality: Alternative reproductive strategies in a polygynous rodent. *Animal Behaviour* 39:1222–24.

———. 1991. Comparative paternal and infanticidal behavior of sympatric white-footed mice (*Peromyscus leucopus noveboracensis*) and deermice (*P. maniculatus nubiterrae*). *Behavioral Ecology* 2:38–45.

Wolff, J. O., and R. Davis-Born. 1997. Response of gray-tailed voles to odours of a mustelid predator: A field test. *Oikos* 79:543–48.

Wolff, J. O., and R. D. Dueser. 1986. Noncompetitive coexistence between *Peromyscus* spp. and *Clethrionomys gapperi. Canadian Field-Naturalist* 100:186–91.

Wolff, J. O., and A. S. Dunlap. 2002. Multi-male mating, probability of conception, and litter size in the prairie vole (*Microtus ochrogaster*). *Behavioral Processes* 58:105–10.

Wolff, J. O., A. S. Dunlap, and E. Ritchhart. 2001. Adult female prairie voles and meadow voles do not suppress reproduction in their daughters. *Behavioural Processes* 55:157–62.

Wolff, J. O., W. D. Edge, and R. Bentley. 1994. Reproductive and behavioral biology of the gray-tailed vole. *Journal of Mammalogy* 75:873–79.

Wolff, J. O., W. D. Edge, and G. M. Wang. 2002. Effects of adult sex ratios on recruitment of juvenile gray-tailed voles, *Microtus canicaudus. Journal of Mammalogy* 83:947–56.

Wolff, J. O., M. H. Freeberg, and R. D. Dueser. 1983. Intra- and Interspecific territoriality in two sympatric species of *Peromyscus* (Rodentia: Cricetidae). *Behavioral Ecology and Sociobiology* 12:237–42.

Wolff, J. O., and R. D. Guthrie. 1985. Why are aquatic small mammals so large? *Oikos* 45:365–73.

Wolff, J. O., and W. Z. Lidicker, Jr. 1981. Communal winter nesting and food sharing in taiga voles. *Behavioral Ecology and Sociobiology* 9:237–40.

Wolff, J. O., and K. I. Lundy. 1985. Intra-familial dispersion patterns in white-footed mice, *Peromyscus leucopus. Behavioral Ecology and Sociobiology* 17:831–34.

Wolff, J. O., K. I. Lundy, and R. Baccus. 1988. Dispersal, inbreeding avoidance and reproductive success in white-footed mice. *Animal Behaviour* 36:456–65.

Wolff, J. O., and D. W. Macdonald. 2004. Promiscuous females protect their offspring. *Trends in Ecology and Evolution* 19:127–34.

Wolff, J. O., and J. A. Peterson. 1998. An offspring-defense hypothesis for territoriality in female mammals. *Ethology Ecology and Evolution* 10:227–39.

Wolff, J. O., and E. M. Schauber. 1996. Space use and juvenile recruitment in gray-tailed voles in response to intruder pressure and food abundance. *Acta Theriologica* 41:35–43.

Wolff, J. O., M. H. Watson, and S. A. Thomas. 2002. Is self-grooming by male prairie voles a predictor of mate choice? *Ethology* 108:169–79.

Wolff, R. J. 1985. Mating behaviour and female choice: Their relation to social structure in wild caught house mice (*Mus musculus*) housed in a semi-natural environment. *Journal of Zoology* 207:43–51.

Wolton, R. J. 1985. The ranging and nesting behaviour of wood mice, *Apodemus sylvaticus* (Rodentia: Muridae), as revealed by radio-tracking. *Journal of Zoology, London* 206:203–24.

Wood, A. E. 1955. A revised classification of the rodents. *Journal of Mammalogy* 36:165–87.

———. 1957. What, if anything, is a rabbit? *Evolution* 11:417–25.

———. 1965. Grades and clades among rodents. *Evolution* 19:115–30.

Wood, S. R., K. J. Sanderson, and C. S. Evans. 2000. Perception of terrestrial and aerial alarm calls by honeyeaters and falcons. *Australian Journal of Zoology* 48:127–34.

Woodard, E. L. 1994. Behavior, activity patterns and foraging strategies of beaver (*Castor canadensis*) on Sagehen Creek, California. PhD diss., University of California, Berkeley.

Woods, B. C. 2001. Diet and hibernation of yellow-bellied marmots: Why are marmots so fat? PhD diss., University of Kansas, Lawrence.

Woods, C. A. 1984. Hystricognath rodents. In *Orders and families of recent mammals of the world*, ed. S. Anderson and J. K. Jones, 389–446. New York: Wiley.

———. 1989. Endemic rodents of the West Indies: The end of splendid isolation. In *Rodents, a world survey of species of conservation concern*, ed. W. Z. Lidicker, Jr., 11–19. Occasional Papers no. 4. Gland, Switzerland: Rodent Specialist Group, Species Survival Commission, IUCN.

———. 1993. Suborder Hystricognathi. In *Mammal species of the world: A taxonomic and geographic reference*, 2nd ed., ed. D. E. Wilson and D. M. Reeder, 771–806. Washington, DC: Smithsonian Institution Press.

Woods, C. A., and D. K. Boraker. 1975. Octodon degus. *Mammalian Species* 67:1–5.

Woods, C. A., and J. W. Hermanson. 1985. Myology of hystricognath rodents: An analysis of form, function, and phylogeny. In *Evolutionary relationships among rodents: A multidisciplinary analysis*, ed. W. P. Luckett and J. L. Hartenberger, 515–48. New York: Plenum.

Woods, H. A. II, and E. C. Hellgren. 2003. Seasonal changes in the physiology of male Virginia opossums (*Didelphis virginiana*): Signs of dasyurid semelparity syndrome? *Physiology and Biochemical Zoology* 76:406–17.

Woodward, R. L., M. K. Schmick, and M. H. Ferkin. 1999. Response of prairie voles, *Microtus ochrogaster* (Rodentia, Arvicolidae), to scent over-marks of two same-sex conspecifics: A test of the scent-masking hypothesis. *Ethology* 105:1009–17.

Woolfenden, G. E., and J. W. Fitzpatrick. 1990. Florida scrub jays: A synopsis after 18 years of study. *Cooperative breeding in birds*, ed. P. B. Stacey and W. D. Koenig, 241–66. Cambridge: Cambridge University Press.

Wrangham, R. W., and D. I. Rubenstein. 1986. Social evolution in birds and mammals. In *Ecological aspects of social evolution*, ed. D. I. Rubenstein and R. W. Wrangham, 452–70. Princeton, NJ: Princeton University Press.

Wright, D. D., J. T. Ryser, and R. A. Kiltie. 1995. First-cohort advantage hypothesis: A new twist on facultative sex ratio adjustment. *American Naturalist* 145:133–45.

Wright, S. 1965. The interpretation of population structure by F-statistics with special regard to systems of mating. *Evolution* 19:395–420.

———. 1969. *Evolution and the genetics of populations. Volume 2: The theory of gene frequencies.* Chicago: University of Chicago Press.

———. 1978. *Evolution and the genetics of populations. Volume 4: Variability within and among natural populations.* Chicago: University of Chicago Press.

Wright, S. L., and R. E. Brown. 2000. Maternal behavior, paternal behavior, and pup survival in CD-1 albino mice (*Mus musculus*) in three different housing conditions. *Journal of Comparative Psychology* 114:183–92.

Wright, S. L., C. B. Crawford, and J. L. Anderson. 1988. Allocation of reproductive effort in *Mus domesticus*: Responses of offspring sex ratio and quality to social density and food availability. *Behavioral Ecology and Sociobiology* 23:357–65.

Wright-Smith, M. A. 1978. The ecology and social organization of *Cynomys parvidens* (Utah prairie dog) in south central Utah. Master's thesis, Indiana University, Bloomington.

Wrigley, R. E. 1972. *Systematics and biology of the woodland jumping mouse, Napaeozapus insignis.* Illinois Biological Monographs 47. Urbana: University of Illinois Press.

Wu, H. M., W. G. Holmes, S. R. Medina, and G. P. Sackett. 1980. Kin preference in infant *Macaca nemestrina*. *Nature* 285:225–27.

Wuensch, K. L., and A. J. Cooper. 1981. Preweaning paternal presence and later aggressiveness in male *Mus musculus*. *Behavioral and Neural Biology* 32:510–15.

Wuerthner, G. 1997. Viewpoint: The black-tailed prairie dog—headed for extinction? *Journal of Range Management* 50:459–66.

Wynne-Edwards, K. E. 1987. Evidence for obligate monogamy in the Djungarian hamster *Phodopus campbelli*: Pup survival under different parenting conditions. *Behavioral Ecology and Sociology* 20:427–438.

———. 2003. From dwarf hamster to daddy: The intersection of ecology, evolution, and physiology that produces paternal behavior. *Advances in the Study of Behavior* 32:207–61.

Wynne-Edwards, K. E., and R. D. Lisk 1984. Djungarian hamsters fail to conceive in the presence of multiple males. *Animal Behaviour* 32:626–628.

———. 1989. Differential effects of paternal presence on pup survival in two species of dwarf hamster (*Phodopus sungorus* and *Phodopus campbelli*). *Physiology and Behavior* 45:465–69.

Wyss, A. R., J. J. Flynn, M. A. Norell, C. C. Swisher III, R. Charrier, M. J. Novacek, and M. C. McKenna. 1993. South America's earliest rodent and recognition of a new interval of mammalian evolution. *Nature* 365:434–437.

Wyss, A. R., J. J. Flynn, M. A. Norell, C. C. Swisher III, M. J. Novacek, M. C. McKenna, and R. Charrier. 1994. Paleogene mammals from the Andes of central Chile: A preliminary taxonomic, biostratigraphic, and geochronologic assessment. *American Museum Novitates* 3098:1–31.

Xia, V., C. Lio, W. Zong, C. Sun, and Y. Tian. 1982. On the population dynamics and regulation of *Meriones unguiculatus* in the agricultural region north of the Yin mountains, Inner Mongolia. *Acta Theriologica* 2:51–72.

Xia, X., and J. S. Millar. 1988. Paternal behavior by *Peromyscus leucopus* in enclosures. *Canadian Journal of Zoology* 66: 1184–87.

———. 1989. Dispersion of adult male *Peromyscus leucopus* in relation to female reproductive status. *Canadian Journal of Zoology* 67:1047–52.

———. 1991. Genetic evidence of promiscuity in *Peromyscus leucopus*. *Behavioral Ecology and Sociobiology* 28:171–78.

Yáber, M. C., and E. A. Herrera. 1994. Vigilance, group size and social status in capybaras. *Animal Behaviour* 48:1301–7.

Yahner, R. H. 1978a. The adaptive nature of the social system and behavior in the eastern chipmunk, *Tamias striatus*. *Behavioral Ecology and Sociobiology* 3:397–427.

———. 1978b. Seasonal rates of vocalizations in eastern chipmunks. *Ohio Journal of Science* 78:301–3.

———. 1980. Burrow system use by red squirrels. *American Midland Naturalist* 103:409–11.

Yahnke, C. J., P. L. Meserve, T. G. Ksiazek, and J. N. Mills. 2001. Patterns of infection with Laguna Negra virus in wild populations of Calomys laucha in the central Paraguayan chaco. *American Journal of Tropical Medicine and Hygiene* 65:768–76.

Yakovlev, F. G., and G. E. Shadrina. 1996. Density and demographic structure of black-capped marmot (*Marmota camtschatica* Pall.) in northeastern Yakutia. In *Biodiversity in marmots*, ed. M. Le Berre, R. Ramousse, and L. Le Guelte, 267–68. Lyon: International Network on Marmots.

Yamashita, J., S-I. Hayashi, and Y. Hirata. 1989. Reduced size of preputial glands and absence of aggressive behaviour in the genetically obese (ob/ob) mouse. *Zoological Science* 6:1033–36.

Yamazaki, K., G. K. Beauchamp, J. Bard, and E. A. Boyse. 1990. Single MHC gene mutations alter urine odor constitution in mice. In *Chemical signals in vertebrates*, ed. S. E. Natynczuk, 255–59. Oxford: Oxford University Press.

Yamazaki, K., G. K. Beauchamp, M. Curran, J. Bard, and E. A. Boyse. 2000. Parent-progeny recognition as a function of MHC odortype identity. *Proceedings of the National Academy of Sciences (USA)* 97:10500–2.

Yamazaki, K., G. K. Beauchamp, I. K. Egorov, J. Bard, L. Thomas, and E. A. Boyse. 1983. Sensory distinction between H-2 *b* and H-2 *bm1* mutant mice. *Proceedings of the National Academy of Sciences (USA)* 80:5685–88.

Yamazaki, K., G. K. Beauchamp, D. Kupniewski, J. Bard, L. Thomas, and E. A. Boyse. 1988. Familial imprinting determines H-2 selective mating preferences. *Science* 240:1331–32.

Yamazaki, K., G. K. Beauchamp, F. W. Shen, J. Bard, and E. A. Boyse. 1991. A distinctive change in odortype determined by H-2D/L mutation. *Immunogenetics* 34:129–31.

Yamazaki, K., E. A. Boyse, V. Mike, H. T. Thaler, B. J. Mathieson, J. Abbott, J. Boyse, Z. A. Zayas and L. Thomas. 1976. Control of mating preferences in mice by genes in the major histocompatability complex. *Journal of Experimental Medicine* 144:1324–35.

Yamazaki, K., M. Yamaguchi, P. W. Andrews, B. Peake, and E. A. Boyse. 1978. Mating preferences of F2 segregants of crosses between MHC-congenic mouse strains. *Immunogenetics* 6:253–59.

Yamazaki, K., M. Yamaguchi, L. Baranoski, J. Bard, E. A. Boyse, and L. Thomas. 1979. Recognition among mice: Evidence from the use of a Y-maze differentially scented by congenic mice of different major histocompatibility types. *Journal of Experimental Medicine* 150:755–60.

Yáñez, J. 1976. Eco-etología de *Octodon degus*. Bachelors thesis, Universidad de Chile, Santiago, Chile.

Yasui, Y. 1997. A "good-sperm" model can explain the evolution of costly multiple mating by females. *The American Naturalist* 149:573–84.

———. 2001. Female multiple mating as a genetic bet-hedging strategy when mate choice criteria are unreliable. *Ecological Research* 16:605–16.

Yates, T. L., J. N. Mills, C. A. Parmenter, T. G. Ksiazek, R. R. Parmenter, J. R. Van de Castle, C. H. Calisher et al. 2002. The ecology and evolutionary history of an emergent disease: Hantavirus pulmonary syndrome. *BioScience* 52:989–98.

Yeaton, R. I. 1972. Social behaviour and social organization in Richardson's ground squirrel (*Spermophilus richardsonii*) in Saskatchewan. *Journal of Mammalogy* 53:139–47.

Yensen, E., and P. W. Sherman. 1997. *Spermophilus brunneus*. *Mammalian Species* 560:1–5.

———. 2003. Ground squirrels (*Spermophilus* species and *Ammospermophilus* species). In *Wild mammals of North America,* 2nd ed., ed. G. Feldhammer, B. Thompson, and J. Chapman, 211–31. Baltimore: Johns Hopkins University Press.

Yin, F., and J. M. Fang. 1998. Mate choice in Brandt's voles (*Microtus brandti*). *Acta Zoologica Sinica* 44:162–69.

Ylönen, H. 1988. Diel activity and demography in an enclosed population of the vole *Clethrionomys glareolus* (Schreb.). *Annales Zoologici Fennici* 25:221–28.

———. 1989. Weasels, *Mustela nivalis*, suppress reproduction in cyclic bank voles *Clethrionomys glareolus*. *Oikos* 5:138–40.

———. 1994. Vole cycles and antipredatory behaviour. *Trends in Ecology and Evolution* 9:426–30.

———. 2001. Predator odours and behavioural responses of small rodents: An evolutionary perspective. In *Advances in vertebrate pest management* 2, ed. H.-J. Peltz, D. P. Cowan, and C. J. Feare, 123–38. Fuerth: Filander.

Ylönen, H., and T. J. Horne. 2002. Infanticide and effectiveness of pup protection in bank voles: Does the mother recognize a killer? *Acta Ethologica* 4:97–101.

Ylönen, H., J. Jacob, M. Davis, and G. R. Singleton. 2002. Predation risk and habitat selection of Australian house mice (*Mus domesticus*) during an incipient plague: Desperate behaviour due to food depletion. *Oikos* 99:285–90.

Ylönen, H., B. Jedrzejewska, W. Jedrzejewski, and J. Heikkilä. 1992. Antipredatory behaviour of *Clethrionomys* voles: "David and Goliath" arms race. *Annales Zoologici Fennici* 29:207–16.

Ylönen, H., E. Koskela, and T. Mappes. 1997. Infanticide in the bank vole (*Clethrionomys glareolus*): Occurrence and the effect of familiarity on female infanticide. *Annales Zoologici Fennici* 34:259–66.

Ylönen, H., and H. Ronkainen. 1994. Breeding suppression in the bank vole as antipredatory adaptation in a predictable environment. *Evolutionary Ecology* 8:658–66.

Ylönen, H., J. Sundell, R. Tiilikainen, J. Eccard, and T. J. Horne. 2003. Hunting preference of the least weasel (*Mustela nivalis nivalis*) and olfactory cues of age and sex groups of the bank vole (*Clethrionomys glareolus*). *Ecology* 84:1447–52.

Ylönen, H., and J. Viitala. 1985. Social organization of an enclosed winter population of the bank vole *Clethrionomys glareolus*. *Annales Zoologici Fennici* 22:353–58.

Yoccoz, N. G., and S. Mesnager. 1998. Are alpine bank voles larger and more sexually dimorphic because adults survive better? *Oikos* 82:85–98.

Yoerg, S. I. 1999. Solitary is not asocial. *Ethology* 105:317–33.

Yoerg, S. I., and D. M. Shier. 2000. Captive breeding and anti-predator behavior of the Heermann's kangaroo rat (*Dipodomys heermanni*). Final report. Sacramento, CA: California Department of Fish and Game.

Yosida, T. H. 1980. *Cytogenetics of the black rat: Karyotype evolution and species differentiation*. Tokyo: University of Tokyo Press.

Young, L. J., R. Nilsen, K. G. Waymire, G. R. MacGregor, and T. R. Insel. 1999. Increased affiliative response to vasopressin in mice expressing the V1a receptor from a monogamous vole. *Nature* 400:766–68.

Young, L. J., J. T. Winslow, R. Nilsen, and T. R. Insel. 1997. Species differences in V1a receptor gene expression in monogamous and nonmonogamous voles: Behavioral consequences. *Behavioral Neuroscience* 111:599–605.

Youngman, P. M. 1975. *Mammals of the Yukon territory*. Ottawa: National Museum of Canada.

Yu, X., R. Sun, and J. Fang. 2004. Effect of kinship on social behaviors in Brandt's voles (*Microtus brandti*). *Journal of Ethology* 22:17–22.

Zahavi, A. 1975. Mate selection: A selection for a handicap. *Journal of Theoretical Biology* 53:205–14.

———. 2003. Indirect selection and individual selection in sociobiology: My personal views on theories of social behaviour. *Animal Behaviour* 65:859–63.

Zahler, P. 1996. Rediscovery of the woolly flying squirrel (*Eupetaurus cinereus*). *Journal of Mammalogy* 77:54–57.

Zahler, P., and M. Khan. 2003. Evidence for dietary specialization on pine needles by the woolly flying squirrel (*Eupetaurus cinereus*). *Journal of Mammalogy* 84:480–86.

Zahm, D. S., S. Grosu, E. A. Williams, S. Qin, and A. Berod. 2001. Neurons of origin of the neurotensinergic plexus enmeshing the ventral tegmental area in rat: Retrograde labeling and in situ hybridization combined. *Neuroscience* 104:841–51.

Zala, S. M., W. K. Potts, and D. J. Penn. 2004. Scent-marking displays provide honest signals of health and infection. *Behavioral Ecology* 15:338–44.

Zammuto, R. M., and J. S. Millar. 1985. Environmental predictability, variability, and *Spermophilus columbianus* life history over an elevational gradient. *Ecology* 66:1784–94.

Zapletal, M. 1964. On the occurrence of the brown rat (*Rattus norvegicus*, Berk.) under natural conditions in Czechoslovakia. *Zoologicke Listy* 13:125–34.

Zarrow, M. X., and J. H. Clark. 1968. Ovulation following vaginal stimulation in a spontaneous ovulator and its implications. *Journal of Endocrinology* 40:343–52.

Zeh, J. A., and D. W. Zeh. 1996. The evolution of polyandry I: Intragenomic conflict and genetic incompatibility. *Proceedings of the Royal Society of London, Series B* 263:1711–17.

Zeitz, P. S., J. C. Butler, J. E. Cheek, M. C. Samuel, J. E. Childs, L. A. Shands, R. E. Turner et al. 1995. A case-control study of hantavirus pulmonary syndrome during an outbreak in the southwestern United States. *Journal of Infectious Diseases* 171:864–70.

Zelley, R. A. 1971. The sounds of the fox squirrel, *Sciurus niger rufiventer*. *Journal of Mammalogy* 52:597–604.

Zeng, A., and J. H. Brown. 1987. Population ecology of a desert rodent: *Dipodomys merriami* in Chihuahuan desert. *Ecology* 68:1328–40.

Zentall, T. R 2004. Action imitation in birds. *Learning and Behavior* 32:15–23.

———. 1988. Experimentally manipulated imitative behavior in rats and pigeons. In *Social learning: Psychological and biological perspectives*, ed. T. R. Zentall and B. G. Galef, Jr., 191–206. Hillsdale, NJ: Lawrence Erlbaum.

Zenuto, R. R., and M. S. Fanjul. 2002. Olfactory discrimination of individual scents in the subterranean rodent *Ctenomys talarum* (tuco-tuco). *Ethology* 108:629–41.

Zenuto, R. R., A. I. Malizia, and C. Busch. 1999. Sexual size dimorphism, testes size and mating system in two populations of *Ctenomys talarum* (Rodentia: Octodontidae). *Journal of Natural History* 33:305–14.

Zeveloff, S. I., and M. S. Boyce. 1980. Parental investment and mating systems in mammals. *Evolution* 34:973–82.

Zhang, J. X., C. Cao, H. Gao, Z. S. Yang, L. X. Sun, Z. B. Zhang, and Z. W. Wang. 2003. Effects of weasel odor on behavior and physiology of two hamster species. *Physiology and Behavior* 79:549–52.

Zhao, Y. J., R. Y. Sun, J. M. Fang, B. M. Li, and X. Q. Zhao. 2003. Preference of pubescent females for dominant vs. subordinate males in root voles. *Acta Zoologica Sinica* 49:303–9.

Zhao, Y. J., F. D. Tai, T. Z. Wang, X. Q. Zhao, and B. M. Li. 2002. Effects of the familiarity on mate choice and mate recognition in *Microtus mandarinus* and *M. oeconomus*. *Acta Zoologica Sinica* 48:167–74.

Ziak, D., and L. Kocian. 1996. Territorial behaviour of bank vole (*Clethrionomys glareolus*) females and its interpretation on the level of relationship between individuals. *Biologia* 51:601–6.

Zielinski, W. J., F. S. vom Saal, and J. G. Vandenbergh. 1992. The effect of intrauterine position on the survival, reproduction and home range size of female house mice (*Mus musculus*). *Behavioral Ecology and Sociobiology* 30:185–91.

Zimen, E. 1976. On the regulation of pack size in wolves. *Zeitschrift für Tierpsychologie* 40:300–41.

Zimina, R. P. 1996. Role of marmots in landscape transformations since Pleistocene. In *Biodiversity in marmots*, ed. M. Le Berre, R. Ramousse, and L. Le Guelte, 59–62. Lyon: International Network on Marmots.

Zimina, R. P., and I. P. Gerasimov. 1973. The periglacial expansion of marmots (*Marmota*) in Middle Europe during Late Pleistocene. *Journal of Mammalogy* 54:327–40.

Zinkernagel, R. M., and P. C. Doherty. 1974. Restriction of in vitro T-cell mediated cyctotoxicity in lymphocytic choriomeningitis within a syngeneic or semiallogeneic system. *Nature* 248:701–2.

Zohar, O., and J. Terkel. 1992. Acquisition of pinecone stripping behaviour in black rats (*Rattus rattus*). *International Journal of Comparative Psychology* 5:1–6.

———. 1995. Spontaneous learning of pinecone opening behavior by black rats. *Mammalia* 59:481–87.

Zuberbühler, K. 2000. Referential labelling in Diana monkeys. *Animal Behaviour* 59:917–27.

Zuk, M. 1992. The role of parasites in sexual selection: Current evidence and future directions. *Advances in the Study of Behavior* 21:39–68.

Zuri, I., I. Gazit, and J. Terkel. 1997. Effect of scent-marking in delaying territorial invasion in the blind mole-rat *Spalax ehrenbergi*. *Behaviour* 134:867–80.

Subject Index

fossorial rodents, 243–54, 291–302, 368–79, 403–15, 427–37

frequency-dependent transmission, 480–81, 483

F-statistics, 163–72

GABA, 193, 308–9

galanin, 191, 266

game theory, 86, 298

GC receptors, 140–41

gene dynamics, 163–71; hierarchical, 166, 168–71

gene flow, 155–56, 160

gene function, 57–67

genera, of rodents, 8–11, 418

genetic: compatibility, 43, 47–48, 57–67; correlation, 115, 164–71; differentiation, 165, 168–71; drift, 64–67, 324, 458; equilibrium, 164, 167; relatedness, 43–44, 180–81, 216–28, 245, 299, 349–54, 360–61, 364, 407–10, 438–44; revolution, 456; structure, 163, 166, 168, 169, 457, 471; variability, 155–56, 167, 172, 444, 466–67

genome sequencing, 381, 392

genomics, functional, 57

geographic isolation, 8–14, 165, 300–302, 420

geographic variation: in behavior, 297–302; in dimorphism, 124–26; in parental care, 239

gestation, length, 387, 407–10

gestational learning, 197, 210

giving up density (GUD), 331–32, 334, 338–39

Glires, 15

global warming, 454, 457

glucocorticoids (GCs), 140–49, 192–94, 198, 200; glucocorticoid cascade hypothesis, 148

gluconeogenesis, 140

gonadal axis, 140–41, 144–45, 148

gonadotropin releasing hormone (GnRH), 145

good genes, 44–47, 57, 60

gradient of fear, 339–41

grooming, 188, 191, 198, 232, 234–35, 312–13

group living, 53–55, 243–54, 356–57, 360–66, 396–97, 399–402, 405–412, 422–26, 431–32, 490; benefits of, 362–63, 410–12, 414; costs of, 247–49, 364–66

growing season length, 346, 349, 354

growth of young, 196–204, 240

H-2 complex, 59–62, 260, 263

habitat: change, 471; complexity, 323–24, 418–20, 425–26, 482; destruction, 454, 457, 467, 473; fragmentation, 156, 457–58, 462, 464–67, 470–71, 482, 485–86; ratio of optimal to marginal, 460; restoration, 475; saturation, 153–54, 179, 366; selection, 335–36, 348–49, 393–402, 406–11, 416–26

habituation-dishabituation technique, 219, 263–64

Hamilton's rule, 217–18, 224, 227, 443

hantavirus pulmonary syndrome, 479, 481–83, 485

haplodiploidy, 432

harsh environments, 349, 358–59, 368–72, 424–25, 432–35

hawks discriminate scent marks, 265

head thumping, 292

hemorrhagic fever, 481, 485

heterochrony, 312

heterothermy, 468–69

heterozygosity, 58, 61, 64–67, 166, 169–71

hibernacula, 348–49

hibernation, 142, 200–201, 204, 349–52, 358–364; group hibernation, 349–52, 363–64

hierarchical structured populations, 164, 166

hippocampus, 139–40

homeostasis stress response, 141

home range, 28, 53, 68–72, 76–80, 84, 347–48, 374–75, 382–83, 394–96, 402

horizontal transmission, 482

hormones, 107–8, 113, 139–49, 185–94, 197, 364, 369

host manipulation, by pathogens, 390–91

HPA axis, 139–40, 198, 200, 308, 314

huddling, 232, 235, 239, 363

human welfare, 453–54, 462

hunting, 398

hypothalamic-pituitary-adrenal axis (HPA), 139–40, 198, 200, 308, 314

hypothalamus, 139–40, 188, 190–91, 198, 200, 308, 314

Hystricognathi, 16–18, 403–4

IgG antibody, 44, 479

IgM antibody, 44, 479

imitation, learning by, 209, 213–14

immune response, 44

immunosuppression, 141–44

imperiled species, 463–67, 476

inbreeding, 47, 62–64, 169–71, 222, 225, 435; avoidance, 62–64, 151–52, 179, 246, 435–37; coefficient, 163–72; depression, 58, 62–64, 151–52, 160, 459

incarceration, of female, 445

incentive circuit, brain, 187–88, 194

index of social complexity, 347–48

indirect selection, 438

individual discrimination, 190–91, 263, 266

individual recognition, 222, 225, 263, 352

inexperience, and predation, 333

infanticide, 40, 42, 49, 52–53, 88, 89, 152, 159, 174, 178, 232–33, 240, 267–79, 308, 351, 389–90, 438–39, 447–50

informants, reliable, 255–60, 320–21

information: center, 382; exchange, 207, 209; theory, 347

intention cues, 313

interconnectedness, of populations, 471

interference competition, 89–90

interhabitat migration, 456

intragenomic conflict, 44

intragroup transfer, 361

intrauterine position (IUP), 153, 197, 398

intromission, 92–93

invasions, 8–13

invasive plants, 470

inverse density dependence, 180, 458–59

irruptive species, 456–57

island endemism, 13, 454, 456, 458, 467

iteroparity, 140–44

Junín virus, 482

juvenile development, 100, 104, 186, 195–206, 220–24, 232–33, 242, 412

kairomones, 433

Kenya rift, 429–31

keystone species, 455, 474–75

killer T cell, 59

kin: cluster, 53–55, 180, 349–50, 353–54; discrimination, 216, 224–25, 227, 263, 349–53, 407; label, 219, 222, 224–25; recognition, 61, 152, 157, 216–28, 349–53; selection, 160, 61, 226, 322, 347, 349–54, 436–37; structure, 245; template, 217, 222

kin recognition: direct, 216–17, 220–28; indirect, 217, 226; prior association mechanism of, 218, 220–22, 226–27; production component of, 217, 222, 225, 227

kinship, 4, 180–81, 216–28, 245, 349–54, 360–61, 364, 407–10, 438–44; effects, on demography, 180–81

kopje, 418

lactation, 232, 448; lactational learning, 210

Laguna Negra virus, 483

landscape variables, 459–462, 474–75

lassa virus, 482–83

learning, 198–204, 207–15, 220–24; of fear, 330, 332–33; by imitation, 209, 213–14; long delay, 384, 392; and memory, 198, 200; observational, 200–203; spatial, 148–49, 310

Lee-Boot effect, 106–8

leishmaniasis, 485

Leydig cells, 145

life history characteristics, 20–22, 68–85, 99–105, 144, 148, 339–41, 347–48, 352, 394, 406, 412–13

limbic system, 139

limestone canyon virus, 482

litter: size, 20–22, 77–79, 82–83, 234–35, 242, 389, 398–99, 408–10; weight, 73–83

littermate recognition, 219–24

local enhancement, 207

luteinizing hormone (LH), 107, 113, 145

lyme disease, 481, 483–84

lymphocytic choriomeningitis virus, 390, 483

Machupo virus, 483

major histocompatability complex (MHC), 47–48, 57, 59–62, 65, 225, 257, 260, 263, 386, 392

major urinary protein (MUP), 61, 225, 257, 262–63

Malagasy rodents, 13

male: -biased dimorphism, 117–19, 126; male-male competition, 42, 46, 59, 89–92, 116, 152, 255, 257–60, 273–79, 385–87; mate choice, 40, 59–61; mating tactics, 27–41, 89–92; spacing behavior, 68–85

male investment. See paternal care

mammae, number of, 235, 241

management, 313

mapped characters, 20–22, 73–85

mast, 71, 485

mate: acquisition hypothesis, 273–74; assessment, 258–60; choice, 42–52, 152, 157, 225, 255, 259–60, 263, 266, 386–88; competition, local, 134–36; guarding, 36–38, 91, 288–89, 396, 445–47, 449; recognition, 190, 193

material benefits, 46

Species Index